AF765296

Handbook of Systems Engineering and Analysis of Electro-Optical and Infrared Systems

There has been a lot of innovation in systems engineering and some fundamental advances in the fields of optics, imaging, lasers, and photonics that warrant attention. This volume focuses on concepts, principles, and methods of systems engineering-related topics from government, industrial, and academic settings such as development and operations (DevOps), agile methods, and the concept of the "digital twin."

Handbook of Systems Engineering and Analysis of Electro-Optical and Infrared Systems: Concepts, Principles, and Methods offers more information on decision and risk analysis and statistical methods in systems engineering such as design of experiments (DOX) methods, hypothesis testing, analysis of variance, blocking, 2k factorial analysis, and regression analysis. It includes new material on systems architecture to properly guide the evolving system design and bridge the gap between the requirements generation and design efforts. The integration of recent high-speed atmospheric turbulence research results in the optical technical examples and case studies to illustrate the new developments is also included. A presentation of new optical technical materials on adaptive optics (AO), atmospheric turbulence compensation (ATC), and laser systems along with more are also key updates that are emphasized in the second edition 2-volume set.

Because this volume blends modern-day systems engineering methods with detailed optical systems analysis and applies these methodologies to EO/IR systems, this new edition is an excellent text for professionals in STEM disciplines who work with optical or infrared systems. It's also a great practical reference text for practicing engineers and a solid educational text for graduate-level systems engineering, engineering, science, and technology students.

Handbook of Systems Engineering and Analysis of Electro-Optical and Infrared Systems

Concepts, Principles, and Methods

Second Edition

William Wolfgang Arrasmith

CRC Press
Taylor & Francis Group
Boca Raton London New York

CRC Press is an imprint of the
Taylor & Francis Group, an informa business

Designed cover image: KanawatTH\shutterstock

MATLAB® and Simulink® are trademarks of The MathWorks, Inc. and are used with permission. The MathWorks does not warrant the accuracy of the text or exercises in this book. This book's use or discussion of MATLAB® or Simulink® software or related products does not constitute endorsement or sponsorship by The MathWorks of a particular pedagogical approach or particular use of the MATLAB® and Simulink® software.

Second edition published 2025
by CRC Press
2385 NW Executive Center Drive, Suite 320, Boca Raton FL 33431

and by CRC Press
4 Park Square, Milton Park, Abingdon, Oxon, OX14 4RN

CRC Press is an imprint of Taylor & Francis Group, LLC

© 2025 William Wolfgang Arrasmith

First edition published by CRC Press 2015

Reasonable efforts have been made to publish reliable data and information, but the author and publisher cannot assume responsibility for the validity of all materials or the consequences of their use. The authors and publishers have attempted to trace the copyright holders of all material reproduced in this publication and apologize to copyright holders if permission to publish in this form has not been obtained. If any copyright material has not been acknowledged please write and let us know so we may rectify in any future reprint.

Except as permitted under U.S. Copyright Law, no part of this book may be reprinted, reproduced, transmitted, or utilized in any form by any electronic, mechanical, or other means, now known or hereafter invented, including photocopying, microfilming, and recording, or in any information storage or retrieval system, without written permission from the publishers.

For permission to photocopy or use material electronically from this work, access www.copyright.com or contact the Copyright Clearance Center, Inc. (CCC), 222 Rosewood Drive, Danvers, MA 01923, 978-750-8400. For works that are not available on CCC please contact mpkbookspermissions@tandf.co.uk

Trademark notice: Product or corporate names may be trademarks or registered trademarks and are used only for identification and explanation without intent to infringe.

ISBN: 978-1-041-03385-1 (hbk)
ISBN: 978-1-041-03510-7 (pbk)
ISBN: 978-1-003-62409-7 (ebk)

DOI: 10.1201/9781003624097

Typeset in Times
by codeMantra

Access the Support Material: Routledge.com/978-1-041-03385-1

This work is dedicated to the late Nitaya Gastin, whose heart was boundless, and whose energy and spirit inspired so many of us. You ran a good race. May you rest in peace sister.

Contents

SECTION II Contemporary Systems Engineering Principles, Methods, and Techniques

Preface

So why do we need this set of books? The short of it is that there has been a lot of innovation in systems engineering as well as some fundamental advances in the field of optics, imaging, and photonics that warrant attention, which have been included in this new version of the book. Some systems engineering-related topics that are receiving increased interest in government, industrial, and academic settings include "Development and Operations (DevOps)," Agile methods, and the concept of the "Digital Twin." Materials related to these topics have been either included as separate chapters or included as expanded content in existing chapters depending on the relevance of the material to the systems engineering discipline and EO/IR systems. We also decided to add new material to provide a more holistic coverage of the systems engineering discipline and tied in some advanced imaging technical and modeling concepts. The result of all this is the addition of six new chapters spread throughout two volumes and significant content updates throughout the remainder of this work. The material is logically split into four sections that are spread out over two volumes. This volume (the first volume of the two-volume set) contains the first two sections that focus on systems engineering fundamentals and provide a holistic set of systems engineering concepts, methods, tools, and techniques. The second volume is broken into two sections wherein the first section of Volume II focuses on the application of systems engineering tools, methods, and techniques to EO/IR systems. Herein, the basic systems engineering waterfall model is used as the framework for illustrating how the fundamental systems engineering principles, methods, and techniques are applied to optical and infrared systems. The second section of Volume II, as the name implies, is, where we are, "Bringing it all together!" Here we show how everything covered in the two volumes of this work comes together to determine and evolve the system's design, and provide insight, and control mechanisms into the three fundamental pillars of program and systems management: technical, cost, and schedule. The technical pillar is addressed by focusing on the collective systems engineering processes and activities for designing, producing, supporting, and eventually retiring and disposing of a generic system. Here, the main focus is on the front end of the systems development process starting with *Problem Definition*. We use the running integrated case study of a fictitious company, Fantastic Imaging Technologies (FIT), to illustrate how the systems engineering principles and methods (including problem definition, requirements definition, functional analysis, requirements allocation, generation of the system's architectures, and model-based systems engineering) are used to generate and evolve the system's design. The technical illustrations are brought to light as FIT conducts an internal review in preparation for an impending systems requirements review (SRR) with their stakeholders from the Department of Homeland Security Customs and Border Protection (DHS CBP). To cover the cost and schedule pillars, lifecycle cost estimation (LCCE) methods are presented to include top-down holistic cost estimation methods, software-dominant estimation methods, and mixed systems (both hardware and software) estimation methods. Among other notable LCCE techniques, function point analysis (FPA) for software-dominant efforts and parametric cost estimation (PCE) methods are presented. As an illustration of these LCCE methods, FIT performs a detailed integrated lifecycle cost and schedule estimate using all of the relevant LCCE methods to estimate the cost and schedule for developing and integrating their Illumination and Sensing System (IaSS) onto a fleet of autonomous and artificially intelligent unmanned aerial vehicles (A/AI – UAV) for their DHS CBP (and special) customers. All of the appropriate lifecycle cost estimates are reconciled with each other to provide a consistent, multi-faceted view of the expected systems cost.

What's different about this work? There are many excellent texts that discuss the field of systems engineering from a general point of view. Similarly, there is much written about imaging, detection, and optical systems from a technical perspective. In the former case, these general systems engineering-oriented books do an excellent job of describing fundamental systems engineering principles, tools, techniques, and methodologies, but often stop short of showing specific discipline-centric, technical examples, and implementations in an integrated fashion. The former general type of systems engineering texts present systems engineering from a holistic and generic point of view, but tend not to show the deeper theoretical, mathematical, and analytical connections between the systems engineering methodology and the technical aspects of the system under development (EO/IR systems in our case). There are many excellent examples of general systems engineering texts from various publishers. One of these excellent general systems engineering texts that this author uses on a regular basis is Benjamin S. Blanchard and Wolter J. Fabrycky's book entitled *Systems Engineering and Analysis*.

On the other end of the spectrum (pardon the pun), many books have been written that thoroughly cover the technical aspects of optical imaging and detection systems. For these types of books, the connections to the systems engineering discipline are mostly tenuous and brief. An excellent (but somewhat dated) example of this kind of book is the book by R. D. Hudson, *Infrared Systems Engineering* wherein abides a wealth of technical information with regard to infrared optical systems, but the systems engineering content is relegated to only a few pages in the beginning of the book. This makes sense since the technical disciplines themselves have a wealth of information and require focus and concentration to understand. Similarly, the field of systems engineering is advancing, evolving, and rapidly maturing, and a similar focus is beneficial in understanding systems engineering principles, concepts, methods, and techniques. Also, the development and description of systems engineering processes have a different flow than the development of technical information in an optics or imaging book. Just as in the previous work, *Systems Engineering and Analysis of Electro-Optical and Infrared Systems* (2015, Taylor & Francis, LLC), in these two volumes we wanted to achieve three goals:

1. We wanted to write an optically oriented systems engineering book that uses fundamental systems engineering principles to drive the optical/imaging technical content.
2. The optical/imaging technical content had to be sufficient to allow the practicing engineer to understand and analyze optical systems and to be able to apply the optical technical content to modern optical systems.
3. We wanted an integrated case study to serve as a running example of how the systems engineering content is applied to optical systems and to provide examples of how to apply the optical systems technical content. This also provided the opportunity to incorporate professional nuances and applications-related discussions that are normally out of place in a purely technical discussion.

In addition to these three goals, we also wanted to provide coverage of enterprise architecture methods, systems-of-systems, family-of-systems, and their connection to systems engineering. This was accomplished largely in the beginning of this volume (Volume I). We decided to introduce the reader to contemporary systems engineering concepts and methods such as Agile Systems Engineering, Digital Engineering (along with the ideas of the Digital Twin, Digital Thread, and executable modeling), DevOps and SECDEVOPS, the importance of various types of architectures in systems engineering (especially the systems architecture), and wicked problems. We provide an expanded treatment of model-based systems engineering (MBSE) methods due to their fundamental importance to the field of systems engineering, and the critical role MBSE plays in the development and evolution of contemporary systems, and the fundamental role MBSE methods will play in the development of complex systems of the future. We also present some examples and applications of MBSE using UML and SysML artifacts for general applications, and also in an Integrated Case Study which applies MBSE methods to a running case study of an Illumination and Sensing System that is being designed for a fictitious Autonomous, Artificial Intelligent Unmanned Aerial Vehicle (A/AI – UAV) program for the Department of Homeland Security, Customs and Border Protection (DHS CBP). Additionally, we also wanted to introduce some new material on research methods in systems engineering which provides an important statistical foundation for hypothesis testing, designing experiments, and statistical process control. Consequently, we developed a new chapter "Research Methods and Experimental Design in Systems Engineering." Since decision making, understanding, and analyzing risk are fundamental activities throughout the systems development lifecycle, we developed another chapter "Decision Making and Risk Analysis in Systems Engineering." This chapter also introduces and describes the useful area of Game Theory, which provides complementary information to decision analysis methods. A short discussion on the utility of artificial intelligence (AI) in systems engineering is also presented.

Finally, we wanted to present some coverage of modern-day electro-optical and infrared systems. We provide material on basic optical system principles, linear and non-linear optical systems, simulating and analyzing optical systems, statistical optics methods including the statistical nature and propagation of light, modeling atmospheric effects, adaptive optics (AO) systems, atmospheric turbulence compensation (ATC) systems, statistical metrics for optical systems, and more.

Why the titles? For Volume I, we chose the title, *Contemporary Optical Systems Engineering Principles, Methods, and Techniques*, because Volume I focuses on modern foundational practices for understanding, developing, and analyzing imaging and sensing systems (e.g., spectral sensing and polarimetric sensing) in the optical part of the electromagnetic spectrum. For the second volume, we chose the title *Applied Systems Engineering in Contemporary Optical Systems*, since the focus of the second volume is on the application of systems engineering principles and methods. We chose the title *Systems Engineering and Analysis of Electro-optical and Infrared Systems, 2nd Edition* for both Volumes I and II together, to show that this is an enhancement and refinement of our previously published book on systems engineering with the same title. The term Systems Engineering leading off the title shows that, first of all, this is a systems engineering-oriented text and that systems engineering is holistically addressed. By emphasizing the word analysis, we state that this book is intended to have enough theory, analytical content, and technical depth, to allow the practicing engineer to analyze optical systems from both a systems perspective and from the technical perspective of an optical systems engineer. We use the words electro-optical and infrared (EO/IR) systems because this indicates that the focus of this work is on electro-optical (EO) and thermal (IR) phenomena. Also, the EO/IR characterization is well known in the US Department of Defense (DOD), and the Intelligence, Surveillance, and Reconnaissance (ISR) communities. We could have used the combination of "Visible and Infrared Systems" instead of "Electro-optical and Infrared Systems" in the title since we predominantly focus on optical systems in the visible and infrared part of the electromagnetic spectrum. However, according to Merriam-Webster's Dictionary, electro-optical means, "relating to or being an electronic device for emitting, modulating, transmitting, or sensing light." This definition has more of a systems interpretation, and for this reason and the name recognition, we chose the "Electro-optical and Infrared Systems" title. Also, the last word in the title is "Systems" to emphasize that we are providing a systems-level technical focus in this book. We should mention at this point that unless otherwise stated, we use the term "optical" to describe applicable parts of the ultraviolet (UV), visible, and infrared (IR) parts of the electromagnetic spectrum. If we need to distinguish between the UV, visible, IR, or other parts of the electromagnetic spectrum, we will make that clear within the text.

For both Volumes I and II of our work, our organizational approach is to start each chapter with a focused discussion that sequentially describes the systems engineering-oriented material in the chapter. These discussions are meant to be not only applicable to general systems development efforts but also directly applicable to EO/IR systems. Our intent here is not to exhaustively cover general systems engineering topics but rather to present the key systems engineering methodologies, techniques, and tools needed to drive the optical analysis and design processes that are presented in this text. We therefore do not intend, nor find it necessary, to replicate large portions of the generic systems engineering texts that are currently prevalent in the literature. We do intend to capture the essential systems engineering methodologies contained in the leading current texts and journals and apply them to EO/IR systems. We will selectively use appropriate processes and tools to outline the fundamental contemporary systems engineering practices and apply these practices to EO/IR systems (mainly imaging, spectral, and polarimetric methods in the visible and infrared domains).

Just as with the systems engineering content of this work, we do not try to exhaustively replicate technical optics books that are widely available on the market. This would be redundant, unnecessary, and require too much space. Instead, we focus on optical systems analysis, metrics, and optical systems design considerations that are needed by the optical systems engineer. Nearly all chapters in both volumes of this work have a section that is devoted to optical systems building blocks. The intent of this part of the chapter is to provide more in-depth theoretical, conceptual, mathematical, and analytical information that is directly relevant to the optical systems engineer and even the optical designer. The approach is still to focus on a level of detail that is appropriate for a practicing systems engineer, optical systems professional, manager, developer, or student who is working on or has interest in EO/IR systems. The intent is for the reader to be able to thoroughly understand, model, and/or analyze optical systems. Mostly, we leave the detailed derivations of the optical technical content to the optical technical books and papers so that we can focus on the systems engineering application of the general results. However, we also provide links to further information in the chapter references, further reading, and bibliography parts of each chapter, as appropriate. Therein, derivations and deeper insights into the optical technical material can be obtained. Said another way, the level of detail in our "Optical Building Blocks" sections would be more in line with a modernization and topical expansion of the classic text from Richard D. Hudson, Jr. entitled, *Infrared Systems Engineering.* Although being light in emphasizing modern fundamental systems engineering principles – an issue that is remedied in this work – the Hudson text has the right level of detail for the optical systems engineer and optical systems professional. We not only provide a similar level of detail but also include an expanded set of topics, modernized material, and an expansion from just infrared to visible and infrared systems. Certain materials such as atmospheric turbulence compensation will, of necessity, receive a deeper treatment than in the Hudson book.

In each chapter, we link the systems engineering content with the optical systems analysis and design content. We call this content "Optical Systems Building Blocks," and we sequentially develop the necessary insights throughout this work. The linking of the driving systems engineering principles to the optical content was challenging since the logical flow of systems engineering content development is different than that of optical systems analysis. For instance, in the early systems engineering conceptual phase, practically all of the optical systems building blocks material is needed to fully conduct an optical systems analysis. We solved this linking issue by abstractly linking the topics covered in the systems engineering content with the optical systems analysis content in each chapter. For instance, in Volume I, Chapter 1, where we *introduce* enterprise architecture (EA), system-of-systems (SoS), family-of-systems (FoS), the acquisition process, and the systems engineering process, we correspondingly *introduce* the optical systems engineering, fundamental units, and radiometric quantities in the Optical Systems Building Blocks section of Chapter 1. Here, the link between this systems engineering content and the optical systems building block content is that, just as the starting point for a system is to define its larger environment (The EA, SoS, FoS, the Acquisition Process, and Systems Engineering development process), the initiating point for the technical aspects of an optical system is the fundamental physics, science, and engineering governing the optical system, as expressed by its fundamental units and radiometric quantities. In this case, the link between the systems engineering content and optical systems technical content is a logical one. In other cases, the link is more direct, such as discussing systems engineering modeling tools in the Systems Engineering section of the chapter and discussing optical systems modeling tools in the Optical Systems Building Block section of the chapter. Another driver in laying out the optical material in each chapter was to properly order the optical systems building block material so that there is a logical progression and flow of the optical contents. For instance, understanding the system of units is of fundamental importance to all of the optical systems building block content, and so, this was introduced right up front. Consequently, the optical systems content can be logically or directly related to the systems engineering content, but it is always sequenced so that there is a logical flow to the material that enhances progressive understanding of the material. Each subsequent chapter has a logical connection between the systems engineering content and the optical systems analysis content, and by the time we get through Chapter 8 and complete Volume I, we cover both the essential systems engineering fundamentals and the general aspects of optical systems analysis to move to the more applied material in Volume II.

We end each chapter with an Integrated Case Study that serves as a running example of how the systems engineering content in that chapter is used to drive the optical systems analysis and design processes. The Integrated Case Study uses the current chapter material as well as previous chapter materials to apply both the systems engineering principles and methods and the optical systems analysis principles and methods in an integrated fashion. We use a fictitious company that bids on, and subsequently wins, a contract with the Department of Homeland Security, Customs and Border Protection (DHS CBP) to provide a high-end optical system for integration on a fleet of unmanned aerial vehicles (UAV) to enhance DHS CBP mission capabilities. This company's piece of development is part of a larger enterprise activity, and so, enterprise architectural issues, systems-of-systems issues, systems engineering issues, and optical systems analysis and design issues can be, and are, addressed in the running Integrated Case Study throughout this work. In essence, the Integrated Case Study ties together the systems engineering content with the optical systems analysis and design material of this work. It also serves to illustrate how to apply the material and provides analytical examples.

We now provide a short description of the contents of our new work. Each volume of the set is complementary to the other volume but is designed to stand on its own, with its own set of interrelated systems engineering concepts, methodologies, techniques, optical systems building blocks, and case studies. Our set consists of two volumes that can be grouped into four major sections. The first major section includes Chapters 1–4 and covers where systems engineering fits within the larger context of an enterprise, systems-of-systems, and family-of-systems. This block of material also describes key aspects of the systems engineering process and ends with the distinct, but fundamentally important topic of establishing a solid systems architecture. Each chapter in this block of material links concepts and processes to the corresponding Optical Systems Building Blocks, and the Integrated Case Study that uses the fictitious FIT and DHS CBP development effort, ties together the systems engineering and optical building block content throughout this work. The second major section of material covers the remainder of Volume 1 (Chapters 5–8) and focuses on holistic and fundamental principles, methods, and techniques of systems engineering such as Model-Based Systems Engineering methods (Chapter 5), Research Methods and Experimental Design in Systems Engineering (Chapter 6), Decision Making and Risk Analysis in Systems Engineering (Chapter 7), and Agile Systems Engineering (Chapter 8).

With regard to the organization of the chapters themselves, chapters are usually broken into three parts consisting of a front section that defines systems engineering principles and methods and applies them to optical systems in general. This general optical systems section is then linked to an Optical Systems Building Block section that provides a technical focus and presents the technical details of optical systems engineering. The Optical Systems Building Blocks are intended to be stand-alone sections as much as possible, but they should be sequentially read for the best effect. In other words, the Optical Systems Building Blocks section in each chapter may draw upon information from previous chapters. The experienced reader may go directly to the chapter of interest, but the learner or trainee should go through the chapters sequentially. Each chapter ends with an integrated case study wherein the principles, tools, and techniques of the chapter are applied to a running case study. We now provide a short overview of what material is in each of the chapters.

Volume I: Contemporary EO/IR Systems Engineering Principles, Methods, and Techniques

Section I: The Larger Context of Contemporary Systems Engineering: Enterprise Architecture, System Architecture, System-of-Systems, Family-of-Systems, Complex, Complicated, and Adaptive Systems, Digital Engineering, DevOps/DevSecOps, and Wicked Problems.

Chapter 1 is an introductory chapter that provides an introduction to systems engineering in the modern age. We provide an introduction to systems engineering along with a short history of systems engineering and introduce the diverse systems engineer. We provide a short introduction to some high-level topics such as enterprise architecture, system-of-systems, and family-of-systems concepts. These topics will be further detailed in a separate chapter. We also discuss how systems are acquired in the U.S. defense sector through the Joint Capabilities Integration and Development System (JCIDS). We introduce the systems engineering development process across the lifecycle and also provide some insight into systems thinking and complex, complicated, and adaptive systems. An introduction to contemporary issues in systems engineering is also provided including topics such as Digital Engineering, the Digital Thread, and the Digital Twin; DevOps and DevSecOps; Agile methods in systems engineering; Systems Architecture, Executable Modeling; Standards; and understanding and dealing with Wicked Problems. Some useful conceptual tools are also presented. In the Optical Building Blocks section, we provide an introduction to optical system units and optical system methodologies and establish some terminology. In the Integrated Case Study section, we introduce our optical systems engineering case study that runs throughout both volumes.

Chapter 2 focuses on understanding enterprises through their enterprise architecture models and methodologies. The systems engineering process is seen to be complementary to enterprise architecture methods. We present the purpose of enterprise architectures, a short historical background, types of architectures, the roles they play in multi-entity development environments, and a short description of some common enterprise architectures. We also discuss some enterprise modeling tools and present a specific description of a useful Enterprise Architectural Framework – the Department of Defense Architectural Framework (DODAF 2.0). Just as the enterprise architecture is the model that unites various entities through its architectural framework, so do linear and nonlinear systems models unite and integrate various aspects of the optical system. In the Optical Building Blocks section, we present linear and nonlinear optical systems models and the integrated case study provides applications of our results.

Chapter 3 focuses on system-of-systems and family-of-systems and shows how the system engineering process is used over the system development lifecycle. In the Optical Systems Building Block section of Chapter 3, we show that the optical system itself can be considered as a system-of-systems, and we define the elements that make up this system-of-systems. The Integrated Case Study section, as will be done in other appropriate chapters of this book, once again integrates the systems engineering content with the optical building blocks content and serves as a section where we apply our results.

Chapter 4 addresses the topic of Systems Modeling and Establishing the Systems Architecture. The chapter starts in Section 4.1 with a brief introduction to systems modeling methods. A little information on historical perspectives on systems architecture and systems modeling is then provided. The processes for architecting systems from concept through verification using the Systems Modeling Language (SysML) are described, and systems engineering modeling methods are presented. Methods and tools for modeling, analyzing, and implementing the system architecture, and a discussion about the systems architect, systems architecture principles, and some case studies round out this section. In Section 4.2, the Optical Systems Building Blocks present modeling, simulation, and analysis methods for optical systems. Section 4.3 applies the concepts and methods in both Sections 4.1 and 4.2 to optical modeling efforts at our fictitious company, Fantastic Imaging Technologies.

Section II: Contemporary Systems Engineering Principles, Methods, and Techniques

Chapter 5 addresses the concepts and methods of MBSE. Section 5.1 provides an overview of the historical, theoretical, and conceptual aspects of MBSE and introduces fundamental MBSE principles, methods, and tools. The shift from document-centric to model-centric systems engineering is presented along with the use of MBSE in the system development process. The economics of MBSE is presented to understand the cost-benefit relationships involved when adopting, implementing, and sustaining an MBSE program in an organization. MBSE methods in relation to Agile Systems

Engineering are then discussed. Some MBSE case studies, MBSE applications, and some challenges of MBSE conclude this section. In Section 5.2, the Optical Systems Building Blocks focus on basic modeling elements for optical systems: the basic optical systems. Herein we cover basic optical principles, the achromatic principle, ray tracing, and optical filters. In Section 5.3, the integrated case study is used to show how MBSE methods are integrated with analytical software like MATLAB® to establish a rapid prototyping capability at our fictitious company, Fantastic Imaging Technologies.

Chapter 6 presents the immensely important topic of Research Methods and Experimental Design in Systems Engineering. This chapter presents useful probability and statistical methods that are prevalent when experimentation and statistical data analysis are important in the Systems Development Lifecycle. These methods are widely used in systems engineering applications such as determining the validity of data, experimental, and/or test results, conducting hypothesis testing, understanding relationships between input design parameters and output response variables through Design of Experiments (DOX), mathematically modeling system effects, applying queuing theory and systems modeling for time and motion studies, conducting Monte Carlo simulations, regression analysis, and statistical process control. Some of the major topics covered are an introduction to probability and statistics fundamentals, hypothesis testing and confidence intervals, analysis of variance (ANOVA), blocking, factorial design (including 2k factorial design), regression analysis, queuing theory, and an introduction to statistical process control (SPC). The Optical Systems Building Block material in Section 6.2 covers the statistical nature and propagation of light, and the integrated case study in Section 6.3 applies these two sections in illustrating the statistical elements of sensor design.

Chapter 7 covers the important topics of how to effectively make decisions and how to identify, quantify, and manage risk. In Section 7.1, theoretical concepts and analytical methods for decision-making and risk analysis are discussed, and the distinction between risk and uncertainty is made. The use of game theory in decision-making is presented and shown to be complementary to decision analysis methods. Some special considerations in modern-day decision-making conclude this section. In Section 7.2, the Optical Systems Building Blocks material focuses on statistical methods for modeling the turbulent atmosphere. Important optical systems models, parameters, and metrics such as the optical transfer function, the point spread function, random phase screens, the atmospheric coherence length (fried parameter), the isoplanatic angle, and the detector signal-to-noise ratio are also provided along with various other useful statistical-based metrics and analytical concepts. The integrated case study in Section 7.3 addresses decision and risk analysis and game theory methods in assessing implementation issues at FIT. The statistically based optical systems metrics are used to assess the performance of the optical system under development at FIT.

Chapter 8 focuses on Agile methods in systems engineering. In Section 8.1, Agile systems engineering is presented including an overview, introduction, historical background, and introduction to fundamental Agile concepts and principles. The potential benefits of Agile in systems development efforts are provided along with an overview of Agile methods, tools, and techniques. We follow this with a discussion of how to implement Agile and Agile assessment metrics. We touch on some of the useful Agile Scrum artifacts and end the section with a discussion on the applicability of Agile methods and some case studies. In Section 8.2, the Optical Systems Building Blocks look at hardware-based AO methods and software-dominant ATC methods, and also hybrid methods, with statistical metrics. The integrated case study in Section 8.3 looks at some new directions that the DHS CBP customer wants to pursue.

We reach the end of Volume I with the completion of Chapter 8. At this point, the reader should have a good understanding of essential systems engineering principles, concepts, methods, tools, and techniques and also have a fundamental technical understanding of EO/IR optical systems. This knowledge will be highly useful in Volume II of this set.

In addition to this material, we also supply some useful appendices at the end of each chapter (if needed), and at the end of each volume, we include important mathematical relationships, useful data, tables, and information needed to understand the technical content within each work.

This book is written for a wide variety of readers. Our intended audience is foremost the practicing systems engineer who is working on optical systems or systems of systems as part of a larger enterprise. The book is also meant for the general engineer who is working on or with optical systems. Optical scientists and specialty engineers will find this book of interest to gain insights into optical systems from a systems engineering perspective. Program and Project Managers who are responsible for optical systems will find useful metrics, analytical tools, and connections between managerial and technical drivers, using systems development methods throughout the systems development lifecycle. This book should also be useful to technicians who are working on optical systems. Additionally, this text should be useful for graduate students in systems engineering, optical systems, Science, Technology, Engineering, and Mathematics (STEM) curriculum, or other multi-disciplinary systems curricula that involve optical systems. This book should also be of interest to international professional colleagues.

Color versions of the figures in this book are available on the Routledge website www.routledge.com/9781041033851

MATLAB® is a registered trademark of The Math Works, Inc. For product information, please contact:

The Math Works, Inc.
3 Apple Hill Drive
Natick, MA 01760-2098
Tel: 508-647-7000
Fax: 508-647-7001
E-mail: info@mathworks.com
Web: http://www.mathworks.com

Acknowledgments

This has been a long journey, and there are many who deserve credit and thanks for helping get this project to completion. Foremost, I want to thank God for giving me peace and strength throughout this project and helping me throughout my life. Nothing would have been possible if not for his grace and support. I also want to thank my family, in particular my wife Lena, whose love and whacky sense of humor put the color in my world and make my days joyful, sunny, and blessed. She has tirelessly supported me throughout this project and has spent countless hours on her own while I stared at a computer screen. I also want to thank my kids Christina (and her husband Carlos), Kari (and her husband Mike), and my grand kids Christian and Tristan, Aria, Adalind, and Ezran, who are all wonderful beings, and bring Lena and me great joy. I also want to thank my late parents, Bill Black, M.D., and Hannelore Black, my brother Gordon, and sister Katrin, for their love, encouragement, and support over the years, and for being great people. Also, thanks to my extended family out there, all my relatives, and great friends (in particular Brian, Darlene and John, Norman and Avie, John and Julie, Eric and Jackie, Barry and Vee, and many more). Also thanks to everyone at Calvary Chapel for your servants' hearts and all you do.

I'd like to thank Dr. Phil Bernhard and Dr. Bhaskar Tenali who have given me the time to work on this project and who both are patient, practical, and kind individuals. Thanks to Barry, Danny, Aldo, Adrian, Arlene, Karen, Juan, and Marivi. I especially want to thank Dr. Barry Webster for many hours of fascinating discussions on game theory and so much else. I also must thank many students over the years who have directly helped me on this project in one form or another. From helping with graphics, to generating figures and equations, to conducting focused, and directed research, to doing related projects, analysis, reviews, and helping with the material itself, many thanks for your hard work and many contributions. There are too many students to individually list, but some deserve special recognition for going above and beyond what was expected: thank you Dr. Erlin He, Anagha Indic, Becky Gregg, Devan Warden, Adam McDonald, Josh Lee, and Gerald Fair. There were many others who rightfully deserve to be listed, and my deepest thanks go out to each and every one of you. Without all of your help, this project could not be accomplished. I consider this our project. Last but not least, I would like to thank Taylor & Francis for their tremendous support and for giving me the opportunity to develop this material.

About the Author

Dr. William Wolfgang Arrasmith is currently a professor in the Department of Mathematics and Systems Engineering at FIT. He has 20 years of experience with government research and development programs and has had extensive exposure to electro-optical, infrared, and laser detection systems. Before his position at Florida Tech, Dr. Arrasmith served as program manager of Physics and Electronics at the Air Force Office of Scientific Research (AFOSR) in Washington, DC. In 1997, he moved to the United States Naval Academy in Annapolis, Maryland to teach courses in Engineering and Linear Adaptive Optics. He was then reassigned to the Air Force Technical Applications Center (AFTAC) at Patrick Air Force Base where he became chief of the Systems and Technology Division. He was later appointed division chief for the Advanced Science and Technology Division of the AFTAC and remained in the position until joining Florida Tech in 2003.

Section I

The Larger Context of Contemporary Systems Engineering

Enterprise Architecture, System Architecture, System-of-Systems, Family-of-Systems, Complex, Complicated, and Adaptive Systems, Digital Engineering, DevOps/ DevSecOps, and Wicked Problems

1 Systems Engineering in the Modern Age

Scientists investigate that which already *is*; Engineers *create* that which has never been.

—**Albert Einstein**

1.1 INTRODUCTION TO SYSTEMS ENGINEERING

The introduction and application of systems engineering (SE) methods, principles, and techniques into the workplace, and their continued support over the years, have made a significant impact on systems development efforts. These efforts can be seen in the government, industrial, commercial, and academic sectors in the United States, as well as its allies, and in various countries throughout the world. In the United States, SE principles and methods are especially prevalent in the Department of Defense (DOD) and its suppliers, partners, and defense contractors. Within the DOD and many of its allies, SE principles are an integral part of their systems development efforts, acquisitions, and support strategies. Outside of the defense sector, many commercial and industrial entities are implementing fundamental SE principles and methods and are making them part of their core business practices. Notable examples are the automotive, transportation, and space industries. Government, corporate, private, and educational sectors have applied SE methods with great success to large, complex programs and small projects alike. This chapter introduces fundamental SE concepts that directly relate and apply to optical systems development activities. This book focuses on optical systems as the subject matter for applying SE principles, methods, and techniques. We will connect SE concepts and methodologies with optical detection methods that are predominantly, but not singularly, optical imaging applications and show how the SE principles and methods drive the optical systems development, support, and eventual retirement and disposal. By optical, unless otherwise stated, we mean systems that operate in the ultraviolet (UV), visible, and infrared (IR) parts of the electromagnetic spectrum. We first start this section with a short history of SE and provide fundamental definitions. Afterward, we follow with a discussion about the modern-day systems engineer and the need for diversity. Afterward, we introduce the overarching concept of an enterprise, its associated enterprise architecture (EA), and its relationship to SE. The EA discussion is followed with an introduction of the systems-of-systems (SOS) and family-of-systems (FOS) concepts. The next section discusses the formal acquisition process in the United States known as the Joint Capabilities Integration and Development System (JCIDS). We end the first section of this chapter with a look at some fundamental SE principles and methods that are applied throughout the systems development lifecycle (SDLC).

The second section of this chapter focuses on optical SE building blocks. These are the fundamental optical systems analytical methods and techniques needed to understand optical SE. The majority of the chapters in this book have an Optical Systems Engineering Building Blocks section; these sections are interrelated and sequential. In this section, we provide introductory information such as units and fundamental concepts that will be needed for later chapters. We introduce the terminology and concepts necessary to lay the foundation for a wide array of optical systems and optical SE applications. The last section of this chapter introduces our optical SE case study that will be used throughout this book. Just as the optical SE building blocks, the case study is intended to be sequential and evolves from chapter to chapter. The intent is to integrate

DOI: 10.1201/9781003624097-2

the SE concepts, principles, and techniques, with the optical SE analytical methods in a running, applications-oriented case study. In the next section, we provide a short history of the field of SE along with some essential definitions.

1.1.1 Short History of SE

In a loose sense, SE can be traced back to ancient days when systems were first engineered. Ancient architectures qualify as systems and so the construction of the pyramids, the Great Wall of China, the Roman aqueducts, and other ancient architectures, and these are all examples of engineered SE. However, from a historical perspective, the bulk of the development of modern-day thinking on SE principles, methods, tools, and techniques occurred relatively recently. SE involves the organized, optimized, and requirements-driven development of systems with an emphasis on the entire system, over its entire lifecycle, and with a focus on integration.

According to the International Council on Systems Engineering (INCOSE), the systems engineering term itself had its origins at Bell Telephone Laboratories in the early 1940s (Schlager 1956; Hall 1962; Fagan 1978), and some of the essential SE concepts within Bell Labs trace back to the early 1900s. The DOD entered into the picture in the late 1940s and applied SE to missile defense systems (Goode and Machol 1957). During World War II, the U.S. Army Air Force formed the RAND Corporation. World War II saw the advent of major SE applications, but it was not until 1950 that the SE discipline, as we know it today, was first taught. This initial attempt at teaching SE began at the Massachusetts Institute of Technology (MIT) by Mr. Gilman who was the director of SE at Bell Labs (Hall 1962). By 1951, Fitts presented the idea of allocating system functions down to the physical elements (Fitts 1951). RAND developed the fundamentally important area of systems analysis in 1956 (Goode and Machol 1957). In 1962, Hall included five phases in his description of SE (Hall 1962). These phases were as follows:

1. Conducting system studies along with appropriate planning activities.
2. Exploratory planning: Well-known SE functions were accomplished.
3. Development planning: A repetition of phase two occurs but at a more detailed level.
4. Development studies.
5. Current engineering: Current engineering takes place during the operational phase of the system when the system gets refined.

According to Eisenberger and G. Lorden, the concept of systems lifecycle costing (LCC) became popular in the 1960s when the DOD began to recognize problems associated with awarding contracts based on price alone. Studies of weapon systems and other procurements revealed that acquisition costs were typically smaller than costs of ownership such as the cost of labor and materials required to operate and maintain the system. Since the 1960s, there has been a robust development in the SE discipline. The development of computers and the subsequent transition to digital technologies gave rise to four system design methods: (1) experimental, (2) simulation, (3) trial and error, and (4) analytic (optimum) design (Faurre 1975).

The latter part of the 1980s saw the development of the Soft Systems Methodology (SSM) that had impacts in solving difficult social problems and change management. According to Checkland, SSM focused on a systematic approach and feasible outcomes and differed from focusing only on the fundamentals of optimization (Checkland 1989). The 1990s saw the rise of systems architecting principles and methods that solved complex problems, mostly in the aerospace community (Rechtin 1991). The concept of the Digital Twin saw early adoption in the 2000s by government agencies, which saw a total system effectiveness increase of 10% (Pettey 2017). The United States Air Force found value added when it modeled a physical system digitally (Stargel 2012).

Some other notable events include the establishment of the Defense Acquisition University (DAU) in 1971 (DAU 2001), the creation of the Air Force Center of Excellence for SE which resides at the Air Force Institute of Technology (AFIT), and the establishment of the INCOSE in 1990. Some significant publications include MIL-STD-499 (DOD 1969) in 1969, Andrew Sage's systems engineering text in 1992 (Sage 1992), IEEE Std 1220-1998 (which looked at the application and management of SE) (IEEE 1998), ANSI/EIA-632-1999 (which focused on SE processes) (ANSI 1999), the systems engineering handbook by INCOSE in 2007, and Blanchard and Fabrycky's books on *Systems Engineering and Analysis* (first edition was in 1981 and fifth edition in 2010) (Blanchard and Fabrycky 2010).

Other recent significant events include the establishment and advancement of EA methods, capability maturity model integration (CMMI) methods, LCC methods, and model-based systems engineering (MBSE) methods. CMMI is a process-centric methodology that is useful in establishing a set of organizational assessment benchmarks and best practices. The CMMI methodology started as a project that involved a productive collaboration between commercial and governmental entities and the Software Engineering Institute (SEI) from Carnegie Mellon. Organizations such as the National Defense Industrial Association and the Office of the Secretary of Defense (OSD) were strong supporters.

In MBSE, the idea is to use integrated models at all levels of the development process and wherever possible to improve integration across disciplines and the integration of diverse SE activities across the systems lifecycle itself. There are various excellent MBSE tools available such as No Magic's MagicDraw™ (now Dassault's CATIA System-of-Systems Modeler) that, among other things, has an EA plug-in that provides a model-based EA framework. It can also integrate with their own (or other) standard requirements management tools such as IBM's Dynamic Object-Oriented Requirements System (DOORS™) and other MBSE tools. Many of the MBSE tools come in an integrated suite that allows modeling of SE methods and techniques such as EA methods, requirements engineering and requirements management methods, business process modeling, and decision and risk analysis modeling.

Looking toward the future, the INCOSE is already developing guidance for its 2035 vision. Adaptability is a priority in this vision, and the aim is to make SE mainstream across the globe. The emphasis is on addressing societal problems, resource management issues, and social inadequacies, focusing on increasing complexity (INCOSE 2021).

EA methods and MBSE are discussed further in Chapter 4. The adoption of "Agile" methods, lean engineering, "Six Sigma" methods, and integrated lifecycle product development tools, along with decision and risk analysis methods, techniques, and tools contributed significantly to the modern understanding and evolution of the SE discipline. These methods will be discussed throughout the text. We next present some of the important definitions and terms that evolved throughout the SE discipline.

1.1.2 Some Definitions of SE

When we think of the word *system*, what comes to mind? According to Merriam-Webster's Collegiate Dictionary, a system is "A regularly interacting or interdependent group of items forming a unified whole" (Merriam-Webster 2003). Taking this definition, we notice that to qualify as being a system, there must be interacting items, and they have to depend on each other in such a way that they are in some fashion unified. This definition is quite broad and fits many items from ordinary to extraordinary. For instance, a coffee cup can be thought to be a system. It has a handle (item 1) that serves as an interface between the hand and the coffee, a flattened surface (item 2) that serves as the interface between the cup and a surface, and a liquid-containing element (item 3) that serves as the receptor for our delightful beverage. On the other extreme, the galaxies themselves can each be thought of as a system that contains billions of stars, dust, planets, black holes, quasars, plasmas, electromagnetic radiation, and more. These elements within the galaxy influence each other through physical interactions such as gravitational and electromagnetic effects and unify to produce the galaxies themselves. Other systems are satellite systems, medical systems, military systems, quality control systems, inspection systems, manufacturing systems, and optical systems, to name a few.

An observation at this point may be that a system is whatever we define it to be, as long as it has interdependent elements that are unified in some fashion. This means that a U.S. guided missile cruiser could be thought of as a system in its role as a maneuverable U.S. Navy surface ship that has a specific role with well-defined functions in the fleet. Equally valid, the cruiser's Aegis combat system itself could also be considered a system, collecting real-time telemetry of targets and providing that information to the ship's command structure. SE involves the principles, methodologies, and techniques involved in engineering a system as we define it. When reading SE texts, the definition for SE varies from one text to another. However, the common elements in the definition of SE are that SE

1. Deals with the entire system;
2. Applies over its entire lifecycle, from conceptual design to retirement/disposal;
3. Focuses on integration.

We can consider these points when we think of engineering a system.

The system must be defined in its environment, so its boundaries and relational context to other systems or entities can be established. The word "*entities*" implies membership in an encompassing EA. Our system may interact with other entities, and if so, EA methods become useful. Once we define what we mean by system, we can further subdivide the system into smaller functional pieces called *subsystems*. This functional decomposition is hierarchical and depends on the system itself and often on who is subdividing the system. The subsystems can then be further divided into *components/assemblies/pieces/parts*. We will use the term *components* to refer to this level of system decomposition.

An example of an optical system as we define it is shown in Figure 1.1, which presents a satellite system as a collection of subsystems. Notice that this satellite has optical components as subsystems (primary mirror, secondary mirror, solar array, sun sensor, and instrument module). Furthermore, other parts of the satellite may be critical components that directly affect the performance of the optical system (baffle, light shield, power, signal conditioning, radiation shielding, environmental conditioning, and data communications).

Therefore, if it suits us, we can define or classify this satellite to be a space-based optical system. On the other hand, this satellite could be part of a larger system wherein the satellite's optical data play a key role. In this sense, the satellite is an entity in a larger enterprise, and the satellite may be considered to be part of an SOS. In essence, we define the system, its constituent parts, and its surrounding environment.

In addition to our understanding and definition of *systems*, we want to distinguish between the words *program* and *project* and also between *service*, *product*, and *process*. *Merriam-Webster's Collegiate Dictionary* defines a *program* as "a plan or system under which action may be taken toward a goal" (Merriam-Webster 2003). In the DOD, the term usually means a large, complex development or acquisition undertaking, and program management then relates to the management activities associated with realizing the program. For example, a high-budget, complex satellite activity in the DOD to develop a new imaging capability would be typically classified as a development or acquisition program. Interestingly, a definition for *project* according to *Merriam-Webster's Collegiate Dictionary* is "a planned undertaking: such as

a. A definitely formulated piece of research;
b. A large usually government-supported undertaking;
c. A task or problem engaged in usually by a group of students to supplement and apply classroom studies."

Item *b* in the definition indicates that projects and programs are synonymous, and some people state as much. From our experience, however, we most often have seen the label of *project* assigned to

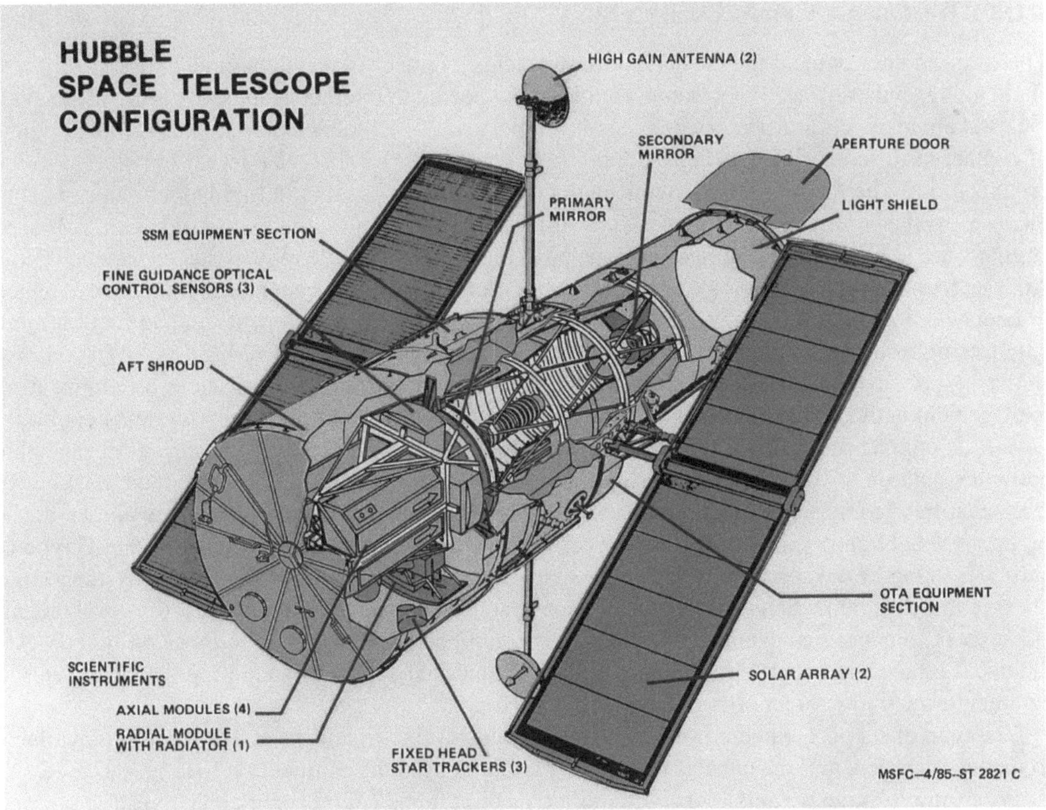

FIGURE 1.1 Hubble space telescope architecture. (Obtained from NASA 2025; https://science.nasa.gov/missions/hubble/30-years-ago-hubble-launched-to-unlock-the-secrets-of-the-universe/. With permission.)

smaller-scale activities. In this text, programs are considered large, complex undertakings, which could possibly consist of a number of projects. Projects are similar undertakings to programs but on a much smaller scale. With this interpretation, you may find a junior engineer or low-level manager working with projects within the confides of a larger program. Project manager, project engineer, and project engineering then have a similar connotation to their larger program counterparts. It must be emphasized that in other disciplines and sectors, the interpretation of the words *project* and *program* may be different.

The SE discipline does not only concern itself with systems, programs, or projects but also concern itself with services, products, and processes. For example, one system may be organized in accordance with a traditional DOD acquisition program paradigm but may also have service or process aspects to it. Another system may be completely service oriented like a help desk support service or technical writing support service system.

In large systems, very often there are components of each present in the activity, and the systems engineer will be involved in all aspects! We use the word product (or material goods) when we are considering systems and activities with commercial aspects to them. For example, cell phones, iPads, and commercial off-the-shelf (COTS) computers are considered products, whereas an optical surveillance system or weapon system is more aptly thought of as a system that may have been implemented through a program or project. Systems engineers may find themselves working with systems, services, products, or processes (SSPPs). We now look at the unique world of the modern-day, multidisciplinary systems engineer.

1.1.3 The Diverse Systems Engineer

The working environment for the current-day systems engineer is very dynamic, multidisciplinary, challenging, and exciting. The systems engineer is expected to interact proficiently with SSPPs, various stakeholders, administrators, managers, support staff, technical experts, and technical points of contact in different fields, officials, and support contractors (whether internal or external to the organization). The modern-day systems engineer needs to be capable in highly technical environments as well as proficient in managerial environments. Sometimes, the expectations for a systems engineer are to have lots of engineering expertise (big "E") and a little managerial proficiency (little "m") such as in a typical junior systems engineer. Often, junior-level systems engineers are assigned to understand and work on lower levels of the functional system components and so require more engineering technical proficiency than managerial proficiency (big "E" and little "m"). In contrast, senior systems engineers usually support upper management and so focus more on engineering management activities (big "M" and little "e"). This does not mean that senior systems engineers have little engineering experience, just that their focus often is more on engineering management activities than on lower-level engineering details. Also, some organizations also have a technical track that has promotion opportunities to senior levels. However, promotion opportunities to the highest levels of an organization in many cases involve a deliberate transition from big E type of activities to big M type of activities. To effectively focus on "big M" activities, senior systems engineers in larger organizations usually have junior systems engineers who provide lower-level technical support to upper management. Senior systems engineers are then free to focus on higher-level technical issues and can effectively provide SE managerial support to the program manager and senior administration of an organization or enterprise.

The type of support expected depends on the needs of the organization, the resources available to the organization, and the capabilities of the particular systems engineer.

For example, large organizations with big pools of systems engineers are able to use senior and junior engineers to support large, complex programs and systems. In smaller organizations, however, there may not be a large pool of systems engineers or even a small pool of systems engineers for that matter, and so the system engineer may be forced to provide both technical and managerial support (big "M" and big "E"). This would be impractical for large organizations dealing with large, complex projects. Smaller companies and organizations tend to use the big "M" and big "E" approaches due to resource constraints. The systems engineer must be able to fulfill multiple roles including both the technical and managerial.

In the case of the optical systems engineer, the expected technical proficiency has to do with optical SSPPs. For proficiency with optical systems, the optical systems engineer needs strong mathematical skills such as familiarity with linear and nonlinear system theory, integral and differential calculus, probability theory, statistics, stochastic processes, complex math, optimization methods, functional analysis, frequency analysis, Fourier transform theory, Laplace transforms, Z-transforms, and wavelet transforms as basic skills. There is great importance given to math since it is the unifying language between different technical disciplines. The optical systems engineer will have to work with optical designers, mechanical engineers, optical testers, software engineers, hardware experts, optical scientists, optical technicians, and a host of other subject matter experts. It is also beneficial to have experience with technical topics such as modern optics, lasers, electromagnetics, electrical circuits, computer architectures, atmospheric turbulence physics with regard to its interaction with light, statistical optics (SO), Fourier optics, nonlinear optics, detectors, material properties, signal/image processing, basic mechanics, system noise analysis, and optical systems modeling methods. We will provide the required basic introductory-level technical understanding in Section 1.2 throughout this chapter.

With regard to systems management, familiarity is expected with requirements management and engineering, modeling and simulation methods, engineering design principles, feasibility analysis, trade studies, functional analysis, requirements allocation, configuration management methods, interface control procedures, technical performance measures and metrics, decision and risk analysis, cost, schedule, performance, risk assessment/management, qualification planning and execution, and system optimization methods. The optical systems engineer must also be proficient in, and aware of, the following representative concepts, processes, and issues:

• Reliability	• Maintainability	• Availability
• Human factors	• Quality	• Logistics
• Engineering ethics	• Value engineering	• LCC
• Concurrent engineering	• Standards	• Benchmarking
• Supportability	• Intellectual property	• Producibility
• Safety	• Security	• Disposability
• Contract management/law	• Flexibility	• Environmental interactions
• Performance	• Liability	• Suitability

The optical systems engineer must ensure that these considerations are properly addressed in the system's requirements development process and throughout the system's development lifecycle. In the next section, we provide a brief overview of the EA concept and its connection to SE.

1.1.4 Introduction to the Enterprise and Its Architectural Description

This section provides a brief introduction to the EA concept and establishes a connection with the SE discipline. A more detailed look at EA methods is provided in Chapter 2. EA methods and their interaction with the SE discipline are becoming critically necessary with the ever-increasing complexity of modern-day technical systems. Although it might be argued that an *enterprise* fits our general definition of a system – and so EA methods should be part of the SE discipline – the reality is that EA methods have evolved on their own. EA methods play a fundamental and complementary role in defining, understanding, developing, and supporting systems that are part of a larger enterprise activity.

According to The Open Group (TOGAF 2012), an *enterprise* is "any collection of organizations that has a common set of goals and/or a single bottom line." One important observation is that we are dealing with multiple organizations. These organizations likely have different motivations, agendas (open and hidden), resources, commitments, and capabilities. They are also likely to have some independence and autonomy but are linked by the common goal(s). The EA methods define the enterprise's strategic goals, establish the framework of cooperation between various organizations within the enterprise, and provide for a governance structure between various *entities* making up the enterprise. An example would be various agencies, components, and organizations making up the DOD. These DOD organizations have separate budgets and a degree of autonomy but cooperate to achieve the strategic goals set by the DOD and U.S. leadership. Note that the DOD itself may be part of a larger enterprise consisting of the U.S. Government and its allies concerned with common defense issues. Each organization in the enterprise is an entity in the DOD enterprise, and their cooperative interactions are structured by the enterprise's architectural framework – in this case, the Department of Defense Architectural Framework (DODAF). Each of these organizations has a key role to play in the overall enterprise. Some organizations provide services, others provide systems, others ensure that processes are followed, still others produce products, and some produce mixtures of SSPPs.

It is the task of the enterprise architectural framework to provide the structure or framework that guides the interactions of the constituent entities. The EA framework does not build the EA architecture itself; it just provides the interaction protocols and tools for jointly building the architecture in an integrated and unified manner. One way this is accomplished is through a collection of predefined views and architectural framework artifacts. For example, many architectural frameworks have viewpoints such as operational views, system views, and service views that are integrated with each other. These EA framework products provide the necessary information required by enterprise stakeholders. The EA typically provides the highest level of integration available to the systems engineer. The exception is when the enterprise is itself part of a larger enterprise and so consequently has a higher-level structure, an enterprise of enterprises. An example would be the U.S. Government structure interacting on the world stage. There would possibly be several different enterprises, each with their own architectural framework, which would be expected to interact efficiently with each other. The enterprise is typically the highest level that the systems engineer has exposure to. In this text, we will assume that the enterprise is the highest level of organizational structure that exists for our applications and needs. If the enterprise were part of a larger enterprise, then the EA methods outlined in Chapter 2 would still apply, and so there is no loss in generality by adopting our "one-enterprise" approach. The EA framework provides the structure for defining the interaction environment of the constituent SSPPs that make up the enterprise.

As an example, one of the key EA artifacts is the concept of operations (CONOPS) for the enterprise. The EA CONOPS shows the critical systems, actors, processes, and the interaction environment itself. Actors are "A person, organization, or system that has a role that initiates or interacts with activities" (TOGAF 2014). We now look at the concepts of SOS and FOS in the context of the EA and also the SE discipline.

1.1.5 Introduction to SOS and FOS

In the previous section, we pointed out that the EA typically contains the highest level of integration that is available to the systems engineer. At the EA level, all the entities in the enterprise are defined as well as the expected interactions between these entities whether we are considering SSPPs or mixtures of these elements. The enterprise then can consist of many separate systems that work together in a unified way. The enterprise can, and often does, contain SOS. We think of the SOS concept as a larger system that is itself made up of a collection of systems, as shown in Figure 1.2.

The larger system then has an overall purpose that can only be achieved by the interaction of the systems that make it up. An example would be a U.S. Navy ship that contains many systems (radar systems, weapon systems, communications systems, damage control systems, defense systems, etc.). All of these systems cooperate harmoniously together to provide the necessary capabilities to the ship commander and crew to fulfill the intended missions assigned by higher command. In this sense, the ship is an SOS, and the constituent systems – HF communications, UHF communications, X-band communications, damage control systems, radar systems, weapon systems, power systems, control systems, optical systems, and as such – are individual systems that make up the SOS. Note that the U.S. Navy ship can be thought of as part of a larger enterprise consisting of early warning satellites, the command structure, other response elements, intelligence services, and support elements. The enterprise itself can be thought of as having SOS as part of its integral components. However, in addition to the SOS component, the enterprise also has services, products, and processes to make up its whole. As an example, the DOD enterprise consists of its collection of individual systems, collaborative SOS, support systems (e.g., weapon systems, communications systems, navigation systems, early warning systems, force protection systems, logistics systems, and maintenance systems), products (e.g., doctrine, tactics, capabilities, military and humanitarian operations, peacekeeping, protection, airmen, soldiers, and sailors), and enabling processes and services (e.g., command and control, intelligence, surveillance, reconnaissance, finance, medical, legal, supply, training, administration, and engineering). This list is not intended to be all-inclusive but just to serve as an illustration how the SOS concept fits into the overall EA paradigm. In other words, we define an SOS to suit our needs and the situation at hand. The EA then defines the interrelationships of the systems and SOS to each other and in relationship to the enterprise itself. "*Family of Systems* (FOS) is defined as a set of systems that provide similar capabilities through different approaches to achieve similar or complementary effects" (ODUSD(A&T) SSE 2008). The difference between the

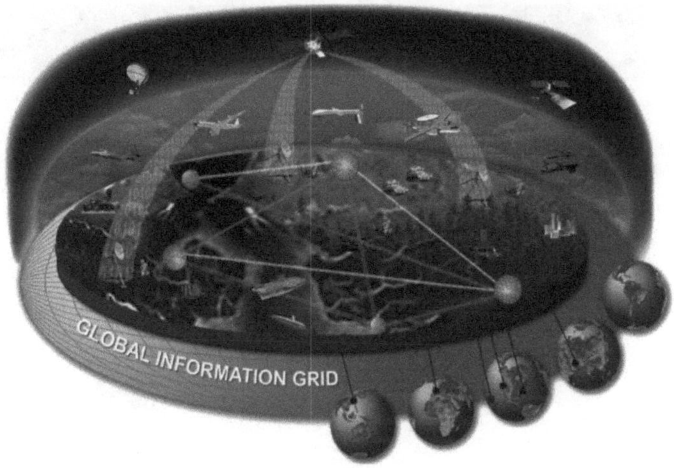

FIGURE 1.2 The SOS concept. (From GIG 2014.)

FOS and SOS definitions is a matter of scope. The FOS provides similar capabilities, whereas SOS provides an increased capability from its constituent systems. In other respects, it would be hard to tell one from the other.

An example of an FOS would be a collection of assets that hunt submarines. There are different assets that all provide the same or similar capability to detect submarines, surface vessels, other submarines, air assets, and specialized sensors. In this scenario, these assets are part of an FOS. In contrast, the U.S. Navy ship we mentioned earlier, working with satellite systems, command-and-control systems, intelligence systems, ground systems, air systems, and integrated communications systems, may be part of an SOS that provides a new, improved, sea-based rapid response capability. The SOS in the latter case is distinct from the FOS since it provides a new capability that is not achievable by the individual systems themselves.

In the following section, the U.S. joint capability integration, development system, and the process of acquiring systems in the DOD are introduced. Most of the jobs relating to SE in the United States are related to the defense industry. A thorough understanding of the DOD acquisition process is relevant and important for those working on or with U.S. defense systems. The next section will provide a brief introduction to the acquisition process used in the DOD.

1.1.6 U.S. JCIDS and the DOD Acquisition System

This section provides a basic overview of the acquisition process to better understand the interrelationship between EA methods, the formal defense capability acquisition process, and the SE discipline. Some good resources for further information are the Defense Acquisition Guidebook (DAU 2025) and material from the Defense Acquisition System (DAU) (DAU 2012). Figure 1.3 shows the relationship between the U.S. Defense Acquisition System (DAS); the planning, programming, budgeting, and execution (PPBE) process; and the JCIDS. From the figure, you can see that there are three distinct entities that have their own oversight structure. For the PPBE process, the oversight comes from the Deputy Secretary of Defense. For the JCIDS process, the oversight comes from the Vice Chairman of the Joint Chiefs of Staff and the Joint Requirements Oversight Council (VCJCS/JROC). The JCIDS process is to a large degree a precursor to the traditional DAS in that it ensures that acquired systems are capability focused and have the required compatibilities with the systems of other services, partners, and allies. There are checks and balances to ensure that the DAS is only engaged after other more cost-effective capability solutions such as doctrine, organization, training, materiel, leadership, personnel, and finance (DOTMLPF) are exhausted.

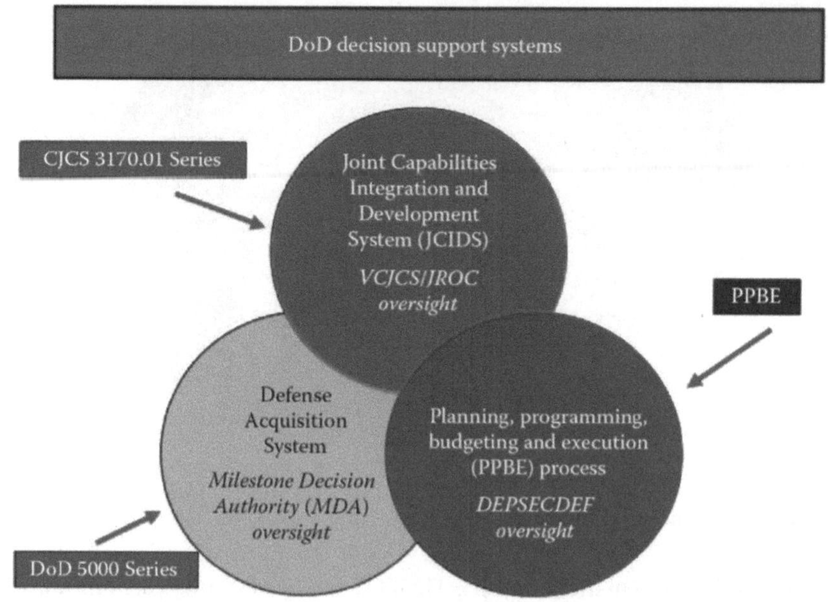

FIGURE 1.3 Relationship of the JCIDS with the defense acquisition system and PPBE process. (From Defense Acquisition Guidebook 2012.)

The JCIDS is governed by a series of instructions from the Chairman of the Joint Chiefs of Staff – CJCS 3170.01 Series instructions. The DAS has oversight from the Milestone Decision Authority (MDA) and is governed by the DOD 5000 Series of directives and instructions. The Federal Acquisition Regulations (FARs) are also prominent as a legal, directive, policy, and process defining entity. While the systems engineer may interact with all of these entities, they mostly are engaged in the DAS itself. SE principles, methods, and techniques are very much a part of defense systems development and acquisition activities.

Figure 1.3 shows the relationship between the JCIDS and DAS. The PPBS is integrated over the entire systems acquisition process and throughout the systems development, support, and retirement phases, which allows optimization of technical performance, development, and support schedule, along with the associated SSPP activities, over all lifecycle phases.

The acquisition process starts with high-level strategic guidance shown at the top left of Figure 1.4. This guidance is itself developed from higher-level strategic guidance such as national policies and is subject to the FARs. A series of joint – multiservice activities that include appropriate representation from air force, army, navy, marine corps, and coast guard stakeholders – concept evaluations, capability assessments, and gap analysis are conducted that lead to an Initial Capabilities Document (ICD). If the MDA is convinced that a material solution is called for in the ICD, then the MDA and appropriate stakeholders make a Materiel Development Decision (MDD). The MDD initiates an analysis of potential materiel solutions – materiel solution analysis and an analysis of alternatives (AoA). In this part of the acquisition process, promising technologies from various sources are investigated to include foreign developments, government laboratories academia, centers of excellence, and industry. Once a materiel solution approach is determined, then the decision is documented in the Acquisition Decision Memorandum (ADM). The ADM and the results of the AoA generate a proposed solution from the lead DOD component that carries out the AoA, and a lifecycle funding profile, generated from the PPBE, marks the end of the materiel solution analysis phase and the start of the technology development phase.

The triangles in Figure 1.4 are formal milestone events that are consistent with the SE development cycle (e.g., conceptual design phase, preliminary design phase, and detailed design phase).

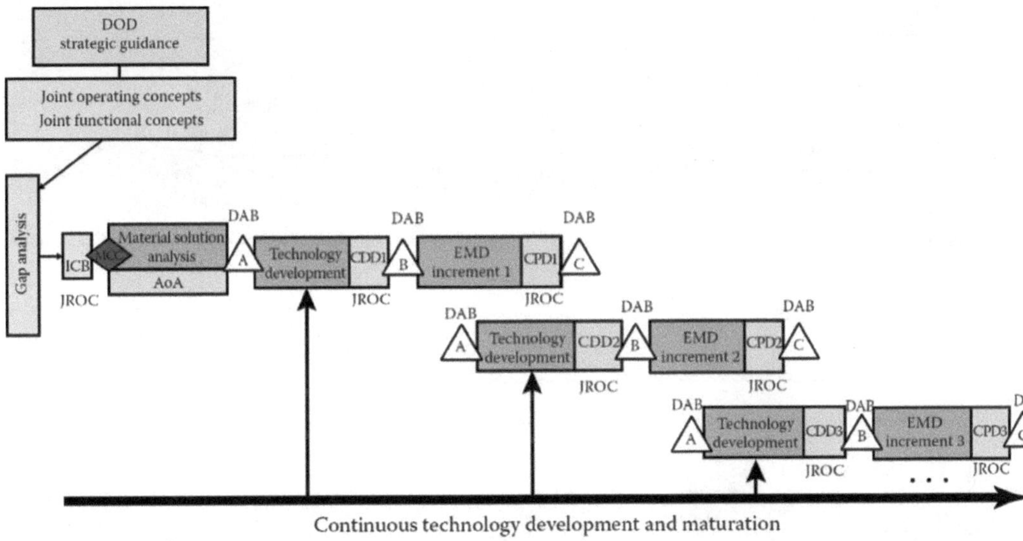

FIGURE 1.4 The requirements and acquisition process flow. (Reprinted from ACQWeb 2014. With permission.)

Depending on program cost levels, the oversight for these milestone events is the Defense Acquisition Board, which is chaired by the MDA. There are different approval levels depending on the overall research, development, test, and evaluation (RDT&E) costs of the program, and the approval levels are based on Acquisition Categories (ACATs) that range from I to IV (with two subcategories for ACATI programs). As an example, ACAT II programs are those programs that do not qualify as an ACAT I system but are still considered major systems. Major systems are those systems in which the RDT&E costs in fiscal year (FY) 2000 in dollars equal to or exceed $140 million. At this stage, it is not as important to delve into the different ACATs, as it is to know that the MDA depends on the RDT&E cost structure of the acquisition program. Where appropriate, the JROC reviews the JCIDS products – Capability Development Document (CDD) and the Capability Production Document (CPD). For major systems development activities, after the JROC reviews the CDD and the MDA approves milestone B – Preliminary Design Review (PDR) – the acquisition proceeds to the engineering, manufacturing, and development (EMD) phase. The next phase in the acquisition process is the CPD review by the JROC and milestone C – Critical Design Review (CDR), approval by the MDA. The overlapping bars in Figure 1.4 illustrate evolutionary development. It is noteworthy to point out that the DOD 5000.02 directive calls out that the DOD EA "shall underpin all information architecture development." Accordingly, any information-related element of an acquisition, development, or activity that must comply with the DOD 5000.02 directive must be subject to the DODAF (current version as of this writing 2.02).

From a business perspective, Figure 1.5 shows the contract process where businesses typically interact and interface with the defense acquisition process. Three phases make up the contracting process: (1) acquisition planning, (2) source selection, and (3) contract administration. Note that these perspectives are from the government side. In the acquisition planning phase, the need for the system is established using the processes discussed in Figure 1.4. The results of these processes are shown in the top block of Figure 1.5 labeled requirement determination. The requirements for the proposed solution that are established are the requirements for a DAS solution. Subsequently, a request for proposal (RFP) is initiated by the government and is posted on government websites such as https://sam.gov/opportunities, https://www.grants.gov, and https://www.sbir.gov, for larger efforts, academic, and small business opportunities, respectively. Once the RFP or broad agency announcement has been released, the business and academic community can choose to respond.

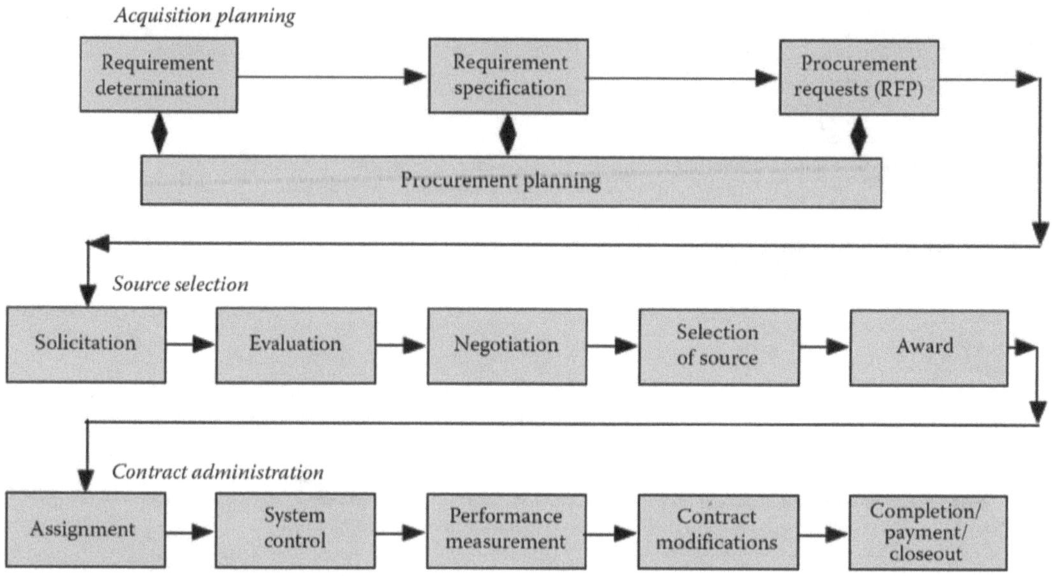

FIGURE 1.5 The contracting process. (Reprinted from Defense Acquisition University Press 2001. With permission.)

The business and/or academic community will use their internal processes to decide whether or not a particular solicitation is worth pursuing. An institution may go through the following representative phases as part of their business development process: opportunity creation, identification, qualification, pursuit, proposal development, proposal review, and proposal submission. Many things are considered in the opportunity creation phase such as evaluating the market need, establishing the business case for a proposed venture, and projecting opportunities for business growth. The identification phase looks at validating the creation opportunities and finding available opportunities that are in line with the strategic vision and core competencies of the business. The qualification phase evaluates the capabilities of the business entity to ensure that there are sufficient technical depth, breadth, and resources to accomplish the work. Collaborations and partnerships are often established at this point to form the "right team" for the effort. Pursuit involves establishing contact with the RFP point of contact (POC) to obtain clarification and available guidance if possible. Once the proposal development phase is completed, there should be a proposal review effort. The proposal review effort consists of independent groups (red team and blue team) that critically review the proposal. The red team tries to challenge the proposal, and the blue team defends it. The purpose of this process is to improve the proposal prior to submitting it to the government/stakeholders for official review. After suggested improvements from the red/blue team reviews are integrated into the proposal, the proposal is then submitted to the contracting office where it becomes a solicitation as shown in the middle block of Figure 1.5.

The government evaluates the submitted solicitation. The government's evaluation process has lifecycle cost estimation as well as competency metrics. The government will use systems engineers and field experts to evaluate the solicitations. Based on the reviews, the government may enter into negotiations with business entities. After the negotiation step, the source selection process determines which business entity or entities are chosen for the work outlined in the RFP. Awarding the contract follows this step. There are several different types of contract vehicles that the government uses such as firm fixed price (FFP), cost plus fixed fee (CPFF), and cost plus incentive fee (CPIF). Which of the contract vehicles is chosen depends on the nature of the task and the amount of risk the government is willing to take on the effort. Once the contract award is made, this ends the source selection phase of the contracting process.

The last row of Figure 1.5 shows the contracting process steps after the contract award has been made. In this contract administration phase, contract oversight agencies such as the Defense Contract Administration Service and the Defense Contract Audit Agency ensure that proper administration and compliance are followed by the business entity. This is done by establishing system control methods and by measuring contract performance. Contract modifications are also conducted if necessary as well as contract completion, payment, and closeout activities at the end of the contract. In this section, we briefly learned about the JCIDS and the DAS. In the next section, we focus on the SE process and look at general SE principles and methods.

1.1.7 Introduction to SE across the Lifecycle

This section provides an introduction to the SE methodology, illustrates some fundamental processes, and defines the SDLC phases. Our approach is consistent with contemporary SE professional practices and consistent with the current academic understanding of the SE discipline. Further, we draw heavily from the applications and methodologies used in the U.S. defense industry. Our approach is general and adaptable to commercial and other applications as detailed in the following. This defense-centric application orientation is adopted since the majority of U.S. SE jobs are in the defense industry or related to it in some fashion. Where appropriate, we focus and tailor the general SE approach to the optical SE application area and discipline. To provide the appropriate context, certain illustrations will remain at the general level, whereas others will be adapted to fit optical SE scenarios throughout the optical SE lifecycle. This approach is used throughout the text and is designed to provide introductory-level information. More detail on the SE methodology, principles, tools, and techniques can be found in Chapters 3–21 of this two-volume set of books.

The SE discipline is fundamentally important in developing large, complex systems, FOS, SOS, and related elements of an EA. SE methods and processes can also be adapted and tailored for other organizations and activities to include small businesses, commercial, industrial, governmental, and academic applications (Arrasmith 2007). The SE discipline consists of principles, methodologies, processes, and techniques for optimally bringing systems into existence, supporting them throughout their lifecycle, and then providing for efficient, safe, environmentally friendly system disposal or retirement. In other words, SE brings complex systems into being and supports them throughout their lifecycle and sees to their disposal in an integrated, optimal fashion.

The evolutionary development of complex SSPPs typically follows an RDT&E-oriented, phased structure such as the traditional SE lifecycle phases: (1) conceptual design phase, (2) preliminary design phase, (3) detailed design and development phase, (4) manufacturing/production/construction phase, (5) operations and support phase, and (6) system retirement/disposal phase. Notice that we are using high-level categories, and these categories will be detailed later in Chapter 3. There are a variety of useful process models that help the systems engineer understand a system and break the system into a manageable set of developmental steps. These models are intended to furnish us with a general view of the SE processes that lead to the successful development of a system. We will briefly discuss some of the more often used process models starting with the generic SE across the lifecycle waterfall diagram and adapt it for optical SE.

This waterfall process model outlines the essential and sequential steps in the SE process. Notice that the steps from optical systems requirements to optical systems design integration are sequential steps but are repeated at each of the three SE design phases (conceptual design, preliminary design, and detailed design). This iteration is shown in the figure by the dashed lines to the left of the individual blocks. We will discuss this more in Chapter 3. As we can see in Figure 1.6, the first step in this waterfall model is to define the problem.

This single block has a number of activities, processes, and events that are associated with it, but for brevity and clarity's sake, we focus only on the high-level detail in this introduction. The next step involves establishing an architectural framework for our system to guide how our system interacts and interfaces with other systems in our enterprise. This step should happen early in the

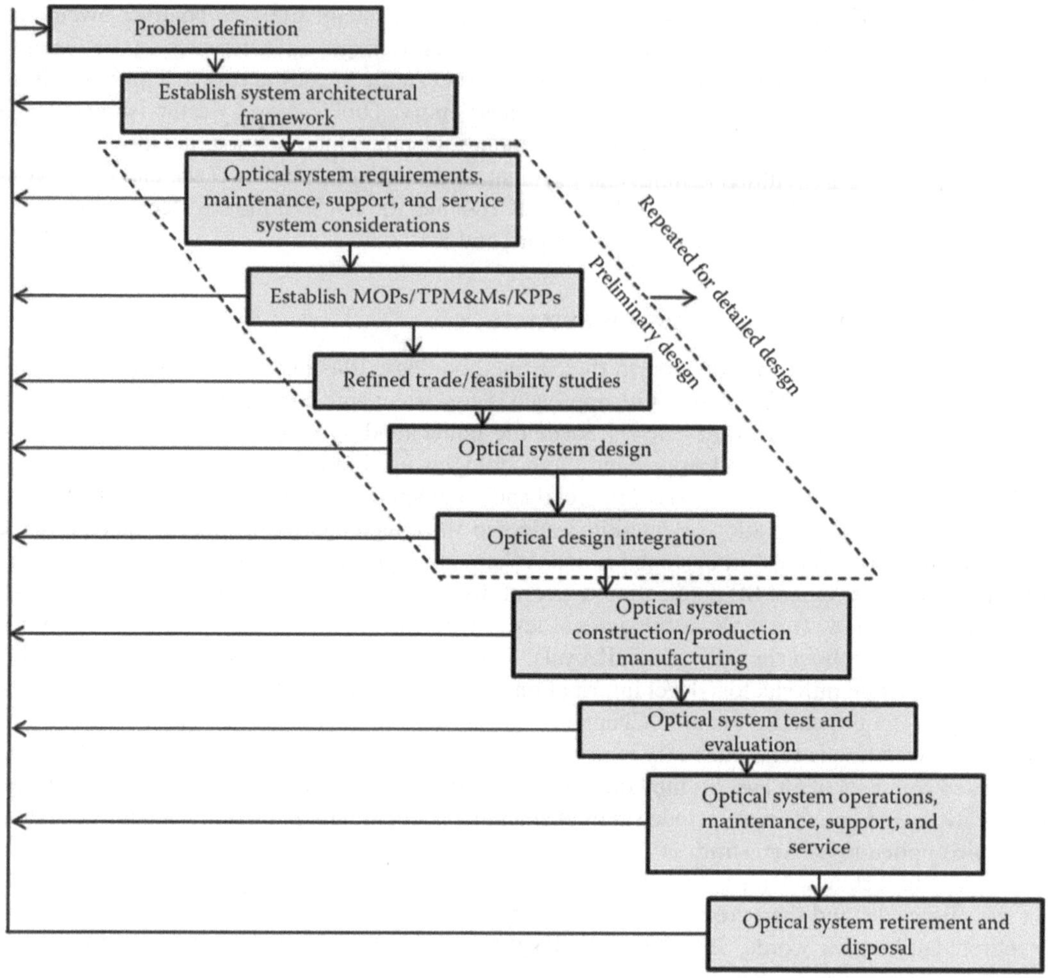

FIGURE 1.6 The optical systems engineering development across the lifecycle waterfall model. (Adapted from Blanchard and Fabrycky 2010.)

SE development lifecycle to ensure that enterprise-related technical issues, systems issues, and governance and strategic drivers are considered in the systems requirements and any derived potential solutions. At this stage, the feasibility studies and trade studies are focused on the stakeholder requirements and are initially assessed at a high and broad level. For example, each stakeholder requirement is assessed not only for technical feasibility but also for organizational, motivational, political, operational, financial, and environmental feasibility. We must also determine whether or not the potential feasible solution to the requirement is sustainable within the enterprise over the systems lifecycle.

The system-level requirements for the candidate optical system are shown in the optical systems requirements block. General SE texts would label this as "Systems Requirements," but since this book is geared toward engineering optical systems, we add the "optical" term as a distinction. Notice that the steps in the waterfall model that start with optical systems requirements and end with optical systems design integration are boxed. These represent core SE processes and are repeated at every design phase. For instance, if we start with system-level requirements, we consider the system-level maintenance, support, and service implications associated with those requirements.

We also determine appropriated measures of performance (MOPs), technical performance measures and metrics (TPM&Ms), and key performance parameters (KPPs) for our system.

These are necessary to effectively test and qualify our system and to effectively communicate the systems performance capabilities to system stakeholders. Afterward, we conduct a functional analysis that is based on the system-level requirements. This functional analysis holistically evaluates the system-level requirements and defines the subsystem functions needed to satisfy these requirements. The system requirements are then allocated into functional partitions (e.g., subsystems) determined by the functional analysis and requirements allocation processes.

The engineering team then designs and integrates the designs based on the functional decomposition. The connection between requirements and design can be divided into two distinguishable steps: (1) generate requirements and (2) design based on the generated requirements. As an example, our optical systems requirements are used to drive the design effort once they have been baselined and put into configuration control. The design process for our optical system starts with a functional decomposition of the optical systems requirements. In the functional analysis process, subsystems are defined and analyzed that functionally partition our technical response to the optical systems requirements. For example, we might choose to partition an optical imaging system into power, control, communications, environmental conditioning, signal processing, image processing, pointing and tracking, and optical subsystems.

The optical systems may be further partitioned into daytime optical subsystems and nighttime optical subsystems, and each of these could be further subdivided into optical telescope, image formation optics, optical detector, detector housing, and detector cooling subsystems. Of course, this decomposition is not unique, and other forms of decomposition are equally valid. The important thing is that the functional decomposition is complete in the sense that all the necessary functions needed to satisfy the optical systems requirements have been defined. The optical system's functional partitions are then grouped together in the requirements allocation process, and the design process continues with further functional analysis, trade studies, modeling, and simulation as needed.

When analyzing the subsystems of our candidate optical imaging system, it is necessary to derive and generate new requirements at the subsystem level. These newly derived and generated requirements along with the allocated system-level requirements form the basis for the subsystem-level requirements and associated requirements documents. This process repeats once the subsystem requirements (B specifications) are baselined. Once all of the requirements generation/design phases are complete, the optical system goes to construction, manufacturing, or production. Subsequently, the complete optical system is tested, accepted, and supported throughout the optical systems lifecycle. At the end of the optical systems lifecycle, it is retired or disposed of in a cost-effective, environmentally friendly, safe, legal, and ethically sound manner.

Figure 1.7 shows another useful SE process model – the "V" process model. This model is effective in showing sequential, hierarchical, and complementary lifecycle SE activities that may be separated in time. We can see the traditional SE design phases in the gray rectangular box. The conceptual design phase starts with problem definition and ends with the preparation of the systems requirements document – the A specification – for our optical system. The systems requirements document is formally approved at the system segment review, and this event starts the preliminary design phase of the SDLC. The preliminary design phase uses the approved A specification to drive the preliminary design. The result of this phase is the production of preliminary design documents and the preparation of subsystem-level requirements documents – B specifications – at the PDR. Upon approval of the B specifications, the last design phase – the detailed design and development phase – is started. In this last design phase, a detailed design, associated specifications, drawings, and detailed requirements based on the A and B specifications are developed. During this stage, prototypes are built and tested. This phase ends with the approval of all detailed design documentation at the CDR. The end of this phase signals the start of the production, manufacturing, and/or construction phase. This phase is shown at the bottom of the V-diagram.

After production, the SE development process continues with the subsequent testing of the "as-built" system as shown on the right side of the V-diagram. Notice that there is a corresponding formal test block for each level of requirements (e.g., system level, subsystem level, and component level).

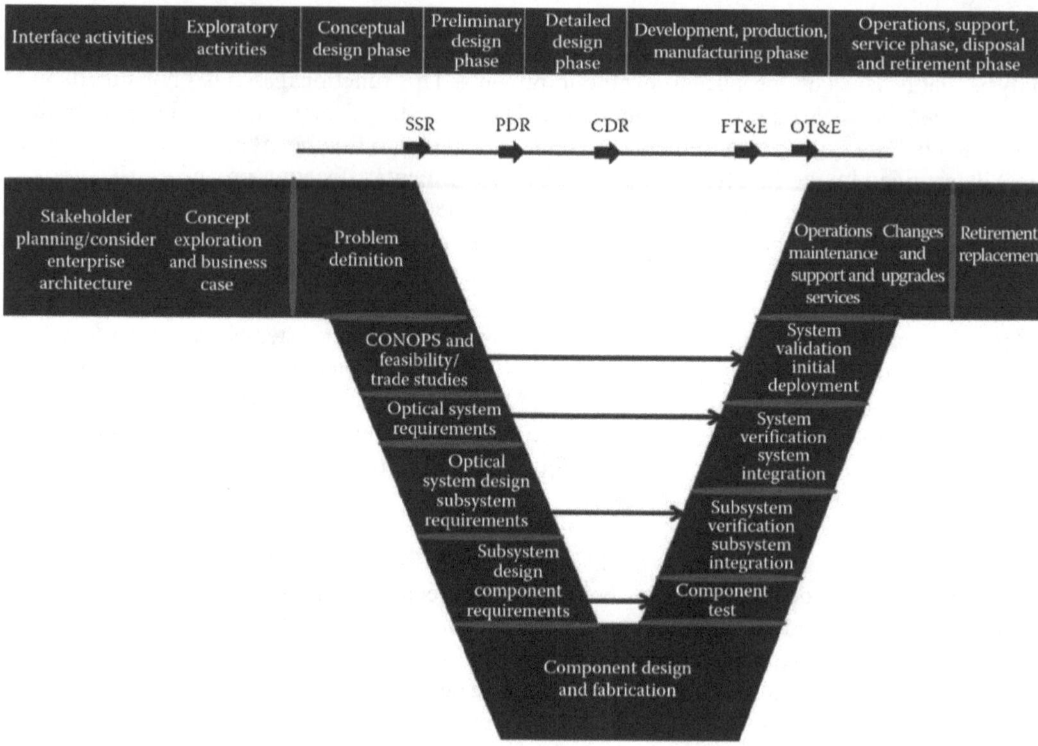

FIGURE 1.7 The V-process model illustrating the connection between system-level, subsystem-level, and component-level requirements and test throughout the traditional systems engineering lifecycle phases.

Establishing a good set of TPM&Ms, MOPs, and KPPs during the requirements generation process is critical in effectively testing the developed system. The system-level formal test is often called functional test and evaluation (FT&E), and the formal test that signals the end of the systems development process is called operational test and evaluation (OT&E). OT&E is often conducted in the field under operational (or simulated) conditions. Once the system passes the OT&E, the development activity is complete, and the system transitions to the operational, maintenance, support, and service phase. In this phase, the systems engineer is involved in evaluating the systems performance, providing preplanned product improvement (PPPI) upgrades for the system, and implementing changes according to a change and configuration control methodology. An example of this would be generating and responding to engineering change proposals (ECPs). The final phase of the SDLC is the system retirement and disposal phase, which should go smoothly if it is properly considered early on in the SDLC.

1.1.8 Systems Thinking and Complex, Complicated, and Adaptive Systems

Our world is interesting! As we look at our surroundings, what stands out? Depending on our location and focus, we might note billowing clouds, majestic mountains, or rushing streams that spill over the sides of a mountain in an endless torrent. Picking up a magnifying glass, we might see the detailed structure in a leaf, or the world that the ant finds itself living in. Further, with a microscope, cellular structures can be observed. On the other end of the spectrum, with a good telescope, we can look at the depths of space and back in time. Whether bodies in our solar system, entities in our galaxy, the galaxies themselves, or objects in deep space, our focus changes and so does the definition of the observed system. Each of these entities can be viewed in terms of a system and so

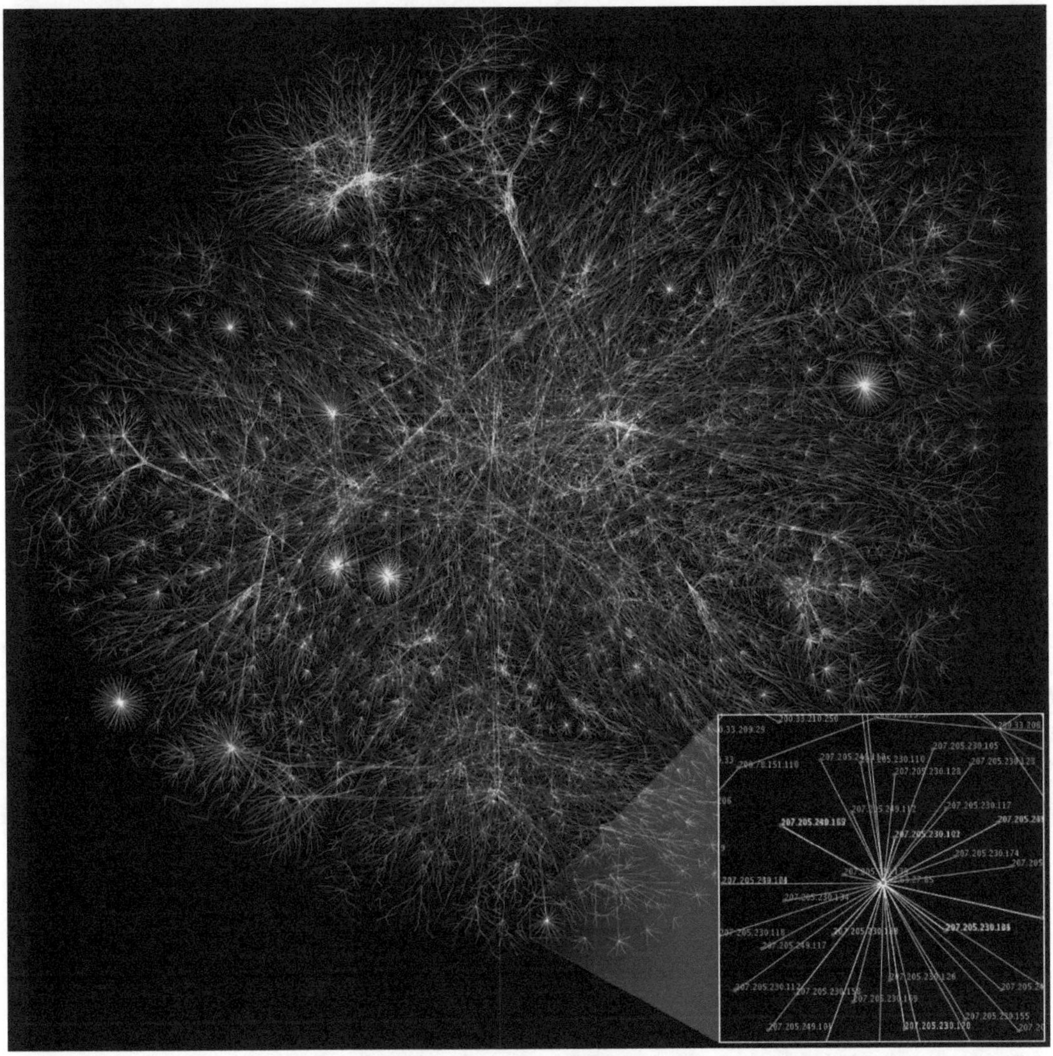

FIGURE 1.8 Internet map. (Courtesy of the Opte Project Wikipedia, 2005, The Internet Map. https://en.wikipedia.org/wiki/File:Internet_map_1024.jpg"jpg, Accessed 9 May, 2017.)

SE methods, tools, and techniques can be applied to understand and characterize them. Whether we have natural or manmade systems, if sufficiently advanced, they can be categorized as complex, complicated, and may also have adaptive properties. But what do we mean by these terms? How do we think about and effectively develop these kinds of systems? We will see that the solution comes from adopting contemporary systems thinking and SE development methods as presented below.

We live in a complicated, complex, and adaptive world. Life goes on around us often at a breathtaking pace. Machines are getting faster and smarter and marvelous scientific and engineering achievements are occurring at an ever-increasing rate. We see from Figure 1.8 that our communications systems are extremely interconnected and are evolving rapidly. Figure 1.8 shows various internet communication nodes and data communication traffic patterns and gives an indication of how connected we are. All it takes is for a storm to take down our electrical grid and internet and life seems to come to a screeching halt. Oh no! I can't communicate with my colleagues, friends, and family, systems are down, and the effects can range from mildly annoying to disastrous. Technology marvels abound as can be seen in Figure 1.9 where we see a virtual reality (VR) robotic surgery

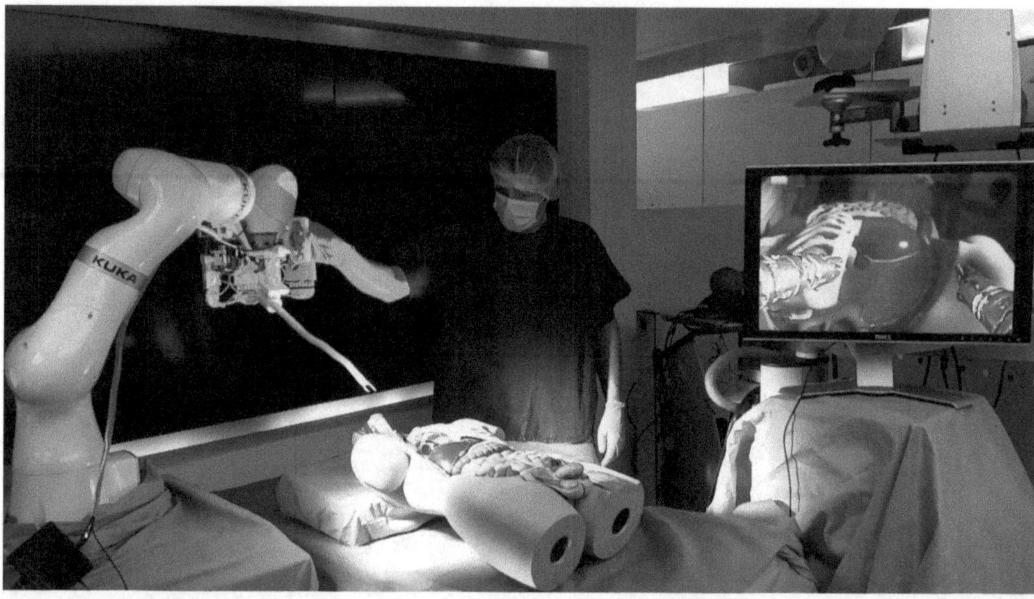

FIGURE 1.9 VR robotic surgery, Davinci robotic surgery system, cambridge science festival 2015, By Cmglee - Own work, CC BY-SA 3.0, https://commons.wikimedia.org/w/index.php?curid=39154360. (accessed 24 January 2025).

FIGURE 1.10 The big data system and its environment. Courtesy of Camelia Boban, Wikimedia 2014, https://commons.wikimedia.org/wiki/File:BigData_2267x1146_white.png (accessed 7 May, 2017).

occurring. Technology advancements have been made across the board at an incredibly fast pace, with fundamental achievements in quantum computing, quantum entanglement, machine learning, artificial intelligence, big data, and cyber-security, for example, regularly making the news. Form a systems thinking/SE perspective, we notice that there are many connected and related systems, parameters, external factors, and issues that can impact contemporary systems. For example, when we consider systems that handle Big Data, we see in Figure 1.10 that it is no trivial matter to define the Big Data system and its environment. In this case, and for other complicated, complex, and/or adaptive systems development efforts, it is essential that systems thinking and contemporary SE methods are used to overcome the challenges we face in our modern systems development environment. Some of these developmental challenges were articulated by INCOSE back in 2014:

1. Mission complexity is growing faster than our ability to manage it … increasing mission risk from inadequate specifications and incomplete verification.
2. System design emerges from pieces, rather than from architecture … resulting in systems that are brittle, difficult to test, complex and expensive to operate.
3. Knowledge and investment are lost at project lifecycle phase boundaries … increasing development cost and risk of the late discovery of design problems.
4. Knowledge and investment are lost between projects … increasing cost and risk: dampening the potential for true product lines.
5. Technical and programmatic sides of projects are poorly coupled … hampering effective project risk-based decision-making.

Interestingly, these developmental challenges still exist today although progress is being made through the refinement and dissemination of SE methods and processes, the wider spread adoption and implementation of Model-Based SE in the commercial and governmental sectors, and SE education and training programs. As will be seen below and throughout the remainder of this work, the SE discipline has methods and processes to deal with these developmental challenges, but these methods and processes must be carefully implemented, used throughout the developmental effort, and effectively communicated throughout the development team. In this vein, one of the major difficulties in effectively implementing modern-day systems, whether complicated, complex, or not, is in correctly understanding, defining, capturing, and addressing the stakeholder's needs, problem to be solved, and the resulting requirements of the system. Establishing clear and precise communications is essential in any sophisticated SE development effort. However, human languages can be inherently ambiguous and full of double meanings, insinuations, and nuance, providing ample opportunities for misinterpretation and misunderstanding. Error mitigation is accomplished through careful consideration of the problem at hand, along with a precise definition of terms, the use of standards and external models where possible and practical, using model-based SE methods, and even natural language processing (NLP) methods where appropriate. At the center of the SE approach lies careful, intentional, and precise wording to mitigate ambiguity, properly define the system, service, project, and/or processes (SSPPs) needed to correctly design, develop, produce/implement, support, and eventually retire SSPPs to the stakeholders' satisfaction. For conciseness, we will use the word "system" to represent the entity that is the focus of the SDLC effort, which itself could consist of various combinations of systems (such as in the case of System-of-Systems development efforts), services, products, or processes. In the next section, we define some terms and provide some fundamental concepts and thoughts related to Systems Thinking when dealing with complicated, complex, and adaptive systems.

1.1.8.1 Systems Thinking

Many original thinkers have made valuable contributions to the discipline of SE. Here, we highlight several important works that address Systems Thinking in the context of the field of SE. Systems Thinking is an essential mindset in the makeup of the contemporary systems engineer and should be embraced, internalized, and practiced in systems development efforts. So, what do we mean by Systems Thinking? To answer this question, we look to what the great thinkers and scholars have to say. These scholars have contributed original and practical discoveries that advance our knowledge of the characteristics and classifications of contemporary complex, adaptive, and/or sophisticated systems.

One of these great thinkers is Andrew Senge. According to Senge (Senge 2006),

> Systems Thinking [is] a way of thinking about, and a language for describing and understanding, the forces and interrelationships that shape the behavior of Systems. This discipline helps us see how to change systems more effectively, and to act more in tune with the larger processes of the natural and economic world.

This definition contains several interesting ideas. First, it is a method of thinking, which means that the system engineer must structure his or her thoughts correctly. It is also a language and thus represents a means of communicating important aspects of the system, such as interrelationships, system description, forces acting on the system, and system behavior. It assists the system engineer in determining how to effectively bring modifications to the system, as well as gain an understanding of the system in a broader environmental and economic context. As a result, both the system's financial features and its underlying physical nature are within scope. Mella also stated,

> Systems thinking is a discipline that recommends that we observe the world as a system of dynamic systems, recursive and repetitive – often with memory – each composed of interacting and interconnected variables.
>
> **Mella (2012)**

In this definition, the ideas of the systems themselves being dynamic and interconnected come to light. Additionally, the systems are composed of variables that interact within the system (intra-connected) and without (interconnected). A key component in this definition is the reference to "memory" which is fundamental to complex systems. Further, we should observe the world from the perspective of systems of dynamic systems with the attributes of recursion and repetition. Mella states that Systems Thinking is "a mental attitude, an approach, a logic, and a language" in addition to a useful technique used in constructing models (Mella 2012). Mella presented five rules that are useful in applying Systems Thinking to development efforts:

- Rule one: in order to understand the world, the Systems Thinker must be able to, "see the trees and the forest;" by developing the ability to "zoom" from the whole system to parts of the system, from the entire system to its components, and vice-versa.
- Rule two: the systems thinker must not limit their focus to just observations of what appears constant but also "search for what varies;" the changing variables are what is interesting to the systems thinker.
- Rule three: to understand change and the underlying reality, effort must be made "to understand the cause of the variations in the variables we observe;" and the chains of causal relationships between the connected variables must be formed.
- Rule four: just searching for the causes of variation is not enough; it is also essential to "link together the variables in order to specify the loops among all the variations."
- Rule five: when making observations, we must "always specify the boundaries of the system we wish to investigate."

The first rule is interesting since it calls for more than just top-down or bottom-up approaches when it comes to systems development and integration. Rather, a top-down and bottom-up strategy is required, sometimes simultaneously. To create a design instance that is coupled to bottom-up design elements, design threads that push the design down to lower levels are created. This method frequently reveals a range of considerations and issues that are beneficial to the overall developing program. It the design thread is implemented digitally; we have the so-called "digital thread" which links the lower design levels to the higher design levels through the digital model.

Another Systems Thinker was Gharajedaghi whose contributions toward Systems Thinking included providing the four foundations of Systems Thinking shown in Figure 1.11.

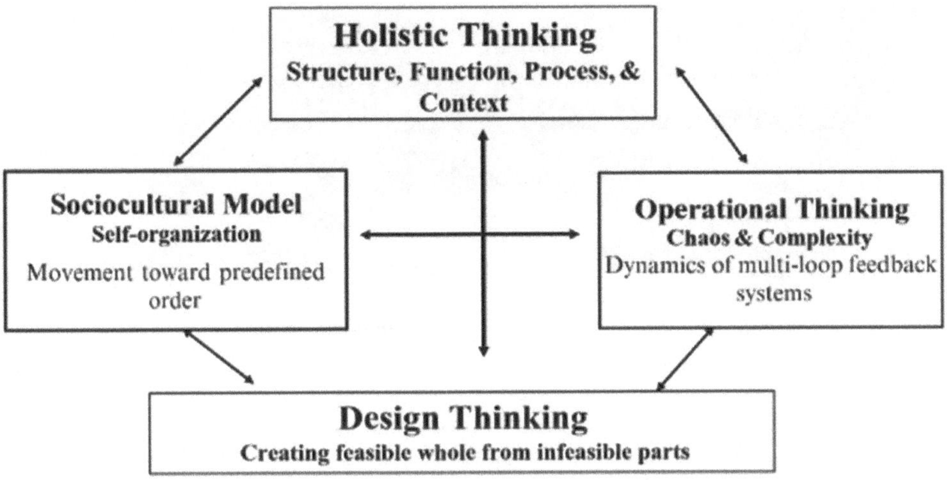

FIGURE 1.11 Foundations of systems thinking according to Gharajedaghi (2011).

At the top of the figure is Holistic Thinking which is a key foundational element of Systems Thinking. What elements make up the entire system? What elements make up its structure? What functions must the system accomplish? These types of questions stem from Holistic Thinking. The contextual elements of the system must also be addressed along with the environmental conditions the system finds itself in, along with the necessary processes that need to be accomplished. The focus is on the "big picture" elements of the system along with how the system is interconnected with other systems and the system's completeness. In the sociocultural model, the idea of self-organization is indicated along with movement toward predefined order. Self-organization has connections to the concept of emergence in complex systems but is separate and distinct. On the right side of Figure 1.11, there is Operational Thinking which involves considering the operational aspects and operational context of the system along with its dynamics and feedback mechanisms including multiple feedback loops. Some interesting and applicable concepts include chaos theory, which has its uses in understanding system stability, or system instability. The idea of system complexity also is considered here. We will provide further detail on systems complexity below. Design thinking, which deals with bringing the system together from its elemental pieces to achieve its larger purpose, is seen at the bottom of Figure 1.11. There is a relationship between holistic and systems thinking. In a way, systems thinking is used to achieve holistic thinking by first focusing on the issue at hand and then iteratively expanding the focus outward to include other areas.

Mugadza stated that,

> System Thinking aims at being holistic by following a method whereby the understanding of a system starts from the apparent issue and widens the system's boundary by expanding the circle to include those other factors that may not be so apparent but have an influence on and are connected to it.
>
> **(Mugadza 2015)**

On the other hand, holistic thinking has a systems thinking element that has a systemic way of looking at a system from nine different viewpoints (Kasser 2010),

1. Big picture	4. Structural	7. Temporal
2. Operational	5. Generic	8. Quantitative
3. Functional	6. Continuum	9. Scientific

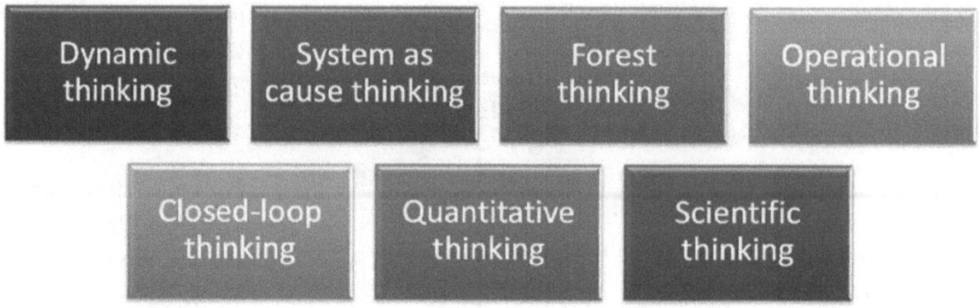

FIGURE 1.12 Seven thinking tracks of systems thinking (Richmond 1994).

The viewpoints shown here provide different perspectives that help the systems thinker gain useful insights into the nature of the system. For example, the big picture perspective provides a macroscopic view that enables the systems engineer to understand the whole system instead of getting stuck in lower-level details. Richmond (Richmond 1994) and Maani (Maani 2004) suggested that there are at least seven thinking tracks that require simultaneous attention (Figure 1.12). In the top-left box entitled Dynamic thinking, we draw a connection with the application of the behavior over time aspects of the system. A problem or issue can be framed in terms of behavior patterns that can vary over time. This type of thinking emphasizes that the concept of time scale needs to be included in the contextual interpretation of the system. We can build further on the dynamic thinking concept in the System-as-cause thinking box wherein this consideration enables plausible explanations for the patterns of behavior observed with the dynamic thinking block. In system-as-cause thinking, the argument is that relationships that can't be controlled by decision makers should be eliminated from consideration (Richmond 1994, p. 6). Forest thinking is what is meant by seeing the "big picture" in that the focus of thought is on macroscopic elements and lesser on fine detail (the trees themselves in this analogy). This type of thinking provides the ability to rise above functional silos and looks at the relationships that tie the system components together. With Operational thinking, the focus is on causality evaluating "how" behavior is generated. Some useful concepts in Operational thinking are to think about how things correlate, or how things influence each other. Maani says that operational thinking considers the structure and/or physics of relationships, and the effects of variables on each other (Maani 2004). Closed-loop thinking focuses on closed-loop structures. With this type of thinking, causality does not just run in a single direction, but includes feedback mechanisms and effects that may impact and change one or more of the causes. The causes can also directly impact each other. Quantitative thinking involves thinking about things in terms of measurable events wherein the ideas of forecasting and prediction come into play. An example would be a statistical description of atmospheric variations permitting predictions and forecasting to be made. Odds and likelihood can play a significant part in this type of thinking. Scientific thinking involves the use of science and scientific principles, investigational processes, and testing to understand the system. Richmond developed a ranking system that placed these different thinking categories in order from the highest (most comprehensive) category starting with 1, and the lowest level (most detailed) category ending in 5. These categories were ranked as follows:

1. Forest thinking,
2. Closed-loop thinking,
3. Operational thinking,
4. System-as-cause thinking,
5. Dynamic thinking.

Systems Thinking offers various benefits that help systems engineers throughout the SDLC. Some of the more significant advantages of systems thinking include the ability to integrate the development team, understand the aim of the developmental effort, its processes, and establish performance expectations. Systems Thinking aids in understanding the system in its environment and its relation to other systems, understanding system complexity and difficult challenges, and optimizing system performance. Furthermore, Systems Thinking aids in avoiding, minimizing, and/or mitigating unintended consequences, managing uncertainty and risk, aligning teams and disciplines, and identifying, understanding, and reducing impacts. All of this saves time and money!

1.1.8.2 Complicated, Complex, and Adaptive Systems

In this section, we consider and define the terms for *complicated, complex*, and *adaptive* systems. We also describe the attributes and characteristics of these types of systems to gain an understanding of some of the additional challenges that the system engineer must face and overcome. Often, the terms complicated, and complex are interchanged and confused with each other, but these types of systems are fundamentally different. Let's start with a preliminary set of definitions to provide a starting foundation for the material that is presented in this section (Table 1.1). Understanding complex systems is important because more modern systems are becoming complex because of the increase in the number of connections within systems. The impact on the system engineer is that instead of the traditional decomposition approach in analyzing individual subsystems, for complex systems the system's behaviors and patterns that the system creates are evaluated and characterized. Additionally, not all systems can be understood. This is especially true if the system is complex. There simply can be too many connections and interdependencies to grasp how the system works. Instead, patterns from the system's behavior are used to understand the system. Some measures of complexity may also be used, such as Logical Depth and Effective Complexity which are complementary quantities (Ladyman and Lambert 2012). Logical Depth is a measure of the complexity of individual strings of a computer algorithm that is used to recreate some information. It was developed by Charles Bennet using computational complexity (Bennett 1988). Effective Complexity attempts to determine the amount of non-random information in a system. Other complexity measures such as Kolmogorov Complexity of an object (Kolmogorov 1965) relate to the shortest computer algorithm that produces that object as an output. Self-dissimilarity relates to patterns of a system wherein observations of the system at different scales provide a "signature" of the complexity of the system (Wolpert and McCready 2004). A variety of other complexity measures exist and will be introduced where and as needed.

We now look at the distinction between the words Complicated and Complex since these words are often misunderstood. Let's first use a piano as a starting example of a complicated system. In its state as a complicated system, the piano has many parts that work together to make beautiful music (hopefully). The piano does this by using a hammer that strikes a string which is set to oscillate at

TABLE 1.1
Useful Definitions in Understanding Complicated, Complex, and Adaptive Systems

Term	Definition
System	A system is a combination of interacting elements integrated to realize properties, behaviors, and capabilities that achieve one or more stated purpose(s) (Dickerson and Mavris 2010)
Systems architecture	The fundamental and unifying system structure defined in terms of system elements, interfaces, processes, constraints, and behaviors (INCOSE Systems Architecture Working Group ND)
Complex system	A system that has the properties of connectedness, interdependence, diversity, and adaptability and that can exhibit properties of emergence, memory, and self-organization
Complicated system	A system with many connections that can be hierarchically structured and understood and does not exhibit the properties of emergence and self-organization
Adaptive system	A system that can change based on external or internal influences and behaviors
Model	An approximation, representation, or idealization of selected aspects of the structure, behavior, operation, or other characteristics of a real-world process, concept, or system (IEEE 1990)

a particular frequency. There are 88 keys on the piano that are all tuned differently. The keys are made of individual pieces of specialized materials. There are pedals for dampening the sound or sustaining it. There are mounts, and brackets, and interfaces, and screws to hold things in place. The cabinet itself which holds the various pieces is also specially designed and structured to give the piano a unique and beautiful sound. Notice that regardless of how many parts there are, we can itemize them, categorize them, and put them in a physical and structural hierarchy. Consequently, we can completely define, and understand the system by conducting a top-down decomposition of the system, a hallmark of complicated systems. Another example of a complicated system is a car which can be broken down into smaller definitive pieces such as wheels, engine, exhaust subsystem, braking subsystem, safety subsystems, and so forth. In turn, each of these subsystems can be decomposed into smaller components pieces or parts such as screws, fasteners, and wiring until the entire car system is fully defined, characterized, and understood.

With regard to complex systems, Poli states that even infinite knowledge of the system does not provide understanding of the system, nor does it provide the ability to model the system (Poli 2013). Let us now look at an example of a complex system by placing the piano inside a concert hall. Let's begin by considering some of the complicated aspects of the system. We have the physical structure of the auditorium with the relative positioning of architectural and acoustic components, the propagation, dampening, reflection, and interference effects of the acoustic waves. All these things can be considered as a finite number of components that can be enumerated and characterized, and to some level, this system can be considered to be complicated and not complex. If we now expand the system definition to include the entire orchestra providing a moving experience to the listener, we now have something that defies simple decomposition, and classification. One might argue that to get an exceptionally moving experience in the concert hall, the musicians play off each other, "feel" the music, and contribute with their own talents and superb skills so that the sum of the individual contributions is greater than the individual parts, and what emerges is a truly exhilarating, musical experience. Arguably, this is even more the case in an interactive concert such as in rock music where the audience participates and provides feedback to the musicians. This interaction can make members in the band, or in the audience do unanticipated things (such as launching themselves into the audience expecting to be caught). The idea of unanticipated emergence, among others, is a key element in complex systems. Complex systems can be thought of as having four characteristics as shown in Figure 1.13. We begin by describing each of these. For the "Connectedness" characteristic, we often consider systems in terms of a hierarchy (top to bottom ordering of the system elements). In Figure 1.13, *connectedness* as the name suggests, means that there exists some sort of connection between the elements of the system. This connectedness can be in the form of a physical connection, like a wire in a circuit diagram, or something else entirely like a behavioral connection when an angry, large dog comes racing across the yard, straight at you, teeth bared, snarling, snapping, and barking up a storm. The rational response being to either jump back in your vehicle, slam the gate in the dog's face, or race up the nearest tree or object that separates you from your not-so-friendly

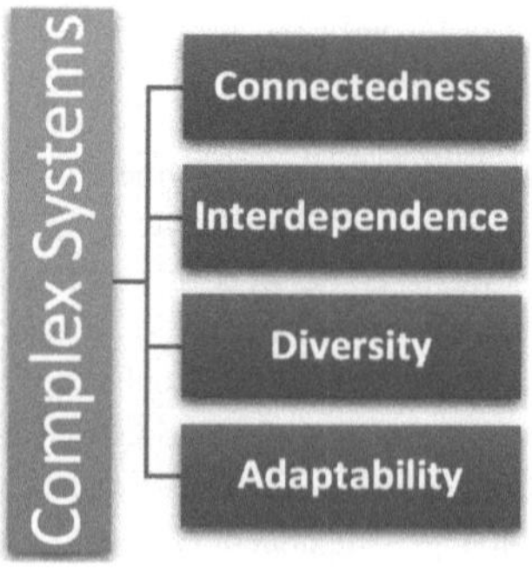

FIGURE 1.13 Four characteristics of complex systems.

new acquaintance. As the number of connections within the system becomes large, the system can become complex. Some examples are the motion of the seas, winds, or the interaction of the billions and billions of stars in the heavens (thank you, Carl Sagan). A more down-to-earth example is the internet itself with its myriad of connected devices and networking hardware. The sheer number of connected elements makes it impossible for anyone to understand the entire system.

For the "Interdependence" characteristic, this is where things become interesting. In addition to having the connections themselves, in the complex system, it is important how those connections interact and depend on each other. For instance, when a vehicle is driven on the road, the driver is aware of his or her surroundings and other vehicles. Each of the drivers is influenced by and reacts to what is going around them. There is an interdependence between the drivers. If they all come up to a four-way stop at the same time, they may all freeze and wait for someone to go. Or, they all go at once, or some combination. Whatever they choose, their actions affect the other drivers and so these entities (drivers in their cars) are interdependent. *Connectedness* and *Interdependence* can be combined into "Connectivity (Chan 2001)." Perhaps the concert hall example would be a good reflection of this combination. Although the artists likely have the largest contribution to the perceived satisfaction of the audience, the audience is connected at some level to the artist, and both influence each other through real-time feedback.

The third element in Figure 1.13 is Diversity. For this concept, we consider the contributions of different elements of the system and how these different elements affect the overall system both as intended, but also in unintended ways. Often, there can be multiple independent control structures within the system itself that can interact in unintended ways. For example, in the concert hall, each member of the audience has their own specific satisfaction criteria and has some control over this satisfaction (sit, stand, move closer, cover your ears, relax with a beverage of choice, ….). This is referred to as decentralized control (Chan 2001). If the system has decentralized control, there is no single entity controlling the system's output. As before, the dynamics of the earth's atmosphere or oceans are functions of innumerable forces and factors that serve as inputs and outputs to these massive, complex systems, wherein the system itself is made up of innumerable, connected, interdependent, diverse elements.

The final characteristic, *Adaptability*, refers to the system's ability to change its state in response to some stimuli. If a hardware node is compromised on the internet or degraded performance is detected, the node in question may often be automatically bypassed in favor of a better (faster, more reliable) node. Regarding complexity, a related concept is that of emergence. Emergence is a signature attribute of complex systems that can result from the four elements of Figure 1.13. According to Ladyman and Lambert, "It is attractive to define emergence in terms of increase in complexity (Ladyman and Lambert 2012)." The field of Game theory has a lot to say about emergence. Game theory can be considered to describe decision-making under competitive circumstances. Each decision-maker tries to optimize conditions so that they maximize their competitive results and minimize the results for their competitor. The assumption is that there are underlying patterns in the decision-making process that are waiting to be found. This pattern may be illusive but can be determined given enough observations. We will discuss Game theory further in the Decision and Risk Chapter of this book. One example of emergence in a complex system is spontaneous order which is order that arises in a complex system which has less than a complete set of rules. In this fashion, the workings of the system are, "not completely random but also not completely ordered (Killingback and Doebeli 1996)." An example of the emergence of spontaneous order is when a distinct pattern is not seen by the observers of a system, but it is in fact there.

Complex systems then can be characterized as having the ability for self-organization and may have emergent properties. Complex systems can also have memory in that even after an obstacle is removed, the system still behaves as if the object was there. An example of this type of behavior is a traffic jam due to an accident. Even after the vehicles involved in the accident are removed, the traffic jam can persist for a while. Complex systems can also give rise to some counter-intuitive outcomes, such as Braess' Paradox wherein adding capacity to a network can actually degrade

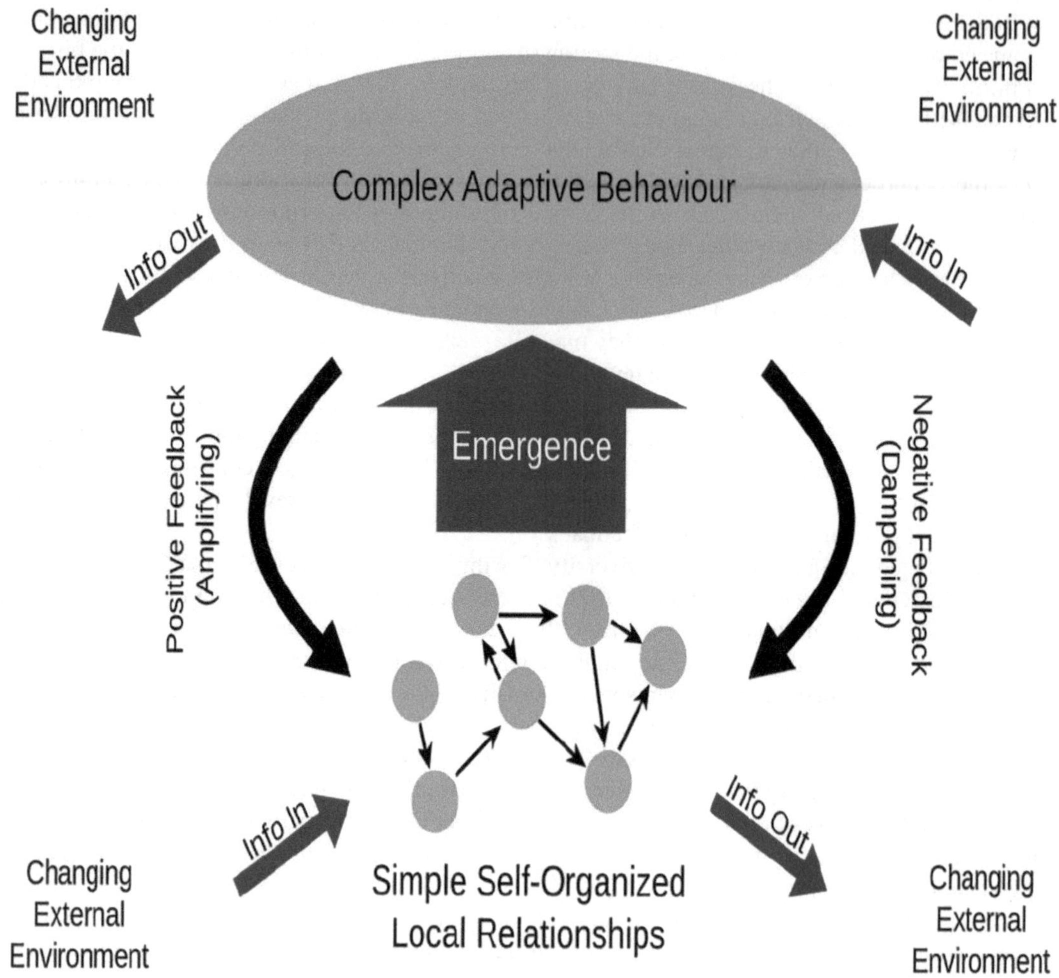

FIGURE 1.14 Complex adaptive systems.

network performance. Much of this can stem from the ideas of the system consisting of lots of diverse, adaptable, interconnected elements that have decentralized control and interdependence. In contrast, a *complicated* system can still have many interacting elements but without emergence or the ability for self-organization. An example of a complicated system is a modern transportation vehicle that can be made up of interacting elements (even systems) such as the fuel system, electrical system, transmission, safety systems, and more. In contrast to a complex system, where a complete system decomposition may not be possible or practical, a complicated system can be repeatedly broken down to lower levels until the system is fully defined and understood. Complicated systems can have aggregate properties such as speed, range, carry weight, and reliability for the transportation system just discussed, but complicated systems do not have emergent properties. Unexpected outcomes can still occur with complicated systems, but their source is different than those for complex systems (i.e., do not result from emergence). It is interesting to note that although the transportation vehicle itself is a complicated rather than a complex system, the larger transportation system itself, made up of many vehicles with decentralized control, can be a complex system as evidenced by the traffic jam example. If the complex system is allowed to be adaptive, then Figure 1.14 shows the interesting dynamics.

We see that the ideas of emergence and self-organization are captured at the bottom center of the figure and are the hallmarks of complex systems. We also see at the left and right of the figure, the interaction between the complex system and its environment is shown by information flow in and

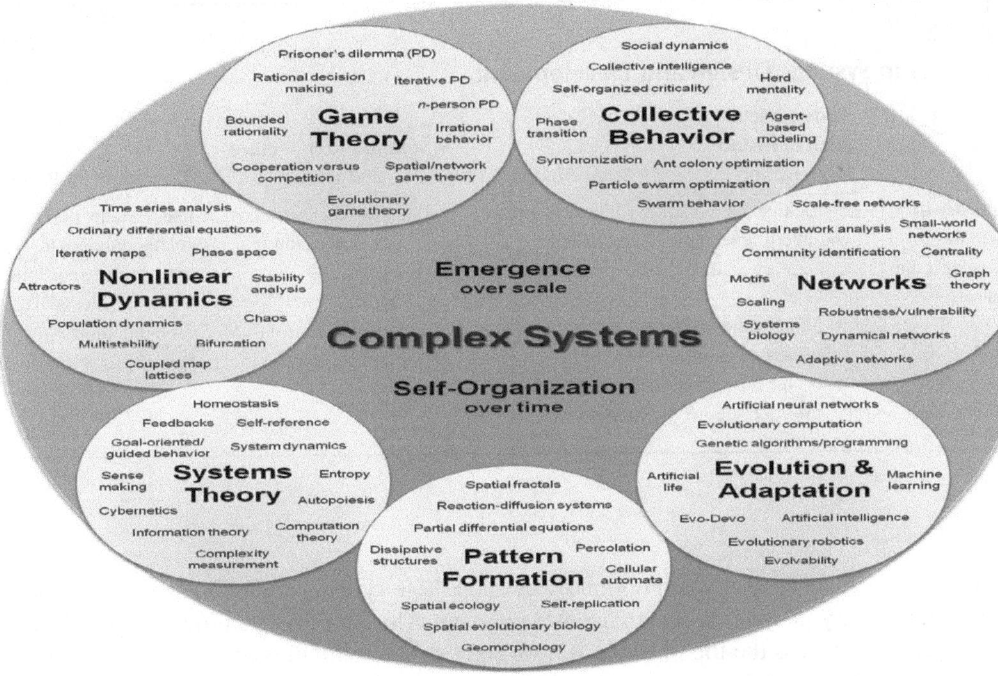

FIGURE 1.15 Complex systems organizational map.

out of the self-organizing elements of the complex system (bottom left and right), and the adaptive complex system itself (top left and right side of the figure). Changes in the external environment influence the complex system and the system reacts to the stimulus through simple self-organization using local relationships. During the self-organization process, unexpected system responses can "emerge" that affect the overall behavior of the complex system. As a result, the complex system can provide positive feedback to the self-organizing elements of the system that amplifies the emergent properties, or negative feedback that dampens these emergent properties. The changes in the complex system are observed through its adaptive behavior and interaction with the environment. Figure 1.15 presents an interesting view of complex systems from an organizational point of view. We see in the center of the figure that the emergence property is seen over the scale of the system whereas the self-organization is seen over time. To successfully deal with complex systems, we see the foundational elements that are necessary to properly understand and develop these kinds of systems. These foundational elements need to be keenly understood by the practicing systems engineer at a theoretical, mathematical, and applied level, and supported with a robust modeling and simulation capability, in a mature system engineering development environment. Some of the expected foundational elements are a strong foundation in systems theory to include such elements as information theory, complexity analysis and measurement, system dynamics, and computation theory among others; nonlinear dynamics including chaos theory, multi-stability, and time series analysis, and Network theory including adaptive networks, dynamical networks, robustness, centrality, scaling, and systems biology. However, equally important, and complementary are other foundational elements such as methods for understanding the formation of patterns, systems evolution and adaptation methods, methods for understanding collective behavior, and decision-making methods such as making decisions in competitive environments as in Game theory. By looking at the sub-elements associated with these foundational elements, it is seen that quite an extensive toolkit is required to understand and successfully manage, design, develop, implement, support, dispose-of, and ultimately understand the adaptive complex systems that are becoming more prevalent in society at large.

TABLE 1.2
Complexity in Systems Design and Development

	Requirements Elicitation and Derivation	Trade Studies	Solution Architecture and Design	Development Process
Complexity in system design & development (general)	Use multi-scale modeling (linking macro- and micro-level models), including exploratory analysis and experimentation: • To generate insight into the implications of derived requirements • As the basis for trade studies and to inform trade-off decision		Emphasize selection of robust and adaptive elements and structures over optimizing to meet current requirements	Use Systems-of-Systems methodologies to synchronize constituent systems; incentivize collaboration. Ensure prototyping and experimentation are used.

How then do we successfully and optimally develop these systems and consequently how do we deal with complexity in systems design and development efforts? Table 1.2 provides some insights into this question. Across the top of Table 1.2, we see essential activities/processes within the system engineering discipline that are strongly impacted by complexity in systems, and for which there are methods to successfully deal with this complexity. In the gray area, there are potential solutions and areas of emphasis presented for dealing with complexity in the corresponding SE activities/processes. For instance, when engaging in the SE activity/process of requirements elicitation the use of multi-scale modeling methods is useful in linking macro-level models with micro-level models. Exploratory analysis, agent-based modeling and experimentation can be used to get insights into requirements and trade studies. When establishing a solution architecture and engaging in design activities, the emphasis for complex systems should be on robustness and adaptive elements and structures rather than optimizing to meet current requirements as in traditional SE development efforts. This does not mean ignoring the requirements, but rather emphasizing different elements in the solution space. Using experimentation and prototyping in the development process along with system-of-systems methods to synchronize efforts and incentivize collaboration is also recommended. There are a wide variety of considerations tools, methods, and solution concepts available to the systems engineer. Figure 1.16 shows some of these considerations and solution concepts and indicates the holistic nature of the SE discipline. For each of these core SE solution concepts, there are corresponding SE principles, methods, tools, and techniques that will be presented and further developed throughout this book, and that need to be mastered as part of the SE practice. Figure 1.16 provides some insight to the holistic set of expected core competencies in the SE profession and highlights the diversity in the SE profession. In addition to the core SE solution concepts, there is a set of building blocks that are necessary for modeling contemporary systems. The building blocks are shown in Figure 1.17 and are illustrations of useful ways that the systems engineer can deal with difficult systems development efforts such as large, dynamic, complicated, complex, and or adaptive systems. The concept of EA deals with "A coherent whole of principles, methods, and models that are used in the design and realization of an enterprise's organizational structure, business processes, information systems, and infrastructure (Lankhorst 2013)." An example of an Enterprise-level modeling tool is CATIA's (NoMagic's) System-of-Systems Modeler which can be used to capture/model/define system-wide artifacts; Incorporate requirements tracking through development phases and Verification and Validation (V&V) during delivery; enable model-based SE (MBSE) themed development of a comprehensive systems model; support system and subsystem functionality development; and allow system-wide functionality modeling and behavioral simulation. There are a variety

- Enterprise Architecture
- Systems Engineering
- Model Based Systems Engineering
- Complex vs Complicated
- Architectural Framework
- Types of Modeling/Models
- Measures and Metrics
- Requirements Engineering
- Standards and External Models
- Functional Analysis
- Requirements Allocation
- Life Cycle Costing
- Reliability Centered Maintenance
- Logistics & Support
- Intellectual Property
- Requirements Leveling
- Dynamic Systems
- Adaptive Systems
- Linear/Non-linear Systems
- Stochastic Systems
- Intelligent Systems
- Fuzzy Systems
- Quality
- Functional/Non-functional requirements
- Configuration Control
- Risk Management
- Systems Thinking
- Holistic Thinking
- Systems Architecting
- Embedded Systems
- System Development Life cycle
- System Architecting
- Modeling (Enterprise, System, Conceptual, Physical, Mathematical, ...)
- Optimization Methods
- Decision Engineering
- Statistical Methods (DoE, SPC, ...)
- Process Control
- Analysis/Analytical Methods
- Design
- Production/Automation
- Decision Engineering/AI/ML
- Human Factors
- Disposal/Retirement
- Metrics
- Systems of Systems
- Digital Transformation (digital twin, digital thread)
- Agile
- DevSecOps
- Verification/Validation
- Big Data
- Safety/Security
- Education/Training
- Environmental Issues
- Leadership/Ethics/Society ...

FIGURE 1.16 Some core systems engineering solution concepts.

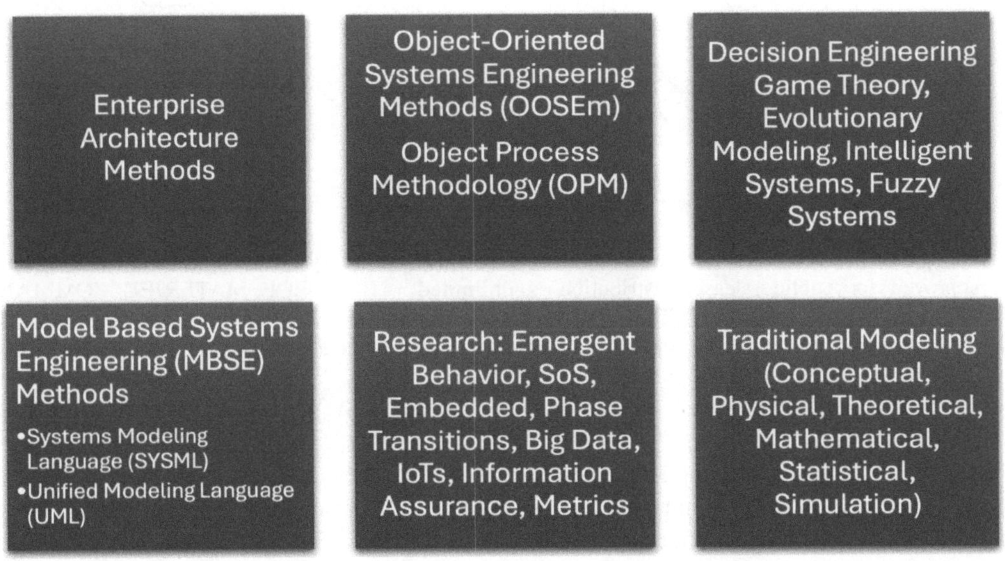

FIGURE 1.17 Building blocks for modeling large, dynamic, complicated, complex, and/or adaptive systems.

of well-developed enterprise, system-of-systems, and systems-level modeling tools that provide essential systems architecting capabilities, and provide necessary artifacts for controlled, optimal systems development efforts. Figure 1.18 shows a conceptual example of how these MBSE modeling tools can be used to integrate SE concepts and activities throughout the SDLC. There are many excellent tools available based on individual and corporate preferences. At the time of this writing, the MagicDraw toolset became part of Dassault's CATIA modeling suite, which is an excellent tool for professional enterprise, modeling, systems modeling and architecting, and can handle a variety of Model-Based Systems Engineering (MBSE) tasks.

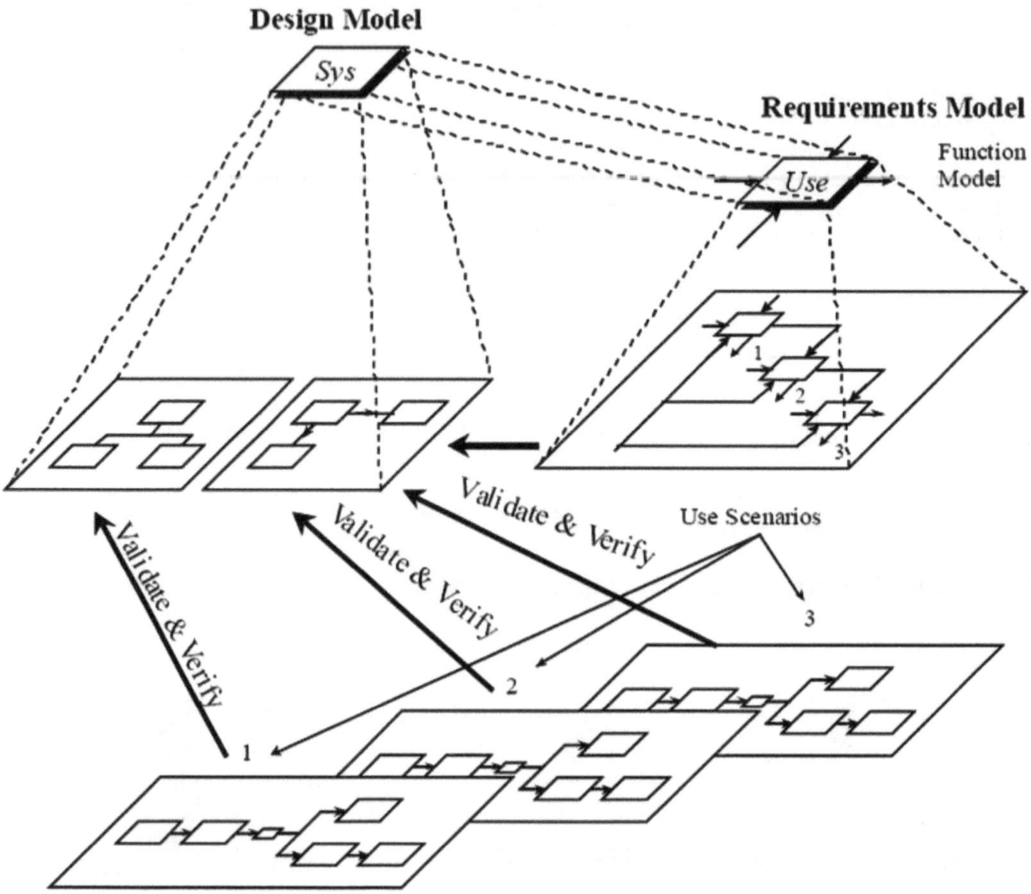

FIGURE 1.18 Conceptual modeling example using model-based systems engineering tools. (Image was approved for public release; distribution is unlimited, AIR FORCE MATERIEL COMMAND, WRIGHT-PATTERSON AIR FORCE BASE, OHIO 45433-7604. Accessed 22 November 2024 at, https://commons.wikimedia.org/wiki/File:Static,_Dynamic,_and_Requirements_Models_for_Sys_Partition.jpg.)

The Object-Oriented Systems Engineering Method (OOSEM) evolved from work done at the Software Productivity Consortium (SPC) in collaboration with Lockheed Martin Corporation. The OOSEM approach is built on various foundations, including SPC's Integrated Systems and Software Engineering Process (ISSEP), and early work related to both structured and Object-Oriented methods (Chism 2004). OOSEM uses a model-based approach to represent the various artifacts generated by these activities. As such, it enables the systems engineer to apply a disciplined approach to the specification, design, and verification of the system, and ensure consistency among the requirements, design, and verification artifacts (Chism 2004). The INCOSE pyramid shows the foundation of OOSEM as seen in Figure 1.19 (Estefan 2008). In the figure, OOSEM has unique aspects that sit on the foundations of OOSE and on the SE foundational base. At the base of the figure, core foundational SE practices such as requirements development and engineering, the SE process, trades, and more provide the backdrop for the OOSE activities in the center of the figure. Some well-known examples at the middle tier of the figure are using UML/SysML to conceptually model the system, generate Use Cases and Scenarios, and Black Box/White Box testing methods. Some unique and useful OOSEM capabilities/activities include conducting causal analysis, developing a model of

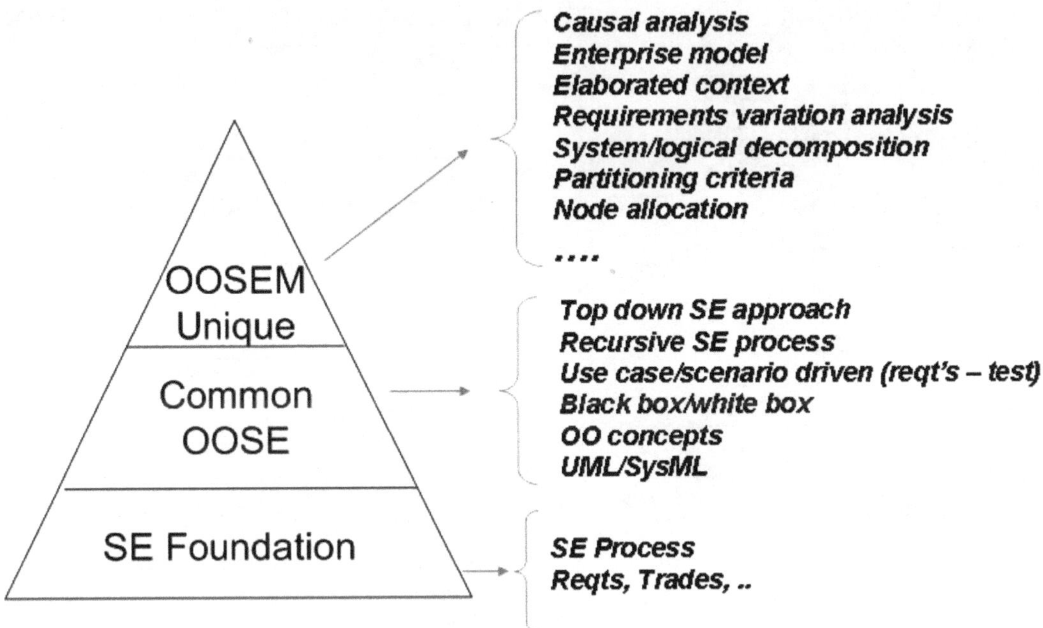

FIGURE 1.19 Foundation of object-oriented systems engineering method.

the entire enterprise, analyzing how requirements change/vary, conducting logical and/or system decomposition, and node allocation for example. The Object Process Methodology (OPM) is an integration approach that unifies the system's lifecycle stages within one single frame of reference, which can be applied in modeling of static, dynamic, and combined systems. According to Dori, OPM is founded on a universal minimal ontology, according to which objects exist, while processes transform them. The transformation includes object creation and consumption, as well as change of the state of an object (Dori 2016).

Model-Based Systems Engineering (MBSE)

> is the formalized application of modeling to support system requirements, design, analysis, verification and validation activities beginning in the conceptual design phase and continuing throughout development and later lifecycle phases (INCOSE 2025).

MBSE is all the aspects of Model-Based Engineering that deal with SE. For example: Behavior Analysis, System Architecture, Requirements Traceability, Performance Analysis, Simulation, and Test. According to Dori (2016), conceptual modeling, which often precedes or is done alongside mathematical and physical modeling, is the primary activity required for engineering systems to be understood, designed, and managed and that modeling is the process underlying MBSE. Dori goes on to show that MBSE is not just about modeling, as some people mistakenly perceive; it is SE that is based on formal modeling of various kinds (conceptual, mathematical, and physical). Thus, MBSE requires a rigorous conceptual modeling methodology that encompasses a universal ontology, a language, a set of principles and guidelines, and a supportive modeling software environment. Figure 1.20 shows some of the advantages for using MBSE as opposed to the the traditional document-based SE methods. First, instead of just capturing/documenting requirements like in a traditional requirements document, the requirements can be modeled. Requirements can be connected to each other, characterized with a set of attributes, and automatic checks can be implemented in

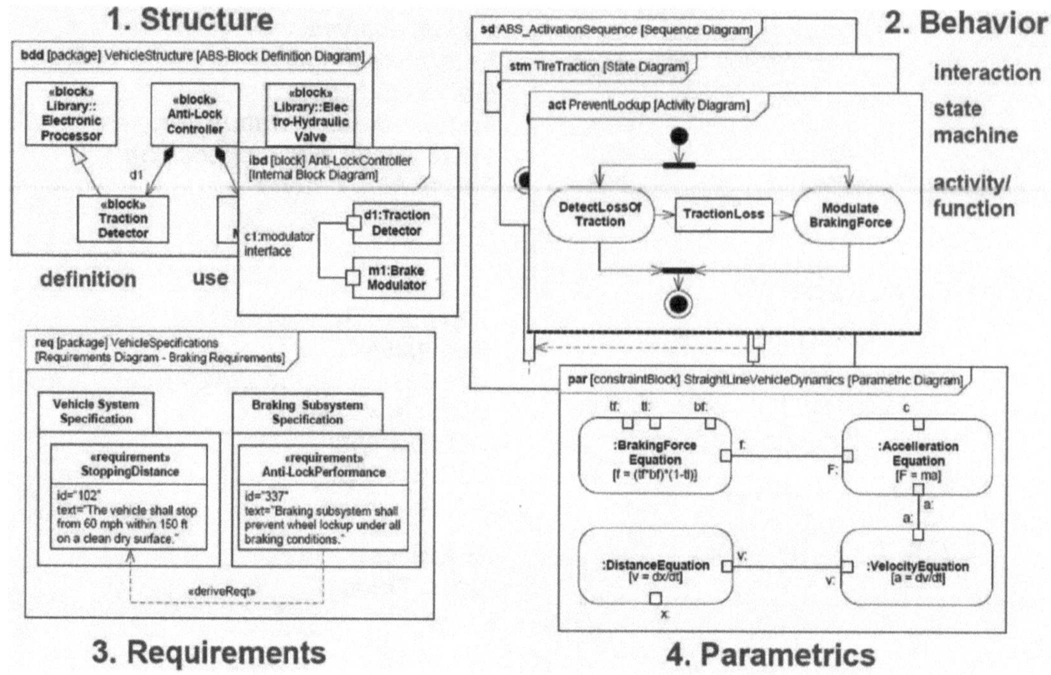

FIGURE 1.20 Conceptual modeling with SysML.

the model to ensure the requirement is enforced. For example, if the system weight cannot exceed 500 kg, an attribute can be set in the model for the requirement such that its maximum is 500 kg. A parametric model can be generated that checks component weights and makes sure the sum is less than or equal to 500 kg. This model can run "under the hood" and will let system developers know if a design element causes the weight budget dictated by the requirement to be exceeded. The requirement itself is documented in the model and is now connected by the model to other requirements and is the basis for the system design. Note also that behavioral aspects of the system can be captured by the model through sequence diagrams, state diagrams, and activity diagrams as shown in the top-right of Figure 1.20. The model can also capture structural elements of the system through block definition diagrams, and interface block diagrams and can capture/describe elements of the system definition and system use. We will develop these ideas further in later chapters of the book dealing with EA methods, systems architecture, and MBSE methods.

Evolutionary Modeling methods also play an important role in understanding complex systems. According to Krzanowski, the paradigm of evolutionary models can change the way we view, explain, and model the most fundamental mechanisms underlying our reality The "old" modeling paradigm required that we first try to guess the structure of the problem, then express it in some form of symbolism, and finally manipulate these symbols using some rules (Krzanowski 2001). Krzanowski continues to point out that in evolutionary modeling, we do the very opposite: we make our representation of the problem model itself. Kraznowski states that evolutionary modeling methods solve complex problems not by telling the algorithm how to solve the problem, but by allowing the algorithm to find the solution itself. Evolutionary algorithms came out of research into complex adaptive systems which, in terms of their structure, resemble closely concepts of natural evolution. These methods are excellent for dealing with the following kinds of problems,

1. [problems] where the interrelationships among the relevant variables are poorly understood,
2. [problems] where finding the size and shape of the ultimate solution to the problem is a major part of the problem,

3. [problems] where conventional mathematical analysis does not, or cannot provide analytical solutions,
4. [problems] where an approximate solution is acceptable (or is the only result that is ever likely to be obtained),
5. [problems] where small improvements in performance are routinely measured and highly prized,
6. [problems] where there is a large amount of data, in computer-readable form, that requires examination, classification, and integration (such as molecular biology for protein and DNA sequences, astronomical data, satellite observation data, financial data, marketing transaction data, or data on the world wide web (Krzanowski 2001)

The idea of having the model find the solution itself is powerful. An excellent example is agent-based simulation methods wherein the answer that is sought evolves from the evolution of the model itself.

1.1.9 Contemporary Issues in SE

In this section, we provide an overview of some contemporary concepts, methods, and techniques that are evolving in the SE discipline, that can have a great and lasting positive impact on future systems development efforts. There is much activity in government, industrial, and academic circles with regard to these topics, and the modern-day systems engineer needs to understand, embrace, and incorporate these concepts, methods, and techniques into their toolkit. The approach we take is to provide an introductory-level overview in the sections below, along with references for additional information and insights, and integrate the core ideas, concepts, methods, and techniques in further detail, where needed in other chapters. The topics presented include (1) the Digital Thread, and the Digital Twin, (2) DevOps and DevSecOps, (3) Agile Methods in SE, (4) Systems Architecture and Architecting the System, and (5) Standards in Design, Modeling, and Architecture. There are of course other topics that come to mind such as MBSE, which in fact has a dedicated chapter all to itself, but these topics promise to have a large impact on the field of SE and so are presented up front. With regards to all these topics, we have a way to go to reach the full potential these topics have to offer, but with proper focus and emphasis, education, and collective work we will get there.

1.1.9.1 Digital Engineering, Digital Thread, and the Digital Twin

The idea of the Digital Twin (DT) has been around for quite a while. Back in 1991, David Gelernter was the first to present this technology in a publication in *Mirror Worlds*. In 2002, the Digital Twin concept was first applied to manufacturing by Michael Grieves from the University of Michigan. Currently, the Digital Twin is seeing extensive use in manufacturing operations, the automotive industry, healthcare, and the power generation industry. There are numerous advantages in implementing the Digital Twin concept such as providing insight into the real system that is modeled, generating real-time data that can be used for research, improve the design and development activities, increase system efficiency and effectiveness, and make product refinements before going into production. Even after production, the Digital Twin can be used to maximize and sustain the efficiency of the manufacturing process. The manufacturing process can be digitally replicated and monitored. Digital Twins can also be used effectively in end-of-product-life considerations such as system retirement, disposal, salvage, and recycling.

The concept of the Digital Twin is to implement a system digitally so that it provides an electronic replica of the system that is useful for modeling, system assessment, analysis, design and development, performance projection, verification, validation, and acts like a digital surrogate for the physical system. Digital Twins virtually detail the physical system from small-scale to large-scale properties to digitally provide a "Twin" of the real-world physical system. Information that is sought from the real-world, physical system can be obtained through the DT. Since the DT is a model of the real system, the adage from George Box comes to mind, "All models are wrong, but some are useful."

The takeaway is that the DT is still an approximation (model) of the system in digital form so it cannot exactly produce results that the real system would produce under, say, testing for instance. However, the DT can still provide useful information in the sense that it provides benefits throughout the SDLC such as the capability to simulate the real, physical system, capture system evolution, and provide real-time data updates. The DT can be used to predict system performance, simulate the real system in different environmental conditions, and optimize system performance. An example would be a wind turbine that has a variety of sensors attached to gather real-time data about important system and environmental parameters related to the wind turbine and feed this information back to its Digital Twin. For example, the sensors can capture real-time performance characteristics that are critical to the turbine operations and send them to update the Digital Twin. The Digital Twin can then be analyzed to assess and/or predict the performance of the actual wind turbine or run what-if scenarios safely on the Digital Twin. The Digital Twin comes in two varieties, (1) the Digital Twin Prototype (DTP), and (2) the Digital Twin Instance (DTI). The DTP is used in the design and development activities to predict the behavior of the product under various expected and unexpected scenarios. Changes in tolerances can be expeditiously modeled and simulated and impacts on the components of the Digital Twin are readily assessed. This gives insight into how the system will perform and behave in the future and can also give insight as to whether or not the requirements of the system are being met. For the DTI, in this case, performance predictions are made on a specific set of DT components for a given realization of the manufactured product/system. The Digital Twin is kept in a Digital Twin Environment (DTE) which provides for a controlled, integrated, multi-domain physics application space in which to conduct the assessments. This digital environment can be used for predictive and interrogative assessments and can be based on past or present instantiations of the Digital Twin. There are different types of Digital Twins that represent different domains of the system. These types of Digital Twins have different magnification levels and include component twins, asset twins, unit twins, and process twins. The most basic magnification level is the component twin which represents the smallest, most detailed level functioning component of the Digital Twin. If two or more component twins are connected to each other, there is an asset twin that provides insight into the interactions between the components. Combination of components is observed at the unit twin magnification level to understand how the components work together in establishing a system with full functionality. At the highest magnification level, we have the process twin which provides insight into how multiple systems work together throughout the production facility. Effectiveness and timing studies are examples of some of the ways process twins are used for understanding and improving manufacturing and production operations.

Even though there are a lot of benefits to employing the Digital Twin concept, it isn't always practical to do so. The system needs to be sufficiently complicated, or complex, to justify the effort of implementing a real-time sensor system to measure critical system parameters. Sometimes, the resource cost to implement the Digital Twin is too high. It can be a substantial effort to develop a complete replica of the physical object and for some smaller development efforts, the cost of generating and sustaining the Digital Twin is not worth it. Some projects, however, benefit greatly from a Digital Twin. If the system is physically large and/or complex such as manufacturing projects, power equipment, automobiles, and aircraft, there is a great benefit, and perhaps even a necessity, in using a Digital Twin. Large-scale systems projects that are developed using Digital Twins often are highly successful. Benefits are also observed in using the Digital Twin to generate timelines for routine maintenance.

The future for the Digital Twin market looks very bright. There is an expected rise in the market growth for the Digital Twin for years to come. In 2020, there were 3.1 billion U.S. dollars spent in the Digital Twin market and speculations are that this number will rise to 73.5 billion by 2027 (IBM 2022). These are promising numbers and have captured the eye of savvy program management offices and business executives interested in establishing and sustaining a competitive edge for their business lines and becoming more efficient and prepared to face the challenges of developing the complex and complicated systems of the future.

1.1.9.2 DevOps and DevSecOps

The term DevOps stems from the idea of merging the development activity (DEV) with operations (OPS) in what is a continuous integration (CI) and continuous delivery (CD) framework that seeks to provide rapid functionality/capability that evolves over time. DevSecOps is an evolution of DevOps that integrates security into the DevOps process (Queen 2025). Instead of having security matters addressed at the end of the development process, security is "moved left" and made an essential aspect of DevOps. Many of the systems development efforts that use DevOps are associated with software (S/W) and information technology (IT) based development efforts, but mixed hardware and software systems developments have also used DevOps and DevSecOps methods. In this section, we describe the concepts behind DevOps and DevSecOps and show how they relate to SE and modern-day development efforts in general.

So, what is DevOps in general and what's the big deal? DevOps consists of practices and an overall cultural mindset, along with tools, that emphasize collaborative, incremental developments. Typically, the intent is to combine the software (possibly systems) development and operational teams to generate fast, high-quality deliveries and service. The capability to provide fast and high-quality products with excellent service can provide a competitive edge that leads to organizational success.

Today, software is becoming more prevalent and is often an integral part of contemporary systems development efforts such as in embedded software systems. Time-to-market can strongly affect production schedules and pressures exist for product development to be faster, better, and cheaper to stay competitive. The importance of IT is increasing with the need to process and secure vast amounts of data to provide timely information needed by the development team. In DevOps, systems development occurs in teams that are self-sufficient as opposed to traditional stovepipe (siloed) organizational structures. Since the development teams are self-sufficient, traditional bottlenecks that would be prevalent in siloed organizational structures are minimized and feedback within the self-sufficient development team is faster. Better feedback also promotes the ability to adapt better to changes and react to problems. The DevOps lifecycle has six steps which are depicted by a continuous, repetitive process represented by the infinity symbol shown in Figure 1.21.

The process starts with the *plan* step that involves engagement with stakeholders to develop a roadmap tailored to minimize problems and optimize the chances of success. The planned functionality is coded, built, tested, and then released. The planned functionality is provided to the stakeholders in the *deploy* step whereafter the operators make use of the functionality. Feedback is provided through the *monitoring* activity and is used in the upcoming *plan* to evolve the provided functionality. Throughout this, there is continuous integration (CI) and continuous delivery (CD) wherein test environments are used to ensure that the code build produces high quality. As a part of CI, central locations are used to collect code to form a code repository. This repository can be used for automated testing and debugging which help speed up the development time. For continuous delivery, automation is used in deploying the code. Once the team is satisfied with the quality, frequent deliveries are made. With continuous testing, functionality can become available at any time

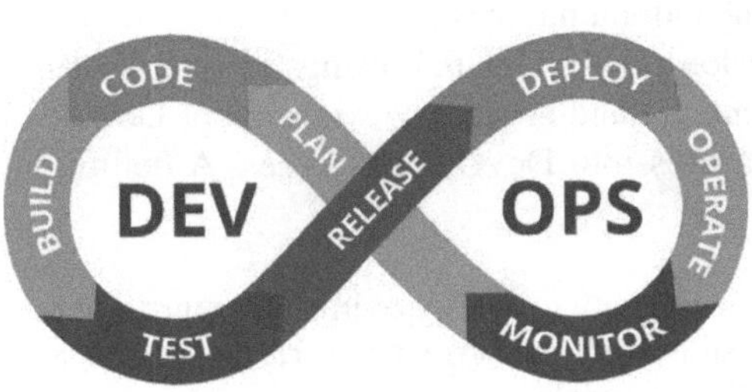

FIGURE 1.21 Model depicting DevOps. (Courtesy of AF DevSecOps 2024).

with automated deployment, which leads to continuous deployment (Cde). Some other important DevOps ideas include,

- *Shift-left*: Here testing/correcting software occurs earlier in the development process.
- *Microservices architecture*: Complex systems are reduced to smaller pieces and apps are built independently for the smaller pieces requiring less coordination.
- *Infrastructure as code*: infrastructure such as computing architecture is serviced through code which can be automated as opposed to manual processes.
- *Environment parity*: the systems environments are kept the same as much as possible between the development, staging, and production environments.

Throughout the development cycle, continuous monitoring occurs which provides for a quick response capability to developing issues.

Besides the integrated development and operations team, DevOps has an interesting view with regard to failures. Instead of attempting to "mistake proof" the development environment, DevOps understands that failures are bound to occur and so accounts for them in their development process. The goal is to engineer things so that failures are manageable and happen early in the development process. The associated culture is that failures aren't punished but embraced and dealt with. Many inventors follow this model often stating that for each mistake, something was learned.

In setting up a DevOps development environment, the DevOps team must deal with the full development cycle, repeatedly. Consequently, the skill set of a DevOps team is more general than siloed. Usually, DevOps team members have varied experiences that help them have a general perspective. For instance, coders may also have operational experience that provides technical depth as well as breadth of experience. This is like the holistic views and generalist nature of many systems engineers that have emerged from a strong engineering, math, or science technical discipline. According to CrowdStrike, DevOp teams should have the following types of specific skillsets present (CrowdStrike 2024):

- ***DevOps Engineer**: Provides for the designs of the infrastructure and also integrates the resources across the project.*
- ***DevOps Automation Architect**: Helps reduce overall workload by automating systems where practical.*
- ***DevOps Architect**: Accomplishes the overall strategic and tool design for the project*
- ***Release Manager**: Responsible for managing smooth deliveries including planning and scheduling.*
- ***Software Developer/Tester**: Adds new functionality and features, writes/updates/fixes code.*
- ***Security and Compliance Engineer**: In charge of overall security across the developmental effort the whole environment.*

These skillsets are the enablers for the DevOps methodology and are founded on the ideas of shared responsibility, collaboration, and mutual trust.

DevSecOps is an expansion on DevOps that brings in security aspects to the DevOps process. This is considered a fundamental and necessary extension of DevOps and involves the continuous integration of security measures into DevOps processes. A definition of DevSecOps is given as (CROWDSTRIKE 2023),

> DevSecOps is the practice of integrating security continuously throughout the software and application development lifecycle to ensure optimal security and performance efficiency. It is considered a necessary extension of the DevOps methodology.

DevSecOps brings in strategic and tactical integration of security methods so, unlike in many traditional development efforts where security is considered at the tail end of the development effort, security is thought about and integrated upfront and continuously. A security strategy is developed, and risk assessments are added to DevOps process steps. The addition of the security element is essential in that cloud environments and platform-agnostic containers (software and their environments) are facing increasing threats from cyber-attacks and need data and information assurance as well as better security tools, plans and procedures, guidelines, and practices. Besides the enhanced security, some added benefit of the security element is in the area of compliance. Many development programs have strict security requirements wherein adherence is mandatory. Security failures are not tolerated and can lead to the shutdown of a program.

DevOps and DevSecOps have a set of metrics that can be used to assess performance (Air Force DevSecOps 2024). According to the US Air Force, some useful DevOps metrics are,

Mean Time To Recovery (MTTRe),
Mean Time To Production (MTTP),
Average Lead Time (ALT),
Deployment Speed (DS),
Deployment Frequency (DF),
Production Failure Rate (PFR).

To these metrics, DevSecOps adds,

The ability to detect security flaws and injections.
The ability to perform fuzzing and static/dynamic source code analysis.
The ability to monitor container security including container base images and libraries.

Note that the context needs to be understood with some of these metrics so that they aren't confused with conventional SE terms. For instance, the term Mean Time To Recovery if presented with the logical acronym of MTTR, could be confused with the common SE metric of Mean Time To Repair (MTTR). For MTTRe, the recovery term means how long it takes an application to recover from a failure. The MTTRe metric then is the average time it takes for all the applications to recover from errors. MTTP is the average time it takes code to get committed to the code repository. ALT relates to a new requirement and the time it takes to get delivered and deployed. Because of the DevOps and DevSecOps emphasis on automation, CI, CD, and Cde, this metric has the potential to become very small and directly captures the potential benefits of the DevOps and DevSecOps methodology. DS relates to the speed between the test and staging areas and production. Similarly, DF represents how many times a new release is deployed to test, quality assurance, and production. PFR represents the rate software fails in production. For the DevSecOps contributions, injections are cyber-attacks where untrusted inputs are used as the basis for the attack. Fuzzing is a testing method where incorrect, invalid, and/or unanticipated inputs are injected into the system to find problems and vulnerabilities. And, as was stated above, containers mean software and its associated operating environment that can be used in a platform-agnostic setting. The container has all the necessary libraries, binaries, executables, and configuration files to function as intended in different computational settings. An example is the virtualization of an operating system. Kubernetes are platforms that run and manage containers and can have their own central processing unit (CPU), memory, process space, file system, and more. Kubernetes can resemble virtual machines.

In connecting DevOps and DevSecOps to SE principles, methods, tools, and techniques, it must be understood that DevOps and DevSecOps have their foundations in Lean and Agile methods. So, the connection to SE must be made with that perspective in mind. Implementing DevSecOps provides a direct benefit to SE activities and applications associated with systems, services, programs, and processes (SSPPs) that are well-suited for an Agile framework. We expand on this suitability in the next section. Conversely, DevOps and DevSecOps teams can benefit greatly from having SE representation. Particularly, the SE propensity toward holistic thinking, understanding complexity, and embracing Model-Based System Engineering (MBSE) concepts and methods can provide great benefit. Some additional guidance for how to integrate DevSecOps into a system-of-systems environment is given by the Software Engineering Institute (SEI) in their Guide to Implementing DevSecOps for System-of-Systems in Highly Regulated Environments (Morales et al. 2020).

1.1.9.3 Overview of Agile Methods in SE

Agile methods have made a positive impact in many modern development efforts, not only for software-dominant systems but also for some hardware-based systems. If schedules are tight and the development effort can be broken into parallel, incremental stages, wherein each stage provides some sort of new functionality to the stakeholder, and further iterations provide increased functionality, then Agile methods may provide a better approach than the traditional waterfall development structure. In this section, we present an overview of Agile methods, some background, and fundamental characteristics of Agile methods, and some of the impacts Agile methods have had on modern-day development efforts. Later, we provide a more detailed treatment of Agile methods in a separate chapter on Agile Methods in SE. Merriam-Webster defines *agile* as, "able to move quickly and easily," or "quick, smart and clever." In systems development efforts, *Agile* is often associated with software development efforts largely due to the inherent nature of software projects to provide early functionality with the possibility of parallel, incremental improvements. In contrast to Agile, the traditional SE waterfall model uses a sequential development path that breaks the system development into three progressively detailed stages, wherein the stakeholders do not see the product until all the features have been designed, developed, built, and tested. The waterfall model appeared in the software literature, though not by its name, in 1970 by Dr. Winston Royce who was investigating the management of developing large software systems (Royce 1970). The credit for the term Waterfall Model is popularly given to a paper by Bell and Thayer in 1976, where the term Waterfall Model was associated with the Waterfall Method as shown in Figure 1.22.

One point that was made by Dr. Royce is that design iterations are not just confined to successive steps, and he concluded that this type of development was the most ideal given the software development cycle of the time. For the DOD, MIL-STD 490A detailed the specification practices, and correspondingly the sequential, staged waterfall development structure of that day. In 1987, it was noted that 30% of development cost was from breakages of the waterfall model (McLaughlin 1987). The upside of the Waterfall Method is that it is driven by requirements and can progressively adapt to changes. Table 1.3 shows a comparison of some development characteristics between the Waterfall and Agile methods (Kamarani 2005). Notice that Agile outperforms the Waterfall method in all categories except the Requirements-Driven category. The last category, however, is an important one since having the development effort driven by requirements lies at the very foundation of the SE discipline. Agile development methods work best for efforts that have a large amount of uncertainty, or where requirements changes abound. However, Agile methods don't work on all development efforts and require certain system development characteristics to be in effect. These characteristics include (1) the development effort must be *partitionable* into parallel complementary efforts, (2) the individual parallel efforts must provide *successive incremental functionality* so that useful functions are provided to the stakeholder early on, and in subsequent incremental development sprints, (3) the functionality provided by the parallel teams must be capable of *progressive functional integration* wherein the functionality provided in earlier parallel sprints can be successfully integrated along the way, (4) the Agile development effort requires good communication between parallel teams, and strong leadership to successfully coordinate development activities, and (5) the Agile development effort must be in a *schedule driven environment* wherein there is strong pressure to deliver the system quickly. Not all development efforts have these conditions in place and not all efforts will benefit

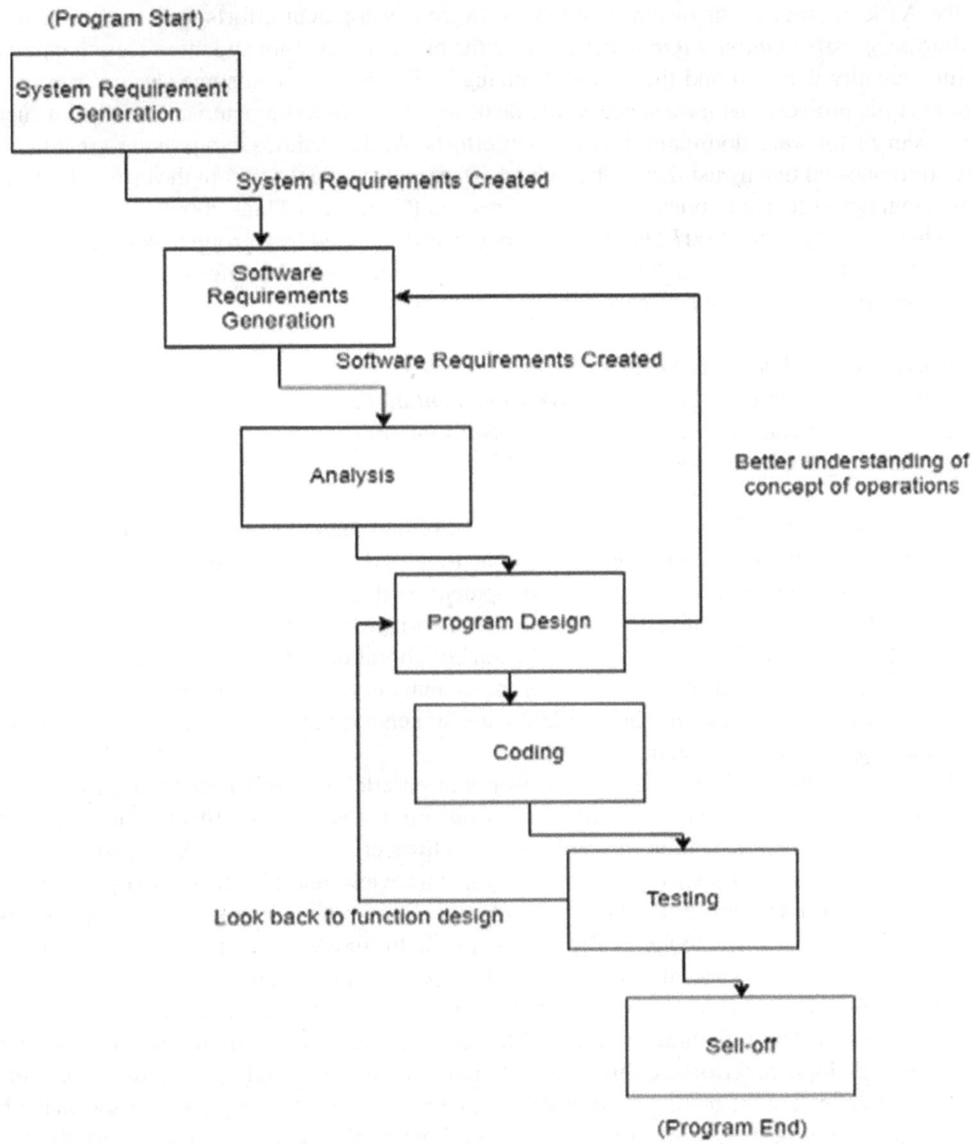

FIGURE 1.22 Illustration of waterfall methodology.

TABLE 1.3
Methodology Comparison

Methodology	Easy to Implement and Prototype	Early Customer Involvements	Ease of Engineering Changes	Meets Customers Schedule	Requirement Driven
Waterfall	– –	+	+	– –	+
Agile	0	++	+	++	–

++, Strong; +, average; 0, neutral; –, low; – –, very low.

from the Agile approach. For instance, some hardware development efforts may have difficulty in providing *progressive functional integration* since the physical structure and interfaces change every time functionality is added and the hardware changes. There have been some success stories with hardware Agile projects that have successfully dealt with these five characteristics, but most success stories occur in software-dominant development efforts. Agile methods can provide great benefits. A case study showed that by using distributed and offshore teams, only 10% of the original estimated full-time manpower to accomplish a project was needed (Cohen and Thias 2009).

So, what is the Agile method? The Agile method was developed by a group of software practitioners in 2001 in their proclaimed Agile Manifesto (2001). The Agile Manifesto emphasized certain key development principles over others and are restated here,

1. *Individual and Interaction over Processes and Tools,*
2. *Working Software over Comprehensive Documentation,*
3. *Customer Collaboration over Contract Negotiation, and*
4. *Responding to Change over Following a Plan.*

Although this seems at odds with the traditional SE, requirements-driven development approach, that encourages the attributes on the right side of the four Agile principles above, this is not directed at all projects. Coming from the software development world, the adoption of Agile methods provides benefits to certain types of system development efforts, where there is high uncertainty, shorter development times, and high customer interaction/collaboration (Huang 2016). Higher customer collaboration results from getting prototypes of the system out early. These prototypes provide early useful functionality to the stakeholder and allow the stakeholder to weigh in and participate in the incremental evolution of the system.

In the Agile method, a development team works in parallel with other teams to provide incremental functionality (features) through *sprints*. *Sprints* are focused efforts that provide system features over defined intervals of time of fixed length (Schwaber and Beedle 2002). Sprints have four cycles, (1) sprint planning, (2) sprint execution, (3) sprint review, and (4) sprint retrospective. In the last step, the sprint team can think about their efforts before proceeding to the next sprint. Agile has a project management framework called *scrum* that helps the Agile teams structure and manage their work. Prototyping is carried out in each of the sprints and an emphasis is placed on testing. This early testing also provides early insights into the viability of the design and ends up saving time in the long run. The early (and continuous) testing mindset may require testers to present much earlier in the development effort (or throughout it) and so adds additional upfront overhead over the Waterfall Method where the testing is done after the development effort. Figure 1.23 shows the fundamental importance of the testing as part of an Agile sprint. We see the sprint start with the design

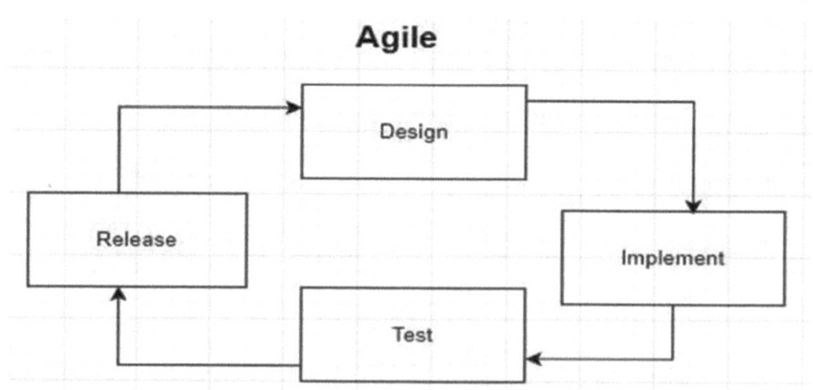

FIGURE 1.23 Release and test cycle in a sprint.

effort associated with providing the feature followed by its implementation. The implemented feature is tested and then released signifying the end of the sprint. The cycle repeats itself in the next sprint. One of the challenges in testing during a sprint is that of providing complete test coverage. Sometimes there is a gap that exists between what the stakeholder expects and what is delivered at the end of the sprint. This gap can be somewhat mitigated by collaboration and adapting test parameters (Mukherjee et al. 2021). Other challenges in implementing Agile methods include (1) reluctance to commit to a decision, (2) conflicting priorities, (3) inconsistent resource availability through the sprint, (4) decisions not acted upon, (5) lack of ownership, and (6) a lack of empowerment as was shown in a case study at the University of Ireland that studied 43 Agile developers and managers (Drury et al. 2011). An Agile technique for mitigating some of these challenges is the implementation of *dojos*, which are periodic meetings with subject matter expert mentors that focus on ways to meet challenges, learn Agile principles, and learn ways to improve the development of the system. Some techniques for increasing learning that are adaptations of Agile are Book Club, Brown Bags, Pair Programming, and Open Spaces which have been validated to increase learning (Aniche et al. 2011).

1.1.9.4 System Architecture, Executable Modeling, and Standards

In this section, we will present three concepts that have provided great utility in systems development efforts, large and small, and deserve special consideration especially when dealing with complicated, complex, and adaptive systems. So, what do we mean by system architecture? According to IEE STD 610.12-1990, "[System] Architecture is the organization of the system components, their relations to each other, and to the environment, and the principles guiding its design and evolution." A graphical view shown in Figure 1.24 is helpful in separating the constituent parts of this definition. Note that the overarching focus of this definition is organization which refers to an activity (the act of organizing). The elements that are being organized are system components, relations, and principles. The relations that are being organized refer to the relations between the system components themselves and how they relate to the environment. The principles that are being organized refer to the systems design principles and the systems evolution. In this definition, we see structural elements (e.g. how the system components are organized), behavioral elements (how the system components relate to each other and the environment), and definitions (guiding principles) and so UML and SysML artifacts are very useful in developing and representing the system architecture. Note that there is some inherent ambiguity in this definition and some inconsistency in terminology. The term *system component* refers here to a general structural element of the system, whereas in some leading system engineering texts, *components, pieces, parts* are relegated to elements of the Detailed Design and Development process. These types of inconsistencies are prevalent in definitions throughout the SE discipline, SE literature, and in standards, and are to be expected in such a young discipline. Where justified, the use of language-based processors can

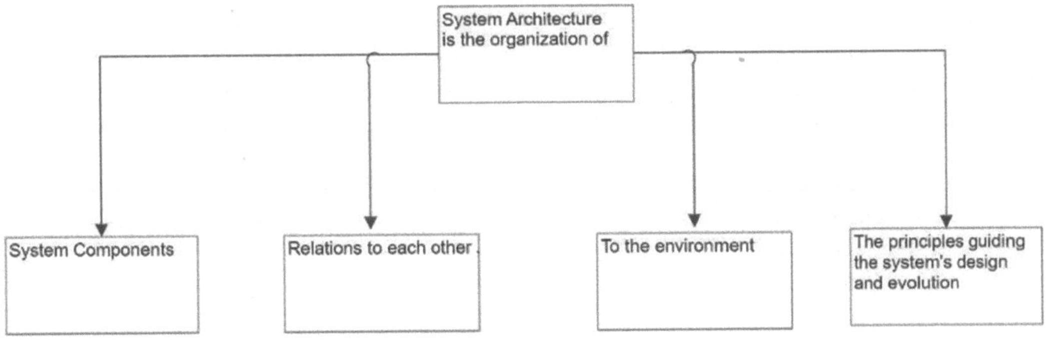

FIGURE 1.24 IEEE system architecture definition decomposition.

mitigate these differences in terminology and definition and provide proper context and definition of terms within a development effort. The system's conceptual architecture is initially established early in the Conceptual Design Phase of the systems development lifecycle (SDLC) when the system's concept has been selected. Initially when the Concept *of Operations (CONOPS)* is developed by the SE team, the system's operational and conceptual aspects and capabilities of the system are presented in the systems environment from the perspective of the stakeholders at the enterprise level. Use Cases, Use Case Scenarios, and Activity Diagrams can be used from the UML and SysML set of artifacts to present behavioral aspects of the system under consideration, still from the Enterprise/Stakeholders point of view. These artifacts can be used to gain understanding of the system's required capabilities, functions, features, and behaviors from the stakeholder's perspective. This collection of artifacts can be used to produce the *Operational Concept* (OpsCon). The OpsCon is similar to the CONOPS but provides the operational and conceptual aspects of the system in its environment from a developer's and system user's perspective. The OpsCon shows the operational details of the specific system(s). From the CONOPS and OPSCON, artifacts like the Structure Diagram and Block Definition Diagrams are useful in capturing the logical architecture of the system. The logical architecture according to IEEE ISO 24748-3 can take several forms, for example the Functional Block/Flow Diagrams wherein higher-level functions/flows can logically be broken down and arranged into lower-level functions/flows. The architecture for the functional decomposition of the system is referred to as the functional architecture. Other diagrams in SysML can be used for interface definition as well, but Block Definition Diagrams (BDDS) in SysML can accomplish this by using an interface block. According to Xia, et al., interface definition provides "insight into controlling factors and resulting behaviors, and overall descriptive conception of states and modes of the system (Xia et al. 2017)." During the Conceptual Design Phase, when the system-level requirements are being established, the system architecture is developed and captures the structure and behavior of the system, along with other useful views of the system. It forms the bridge between requirements and design and constrains the evolution of the design. The system architecture incorporates insights from the conceptual and logical architectures and helps establish and contain the physical architecture. The physical architecture goes one step further than the logical architecture in that it associates physical components (hardware and software) with the functional decomposition of the system. All these architectures (conceptual, logical, functional, system, and physical) are important in constraining and guiding the design process. Typically, there is frequent interchange between the requirements development process and the design process wherein the system architecture provides the structure, boundaries, and guidance for the design evolution. For instance, if the stakeholder wants a transportation system to go between one city and the next, and after some conceptual development the decision is to accomplish this with a car, the system architecture would reflect that the solution space for the evolving design for this transportation system is restricted to cars. This constrains the design space in that airplanes, trains, boats, and skateboards are not permitted to be considered in the solution space after this architectural decision. The logical architecture (expressed in terms of functions – functional architecture) would capture the functional elements and flows of the car (e.g., engine function, control function, power function, breaking function, steering function, ….) and the physical architecture would provide the decomposition for the physical elements of the system (e.g., from car to fuel subsystem to carburetor to assembly/pieces/parts of the physical elements of the carburetor). The importance of the systems architecture cannot be understated in that implementing an effective systems architecture can make all the difference in the long-term success or failure of the system. It might be argued that establishing the system architecture is just regular "design," but the design decisions made in establishing the system architecture are far-reaching, and dramatically impact the performance and flexibility of the resulting system. Bad system architecture decisions can lead to impractical implementations, loss of flexibility, and even unrecoverable system failure.

In addition to having an effective systems architecture, establishing an executable modeling capability is essential in modern systems development environments. One approach in developing this capability is to transform static models. Usually, this approach is associated with model-driven architecture methods (Ge et al. 2012). Within the Department of Defense Architecture Framework (DoDAF 2.0) there is a DoDAF Meta-model (DM2) that uses a data-centric approach. The conventional ontology that is used in DM2 stems from the International Defense Enterprise Architecture Specification (IDEAS). The combination of DM2 with the data-centric approach provides a powerful enhanced capability in developing strong architectural descriptions. These architectural descriptions can be shared between systems and used in a methodical way. However, there are many static models that do not use DM2 or the data-centric approach, and so may introduce semantical inconsistencies into the modeling framework. These inconsistencies can result in misinterpretation and errors and degrade the usefulness and effectiveness of the model. The most widely used framework for developing architectural descriptions is UML. Some variants of UML are executable UML ($_x$UML), and executable and translatable ($^x{}_T$UML). SysML borrows from and builds on UML providing a modeling capability tailored for the SE discipline. However, both UML and SysML lack formal semantics useful in executing programs (Ge et al. 2012), have no provisions for accommodating the formal specification of executable models, nor provide a means for verifying and validating these models. Another type of executable model that is based on the OMG Model-Driven Architecture (OMG MDA) is computation independent model (CIMS) which operates at a high level of abstraction. These executable models can be used to capture business knowledge like organization, functions, processes, and system requirements and are designed to work with various modeling languages on independent platforms. OMG MDA also recognizes platform independent models (PIMS) which are developed independent of the technical consideration of the underlying platform, and Platform Specific Models (PSMS) which do consider the technical details of the underlying platform. Besides the semantic issue mentioned earlier, transforming static models is an automatic process and requires bidirectional traceability between both the static and executable models (Wagenhals et al. 2009). The use of standardized terminology and recognized defined semantics are required for machines to have the capability for automatic interpretation. To get to the standardized terminology, there are a variety of important SE standards used in the discipline. Some of these major standards are:

1. ISO/IEC/IEEE 15288– System lifecycle processes (2023).
2. IEEE 15288.1 – IEEE Standard for Application of SE on Defense Programs (2014).
3. IEEE 15288.2 – IEEE Standard for Technical Reviews and Audits on Defense Programs (2014).
4. ISO/IEC/IEEE 15289 – Systems and software engineering – Content of lifecycle information items documentation (2019)
5. ISO/IEC TR 24748-1 – Lifecycle management (2018).

Even though these and other SE standards provide a solid basis for standardized terminology, there are still ambiguities, inconsistencies, and imprecisions present in the standards that need to be rooted out over time. The collection of standards prevalently used in the SE discipline doesn't provide the level of precision and consistency in terminology and semantics necessary for pervasive, consistent automated natural language-based processors needed for broad adoption and implementation of machine-based automatic interpretation. Even so, strides are being made daily in this area and the horizon looks promising.

1.1.9.5 Wicked Problems, Solution Approaches, and Useful Conceptual Tools

When looking into the crystal ball for future systems developments, the expectation is that systems will become more connected, interdependent, diverse, and adaptive, and consequently more complex. The inherent properties of emergence, self-organization, and adaptation found in adaptive complex systems can lead to system development efforts that face new challenges and require new techniques, methods, and approaches to succeed. In this section, we will present an overview of associated *wicked problems* that can be expected when developing complex systems, and describe some solution approaches for successfully dealing with them. We also present some interesting and useful conceptual tools (the Conceptagon and Systemigram) which can be used to holistically think about the system and help in addressing these wicked problems. Systems acquisition and development efforts face continual scrutiny across the globe by entities such as the media, governmental agencies, and audit organizations [such as the Government Accountability Office (GAO) in the United States]. Among other reasons, such as technical oversite and security, this scrutiny stems from the rather large number of resources associated with these acquisition efforts. For instance, in the United States, the fiscal year 2024 President's Budget request *once again includes a record investment in research, development, test, and evaluation (RDT&E) of $145 billion. This budget also funds $170 billion for procurement, the largest in history* ((Department of Defense 2023). Results of this scrutiny come in a variety of forms from inspections, audits, increased internal and external reviews, establishment of governmental and corporate oversight structures, required documentation, processes, and modeling artifacts, demanded program changes, strategic partnering and community relations building, and acquisition innovation initiatives. In the US, a wide range of strategic planning documentation exists to provide strategic guidance for the evolution of our defense systems. Some of the notable documents that reflect this strategic guidance and vision include the National, Joint, and Service-oriented Strategy and Vision documents (e.g., National Defense Strategy 2022, National Defense Science and Technology Strategy 2023, Joint Warfare 2040, Joint Operational Environment 2040, U.S. Space Force Vision for a Digital Service 2021, The Army's Vision and Strategy, Chief of Naval Operations Navigation Plan 2022, Marines – Force Design 2030, and U.S. Army Unmanned Aircraft Systems Roadmap 2010–2035.

Even though there is heightened scrutiny of acquisition programs and processes, little improvement in terms of cost savings, performance expectations of the delivered systems, or in the ability to meet schedule has been achieved. Some reasons for this lackluster performance stem from several sources. According to Dickerson and Mavris, there are four key concepts that introduce difficulties/problems in defense acquisition development efforts (Dickerson and Mavris 2010). First, there are difficulties in accurately assessing and translating technical requirements from stakeholder-based capabilities perspectives. This abstracted capabilities perspective, which is ingrained in the front end of the DOD acquisition process in the United States, is great for showing operational needs, capability gaps, and shortfalls, however, the translation from the operational and capability perspective to the technical requirements of the system(s) that address the needed capabilities, is not straight forward, precise, or robust. Consequently, there can be misinterpretations and misunderstandings between the stakeholders and developers because of this difference in perspective. Second, with multiple stakeholders, there is a need for balanced multi-perspective integration wherein compromise between stakeholder needs and expectations is a necessity. Third, there can be a lack of a definitive formulation of the problem space (or the problem space can be malformed and erroneous). This can happen with competing priorities, individual or shared knowledge gaps, on both the stakeholders' and developers' side, stakeholder and developer predilections and biases, and political interference and restructuring. Fourth, ambiguity in closure criteria can impact the proper assessments of the delivered systems. Additionally, because of the increased interconnectedness and interdependence of modern-day systems, complex systems development efforts are having to deal with wicked problems on an increasing basis.

What are wicked problems? *Wicked Problems* were brought to light originally by Horst W. J. Rittel and Melvin M. Webber in 1973 from their statement in their Policy Sciences article, *Dilemmas in a general theory of planning*, "… designing systems today is difficult because there is no consensus on what the problems are, let alone how to resolve them (Rittel and Webber 1973)." Therefore, wicked problems can be essentially unsolvable. This definition rightfully raises the hackles on any good engineering development team that chomps at the bit to solve the great problems of the day. Don't tell me what I can't do! But, as the famous actor Steve Martin once said,

"You can't put a piano up your nose!" Not that anyone would want to, but the reality is that there are certain problems that are just…wicked! According to Rittel and Webber, Wicked Problems have the following characteristics,

1. They cannot be definitively formulated.
2. No stopping rule exists (there are no criteria to know if you solved the problem or not, or stakeholders can't agree whether the solution is sufficient/acceptable or not).
3. Solutions are not true or false, but qualitative such as in good or bad.
4. There is no available current or future test of the solution.
5. No chance exists to learn from trial or error. Every attempt at a solution is significant, and every attempted solution only gets one chance.
6. There is no enumerable set of possible solutions, neither is there a collection of allowed operations that can be captured in the plan.
7. For all practical purposes, they are unique.
8. They can be a symptom of another problem.
9. They have numerous explanations and stakeholders can have differing ideas about what the problem is, what is causing it, and how it gets resolved.
10. The planner or developer is not allowed to get it wrong.

Dickerson and Mavris sum these points well with, "In some sense, these characteristics can be considered as expressing an ill-posed question with an open-ended set of interrelationships and dependencies requiring a formulation of constraint mathematics that might preclude a solution space in the first place (Dickerson and Mavris 2010, p. 357)." Wicked Problems can

- Have many causes,
- Be hard to describe,
- Produce undesirable effects,
- Have no "right" answer,
- Have no simple solution.

In fact, some attempted solutions to wicked problems can sometimes make the problem even worse. The best approach then may be not to solve the problem directly but to mitigate its negative effects. Wicked problems abound and are an inevitable part of life. Many wicked problems are unsolvable because there isn't a simple solution and/or the solution causes unanticipated, and perhaps unseen effects. This is akin to the principle of emergence in complex systems theory where there are many interdependencies and solutions that may change the effects of the system. Wicked problems have many causes and solutions may make problem worse.

So, what are some of the solution approaches to effectively deal with wicked problems? A holistic response is required that takes advantage of modern SE practices, tools, and techniques, and understands the diversity in program/project styles/systems development efforts (e.g., basic, complicated, complex, and/or adaptive). Visualization tools and techniques such as augmented reality, virtual reality, and 3D and 4D modeling tools are helpful in representing the evolution of concepts or system dynamics. Parallel (perhaps Agile) analytical methods and teams are necessary to rapidly process changing states in the system and evaluate emergent system behaviors and properties. Virtualization of the system is needed to create its digital twin to rapidly analyze, simulate, and study system evolution, adaptation, emergent behavior, and/or the effects of external actors, and the environment on the system. Model-Based SE methods must be implemented to capture, understand, control, communicate, and evolve structural, behavioral, definitive, and contextual elements, system requirements, information and data flows, interactive models, and develop the system architecture. The development team must separate the problem domain, from the solution domain and have mechanisms to iterate back and forth between these domains. Decision and risk models with game theoretic methods, artificial intelligence and machine learning methods, and agent-based simulation must be developed and used along with statistical, heuristic, and optimization methods to capture the impacts of adversity, variation, and uncertainty. Solid configuration control and change management capable of rapid response is also essential along with a culture of expected change and volatility in the evolution of the system.

Some of the attributes of the organization, workforce, and individuals that are necessary for successfully overcoming wicked problems are now presented. The development team must be capable of successfully reconciling the abstract capabilities perspective from stakeholders with the different, commercial product orientation in the industry. Early in the conceptual development, stakeholders and developers need to work together to overcome the "valley of death" where the time from initial prototype to adoption can take several years (typically two in the US defense sector). The system planners and workforce must have an ingrained awareness of systems thinking methods and understand complex systems. This holistic view encourages comprehensive evaluation of the development effort and searches for areas that require special processes and considerations such as the employment of nonlinear, parallel, or recursive methods. The systems thinking approach fosters the ability to connect abstract ideas and concepts and deal with uncertainty and change. The team members must have a heightened sense of working in a supportive environment wherein everyone is valued and depended on. Teammates must also be intuitive and have strong interpersonal skills. The development team must expect that the organization in which they work may adapt their structure, behaviors, and values as the system evolves and must be able to use the appropriate development style for the project at hand. Some models, such as the four quadrant model shown in Table 1.4, help in understanding different types of development styles in projects, programs, and systems development efforts. Table 1.4 shows an example of different development styles (shown in parentheses) based on the degree of intricacy and uncertainty in the development effort. In this table, the vertical axis deals with intricacy of the effort, and the horizontal axis deals with the uncertainty of the effort. If an acquisition effort has low intricacy and low uncertainty, then it is considered a basic development effort and a standard waterfall process can be successfully employed. If intricacy is high, but uncertainty is low, then this conforms to a complicated project and the standard V-shaped SE effort is highly applicable. For a project that is low in intricacy but high in uncertainty, then options models are beneficial wherein different types of options are laid out corresponding to the different uncertain outcomes. Statistical and/or heuristic methods are also often highly applicable. If both intricacy and uncertainty are high, this is the space of wicked problems, and an emergence model is necessary.

Another useful tool for understanding a system under investigation in a holistic way is the Coneptagon. The motivation for the Conceptagon was to make an important addition to the Systems Thinking toolkit (Boardman et al. 2009). The Conceptagon concept is shown in Figure 1.25 and provides seven distinct ways to think about a system of interest. The approach is intended to provide

TABLE 1.4
Development Styles for Levels of System Intricacy and Uncertainty

Four Quadrant Development Style

High intricacy	Complicated system (SE V-model)	Complex system (emergence model)
Low intricacy	Basic system development (Waterfall model)	Probability/heuristics-based (different options model)
	Low **uncertainty**	High **uncertainty**

Source: Adapted from Dickerson and Mavris (2010).

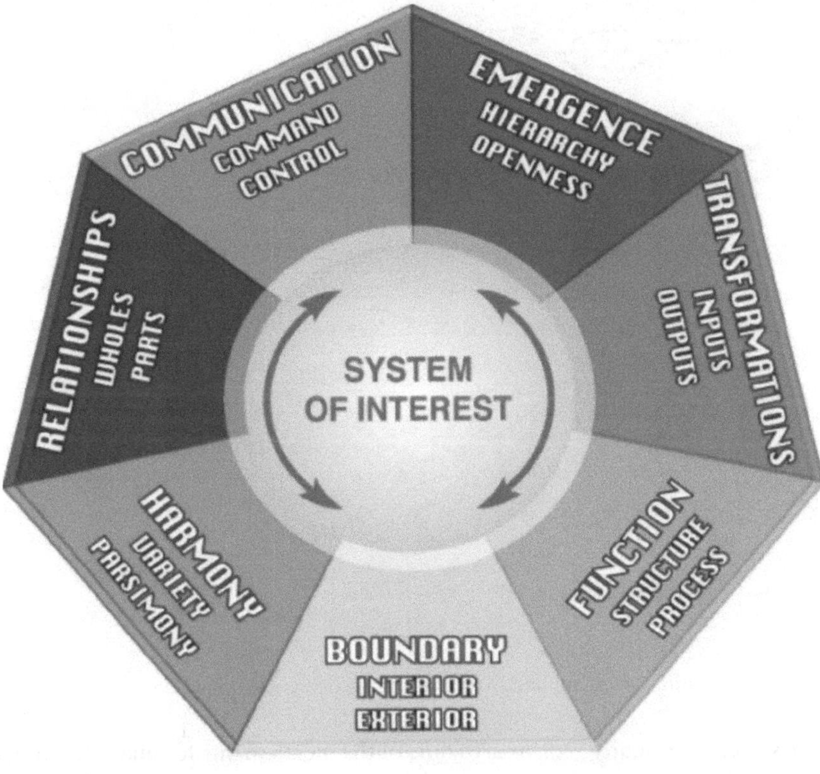

FIGURE 1.25 The Conceptagon as part of the systems thinking toolkit.

a holistic view of the system and provide understanding about the system at several levels. Familiar to the systems engineer is the orange area of the Conceptagon which involves thinking about the system's function(s). What is the system supposed to do? What functionality does it provide? What is the structure of the system and what are the underlying processes? The bottom-yellow layer considers where the boundaries of the system are. Where does the system stop and where does the system's environment begin? The blue area deals with harmonious and parsimonious aspects of the system. The purple area involves relationships between lower elements of the system (subsystems, components, assemblies) and how they relate to each other. What are the parts of the system? How do they combine? How the system communicates is shown in the gray area and system command and control considerations are thought of here. From complexity theory, the idea of emergence is addressed in the red area along with how open the system is and system hierarchy. Finally, transformational aspects of the system, such as how inputs are transformed into outputs are addressed in the green area. Because there are no prescriptive methods for using the Conceptagon, domain experts enjoy the freedom to think in new ways about the System of Interest, choosing their own navigation of the set of triples as it becomes obvious and intuitive how to do so. The systemigram shown in Figure 1.22 is another useful systems thinking tool that permits the visualization of complex systems. The top-left of the systemigram is the system of interest and the diagram proceeds from top-left to bottom-right. At the bottom-right part of the systemigram, the goal of the system of interest is articulated. The circular aspects of the diagram are nodes of the system of interest, and they are connected by relationships that are described in the square boxes. Lines are not allowed to cross. As an example, we show a systemigram for a system in Figure 1.26 that consists of a collection of infrasound signals (shown at the top-left). The goal is to have an optimized detection system for these infrasound signals (shown at the bottom-right part of the figure). The important characteristics

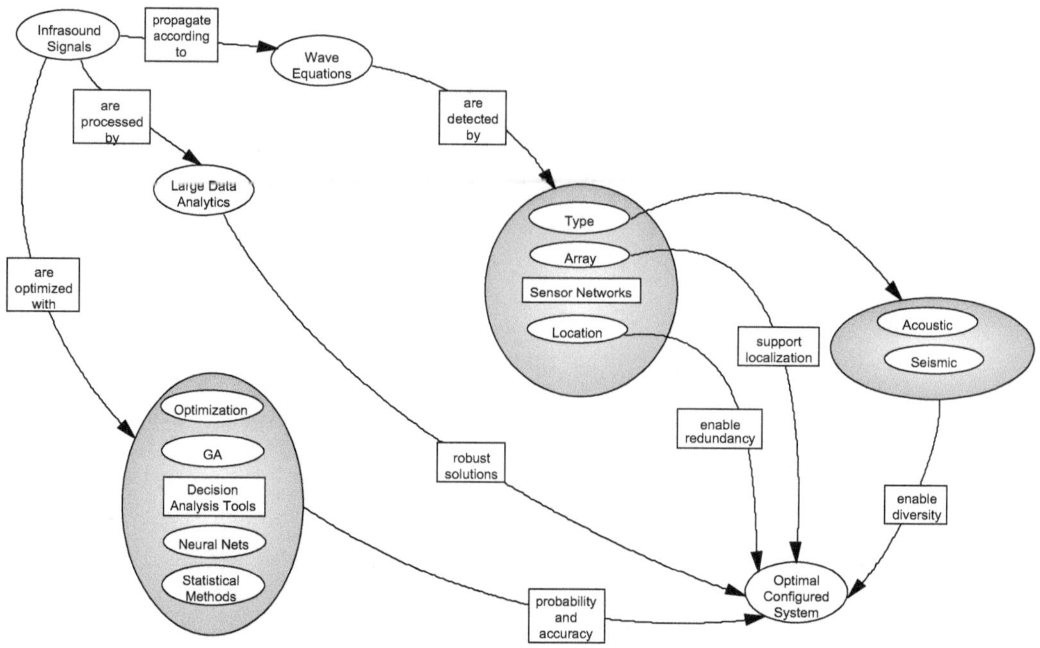

FIGURE 1.26 The Systemigram as applied to an infrasound detection system.

of the detection system are clearly seen as are some of the mechanisms for understanding the system (e.g., wave physics, large data analytics, probability theory, and sensor network analysis). Notice how the Systemigram gives good insight into the overall system, regardless of whether the system is complicated, complex, or even basic. It is easy to see what the main elements of the system are by looking at the internal elements of the ovals. How the ovals relate to each other is shown by the arrows and referenced by the rectangular boxes. The Systemigram is a great tool for interacting with the systems stakeholders because of its simplicity and ability to capture the important system elements and relationships. In the next section, we provide some optical systems building blocks that are essential to understand optical systems from a technical perspective.

1.1.10 Transition to Optical Systems Building Blocks

We provided an introduction to SE, modern-day systems and EA methodologies and processes. In the next section, we will begin to build the optical systems building blocks that will be necessary to understand and successfully analyze optical systems. Most chapters of this book will have an optical building blocks section that relates to the chapter content in some way but also is largely independent. The intent of the optical building blocks sections in each chapter is to provide the technical details needed to understand and analyze optical systems. They are intentionally designed as related, but stand-alone, sections that are sequentially progressive. By this, we mean that subsequent optical building block sections in later chapters may depend on the principles, methods, and understanding developed in previous sections.

We have attempted to make each section stand-alone as much as possible, but due to the complexity of optical systems in general, some progressive material development and interdependence of sections are unavoidable. Nevertheless, the technical content in these optical building block sections is essential for the practicing systems engineer, working professional, or student who is working with optical systems. Just as we provided an introduction to SE in the Modern Age section, we now provide introductory material needed to understand and analyze optical systems in the following section.

1.2 OPTICAL SYSTEMS BUILDING BLOCKS: INTRODUCTION, SYSTEMS OF UNITS, OPTICAL SYSTEMS METHODOLOGIES, AND TERMINOLOGY

In this section, we start with a short introduction and background of important discoveries associated with the technical aspects of optical SE. We then provide an overview of our adopted optical SE systems of units. Subsequently, we introduce optical systems methodologies, which are useful in understanding, describing, modeling, and analyzing optical systems. We also want to provide different perspectives in modeling optical systems and discuss when particular approaches and methods apply. For instance, we want to introduce terms that are essential to the reader's understanding of different types of propagation models. We can model and analyze the propagation of visible and IR light using several different approaches. The chosen approach depends on the physical situation. In the appropriate chapter, we discuss the means for deciding which approach to take and also key results from each approach.

A full treatment of just one of these areas could easily fill this text, so we focus on results that can be directly applied by the practicing engineer and optical systems manager. Some examples of different approaches to modeling and analyzing the propagation of optical waves in the visible and IR part of the electromagnetic spectrum are as follows: vector electromagnetic theory, scalar electromagnetic theory, SO, geometrical optics, Fourier optics, or radiation transport analysis. Each approach has a conceptual framework that needs to be understood by the practicing optical systems engineer or optical systems manager, and we attempt to provide introductory-level details in this section. We end this section with a description of essential terminology and concepts that will be needed in later chapters. We will use International System of Units (SI) throughout the text. We first start with a little background information on key developments in the optics field.

1.2.1 Introduction to Optical Systems

For millennia, the only known portion of the electromagnetic spectrum was that of visible light. The IR region of the electromagnetic spectrum was discovered when early scientists and scholars began to investigate what was beyond the red end of the rainbow of colors that were identified with light. In 1665, Sir Isaac Newton determined the basic principles of light and color and the use of the prism. He also discovered that without any correction, *diffraction-limited* performance could not be achieved by an optical system that is imaging at visible wavelengths with an aperture greater than a few tens of centimeters (Newton 1952).

Diffraction-limited performance, in the context of the spatial resolution of a given imaging system, refers to the best possible spatial resolution attainable using classical electromagnetic theory In essence, it is the resolution of a perfect imaging system – no optical systems aberrations or noise, no atmospheric effects – and so represents a benchmark in analyzing the performance of an optical system. During Newton's time, it was also known that when viewing a point source – an object that is smaller than the spatial resolution of a given optical system – from a distance, the spreading of the light is broader than if the point source were closer. Newton looked at starlight using a telescope and discovered that the spreading of the light – the observed point spread function – was broader than when telescope was used to look at a similar point source in the laboratory. This was an early indication of the effects of atmospheric turbulence on optical imaging systems. Unfortunately, scientific equipment, photographic, and camera technology did not evolve sufficiently to do anything about this until the 1900s.

During the 1800s, Sir Friedrich Wilhelm Herschel discovered a region of the spectrum beyond the region of visible light that emitted heat. Later on, this part of the electromagnetic spectrum was dubbed "invisible rays" or "dark heat" and is today known as the IR part of the electromagnetic spectrum. Herschel set out to investigate the colors of the spectrum that were accountable for heating objects. Hershel knew that sunlight produced the colors red, orange, yellow, green, blue, indigo, and violet (ROYGBIV) with the use of a prism, so he ventured to measure the temperature associated with each of these colors. By means of a prism and thermometers, he measured the temperatures of the different colors to observe where the increase in temperature occurred as he moved the thermometer throughout the prism. He discovered that the temperature reached its highest levels when it was past the red light of the prism in a region where there was no color. The radiation causing this heating was not visible; as a result, he concluded that these invisible rays had some kind of "dark heat." In 1860, Gustav Kirchhoff found that good absorbers are also good radiators, which led to the idealized blackbody model. In 1879, Jozef Stefan discovered that radiated power from a blackbody source is related to its temperature. In 1884, Ludwig Boltzmann derived the same result from theoretical thermodynamic calculations leading to the Stefan-Boltzmann law. As will be seen later, this law is fundamentally important in estimating the expected spatial power density (W/m^2) emitted from an ideal (blackbody) source. In 1893, Wilhelm Wien derived Wien's law using thermodynamic principles showing that the shape of the thermal radiation that is emitted from a blackbody, as a function of wavelength, is essentially the same for all temperatures. He also showed that the peak of this thermal distribution moves toward lower wavelengths as temperature increases. Lord Rayleigh (aka John William Strutt) and Sir James Jeans developed the Rayleigh-Jeans law that attempted to analytically describe thermal radiation from a blackbody as a function of wavelength. Their law was shown to work for higher wavelengths but diverged from physical observations in the lower-wavelength regime. In 1900, Max Planck empirically found the correct blackbody radiation law that works for all wavelengths. Also in 1900, Heinrich Rubens and Ferdinand Kurlbaum experimentally verified Planck's work which led to the establishment of quantum theory. In 1941, Andrey Nikolaevich Kolmogorov published a series of mathematical papers on atmospheric turbulence (Kolmogorov 1941a,b). In 1961, Valer'ian Il'ich Tatarski developed a model for wave-front perturbations that were based on Kolmogorov's work. In 1965, Dave Fried developed a parameter r_o using statistical methods that presented a metric related to the quality of "seeing" through the atmosphere. His parameter is called the Fried parameter and represents the diameter of a patch of atmosphere over which the atmosphere can be thought of as being somewhat the same. Effectively, for telescope diameters that are larger than r_o, adaptive optics and/or atmospheric turbulence compensation methods are required to obtain the best resolution out of the imaging system. Figure 1.27 presents a timeline of key technical contributions to the optical systems field relating to remote detection systems that are affected by the Earth's atmosphere.

After the 1900s, we only show a few contributions that relate to remote optical detection systems that are affected by atmospheric turbulence. We intend this to be an introductory-level timeline that is relevant to remote optical detection systems such as ground-to-ground, ground-to-air, air-to-air,

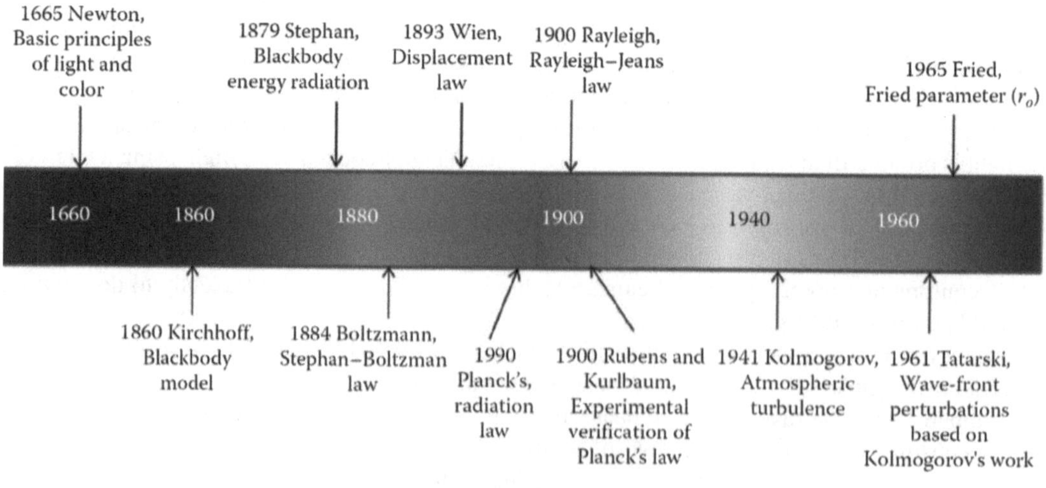

FIGURE 1.27 Timeline of important optical systems developments.

and air-to-ground imaging systems found on a variety of optical sensing platforms. A more complete list of key contributions is included with each of the subsequent chapters. Also, in the later part of this timeline, we focus on key developments that impact optical remote detection systems and also key results in understanding and modeling the effects of atmospheric turbulence.

Since 1965, a tremendous amount of activity and research has been done in adaptive optics (real-time, hardware-dominant solutions), atmospheric turbulence compensation methods (traditionally non-real-time, software-dominant solutions), and hybrid methods (a mix of the two approaches) to correct for the effects of atmospheric turbulence in optical images. For many well-designed optical imaging systems, the effects of atmospheric turbulence are the limiting factor in the optical systems spatial resolution.

1.2.2 Overview of Adopted Systems of Units

In this section, we provide an overview of our adopted system of units. It is important to focus on terminology early on since many terms can be confusing when speaking to someone in another technical field. Some terms have different meanings depending on which technical discipline uses the term. Much confusion can be avoided by using optical systems-related terms correctly. For example, the word intensity is used in a variety of ways in literature and in practice. The "intensity of the target" is often used in practice and in literature when the target is fully resolved by the optical system. In other words, the target system has observable dimensions of length, width, and possibly depth. In this case, the term intensity is not appropriate. The actual units for intensity are Watts/steradian, and so the term *intensity* technically only applies to point sources. As mentioned earlier, for our purposes, a point source is a source of optical radiation whose dimensions are smaller than the spatial resolution limits of the observing optical system.

In the case of a point source, the source object can be thought of as a radiating "point" that emits an expanding spherical wave. The intensity is then that portion of the optical power that is radiated into a steradian (sr). The steradian is determined by dividing a part (or whole part) of the spherical surface area of the expanding optical wave by its separation distance squared. For instance, if we wanted to know how many steradians our point source radiates into over the entire spherical surface, we would divide the surface area of the sphere ($4\pi r^2$) by the square of the separation distance of the spherical surface and the point source (r^2) and obtain 4π sr. The intensity is the optical power in Watts that radiates into this sphere. Similarly, if we were interested in finding the intensity in a hemisphere about the point source, we would want to know how much optical power is being radiated into the 2π sr of the hemisphere. In essence, the word intensity applies to point sources. As previously stated, this term is often misused in conversations and even in published works in the sense that the word intensity is used to also describe optical power associated with *extended objects* – objects that have larger spatial dimensions than the resolution limits of an observing optical system. This abuse is so frequent that many optical scientists and engineers have learned to turn a blind eye to it. A more appropriate term in the extended object case is radiant emittance, which has units of W/m^2. To avoid these issues, it is helpful to use an established system of units and define terms early on.

There are different systems of units for describing properties and characteristics of observable physical phenomena like electromagnetic, seismic, acoustic, and infrasound waves. Which system is used depends on the convention of the technical discipline and also the problem at hand. For convenience, and in consideration of our international audience, we will use SI units throughout this book. With regard to optical systems, there are two SI systems of units that are particularly useful when it comes to the measurement of electromagnetic radiation – photometric units and radiometric units. Photometry typically involves measuring light at individual wavelengths and includes the response of the human eye as a filtering function within its formulation. Radiometry usually deals with measuring radiation in terms of total power (e.g., integrated over all wavelengths). The radiometric units also have single-wavelength elements, but these are distinguished by adding qualifiers

TABLE 1.5
Photometric Measurement Elements

Photometric Name	Symbol	SI Units	Dimensions
Luminous flux (power)	Φ_v	Lumen	J
Luminous energy	Q_v	Lumen second	T J
Luminous intensity or candlepower	I_v	Candela (lm/sr)	J
Luminous emittance	M_v	Lux (lm/m^2)	$L^{-2}.J$
Luminous incidence or illuminance	E_v	Lux (lm/m^2)	$L^{-2}.J$
Luminance, luminous sterance, or photometric brightness	L_v	Candela per square meter (cd/m^2)	$L^{-2}.J$
Luminous exposure	H_v	Lux second (lx· s)	$L^{-2}.T.J$
Luminous energy density	ω_v	Lumen second per meter^3 (lm·s/m^3)	$L^{-3}.T.J$
Luminous efficacy	η	Lumen per Watt (lm/W)	$M^{-1}.L^{-2}.T^3.J$
Luminous efficiency	V		1

such as *spectro* or *spectral* in front of the radiometric term. For instance, a *radiometer* measures the total optical power across a spectrum of wavelengths, whereas a spectroradiometer measures the optical power within a narrow band of wavelengths (or frequencies) surrounding a particular center wavelength. The spectroradiometer can be tuned to different center wavelengths to provide narrowband optical power readings at differing center wavelengths.

We use the Si-based photometric units when the quantities of interest contain the response of the human visual system, which attenuates or filters light differently at separate wavelengths. Table 1.5 shows the elements for the photometric measurement system.

This set of units is used quite often in specification sheets for optical imaging systems, camera systems, and some detectors. Column 2 shows the corresponding SI symbols. Authors often define and use different symbols in their work, so the practitioner needs to check the units and dimensions of symbols to understand their proper usage. For example, if an author uses the words *luminous intensity* and the symbol I for a source of optical radiation, but the units end up being Lumens/m^2, then we know that what the author meant was luminous exitance and not luminous intensity. Unit analysis is critical in correctly interpreting the technical context.

When conducting requirements analysis, one step should be to consider the best consistent system (or systems) of units to use in the application and then rewrite the requirements in that set of units. The dimension column entries in Table 1.5 are useful when comparing table elements in different systems of units. When it comes to radiometric analysis, the system of units in Table 1.6 is quite common.

We will adopt this system of units throughout the majority of the text since these units are well understood and are applicable to a variety of optical systems analysis applications. We will use Si-based photometric units where necessary, but because of their familiarity, we will predominantly use Si-based radiometric units. Table 1.7 shows the symbols and units of some important physical quantities. The fundamental unit of energy in the Si-based radiometric system of units is the *joule*, which can be thought of as the amount of work that is done by applying 1 N of force over a distance of 1 m – a Newton-meter. Recall that force is given by a mass times its acceleration and 1 N of force is equal to a 1 kg mass experiencing and acceleration of 1 m/s^2. The units of 1 J of energy are then kg m^2/s^2. In terms of electrical current, this equates to 1 A of current running through a 1 Ω resistor for 1 s. *Radiant flux* is radiated optical power and is given in *Watts*. A Watt is given as the time rate of change of energy and so has units of J/s. The spatial power density for an electromagnetic wave has units of W/m^2 and has the term *radiant exitance* or *radiant emittance* when applied to a radiometric source with finite extent.

TABLE 1.6
Radiometric Measurement Elements

Radiometric Name	Symbol	SI Units	Dimensions
Radiant energy	Q_e	Joule	$M{\cdot}L^2{\cdot}T^{-2}$
Radiant flux	Φ_e	Watt	$M{\cdot}L^2{\cdot}T^{-3}$
Spectral power	$\Phi_{e\lambda}$	Watt per meter	$M{\cdot}L{\cdot}T^{-3}$
Radiant intensity	I_e	Watt per steradian	$M{\cdot}L^2{\cdot}T^{-3}$
Spectral intensity	$I_e\lambda$	Watt per steradian per meter	$M{\cdot}L{\cdot}T^{-3}$
Radiance	L_e	Watt per steradian per square meter	$M{\cdot}T^{-3}$
Spectral radiance	$L_{e\lambda}$ or L_{ev}	Watt per steradian per cubic meter or Watt per steradian per square meter per hertz	$M{\cdot}L^{-1}{\cdot}T^{-3}$ or $M{\cdot}T^{-2}$
Irradiance	E_e	Watt per square meter	$M{\cdot}T^{-3}$
Spectral irradiance	$E_{e\lambda}$ or E_{ev}	Watt per cubic meter or Watt per square meter per hertz	$M{\cdot}L^{-1}{\cdot}T^{-3}$ or $M{\cdot}T^{-2}$
Radiant exitance/radiant emittance	M_e	Watt per square meter	$M{\cdot}T^{-3}$
Spectral radiant exitance/ spectral radiant emittance	$M_{e\lambda}$ or M_{ev}	Watt per cubic meter or Watt per square meter per hertz	$M{\cdot}L^{-1}{\cdot}T^{-3}$ or $M{\cdot}T^{-2}$
Radiosity	J_e or $J_{e\lambda}$	Watt per square meter	$M{\cdot}T^{-3}$
Radiant exposure	H_e	Joule per square meter	$M{\cdot}T^{-2}$
Radiant energy density	ω_e	Joule per cubic meter	$M{\cdot}L^{-1}{\cdot}T^{-2}$

TABLE 1.7
Symbol and Units of Physical Quantities

Quantity	Symbol	SI Unit
Length	L	1 meter (m)
Mass	m	1 kilogram (kg)
Time	s	1 second (s)
Force	F	1 Newton (N)
Frequency	v	1 hertz (Hz)
Energy	U	1 Joule (J)
Power	P	1 Watt (W)
Current	I	1 Ampere (A)
Potential	V, ϕ	1 volt (V)

This term can be interpreted as the amount of optical power emitted from 1 m^2 area from the source. If the radiant exitance is constant across the emitting surface, it is said to have a *uniform* spatial power density across the surface. Otherwise, it has a *nonuniform* spatial power density distribution or a *spatially varying power density* across the emitting surface. Notice that the units for *irradiance* are the same as those for *radiant emittance.* That is because irradiance refers to the spatial power density received in a distant observation plane, whereas radiant emittance refers to the spatial power density emitted from the source.

In our system of units, radiant intensity refers to the optical power radiated into a solid angle of 1 sr. The term radiant intensity, therefore, applies only to radiating sources that can be properly characterized as point sources and not to radiometric quantities that are in a distant observation plane. For instance, it does not make sense to say the radiant intensity received on an optical detector in a remotely located detection plane was 1 W/m^2 since (a) the units of intensity are W/sr and (b) radiant intensity is an emitted property and not a received property. The term *radiance* is given to characterize a source that has both point source and extended source attributes. For example, the source can have a known finite extent but still be unresolved by an imaging system and so appear like a point source. Radiance is the spatial power density per unit solid angle and so has units of W/(m^2 sr). Note that if the radiance is integrated with respect to the source's surface area, then we arrive at the source's intensity. Likewise, if we integrate the source over the solid angle, we arrive at the source's radiant emittance. If we add the term spectral in front of the radiometric term, this means that we are referring to the radiometric quantity over a narrow spectral band of frequencies. For instance, if we were interested in measuring optical power radiated from a source over a narrow spectral range of frequencies, we would use the term spectral radiant flux. Other non-SI systems of units exist and include luminance units such as the footlambert, millilambert, and stilb and illuminance units such as the footcandle and phot. As previously stated, we adopt Si-based units throughout this book. In the next section, we briefly introduce some optical systems methodologies that are useful in analyzing optical systems. These will be further detailed in subsequent chapters.

1.2.3 Optical Systems Methodologies

There are a variety of very useful analytical methodologies that help us understand optical systems and their performance. In this section, we provide an overview of these methods and discuss in a general sense when it is appropriate to use them. However, before we present these methods, we want to provide some fundamental physical concepts and, from a systems perspective, describe some important characteristics of the electromagnetic spectrum. Many types of radiation are familiar in everyday life, including sunlight, heat, radio waves, X-rays, and microwaves. These types of radiation appear to be very different from one another, but they are all considered to be part of the *electromagnetic spectrum*. In using the term *optical*, we mean electromagnetic radiation that is radiating in the UV, *visible*, or IR part of the electromagnetic spectrum.

Since we will most often be working with visible and/or IR systems, our usage of *optical* typically refers to these parts of the electromagnetic spectrum either individually or together. The correct meaning will be made clear by the context of the discussion. The type of electromagnetic radiation depends on the wavelength or equivalently the frequency of the radiating *electromagnetic wave*. The following equation provides a fundamental physical relationship between the electromagnetic wave's wavelength and its corresponding frequency:

$$\lambda \upsilon = c. \tag{1.1}$$

The frequency is the number of electromagnetic wave cycles per unit of time(s) measured at a point in hertz (Hz). The term c on the right side of the equation is the speed of light in a vacuum (299,792,458 m/s). Equation 1.1 applies at every frequency. In our adopted system of units, the unit of length that applies to the wavelength is meters (m). Because of the extremely short length of an optical wavelength, it is convenient to further characterize and define our unit of length. The optical wavelength is often specified in terms of microns, aka micrometers (μm), or nanometers (nm), for optical radiation, with occasional reference to units of millimicrons (mμ), inverse centimeters, or Angstroms (Å). The following relationship helps in converting between these units, the unit of an inverse centimeter is called a wave number, and it is measured, as the name implies, in terms of an inverse of a centimeter:

$$1\text{ m} = 39.37\text{in.} = 10^2\text{cm} = 10^3\text{mm} = 10^6\mu\text{m} = 10^9\text{m}\mu\text{m} = 10^9\text{nm} = 10^{10}\text{Å}. \tag{1.2}$$

For example, a 1 cm length is equivalent to 1 wave number (e.g., inverse centimeter). Similarly, a 10 cm length is equivalent to 0.1 wave numbers. The wave number unit is obtained by first converting the length unit into centimeters and then taking the reciprocal. A 1 μm wavelength is equal to 10^{-4} cm or equivalently 10,000 wave numbers. Wave number units are often used in fields that perform spectral analysis on materials such as in the field of spectroscopy. Spectroscopy is the study of the interaction of matter with radiated energy (Skoog et al. 2006). As the units for the frequency in Equation 1.1 are in cycles per second, the inverse centimeter unit can be thought of as one cycle per centimeter – a spatial frequency. In some cases, where lambda is part of a narrow band of frequencies or wavelengths, lambda is thought of as the center wavelength of the band. The following equation shows the relationship between a change in frequency and the corresponding change in wavelength:

$$|\Delta \upsilon| = \frac{c}{\lambda^2}|\Delta \lambda|. \tag{1.3}$$

Solving Equation 1.1 for the frequency, introducing a perturbation on both sides of the equation, and taking the absolute value of the result to obtain this expression. Equation 1.3 can then be used to relate a change in frequency to a change in wavelength.

The electromagnetic spectrum can be characterized either by wavelength, frequency, or the wave number of a radiating electromagnetic wave. Figure 1.24 shows useful properties of radiation throughout various parts of the electromagnetic spectrum. The top part of the figure shows the parts of the spectrum for which the Earth's atmosphere is relatively transparent. These transparent regions are good for observing distant objects either within or outside of the atmosphere. The transparent regions are shown with a Y at the top of Figure 1.28, and nontransparent regions are shown with an N. The gray areas are partially transparent. Notice that the UV part of the electromagnetic spectrum is not transparent and so that is a typical reason why we do not build long-range communication and/or detection systems in the UV part of the electromagnetic spectrum. Underneath the top bar, the relative size of the wavelength is shown. On the left side of the figure, there are very long wavelengths, and on the right side, the wavelengths are very short. The wavelength, and frequency component of the electromagnetic wave, is determined by Equation 1.1.

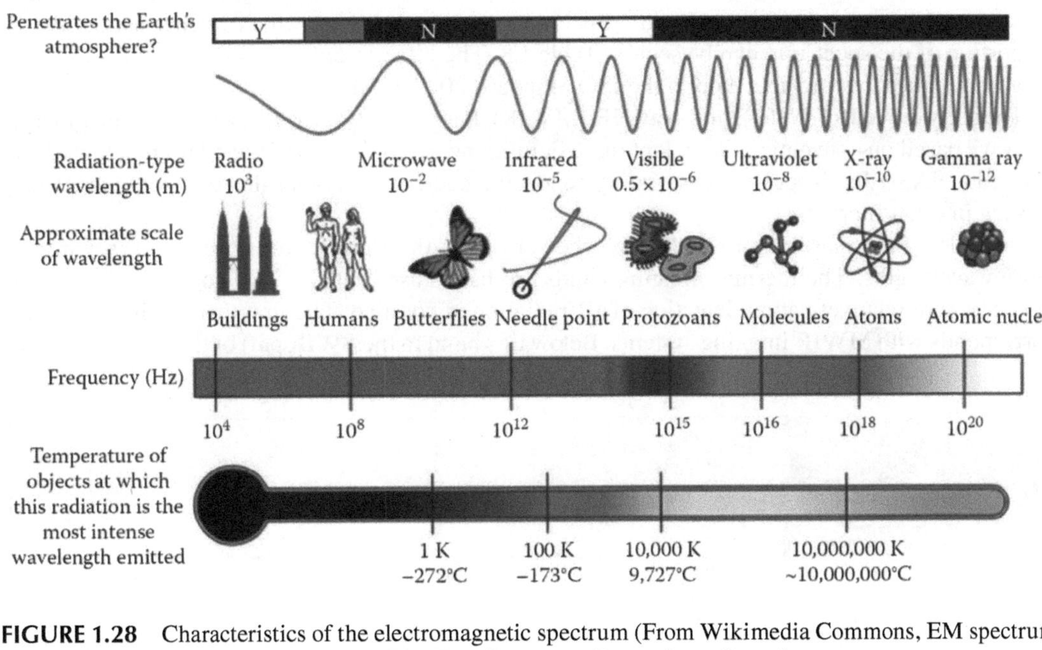

FIGURE 1.28 Characteristics of the electromagnetic spectrum (From Wikimedia Commons, EM spectrum properties, http://en.wikipedia.org/wiki/File:EM_Spectrum_Properties_edit.svg).

Underneath the qualitative picture of the wavelength, the corresponding region of the electromagnetic spectrum and the numeric value of the wavelength are shown along with a picture of something familiar that represents the size of the wavelength. For instance, at the far left of the figure, we see a transparent region of the atmosphere that contains radio waves. A representative wavelength in this part of the electromagnetic spectrum can range from the size of buildings to over 1,000 m long. Wavelengths in the microwave part of the electromagnetic spectrum are strongly absorbed and scattered by the Earth's atmosphere, and the wavelengths are approximately the size of a butterfly. In the visible region, the size of the wavelength is on the order of protozoans, and at the far right of the chart, for gamma rays, the wavelength is on the order of the radius of atomic nuclei. Underneath the approximate scale of wavelength, we see the corresponding frequency of the electromagnetic wave according to Equation 1.1. The different types of electromagnetic radiation share common characteristics and obey similar laws of refraction, reflection, diffraction, and polarization.

The different types of radiation also propagate at the same speed, often referred to as the "speed of light." The significant difference is in their frequency or wavelength. Finally, underneath the frequency bar, we see a temperature bar given in kelvins. This bar shows the equivalent temperature of an object that radiates most intensely at the peak of the wavelength shown. The temperature bar at the bottom of Figure 1.28 is important because energy can also be quantified in terms of heat. This is particularly useful in the IR part of the electromagnetic spectrum where the electromagnetic wave is not visible to the human eye. Thermal radiation is a type of electromagnetic radiation emitted by objects because of the object's temperature and material properties. This type of radiation is readily observable in the IR part of the electromagnetic spectrum because the atmosphere is relatively translucent in the IR. Consequently, depending on the temperature of the object, the emitted light due to thermal radiation may dominate over reflected and scattered light in the IR part of the electromagnetic spectrum. This is especially true at night when the reflected component of light is typically minimalized due to the absence of the sun.

Figure 1.29 shows a further subdivision of the IR part of the electromagnetic spectrum and its relationship to other parts of the electromagnetic spectrum. We can see that the IR radiation appears on the electromagnetic spectrum from near infrared (NIR) at 0.7 μm through the extreme IR at 100 μm. This is bounded between visible radiation on the shorter-wavelength side and microwave radiation on the longer-wavelength side. Specific wavelength limits, which further subdivide the IR portion of the spectrum, can be seen in Table 1.8. The NIR, middle-IR (MIR), and far-IR (FIR) ranges are those that are specified in the ISO Standard 20473:2007.

The short-wave IR (SWIR), mid-wave IR (MWIR), long-wave IR (LWIR), and others in Table 1.8 can vary based on convention. We adopt these definitions based on the ISO standard mentioned earlier and a NASA Jet Propulsion Laboratory report but caution the reader that different ranges may be seen in other literature.

In Table 1.8, the last column indicates whether or not the Earth's atmosphere is transparent at those wavelengths. The thermal imaging bands are bands associated with a common IR imaging system where relatively good detection of IR radiation is possible. The first thermal imaging band corresponds with MWIR imaging systems. Below this band in the SWIR part of the electromagnetic

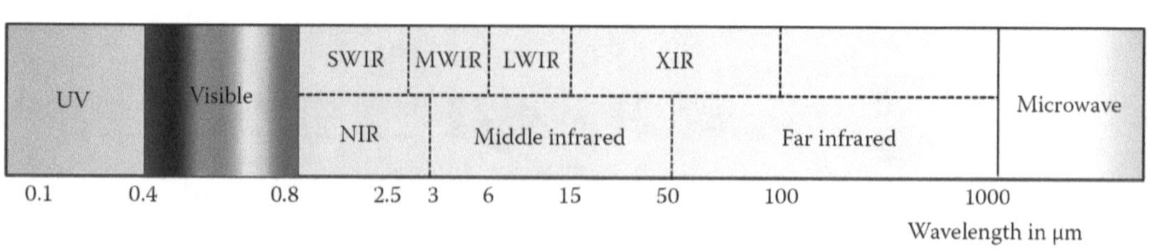

FIGURE 1.29 Expanded view of the visible and IR parts of the electromagnetic spectrum.

TABLE 1.8
IR Radiation Band Subdivisions

IR Segment	Abbreviation	Range (μm)	Relative Atmospheric Transparency
Near infrared	NIR	0.7–3.0	Transparent
Short-wave infrared	SWIR	1.1–2.5	Transparent
Middle infrared or	MIR	3.0–50	Transparent
mid-wave infrared	MWIR	2.5–6	
First thermal imaging band		3.0–5.5	
Blue spike plume		4.1–4.3	
Red spike plume		4.3–4.6	
Far infrared or	FIR or	50–1,000	Transparent
Long-wave infrared	LWIR	7.0–15.0	
Second thermal band		8–14	
Very long-wave infrared	VLWIR	>15	Opaque
Extreme infrared	XIR	15–100	
Near millimeter		100–1,000	
Millimeter		1,000–10,000	

spectrum, IR imaging systems are typically dominated by reflected light. In the first thermal imaging band, both reflected light and emitted light are important. Therefore, in the MWIR, imaging systems can take advantage of low light levels if they are present, but these imaging systems can also work in complete darkness by detecting a target's emitted radiation. The latter of course depends on the potential target having the right temperature and emission properties to be detected by the MWIR imaging system. The spectral radiant emittance over all wavelengths increases, as the temperature of the system (object) increases. Stefan and Boltzmann found the relationship between an object's radiant emittance and its temperature. Wien found that the peak of the radiant emittance for a given object moves to shorter wavelengths, as its temperature increases. This explains why we can see the hot exhaust of a jet engine with our naked eyes in the visible band of wavelengths, but we cannot see the radiation emitted by the "relatively cool" temperatures of people. For example, a person's temperature and physical properties are such that the peak radiant emittance for a human occurs at about 9.4 pm. Consequently, an LWIR imaging system would be better matched to detect this peak radiant emittance than an MWIR imaging system. We will discuss the Stefan-Boltzmann law and Wien's law more in Chapter 9. The second thermal band is in the long-wave part of the electromagnetic spectrum. Here, imaging systems are designed to detect emitted radiation since emitted light dominates over reflected light in this region of the electromagnetic spectrum. The blue spike plume and the red spike plume have to do with gaseous emissions. The Earth's atmosphere is relatively transparent within the NIR, MIR, and FIR regions; in the extreme IR region, the atmosphere is opaque for paths exceeding a few meters in length.

In order to grasp and properly analyze optical systems concepts, some general understanding of the behavior and properties of electromagnetic waves is necessary – especially in the optical part of the electromagnetic spectrum. Different mathematical models and physical concepts have been developed throughout the years that help in analyzing and modeling optical waves. We introduce these methods and provide a short overview along with some key physical concepts and considerations. We focus on the key concepts, results, assumptions, and limitations from a SE perspective. Further detail can be found in some of the discipline-centric texts in the reference section.

One physical concept that is important when working with optical systems is the *duality of light*. Sometimes, it is convenient to think of light as a propagating electromagnetic wave. At other times, it is more convenient to think of light as particles. An interesting fact is that light has both particle and wave properties. For our purposes, we work with what is most convenient for a given application. For example, Equation 1.1 shows that the frequency of an electromagnetic wave can be given by the speed of light c divided by the wavelength of the electromagnetic wave. In this case, we are thinking of an electromagnetic *wave*. When the wavelength is in the optical part of the electromagnetic spectrum (UV, visible, and IR), it is referred to as an *optical wave*. We also know that the energy of a *photon* is given by

$$E_{\upsilon} = h\upsilon, \tag{1.4}$$

where h is Planck's constant, $h = 6.626 \times 10^{-24}\,\text{m}^2\,\text{kg/s}$, and is a fundamental physical parameter.

The photon can be thought of as a quantum of optical energy that behaves like a particle. In some optical systems analysis applications, it is more convenient to think in terms of particles than waves. Some examples would be in understanding low light detection instruments such as *photon detectors* and *photomultiplier tubes*. Photon detectors can detect very low levels of light – down to a few photons – and photomultiplier tubes can effectively amplify photons that are converted into electrons. In this text, we will adopt either the wave or the photon approach where necessary to conveniently analyze optical system properties.

There are several useful analytical methodologies when it comes to understanding, analyzing, and modeling electromagnetic phenomena in the optical part of the spectrum. Which method applies depends on the application at hand. We will provide a brief introduction to these methods and some guidelines for when to apply them, but we will reserve more detailed comments and mathematical developments for later chapters as necessary. The most rigorous analysis in the subject of optical waves and photons can be accomplished through the application of classical electrodynamics (Jackson 1975). Many of the significant results that will be used in our analyses and modeling activities can be derived from fundamental principles using this methodology. Electrodynamics works well for understanding the interaction of light with materials and for relatively well-defined and simple objects such as planes, spheres, crinkled spheres, hemispheres, slabs, cylinders, and combinations of basic shapes. An electrodynamics approach becomes analytically intractable when dealing with very rough, macroscopic objects, and in this regime, computer simulations and other analytical methods become more useful. For manageable boundary conditions, electrodynamics treats electrostatics, magnetostatics, time-varying electromagnetic fields, radiating systems, wave propagation, scattering, diffraction, the interaction of the electromagnetic wave with particles and matter, plasma physics, and more. Electrodynamics applies to homogeneous and nonhomogeneous media, can be applied on extremely small scales – *quantum electrodynamics* – and can also deal with physical problems that involve interactions occurring near the speed of light – *relativistic electrodynamics*. For our applications, we are typically most interested in the radiating, propagating, scattering, diffracting, absorbing, reflecting, transmitting, and material interactive aspects of the electromagnetic field. Additionally, electrodynamics is also useful in describing the electromagnetic wave in a waveguide or in a resonating cavity like in a laser. The fundamental physical equations that describe the behavior of the electromagnetic field are *Maxwell's equations*, and they have both a differential and integral formalism. Additionally, the Helmholtz wave equation can be used to describe electromagnetic wave behavior. Electrodynamics provides a means for understanding the fundamental physical properties of the electromagnetic field and is useful in explaining, deriving, and understanding key characteristics of optical waves.

Fourier optics methods provide a powerful set of analytical tools in the spatial frequency domain that relate the electromagnetic field spectrum, intensity spectrum, or radiant emittance spectrum to corresponding elements in some distant plane such as the imaging system's image plane. Fourier optics methods are a great simplification over the fundamental Maxwell's equations and the Helmholtz equation. The Fourier optics relationships are obtained by simplifying the Rayleigh-Sommerfeld diffraction integrals using a pair of distance constraints. These constraints ensure that the *Fraunhofer diffraction* conditions are established. Fourier optics methods can be derived using classical electrodynamics for the case that the dimensions of the source of the electromagnetic wave d are small compared to z, the separation distance of the source from the observation plane. More specifically, we must look at the following conditions in order to fully apply Fraunhofer diffraction and consequently Fourier optics methods:

$$\frac{kd^2}{2z} \ll 1, \tag{1.5}$$

and,

$$d \ll z, \tag{1.6}$$

where

k is given by $2\pi/\lambda$;

d is the radial size of the diffracting system;

z is the separation of the diffracting system from the observation plane.

As an example, if we were looking at light that is reflected from a 1 m circular plate satellite surface with an optical system that is located 1×10^6m away, we would have 0.785, which is not much smaller than 1, and so Fraunhofer conditions may not apply. However, in many practical situations, Fraunhofer diffraction assumptions and Fourier optics methods apply for the less stringent case $d/z \ll 1$ (Jackson 1975). As an example for the second criterion, a 15 m diameter satellite at 1×10^6m separation distance easily satisfies this condition but fails the conditions of Equation 1.5. Nevertheless, Fourier optics methods are routinely applied using the less stringent criterion of $d/z \ll 1$ with excellent results. We advise a somewhat lenient interpretation of Equation 1.5 as a discriminating factor for the applicability of Fourier optics but advise caution in applying Fourier optics methods if this equation is grossly violated. We will consider the second condition – $d/z \ll 1$ – as the primary condition and Equation 1.5 to play an important supporting role. The result is useful collections of 2D Fourier transform relationships between an emitting and/or reflecting source and relatable electromagnetic elements in a distant observation/image plane. If Equation 1.5 does not hold, then Fresnel diffraction conditions apply and a somewhat more complicated set of relationships exist. For now, we will proceed with Fourier optics methods where appropriate and address Fresnel propagation where needed.

SO is another powerful analytical tool that is very useful in analyzing and understanding optical systems. With SO methods, we are no longer limited to simple scattering models that are required to make classical electrodynamics calculations tractable. We can use statistical techniques to analyze the properties and characteristics of the optical wave. These methods can be combined with Fourier optics methods to provide a powerful and rigorous mathematical framework for understanding optical wave propagation and the overall detection process. For low light levels, Poisson's statistics are useful in describing photon-counting processes. However, at high levels of light, Gaussian statistics are more commonly used. We will provide more insight into this powerful approach in Chapter 6 where we discuss the propagation of radiation in more detail.

There are a variety of other methods that must be mentioned. Modern optics methods allow us to understand and model optical components such as lenses, filters, optical mounts, coatings, optical wave propagation through these components, and polarization effects to mention a few. These methods are important in designing optical components. *Physical optics methods* include diffraction effects in their analytical approach, whereas *geometrical optics* use simplified ray-tracing methods to approximate optical propagation phenomena. Nonlinear optical (NLO) methods provide an approach for understanding the behavior of optical waves when traditional assumptions of linear, shift-invariant systems do not apply. We will discuss the linear, shift-invariant assumption and NLO models in Chapter 2. We will use results from these methods as needed to help us understand, characterize, and model optical waves and optical systems. As optical systems engineers and optical systems managers, we will focus on fundamental results and depend on our optical engineer for detailed analysis. More detailed information may also be found in the references.

We want to make one last comment about our chosen system of units with regard to electromagnetic applications. There are several systems of units that are useful in working with electromagnetic physical phenomena. These are electrostatic units (esu), electromagnetic units (emu), Gaussian units, Heaviside-Lorentz units, and the rationalized MKSA units (SI). The most common two of these units are the Gaussian units and the rationalized MKSA units. The Gaussian units are more practical when dealing with microscopic issues, and the rationalized MKSA units – our adopted SI units – are more practical for macroscopic problems such as in large-scale engineering applications. We now present some definitions of important terms that will be used throughout the text.

1.2.4 Fundamental Optical Systems Concepts and Terminology

This section contains definitions that will become important in understanding concepts throughout the book. We would like the reader to be able to skip to any subsequent section of the book and be able to understand the content. Of course, it is better to sequentially go through the material since each chapter builds on the preceding one. But we wanted to enable those readers who want to jump straight to a particular topic. For instance, if someone wanted to learn about optical detectors, after reading and understanding this section, they should be able to jump straight to Chapters 8–10 and understand the majority of material found there. The reader would not know about any of the preceding material (e.g., sources, optical radiation, basic optical systems), but the material in the detector sections should be understandable. We hope to make each chapter as stand-alone as possible. In order to facilitate this, we need to define some common terminology. In addition to the terminology definition, we provide some additional descriptive detail such as equations and/or insight where appropriate.

Absorptance (α): The ratio of radiant flux absorbed by a body to radiant flux incident on the body:

$$\text{Absorptance} = \frac{\varnothing \text{ absorbed}}{\varnothing \text{ incident}}. \tag{1.7}$$

Acousto-optics: The interaction of light with sound. Acousto-optical devices manipulate light with sound waves.

Active imaging: The imaging system uses a coherent imaging model under manmade illumination conditions such as laser illumination.

Atmospheric turbulence-limited spatial resolution: For most well-designed optical imaging systems that have a large entrance pupil aperture compared to the Fried parameter, r_0, and are in the Earth's atmosphere, atmospheric turbulence effects limit the spatial resolution of the optical imaging system. For this case, the atmospheric turbulence-limited spatial resolution of the imaging system is given by

$$\Delta x_{\text{atm}} = 1.22\frac{\lambda}{r_0}z, \tag{1.8}$$

where

1.22 results from a circular entrance pupil aperture

λ is the center wavelength of the optical wave

z is the separation distance of the observation point from the imaging system's entrance pupil

Adaptive optics: A mostly hardware-based system for real-time correction of atmospheric turbulence effects.

Atmospheric turbulence compensation: Software-dominant methods that correct for atmospheric turbulence in collected data (usually images). These methods are typically much slower than adaptive optics methods and are performed in a postprocessing environment.

Brightness: Spatial power density that is independent of location (e.g., can be either at the source or detector). The units for brightness are W/m^2.

Coherent wave: The autocorrelation of the electromagnetic wave is nonzero at more than one point.

Diffraction-limited spatial resolution: This is the theoretically best possible spatial resolution from an imaging system. It is given by

$$\Delta x_d = 1.22\frac{\lambda}{D}z, \tag{1.9}$$

where

1.22 comes from using a circular aperture

λ is the center wavelength of the light in meters

D is the diameter of the optical system's entrance pupil (typically the primary mirror of a telescope) in meters

z is the separation distance between the observed object and the imaging system's entrance pupil in meters

The left side of the equation represents a size scale for which smaller spatial details cannot be observed with the given imaging system. It can be considered as the spatial resolution of a theoretically perfect imaging system that has no optical or atmospheric aberrations and no system noise effects.

Electro-optics: These are mostly optical devices that are influenced by electronics. An example is an optical switch.

Emissivity (ϵ): It is the ability of a surface to emit radiation. It is equal to the radiated energy from a given source (across all wavelengths) divided by the radiated energy of a blackbody source (at the same temperature).

Extended source: If an object that is viewed by an imaging system has spatial dimensions that are larger than the spatial resolution of the imaging system in question, then, ideally, the imaging system can resolve the spatial features of the object. In this case, the source object can be thought of as an extended source.

Fourier optics: A methodology that uses the spatial frequency domain to analyze and characterize optical systems.

Fried parameter (pronounced freed): The Fried parameter, r_o, was developed by Dave Fried and represents the diameter of a patch of atmosphere over which the atmosphere can be considered spatially coherent. Another name for the Fried parameter is the atmospheric coherence length. For imaging systems that have an entrance pupil diameter equal to or smaller than the Fried parameter, adaptive optics/atmospheric turbulence compensation (AO/ATC) methods have little effect. For imaging systems that have a larger diameter than r_o, AO/ATC methods have some observable benefits.

Incoherent wave: The autocorrelation of the wave is nonzero at only one point.

Photonics: A discipline that combines optics with electronics. It involves using electronic devices to control light in either its wave or particle form (photons). The discipline includes optics, optoelectronics, electro-optics, acousto-optics, optical information processing, lasers and quantum electronics, optical electronics and communications, lightwave devices, fiber-optics communications, or lightwave systems (Saleh and Teich 1991).

Optoelectronics: Devices that are mostly electronic but also have optical aspects to them. Some examples include television displays and photodiodes.

Statistical optics: A methodology wherein the interaction of light with matter is considered a fundamentally statistical phenomenon (Goodman 1985). This methodology provides a powerful set of statistically based analytical tools for understanding light propagation, the interaction of light with macroscopic objects, and spatial and temporal coherence effects of the optical wave.

Lightwave optics: A field wherein optical methods are merged with optical signal processing and optical communications devices.

Nonlinear optics: The interaction of light with nonlinear media. Some examples include phase conjugation, two-wave mixing, four-wave mixing, self-focusing, third harmonic generation, and optical amplification.

Passive imaging: The imaging system uses an incoherent imaging model under naturally illuminated conditions (e.g., sunlight, starlight, and emitted light in the IR part of the electromagnetic spectrum).

Photometry: The measurement of light's brightness as perceived by the human eye. The sensitivity of human eyes varies with the wavelength of the incident energy, so in photometry, the brightness is measured in terms of radiant power at each wavelength.

Photon flux (Φ_p): The number of photons emitted per second, measured in *photons/s*.

Point source: If the spatial resolution of a given imaging system is larger than the physical dimensions of a source object viewed by an imaging system, then the imaging system cannot resolve any spatial details of the object. In this case, we can consider the object to have properties of a point source with respect to the imaging system even though the object may have some physical dimensions.

Radiant energy (U, E, Q_e): This is the energy that is carried by an electromagnetic wave. The units of energy are given in *joules* (*J*). One joule is equivalent to 1 N of force being applied over 1 m. It is also equivalent to 1 W of power over 1 second.

Radiant flux or radiant power (P, Φ_e): This is a measure of the change in energy as a function of time rate and has units of *Watts* (W):

$$1 \text{ W} = 1 \text{ J/s}. \tag{1.10}$$

Radiant emittance or radiant exitance (W, M_e): Radiant flux emitted from a source per unit area. The units of radiant emittance are W/m^2 or W/cm^2. This quantity is called radiant incidence if measured at the imaging system.

Radiant intensity (J, I_e)*:* Radiant flux emitted per unit solid angle, measured in *W/sr.*

Radiance (N, L_e): Radiant flux emitted per unit solid angle per unit area of source, measured in $W/(m^2 \cdot sr)$.

Radiant photon emittance (Q_p): The number of photons emitted per second per unit area, measured in $photons/(s\ m^2)$.

Radiometry: Measurement of electromagnetic radiation. Measurements are made over a range of wavelengths or equivalently frequencies.

Reflectance (ρ): The ratio of radiant flux reflected by a body to radiant flux incident on the body,

$$\text{Reflectance} = \frac{\varnothing \text{ reflected}}{\varnothing \text{ incident}}. \tag{1.11}$$

Steradian: An angular measurement obtained by dividing the surface area of a portion of a sphere by its radius squared. For instance, if the surface area under consideration is the entire sphere, then the associated angular measurement is 4π sr.

Spectral radiant flux (P_λ): The radiant flux per unit wavelength interval, measured in $W/\mu m$.

Spectral radiant emittance (W_λ): Radiant emittance per unit wavelength interval, measured in $W/(m^2 \cdot \mu\text{m})$.

Spectral radiant intensity (J_λ): Radiant intensity per unit wavelength interval, measured in *W/(sr* μm).

Spectral radiance (N_λ)*:* Radiance per unit wavelength interval, measured in *W/(m² sr*·μm).

Spectral irradiance (H_λ): Irradiance per unit wavelength interval, measured in *W/m²*·*μm*.

Total power law: The sum of the absorptance, reflectance, and transmittance of a body is unity Or, in other words, the radiant flux φ incident on a surface is equal to the sum of the radiant flux absorbed, radiant flux reflected, and radiant flux transmitted:

$$\alpha + \rho + \tau = 1. \tag{1.12}$$

or

$$\varnothing \text{absorbed} + \varnothing \text{reflected} + \varnothing \text{transmitted} = \varnothing \text{incident}. \tag{1.13}$$

Transmittance (τ): The amount of radiant flux that is transmitted divided by the total radiant flux that is incident on the body,

$$\text{Transmittance} = \frac{\varnothing \text{transmitted}}{\varnothing \text{incident}}. \tag{1.14}$$

Some optical SE building blocks have been introduced in this chapter that present a conventional system of units and introduce standard optical systems terminology. This chapter provides some basic concepts and methodologies that should help those readers who find it necessary to jump ahead to later chapters. The best approach is to sequentially read the material in this text. That way, the proper foundation and contextual elements can be put in place. The intent of this section was not to provide an all-inclusive set of definitions but just the key definitions needed for general understanding. The next section provides an integrated case study that applies some of the concepts learned in this chapter.

1.3 INTEGRATED CASE STUDY: INTRODUCTION TO OUR OPTICAL SE CASE STUDY

In this section, a case study is included that relates to the topics discussed in the previous two sections. This section demonstrates how the SE approach is applied in the optics field. The intention behind this case study is to help the reader understand how SE actually works in the professional environment and illustrate some of its advantages. Also, we want to use a single integrated case study to show how different aspects of the SE discipline are applied to the same scenario. We adopt a fictitious, mid-size, expanding company called Fantastic Imaging Technologies (FIT) to demonstrate both the higher-level SE concepts and principles and their connection to SE-oriented optical technical analysis, modeling, feasibility, and trade studies. Table 1.9 provides a list of the fictional main characters and their roles in this case study.

FIT is a mid-size imaging company located in Melbourne, FL, in the United States. FIT develops imaging systems that operate in the visual and IR regions of the electromagnetic spectrum. The company has distinct techniques for ground-to-ground, ground-to-air, and air-to-ground imaging scenarios. FIT also has a dedicated research and development team concentrating on improving the performance of long-distance imaging systems with atmospheric turbulence compensation and adaptive optics methods.

Now, let me introduce you to some of FIT's leadership. The chief executive officer (CEO) of the company is Dr. Bill Smith. He is a retired U.S. Air Force officer with over 25 years of experience with optical detection methods, SE, and the defense acquisition process. The chief technical officer (CTO) of the company is Dr. Tom Phelps who is an expert in optics with an emphasis on optical imaging systems. The chief information officer (CIO) of the company is

TABLE 1.9
Cast of Main Characters for Integrated Case Study

Actor	Role	Affiliation
Dr. Bill Smith	CEO	FIT
Dr. Tom Phelps	CTO (optics expert)	FIT
Mr. Garry Blair	CIO (EA)	FIT
Mr. Karl Ben	Senior systems engineer	FIT
Elana D.	Contracting officer	FIT
Jennifer O.	Systems engineer	FIT
Ron S.	Systems engineer (new employee)	FIT
Barry W.	Artificial intelligence and computer science	FIT
Ginny R.	UAV program manager	FIT
Arlene G.	Business development	FIT
Wolf B.	Facility security officer (FSO)	FIT
Marie C.	Requirements management/CM	FIT
Carlos R.	Quality manager	FIT
Lena A.	Logistics analyst	FIT
Phil K.	Software engineering	FIT
Erik S.	Digital engineering specialist	FIT
George B.	Hardware specialist	FIT
Serana V.	Machine learning and signal processing	FIT
Coralyn A.	Computer Networking Expert	FIT
Daegon M.	Mathematician	FIT
Amanda R.	Optical engineer (optics lead)	FIT
Christina R.	Optical technician	FIT
Kari A.	Test manager	FIT
Malcolm P.	Production manager	FIT
Christian A.	Analyst	FIT
Rodney B.	Field service	FIT
Julian F.	Product service	FIT
Tristan R.	MBSE Modeler	FIT
Steven E.	Technical support and IT	FIT
Andy N.	Maintenance/support	FIT
Warren M.	Agency liaison	FIT
Doris	Executive administrative assistant	FIT
Wilford Erasmus	Chief of operations and acquisitions U.S. Customs and Border Protection Primary stakeholder	DHS
Dr. Glen H.	Technology specialist	DHS
Jean H.	Operations management	DHS
Aria R.	Mission planner	DHS
Kyle N.	User	DHS
Adalind R.	UAV operator	DHS
Ben G.	Depot maintenance	DHS
Simon S.	Chief, acquisitions, and procurements	Central Intelligence Agency
Rebecca H.	Special customer	Unnamed organization

Mr. Garry Blair. Dr. Bill Smith is a strong proponent of the SE discipline and is insistent about always following SE methods in all the projects within the company. From the beginning, FIT has applied state-of-the-art SE principles, methods, tools, and techniques in all internal and external programs and projects. The focus on SE is never limited to the products, projects, programs, and systems of the company but also includes FIT's processes and services. Systems thinking and the SE methodology are embedded in the company's culture and core values.

FIT takes a holistic approach to product development and support. From the start, FIT has been actively engaged in an optical system's development over its entire lifecycle (RDT&E, production/manufacturing, operational support, maintenance, service, and system retirement/disposal). MBSE methods are practiced for requirements management and the design phases of the system. The company uses both DOORS and CATIA System-of-Systems Modeler as its customer-specific requirements management and SysML-based modeling tools and uses Zemax™ as its optical design suite. The company also has some internally developed and patented algorithms in a MATLAB®-based rapid prototyping environment that do high-speed atmospheric turbulence compensation. These algorithms have been ported to various high-speed hardware architectures such as the field-programmable gated arrays (FPGAs) and have been developed using modular design methods and state-of-the-art security and service protocols.

As the company started growing and handling multiple projects, the normal information flow between the top-level management team and the technical staff in the company became noticeably less efficient and strained. The normal budgeting and accounting processes also became more difficult to manage. This resulted in an increase in budgeting and scheduling problems within the company. To handle this situation, the CIO proposed adopting an EA framework for the management, integration, and documentation of the company's strategic vision, core business processes, technical processes, information flows, systems, operations, and standards. The CEO and board of directors approved this EA approach, and a baseline EA was established along with a projected development path and review cycle. The EA made it possible to clearly identify the goals and objectives of the company. It helped to define the organizational structure, the information flow, the decision flow, the interactive structure of the people, and the products and processes that are critical to the company's operations and growth. With the implementation of an EA, it became easier to follow up on each of the projects in the company, and also the decision-making process was much simplified and more effective.

FIT recently received an RFP from the U.S. Department of Homeland Security (DHS). The project is to develop an optical subsystem for an unmanned aerial vehicle (UAV) in order to identify human bipedal traffic crossing the United States-Mexico border. This system was required to have a standoff ground range of at least 5 miles for daytime and nighttime conditions. The system also needs to have the capability to do near-real-time image processing and send images to a remotely located mission control center (MCC) within 5 s of the image capture. The optical system must also be capable of bidirectional communication and control for local as well as over long distance. Here is a scene from the discussion that happened between the CEO, Dr. Bill Smith; CIO, Mr. Garry Blair; and CTO, Dr. Tom Phelps, on the project.

Bill: "Hello everybody, I appreciate that you both could make it for the meeting even though this is such short notice. Before deciding about whether to go forward with the proposal effort for the DHS RFP, we need to take a look at the current obligations and resources of the company. We need to discuss where our resources are currently allocated. Also, do we have the proper resources and infrastructure to take on this project? Do we have the staff in house with the needed skill sets? How many employees will we need to work on this project? Also, what will be our strengths and weaknesses for winning the proposal? This is a huge project for our company, so we should be aware of the associated risks."

Tom: "Yes I agree with you Bill. Before jumping into it, we must do some homework and make sure we are ready and capable of taking up this huge new task. When I saw this opportunity, I got excited feeling that this is the break that would propel our company to the next level. Our company has been in the imaging industry for a while now focusing largely on commercial projects. These projects were fitting for the company as we were focusing on optimizing our atmospheric turbulence compensation methods. Now we have a leadership position in atmospheric turbulence compensation methods, and some new doors are beginning to open for us. So, if we are all in agreement that we have enough resources to support this effort, it is time for us to take the next step in becoming the leaders in our industry. This project will be a major step in that direction and, if successful, will open up a new market base.

For this project, we have the right technology for the imaging system. This project presents an opportunity for us to take our system to the next level and integrate it onto a UAV. I understand that we are involved in other projects, but this represents a growth opportunity we can't ignore. The proposal team will put together a very competitive package. The proposal activity itself should only take 2–3 months of effort, so by the time we start the project, some of our other projects will be over, and we will have people to work on the DHS project. In parallel, we can also put some feelers out for potential hires should we win the bid. There's a good university in the local area; we could make some starting inquiries there."

Garry: "I agree with you Tom. Now that we have our enterprise architectural framework set up, and all the goals, objectives, policies, processes, systems, operations, and standards of the company are being laid out clearly, it is easier to see what we have and make the choice. This project clearly matches with the transition we planned for the company while setting up the EA. The company has aligned itself now to take on a project for a government organization. In 1996, the U.S. government passed the Cohen-Clinger Act, which makes it mandatory for all the federal agencies to follow EA methods for their activities. This requirement trickles down to contractors that support these government agencies such as us. Since we already have implemented an enterprise architectural framework, it will be an advantage on our side when submitting our solicitation to the DHS. From a technology standpoint, FIT will also have an advantage that none of our competitors have – we have the best high-speed, atmospheric turbulence compensation technique in the industry! This project calls for long-distance imaging, so the effect of the atmosphere on the image is unavoidable for sufficiently large-aperture optical systems. Our systems will have longer standoff ranges or higher spatial resolution at equivalent standoff ranges than our competitors. We have the technology edge! We should have a better proposal than our competition; now we will just have to keep the lifecycle costs down."

Bill: "Since we all agree that we should go for the DHS project, let's start working on the proposal as soon as possible. This is a new step for us, and so we may need some extra time to ramp up. Tom, you are in charge of the proposal. You and Garry need to make sure that we have all the necessary architectural framework artifacts in the proposal. The proposal will have to be completed within 2 months from today. Should you need any additional details contact Mr. Wilford Erasmus. He is our DHS POC for this project. For proposal discussions, I want you to have two teams: red and blue. The blue team will discuss the pros, and the red team will discuss the cons. I would like to see a weekly status report. We will meet every 2 weeks to discuss our progress. Let's do this! Good luck!!"

After the meeting, Tom set up a team to work on the proposal effort for the DHS project under Mr. Karl Ben, a senior systems engineer in the company. They started the proposal effort by first identifying the stakeholders. Here is another discussion that happened 2 weeks later between Bill and Tom during the first review of the proposal effort.

Tom: "We started the effort of identifying the stakeholders for this project. The team identified many potential stakeholders apart from DHS – like CIA, FBI, and other foreign customers – but we decided to focus solely on the DHS requirements for this proposal effort."

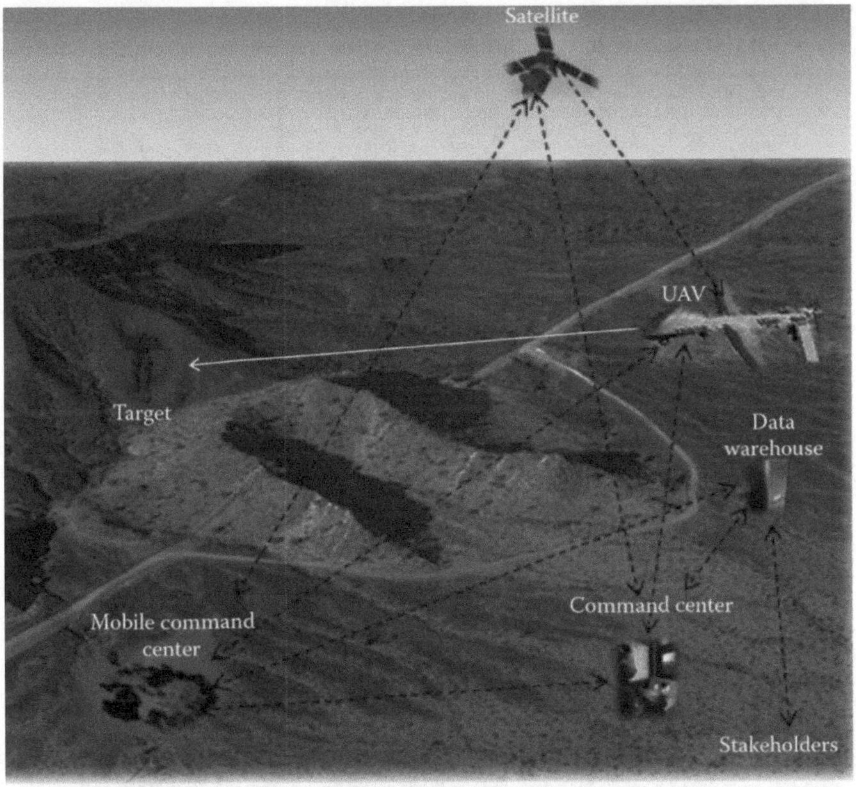

FIGURE 1.30 CONOPS of UAV border patrol mission.

Bill: "That's a good decision. Let us focus only on the DHS requirements right now, and we can think about secondary stakeholders after winning the proposal. If we win, we can use our modular design principles and our rapid prototyping methodology to efficiently pursue opportunities with the secondary stakeholders. Have you come up with an idea of how we are going to address the primary stakeholder requirements in the RFP?"

Tom (pointing to the presentation slide containing the CONOPS): "Yes, we have come up with a CONOPS (Tom points to Figure 1.30 on the screen). Our electro-optical and infrared imaging system within the UAV will be searching for human traffic by using the UAV's conventional scanning mechanism for the imaging solution. Our images will be processed for atmospheric turbulence compensation onboard the UAV and also to make them compatible with the UAV's standard communications protocol. The processed images will then be sent to the mission control center via the communication mechanism available on the UAV. The integrated imaging system will include two different subsystems: a telescopic camera system that works in the visible region of the electromagnetic spectrum for daytime imaging and a telescopic IR camera for nighttime imaging."

Bill: "Can you please explain how each of the systems mentioned above are supposed to work?"

Tom: "OK. Let me start with our nighttime optical imaging solution. All materials radiate and have a radiation profile that peaks at a particular wavelength in the electromagnetic spectrum. Since the target of our optical imaging system is human, humans will be the source of radiation. (Tom shows Figure 1.31 of a full human body.)

The human body emits radiation in the IR part of the electromagnetic spectrum. So, if we know the peak wavelength at which human bodies emit IR radiation, we can use an

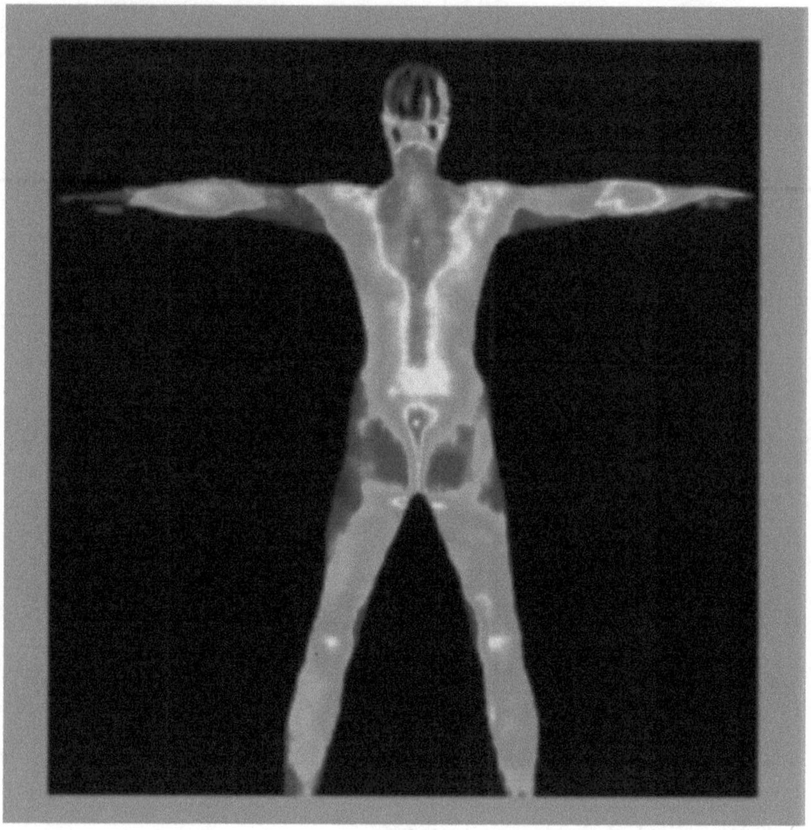

FIGURE 1.31 IR image of human body. (Reprinted from NASA 2014. http://commons.wikimedia.org/w/index.php?title=File:Atmospheric_window_EN.svg&page=l. With permission.)

optical detector that is optimized for detecting humans around that wavelength. There will be risk factors to consider, but assuming that there isn't a lot of absorption, noise from the optical system, and clutter from other objects at, or close to, this wavelength, then we should be able to design our optical system to take full advantage of the peak radiation emitted from the human target of interest. One important thing here is to make sure that the optical detector captures signals only from the desired spectral region in the electromagnetic spectrum. There is a trade-off between the bandwidth of the system, the detector noise, the received irradiance on the detector, and background noise. This should be done as part of our design process where we construct a radiometric model, analyze the imaging scenario, determine the optimal detector characteristics, and then design appropriate optical or signal filters for our CONOPS-driven scenarios.

Another thing to be taken into consideration is the effect of the atmosphere in the region in which the human body emits radiation. If we use a narrowband detector in the wrong part of the electromagnetic spectrum, we may lose a good portion of the available signal from our target of interest due to atmospheric absorption."

[Tom then jumped to the following slide that shows how the atmosphere and different compounds in the atmosphere affect transmitted light, as shown in Figure 1.32. He then continued his explanation.]

Tom: "We know that the human body has its peak radiation at a wavelength of about 9.5 μm. Here, we see that there is a loss in the optical power being transmitted at, and around, this

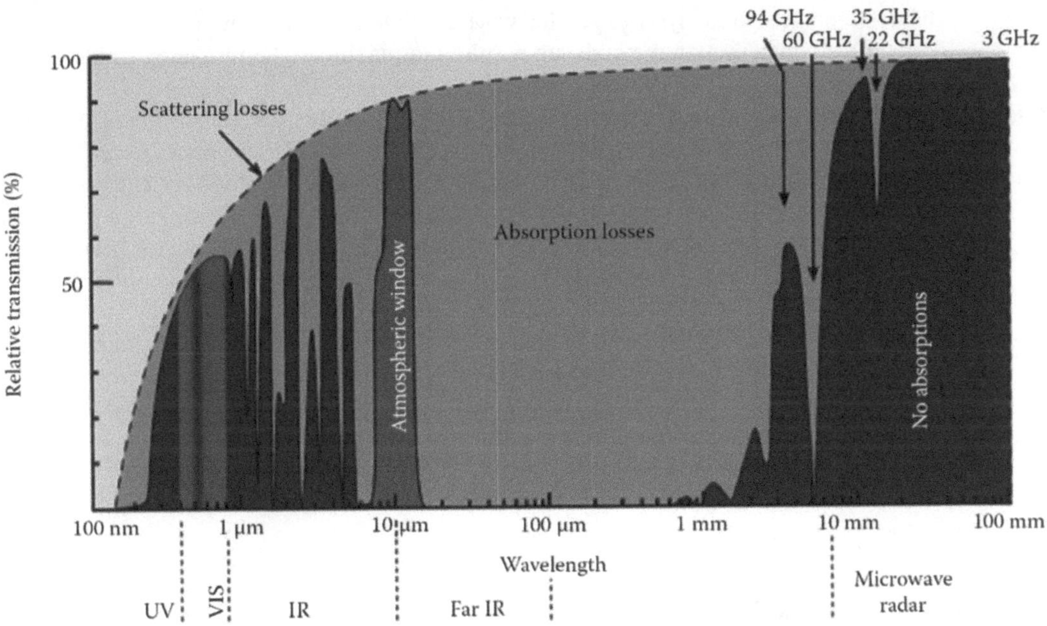

FIGURE 1.32 Atmospheric effects on transmittance. (Reprinted from Cepheiden 2009. With permission.)

wavelength – a little over 20% from this chart. We will have to get a more detailed chart of the atmospheric transmission in the vicinity of 9.5 μm, but this gives us a ballpark result for now. There is also notable absorption to the right of the wavelength for the peak radiant emittance of our prospective human emitter, so we should avoid this part of the electromagnetic spectrum. Now, a major concern we have is to make sure that our optical detector receives enough signals from the source to put out a detectable response. For that, the radiant emittance from the source and the corresponding irradiance at the detector is calculated. If the signal strength is low but within detectable limits, appropriate amplification, noise reduction, and signal processing methods need to be used. Since the radiant emittance produced by humans (and warm-blooded animals) is typically far above the background in nighttime desert environments, we can go with a wide instantaneous field of view (IFOV) imaging system as a queuing sensor. We can then zoom our imaging system to get higher-spatial-resolution images once the queuing sensor tells us where to look."

Tom: "A similar radiometric argument can be made for our daytime solution. Here, the detector has to be matched to the visible part of the electromagnetic spectrum since the dominant available optical signal during the day is from scattered and reflected visible light from the target and from the scene. The queuing sensor idea may not work since there is so much background clutter during the day. There may be some potential with the queuing approach if we can adapt and incorporate some of the latest automatic target recognition algorithms, but we may need to go to a scanning approach to get acceptable spatial resolutions as our primary method. We are still evaluating this scenario. Also, we can use our real-time atmospheric turbulence compensation methods to enhance the spatial resolution of our optical system or provide a further standoff range than our competitors. I think DHS will like that."

Bill: "How much spatial resolution do you need for the DHS CONOPS?"

Tom: "A rule of thumb is that we need to have at least ten picture elements (pixels) across our target of interest to properly distinguish whether we are looking at a human or something else. Remember that this is actual resolution across the target of interest (TOI), and so we will have to conduct an analysis to see for what CONOPS-driven scenarios our atmospheric turbulence compensation methods are required to maximize the resolution of our optical system."

Tom: "All I have talked about so far are the technical basics for this project. In the next part of the effort, we will have to develop an optical systems model and begin to analyze our optical scenario in more detail. We also have to do some preliminary calculations to see what effects that we can expect on our imaging system spatial resolution due to the atmosphere. As you know, for many well-designed optical systems, atmospheric effects are often the principal limiting factor in the optical systems spatial resolution. We also have to model the imaging scenarios that we expect from our CONOPS and start to get a handle on look angles, expected slew rates for the pointing and tracking system, and the basic optics needed for our design models.

I was talking to Garry earlier today, and he was mentioning that we should consider some actions relative to our EA. He said it wasn't anything that couldn't wait until the next meeting, but he seemed pretty excited."

Bill: "This looks like a great start; I am quite happy with the work so far. I will see you and Garry at the next review."

1.A APPENDIX: ACRONYMS

AFIT	Air Force Institute of Technology
CDR	Critical Design Review
CEO	Chief executive officer
CIO	Chief information officer
CMMI	Capability Maturity Model Integration
CONOPS	Concept of operations
COTS	Commercial off-the-shelf
CPFF	Cost plus fixed fee
CPIF	Cost plus incentive fee
CTO	Chief technical officer
DAS	Defense Acquisition System
DAU	Defense Acquisition University
DHS	Department of Homeland Security
DOD	Department of Defense
DODAF	Department of Defense Architectural Framework
DOORS	Dynamic Object-Oriented Requirements System
DOTMLPF	Doctrine, organization, training, materiel, leadership, personnel, and finance
EA	Enterprise architecture
ECPs	Engineering change proposals
emu	Electromagnetic units
esu	Electrostatic units
FAR	Federal Acquisition Regulations
FFP	Firm fixed price
FIR	Far infrared
FIT	Fantastic Imaging Technologies
FoS	Family-of-systems
FPGAs	Field-programmable gated arrays

FT&E	Functional test and evaluation
IFOV	Instantaneous field of view
INCOSE	International Council on Systems Engineering
JCIDS	Joint Capabilities Integration and Development System
JROC	Joint Requirements Oversight Council
KPPs	Key performance parameters
LCC	Lifecycle costing
LWIR	Long-wave infrared
MBSE	Model-based systems engineering
MCC	Mission control center
MDA	Milestone Decision Authority
MIR	Middle infrared
MIT	Massachusetts Institute of Technology
MOPs	Measures of performance
MWIR	Mid-wave infrared
NIR	Near infrared
NLO	Nonlinear optical
OSD	Office of the Secretary of Defense
OT&E	Operational test and evaluation
PDR	Preliminary Design Review
POC	Point of contact
PPBE	Planning, programming, budgeting, and execution
PPPI	Preplanned product improvement
RDT&E	Research, development, test, and evaluation
RFP	Request for proposal
SDLC	Systems development lifecycle
SEI	Software Engineering Institute
SI	International System of Units
SI	Rationalized MKSA units
SO	Statistical optics
SOS	Systems-of-systems
SSPPs	Systems, services, products, or processes
SSR	System Segment Review
SWIR	Short-wave infrared
TPM&Ms	Technical performance measures and metrics
UAV	Unmanned aerial vehicle
VCJCS	Vice chairman of the joint chiefs of staff

REFERENCES

ACQWeb. 2014. Requirements and acquisition process flow. Adapted from: https://acqnotes.com/acqnote/acquisitions/dodi-5000 (accessed March 3, 2025).

Agile Manifesto. 2001. Manifesto for Agile software development. https://Agilemanifesto.org/ (accessed February 18, 2022).

AF DevSecOps. 2024. DevOps and DevSecOps metrics https://software.af.mil/resources/devops/ (accessed on February 8, 2024).

Queen, C. 2025. What is DevOps. https://www.crowdstrike.com/cybersecurity-101/cloud-security/devops/ (accessed March 3, 2025).

Aniche, M., A.M. Finavaro, and G.A. Silveira. 2011. Increasing learning in an Agile environment: lessons learned in an Agile team. *2011 AGILE Conference, 2011*, São Paulo, Brazil. https://ieeexplore-ieee-org.portal.lib.fit.edu/document/6005834.

ANSI/EIA-632-1999 (R2003). 2003. Processes for engineering a system. American National Standards Institute and Electronic Industries Alliance. Washington. https://webstore.ansi.org/ (accessed November 10, 2014).

Arrasmith, W.W. 2007. A systems engineering entrepreneurship approach to complex, multi-disciplinary university projects. Paper presented at *National Conference of the American Society of Engineering Education*, Honolulu, HI, June 23–27, 2007.

Bennett, C.H. 1988. Logical depth and physical complexity. In: Herken, R. (ed.), *The Universal Turing Machine: A Half-Century Survey*. Oxford: Oxford University Press, pp. 227–257, CiteSeerX 10.1.1.70.4331

Blanchard, B.S. and W.J. Fabrycky. 2010. *Systems Engineering and Analysis,* 5th edn. London, UK: Pearson.

Boardman, J., B. Sauser, L. John, and R. Edson. 2009. The conceptagon: a framework for systems thinking and systems practice. In: *Proceedings of the IEEE Conference on Systems, Man and Cybernetics*, San Antonio, TX, October 11–14.

Camelia Boban, Wikimedia. 2014. https://commons.wikimedia.org/wiki/File:BigData_2267x1146_white.png (accessed May 7, 2017).

Cepheiden. 2009. Atmospheric window. Wikimedia commons website: https://commons.wikimedia.org/w/index.php?title=File:Atmospheric_window_EN.svg&page=l (accessed November 10, 2014).

Chan, S. 2001. Complex adaptive systems. ESD.83 Research Seminar in Engineering Systems, MIT. https://web.mit.edu/esd.83/www/notebook/Complex%20Adaptive%20Systems.pdf.

Checkland, P.B. 1989. Soft systems methodology. *Human Systems Management*, Vol. 8, No. 4, pp. 273–289. https://journals.sagepub.com/doi/abs/10.3233/HSM-1989-8405.

Chism, J. 2004. Overview and status of the object-oriented systems engineering methodology OOSEM. *Insight*, Vol. 7, No. 2, pp. 31–33.

Cohen, B. and M. Thias. 2009. The failure of the off-shore experiment: a case for collocated Agile teams. *2009 Agile Conference*. https://ieeexplore-ieee-org.portal.lib.fit.edu/document/5261076, Chicago, IL. ISBN: 978-0-7695-3768-9.

CROWDSTRIKE. 2024. What is DevSecOps. https://www.crowdstrike.com/cybersecurity-101/what-is-devsecops/ (accessed February 8, 2024.)

DAU. 2025. *Online Defense Acquisition Guidebook.* Defense Acquisition University website. https://aaf.dau.edu/guidebooks/ (accessed March 03, 2025).

Defense Acquisition Guidebook. 2012. *Defense Acquisition Guidebook PDF Created 1 November 2012.* https://www.acqnotes.com/Attachments/Defense-Acquisition-Guidebook-11-01-2012.pdf (accessed March 03, 2025).

Defense Acquisition University Press. 2001. *Systems Engineering Fundamentals.* Fort Belvoir, VA: Defense Acquisition University Press.

Department of Defense. 1969. MIL-STD-499 Military standard system engineering management. https://segoldmine.ppi-int.com/node/45238 (accessed March 03, 2025).

Department of Defense. 2023. Department of Defense Releases the President's Fiscal Year 2024 Defense Budget. https://www.defense.gov/News/Releases/Release/Article/3326875/department-of-defense-releases-the-presidents-fiscal-year-2024-defense-budget/#:~:text=The%20FY%202024%20budget%20request,credible%20forces%20across%20all%20domains (Accessed July 20, 2023).

Dickerson, C.E. and D.N. Mavris. 2010. *Architecture and Principles of Systems Engineering*. Boca Raton: CRC Press.

Dori, D. 2016. *Model-Based Systems Engineering with OPM and SysML.* New York: Springer.

Drury, M., K. Conboy, and K. Power. 2011. Decision making in Agile development: a focus group study of decisions and obstacles. *2011 AGILE Conference.* https://ieeexplore-ieee-org.portal.lib.fit.edu/stamp/stamp.jsp?tp=&arnumbe r=6005504.

Estefan, J. 2008. Survey of Model-Based Systems Engineering (MBSE) Methodologies. https://www.omgsysml.org/MBSE_Methodology_Survey_RevB.pdf.

Fagan, M.D. 1978. *A History of Engineering and Science in the Bell System: National Service in War and Peace.* Murray Hill, NJ: Bell Laboratories.

Faurre, P.L. 1975. Systems engineering methodology. *IFAC Proceedings Volumes*, Vol. 8, No. 1, pp. 660–63. Part 3.

Fitts, P.M. 1951. *Human Engineering for an Effective Air Navigation and Traffic Control System.* Washington, DC: National Research Council. https://apps.dtic.mil/sti/tr/pdf/ADB815893.pdf (accessed March 3, 2025).

Ge, B., K.W. Hipel, L. Li, and Y. Chen. 2012. A data-centric executable modeling approach for system-of-systems architecture. *2012 7th International Conference on System of Systems Engineering (SoSE)*, Genoa, Italy, pp. 368–373, https://doi.org/10.1109/SYSoSE.2012.6384149.

Gharajedaghi, J. 2011. *Systems Thinking: Managing Chaos and Complexity: A Platform for Designing Business Architecture*. Morgan Kaufmann, MA: Burlington.

GIG. 2014. Global information Grid OV-1. Wikimedia commons website. https://commons.wikimedia.org/wiki/File:Gig_ov1.jpg (accessed Mar 3, 2025).

Goode, H.H. and R.E. Machol. 1957. *Systems Engineering*. New York: McGraw-Hill.

Goodman, J.W. 1985. *Statistical Optics*. New York: Wiley-Interscience.

Grants.Gov. 2014. https://www.grants.gov (accessed November 8, 2014).

Hall, A.D. 1962. *A Methodology for Systems Engineering*. Princeton, NJ: Van Nostrand.

Huang, P. 2016. Agile development projects-building the fast track. In Ann Garrison Darrin, M. and J.A. Krill (eds), *Infusing Innovation into Organizations: A Systems Engineering Approach*. CRC Press, Boca Raton, FL, pp 295-313.

IBM. 2022. Digital Twin Market by Enterprise, Application (Predictive Maintenance, Business optimization), Industry (Aerospace, Automotive & Transportation, Healthcare, Infrastructure, Energy & Utilities) and Geography (https://www.ibm.com/think/topics/what-is-a-digital-twin)—Global Forecast to 2027, Digital Twin Market.

IEEE. 1990. IEEE standard glossary of software engineering terminology. *IEEE Std 610.12-1990*, pp.1–84, 31 December 1990. https://doi.org/10.1109/IEEESTD.1990.101064.

IEEE. 1998. Standard for application and management of the systems engineering process—description. IEEE Std 1220-1998. https://standards.ieee.org/findstds/standard/1220-2005.html (accessed April 24, 2014).

INCOSE. 2021. INCOSE Board of Directors, "Systems Engineering Vision 2035 Executive Summary." https://www.incose.org/docs/default-source/se-vision/incose-se-vision-2035-executive-summary.pdf?sfvrsn=ff2063c7_2.

INCOSE 2025. INCOSE MBSE Initiative Homepage, "MBSE Initiative." https://www.incose.org/communities/working-groups-initiatives/mbse-initiative#:~:text=MBSE:%20A%20formalized%20application%20of,to%20a%20model%2Dbased%20practice.

Jackson, J.D. 1975. *Classical Electrodynamics*. New York: John Wiley & Sons.

Kamarani, A.K. 2005 (August 5). Systems engineering and integration for advanced life support system and HST." *NASA Summer Faculty Fellowship Program 2004* 1 and 2, pp. 1–17. https://ntrs.nasa.gov/citations/20050202029.

Kasser, J. 2010. Holistic thinking and how it can produce innovative solutions to difficult problems. *Proceedings of 7th Bi-annual European Systems Engineering Conference (EuSEC)*, Stockholm, Sweden. INCOSE, pp. 1–17.

Killingback, T. and M. Doebeli. 1996. Spatial evolutionary game theory: Hawks and Doves revisited. *Proceedings: Biological Sciences*, Vol. 263, No. 1374, pp. 1135–1144.

Kolmogorov, A.N. 1941a. Dissipation of energy in the locally isotropic turbulence. *Proceedings of the USSR Academy of Sciences*, Vol. 32, pp. 16–18.

Kolmogorov, A.N. 1941b. The local structure of turbulence in incompressible viscous fluid for very large Reynold's numbers. *Proceedings of the USSR Academy of Sciences*, Vol. 30, pp. 301–305.

Kolmogorov, A.N. 1965. Three approaches to the quantitative definition of information. *Problems of Information Transmission*, Vol. 1, No. 1, pp. 1–7.

Krzanowski, R. and J. Raper. 2001. *Spatial Evolutionary Modeling*. Oxford: Oxford University Press.

Ladyman, J. and J. Lambert. 2012. What is a complex system? *European Journal for Philosophy of Science*. https://doi.org/10.1007/s13194-012-0056-8.

Lankhorst, M. and SpringerLink (Online service). 2009; 2012; 2013. *Enterprise Architecture at Work: Modelling, Communication, and Analysis*, 2nd ed. Heidelberg; Berlin: Springer-Verlag. Aufl. 2; 3.

Maani, K.E. and V. Maharaj. 2004. Links between systems thinking and complex decision making. *System Dynamics Review*, Vol. 20, No. 1, pp. 21–48.

McLaughlin, L.L. 1987. User engineering: a new look at system engineering. *NASA. Lyndon B. Johnson Space Center, Houston, Texas, First Annual Workshop on Space Operations Automation and Robotics (SOAR 87)*, October 1, 1987. https://ntrs.nasa.gov/citations/19880007847.

Mella, P. 2012. *Systems Thinking: Intelligence in Action*. Aufl.;1; ed. Vol. 2;2. New York: Springer.

Merriam-Webster. 2003. *Merriam-Webster's Collegiate Dictionary*, 11th edn. Springfield, MA: Merriam-Webster.

Morales, J., R. Turner, S. Miller, P. Capell, P. Place, and D. Shepard. 2020. Guide to implementing DevSecOps for a system of systems in highly regulated environments. (CMU/SEI-2020-TR-002). *Carnegie Mellon University, Software Engineering Institute's Digital Library*. Software Engineering Institute, April 8, 2020. https://doi.org/10.1184/R1/12363770.v1.

Mugadza, G. 2015 (February 9). Systems thinking and design thinking: complementary approaches. Research Gate: https://www.researchgate.net/publication/274071837_Systems_Thinking_and_Design_Thinking_Complimentary_Approaches, 5 pages.

Mukherjee, S., S. Roy, and P.P. Acharjya. 2021. Test evaluation metrics and test case prioritization in the dynamics of software testing. *2021 Fourth International Conference on Electrical, Computer and Communication Technologies (ICECCT)*, Erode, Tamil Nadu, pp. 1–8, https://doi.org/10.1109/ICECCT52121.2021.9616658.

NASA. 2007. EM spectrum properties. Wikimedia Commons website: https://en.wikipedia.org/wiki/File:EM_Spectrum_Properties_edit.svg (accessed November 10, 2014).

NASA. 2014. Hubble space telescope diagram. NASA Quest website: https://science.nasa.gov/missions/hubble/30-years-ago-hubble-launched-to-unlock-the-secrets-of-the-universe/ (accessed March 4, 2025).

Newton, I. 1952. Optics. In: R.M. Hutchins (ed.), *Great Books of the Western World*. Chicago, IL: Encyclopedia of Britannica, pp. 377–550.

Opte Project, Wikipedia. 2005. The Internet map. https://en.wikipedia.org/wiki/File:Internet_map_1024.jpg (accessed May 9, 2017).

Pettey, C. 2017. *Prepare for the Impact of Digital Twins*. Stamford, CT: Gartner.

Poli, R. 2013. A note on the difference between complicated and complex social systems. *Cadmus*, Vol. 2, 142–147. https://www.cadmusjournal.org/files/pdfreprints/vol2issue1/reprint-cj-v2-i1-complex-vs-complicated-systems-rpoli.pdf.

Rechtin, E. 1991. *Systems Architecting: Creating and Building Complex Systems*. Englewood Cliffs, NJ: Prentice Hall.

Richmond, B. 1994. Systems thinking/system dynamics: let's just get on with it. *System Dynamics Review (Wiley)*, Vol. 10, No. 2/3, pp. 135–157.

Rittel, H.W.J. and M.M. Webber. 1973. Dilemmas in a general theory of planning. *Policy Sciences*, Vol. 4, pp. 155–169.

Royce, W.W. 1970. Managing the development of large software systems, pp. 1–11. https://Agileconsortium.pbworks.com/w/file/fetch/52184636/waterfall.pdf.

Sage, A.P 1992. *Systems Engineering*. New York: Wiley.

Saleh, B.E.A. and M.C. Teich. 1991. *Fundamentals of Photonics*. New York: Wiley-Interscience.

Schlager, J. 1956. Systems engineering: key to modern development. *IRE Transactions*, EM-3, pp. 64–66.

Schwaber, K. and M. Beedle. 2002. *Agile Software Development with Scrum*. Upper Saddle, River, NJ: Prentice-Hall.

Senge, P.M. 2006. *The Fifth Discipline: The Art and Practice of the Learning Organization*. Rev. (updated). New York: Doubleday/Currency, available at: https://www.google.com/books/edition/The_Fifth_Discipline/b0XHUvs_iBkC?hl=en&gbpv=1&printsec=frontcover.

Skoog, D.A., F.J. Holler, and S.R. Crouch. 2006. *Principles of Instrumental Analysis*, 6th edn. Wadsworth, Australia: Thomson Brooks/Cole.

Stargel, D., E. Glaessgen, et al. 2012. The digital twin paradigm for future NASA and U.S. Air force vehicles: 53rd AIAA/ASME/ASCE/AHS/ASC structures, structural dynamics and materials conference. *Structures, Structural Dynamics, and Materials and Co-located Conferences*, ARC, Honolulu, Hawaii, June 14, 2012. https://arc.aiaa.org/doi/abs/10.2514/6.2012-1818.

TOGAF. 2012. The TOGAF welcome screen. https://pubs.opengroup.org/architecture/togaf8-doc/arch/ (accessed April 24, 2014).

TOGAF. 2014. Definitions. The Open Group Definitions web page. https://pubs.opengroup.org/architecture/togaf9-doc/arch/chap03.html (accessed November 10, 2014).

U.S. Government. 2014. SBIR/STTR. https://www.sbir.gov (accessed November 8, 2014).

Wagenhals, L.W., S.W. Liles, and A.H. Levis. 2009. Toward executable architectures to support evaluation. *2009 Proceedings International Symposium on Collaborative Technologies and Systems*, Baltimore, MD, pp. 502–511.

Wolpert, D.H. & W. Macready. 2005. Self-dissimilarity as a high dimensional complexity measure. https://ntrs.nasa.gov/api/citations/20050185522/downloads/20050185522.pdf (Accessed: March 4, 2025).

Xia, X., K. Qu, J. Shi, Z. Fan, and L. Xu. 2017. The construction of effectiveness evaluation model based on system architecture. *2017 IEEE International Systems Engineering Symposium (ISSE)*, Vienna, Austria, pp. 1–4, https://doi.org/10.1109/SysEng.2017.8088319.

2 Enterprise Architecture Fundamentals

Some regard private enterprise as if it were a predatory tiger to be shot. Others look upon it as a cow that they can milk. Only handfuls see it for what it really is—the strong horse that pulls the whole cart.

—Winston Churchill

2.1 HIGH-LEVEL INTEGRATED MODEL

Enterprise architecture (EA) methods are becoming an essential element in forging successful partnerships and collaborations, between multiorganizational entities working on complex programs, projects, and efforts. These methods are used to apply architecture principles and practices needed to guide large and/or expanding organizations to evolve from simple to a complex organizational structure through business, information, process, and technology changes to execute core enterprise strategies.

The Institute of Electrical and Electronics Engineers (IEEE) defines architecture as "architecture is the fundamental organization of a system embodied in its components, their relationships to each other, and to the environment, and the principles guiding its design and evolution." The main aim of EA is to capture/establish the core business, organizational and technical integration, standards, and governing aspects of an enterprise and focus on the planning, design, and evolution toward a future state of the enterprise. The EA methodology also provides a strategic and business-driven framework for making systems and infrastructure development decisions, provides an enterprise modeling capability that integrates with model-based systems engineering (MBSE) tools, and captures the vision and direction of an enterprise. The main three perspectives of EA are the regulation-oriented perspective, design-oriented perspective, and pattern-oriented perspective.

The "blueprint analogy" is often used in describing and understanding EA. Imagine building a room without blueprints. This can be compared to developing business resources and systems without EA (Bernard 2012). Systems have become more complex over the past century presenting organizations with new opportunities and challenges. Stakeholders use systems and architectural concepts, principles, and procedures to understand and manage the composition and evolution of the enterprise. Stakeholders can use architectural descriptions to improve communication and cooperation regarding business strategies to achieve desired goals. Architectural frameworks and description languages are used to organize conventions and common practices of architecture in various communities and application domains. EA acts as the "collaboration force between the aspects of business planning, business operations, automation and enabling technological infrastructure of the business" (de Vries 2010, pp. 17–29).

EA outlines organizational mission, goals, and objectives and translates business vision and strategy by improving and communicating key requirements, processes, and models to enable change throughout the enterprise. This includes information on how the architecture should adapt in response to the change in the mission.

EA provides a structure for management (Bernard 2012) and aims to define clear relationships from business and/or strategic goals and objectives to performance improvements for the enterprise. To be able to achieve the desired performance improvements, the EA must evolve and be progressively applied to the entire enterprise. The EA and its principles and methods must be supported

DOI: 10.1201/9781003624097-3

by top management and must be progressively implemented throughout the entire enterprise. EA methods help facilitate global thinking in technology and systems selection and provide control mechanisms, standardization, and governance mechanisms to the enterprise. The EA should also project and capture the impact of performance improvements. Metrics should be established that incorporate stakeholder feedback, capture the value added, and improve the quality of EA products, services, and processes (Schekkerman 2008).

In the previous analogy of the blueprints for a house, we described the importance of knowing the "big picture" (the blueprints) before embarking on a venture (building the house). To extend this analogy, a little further, if building a house with blueprints represents building a product, process, system, or service with systems engineering (SE) principles, then EA methods would be equivalent to urban planning activities. The house (system) that is developed by the blueprints (SE methods) is part of a greater enterprise (the city/state/country subject to urban planning) that has its own governance structure, processes, systems, and standards (the EA). The EA captures the governance, policies, strategic direction, technical systems, standards, and processes (the urban plan and associated actors and processes) and integrates the activities of the entities making up the enterprise (individual architects/builders). In essence, the EA provides a high-level integrated model for its constituent members (entities). These principles and methods work for massively large organizations such as the Department of Defense (DOD), federal agencies, and even international cooperative efforts. They also apply to smaller organizations that integrate into larger organizations or that plan on evolving into large organizations.

2.1.1 Purpose of EA

Initially, EA tries to improve the overall efficiency of an organization or business by considering and implementing structural improvements within an organization, the centralization and/or integration of business processes, and the data flow and ensuring the justification of information technology (IT) investments. The EA approach can be used at multiple levels of an organization by using strategic drivers to align business practices and processes, standardize and integrate ITs and systems, streamline decision making, and manage business transformation activities (GAO 2010). EAs are built on an architecture framework. There are many EA frameworks to choose from (e.g., The Open Group Architecture Framework [TOGAF]). TOGAF describes a method for developing an "architectural vision" also referred to as a business with an objective. If a vision is articulated, a road map in the form of transitional architectures is constructed to provide guidance for changing from the present to the future state. Similar methods are described in other EA frameworks. Each of the enterprise architectural frameworks exhibits uniqueness. TOGAF, for instance, has strong defining methods for architectural development, while others have a strong definition of EA artifacts and taxonomy (Zachman 1987, pp. 276–292).

Enterprises will face changes and challenges, which may include mergers, acquisitions, innovations, new technologies, new business models, deregulation, and increased global competition. Advances in e-commerce, globalization of networks, virtual enterprises, and resource availability put a further strain on corporations to manage informational technology and data flow. As technology evolves at an ever-increasing rate, the business environment becomes increasingly dynamic. In order to improve the chance of business survival in today's competitive global environment, enterprises need to evolve and become more agile to allow them to adapt to change quickly and effectively.

2.1.2 Historical Background

In his article entitled *A Framework for Information Systems Architecture* released in 1987, J.A. Zachman (Zachman 1987, pp. 276–292) describes the challenges and the vision of EA as a way to manage the complexity of disturbed systems as they evolve and the addition of needed capabilities into these entities. He envisioned a holistic approach to systems architecture that valued each important aspect that would allow businesses to adapt to change. The costs and the success of a business that depends on its information system require a disciplined approach to managing those systems. This broad view method for architecture systems was first called an "information systems architectural framework" and later coined "enterprise architecture framework" (Zachman 1987, pp. 276–292).

In 1994, the U.S. DOD used Zachman's work in creating an approach known as the Technical Architecture Framework for Information Management (TAFIM) (DOD 1994) that promised to align business needs with technical projects to ensure the achievement of strategic goals and objectives. In 1996, the U.S. Congress issued a bill known as the Clinger and Cohen Act of 1996 or the Information Technology Management Reform Act (ITMRA) (Clinger and Cohen Act 1996). This act established chief information officers (CIOs) for government agencies. The federal agencies were required to focus on the results and the streamlining of IT procurement processes and placed emphasis on the selection processes and management of IT projects to improve business value. Simplified acquisition procedures were allowed if project costs were under $5 million.

The CIO Council began to work on the Federal Enterprise Architecture Framework (FEAF) in 1998 and was able to release Version 1.1 by September 2009 (Hite 2004). FEAF discussed segment architectures as a starting point for smaller parts of an organization. Transitioning from the CIO, the Office of Management and Budget (OMB) became responsible for the FEAF in 2002. OMB changed the name of FEAF to Federal Enterprise Architecture (FEA), and its function will be discussed further later. Though the Federal Government placed significant emphasis on EA, the progress was rather slow, and high-profile failures were documented. The General Accountability Office (GAO) reported that in 8 years after the implementation of the Clinger and Cohen Act, only 20 out of 96 agencies had started establishing the groundwork necessary for managing architectures. Furthermore, the GAO determined that a number of failures occurred within agencies' application of EA methods, including the Federal Bureau of Investigation (FBI) (GAO 2005a), DOD (Hite 2006), and the National Aeronautics Space Administration (NASA) (GAO 2005b).

TAFIM was revoked 4 years after its introduction, and The Open Group assumed the responsibility of documenting a new standard known as "The Open Group Architecture Framework (TOGAF)" released in 2003. Around this time, the DOD also created their own architectural framework named the "Department of Defense Architecture Framework (DODAF)." These two standards continued to expand and were improved for years along with the FEA. Figure 2.1 shows the evolution of the EA methodology over the years.

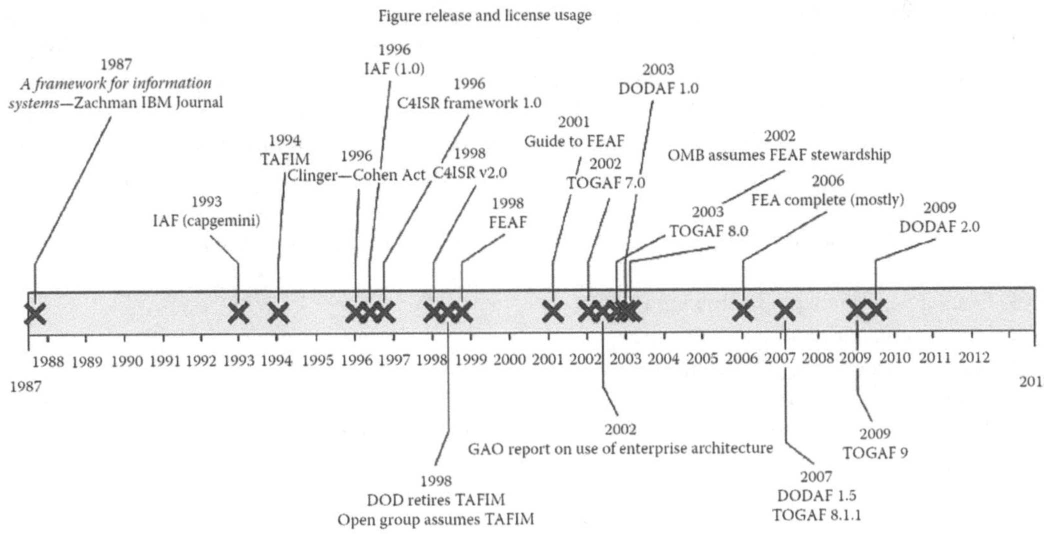

FIGURE 2.1 EA timeline. (Courtesy of Chris Lannon.)

2.1.3 Types of Architectures

Many varieties of EA frameworks are in use today such as Gartner, TOGAF, the FEAF, and the Zachman EA framework. Each of these frameworks has advantages and disadvantages that need to be considered in certain scenarios. Each type is designed to address the problems of all businesses including the relationship of business strategy to IT. We start with providing some of the key features for each of the major architecture frameworks.

The first EA type is the Gartner approach created by the Gartner Group. The Gartner Group is known for its IT research as a consulting organization (de Vries 2010, pp. 17–29). Their approach to EA is considered more of a practice than a methodology or process. This EA practice is highly collaborative and promotes communications among business owners, IT specialists, and technology engineers.

The second EA framework is TOGAF. This EA framework integrates four different architectures into its approach. These architectures are:

- *Business architecture:* Describes the processes able to meet business goals;
- *Application architecture:* Deals with the how applications interact;
- *Data architecture:* Concerned with the organization of data;
- *Technical architecture:* Concerned with the infrastructure of hardware and software in the enterprise (Hornford 2011).

The third type of architecture is the FEA. This architecture unites a number of federal government agencies under one common architecture. FEA has a well-defined architectural process and taxonomy that is usable as a possible methodology or as an evaluation tool of results achieved by executing the process. It has five reference models to measure performance: business, components, service, technical, and data. Furthermore, it gives perspectives in viewing the EA, processes creating EA, taxonomy, and approaches for measuring success (Whitehouse 2007). Finally, the Zachman framework mostly contains taxonomy on how EA is structured and organized. Zachman uses a catalog approach by organizing items into classifications that allow stakeholders to make better decisions (O'Rourke et al. 2003).

2.1.3.1 Relative Comparisons

In this section, we make a subjective comparison between the TOGAF, FEA, Zachman, and the Gartner EA frameworks. The intent is to highlight the relative strengths and weaknesses of these major EA frameworks. The categories and rankings come from Sessions (2007) from a posting in the Microsoft Developer Network (MSDN) library (Sessions 2007). This comparison is a useful example, and the results are shown practically verbatim as follows. Some of these criteria may not be directly relevant for your organization or enterprise, but these evaluations provide a place to start when considering an architectural framework for your own organization. The comparative analysis is as follows:

Each EA framework methodology is ranked according to the following criteria:

1. Does a very poor job in this area;
2. Does an inadequate job in this area;
3. Does an acceptable job in this area;
4. Does a very good job.
   1. *Taxonomy completeness:* Refers to the efficiency of using the indicated EA framework methodology to classify various architectural artifacts. This is almost the entire focus of Zachman. None of the other methodologies focuses as much on this area.

Ratings:

Zachman:	4	TOGAF:	2	FEA:	2	Gartner:	1

2. *Process completeness:* Refers to the efficiency of a methodology's guidance through a systematic process in creating EA. This is almost the entire focus of TOGAF, with its Architecture Development Method (ADM).

   Ratings:

Zachman:	1	TOGAF:	4	FEA:	2	Gartner:	3

3. *Reference model guidance:* Refers to the usefulness of the methodology in building a relevant set of reference models. This is almost the entire focus of FEA. TOGAF also provides support; however, the TOGAF reference model guidance is not as complete as the FEA reference model guidance.

   Ratings:

Zachman:	1	TOGAF:	3	FEA:	4	Gartner:	1

4. *Practice guidance:* Refers to how much the methodology facilitates assimilating the mindset of an EA approach into the organization and develops a culture in which the EA is valued and used. This is a primary focus of Gartner's architectural practice.

   Ratings:

Zachman:	1	TOGAF:	2	FEA:	2	Gartner:	4

5. *Maturity model:* Refers to how much guidance the methodology gives in assessing the effectiveness and maturity of different organizations within the enterprise in using EA.

   Ratings:

Zachman:	1	TOGAF:	1	FEA:	3	Gartner:	2

6. *Business focus:* Refers to the methodology's focus on using technology to drive business value specifically defined as either reduced expenses and/or increased income. Ratings:

Zachman:	1	TOGAF:	2	FEA:	1	Gartner:	4

7. *Governance guidance:* Refers to the methodology's understanding and creating an effective governance model that guides stakeholder interactions.

   Ratings:

Zachman:	1	TOGAF:	2	FEA:	3	Gartner:	3

8. *Partitioning guidance:* Refers to how well the methodology will guide effective autonomous partitions of the enterprise, which is an important approach to managing complexity.

   Ratings:

Zachman:	1	TOGAF:	2	FEA:	4	Gartner:	3

9. *Prescriptive catalog:* Refers to how well the methodology guides in setting up a catalog of architectural assets that can be reused in future activities.
   Ratings:

Zachman:	1	TOGAF:	2	FEA:	4	Gartner:	2

10. *Vendor neutrality:* Refers to how likely you are to get locked in to a specific consulting organization by adopting this methodology. A high rating here indicates low vendor lock in.
    Ratings:

Zachman:	2	TOGAF:	4	FEA:	3	Gartner:	1

11. *Information availability:* Refers to the amount and quality of free or inexpensive information about this methodology.
    Ratings:

Zachman:	2	TOGAF:	4	FEA:	2	Gartner:	1

12. *Time value:* Refers to the length of time you will likely be using this methodology before you start using it to build solutions that deliver high business value.
    Ratings:

Zachman:	1	TOGAF:	2	FEA:	1	Gartner:	4

Sessions (2007)

As can be seen by these ratings, various EA frameworks have different strengths and weaknesses. Which EA framework should be used depends on the relative importance of the evaluation categories to the organization/enterprise. The choice also often depends what EA framework is used by the stakeholders. For instance, the DOD uses the DODAF.

2.1.4 Architecture Type Roles in Multiple-Entity Development Process

The goal of each architecture type is to determine a framework that allows the flow of information, provides understanding how change affects the complete system or enterprise, and helps manage that change with minimal impact or consequences. While each architecture type takes a different approach, they all have the same goals. The role of these architectures is to effectively integrate systems from multiple entities across the entire enterprise and provide a holistic, optimized approach in entity interactions. An EA aids in effectively incorporating and integrating new capabilities into the enterprise. In turn, new capabilities affect the overall enterprise with respect to data flow, IT infrastructure, and system/company resources.

2.1.5 Architectural Framework

Architectural frameworks define the organization or structure of an organization (Minoli 2008). These frameworks should be used as guiding principles for an EA that defines the procedures for creating, analyzing, and applying architecture descriptions within an organization utilizing business processes, technology, and IT infrastructure to determine the ways the new capabilities affect the organization. Architectural frameworks typically contain three components (Minoli 2008):

1. *Views:* Used for information flow and to allow stakeholders to visualize relationships between subsystems or other systems within the organization.
2. *Methods:* Supplies the necessary steps a user needs to obtain a specific goal. Multiple methods can be present within a framework and can consist of processes, tools, or rules.
3. *Approaches:* Defines how to accomplish a desired result.

Many open-source and proprietary frameworks are available for general and specific uses. Each architectural framework should be considered for applicability and fit with respect to the enterprise. The industry applies these frameworks to build EAs that facilitate evolving the business vision, mission, goals, and new desired capabilities to meet stakeholder needs. Architectural frameworks provide the components, processes, and procedures necessary to support the increase in complexity of system of systems architectures (Minoli 2008) and define the interrelationships of these systems, allowing the definition of business goals and a more holistic view of the enterprise. This provides a mechanism for understanding the capabilities of the enterprise and responding to new opportunities to meet business needs. Several types of EA frameworks are described in the following sections.

2.1.5.1 Enterprise Architecture Cube

The Enterprise Architecture Cube (EA3) is a procedure to document the enterprise with different levels of detail. There are two opposing views that formulate this approach: the first view, called "AS IS," shows the EA as currently defined (Bernard 2012), and the second view, called "TO BE," seeks to identify the future state of the enterprise. The EA3 framework is built around a cube structure that defines the architecture management and transition plans between the "AS IS" and the "TO BE" views (Bernard 2012). When applying the EA framework (EA3), there are six core elements that are necessary for the completion of the framework (Bernard 2012):

1. *Enterprise architecture documentation framework:* Identifies the scope and relationships between various documentation levels;
2. *Implementation methodology:* Details the current (AS IS) and the future (TO BE) views of the architectures and also includes specific steps to maintain the architecture;
3. *Processes for architecture governance:* Manage planning, decision-making, and process oversight for changes to the architecture;
4. *Documentation artifacts:* Details the types and methods of documentation used in each subsystem, which includes strategic analysis, security, plans, and controls;
5. *Architecture repository:* Examples are databases, websites, or other software that stores the documentation necessary for the framework;
6. *Proven methods:* The best practices that are implemented for the entire architecture.

Each of these core elements defines the EA3 and together allows the EA framework to be used to identify new capabilities that move the enterprise from the current to the future state. Figure 2.2 shows the EA3 framework as it relates to business objectives.

2.1.5.2 Department of Defense Architectural Framework

The DODAF is intended to ensure that the architecture descriptions are relatable between and among each organization's operational, systems, and technical architectural views and are comparable across organizational boundaries. The framework provides both the rules and guidance for developing and presenting architecture descriptions. The framework provides direction on how to describe architectures; it does not provide guidance in how to construct or implement a specific architecture.

The DODAF Version 2.0 is the encompassing, comprehensive framework and conceptual model that enables the development of architectures to facilitate the ability of DOD managers at all levels

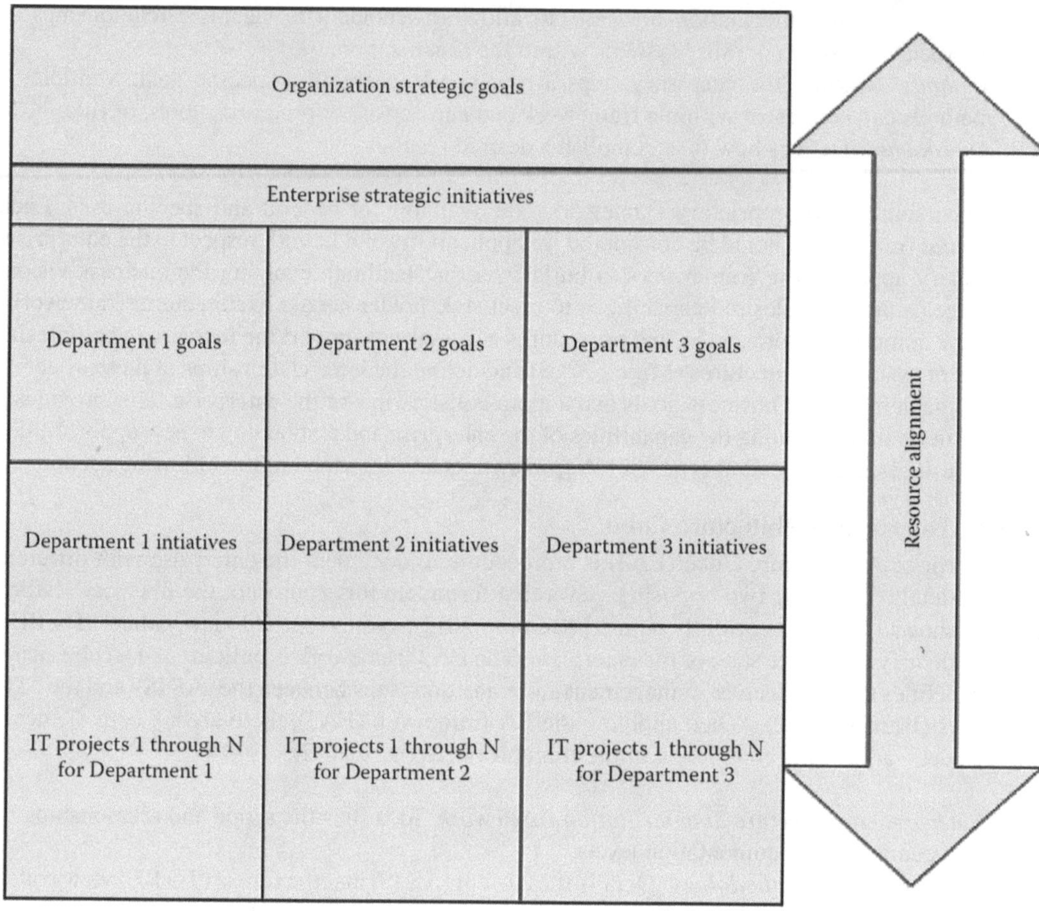

FIGURE 2.2 EA3 framework as it relates to business field. (Courtesy of Chris Lannon.)

to make key decisions more effectively through organized information sharing across the department, joint capability areas (JCAs), mission, component, and program boundaries (DODAF 2014). The DOD created this framework in response to the Clinger and Cohen Act of 1996 to regulate the CIO's responsibilities with that of developing EAs for the DOD (DOIDAF 2014). Conformance with DODAF processes allows for the reuse of information and defining the enterprise as a whole, in providing common artifacts across the enterprise for common understanding.

DODAF's latest version (2.0) switched focus from individual products to the process of data flows through the enterprise's systems. This change in focus was meant to facilitate the understanding of data flows and guide the collection and organizing of data. The enterprise is defined in different viewpoints provided by the architectural framework. The DODAF viewpoint allows users to understand the architectural description that is consistent with a particular purpose, mission, or objective (DODAF 2014). Each viewpoint provides a different purpose with different content, structure, and level of detail. These varying viewpoints allow the architectural descriptions to address the specific needs and purposes of the stakeholders.

DODAF supports six core processes that are designed to provide uniform processes in critical defense-related decision-making areas (DODAF 2014). The six core processes are as follows:

1. Joint capability integration and development system (JCIDS);
2. Defense acquisition system (DAS);

3. Systems engineering (SE);
4. Planning, programming, budgeting, and execution (PPBE) system;
5. Operations planning;
6. Capabilities portfolio management.

JCIDS focuses on the joint capabilities needed to meet DOD mission requirements (DoDAF 2014). This core process identifies the necessary capabilities that must be prioritized along with the mission scope, description, training, and personnel to meet established capability gaps. Some EA artifacts include the Capabilities-Based Assessment (CBA), Initial Capabilities Document (ICD), Capability Development Document (CDD), and Capabilities Production Document (CPD).

The DAS manages the investment in technologies, program, and product support for the National Security Strategy (DoDAF 2014). Processes and artifacts needed for DAS support the framework for translating mission needs and technology opportunities into stable, affordable, and well-managed acquisition programs. Acquisition programs are advanced through the enterprise based on completing milestones that are significant to program phases.

DODAF artifacts for DAS include road maps and integrated plans that are utilized to provide capability assessments, define investment plans, and guide systems development for resource allocation and input to the defense planning and budgeting reviews (DoDAF 2014).

Many DOD programs are required to use a robust SE approach per policy directives like the Sambur memo dated January 6, 2003 (Sambur 2003). The SE approach aligns with the DAS processes to provide an optimized approach to systems development. A systems engineering plan (SEP) is created to describe the SE technical approach including activities, resources, measures (and metrics), and applicable performance incentives. SE processes are used throughout the systems development lifecycle.

DODAF provides guidance, direction, and support for program planning, budgeting, and execution by capturing and identifying military capabilities and resource allocations and providing a structure for the program planning process that allows agencies to develop strategies for prioritization and affordability (DoDAF 2014). The DODAF also supports portfolio management for IT infrastructure investments by establishing the framework for capturing and identifying risks and performance measures to meet agency mission and goals. The collective features of an EA establish a knowledge-based approach that allows stakeholders to obtain critical information at the right times to make informed decisions (DoDAF 2014). EA processes continue to evolve with emphasis on a holistic view of the enterprise rather than individual program efforts.

At the core of DODAF 2.0 are its viewpoints that provide a mechanism for stakeholders to view the EA from a particular perspective. The different viewpoints present data at varying levels of detail and from different perspectives. Some viewpoints offer summary information regarding the enterprise as a whole, such as a high-level concept. Other viewpoints offer a deeper dive into a particular subsystem at a more technical level. Yet others indicate the way various subsystems, or individual systems, are interconnected, providing data and control flow diagrams. Different viewpoints are depicted in Figure 2.3, which also shows the relationship of DODAF version 1.5 to the new version 2.0.

2.1.5.3 The Open Group Architectural Framework

TOGAF is used by organizations to improve business efficiency (Opengroup 2014). It focuses on the business' vision and current capabilities to decide the best approach for designing, planning, and implementing information throughout an organization. Practitioners of this approach are open to processes that allow the efficient use of resources (Hornford 2011). As stated in Section 2.1.2, TOGAF had its roots in the DOD as the TAFIM, which was turned over to The Open Group and transformed into an Architectural Development Framework in 2003 (Hornford 2011). TOGAF is typically used for developing multiple interrelated architectures in conjunction with other frameworks. A major focus is on its ADM that aids in developing EA.

DoDAF V2.0 / DoDAF V1.5	Operational Viewpoint	Systems Viewpoint	Services Viewpoint	All Viewpoint	Standards Viewpoint	Data & Information Viewpoint
AV-1				AV-1		
AV-2				AV-2		
OV-1	OV-1					
OV-2	OV-2					
OV-3	OV-3					
OV-4	OV-4					
OV-5	OV-5a, OV-5b					
OV-6a	OV-6a					
OV-6b	OV-6b					
OV-6c	OV-6c					
OV-7						DIV-2
SV-1		SV-1	SvcV-1			
SV-2		SV-2	SvcV-2			
SV-3		SV-3	SvcV-3a, SvcV-3b			
SV-4a		SV-4				
SV-4b			SvcV-4			
SV-5a		SV-5a				
SV-5b		SV-5b				
SV-5c			SvcV-5			
SV-6		SV-6	SvcV-6			
SV-7		SV-7	SvcV-7			
SV-8		SV-8	SvcV-8			
SV-9		SV-9	SvcV-9			
SV-10a		SV-10a	SvcV-10a			
SV-10b		SV-10b	SvcV-10b			
SV-10c		SV-10c	SvcV-10c			
SV-11						DIV-3
TV-1					StdV-1	
TV-2					StdV-2	

FIGURE 2.3 DoDAF-V2.0-Viewpoints. (From DoD, DoD Architecture Framework Version 2.0. http://en.wikipedia.org/wiki/File:DoDAF-V2.0-Viewpoints.jpg#mediaviewer/File:DoDAF-V2.0-Viewpoints.jpg, 2014.)

TOGAF helps define architectures by providing a description of a system, with details on its components including interrelationships and guidelines for governance, along with the EA evolution over time. TOGAF offers guidelines for the development of four related types of architectures. The four types of architecture are shown in Table 2.1 (Hornford 2011).

As explained earlier, one of TOGAF's essential features is the AMD. The AMD is segregated into several sequential phases. Each phase allows the enterprise architect to understand what is needed within that phase. The phases are defined as follows (Hornford 2011):

TABLE 2.1
Architecture Types Supported by TOGAF

Architecture Type	Description
Business architecture	Strategy of the business, governance, processes, and organization
Data architecture	Organization's data assets and data management resources
Application architecture	Organization's applications in use, their interfaces, and relationship to core business processes
Technology architecture	SW and HW capabilities required to support the business, data, and application services; IT infrastructure, networks, communications, etc.

Source: Adapted from Barroero et al. (2010).

1. *Preliminary phase:* Defines the architecture principles. It is also used to identify specific frameworks that can/should be used;
2. *Phase A:* Defines the scope, vision, stakeholders, and mission for creating the architecture;
3. *Phase B (business architecture):* Describes the product and service strategies and processes based on the scope and vision from Phase A;
4. *Phase C (information systems architecture):* Develops the data/application systems services;
5. *Phase D (technology architecture):* Forms the basis of implementation;
6. *Phase E (the opportunities and solutions):* Allows the architect to evaluate the implementation to determine and identify strategic parameters for future capabilities;
7. *Phase F (migration planning):* Assesses the parameters from Phase E and defines cost, benefits, and migration strategies;
8. *Phase G (implementation governance):* Makes recommendations for implementation and provides for change management and deployment once it is completed;
9. *Phase H (architecture change management):* Continues to monitor new capabilities or changes to technology and business environments. Figure 2.4 shows the complete flow of these processes.

TOGAF also defines a Content Metamodel (displayed in Figure 2.5) that illustrates the existing architecture building blocks and details the interrelationship (Hornford 2011) between vision and drivers to visualize business capabilities and vice versa. Together with the Architectural Development Process and the Content Metamodel, enterprises can determine how new capabilities affect the overall enterprise and drive business needs (Hornford 2011). Figure 2.5 displays a Content Metamodel diagram.

2.1.5.4 Federal Enterprise Architectural Framework

The OMB of the U.S. Federal Government created the FEAF, which defines a common approach in coordinating business drivers and goals to formulate a clear vision and strategic direction across the enterprise. This includes reducing waste and duplication methods along with assessing technological impacts and fostering advancements. A lot of money is being spent to support and develop IT infrastructures throughout the federal agencies. FEAF was developed to ensure that programs are taking a more holistic approach in developing various architectures, with emphasis on interrelationships and cost reductions. There are four primary outcomes consistent with using the FEAF:

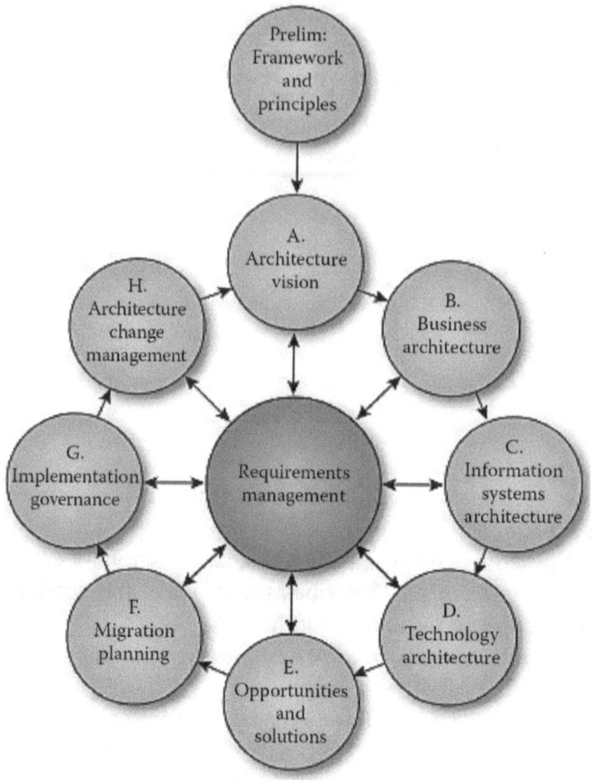

FIGURE 2.4 TOGAF architectural development process. (From Wikipedia, TOGAF Architectural development process is located, on April 13, 2014, on the website http://en.wikipedia.org/wiki/File:TOGAF_ADM.jpg and is considered public domain, 2014.)

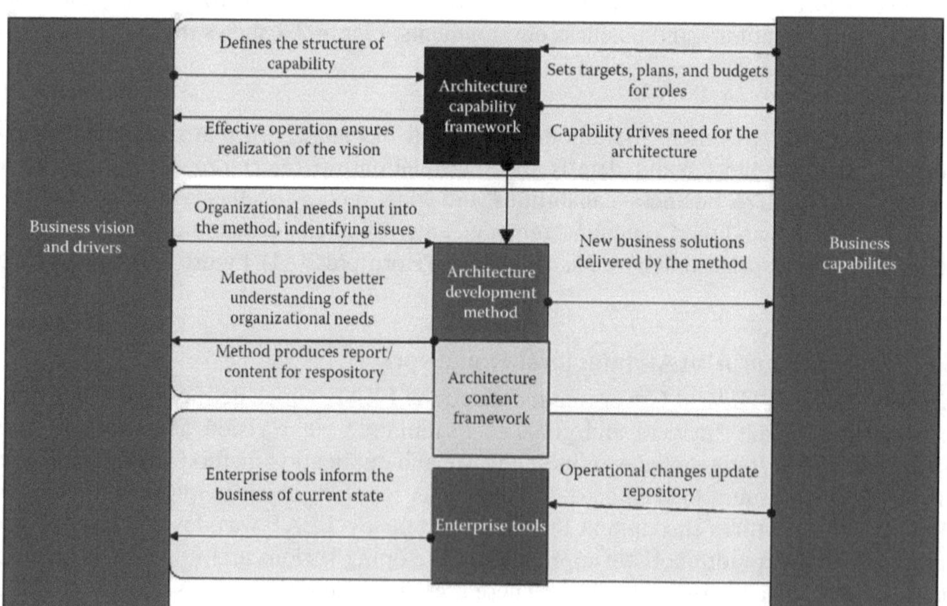

FIGURE 2.5 TOGAF content metamodel. (Courtesy of Chris Lannon.)

1. *Service delivery:* Concerned with responding to agency missions and allowing the architecture to meet those needs (Whitehouse 2007).
2. *Functional integration:* Functional integration means interoperability between programs, systems, and services, which requires metacontext and standards to be successful. EA can provide both metacontext across all functional domains (strategic, business, and technology), as well as related standards for the full lifecycle of activities in each domain. Program, systems, and services interoperability is foundational for Federal Government organizations to be able to successfully partner in new shared service models that may involve outside providers and new roles for participation (e.g., consumer, developer, or provider). The EA should provide context and be the source of standards for all levels of interoperability (Whitehouse 2007).
3. *Resource optimization:* Defining the best way to use the available resources including asset management (e.g., hardware inventory and software licenses) and configuration management (maintaining and monitoring a documented baseline of users, processes, hardware, and software) that are important elements of resource optimization that an EA also enables (Whitehouse 2007).
4. *Authoritative reference:* Just as the blueprints of a building are the authoritative reference for how the structure will function, the organization's EA provides an integrated, consistent view of strategic goals, mission and support services, data, and enabling technologies across the entire organization, including programs, services, and systems. This allows the EA to document ownership, goals and objectives, and performance metrics (Whitehouse 2007).

Based on the outcomes stated previously, FEAF outlines principles and standards to develop the business, technology, and data using the following models:

- *Performance reference model:* Standards to measure the performance and success of IT investments and the impact to agency outcomes (GAO 2001).
- *Business reference model:* Functional view of the business and its services with a goal to promote collaboration across agencies (GAO 2001).
- *Service component reference model:* Used to identify and classify services that can aid in new capabilities that can be used for reuse in support of new services (GAO 2001).
- *Data reference model:* Provides for information sharing and collaboration across agencies by providing uniformity to describing data, facilitation of data discovery, and data sharing (GAO 2001).
- *Technical reference model:* A technical model that supports the Service Component Reference Model to enable its capabilities. It aims to reuse technology assets and reclassify existing architectural elements to meet agency mission needs (GAO 2001).

FEAF is similar to other framework architectures; however, it uses architecture or business drivers as input to promote new capabilities, objectives, visions, and principles. It aims to reuse and reduce costs across agencies. The FEAF process is shown in Figure 2.6.

2.1.5.5 Ministry of Defense Architectural Framework

The Ministry of Defense Architectural Framework (MODAF) is an internationally recognized architecture framework developed by the Ministry of Defense (MOD) in the United Kingdom to promote defense planning and change management activities (MoD 2012) and to provide structure and definition to equipment capability. MODAF provides rules and templates that provide visualizations of the business area from different views. These views were defined in seven categories as shown in Table 2.2 (MoD 2012).

These views interact within the MODAF model detailing the interrelationships between the what's, why's, and how's determined by the stakeholders in trying to accomplish specific tasks (MoD 2012).

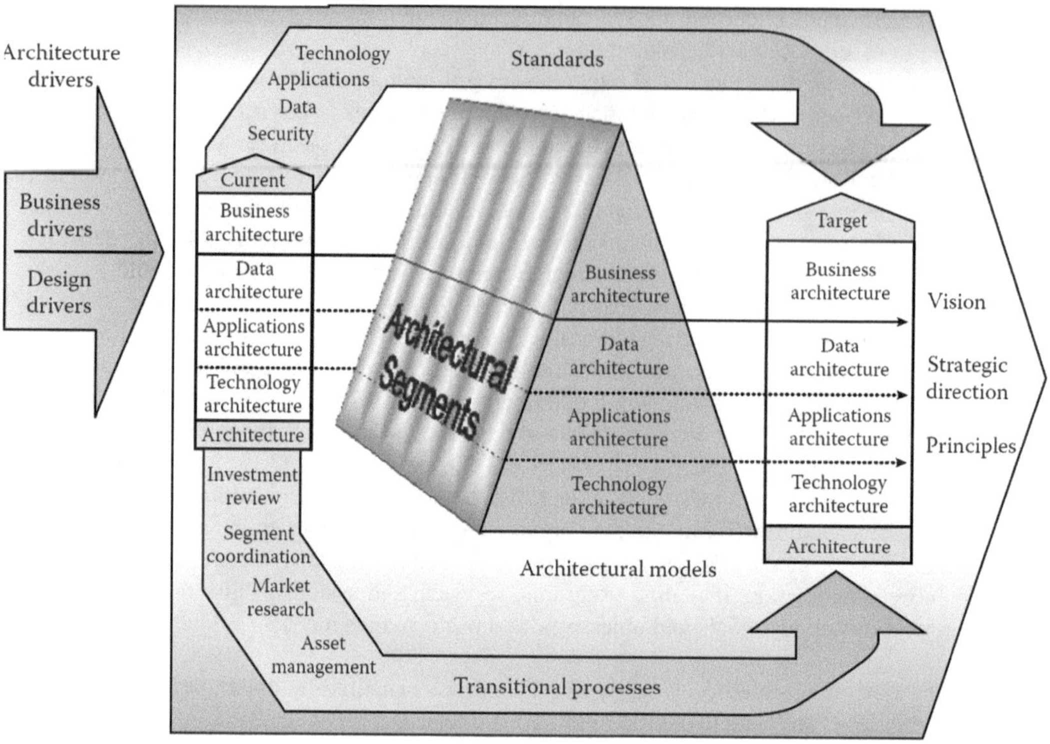

FIGURE 2.6 Structure of the FEAF Components. (From Chief Information Officer Council - A Practical Guide to Federal Enterprise Architecture. Licensed under Public domain via Wikimedia Commons - http://commons.wikimedia.org/wiki/File:Structure_of_the_FEAF_Components.jpg#mediaviewer/File:Structure_of_the_FEAF_Components.jpg, 2014.)

TABLE 2.2
MODAF View Descriptions

View Name	Description
Strategic view	Shows desired business outcome and steps to achieve it
Operational view	Processes and information needed to achieve capability
Service-oriented view	Supports the process of operational view and describes the services by providers
System views	Physical implementation of operational and service-oriented view
Acquisition view	Timelines and milestones for projects to deliver the capability
Technical view	Describes the technical requirements to be applied
All views	Overview description

Source: Adapted from MoD (2012).

2.1.6 Modeling Tools

This section provides a brief overview of architecture development and modeling tools. Architectural tools support the development, presentation, storage, and modeling of the enterprise views of the selected framework. Tools generally provide a graphical view of the system to establish the current (AS IS) model and the future (TO BE) models including new capabilities that affect the organization as a whole.

The uses of automated tools are becoming more extensive due, in part, to the sheer complexity of systems in design and the easy, accessible repository of inherent artifacts and EA building blocks. Some of these tools come in the form of architectural suites, providing modeling tools and repository management, along with multifunctional EA tools. When selecting a tool, organizations should consider the following (van den Berg and van Steenbergen 2006):

- Functionality;
- Tool architecture;
  - Central or distributed repository;
  - Version control;
  - Accessibility (web, PC, Mac, and Unix);
- Cost of ownership.

There are several commercially available and open-source EA tools available on the commercial market. Some representative examples include Rhapsody, IBM's Systems Architect, Essential Project, Modelio, and MagicDraw. Figure 2.7 is an example screenshot from IBM Rhapsody using DODAF. The main aim of choosing a comprehensive EA modeling tool must be discussed and agreed upon by the enterprise stakeholders. Once the tool selection is made, training will typically be required to get the full utility from the tool.

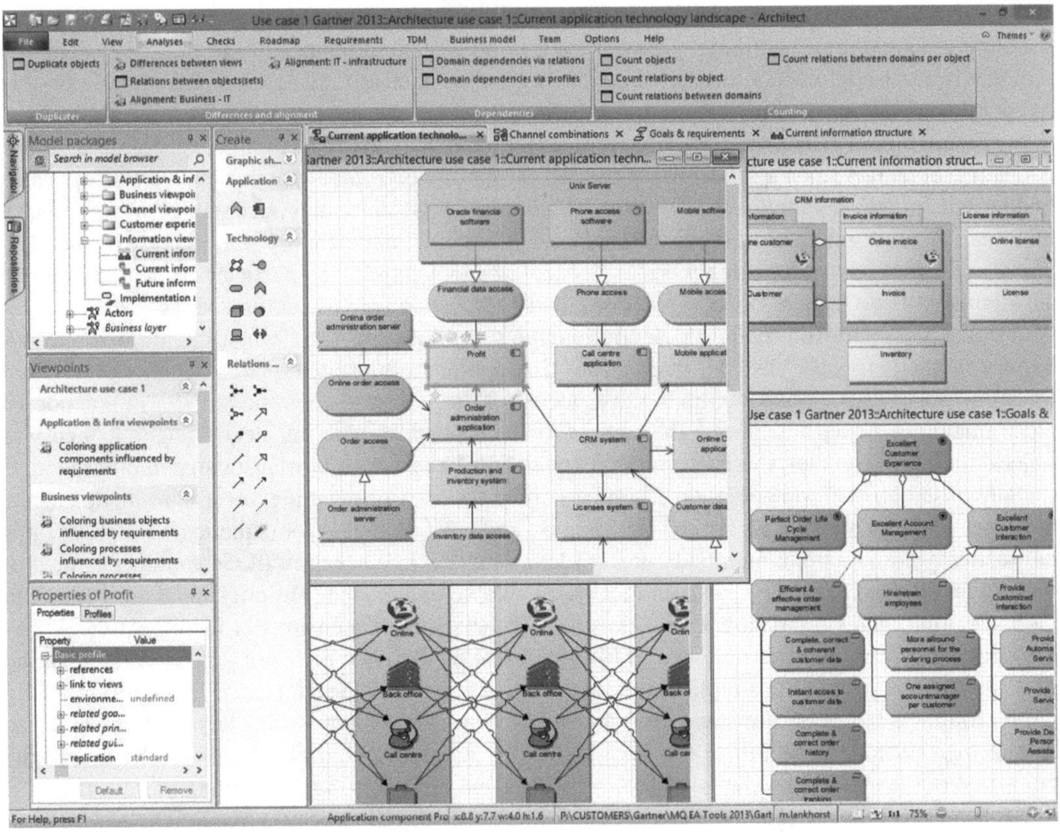

FIGURE 2.7 Screenshot of enterprise architecture tool showing several views of the enterprise architecture model. Courtesy of marklankhorst, obtained from Wikimedia Commons at https://commons.wikimedia.org/wiki/File:BiZZdesign_Architect_4.3.1_screenshot.jpg (accessed on 6 February, 2025).

2.2 OPTICAL SYSTEMS MODEL

In this section, we provide models and an analytical framework for understanding the technical aspects of optical systems. In a sense, these frameworks and models have a similar role as the EA has in assimilating, integrating, and defining the relationships between enterprise entities and their artifacts. The analytical frameworks and models describe the technical relationship between key system elements and their observables. By observables, we mean observable quantities of interest such as photometric or radiometric elements of interest, for example, the irradiance on the detector element. The models describe and capture all the essential enterprise or system components and their necessary interactions to properly address the analytical problem at hand. In this chapter, we develop a variety of useful mathematical models and analytical methods that help in our understanding of optical systems. These models are useful for analyzing and simulating the physical phenomena associated with our optical system. Each model has its strengths and weaknesses as discussed later.

Radiometric and photometric models are useful in determining whether or not there is sufficient signal on the detector, as compared with the noise that affects the detector, to successfully conduct an observation. By noise, we mean any unwanted signal from the source, medium between the source and optical system, and optical system itself. Often, these are separated into background and clutter effects, scintillation effects, scattering, atmospheric turbulence effects, aberrations of the receiver optics, detector noise effects, detector readout electronics effects, shock, jitter, environmental effects (pressure, temperature, humidity, radiation, and dust), and statistical uncertainty from the random nature of the light itself. We will look at radiometric and photometric models in Chapter 6 and address each of these effects in more detail as needed.

Although good at determining various signal-to-noise characteristics of our optical system, the radiometric and photometric models are not so good at determining spatial effects and possibly temporal correlations that may become significant in some imaging applications. If we are looking at the interaction of light with very simple or idealized objects such as a plane, sphere, or a simple geometric boundary, then a model using the full power of electromagnetic wave theory is workable and most accurate. A simple geometric model assumption is sometimes acceptable for preliminary or rough order of magnitude types of analyses. Electromagnetic theory and derivative methods can also be applied to exceedingly small physical phenomena such as particles; molecules; atoms; subatomic particles such as bosons, fermions, and hadrons; and nanostructures using quantum electrodynamics. Relativistic quantum electrodynamics methods are used when quantum electrodynamics methods involve physical interactions that approach the speed of light. Although modern electromagnetic theory and methods most accurately represent the physics of the interaction of electromagnetic waves and matter, the computational aspects of applying electromagnetic theory to objects that have complex spatial structures and details become too burdensome, and other models that are based on electromagnetic methods are more useful. For example, most man-made objects are "rough" on the scale of an optical wavelength, and a detailed electromagnetic treatment of optical scattering from a large manmade object like a satellite, or a scene in the instantaneous field of view of an imaging system, is mathematically intractable unless some simplifying assumptions can be made. One such simplifying approach results if the spatial dimensions of the object viewed by the imaging system are small compared to the separation distance of the object from the imaging system. We will discuss the simplification of spatial dimension in Section 2.2.3. Another modeling approach that is based on statistical analysis works well if the object is very large compared to the size of the optical wavelength of the imaging/observing system and also if the object is "rough" compared to the optical wavelength. Since most man-made objects in typical imaging scenarios fall into this category (apart from mirrors or reflective surfaces that are treated separately), the statistical optics models are very useful and will be discussed in Section 2.2.4.

This chapter is organized as follows: we will first start with an overview of basic electromagnetic propagation theory to show the basis and general assumptions for the linear (and nonlinear) systems models. We then present some fundamental concepts of linear and nonlinear systems along with some useful properties such as superposition and shift invariance. We then develop a linear systems model and list some useful properties. We extend the model to two spatial dimensions (2D) for visible and infrared imaging applications and three dimensions, two spatial dimensions and one wavelength (or frequency) dimension, for spectral imaging applications such as ultraspectral, hyperspectral, and multispectral imaging systems. These will be discussed in more detail later. We then provide some useful relationships between the spatial domain and spatial frequency domain for both coherent and incoherent imaging systems. We discuss the energy spectrum, power spectrum, generalized pupil function (GPF), impulse response, point spread function (PSF), optical transfer function (OTF), modulation transfer function (MTF), coherent transfer function (CTF), and some useful associated properties. We finish our presentation of coherent and incoherent imaging models by discussing some sampling effects and illustrating how to properly represent a continuous signal or image on a computer. We then move on to a statistical optics analytical and modeling approach and start with some important terminology, concepts, and results from statistics and probability theory. As previously stated, this methodology is very useful in modeling very large optical systems with respect to the size of the optical wavelength. As before, our approach will be focused on the optical systems engineer and optical systems manager, so we will focus on the key points and results but will provide references to more detailed supporting information. A good understanding of the terminology and concepts presented in Chapter 1 is necessary to get the most out of this chapter and the rest of this text.

2.2.1 Fundamentals of Electromagnetic Propagation Theory

For all of the models that we will present, the characteristics of the electromagnetic field, its derivative components, its propagation effects, and interactions with matter are central considerations in their formulation. The differential form of Maxwell's equations describes how electromagnetic waves interact with matter (or propagate through free space) at a given point in space and time. The differential form of Maxwell's equations is given by

$$\nabla \times H = J + \frac{\partial D}{\partial t}, \quad \nabla \cdot D = \rho,$$
$$\nabla \times E = -\frac{\partial B}{\partial t}, \quad \nabla \cdot B = 0, \tag{2.1}$$

where

E represents the electric field in V/m
B is the magnetic induction in teslas
J is the current density in A/m^2
D is the displacement in C/m^2
H is the magnetic field in (ampere turns)/m
ρ is the charge density in C/m^3
∇ is the gradient operator given by

$$\nabla = \frac{\partial}{\partial x}\hat{x} + \frac{\partial}{\partial y}\hat{y} + \frac{\partial}{\partial z}\hat{z}. \tag{2.2}$$

Here, the "hatted" (^) coordinates are unit vectors in the x, y, and z directions, respectively.

Notice that these are a set of first-order, coupled, electromagnetic equations. The top right of Equation 2.1 is Coulomb's law, the top left is Ampere's law, the bottom right shows that there are no free magnetic poles, and the bottom left is Faraday's law. Although these equations can be solved directly, it is mathematically preferable to decouple these equations and deal with separate scalar and vector quantities. This can be done by defining a scalar potential Φ and a vector potential A and relating these terms to the electric field and magnetic induction by

$$\boldsymbol{E} = \nabla\Phi - \frac{\partial \boldsymbol{A}}{\partial t}, \qquad \boldsymbol{B} = \nabla \times \boldsymbol{A}. \tag{2.3}$$

By making an additional substitution called a gauge transformation for mathematical convenience, we can substitute these results into Maxwell's equations and solve for two decoupled inhomogeneous wave equations:

$$\begin{aligned} \nabla^2\Phi - \mu_o\epsilon_o\frac{\partial^2\Phi}{\partial t^2} &= -\frac{\rho}{\epsilon_o}, \\ \nabla^2 A - \mu_o\epsilon_o\frac{\partial^2 \boldsymbol{A}}{\partial t^2} &= -\mu_o J. \end{aligned} \tag{2.4}$$

In these equations, the ∇^2 symbol is the Laplacian operator given by

$$\nabla^2 = \frac{\partial^2}{\partial x^2} + \frac{\partial^2}{\partial y^2} + \frac{\partial^2}{\partial z^2}. \tag{2.5}$$

This set of wave equations, in Equation 2.3, is mathematically equivalent to Maxwell's equations but has the advantage of being decoupled from each other. Notice that these wave equations have the same structural form as

$$\nabla^2\psi_{(\bar{x},t)} - \mu_o\epsilon_o\frac{\partial^2\psi_{(\bar{x},t)}}{\partial t^2} = -\rho f(\bar{x},t), \tag{2.6}$$

where the term on the right is a driving term that has to do with the presence of sources such as charge densities or current densities. We can solve this equation by removing the explicit time dependence in expressing the solution in terms of its Fourier transform pair

$$\Psi(\bar{x},t) = \int_{-\infty}^{+\infty} \psi(\bar{x},\omega)e^{-j\omega t}d\omega,$$

$$F(\bar{x},t) = \int_{-\infty}^{+\infty} f(\bar{x},\omega)e^{-j\omega t}d\omega,$$

and the corresponding inverse

$$\begin{aligned} \psi(\bar{x},\omega) &= \int_{-\infty}^{+\infty} \Psi(\bar{x},t)e^{j\omega t}dt, \\ f(\bar{x},\omega) &= \int_{-\infty}^{+\infty} F(\bar{x},t)e^{j\omega t}dt. \end{aligned} \tag{2.7}$$

Substituting Equation 2.7 into Equation 2.6, we end up with the inhomogeneous Helmholtz wave equation:

$$\left(\nabla^2 + k^2\right)\psi(\bar{x},\omega) = -f(\bar{x},\omega) \text{ in } \mathbb{R}^n. \tag{2.8}$$

In this equation, k is a term related to the velocity of the wave in the medium in which it is propagating. If we now consider the propagation of the electromagnetic field through free space, then we have no current density J or charge density ρ, and the form of the wave equation in Equation 2.6 becomes

$$\nabla^2\psi_{(\bar{x},t)} - \mu_o\epsilon_o \frac{\partial^2\psi_{(\bar{x},t)}}{\partial t^2} = 0. \tag{2.9}$$

Since the driving terms shown explicitly on the right side of Equation 2.4 are zero in this case, the corresponding homogeneous Helmholtz equation is given by

$$\left(\nabla^2 + k^2\right)\psi(\bar{x},\omega) = 0, \tag{2.10}$$

where k is given by

$$k = \frac{\omega}{c} = \frac{2\pi}{c}\nu = \frac{2\pi}{\lambda}, \tag{2.11}$$

where

λ is the electromagnetic wavelength in meters
ν is the corresponding frequency in Hz

The Helmholtz equation has several solutions that propagate. The simplest of these solutions for the scalar, homogeneous Helmholtz equation is the plane wave given by

$$\begin{aligned} \underline{u}(P,t) &= \text{Re}\left\{U(P)e^{j\phi(P)}e^{j2\pi\nu t}\right\}, \\ \underline{u}(P,t) &= U(P)\cos\left(2\pi\nu t + \phi(P)\right). \end{aligned} \tag{2.12}$$

In these equations, the symbol u represents either the electric field or the magnetic induction since both these quantities satisfy the wave equation. The parameter P indicates position, $U(P)$ is the amplitude of the wave at position P, $\phi(P)$ is the phase at position P, and indicates taking the real part of the term inside the brackets. The phasor notation in Equation 2.12 is quite useful and will be adopted throughout the book. A useful consequence of using the phasor notation is that the phasors can usually be manipulated through a wide range of analytical calculations and the determination of the real physical result, obtained by taking the real part of the solution, can be deferred until the last step of the calculation. In using u, we will make it clear by the context of our discussion whether we are referring to the electric field or the magnetic induction. The plane wave solution to the homogeneous Helmholtz equation is itself quite useful since expanding waves that are far enough away from their source point are well approximated by plane waves. A physical example of this would be starlight outside of our solar system. Stars are so far away that they are well approximated as point sources. By the time the light from the star reaches us, the electric field that is carrying the information from the star is well approximated by a plane wave.

The solutions to Maxwell's equations and the Helmholtz wave equation accurately describe the behavior of the electromagnetic field at a particular point in space and time; however, we need to get expressions that relate the interactions of the electromagnetic field with extended objects. We also need to understand how these fields propagate to, and interact with, some remote observational systems. An example of this scenario is an imaging system that is observing a resolvable object, such as an earthbound telescope viewing a lioness on the plains of the Serengeti. For this case, an integral representation based on the Helmholtz wave equation is most practical. We will briefly describe this approach and provide some references for a more detailed explanation. Our aim is to arrive at the main results and show the connection to our modeling methodology. In transforming the Helmholtz wave equation into an integral expression, Green's theorem is very useful (Goodman 2005):

$$\int\int\int_V \left(\underline{G}(P)\nabla^2\underline{U}(P)-\left(\underline{U}(P)\nabla^2\underline{G}(P)\right)\right)dV=\iint_S \underline{G}(P)\frac{\partial \underline{U}(P)}{\partial n}-\underline{U}(P)\frac{\partial \underline{G}(P)}{\partial r}dS, \tag{2.13}$$

where the function *G(P)* is the well-known Green's function at spatial position *P*. Notice that we have dropped the time dependency in Equation 2.12 since we are interested in spatial effects and since this is a separable function in its phasor form. We will bring in this dependency later when necessary. In Green's theorem, *G(P)* and *U(P)* can be any two complex-valued functions. The integration is over a volume *V* that is contained within a surface *S*. The partial derivative on the right side of Equation 2.13 is in the outward normal direction of the surface *S*. For our purposes, a free-space Green's function is given by

$$G(P_o)=\frac{\exp(jkr_{os})}{r_{os}}, \tag{2.14}$$

where r_{os} is the distance from a source point s to an observational point P_o. This Green's function describes a spherically expanding wave that originates at point s. Note that an assumption here is that the observational point is not directly on the source point since this term would tend toward infinity. In remote observation applications, the term r_{os} is never zero. The choice of this spherically expanding Green's function is motivated by the work of the Dutch physicist Christiaan Huygens who noted that the electromagnetic field at some distant observational point from an extended object can be determined by superimposing spherically expanding wavelets from each source point in the observational plane (Born and Wolf 1970). By substituting Green's function in Equation 2.14 into Green's theorem along with the spatial component of our phasor in Equation 2.12, evaluating the partial derivatives for Green's function, and simplifying, we get the integral theorem of Helmholtz and Kirchhoff, an intermediate result:

$$\begin{aligned} U(P_o) &= \iint_s \left\{\frac{\partial \underline{U}(P)}{\partial n}\left[\frac{\exp(jkr_{os})}{r_{os}}\right]-\underline{U}(P)\frac{\partial}{\partial n}\left[\frac{\exp(jkr_{os})}{r_{os}}\right]\right\}ds \\ &= \iint_s \left\{\frac{\partial \underline{U}(P)}{\partial n}\underline{G}-\underline{U}(P)\frac{\partial \underline{G}}{\partial n}\right\}ds. \end{aligned} \tag{2.15}$$

This equation was used by Kirchhoff to describe the observed electromagnetic field in a distant observational plane due to diffraction effects from an aperture. The only nonvanishing source term in Equation 2.15 was in the actual aperture itself. Consequently, the limits of integration in Equation 2.15 are confined to just the area of the aperture. Equation 2.15 was used to describe the diffraction effects of a plane screen, and the Fresnel and Kirchhoff diffraction formula was developed. Although surprisingly accurate experimentally, this formulation had some theoretical problems in satisfying the Kirchhoff boundary conditions near the aperture. To solve this problem, Rayleigh and Sommerfeld solved Equation 2.15 using a different Green's function:

$$G_{rs}(P_o)=\frac{\exp(jkr_{os})}{r_{os}}-\frac{\exp(jk\tilde{r}_{os})}{\tilde{r}_{os}}, \tag{2.16}$$

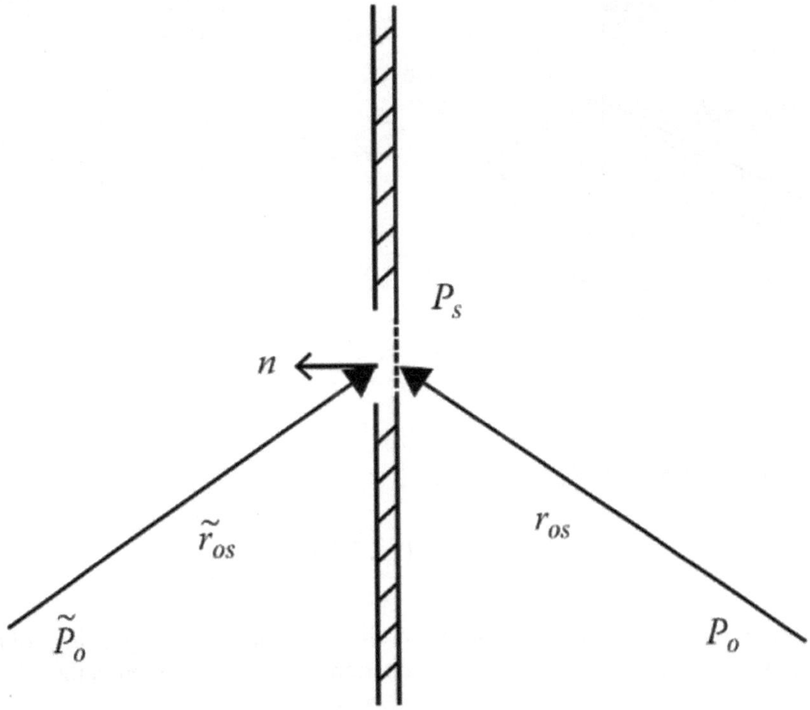

FIGURE 2.8 The geometry for the Rayleigh-Summerfield formulation.

where the addition of the second term was found by Sommerfeld to satisfy the appropriate boundary conditions. Figure 2.8 shows the geometry for the Rayleigh-Sommerfeld formulation of diffraction from a plane screen.

The point source on the left is a mirror image of the point source on the right and is 180° out of phase. This choice of Green's function, in addition to satisfying the Kirchhoff boundary conditions, permits expressing the electromagnetic field in a distant observation plane in terms of integration over the clear aperture of the diffracting screen (*Sa*). The point in the clear aperture of the screen is P_s, and the point at the observer is given by P_o. Substituting Equation 2.16 into Equation 2.13 and once again evaluating the partial derivative on the new Green's function in Equation 2.16, and simplifying the result, lead to the Rayleigh-Sommerfeld diffraction integral:

$$u(P_o) = \frac{1}{j\lambda} \iint_{Sa} \underline{U}(P_o) \left[\frac{\exp(jkr_{os})}{r_{os}} \right] \cos(\boldsymbol{n}, \boldsymbol{r}_{os}) dSa. \tag{2.17}$$

This equation is very general and useful in describing the electromagnetic field in a distant plane, due to a source with finite extent, represented by the field within the aperture of the plane screen, to a remote observational point. Let us make some variable substitutions to make the notation of the result in Equation 2.17 more convenient with the notation of our linear systems model presented later. Let the point P_s in the clear aperture of the plane screen be represented by the 2D position vector $\bar{x}_o$ where the subscript "*o*" represents the object (instead of the observer as was previously defined earlier). Let the subscript "1" represent the observational plane so that $\bar{x}_1$ represents a 2D position vector in the plane of the observer. Let the lowercase u represent the complex electromagnetic field, and let θ be the angle between the normal surface unit vector $\boldsymbol{n}$ and $\boldsymbol{r}_{os}$. The Rayleigh and Sommerfeld diffraction formula then has the following form (Roggemann and Welsh 1996):

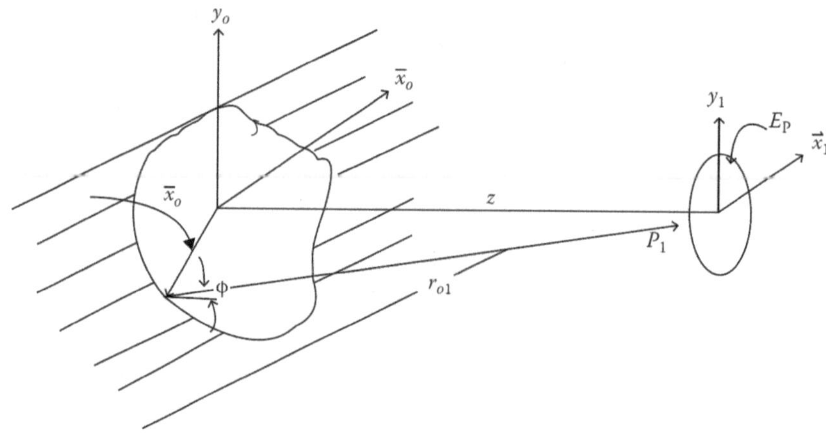

FIGURE 2.9 Superposition integral geometry.

$$u_1(\bar{x}_1) = \frac{1}{j\lambda}\int d\bar{x}_o u_o(\bar{x}_o)\frac{\exp(jkr_{o1})}{|\vec{r}_{o1}|}\cos\theta. \tag{2.18}$$

We adopt the notation from Roggemann and Welsh since it is intuitive and practical for a number of imaging concepts and applications. Notice that Equation 2.18 can be written in the following form:

$$u_1(\bar{x}_1) = \int d\bar{x}_0 u_o(\bar{x}_o) h_d(\bar{x}_1, \bar{x}_o), \tag{2.19}$$

where

$$h_d(\bar{x}_1, \bar{x}_o) = \frac{1}{j\lambda}\frac{\exp(jkr_{o1})}{|\vec{r}_{o1}|}\cos\theta. \tag{2.20}$$

Equation 2.20 is the *impulse response* of the imaging system, and Equation 2.19 is a superposition integral. The corresponding geometry to Equations 2.19 and 2.20 is shown in Figure 2.9. Notice that the clear aperture on the left side of the diagram contains the object plane complex field distribution.

This complex field distribution can be considered as an extended object source term and is applicable as a model for various physical interpretations. For instance, just as in the derivation earlier, we can assume that we have a source on the left side of a very large screen with an opening at the center. In this case, we know that the complex electromagnetic field distribution in the clear aperture is due to the illuminating source. In addition, we also know that the value of the field is zero outside the clear aperture along the infinite plane. Alternatively, we can think of the complex electromagnetic field in the aperture as consisting of monochromatic emitted or reflected light from an extended object. In both these alternate cases, the source electromagnetic field is zero beyond the spatial extent of the object, and the complex field itself is fully represented across the spatial regions of the object (represented by the finite extent of the aperture). Stated in another way, once we have a spatial description of the field within the source aperture (the object plane), it is irrelevant what generated the field within the aperture. By knowing the field within the aperture, we have the ability to describe the complex electromagnetic field in a separate, distant observation plane. From here on, we use the subscript "o" to denote items of interest in the object plane and the subscript "1" for any items of interest in the observation plane. The position vector $\vec{r}_{o1}$ connects an arbitrary point P_o in the observation plane to an arbitrary point P_1 in the observation plane, and its magnitude is given by

$$r_{o1} = \left[z^2 + (x_o - x_1)^2 + (y_0 - y_1)^2\right]^{1/2}. \tag{2.21}$$

We now want to look at some special cases of the Rayleigh-Sommerfeld diffraction formula considering Equation 2.21. It turns out that it is difficult to work with Equation 2.20 because of the square root term in Equation 2.21. However, if the distance z is much larger than the maximum finite extent of the source term $|\bar{x}_o|_{\max}$ and $\lambda \ll |\bar{x}_o|_{\max}$, then the paraxial approximation can be made, and only the lower-order terms in the binomial expansion of Equation 2.21 need to be kept. This condition is satisfied if (Bahaa and Teich 2007)

$$F = \frac{|x_o|^2_{\max}}{\lambda z} \geq 1. \tag{2.22}$$

In this equation, F is the Fresnel number. If this condition is met, then we can approximate the Rayleigh-Sommerfeld diffraction formula with the Fresnel diffraction integral given by

$$u_1(\bar{x}_1) = \frac{\exp(jkz)}{j\lambda} \frac{\exp\left[jk|\bar{x}_1|^2\right]}{2z} \iint_{S_A} dx_o u_o(\bar{x}_o) \exp\left[j\frac{2\pi}{\lambda z}|\bar{x}_o|^2\right] \exp\left[j\frac{2\pi}{\lambda z}(\bar{x}_o \cdot \bar{x}_1)\right]. \tag{2.23}$$

Notice that Equation 2.23 can be written in the compact form:

$$u_1(\bar{x}_1) = \iint_{S_A} dx_0 u_o(\bar{x}_o) h_d(\bar{x}_1 - \bar{x}_o), \tag{2.24}$$

where

$$h_d(\bar{x}_1 - \bar{x}_o) = \frac{\exp(jkz)}{j\lambda z} \exp\left\{j\frac{k}{2z}\left[|\bar{x}_1 - \bar{x}_o|^2\right]\right\}. \tag{2.25}$$

The impulse response shown in this equation is the free-space impulse response (Roggemann and Welsh 1996; Goodman 2005). We have adopted Goodman's assumptions that the distance z is much larger than the angular extent of the object and the distance off of the optical axis and so the cosine term in Equation 2.20 is approximately one and the r_{o1} term in Equation 2.21 is approximately equal to z. Equation 2.24 is a 2D spatial convolution integral, and Equation 2.22 can be interpreted as a 2D spatial Fourier transform as follows:

$$u_1(\bar{x}_1) = \frac{\exp(jkz)}{j\lambda z} \frac{\exp jk|\bar{x}_1|^2}{2z} \Im\left\{u_o(\bar{x}_o) \exp\left[j\frac{k}{2z}|\bar{x}_o|^2\right]\right\}\Bigg|_{\bar{f}=\frac{\bar{x}_1}{\lambda z}}, \tag{2.26}$$

where J represents the 2D Fourier transform given by the relation

$$G(\bar{f}) = \Im[g(\bar{x})] = \int_{-\infty}^{+\infty}\int d\bar{x} g(\bar{x}) \exp(-j2\pi\bar{f} \cdot \bar{x}), \tag{2.27}$$

and the associated 2D inverse Fourier transform

$$g(\bar{x}) = \Im^{-1}\left[G(\bar{f})\right] = \int_{-\infty}^{+\infty}\int d\bar{f} G(\bar{f}) \exp(j2\pi\bar{f} \cdot \bar{x})\Bigg). \tag{2.28}$$

The parameter $\bar{f}$ is a 2D spatial frequency variable that has units of cycles per meter. We use the convention here and throughout the book that lowercase letters represent parameters in the spatial domain and uppercase letters represent parameters in the frequency domain. Equations 2.23 through 2.28 apply if Equation 2.22 is satisfied, and if $z \gg \lambda$, and if the source aperture maximum spatial extent $|\vec{x}_o|_{max} \gg \lambda$. Notice that z cannot be too large; otherwise, the Fresnel condition in Equation 2.22 no longer holds. In such a case, we need to look at our second case of interest, which is the Fraunhofer diffraction formulation of the Rayleigh-Sommerfeld diffraction integral.

In the case of Fraunhofer diffraction, we have an additional assumption on the exponential term inside the Fourier transform operator in Equation 2.26. We need to ensure that this exponential term is practically constant in the object plane aperture. In this case, the argument of the exponential term inside the Fourier transform integral tends to zero and the exponential term itself tends to one. This condition is satisfied when the Fresnel number is much less than one:

$$F = \frac{|x_o|^2_{max}}{\lambda z} \ll 1. \tag{2.29}$$

Notice that this condition is satisfied if the product of λ and z is much larger than the square of the maximum spatial extent of the source aperture in the object plane. If Equation 2.29 holds, we have the Fraunhofer diffraction integral

$$u_1(\vec{x}_1) = \frac{\exp(jkz)}{j\lambda z} \frac{\exp jk|\vec{x}_1|^2}{2z} \iint_{SA} d\vec{x}_o u_o(\vec{x}_o) \exp\left[j\frac{2\pi}{\lambda z} \vec{x}_o \cdot \vec{x}_1 \right], \tag{2.30}$$

that has the compact notation

$$u_1(\vec{x}_1) = \frac{\exp(jkz)}{j\lambda z} \frac{\exp\left(jk|\vec{x}_1|^2\right)}{2z} F\{u_o(\vec{x}_o)\}_{\vec{f}=\vec{x}_1/\lambda z}. \tag{2.31}$$

At this point, we should mention something about extending this monochromatic result to multiple wavelengths, or over a spectrum of wavelengths. For multiple wavelengths, or over a broad range of wavelengths, the results are equally valid at each separate wavelength, and the superposition principle still holds. The resulting field in the observation plane would be the superposition of the separate fields at each wavelength or, by extension, the superposition of the fields across the spectrum of wavelengths.

Equations 2.22, 2.25, 2.26, 2.29, and 2.31 are the main results of this section and will be the basis for our linear systems model in the next section. It is worth repeating that these results originate from Maxwell's equations and require only the Huygens superposition principle to establish linearity. Additionally, in order to be valid, some constraints on aperture size with respect to the wavelength and aperture size with respect to the object plane to observation plane distance z must be met. These constraints are typically easily met for most remote imaging applications. In the next section, we present a linear systems imaging model that is based on these results.

2.2.2 Linear and Nonlinear Systems Models

In the previous section, we presented a short development of how the electromagnetic field in a distant observational plane can be related to the electromagnetic field distribution in the object plane. We showed that if the Fresnel conditions are met, the complex electromagnetic field in the observer plane could be obtained by a 2D spatial convolution of the complex electromagnetic field in the object plane with a shift invariant impulse response. We also showed that for both Fresnel and Fraunhofer imaging conditions, the electromagnetic field in the observer plane is determinable by 2D Fourier transforming the spatially modulated object electromagnetic field, in the case of Fresnel diffraction, or by the 2D Fourier transform of the object plane electromagnetic field itself, as in the case of Fraunhofer diffraction. In essence, the 2D Fourier transform plays prominently in the determination of the observed field for both the Fresnel and Fraunhofer diffraction cases.

In order to determine our linear systems model, we need one additional piece of information. We know that since the object plane field and observation plane field are related through a Fourier transform, then in order to recover the image, we need to determine the inverse Fourier transform (or something somewhat equivalent) of the observed field. This can be done using a thin lens. To see how this is accomplished, we first need to describe the effect that a thin lens has on an optical field. The effect of a thin lens on an incident electromagnetic field can be thought of as multiplication by a transparency function

$$u_t(\bar{x}) = u_i(\bar{x})t_l(\bar{x}), \tag{2.32}$$

where the transparency function itself is given by

$$t_l(\bar{x}) = \exp\left(-j\frac{k}{2f}\,|\,\bar{x}\,|^2\right), \tag{2.33}$$

$$u_i(\bar{x}) = \int_{-\infty}^{+\infty} d\bar{x}u_o(\bar{x}_o)h(\bar{x}-\bar{x}_o)d\bar{x}_o. \tag{2.33a}$$

The subscript "i" in Equation 2.32 refers to the electromagnetic field that is *incident* on the thin lens. The vector $\bar{x}$ is a position vector in the plane of the lens, and f is the focal length of the thin lens. The subscript "t" represents the transmitted electromagnetic wave just after it has passed through the thin lens. If we now apply this concept to the electromagnetic field in the observer plane given by Equation 2.31 by placing a thin lens in the observer plane, assuming that the electromagnetic field in the observer plane is now the source, and propagating a distance equal to the focal length of the thin lens, it is seen that the resulting image will be a scaled version of the original object in the object plane. Working out the details is a fun exercise for the reader. The effect of the lens is to effectively take the 2D Fourier transform of the field in the observer plane. From now on, we will refer to the previously discussed observer plane as the entrance pupil plane of our imaging system. As a point of reference, for a well-designed telescopic imaging system, the primary mirror of the telescope is placed in the entrance pupil of the imaging system. Working through the details just discussed, we find that the electromagnetic field in the focal plane of the thin lens is, apart from a complex scale factor and diffraction effects, a scaled version of the electromagnetic field in the object plane. In effect, the thin lens cancels out the quadratic phase factor in the entrance pupil and takes the 2D Fourier transform of the result. We, therefore, have a scaled version of the Fourier transform, of the Fourier transform of the object field, in the image plane of the optical system. The effect of the second Fourier transform, as opposed to an inverse Fourier transform, is nothing more than inverting the image of the object in the focal plane. We can generalize the thin lens approach to a collection of arbitrary optical elements by using the systems model shown in Figure 2.10.

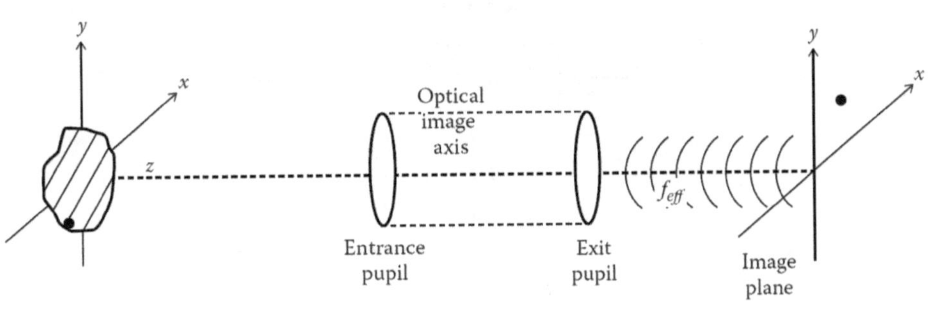

FIGURE 2.10 Optical systems model.

The left coordinate system in Figure 2.10 is the plane where the object resides. The first aperture to the right of the object is the entrance pupil of the imaging system. The optics of the optical imaging system are treated as a "black box" and can consist of lenses, relay optics, optical filters, polarizers, beam splitters, and a variety of other beam-shaping and beam-forming optics. We will discuss how to characterize and mathematically describe the optics within the "black box" in Chapter 7. The next point of interest in Figure 2.10 is the exit pupil of the imaging system. The distance from the exit pupil to the image plane is the effective focal length f_{eff}. Notice that we have dropped the subscripts distinguishing the object plane from the entrance pupil plane and the image plane. The reason for this is that, in an aberration-free environment, the image is a scaled version of the object, and so the relative image scaling issues that are addressed by the optical systems transverse and lateral magnification are typically addressed separately from the image formation process. For convenience, we then use the same coordinate system in all three planes, object, entrance pupil, and image plane, and separately address magnification. We will make it clear by the context what plane is being discussed. We address magnification issues in Chapter 4.

We now have sufficient information to present our linear, shift invariant systems model. The linearity in our model results from the superposition of the Huygens wavelets and the shift invariance results from the geometrical simplifications resulting from the Fresnel or Fraunhofer approximations. If the shift invariance assumption does not hold, then we would have to use Equation 2.19 as the governing relationship between the object plane electromagnetic field and the observation plane electromagnetic field with Equation 2.20 as the impulse response. Luckily, the shift invariance assumption is valid for a large number of practical imaging applications, and so we can benefit from some mathematical simplifications in both the spatial domain and in the spatial frequency domain. In this section, we develop two types of linear, shift invariant models that have to do with the spatial coherence properties of the electromagnetic field. If the electromagnetic field has a high degree of spatial coherence, as is the case for a laser, we will adopt a model that is linear and shift invariant at the field level. On the other hand, if the electromagnetic field is highly incoherent, as is the case for emitted or reflected electromagnetic waves from optically rough surfaces, then we adopt a model that is linear and shift invariant at the optical power level (e.g., magnitude squared of the electromagnetic field). We first start with the assumption that the electromagnetic field is highly spatially coherent. In this case, the linear, shift invariant imaging model that we will use is illustrated in Figure 2.11.

The relationship between the electromagnetic field in the object plane $u_0(\vec{x})$ and that in the image plane $u_i(\vec{x})$ is given by a 2D spatial convolution of the object electromagnetic field with the impulse response of the optical imaging system $h(\vec{x})$. The asterisk represents convolution, $\Im_{2D}$ represents the 2D Fourier transform operator, $\Im s_{2D}^{1}$ is the 2D inverse Fourier transform, and $H\left(\vec{f}\right)$ is the CTF. Notice that these relationships hold true at every spatial point $\vec{x}$ and every spatial frequency $\vec{f}$. A physical example wherein this model applies would be a laser-illuminated circular mirror wherein the reflected field is detected at some far distant point (also in space) that satisfies the Fresnel or

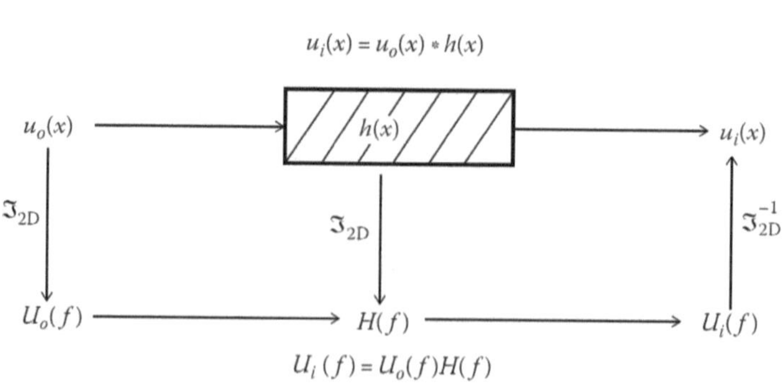

FIGURE 2.11 Linear systems model.

Fraunhofer criterion previously specified. In this case, the reflected field would be highly coherent and bounded by the finite extent of the mirror. Let us say that the detector is within the Fresnel region of the reflected electromagnetic wave, and no lens is used (the detector is placed directly in the entrance pupil instead of the image plane); then, the impulse response would be given by Equation 2.25, and the electromagnetic field in the entrance pupil can be obtained by either using Equation 2.24 directly or solving the problem in frequency space and then taking the 2D inverse Fourier transform of the result. This latter approach is shown at the bottom of Figure 2.11. The computational burden is typically much less cumbersome by taking the Fourier transform and solving in frequency space than by solving the problem in the spatial domain. However, with the advent of parallel processing hardware architectures, especially 3D hardware architectures and clever image processing algorithms, this may not always be universally true (Arrasmith 2010). Extending our simple example, let us say that the illuminating laser light is constant and the reference phase is zero everywhere. Then, the spatial component of the reflected field is modeled by

$$u_o(\bar{x}) = A\,\text{circ}\left(\frac{|\bar{x}|}{\rho}\right). \tag{2.34}$$

The circ(ax) function is defined as 1 where the quantity ax is less than 1 and zero elsewhere. In effect, we have a constant value A within the clear aperture of a circular mirror with radius ρ. Equation 2.25 gives the corresponding free-space impulse response. This impulse response has a Fourier transform given by (Goodman 2005; Roggemann and Welsh 1996)

$$H_d(\bar{f}) = \begin{cases} \exp(jkz)\exp\left[-j\pi\lambda z\,|\bar{f}|^2\right], & |\bar{f}| < \dfrac{1}{\lambda} \\ 0, & \text{elsewhere} \end{cases}. \tag{2.35}$$

The "d" subscript denotes that this is the CTF for free-space diffraction. The Fourier transform of Equation 2.34 is seen from Fourier transform tables to be

$$U_o(\bar{f}) = A\frac{\rho^2}{(\lambda z)^2}\frac{J_1\left(2\pi\rho|\bar{f}|/\lambda z\right)}{\left(\rho|\bar{f}|/\lambda z\right)}, \tag{2.36}$$

where the symbol J_1 means the Bessel function of the first kind and of order 1. The field spectrum in the entrance pupil plane is obtained by pointwise multiplying Equation 2.36 by Equation 2.35 everywhere within the clear aperture of the detector in the entrance pupil and then inverse Fourier transforming the result. This procedure is much simpler computationally than calculating the 2D spatial convolution using Equations 2.24, 2.25, and 2.34.

We now want to consider the case for imaging a coherent electromagnetic field. In this case, the detector is placed in the image plane and not in the entrance pupil plane. By making use of the lens function in Equation 2.33 and ensuring that we have paraxial imaging conditions, we can determine the impulse response of the coherent imaging system as (Goodman 2005)

$$h(\bar{x}) = \int\int_{-\infty}^{+\infty} d\bar{f}\,W\left(\bar{f}\lambda f_{\text{eff}}\right)\exp\left[-2\pi\bar{f}\cdot\bar{x}\right], \tag{2.37}$$

or

$$h(\bar{x}) = \mathcal{F}\left\{W\left(\bar{f}\lambda f_{\text{eff}}\right)\right\}. \tag{2.38}$$

The function $W\left(\bar{f}\lambda f_{\text{eff}}\right)$ is the pupil function and, for the coherent imaging case, takes on the value of 1 inside the clear aperture of the imaging system's entrance pupil and zero elsewhere. The impulse response is then the 2D Fourier transform of the pupil function. The parameter is the effective focal length and is given by the distance from the imaging system's exit pupil to the focal plane. Equation 2.38 is very useful. If we know the shape of the entrance pupil aperture (e.g., thc primary mirror of an imaging system's telescope), then we can determine the pupil function by scaling the spatial coordinates x in the entrance pupil plane by

$$\bar{x} = \bar{f}\lambda f_{\text{eff}}. \tag{2.39}$$

We can then determine the impulse response for the imaging system by simply taking the 2D Fourier transform of the result. As an example, many telescopic imaging systems have a circular primary mirror. As before, we can describe the entrance pupil using the circ(ax) function and then scale the entrance pupil coordinates appropriately according to Equation 2.39:

$$H\left(\bar{f}\right) = \text{circ}\left(\frac{\left|\bar{f}\right|\lambda f_{\text{eff}}}{\rho_{\text{ep}}}\right). \tag{2.40}$$

This scaled entrance pupil function is the CTF. The quantity ρ_{ep} is the radius of the aperture in the entrance pupil. Since the CTF has radial symmetry across the circular pupil, we show the spatial frequency coordinate as a magnitude. The impulse response is found by taking the 2D inverse Fourier transform of Equation 2.40:

$$h(\bar{x}) = \frac{\rho_{\text{ep}}^2}{\left(\lambda f_{\text{eff}}\right)^2} J_1 \frac{\left(\left(2\pi\rho_{\text{ep}}^2|\bar{x}|\right)/\left(\lambda f_{\text{eff}}\right)\right)}{\left(\left(\rho_{\text{ep}}^2|\bar{x}|\right)/\left(\lambda f_{\text{eff}}\right)\right)}. \tag{2.41}$$

This equation is, as before, a Bessel function of the first kind with order 1. Notice that if we had a rectangular entrance pupil aperture instead of a circular aperture, then the CTF would be

$$H\left(f_x, f_y\right) = \text{rect}\left(\frac{f_x\lambda f_{\text{eff}}}{D_x}\right)\text{rect}\left(\frac{f_x\lambda f_{\text{eff}}}{D_y}\right), \tag{2.42}$$

where

- f_x and f_y are spatial frequency components in the x and y directions, respectively;
- D_x and D_y are the respective diameters of the rectangular entrance pupil aperture in the x and y directions.

The function rect(x/b) is defined as 1 for $-b/2 < x < b/2$ and 0 elsewhere. As mentioned earlier, the impulse response is found by taking the 2D inverse Fourier transform and is given by

$$h(x, y) = \frac{D_x D_y}{\left(\lambda f_{\text{eff}}\right)^2}\text{sinc}\left(\frac{xD_x}{\lambda f_{\text{eff}}}\right)\text{sinc}\left(\frac{yD_y}{\lambda f_{\text{eff}}}\right). \tag{2.43}$$

The sine(ax) function is given by sin(ax)/ax, and the $x=0$ point is defined as 1. We will now provide a quick example to show the utility of our modeling system shown in Figure 2.10. Let us assume that we have a highly polished, mirrorlike, circular surface that is illuminated by a laser beam. As before, we can model the source term (e.g., the mirror) with a "circ" function with amplitude "A" as in Equation 2.34. The Fourier transform of our electromagnetic source field is the electromagnetic field spectrum and is given by Equation 2.36. Equation 2.40 gives the CTF for an imaging system with a circular entrance pupil with radius ρ_{ep}. The product of the electromagnetic field spectrum and the CTF gives the electromagnetic field spectrum in the image plane. The 2D inverse Fourier transform gives the electromagnetic field in the image plane, and the magnitude squared of the image plane electromagnetic field gives the irradiance in the image plane.

We will now take a look at the opposite of the coherent imaging case, incoherent imaging. For incoherent imaging systems, we assume that the object is spatially incoherent that means that neighboring points further apart from each other than an optical wavelength do not constructively interfere with each other. The effect of this assumption is that a 2D spatial autocorrelation of the object would be nonzero only for an offset of zero, a direct overlap of the object with itself. This of course is strictly speaking not true, but for most man-made objects, this assumption works very well. The reason for this validity is that most man-made objects, which are not highly polished, tend to be rough on the scale of an optical wavelength, and so neighboring points on the object's surface do not constructively interfere with each other. We will make good use of this fact in determining the mathematical model that governs the relationship between the object's brightness (watts/m^2) and the irradiance (watts/m^2) in the image plane. To determine this relationship, we must recall that brightness in the object plane is proportional to the magnitude squared of the electromagnetic field in the object plane:

$$i_o(\bar{x}) = u_o(\bar{x})u_o^*(\bar{x}). \tag{2.44}$$

We can relate the object brightness to the image plane irradiance by making use of Equation 2.33a and substituting into Equation 2.44. Taking the expected value of the result, we get the following relationship (Roggemann and Welsh 1996):

$$\langle i(\bar{x})\rangle = \int_{-\infty}^{\infty} d\bar{x}_o' \int_{-\infty}^{\infty} d\bar{x}_o'' u_o(\bar{x}')u_o^*(\bar{x}'')h_i(\bar{x}-\bar{x}_o')h_i^*(\bar{x}-\bar{x}_o''). \tag{2.45}$$

The expected value produces the average irradiance in the image plane. At this stage, we make use of the incoherent imaging assumption. As previously stated, for incoherent imaging conditions, the only nonzero contributing factor to the expectation is when the fields overlap. Mathematically, we can write this as (Goodman 2005)

$$\langle u_0(\bar{x}_0')u_0^*(\bar{x}'')\rangle = k\langle o(\bar{x}_0')\rangle\delta(\bar{x}_0' - \bar{x}_0''). \tag{2.46}$$

The factor k is a proportionality constant that provides equivalence between both sides of this equation. The symbol 6 is the Dirac delta function. For this equation, we have annotated the integration variables to show clearly in what plane the integration takes place. Substituting Equation 2.46 into Equation 2.45 and carrying out the integrals, we find that

$$i(\bar{x}) = \int_{-\infty}^{+\infty} d\bar{x}_0 o(\bar{x}_0)\left|h_i(\bar{x}-\bar{x}_0)\right|^2, \tag{2.47}$$

or simply written as

$$i(\bar{x}) = o(\bar{x}) * \left|h_i(\bar{x}-\bar{x}_0)\right|^2. \tag{2.48}$$

We have dropped the expectation values on the left side of the equation for convenience and with the understanding that, unless otherwise stated, we are referring to the average quantities, in this case, the average image irradiance on the left side of Equation 2.47. Likewise, the object brightness within the integral is also an averaged quantity. It is interesting to note that this equation has the

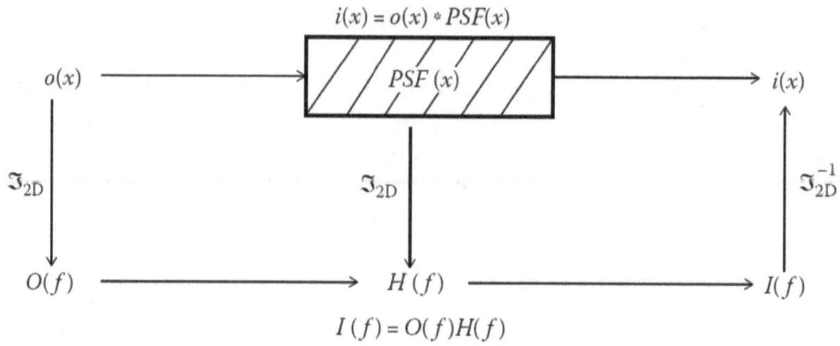

FIGURE 2.12 Incoherent imaging model.

same form, 2D spatial convolution, as what we had in the coherent imaging case. The difference is that, for the coherent imaging case, the 2D spatial convolution was at the field level, whereas for the incoherent imaging case, the 2D convolution is at the magnitude squared of the electromagnetic field. At this point, we can present in Figure 2.12 our model for the incoherent imaging case.

Just as before, there is a strong connection with the 2D Fourier transform and its inverse. At the top left of Figure 2.12, we have the object brightness denoted by a lowercase "*o*." The average irradiance in the image plane is obtained by convolving the PSF with the object brightness. Similar to before, a conventionally, more computationally efficient approach to determining the average image plane irradiance is to first take the 2D Fourier transform of the object brightness to obtain the object brightness spectrum. Then, multiply the object brightness spectrum by the OTF to obtain the image spectrum. As before in the coherent imaging case, we take the inverse Fourier transform of the image spectrum to obtain the average image plane irradiance. Modeling the propagation physics of the object brightness in the spatial frequency domain is most often computationally more efficient; however, the advent and further development of high-speed, parallel processing devices, such as field programmable gated arrays, 3D neural network architectures (Irvine Sensors Corporation n.d.), cellular neural network chips (Arrasmith 2010), and advancements in high-speed, parallel processing, image processing, and atmospheric turbulence compensation methods, offer promising potential in up-and-coming high-speed imaging solutions in both the spatial and frequency domains.

The incoherent imaging model, as depicted in Figure 2.12, applies to a wide variety of *passive* imaging scenarios and applications. By passive imaging, we mean imaging that occurs under natural conditions such as ambient light or natural emissions when operating in the infrared part of the electromagnetic spectrum. In contrast, an *active* imaging system is one in which man-made sources such as lasers and/or radars are used to illuminate the target of interest (TOI). If a high, spatially coherent source is used to illuminate an *optically flat*, smooth on the scale of the optical wavelength, TOI, then the coherent imaging model shown in Figure 2.12 should be used to model the basic propagation and detection physics of the imaging system. An example of this scenario would be a spatially coherent laser that illuminates a mirrorlike surface and is detected in some distant imaging plane under Fresnel or Fraunhofer conditions. It should be noted that in the event a source that has a high degree of spatial coherence illuminates an object that is rough on the scale of an optical wavelength, the reflected light will be spatially incoherent, regardless of the source, and a mixed model should be employed (a coherent propagation model from source to target and an incoherent imaging model for the reflected brightness from the target to the imaging plane).

For the incoherent imaging scenario, since the surface granularity of the TOI is often rough on the scale of the optical wavelength, or the illuminating source itself is spatially incoherent, the incoherent imaging model works exceedingly well in describing the essential characteristics of passive imaging systems or even actively illuminated objects that are rough on the scale of the illuminating wavelength. We will now describe some essential aspects of the incoherent imaging system that are needed in modeling and understanding the performance of the optical imaging system.

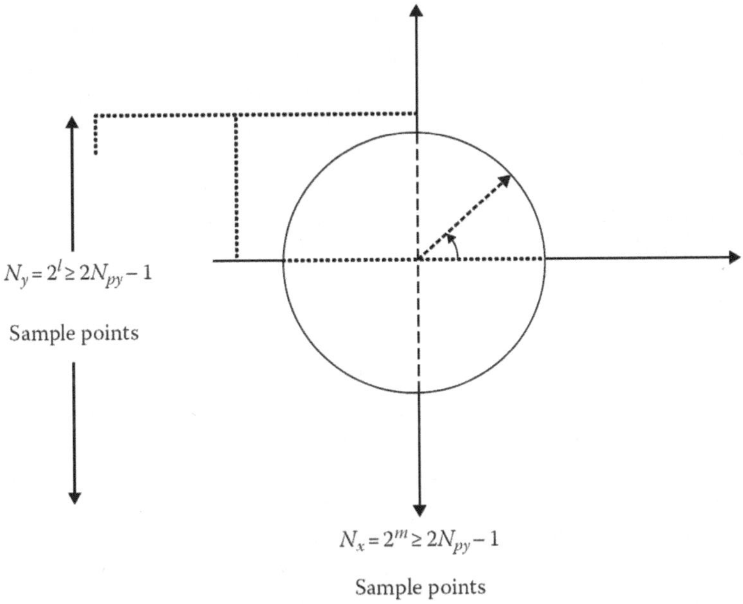

FIGURE 2.13 Sampled circular entrance pupil model.

Just as in the coherent imaging case, the entrance pupil plays a dominant role in modeling imaging system and atmospheric turbulence effects. The imaging model depicted in Figure 2.12 still applies for the incoherent imaging case. Recall that for many well-designed optical, telescopic imaging systems, the primary mirror of the telescope is placed in the entrance pupil of the imaging system. Due to its importance and its connection to other essential elements of the optical system, we will develop a separate model for the entrance pupil plane of the imaging system. Figure 2.13 provides an illustration of our assumed circular entrance pupil. We use a circular entrance pupil in this illustration since most telescopes have circularly shaped primary mirrors. Although the real and imaginary parts of the electromagnetic wave are continuous within the telescope entrance pupil, to properly model the imaging scenario on a computer, the electromagnetic field within the imaging system's entrance pupil must be correctly spatially (and temporally) sampled. We focus on the spatial sampling requirements for the moment and bring the temporal aspects in the statistical optics section later. The discrete points in the entrance pupil aperture are sample points, in the $\hat{x}$ and $\hat{y}$ directions, which are chosen to obey Nyquist sampling conditions. We will discuss Nyquist sampling conditions further in this chapter.

We are keeping with the convention in dropping the subscripts on the position variables and identifying the proper plane, in this case the entrance pupil, within the context of the dialog supporting the figure. The sampling shown in Figure 2.13 is uniform in both spatial directions. This is not a limiting restriction; however, it is convenient, and so we will continue with this entrance pupil uniform spatial sampling model unless otherwise needed. We have also found it to be convenient to use an odd number of points across the entrance pupil clear aperture so that both the physical center point of the aperture and the aperture edge points along the main coordinate axes coincide with a sample point. In our model, we embed this sampled aperture in a matrix of zeros that has at least $(2N_{px} - 1)$ points in the x direction and $(2N_{py} - 1)$ points in the $\hat{y}$ direction. Additionally, we increase the number of sample points in each linear direction to the smallest power of 2 that is larger than the preceding sampling requirements to aid in the efficiency of doing the 2D Fourier transform. In other words,

$$N_y = 2^l, \quad l = \min\{l = 1,2,3,\ldots\} \ni 2^l \geq 2N_{py} - 1, \tag{2.49}$$

and

$$N_x = 2^m, \quad m = \min\{m = 1,2,3,\ldots\} \ni 2^m \geq 2N_{px} - 1, \tag{2.50}$$

where

N_{px} and N_{py} are the number of sample points along the diameter in the clear aperture of the entrance pupil in the x and y directions, respectively.
l and m are integers that are chosen to be as small as possible and that satisfy the right side of Equations 2.49 and 2.50, respectively.

N_x and N_y are then the respective total number of entrance pupil sample points in a given linear direction in our model. For example, say that we had 127 sample points across the diameter in the $\hat{x}$ direction. The center point corresponding to $x=0$ is the 64th point. There are an equal number of points to the left and right of the center point, 63 points. The 1st and 127th point would rest on the edge of the clear aperture. Since we are using uniform sampling, the $\hat{y}$ direction would be identical to the $\hat{x}$ direction for our circular aperture. The number of sample points along the diameter within the clear aperture, in the $\hat{x}$ and $\hat{y}$ directions, respectively, is $N_{px}=N_{py}=127$. The minimum number of required points in our model is $(2N_{px}-1)=253$ points in the x direction and $(2N_{py}-1)=253$ points in the $\hat{y}$ direction. The smallest l and m that satisfy Equations 2.49 and 2.50 are l and $m=8$, leading to $N_x=N_y=256$. To model our entrance pupil aperture, we would have a 256×256 array that is 0 everywhere except within a circular aperture that has 127 points along the $\hat{x}$ and $\hat{y}$ diameter. The circular aperture is centered within the array with the center point located on the (129, 129) sample point measured from the top left of the array. Note that if we assign a length to the diameter of the circular entrance pupil diameter, then we can determine sample spacing Δx and Δy. For example, if we had a telescope with a 20 cm diameter primary mirror, then Δx and Δy would be $20/(129-1)=0.15625$ cm. Within the clear aperture of the entrance pupil, we define a GPF as

$$W(\vec{x}) = W_p(\vec{x})e^{j\Phi(\vec{x})}, \tag{2.51}$$

where

$$W_p(\vec{x}) = \begin{cases} 1 & r(\vec{x}) \leq D/2 \\ 0 & \text{e.w.} \end{cases}, \tag{2.52}$$

where Equation 2.52 is an aperture function that describes the physical aperture of the entrance pupil, in this case 1 inside our circular aperture of diameter D meters. It should be mentioned that the physical aperture can take on values other than 1 for physical situations that have strong amplitude effects, such as in scintillation, but for a broad range of imaging conditions, the constant amplitude assumption is acceptable. If it becomes necessary to include amplitude effects in our imaging model, we would just replace the "1" on the right side of Equation 2.52 with a spatially varying amplitude. The complex exponential in Equation 2.51 is used to describe entrance pupil phase aberrations to include phase aberrations due to the atmosphere. The GPF is shown as a continuous function of spatial position vector $\hat{x}$ but computationally is evaluated at the discrete sample points shown in Figure 2.13. As will be discussed later in the section discussing spatial sampling, if proper sampling conditions are established, it can be shown that an exact copy of the continuous function can be reproduced from the sampled data. For now, we will show the continuous variable $\hat{x}$ but understand that when implemented on the computer, sampled points of our GPF, as well as other functions, will be used to represent their continuous counterparts. We now want to develop some fundamentally important optical SE functions to help characterize and understand our optical system's performance. As will be seen, our approach has many similarities and parallels with the coherent imaging case mentioned earlier.

If we make the variable substitution from Equation 2.39 into Equations 2.50 and 2.51 and take the 2D Fourier transform of our GPF just as in Equation 2.38, we obtain the impulse response for our optical system. By then taking the magnitude squared of the resulting impulse response, we get the PSF for our incoherent imaging system:

$$s(\bar{x}) = |h(\bar{x})|^2 = \left|\mathrm{F}\left\{W\left(\bar{f}\lambda f_{\text{eff}}\right)\right\}\right|^2. \tag{2.53}$$

The PSF describes how the diffraction, atmospheric effects, and system effects "spread" the light across the detector surface in the image plane and thereby decrease the spatial resolution of the imaging system. Even if we had a perfect optical system with no aberrations and we were imaging in a vacuum (no atmospheric aberrations), the diffractive effects of the entrance pupil aperture would still impact our imaging system to some degree. Superresolution methods would need to be employed to potentially mitigate the spatial resolution loss due to diffraction. For most optical systems in the earth's atmosphere, optical systems aberrations and atmospheric turbulence are the dominant effects in the imaging systems loss of spatial resolution. For well-designed optical systems, aberrations due to the optical system itself are minimized, and atmospheric turbulence effects dominate the loss in spatial resolution of the imaging system.

If we take the 2D Fourier transform of the PSF and normalize the result so that the maximum value of the resulting function is 1, for example, by dividing the result of the 2D Fourier transform by its maximum, we obtain the OTF

$$H(\bar{f}) = \frac{\mathcal{F}\{s(\bar{x})\}}{\mathcal{F}\{s(\bar{x})\}|_{\bar{f}=0}}. \tag{2.54}$$

The maximum value of the 2D Fourier transform of the PSF is always the zero spatial frequency component of the resulting 2D Fourier transformed function. Another way that the OTF can be determined is by making use of two convenient properties of the Fourier transform. The first property is that the Fourier transform of the product of two functions is the same thing as the spatial convolution of the Fourier transform of each function. The second property is that the complex conjugation of a function has the effect of time/space reversal in the Fourier-transformed space. Combining these two properties and noting that the PSF is the product of the optical systems impulse response with the conjugate of the impulse response, we see that the OTF can also be obtained by an autocorrelation of the GPF:

$$H(\bar{f}) = \frac{\mathcal{F}\{h(\bar{x})h^*(\bar{x})\}}{\mathcal{F}\{h(\bar{x})h^*(\bar{x})\}|_{\bar{f}=0}} = \frac{W\left(\bar{f}\lambda f_{\text{eff}}\right) \otimes W\left(\bar{f}\lambda f_{\text{eff}}\right)}{W(0) \otimes W(0)}, \tag{2.55}$$

where the symbol ⊗ means 2D autocorrelation. The method of determining the OTF leading up to Equation 2.54 is typically preferred over Equation 2.55 since the former approach tends to be computationally faster than Equation 2.55 for most conventional applications. From Figure 2.5, and using Equation 2.54 or 2.55, we can get the image spectrum by

$$I(\bar{f}) = O(\bar{f})H(\bar{f}). \tag{2.56}$$

The image is then obtained by taking the 2D inverse Fourier transform of the image spectrum:

$$i(\bar{x}) = \mathcal{F}^{-1}\left\{I(\bar{f})\right\}. \tag{2.57}$$

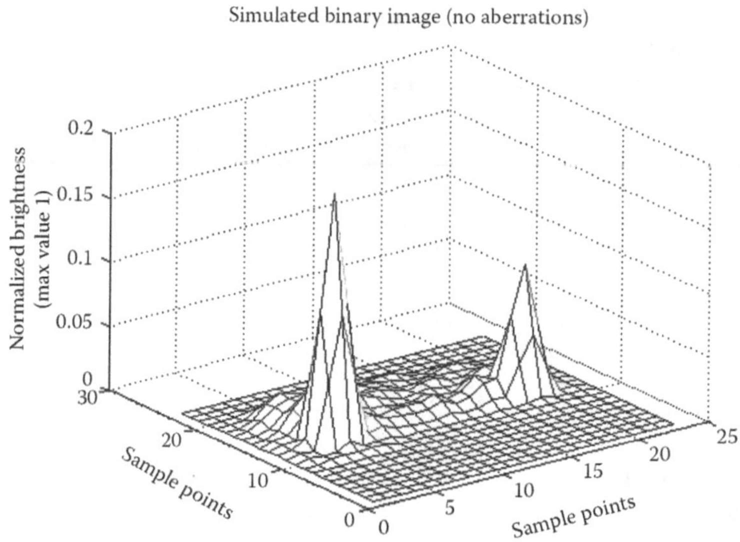

FIGURE 2.14 Object model of a visible binary star system such as a-Centauri A and α-Centauri B.

We now provide a simple example to illustrate this approach. Let us say that we have a telescopic imaging system that has a diameter of 1 m and an effective focal length f_{eff} of 1 m. Say we want to model the imaging systems performance when it is looking at a binary star system, one with twice the magnitude of the other star. Because of the geometries involved, our earthbound telescope cannot resolve the individual stars themselves, so we can represent them in our model as delta functions. We normalize one star to have a maximum brightness of 1 W/m^2 and the other star to a relative brightness of 0.5 W/m^2. Figure 2.14 shows a MATLAB® model of our TOI.

Experimentally, the PSF can be obtained by looking at a single star. It can be seen from Equation 2.47 that if the object is a point source, we recover the PSF. This PSF would incorporate the effects of the atmosphere, the entrance pupil aperture, and any optical system effects. If the imaging system were to observe a distant point source object, the effect in the image plane would be to spread the irradiance due to the point source, hence the name PSF. For now, let us assume that we want to determine the effects of the entrance pupil aperture by itself without any atmospheric turbulence or optical system effects. A physical example would be looking at a binary star system with a telescope outside of the Earth's atmosphere. Let us also assume, as we have been so far, that we can ignore temporal effects; we will model temporal effects later. Under these assumptions, our modeling results would give us an upper bound on the expected conventional spatial resolution performance of our optical system. As an example, if we use uniform entrance pupil sampling of 127 points across the x and y directions, then, as previously discussed, we have $N_x = N_y = 256$ samples in our entrance pupil model. Figure 2.15 shows the model for our entrance pupil plane aperture. Notice that by setting the phase term to zero everywhere within the aperture, we effectively eliminate optical system phase effects and atmospheric turbulence effects from our model. By temporarily ignoring system and signal noises, we can isolate the effect of the entrance pupil aperture in our model. The GPF only has real values when the phase is identically zero everywhere within the clear aperture of the entrance pupil. The approximation of a constant phase throughout the entrance pupil only applies to a well-designed optical system that is imaging through a vacuum, like space. In the presence of atmospheric turbulence, the constant entrance pupil plane phase assumption is not valid and the GPF is a complex quantity. System noise and atmospheric turbulence effects can readily be included in our model, but, for now, we want to isolate the effects of the entrance pupil aperture on our imaging system. Notice in Figure 2.15 that our GPF values are nonzero only within the clear aperture of the entrance pupil itself.

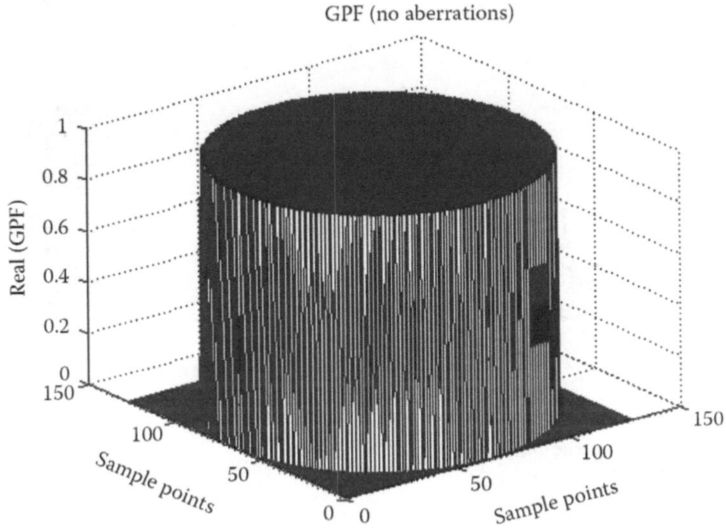

FIGURE 2.15 Model of GPF for no system and atmospheric aberrations.

The values outside of the entrance pupil clear aperture are set to zero since only the irradiance within the clear aperture of the entrance pupil contributes to the formation of the image. Notice also that the maximum value of GPF is one since we have assumed no amplitude effects are present in our imaging system.

Taking the 2D Fourier transform according to Equation 2.38, we obtain the impulse response of the optical system. Taking the pointwise magnitude squared of the impulse response gives us the PSF. In the spatial domain, the PSF is an indicator of our imaging system's loss of resolution. For no aberrations present in the imaging system and an infinitely large entrance pupil aperture, the PSF would approach a delta function and, according to Equation 2.47, would exactly replicate the object brightness. In the absence of aberrations in our imaging system, the presence of an entrance pupil with finite spatial extent would effectively spread the received irradiance in the image plane. Therefore, apart from superresolution methods, even with a perfect imaging system with no aberrations or system noises, we cannot exactly reproduce the object brightness because of the presence of the entrance pupil with finite extent. Said another way, the entrance pupil acts as a low-pass filter that effectively removes higher spatial frequency content from our image. Taking the 2D Fourier transform of the PSF and then normalizing by its maximum value yield the OTF.

We now have all the elements necessary to model the performance of our optical imaging system that is viewing a representative, visible binary star system like α-Centauri A and α-Centauri B. The combination of α-Centauri A and α-Centauri B – α-Centauri AB – appears as a single star to the naked eye but are easily resolved into a binary star system with a small telescope. Together, they are the third brightest objects among stars, when excluding our sun. We model each of the binaries as scaled Kronecker delta functions to represent point source like objects with differing brightness and a variable separation between 2 and 22 as. The brightness for α-Centauri A is about 3.038 times that of α-Centauri B, so we can normalize the brightness of α-Centauri A to 1 and α-Centauri B to approximately 1/3. We assume that the telescope is aligned to the center of the binary star system and that the image of the stars lies on the *x*-axis. We assume that the binaries are separated from each other by 2 as (equivalently 9.69 μrad). The sample point spacing in the object's spatial domain is 0.05 as, and the spatial sampling in the entrance pupil plane is 0.0794 m. We use 500 nm as the center wavelength of our narrowband, filtered, imaging system. Figure 2.16 shows the results for our diffraction-limited optical imaging systems model with a 1 m circular entrance pupil. The modeling was done in MATLAB R2012a.

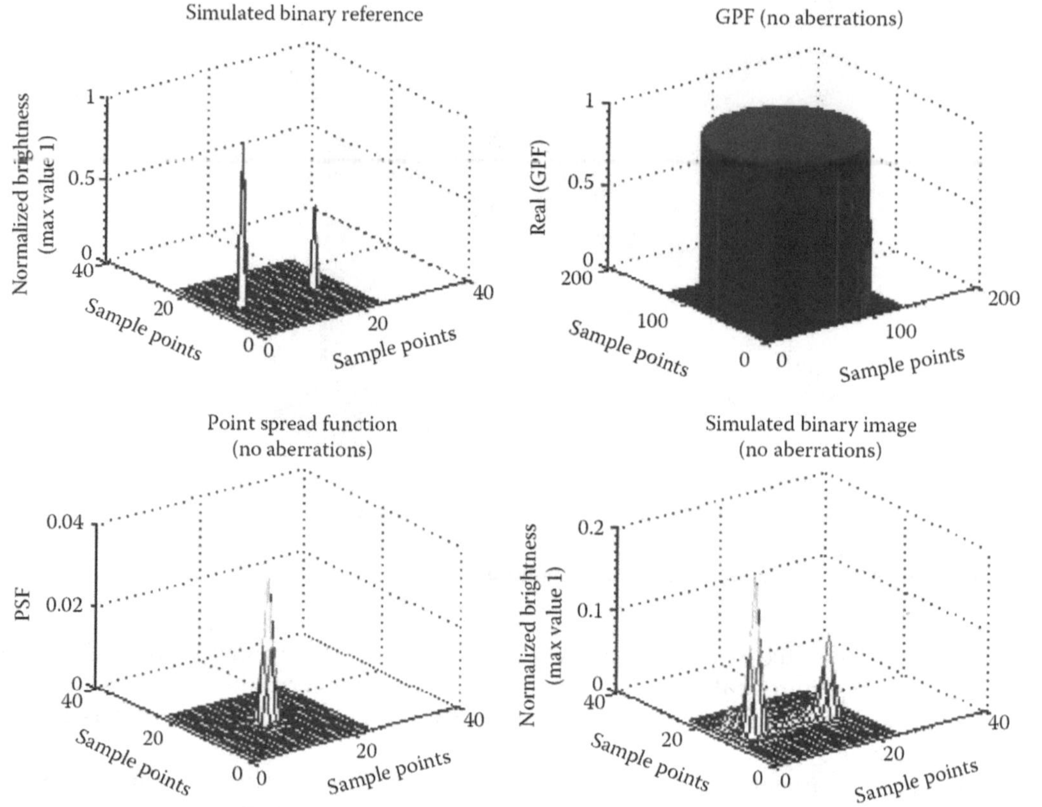

FIGURE 2.16 Simulated binary star system.

The top-left figure is a plot of our binary star system wherein the maximum brightness has been normalized to one. The brighter point source object (star) on the left represents α-Centauri A, and the dimmer object on the right represents α-Centauri B. The top-right figure is that of our generalized pupil plane assuming no system noises or atmospheric aberrations. The bottom-left figure is that of our PSF obtained by implementing Equations 2.51 through 2.53. The image obtained by our diffraction-limited imaging system is shown on the bottom right. This simulated image was obtained by first determining the OTF in accordance with Equation 2.54. We then solve for the image by implementing Equations 2.56 and 2.57. We can see that the image is a blurred version of the original object even though we have assumed a noise-free imaging system with no aberrations. If we multiply the image irradiance that falls on the detector by the detector area, we end up with the optical power on the detector. The 2D Fourier transform of the optical "power" gives the *power spectrum* of the image. Likewise, the 2D Fourier transform of the product of the object brightness with the object area produces the power spectrum of the object. In determining the *energy spectrum* of an object or image, the optical power must first be converted into energy units by integrating the optical power over an appropriate time scale. If the total energy spectrum is desired, the power spectrum must be integrated with respect to time from minus to plus infinity. This can only be done if the integrated energy over time is finite. Mathematically, we can define the energy spectral density as

$$\varepsilon(\bar{f}) = \left|u(\bar{f})\right|^2, \tag{2.58}$$

that is only applicable if the temporal integral of the magnitude of the electromagnetic field is finite when integrated from minus infinity to plus infinity (Goodman 1985). If the integrated energy is

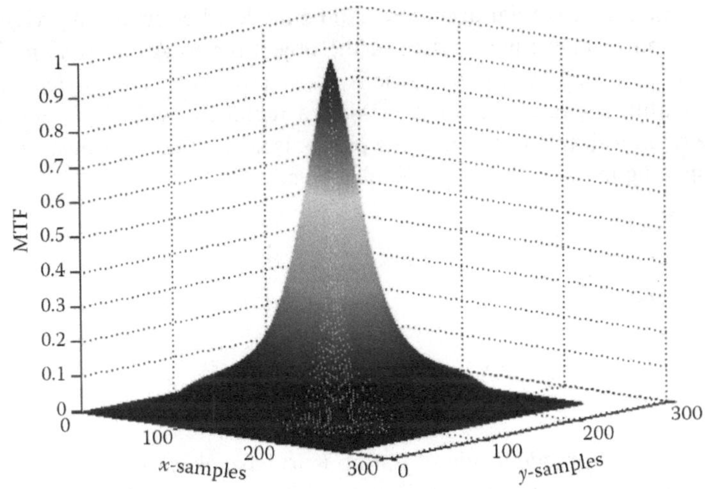

FIGURE 2.17 MTF of imaging system viewing the binary star system.

not finite, then it may be possible to integrate the optical power over a shorter period T and obtain a finite result. If this can be done, we can determine the power spectral density by

$$P(\bar{f}) = \frac{\left|u(\bar{f})\right|^2}{T}. \tag{2.59}$$

Another useful metric is the MTF. This is the magnitude of the OTF and describes the magnitude effects of the OTF on the object spectrum at every applicable spatial frequency. Figure 2.17 shows the MTF for the imaging system viewing our binary star system.

2.2.3 Sampling Considerations

To properly represent a spatially continuous object on a computer, the object is often spatially sampled with the assumption of continuity between sample points. This idea is nothing new to most people. Anyone who has watched television (TV) "sees" a continuous picture but knows that the picture is made up of lots of small picture elements (pixels) that each put out a sampled brightness. Move inches away from the TV display, and often, individual pixels can be seen with the naked eye. However, move a short distance away from the screen, and the picture looks smooth and continuous (for a properly working, good TV). In the previous example of the binary star system, we provided a sampling scheme to facilitate our illustration. But how is the sample spacing determined? To properly use the linear systems methods presented in the previous section, it is necessary to understand spatial and temporal sampling methods. In this section, we provide a short discussion on basic sampling considerations. We will provide both the theoretical sampling requirements dictated by mathematics and also some practical sampling guidelines dictated by experience.

A good point to start in a discussion about spatial sampling is to recall that the relationship between a distant source object and the entrance pupil plane electromagnetic field is often expressible as a 2D Fourier transform relationship. Equation 1.7 gives the diffraction-limited spatial resolution of the optical imaging system for a circular entrance pupil. Conventional telescopes cannot see spatial details smaller than the diffraction-limited resolution. Additionally, it takes a well-designed optical system to provide spatial resolutions that approach the diffraction limit. Consequently, for spatially resolvable objects, the spatial sampling in the object plane, obtained by projecting the dimensions of the imaging system's pixels onto the object, should not exceed the diffraction-limited resolution of the telescope. A similar argument can be made when imagining viewing the imaging system's entrance pupil from the perspective of the object plane. A question might be as follows: what sample spacing in the entrance pupil plane is needed to resolve spatial details due to an object that is contained within a circle of diameter D_o or a rectangle with dimensions D_{ox} and D_{oy} in meters? Since the relationship between the two planes is still a Fourier transform relationship, the required sample spacing in the entrance pupil must be smaller than

$$\Delta X_p = 1.22\frac{\lambda}{D_o}z, \tag{2.60}$$

for the circular case and

$$\Delta X_{p_x} = \frac{\lambda}{D_{ox}}z,\ \Delta X_{p_y} = \frac{\lambda}{D_{oy}}z, \tag{2.61}$$

for the rectangular case. A practical rule of thumb is to have at least two samples in the entrance pupil per required sample spacing given by Equations 2.60 and 2.61. The value D_o is the diameter in meters of a circle that circumscribes the object. The values D_{ox} and D_{oy} are the diameters of a rectangle that just contains the object. The intent is to provide a simple object such as a square, rectangle, or circle that just contains the object. In this manner, the sampling requirements in either the object plane or the entrance pupil plane can be set up, to ensure that available spatial information is captured by the optical imaging system. Because of the nature of the 2D Fourier transform that relates the electromagnetic field from the object plane to the entrance pupil plane, it turns out that the largest spatial dimension of the object to be imaged determines the sampling requirements in the entrance pupil plane (see Equations 2.60 and 2.61) and the largest spatial dimension of the entrance pupil determines the smallest sampling requirements in the object space (see Equation 1.7).

A more formal treatment of sampling involves understanding the Whittaker and Shannon sampling theorem (Goodman 2005). It can be shown that for a band-limited, continuous 2D function, $g(x, y)$, the sampled function

$$g_s(x,y) = \text{comb}\left(\frac{x}{X}\right)\text{comb}\left(\frac{y}{Y}\right)g(x,y), \tag{2.62}$$

where the "comb" functions are given by

$$\text{comb}\left(\frac{X}{T}\right) = \sum_{n=-\infty}^{\infty}\delta\left(X - \frac{n}{T}\right). \tag{2.63}$$

The Kronecker delta function on the right side of this equation has the value 1 at sample separations of T units. This effectively looks like an infinitely long "comb" that has "teeth" with a height of 1 and teeth separation of T. For a ID signal with finite extent, the comb function in Equation 2.63 is multiplied by a "rect" function:

$$\text{rect}\left(\frac{X}{L}\right) = \begin{cases} 1, & |X| \le \frac{L}{2}. \\ 0, & \text{e.w.} \end{cases} \tag{2.64}$$

In two dimensions, each "comb" function can be multiplied by its own "rect" function to define the maximum extent of the object. In this way, different sampling in the x and y directions can be accomplished if desired. For uniform sampling, the parameters X and Y in Equation 2.62 are set equal to each other. Figure 2.18 shows the case of a 2D, continuous function, $g(x, y)$, being sampled in the x and y directions with sample spacing of X and Y, respectively.

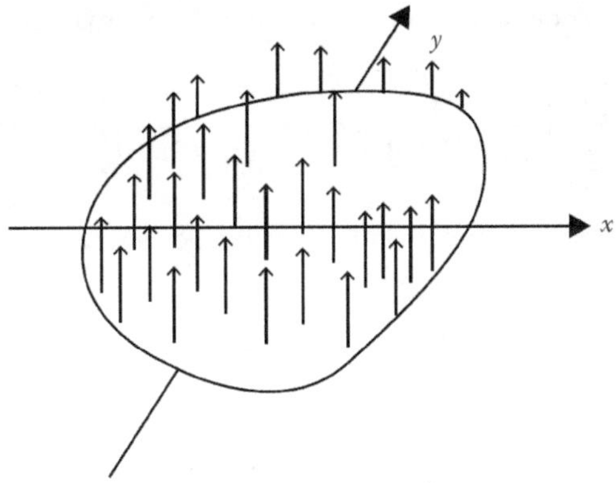

FIGURE 2.18 Two-dimensional, continuous sampling functions.

As can be seen, lower spatial frequencies of the object spectrum are affected less than higher spatial frequencies. Since the OTF pointwise multiplies the object spectrum, the effect of the OTF is to attenuate the higher spatial frequency information in the object spectrum. This higher spatial frequency information contains the fine details in the image. The MTF shows this attenuation and so is a good metric for image quality.

Before leaving this section, we should say a few words about nonlinear optical systems models. The field of nonlinear optics (NLO) is quite rich, interesting, and beyond the scope of this book. Topics such as optical phase conjugation, three-wave and four-wave mixing, second harmonic generation, and optical parametric amplification are examples of nonlinear optical methods that produce intriguing results (Boyd 2008). In NLO, the super-position principle no longer holds, and our linear systems models just mentioned break down. Often, nonlinear optical processes occur in materials where the material properties vary as function of the strength of the electromagnetic field. Typically, very high electric field strengths are required to generate the NLO effect. We consider NLO an advanced and specialized topic that is well treated in the following references (Shen 1998; Anderson and Erdogan 2014). For most practical optical imaging and optical detection scenarios, our linear systems model is sufficiently accurate. However, if high-power lasers interact with nonlinear material such as a barium borate (BBO) crystal, then NLO analytical methods need to be employed to accurately model the optical phenomenon. In this book, we employ NLO methods where needed and space allows and refers the reader to more detailed literature where needed.

The resulting sampled function, $g_s(x, y)$, is given by Equation 2.62. It can be shown that for Nyquist sampling conditions given by

$$X \leq \frac{1}{2B_x}, \tag{2.65}$$

and

$$Y \leq \frac{1}{2B_y}, \tag{2.66}$$

the original continuous function can be reproduced *exactly* from the sampled data (Oppenheim and Schafer 1975; Goodman 2005). For the ID case, a simple reconstruction equation is given by (Oppenheim and Schafer 1975)

$$g(t) = \sum_{k=-\infty}^{\infty} g(kT)\frac{\sin\left[(\pi/T)(t-kT)\right]}{(\pi/T)(t-kT)}, \tag{2.67}$$

and for the 2D case, the interpolation formula is given by (Goodman 2005)

$$g(x,y) = \sum_{n=-\infty}^{\infty}\sum_{m=-\infty}^{\infty} g\left(\frac{n}{2B_x},\frac{m}{2B_y}\right)\operatorname{sinc}\left[2B_x\left(x-\frac{n}{2B_x}\right)\right]\operatorname{sinc}\left[2B_y\left(y-\frac{m}{2B_y}\right)\right], \tag{2.68}$$

and the parameters B_x and B_y are the bandwidths in frequency space. The "sinc" function is given by

$$\operatorname{sinc}(ax) = \frac{\sin(a\pi x)}{a\pi x}, \tag{2.69}$$

where Equation 2.69 can be used in concert with the shift theorem to evaluate Equation 2.68.

The critical assumptions in obtaining the interpolation formulas in Equations 2.67 and 2.68 are that the object can be Fourier transformed and that the object is band-limited. For the first assumption, the conditions that lead up to Equations 2.58 and 2.59 need to be satisfied. In essence, in order to be Fourier transformable, the source must be able to be absolutely integrated

$$\int_{-\infty}^{+\infty}\int_{-\infty}^{+\infty} |g(x,y)|\,dx,dy < \infty, \tag{2.70}$$

or the source must be able to be integrated over a finite period

$$\lim_{T_y\to\infty}\frac{1}{T_y}\int_{-T_y/2}^{T_y/2}\lim_{T_x\to\infty}\frac{1}{T_x}\int_{-T_x/2}^{T_x/2} u^2(x,y)\,dxdy < \infty. \tag{2.71}$$

In practical cases for real objects with finite extent, this is typically the case. For the temporal component, equivalent to the discussion of the energy spectrum and/or power spectrum earlier, the conditions in order to be Fourier transformable are

$$\int_{-\infty}^{\infty} |g(t)|\,dt < \infty, \tag{2.72}$$

and

$$\lim_{T\to\infty}\frac{1}{T}\int_{-T/2}^{T/2} g^2(t)\,dt < \infty. \tag{2.73}$$

The second assumption is the basis for the definition of the *bandwidth*, which for our optical system is just the maximum spatial frequency in a given linear direction. The *Nyquist rate* is twice the bandwidth and is the spectral extent of our band-limited signal as shown in Figure 2.19.

For most practical imaging scenarios, the object has finite extent and finite power/energy, so Fourier transform methods in modeling the optical system are very practical and computationally efficient. To aid in the Fourier-based analysis of optical systems, some useful properties and transform relationships have been developed over time. For convenience, we list the more notable properties and transforms in Table 2.3.

By using the shift theorem and scaling property of the Fourier transform in combination with the basic functions earlier, spatially complex functions can be modeled and analyzed using a combination of these simple relationships. As an easy example, our binary star system was modeled using two scaled and shifted Kronecker delta functions. A more complex object could be constructed by using individually scaled and shifted Kronecker delta functions that were generated to satisfy the Shannon and Whitaker sampling theorem discussed earlier.

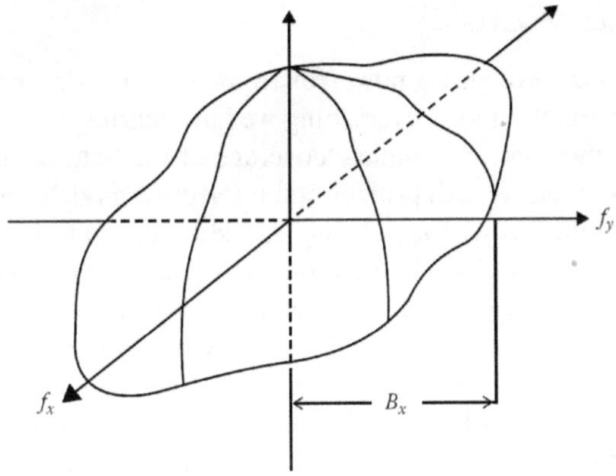

FIGURE 2.19 Band-limited function.

TABLE 2.3
Principles/Theorems

Linearity (1D and 2D)	$\mathcal{F}(c_1u_1 \mp c_2u_2) = c_1U_2 \mp c_2U_2$
Shift (1D)	$\mathcal{F}\{u_1(x-a)\} = e^{j2\pi fa}G(f)$
Shift (2D)	$\mathcal{F}\{u_1((x-a, y-b))\} = e^{j2\pi(f_xa+f_yb)}G(f_x, f_y)$
Scaling (1D)	$\mathcal{F}\{u_1(ax)\} = \frac{1}{\lvert a\rvert}G\left(\frac{f_x}{a}, \frac{f_y}{a}\right)$
Scaling (2D)	$\mathcal{F}\{u_1(ax, by)\} = \frac{1}{\lvert ab\rvert}G\left(\frac{f_x}{a}, \frac{f_y}{b}\right)$
Parseval's theorem (1D)	$\int_{-\infty}^{\infty} \lvert u_1(x)\rvert^2 dx = \int_{-\infty}^{\infty} \lvert U_1(f)\rvert^2 df$
Parseval's theorem (2D)	$\iint_{-\infty}^{\infty} \lvert u_1(x, y)\rvert^2 dxdy = \iint_{-\infty}^{\infty} \lvert U_1(f_x, f_y)\rvert^2 df_xdf_y$
Autocorrelation theorem (1D)	$\mathcal{F}\left\{\int_{-\infty}^{\infty} h(\lambda_1 - x)h^*(\lambda_1)d\lambda_1\right\} = \lvert H(f)\rvert^2$
Autocorrelation theorem (1D)	$\mathcal{F}\iint_{-\infty}^{\infty} h(\lambda_1, \lambda_2)h^*(\lambda_1 - x)(\lambda_2 - y)d\lambda_1\lambda_2 = \lvert H(f_x, f_y)\rvert^2$
Convolution theorem (1D)	$\mathcal{F}\left\{\int_{-\infty}^{\infty} u(\lambda_1)h(x-\lambda_1)d\lambda_1\right\} = U(f)H(f)$
Convolution theorem (2D)	$\mathcal{F}\iint_{-\infty}^{\infty} u(\lambda_1, \lambda_2)h(x-\lambda_2)(y-\lambda_1)d\lambda_1\lambda_2 = U(f_x, f_y)H(f_x, f_y)$

So far, we have presented some linear systems models that deal with fully coherent imaging systems and fully incoherent imaging systems. In addition, we have discussed some concepts related to spatial coherence so far. What models do we use if our imaging model lies somewhere in between these two extremes? How do we model coherence properties in time as well as space? The field of statistical optics provides some relevant and useful methods to answer these questions.

2.2.4 Statistical Optics Models

The field of statistical optics provides a powerful framework for describing, characterizing, and analyzing optical systems. Furthermore, everything we have learned so far still applies and extends to the statistical optics framework. A complete coverage of the field of statistical optics is beyond the scope of this book; however, we will provide some selective models and principles that will add to our present understanding. We will also bring in additional material as needed in subsequent sections. In this section, we predominantly focus on statistical optics methods for describing spatial and temporal coherence. In addition to the previous limiting cases of fully coherent and fully incoherent light, we present a means for treating and modeling partially coherent light. We also present some useful tools that address the coherence properties of light such as the mutual coherence function and the van Cittert and Zernike theorem. We also present some expressions for the limits on the expected signal on a photoelectric detector. A good text for additional details on statistical optics methods can be found in the references (Goodman 1985). We first start with a few mathematical preliminaries. A basic understanding of probability and statistics is assumed.

For high signal levels, a Gaussian density function is often used in modeling, analyzing, and understanding signals of interest. The ID and 2D probability density functions for a Gaussian signal are given by

$$P_u(u) = \frac{1}{\sqrt{2\pi}\sigma} \exp\left\{-\frac{(u-\bar{u})^2}{2\sigma^2}\right\}, \tag{2.74}$$

and

$$P_{uv}(u,v) = \frac{\exp\left[\dfrac{-\left(u^2+v^2-2\rho uv\right)}{2\left(1-\rho^2\right)\sigma^2}\right]}{2\pi\sigma^2\sqrt{1-\rho^2}}. \tag{2.75}$$

The parameters U and V are random variables that represent random signals of interest. The parameters $\bar{u}$ and σ represent the signal mean and standard deviation, respectively. The parameter ρ in Equation 2.75 is given by

$$\rho \triangleq \frac{\overline{uv}}{\sigma^2}, \tag{2.76}$$

and $\overline{uv}$ is the joint moment of the jointly distributed Gaussian random variables U and V, and the parameter ρ ranges from 0 to 1. In Equation 2.75, we have made a common assumption that the random variables have zero mean and equal variance. If this is not the case, a more general expression for the jointly distributed Gaussian probability density function is given by

$$P_U(\underline{u}) = \frac{1}{2\pi^{n/2}\,|\underline{C}|^{1/2}} \exp\left\{-\frac{1}{2}\left(\underline{u}-\underline{\bar{u}}\right)^t |\underline{C}|^{-1}\left(\underline{u}-\underline{\bar{u}}\right)\right\}. \tag{2.77}$$

Here,

$\underline{C}$ is an $n\times n$ covariance matrix
$\underline{u}$ is a column vector of length n that consists of the random values u
The superscripts "t" and "–1" stand for transpose and inverse, respectively

The covariance matrix consists of "i" rows and "k" columns of covariance values given by

$$\sigma_{ik}^2 = E\left[\left(u_i-\overline{u_i}\right)\left(u_k-\overline{u_k}\right)\right], \tag{2.78}$$

where the "E" is the expectation operator given by

$$\sigma_{ik}^2 = \int_{-\infty}^{+\infty}\int \left(u_i - \overline{u_i}\right)\left(u_k - \overline{u_k}\right) P_{u_i u_k}\left(u_i, u_k\right) du_i du_k. \tag{2.79}$$

In general, expanding the random variables U_1 and U_2 in terms of two parameters t_1 and t_2, such as time t_1 or time t_2, we can write the joint expectation as

$$\overline{u_1 u_2} = E\left[u_1 u_2\right] = \int_{-\infty}^{+\infty}\int u_1 u_2 \rho_u\left(u_1, u_2; t_1, t_2\right) du_1 du_2, \tag{2.80}$$

and, in general,

$$\overline{u_1^n u_2^m \ldots u_L^p} = E\left[u_1^n u_2^m \ldots u_L^p\right] = \int_{-\infty}^{+\infty}\int \ldots \int u_1^n u_2^m \ldots u_L^p \rho_{u_1 u_2 \ldots u_L} \left(u_1, u_2, \ldots, u_L; t_1, t_2, \ldots, t_L\right) du_1 du_2 \ldots du_L. \tag{2.81}$$

Equation 2.81 is useful in calculating higher-order moments of random variables U_1 and U_2. At this point, it is useful to discuss stationary random processes. A random process is strictly *stationary* if the Lth order probability density function is independent of the time origin (Goodman 1985). It is considered wide-sense stationary (WSS) if the expected value of the random process is independent of time and the correlation of the random process at two different points in time is only a function of the difference in time. In other words,

1. $E[u(t)]$ is independent of time
2. $E[u(t_1)u(t_2)]$ is a function of $\tau = t_2 - t_1$

In Figure 2.20, we see a visual representation of these ideas. The horizontal axis is the temporal axis, and each curve is a particular realization of the random process U. A random process is said to be stationary in increments if the random process, $\Upsilon(t) = U(t) - U(t-\tau)$, is stationary for all values of τ. Also, a stationary random process is ergodic if the temporal average is equivalent to the statistical average. This last property is useful if we want to infer statistical attributes of a random process from temporal data. The stationarity of the random process is important since, for nonstationary random processes, we may have different statistical results based on different starting times. Fortunately, there are a variety of optical systems applications where one or more of our stationary assumptions are valid. However, consideration must be given to each application to examine whether or not an assumption of stationarity is warranted.

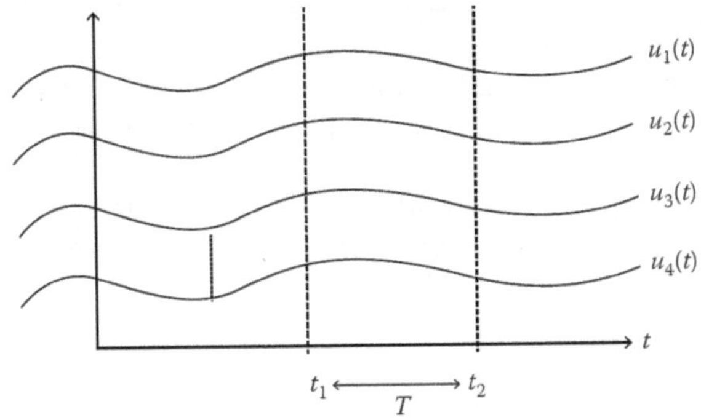

FIGURE 2.20 Random process.

Notice that every stationary random process is also WSS, but the same cannot be said for the other way around, and every WSS process is not stationary. Additional details and a discussion of other forms of stationarity can be found in the book *Random Signals, Detection, Estimation, and Data Analysis* by K.S. Shanmugan and A.M. Breipohl (Shanmugan and Breipohl 1988). A major reason for mentioning these distinctions is that there are a host of useful analytical properties for random process that satisfy one or more of our stationary and ergodic assumptions. A summary of some of these useful properties is presented in Table 2.4. In this table, $R_{xx}(t)$ is the autocorrelation of the random signal $X(t)$, $R_{xy}(t)$ is the cross-correlation of the random processes $X(t)$ and $Y(t)$, $S_{xx}(f)$ is

TABLE 2.4
Useful Attributes of Random Processes

Name	Property	Real/Complex	Stationary
Autocorrelation function	$R_{xx}(\tau)=E[X(t)\,X(t+\tau)]$	Real	WSS
	$R_{xx}(0)=E[X^2(t)]$	Real	WSS
	$R_{xx}(t)=R_{xx}(-\tau)$	Real	WSS
	$R_{xx}(0)\geq \lvert R_{xx}(\tau)\rvert$	Real	WSS
	$\lim\limits_{\tau\to\infty} R_{xx}(\tau)=\mu_x^2$	Real	WSS
Cross-correlation function	$R_{xy}(\tau)=E[X(t)\,Y(t+\tau)]$	Real	WSS
	$\sqrt{R_{xx}(0)R_{yy}(0)}\geq \lvert R_{xy}(\tau)\rvert$	Real	WSS
	$\lvert R_{xy}(\tau)\rvert\leq \frac{1}{2}\left[R_{xx}(0)+R_{yy}(0)\right]$	Real	WSS
	$R_{xy}(\tau)=0$, orthogonal processes	Real	WSS
	$R_{xy}(\tau)=\mu_x\mu_y$, independant processes	Real	WSS
Power spectral density	$s_{xx}(f)=\mathcal{F}\{R_{xx}(\tau)\}=\int_{-\infty}^{\infty}R_{xx}(\tau)e^{-j\pi f\tau}d\tau$	Real	WSS
	$R_{xx}(\tau)=\mathcal{F}^{-1}\{s_{xx}(f)\}=\int_{-\infty}^{\infty}s_{xx}(f)e^{-j\pi f}df$	Real	WSS
	$R_{xy}(\tau)$ is even function	Real	WSS
	$s_{xy}(f)$ is real and non-negative	Real	WSS
	$s_{xy}(f)$ is even function	Real	WSS
Cross-spectral density	$s_{xy}(f)=\mathcal{F}\{R_{xy}(\tau)\}=\int_{-\infty}^{\infty}R_{xy}(\tau)e^{-j\pi f\tau}d\tau$	Real	WSS
	$R_{xy}(\tau)=\mathcal{F}^{-1}\{S_{xx}(f)\}=\int_{-\infty}^{\infty}S_{xy}(f)e^{-j\pi f\tau}df$	Real	WSS
	$[(\mathrm{Re}\{S)]_{xy}(f)\}$ is an even function	Complex	WSS
	$[(\mathrm{Im}\{S)]_{xy}\,(f)\}$ is an odd function	Complex	WSS
	$S_{xy}(f)=0$, orthogonal processes	Complex	WSS
	$S_{xy}(f)=\mu_x\mu_y$, independent processes	Complex	WSS
	$\Gamma_{11}(\tau)=E[u(P_1,t+\tau)u^*\,(P_1,t)]$	Complex	WSS
	$\Gamma_{11}(0)=I(P_1)$	Real	WSS
	$J_{12}=E[u(P_1,t)u^*\,(P_2/t)]$	Complex	Stationary
	$\Gamma_{12}(\tau)=E[u(P_1,t+\tau)u^*\,(P_2,\,t)]$	Complex	Stationary
	$\mu_{12}=X_{12}(0)=\dfrac{J_{12}}{\left[J_{11}J_{22}\right]^{1/2}}$		
	$X_{12}(\tau)=\dfrac{\Gamma_{12}}{\left[\Gamma_{11}(0)\Gamma_{22}(0)\right]^{1/2}}$	Complex	Stationary

the power spectral density of the random process $X(t)$, and $S_{xy}(f)$ is the cross-spectral density function (Goodman 1985; Shanmugan and Breipohl 1988).

In Table 2.4, the temporal relationship between the autocorrelation function and its power spectral density is known as the Wiener-Khinchin theorem. Simply put, the power spectral density is the temporal Fourier transform of the autocorrelation of a WSS random process. For example, by taking the temporal Fourier transform of an electromagnetic field that is at least WSS, we end up with its power spectral density. The functions shown in Table 2.4 are quite prevalent and useful in describing, characterizing, and analyzing important optical system properties such as the coherence properties of the light. Both spatial and temporal coherence properties of light are of interest, and these can be described by the relationships shown in Table 2.4. Temporal coherence can be thought of as shifting the electromagnetic wave in the observation plane with respect to itself in time, but not in space, and looking at the correlation properties between the shifted and not shifted versions of the electromagnetic field. Temporal shifts that are within the coherence time of the light are expected to produce sinusoidal variations in the observed intensity pattern at a point in the observation plane. As the temporal shift τ becomes progressively larger, the variations in the observed intensity at the observation point become progressively smaller approaching a constant value. Further increasing τ beyond this point has no effect, and the observed intensity at the observation point remains constant. The temporal separation τ of the electromagnetic field where the observed intensity becomes constant is the temporal boundary where the electromagnetic field becomes temporally incoherent. Experimentally, temporal coherence properties are measured by using a Michelson interferometer as shown in Figure 2.21. The source of optical radiation is shown at the bottom of the figure as S_1. The light propagates to the collimating lens L_1 and then to a partially reflecting beam splitter shown as the tilted rectangle in the center of Figure 2.21. The beam splitter consists of a layer of glass that is shown as the clear area in the figure and a partially reflecting surface that is shown by the cross-hatching.

Notice that the back end of the beam splitter has the partially reflecting optical coating and the front end is an optically transparent material like glass. Half the light is reflected from the back surface of the beam splitter to mirror M_1. The beam splitter transmits the other half of the light toward mirror M_2. The compensator C adjusts the optical path length so that the light in both directions (M_1 and M_2) experiences the same amount of propagation through the glass. The reflected light from M_1 and M_2 is directed toward the collecting lens L_2 that focuses the light from both directions onto the point detector D. The compensator C is designed so that the optical path lengths in both directions are initially exactly the same. As M_1 is moved, the relative optical path length changes as a function of τ. It can be seen that the temporal shift τ is given by

$$\tau = \frac{2d}{c}, \tag{2.82}$$

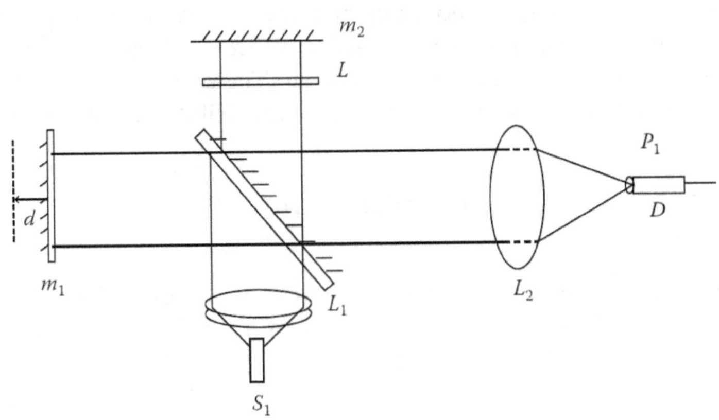

FIGURE 2.21 Source of optical radiation.

where

d is the relative separation of the electromagnetic fields in meters;
c is the speed of light.

Dividing d by the speed of light converts the spatial separation into a temporal separation. The expression for the self-coherence function in Table 2.5 provides the observed intensity at observation point P_1, and the expression for the complex degree of coherence can be evaluated with $P_2 = P_1$ for a normalized result with a maximum value of 1 and a minimum value of 0. A technical performance measure that is associated with the temporal coherence of an optical field is the classical visibility given by

$$V = \frac{I_{\max} - I_{\min}}{I_{\max} + I_{\min}}, \tag{2.83}$$

where $I_{\max}$ and $I_{\min}$ are the maximum and minimum intensity values at the observation point P_1, respectively. Note that the complex degree of coherence and the classical visibility will both be one for full temporally coherent fields and zero for full temporally incoherent fields. Intermediate values result from partial temporally coherent fields. A plot of the self-coherence function with respect to τ from a Michelson interferometer is called an inter-ferogram. According to Mandel, the coherence time τ_c can be given by (Goodman 1985)

$$\tau_c \triangleq \int_{-\infty}^{+\infty} |\gamma(\tau)|^2 \, d\tau, \tag{2.84}$$

where the integration is over all possible temporal separations τ. The interpretation of the coherence time is that for time scales equal to or less than the coherence time, the fields are expected to interfere constructively and fringes will be observable in an inter-ferogram. For time scales larger than the coherence time, the interferogram will produce a constant value.

For spatial coherence, an electromagnetic field (reference field) is spatially shifted with respect to itself (shifted field), and the temporal component of the electromagnetic field is kept identical for both the original and space-shifted versions of the electromagnetic field. The spatial correlation properties of the reference field and shifted field can be characterized by the mutual intensity or the complex coherence factor shown in Table 2.4. Experimentally, spatial coherence can be observed in Young's double-slit experiment.

A source on the left of Figure 2.22 is used to illuminate an opaque screen with two small holes cut into it at P_1 and P_2. The optical path from P_1 to the observing point Q_1 is r_1 and from P_2 to Q_1 is r_2. If the observation point Q_1 is moved up and down the observation plane, fringes will be observed. As the observation point moves further away from the projected center of points P_1 and P_2, the fringes become smaller until they disappear. The point where the fringes disappear is where the fields become spatially incoherent. The classical visibility for the spatial coherence case is given by

$$V = \frac{2\sqrt{I_1 I_2}}{I_1 + I_2} \gamma_{12}(0), \tag{2.85}$$

where I_1 and I_2 are the intensities observed at Q_1 due to P_1 and P_2, respectively. The complex degree of coherence in Equation 2.85 is due to the source field at the diffracting plane and is evaluated at points P_1 and P_2. The intensity at the observation point Q_1

$$I(Q_1) = I_1(Q_1) + I_2(Q_1) + K_1 K_2^* \Gamma_{12}\left(\frac{r_2 - r_1}{c}\right) + K_1^* K_2 \Gamma_{21}\left(\frac{r_1 - r_2}{c}\right), \tag{2.86}$$

TABLE 2.5
Useful Temporal and Spatial Coherence Properties

Name	$\Gamma_{11}(\tau)$ Self-Coherence Function	$\gamma_{11}(\tau)$ Complex Degree of Self-Coherence	$\Gamma_{12}(\tau)$ Mutual Coherence Function	$\gamma_{12}(\tau)$ Complex Degree of Coherence	J_{12} Mutual Intensity	μ_{12} Complex Coherence Factor
			Source Plane			
Narrow band	$\langle u(P_1,t+\tau)u^*(P_1,t)\rangle$	$\frac{\Gamma_{11}(\Gamma)}{\Gamma_{11}(0)}$	$\langle u(P_1,t+\tau)u^*(P_2,t)\rangle$	$\frac{\Gamma_{12}(\tau)}{\left[\Gamma_{11}(0)\Gamma_{22}(0)\right]^{1/2}}$	$\langle u(P_1,t)u^*(P_2,t)\rangle$	$\gamma_{12}(0)$
Coherent case					$A(P_1)A^*(P_2)$	$e^{\{j[\phi(P1)-\phi(P2)]\}}$
Incoherent case			$\|\Gamma_{12}(\tau)\|=0$ $P_1 \neq P_2,\ \tau \neq 0$		$kI(P_1)\delta(\Delta x,\Delta y)$	$\delta(\Delta x,\Delta y)$
Quasimonochromatic case	$I_1 e^{-j2\pi\upsilon\tau}$	$e^{-j2\pi\upsilon\tau}$	$J_{12}e^{-j2\pi\upsilon\tau}$	$\mu_{12}e^{-j2\pi\upsilon\tau}$	J_{12}	μ_{12}
			Image Plane			
Narrow band	$\langle u(Q_1,t+\tau)u^*(Q_1,t)\rangle$	$\frac{\Gamma_{11}(\tau)}{\Gamma_{11}(0)}$	$\langle u(Q_1,t+\tau)u^*(Q_2,t)\rangle$	$\frac{\Gamma_{12}(\tau)}{\left[\Gamma_{11}(0)\Gamma_{22}(0)\right]^{1/2}}$	$\langle u(Q_1,t)u^*(Q_2,t)\rangle$	$\gamma_{12}(0)$
Incoherent source (I_s) case					$\frac{ke^{-j\psi}}{(\bar{\lambda}z)^2}\Im_2(I_s)\Big\vert\ f_x=\Delta x/\bar{\lambda}z,\ f_y=\Delta y/\bar{\lambda}z$	$\frac{e^{-j\psi}}{(I_s)^2}\Im_2(I_s)\Big\Vert_{f_x=\Delta x/\bar{\lambda}z,\ f_y=\Delta y/\bar{\lambda}z}$
Partially coherent source (I_s) case					$\frac{k(\bar{x},\bar{y})e^{-j\psi}}{(\bar{\lambda}z)^2}\Im I_2(I_s)\Big\vert_{f_x=\Delta x/\bar{\lambda}z,\ f_y=\Delta y/\bar{\lambda}z}$ $k(\bar{x},\bar{y})=\Im_2\left(\mu(\Delta x_s,\Delta y_s)\right)\Big\vert_{f_x=\bar{x}/\lambda z,\ f_y=\bar{y}/\lambda z}$	$\frac{e^{-j\psi}}{I_s}\Im_2(I_s)\Big\Vert_{f_x=\Delta x/\bar{\lambda}z,\ f_y=\Delta y/\bar{\lambda}z}$

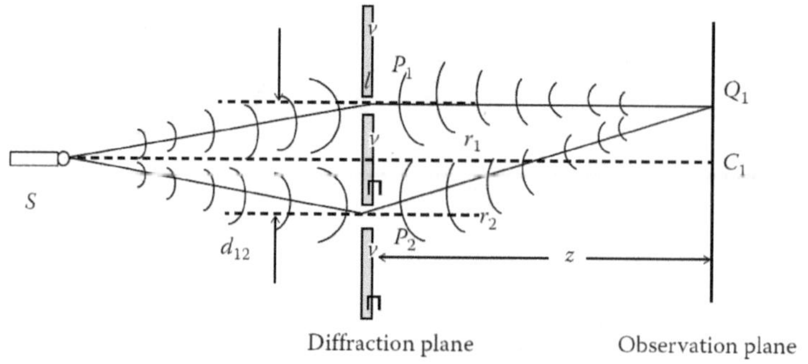

FIGURE 2.22 The experimental details of Young's double-slit experiment.

where K_1 and K_2 are complex scaling factors arising from geometrical considerations and the intensities observed at Q_1 due to the fields at P_1 and P_2 and are given by

$$I_1(Q_1) = K_1^2\Gamma_{11}(0), \quad I_2(Q_1) = K_2^2\Gamma_{22}(0). \tag{2.87}$$

The assumption required for Equations 2.86 and 2.87 to hold is that the pinholes at P_1 and P_2 are small enough so that the electromagnetic field does not vary across each opening and the illuminating light from the source S is narrow band. Additionally, we assume that the paraxial conditions hold as was done in the development of the Fresnel and Fraunhofer electromagnetic fields earlier in this chapter. The fringe spacing in the observation plane is given by

$$L = \frac{\bar{\lambda} z}{d_{12}}, \tag{2.88}$$

where

d_{12} is the distance between the two pinholes
z is the separation of the observation plane from the diffracting plane
$\bar{\lambda}$ is the center wavelength of the narrowband light

Note that the bottom entries of Table 2.5 apply to both spatial and temporal coherence properties of the light. The number of fringes that are observable is given by

$$N = 2\frac{\bar{\nu}}{\Delta\nu}, \tag{2.89}$$

where

$\Delta\nu$ is the half-power bandwidth
$\bar{\nu}$ is the center frequency of the narrowband light

Analogous to the coherence time τ_c, the coherence length L_c can be thought of as the distance in the observation plane over which the electromagnetic fields are coherent and, therefore, produce fringes. The width of the packet that contains the fringes in the observation plane is given by

$$\Delta l = \frac{2zc}{\Delta\nu d_{12}}. \tag{2.90}$$

Some remarkable simplifications occur in the mutual intensity and the complex coherence factor when *quasimonochromatic* conditions prevail. For quasimonochromatic light, the half-power bandwidth $\Delta\nu$ must be much smaller than the center frequency of light $\bar{\nu}$, and the difference in the optical path length from the source to the observation point through pinholes P_1 and P_2 must be much smaller than the coherence length of the observed light. An example of this kind of light may be a low-grade laser. For quasimonochromatic conditions, the mutual intensity becomes

$$\Gamma_{12}(\tau) \simeq J_{12}e^{-j2\pi\bar{\nu}\tau}, \tag{2.91}$$

and the complex coherence factor is given by

$$\gamma_{12}(\tau) \simeq \mu_{12}e^{-j2\pi\bar{\nu}\tau}. \tag{2.92}$$

Under paraxial conditions, the observed intensity at the observation point $\vec{Q}_1$ is given by

$$I(\vec{Q}_1) = I_1 + I_2 + 2K_1K_2J_{12}\cos\left[\frac{2\pi}{\lambda z}(\vec{d}_{12}\cdot\vec{Q}_1) + \phi_{12}\right]. \tag{2.93}$$

Here, we have generalized our expression slightly so that $\vec{Q}_1$ is a 2D position vector in the observation plane and $\vec{d}_{12}$ is a vector in the diffraction screen plane that corresponds to the difference in positions of the pinholes P_1 and P_2. The phase term ϕ_{12} is associated with the mutual intensity term.

We now have a basic understanding of spatial and temporal coherence. Note that the mutual coherence factor and the complex degree of coherence address both spatial and temporal coherence simultaneously. We want to introduce another useful technical performance metric related to both spatial and temporal coherence of an electromagnetic field. This technical performance metric is very useful in characterizing the effects of the atmosphere on optical systems. The technical performance measure is the structure function given by (Roggemann and Welsh 1996)

$$\begin{aligned} D(\Delta x, \Delta y) &= \overline{\left(\chi(\vec{x}) - \chi(\vec{x}-\Delta\vec{x})\right)^2} + \overline{\left(\psi(\vec{x}) - \psi(\vec{x}-\Delta\vec{x})\right)^2} \\ &= D_x(\Delta\vec{x}) + D_\psi(\Delta\vec{x}). \end{aligned} \tag{2.94}$$

where $D_x(\Delta\vec{x})$ is the log-amplitude structure function, and $D_\psi(\Delta\vec{x})$ is the phase structure function associated with the pupil plane field, and as before, the bar in the expression infers statistical expectation. The mutual intensity and the complex coherence factor deal with spatial coherence, and the self-coherence factor provides information about the temporal coherence of a given electromagnetic field. We also notice that for the complex degree of coherence and the complex coherence factors, fully coherent fields produce a value of 1, fully incoherent fields produce a value of 0, and partially coherent fields produce an intermediate value.

We now want to look at two limiting cases of the mutual coherence function. The first is a fully incoherent source term, and the second case is a fully coherent source term. For the first case, most sources apart from a laser fall into this category. Additionally, even if a source is used with a high degree of spatial coherence, if it is transmitted through, or reflected from, an object that is rough on the scale of an optical wavelength, then the transmitted or reflected light becomes spatially incoherent. A majority of imaging scenarios fall under these conditions, and so it is essential to understand the nature of the mutual coherence function under an incoherent source approximation. Figure 2.23 provides the geometry for visualization. We again assume paraxial imaging conditions, such as those that were used to derive the Fresnel and Fraunhofer diffraction formulas. The source plane is shown at the left of Figure 2.23 with the clear area showing the projection of the source onto the 2D plane. In the case of a transmitting system, the open area on the left side of the figure is the clear aperture of the transmitted electromagnetic field. The distance from the source plane to the observation plane is given by z. A point in the source plane is denoted by P_1, and two points in the observation plane are labeled Q_1 and Q_2. The distance from P_1 to Q_1 is given by r_1, and the

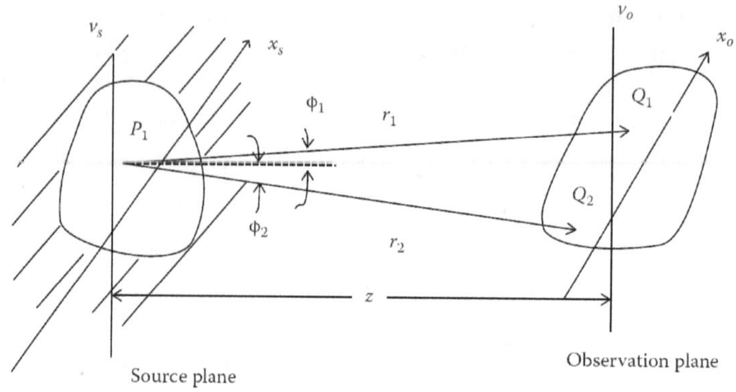

FIGURE 2.23 Transmitting system case.

distance from P_1 to Q_2 is given by r_2. The crosshatched area on the left of Figure 2.23 is the region where the source fields are zero.

We want to take a look at the propagation of the mutual intensity. We will assume that quasimonochromatic imaging conditions hold. In general, the propagation of the mutual intensity is given by (Goodman 1985)

$$J(Q_1,Q_2)=\iint_{\Sigma}\iint_{\Sigma} J(P_1,P_2)\exp\left[-j\frac{2\pi}{\bar{\lambda}}(r_2-r_1)\right]\frac{\xi(\theta_1)}{\bar{\lambda}r_1}\frac{\xi(\theta_2)}{\bar{\lambda}r_2}ds_1ds_2, \tag{2.95}$$

where the integration is over the clear aperture shown on the left side of Figure 2.23, and the angular terms are associated with the imaging geometry. By assuming that paraxial conditions hold just as in the development of our Fresnel and Fraunhofer diffraction formulas, these angular terms approach one, and we get what is known as the van Cittert-Zernike theorem:

$$J(x_1,y_1;x_2,y_2)=\frac{K\left(e^{-j\psi}\right)}{\left(\bar{\lambda}z\right)^2}\int\int_{-\infty}^{+\infty} I(x_s,y_s)\exp\left[j\frac{2\pi}{\bar{\lambda}z}\left((x_2-x_1)x_s+(y_2-y_1)y_s\right)\right]dx_s\,dy_s. \tag{2.96}$$

In Equation 2.96, the phase term ψ in front of the integrals is given by

$$\psi=\frac{\pi}{\bar{\lambda}z}\left[\left(x_2^2+y_2^2\right)-\left(x_1^2+y_1^2\right)\right], \tag{2.97}$$

and K is a volume normalization term given by

$$K=\frac{\left(\bar{\lambda}\right)^2}{\pi}. \tag{2.98}$$

The van Cittert and Zernike theorem is a remarkable result in that we can determine the mutual intensity in the observation plane by taking the 2D Fourier transform of the object brightness. Normalizing the mutual intensity, we can eliminate a majority of the scale factors in the mutual intensity and arrive at the complex coherence factor:

$$\mu(x_1,y_1;x_2,y_2)=\frac{e^{-j\psi}\int\int_{-\infty}^{+\infty} I(x_s,y_s)\exp[j\left(2\pi/\bar{\lambda}z\right)\left(\Delta x x_s+\Delta y y_s\right)dx_s\,dy_s]}{\int^{+}\int_{-\infty}^{\infty} I(x_s,y_s)dx_s dy_s}, \tag{2.99}$$

where $\Delta x = x_2 - x_1$ and $\Delta y = y_2 - y_1$. Equations 2.98 and 2.99 are applicable to most sources in the visible and infrared parts of the electromagnetic spectrum except for direct illumination of a highly spatially coherent source such as a laser. For the case of a source with a high degree of spatial coherence, the analytic electromagnetic field can be represented as

$$u(P,t) = A(P,t)e^{-j2\pi\bar{\nu}t}, \quad (2.100)$$

where the amplitude term is a complex quantity. This analytic form of the electromagnetic field is often useful in analyzing the electromagnetic field, its propagation, and its interaction with the optical system. It can be shown that the mutual intensity for a coherent source field is given by (Goodman 1985)

$$J_{12} = A(P_1)A^*(P_2), \quad (2.101)$$

where the complex amplitude terms are evaluated at two spatial points on the coherent source, P_1 and P_2. The associated complex coherence factor is given by

$$\mu_{12} = \exp\left(j\left[\varphi(P_1) - \varphi(P_2)\right]\right), \quad (2.102)$$

where the phases are obtained by taking the argument of the complex amplitude of Equation 2.100. In the case of Young's experiment, the fringes that would be observed in the observation plane are given by

$$I(Q_1) = I_1(Q_1) + I_2(Q_1) + 2\sqrt{I_1(Q_1)I_2(Q_2)}\cos\left(\frac{2\pi(r_2 - r_1)}{\bar{\lambda}} + \varphi(P_2) - \varphi(P_1)\right), \quad (2.103)$$

where Figure 2.24 is again used for reference of Young's double-slit experiment. To generalize the geometry for the mutual coherence function and its derivate products, we use the following four-point figure.

The angles θ_1 and θ_2 are measured from the normal to the surface at the source point and the line connecting the source point to the corresponding exploratory point in the observation plane. To propagate the mutual coherence function or, when appropriate, the mutual intensity function, from the source plane to the observation plane, several special cases are of interest: (1) narrowband light, (2) narrowband light with quasimonochromatic assumption, (3) coherent light, (4) incoherent light, and (5) partially coherent light. We summarize the main propagation results in Table 2.5 (Goodman 1985).

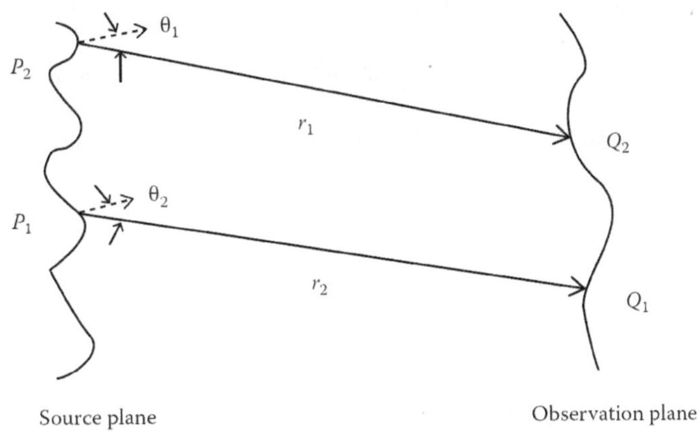

FIGURE 2.24 Young's double-slit experiment.

We see that for both the spatially incoherent and partially coherent image plane cases, the primary propagation integrals obey the van Cittert-Zernike theorem and are 2D spatial Fourier transforms. Of course, the "k" factor out front of the partially coherent case complicates things somewhat. The condition for which the partially coherent case applies is

$$z > \frac{2D_{\max} D_c}{\bar{\lambda}}, \tag{2.104}$$

where

$D_{\max}$ is the maximum physical extent of the source object
z is the separation of the source plane from the observation plane
D_c is the maximum coherence length in the source plane
$\bar{\lambda}$ is the mean wavelength

Since many sources are well modeled by assuming an incoherent source, we focus some attention on this case. Notice from the equation for the mutual intensity in Table 2.5, that if we are interested in describing the intensity in the observation plane for an object that can be modeled as an incoherent source (which is often the case), the observation points Q_1 and Q_2 coincide, and the exponential phase term with Ψ reduces to 1. The relationship between the source intensity and the observed intensity is just a 2D spatial Fourier transform that is readily solvable with either analytical or numerical methods.

We want to end this section of statistical optics models with an overview of some useful probability density functions that can be used to model useful statistical quantities such as the mean, covariance, and even higher moments. An interesting fact, according to the *central limit theorem*, is that for any set of N independent random variables with arbitrary probability density function(s) and finite variance, the summation of a sufficiently large number of these zero mean, unit variance random variables produces a probability density function that approaches a normal distribution with zero mean and unit variance (Craig et al. 1995). For instance, a source that produces a large number of independent photons, high-light-level sources can be modeled as having a Gaussian probability density function such as presented in Equations 2.74–2.77. For low light levels, analysis and experimental results indicate that photon arrivals obey Poisson statistics wherein the mean and the variance have the same value. Circularly complex Gaussian statistics have been used to model certain aspects of lasers, while low levels of photons arriving on a detector have been shown to obey a doubly stochastic Poisson process (Andrews and Phillips 1998). We will treat these special cases in more detail later in the book.

A useful tool in determining moments of a known probability density function is the *characteristic* function. The characteristic function is defined as the expected value of the complex exponential of the weighted sums of the random variables:

$$M_u^{(n)}(w_1, w_2, \ldots, w_n) \triangleq E\left[\exp\left\{j(w_1 u_1 + w_2 u_2 + \cdots + w_n u_n)\right\}\right], \tag{2.105}$$

where n is the highest number of random variables. This expectation takes the form of an n-dimensional Fourier transform

$$M_u^{(1)}(w_1) = \int_{-\infty}^{+\infty} \exp\left\{j(w,u)\right\} P_{u1}(u_1)\,du_1, \tag{2.106}$$

or

$$M_{u_1 u_2}^{(2)}(w_1, w_2) = \int\int_{-\infty}^{+\infty} \exp\left\{j(w_1 u_1 + w_2 u_2)\right\} P_{u_1 u_2}(u_1, u_2)\,du_1 du_2, \tag{2.107}$$

respectively.

Conversely, the corresponding probability density function can be determined from the characteristic function by inverse Fourier transform. If the moments exist, the characteristic function can be used to generate higher-order moments (Goodman 1985). For example,

$$\overline{u_1^n u_2^m} = \frac{1}{j^{(n+m)}} \frac{\partial^{n+m}}{\partial w_{u1}^n \, \partial w_{u2}^m} M_{u_1 u_2}\left(w_{u_1}, w_{u_2}\right)\Big|_{w_{u1}=0, w_{u2}=0}, \tag{2.108}$$

where, as before, the bar means statistical expectation. This ability to generate higher-order moments from the characteristic function is very useful, especially in the case of Gaussian random variables, since all the higher-order Gaussian moments can be written as linear combinations of the joint moments. For the case of Gaussian and Poisson probability density functions, the first-order characteristic functions are given by

$$M_{u_1}(w_1) = \exp\left(-\frac{\sigma^2 w_1^2}{2} + j w_1 \overline{u_1}\right) \quad \text{Gaussian}, \tag{2.109}$$

and

$$M_{u_1}(w_1) = \sum_{k=0}^{\infty} \frac{\overline{k}^k}{k!} e^{\left(-\overline{k} + j w_1 k\right)} \quad \text{Poisson}. \tag{2.110}$$

The nth-order Gaussian characteristic function is given by

$$M_{u_1 \ldots u_n}(\underline{w}) = \exp\{j \underline{\overline{u}}^t w - \frac{1}{2} \underline{w}^t \underline{\underline{C}} \underline{w}, \tag{2.111}$$

where the same notation is used as in the case of the multiple-dimensional Gaussian probability density function in Equation 2.77. We end with the 1D probability density function for the Poisson case given by (Goodman 1985)

$$P_{u_1}(u_1) = \sum_{k=0}^{\infty} \frac{\left(\overline{k}\right)^n e^{-\overline{k}}}{k!} \sigma(u_1 - k), \tag{2.112}$$

that has mean and variance given by

$$E[u_1] = \overline{u}_1, \tag{2.113}$$

and

$$\sigma_{u_1}^2 = \overline{u}_1, \tag{2.114}$$

respectively.

This section has provided some fundamental results from the field of statistical optics. These results will be used throughout this book and are complementary to and consistent with previous results. We started this chapter section with Maxwell's equations and showed that the electromagnetic field satisfies either the inhomogeneous or homogenous Helmholtz wave equation. We then showed that the electromagnetic field under paraxial conditions satisfies the Fresnel or Fraunhofer diffraction integrals and that the relationship between the object plane and observation plane is relatable to a 2D Fourier transform. We provided a linear systems model and defined some important optical linear systems terms such as the GPF, impulse response, PSF, and OTF and provided some simple examples. We presented the energy spectrum and power spectrum and discussed sampling requirements in both the object plane and the entrance pupil plane of an imaging system.

Lastly, we provided a short overview of some key statistical optics results. We provided probability density functions for two frequently occurring random processes encountered in statistical optics, Gaussian and Poisson. We showed that for high light levels, the Gaussian distribution is a good model to use and for low light levels, Poisson statistics are applicable. We talked about ergodic, stationary, WSS, and stationary in increments processes and their relations to statistical analysis. We discussed temporal and spatial coherence and showed how these properties can be measured using a Michelson interferometer (temporal coherence) and Young's double-slit experiment (spatial coherence). We provided associated analytical expressions for the expected intensity in the observation plane. For Young's double-slit experiment, we presented expressions for the intensity in the observation plane for a fully incoherent source and a fully coherent source. The van Cittert and Zernike theorem was then presented which showed that for a fully incoherent source, the observed mutual intensity in the observation plane is the scaled, 2D spatial Fourier transform of the object brightness. We presented some technical performance metrics such as the correlation coefficient, classical visibility, structure function, coherence time, and coherence length. We also showed the importance of correlation functions to optical systems analysis. We presented correlation functions like the mutual coherence function, mutual intensity, self-coherence factor, complex coherence factor, and the complex degree of coherence and showed how to relate these correlation functions in the source plane to the observation plane. We presented results for these correlation functions under (1) narrowband light; (2) narrowband, quasimonochromatic light; (3) broadband light; (4) incoherent light; and (5) partially coherent light assumptions. We provided expressions for the mean, variance, and a means to generate higher-order moments using the characteristic functions. This section is not intended to provide a comprehensive view of these topics but, instead, to present important and useful optical analysis concepts and methods. We will build upon these basics in later chapters. For a more detailed and comprehensive treatment, please see the provided references.

In the next section, we continue our case study from Chapter 1 and illustrate the tools, techniques, and methods we learned in this chapter. To set the stage, our imaging company, Fantastic Imaging Technologies, has just been informed that they won the Homeland Security contract to build imaging systems for detecting illegal bipedal border crossings into the United States using an unmanned aerial vehicle (UAV). We continue with our integrated case study.

2.3 INTEGRATED CASE STUDY: INTRODUCTION TO EA

This case study illustrates the process of identifying the EA for FIT and how it is used to understand the effects that a new business has on the rest of the company. It also illustrates some key optical system technical considerations. The company reported that with the implementation of EA, decisions became more clear, simplified, and more effective. With FIT being awarded the Homeland Security's UAV optical system effort, Dr. Bill Smith called an executive committee meeting to congratulate the team and discuss plans for the future of the business. This scenario has the following characters: Dr. Bill Smith, FIT chief executive officer (CEO); Dr. Tom Phelps, FIT chief technical officer (CTO) and optics expert; Mr. Garry Blair, CIO and chief enterprise architect; Karl Ben, FIT lead systems engineer; and Ron, S., FIT systems engineer (new hire).

2.3.1 Defining the EA for FIT

Bill: "Good Morning everyone, I hope everyone is as excited as I am. Let me open the meeting with my congratulations to the team for a job well done! Winning this program will position the company for many future endeavors in this industry. As you know, we have been mainly focused on providing optical imaging for the commercial sector, and this win is a great opportunity for us. One of the ideas in our proposal that stood out was our enterprise architecture model. During the proposal process, we were able to identify new capabilities that would be needed to support this new program and how it would affect our business.

We were able to persuade the Department of Homeland Security that we would be able to complete the program on time and within budget. Now we need to deliver! With that in mind, I called this meeting to discuss what the next steps are, and I want to discuss how we can successfully execute this program. I want to open the floor for discussion."

Garry: "I would also like to offer my congratulations to the team on a job well done. During the proposal process, we were able to use our enterprise architecture to map how changes in the proposal would affect the overall company business. This included current projects, staffing resources, infrastructure, information flow, and processes already in place. Winning this effort is largely due to how well we all worked together. Our enterprise architecture is still in its infancy, some systems still need to be implemented, and since being awarded the program, we have noticed some deficiencies that need to be addressed with respect to resources, infrastructure, and tools needed to be successful."

Tom: "A good introduction Garry to the items for discussion. I have put together some slides on the new program and what it means for our company. As we all know, our mission is to provide the best optical imaging products for our stakeholders. We also strive to provide our clients with solutions designed to execute their missions effectively by providing reliable, efficient, and rapid solutions to their needs. This first slide shows our goals and objectives. We have also decided to do a feasibility analysis on various aspects of this project, so that we can be confident on the staffing resources, information flow, and various processes already in place" (see Table 2.6).

Bill: "Mr. Phelps, at this stage of the project, why do you want the feasibility study?"

Tom: "That is a good question. Once the requirements for an effort are understood, it is important to conduct a complete feasibility study to determine whether or not the intended effort is doable. The feasibility study should address all the requirements from a holistic perspective."

Bill: "That is a good idea. Can you explain to us what you mean by holistic? Do you mean evaluating the feasibility of the project from more than just the technical perspective?"

Tom: "Yes, exactly! Although other categories can be considered, our feasibility study will target the following six items:

1. Technical feasibility
2. Economic/financial feasibility
3. Organizational feasibility
4. Operational feasibility
5. Motivational feasibility
6. Managerial/political feasibility"

TABLE 2.6
FIT Goals and Objectives

Goals	Objectives
Develop and integrate an electro-optic subsystem for a UAV	Design and manufacture an optical subsystem that integrates with other subsystems becoming a system of systems
Train U.S. Military personnel on use and maintenance of system	Provide support for optical subsystem to allow for users to execute missions successfully
Promote sales of optical subsystems to other government agencies	Increase sales and revenue for the company by 20% over the next 2 years
Expand company assets with respect to personnel resources, infrastructure, and intellectual property	Become the industry leader in optical imaging with an emphasis on atmospheric turbulence compensation and adaptive optics methods

Bill: "Hmmm, doing all those feasibility studies sounds costly."

Tom: "True, this takes some effort. However, it is more costly to miss some critical aspects of the project and have it bite us later on in the development cycle. Also, for now, we have been focusing on the technical and organizational feasibility areas."

Bill: "Ok, so how are things going on the technical front?"

Tom: "We have run all of our models for the typical scenarios and parameters identified in the DHS proposal. We have done a complete radiometric analysis using humans as the target. We spent the majority of the time looking at nighttime operational scenarios and evaluated both a low-light-level and thermal-based infrared imaging system. At this point, a full-blown computational electromagnetic analysis is not warranted. For the daytime visible imaging system solution, we ran our atmospheric turbulence models to see what spatial resolution our imaging system would provide. For the size of the imaging apertures involved in the daytime system, the atmospheric turbulence compensation systems may not strictly be required. However, we may get a further standoff range, which will increase our sensor footprint on the ground. In this way, we may be able to cover more area than a system without the atmospheric turbulence compensating system."

Bill: "Good! What kind of models are we setting up?"

Tom: "Well, we have a radiometric model in SI units that provides the signature levels for expected border crossers. We also have the atmospheric turbulence models that we can turn on or off that provide insight into the spatial resolution of our optical system. We are using statistical optics models to describe the spatial correlation properties of the detected image. We also have generated some basic optical models for the imaging sensor on the UAV host, using their gimbal system. We can model both Fresnel and Fraunhofer propagation. For now, we are assuming that we are integrating into their gimbal system, but we can build our own collection and relay optics if necessary. We also have a geometrical optics model ready for some quick calculations and ray tracing. We also have developed a linear systems model with a circular entrance pupil for incoherent imaging scenarios. We have computer models based on the UAV collecting aperture, the Fresnel zone free-space impulse response, point spread function, and optical transfer function. These models allow us to respond to variations in their CONOPS and allow us to quickly evaluate changes in their requirements."

Bill: "What if we need to use some active imaging methods?"

Tom: "If that becomes necessary, we can quickly generate a coherent imaging model for our optical imaging system on the UAV platform. At this stage, our incoherent imaging models supported by Fourier optics and statistical optics methods are sufficient."

Bill: "Wait, you said you were using a free-space impulse response. That applies to a vacuum and doesn't account for atmospheric effects."

Tom: "That's true. We are using the free-space impulse response to determine an upper limit on the performance of our imaging system. Basically, we want to isolate the entrance pupil effects to provide diffraction-limited, spatial resolution performance estimates. If we build an excellent imaging system, like we always do, we should be able to approach this level of performance if we implemented our atmospheric turbulence compensation system. By having this estimate, we can show how much of an improvement their current optical system would have if they added the atmospheric turbulence compensation subsystem."

Bill: "That makes sense. Mr. Blair, you were mentioning something about deficiencies regarding our enterprise architecture."

Garry: "Yes. Nothing unexpected. However, in order to achieve our business goals and objectives, we needed to find a way to have the ability to unify and integrate business processes across the enterprise and link with external partners. Our documentation needs to be readily available. We need to reduce delivery time and development costs by maximizing

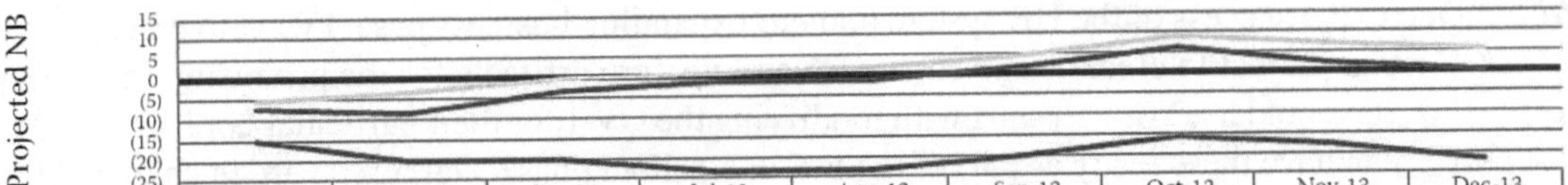

	Apr-13	May-13	Jun-13	Jul-13	Aug-13	Sep-13	Oct-13	Nov-13	Dec-13
Pastdue	0.00								
Worst case	(5.20)	(3.20)	(0.20)	0.80	1.80	4.80	8.99	7.09	5.29
Nom case	(7.20)	(8.20)	(3.20)	(1.20)	(1.20)	1.80	5.99	2.09	0.29
Best case	(15.20)	(20.20)	(20.20)	(23.20)	(23.20)	(20.20)	(16.01)	(17.92)	(21.71)

FIGURE 2.25 Staffing forecast versus new business. (Courtesy of Chris Lannon.)

reuse. Finally, a common vision was needed for both the business and IT departments. Now that we have won the contract, we need to look at how this win affects the company at an enterprise level and not just at the program level. The new program will put a strain on personnel resources, IT infrastructure, sales and inventory, and so forth."

Bill: "Garry, Do you have specific data that reflect how the new program will affect the enterprise?"

Garry: "When we started the proposal for Homeland Security, we were concerned with possible technical staffing shortages; some of our current programs are running a little behind schedule. Utilizing the reports from the HR system that tracks staffing resources, we show a staffing shortage of 25 heads over the next few months. Figure 2.25 shows the deficiency."

Bill: "Garry, is this staffing issue going to impact this program?"

Garry: "No. Fortunately, we have the HR system integrated with our manpower tools that allowed us to be aware of the situation and act accordingly. We have already begun hiring to meet the needs, and I have to say, there are some really great candidates."

Bill: "But it will also take time to train the new staff for the project. Have we accounted for the training time?"

Garry: "Yes. As part of the organizational feasibility study, analysis was done on our projected staffing needs. The manpower tool accounts for training and also the learning curve. Our staff did an amazing job of forecasting our resource needs based on our existing and expected workload. The results of the organizational feasibility analysis showed that we could handle the needs of this project."

Tom: "With the success of the HR system integrating with our manpower tools, which allows us to strategically plan for new business, we began researching other areas that would benefit the company and allow us to maintain perspective on our business goals and objectives. Enterprise architecture techniques can be used to help our company in areas such as sales, our inventory system, our cost accounting system, and our training systems, to mention a few. We have begun talking with the various program managers, as stakeholders in the cost accounting system, regarding their needs and the capabilities of the tools needed to support how we do business. In addition, we began looking at the training needs, not only internally but also for the stakeholder. When we looked at the Homeland Security project during the proposal phase, we noticed there were some required integrated logistic capabilities that were needed to win. Part of our response involved stakeholder training with our new optical system in the operational and maintenance phases of the systems development lifecycle. We also are getting some interest from other government agencies with respect to our capabilities and how we may benefit them. In our design, we are considering these other government agencies as secondary stakeholders. With that in mind, we should consider training for both internal and external stakeholders. This will be a big undertaking since many of the external stakeholders are using localized tools that range from spreadsheets, to homegrown databases, to sophisticated enterprise architecture suites to store and track data."

Bill: "Due to the success of the HR system and our expanding business base, I want to create a special project to create basic enterprise architecture artifacts for the promising secondary stakeholders. You can start with modifying the OV-1 for their particular applications. We can have these ready as part of the next EA rollout phase. Garry, you are in charge. After the meeting, I am going to need you, Garry, to set up a meeting with Mr. Karl Ben as the lead systems engineer on the new Department of Homeland Security (DHS) program."

(They started evaluating the current systems and processes implemented in the company and how new capabilities would affect it. The meeting discussion started with the following.)

Karl: "We started evaluating the new configuration items the DHS program was going to need along with new secondary stakeholders, and we determined we may have to implement some new capabilities into our enterprise architecture. I will start with an overview of the program so we can all be on the same page with respect to the concept of operations (CONOPS) and operational view of the system as shown in Figure 2.26."

Karl: "I will explain some of these systems to provide a more detailed picture. The capabilities I wanted to discuss today are the ground control station (GCS), mobile control station (MCS), and the optical imaging system, as it pertains to maintenance and support. As you know, we have a requirement to provide support including training and maintenance of the system we are delivering. In the past, we usually created training modules and did on-site or off-site training for the operators of the system and maintenance personnel. This requires several man-hours, personnel, and travel expenses. We have also had interest in the system from other stakeholders including the FBI, CIA, and some other foreign stakeholders. We are in the process of generating new enterprise architecture artifacts for our Homeland Security stakeholder and will adapt these for our most promising secondary stakeholders. Since many of our prospective customers fall under the DOD, we

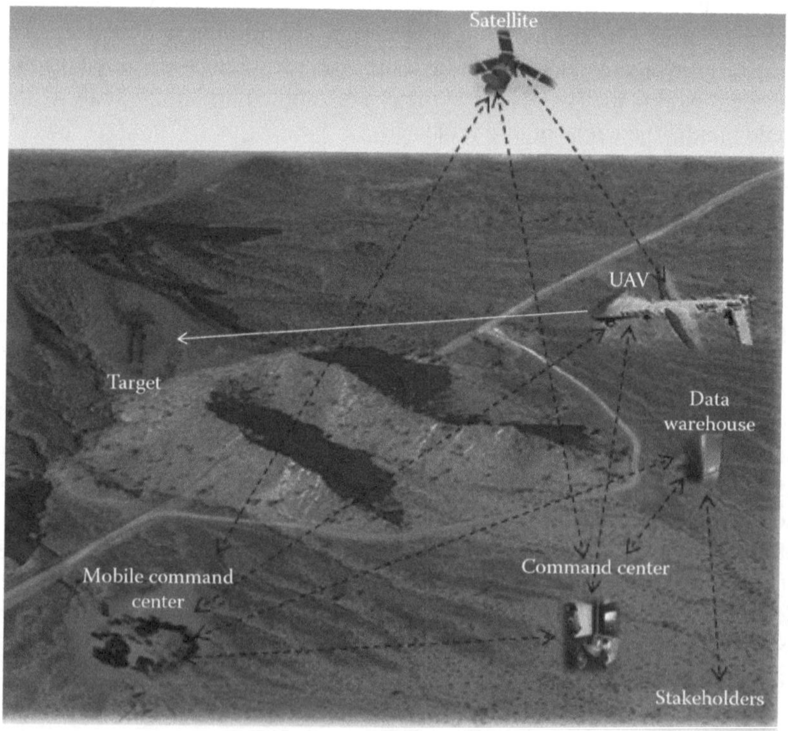

FIGURE 2.26 CONOPS of UAV border patrol mission.

need to ensure that our EA is compatible with what they expect. Are you familiar with what the DOD expects with regard to EAs?"

Garry: "Yes. The DOD has some specific goals and expectations when it comes to enterprise architecture. The EA should

- Eliminate duplication, incompatibility, and redundancy of systems and business processes
- Manage knowledge as a corporate asset using standard shared information as a driver
- Facilitate effective business decisions
- Capture information once and reuse across the enterprise
- Provide security and protection of sensitive information
- Affect changes through increasing efficiency and economy of scale
- Create consistent EA products with sufficient detail for implementation

We need to make sure our EA accomplishes these items."

Karl: "Is there some guidance for how to accomplish this?"

Garry: "Yes. The DOD has an architectural framework known as the DODAF that facilitates implementing an EA. The DODAF provides the guidance and rules for developing, representing, and understanding architectures based on a common denominator across DOD, joint, and multinational boundaries. It provides insight for external stakeholders into how the DOD develops architectures. The DODAF is intended to ensure that architecture descriptions can be compared and related across programs, mission areas, and ultimately the DOD enterprise, thereby establishing the foundation for analysis that supports decision-making processes throughout the DOD. The DODAF is organized into several 'views/These views consist of operational views that show the major entities, or nodes, upon which the system is based, system views that describe the systems that support data communication between or within the nodes, and technical standards views that show the standards and conventions upon which the whole system is based. One of the feedback points from DHS as to why we were awarded the contract was that they were looking for a company that had sound EA principles."

Karl: "We have been focused on hiring technical staff, and I believe we will be well positioned for the design and development of the new optical system per the new numbers that came out today from the HR system. With the new talent coming in though, we also should plan for training internally as well."

Karl: "Actually, that is what I wanted to discuss. We are considering new training software to implement across our enterprise. We have been working with your enterprise architecture team in determining the best system for our needs."

Ron: "I am not sure I get the connection of this training software to our EA. Can you explain that please?"

Karl: "Sure, Ron. I will start with an overview of FIT's enterprise architecture process and how we have implemented our enterprise architecture. Remember, in simple terms, an enterprise is an organizational structure of people, processes, and resources acting together to achieve a set of objectives that are expressed as a mission or a set of goals. This organizational structure collects, produces, and uses information to support performance of the business processes as they develop products. The organizational structure uses business processes to achieve objectives. Portions of the business processes are performed by different parts of the organizational structure. Enterprise architecture is a representation of

- The business architecture and the technical architecture of an enterprise
- How the business and technical architectures align to achieve the objectives of the enterprise

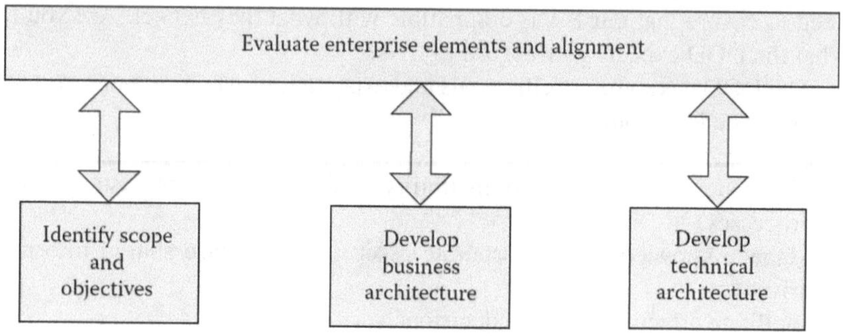

FIGURE 2.27 FIT top-level EA process. (Courtesy of Chris Lannon.)

- How the technical and business architectures will transform to more effectively achieve current or future mission objectives

While this description of an enterprise uses terms that are business oriented, an enterprise is not necessarily a business, but instead could be an instance of a commercial entity, civil government entity, military entity, an intelligence community entity, or any other stakeholder enterprise representation. The enterprise business architecture is a model that depicts the enterprise business processes, the organizational structure, the information needs, and the products and their interrelationships. The enterprise technical architecture is a model that depicts the enterprise technology and how it supports the business architecture. This technology includes applications and support software and computing and infrastructure hardware. Since performing the enterprise architecting tasks is based on heuristics and experience, the process does not describe "how to" do the tasks. We placed emphasis on the set of products needed to adequately define our EA. The slide I am showing now depicts FIT's top-level EA process.

As depicted in Figure 2.27, the general concept of operation for the enterprise architecting process is as follows:

- *Identify scope and objectives:* Develop an understanding of the vision, mission needs, and key requirements of the EA; validate FIT's understanding with stakeholders and determine the scope of the EA effort. The result of this activity is the definition of the scope of EA activities, ranging from validating our understanding of an existing stakeholder EA to generating an EA for a stakeholder.
- *Develop business architecture:* Validate existing or create new process descriptions, information flows, and enterprise security and show how these operate to achieve enterprise objectives.
- *Develop technical architecture:* Validate existing or define the technology (software and hardware) used by the enterprise and show how this technology supports the business architecture.
- *Evaluate enterprise elements and alignments:* Evaluate alignment of the business and technical architecture to enterprise objectives and define improvement projects. The improvement projects can be for current objectives or achievement of future objectives.

At each step of the EA process for FIT, we have identified activities and products that will be created. A high-level table is shown in Figure 2.28 on the next slide. With this process, we have developed a proposal for the support, training, and maintenance of the DHS optical system. At the high level, we identified the scope and objectives, stakeholders, and governance. Using those, the EA team identified improvement opportunities both internally and for stakeholder use. I have put together a proposal for the support, training, and maintenance subsystem that meets the enterprise objectives and goals. The EA team was able to identify

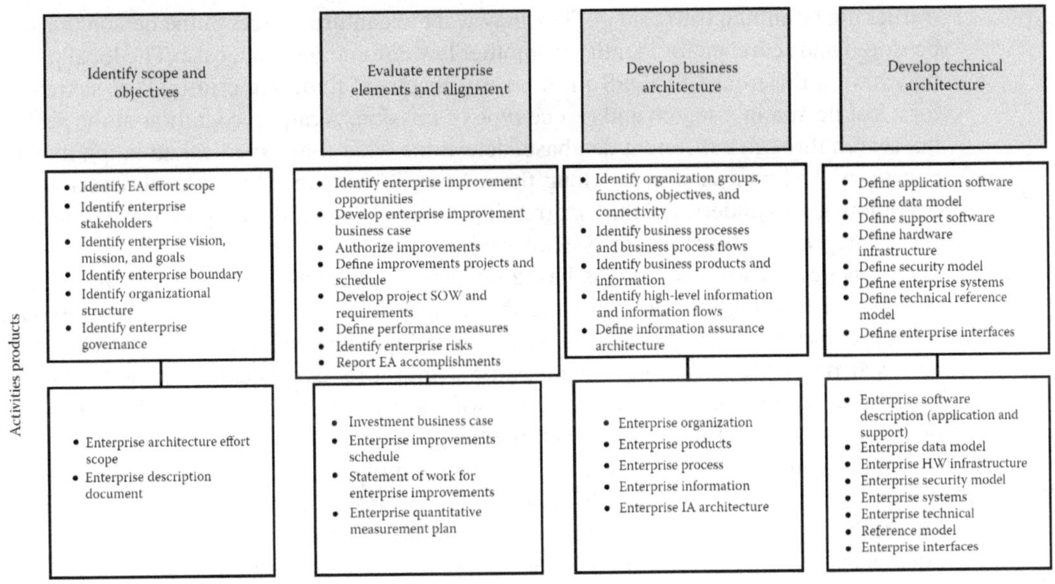

FIGURE 2.28 FIT EA process details. (Courtesy of Chris Lannon.)

the enterprise owner for this subsystem allowing us to identify the needs early and help meet our own business needs as well. The mission for this particular subsystem is to support, train, and maintain the optical system. This includes training of the operators, maintenance personnel, our own manufacturer staff, technical staff, and field support.

We conducted technical feasibility analysis to evaluate and understand the benefits of the present technical resources of the organization and their applicability to the present and future needs of the proposed system. We conducted the evaluation of the hardware and software packages to see how they will meet our proposed and future needs. We have done some research on training software packages that allow for training documentation to be created and tracked online. The training can run on mobile devices, computers, and printed to hardcopy. Users can ask questions that are sent to an expert, and the answer is then stored with the question online for easy searching later. The training package also allows for curriculums to be created so the user is able to expand upon their knowledge in an easy-to-follow manner. This software also contains "snippets" or modules of information that allows the user to easily find answers to questions. Simply send an email to the training system account; the system searches the database and provides training that is associated to your desktop or mobile device. The training packages are easily created with the software installed on an expert's computer that allows them to be video recorded or capture input from the keyboard, microphone, or other input device. The training software is similar to YouTube."

Garry: "Sounds too good to be true. What are our current capabilities and what needs do we have for the future?"

Karl: "The training we currently have is in PowerPoint slides and manuals. It is not easy to search or select parts of training that are needed on the fly. For our commercial stakeholders, we travel to the site to help with installation, training, and maintenance. For future government projects, we will be dealing with multiple agencies that will view the training differently. For example, DHS will be primarily using it for border patrol, looking for people crossing the border illegally. They are primarily concerned with immigration issues such as, did someone cross the border illegally? Where the DEA may want to use it for drug trafficking or human trafficking. They may want to capture images of the person that can be stored and searched for identity in another law enforcement database. The local police may have a different CONOPS in mind, for instance, finding a criminal, or searching for a lost person in a search and rescue type of mission. Again, recognition of the person but search through a different database, determine what if any medical attention may be necessary, and answer queries along those lines. These are all high-level descriptions, but you begin to understand that the training scenarios can quickly spiral out of control."

Garry: "Does it integrate with any of the systems we have now?"

Karl: "Yes, the HR system we now have is based on service-oriented architecture (SOA), which is a design methodology for software that allows for collections of separate software modules or services that provide higher functionality when integrated. The SOA technology is also quickly becoming a standard or sought-after enhancement in the DOD area because of the modularization of discrete software components. Sort of like a plug and play of systems, not fully realized yet, but the goal of SOA."

Garry: "What is the return on investment (ROI) for this technology?"

Karl: "The cost of the system is over 3 million dollars. However, this software will not only be used internally. As we design the system for DHS, all drawing packages, ICDs, and any other documents associated with the design, implementation, and production of the optical system can be uploaded. This allows for the developed information to be incorporated into stakeholder training packages as well. The stakeholder can login to the system and be able to view and/or download training material that is current to the system. They will also have the ability to ask an expert similar to the way our own employees are trained."

Garry: "So this could reduce travel costs and field staffing needs?"

Karl: "Yes, we estimate an ROI in 1.5 years and believe it fits with our company objective of providing high-quality support for our optical systems to our stakeholder base."

Garry: "Ok Karl, I think you have a good start for a package that we can present to Dr. Smith. Let's spend the next couple of weeks with the enterprise architecture team and run through the defined enterprise architecture process. We should identify each of the activities and processes as it pertains to the support and maintenance subsystem and the overall enterprise. These should include the stakeholders, goals and objectives, description, how it aligns with the business and program objectives, and improvement opportunities. We should understand our current capabilities and how the new system will expand upon them to meet and/or exceed our goals."

(Two weeks later, the team had set up a meeting with the CEO [Dr. Smith], CTO [Dr. Phelps], and the CIO [Mr. Blair]. This meeting was to obtain authorization to implement the support and maintenance training module that was identified in the previous section of the case study. The meeting began with a brief overview of the EA process as defined in the previous meeting with Mr. Ben and Mr. Blair. We start now with the overview of where the training module fits into the system for the organization.)

Karl: "As seen in the previous overview section of our enterprise architecture for the company, the programs we support, and the infrastructure capabilities we currently have established, we are lacking in some areas that need to be addressed if we are to make the current and future programs successful and allow us to meet our company objectives. The U.S. Government and the agencies we are supplying our system to have placed a high value on the integrated logistics and support components of the optical system. This was seen during the proposal and in conversations with the stakeholders of the contract through our usage of the House of Quality, shown in Figure 2.29. Some of the highest ratings were for ease of use, maintainability, and serviceability. With our current system of documentation control, training design, maintenance, and support, we will have a difficult

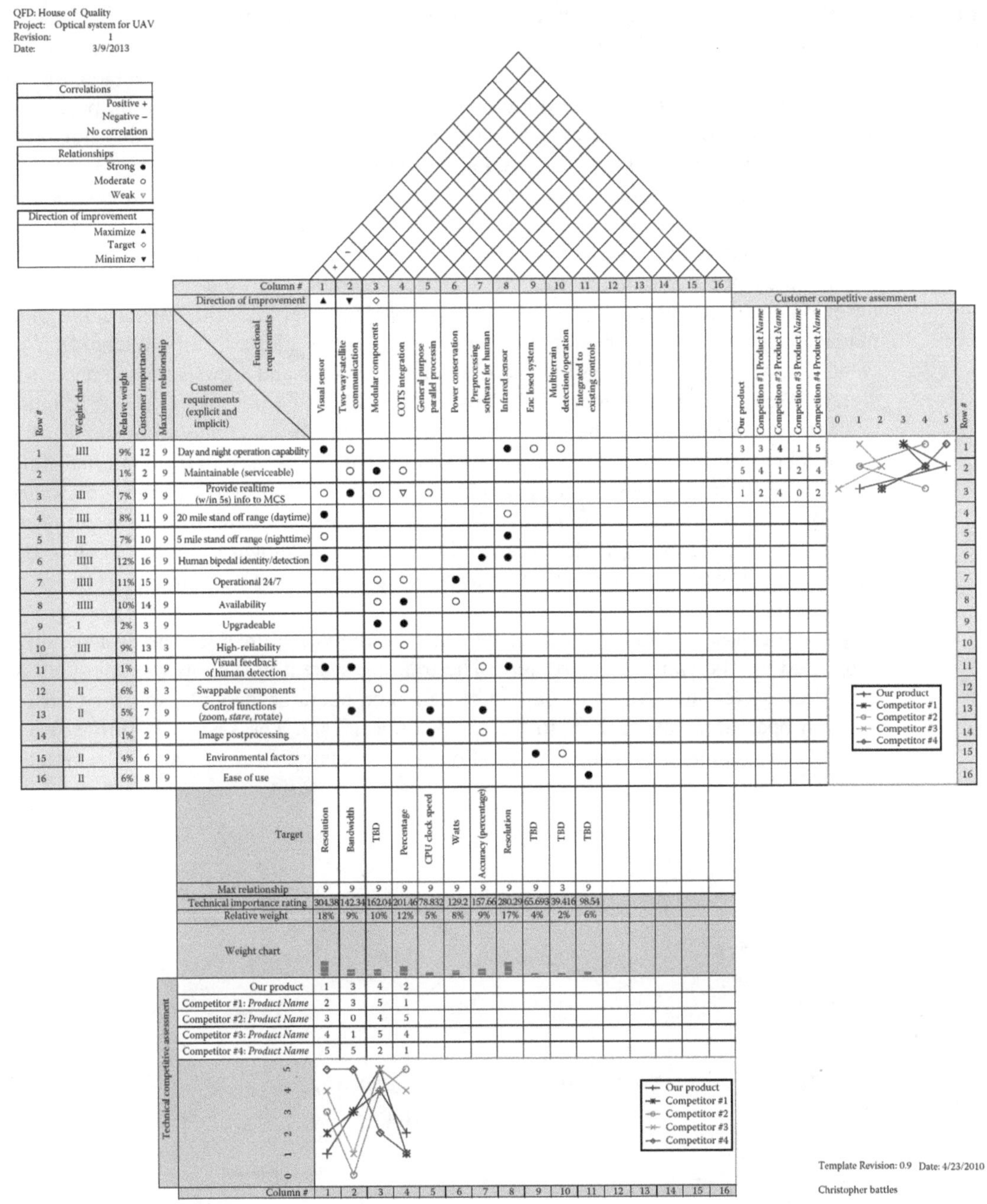

FIGURE 2.29 House of quality for UAV optical system. (Courtesy of Chris Lannon.)

time responding to the various agencies that are now engaged in discussions with our company for additional enhancements and new programs. We will need to add additional capabilities to our training system to support future company growth and improve the overall training, support, and maintenance regimen.

I am proposing that we incorporate a new training module that allows training to be created and viewed in such a manner that will reduce our on-site travel expenses by 35% and technical staff requirements by 20% and increase communications efficiency throughout the company by 25%."

Bill: "Those are some large returns; how have you been able to identify them?"

Karl: "We identified several areas of training that could be modularized and centralized that we currently do on-site. This frees up our technical staff so that they can focus more on the design activities and other core duties. We also identified several redundant processes and documentation. We were even able to eliminate some repetitive steps and standardize the information flow between functional departments."

Tom: "Karl, you have discussed internal training as well as training for the optical system operators. Who are we targeting as the stakeholders of this system?"

Karl: "The stakeholders of the new systems include DHS, border patrol; this includes the operators and maintenance personnel at the site, technical members of other agencies, business development, procurement and acquisitions people, our internal technical staff, program management, and human resources. I know that sounds like a lot of stakeholders; as we all know, each stakeholder has their own requirements and views into and about the system. We are targeting internal staffing and the DHS, border patrol as our primary stakeholders at this time. The enterprise boundary will be to create and establish the initial set of training and maintenance procedures that can be used to train internal staff. Developing these modules in steps will have the impact of allowing training packages to be developed modularly and at the same time reduce cost of training as we hire new employees. The following is the organizational structure, shown in Figure 2.30, of the new system. The enterprise description also includes how we plan to manage change to this system.

As you can see from the picture, the training module allows for the system to be connected via the cloud infrastructure. The training system will allow the operators of the optical system to have up-to-date information on operator manuals, specific instructions on operations, and access to technical support 24/7. The system is fully searchable. If a user enters a question, the question and the answer become part of the system.

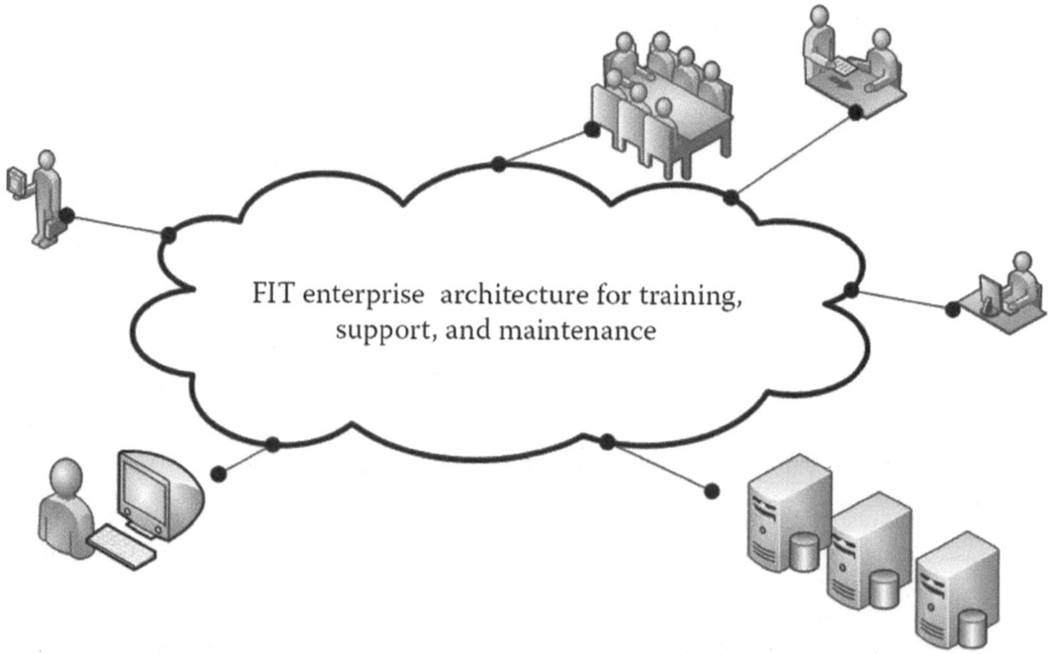

FIGURE 2.30 Organizational view of training support module. (Courtesy of Chris Lannon.)

This feature is similar to what users experience now on discussion forums and behaves the same way. For the maintenance operator, maintenance manuals are also up to date with the latest information. Videos are encoded with keyword searches to identify parts, how-to illustrations, and specifications for various parts. The system keeps track of various training a user has completed and assigns the user to a category, novice, intermediate, or expert based on the training completion and the user's competency on assessments."

Bill: "Is it possible to do this internally for training on systems design processes, coding standards, HR policies, and such? Does it integrate with our current Human Resource system?"

Karl: "The system is actually another module that has been developed by ACME. This is the same vendor that we bought the HR system from and still support. They are developing service-oriented applications that allow for discrete software packages to interact with each other."

Tom: "So what type of reports can we expect to see from this system, and how can we measure the effectiveness of the training?"

Karl: "Each training package is scored by the user. These scores are then tallied by number of users and given an effectiveness rating. If a rating falls below a certain score, the training package is flagged for review and sent back to the originator or a dedicated member of our technical staff for update. The system will also provide metrics on how many questions were asked that could have been alleviated by completing a training module. These metrics are then calculated to formulate decisions on suggested training to the user. There are also more standard metric reports of how many times a user has logged in to the system, how many times a particular training item has been accessed, number of users in the system, and so forth. We could use these metrics to determine types of training users need the most, whether it is video, documentation, or discussion type threads."

Bill: "Ok, I think we all agree that this new system provides us some needed capabilities. These seem to be in line with our requirement to deliver support, training, and maintenance components of the optical system we are developing for the DHS. It integrates with the current HR system and fits in nicely with the overall enterprise concept. What is the cost of the new system and what is the return on investment?"

Garry: "The cost of the new software system is 3 million dollars; however, because we have provided the ACME company with good references and feedback on the current human resources tool, they have offered us a 10% discount. We have also had the opportunity to set up temporary licenses internally and begin using the system in a small pilot study. The study team had some of our technical employees, Human Resources members, program management, and some business development folks. We also recently had one of the Department of Homeland Security personnel in house for a systems requirements meeting and showed it to them. They were impressed with the layout and ease of use. They also were intrigued by the SOA technology as they have other systems using that functionality. The return on investment we estimate to be 1.5 years. This includes acquisition costs, support costs, labor costs for training internal employees, cost of generating training packages, travel expenses, and securing the data on our data encrypted servers."

Bill: "ACME seems like a reputable company and has provided a system that has allowed us to help win programs by providing sound metrics and reporting capabilities providing an overview of the enterprise and how new capabilities or changes will affect the company's strategic planning. I think we will go ahead and buy the new capabilities and begin implementing it into our DHS optical system program as part of our solution to the support, training, and maintenance requirements. In addition to helping Homeland Security, this will allow us to be a step ahead of our competition and win new business. Karl please go forward with this acquisition."

2.4 SUMMARY

Fantastic Imaging Technologies has standardized a process for evolving their EA. FIT's EA process is anchored to a 25-year legacy of SE discipline in requirements analysis and management using industry standard tools. All architecture and system-level elements are contained in a common repository and linked to requirements, thus ensuring long-term maintainability of the complete technical baseline. The system is supported by a skilled support staff and training for internal and external stakeholders. FIT used a tailored EA framework based on DODAF 2.0. FIT systems engineers follow a standard process and workflow to capture, analyze, and model requirements and efficiently interact with other enterprise actors.

Investment decision-making is simplified: FIT's E A uses an E A process to simplify the evaluation of proposed solutions for program or IT infrastructure that provides a common language linking program solutions and IT initiatives with the company's strategic goals, objectives, and core business functions. FIT's holistic approach provides the executive-level staff with a strategic, enterprise view of the complete portfolio.

System implementation is accelerated: FIT's EA practice applies existing processes to accelerate systems design and development and define core business practices, common data elements, applications, and standard system platforms. The processes are used to verify system needs that cross between program areas and internal areas and facilitate communication between program areas and technical staff to define custom requirements.

IT system diversity is reduced: FIT's EA practice uses processes to streamline FIT's IT environment that define core business processes, common data elements, and standard applications and platforms. Standardization allows FIT to reduce cost associated with IT products, for systems maintenance and operation costs, and simplifies training requirements for both internal employees and stakeholder training. IT system diversity is reduced due to the standard-based approach, utilizing SOA technology and techniques.

EA allows entities to respond to the rapid changes in a complex, dynamic, and evolving development environment. Economic and social impacts can be analyzed, and responses to these impacts can be formulated to allow for new capabilities. A study was conducted in 2005 by the National Association of Chief Information Officers (NASCIO), which concluded that since 1999 states have made significant progress in adopting EA processes (NASCIO 2005). While most of these are in the area of technology, other areas of interest are beginning to adopt EA methodologies. These include performance management, process architecture, and business architecture. Many vendors are creating and refining modeling tools to support the usage of EA frameworks and processes.

2.A APPENDIX: FOURIER TRANSFORM PAIRS

$u(x)=\int_{-\infty}^{\infty} u(f)e^{j2\pi xf}df$	$U(f)=\int_{-\infty}^{\infty} u(x)e^{-j2\pi fx}dx$
$u(x,y)=\iint_{-\infty}^{\infty} u(f_x,f_y)e^{j2\pi(f_x x+f_y y)}df_x df_y$	$U(f_x,f_y)=\iint_{-\infty}^{\infty} u(x,y)e^{-j2\pi(f_x x+f_y y)}dxdy$
$\delta(x)$	1
1	$\delta(f)$
Tri(x)	$\text{sinc}^2(f)$
$e^{-\lvert f\rvert}$	$\dfrac{2}{1+(2\pi f)^2}$

(*Continued*)

(Continued)

$\frac{2}{1+(2\pi x)^2}$	$e^{-\lvert f \rvert}$
$\cos(\pi x)$	$\frac{1}{2}\delta\left(f-\frac{1}{2}\right)+\frac{1}{2}\delta\left(f+\frac{1}{2}\right)$
$\sin(\pi x)$	$\frac{1}{2}\delta\left(f-\frac{1}{2}\right)-\frac{1}{2}\delta\left(f+\frac{1}{2}\right)$
$e^{-\pi x^2}$	$e^{-\pi f^2}$
$J_0(2\pi x)$	$\frac{\mathrm{rect}\left(f/2\right)}{\pi\left(1-f^2\right)^{1/2}}$
$\frac{J_1(2\pi x)}{2x}$	$(1-f^2)^{1/2}\,\mathrm{rect}(f/2)$
$\mathrm{rect}(x)\,\mathrm{rect}(y)$	$\mathrm{sinc}\,fx\;\mathrm{sinc}\,fy$
$\mathrm{circ}\,(r)$	$\frac{J(2\pi\rho)}{\rho}$
$\delta(r-a)$	$2\pi a J_0\,(2\pi a\rho)$
$\mathrm{Tri}(x)\mathrm{Tri}(y)$	$\mathrm{sinc}^2\,(f_x)\;\mathrm{sinc}^2\,(f_y)$
$e^{j\pi r^2}$	$je^{-j\pi\left(fx^2+fy^2\right)}$
$\frac{1}{r}$	$\frac{1}{\rho}$
$\left[\frac{J_1(2\pi r)}{r}\right]^2$	$2\left[\cos^{-1}\left(\frac{\rho}{2}\right)-\frac{\rho}{2}\sqrt{1-\left(\frac{\rho}{2}\right)^2}\right]\mathrm{rect}\left(\frac{\rho-1}{2}\right)$

REFERENCES

Anderson, N. and T. Erdogan. 2014. Multimodal nonlinear optical (NLO) imaging. https://www.semrock.com/multimodal-nonlinear-optical-nlo-imaging.aspx (accessed November 14, 2014).

Andrews, L. and R. Phillips. 1998. *Laser Beam Propagation through Random Media*. Bellingham, Washington: SPIE Press.

Arrasmith, W.W. 2010. Novel wavelength diversity technique for high speed atmospheric turbulence compensation. *SPIE's Defense and Security Symposium,* Orlando, FL, May 5–9.

Bahaa, E.A.S. and M.C. Teich. 2007. *Fundamentals of Photonics*, 2nd edn. New York: Wiley.

Barroero, T., G. Motta, and G. Pignatelli. 2010. *Business Capabilities Centric Enterprise Architecture*. Berlin: Springer.

Bernard, S.A. 2012. *An Introduction to Enterprise Architecture EA3*, 3rd edn. Bloomington, IN: AuthorHouse.

Born, M. and E. Wolf. 1970. *Principles of Optics: Electromagnetic Theory of Propagation, Interference and Diffraction of Light*, 7th edn. Cambridge: Cambridge University Press.

Boyd, R.W. 2008. *Nonlinear Optics*, 3rd edn. San Diego, CA: Academic Press.

Clinger and Cohen Act. 1996. Information Technology Laws. U.S. Chief Information Officers Council: https://www.cio.gov/handbook/it-laws/clinger-cohen-act/ (Accessed on March 4, 2025).

Craig, A.T., J. Hogg, and R.V. Hogg. 1995. *Introduction to Mathematical Statistics*, 5th edn. Upper Saddle River, NJ: Prentice Hall.

de Vries, M. 2010. A framework for understanding and comparing enterprise architecture models. *Management Dynamics: Contemporary Research Journal of the Southern Africa Institute for Management Scientist*, Vol. 19, No. 2, pp. 17–29.

DOD. 1994. *Technical Architecture Framework for Information Management* (TAFIM), vols. 1–8, version 2.0. Reston, VA: DIS A Center for Architecture.

DoD AFv2.02. 2014. DoD architecture framework Version 2.02. DOD Deputy Chief InformationOfficer. U.S. Department of Defense. https://dodcio.defense.gov/TodayinCIO/DoDArchitectureFramework.aspx (accessed November 14, 2014).

GAO. 2001. A practical guide to federal enterprise architecture, Version 1.0. CIO Council. https://www.gao.gov/products/P00201 (Publicly Released: 9 February 2001).

GAO. 2005a. FBI is taking steps to develop an enterprise architecture, but much remains to be accomplished. Information Technology. https://www.gao.gov/products/GAO-05-363 (Publicly Released: 9 September 2005).

GAO. 2005b. Some progress made toward implementing GAO recommendations related to NASA's integrated financial management program. Business Modernization. https://www.gao.gov/products/GAO-05-799R (Publicly Released: 27 October 2005).

GAO. 2010. Federal enterprise architecture program EA assessment framework, Version 2.0. Organizational Transformation. https://www.gao.gov/products/GAO-10-846G (Publicly Released: 5 August 2005).

Goodman, J.W. 1985. *Statistical Optics*. New York: Wiley.

Goodman, J.W. 2005. *Introduction to Fourier Optics*, 3rd edn. Greenwood Village, CO: Roberts & Company.

Hite, R.C. 2004. The federal enterprise architecture and agencies' enterprise architectures are still maturing. Information Technology: [GAO Testimony]. https://www.gao.gov/products/GAO-04-798T (Publicly Released: 19 May 2004).

Hite, R.C. 2006. Progress continues, but challenges remain on department's management of information technology. Homeland Security: [GAO Testimony]. https://www.gao.gov/products/GAO-06-598T (Publicly Released: 29 March 2006).

Hornford, D. 2011. *TOGAF, Version 9.1*. Zaltbommel: Van Haren Publishing.

Irvine Sensors Corporation. n.d. Development & sales: miniaturized infrared and electro-optical devices. https://www.irvine-sensors.com (accessed April 26, 2014).

Minoli, D. 2008. *Enterprise Architecture from A-Z: Frameworks, Business Process Modeling, SOA, and Infrastructure Technology*. Boca Raton, FL: Auerbach Publications (CRC).

Mo, D. 2012. MOD architecture framework. https://www.gov.uk/mod-architecture-framework (accessed November 14, 2014).

NASCIO. 2005. NASCIO enterprise architecture assessment. NASCIO Online. https://www.nascio.org/publications/documents/NASCIO-eaAssessment.pdf (accessed November 14, 2014).

Opengroup. 2014. The open group architecture framework (TOGAF). TOGAF Online. https://www.opengroup.org (accessed November 14, 2014).

Oppenheim, A. and R. Schafer. 1975. *Digital Signal Processing*. Englewood Cliffs, NJ: Prentice Hall.

O'Rourke, C., N. Fishman, and W. Selkow. 2003. *Enterprise Architecture Using the Zachman Framework*. Boston, MA: Course Technology.

Roggemann, M.C. and B. Welsh. 1996. *Imaging through Turbulence*. Boca Raton, FL: CRC Press.

Sambur, M.R. 2003. "Incentivizing Contractors for Better Systems Engineering, Assistant Secretary of the Air Force (Acquisitions), SAF/AQ Policy Memorandum dated January 6, 2003. https://apps.dtic.mil/sti/tr/pdf/ADA437444.pdf (Accessed Mar 4, 2025).

Schekkerman, J. 2008. *How to Manage the Enterprise Architecture Practice*. Bloomington, IN: Trafford Publishing.

Sessions, R. 2007. A comparison of the top four enterprise architecture methodologies. Microsoft Developer Network. https://msdn.microsoft.com/en-us/library/bb466232.aspx (accessed November 14, 2014).

Shanmugan, K.S. and A.M. Breipohl. 1988. *Random Signals Detection, Estimation, and Data Analysis*. New York: John Wiley & Sons.

Shen, Y.R. 1998. *The Principles of Nonlinear Optics*. Chichester: John Wiley & Sons.

van den Berg, M. and M. van Steenbergen. 2006. *Building an Enterprise Architecture Practice: Tools, Tips, Best Practices, Ready-to-Use Insights*. Dordrecht: Springer.

Whitehouse. 2007. FEA consolidated reference model document, version 2.3. https://www.reginfo.gov/public/jsp/Utilities/FEA_CRM_v23_Final_Oct_2007_Revised.pdf (accessed April 10, 2025).

Zachman, J.A. 1987. A framework for information systems architecture. *IBM Systems Journal*, Vol. 26, No. 3, pp. 276–292.

3 System-of-Systems, Family-of-Systems, and Systems Engineering

What is the use of a fine house if you haven't got a tolerable planet to put it on?

—**Henry David Thoreau**

3.1 OVERVIEW

Systems are everywhere! From small, familiar objects like a coffee mug to complex objects like a satellite, we are continuously surrounded by systems. Some systems occur naturally, like planet earth, while others are man-made, like an automobile. When classifying each of these items as a system, it is important to remember that a "system" is defined as a combination of things that work together to achieve a common goal. With this definition, it becomes clear that many things can be classified as a system with the proper reference applied.

In early-recorded human history, system principles were recognized but not fully understood, as evidenced by the use of mysticism to interpret some natural interactions such as seasons, phases of the moon, sunrises, and herd migrations. As human thought developed and advanced, a broader interpretation of natural systems started to develop, as seen by the creation of theories of orbital mechanics, relativity, and quantum mechanics. Eventually, man-made systems interacted with natural systems in significant and observable ways (National Aeronautics and Space Administration Goddard institute for Space Studies 2012). The impacts of some man-made systems on these natural systems became destructive, giving start to the depletion of nonrenewable and limited natural resources. On many occasions, studies into system interactions were initiated to avert further resource exhaustion and damage. For example, the North American Whitetail Deer, Elk, and Bison were once almost hunted into extinction until cattle ranching, as a renewable food source, gained popularity (Evans 1996). Another example is how the clothing industry relied heavily on North American beaver, mink, and cougar, causing them to be hunted and trapped almost into extinction until renewable resources such as sheep ranching, silk worm production, and cotton crops gained popularity (Saundry and EH.net. 2012). However, one of the most recognized and significant impacts of man-made systems on natural systems occurred with the advent of the industrial revolution at the turn of the century. Massive factories' increased smog, soot, and pollution were impacting water and air quality, spurring groups and governments into action to limit industry pollution on the population and the environment. These actions started the modern-day environmental movements and man-made climate change research (Robbins 2007; Lorenz 1972). With better human understanding of the effects that systems have on other systems, insights into the nature and behavior of systems, system-of-systems (SoS), and family-of-systems (FoS) began to emerge.

The earth can be used to examine more closely the definition of a system. The earth is a natural system that contains many elements within it: water, land, gravity, atmosphere, plants, people, and animals. Each of the aforementioned elements brings with it certain characteristics, and all of these elements support each other to achieve harmony within the system, maintaining the ultimate goal of existence. Perturbing one or more of these elements will cause the system to experience a shift that may or may not impact other elements of the system or endanger the system itself. Nevertheless, the degradation or removal of an interacting element within the system would cause turmoil within

DOI: 10.1201/9781003624097-4

the system until a new balance is again obtained. An example of this would be a shift in the atmosphere causing a shift in sustainable plant life within a specific region. Eventually, a group of elements affected by it, for example, the plants within that region, would balance with the atmospheric change, whether by adapting or by a different species becoming more prominent.

The earlier example of the automobile as a man-made system also contains many elements within itself: body, tires, engine, seats, steering, fuel, battery, and computer. Although each element will bring certain properties to bear within the system, all of the elements must work together to achieve the ultimate goal of transporting something somewhere. Without the engine, the system will fail; without the tires, the system will fail; and without the fuel, the system will fail. The engine needs fuel, the computer needs energy, the wheels need steering, and the body needs seats; therefore, it takes multiple elements working together to achieve the system goal.

Blanchard said, "Systems are as pervasive as the universe in which they exist. They are as grand as the universe itself or as infinitesimal as the atom. Systems appeared first in natural forms, but with the advent of human beings, a variety of human-made systems have come into existence. In recent decades, humans have begun to understand the underlying structure and characteristics of natural and human-made systems in a scientific way" (Blanchard and Fabrycky 2011). Systems are made up of individual elements. For example, our galaxy as a system has individual elements. These elements are the Sun, planets, and moons. The atom, as a system as well, has individual elements: the electrons, nucleus, protons, and neutrons. Each individual element of a system brings its own contributions to the system as a whole, and the interactions within each of these elements are what determine how a system works to achieve a goal. In the end, the system is as we define it. It can be larger or small and it can, itself, be made up of numerous systems – an SoS. As systems engineers, we need to define the system in context to its relation to other systems and its environment.

The remaining sections will discuss in detail the differences between an SoS and an FoS. Also, it will look into the specific example of an optical subsystem on an unmanned aerial vehicle (UAV) and how different components can be treated as either an SoS or FoS. Furthermore, a case study will be presented, integrating the systems engineering (SE) and the UAV technical aspect of this chapter.

3.1.1 System-of-Systems

A system is a group of elements that work together to accomplish a common goal; similarly, an SoS is a group of systems where all systems work interdependently to accomplish a higher-level goal. One formal definition of an SoS is "a collection of task-oriented or dedicated systems that pool their resources and capabilities together to create a new, more complex system, which offers more functionality and performance than simply the sum of the constituent systems" (Wikipedia 2025). Another definition for SoS, as defined in the 2008 Department of Defense (DOD) Defense Acquisition Guidebook (DAG), is "a set or arrangement of systems that results when independent and useful systems are integrated into a larger system that delivers unique capabilities" (ODUSD 2008). Applying these definitions, a quick example could be found in an aircraft navigation system. Sensor suites that take advantage of an existing basic navigation system can be added to the aircraft while also incorporating additional stand-alone sensor systems in such a manner that they complement each other's strengths/weaknesses. All of these systems would then collectively be recognized as one SoS providing its host with robust navigation capabilities. Each system within the SoS has interdependency on the other systems within the system. That is, the inputs and outputs of a system in an SoS have a cause-and-effect relationship with the system's operation as a whole.

Using the previous example of the earth's natural system, the element considered as people could be further defined as a subsystem with elemental composition and functions allocated at the lowest levels. Many distinct societies come together to form this element, and furthermore, these societies are comprised of smaller, independent operating groups of people that work together. The people of the earth are varied by many characteristics: geographic regions, ethnic

FIGURE 3.1 Automobile subsystems. (Courtesy of Richard Qualis. With permission.)

backgrounds, socioeconomic backgrounds, religious backgrounds, and political backgrounds. However, they all reside on and rely upon the earth system; they have a symbiotic relationship with the higher-level system — the earth. As mentioned before, the impacts that people have on the earth system can directly impact all of the other systems within it, notably the depletion of resources, impacts on other living systems (species), impact on environmental systems, and the climate system.

Looking back, the previous automobile example could be considered an SoS as well. The automobile is composed of smaller systems, each made up of individual parts, altogether functioning as a system. As illustrated in Figure 3.1, these smaller systems include the body system, wheel system (tires), engine system, seating system, steering system, fuel system, power system, and computer system. Each of these smaller systems, with adaptation, could conceivably provide a capability on their own and are themselves made up of smaller components, thereby satisfying our concept of a system. The interdependency of each of these smaller systems within the automobile system is vital to the system as a whole. They operate through working together, thereby inputs and outputs of one system can affect the inputs and outputs of another. For example, if the power steering subsystem suffers a leak in the power steering liquid tank, the car system's wheel (steering system) will become very stiff, thus impeding proper functionality of the wheel system. Both failures will have negative effects on the car's ability to function as a whole.

Without the ability to steer, tires will not be able to turn; thus, no direction may be given accurately to the car when in motion.

Using the DOD's definition for software SoS, they describe the four primary types of SoS in use today as virtual, collaborative, acknowledged, and directed (ODUSD 2008). The virtual SoS does not use central management, nor does it operate with an agreed-upon purpose for the overall SoS. Collaborative SoS have constituent systems that interact only voluntarily as needed to satisfy a purpose. An acknowledged SoS has common objectives and resource management, but the systems are maintained independently, and they collaborate together to achieve the system's purpose. The last type of system is the directed SoS; it is fully integrated, and its engineered system is geared toward being managed over the long-term operation of the system. However, with the constant desire (and effort) to link systems with information-sharing purposes, new approaches are being considered to allow this linkage, without the need for common goals among systems (ODUSD 2008).

Just as a system is a group of elements working in conjunction to accomplish a goal, an SoS is a group of constituent systems working in conjunction to accomplish a larger goal than the individual systems could accomplish on their own. A defining characteristic of the SoS's architecture is the interdependency that exists among the inputs and outputs of the constituent system elements to accomplish the overall SoS goals. The U.S. military recognizes that combining some systems to form larger systems can expand their capabilities, but managing all of those constituent systems to achieve the desired SoS goal would require developing new capabilities.

3.1.2 Family-of-Systems

Since a system is a group of elements that work together to accomplish a goal, an FoS can be viewed as a group of systems that work independently to accomplish the same goal. The manner in which the subsystems are arranged will determine the output of the overall system (Office of the Under Secretary of Defense for Acquisition, Technology and Logistics). The Office of the Deputy Under Secretary of Defense (ODUSD 2008, p. 4) stated that the FoS can also be defined as a "set of systems that provide similar capabilities through different approaches to achieve similar or complementary effects." Each system within the FoS operates independently from one another, though they all work to accomplish the overall system goal.

Revisiting the automobile example for the FoS concept, it could be considered to be part of an FoS by looking at the overall system (the car) as part of a larger system, that is, a short-range personal transport system. This system has the ability to provide a means for passengers to commute to work and school. As part of its family, there are various systems, including cars, buses, motorcycles, and even bicycles. Comparing the automobile system to the other family members, they are very distinct in capabilities and do not operate with synergy, yet at the same time, all of these systems come together to form the overall personal transport system. These individual systems also have the same goal – to provide a personal transport capability for an individual. The system is not paralyzed if one system or vehicle breaks down thanks to each system's independence, and another vehicle may step in to provide the short-range transport capability for the family in need.

Just as a system is a group of elements working in conjunction to accomplish a goal, an FoS is a group of systems that can provide similar or complimentary capabilities through different means. The FoS is different from the SoS concept since the former does not provide expanded functionality or capabilities larger than the sum of its parts.

3.1.3 SE Processes across the Systems Development Lifecycle Phases

The SE approach takes the long view of how the system under study will be designed and operated over that system's life expectancy The SE lifecycle is commonly divided into phases from the conceptual design to disposal of the system. As depicted in Figure 3.2, the systems development lifecycle (SDLC) phases are the conceptual design phase, the preliminary design phase, the detailed design and development phase, the production phase, the system operations and support phase, and the system retirement and disposal phase. Each of these phases has its own set of rules, best practices, tools, techniques, and methodologies, which we will discuss further ahead.

3.1.3.1 Overall SE Development Process

The conceptual design phase of Figure 3.2 is designed to give the system shape by defining the system-level requirements. Block one begins with problem definition and needs identification, both of which are critical processes of this phase. Here, the problem that the system will solve is thoroughly discussed with the stakeholders and the development team. The essential motive is to clarify and take into account the needs and wants of the stakeholders as much as possible and develop the system concept. The needs, and agreed wants, are expanded into a set of stakeholder requirements that are subsequently evolved into systems requirements by the end of the phase. The systems requirements are captured in a system-level requirements document. This requirement document contains both functional and nonfunctional operational requirements. The system operational requirements and the maintenance and support concepts tell us how the system will perform and what is required to keep the system operational through the operations and support phase of the SDLC. Trade studies will be completed to begin architecting the system. Feasibility studies are conducted, typically on every requirement one-on-one to see if they can be accomplished or not. It is important that we have a solid system specification (A spec) before advancing to the next system lifecycle phase.

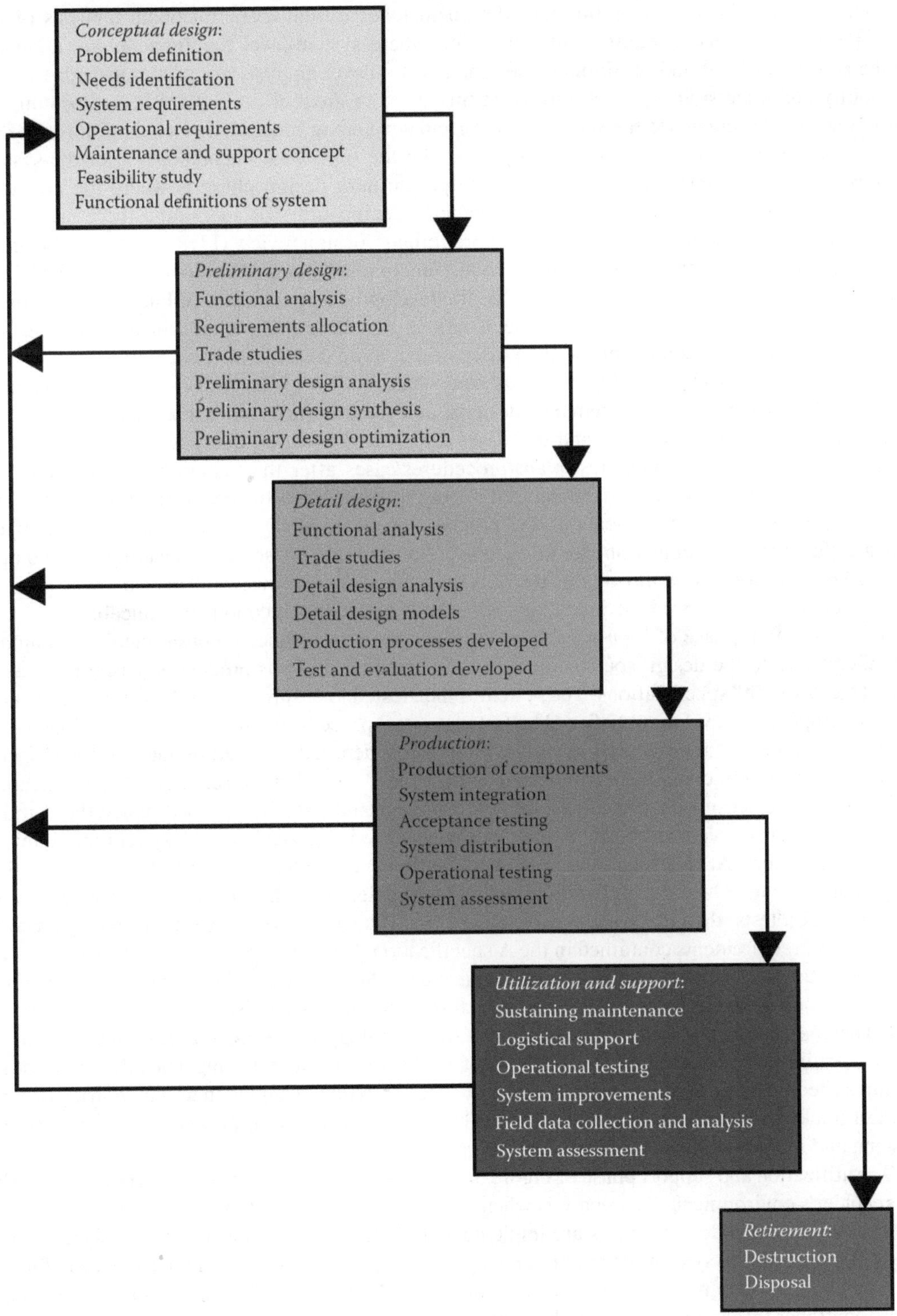

FIGURE 3.2 The SE development process. (Courtesy of Jeffe Baudek. With permission.)

Block two of Figure 3.2, preliminary design includes a high-level functional analysis of the system and a subsequent requirements allocation where system-level requirements are allocated to the subsystem level and preliminary designs and midlevel analysis are accomplished. Refined feasibility and trade studies are performed at this stage. A great effort is placed on allocating the system-level requirements to the subsystem level and developing functional block diagrams (FBDs)/functional flow block diagrams (FFBDs) to describe the functional decomposition of the system into its constituent subsystems. The end of the preliminary design phase is where the allocated requirements are used to generate the development specifications (B specs) and these are placed under configuration control (baselined) at the preliminary design review (PDR). There can be many B specs for a given system. These often match the functional decomposition of the system itself and usually are separated into hardware and software development specifications. Further, preliminary specifications such as the C, D, and E specifications may be developed concurrently during this phase as well. These specifications will be discussed in more detail in later chapters.

The detail design block of Figure 3.2 is somewhat similar to the previous one; in this case, higher-level products (e.g., requirements, designs, analyses) are taken to a more detailed and final level. Test plans, test procedures, and test cases are created covering every requirement. The test team eventually executes these test plans/procedures/cases after the system has been produced. In some cases, prototypes are manufactured during this phase, to further define the system. The product specification (C specification), the process specification (D specification), and the material specification (E specification) are all completed during this phase before heading into the next phase. The milestone event during this phase is the critical design review (CDR) where the product specifications (C specs) are baselined along with the D and E specifications if applicable.

The production phase of Figure 3.2 is where the system is produced, constructed, or manufactured. Changes to the design specifications are documented in a postproduction baseline revision called the "as-built" specifications. The system is then tested to ensure that all of the requirements at each developmental level are satisfied. The test plans, test procedures, and test cases developed during the design phases are executed starting with the most detailed level requirements – the C specifications. The system components are tested against the C specification requirements. Verification testing is used to ensure that each requirement has been individually satisfied. Once the component requirements have been verified, the components are integrated into the system and integration testing occurs. At this stage, the integrated components are tested against the B specification level requirements. Once the integration testing is complete and the integrated components pass the integration tests, then the system-level tests begin. The system-level tests are tested against the system-level requirements contained in the A specification. Once the system passes the system-level tests, operational test and evaluation (OT&E) take place. During OT&E, the stakeholder evaluates if the system performed as envisioned. Validation testing occurs during OT&E wherein the system is tested against real-world/operational conditions. As an analogy, checking a racecar engine's performance requirements on a stand in a laboratory can satisfy verification testing, but validation testing requires checking requirements or groups of requirements with the engine in the car, roaring around the test track! At a successful completion of OT&E, the stakeholder typically takes charge of the system, and the developmental effort is considered at an end.

The utilization and support phase of Figure 3.2 block five consists of the system's operation in the stakeholder's environment. This phase is when the maintenance and support plans that were developed in the system's design phases are implemented. Field data are collected and analyzed in the areas of maintenance, system operations, and logistical support for assessment purposes and for use in systems reliability, maintainability, and availability (SRMA) and lifecycle support models. This phase continues throughout the life of the system until the system is ready to be retired from service.

The retirement phase of Figure 3.2 block six consists of the removal from service and satisfactory disposal of the system. The system may have critical components disassembled and destroyed, whereas other components may be salvaged. The space shuttle, for instance, may find a sanitized version of itself on display in a museum somewhere. Component-level disassembly may be required to meet federal, state, or local environmental recycling/disposal requirements or security requirements. Disposal is a crucial design criterion that must not be ignored since eventually all systems face retirement and disposal. Proper planning that includes logistical considerations must be developed early on in the SDLC. These plans must detail how to handle disposal, including the transportation of hazardous materials during disassembly and the safeguarding, handling, and disposal of sensitive/classified information.

3.1.3.2 SE Waterfall Process

The spiral model represented in Figure 3.3 is generally intended for large complex programs where it is beneficial to have incremental releases and refinements for each successive loop around the circle (Boehm and Hansen 2000). In the spiral model, a system development effort is broken into a set of progressive, spiral activities that have clearly defined exit criteria at the end of each spiral.

For instance, in Figure 3.3, some precursors to the first spiral starting with requirements plan include conceptual formulation activities such as stakeholder requirements generation and the formation of the concept of operations (CONOPS) for the system. Upon successful review of these products, the first spiral starts with the generation of the requirements plan. The requirements plan captures the essential elements of the CONOPS and stakeholder needs and wants. The requirements plan also determines the objectives that need to be satisfied in the requirements generation process. In the top right quadrant of Figure 3.3, system/service/product/or project risks are identified and resolved, and high-level prototypes may be generated as part of risk reduction activities. The main products of this quadrant of this part of the spiral development are a well-defined requirements plan and a high-level risk analysis with acceptably mitigated risks. The next quadrant of this spiral involves the generation, verification, and validation of the system requirements. This quadrant of the spiral development method is about development and test. Once the requirements are verified, validated, and accepted, planning for the next spiral evolution is accomplished in the lower left quadrant of Figure 3.3. In this case, the development plan is generated. The end of a particular spiral occurs

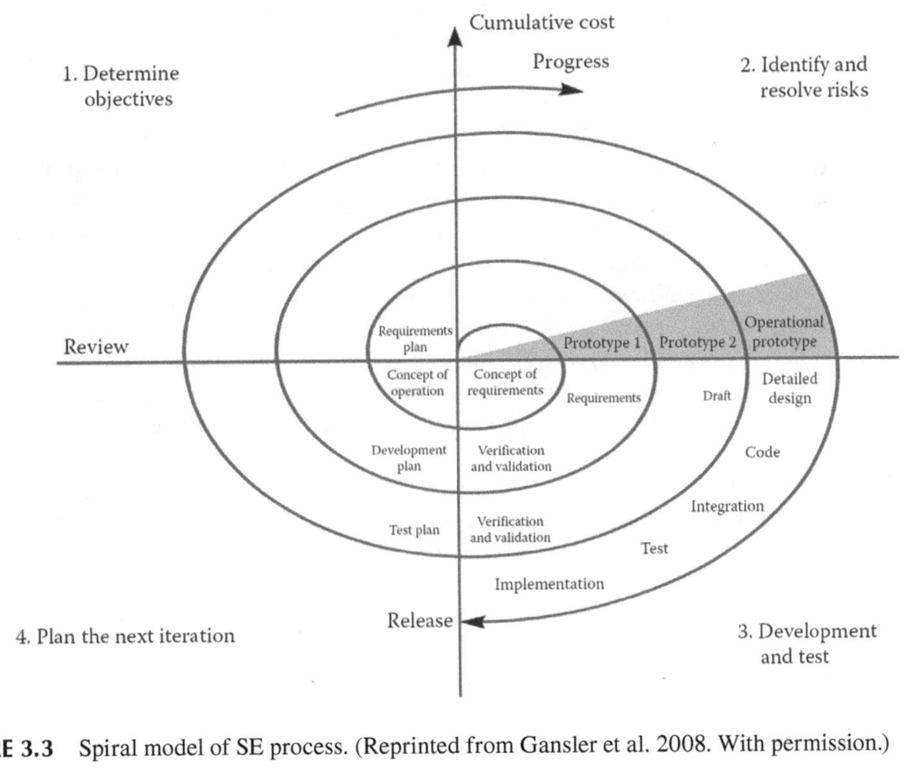

FIGURE 3.3 Spiral model of SE process. (Reprinted from Gansler et al. 2008. With permission.)

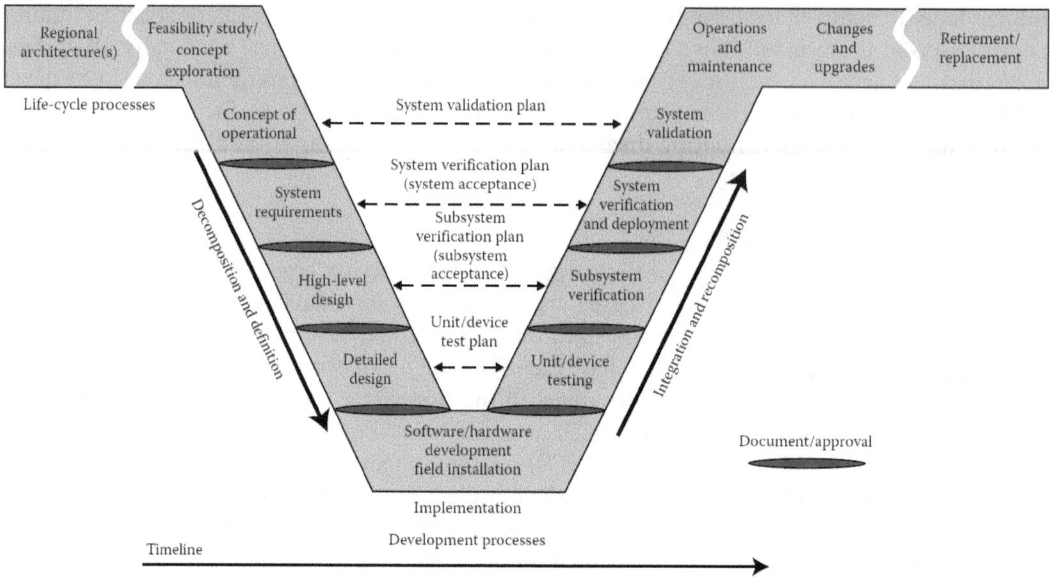

FIGURE 3.4 V diagram of SE lifecycle process. (Reprinted from Rausch et al. 2007. With permission.)

at a review that evaluates the results of the spiral activities. A set of exit criteria are established that must be satisfied in order to proceed to the next spiral. In this case, the results are acceptable versions of the requirements plan, risk analysis, and plan for risk mitigation, any high-level prototypes, the requirements themselves, and the test plan. Assuming that the review is positive, progression to the next spiral is authorized. The spirals continue through successive steps of determining objectives, evaluating and reducing risk, evolving prototypes, developing and testing, and then planning for the next spiral. This model is very effective in projects that involve incremental funding. Funding decisions can often be linked to demonstrated progress at the review points of the spirals.

Figure 3.4 depicts another common SE tool called the V diagram (U.S. Department of Transportation, Federal Highway Administration 2007). The left side of the V shows the systems design process and the evolution from the systems requirements downward to the detailed design. At the bottom of the V diagram, the system is produced. The right side of the V illustrates the testing process starting with verification testing at the unit level (Rausch et al. 2007, p. 5).

Notice that the unit is tested against the detail design level requirements. Next, the units are integrated into subsystems and integration testing occurs. These subsystems are tested against the high-level design (e.g., B specification level). After integration testing, system-level testing occurs. For system test, the system is tested against the operational requirements given in the A specification. System validation occurs during OT&E, and the system is subsequently fielded.

Figure 3.5 shows the so-called waterfall model. The spiral model, the V diagram, and the waterfall model are all standard models for depicting the SE design process over a product lifecycle (Harris Corporation 2012). Each model has broad applications; however, the waterfall model is the methodology that will be used for depicting the iterative design process here. Because the waterfall model has a clear, practical, and sequential process flow, this model will also serve as the organizational model for the rest of this text.

The waterfall model of Figure 3.5 displays the major steps followed in the SDLC. The SE process begins with the problem statement and its definition. The problem definition is the most important step in the design process because it is at this point in the process where the development team needs to listen carefully to the stakeholder to clearly identify the problem that the system is intended to solve. This step is accomplished by identifying stakeholder needs and wants through interviews and conversations with the stakeholders and the engineering team. Multiple sessions are usually required in meetings between engineering and stakeholders with each meeting being

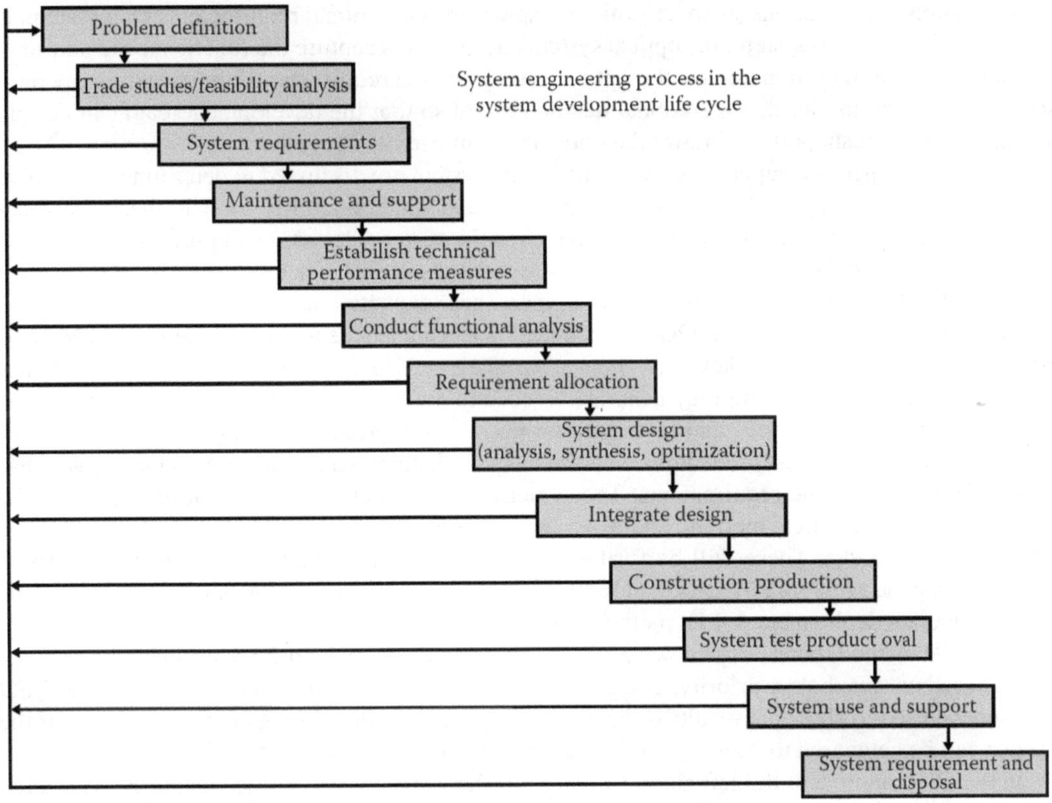

FIGURE 3.5 The waterfall SE process in the system lifecycle.

carefully dissected by the engineering team for further problem definition refinement. A CONOPS is developed to understand the role the system plays in its environment and its relationship to other entities within a given enterprise.

Stakeholder requirements are defined from the stakeholder needs and CONOPS. The systems requirements at this level are very broad, usually with high-level technical performance specifications that focus on "what" the system must do and not "how" to do it. This level of requirement is usually written in the stakeholder's language with engineering intent to refine the requirements at a later date (e.g., feels smooth, lasts long time, runs fast, and works well). In the problem definition phase, a set of goals and objectives are typically developed. The goals are measurable "big deal" type of events and projections that have business or technical focus. An example of a business goal is, "We will capture 10 percent of the local high-resolution imaging market, within 3 years of the delivery of our first sensor." Another might be, "We will achieve an average of 10 percent sales growth annually over the next three years." An example of a technical goal could be, "Our optical imaging systems will provide the highest spatial resolution imaging systems available on the market." Any ambiguities in the high-level stakeholder requirements must be removed in subsequent requirements levels (e.g., system-level requirements and below) by including measurable metrics in the lower-level requirements. For instance, from the previous example, "lasts a long time" would be further refined at the system specification level using reliability-centered maintenance metrics such as the mean time between failure, mean time between repair, and operational availability to name a few.

As an entity in an enterprise, the enterprise architecture that is to be used needs to be established and understood. For instance, if the system is a DOD acquisition, then the DOD Architectural Framework (DODAF) is mandated. The system architecture needs to be developed so that it is in compliance with the guidance in the higher-level enterprise architecture framework.

The system specifications (A specifications) capture the operational requirements for a system. If the system is an optical system, the optical system requirements capture the functional and nonfunctional operational requirements for the optical system. These requirements are written in operator language and capture all the operational details needed so that the development team can design, build, test, deliver, support, and retire the optical system over its lifecycle.

The maintenance and support expectations for the system are discussed to determine what infrastructure, logistics, repair, factory sparing models, and maintainability models the stakeholder needs for their system. This is also the moment when the stakeholder has an opportunity to communicate the operational flow and tempo expected from the system. This helps the engineers determine the reliability the system must have in order to meet these expectations.

Measures of performance (MOPs) and technical performance measures and metrics (TPM&Ms) begin to be established, as are key performance parameters (KPPs). The engineering team evaluates the systems requirements, and they generate associated measures and metrics to indicate systems performance. These measures and metrics are especially important in testing since they are typically used in test plans, test procedures, and test cases to demonstrate whether or not the system has successfully satisfied the requirements. These measures and metrics are also prioritized with the more important measures, metrics, and associated parameters being separately tracked as MOPs, TPM&Ms, and KPPs. These will be further discussed in a later chapter of this text. In order to communicate the MOPs, TPM&Ms, and KPPs to the stakeholders and to the development team, the quality function deployment (QFD) method is often effective.

The QFD method, otherwise known as the house of quality, can be used to evaluate the requirements based on stakeholder priority. The generic form of the QFD is shown in Figure 3.6. The QFD can also be used to show the results of benchmarking studies wherein performance aspects of the system can be compared to "best in class" systems available on the market.

In the QFD approach, the left side of Figure 3.6 shows the "voice of the customer." Typically, you would list stakeholder requirements or high-level MOPs, KPPs, or TPMs to illustrate what

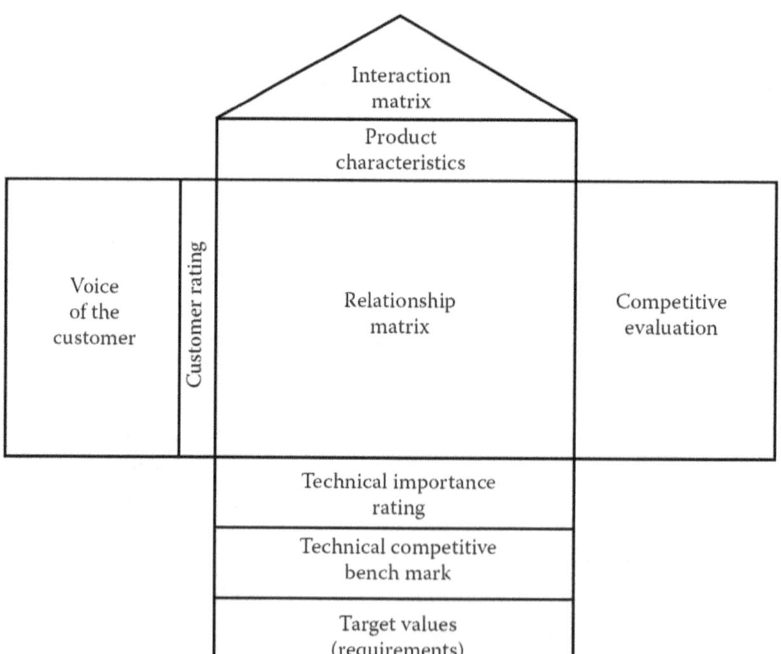

FIGURE 3.6 Sample QFD aka house of quality. (Reprinted from Sherif and Tran 1994).

is important to the stakeholder. The customer rankings are accomplished by the stakeholder to show the relative importance of each of the items in question. The top of the QFD shows the technical response to the voice of the customer entries. In this case, product characteristics would contain columns of attributes that would show how product characteristics would address the row entries in the voice of the customer section (e.g., the stakeholder requirements). The right side of Figure 3.6 addresses the competitive evaluation such as benchmarking results. The relationship matrix contains scores that relate the voice of the customer to the product characteristics. For example, if the optical system voice of the customer entry were "ease of use," then the product characteristic response could be a "GUI menu for operator." The relationship matrix would score the overall importance of the product characteristic response relative to the weighted "ease of use" row entry. The weighting comes from the stakeholder relative preference rating. At the bottom of the house of quality are the technical importance ratings, which are the column sums. A technical competitive analysis along with requirements-based target values for each technical response (product characteristic column entry) can also be provided. In this fashion, the QFD provides quite a lot of useful information that can be quickly communicated to the stakeholders or the development team.

Once the MOPs, TPM&Ms, and KPPs have been established, trade studies and/or feasibility studies are performed. The system engineer should focus on at least three fundamental feasibility studies: cost, technical, and schedule. Feasibility can be considered the complement of risk. The more feasible something is, the less risky it is and vice versa. Cost feasibility is an assessment of the capability to complete the requirement, task, or activity given the allocated funds. Technical feasibility assesses the state of technology and determines if the requirement, task, or activity can be accomplished with the available technologies. Schedule feasibility is an assessment if the requirement, task, or activity can be completed within the allowed timeframe. If a requirement, task, or activity is found to be unfeasible or have low feasibility, a risk assessment should be accomplished to assess and document the risk. Risk management methods should be employed to mitigate the risk if possible. Trade studies occur when there is a trade-off decision to be made. For example, choosing the type of sensor used in an imaging application, determining the type of data exfiltration mechanism, or choosing the best imaging system design can be accomplished by a trade study.

As part of the optical systems design block, a functional analysis is conducted. The functional analysis identifies the necessary functions of each subsystem within the system. FBDs and FFBDs are used to describe the necessary subsystems that are generated from analyzing the system-level requirements. The FBDs are used to define and expand the next lower design level, in this case, the subsystem level. FFBDs are used to illustrate processes or activities wherein sequencing and order are important. The functional analysis includes the FFBD where the functional blocks are linked showing how the inputs and outputs of each system element flow within the system design (USAF 1969). For FBDs and FFBDs, functions should have a high-level evaluation of the necessary inputs, the function's constraints, the resources required for the function, and the function's output as illustrated in Figure 3.7.

Each block in a given FBD should be evaluated to ensure that interfaces between functional blocks are properly considered, resources are available to implement the function, all constraints are considered, and there are no missing requirements for the function. As an example, a function for a spaceborne imaging system would require special testing facilities such as shock and vibration testing, thermal cycle testing, radiation testing, and vacuum testing. After the functional analysis is completed, the requirements allocation process commences wherein the functional blocks generated during the functional analysis decomposition are partitioned and requirements are allocated, apportioned, or derived into the lower-level functional blocks and captured in the lower-level requirements document (B specifications).

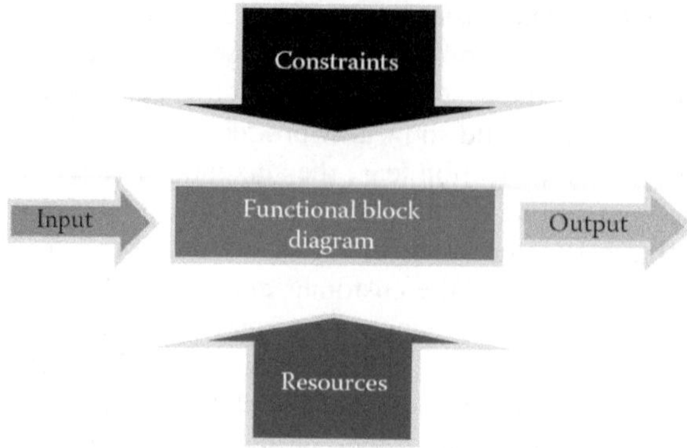

FIGURE 3.7 Analysis mechanism for functional block diagrams (FBDs) and functional flow block diagrams. (Courtesy of Jeffe Baudek. With permission.)

Further along, in the system design and analysis, the individual system components are developed through an iterative process of model/design, analysis, trade-off, and test. This design process seeks to design and evaluate all alternatives to the design with respect to all of the "ilities": reliability, maintainability, usability, supportability, transportability, flexibility, availability, survivability, testability, producibility, disposability, affordability, and so forth. The result of the system design is a set of components or elements that are ready to be integrated into a system.

In the optical design integration block, the individual subsystem designs are integrated into the overall system design. The integrated design is reviewed and expressed in a lower set of requirements documents, the B specifications. A formal review, the PDR, evaluates the preliminary design and associated requirements documents, and, after a successful review, this marks the end of the preliminary design phase. The "boxed" process shown in Figure 3.5 repeats for the next, more detailed, level of the design, the detailed design phase. This phase starts with the B specification requirements and proceeds in the same fashion as in the preliminary design phase. Namely, the B specifications are used to drive the functional analysis for the lower-level design. After the functional decomposition, requirements allocation is used to derive, allocate, or apportion B spec level requirements to the most detailed level — the component level. The allocated requirements are captured in the product specifications (C specs), process specifications (D specs), and material specifications (E specs), and all are baselined and reviewed at the CDR. Once the design documents are all completed, the optical system is built in the optical system production, construction, and manufacturing block. The optical system then undergoes verification testing for the components and integrated components. Testing starts small with the individual components to meet the lower-level detailed design requirements (C specs). Once component testing is completed and all the components passed the tests, then the components are integrated into higher assemblies (subsystems) and tested against the preliminary design requirements (B specs). The optical systems test block completes with the completed optical system being tested against the A specification requirements and the optical system passing validation testing in the field at OT&E.

In the optical systems operations, maintenance, support, and service block of Figure 3.5, optical systems performance is linked to the reliability, maintenance, and logistical models developed during the conceptual, preliminary, and detailed design phases. During this block, the maintenance infrastructure is implemented and field testing data are compiled and analyzed against the expected models to ensure that the system design is behaving as expected (Rausch et al. 2007). If the maintenance and support design is not meeting requirements, then changes may have to be made to the system design or the maintenance and support models to ensure that system requirements are being satisfied. This too is an iterative process that consists of evaluating field data to ensure that systems requirements are being satisfied.

Finally, the system retirement and disposal process can often be overlooked within the system design, but it is extremely important. Today's environmental concerns and regulations force good stewardship of the materials used in manufacturing. The technicians, material, and logistic support required to safely disassemble, transport, and dispose of the system components can be a significant cost driver that must not be overlooked in the overall system design.

The SDLC process model depicted in Figure 3.5 is a highly iterative and multidisciplinary process that forces the system engineer to contemplate and plan for many contingencies. The SDLC process starts at the conceptual 30,000 ft level of what problem the system is intended to solve and finishes with a fully functional system that may or may not suffer eventual decommissioning and disposal.

3.2 OPTICAL SYSTEMS BUILDING BLOCK: OPTICAL SoS MODEL

In the following section, an SoS model will be described; this encompasses the overall technical components that interact to form an optical system. Similarly to how the previous section was able to fashion the SoS model, and through its use depict the interactions of the constituent members found within the enterprise architecture model, the intent of the technical SoS model will be to unravel the interactions found within a specific optical system. Shown within Figure 3.8 is a representation of a general-purpose imaging system, depicting its technical-oriented SoS model.

This model may be refined and modified or replaced with other models as necessary to be used with other optical applications such as the following: active imaging (e.g., laser-based imaging), space-based imaging, medical imaging (e.g., magnetic resonance imaging), airborne imaging, optical tomography, spectroscopy, and underwater imaging. An example of a modification to the model required to represent space-based imaging would be to change the intervening medium to a vacuum, instead of atmosphere. Similar modifications become necessary to properly model underwater imaging, with the intervening medium being again changed, this time to water, with the collecting aperture located either in or outside of the water. For an airborne imaging system that is looking at an object in the water, the intervening media would be two-layered. The target would be immersed in the water and the immediate intervening media surrounding the target would be water. However, between the surface of the water and the airborne collecting aperture, the intervening media would be air. For a wide variety of optical detection scenarios, the object is at a remote location from the optical detection system and the intervening media is air. As will be seen in subsequent chapters — especially the chapter discussing the effects of the atmosphere on optical systems — the properties of the intervening media can dominantly affect the performance of the

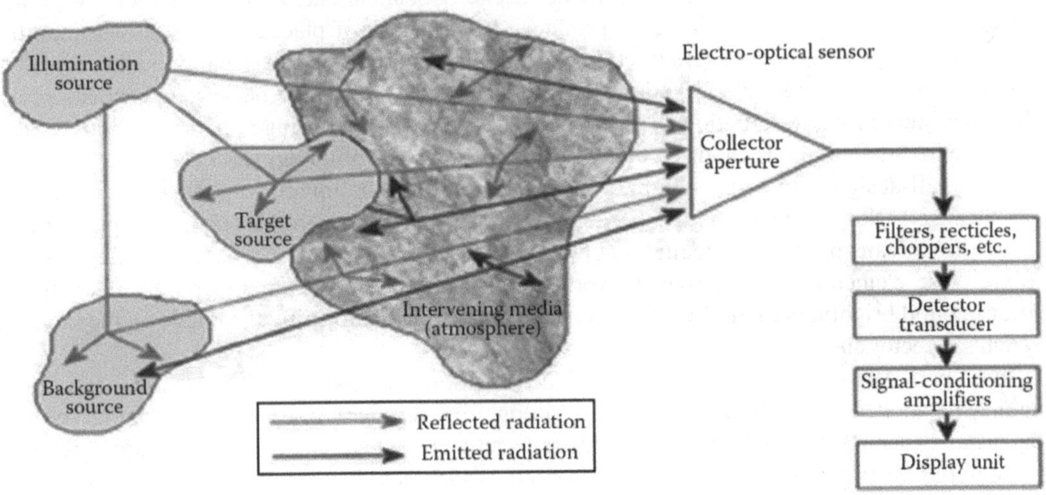

FIGURE 3.8 General-purpose physical SoS model for analyzing an optical system.

optical system. For instance, in well-designed optical imaging systems that operate in the visible part of the electromagnetic spectrum, the atmosphere is often the dominant effect that contributes to the loss of spatial resolution of the optical imaging system (Roggemann and Welsh 1996).

We now describe the essential elements of our optical SoS. Why SoS? We use this terminology to illustrate the multidisciplinary nature of the optical detection problem. For instance, there are a variety of excellent imaging system manufacturers that develop complete imaging systems but do not overly concern themselves with the effects of the atmosphere on the resulting optical image that is captured by their respective imaging system. For example, nearly all camera systems sold to the average consumer do nothing to remove the effects of higher-order atmospheric turbulence from the captured image even though atmospheric turbulence is often the dominant factor affecting spatial resolution in well-designed consumer cameras with sufficiently large apertures. Some cameras come with built-in image stabilization software that corrects low-order atmospheric turbulence – specifically atmospheric tilt. These imaging systems can be considered as "systems" in their own right. We have atmospheric turbulence compensation/adaptive optics systems; beam shaping, beam forming, and relaying optical systems; environmental control and conditioning systems; data communication systems; image processing and analysis systems; signal conditioning systems; command and control systems; power systems; data exfiltration systems; mission planning systems; support systems; and service systems. Also, the sources of optical radiation and the intervening media can often be viewed as systems. The interrelationships of all of these systems are quite complex and require multidisciplinary knowledge and expertise. Additionally, when all of these systems act together, they provide a capability that is greater than the sum of its parts (e.g., a nearly diffraction-limited image that provides higher than conventional spatial resolution to the interested observer). For these reasons, we think of the physical scenario in terms of interacting systems — or an SoS.

In Figure 3.8, the illumination source at the top left can either be a passive source such as the Sun, Moon, starlight, or infrared (IR) heat emitted from a target, or it can be an active source like a laser, flood light, thermal lamp, or other man-made source. Although the illumination source is shown at the top-left corner, this does not mean that there are geometrical restrictions on the relative position of the illumination source to the target or detector. Rather, the illumination source block of the figure illustrates the possible need for a source of illumination. The green lines in Figure 3.8 show reflected radiation, and the red lines show emitted radiation. The background radiation is radiation that is detected by our optical detection system that is generally from sources other than the one we are interested in. For instance, in an IR imaging system, other heat-producing sources in the imaging system's field of view such as hot pipes, walls of a heated structure, motors, or other living beings produce IR radiation that provides scene-dependent background radiation. As shown in Figure 3.9, the dog acting as a target produces IR radiation; if placed in a scene with another target, the optical detection system would pick up both signals. The intervening media will often scatter, absorb, and refract the radiation from the source as well as the background radiation and will sometimes emit new radiation at different wavelengths. The resulting radiation — both wanted and unwanted radiation – is captured by the collecting aperture of the optical system.

For a well-designed imaging system, the entrance pupil to the imaging system is the collecting aperture itself. Next in the system, subsequent to its entrance through the pupil, the optical beam travels through an application-specific set of optical elements and components. In an imaging application, these elements and components work to shape the optical beam, essentially relaying the entrance pupil brightness to the optical system's image plane and forming the image on the imaging system's detector elements.

To properly shape the spectral response of the radiation prior to it arriving at the detector, filtering may be applied to the incident radiation, manipulating its characteristics throughout its spectral regions and shaping it to include only the desired sections while attenuating the unwanted regions, by applying filter elements such as narrowband, wideband, band-pass, band-reject, low-pass, high-pass, notch, and adaptive filters. Optical hardware such as turning mirrors, corner cubes, beam splitters, lenses, and prisms are used to apply 2D Fourier transformations of the radiation in

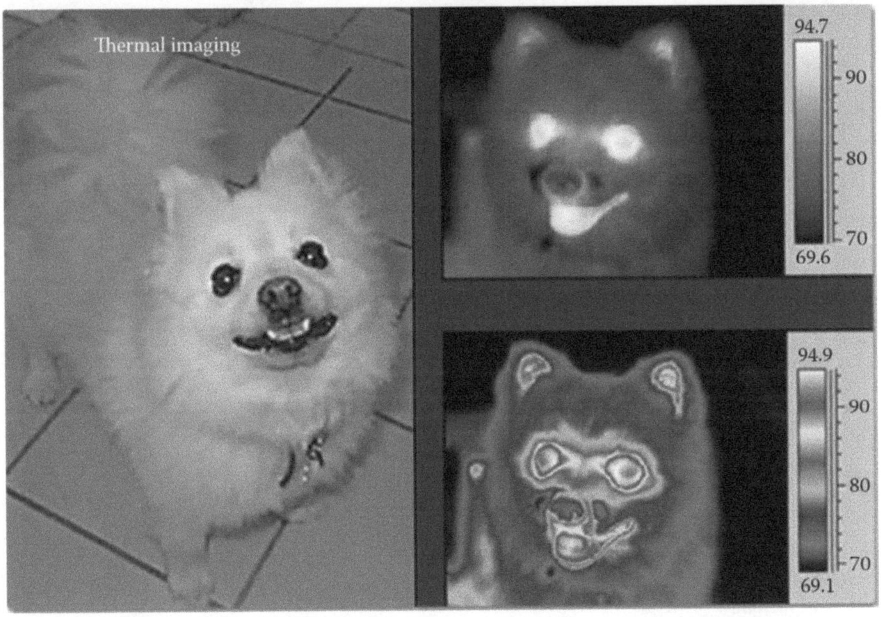

FIGURE 3.9 IR images of a dog. (Reprinted from NASA. With permission.)

the entrance pupil and map the result upon an image plane of the small detector found within the optical system. Neutral density filters can be used to control the amount of radiation that falls on the detector to ensure that the detector does not become saturated. This is similar to putting on a pair of sunshades when first walking outdoors to reduce the brightness on the eyes. Items such as reticles and choppers are used to help identify targets of interest (TOIs) and will be discussed in a later chapter. The detector in Figure 3.8 is shown as an electro-optical detector, but the model is not restricted to electro-optical detectors alone since other transducers are equally valid. Electronic transducer hardware is used to detect the change in a specific detectable physical quantity (e.g., pressure, brightness, temperature, moisture, and electromagnetic waves), by converting the energy from the physical change into a much-easier-to-quantify electrical signal. Hence, photoelectric, chemical, mechanical, and thermal optical detectors all fit within our optical SoS paradigm. The output of the detector undergoes various application-specific signal processing prior to appearing on an image/data display. The following tasks would be completed during signal processing as needed: signal conditioning, encryption or decryption, filtering, imagery, and/or signal analysis. Roles and specifics for each of these model elements will be explored deeper within future chapters.

Next, let us further evaluate the radiation in our optical system, in terms of the radiation's source. To better illustrate this, a space-based optical system is depicted in Figure 3.10 as an example of an optical system, along with the various corresponding sources of radiation that would arrive at its entrance pupil plane.

We are assuming that the sun is the illuminating source. The TOIs are the small objects in the water (fish). The signal of interest (SOI) is indicated by the line that represents the reflected sunlight off of our TOI that is captured by the field of view of our optical system. Part of the energy of our illuminating source is transmitted into the water and is subsequently reflected off of our TOI in the direction somewhat toward our optical system. At the water/air interface, the reflected light from our TOI is refracted in such a fashion that the propagation direction of the electromagnetic field associated with it is within the instantaneous field of view (IFOV) of the space-based optical system. Notice that we are focusing on the radiation that is captured by our optical system and are not showing reflected or emitted radiation that is outside our optical system's IFOV. All other radiation sources are considered background radiation, such as the segment that shows the reflected light off

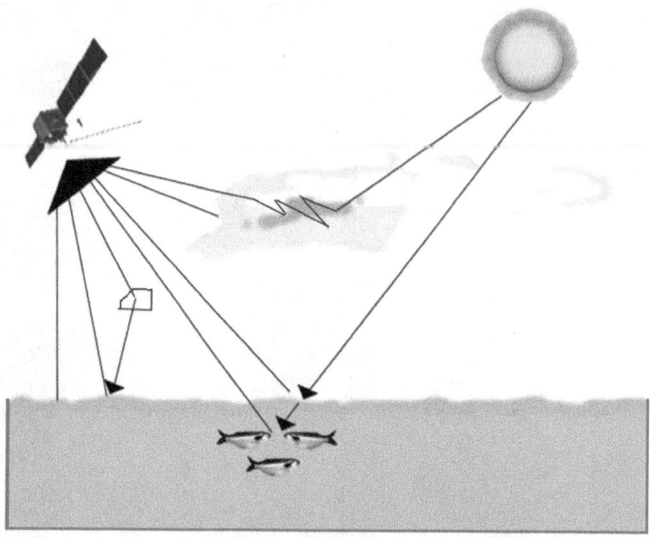

FIGURE 3.10 Sources of radiation. (From California Department of Fish and Wildlife, Marine sportfish identification: Other fishes 2014.)

the surface of the water that is reflected into the IFOV of our optical system. Another line segment shows sunlight that is scattered by the earth's atmosphere into the IFOV of our optical system. As will be seen later, the atmosphere itself can generate radiation through absorption and emission processes. Energy at one wavelength can be absorbed and reradiated at another wavelength. A line segment shows this generated radiation that is emitted in the direction of our optical system. Clouds can often be an issue with optical imaging systems since they generate radiation that can sometimes interfere with the optical systems' ability to detect the desired TOI. As will be seen later, optical modulation techniques and clutter reduction algorithms are used to often mitigate the effects of unwanted radiation on an optical system. Another segment in Figure 3.10 represents the emitted radiation from clouds intercepted by our optical system. There is also a segment that shows the energy from the clouds that reflect off the surface of the water into the IFOV of our optical system. Finally, there is a segment that shows the radiation that is emitted from the water itself and that is intercepted by the space-based optical system. As can be seen in Figure 3.10, there are a variety of unwanted signals that clutter the IFOV of our space-based optical system example. Additionally, as we learned in the statistical optics section of the preceding chapter, the desired signal can itself have a random component due to the random arrival rates of the photons themselves. These effects cannot be disregarded and instead must be taken into account and incorporated accordingly into the optical system's model. Furthermore, the models depicted within Figures 3.8 and 3.10 could be used to represent other scenarios; with some easily made adaptations, they can be made to apply as general use. A possible adaptation would be to consider a ground-to-ground remote imaging example, depicted by the following perimeter surveillance OV-1 (Figure 3.11).

In the depicted perimeter surveillance mode OV-1, there is a telescope with a digital camera connected to it, as well as a laptop with software installed to handle atmospheric turbulence compensation algorithms.

An example application could be illustrated using a U.S. Embassy located within foreign borders; the telescope would then be focused via local or remote means so that clarity is maximized for a particular *area of interest* (AOI). The AOI would be defined such that any relevant data produced from the telescope are able to satisfy its intended mission, for example, in the case of the U.S. Embassy example, the AOI would be suspicious people outside of the embassy. To provide the desired coverage of this AOI, the ground-to-ground optical imaging system would capture photos

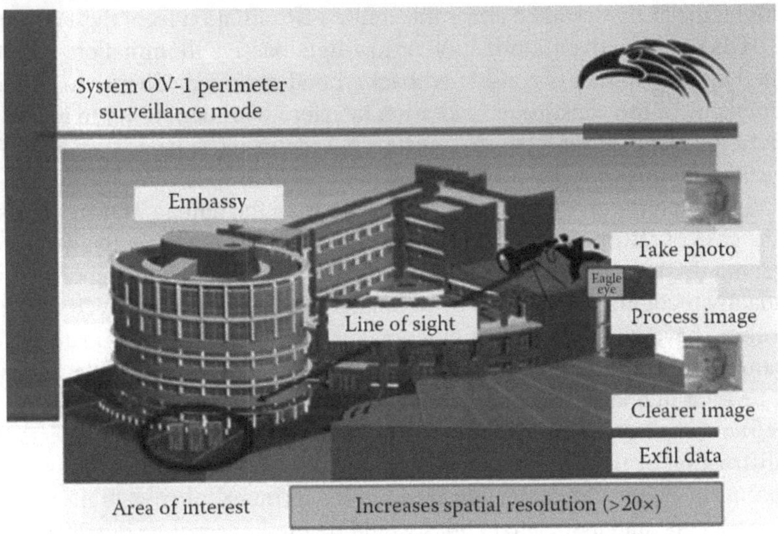

FIGURE 3.11 Perimeter surveillance mode OV-1 with atmospheric turbulence compensating imaging system and data exfiltration.

of the AOI using the telescope. However, for sufficient aperture sizes, the direct resulting image would be much too blurry, due to atmospheric turbulence causing persons within the image to have insufficient spatial detail for subsequent characterization and/or identification purposes, thus not providing usable information for the U.S. Embassy's intended surveillance. Atmospheric turbulence compensating software resident on the laptop computer removes the atmospheric turbulence and produces an image closer to the best possible image — the diffraction-limited image. The amount of improvement, measured in terms of spatial resolution, increases proportionally with the aperture size of the telescope used with the imaging system. For example, if a telescope was built with its primary mirror size being 8.5 in. (e.g., as viewed through an 8.5 in. telescope), the resulting resolution of the image after atmospheric turbulence compensation is applied would be improved by at least 20 times in comparison to the original uncompensated image.

When this ground-to-ground imaging scenario in terms of Figure 3.8 occurs, we see that the illumination source during the day is the sun. At night, the illumination source could be lights that are strategically located around the embassy. If there are no such lights, then low-light visible imaging technologies, active imaging methods, or IR methods could be used. It should be noted that switching to an IR imaging technology increases the wavelength of the observed light and consequently reduces the spatial resolution of the imaging system. Careful analysis needs to be accomplished to determine whether or not the gains in resolution made by implementing an adaptive optics or atmospheric turbulence compensation methodology offset the loss of resolution due to the increased wavelength from switching from a visible sensor to an IR sensor. All significant sources of radiation that emit in the spectral band of the detector need to be accounted for in developing the imaging model. The target sources in Figure 3.8 are the individuals in the AOI. The source radiation would be considered incoherent whether it was reflected light in the visible part of the electromagnetic spectrum or IR radiation in the IR part of the electromagnetic spectrum. Recall that even if an active source (such as an eye-safe laser) is used to illuminate the AOI – in order to provide a controllable amount of measurable light at the optical detector — the reflected light is considered spatially incoherent since the reflecting surface is optically rough. Therefore, an incoherent imaging model is appropriate for this imaging scenario. The background radiation in Figure 3.8 is all the radiation that is within the imaging system's IFOV that is not due to our intended objects — for example, the suspicious people in front of the embassy. In daylight, our SOIs would be the reflected light from our TOIs. At night, the SOIs would be either the emitted IR radiation from the TOIs or the reflected light from the TOIs due to active illumination or low-light passive illumination. All other radiation captured by the imaging system is considered background radiation. The term clutter can be used when the components of the background radiation interfere with our ability to locate, isolate, classify, characterize, or identify our TOIs. For instance, identifying a person wearing forest camouflage in foliage with an imaging system operating in the visible spectrum, the foliage is considered clutter since it visibly interferes with the ability of separating the human TOI from its surroundings. Similarly, in the case of IR imaging, heat sources that radiate at the same temperature as a human and are in the same IFOV could be considered clutter. Clutter is especially important when attempting to automatically recognize the TOI as in automatic target recognition (ATR) algorithms. Many of these algorithms have clutter rejection components that help them eliminate false targets and help, as the name implies, to reject the clutter in the scene. In Figure 3.8, the intervening media is air. As will be seen in a later section, the atmosphere scatters, absorbs, emits at different wavelengths, and refracts an electromagnetic wave. The scattering and absorption effects, in addition to the ordinary diffraction of the electromagnetic field itself, lead to the attenuation of our SOI. The collecting aperture in Figure 3.8 is the entrance pupil – primary mirror — of our telescope. The filters, reticles, choppers, and associated optics would be used to separate our TOIs from the clutter and place the resulting image on the detector in the optical system's image plane. The output of the detector is typically a low-level current or low voltage that has to be amplified, possibly denoised, and further conditioned to interact with processing and/or display hardware. In our example case, the detector output must also be converted to digital form to be used by the atmospheric turbulence compensating algorithms that reside on the laptop computer. The signal amplifiers from the detector output must amplify the signal so that it is in the range of the analog-to-digital (A/D) converters. The image is then processed to remove the atmospheric turbulence and provide a high-resolution atmospheric turbulence-compensated image. The image may then be encrypted and sent to a separate facility (e.g., exfiltrated) where it is decrypted and evaluated for possible further action.

Our perimeter surveillance OV-1 can also be evaluated from the perspective of an adapted Figure 3.10. In our case, we do not have a TOI in the water and neither is our detector in space. Even so, the different segments can still be evaluated with some adaptations as follows. During the day, the sun is still the primary passive illumination source. Unless there is a pool of water in the AOI of the imaging system and the sun is positioned in such a way that the specular reflection off the water is in the IFOV of the optical imaging system, the direct reflection of the water into the camera as indicated by the ray from the sun, to the water, and up through the path to the imaging system does not apply in the perimeter surveillance scenario. Instead, the equivalent path in our perimeter surveillance application involves the reflection of sunlight off of the ground and all the structures that are subsequently captured by the imaging systems IFOV. Since this application has relatively short range and the angle of the optical system is toward the ground and not the sky, refracted direct illumination from the sun, direct emission from the atmosphere and clouds, and reflected radiation from the ground and from the clouds, as in all corresponding line segments in Figure 3.11, are not significant factors. The last segment we discussed shows direct emissions from the water, which in our case correspond to the ground and surrounding structures that are captured by the IFOV of the optical imaging system. At night, depending on the thermal emission properties of the structures and emitting objects in the imaging systems IFOV, this may or may not be a significant factor. During the day, however, the solar reflection off the background structures and TOIs alike typically dominate over emitted radiation for our typical perimeter surveillance scenario. The segment that reflects light from our TOI that is captured by the imaging system is the radiation that is of most interest. The reflected sunlight/moonlight/starlight or thermal emissions off of the background structures are also of consequence and must be considered in developing the imaging systems model. Further, we are interested in the refractive properties of the atmosphere that affect the TOI reflected electromagnetic fields and that are received by the optical system. At this point, an example may be helpful.

If our optical system in the perimeter surveillance scenario had an 8.5 in. telescope, the entrance pupil would measure 8.5 in. in diameter. The best possible resolution that an optical imaging system attached to this telescope could provide would be the classical diffraction limit

$$\Delta x_d = 1.22\frac{\lambda}{D}z, \tag{3.1}$$

or

$$\Delta x_d = \frac{\lambda}{D}z, \tag{3.2}$$

where

Factor 1.22 in Equation 3.1 is due to a circular aperture

λ is the center wavelength of the light in meters received by the optical system

D is the diameter of the entrance pupil telescope in meters (in this case 8.5 in. or 0.2159 m)

z is the separation distance of the entrance pupil in meters from our TOIs

Equation 3.2 holds if the entrance pupil of the telescope was square instead of round. For the case of narrowband optical illumination centered at 500 nm and a separation of 1.5 km using a circular primary mirror on the telescope, we would arrive at a diffraction-limited resolution of about 4.24 mm. In the case of a square primary mirror with an 8.5 in. edge length, the diffraction-limited resolution of a well-designed optical system would be about 3.47 mm. These spatial resolution expressions can be considered as an upper limit on the performance of a well-designed optical system in vacuum. Unfortunately, in the case of our example optical system, we are imaging through the atmosphere, and without atmospheric turbulence compensation, the atmosphere is often the limiting factor on an optical system's spatial resolution performance. To quantify the effects of the atmosphere on our imaging system, Dave Fried developed the atmospheric coherence length, or Fried parameter (r_0), that provides a metric as to the seeing quality of the atmosphere (Quirrenbach 2000, pp. 76–78). The Fried parameter can be considered the diameter for which atmospheric turbulence becomes a consideration in the optical systems performance. Entrance pupil diameters that exceed the Fried parameter have their spatial resolution affected by the atmosphere, whereas entrance pupil diameters that are smaller than the Fried parameter have their spatial resolution relatively unaffected by the atmosphere. A set of equivalent expressions for the spatial resolution of the optical imaging system that use the Fried parameter are given by

$$\Delta x_a = 1.22\frac{\lambda}{r_0}z, \tag{3.3}$$

or

$$\Delta x_a = \frac{\lambda}{r_0}z. \tag{3.4}$$

In these equations, the diameter of the entrance pupil of the telescope is replaced with the Fried parameter. Typical values of the Fried parameter for ground-to-ground imaging systems are on the order of 1–4 cm. In our scenario, choosing a Fried parameter of 4 cm for good viewing conditions, the actual spatial resolution for our optical imaging system in the presence of atmospheric turbulence is 0.022875 and 0.01875 m for the square telescope primary mirror. Spatial features that are smaller in scale than the calculated spatial resolutions cannot be resolved by our imaging system. Using the example for the circular telescope aperture discussed previously, we see that facial features that are smaller than 2.2875 cm practically cannot be distinguished by our optical system. The face would be blurry and features such as the eyes and other notable features that are smaller than 2.2875 cm would not be observable. On the other hand, if the atmospheric turbulence was removed from the image, then the spatial resolution for the circular entrance pupil would be on the order of 4.24 mm and much more of the face would become resolvable. If we look at the case for bad viewing, as in a Fried parameter of 1 cm, the resolution increase is even more pronounced. The actual resolution of the optical imaging system is of 0.0915 m for the circular telescope entrance pupil aperture and 0.075 m for the square entrance pupil aperture. The upper limit for the spatial resolution improvement achievable by removing the effects of atmospheric turbulence from the image is given by dividing the expression for the spatial resolution with atmosphere present, Equation 3.3 or 3.4, and by the diffraction-limited resolution of the optical imaging system, Equation 3.1 or 3.2. The result for telescopes with either circular entrance pupil or square entrance pupil is given by

$$\Gamma = \frac{\Delta x_a}{\Delta x_d} = \frac{D}{r_0}. \tag{3.5}$$

For our 8.5 in. telescope, the spatial resolution improvement for good seeing is over a factor of 5, whereas for bad seeing, it is upward of 22. Note that in the presence of atmospheric turbulence, without atmospheric turbulence compensation, telescope diameters larger than the Fried parameter do not produce higher spatial resolution. With atmospheric turbulence compensation methods applied, the upper bound on spatial resolution scales with the telescope entrance pupil diameter.

A crucial limiting factor that must be heeded during optical imaging system design is the spatial resolution attained by the optical imaging system, which must be larger than the detector's pixel size when projected onto the object plane. For instance, in our perimeter surveillance example with a circular aperture, assuming square pixels and in the case of uncompensated atmospheric turbulence with bad viewing (Fried parameter is 1 cm), the projected pixels must be lesser than 0.0915 m by 0.0915 m. In the case of compensated atmospheric turbulence, the projected pixel size must be less than 0.00424 m by 0.00424 m. If this is not the case, the spatial resolution performance of the optical imaging system will be limited by the pixel size instead of attaining the near diffraction-limited imaging performance of the well-designed optical system (Motorola White Paper 2012). Figure 3.12 is useful in understanding the projected pixel size requirement for our perimeter surveillance scenario. This figure is a general model of an optical system. Shown on the left is the IFOV, which consists of all the pixels of the optical detector, projected onto the object plane. The half angles of the IFOV are given by $m_l * \alpha_l$ and $m_t * \alpha_t$ corresponding to the longitudinal and transverse directions. The parameters m_l and m_t are the longitudinal and transverse magnifications, respectively, and are determined by the specific optical system. We will discuss this more in Chapter 4. For now, these are treated as selectable parameters over a certain range (say 1–25). In typical cases, the IFOV is determined with the assumption that the lateral and transverse magnifications are set to one and then adjusted with respect to the specific magnification, within the acceptable range of possibilities that is required for the specific imaging scenario.

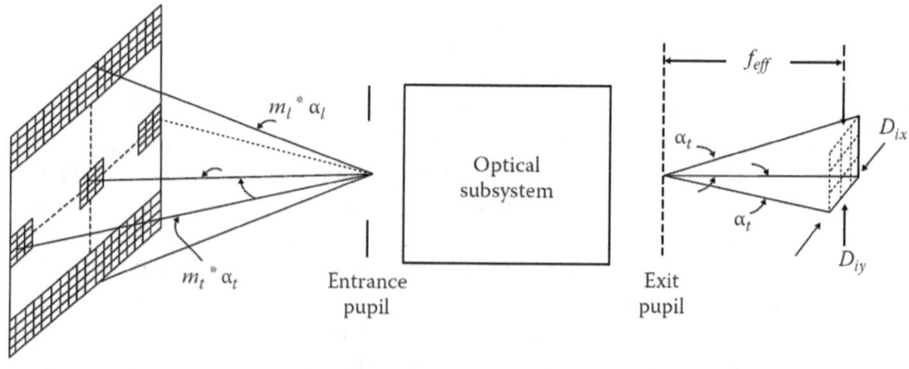

FIGURE 3.12 General model for an optical subsystem as part of an imaging system.

For instance, if the imaging scenario requires the highest resolution possible, then the magnifications are both set to one and the IFOV becomes as small as possible. If the imaging scenario requires wide-area surveillance, then the magnification is set to the highest value (say 25) and the IFOV becomes as wide as possible. The entrance pupil of the telescope is shown to the left of the box labeled "Optical System." The entrance pupil plane is of course defined by the optical system itself, but it is shown in Figure 3.12 as a point of reference. The "Optical System" box treats the optical system as a black box, representing a general optical system. We will use the techniques in our "Basic Optics" chapter section later in the book to describe the specifics of the optical system. For now, we treat it as a "black box" and are only concerned with the entrance pupil and the exit pupil. From the exit pupil, we have the effective focal length given by f_{eff}, which gives the distance from the exit pupil to the image plane. The detector element is placed in the image plane. The backplane transverse and longitudinal angles are given by, respectively

$$\alpha_t = \tan^{-1}\frac{D_{ix}}{2f_{\text{eff}}}, \tag{3.6}$$

and

$$\alpha_l = \tan^{-1}\frac{D_{iy}}{2f_{\text{eff}}}, \tag{3.7}$$

where

D_{ix} and D_{iy} are the transverse and longitudinal dimensions of the detector element itself, respectively

f_{eff} is the effective focal length

For a fill factor of 100%, the entire face of the detector area is filled with pixels. This is not always the case since some detectors have their pixel readout electronics connected to the edge of the pixels. Other manufacturers place the readout electronics at the bottom of the pixel and consequently are able to reach very high fill factor percentages. The actual dimensions of the pixels and associated fill factor are determined by the detector element manufacturer, and the optical system is often designed to work with the given detector element. In terms of the detector-limited spatial resolution, the pixel density is an important parameter. If a given detector has n by m pixels in the transverse and longitudinal directions, respectively, in the case of detector-limited spatial resolution, the pixels end up limiting the performance of the optical system. Since this is a consequence of inadequate optical system designing, it can be countered with the application of good SE coupled with effective optical systems design practices. An example of this kind of design error is if the projected pixel size on the target plane is larger than the diffraction-limited resolution (in the atmospheric turbulence-compensated imaging case) or larger than the Fried parameter-based resolution obtained when imaging in the presence of atmospheric turbulence. In both cases, the spatial resolution is determined by the pixels and not by other aspects of the optical system or atmosphere itself. In proper optical systems design, the optical system is designed and applied in a fashion that the pixel sampling in the object plane is smaller than the intended spatial resolution given by the optical system. Figure 3.13 provides a 2D view of the projected pixel in the object plane.

In order to calculate spatial sampling in the object plane, both the transverse and longitudinal directions' projected pixel size needs to be determined. In Figure 3.12, we look at one dimension of Figure 3.13 with an expanded view of the pixels in the image plane (right side of the image) and also the plane of the TOI (left side of the image). Let us assume that in this case, the pixels are located along the longitudinal axis of our optical system. If the detector element manufacturer had 4096 pixels in the longitudinal direction and 4096 in the transverse direction with a 100% fill factor, then we would have square pixels. If the detector element has an active area of 1 cm by 1 cm, then the 4096 by 4096 pixels would be uniformly spread across this area. In the longitudinal direction, we would have 4096 pixels as illustrated at the right of Figure 3.13. The figure is exaggerated to show

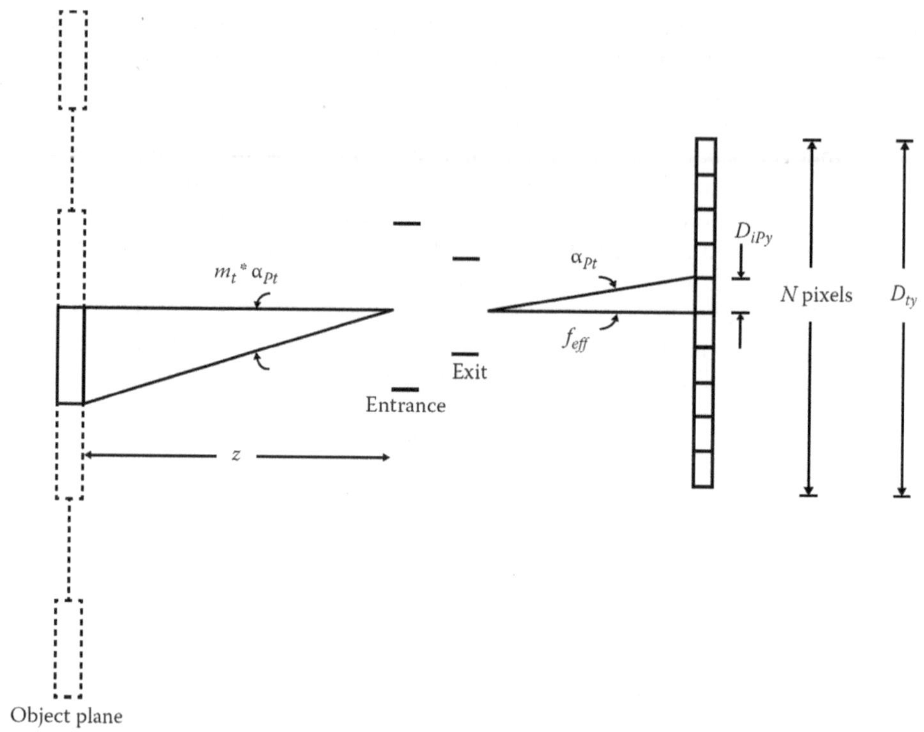

FIGURE 3.13 Projection of pixel size into object space in one linear direction (e.g., horizontal).

the relevant geometry. In our example, the linear extent of the detector element in the longitudinal direction is 1 cm. We would have all the pixels adjacent to each other across the 1 cm active area of the detector. Each pixel would have a length of 1/4096 cm per pixel or about 2.44 μm per pixel in the longitudinal direction. The angle α_{pl} can be determined from

$$\alpha_{pl} = \tan^{-1}\frac{D_{iPy}}{2f_{eff}}, \tag{3.8}$$

where

D_{iPy} is the length of the pixel in the longitudinal direction

f_{eff} is the effective focal length

Once we know the angle α_{pl}, we can determine the size of the projected pixel in the object plane by noticing that the angle subtended by the projected pixel is just the angle α_{pl} multiplied by the optical systems longitudinal magnification, m_l. We can determine the length of the projected pixel in the object plane from

$$D_{opy} = z\tan\left(m_l * \alpha_{pl}\right), \tag{3.9}$$

where D_{opy} is the length of the projected pixel in the object plane, z is the separation of the object plane from the entrance pupil, and the other parameters are as before. This analysis has to be repeated for the transverse direction. In our example, we have the same number of pixels in the transverse direction as in the longitudinal direction, so the analysis will be identical to what we have just done. In general, we would replace the subscripts l and y with t and x, respectively, in Equations 3.8 and 3.9 and Figure 3.13 and repeat the preceding analysis. We can see that the further away the object plane is from the entrance pupil, the larger the projected pixel will be. If we want to project the smallest pixel size possible, then we use a longitudinal magnification of one. We can also decrease α_{pl} by increasing the effective focal length and/or, if possible, choosing a detector that has a smaller pixel size. Typically, once chosen, the effective focal length and the pixel size are fixed for a generic, simple optical system. As an example, if we used the pixel length of 2.44 μm with a focal length of 0.5 m, the angle α_{pl} would be 2.44 μrad. Using a longitudinal magnification of 1 in order to get the smallest resolution possible, we would get a projected pixel size of about 3.7 mm for objects that are 1,500 m away. We see that this result is less than the previous 4.24 mm upper limit needed for the turbulence-compensated imaging example previously discussed. In this case, the projected pixel size is sufficient to properly sample the TOIs in the object plane. It should also be noted that if we did not have an optical system that was capable of compensating for atmospheric turbulence, then the resolution achievable for our optical system with a circular aperture would remain at 91.5 mm, and the additional sampling capability of our optical system would be of no benefit. In this later case, we would have an atmospheric turbulence-limited spatial resolution, as is often the case.

In summary, we presented a general model for an optical system that, similar to the SoS concept in high-level systems modeling methods, provides a comprehensive model for a system's oriented decomposition of a general optical system. We showed how this model could be applied to a generic optical system to evaluate different types of radiation that may be captured by the optical system. We also defined and distinguished between SOIs, background radiation, TOIs, and clutter. We provided an example using optical imaging perimeter surveillance CONOPS and showed a corresponding OV-1. We described how Equation 3.5 could be used to determine whether or not the optical system would benefit from atmospheric turbulence compensation methods. If the result of evaluating Equation 3.5 is larger than one, then atmospheric turbulence methods would provide some benefit in spatial resolution, whereas if it is less than one, then the optical system's spatial resolution would not benefit from atmospheric turbulence compensation. If atmospheric turbulence compensation is required, then Equations 3.1 and 3.2 provide the best possible resolution attainable by a well-designed optical system. The optical system must be designed so that the projected pixel size in the object plane is slightly smaller than that given by Equation 3.1 or 3.2. No additional benefit will be gained by having projected pixels that are smaller than the diffraction-limited spatial resolution size of the optical system. Doing so would just cause additional processing with no spatial resolution benefit. If atmospheric turbulence compensation methods are not used, then the projected pixel size in the object plane must be slightly less than the size dictated by Equations 3.3 and 3.4. To resolve features in the object plane, the projected pixel size and the spatial resolution of the optical system should be a factor of 10 or less than the size of the feature that needs to be resolved. For instance, to resolve a human object sufficiently to tell that it is human and not some other animal or object, a guideline is that 10 resolution cells should be placed across the human object. For a 2 m tall human object, this means that the spatial resolution required to tell that the object is human should be on the order of 20 cm or less. On the other hand, if we wanted to resolve the color of someone's eyeball, we would need resolution cells on the order of 10 across the approximately 1 cm colored portion of the eye (e.g., a required spatial resolution and object plane projected pixel size of 1 mm). The optical system must be designed to address the required resolution for the particular application, as well as the spatial sampling requirements, and the effects of atmospheric turbulence.

In the next section, we provide an example of the SE and optical systems principles methods and techniques learned in this chapter. We see FIT analyzing the particular design parameters required for the UAV example developed in the preceding section.

3.3 INTEGRATED CASE STUDY: IMPLEMENTING A SoS OPTICAL MODEL

In this section, we continue with our integrated case study and show how the SE principles learned earlier are used in a simulated, realistic technical setting. In this chapter, we see FIT coming to grips with deciding and using the SoS and FoS ideas and applying SE practices to evolve their concepts and technical approach for their primary stakeholder, the Department of Homeland Security (DHS), U.S. Customs and Border Patrol. They are in the midst of an optical propagation analysis of the optical detection scenario. This effort triggered off a series of interactive meetings.

During one of these interactive meeting with the stakeholders, the team identified the problem statement and focused on obtaining and understanding the problem definition. This definition identified what problems DHS CBP was seeking to solve with the system. Along with detailed information gathered at these meetings and from other discussions, the team was able to produce a stakeholder requirements document, which was further analyzed to understand the operational, maintenance, and functional details of the intended optical system.

With a solid understanding of the problem, the team proceeded to perform the needs identification process for the primary stakeholder. The primary stakeholder's need was identified to be a requirement for a general-purpose optical imaging system that was intended to detect illegal bipedal border crossers from an unmanned aerial platform along the U.S.-Mexican border. The intended UAV optical system was in itself part of a larger SoS that included direct line-of-site communications, over-the-horizon communications, a Mission Command Center (MCC), Mobile Control Center (MobCC), distributed information, and deployable border patrol and other assets.

The team worked aggressively to conduct additional interviews with DHS CBP, which yielded some essential information about the UAV platform so that the interfaces between the optical system and the UAV and the UAV capabilities (e.g., pointing and steering, communications, power, size, and weight restrictions) could be identified. The stakeholder requirements have been identified, and preliminary models are in progress and in various states of completion. These models include a radiometric model, a statistical optics model, a Fourier optics model, a linear systems model, and a geometrical optics model. One immediate use of the geometrical optics model was to determine the projected pixel sizes in the object plane for representative imaging sensors. These representative pixel sizes were used in spatial resolution analysis.

The FIT engineering team has also worked on the initial feasibility study of the stakeholder requirements to make sure that the intended optical system meets the system design criteria and expectations envisioned by the DHS. An FBD of the optical system has been generated, FFBDs are in process, and a secondary examination of the optical system has yielded a good look at what the conceptual design should be for the imaging system.

In one of the internal meetings Jennifer O., systems engineer (Jen), and Mr. Tom Phelps, optics expert (Tom), met to discuss a trade study on the potential detectors to be used for the primary sensors of the UAV optical system.

(Monday morning, Jen walks into Tom's office for a technical meeting.)

Jen: "Morning Tom."

Tom: "Good morning, Jen."

Jen: "Thanks for meeting me on such a short notice. Karl asked me to put together a trade study for the detectors, and I figured I could use your knowledge and expertise."

Tom: "Sure, no problem. What you got so far?"

Jen: "So far, I've got a list of sensors for object detection. During my studies, I have learned that every imaging system must include a detector of some form. Most detectors exhibit some degree of nonlinearity in their response to incident radiation. Some detectors, such as photographic film, are intrinsically very nonlinear, while others, such as silicon photodiodes, are quite linear over several orders of magnitude if operated properly. All detectors, however, eventually saturate at high radiation levels or display other nonlinearities which may be either global or local (Barrett and Myers 2004, p. xix)."

Tom: "That's right. Let's go over each list so we can eliminate the obvious ones and then start the trade study from there."

Jen: "Perfect, you read my mind. On the passive list we have…."

Tom: "Sorry to interrupt you, but passive sensors? Although most of these options cost less, they also provide fewer features. We are also at the mercy of the objects themselves, in the case of thermal radiation, providing enough optical power to get a detectable signal. In the visible region, we are dependent on passive sources like the sun to provide enough illumination on the target to be able to see it. We will have to see if we have enough optical power from our Targets of Interest (TOIs) for both day and nighttime operating scenarios. If not, the added complexity of an active source may be justified. Hmm, an interesting one from this category is the stereo imaging camera due to its ability to work well in different types of terrain and its ability to estimate distances given some reference information."

Jen: "Ok. How about optical flow?"

Tom: "Technology is too new to tell. We need something proven and robust for this stakeholder. Leave it in there to see what results we obtain."

Jen: "Alright. As for the active sensors, we have lasers, which have amazing precision and accuracy but add complexity to the system, and may provide size, weight, and power issues (SWaP). Radar is less susceptible to atmospheric effects but is more expensive and doesn't provide comparable spatial resolution. Spectral systems are too costly and complex and may be overkill for this application."

Tom: "I agree."

Jen: "Thanks. At this point we want to evaluate options based on size and weight, cost, range and accuracy. I've taken the liberty of assigning weights to all. Based on our radiometric analysis that is in progress, we can temporarily defer the discussion on passive sensors for a later date and address the more complex active sources now. That way, we will be prepared for the worst-case scenario of not enough optical power from passive sources at our optical receiver to register a detection."

Tom: "Ok. What does it come down to?"

Jen: "It comes down to Optical Flow."

Tom: "It can't be."

Jen: "Yup, gives us the highest score for size, cost, range and accuracy."

Tom: "I'd say we add a reliability column to this trade study and see how long these sensors have been in the market and what's their sustainability has been like."

Jen: "Makes sense. I'll work on adding that into the trade study and schedule another meeting with you. We should be prepared to discuss the results of the radiometric study and answer questions in case we have a meeting this week. Karl suspects Wilford Erasmus from DHS CBP will contact us soon."

Tom: "Hmm, we need some time to finish the radiometric study. Hopefully, it won't come up until we have a chance to complete it. We can't even decide if we can use passive sensors or if we need an active sensing solution until we complete this study. If I find any relevant information I will send it to you."

Jen: "Thanks! Talk to you soon."

(Wednesday morning)

Bill: **(Thinking to self)**, "I don't completely understand our technical approach and how it relates to the larger DHS CBP enterprise… What if our advisory board asks questions about this project? I know, I'll get my experts together and have them explain our proposed system in terms of how it addresses our Stakeholder's requirements."

Bill: "Can you please set up a meeting for today at 15:30 with our Systems Engineers, CTO and CIO. Tell them to bring specifics regarding the technical approach used in our UAV contract. Oh, and have them bring their results of the stakeholder requirements analysis."

(Meeting time, Dr. Smith shows up at 15:45)

Attendees to the meeting include Dr. Bill Smith, CEO (Bill); Mr. Karl Ben, Senior Systems Engineer (Karl); Jennifer O., Systems Engineer (Jen); Mr. Tom Phelps, Optics Expert (Tom); and Mr. Gary Blair, Engineering Architecture (Gary).

Bill: "My apologies for being late. I just got off the phone with our sponsor. Thank you everyone for attending. As you may already know, our Stakeholder is the Department of Homeland Security Customs and Border Protection (DHS CBP). They have identified their need for a system to monitor the US-Mexican border (U.S. Customs and Border Protection 2012). I would like this meeting's focus to be on the overall approach to meet the early-established stakeholder requirements for our UAV contract. First, Karl could you please tell us about the highest-level system breakdown?"

Karl: "Absolutely Sir! As you know, the major systems that concern this project are the sensor systems and communications systems, and the other major systems that also affect the UAV operation are the flight system and power system. The UAV, with our optical system in it, is part of a larger SoS that includes our optical systems, other UAV sensors on multiple UAVs, communications satellites, a Mission Command Center, a Mobile Control Center, deployable assets like helicopters, cars, airplanes, boats, All Terrain Vehicles (ATVs), and assets from other organizations. These major systems work together and are very interdependent and come together to provide a much larger output than just the sum of each one's individual contribution, thus they are a SoS. Working together they enable the UAV to operate and fulfill its mission."

Bill: "Karl, before we get into the details of the stakeholder requirements, can you tell everyone how much of the border the optical system is monitoring? What is the typical climate and terrain in the monitored area? What is the typical population expected to be in the monitored area? Why is this border being monitored?"

Karl: "Sure, no problem. We will monitor at 10 'hot spots' along the border that are each 10 km by 10 km with the border in the middle. These are all away from population centers with variations in terrain indicated by the different landscapes."

Bill: "Will the hot spots being monitored have a physical representation of the border, such as a river or a fence? Are the hot spot locations expected to change over time or are they fixed locations?"

Karl: "The hot spots will likely change over time. There are fence lines, rivers, or natural barriers that separate the countries. We should be prepared to deal with all terrain types along the border (away from population centers)."

Bill: "Tom, I would like to get some background information on the intended optical system. What is the make-up of the optical system, that is, what does and does not lie within our development responsibility for this optical system?"

Tom: "The optical system will consist of the telescope, which is used to change the instantaneous field of view (IFOV), relay optics to image the radiation in the entrance pupil (the telescope irradiance (W/m^2) on the primary mirror) onto the detector focal plane. The detectors themselves cover the visible and IR spectral regions. If we want to passively detect a human's thermal radiation in the IR, the peak wavelength should be roughly 9.5 μm. If we need an active system (say a laser), we will need to choose/build the laser and provide the power stabilization, power, and pointing and tracking circuitry and optics. Alternatively, we may see if we can leverage existing assets that may be on board the UAV. We may possibly need cooling systems for the detectors to reduce noise. We also need filter wheels to apply notch filters, high-pass filters, low-pass filters, perhaps adaptive filters, and optical modulators (for background suppression), vibration and shock isolators. We also need an Environmental Conditioning unit with temperature regulators, moisture and humidity control, dust mitigation, pressurization, and light shields. We will need to develop anti-reflection coatings for the optical surfaces. Also, the atmospheric turbulence compensation (ATC) system or adaptive optics (AO) system, if we need it, requires light captured at least two wavelengths simultaneously - such as an RGB digital camera. The ATC system also requires a general-purpose parallel processing device (GPPP) and these can be miniaturized. If we go with the AO system, we don't need the GPPP but instead need a wave-front sensor, deformable mirror, or microelectromechanical mirror system (MEMS) along with a controller. We also need a controller for the telescope, a controller for the sensors, an on-board signal-processing unit, and UAV communications unit. We also need an on-board image processing capability that has, at a minimum, a means to select the size of the image frames to transmit back to the MCC or MoCC. We also need to design the interface circuitry for our data with the UAV and hook into the UAV power grid. We assume the gimbal pod is supplied as Government Furnished Equipment (GFE) and we need to interface with that. Raytheon manufactures their Gimbal Pod and that should handle our pointing and tracking needs. Finally, we need to encrypt our data and make sure our data and images are formatted to integrate in the DHS/DOD enterprise architecture."

Tom: "For this project an important aspect is the telescope. 'Telescopes may be broadly classified as refracting or reflecting, according to whether lenses or mirrors are used to produce the image. There are, in addition, catadioptric systems that combine refracting and reflecting surfaces. Telescopes may also be distinguished by the erectness or inversion of the final image, and by either a visual or photographic means of observation (Pedrotti and Pedrotti 1987, p. 135)."

Bill: "Great overview of the system. Does the optical system require an on-board storage of the raw optical data?"

Gary: "This is a good idea for high value data. We will include that in our technical approach."

Bill: "Tom, is a separate transmission system required for the optics system or can it piggyback on the UAV's communication system? If it can piggyback, what is the bandwidth available for the optics communication system?"

Tom: "I'm an optics guy. I don't know the communications piece. Karl, any ideas or maybe we need to research to see what is doable."

Karl: "We can interface with the existing communications package on the UAV. We will have to build the interface circuitry for our sensor. Also, the UAV has both C-band for line of site and X-band for long-haul communications so we will have to interface with both these systems. We also will need to adjust our data stream on the fly to meet the bandwidth requirements of the UAV. We will also need to share the bandwidth with other systems and so we will need to work closely with DHS's UAV communications engineers."

Bill: "After the video and images are captured, possibly stored, and transmitted 'real time,' will there be a data reduction/evaluation process?"

Jen: "No, there are no plans for that. Raw data and 'interesting' mission critical data will be passed on to interested users (e.g. DIA, CIA, FBI) in both raw data form and real-time processed form (either video or still images) if we use the ATM/AO system."

Bill: "Do we have enough detectable optical power, or do we need some image intensifiers?"

Tom: [winces], "We are currently doing some research on that, Bill. A little history: image amplification by means of an electronic device was first indicated by Holst et al. in 1934. At that time, vacuum-tube technology did not permit the realization of the required devices. We had to wait until the late 1940s when simulated by Chamberlain through his paper on fluoroscope and fluoroscopy in 1942, image intensification was achieved by Coltman, Tol and Oosterkamp, and Teves with the image intensifier as we know it in its present form (Biberman and Nudelman 1999, p. 150)."

Bill: "All right. Thanks for the image intensifier history, but we need to first understand the radiometry to see whether we even need them. Let's expedite our modeling efforts to see how things look. Do we have any insights on the specific hardware and software that we want to use?"

Karl: "That is a design question, and we aren't there yet. This could be an internally developed system or parts could be commercial-off-the-shelf (COTS). We will have to do a trade study to see which is the best way to proceed."

Bill: "Is the system expected to deliver an image and video independently or can we assume that the system will send either an image or a video at any given time?"

Jen: "Predominantly video. We can select images from the video stream to display. Alternatively, we can go into a mode where it only sends an image at selectable time increments. The adaptive optics system will process video data on streaming data at video rates (30 Hz), or on individually selected images, or not at all."

Bill: "Can we assume that the UAV will predominantly communicate to the command center by means of satellite communications?"

Gary: "Yes. However, it will also communicate with a mobile command center, or both simultaneously (one has control — either mobile or command center depending on mission requirements), the other gets an image stream."

Bill: "Thanks Gary. Now I have a high level of understanding of the technical approach. What methodology are we using to manage the process?"

Karl: "We are using our standard systems engineering processes for developing the system. We are also building the enterprise architecture models so that we can interact with the other entities involved in the DHS Border Patrol CONOPS. In this sense, our optical sensor system is just one of many systems that operate together to provide the DHS with the ability to achieve its border patrol mission. In this way, we are part of DHS Border Patrol's SoS.

Bill: "Are there any constraints on the operating environment?"

Karl: "The stakeholder would like our system to be very adaptable to a variety of different environments. This may be very important for our marketing strategy for secondary stakeholders like downstream customers"

Bill: "Did we receive sufficient high-level technical information from the stakeholder to understand the operational scenario?"

Karl: "Yes, we have sufficient information to refine our feasibility studies and start developing the system-level requirements. Of course, we still need to do some analysis like the radiometric study, but we are ready to proceed. We will schedule some regular technical interchange meetings with the stakeholder to make sure we stay on track."

Bill: "I see there is a need for encryption, what kind of encryption are we talking about?

Karl: "That is correct. We think that 128-bit encryption would suffice, but we will need to confirm this with the client."

Bill: "Let's do this ASAP. We need to tie up some of these loose details before we proceed."

Karl: "Absolutely Bill. I will make a note to contact the client today."

Bill: "How do we plan on measuring our performance?"

Karl: "Well, we have sample values for key metrics provided by the stakeholder's technical team. Also, the stakeholder did provide some expected results in the form of Measures of Performance (MOPs), Key Performance Parameters (KPPs), and Technical Performance Measures and Metrics (TPM&Ms). We need to ensure that our system meets the baselined requirements and these performance metrics."

Bill: "I see here that there is a requirement that our system may be used in other scenarios and that it should be adaptable. Do we have some insight into these other CONOPS?"

Karl: "We were given some information on their Perimeter Surveillance OV-1, and we were able to see one of their systems in operation, and have a good idea what the requirements mean. Our optical sensor system will be designed to easily be implementable as a stand-alone ground-to-ground imaging system. We will have to take this into account during our design process and use modular design principles."

Tom: "Also, we were provided with Automatic Target Recognition algorithms by our primary stakeholder. With these we can simulate the various alternatives and directly estimate our optical system performance."

Bill: "Team, what about the maintenance, upgrades, pre-planned product improvement, service, availability, reliability, and things along that line. We need to understand those requirements. Karl, can you head the effort to identify these for our next meeting, please?"

Karl: "Yes Bill."

Bill: "For our next meeting, let's also have a list of all the risks, and the corresponding planned mitigation strategies, for the medium to high-risk items. I also want to see the preliminary modeling results, especially the radiometric models. That will tell us if we need an active imaging approach or if we can go with the lower cost passive imaging solution. Based on the radiometric modeling results and spatial resolution analysis, we need to identify alternatives to our approach and conduct a trade study."

Bill: "Thank you all for all the feedback at such short notice. Are there any concerns? Is there anything that we need to address? If not, I will have my secretary schedule a follow up meeting next week same time."

3.4 SUMMARY

Through this chapter, we have provided an introduction to SoS and FoS and have described a high-level overview of the SE process. We also provided an introduction to an optical systems model that can, itself, be considered an SoS model. We provided some simple equations for determining the spatial resolution of a circular or square well-designed optical system with or without atmospheric turbulence. We also showed how to determine the projected pixel size in object space for an arbitrary pixelated sensor. Through the integrated case study, we were also able to see some specific considerations involved in understanding high-level attributes of a UAV optical system and how some integral SE concepts relate to the technical understanding and development of the optical system. In the next chapter, we will present some concepts related to model-based SE.

REFERENCES

Barrett, H.H. and K.J. Myers. 2004. *Foundations of Image Science.* Hoboken, NJ: John Wiley & Sons, Inc.

Biberman, L.M. and S. Nudelman. 1999. *Photoelectronic Imaging Devices,* Volume 2: Devices and Their Evaluation. New York: Plenum Press.

Blanchard, S. and W.J. Fabrycky. 2011. *Systems Engineering and Analysis,* 5th edn. Englewood Cliffs, Upper Saddle River, NJ: Pearson Education, Inc.

Boehm, B. and W.J. Hansen. 2000. Spiral development experience, principles, and refinements. *Spiral Development Workshop,* University of Southern California, Los Angeles, CA, February 9, 2000. https://www.sei.cmu.edu/reports/00sr008.pdf (accessed March 30, 2014).

California Department of Fish and Wildlife. 2014. Marine sportfish identification: other fishes. https://www.dfg.ca.gov/marine/mspcont7.asp#tomcod (accessed March 30, 2014).

Evans, S. 1996. *The Deer Hunter's Almanac.* New York: The Hearst Corporation.

Gansler, J.S., W. Lucyshyn, and A. Spiers. 2008. Using spiral development to reduce acquisition cycle times, UMD-AM-08-128. College Park, MD: Center for Public Policy and Private Enterprise. https://apps.dtic.mil/sti/citations/tr/ADA494266 (accessed March 4, 2025).

Harris Corporation. 2012. *Systems Engineering Manual, S-401-001,* Revision 47, Revised November 12, 2012. New York: Harris Corporation, Government Communications Systems Division.

Lorenz, E. 1972. Predictability: does the flap of a butterfly's wings in Brazil set off a tornado in Texas? *139th Meeting of the American Association for the Advancement of Science,* Boston, MA, 29 December 1972. https://mathsciencehistory.com/wp-content/uploads/2020/03/132_kap6_lorenz_artikel_the_butterfly_effect.pdf (accessed March 4, 2025).

Motorola White Paper. 2012. Video surveillance trade-offs, a question of balance: finding the right combination of image quality, frame rate, and bandwidth. https://www.motorolasolutions.com/content/dam/msi/docs/business/solutions/business_solutions/incident_scene_and_event_management/intelligent_video_surveillance_and_control__ivsc_/_documents/staticfiles/ro-99-2175a_video_surveillance_tradeoffs_wp_412.pdf (accessed March 4, 2025).

National Aeronautics and Space Administration Goddard Institute for Space Studies. 2012. Earth's energy budget remained out of balance despite unusually low solar activity. https://www.giss.nasa.gov/research/news/20120130b/ (accessed November 18, 2014).

Office of the Deputy Under Secretary of Defense (ODUSD) for Acquisition and Technology Systems and Software Engineering. 2008. *Systems Engineering Guide for Systems of Systems*, Version 1.0. Washington, DC: ODUSD(A&T)SSE. https://www.cto.mil/wp-content/uploads/2024/06/DoD-SE-for-SoS-2008.pdf (accessed March 4, 2025).

Pedrotti, F.L. and L.S. Pedrotti. 1987. *Introduction to Optics*. Englewood Cliffs, NJ: Prentice-Hall, Inc.

Quirrenbach, A. 2000. Observing through the turbulent atmosphere. In: Lawson, P.R. (ed.), *Principles of Long Baseline Stellar Interferometry*. Pasadena, CA: JPL Publications, pp. 76–78.

Rausch, R., Benevelli, D., & M. Serell. 2007. Testing programs for trqansportation management systems: A technical handbook (Report No. FHWA-HOP-07-088). Washington, DC: U.S. Department of Transportation. Pp 5-14.

Robbins, P. (ed). 2007. *Encyclopedia of Environment and Society*. London, GB: SAGE Publishing.

Roggemann, M.C. and B. Welsh. 1996. *Imaging through Atmospheric Turbulence*. Boca Raton, FL: CRC Press.

Saundry, P. and EH.Net. 2012. Economic history of the North American fur trade, 1670 to 1870. https://editors.eol.org/eoearth/wiki/Economic_history_of_the_North_American_fur_trade,_1670_to_1870 (accessed March 4, 2025).

Sherif, J.S. and T.-L. Tran. 1994. An overview of the quality development (QFD) technique. Software Assurance Group, Jet Propulsion Laboratory, Pasadena, CA, https://trs- https://dataverse.jpl.nasa.gov/dataset.xhtml?persistentId=hdl:2014/33621 (accessed March 4, 2025).

U.S. Customs and Border Protection. 2025. Vision Statement, Mission Statement, and Enduring Mission Priorities. https://www.cbp.gov/about (accessed March 4, 2025).

U.S. Department of Transportation, Federal Highway Administration. 2007. Testing programs for transportation management systems: a technical handbook, FHWA-HOP-07-088. Washington, DC: U.S. Department of Transportation, Federal Highway Administration, May 10,2007. https://rosap.ntl.bts.gov/view/dot/41907 (accessed March 4, 2025).

USAF. 1969. MIL-STD-499A - system engineering management. Systems Engineering Goldmine: https://segoldmine.ppi-int.com/node/45238 (accessed March 4, 2025).

Wikipedia. 2025. Definition of System of Systems (SoS). https://en.wikipedia.org/wiki/System_of_systems#:~:text=System%20of%20systems%20is%20a,to%20system%20of%20systems%20engineering. (Accessed 4 March, 2025).

4 Systems Modeling and Establishing the Systems Architecture

All models are wrong, but some are useful.

—**George E. P. Box**

4.1 OVERVIEW

In this chapter, we focus our attention on systems modeling activities to understand the nature and importance of modeling in the holistic development of systems in general. We also focus on some analytical aspects of systems engineering to show how these models are used in the overall systems development process. We provide a brief discussion on general modeling methods in systems engineering but address many of the specific methods in greater detail in subsequent chapters where they are more directly relevant. For example, we address statistical modeling methods such as design of experiments (DoE) methods in Chapter 6 where we present probability and research methods in systems engineering. In this chapter, apart from a brief introduction to general modeling concepts and methods, we quickly focus on modeling efforts that support the development of the systems architecture. Establishing a solid, holistic systems architecture has a major impact on the successful design evolution and ultimate success of systems development efforts. Consequently, the bulk of this chapter is dedicated to modeling efforts that support the development of the system/systems architecture.

This chapter is organized much like other chapters in this book; in that, there are three major sections. The first part of this chapter (Section 4.1) deals with systems engineering-oriented material such as, in this case, focusing on systems modeling elements with a major emphasis on generating and analyzing the systems architecture. The second part of this chapter (Section 4.2) presents analogous and relevant optical technical material such as modeling and analyzing optical systems in general. The last part of this chapter (Section 4.3) provides our running case study that applies the material in Sections 4.1 and 4.2 to a fictitious US Department of Homeland Security, Customs and Border Protection program.

In the first part of this chapter, Section 4.1.1 starts with an introduction to systems modeling methods with an emphasis on modeling methods that involve establishing and evolving the systems architecture throughout the systems development lifecycle. In Section 4.1.2, we provide some historical perspectives related to systems modeling and modeling the systems architecture. Section 4.1.3 focuses on architecting systems from concept to verification and then addresses general systems modeling methods in Section 4.1.4. In Section 4.1.5, we discuss methods for modeling and analyzing the systems architecture and introduce the Unified Modeling Language (UML) and Systems Modeling Language (SysML). Next, in Section 4.1.6, we introduce mathematical concepts that directly relate to modeling the systems architecture, namely propositional calculus and predicate calculus. In Section 4.1.7, we present useful systems architecture methods and tools and follow this in Section 4.1.8 with systems architecture implementation methods and procedures. In Section 4.1.9, we have a detailed section on the System Architect and provide an overview of this important position as well as present the roles and responsibilities of the system architect. Section 4.1.10 presents 26 systems architecting principles which are fundamental in establishing a solid

DOI: 10.1201/9781003624097-5

system architecture (Crawley et al. 2016). These principles provide a holistic foundation for systems developments and present an excellent framework for developing the systems architecture. Finally, Section 4.1.11 presents national and international case studies that demonstrate the systems modeling or systems architecture concepts, principles, or methods presented in this chapter. Additionally, because of their theoretical, conceptual, and ultimate practical importance, we also present the 26 systems architecture principles in detail, as originally stated from their source, along with a short interpretive and application-oriented synopsis in Appendix 4.B, to guide those involved in the development of the systems architecture, especially the system architect. In Appendix 4.B, we also include some SysML-based examples to help illustrate these principles.

4.1.1 Introduction to Systems Modeling Methods

This chapter addresses the very important topic of systems modeling and establishing the system architecture. We use the terms system architecture and systems architecture interchangeably since the difference is only dependent on whether there are one or many systems that are under consideration. INCOSE defines systems architecture as a "solution based on principles concepts and properties logically related to and consistent with each other" (Walden et al. 2015). Further refinement provides that the "solution has features, properties and characteristics which satisfy, as far as possible, the problem or opportunity expressed by a set of requirements… and lifecycle concepts" (Walden et al. 2015). One of the primary professional bodies for systems engineering is the International Council on Systems Engineering (INCOSE) which according to their website:

> …is a not-for-profit membership organization founded to develop and disseminate the transdisciplinary principles and practices to enable realization of successful systems. INCOSE connects systems engineering professionals with educational, networking, and career-advancement opportunities in the interest of developing the global community of systems engineers and systems approaches to problems. INCOSE creates state-of-the-art work products that support and enhance this discipline's visibility in the world, including documenting best practices
>
> *INCOSE Homepage (2023)*

INCOSE is one of the most prevalent and active organizations for systems engineering in the United States and is a great source for current information of modern and evolving systems engineering practices. There are many other professional and regulating bodies that are concerned with systems engineering and/or are involved with standards that are widely used in the systems engineering community. Some of these organizations include the American National Standards Institute (ANSI), National Institute of Standards and Technology (NIST), International Electrotechnical Commission (IEC), International Organization for Standardization (ISO), and Institute of Electrical and Electronic Engineers (IEEE), to name a few. There are many other standards organizations, too many to list, that play an important role in the systems engineering discipline, but the ones here play a prominent and regular role in the evolution of contemporary systems engineering practices. Consequently, these organizations strongly influence the evolution of systems engineering-oriented topics such as model-based systems engineering (MBSE), the definition and characteristics of the systems architecture, and systems modeling in general. It is interesting to note that some of these standards organizations have joined together to release joint standards such as ISO/IEC/IEEE 15289:2015 – Content of systems and software lifecycle information products.

There are many things said about the nature and importance of systems architecture. According to Jaakola and Thalheim, system architecture is a form of a conceptual model that provides insight into the structure, behavior, and views of a system along with details on how a system satisfies requirements (Jaakola and Thalheim 2011).

The systems architecture in essence is a conceptual model that addresses structural and behavioral system characteristics and provides other ways of looking at critical elements of the system. The systems architecture is instrumental in organizing these critical elements and strongly relates to the system requirements. The physical architecture emerges from the systems architecture and details the hardware architecture, software architecture, data architecture, and key operational aspects/procedures of the system. Over time, technology advances and systems are becoming more complex, exhibiting emergent behavior and becoming more interconnected. System architectures consequently become increasingly important and need to be more robust (Mandel et al. 2022). When engaging in modeling the systems architecture, the focus is on capturing elements of the problem, and not so much on the details of the solution. Very similar to requirements, the systems architecture describes *what* needs to be accomplished and not *how*. The system architecture is established early in the development process and provides the means to inject realism into the concept and design (NASA 2007). Establishing an early system architecture can sometimes be different due to dissimilar expectations and vested interests among the stakeholders. Consequently, significant compromise can occur in establishing the systems architecture that satisfies the primary needs of the program and satisfies the interests of the diverse set of stakeholders (Rechtin 1992). The more mature the system architecture model becomes, the more resistance to change occurs. This resistance to change can manifest in increased system costs and even compromised system functionality. Even though the system architecture should be resolved early, it evolves continuously and drives the system design process. It is also an essential link between requirements and design and contributes to eliminating non-feasible design options. A brief example is as follows. Say that we are interested in developing a transportation system between two cities. Well, this concept could be achieved in many ways. We could fly from city A to city B. We could take a car, a bus, roller-skates, or wear a fine set of sneakers and trek there. If the stakeholders state that the distance between cities is 50 miles, that we need to be able to go back and forth between these cities often, that we need to be able to take luggage and materials, and that we would have many activities the day we get there, and, finally, that cost is important. After conducting some trade studies, we may decide that a car is the best way to accomplish these needs. We now have enough information to specify the system architecture – namely, the transportation system is a car. Notice, that once this is decided, all the other non-solutions (sneakers, airplane, bicycle, ….) are eliminated, and the requirements generation and associated design processes can focus strictly on the vehicle described by the system architecture (the car). Note the car has software aspects, data aspects, and hardware aspects that can be detailed in the software architecture, data architecture, and hardware architecture. We also know that we now need to flesh out the requirements for the car in our operational procedure of driving back and forth between cities A and B. We need to address questions such as, how often does our system need to go back and forth between cities A and B? Are there constraints on arrival times? Safety consideration, reliability, and a slew of other operational considerations need to be determined and articulated as a set of system requirements. Once these are specified, then new questions about the system architecture emerge and need to be addressed. Some examples of these questions include, do we need two doors, or four doors? How much capacity do we need in the trunk? How big of a fuel tank is needed? As these details are captured in the systems architecture and reflected in the physical architecture, this will generate more details that need to be specified in the requirements. What ensues is a steady back and forth between the design space and requirements space wherein the systems architecture becomes an essential link. This concept is well summarized by Dickerson and his associates when they state that the system architecture describes the "entire system based on the physical architecture and the definition of supporting products and services across all phases of the lifecycle" (Dickerson and Mavris 2016, p. 282). There are a variety of uses of the system architecture, including establishing the framework for the design effort itself. The base system architecture can be implemented on a program using well-vetted architectural frameworks such as the Department of Defense Architecture Framework (DODAF) for largely US (and allies) defense applications, The Open Group Architecture Framework (TOGAF) for commercial applications, the Ministry of Defense Architecture Framework (MODAF) for defense applications in the United Kingdom, and many others. Each architecture framework has its advantages and disadvantages but serves to establish the systems architecture for systems, services, processes, or products (or combinations thereof). These architecture frameworks have associated modeling languages such as UML or SysML, which are used to implement various architectural products and artifacts for the system at hand. For instance, the OV-1 diagram in the DODAF and MODAF can depict a graphical rendering of the concept of operations (CONOPS) for the system, or the Use Case diagram can illustrate how a system that is captured in the OV-1 is used to accomplish a stated purpose or stakeholder requirement. These two examples of artifacts, when applied to the system under consideration, provide useful views of the system, its unique interrelations and interactions with other entities, system elements, and its environment. These artifacts begin to define the salient features of the system architecture. The collection of artifacts that are generated using the architecture framework provides various complementary views and insights into the behavior, structure, requirements, and parametric aspects of the system, such that the system architecture becomes defined and emerges. The logical and scientific approach in developing the system architecture provides the foundation for MBSE methods, which are addressed in further detail in Chapter 5. The systems architecture is an essential element in general systems developments and has seen application in complicated and/or complex systems modeling efforts such as in modeling warfare systems illustrated in the CONOPS or Operational Concept (OpsCon). The system architecture finds applications as a fundamental feature in Capability-Based Acquisition Policy, wherein it plays prominently in transitioning from the problem domain to the solution domain as we now will see. The system architecture is also useful in improving system planning activities and aids the development of model representations at various levels of the system.

Problem & solution domains and systems architecting: There are two major domains that can be used to abstract the system architecting process: the Problem Domain, and the Solution Domain. In general, the Problem Domain is for understanding stakeholders' needs, requirements, and critical aspects of the system in its environment, whereas the Solution Domain provides the design and development products and activities needed to realize the system. The well-known standard, ISO/IEC 15288, which is an internationally recognized standard for systems engineering, describes the *problem domain* as the "environment in which the system will be used and the constraints that the system must operate under." The *solution domain* is defined by the same standard as, "set of technologies and techniques that will be used to create the system" (ISO/IEC/IEEE 2015). The domains just mentioned are fundamental abstractions of the systems design process that help determine which SysML artifacts need to be produced and in what sequence, so that an efficient, compliant, and comprehensive system may be developed using contemporary systems engineering methods such as MBSE.

4.1.2 Historical Perspectives on Systems Architecture and Systems Modeling

The concept of system architecture developed over time through application versus formal definition. The current understanding of system architecture emerged through necessity, application, and the interactions of many diverse groups. Multiple perspectives were taken in arriving at a formal definition for system architecture. These perspectives are, (1) the large system perspective, (2) product-based perspective, and (3) the software-based perspective. When combined, these perspectives provide the proper insights into understanding the modern view of system architecture. We now describe these perspectives.

Large systems perspective: **Systems are becoming more intricate and complex as time progresses.** The ever-growing complexity of systems is a major factor behind the development of more structured Systems Architecture approaches. Because of the increasing likelihood of problems that come with large systems development efforts, the necessity of putting formalized systems architecture methods into practice scales with system size and complexity. Large systems carry inherent risks that often lead to high-impact problems, which consequently result in greater systems development and support costs. The defense industrial base provides an excellent source for examples of large and sophisticated development programs that require and benefit from the usage of systems architecture principles, methods, tools, and techniques. System architecture principles and methods had to grow in step with the development of large-scale military systems like air defense and missile defense systems. Even large commercial space exploration systems require sound systems architecture methods to ensure the necessary organization and structure are put in place for a successful system launch (Leonard 1999). Major advancements in Systems Architecture, concerning both the system definition and the method by which the system definition is generated, were instrumental in the first successful United States mission to carry a person to the moon. Large-scale initiatives like this one led to the development of new methods and instruments for large-scale system acquisition. Figure 4.1 illustrates one of the novel techniques, the intersecting waterfall model, which explains both the acquisition product waterfall, and systems architecture process from needs identification to operations (Fixson 2000). The evolution from the top-center to the bottom shows a particular rendition of the traditional *waterfall* development method with the general developmental steps to get to production and beyond. The *process* flow shown from the top-left of Figure 4.1 to the bottom-right clearly shows the modeling step after the identification and establishment of resources, wherein the systems architecture gets defined as part of the initial steps of the process. Both paths intersect at the production block, which is common to both views, but the importance of modeling, prior to engineering is prominently displayed.

Systems architecture moved away from historical models like the traditional Waterfall Model and toward heuristics as technology advanced and systems grew increasingly integrated with software (Maier and Rechtin 2000).

Product-centric perspective: The increase in large, complex systems offered ample motivation for the development of Systems Architecture as an engineering tool. However, systems architecting also has origins in the process for how products are realized, especially with respect to physical systems. This product-based perspective tries to provide a simple view of why architecting is critical to system success. The fundamental reason behind product-centric systems architecture is that a basic goal in the systems development effort is to segregate the system's pieces into physical chunks (Ulrich and Eppinger 2000). These chunks are modular by nature and include interface characterization to guarantee that each element's interaction is fully specified. Modularity provides a straightforward function and form description, resulting in a more favorable interchange and scalability factor inherent in each physical piece (Eppinger et al. 1994).

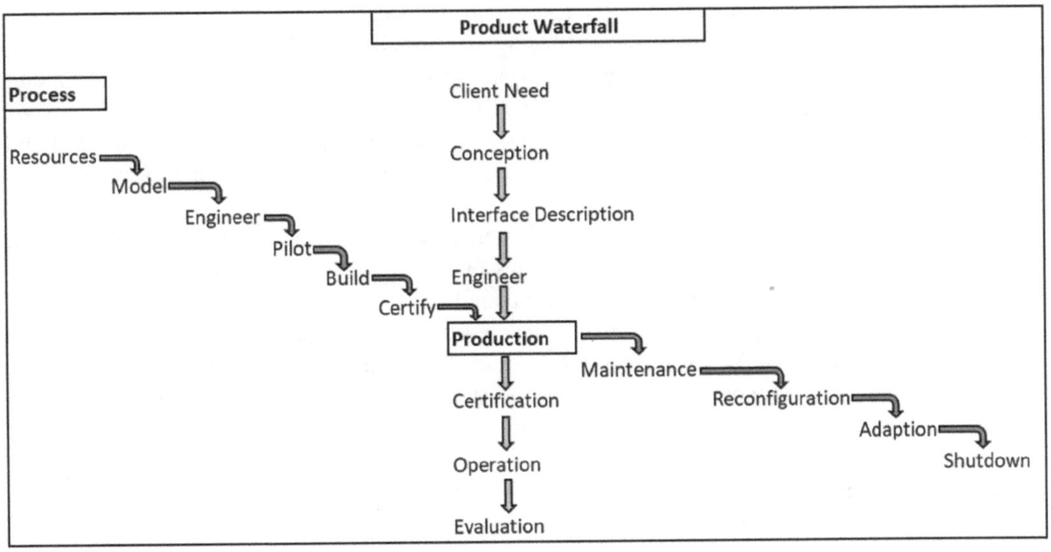

FIGURE 4.1 Intersecting Waterfall Model.

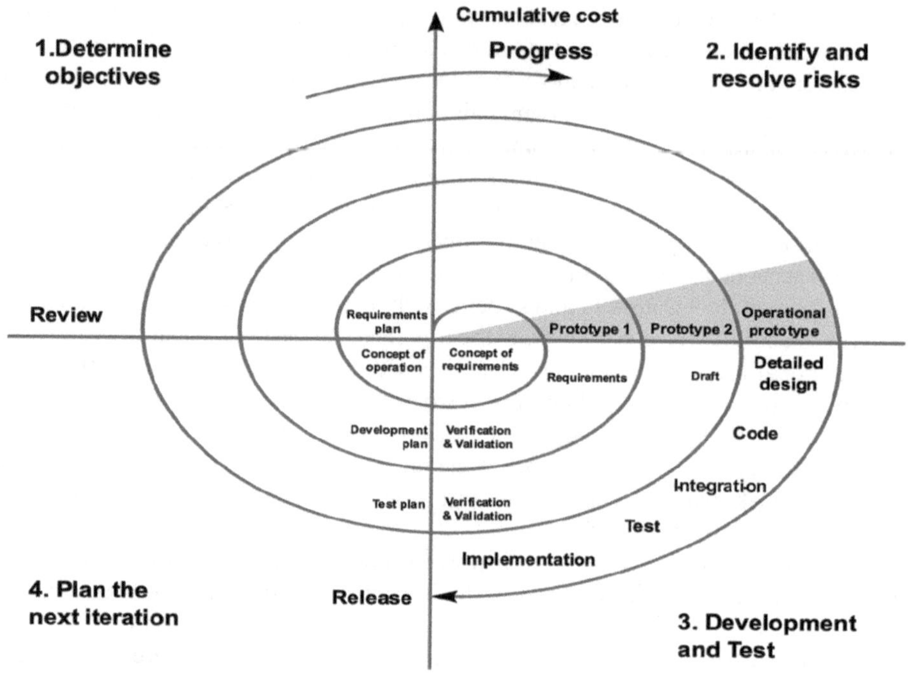

FIGURE 4.2 Spiral model. (Source: https://commons.wikimedia.org/wiki/File:Spiral_model_(Boehm,_1988).svg. Accessed December 18, 2023).

Software-based perspective: This perspective provides a marked difference from the product-centric approach in that the software-based perspective transcends the bounds of physical limitations such as those introduced by hardware and provides for more abstract possibilities such as designing with object-oriented systems design methods. From this modern perspective, more layered designs are possible, with the differences narrowing the gap between prototypes and production items. This can be seen in Figure 4.2 where the items in gray show the various prototype realizations in a respective spiral development ring. In the outer ring, an operational prototype is evident that likely exhibits many of the desired attributes of the released system.

The spiral development model fits in well with concepts like Agile development wherein the development of prototypes can represent points of provided features/capabilities. There is a notable difference between the spiral model and models based on the traditional systems engineering development cycle in that the spiral model is iterative and evolutionary wherein the functionality/capability of the system expands with each new completion of the spiral as the system matures (Barroca et al. 1999).

Systems architecture controls/constrains the elements that make up a system, how those elements communicate with one another, and the limitations placed on the system (Luckham et al. 1995). A well-designed architecture provides for a guaranteed set of system outputs, in accordance with the requirements, while also establishing the expected behavior of the system. Although the concepts used in developing the Systems Architecture can differ, the end results are usually consistent regardless of the chosen concept. Three notions of systems architecture that are frequently used include: Plug and socket, object connection, and interface connection.

Interface connection concept: The idea behind the Interface Connection Concept is to establish the architecture of a system in such a way that the interfaces between system elements, rather than the modules themselves, define the connections between them (Luckham et al. 1995). Both the needed requirements for the interface and a set of holistic features that will be provided by the interface are captured in the interface definitions. Between the system elements themselves, the stability of element inter-communications is ensured by the Interface Connection Concept so that only the System Architecture-level information is required to manipulate the system elements.

Object connection concept: Object Connection aims to define the architecture through supplied features of each system element rather than the interactive aspects of the system elements as presented in the Interface Connection Concept, which strives to build the architecture of a system through those interactions (Luckham et al. 1995). The most prevalent use of object-connected, or object-oriented, architectures is in software development, where they serve to enhance system evolution. As new requirements are imposed on a system over its existence, this evolution enables smooth system transitions (Ehrig et al. 2005).

Plug and socket concept: A "plug and socket" architecture is one in which a central system is responsible for housekeeping and additional systems are built to connect to it through service connections made using predefined interface connections. By serving as a channel for data transfer and consolidating numerous connections into one long one, the service connection streamlines the communication pipeline between the system's core and its subsystems (Luckham et al. 1995).

4.1.3 Architecting Systems – Concept through Verification in SysML

An important aspect of this chapter is that the system architecture artifacts, particularly the nine diagrams-as-artifacts specified by the SysML specification, are created during the system lifecycle stages of problem definition, system design, development, and test (verification and validation). In addition, sophisticated tools like CATIA System-of-Systems Modeler automatically update artifacts when they are correctly configured and scoped to model system elements, behaviors, and other essential system attributes. It is critically important to properly scope an artifact. The process of scoping typically brings the systems-thinking idea of abstraction to systems modeling allowing for model creation and evolution that produces useful and simplified representations of system problems, needs, definitions, and analyses for the development team and stakeholders alike.

Everyone from end users to management to engineers to acquisition stakeholders should be able to understand the system and essential system concepts when the system is represented in a systems modeling tool, especially those tools that employ SysML as their primary modeling language. These SysML artifacts provide a highly visual, structured, and methodical way to capture essential aspects of systems such as the system's requirements, system architecture, the behavior(s) of the system, and system interconnections, intra-connections, and relationships. In a nutshell, SysML artifacts are a tremendous aid in understanding complex systems and are fundamentally important in the systems development process. Some of the provided benefits of the SysML artifacts include:

- *Improved communication and collaboration:* An important benefit of generating system artifacts with a SysML tool is the improved communication that results between the development team and program stakeholders. SysML provides for a common language and structured framework that facilitates understanding and communication among stakeholders, developers, program facilitators, and collaborators.
- *Early identification of problems:* To avoid expensive mistakes, the SysML artifacts that are covered below help to find possible system issues earlier in the design phase. Fixing these issues later in the development process can be costly and time-consuming. Developing SysML artifacts can assist in avoiding these negative outcomes. Multiple case studies have shown that, when compared to conventional, document-centric methods of design and verification, SysML tools significantly reduce total costs by allowing for the early detection and correction of design mistakes.
- *Improved traceability:* To enhance traceability, it is helpful to use relational and metrics diagrams like SysML Requirements Diagrams to keep track of the connections between various parts of the system, including requirements, design, and implementation. An important aspect of SysML and associated modeling tools in general is that changes made

to one element of the system are reflected in other parts of the system. CATIA System of Systems Modeler and IBM Rhapsody are two examples of SysML tools that are specifically made to work with more conventional requirement management databases that mainly focus on documents.
- *Improved documentation:* By employing systems modeling tools and SysML diagrams, the requirements, architecture, and behaviors of the system can readily be documented. In the long run, this can aid with the system's upkeep and streamline the design and development processes. When new system models become necessary, several industries employ tools like IBM Rational DOORS, IBM Rhapsody, or CATIA Enterprise Architect to keep track of requirements and necessary/beneficial systems design and development artifacts.

Expected deliverables of the system architect: Whether a document-centric systems engineering approach is used or an MBSE approach with languages like SysML is employed, a set of expected deliverables include:

- Project or program goals, with a focus on their functional aspects. Clear, succinct, consistent, and comprehensive goal setting is essential, as is a level of certainty (80%–90%) that the goals will be met. Even if a decision to acquire a system is still in the Analysis of Alternatives (AoA) phase in the DoDAF framework, following the Joint Capabilities Integration and Development System (JCIDS) process, the project or program goals align most closely with stakeholder demands and the problem definition phase (Hause 2010).
- How the system is situated within its natural environment (OV-1 using context diagram in SysML). The description ought to cover the goals of the system's operation(s), the operating system(s), the standards that were used, and any regulations that are applicable.
- The system's CONOPS. Similarly, SysML may be used to implement the CONOPS. The CONOPS describes the capabilities of the system and how they will cooperate to meet mission objectives. It must include emergency preparation for unforeseen circumstances as well as contingencies for various scenarios.
- The decomposition and analysis of the system in functional, logical, and physical forms. Block Definition Diagrams and Package Diagrams are two ways in which these decompositions can be implemented in SysML. These schematics illustrate relationships based on logic and abstraction. At least two levels of decomposition should be used, with each level describing a service and its relationship to the layers above and below it.
- A description of the interfaces. Interactions between internal system elements (intraoperability) and external systems, actors, and entities (external interfaces) should be considered (interoperability). Software based on SysML can automatically generate allocation tables describing interfaces and their properties connected to model elements, and ports on model element blocks can be used to display interfaces in SysML and CATIA System of Systems Modeler (Herzig et al. 2018).
- A development plan. Time, money, risk, design, and execution all factor into this plan. To make sure the system satisfies the needs of stakeholders, it is essential to have an initial plan that details the development timetable, resources, budget, and other important factors. Automating the tracking of metrics and risk in requirements management is one of several compelling techniques in contemporary risk management (Hayat et al. 2020).

Fundamentally, systems architecture is a set of beliefs and practices that involves thoroughly defining a system or systems of systems, including all its parts, how they interact with one another, the results that should be expected from the finished product, and the resources that will be needed to make it happen (Wilkinson et al. 2010). Pursuing a system architecture primarily aims to simplify the design while providing clarity through a holistic system view (Maier and Rechtin 2000). When the fields of architecture and systems engineering came together, the resulting philosophy was known as systems architecting. The rising complexity of systems, caused by the increased integration of software and computers into systems with expanded functionality and capabilities, was the primary motivator for this merging and adaptation of the design process (Rechtin 1992). An issue arose in the field of system design with the emergence of these increasingly complex systems: as the size of individual systems increased, the method of architecting became too laborious and impractical. Systems architecting emerged as a discipline to address this issue by developing and disseminating best practices for streamlining complex systems into more manageable and, above all, fruitful undertakings that deliver as promised. The systems engineering community has embraced systems architecting, and it is now well established in the system design process. We also recognize the tireless efforts of many system architects in the past for their fundamental efforts in bringing many modern-day systems into existence.

4.1.4 Systems Engineering Modeling Methods

In this section, we provide a short discussion on some general systems engineering-oriented modeling methods and activities that are an essential part of a systems engineer's toolkit. Herein, we provide brief overviews of these methods and defer detail to subsequent sections of this book where the modeling method is more directly relevant. The bulk of this section focuses on modeling activities that are relevant to the generation of the systems architecture.

General modeling concepts: The following summarizes some important modeling concepts within the field of systems engineering. The first point to consider is what do we mean by *modeling* in the first place. This word, like many others, can mean different things to different people. For example, considering Figure 4.3, we see that there are a variety of different legitimate considerations when contemplating modeling systems. Depending on the topic at hand, one or more of these considerations must be applied in the modeling process. For instance, early in the development of a system, conceptual modeling is prevalent and quite useful. These conceptual models, among other things, can be abstract and involve both static and dynamic aspects of the system. They can be descriptive but also have mathematical underpinnings that can be implemented in MBSE environments. The idea of modeling can be different in the minds and hearts of different stakeholders and members of the development team so, just as in establishing good requirements, proper care must

Considerations in Modeling Systems

- Physical vs Abstract
- Mathematical vs Descriptive
- Static vs Dynamic
- Steady State vs Transient
- Lumped vs Distributed
- Open vs Feedback
- Deterministic vs Stochastic
- Continuous vs Discrete
- Observability
- Linear vs Non-linear
- Model Order Reduction
- Classification by,
    - Timeframe
    - Complexity
    - Interactions
    - Nature/Type of Components
    - Uncertainties
- Hard vs Soft
- State Space Formulation
- Systems Analysis
- Conceptual Models
- Controllability
- Interpretive Structural Modeling
- System Dynamics
- Simulation
- Discrete Event
- Chaotic Systems
- Fuzzy Systems
- Artificial Neural Nets
- Sensitivity
- Liapunov Stability

FIGURE 4.3 Necessary things to consider when modeling systems.

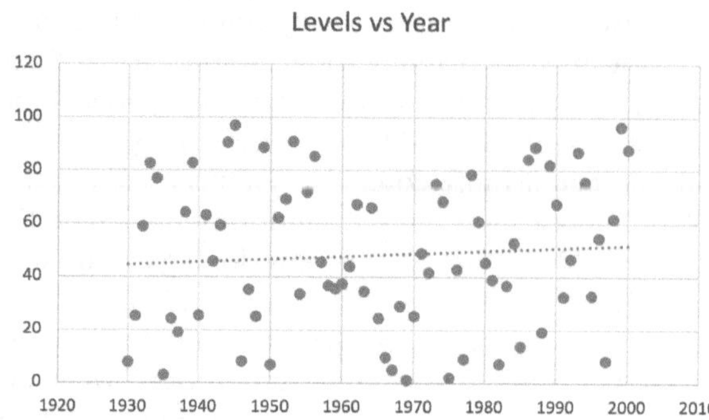

FIGURE 4.4 Parameter levels vs years linear regression plot.

be taken in defining this term. With this in mind, we provide some general modeling concepts, methods, and activities that are commonly used by systems engineers throughout the systems development effort.

Statistical modeling: Statistics is very important to systems modeling in several ways. For example, it is important to properly account for statistical variations in the modeling process, and sometimes optimize systems performance. Some optimization examples include accounting for statistical noise variations in detectors, to optimizing data routes or delivery routes, to determining the quantity of products that can be made with the components on hand. There are two especially noteworthy types of models that often occur in system modeling efforts, and that are generally useful in a wide range of applications. These models are regression models and classification models.

Regression models are predictive and help analyze the correlation between dependent and independent variables. Most often, linear regression models are the most generally applicable and easiest to analyze. A basic example of a linear regression is shown in Figure 4.4 wherein a set of data is "fit" with a straight line. This type of modeling plots the points graphically in a scatter plot and then fits a straight line across the data set so that, for instance, the variances of the points relative to the straight line are minimized. In this example, the fit is not all that good since the variation of the points with respect to the fitted line (standard deviation, or variance of the points) is relatively high. Consequently, some statistical metrics that reflect the *goodness of fit*, such as the R^2 value, would be low, indicating a poor fit. We present more detailed information about relevant statistical metrics in our chapter about research methods, Chapter 6 of Volume 1.

Another interesting example of a widely used statistical modeling method is DoE. DoE provides a means for understanding data that has inherent random characteristics. DoE is structured and mathematically rigorous and is useful in applications such as hypothesis testing, and random data analysis methods such as Analysis of Variance (ANOVA), or Analysis of Covariance. As an example of DoE modeling, two data sets (say one from laboratory testing and another from field data) can be compared to each other to determine whether or not they are the same in a statistical sense, and that there are statistically significant results. DoE can also be used to ensure the optimal set of parameters are compared in a multi-factor experiment so that statistically meaningful results are obtained and that an optimal number of experiments are conducted in making the statistical comparison. Correct application of DoE methods ensures that there is a completely randomized design (CRD) in the analytical process, whether the results are correctly analyzed and processed, and whether they are believable and defensible in a statistical sense. Table 4.1 shows a tabular layout for DoE with three factors that indicate the number of experimental runs that need to be accomplished to properly compare the factor results with each other. The plus sign indicates the possible permutations with the three factors A, B, and C. We present a very detailed explanation of DoE methods in our chapter on research methods. Therein, we provide the mathematical formalism of DoE along with implementation methods and examples. Other prevalent statistical methods such as statistical process control (SPC) methods are covered in detail in our chapter dealing with production.

TABLE 4.1
2^3 Factorial Layout Matrix

		Factors			
		A	**B**	**C**	**Label**
Runs	1	–	–	–	1
	2	+	–	–	A
	3	–	+	–	B
	4	–	–	+	C
	5	+	+	–	AB
	6	+	–	+	AC
	7	–	+	+	BC
	8	+	+	+	ABC

Risk and decision modeling: Although they are linked, risk and decision modeling are represented as distinct concepts. They are connected by the fact that the risk compels program leadership to decide what to do depending on the likelihood of an event occurring and how it would affect the system (consequences). Risk management involves striking a balance between potential hazards and risks, and cost. We then monitor the risks and offset their impact with planned and feasible risk mitigation methods. This is the point at which the decision-making element emerges and affects every facet of our program (Haimes 2004). Our later chapter on decision and risk analysis goes into great length about risk and decision models and approaches.

Reliability modeling: System failure-related elements can be estimated with the help of reliability modeling. As a result, we may develop mathematical formulas to calculate the system's availability, mean time to failure (MTTF), mean time between failures (MTBF), and the likelihood of failure. This is accomplished by applying mathematical formulas that assess each systems element's reliability at the element level, and in relation to the system, providing an overall reliability. Predicting maintenance cycles with the use of maintainability and availability metrics can help save needless downtime. Reliability block diagrams can be made using MBSE tools like CATIA Systems-of-Systems Modeler. According to Trivedi and Bobbio (2017), "this method is a pictorial representation of the systems reliability performance." We may use the model to determine the overall reliability of the system by applying reliability factors to each individual component and running the model. Another important thing to think about is maintainability; if maintenance can be accomplished, the system will remain viable longer than if it is poorly maintained.

The predicted mean time it takes for a system to fail is known as the mean time between failure (MTBF). Related to this metric is the MTTF which indicates the average time, measured from the current moment, it takes to get to the point in time determined by the MTBF. The MTBF is a widely used measure that forms a major component in six sigma. Because the MTBF can be used to determine the expected longevity of the system and aid in the determination of when each system will require maintenance, these concepts provide great benefits to systems modeling efforts (Trivedi and Bobbio 2017).

The system's availability is another interesting measure to consider. Although the availability and reliability are calculated differently, they are related. The distinction is that availability is typically expressed as a higher percentage than many reliability metrics since availability represents the portion of time the system spends operating normally (without failing), versus the probability that the system works, as is the case for reliability. In reliability modeling, metrics such as MTBF and MTTF, decrease as time goes on. There are several different versions of the availability measure, but in one instantiation, Operational Availability; this measure assumes that if the system is in a standby condition, there is no stress on the system and there is no decrease in operational availability during this time.

MBSE: One of the most popular methods in System Engineering is the application of Model-Based System Engineering or MBSE for short. MBSE is applicable and beneficial across all system development lifecycle phases (SDLC). Specifically, the SDLC includes the phases for, Conceptual Design, Preliminary Design, Detailed Design and Development, Production, Deployment/Operation/Support, and System Retirement/Disposal. We will use various tools during these systems development lifecycle phases, most of which will cross over from one phase into others. Examples of these tools include tools for decision and risk analysis, architectural modeling, systems modeling, requirements verification and validation, analysis tools, and system modeling and process optimization.

Programs in the public sector, such as the Department of Defense, are subject to specific laws and regulations that are not often applicable in the private sector due to the unique nature and requirements of public sector problems. When it comes to standards in the private sector, it is entirely up to the individual company which standards to use, or not, unless there is a specific law that applies to the job being done in that field. Using the adaptive acquisition framework (AAF) is one way the public sector functions differently. To guarantee that systems are delivered in a fashion that provides effective, acceptable, survivable, sustainable, and cheap solutions to the end user in a timely manner, the Department of Defense in part established program milestones in accordance with the AAF.

Modeling using natural language processing: The study of programming computers to understand human language data is known as natural language processing. There have been some early successes in attempts to build MBSE model elements and diagrams using text-based statements within the context of digital engineering and SysML. Using artificial intelligence, Chami et al. developed a text-to-model framework and method for automatically generating actors, Use Cases, blocks, and associations between them (Chami et al. 2019). An important thing to note is that the text-based inputs, including formulaic text, did not have to be completely structured to produce useful outputs. Their reported outputs did, however, require some minor alteration and clarification. The trained AI was able to generate basic model elements, such as Use Cases and relationships to actors and blocks, even with unformatted inputs.

Zhong et al. were able to demonstrate the usefulness and applicability of using natural language processing – which automatically extracts relationships and entities from text – as a foundation for a SysML system model. Once most of the work has been completed by machines, the system model can then be manually adjusted and improved (Zhong et al. 2023). Here, AI natural language processing was used to automatically create SysML diagrams. The results were compared favorably to case studies and also manually created artifacts where the outcomes were compared. This is quite interesting because a trained AI might automatically scan in and convert existing design documents, interface control documents, and operating manuals for legacy complex systems in order to develop a system model and requirements. Zhong et al. in particular demonstrated that criteria may be constructed semi-accurately without the need for a relational MBSE model to be established first.

Fundamentally, systems modeling efforts need to follow certain guidelines to be effective. Normally, this is known as establishing constraints for the model to follow. Since mathematical models are the best in terms of precision, the question of why we need to construct models using written language still must be answered. At first glance, it may seem that generating written language models presents more problems than it's worth. Mathematical models are global in scope, in contrast to written language, which might differ even among nations that speak the same mother tongue. Ambiguities in a written language can lead to many failures. For example, a significant safety risk, through inadequate communications, has been found to be the primary cause of over 70% of problems in the aviation industry (Buarqoub 2019). When we compare the many kinds of languages that are spoken around the world, precise communication becomes even more of a concern. German is ordered by a small number of syntactic limits in conjunction with a set of fundamental ordering criteria. Chinese is based on linear patterns of slots that are described by a combination of structural semantic and pragmatic conditions, and English is based on a set of 26 characters (Dietrich and Zheng 1993).

So why do we persist in proceeding to develop natural language-based models if there are so many variations and ambiguities when using language as the medium for constructing the model? One reason is that it is unrealistic to expect everyone to comprehend the system's underlying mathematics. Different stakeholders have different needs; therefore, if we limit our models to mathematical and algorithmic ones, we will only be able to communicate the models to mathematicians and software developers. To demonstrate a model and highlight the key model elements to a general audience, we require a language-based modeling system.

It is actually very easy to understand why language processing is used in the first place. Our system has requirements, and those requirements have associated criteria that must be fulfilled. We must own the requirements if we are to meet them, and the best way to demonstrate that requirements can be met is to prove that they can be met. One of the main reasons we need to use natural language processing is that our requirements are written in natural language phrasing rather than being expressed as mathematical terms.

Steven Hawkings, one of the greatest minds of the 21st century, talked about good physics models. According to his criteria, a successful physics model "had to be simple, mathematically correct, and scientifically verifiable." This prompted him to make extensive use of the scientific paradigm known as model independent realism (Murphy 2023). We should integrate this theory into our natural language model using Hawkings' methods.

Our models must be clear-cut, consistent, logical, and verifiable. It is easier to diagnose and analyze the findings of a simpler model. To facilitate the integration and adoption of other languages, simple language usage is necessary. This kind of cross-language adjustment may best be achieved with a straightforward word lookup and replace function. To lessen the chance of misunderstandings, a glossary of terminology is another essential but frequently disregarded modeling concept/tool. For reviewers of the model, they must be able to understand and accurately interpret the model. The model must make sense to those who use it. Physics-based models must be repeatable and have consistent logic, and our natural language-based models also require this kind of consistency in repetition; otherwise, we risk inaccuracies and errors in interpretation that make our natural language-based models unusable.

Ultimately, the models must be able to be verified and validated. This can be accomplished by either first manufacturing the system and then testing it or testing the system through simulation. The simulation option is usually much cheaper.

Parametric modeling: "Parametric modeling constrains the properties of a system which can then be evaluated by an appropriate analysis tool" (Friedenthal et al. 2015). This indicates that system studies involving physical attributes such as the predicted weight and area of a system can be carried out by using parametric modeling methods, which can then be further assessed with other useful analytical tools and methods. Both the Ministry of Defense Architecture Framework (MODAF) and the Department of Defense Architecture Framework (DODAF) explicitly employ parametric modeling, mostly through trade-off analysis (Hause 2010).

Parametric modeling is a computer-aided design (CAD) approach that system engineers employ on a regular basis. CATIA System-of-Systems Modeler is one CAD tool that uses parametric modeling, and that has seen extensive use in the systems modeling community. Constraints can be applied to our systems model to give it the structure and behavior it needs to effectively represent the system. Logistic regression, exponential distributions, and linear regression are examples of parametric analysis. Of these, linear regression is most widely used. In algebra, linear regression is the process of evaluating two variables to see whether they have a relationship (Ranganathan and Aggarwal 2017).

4.1.5 Modeling and Analyzing the System Architecture

Modeling and analysis (M&A) of the system, or systems architecture, depending on whether there are one or more systems in question, is conventionally accomplished using modeling tools that are based on a standard ontology that uses a language like the UML or the SysML. The M&A of the system's defined Use Cases, which are represented by scenarios and flow charts that implement Use Case diagrams, Activity Diagrams (also known as flow diagrams), and Sequence Diagrams, is the focus of the M&A activity. The design diagrams that mathematically model the blocks of the system and the structural aspects of the system are typically produced using Parametric Diagrams in SysML. In design, when design diagrams are appropriately scoped to the model elements in the Containment Tree, they can even be produced automatically in CATIA System of Systems Modeler and other MBSE SysML tools – making them a kind of artifact in and of themselves (Friedenthal et al. 2015). To generate the SysML and system architecture diagrams required to define and construct complex systems, combining MBSE tools, such as CATIA System of Systems Modeler, with other Multi-Disciplinary Analysis and Optimization (MDAO) tools/techniques is also a detailed and effective approach (Saqui-Sannes et al. 2022). These SysML artifacts are useful throughout the entire system development lifecycle, from the system's inception through system retirement, and in between.

Blocks (SysML) vs classes (UML): It is important to mention the distinctions and similarities between SysML and UML ontologies for systems engineers, even though blocks and classes in both systems are not System Artifacts per se. SysML is an extension of UML with application tailored for the systems engineering discipline. Data and objects, rather than functions and logic, serve as the organizing principles in both languages, which are known as Object-Oriented principles.

Within the framework of a system that has many object kinds, classes (UML) are described as blueprints for a certain object type. An object's class specifies its characteristics, actions, and connections to other parts of the model. Classes, modeled with state information as objects, can represent real-world items, concepts, and even entire systems, even though UML was originally intended for software architecting (in the system-of-systems context).

Like Classes, Blocks (SysML) represent actual objects; but, unlike Classes, they are defined by their interfaces, actions, and properties rather than their attributes and operations. Additionally, in the context of the structure diagram, they share comparable relationships with other blocks using SysML. In contrast to classes, which are only blueprints for objects, blocks are actual objects or model elements in their own right. Finally, in contrast to UML Classes, which can only have relationships between other classes, SysML Blocks can have relationships and various types of ports. This is an attempt by SysML to accommodate systems engineers' more abstract ideas regarding interface definitions, such as flows, ports, and the energy, information, or other simulated parameters that flow across these interfaces between the abstracted components of a system.

The four pillars of SysML: SysML is a modeling language that is central to many modern-day systems development efforts. SysML is implemented on various commercially available tools such as CATIA System-of-Systems Modeler. SysML provides a series of diagrams (artifacts when applied to a system) that provide different views of the system under development, which are useful in understanding the system, analyzing the system, modeling the system, and establishing and evolving the system's requirements, system's architectures, and design. Before taking a closer look at the available SysML diagrams that can be generated using SysML, it is first necessary to identify the four different pillars of SysML that contain these diagrams and provide an overall categorization scheme for these diagrams.

4.1.5.1 Structure Pillar

An object-oriented design of a system can be represented and baselined by structure diagrams, which are mostly concerned with the system's physical components and use SysML block elements (Hampson 2015). Structure diagrams are common to both SysML and UML (Delligatti 2013). Because UML originated as a language and tool for software engineering, structure diagrams are often used to record software architecture and system decompositions. Put differently, UML structure diagrams often indicate how software components are applied to the system hierarchy, while both UML and SysML structure diagrams typically display the internal structure and system architectural hierarchy (Douglass 2014).

Structure diagrams cover more than just the configurations of model parts or objects. The definition and representation of system properties is another function of structure in modeling; parts and properties can be allocated to the system and/or its components (Bouabana-Tebibel et al. 2012).

4.1.5.2 Behavior Pillar

Behavior diagrams show interactions and changes between the model's items and components. They are ideal for summarizing how the architecture or system works. The purpose of the SysML behavior diagrams, to capture behavioral aspects of the system, is implemented by the four types of behavior diagrams. These behavior diagrams are capable of defining activities and information/energy/mass flows; tracking system state and transitions; defining system controls and constraints; defining interactions; and assigning responsibilities to instantiated model elements (Zdanis and Cloutier 2007).

4.1.5.3 Requirements Pillar

Requirements Diagrams are used to link Functional and Non-Functional Requirements to other model elements that satisfy them. They are also used to link the requirements to Test Cases that verify and validate them. Both Functional Requirements and Non-Functional Requirements can be expressed within the model. Requirements Diagrams can provide requirement hierarchies that reflect the regular decomposition of system requirements. For example, Requirements Diagrams can be used to capture requirements-related information, from stakeholder demands of what the system must do to be successful (functional requirements), to derived requirements and Key Performance Metrics (KPMs), to non-functional requirements that indicate how *well* the system does what it needs to do. The stereotypes of <<derive>> and <<extend>> are used in SysML modeling tools to indicate the derived requirement and additional non-functional aspects of a requirement respectively. Additionally, the <<satisfy>> stereotype can be used to link requirements to model objects/elements that have satisfaction criteria, or the <<verify>> stereotype connections can be used to link a test case element to a model element, indicating that the model element will be tested by the specific test case that will verify the requirement.

4.1.5.4 Parametric Pillar

The SysML parametric pillar can facilitate engineering analysis of important system parameters. SysML implements parametric analysis by appending parameters to model blocks that can be changed during simulation. Constraints can be used to implement relations, which are equations that model many types of physical and mathematical aspects of the system. Some examples include physical attributes (e.g., weight, speed, size, force), temporal behaviors, and non-functional performance features (e.g., reliability, availability, maintainability equations) for instance.

The non-standardized engineering analysis models are often kept apart from the SysML-based systems architectural models, which define the requirements and structural, behavioral, and parametric characteristics of the system. A way to link these external models to the SysML-based models is provided by parametric modeling. Parametric modeling combines behavior and structural design models with engineering analytical models and tools, and system requirements (Johnson et al. 2007). Additionally, since parametric modeling has a constraint representation, limitations on a system or its internal components and attributes, such as voltage limits, minimum temperatures, weight, or reliability constraints, for instance, are easily rendered.

SysML diagrams by pillar: Structure diagrams (apart from the parametric diagram), Behavior diagrams, Requirements Diagrams, and the Parametric diagram (which is contained within the Structure block) are the four *types* of SysML diagrams that the Systems Engineer can use to conceptually represent and relate the four SysML pillars, as shown in Figure 4.5 below (Hause and Francis 2008). Requirement matrices, which are simply a subclass of Allocation Tables that relate requirements to other model elements (e.g., Blocks, Test Cases, Use Cases, and Packages) or between types of requirements (such as functional to non-functional requirements traceability and metric

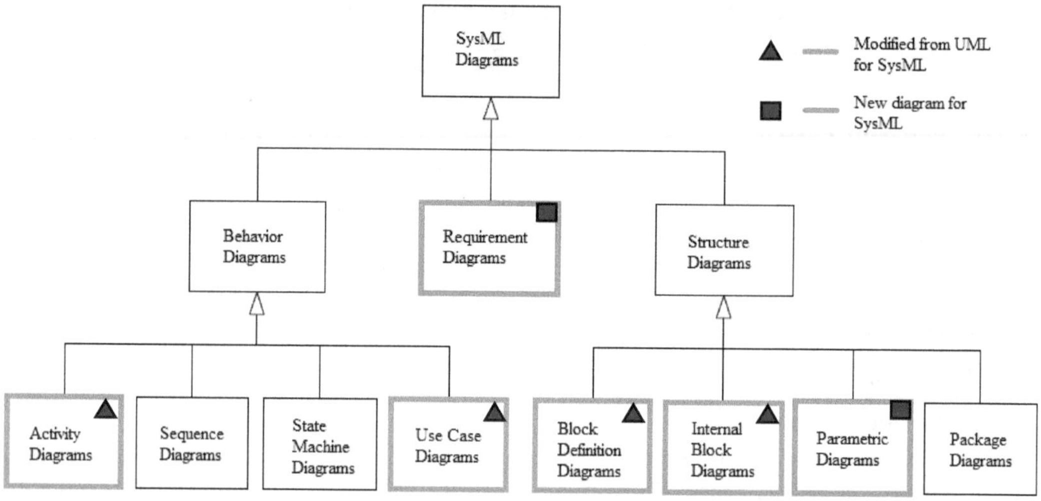

FIGURE 4.5 SysML diagrams and their changes and additions from UML.

reporting), should be distinguished from Requirements Diagrams. An Allocation Table is a particular kind of relationship matrix that is just as helpful for mapping system modeling elements – that is, system components as blocks or parts – across hierarchical groupings of elements.

The systems architecting process, from the problem domain to the solution domain, may be applied using these four SysML pillars and related diagram artifacts. Requirements Diagrams, for instance, can be used to record stakeholder demands, system-level A-spec requirements, and even component-level requirements in the solution domain. Use Case diagrams, in the Behavioral Pillar, can be used to establish contextual information about the system from program information like the CONOPS or the OpsCon. Activity Diagrams can be useful in functional flow analysis, and State Machine Diagrams apply well in documenting dynamic system behaviors. Sequence Diagrams, are also in the Behavioral Pillar and, can provide information about the interactions between objects and/or distributed services and so aid in capturing dynamic interactions of the system elements. On the Structure Pillar side, Block Definition Diagrams (BDDs) are good for showing structural details such as the functional or physical decomposition of the system. Internal Block Diagrams, also in the Structure Pillar, can show the inner workings of a static block. Parametric Diagrams can aid in the simulation of system internals by using Parametric Diagrams from previously modeled parts, binding connectors, and constraint properties (Wagner et al. 2012). Parametric Diagrams can also aid in the early generation and maintenance of requirements by using constraint blocks to provide relations that check and enforce the requirement's technical performance measures such as weight limits, size constraints and the like. Package Diagrams are static diagrams that are part of the Structure Pillar and are useful in illustrating the contents of packages and the relationship between the packages.

4.1.6 Mathematical Considerations

This section aims to explain some important mathematical concepts concerning mathematical logic and its formal integration into our system models. Propositional Calculus and Predicate Calculus are the two formal languages of mathematical logic. Sentential calculus is another name for propositional calculus. This kind of calculus deals with the terms NOT, AND, OR, and IMPLIES. To attain "consistency completeness and independence of axioms" was the reason this type of calculus was developed (Wolfram Homepage 2023). With this form of calculus, letters, and symbols can be used for NOT, AND, OR and IMPLIES functions, and logical problems can be addressed. Propositional calculus also gives rise to several rules, including Modus Ponens, which eliminates a conditional statement in a logical argument (or proof) and so shortens and clarifies evidence. An example of a Modus Ponens rule symbol is $F \rightarrow G$ wherein the arrow symbol means IMPLIES. In this case, *F implies G*. Furthermore, this rule provides the foundation for Hilbert-type proof theories, whereas Gentzen-type proof theories are built on introduction and elimination rules. In this kind of calculus, these all-formal theorems are known as "tautologies, and every one of them is formally provable" (Wolfram Homepage 2023). Because we are now able to implement parameters using a formal language, propositional calculus could be used in our systems modeling process. For instance, we can specify that the camera on our AUAV must be able to activate in 5 seconds OR fewer.

A formal language for expressing propositions constructed from atomic formulas is called predicate calculus. Compared to the preceding calculus kind, this one is more general. We can use variables, quantifiers, and relations with this type of calculus. A phrase expressed in terms that turn into a proposition is called a predicate. This is precisely how we would set limits on our model, which addresses, for instance, the amount of time needed to achieve a goal. Interesting fact: predicate calculus and propositional calculus are also known as first-order logic and zeroth-order logic, respectively (Goldmakher 2023). One way to understand the difference between these two types of calculus is that propositional calculus is the study of Boolean algebra without the use of quantifiers, variables, or relations – that is, without dealing with predicates.

4.1.7 Systems Architecture Methods and Tools

The terms *science* and *art* have been used to describe systems architecture development, and systems architecting methods in general, with systems developments that reflect a compromise of these two terms resulting in the ultimate product (Maier and Rechtin 2000). Since most systems are developed using different methodologies, systems architecting, and developing/characterizing the systems architecture in particular, tend to fall into one of two categories: science or art. While the Normative and Rational methods encompass common scientific approaches, the Participative and Heuristic methods are more suited for an artistic approach (Lang 1987).

4.1.7.1 Normative Method

The Normative Method in systems architecting is solution-focused and mostly centered on the codification of rules and stringent requirements for a system. It is largely grounded in science and standards (Maier and Rechtin 2000). Systems created using this methodology often emphasize analysis and system obligations while providing a logical description of element interactions (Boella and Torre 2006). When applied rigorously enough, this method produces systems whose success is almost certain if the system's scope and design don't vary too much over time. A significant drawback of this method is the rigidity needed for it to function at its best. Changes in requirements, desires, or circumstances lead to a breakdown of the normative method and pave the way for the rational method, as is normal with big and complex systems (Maier and Rechtin 2000).

4.1.7.2 Rational Method

The Rational Method in systems architecting is based on the scientific foundation of the normative method, but it shifts the emphasis from solution-based to method-based, allowing for greater flexibility and adaptability to changing system priorities and requirements (Maier and Rechtin 2000). This rule-based method is analytical, easily validated, and well-understood, much like the normative method. The main benefit of the Rational method is that it can be applied in a modular fashion, making system changes easier to implement because the rules governing the system are given priority over the originally planned solution.

4.1.7.3 Participative Method

With the Participative Method of systems architecting, a more creative approach is implemented. The aim is to use insight and vision to develop the systems architecture with stakeholder consensus.

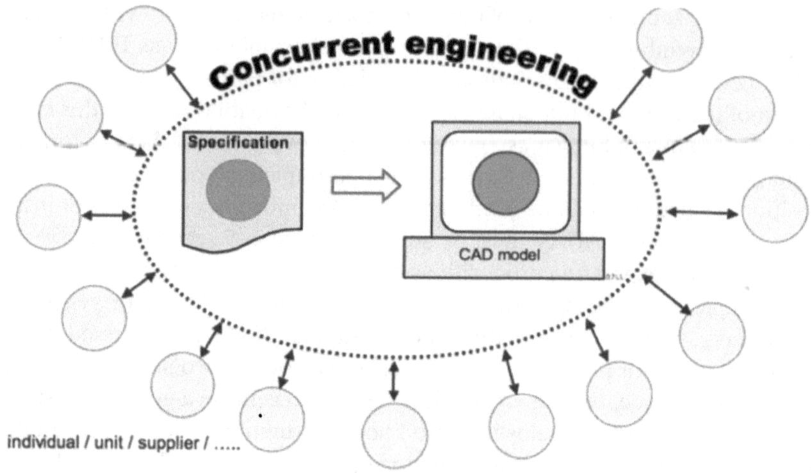

FIGURE 4.6 Concurrent engineering. (Source: https://commons.wikimedia.org/wiki/File:Concurrent_engineering.png).

This is in stark contrast to science-based normative and rational techniques (Maier and Rechtin 2000). Said another way, participatory design fosters a collaborative atmosphere in which researchers, end users, and designers all contribute to the design methods that are most likely to meet the system's requirements as well as the intended system outputs (Sanders and Stappers 2008). Close collaboration and communication are encouraged by this method, as seen in Figure 4.6, Concurrent Engineering.

The participatory method is inductive rather than deductive, and it is always changing with each system it is used to build, much like the study of medicine. This method works especially well for novel systems, where it may be challenging to monitor the system's key outputs and there is little to no a priori data available. By using a participative architecture development method, it is possible to achieve success measures that are difficult or impossible to measure, and results in a system that is based more on recommendations and assumptions than on strict regulations (Maier and Rechtin 2000). These positive outcomes are frequently achieved through a contemporary development process known as concurrent engineering, which is included in the participative category. The focus of concurrent engineering is on designing a process where development activities occur simultaneously and within the same time window (Sawhney 2018). The goal of this technique is to guarantee broad agreement and collaboration between the system design team and stakeholders by implementing simultaneous engineering and real-time feedback throughout a system's lifecycle (Winner et al. 1988). Although the Participative Method is very powerful and economical in theory, difficulties can occur in real-world applications where exchanging computer models may prove challenging and impede progress (Rosenblatt and Watson 1991).

4.1.7.4 Heuristic Method

The Heuristic Method is the last and least scientific strategy used in system architecting. Heuristics, commonly known as "common sense," are tactics learned from the past that are then used in a comparable scenario that arises in the present (Pearl 1984). One of the main principles of the heuristic problem-solving strategy is simplicity, which holds that the most important instrument for assisting in the solution process is the simplification of a problem (Polya 1945; Maier and Rechtin 2000). The best way to lower costs and reduce schedule overruns while improving system reliability is to simplify the system. "Don't assume that the initial formulation of the problem is necessarily the best, or even the appropriate one," is another popular heuristic approach used in systems development (Maier and Rechtin 2000). When combined with other strategies, heuristics can be used to solve a wide range of systems development problems.

Systems architecting and developing the systems architecture is not so much a science as it is a philosophy and an art (Rechtin 1992). Considering this, it is to be expected that many of the methods and guidelines governing the creation of the systems architecture are based more on ideas and context than on specific tools. Therefore, while discussing the origins of a given system architecture, it is often heard that common sense was applied and nothing more (Rechtin 1992). This is not to argue that there are no available tools for systems architecting; rather, there are only a fixed number of these tools that are accessible, and many of these tools are taken directly from the realm of Enterprise Architecture. A list and a short description of some of the contemporary systems architecting tools are provided as follows.

4.1.7.5 Lucidscale

The purpose of this tool for systems architecting is to arrange cloud environments and give organizations clarity so that design teams can work efficiently to create and implement cloud-based systems architectures. Diagrammatically presented, these architectures offer uniform graphics and automatically updated schematics for cross-device real-time collaboration. The process of approving architectures is streamlined since the clarity offered enables audiences like executive management and stakeholders to understand the material better (Lucid Software Inc. 2022).

4.1.7.6 Arkitect

System safety, a sometimes-overlooked aspect of system design, was taken into consideration when creating this systems architecting tool. In contrast to many systems development efforts, which create safety plans either during or after the late stages of systems modeling, Arkitect aims to put safety first to guarantee that the system is built with good safety principles, rather than safety coming as an afterthought. The tool's type-based, graphical interface makes it easy for system designers to use by displaying the architecture in a hierarchical manner. Like SysML, Arkitect can interface with other popular tools used in systems engineering, like Excel, DOORS, and Simulink (Langheim et al. 2010).

4.1.8 Implementation Methodology and Procedures

Whether using SysML tools or basic flow diagram software like Microsoft Visio, it is important to make sure that all relevant stakeholders are identified early and included in the OV-1. The stakeholder's needs (and approved wants), through the stakeholder requirements, must be properly assessed, and the resulting capabilities and features included in the OV-1 context diagram. In this manner, any upcoming OpsCon (OV-2) diagrams and model components need to link to and trace back to the OV-1 stakeholders. This guarantees that the SysML-based modeling process considers the requirements of all stakeholders and that these requirements are properly addressed (Anyanhun et al. 2020). When designing, creating, confirming, verifying, and validating a system, it is equally important to pay attention to stakeholders, particularly the system's end users (Crawley et al. 2016). Correctly evolving the OV-2 into Use Case Diagrams and Activity Diagrams so that there are fewer functional and interface flaws later in the systems engineering development process depends on correctly interpreting and understanding stakeholder requirements (with context) and ensuring there is thorough traceability later in the verification and validation stages (Haidrar et al. 2017).

The OV-2 should then be used to produce Use Case diagrams. After generating Use Case diagrams from the OV-2, Activity and State Diagrams are made to include information flows and diagrammatically represent system operations and activities with more specific details and limitations. Concurrently, the requirements can be broken down from functional system-level requirements into non-functional requirements (e.g., performance requirements), and the system logical decomposition from the high-level Package Diagram can start to iteratively take shape. To guarantee requirements coverage and full requirements satisfaction, these non-functional requirements must then be coupled (linked) to model elements (blocks and Internal Block Diagrams) and Use Cases.

TABLE 4.2
Comparison of OV-1 and OV-2 Artifacts/Views

Characteristic	OV-1 (CONOPS)	OpsCon (Implemented as OV-2)
Purpose	Shows how system(s) will be used in the organizational context. Shows how leadership intends to operate the organization. Useful in strategic planning Describes what a system will do (not how it will do it) along with why (rationale). Provides a high-level overview of the system's operations and technologies/phenomenologies	
Audience	Stakeholders, organizational level	Users, acquirer, supplier, other organizational elements
Content	*Stakeholder-organizational focus:* System capabilities, actors, entities, and how they are used to achieve the organizational goals and objectives	*Operations-based:* Describes system characteristics of the to-be-delivered system from the user's viewpoint. Communicates overall quantitative and qualitative system characteristics
Level of detail	High-level	High-level
Use Cases	Depiction of the systems as black boxes. Show system's role in achieving the goal(s) of the organization	Systems operational nodes, activities performed at each node, connections and flows

Implementing the OV-1 and OV-2 SysML artifacts: The Department of Defense Architecture Framework (DoDAF), like the logical System Theoretic Process Analysis (STPA) technique, logically delays requirements capture until the system's intended application and stakeholders are known and defined. This entails specifying the system's goal(s), its users and stakeholders, as well as providing information about the operational details of the system in its environment. The OV-1 (Concept of Operations) and OV-2 (Operational Concept) views, which are systems architecture artifacts themselves, can be used for this. Table 4.2 shows a comparison of OV-1 and OV-2 details such as purpose, audience, content, level of detail, and related Use Cases.

It is important to understand the differences between the OV-1 and OV-2 views/artifacts in SysML/CATIA System of Systems Modeler, as well as how they serve to define the stakeholders and enterprise-level system description in the model. There is a significant difference between the CONOPS and the OpsCon and according to ISO/IEC 29148. The ConOps document, which in SysML models should be completed as an OV-1 before any document-based artifacts, is defined by ISO/IEC 29148 as follows:

> The ConOps, at the organization level, addresses the leadership's intended way of operating the organization. It may refer to the use of one or more systems, as black boxes, to forward the organization's goals and objectives. The ConOps document describes the organization's assumptions or intent in regard to an overall operation or series of operations of the business with using the system to be developed, existing systems, and possible future systems. This document is frequently embodied in long-range strategic plans and annual operational plans. The ConOps document serves as a basis for the organization to direct the overall characteristics of the future business and systems, for the project to understand its background, and for the users of this International Standard to implement the stakeholder requirements elicitation
>
> *ISO/IEC (2018)*

The operationally-oriented OpsCon document, which should be finished second as an OV-2 diagram rather than a document-based artifact in SysML models, is defined separately by ISO/IEC 29148 as follows:

> A system OpsCon document describes what the system will do (not how it will do it) and why (rationale). An OpsCon is a user-oriented document that describes system characteristics of the to-be-delivered system from the user's viewpoint. The OpsCon document is used to communicate overall quantitative and qualitative system characteristics to the acquirer, user, supplier, and other organizational elements.

The key distinction is that the creation of the ConOps comes first, ensuring that the environment, capabilities, core functions and needs, and stakeholders are all identified. The ConOps serves as the foundation for the creation of OpsCon and is centered on the enterprise and end-user level of detail. The OpsCon describes functional aspects of the system in the sense that it describes what the system must do, and so indicates some understanding of early aspects of the systems architecture and consequently having completed some early architecting. The OpsCon follows the initial ConOps. SysML diagrams are used in the initial SysML system architecting process to generate the ConOps and the ensuing OpsCon diagrams.

In creating a new project or program, the first consideration is who are the stakeholders that must be included in the effort. This is because a decision must be made at the program's inception regarding the model framework that is to be used: DoDAF, MoDAF, or another architecture framework. The E-Government Act of 2002 and the Clinger-Cohen Act of 1996 were two key legislations that gave rise to the Department of Defense Architecture Framework, or DoDAF. The following are the ways in which these laws compelled the defense sector to adapt to the 21st century. The Clinger-Cohen Act of 1996 stated, "With respect to an executive agency, information technology architecture means an integrated framework for evolving or maintaining IT and purchasing new IT to achieve the agency's strategic goals and information resources management goals," (Clinger Cohen Act of 1996, Title 40 1996). In essence, it compelled all federal agencies to alter how they selected and managed their IT (Information Technology) resources. The E-Government Act mandated that federal agencies adopt architecture methods to enhance the administration of electronically supplied services (Public Law 1007–347 2002).

As a sidenote, the CATIA System of Systems modeler launches with the DODAF selectable out-of-the-box because this architectural framework is so fundamental, pervasive, and essential to modeling. We describe some summary details about starting the modeling process. After loading the desired modeler's perspective (e.g., System Architect, Systems Engineer) or the architectural framework (e.g., DODAF, MODAF, ….), the CATIA System of Systems Modeler tool opens a screen to either import an existing modeling project or start a new project.

Setting up appropriate *packages* is one of the first things that needs to be determined. To accomplish this, choose an element and define a package by left clicking in CATIA SOS Modeler's containment field. Packages will assist in providing further clarification on important aspects of the model and are "used to arrange the model" (Holt and Perry 2018). Block Definition Diagrams, Internal Block Diagrams, Parametric Diagrams, Activity Diagrams, Use Case Diagrams, and Requirements Diagrams are systems artifacts in these packages that are commonly used. A glossary table is essential in defining terminology that can be unclear or misunderstood so stakeholders and developers alike can understand the model appropriately. We will go through each diagram in the parts that follow.

The blocks that we include inside our modeling software packages are composed of Block Definition Diagrams, or BDDs. A block is a versatile feature that can be used to explain an essential aspect of the system such as its structure, and linked together with other blocks can give an indication of the system's architecture. Blocks can also depict the relationship between and among systems as well as subsystems. A BDD can therefore be extensive and intricate. It is possible to add components, add constraints, add values, reference other blocks, and establish flows between blocks in a BDD. Because of its role in defining our system, this kind of diagram is essential to every model. We can use CATIA System of Systems modeler to generate blocks by selecting block with a left click in the containment field. After the blocks are constructed, we can start defining their surrounding environment and subsequently display them graphically in a BDD. This serves as the foundation of the model and the source of many of the links to other model elements.

Another useful modeling diagram is the Internal Block Diagrams (IBD). An internal block diagram, which contains components, connections, ports, and interfaces, belongs to a block. In the CATIA System of Systems modeler, it appears as a white box. The model designer can illustrate activities and dependencies between these blocks inside an IBD. These complex diagrams aid in the system designers' understanding of the interactions between the blocks (Weilkiens 2008).

Activity Diagrams are a vital element of our CATIA System of Systems modeling software and are indispensable in modeling the system's flow and behavior. A useful analogy is that the Activity Diagram can be like a flowchart with the ability to have more precise linkages between elements. This means that the relationships between system elements can be further defined to describe the precise events or inputs being added to the system that drives the activity (Paech 1999).

To meet and monitor the system's requirements, Requirements Diagrams are generated. CATIA System of Systems Modeler can be used to generate a requirements table, which can then be used to link the requirements with other important system elements, such as test elements and other useful attributes needed for qualification purposes.

4.1.9 The System Architect

The systems architect is a central figure in the creation of systems, especially ones that are large and complex. Rather than being overly specialized like a discipline engineer, the systems architect is often more of a function within a corporation, with the architect being a systems engineer with a thorough understanding of how systems interact and the consequences of those interactions (Firesmith et al. 2008). From one angle, a systems architect can be seen either as a specialist who concentrates on creativity, specialized knowledge and analysis or as a generalist whose wide knowledge aids in uniting cross-functional groups under a common framework. The onus of developing the system's architecture or framework rests with the system architect in either scenario. Since "the success or failure of many civil and defense systems depend mainly on their design," this architecture's creation is essential (Rechtin 1992).

Over 100,000 professionals work as systems architects; however, there are many more who go by other names (Crawley et al. 2016). One might wonder, what is this relatively new expanding discipline? The increased complexity of modern systems is mirrored in the growing demand for systems architects. "Modern society tends to use more and more complex electronic equipment" even from the perspective of the consumer (Herrmann et al. 2005). In addition to being necessary from a technical standpoint, architecturally unique systems are more profitable from a financial standpoint (Rechtin 2002). U.S. Government acquisitions, especially those in the DoD, are often more complex than their civilian counterparts. The complexity of systems means that there is no mathematical method to find the "correct" response (Crawley et al. 2016). Consequently, a systems architect is required to do much more than just establish a framework and evaluate risk and opportunity (Crawley et al. 2016). Figure 4.7, below, provided by Dam, provides a humorous illustration of different discipline perspectives of optimality when it comes to designing an aircraft (Dam 2019). Streamlining is essential and other considerations are noticeably absent from the aerodynamics standpoint. Large intakes, fuel, and engines may be considered ideal for propulsion. Although manufacturing would like something as simple to construct as putting two pieces of wood together, structural engineers would prefer a steal beam with wings. Even though the illustration uses lighthearted comedy, this is regarded as a classic example of how different design teams see "optimal" solutions. In this instance, the systems engineer/architect is regarded as the functional design engineers' "Honest Broker" (Dam 2019). The systems architect needs to have a comprehensive understanding and holistic perspective of the fundamental stakeholder needs, wants, and requirements to be "viewed as someone who is knowledgeable about each of the disciplines, seen as valuing the design engineer's points of view, and can communicate and translate between other disciplines" (Dam 2019).

From one angle, the architect possesses a wide variety of information, and so they are viewed by many to be generalists. Even so, if the systems architect possesses a broad range of knowledge, they are still typically regarded as specialists in the field of system engineering.

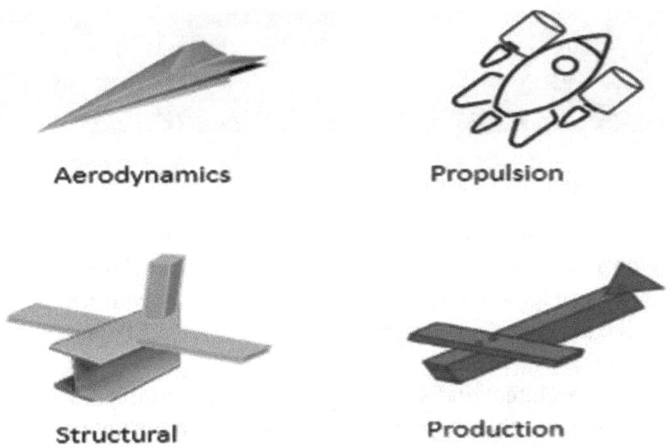

FIGURE 4.7 How different disciplines may see optimal design.

The systems architect needs to hone (and become proficient in) a very specialized skillset. Apart from understanding the requirements of the design disciplines, they also need to develop an architecture that meets the demands of all the disciplines, limits the solution, but does not unduly restrict designers when it is not required. Systems architects, according to Rechtin, are specialists in systems; as such, their models need to be refined into the technology-specific models of the domains in which their systems are to be implemented (Rechtin 2002). Systems architects are not generalists. They are "specialists at simplifying complexity, clarifying ambiguity, and focusing creativity" (Crawley et al. 2016).

Characteristics of the system architect: Even with a specialized skillset, system architects nevertheless need to maintain a strong technical foundation, possess a wide variety of knowledge, and prioritize practicality. It is a common risk for a systems architect to become so cut off from peers or the system that effectiveness may wain. Care needs to be taken by the system architect not to fall into some traps that can mitigate their effectiveness. These traps are characterized by Buschmann as the "techno-geek," the "architecture dwarves," and the "architect astronauts" (Bushmann n.d.).

The "architect astronaut" is a person who demands high standards and goes too far. Often, their guidance is in the stratosphere and is, "So distant from systems they build that much of what they detail lacks technical depth and practicality" (Bushmann n.d). The architectural products in this case end up being relatively ineffective.

Another kind of architect is the "techno-geek." These types of system architects disregard the real *commercial* needs of the organization, since they are so technically busy. These types of system architects are the ones that program management, executive leadership, and business development lament because of the perception that, "systems architects are *techno-geeks* and seemingly don't care about the business or financial health of the organization" (Bushmann n.d.).

The "architecture dwarves" are a third kind of trap to stay away from (Bushmann n.d.). At the opposite end of the "astronaut architects," these are system architects who enjoy delving deeply into the specifics of technological advancement. This may lead to too complex, overly defined, and challenging to implement systems architectures.

It is essential that the system architect maintains their pragmatism to avoid these extremes of having either too high level or too low level of a focus. This is not to argue that technical zeal, abstraction, or in-depth detail-oriented thinking are undesirable behaviors. All are occasionally required. For example, a preference for abstraction can facilitate high-level communication with a wide range of stakeholders. Technical competence and a wide grasp of technical intricacies are required in addressing complex problems. Sometimes having a keen interest in understanding the details of a system or problem might even make one an "architecture dwarf." The pragmatic architect embodies

FIGURE 4.8 Typical development of a systems architect.

the finest qualities of these varied styles. These architects need to be able to maintain a high-level perspective and develop abstractions, akin to an "architect astronaut," but also be able to change into a "techno-geek" with a thorough comprehension of the solution's technical details. At times, going almost too far, much like the "architect dwarf." The key is to have a balanced approach and employ these styles as necessary.

The pragmatic systems architect and successful architectural implementations are made of balance. The pragmatic systems architect "rarely fails, owing to their knowledge, skills, experience, and everyday practice," even though no systems architect is flawless (Bushmann n.d.).

Muller asserts that genuine systems architects are "scarce." To be effective, they require a "broad range of knowledge, abilities, and experience" (Muller 2011). A typical development path for a systems architect is depicted in Figure 4.8.

Whether it's mechanical, software, firmware, or electrical engineering, most architects begin with a particular technical specialty. This technical depth is important in developing one's character, problem-solving skills, and teamwork in addition to being essential for early career advancement. A "deep comprehension of a single technological subject is a necessary underpinning," as Muller states (Muller 2011).

A systems architect's attributes then progress from a single technical knowledge base to an understanding of several technologies that are relevant to a system, resulting in a growth in general knowledge and an expansion of expertise with various technology areas. This is a logical progression since more seasoned designers are given greater authority, bigger projects, and development possibilities. This growth in depth and experience frequently results in additional inputs across the entire system and could signal the start of a specialized career in systems engineering.

It is common for an engineer to obtain insight into business needs and operations after broadening their technical knowledge base. The "market, competition, value, logistics, and service factors" start to become clearer (Muller 2011). Working with various stakeholders and knowing "who does what and why" are better understood from a process perspective. This process perspective and analytical/technical background can be of great benefit to those engineers who also have business acumen and have interest in applying their unique skillset to providing comprehensive business solutions.

A blossoming systems architect's awareness of the social and psychological aspects of a system's lifecycle is what is meant by building psychosocial abilities, which might seem a bit sinister. Diverse parties involved bring unique goals and points of view to the table. Making better social connections when working with stakeholders, teams, engineers, and company executives is important. The social and political factors in question "interact heavily with the architecting process, directly impacting the missions and designs of large scale, complex systems" (Rechtin 1992).

According to Muller, a systems architect is more of a generalist than a specialist. A broad understanding of several technical domains is essential for architects "because one of their core functions is to generate top-level specifications and designs of the system" (Muller 2011). Figure 4.9 illustrates the necessity of both the generalist and the specialist. Generalists are needed for their capacity to integrate or span areas, whereas specialists are needed for their extensive expertise. In most cases, a program will employ a far higher number of experts than generalists.

Rechtin emphasizes that system architects are "not generalists; they are experts," in contrast to Muller's unequivocal assertion that they should be generalists (Rechtin 1992). Between these two accomplished, well-known, and successful systems architects, this sounds somewhat contradictory.

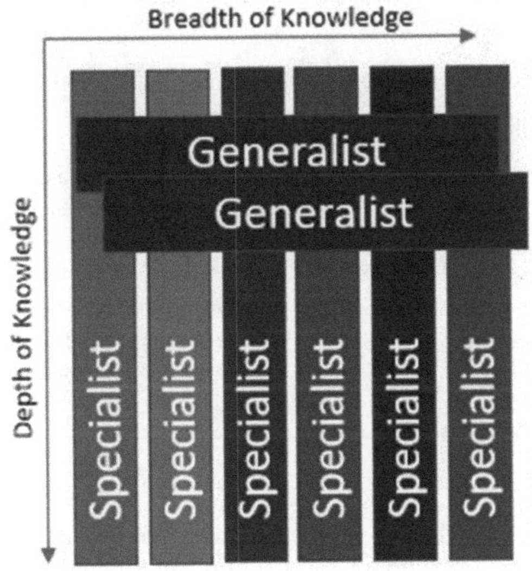

FIGURE 4.9 The specialists and generalist have complementary skills.

More importantly, people wishing to recruit or assess systems architects as well as those aspiring to become systems architects may find this advice perplexing.

Like many things, though, context is key. Muller is referring to the technical skill base when he considers system architects as generalists. It is true that to understand, interact with, and design architectures for the experts, architects, like other system engineers, require a broad skill set as generalists. Given that architects have specific knowledge within the field of system engineering, Rechtin's observation that they are specialists is likewise accurate.

4.1.9.1 Roles of the System Architect

It is often unclear what is expected of the systems architect. System architects design the systems architecture, of course, but how can this expectation be pragmatically scoped? The role of the systems architect can be approached from a more practical perspective. Bass and others describe the necessary steps involved in creating a sustainable architecture as follows (Bass et al. 1998):

- Establish the system's business case;
- Gain understanding of the requirements;
- Develop or choose the architecture;
- Communicate and articulate the architecture;
- Assess and analyze the architecture;
- Develop/implement the system using the architecture;
- Confirm that the implementation is consistent with the architecture.

Another way that Crawley et al. attempt to define the systems architect's role is in terms of three primary, yet not exhaustive, functions of a system architect. These primary functions are:

- Reduce ambiguity
- Use creativity
- Deal with complexity

Surprisingly, both these views fit rather nicely into another characterization of a system architect's roles, namely the roles of creator, moderator, and cat herder. Table 4.3 presents

TABLE 4.3
Concise Definition of Architects Roles from Three Different Sources

Roles of the Architect According to		
Clark	**Crawley et al.**	**Bass et al.**
Creator		• Creating or selecting the architecture • Analyzing and evaluating the architecture
Moderator		• Creating a business case for the system • Understanding the requirements
Cat Herder	Manage complexity	• Ensuring that the implementation conforms to the architecture • Representing and communicating the architecture

Source: Courtesy of Jason Clark.

Moderator/reduce ambiguity: Despite not hearing the title of *moderator* very often, the systems architect does indeed serve as a moderator for a significant amount of time. Different agendas and interactions between stakeholders, which include those between the technical team and business development, the client (requirements) and the development team, and the distinctions between functional designers and systems architects, must all be carefully managed and controlled. These stakeholder interactions are required to make the system less confusing and often require moderation.

Gaining a thorough understanding of the needs (and wants) of the stakeholders is usually the first, and possibly most critical step in reducing ambiguity. To do this, the systems architect will often collaborate with a requirements/systems engineer.

Many of the stakeholder's requirements are simple to understand. However, some requirements can be unclear, ambiguous, and/or incomplete. For example, terms like "quick" and "little" are frequently used, yet they need more thorough definitions. Even something as basic as the color "green" might be confusing. Various people would perceive green in different ways – for example, as mint, olive, or forest green. Even something that is clearly defined can seem to be in conflict with something else.

One responsibility of the systems architect is to understand these ambiguities, make sense of them, and create an architecture that is consistent with any fixes. If getting input from stakeholders isn't feasible, the systems architect must overcome problems using precedent, discretion, standards, ingenuity, and – above all – experience. In practical terms, the systems architect serves as a mediator/moderator between the design/development team and the stakeholders in relation to the stakeholder requirements.

The systems architect must understand the needs of various stakeholders in particular, those of the user. For example, as Bass notes, the systems architect must "make a business case for the system" (Bass et al. 1998). Even a superb technological solution will fail if it cannot demonstrate that it supports the needs of the company and obtains support from executives, business development, and other relevant stakeholders.

Creator/employ creativity: The "creative task of establishing a concept for the system" comes next once the requirements, goals, and objectives for the system have been established (Crawley et al. 2016).

The function of creator is arguably one of the most distinctive that a systems architect may have. The systems architect has more creative freedom than just about any other position. The systems architect studies the specifications before developing an idea. The overall system concept is also influenced by internal business or political requirements. The systems architect navigates these waters and synthesizes a variety of concepts from seemingly a blank slate at times. These concepts are rapidly narrowed down, usually with feedback from the design team. Subsequently, the systems architect develops a framework to enable the system to be developed.

One of the greatest gifts and greatest challenges of being a systems architect is having a creative spirit. While most engineers love to be creative, they frequently do not have the same opportunities as those that are available to the systems architects. On the other hand, a system may fail if its architecture is inadequate or the systems concept does not adequately align with the requirements, both internal and external. Thus, there is a great deal of responsibility involved in this creative process.

Because of this great responsibility and the great effect the systems architecture has on the entire system and also the development team, the creative skill set is widely acknowledged as a critical attribute of system architects. This is the highest level of skill in Bloom's taxonomy shown in Figure 4.10. Even so, it is necessary to learn and become proficient in lower-tier skills first. Muller states that proficiency in the upper levels is critical in gaining competency in systems architecting (Muller 2018).

Cat herder/manage complexity: Now it is time to present the Cat Herder, shown in Figure 4.11, which is perhaps no one's favorite role, to step in once the systems architect has developed a plan

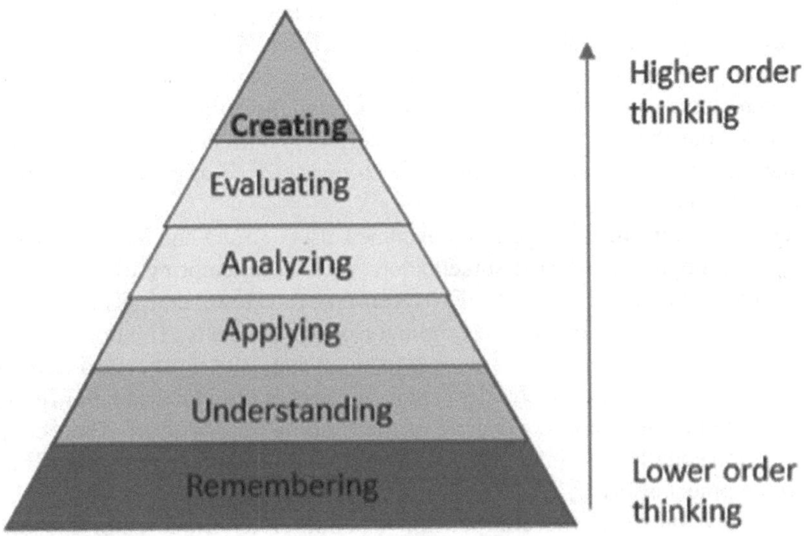

FIGURE 4.10 Lower and higher-level skills in Bloom's taxonomy.

FIGURE 4.11 Herding cats in the world of the systems architect. (Adapted from EDS SB 2000 Commercial, modified as 'Architect' herding 'cats' (EDS SB 2000).

and an architectural framework. Awareness of the systems architecture needs to be spread throughout the organization. Additionally, "that implementations conform to the architecture" is also the responsibility of the systems architect (Bass et al. 1998). Along with project engineers and designers, these people might also include representatives from manufacturing, quality, safety, and many other parts of the organization that are involved in the systems development lifecycle. The result is required interactions with a sizable group of experts, each with their own goals, due dates, and viewpoints. It can feel like herding cats at times to try to keep everyone moving in the same direction.

A great way of summarizing the duties of a systems architect is given by Crawley et al. as, "Identifying the essential, consistent, and relevant information by eliminating ambiguity; adding new information through innovation; and managing the explosion of information in the final design are the three major duties of the architect that are all centered around information" (Crawley et al. 2016).

4.1.9.2 Responsibilities and Activities of the System Architect

According to Muller's "Systems Architecting: A Business Perspective," the systems architect dedicates a significant amount of time to simple communication. This entails first gaining a written and verbal grasp of the requirements, goals, and objectives of internal and external stakeholders. Maintaining open communications with stakeholders throughout the creation of the systems architecture and throughout the solution domain is essential. The architectural framework constrains and evolves the system design, and it is important to maintain communications with stakeholders throughout its evolution.

The competition, governmental regulatory agencies, users, direct and indirect clients/customers, and suppliers are examples of external stakeholders. It is quite tempting to call requirements that are specified in the contract source "customer" requirements. However, it is an important observation that the end user is frequently not the originator of the contracts. The documents that include the request for proposals, for instance, can come from an acquisition office and development plans might be assessed by some other stakeholder office. Soldiers or unnamed end users could be the end user and may be the actual *customers* instead of the funding agency itself. Their cooperation and satisfaction are often equally (and sometimes even more) important than the organization that funds the systems development effort.

The need to stay ahead of the competition is another external factor that drives needs, wants, goals, and objectives. In other words, to come up with a proposal that delivers a better solution, it is essential to know what the competition is offering. If systems architects want to be "world-class competitive," Franke says they need to know "how and what other internationally successful programs are doing" (Franke and Levis 2000). This enables the supply of a superior product and is critical to the company's successful operations. Various factors such as price, features, and performance for instance can constitute a superior proposition. A best "value," rather than a particular feature, is frequently desired, that is an optimal compromise among dependability, performance, and cost, rather than the lowest or greatest performance in and of itself.

Other external stakeholders that have the power and responsibility to impact the system are regulatory authorities. The Underwriters Laboratory (UL) is one example of an independent, non-governmental organization that verifies the safety of consumer goods. Among the many services offered by UL are consulting, risk management, inspection, and certification. The FAA oversees certifying and regulating avionics safety. In the United States, the Federal Communications Commission (FCC) establishes rules for electromagnetic emissions. "Intentional, Unintended, and Incidental" are the three ways they classify emissions (Compliance Testing Corporation 2021). In addition, there are emission rules like Class A for business and industry and Class B for homes. The main goals of the regulations are safety of individuals exposed and non-interference with other items. In a similar vein, the European Union (EU) Radio Emissions Directive (RED) program offers comparable regulations in Europe.

Compliance with these standards is essential both throughout operations and, more specifically, at system retirement or end of life (EOL), when several environmental protection regulations become applicable. When it comes to electrical and electronic equipment, one example is the EU's Restriction of Hazardous Substances (RoHS). The purpose of the RoHS directive is to lessen the likelihood of harmful substances. For example, restrictions on the handling of lead, since it may leak into groundwater during product disposal. The amount of Waste Electrical and Electronic Equipment (WEEE) has risen dramatically due to the increasing number of devices used by customers and their ongoing electronics upgrades to many of them. An approximate estimate of 1.1 million ton of WEEE was produced in Germany in 2005, according to (Herrmann et al. 2005). To increase product recyclability and to shoulder responsibility for recycling and end-of-life expenses, new EU laws like the WEEE directive and RoHS "force[s] producers to take action" (Herrmann et al. 2005). This means that when designing a system, systems architects must think about retirement and end-of-life scenarios.

The success of a system depends on communicating with internal stakeholders as much as external ones. The design and development team, program management, manufacturing, and other associated team members are all considered internal stakeholders.

The systems architect spends considerable effort in collaborating with internal design groups to formulate and implement viable architectural solutions. According to Muller, bridging different contributing engineering disciplines, including mechanical, electrical, and software engineering, is a major issue for systems architects working on traditional systems (Muller 2011). It is common practice for the systems architect to collaborate with system, subsystem, and component-level designers to integrate lower-level design details into the overall systems architecture (Muller 2011). The systems architect's interactions with the design team also result in the establishment of other useful architectures such as conceptual architecture, functional architecture, physical architecture, and logical architecture for instance.

Provide framework: A well-designed architecture is incredibly important to the success of a systems development effort. A well-designed architecture is especially important in light of demands from distributed teams, rapid development, and concurrent work (Rechtin 2002). An order of magnitude increase in technical capabilities occurs every 4–6 years when applying Moore's law to engineering. The time that many modern aerospace defense systems firms can create a single system is about equivalent to this. To preserve a decisive, competitive edge, evolutionary systems architecting and systems engineering methodologies are needed due to the ever increasing complexity of modern systems, and the rapid change expected in the contemporary systems development environment.

Thankfully, the systems architect has access to pre-existing architectural frameworks that can provide guidance. When the systems architect needs to create an architecture, an architectural framework can provide the necessary elements to accomplish this task. Numerous well-known architecture frameworks are available and can be customized. Some representative examples are the Zachman Framework, the NATO Architecture Framework (NAF), the European Space Agency Architecture Framework, the Department of Defense Architecture Framework (DoDAF), the Ministry of Defense Architecture Framework (MoDAF), the Treasury Enterprise Architecture Framework (TEAF), the Federal Enterprise Architecture Framework (FEAF), and the Open Group Architecture Framework (TOGAF). Many of these frameworks have been covered in our previous chapter on Enterprise Architecture Fundamentals.

As an illustration of an architectural framework used extensively by the Department of Defense (DoD) in the United States, we will give a succinct overview of the DoDAF and examine it from the standpoint of the systems architect. As you may remember, the purpose of DoDAF was to create a uniform vocabulary for comparing architecture frameworks for Department of Defense systems across the nation (Dickerson and Mavris 2016). Its foundations were laid in the early 1990s with the goal of promoting interoperability and the development of more affordable military equipment. The C4ISR Architecture Framework is the product of the Clinger-Cohen Act of 1996, which promoted the use of architectural frameworks. DoDAF 1.0 superseded the C4ISR framework in 2003.

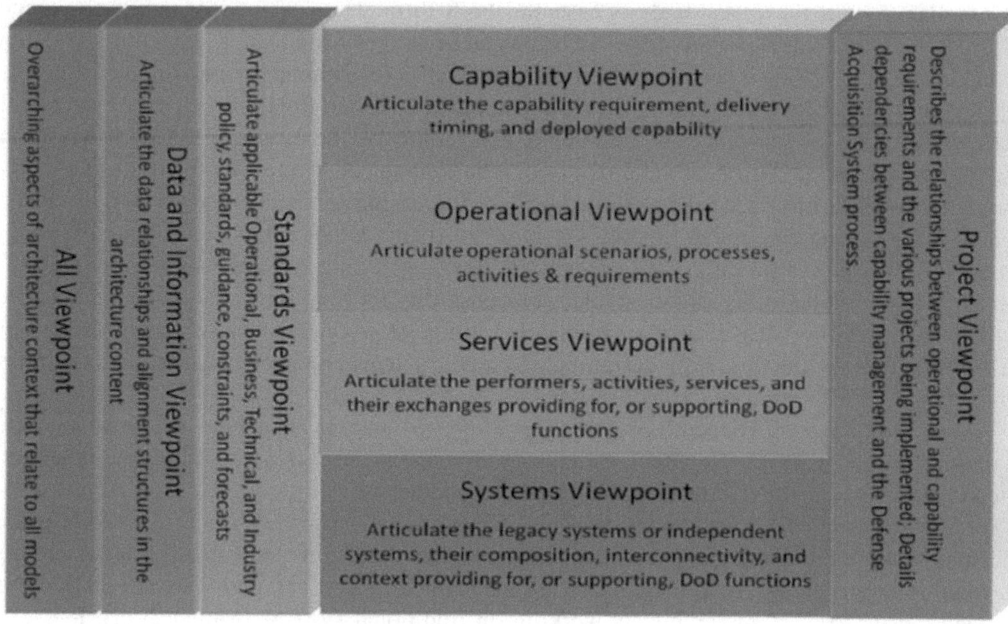

FIGURE 4.12 DoDAF views and relationships (CIO 2024).

The year 2007 saw the release of version 1.5 which was based on different views. Version 2.0 of DODAF was released in 2009, and the current version DODAF 2.02 was released in 2010. The current DODAF 2.02 version has a set of viewpoints (versus views) that provide the organizational structure for the architecture framework. These viewpoints come in text, tabular, or graphic formats and are illustrated in Figure 4.12. Viewpoints are divided into several categories: All Viewpoint (AV), Data and Information Viewpoint (DIV), Standards Viewpoint (StdV), Capability Viewpoint (CV), Operational Viewpoint (OV), Services Viewpoint (SvcV), Systems Viewpoint (SV), and Project Viewpoint (PV). Note that these different viewpoints provide

A comprehensive set of viewpoints in which to organize, understand, and define the system and its evolution through the systems development lifecycle phases. Cross-cutting data about every viewpoint in the architecture framework is provided by the All Viewpoint. The scope, purpose, intended uses, surroundings, and analytical results of the architecture development process can all be represented. Additionally, the AV includes a glossary with definitions for every term used in every architecture description product (CIO 2024).

The functional relationships between the various components of the architecture are detailed in the Operational Viewpoint. It provides a set of artifacts that describe operational aspects of the system that are essential in its evolution. One high-level picture that illustrates the main system elements and how they work together to accomplish the operation tasks/mission is the OV-1.

The system, its services, and the connections that support system functions are all described in the System Viewpoint and consist of a collection of mostly graphical and tabular products. The actual architecture, along with its connections to the OV, is described in the SV. Most products are designed to target physical system aspects such as geographic locations or operational domains (CIO 2024).

The Technical Viewpoint outlines both established and developing standards. The arrangement, interaction, and dependency of system components or pieces are governed by this minimal set of rules. There are numerous artifacts (such as SV-1 to SV-11) within each of the major viewpoints and their contents can be seen in Table 4.4.

TABLE 4.4
Overview of DoDAF v2 Products

All Viewpoint	Operational Viewpoint	Capability Viewpoint	System Viewpoint	Data & Information	Project Viewpoint	Services Viewpoint	Standards Viewpoint
AV-1: overview and summary	OV-1: High-level operational concept graphic	CV-1: Vision	SV-1: Systems interface description	DIV-1: Conceptual data model	Pv-1: Project portfollo relationships	SvcV-1: Services context description	StdV-1: Standards profile
AV-2: Integrated dictionary	OV-2: Operational resource flow	CV-2: Capability taxonomy	SV-2: Systems resource flow	DIV-2: Logical data model	PV-2: Project timelines	SvcV-2: Services resource flow desc.	StdV-2: Standards forecast
	OV-3: Operational resource flow matrix	CV-3: Capability phasing	SV-3: Systems-systems matrix	DIV-3: Physical data model	PV-3: Project to capability mapping	SvcV-3a: Systems services matrix	
	OV-4: Organizational relationship chart	CV-4: Capability dependencies	SV-4: Systems functionality			SvcV-3b: Services systems matrix	
	OV-5a: Operational activity dec. tree	CV-5: Capability to org. dev. map	SV-5a: Operational activity to systems function trace			SvcV-4: Services functionality desc.	
	OV-5b: Operational activity model	CV-6: Capability to operational activities mapping	SV-5b: Operational activity systems trace			SvcV-5: Operational activity to services traceability matrix	
	OV-6a: Operational rules model	CV-7: Capability to services mapping	SV-6: Systems resource flow matrix			SvcV-6: Services resource flow matrix	
	OV-6b: State transition desc.		SV-7: Systems measures matrix			SvcV-7: Services measures matrix	
	OV-6c: Event-trace description		SV-8: Systems evolution description			SvcV-8: Services evolution description	
			SV-9: Systems technology and skills forecast			SvcV-9: Services technology and skills	
			SV-10a: Systems rules model			SvcV-10a: Services rules model	
			SV-10b: Systems state transition			SvcV-10b: Services state transition	
			SV-10c: Systems event-trace			SvcV-10c: Services event trace desc	

These viewpoints can be tailored to the systems architect's and development team's satisfaction based on the nature of the system, and available time and resources. The artifacts shown in Table 4.4 provide a common set of artifacts that provide unique insights and perspectives that, once tailored to the program/project, are essential in successfully developing the architecture for the system and properly framing/constraining the development effort.

The persona, duties, and activities of a systems architect are what make them unique and valuable. System architects should be well-versed in a variety of fields and have a generalist background. The systems architect should also have specific knowledge with respect to the field of systems engineering. Their expertise lies in creatively coming up with novel solutions to problems and defining and implementing viable architectures for the system, which is what really distinguishes them from other engineers. One way to think of the duties of a systems architect is as a moderator, cat herder, or creator. Also, among the many important tasks and duties that the systems architect must accomplish, some of the more notable ones include facilitating effective communications, providing viable architectures, generating and analyzing requirements, modeling systems, interacting with the design, development, and testing (verification and validation) teams. As a system ages, the influence of the systems architect's decisions becomes increasingly apparent and profoundly affects the system's viability, features, capabilities, and lifespan.

4.1.10 Systems Architecture Principles

A "collection of standards that allow meaningful communication among the components" is the definition of architecture (Maier 1998, p. 1). The systems architecture reflects a system's requirements at every stage of its lifecycle and describes the system as a whole. "The allocation of physical/informational function to the elements of form, and the specification of interactions among the elements and with the surrounding context" is another definition of systems architecture (Crawley et al. 2016, p. 124). Form is what a system is, "a system property comprised of parts and structure," whereas function is what a system accomplishes such as, "actions, consequences, and outputs" (Crawley et al. 2016, pp. 24, 124). To create an efficient systems architecture, form and function must be understood. An essential component of systems architecture is the context or the environment in which the system operates. Systems architecture requires a thorough understanding of the system's requirements, its purpose, and the environment in which it will operate. How are functional interactions supported? What are the influences of emergence? What are the supporting functions? What are the internal and external system boundaries? How are things processed? These are some of the questions a systems architect should be asking to aid in the definition of the systems architecture.

A principle is characterized as "a fundamental truth or statement acting as the foundation for belief or behavior" and is intended to direct the design of the system (Woods 2016. p. 1). Figure 4.13 shows 26 fundamental system architectural principles that need to be understood for proper architectural design and implementation. These principles must be ingrained in the systems architect's worldview and must be applied in developing architecture and architectural products.

A principle that "clearly outlines the essential limitations on a system's design without prescriptively defining all the design specifics" is referred to as an architecture principle (Woods 2016, p. 1). The limits and priorities of a system can also be established with the help of architecture principles. Constructive (guiding decisions), reasoned (rational and logical), well-articulated (comprehensible by stakeholders), tested, and significant are qualities that a good architecture principle should possess (has value). In general, principles are meant to offer practical and beneficial advice on system architecture. For brevity's sake, in Appendix 4.B, we provide the definition of each of these principles verbatim from the source (Crawley et al. 2016), and a discussion for understanding and applying the principle. We also provide a SysML-based example relating to the principal at hand, that is focused on our UAV-based case study presented in Section 4.3.

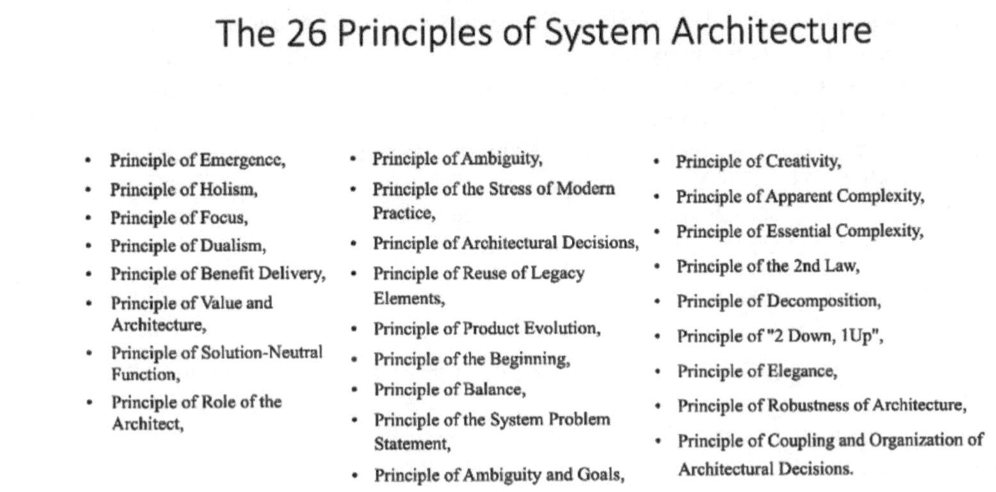

FIGURE 4.13 Twenty-six principles for developing a successful systems architecture.

4.1.11 Case Studies

Since its inception, systems modeling has seen a lot of attention in the literature. The value of systems modeling and object-oriented languages like UML and SysMLunified modeling languages in research was highlighted in a recently published case study from the United States. The proliferation and expansion of powerful internet features has led to an increase in database and data mining research and capabilities. Due to the exponential growth of databases and related technologies, vast quantities of data now reside on the cloud. As the number of users of cloud storage continues to rise, data management has gained prominence. It has been pointed out that there are many ways in which data management modeling could improve efficiency and produce a more effective system (Kennedy 2023; Erickson and Siau 2013).

4.1.11.1 National Case Study

4.1.11.1.1 MBSE Methodology for Implementing Networked Aircraft Control System on Integrated Modular Avionics

There is a growing demand in the aerospace industry to assure stakeholders that all their requirements are satisfied in the final deliverable. Using the Architecture Analysis & Design Integrated Approach (ARCADIA) methodology and the open-source tool Capella, one case study investigates a novel approach to MBSE (Mathew et al. 2018). ARCADIA is a methodology that defines the system's processes on four different levels. The first step in the process is to perform an operational analysis that specifies the operational functions, processes, and goals of the system. The second step is to conduct a systems analysis to define the requirements and features of the system. Establishing the logical architecture (LA) follows next and specifies the steps the system will take to accomplish its objectives. This goes a long way toward figuring out what are the essential parts for the system to work. The last step is to establish the Physical Architecture (PA), which details the construction process and specifies the hardware components that will be required. Establishing a risk-free internal environment is the primary focus of the case study's environmental control system (ECS) which focused on the air conditioning systems with cabin pressurization systems.

Based on the case study, aviation systems often rely on document-based processes, however implementing ARCADIA with MBSE was the adopted approach. Aerospace systems design guideline ARP4754A and MBSE methods were adopted as fundamental building blocks in this case study. Capella was selected as the MBSE tool to create requirements and document the system models. The case study's methodology was multi-tiered, beginning with the foundational level and working its way down to the subsystems and, finally, the items that specify the components required by each subsystem. A multi-tier approach was adopted that focused on all levels of design, from the system level to components, assemblies, pieces, and parts. The case study demonstrated that combining MBSE with the ARP4754A guideline was beneficial in the development effort. By proposing this new MBSE-centric approach, the resulting system development effort was pushed forward, and the entire system was properly, and usefully represented.

4.1.11.2 International Case Study

4.1.11.2.1 China's Modern Electric Multiple Unit Train System

China has strict standards for the reliability and safety of its modern Electric Multiple Unit (EMU) train system, which is a complex system (Ferrogalini 2015). A departure from the conventional document-based method of system engineering was made to construct the EMU using the SysML language and many SysML-based MBSE tools. It was the China Academy of Railway Sciences that provided funding for the case study (Chen et al. 2018). By tying together conventional design loops, this case study consolidated the systems development process into just three stages, each of which acknowledged the inherent iterative steps in system design.

For the first design process stage, the "user requirements" and "external constraints" that were part of the overall requirements analysis process (or requirements loop) seemed to overlap with the steps taken in the problem definition activity to identify problems and stakeholder needs/requirements, as outlined in DoDAF and other Systems Engineering documents (Chen et al. 2018). Some examples of activities in this stage included the analysis of systems settings and tasks (functions), the definition of performance and design limitations (non-functional requirements and constraints), and the identification (derivation) of functional requirements.

During this stage, three out of the nine SysML diagrams – the Requirements Diagram, the Use Case diagram, and the Package Diagram – were used. Use Case diagrams document interactions, functions, and activities; and Package Diagrams contain them in the logical structure. These diagrams help capture and define the requirements and their attributes and provide insight into their interactions and packaging/containment structure.

In their second design process stage, decomposing functional system requirements into EMU subsystem requirements and associated performance requirements (non-functional requirements) was the first step in the process loop known as "subsystem requirements refining and function breakdown (requirements decomposition)." By following the typical procedure for breaking down system Use Cases into activities using three more types of SysML diagrams – Sequence Diagrams, Activity Diagrams, and State Machine Diagrams – the system's states, behaviors, interactions, and logical processes were also evaluated.

Subsequently, in their third and final design process stage, iterations of the architecture (the design loop for subsystems) were accomplished providing for: the definition of the system logical architecture, the definition and configuration of the systems architecture, the selection of the preferred physical components and process plan, and the definition of the internal and external interfaces. Block Definition Diagrams (BDD), Internal Block Diagrams (IBD), and Parametric Diagrams were the last three of the nine SysML diagrams used in this stage to describe and explain the EMU SysML model.

To summarize, this case study did not detail the project's implementation, but it did cover the Chinese railway SysML modeling process and its associated artifacts. It seemed like the three-step procedure outlined in MIL-STD-499B (EIA/IS-632) systems engineering process was quite like the three-stage procedure that the B. Chen EMU case study offered for the Chinese EMU system design using MBSE tools. The three similar steps of the MIL-STD-499B process are requirements analysis, functional analysis/allocation, and synthesis, as shown in Figure 4.14.

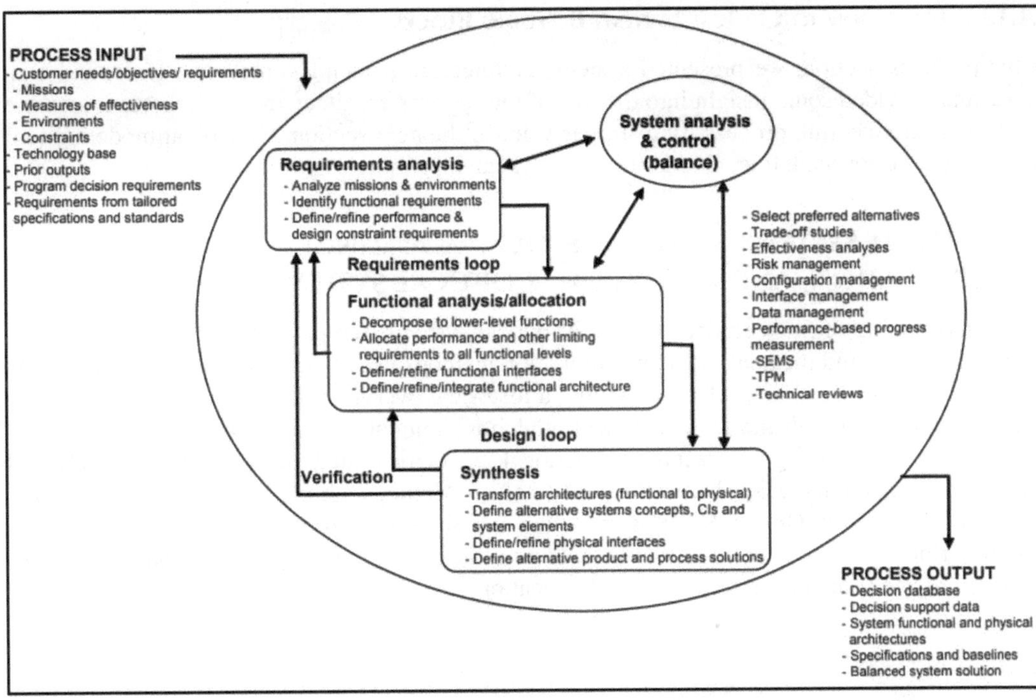

FIGURE 4.14 The MIL-STD-499B (EIA/IS-632) systems engineering process.

This case study demonstrated that SysML and MBSE can be used as the foundation to effectively implement a highly regulated complex systems development effort, the Chinese Electric Multi-Unit Train System. In this project, conventional systems engineering design loops were effectively consolidated into three steps using SysML-based MBSE methods.

4.1.12 Conclusions

The systems engineering methods and resulting artifacts needed to build complex systems are increasing in popularity, but MBSE methods haven't completely supplanted traditional, document-centric systems engineering (DCSE) frameworks and acquisitions just yet. This is largely due to the relatively slow and inconsistent implementation of MBSE concepts and methods throughout the industrial and governmental sector, issues with integrating MBSE and design/development tools, and the lack of long-term consistent metrics on the value of MBSE and its effect on return of investment (ROI). Also, according to Madni and Sievers, there aren't many software tools and supporting languages, and broad adoption has been slow (Madni and Sievers 2018). Nevertheless, in many real-world applications, case studies, and evaluation criteria, it has been demonstrated that MBSE advantages can be quantified and do in fact demonstrate a superior return on investment compared to DCSE methods (Goodwin et al. 2022). Future system capabilities, their quality, and their development time will undoubtedly improve with the continued standardization of languages like SysML, reusable model building blocks, artificial intelligence, and stakeholder model-centric frameworks. The SysML diagram examples and debate on auto-generation of such systems development products show that the artifacts of SysML models may be used with automation to build more complicated, and even complex systems. Beyond a shadow of a doubt, MBSE methods are integral to contemporary systems design practices since MBSE artifacts, such as those produced using UML and SysML, enhance communications, facilitate system assessments, and better prevent expensive design mistakes than conventional DCSE methods.

4.1.13 Transition to Optical Systems Building Blocks

In the previous section, we presented systems architecture principles, methods, tools, and techniques and provided some insight into the role of the systems architect and the important nature of the systems architecting process. In a similar vein, in the next section, we look at modeling tools that can be used for modeling, simulating, and/or analyzing optical systems.

4.2 OPTICAL SYSTEMS BUILDING BLOCK: MODELING, SIMULATION, AND ANALYSIS FOR OPTICAL SYSTEMS

The understanding of the technical aspects of optical systems, the physics of the propagation and detection of light, and the analytical and simulation methods used today have their roots in ancient times. We kick off Section 4.2.1 by providing a historical overview of the development of optical systems, as seen through the lens of optical modeling, simulation, and analysis. The Purpose of Optical Systems Modeling is presented in Section 4.2.2. Optical modeling and simulation methods are covered in Section 4.2.3, whereas Optical Analysis Methods are presented in Section 4.2.4. In Section 4.2.5, we cover Optical Systems Modeling and Simulation Tools. Here we present a list and short description of some useful available tools for optical modeling in general. Finally, in Section 4.2.6 we provide some interesting Optical Applications.

4.2.1 Overview

Ancient civilizations are the starting point for the history of optical systems. Simple lenses were the first optical components that mankind created. The oldest known lenses date back to around 4,600 years to the ancient Egyptian culture (Enoch and Lakshminarayanan 2000). The Nimrud lens, shown in Figure 4.15, dates back to the 7th century BCE and is the earliest lens item that is still in existence. It is currently on display in the British Museum. Although primitive, the first lenses made of polished crystal were able to focus light to a certain extent. Relics from ancient Babylon, Egypt, and Greece show more refined polished crystal lenses. No one knows the function of these lenses

FIGURE 4.15 The Nimrud lens.

or whether they served only as decoration. Commonly referred to be "burning glasses" for starting fires, the ancient Greeks and Romans also made use of water-filled glasses as lenses.

Optics as a scientific discipline has its roots in classical Greece. Optica, written by Euclid of Alexandria, was published about 300 BCE. Euclid wrote a set of geometric postulates about sight. After describing the law of reflection, Euclid concluded that light travels in straight lines. The idea of perspective was also brought up by Euclid. The foundations of geometrical optics can be found in Euclid's writings.

The theory and application of mirrors were discussed in Hero of Alexandria's Catoptrica circa 10–70 AD. Refraction, in which light undergoes a change in direction at the interface between two media, was first described by Ptolemy around 140 AD.

While the rest of the globe froze in the Middle Ages, the Islamic world advanced further during the subsequent 1,000 years. The law of refraction, which is today known as Snell's Law, was described by the Persian mathematician Ibn Sahl in his work On Burning Mirrors and Lenses in 984 CE. Sahl described aspheric lenses using the law of refraction (Rashed 1990). The seven-volume Book of Optics, authored later by Ibn-al-Haitham (often called Alhazen), was groundbreaking because it was based on empirical evidence rather than theoretical speculation. The 965–1040 CE scholar Al-Haitham studied atmospheric refraction, lens magnification, spherical and parabolic mirrors, and more (he estimated the atmosphere was 15 km thick based on observations of atmospheric refraction). A title bestowed upon Al-Haitham is "the father of modern optics." Light obeyed the rules of geometry and mathematics, according to Robert Grosseteste of Oxford University in 13th-century England (Krebs 2004). Although Grosseteste's effort was comparable to that of Al-Haitham, it is significant since it increased the work's visibility in Europe. Following in the footsteps of Grosseteste and Oxford, Roger Bacon compiled findings from Islamic scientists and Grosseteste's work in his magnum opus, Opus Majus. Figure 4.16 shows Bacon's description of light refraction in a water-filled cylinder; he also proposed that lenses could improve nearsightedness.

Germany saw the publication of Johannes Kepler's Ad Vitellionem Paralipomena in 1604. Kepler provided the first accurate description of the eye's operation, describing how an image is created on the retina and how vision is the result. Nearsightedness and farsightedness were both described by him. In his subsequent work Dioptrice, Kepler detailed the inner workings of the telescope and convergent/divergent lens microscope.

It was in 1608 that the Dutchman Hans Lippershey built the first telescope. The year after that, Galileo Galilei built his very own telescope. The discovery of Jupiter's moons was one of the discoveries that Galileo announced in 1610.

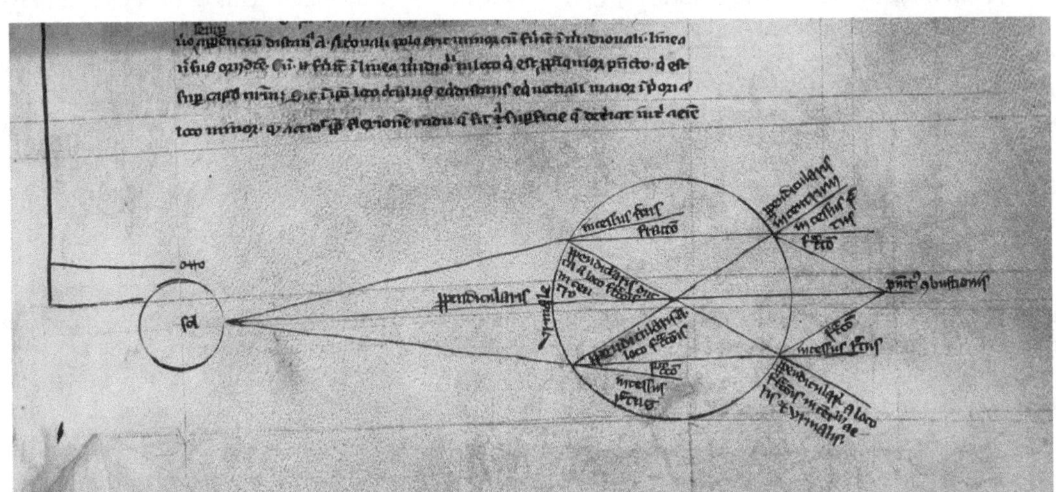

FIGURE 4.16 Roger Bacon's drawing of light refracting through a water-filled glass, c. 1267.

In 1621, the law of refraction – now named after its creator, Willebrord Snellius – was presented. In 1637, the law of refraction was mathematically described by René Descartes. Pierre de Fermat demonstrated in 1657 that the law of refraction agrees with the "principle of least time," which states that light travels more slowly through denser media.

Originally, diffraction was used in tests conducted by Francesco Maria Grimaldi. Shadows cast by objects in a light beam have fuzzy edges and colored fringes, according to Grimaldi's 1665 work Physico-mathesis de Lumine. What's more, the sizes of the shadows don't match up with the sizes of the objects when the shadow is calculated using geometrical optics. The term "diffraction" was first used by Grimaldi to characterize this occurrence (Cajori 1899).

It was in 1678 when Christiaan Huygens initially proposed a theory of light waves. His treatise Traité de la lumiere, published in 1690, elaborated on this idea. But others didn't seem to agree with this theory (Huygens 1690). In his 1704 publication Opticks, skeptic Isaac Newton upheld the corpuscular hypothesis of light (Newton 1704). Thomas Young's work in 1801 proved that light interfered, and the wave theory of light didn't get any traction until then.

A paraxial approximation analysis of image creation was published by Carl Friedrich Gauss in 1840. The study of lenses is substantially simplified by this approximation, which reduces their properties to relatively simple equations, since rays maintain a modest angle with respect to the optical axis of the system. It is also possible to express the properties of a lens system using straightforward explicit mathematics. This approximation is still used today and is known as Gaussian optics.

Light, according to James Clerk Maxwell's 1865 theory, is an electromagnetic radiation. After expanding his work on electromagnetics to optics, Maxwell established a unifying theory that governs all three disciplines. Nowadays, most physical optics problems are solved using Maxwell's equations.

Quantum optics emerged in the middle of the 20th century. The quantum mechanical analogue of classical electrodynamics is the theory of quantum electrodynamics. Quantum optics is necessary to understand the operation of laser cavities and other niche applications. Maxwell's equations most often provide a good enough description.

4.2.2 Purpose of Optical Modeling and Simulation

According to the Optical Society of America (OSA), the goal of optical modeling and simulation has traditionally been to provide a mathematical description for meeting the specified requirements of an optical system's elements (OSA 2009). To accomplish this, the "first-order" layout of the optical system must be established before any optical system can be modeled. This optical element layout satisfies the system's first-order requirements such as the required image size, its orientation, its position on the detector, and detector signal-to-noise ratio considerations. The requirements for the first-order layout, of course, will be different for non-imaging applications but still have a first-order set of parameters that need to be determined, just like imaging systems. These first-order layout requirements also impact optical system performance parameters such as physical size constraints, aperture sizes, *f*-numbers, and fields of view. Higher-order layout requirements are determined after the first-order layout has been completed.

For imaging applications, determining the optical system's configuration and settings to get the best image possible was traditionally the end objective of modeling and simulation. This was accomplished by using lenses and other optical elements in a system configuration that balanced the aberrations of the individual elements. The goal for most high-end imaging systems is to get as close to the so-called diffraction-limited spatial resolution of the optical imaging system as possible. However, this is not an easy task for optical systems that are within the Earth's atmosphere, and that have entrance pupil plane aperture sizes (primary mirror sizes in telescopes for instance) that are greater than the atmospheric coherence length (Fried parameter), the atmosphere will destroy the spatial resolution of the optical system. This atmospheric turbulence effect is so pronounced, that a 3-m class observatory telescope will have no better spatial resolution than a 4-in. telescope bought at a toy store, if the telescope is on the Earth's surface and is looking through atmospheric turbulence. Atmospheric turbulence and/or adaptive optics methods need to be employed to mitigate atmospheric turbulence effects on optical systems. The atmospheric turbulence, its effect on optical systems for both imaging and non-imaging applications, and adaptive optics/atmospheric turbulence compensation methods can all be modeled and simulated to gain an understanding for how the optical system works in the presence of atmospheric turbulence and to know how best to configure and optimize the optical system's performance.

Cost has emerged as a key factor in the modeling and simulation of optical systems due to the recent development of consumer devices that use sophisticated optical elements like lasers. Optoelectronic component sales are currently being driven by consumer electronics applications. New cost constraints on optical systems design emerged because of the price sensitivity of consumer goods (Kincade and Anderson 2005). "Good" and "affordable" optical systems are now perceived necessities.

Simultaneously, optical systems have shrunk in size due to miniaturization, rendering conventional geometric analysis sometimes inadequate and physical optics methods to get accurate modeling results. Advances in computer and software technology have thankfully made it possible to optimize designs for more than just basic applications. With the help of these new sophisticated and capable modeling and simulation tools, it is possible to not only improve the performance of optical systems but also reduce overall optical systems cost and increase quality. For an optical system to be reliably produced, it is important to perform tolerance analysis early in the design phase. Incorporating alignment features into the design during early stages allows for faster manufacturing and shorter time to market by mitigating potential manufacturing-related misalignments. Finally, analysis tools enable cost optimization and trade-off studies between optical system components such as lenses, mirrors, beams, splitters, filters, and more to ensure performance while also lowering the overall cost of the system. Overall cost needs to be at the forefront of design and development considerations since "a fantastic design doesn't help much if you can't make it cost efficient" (ORA 2013).

4.2.3 Optical Modeling and Simulation Methods

Modeling an optical system has traditionally begun with the design of the lens. Optical systems were manually modeled with geometrical optics approximations before the calculator was invented. Extensive manual calculations involving logarithm and trigonometric tables were performed to determine the dimensions, placement, and powers of the lenses. This approach was very time-intensive and prone to human error. Once calculators became widely available, many of the computations were done with them, but modeling the optical system was still a laborious and error-prone process. Consequently, optical systems modeling was typically done using only a limited number of individual optical elements.

Digital computers have completely changed the way optical system modeling is done, and they have quickly supplanted calculators in professional settings. A sophisticated optical design program is now the starting point for most optical systems models. The optical elements and their layout are typically combined with the mechanical elements of the optical system using specialized software for optical engineering. It is common practice to use paraxial ray tracing or a known-good design to estimate the approximate "first-order" locations of optical elements, even if computers and specialized software are becoming more commonly available and are capable of higher-order modeling.

Instead of being distinct processes, the modeling, simulation, and analysis of optical systems are now closely related activities. Thanks to the dramatic improvement in computer processing power, simulations that used to take weeks to finish now take only seconds (Savage 2007). The surge in computational processing power has acted like a springboard for more powerful optical modeling and simulation software products and so facilitates the rapid modeling, simulation, and analysis of candidate optical systems designs. Faster evaluation of the new design is also possible after the analytical results are incorporated in the design. This optimization loop can be executed rapidly by using the optimization features included in several contemporary optical systems software programs.

Geometrical optics, often known as ray optics, and physical optics are the two main approaches to simulating optical systems. Up until recently, geometrical optics was the go-to method for simulating macro-scale applications, while physical optics was the way to go for nanoscale applications. Although physical optics is useful for certain specialized macro-scale applications (such as laser system design), the two domains rarely interacted (Sinclair 2000). Due to the proliferation of laser-based technologies, physical optics analysis has become increasingly prevalent.

When imaging a distant object, the light ray is the focus of geometrical optics (also known as ray optics). Light travels in waves, but the geometrical optics approximation doesn't take that into account. This means that geometrical optics can't be used to mimic diffraction or any other such wave-based optical phenomenon (Hecht 1998). The geometrical optics approximation has some shortcomings, but its simplicity makes up for them in many cases. While dealing with reflection, one applies the "law of reflection," which states that the angle between the incoming ray's angle of incidence and the surface normal is equal to the angle of reflection relative to the same surface normal. When dealing with refraction, Snell's Law is applied. The term "ray tracing software" is commonly used to describe programs that use the geometrical optics method. "Sequential" and "non-sequential" are the two main classifications for this software. It is presumed that light rays interact with system elements in a certain order in a sequential ray tracing algorithm. The path of light from its origin to the final image is determined by the optical system's constituent parts. Each incident ray of light is converted into one reflected, refracted, or absorbed ray when it interacts with optical components in its path. However, there is no set direction for a non-sequential code. There are three ways in which light can interact with a surface: reflection, refraction, and absorption. Put simply, when one ray hits a surface, it has the potential to generate two additional rays, the *reflected ray*, and the *transmitted ray*.

Even though geometrical optics methods are simple and convenient, it is impossible to disregard phenomena brought about by the wave character of light in certain applications. Among these effects include polarization, interference, and diffraction. Such effects are not inherently modeled by the geometrical optics approach and belong instead to the realm of physical optics. Physical optics phenomena caused by light waves interacting with their surroundings become noticeable when the size of the wavelength of light becomes significant (on the same scale) compared to the dimensions of what the light is interacting with. Physical optics, like geometrical optics, is still an approximation; however, it is a more precise approximation of the interaction between light and matter, and it describes the propagation of light as a wave phenomenon. Physical optics-based modeling software can be implemented in various ways from rigorous finite difference solutions of Maxwell's equations to various extensions and modifications of the geometrical ray tracing approach.

4.2.4 Optical Analysis Methods

Image quality metrics are used to evaluate the results of simulations of optical systems applications. Spot diagrams, fan plots, ringed energy plots, and the optical transfer function are some of the common means to assess image quality (Irving 2013). When evaluating non-imaging systems, one way to measure efficiency is by looking at the quantity of energy transferred from the imaging system's entrance pupil to the exit pupil aperture.

Spot diagrams, created by ray tracing analysis, offer a visual means of quickly determining the quality of an optical system or lens. With spot diagrams, the points where the light through an optical system meets the focal plane can be mapped. Spot diagrams are based on geometrical optics (ray tracing optics) and so do not include diffraction effects. They do give a good indication of how optical rays trace through the optical system elements and the pattern that is produced (barring diffraction) in the imaging system's focal plane. Figure 4.17 illustrates a spot diagram as an example.

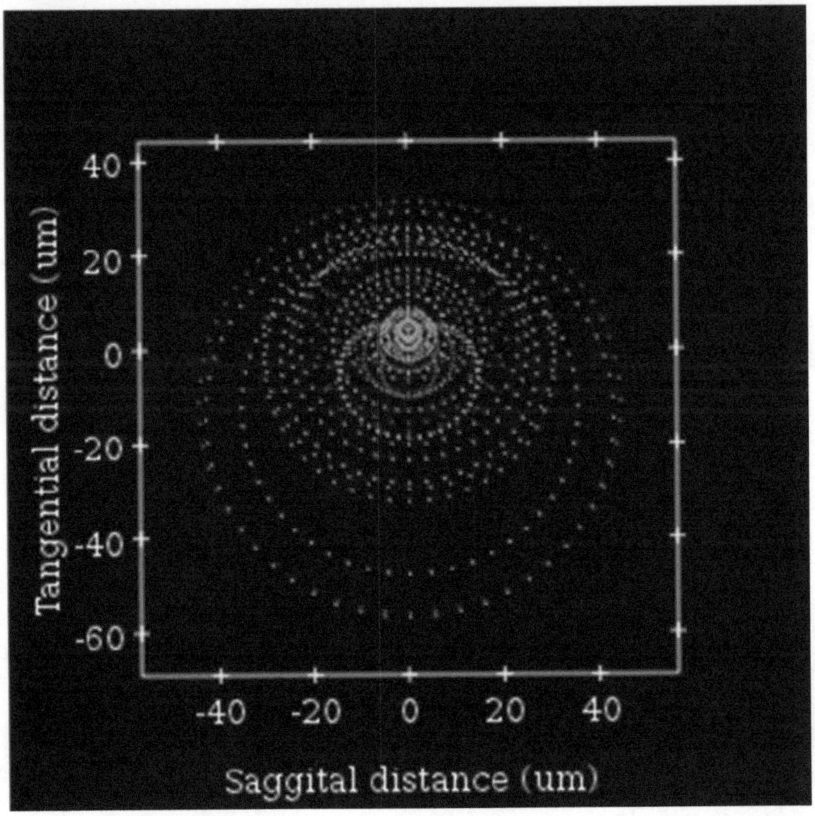

FIGURE 4.17 Spot diagram.

It is common practice to make the plots at various wavelengths to measure the system's chromatic aberration.

Plots of encircled energy show the spot diagram's energy distribution versus radius. The size of the focal spot can be estimated in this way. You can see an example of an encircled energy diagram in Figure 4.18.

Differential ray path lengths through the optical system are another important metric to determine. Many times, these are shown as Ray Fan Plots. Figure 4.19 is an example of such a plot. Dissimilar optical paths lead to focal plane wavefront distortion and phase mismatches.

The modulation transfer function, or MTF for short, is another optical system metric that is useful in characterizing the quality of an imaging system. The MTF is the magnitude of the Optical Transfer Function (OTF) plotted at every non-zero spatial frequency. With imaging system aberrations present, higher spatial frequencies fall off rapidly with respect to the MTF for a diffraction-limited system, and this effect can be seen in the three-dimensional frequency response characteristics of the MTF.

4.2.5 Optical Systems Modeling and Simulation Tools

There are many software packages that are beneficial to optical modeling and simulation are readily available (Savage 2007). For those in need of optical design and analysis supplies (hardware and software), the Photonics Buyers Guide, which is accessible in print (Photonics Media Print 2024) and online (Photonics Media Online 2024), features listings from more than 40 businesses. There are many software design programs available on the market nowadays and the casual observer may initially find the task of navigating these tools a bit daunting (Jones-Bey 2005). However,

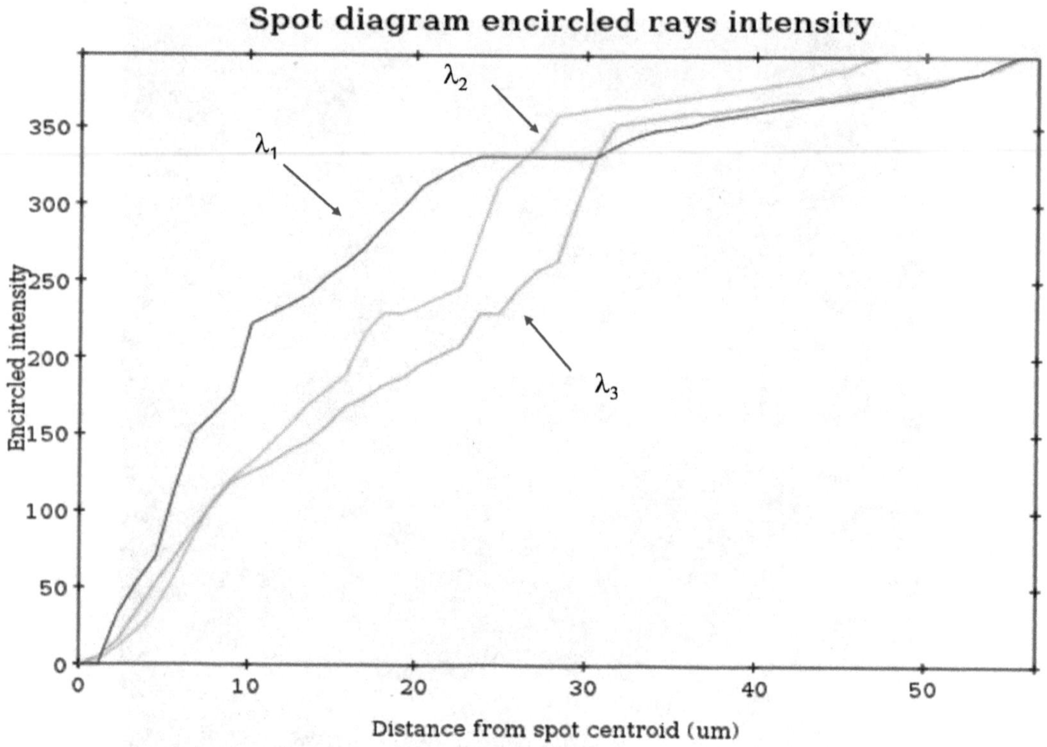

FIGURE 4.18 Encircled energy diagram.

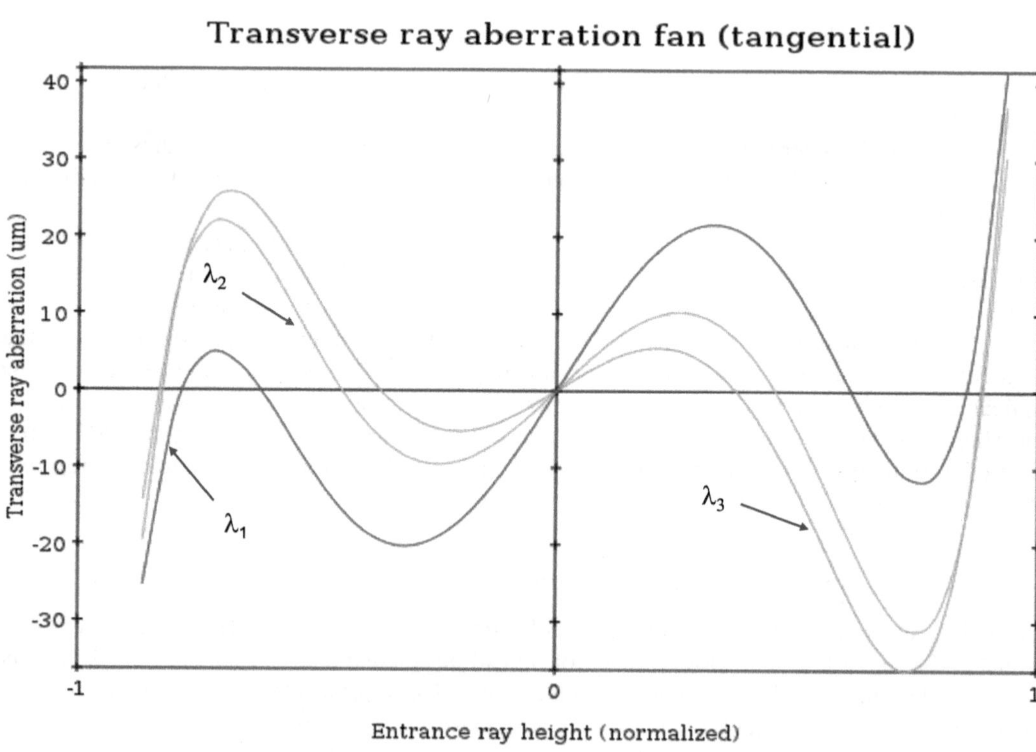

FIGURE 4.19 Example ray fan plot.

it is possible to classify the programs into multiple broad groups according to the modeling and simulation method that is used. Lens design is the forte of certain software packages; they include lenses for cameras, telescopes, microscopes, and other imaging devices. Commonly, these programs employ sequential ray tracing. Sequential ray tracing may be used to build complicated optical systems and undertake extensive performance characterization, despite its apparent simplicity. A high degree of design optimization is also within reach, with methods like dethermalization and tolerancing analysis that aid in determining sensitivities and optimizing thermal and tolerancing effects in optical systems design.

As computing power increased, certain software packages that initially only supported sequential ray tracing now also support non-sequential ray tracing (the number of rays generated in a non-sequential analysis can be many times that of a sequential analysis). Even with relatively simple designs, non-sequential ray tracing codes can carry out more complicated studies (such as stray light analysis) as compared to sequential ray tracing software. A non-coated lens will reflect around 4% of the light that hits it instead of letting it pass through. To determine how this unwanted light affects the optical system's performance, a non-sequential ray tracer can model its route. Illumination modeling is another domain where non-sequential ray tracing codes find use. The outcome of these types of analyses is not a picture but rather the distribution of light over a targeted area.

The last group of applications models light propagation using physical optics techniques. These applications come in handy when the size of the optical system being studied decreases to a magnitude close to that of the wavelength of the light. Approximations from geometrical optics are inadequate at this relative scale. Specialized laser design packages are also available in software; these make use of physical optics solutions as well. When it comes to physical optics, there are a variety of modeling tools available.

A quick overview of some of these software programs that can be used to model optical systems is given below. Here you can find codes for generic physical optics, laser design, sequential and non-sequential ray tracing, and even some tools for thin film analysis. Beyond the scope of this list, there are other software tools available for modeling and controlling the manufacturing of optical elements.

4.2.5.1 APEX/ASAP/ReflectorCAD

Breault Research Organization, Inc. released these applications as part of their optical design software bundle. One such ray tracing add-in for SolidWorks is APEX. The ability to build and analyze intricate optical and illumination systems within SolidWorks is made possible by APEX's built-in ray tracing analysis.

An example of a non-sequential ray tracing program is ASAP, which stands for Advanced Systems Analysis Program. Some physical optics capabilities are also a part of ASAP. A wide range of imaging and non-imaging applications currently make use of ASAP, despite its initial design being for illumination design.

ReflectorCAD is a CAD software that is useful for ray tracing applications such as headlight reflector design and other similar applications.

4.2.5.2 ATMOS

Telescopes built by amateurs are often designed using the Amateur Telescope Maker Optical Design and Analysis Software or ATMOS. Roughly 30 different optical combinations are available for analysis in the program. The user can conduct tolerance analysis with the help of the optimization tools included in the software.

4.2.5.3 Beam Four

Stellar Software developed the Java program known as Beam Four. Lenses, gratings, irises, mirrors, and prisms are just some of the optical elements that can be modeled in this sequential ray tracing software. An optical system's image quality can be more readily assessed with the use of the program's analysis tools.

4.2.5.4 CODE V

For creating imaging optics, Synopsys, Inc. produced the software package known as CODE V. Designing lenses for cameras, microscopes, telescopes, projectors, and laser scanning systems are all examples of some practical applications of this software package. NASA has been using CODE V for a long time. As an example, CODE V was used to assess the Cassini probe's spectrometer (Crooke and Hagiopian 1998). The James Webb Space Telescope's scientific equipment, which include a fine guidance sensor, two infrared spectrometers, and two infrared cameras, was more recently modeled using CODE V (Davila 2006). CODE V has expanded its feature set to include physical optics from its original focus on geometrical ray tracing lens construction.

4.2.5.5 dBOptic

Sky Scientific, Inc. developed their useful tool, dBOptic for studying engineering problems, regardless of the user's proficiency in optical design. It is useful for anyone interested in engineering and is very applicable for use as a teaching tool.

4.2.5.6 FDTD Solutions

Lumerical Solutions' FDTD software is a 3D Maxwell equations solution package for physical optics problem solving in the finite difference time domain. Initial releases of the program occurred in 2003. Microscale and nanoscale optics and photonic devices, where the device size is on par with the wavelength of light, are the ideal candidates for this software's analysis capabilities. The software has found utility in the modeling of many electronic devices, such as CMOS imaging sensors, thin film photovoltaic cells, and LED/OLED displays. Optimizing nanophotonic devices is relatively straight forward using the software's built-in optimization tools. Globally, both academic institutions and businesses rely on FDTD Solutions.

4.2.5.7 FilmStar^{tm}

Developed and released by FTG Software Associates, FilmStar^{tm} is a thin film coating design and optimization application.

4.2.5.8 Film Wizard^{tm}

The FilmWizard^{tm} software was developed by Scientific Computing International (SCI) and is useful in optical thin film design.

4.2.5.9 FRED

The non-sequential ray tracing software package FRED is offered by Photon Engineering, LLC. FRED sounds like an acronym but isn't one. Among FRED's many potential uses are imaging systems, lighting, non-imaging optics, stray light, and laser applications. FRED can be used to construct lenses or import lens definitions from another application. By combining the optical components with imported CAD geometry, FRED can generate a comprehensive model of the optical system. The Goddard Space Flight Center, for instance, uses FRED for the purpose of simulating complete optical systems and assessing issues like stray light, ghosting, and image quality (Hagopian et al. 2010).

4.2.5.10 General Laser Analysis and Design

Applied Optics Research publishes GLAD, which stands for General Laser Analysis and Design. The program's main emphasis is on laser applications, although it also covers physical optics in general. Laser cavities can be analyzed and designed using GLAD. Complete optical systems, including intricate configurations with numerous lasers, can also be modeled. Apertures, lenses, mirrors, aberrations, active media, and diffraction can all be modeled with this software tool.

4.2.5.11 LASCAD

Laser resonator design and analysis software is LASCAD (LASer Cavity Analysis and Design) developed by LASCAD Gmbh. This software bundle incorporates laser crystal structural and thermal finite element analyses. To simulate beam propagation, the ABCD ray tracing method is employed. Three-dimensional wave optics beam propagation is used to analyze diffraction and gain saturation, and the program also contains analysis for multi-mode and Q-switched lasers.

4.2.5.12 LensForge

Only Mac OS X users can use LensForge, a lens design application created by Ripplon Software. Simplicity is the program's main goal. It is possible to read Zemax lens files and use the program's extensive lens library. Additional features are also available for optimizing lens systems.

4.2.5.13 LightTools

Synopsys, Inc. released LightTools, a non-sequential ray tracing program. Illumination system design is the main emphasis of this software package. Users may model, simulate, optimize, and visualize illumination systems with this application.

4.2.5.14 MODAS and JODAS

Telescope design packages include MODAS (Modern Optical Design and Analysis Software) and JODAS (Java-based Optical Design and Analysis Software). Both are well-liked by telescope manufacturers and amateur astronomers. The software is offered as freeware.

4.2.5.15 Newt

A program for designing Newtonian telescopes is called Newt. As freeware, it is offered for download. Also available at http://stellafane.org/tm/newt-web/newt-web.html is a Java-based web version as well. This ray-tracing software will print drawings and optimize telescope designs.

4.2.5.16 OpTaliX

Optenso Optical Engineering Software is the manufacturer of OpTaliX. Thin film multilayer coatings, illumination systems, and optical imaging systems can all be designed with this software. Thin film multilayer analysis and refinement, optimization, tolerance analysis, non-sequential ray tracing, and physical optics propagation are all included.

4.2.5.17 Optica

Developed by Barnhart Optical Research LLC, Optica is a 3D optical design package. For the software to function properly, Mathematica must be installed. Unlike other commercial programs, this one comes with source code. The company also offers LensLab, a less feature-rich version of Optica that is a lens design application.

4.2.5.18 Optics Lab

Science Lab Software created the optical ray tracing software suite known as Optics Lab. Optics Lab is a three-dimensional non-sequential ray tracing application. It can accommodate a wide range of lens types as well as optical elements like prisms, mirrors, beam splitters, gratings, apertures, and more. Basic optimization capabilities are offered.

4.2.5.19 Optiwave Products

Optiwave Systems, Inc. develops several programs that help in the design of optical communication systems and components. These programs include OptiFDTD, OptiBPM, OptiFiber, OptiGrating, OptiSystem, OptiSPICE, and OPTIPerformer. Applications involving fiber optics and optical circuit design are the primary emphasis of these tools. The applications' simulation engines are finite difference time domain physical optics solvers.

4.2.5.20 OptiCAD

OptiCAD Corporation publishes OptiCAD which is a tool for optical system analysis and design. The program is set up similarly to a conventional CAD program.

4.2.5.21 OptisWorks

Several products made by OPTIS, Inc. are directly compatible with well-known 3D CAD applications, such as Pro/Engineer, SolidWorks, and CATIA V5. The program makes it possible to model the optical design without requiring data exportation or translation into a stand-alone application. OPTIS software has functionality for optical imaging systems and laser applications. Its primary usage is modeling and simulating illumination systems.

4.2.5.22 OSLO

Lambda Research Associates publishes OSLO (Optics Software for Layout and Optimization). OSLO's main application is finding the ideal shapes and sizes for the components in an optical system. Ray tracing, analysis, and optimization features are all included in OSLO, which was initially developed as a lens design program (Lambda Research Associates 2024). OSLO is applicable to imaging and non-imaging optics, laser systems, and visual systems and is widely used. For example, OSLO was used to evaluate how thermoelastic distortion affected the James Webb Space Telescope (Johnston et al. 2004).

4.2.5.23 Paraxia-Plus

A software specifically developed for laser design is called Paraxia-Plus from SCIOPT. The user can design and examine laser resonator properties with this software. Paraxia-Plus can also be used to model the interactions between components outside of the resonator, and the propagation of laser beams.

4.2.5.24 Photon Design

A software suite for designing and analyzing optical waveguides and other photonic devices is available from Photon Design. Depending on the application, several physical optics solvers are available, including mode solvers, finite element frequency domain, and finite difference time domain solvers.

4.2.5.25 Photonics Workbench

Photonics Workbench, published by Inspired Photonics, is a tool for optical fiber system analysis. Numerous Laser sources and fiber effects, including noise effects, scattering, amplification, and attenuation, can be simulated using this software.

4.2.5.26 Photopia

Photopia is a rapid and precise optical design and analysis application for non-imaging optical systems, developed by Colorado-based LTI Optics (Denver). A non-sequential ray tracing code, Photopia is widely used in the architectural lighting sector.

4.2.5.27 Radiance

The ray tracing software program Radiance is used for illumination system design and simulation. The program has a large user base and is available for free download.

4.2.5.28 RayCAD

RayCAD, Inc. launched the first edition of RayCAD in 1993. The raytracing capabilities of AutoCAD and Autodesk Inventor are enhanced by this application. The software can import ZEMAX lens data.

4.2.5.29 RSoft Product Suite

The RSoft Product Suite includes programs for both design and simulation and was purchased by Synopsys, Inc. This application comes in bundles of physical optics-based packages. From optical communication devices to optoelectronics, RSoft provides a suite of software tools for simulating all kinds of photonic devices.

4.2.5.30 The SYNOPSYS

Optical Systems Design, Inc. publishes SYNOPSYS (SYNthesis of OPtical SYStems). The program's development got underway in 1962. The main uses of this program are for multi-lens system design and lens design. The application comes with a full range of optimization and tolerancing capabilities in addition to traditional lens design elements. Using these methods, the lens designer can lower the cost of the lens by performing "tolerance desensitization" (Dilworth 2008).

4.2.5.31 TracePro

A non-sequential ray tracing tool called TracePro is offered by Lambda Research Associates. Although it can be used for both imaging and non-imaging optical systems, this software is mainly used for modeling illumination systems. It is used to evaluate geometry created in typical 3D CAD products (that use the Initial Graphics Exchange Standard – IGES, for surface modeling; or the Standard for the Exchange of Product data – STEP, for solid models), and optical elements designed in classic optical design applications, like OSLO, CODE V, or ZEMAX. It has fewer lens design features than OSLO. Additionally, it has an interface with SolidWorks that enables direct application of optical attributes to the SolidWorks model (Lambda Research Associates (Tracepro) 2024). At NASA, TracePro is widely used. For instance, it was used to assess the long-range imaging sensor on the New Horizons Expedition (Cheng et al. 2007).

4.2.5.32 VirtuaLab

LightTrans has developed a physical optics package called VirtuaLab. LightTrans employs a novel modeling approach known as field tracing, which blends electromagnetic systems modeling and ray tracing. This makes it possible to employ finite element methods when needed and geometrical optics when appropriate (lens design) (Wyrowski and Kuhn 2011).

4.2.5.33 ZEMAX

Radiant Zemax Corporation has been publishing ZEMAX since its 1988 debut. In its early days, ZEMAX was mainly used for lens design and ran on personal computers for ray tracing. A non-sequential ray tracing algorithm with physical optics analysis of beam propagation through an optical system is the result of the program's evolution over time. Included are elements that can be put to good use in many different contexts, including but not limited to physical optics, basic lens design, illumination systems, microscopes, and telescopes. Thermal compensation calculations and optimization capabilities are also included. ZEMAX has been integrated into multiple NASA initiatives, including multiple Mars rovers.

4.2.6 Optical Applications

While the "classic" uses of optics – such as cameras, telescopes, microscopes, eyeglasses, contact lenses, and such are ever present and popular, the exponential growth of optical technology has led to an explosion of new optical applications.

The optical mouse is a commonplace example. Mechanical mice have been mostly superseded by optical mice. The optical mouse has a tiny optical sensor and its own miniature imaging system. While it is on its mouse pad, the system snaps many images of the same spot. The surface can be illuminated using either a laser or an LED light source. The mouse can detect its own movement direction by comparing photos taken at different times.

With the advent of ray tracing algorithms and powerful computers, the optimization and modeling of illumination systems have seen a remarkable transformation. With the advent of non-sequential ray tracing codes, it is now possible to simulate the spread of light from many sources, optimize its distribution, and design lighting systems for both indoor and outdoor areas, and even inside or outside of vehicle (like the previous headlight reflector example).

Optical systems are also prevalent in optical scanning systems for residential, business, and workplace applications. The optical systems are what allow modern fax machines, scanners, and copy machines to accomplish their functions. Another device that uses optical components to print is the laser printer. The bar code scanner reads the Universal Product Code (UPC) that is written on packaging by means of optical components and laser beams.

Laser applications have been one of the most rapidly expanding markets. Although the computer and laser were both invented in the 1950s, optics was truly transformed in the late 20th century by the invention of the latter. It was said that lasers were "a solution looking for a problem" when they were initially introduced (Townes 2003). Lasers were initially used in scholarly and commercial contexts. Nowadays, lasers seem to be everywhere. To the tune of over $5.3 billion, 131,000 lasers and 733 million diode lasers were sold in 2004 (Kincade and Anderson 2005). A wide range of consumer goods use lasers as fundamental parts of their functionality. Among the many examples, laser pointers stand out.

Another common yet noteworthy application of laser optics is laser storage media. The LaserDisc, which came out in the 1970s, was the first consumer product to use this technology. The LaserDisc\ was a commercial flop, but the Compact Disc (CD) that came out in the early 1980s was a smashing success. Media players like the DVD and Blu-ray players came later. Each of these devices essentially works in the same way. Using an optical system, a beam from a laser diode is concentrated on the surface of a disk. A photodetector takes readings from the disk's reflected energy and makes use of the fluctuating light levels.

The commercialization of the laser has completely transformed the discipline of metrology. In the early 1990s, the first laser tracker came out. This replaced theodolites and photogrammetry as the primary tools for measuring three-dimensional data (both of which are also applications of optical systems). Accurate mechanisms enable precision measurement of azimuth and elevation angles, and the laser tracker uses a laser and interferometer to generate three-dimensional coordinate data. Data may be acquired more precisely and rapidly using the laser tracker. Following suit are other laser-based metrology technologies, such as laser scanners and laser radars.

Recently, laser-assisted surgeries have grown in popularity. The most basic methods eliminate the need for a scalpel by cutting tissue using laser beams. Cauterization, which slows bleeding, is another usage. Nearsightedness correction via laser-assisted selective intraocular lens implantation (LASIK) is probably the most well-known application. This technique requires micron-level precision in concentrating a laser beam to excise eye tissue.

Fiber optic communications is another field that has seen tremendous growth. The 1970s saw the initial development of this technology. Nowadays, optical fibers are the norm for high-speed data transmission instead of copper cables because of the increased data capacity of optical fibers. Nowadays, optical fiber is the material of choice for most underwater communications cables. Transmitting light across optical fibers and the optical interface components at their ends is an expanding area of study.

In the next section, we apply some of the methods developed above to our fictitious program development where Fantastic Imaging Technologies (FIT) is working on a sensing system for the US Department of Homeland Security Customs and Border Protection (DHS CBP). We focus mainly on optical M&A methods and how they benefit the program development.

4.3 INTEGRATED CASE STUDY: OPTICAL MODELING AT FIT

In this section, we show how the information in the previous two sections work together in developing a fictitious sensing suite for an airborne unmanned aerial vehicle (UAV) for the US Department of Homeland Defense, Customs and Border Protection. The intent is to provide a multi-sensing capability that can be mounted in the UAV to help find illegal border crossers, and conduct search and rescue, and other US CBP missions. Here we find FIT needing to understand some of the US DHS CBP stakeholder requirements, so optical modeling is being considered. We start with a meeting between FIT's CEO, Dr. Bill Smith, FIT's Senior Systems Engineer, Karl Ben, FIT's Chief Technical Officer (CTO), Dr. Tom Phelps, FIT's Facility Security Officer (FSO), Wolf B., FIT System Engineer, Jennifer O, and FIT Optical Lead, Amanda R.

[The Scene*: Dr. Bill Smith calls a meeting to gain a better understanding of the required capabilities for the optical sensing systems and to throw in a little twist.]*

Karl: "Hi Jen, it's good to see you again. Were you able to finish the analysis on which MBSE tool we should use to evaluate the optical modulator?"

Bill: "Hello everyone, thanks for coming. I just got off the phone with our customer at US DHS CBP and there have been some interesting developments. We will need to move to the sensitive compartmented information facility (SCIF) to continue our discussions."

[A few minutes later in the SCIF, Bill continues.]

Bill: "Okay, I just got off the horn with Wilford over at Department of Homeland Security Customs and Border Protection. It seems that there is a big interest by a classified *three-letter agency* in our project. We will need to address some additional capabilities with their interests in mind and integrate that seamlessly into our current project with US CBP. According to the three-letter office's security classification guide, we cannot mention that there is any interest by this three-letter agency in our system at all outside of this SCIF, and only cleared individuals with a need to know will be aware of their involvement. All of you here have the clearances and need to know and have been vetted to participate. Outside of this room, this topic or this special customer will never be mentioned, discussed at all. Clear? Good. We do not want a security incident."

Karl: "Okay, no mention of the special customer or any details of their interests, or their impact on the overall sensor system design. Got it. But how will we engage in the day-to-day program development and technical discussions outside of this room?"

Bill: "Good question Karl. This means we need to look at the entire US CBP mission and generalize our optical sensor suite to address ALL their required capabilities, not just the human detection problem. We will basically use our US CBP program to develop a dual use system that is of interest to this three-letter agency. Outside of this room, we only talk about the US CBP program and emphasize our current primary mission to detect bipedal human traffic. However, we can add in the resto of the CBP missions and talk about addressing them outside of this room. For our three-letter customer, the design will allow for much more than that. For instance, our three-letter agency (now our new special customer) has a worldwide deployment requirement. We can design our system to be functional in all the different terrain types in the world (mountainous, desert, forest, tropical, subtropical, plains, and even urban environments for instance). For our US CBP program, we can address this requirement by noting that for US border protection, our sensors are likely to encounter each of these environments along our border, and so must be designed to function in all of them. In this way, we make sure our optical sensor suite is worldwide deployable without stating the requirement directly. Someone would surely wonder why US CBP has a worldwide deployable requirement if their responsibility begins and ends with the borders of the United States."

Karl: "Clever."

Bill: "Thanks. I have my days. So, outside of this room, the primary discussion and sensor development effort will be focused on the extended DHS CBP problem set. This will require some additional capabilities that will need to be addressed and included in our optical system development effort. Before we get into that though, I would like our FSO Wolf to fill us in on what can and what cannot be said outside of this room."

Wolf: "Thanks Dr. B! The safest rule to use is to say nothing outside of this secure environment. Don't write anything down, don't whisper things to each other, don't act things out, don't even think about our three-letter customer outside of this vault. That is the safest policy. All technical information in any form (written, verbal, any media) is generated here and stays here. No phones, communication devices, thumb drives, external drives, or anything that associates the three-letter customer with this development project leaves this secure environment. All of our activities, communications, and documentation stays in here where properly approved communications gear and processes are used. Now, that said, the security guide states that the "interest" of "a" three-letter agency in this project is unclassified. So, it is okay that there is interest by someone. But the actual name and/or designation of the three-letter agency is classified. Feel free to state that a three-letter agency is interested in your project outside of the SCIF but leave it at that. We can't say who it is. Everything else happens in here. We have a copy of the security classification guide here in the vault so please get familiar with it. You will need to review it by the end of the week and sign off that you did."

Bill: "Thanks Wolf. We appreciate the run down. So, we will start to broaden the scope of our current project to include the capabilities that are needed by the three-letter agency, but we will tie the scope change to addressing more and more of the US CBP mission requirements. For example, to help US CBP in anti-terrorist mission activities, they require the ability to *identify* their Target of Interest (TOI) and not just *detect* it. They also need to be able to *track* TOIs. Additionally, the TOI set will need to be expanded beyond just bipedal human border crossing traffic and now include animals, cars, trucks, motorcycles, helicopters/airplanes, boats, and concealed targets. There is also a material characterization requirement such as in identifying hazardous chemicals and materials, an anti-glare capability such as when you try to identify a person in a car with high glare coming from the reflected light in the windshield, a camouflage detection capability, such as when illegal aliens are covering themselves with camouflage netting or a tarp that matches their surroundings to avoid our EO sensors. There also needs to be a low-light capability for cases where there is not enough reflected light, and an all-weather capability when the weather is stormy and/or cloudy days, hindering the optical assets. Finally, there is a requirement that our optical sensor suite needs to be capable of working autonomously with artificial intelligence (A/AI) that is resident on the UAV. The optical sensor suite must have its own AI package that can autonomously work with the AI on the UAV to perform/adapt the mission, avoid harm, protect itself, and secure all data. There should be onboard capabilities to learn and adapt using state-of-the-art methods like machine learning and computer vision, self-healing and adapting to local environmental conditions (including hostile actions), independent decision-making, dealing with big data, atmospheric turbulence compensation, and automatic target recognition with learning features. All these things can (and must) be linked to the current US DHS CBP mission set and our whole development process outside of this room will be based on this broadened set of US DHS CBP mission requirements. The human detection problem is still very much at the forefront but we now will have these other things that need to also be accomplished. Questions?"

Jennifer: "Interesting, so the rest of the development team thinks this is a US DHS CBP program and that's it. Only those of us here right now have the need to know and are read-in and briefed?"

Bill: "Yes, Jennifer. It will only be us for now. If we need to read anyone else in, we will address that on a case-by-case basis and work with our FSO and customer for prior approval."

Jennifer: "Got it, thanks."

Tom: "Wow, very interesting. This brings up a ton of questions on the impact on our current program. Are the current spatial resolution requirements still the same? Does our new special customer have any different requirements with regard to stand-off ranges? We will need to nail this down immediately and rerun our optical models to determine the impact on the current design. Do we have any relaxation in our schedule to deal with this?"

Bill: "Right on top of it as usual, Tom. Unfortunately, our special customer needs this yesterday and so they are wanting to piggy-back off our current effort. There will be additional money to bring in these additional capabilities, but we are not getting any schedule relief for now."

Tom: "That makes it even more important to make sure that there are no major requirements changes from what we have already done. I think it is a very good idea to verify this with both customers. Who at US DHS CBP is read into this?"

Bill: "For now, just talk to Wilford at US DHS CBP and Rebecca, our Special Customer POC over the secure line in this SCIF. It will only be us that knows about this."

Tom: "Okay. Anything else? I am itching to talk to Wilford and Rebecca."

Bill: "That's it for now. Let's reconvene in here after you had a chance to investigate the requirements changes. Wolf, you are welcome to join us next time, but I imagine it may be a technical discussion about the optical system requirements. You may have other things that you need to take care of."

WOLF: "I do indeed. We are preparing for our security inspection and need to be involved there. If you need me, I can pop in as needed."

Bill: "Sounds good! Thanks Wolf."

Wolf: "No problem!"

[The meeting breaks up and Tom heads for the secure communications in the cubby outside the meeting room. A few days later, in the SCIF, the team convenes again.]

Bill: "Hi everyone! Tom, any good news?"

Tom: "Actually, yes. We had quite the discussion regarding mission parameters, and nominal, maximum, and minimum flight altitudes and the like and there were several cultural differences that needed to be overcome. In the end, we managed to sort things out."

Karl: "Cultural differences? Like what?"

Tom: "Well, our new special customer, as you can imagine, has a different set of missions than US DHS CBP with different strategic and tactical objectives. Consequently, the expected flight profiles were initially different. For example, the nominal altitude for US DHS CBP conventional applications is 25,000 ft, whereas the nominal altitude for our special customer is…lower…let's just say. They also have some missions that fly at higher altitudes but there is flexibility in the actual altitude. When we discussed the reasons for the choice in altitude, platform safety was at the top of the list. There was also mention of the risk of being heard in some situations where the flight path is too low. The new customer also mentioned a switch to the MQ-9 Reaper drone that may have offensive capabilities. This is a bigger, more capable UAV platform and provides for the possibility of using the payload bay for additional specialized sensors."

Karl: "Wow, I have always wanted to work with those. The Raytheon MTS-B gimbal can easily be mounted on the Reaper, so we won't have to redesign the gimbal housing and optics. If the stand-off distance requirements haven't increased, with regards to spatial resolution requirements, "lower" is not a problem since the spatial resolution goes up, the closer we get to our TOI."

Tom: "Exactly. I put that question to our new customer, and after discussing the threat analysis to the Reaper platform and the possibility of being heard, there was some mutual agreement between US DHS CBP and our three-letter customer with regard to acceptable nominal, maximum, and minimum stand-off ranges. Again, the minimum stand-off range is not a concern for us since our optics and methods work close in."

Jenn: "So, how did you resolve the safety issue?"

Tom: "Ah, good question Jenn. It turns out that both customers were worried about small arms fire, rifles, and the occasional stinger missile. Fortunately, we already had run these scenarios in our Integrated Modeling Software and found that small arms fire is typically only good up to 500 m, way below the nominal and minimum flight profile of both of our customers. Stinger missiles, even though they have a range of about 5 miles, can only hit targets at an altitude of about 3,500 m (11,500 ft). With sound propagation, sound moves at roughly 346 m/s in air at the Earth's surface. At 15,000 ft, it is 322 m/s and is even slower. It takes about 43 seconds for a drone's engine noise to propagate from directly overhead. Considering, that the Reaper can move at 134 m/s, it can cover about 5,809 m (conservatively using the faster 346 velocity) before the sound can reach the ground from directly overhead. After going over this, both customers agreed on a nominal 15,000 ft (4.572 km) flight path altitude with a minimum of 13,123 ft (4 km), and a maximum of 50,000 ft (15.24 km) altitude for their flight path. The nominal stand-off distance (slant range from sensor to TOI) is 9.543 km. We are currently designing our optical system to these specifications and so there isn't any expected change to what we are currently doing. Also, we were already running our models to determine the resolution for *detection* and *identification* of the human bipedal TOIs and, apart from the new material characterization and concealed/camouflaged object detection requirements, and the all-weather capability, the TOIs are all bigger and so our current spatial resolution requirements will still work. The new tracking requirement also doesn't impact us very much. In a nutshell, we are well situated with our current set of analyses. We just need to rerun our models for some of the new features/capabilities that our special customer needs."

Bill: "That is great news! I did have my concerns about the impact of the potential requirements changes, but this is good news indeed! What about the impact of moving to the MQ-9 Reaper from the RQ-1 Predator?"

Tom: "That is also good news for us. Let me bring up the Reaper on the screen."

[Tom clicks the intelligent mouse and Figure 4.20 comes into view.]

Tom: "This is the Reaper that is currently being used by US DHS CBP.

It has a Raytheon MTS-B gimbal that is twenty-two inches in diameter and weighs 230 pounds. You can see the gimbal pod at the front bottom of the picture. We can fit a 20-inch (0.508 m) mirror for both our EO and IR systems into this gimbal. We can also design and build some specialized sensors to fit into the payload bay for some of the other new requirements needed by our new special customer."

Tom: "Here is another picture (Figure 4.21) of a weaponized version of the Reaper. Again, the optical sensor pod can be seen at the front bottom of the picture."

[***Next scene****: Internal status review of UAV optics program.*]

Dr. Bill: "How's it going? Do you have any interesting developments to report? How are the updated requirements impacting our system?"

Tom: "As you know, the range was lowered to around 5 miles for the daylight, visible portion of the mission, making the range the same for both missions, visible and infrared. The altitude was also lowered, to a nominal value of 15,000 ft for both mission profiles. Interestingly, with these changes, we are good with our daylight detection requirements. The nighttime identification mission, however, is causing us a few issues. As you know the identification

FIGURE 4.20 U.S. Customs and Border Protection Predator MQ-9.

FIGURE 4.21 MQ-9 Reaper with optical sensor package (front bottom). (By U.S. Air Force photo/Staff Sgt. Brian Ferguson - This image was released by the United States Air Force with the ID 071104-F-2185F. Obtained from http://www.af.mil/News/Photos/igphoto/2000694430/ (Direct Link), Public Domain, https://commons.wikimedia.org/w/index.php?curid=4880775. Accessed March 5, 2024).

requirements are much more stringent when it comes to required spatial resolution and we seem to be operating on the hairy edge."

Bill: "Well, this sounds like we are going to have a fun discussion. What is the problem?"

Tom: "The problem, as I said, only shows up with the limiting form of our identification mission. For identification purposes, we need to resolve the smallest distinguishing feature on the face, which for our purposes is the eye, …, actually the colored part of the eye. This colored part has roughly a 1 cm diameter. To be able to resolve that, conservatively, we need ten spatial resolution cells across the smallest feature we want to discriminate, the colored part of the eye in this case, which gives a spatial resolution requirement of around 1 mm. From our models, using the thermal wavelength of 9.5 μm, we need to be within roughly 43 m of our TOI to get that kind of spatial resolution. This of course is using the conservative resolution cells across the TOI. We can relax that a bit and maybe get to a requirement to be within 50 m or so, but we can't move the center wavelength too much since this is where the peak of the human radiant emittance curve is located. Also, we can't increase the diameter of the imaging system's entrance pupil since the Raytheon MTS – B gimbal is limited to a maximum 20-in. optic."

Bill: "This is for the nighttime identification mission, right?"

Tom: "Yes, that's right."

Bill: "Well, we likely will be able to lower the center wavelength by using a laser illuminator at a lower range that is still in the eye-safe region of the electromagnetic spectrum. I also recall we worked on a special program for another customer called Slip Stream, that employed a bunch of smaller sensors from a mother ship (launched from the payload bay of the Reaper drone), that could fly in close to ranges like that virtually undetected. I think we can provide some supplementary solutions for some of the harder cases. The atmospheric turbulence compensation elements though will be a must. Thanks Tom. I think overall we appear to be in pretty good shape to absorb the change of scope and address the added requirements for new features and capabilities for our special customer."

Karl: "There is another issue that has surfaced with regard to the overall cost."

Bill: "Okay, what is the issue?"

Karl: "It has to do with the scaling of the cost for the larger mirrors that need to be used. We originally were planning on using 16-in. mirrors for the pure daylight detection mission. However, the spatial resolution requirements are forcing the design to a 20-in. entrance pupil plane optic and that is driving up our cost estimates for the primary telescope mirror for both the EO and IR imaging systems."

Bill: "The 16-in. primary mirror solution was limiting our flexibility though if I remember correctly, right Jenn?"

Jenn: "Bill, that's right. It was also limiting the full potential of our atmospheric turbulence compensation (ATC) system."

Bill: "How so?"

Jenn: "The relative gain, or benefit we get from atmospheric turbulence compensation systems with regards to the expected increase in spatial resolution is given by dividing the entrance pupil plane diameter of the imaging system D_p by the atmospheric coherence length r_o. If we take a 16-in. (0.406 m) diameter entrance pupil plane optic and divide it by an r_o value of 0.15 m, since the atmospheric coherence length r_o is larger at altitude than it is on the ground, we get a potential increase in expected spatial resolution of about a factor of 2.71. If instead, we used 20-in. entrance pupil plane diameter optics, that same ratio would be a potential spatial resolution increase of about 3.4."

Bill: "Ah, I see. So how much of an additional cost is it to go from a 16-in. primary mirror to a 20-in. primary mirror?"

Karl: "We checked with two of our best suppliers. Normalized costs are shown in this next figure."

[Karl brings up Figure 4.22]

Karl: "We are using normalized costs as a function of entrance pupil diameter from both vendors plotted as squares and diamonds from previous purchases from them. You can see that the increase from 16 to 20-in. increases the cost on the order of 50%–70%."

Bill: "Ouch! That is noticeable. However, we need to have the larger mirrors to deal with the harder requirements for our special customer, right?"

Karl: "That is correct Bill."

Bill: "Okay, let's go with the larger mirrors then. That will also provide for more flexibility when it comes to other problem spaces. Anything else?"

Karl: "Well, there will be a ripple-down effect to some of the other components, and the vibration and shock isolation elements will need adjusting, but that shouldn't be a problem."

Jenn: "Oh, I have something to add. If the UAV is being manually controlled with a real human analyst that is analyzing the images for identification, there have been some recent studies that show human beings are much better when it comes to pattern recognition than automatic target recognition. The studies showed that humans can do identification with low resolution images with only 9-to-17-pixel images across the face, versus 40×40 minimum, and more realistically 80 by 80 pixels across the human face for automatic target recognition algorithms. That is a much easier spatial resolution requirement to meet than what we are currently using."

Bill: "Jenn, that is a good thing to keep in our hip pocket. However, let's not relax anything until we need to. We want to have some margin if possible. Let's get that study and run our simulations with your data and see how well things performs."

Tom: "Great idea!"

Bill: "All right everyone, great job! Let's get to incorporating the new capabilities and features into our sensing system models. Remember, outside this room. US DHS CBP is our only customer. We have a new secondary set of objectives to address and cast into the US CBP mission areas, including search and rescue, and assisting other agencies, but publicly, our main mission is still the total set of US DHS CBP missions."

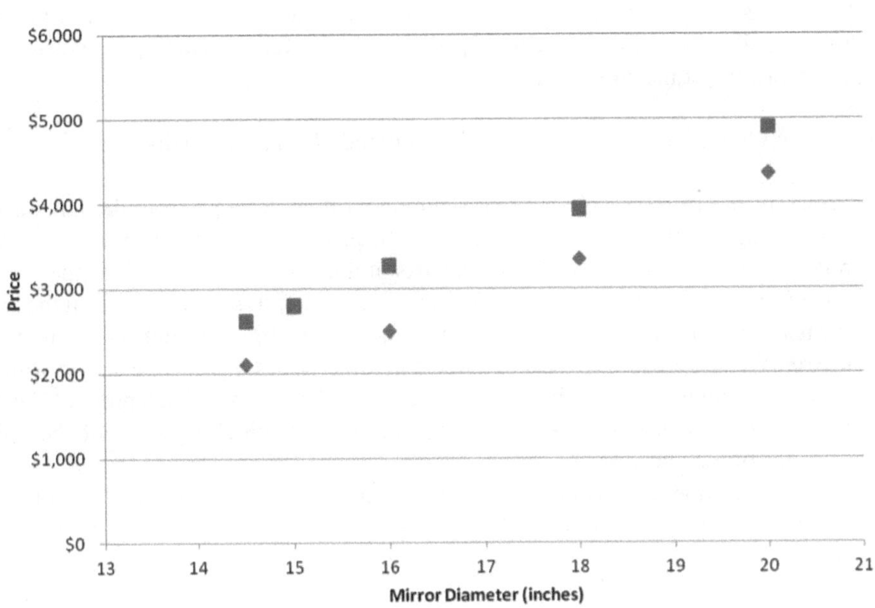

FIGURE 4.22 Mirror cost increase as a function of primary mirror diameter (in entrance pupil plane).

[A few days later in FIT's Optics Lab.]

Jenn: "Okay Amanda, I am bringing up our Zemax model of the optical train. What are some of the parameters that we should be varying to incorporate the changes we got from US DHS CBP?"

Amanda: "Well, we need to run a ray trace through our existing optical train with the new larger 20-in. mirrors to see if we are getting any vignetting. The new mirror has a different nominal focal length, so we will need to make sure our projected pixels still have the right angular resolution and that we can vary the focal length to give us enough control on the Instantaneous Field of View and TOI sampling. We will eventually also want to estimate the detector signal-to-noise ratio across our target sets and for the various Use Cases that we need to address but let's save that for another day. Have you already come up with the CONOPS-based Use Cases for the new features and capabilities yet?"

Jenn: "We are still in the process of generating those, but we have a good start. It makes sense to delay the detector SNR analyses until we get those ironed out. So, for now, are we focusing on simulating the spatial aspects of the imaging process."

Amanda: "That's right. We will also use our multi-physics tools to model the shock, vibration, and other mechanical effects on the imaging system to assess the pointing stability and estimate the amount of boresight jitter we have. That directly relates to the amount of variation we expect in our projected pixels."

Karl: "We have that analytical capability in Matlab™ don't we, or are we using the stand-alone Comsol™ Multi-Physics tool for that?"

Amanda: "We have both. In this case though, we are using our Matlab™ tools to do the analysis. In this simulation, we are going to plot the achievable spatial resolution in the form of projected pixel dimensions at the location of the TOI, as a function of stand-off distance, wavelength, and focal length. We are using the new 20-in. mirror in our model of the Raytheon MTS – B gimbal, with our optics. We are first going to adapt our telescope model to account for the larger primary mirror and look at ways to vary the telescopes effective focal length."

Tom: "What type of telescope are we using, and how will you change the effective focal length?"

Amanda: "We are going to simulate some existing designs that worked well on similar projects in the past. We are going to use a reflective design, likely a catadioptric design to reduce errors. Here, let me show you."

[Amanda brings up Figure 4.23 on the lab's, wall-mounted 86-in., OLED display.]

Amanda: "You can see from our model, that light enters the telescope from the left through two converging lenses that focus the light onto the primary mirror, which is shown by the two reflectors on the right top and bottom parts of the telescope. The light is then reflected to the left onto a secondary mirror that is shown in between the two lenses on the left side of the telescope. This secondary mirror then focuses the light through a hole in the primary mirror. We can get longer focal lengths from this type of a telescope design which allows us to get a smaller IFOV which, in turn, gives us better spatial sampling of the TOI.

Karl: "Is this the way it is usually done? It seems like you are blocking some of the light at the entrance of the telescope."

Amanda: "This a common way. However, there are other telescopic methods that provide for a means not to block the light such as in an off-axis telescope design. Here, let me show you."

[Amanda brings up Figure 4.24 on the lab's monitor.]

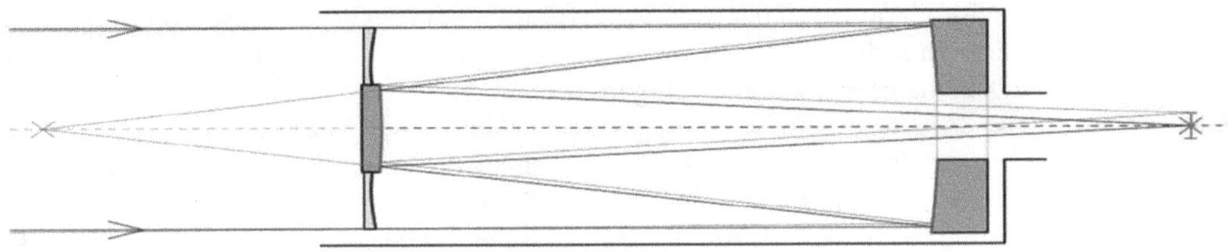

FIGURE 4.23 A Schmidt-Cassegrain telescope. (Image in public domain. https://commons.wikimedia.org/wiki/File:Schmidt-Cassegrain-Telescope.svg. Accessed March 7, 2024).

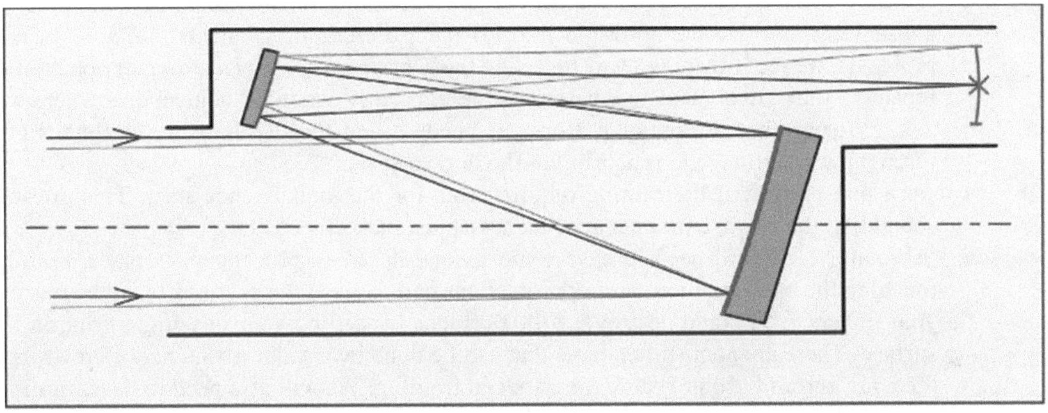

FIGURE 4.24 Off-axis telescope example. (By User:Eudjinnius - Own work, CC BY-SA 3.0, https://commons.wikimedia.org/w/index.php?curid=2116871. Accessed March 7, 2024).

Amanda: "Notice that in this off-axis design, we don't have the obscuration at the entrance of the telescope. However, depending on the size and space available, we may not be able to implement this. There are folding methods that use a combination of lenses and mirrors to effectively fold the light and increase the imaging system's effective focal length. We are investigating those designs too."

Tom: "There are also advantages going with a reflective design with a longer focal length since it is easier to make a reflective mirror that spans a broad range of wavelengths than a similar transmissive object like a lens. There are also some weight advantages."

Karl: "We also need to get a handle on any special maintenance and logistics issues that we will need to deal with."

Amanda: "There is the possibility for misalignment of the optical components. There is a lot of turbulence that causes shock and vibration that can be coupled into the optics and potentially cause some misalignments."

Karl: "Is there something we can do to mitigate those effects?"

Amanda: "Not really. We can, and do, build vibration isolation mechanisms into our system but things will still become misaligned over time. What we can do is to build in some in flight calibration capabilities that allow the system to be adjusted in flight. There are built-in laser-based methods for doing this. That way, if there is some minor drift of the optical axis through the optical system, it can be adjusted to within acceptable mission parameters. We also need to make sure that there is regular maintenance on the ground with precise alignment tools to re-calibrate the optics and make sure everything is in alignment. This should be part of the regular maintenance function while the optical system is in the hanger and not being used."

Jenn: "How precise does the alignment of the optical elements have to be?"

Amanda: "We are going to perform a tolerance analysis of the optical components to determine those requirements. It isn't unusual to have alignments within hundredths, or even thousandths of an inch. We can use Zemax to do the tolerance analysis and investigate all the optical elements in their layout for misalignment and image quality effects. We can simulate the effects of jitter, drift of the optical axis, decentering, tilt, and changes in optical element positioning."

Karl: "You then fold back those results back into the optical design."

Amanda: "Exactly! We look at ways to isolate unwanted vibrations, or shield against some shocks, and try to minimize the boresight jitter. If there are certain elements that are causing problems, we see if there is a way to design those problems out, by moving, or potentially replacing the optical element where possible. Basically, we make adjustments where we can, ensuring that the optics perform as needed, and that we have the optimal light throughput, stability, and reliability of the optical system."

Karl: "I take it that there will be training requirements for the maintenance staff. This doesn't sound like the average maintenance tech will just pick this up."

Amanda: "Absolutely! We will need to have some special handling procedures in place. Simply touching the optics without proper protections can damage the coatings from the oils on your fingers. You could destroy a $100,000 lens by getting your oily fingerprint on its surface. There are some procedures that can be done by regular maintenance crews, but there are certainly some that require special training. We will also need to have training plans and procedures, and specialized testing equipment on hand for the re-calibration of the optics."

Tom: "It might also be good to periodically recertify the system."

Amanda: "There should also be provisions for regular inspection, cleaning, and regular maintenance. This is a complicated piece of equipment, and it needs to be well maintained."

Karl: "Yes, those are good ideas. We will include those in our support planning. Perhaps we can tie the frequency of the maintenance to the reliability analysis of the optical sensor suite. We can set up a reliability-centered maintenance program and adjust the maintenance cycles based on field failure rate data."

Tom: "What about spares?"

Karl: "We are going to evaluate our supply chain for parts availability, cost, number of sources, lead times, and surge capability. We will look at our recent failure rate data and determine the expected mean time between failure (MTBF) for the optical system. Based on the failure data, we can also estimate what parts are likely to be in demand and so we will plan accordingly."

Jenn: "Why do you need the number of sources?"

Karl: "We don't want to get caught in a single supplier situation where the supplier decides not to continue a part we need. That can get very costly over the long run."

Amanda: "We also try to use commercial off the shelf parts, where it makes sense and use standard parts where possible. By the way, Zemax has a cool feature when it comes to common optical parts. It has a built-in lens catalog. We will have to use a custom build for the primary and secondary mirror but for some of the standard optical components in the optical system, this catalog is quite useful. Here, let me show you."

[Amanda brings up Figure 4.25 on the lab monitor.]

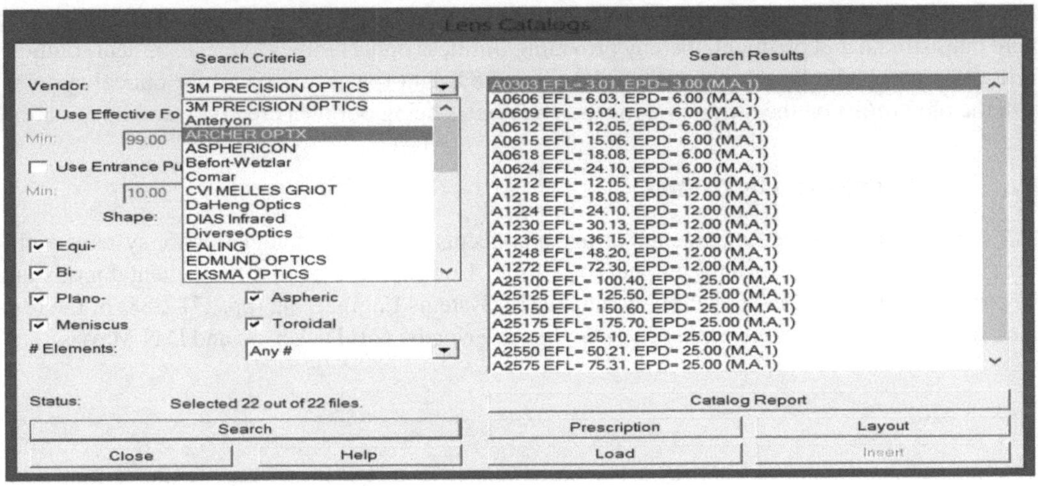

FIGURE 4.25 Zemax includes a catalog of off the shelf lenses.

Karl: "Amanda, anything else going on?"

Amanda: "Yes, we are going to use our modeling tools to do a stray light analysis."

Karl: "What is that for?"

Amanda: "Stray light analysis uses non-sequential ray tracing through all the optical elements. At each optical surface, the ray can be broken into a reflecting component, a transmitting component, or an absorbed component as in the case of the light ray hitting a blocking element. All the possible rays are projected through the optical system to see if light is going where it shouldn't (stray light). For instance, we might get unwanted reflections that end up on our detector resulting in an unwanted signal (noise). Through this analysis, we can design in anti-reflection coatings, and/or adjust the optical layout to mitigate the stray light effects. Given the choice, if we can eliminate the stray light by adjusting the optical layout, that is preferred."

Karl: "Why is that?"

Amanda: "Optical coatings are expensive. They are also fragile and require specialized handling. Also, they are not regularly stocked and require an expensive process to apply. The imaging system is much more robust if we don't need them. Of course, in low-light situations, it is not always possible to avoid coating the lenses since the imaging system can lose about 4% of its signal strength at each glass interface for instance."

Karl: "Makes sense to me."

Tom: "Nice job everyone. I see the optical modeling is coming along well. Let me know when you get done with your telescope and focal length analysis and what results you get. Karl, please let me know when the Use Cases are ready for the new features requested by US DHS CBP. We will set up a meeting with Bill to go over everything."

Karl: "Will do!"

Amanda: "I should have something within the next few days."

4.4 CONCLUSION

Optical systems are everywhere. We see optical devices in all segments of society, from medical, scientific, industrial, educational, and military communities. They have become commonplace and are essential elements in modern systems. Computers rely on optical drives for high-speed storage and surgeries such as LASIK provide clear eyesight for millions who had difficulty seeing before.

With the prevalence of high-speed computers, high-speed optical design software is also becoming more mainstream and practical, thereby providing limitless opportunities to create optical solutions for today's problems. It seems that the future is quite bright (pardon the pun) for optical systems, with the only limits on the utility of optical technologies being our own creativity and imagination.

4.A APPENDIX: DEFINITIONS

This section discusses the fundamental terms or concepts that are the foundation for systems architecture and principles in systems engineering. Table 4.A1 lists and defines fundamental terms and concepts identified in Chapter 17: Fundamentals of Systems Engineering (pp. 271–288) of the textbook *Architecture and Principles of Systems Engineering* by C.E Dickerson and D.N Mavris.

TABLE 4.A1
Fundamental Terms and Concepts of Systems Architecture and Principles in Systems Engineering

Term or Concept	Definition
Baseline	"the complete documentation of the system at some level of detail and design definition, and serves as reference for subsequent development phases" (Dickerson and Mavris 2016, p. 282)
Configurations baseline	"defines the system at different levels within the system hierarchy and is generated during different phases" (Dickerson and Mavris 2016, p. 283)
Customer requirements	"focuses on the operator, state expectations about the system in terms of measures of effectiveness, mission objectives, operational environment, and design constraints" (Dickerson and Mavris 2016, p. 278)
Design loop	"the iteration between Design Synthesis and Functional Analysis and Allocation, whose objective is the evaluation of consistency between the physical architecture and the functional and performance requirements" (Dickerson and Mavris 2016, p. 281)
Design synthesis	The objective "is to generate a physical architecture of the system that performs all the functions with the prescribed required performance as established in the functional architecture" (Dickerson and Mavris 2016, p. 281)
Functional architecture	"embodies the structure of allocated performance and functional requirements" (Dickerson and Mavris 2016, p. 277)
Functional baseline	"created in the conceptual phase of design, is a system-level technical description based on top-level system concept functions and performance requirements, which in turn are derived from customer requirements" (Dickerson and Mavris 2016, p. 283)
Functional decomposition	"a function may be decomposed into a set of functions that comprise it" (Dickerson and Mavris 2016, p. 272)
Functional requirements	"States the required actions that must be performed by the system to meet the aforementioned expectations, and embody top-level functions used in the functional analysis step" (Dickerson and Mavris 2016, p. 278)
Functional view	"addresses what the system must do to attain a prescribed operational behavior, and couples functional, performance, and physical requirements, which shape design specifications downstream in the systems engineering process" (Dickerson and Mavris 2016, p. 278)
Hierarchical structure	"a system exists at a certain level of abstraction, and the elements that are part of that system are at a lower level of abstraction, or at a lower level in the hierarchy" (Dickerson and Mavris 2016, p. 272)
Integration	"it can be conceived as the reversal of the decomposition process within the same hierarchical relationship between systems and elements" (Dickerson and Mavris 2016, p. 272)
Operational view	"embodies in an operational concept document, describes the context in which the system is expected to operate (DoD 2001)" (Dickerson and Mavris 2016, p. 278)

(Continued)

TABLE 4.A1 (*Continued*)
Fundamental Terms and Concepts of Systems Architecture and Principles in Systems Engineering

Term or Concept	Definition
Performance requirements	"States the extent to which actions must be performed to meet expectations, and are expressed in terms of quality, quantity, scope, and availability" (Dickerson and Mavris 2016, p. 278)
Physical architecture	"Provides the breakdown structure of the physical system into various levels of subsystems, components, and parts" (Dickerson and Mavris 2016, p. 277)
Physical decomposition	is the decomposition based on the physical descriptions of the system
Physical view	"addresses the construct of the system by establishing the configuration of operator–system interfaces, characterizing the different users and establishing physical limitations" (Dickerson and Mavris 2016, p. 278)
Requirement analysis	"consolidates the process inputs into a cohesive set of customer requirements, and then translates them into functional and performance requirements" (Dickerson and Mavris 2016, p. 278)
Requirement analysis output	"defines what the system must do and how well it must do it" (Dickerson and Mavris 2016, p. 278)
Specifications	"are based on the functional architecture; they describe what the system must do by documenting functional requirements, describe how well it must do it by documenting performance requirements, and describe how to verify that these requirements are met by documenting the verification process" (Dickerson and Mavris 2016, p. 282)
System	"defined by the elements that comprise it, and implies that there are some criteria that differentiate these elements from others outside the system" (Dickerson and Mavris 2016, p. 272)
System architecture	"a description of the entire system based on the physical architecture and the definition of supporting products and services across all phases of the lifecycle" (Dickerson and Mavris 2016, p. 282)
System engineering	"encompasses a collection of activities, processes, and techniques" (Dickerson and Mavris 2016, p. 273)
	"enables the generation of a system that represents a design solution and meets certain needs or requirements" (Dickerson and Mavris 2016, p. 273)
	"enables the transformation of information from requirements to system definition" (Dickerson and Mavris 2016, p. 274)
	"characterized as being iterative, recursive, and interdisciplinary, and takes a multitude of aspects of the system into consideration" (Dickerson and Mavris 2016, p. 274)

4.B APPENDIX: PRINCIPLES OF SYSTEM ARCHITECTURE WITH SysML EXAMPLES

4.B.1 Principle of Emergence

Source: System Architecture, Strategy and Product Development for Complex Systems by E.C. Crawley, B. Cameron, and D. Selvam Page 26 (Box 2.2):

> As the entities of a system are brought together, their interaction will cause function, behavior, performance, and other intrinsic properties to emerge. Consider and attempt to predict the anticipated and unanticipated emergent properties of the system.
>
> - The interaction of entities leads to emergence. Emergence refers to what appears, materializes, or surfaces when a system operates. It is this emergence that can give systems added value.
> - As a consequence of emergence, change propagates in unpredictable ways.
> - It is difficult to predict how a change in one entity will influence the emergent properties.
>
> System success occurs when the anticipated properties emerge. System failure occurs when the anticipated emergent properties fail to appear or when unanticipated undesirable emergent properties appear.
>
> *Crawley et al. (2016, p. 26)*

The term emergence refers to what is created when the system is operational. Emergence is based on the system's functional interactions. The concept of emergence is "considered within both the framework of Composite and System of Systems (SoS) and it is linked to the problem of defining functions on a given system and evaluating their values" (Karcanias and Hessami 2011, p. 1). Within a SoS, emergence is manifested and considered inevitable. In the paper *Self-organization, Emergence and the Architecture of Complexity* by Francis Heylighen, it was proposed to use dynamic and evolutionary viewpoints to deal with emergent properties (irreducible properties). The approach for dynamical emergence is developed through self-organization models, which are "defined as a spontaneous (i.e. not steered or directed by an external system) process of organization, i.e. of the development of an organized structure" (Heylighen 1989, p. 1). The main takeaway is to use emergence and self-organization models to understand the complexity of a system.

The paper *System principle on emergence of Mobiligence and its engineering realization* by a group of students in Japan discussed mobiligence which emerged from the mobility of robots around an environment. Essentially, mobiligence is the "mechanism of intelligence to behave adaptively even in an unknown environment, and to discover the design principle of artificial systems that have such intelligence" (Asama et al. 2003, p. 1720). Based on the interactions between the system and the environment, there was the emergence of intelligence. The first part of this paper defines mobility as a function that "brings diverse interaction with the environment by changing self-location, that increases experience in spatial properties, as well as dynamic interaction with the environment by movement that increases the experience in temporal properties" (Asama et al. 2003, p. 1716). The interactions based on mobility led to the development of intelligent cognitive functions, such as adaptable behaviors and observations that lead to signals for change in motion.

This idea of adaptive intelligence based on interactions with the environment can be applied to Autonomous Unmanned Aerial Vehicles (AUAVs). The AUAVs have to constantly adjust their location based on mission, environment, weather, etc. The emergence of "mobiligence" for AUAVs will be based on that specific mobility and interactions in the air. This concept will assist with enhancing decision-making skills, threat-aversion skills, search and find maneuvers, and Object of Internet (OoI) tracking of the AUAV.

The paper *On the Emergent Behavior Oxymoron of System-of-Systems Architecture Description* by F. Oquendo discussed the oxymoron of designing system architecture for expected unexpected behaviors (emergent behaviors). Emergence is characterized as unexpected ("characterizes that emergent properties are not predictable") and irreducible (characterizes how emergent properties cannot be explained in relation to the properties of its part) (Oquendo 2018, p. 418). There are two types of emergence: strong emergence, which is an unexpected property that cannot be deduced, and weak emergence, which is an unexpected property that can be deduced. The paper also focused on how complex systems pose the need to separate concerns between architecture and engineering. Architecture is for focusing on the "designing and reasoning about interactions of parts and their aggregative or emergent properties" and engineering for focusing on the "designing and constructing such parts and integrating them as architected" (Oquendo 2018, p. 419). This paper uses the example of a UAV-based Reconnaissance SoS with the purpose to fly an autonomous reconnaissance mission. One of the emergent behaviors of the UAV-based Reconnaissance SoS is flocking. A flock is "defined as a clustered group of individuals, moving with a common velocity" (Oquendo 2018, p. 421). The emergent behavior of flocking resulted in three behaviors: separation behavior (collision avoidance), alignment behavior (velocity matching), and cohesion behavior (flock centering).

The idea of the emergent behaviors of flocking can be used in the AUAV project. Figure 4.B1 displays the three emergent behaviors: separation, alignment, and cohesion.

Figure 4B2 is the SysML Package Diagram for the three flocking emergent behaviors, which was created using No Magic Cameo Systems Modeler.

4.B.2 Principle of Holism

The textbook *System Architecture, Strategy and Product Development for Complex Systems* by E.C. Crawley, B. Cameron, and D. Selva discussed the Principle of Holism and supplied the following information on p. 34 in Box 2.5:

> Every system operates as part of one large system or several larger systems, and each is itself composed of smaller systems. Think holistically about all of these relationships, and develop architectures that are in harmony with the larger, smaller, and peer systems.
>
> - Holism holds that all things exist and act as wholes, not just as the sum of their parts. Its sense is the opposite of that of reductionism, which suggests that things can be understood by carefully explaining their parts.
> - To think holistically is to encompass all aspects of the system at hand, taking into account the influences and consequences of anything that might interact with the system.

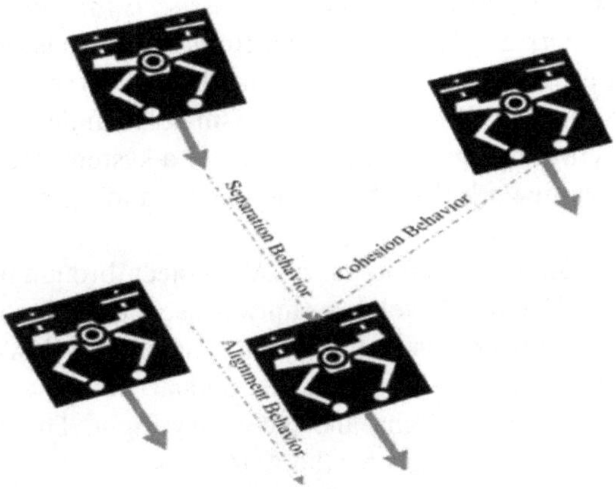

FIGURE 4.B1 A AUAV example of cohesion, separation, and alignment behaviors for SoS platooning. (Drone image from https://commons.wikimedia.org/wiki/File:198_Senpilota_aviadilo.svg, which was accessed May 26, 2022).

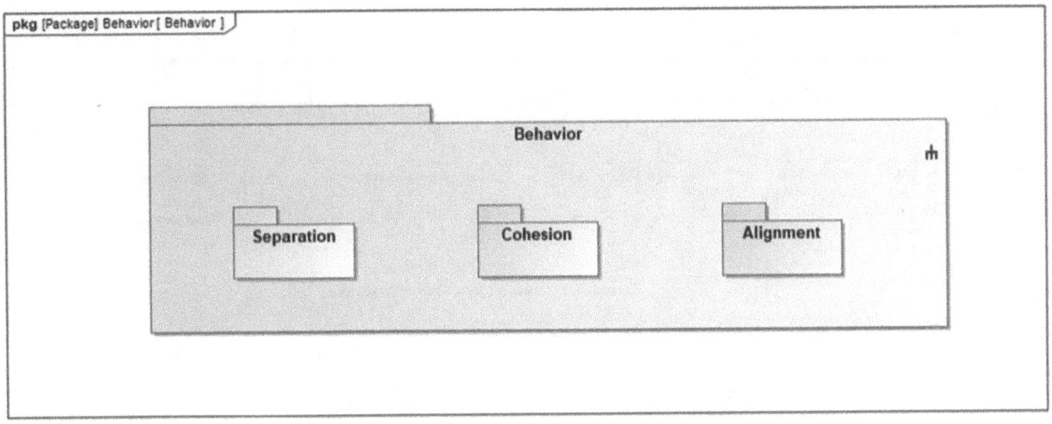

FIGURE 4.B2 Package diagram for the emergent behaviors of the AUAV.

Figure 4.B2 is the SysML package diagram for the three flocking emergent behaviors, which was created using No Magic Cameo Systems Modeler

- In more simple terms, to think holistically is to think about all the things (entities, relationships, and so on) that may be important to the question, circumstance, or problem at hand.
- Methods to stimulate holistic thinking include structures and unstructured brainstorming, frameworks, thinking from different perspectives, and thinking about context."

Crawley et al. (2016, p. 34)

Holism is the modern-day version of the idea that the whole is greater than the sum of its parts. Holistic thinking's purpose is to identify all aspects and entities of a system that might be important. Understanding the holistic view will assist with grasping the context and scope of a system. Holism is referred to as the "looking at any object or a system from multiple viewpoints as a whole unit within a larger system called environment" (Kancherla and Kummamuru 2011, p. 2). The point of holism is to look at the system, components, and problems from multiple angles. Observe the various interactions and create a list of stakeholders and influential attributes from those aspects. This provides a big picture view of the system and assists in understanding all influences of the system. The paper *Advances in the Prospects for Realizing a Scientific General Theory Underpinning Systems Engineering* by D. Rousseau discussed how there is a shift of emphasis to the concept of holism to improve Systems Engineering (SE). The increasing rise of system complexity is a challenge of SE and using holism helps understand the system as a whole and contributes to identifying the emergent properties of a system. Holism is an important part of developing a system's framework that shows the understanding of structure and comprehensiveness of a system.

The Principle of Holism can be applied to the AUAV project through the use of a Top-Level diagram, shown in Figure 4.B3. The Top-Level diagram was used to think about the system as a whole and come up with the various subsystems that will be needed in the AUAV system. The subsystems identified in the Top-Level diagram are for guidance, optical, power, communication, propulsion, controls, structural, payload, primary flight, and secondary flight. The diagram also contains the system requirements and the legacy models for the AUAV system.

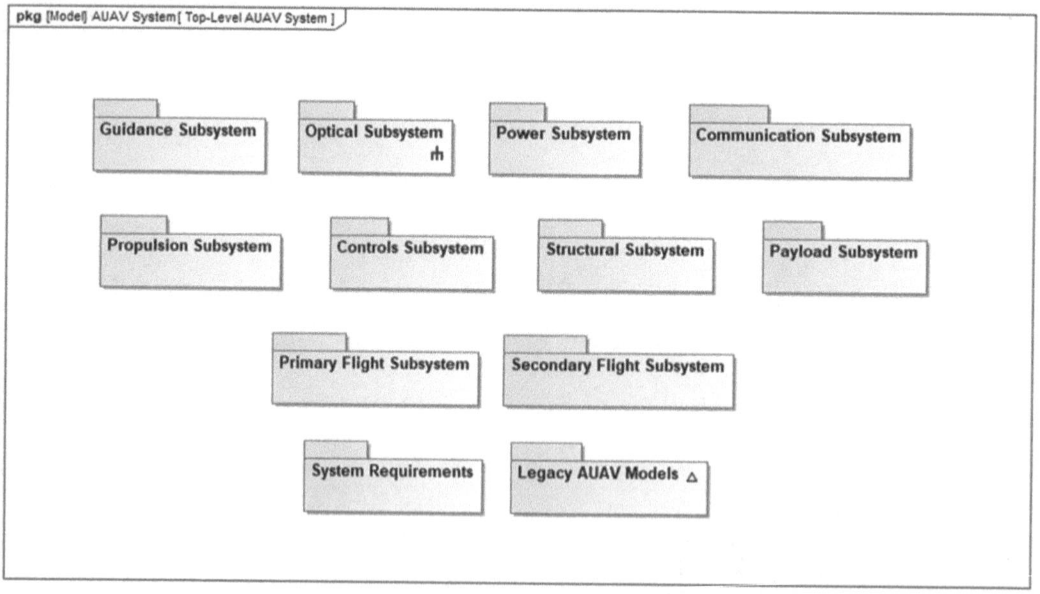

FIGURE 4.B3 Top-level package diagram for the AUAV system.

4.B.3 Principle of Focus

The textbook *System Architecture, Strategy and Product Development for Complex Systems* by E.C. Crawley, B. Cameron, and D. Selva discussed the Principle of Focus and supplied the following information on p. 36 in Box 2.6:

> The number of identifiable issues that will influence a system at any point is beyond one's ability to understand. One must identify the most critical and consequential issues, and focus on them.
>
> - At any given time, there are tens or even hundreds of issues identified by holistic thinking that could impact the system under consideration. This is too many for any individual or small team to simultaneously understand.
> - To sustain close consideration of the important issues at any moment, one must be prepared to leave others behind.
> - Failure rarely occurs in aspects on which you focus.
> - Process or filter this larger set of issues to identify those that are important to that day or activity. Focus on the hard issues, and avoid the temptation to address the easier ones first.
>
> *Crawley et al. (2016, p. 36)*

The Principle of Focus is to identify the important factors to question on a system. This principle narrows the list of potential impacting entities identified during holistic thinking of the system. The goal is to choose and focus on the entities of a system that are truly consequential. Questions a system architect should ask are: What is important for the stakeholders? What are possible emergent behaviors?, What are the important outcomes of the system? The answers to those questions shorten the initial list of impacting factors developed using the Principle of Holism. The final aspect of the Principle of Focus is to verify the focused entities are large enough to cover the overall problem, but still be able to be examined with the available resources. The paper *Non-functional refinement of computer-based systems architecture* by M. Denford, J. Leaney and T. O'Neill suggested using the systems requirements specification (SRS) containing the functional and non-function requirements to think about the system holistically, and then refine those system requirements to help focus the systems architecture for the important factors (Denford et al. 2004). The paper *Implications of systems of systems on system design and engineering* by J. Dahmann and K. Baldwin also stresses defining the system operational requirements to then put the systems architect's focus on understanding and addressing the important factors of the system (Dahmann and Baldwin 2011). The paper *Architecture Design for the Large-Scale Software-Intensive Systems: A Decision-Oriented Approach and the Experience* by X. Cui, Y. Sun, S. Xiao and H. Mei suggested focusing on significant design issues in the system architecture to help accommodate and capture key decisions for large complex systems (Cui et al. 2009). The paper also suggested focusing on decision-making and decomposition of the system to provide support to handle the hard issues of the system.

The AUAV system focus for our example is the optical subsystem. Progressing from the Top-Level diagram shown for the Principle of Holism (Figure 4.B3), a Package Diagram for the focused subsystem, the optical subsystem, was produced. Figure 4.B4 is the Optical Subsystem Package Diagram which identifies the components of the subsystem as hardware, software, Use Cases, behavior, requirements, and analysis.

4.B.4 Principle of Value and Architecture

The textbook *System Architecture, Strategy and Product Development for Complex Systems* by E.C. Crawley, B. Cameron, and D. Selva discussed the Principle of Value and Architecture and supplied the following information on p. 125 in Box 6.2:

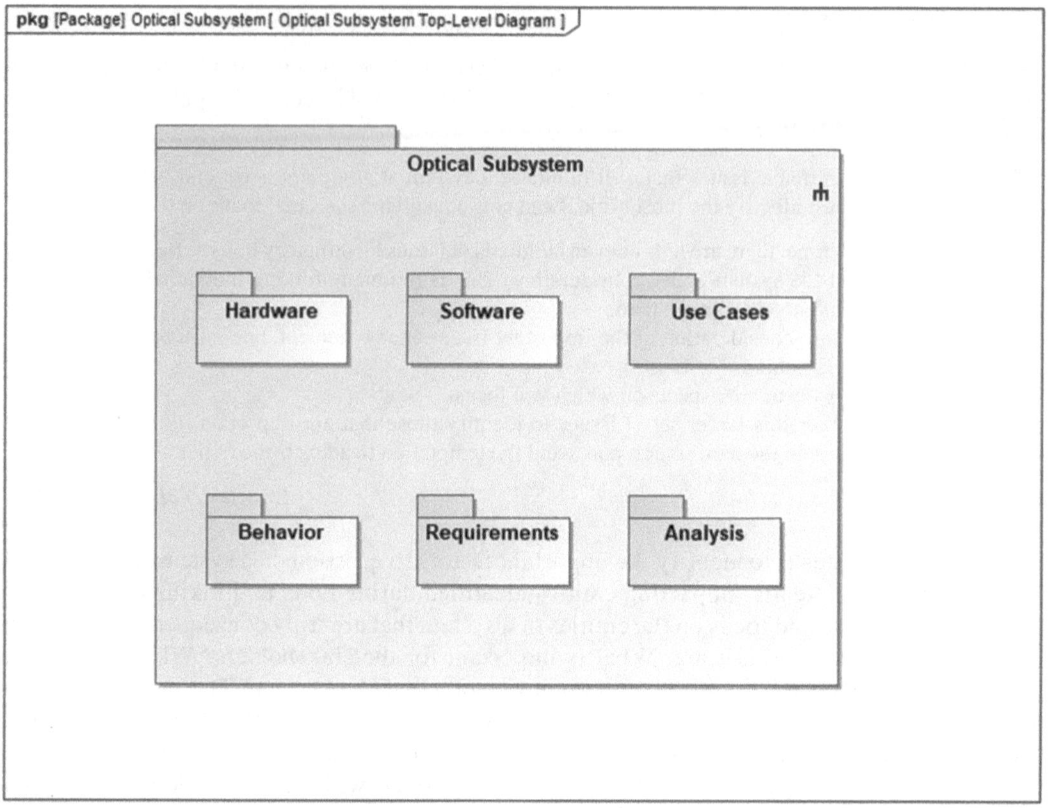

FIGURE 4.B4 Package diagram for the optical subsystem of the AUAV.

> Value is benefit at cost. Architecture is a function enabled by form. There is a very close relationship between these two statements, because benefit is delivered by function, and form is associated with cost. Therefore, developing good architectures (desired function for minimal form) will be nearly synonymous with the delivery of value (benefit at minimal cost).
>
> - Benefit derives from the emergence of primary and secondary value-related functions.
> - An axiom of lean manufacturing is that parts (form) attract cost.
>
> *Crawley et al. (2016, p. 125)*

The Principle of Value and Architecture displays the importance of architecture. Essentially if value is delivered with the system that was built, then there must have been good architecture. A system architect can use the value-based approach to conduct a value assessment for the systems architecture. The value-based approach "uses economic-driven tools, which are based on economic studies including, for example, the areas of cost estimation, cost calculation, and investment calculation" (Ojala 2008, p. 1). This approach identifies any dependencies between the product and process and is used to justify the need to have optimization processes. Based on the value assessment, the system architect can improve on the proposals of the system and add an additional element to the decision thinking process for the designed system architecture. A system architect can also use the value chain approach, which "is based on the major principle of the market value of a company: transformation of operating results into financial ones" (Karlik et al. 2016, p. 1). The analysis of the value chain can be used in system architecture to improve upon the value of the system. The paper *Value-Based Design Decision Rationale Documentation: Principles and Empirical Feasibility Study* by D. Falessi, G. Cantone, and P. Kruchten stresses the importance of the system architect documents reasoning, rationale, and justification for the design decisions, especially ones that impact the system's value (Falessi et al. 2008). A thought-out and documented value-based design can assist the system architect to make good decisions in the architecture.

The Principle of Value and Architecture can be applied to the AUAV system by encouraging the system architect to elaborate on the system. The system architect can use the Model-Based System Engineering (MBSE) tools to create traceability matrices between the requirements and processes that are going to be used. The system architect can also create requirement matrices and dependency matrices to elaborate on the system. By understanding the requirements and the dependencies of the system, along with thinking about emergent properties from these dependencies, the system architect can produce good system architecture. Good architecture will produce the value of the AUAV.

4.B.5 Principle of Architectural Decisions

The textbook *System Architecture, Strategy and Product Development for Complex Systems* by E.C. Crawley, B. Cameron, and D. Selva discussed the Principle of Architectural Decisions and supplied the following information on p. 211 in Box 10.1:

> Architectural decisions are the subset of design decisions that are most impactful. They relate to form-function mapping, they determine the performance envelope, they encode the key tradeoffs in the eventful product, and they often strongly determine cost. Separate these architectural decisions from other decisions, and take the time to carefully decide them up front, because they will be very expensive to change later on.
>
> - Architectural decisions lead to architectures that are fundamentally different from each other — the number of driven wheels on a car, whether or not an artifact has a tail, whether an algorithm runs in real time or not, and the like.
> - Architectural decisions are different from design decisions. The choice between leather and cloth for the seats in a car is not an architectural decision but a design decision. It does not materially impact technical parameters or important metrics. The choice between leather and cloth could affect the cost, but it will do so predictably. It is less likely to have emergent consequences.
> - Identify architectural decisions early and study them carefully, because making the wrong choices can irrevocably hinder realization of the eventful product, where no amount of detailed design or component optimization will fix the fundamental issues.
>
> *Crawley et al. (2016, p. 211)*

The Principle of Architectural Decisions is to understand the influences of a system and to decide on how to handle those influences. Identifying architectural decisions early and understanding them have a huge influence on the system. Poor architectural decisions can hinder the system and impede cost, which will disappoint the stakeholder. Architectural design decisions include "design alternatives, stakeholder concerns, decisions and the rationale for making such decisions and enhancing the decision-making process" (Rekhav and Muccini 2014, p. 1). System architects make decisions to resolve any issues that arise during the architecture creation/process and have to take time and budget into consideration when making design decisions. Architects can use trade studies to "determine how to allocate their limited budgets and identify the options that are most likely to succeed" (Min and Noguchi 2016, p. 1). Architectural decisions in any part of the system architecture development can make a large impact on the system outputs, including overall value/benefit.

The system architect can apply the Principle of Architectural Decisions to the AUAV system by using a block definition diagram (bdd) for stakeholder concerns to influence the architectural design decisions, as shown in Figure 4.B5. Understanding the stakeholder concerns can help the system architect with thinking about how to resolve these concerns, possible design alternatives, and influences on the system as a whole.

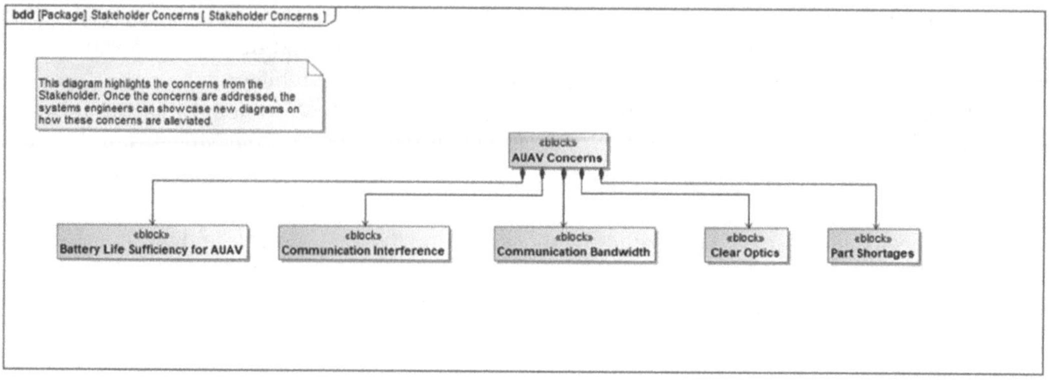

FIGURE 4.B5 BDD for stakeholder concerns for the AUAV system.

The bdd highlights the concerns from the stakeholders of the AUAV system. The concerns shown are battery life sufficiency, communication interference, communication bandwidth, clear optics, and part shortages. All of these concerns impact how the AUAV does on missions. The concerns also help the system architect break down the influences that would cause these concerns. Once the concerns are addressed, the system engineers can showcase new diagrams on how these concerns are resolved.

4.B.6 Principle of Beginning

The textbook *System Architecture, Strategy and Product Development for Complex Systems* by E.C. Crawley, B. Cameron, and D. Selva discussed the Principle of Beginning and supplied the following information on p. 243 in Box 11.1:

> The list of stakeholders (internal and external to the enterprise) that are included in the early stages of product definition will have an outsized impact on the architecture. Be very careful about who is involved at the beginning and how they shape the architecture.
>
> - The initial list should consider which stakeholders have a strong need for your project outputs, as well as stakeholders whose outputs are important to the project.
> - The architect should balance the need to engage stakeholders for their views against the decision rights of the stakeholders, such as whether they are informed, consulted, or voting members and whether-or-not they have a veto.
>
> *Crawley et al. (2016, p. 243)*

The Principle of Beginning is for stakeholder identification in the early stages of developing system architecture and managing those stakeholders. There are internal stakeholders and external stakeholders that have views on what is needed from the system. It is important to balance their views against their decision rights. The paper *Identification and prioritization of stakeholders in the planning process of sustainable development of the smart city* by I. Ielite, G. Olevsky, and T. Safiulins stressed that long-term success of the system can be dependent on the support and input of the key stakeholders (Ielite et al. 2015). The system architect should identify all potential stakeholders, conduct a stakeholder engagement analysis, conduct stakeholder interviews, prioritize the different stakeholders, create and agree on an engagement plan for the stakeholders, and implement the engagement plan. Stakeholder analysis involves identifying stakeholders, prioritization of the stakeholders, managing and understanding the stakeholders, setting goals of the stakeholders, identifying cost of stakeholder engagement, and regularly updating the stakeholder plan as new stakeholders are identified or if the stakeholders' level of influence changes. The paper *Smart Manufacturing Stakeholders and Their Requirements* by A. Weber discussed the use of stakeholder questionnaires and answers to drive other components of the system (Weber 2018). The questionnaires ask the stakeholder about issue areas, success criteria, expectations, and if there are any areas of concern that can be used to create a specification outline. The paper *A Conceptual Model for Capturing Stakeholders' Wish List* by V. Kulkarni also recommended capturing the stakeholders' needs through the use of questionnaires, as well as interviews, workshops, and brainstorming sessions. Kulkarni suggested the system architect should identify and organize the systems' stakeholders in a table for the following criteria: Requirement #, Stakeholder Classification (Internal or External), Requirement Name, Requirement Description, Type or Requirement (Functional, Non-Functional, or a Constraint), Related to Requirements, Priority of the Requirement, Cost Estimation (for implementation), Quality Impact (Positive, Negative, or Neutral), Necessary for Product (Yes or No), and Approved or Conflict (Kulkarni 2008). There is also the idea of stakeholder theory which is "widely used to measure stakeholder influence in terms of power, urgency, and legitimacy" (Balta et al. 2015, p. 1). The proposal of stakeholder analysis based on the stakeholder theory is the integration of ideas and design which are derived during the identification of stakeholders, determining the level of interest, and overall stakeholder evaluation. This supports the principle of beginning by putting emphasis and importance on identifying and analyzing stakeholders early in the system architecture process. The system architect should also take the influence and importance of stakeholders on the corporate social responsibility (CRS) of a company into account, as discussed in the influence and importance of stakeholders on the corporate social responsibility (CRS) of a company. The CRS "refers to context-specific organizational actions and policies that take into account stakeholders' expectations and the triple bottom line of economic, social, and environmental performance" (Oleksiv et al. 2020, p. 1).

The Principle of Beginning is for stakeholder identification in the early stages of developing system architecture and managing those stakeholders. There are internal stakeholders and external stakeholders that have views on what is needed from the system. It is important to balance their views against their decision rights.

The system architect can identify the stakeholders of the AUAV system and create a bdd (block definition diagram) to view the various stakeholders. Figure 4.B6 displays the stakeholder bdd where the category of stakeholder and the specific type of stakeholder for each category is seen. The diagram is from the perspective of FIT because they are the prime contractor of the AUAV. The model can be used to track everything the stakeholders make comments on by adding notes to the diagram. By identifying the stakeholders, the flow down of stakeholder needs to system requirements to component requirements can start.

The key stakeholder is the three-letter agency, under government associations, and should be engaged regularly by the system engineers, considered the main focus, and involved in decisions. The stakeholders that should be considered are the operators, maintenance personal, and first responders. These stakeholders should be engaged in their area of interest and kept informed on related information. Target of interest represents all of the targets of interests under that block.

4.B.7 Principle of the Stress of Modern Practice

The textbook *System Architecture, Strategy and Product Development for Complex Systems* by E.C. Crawley, B. Cameron, and D. Selva discussed the Principle of the Stress of Modern Practice and supplied the following information on p. 206 in Box 9.3:

> Modern product development process, with concurrency, distributed teams, and supplier engagement, places even more emphasis on having a good architecture. Recognize this trend and the impact it has on the architecting process.
>
> - Accelerating the development process by concurrency increases the importance of initial conceptual decisions.

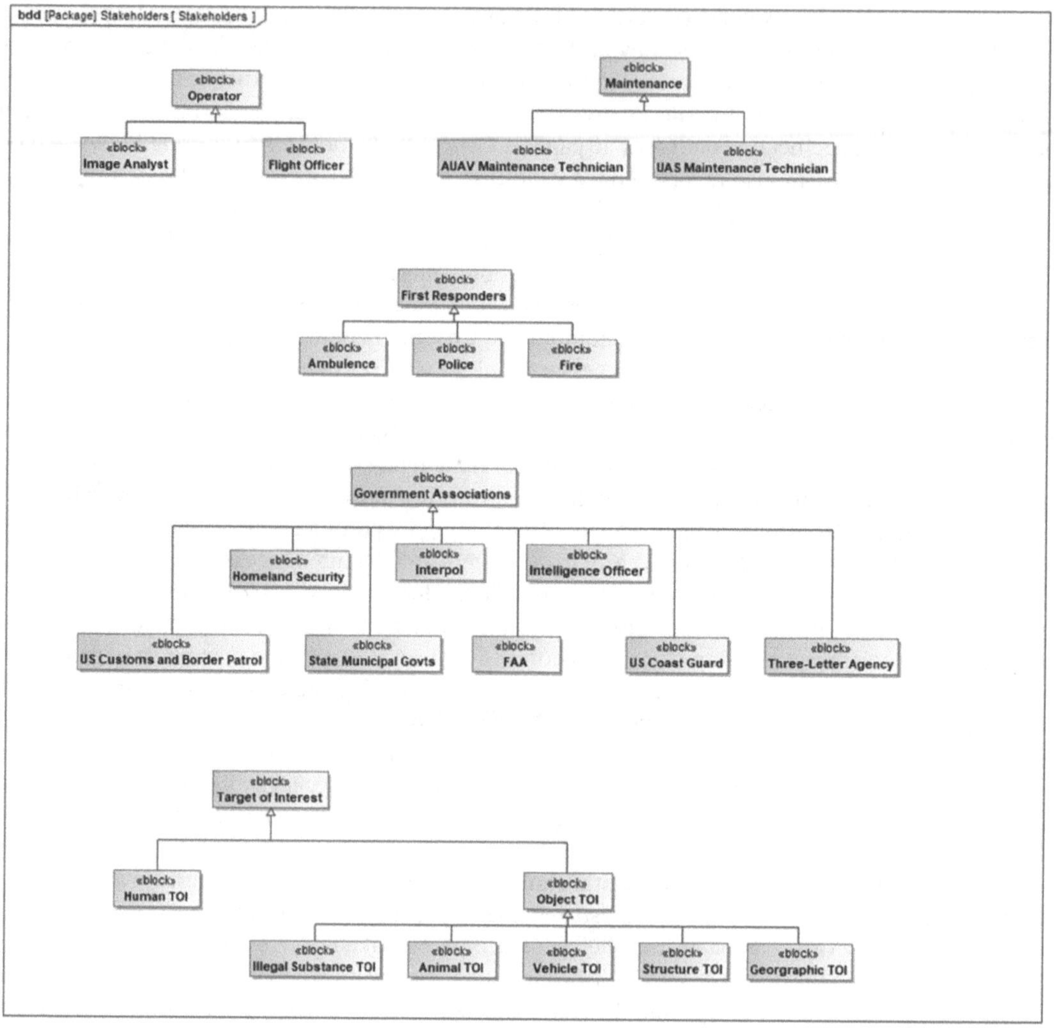

FIGURE 4.B6 BDD for the stakeholders of the AUAV system.

- Driving decision-making to the lowest appropriate level (empowerment) and the use of distributed or non-collocated teams increase the importance of well-coordinated and visible high-level design guidance.
- Involving suppliers early in the process increases the importance of having an architectural concept and baseline; suppliers can define or defy decomposition.

Crawley et al. (2016, p. 206)

The Principle of the Stress of Modern Practice puts emphasis on good system architecture for the accelerating development process, driving decision-making to the lowest level, and involving suppliers early on for architectural concept and baseline. This is a main principle for system architects to focus on and take into consideration. The article *MBSE: Simple, Complicated, or Complex* by D. Long discussed how the model-based system engineering (MBSE) approach for a system is seen. MBSE is described as an approach for "making the descriptive and analytical system models explicit, coherent, consistent, and actionable" and that it's "about an evolution in the fidelity of the representation of information for both communication and analysis" (Long 2020). Depending on

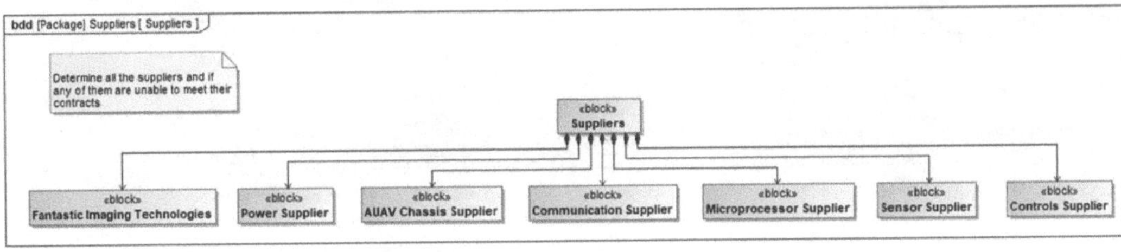

FIGURE 4.B7 BDD for the suppliers of the AUAV system.

the system and the situation, MBSE can be seen as simple, complicated, or complex. Overall, MBSE enables the understanding of the problem and the comprehension of the solution, which addresses the descriptive architecture that drives the systems performance.

The Principle of the Stress of the Modern Practice can be applied to the AUAV system using a bdd (Block Definition Diagram) for the suppliers, shown in Figure 4.B7. This diagram is to identify various suppliers of the system and encourages the system architect to think about how changes with the suppliers impact the overall system. The understanding of suppliers and their impact leads to important conceptual decisions and provides guidance on system components.

The diagram identifies the following suppliers: FIT, Sensor Supplier, Microprocessor Supplier, Communication Supplier, AUAV Chassis Supplier, and Power Supplier. A change in any of these suppliers can lead to a change in design. An example is if the power supplier changes the battery, which could impact how long the AUAV can operate before needing to charge. Battery life sufficiency is one of the stakeholders' concerns, so a change in supplier could lead to further concern and opinions on a design change.

4.B.8 Principle of Dualism

The textbook *System Architecture, Strategy and Product Development for Complex Systems* by E.C. Crawley, B. Cameron, and D. Selva discussed the Principle of Dualism and supplied the following information on p. 92 in Box 4.7:

> All built systems inherently and simultaneously exist in the physical domain and the informational domain. It is sometimes useful to explicitly consider both the physical and the informational views of a system.
>
> - Dualism holds that things can be usefully considered in two domains simultaneously –in case, physical and informational domains.
> - Systems that we usefully think of as information systems are just abstractions of physical objects that store and process the information.
> - Systems that we think of as physical store all the information about their form, but not necessarily about their function.
>
> *Crawley et al. (2016, p. 92)*

The Principle of Dualism is about the architect considering the system in both the physical form and the informational form. This principle puts emphasis on how the system's informational form is encoded in the physical form. For a computer-based information system, the architect can focus the dualism of information through the use of Dual Information Systems (DIS), as suggested in the paper *Redesigning work with dual information systems: the work process benchmarking service* by T. K. Kakola and K. I. Koota (Kakola and Koota 1996). Where DISs store, process, and utilize the data of a system. For a human-to-machine interaction system, the architect should focus on the dualism of information between human and machine for intelligent manufacturing, as suggested in

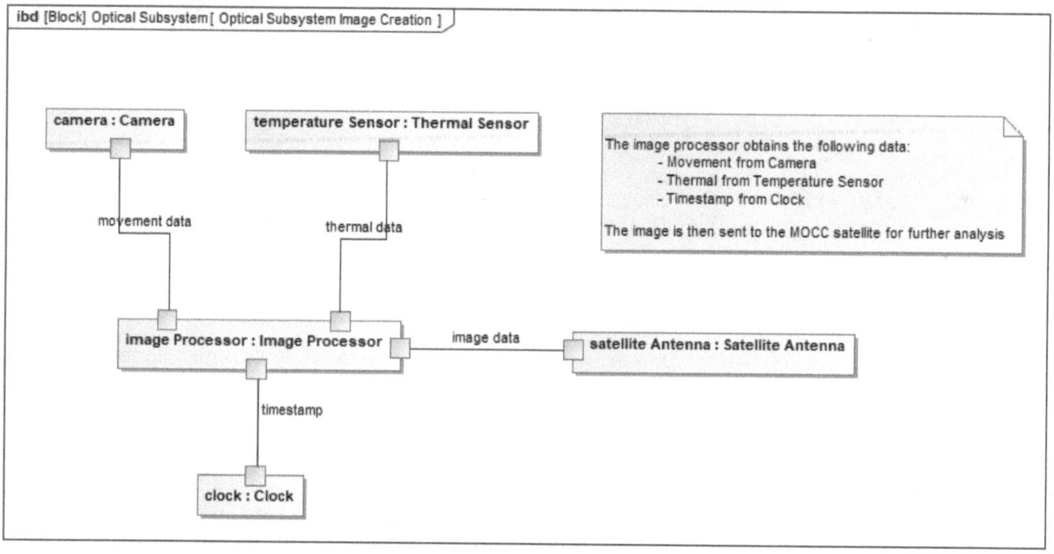

FIGURE 4.B8 IBD for the image creation of the AUAV optical subsystem.

the paper *Dualism of Knowledge Creation for Human-Machine Interactive Processing* by R. Liu, Z.-Q. Ren, and Z.-Y. Wang (Liu et al. 2018). The data and information is transferred from humans to machines and the physical form is represented in the manufacturing process.

The Principle of Dualism can be used for the AUAV through the internal block diagram (ibd) for the image creation of the AUAV optical subsystem, shown in Figure 4.B8. This diagram shows the interaction between the physical and informational of the optical subsystem. The image processor obtains data for camera movement, the temperature sensor, and the clock timestamp. This data is noted that the image is sent to the MOCC satellite for further analysis.

4.B.9 Principle of Benefit Delivery

The textbook *System Architecture, Strategy and Product Development for Complex Systems* by E.C. Crawley, B. Cameron, and D. Selva discussed the Principle of Dualism and supplied the following information on p. 105 in Box 5.5:

> Good architectures deliver benefit, first and foremost, built on the primary externally delivered function of the systems by focusing on the emergence of functions, and their delivery across the system boundary at an interface.
>
> - The benefit delivered by systems is provided by their externally delivered function.
> - There is usually a single externally delivered function for which the system was originally defined, the absence of which will cause failure of the system.
> - The benefit is delivered at the boundary of the system, most likely at an interface.
>
> *Crawley et al. (2016, p. 105)*

The Principle of Benefit Delivery is for the architect to focus on good architecture which delivers benefit at the boundary of the system (most likely at an interface). The value of benefit is subjective to the end user or customer or beneficiary. The focus in the system architecture should be the attributes of the system that will produce the most benefit for the end user. The article *Economic Analysis of Model-Based Systems Engineering* by A. Madni and S. Purohit discussed the economic benefits of MBSE (Madni and Purohit 2019). The focus was on a proposal of framework for the analysis of investment and potential earnings from the implementation of MBSE. The implementation of

MBSE early in systems architecture leads to better standards conformance, an improved definition of the product line, an increased efficiency of supply, a reduction of risk, and the early detection of issues. All of these positive contributions of MBSE saves money in a system's lifecycle and can lead to a better functioning system that can bring profit down the line. The article had an example of the MBSE framework for an economic analysis, where the inputs are from data sources (MBSE tools, information from experts, and literature reviews), the methodology is data driven, and contains the overall evaluation result. The data sources and methodology part of the framework is where the information about effort, cost, required talent, and investment for the implementation of MBSE comes into play. The article determined the investment in MBSE implementation is beneficial for a system to produce cost-benefits. A system architect can use the MBSE approach to think about the principle of benefit delivery.

The Principle of Benefit Delivery for the AUAV system can be seen with a Use Case diagram, shown in Figure 12. "A Use Case describes the goals of a system from a perspective of the users of the system" (Friedenthal et al. 2015, p. 297). This is useful for the Principle of Benefit Delivery because it captures the boundary of the system and the functions that will be occurring, which supports the idea of benefit being provided by the delivered function. Figure 4.B9 displays the AUAV system boundary, the actors outside of the boundary, and the actions within the boundary. The actors are tied to specific actions that can occur, essentially saying what each actor is responsible for. This is to understand how each actor can interact with the system and benefit the system.

4.B.10 Principle of Solution-Neutral Function

The textbook *System Architecture, Strategy and Product Development for Complex Systems* by E.C. Crawley, B. Cameron, and D. Selva discussed the Principle of Solution-Neutral Function and supplied the following information on p. 153 in Box 7.1:

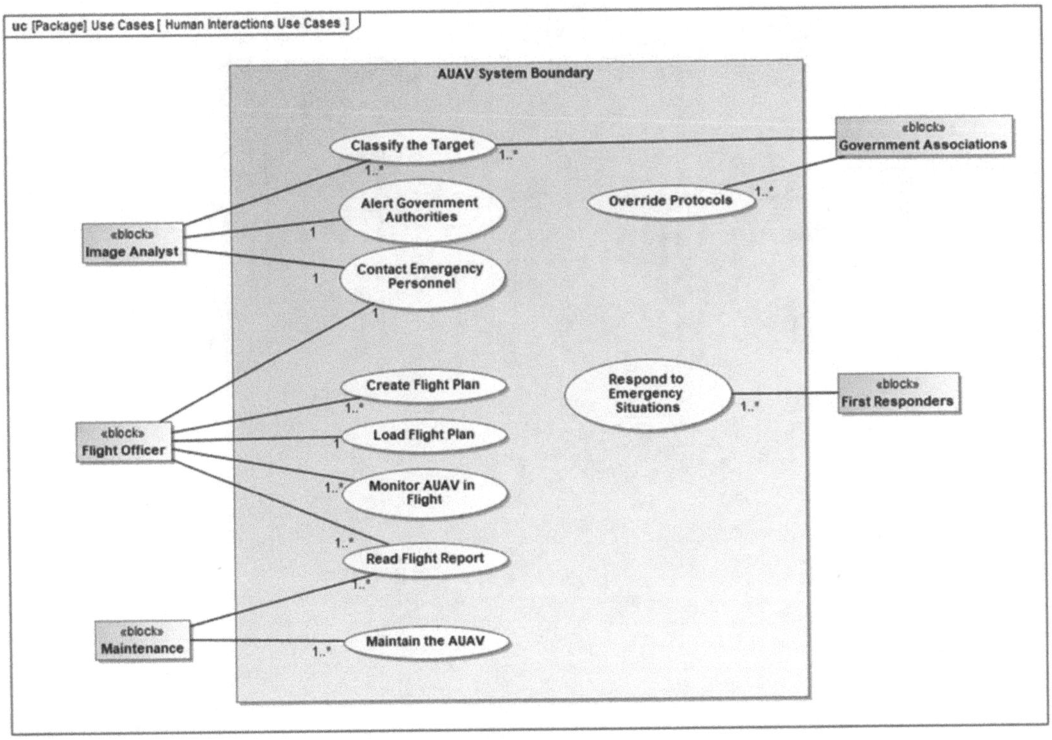

FIGURE 4.B9 Use case diagram for the stakeholders of the AUAV system.

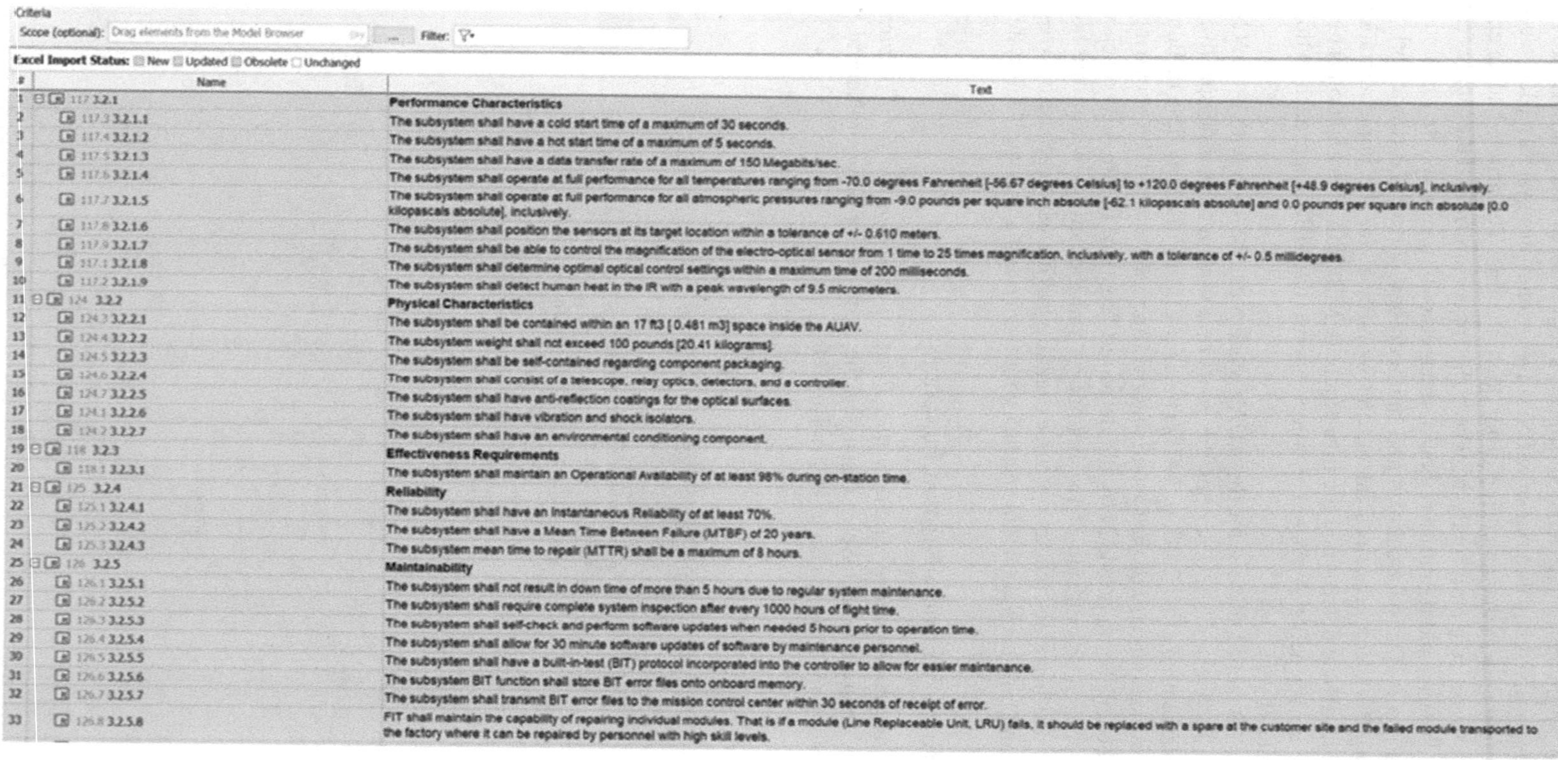

Criteria

Scope (optional): Drag elements from the Model Browser ... Filter:

Excel Import Status: New Updated Obsolete Unchanged

#	Name	Text
1	117 3.2.1	**Performance Characteristics**
2	117.3 3.2.1.1	The subsystem shall have a cold start time of a maximum of 30 seconds.
3	117.4 3.2.1.2	The subsystem shall have a hot start time of a maximum of 5 seconds.
4	117.5 3.2.1.3	The subsystem shall have a data transfer rate of a maximum of 150 Megabits/sec.
5	117.6 3.2.1.4	The subsystem shall operate at full performance for all temperatures ranging from -70.0 degrees Fahrenheit [-56.67 degrees Celsius] to +120.0 degrees Fahrenheit [+48.9 degrees Celsius], inclusively.
6	117.7 3.2.1.5	The subsystem shall operate at full performance for all atmospheric pressures ranging from -9.0 pounds per square inch absolute [-62.1 kilopascals absolute] and 0.0 pounds per square inch absolute [0.0 kilopascals absolute], inclusively.
7	117.8 3.2.1.6	The subsystem shall position the sensors at its target location within a tolerance of +/- 0.610 meters.
8	117.9 3.2.1.7	The subsystem shall be able to control the magnification of the electro-optical sensor from 1 time to 25 times magnification, inclusively, with a tolerance of +/- 0.5 millidegrees.
9	117.1 3.2.1.8	The subsystem shall determine optimal optical control settings within a maximum time of 200 milliseconds.
10	117.2 3.2.1.9	The subsystem shall detect human heat in the IR with a peak wavelength of 9.5 micrometers.
11	124 3.2.2	**Physical Characteristics**
12	124.3 3.2.2.1	The subsystem shall be contained within an 17 ft3 [0.481 m3] space inside the AUAV.
13	124.4 3.2.2.2	The subsystem weight shall not exceed 100 pounds [20.41 kilograms].
14	124.5 3.2.2.3	The subsystem shall be self-contained regarding component packaging.
15	124.6 3.2.2.4	The subsystem shall consist of a telescope, relay optics, detectors, and a controller.
16	124.7 3.2.2.5	The subsystem shall have anti-reflection coatings for the optical surfaces.
17	124.1 3.2.2.6	The subsystem shall have vibration and shock isolators.
18	124.2 3.2.2.7	The subsystem shall have an environmental conditioning component.
19	118 3.2.3	**Effectiveness Requirements**
20	118.1 3.2.3.1	The subsystem shall maintain an Operational Availability of at least 98% during on-station time.
21	125 3.2.4	**Reliability**
22	125.1 3.2.4.1	The subsystem shall have an Instantaneous Reliability of at least 70%.
23	125.2 3.2.4.2	The subsystem shall have a Mean Time Between Failure (MTBF) of 20 years.
24	125.3 3.2.4.3	The subsystem mean time to repair (MTTR) shall be a maximum of 8 hours.
25	126 3.2.5	**Maintainability**
26	126.1 3.2.5.1	The subsystem shall not result in down time of more than 5 hours due to regular system maintenance.
27	126.2 3.2.5.2	The subsystem shall require complete system inspection after every 1000 hours of flight time.
28	126.3 3.2.5.3	The subsystem shall self-check and perform software updates when needed 5 hours prior to operation time.
29	126.4 3.2.5.4	The subsystem shall allow for 30 minute software updates of software by maintenance personnel.
30	126.5 3.2.5.5	The subsystem shall have a built-in-test (BIT) protocol incorporated into the controller to allow for easier maintenance.
31	126.6 3.2.5.6	The subsystem BIT function shall store BIT error files onto onboard memory.
32	126.7 3.2.5.7	The subsystem shall transmit BIT error files to the mission control center within 30 seconds of receipt of error.
33	126.8 3.2.5.8	FIT shall maintain the capability of repairing individual modules. That is if a module (Line Replaceable Unit, LRU) fails, it should be replaced with a spare at the customer site and the failed module transported to the factory where it can be repaired by personnel with high skill levels.

FIGURE 4.B10 Legacy requirements list for the AUAV system.

> Poor system specifications frequently contain clues about an intended solution, function, or form, and these clues may lead the architect to a narrower set of potential options. Use solution-neutral functions where possible, and use the hierarchy of solution-neutral statements to scope how broad an exploration of the problem is to be undertaken.
>
> *Crawley et al. (2016, p. 153)*

The Principle of Solution-Neutral Function is for the system architect to think about "the function of a system stated without reference to how the function is achieved" (Crawley et al. 2016, p. 153). Using solution-neutral functions stimulates the system architect to think about alternative specialized functions. This is a method to creatively think about a system and expand on possible functions of a system. The paper *How function analysis of manufacturing processes can stimulate technology intelligence processes* by M. Vollmer, M. Schmitz, and J. Warschat discussed the function and functional analysis for technology intelligence in a solution-neutral and problem-oriented way (Vollmer et al. 2013). The overall goal was to address the requirements without a preconceived notion of what the conclusion should be. The functional aspect focuses on the main objective/purpose of the system and interprets the requirements in a solution-neutral way, so it does not pronounce how the solution should be. The solution-neutral thinking aspect is to liberate the researcher of stereotypical solutions, preconceived thoughts, obsolete ideas, and mental blockades. The methods for information collection and data sources should be based on the function's purpose.

The AUAV system can use the Principle of Solution-Neutral Function through identifying and listing requirements of the system. Requirements identify the needed functions of the system and it's up to the system architect to think about those needs in a solution-neutral way. Figure 4.B10 displays the legacy system's requirement list. This initial list of requirements can be added as the basis to think about alternative specialized functions and for any new functions that might be needed.

4.B.11 Principle of the Role of the Architect

The textbook *System Architecture, Strategy and Product Development for Complex Systems* by E.C. Crawley, B. Cameron, and D. Selva discussed the Principle of the Role of the Architect and supplied the following information on p. 194 in Box 9.1:

> The role of the architect is to resolve ambiguity, focus creativity, and simplify complexity. The architect seeks to create elegant systems that create value and competitive advantage by defining goals, functions, and boundaries; creating the concept that incorporates the appropriate technology; allocating functionality; and defining interfaces, hierarchy, and abstractions to manage complexity.
>
> - In view of the ambiguity and complexity present in most systems, and the need for balance, it is often desirable to have the architecture created by a single individual or a small group.
> - The architect maintains a holistic view but always focuses on the small number of issues critical to the design.
> - The architect doesn't follow a single method but adopts different frameworks, views, and paradigms as appropriate.
>
> *Crawley et al. (2016, p. 194)*

The Principle of the Role of the Architect puts emphasis on how the system architect should be a specialist in reducing/resolving ambiguity, simplifying complexity, focusing on creativity, and interpreting new product development initiatives. The system architect should also act as the link between the functions of the system and the design environment the system will be operating in. A system architect is defined as "the key solution leader who designs, devises and creates the outcomes of a project that will meet the expectations of the system owner," and who is responsible "for the quality of the outcomes and the knowledge management of the processes that produced these outcomes" (Gore 2003, p. 2). Eberhardt Rechtin, who was a pioneer in the world of system architecture, listed the following characteristics a system architect should have in his article *The art of systems architecting*: be an expert in relationships (of interfaces, factors, etc.), have knowledge of the overall system, the capability to learn details about the system, and have the drive to stay current in the field of research. Rechtin also stressed that system architects should continue learning techniques to evaluate architectures, about the impact of the government (standards, policies, etc.), about how to form effective architecture teams, and how to design intelligent systems. The paper *The challenge of increasing heterogeneity in Systems of Systems for architecting* by G. Muller discussed the system architect's role to connect the organizational context and operational lifecycle context to the technical system requirements and designs, which assists in the management of the ever increasing diversity of systems (Muller 2018). The technology based on the systems design enables the system requirements, which then enable customer value propositions and business propositions. System architects also have the task of achieving the systems functionality and performance by combining the contributions from various engineering disciplines.

The Principle of the Role of the Architect can be applied to all aspects of the AUAV system. The system architect is responsible for reviewing the legacy system, thinking about the system holistically, identifying stakeholders and influences, identifying and resolving ambiguity, creating SysML models, and etcetera.

4.B.12 Principle of Ambiguity

The textbook *System Architecture, Strategy and Product Development for Complex Systems* by E.C. Crawley, B. Cameron, and D. Selva discussed the Principle of Ambiguity and supplied the following information on p. 196 in Box 9.2:

> The early phase of a system design is characterized by great ambiguity. The architect must resolve this ambiguity to produce (and continuously update) goals for the architect's team.
>
> - Development is possible only with the acceptance of uncertainty.
> - In general, no one designs or rigorously controls the process upstream of architecting, so there should be no expectation of unambiguity. There will be uncoordinated, incomplete, and conflicting inputs.
> - Uncertainty can create opportunities; it is not always bad.
> - Ambiguity contains known unknowns and unknown unknowns, as well as conflicting and false assumptions.
> - Ambiguity is especially apparent at the interface to the upstream influences, because no one designs the upstream processes.
> - Uncertainties should be identified and prioritized so that they can be managed.
>
> *Crawley et al. (2016, p. 196)*

The Principle of Ambiguity is for the system architect to determine, characterize, analyze, and reduce the ambiguity of the system. The system architect must first take into account the various influences on a system, identify the sources of ambiguity, and then set constraints and make assumptions to reduce the identified ambiguity. Ambiguity is made up of two ideas: fuzziness (event with numerous connotations) and uncertainty (event with an ambiguous outcome). Ambiguity can lead to unknown information (unavailable/not determined information), conflicting information (information is contradictive), and false information (presented with incorrect information). A systems architect will be responsible for dealing with ambiguity from the beginning of a project. Corporate functions such as customers, operations, research and development, manufacturing, strategy, marketing, regulations, and standards are examples of upstream influences that can cause ambiguity. Possible ways to reduce ambiguity are through analysis, imposing constraints to simplify the problem, or by making assumptions. Engaging upstream influences can reduce ambiguity and support envisioning/planning a successful system. One of the main roles of the system architect is to remove, reduce, and resolve ambiguity within the system. The paper *Complex systems architecting using design primitives, comparative analysis and fuzzy analytical feedback* by J. P. Dauby and C. H. Dagli discussed how the initial system architecture contains ambiguity and how understanding and characterizing the architecture early on results in a better managed and controlled system (Dauby and Dagli 2010). During the systems architecting process, system concepts evolve from fuzzy descriptions from stakeholders to more detailed and precise requirements. The architecting process starts at the functional level, advances in detail to the system architecture level, and then produces a fully developed physical architecture description. Ambiguity continues after the physical architecture is defined due to unknown/undeveloped detailed design work needed for the physical artifacts that have yet to start. A solution to this ambiguity is the Canonical Decomposition Fuzzy Comparative (CDFC) method, which is defined as "an analytically rigorous assessment method that is currently being studied at the physical architecture level" (Dauby and Dagli 2010, p. 1). The four elements to the CDFC method are extensible modeling (interchange data between the different model resolution levels), canonical design primitives (class representation of components), comparative analysis (substantiates the ambiguity of candidate architectures), and fuzzy interference (results in a way to map the systems response to fuzzy features describing in the architecture assessment). Another way to handle ambiguity is with the Fuzzy Voronoi Neural Network (FVNET), which is defined as an approach that "facilitates the handling of ambiguity in system tracking, and can lead to the development of decision aides which systematically process subjective information" (Stadter and Garga 1997, p. 2444). The FVNET uses fuzzy logic as a way to process ambiguous information and the fuzzy classification provides a measure for a given input feature vector. A third way to handle ambiguity is with the Fuzzy Theory and the System of Systems Architecture Description Language (SoSADL). To address uncertainty, the Fuzzy Theory and SoSADL are used together to form the Fuzzy SoSADL. Uncertainty is defined as "an intrinsic characteristic of complex systems, may it be human, natural, or engineered that is in the architectural design of SoS intrinsically" (Oquendo 2020, p. 1). There are two classifications of uncertainty: aleatory and epistemic. Aleatory uncertainty is due to the inherent randomness and cannot be reduced, while epistemic uncertainty is due to having partial knowledge/information and can be reduced by gathering more detailed information to fill in that lack of knowledge. The Fuzzy Theory is a mathematical theory "to represent partial information on the SoS operational environment" (Oquendo 2020, p. 3), and is used to explicitly represent uncertainty in the description of SoS Architecture. The SoSADL is enhanced with concurrent constraints to deal with uncertainty.

The paper *Coping with Uncertainty in Systems-of-Systems Architecture Modeling on the IoT with SosADL* by F.Oquendo used an Unmanned Aerial Vehicle-based (UAV) Reconnaissance SoS as the example to expand on SoS (System of Systems) Architecture under uncertainty (Oquendo 2019). A SoS architect will have to deal with several uncertainties from the architecture description of the UAV-based Reconnaissance SoS. The SoS architect will first have to deal with the uncertainty of not knowing whether or not the UAV will take part in the mission, the type of the model being used, or all the needed capabilities for the mission. The second uncertainty the SoS architect has to deal with is not knowing the position of the UAV during flight while describing the design and functionality. The UAV must adapt as its position and environment changes during flight. The UAVs in this paper's scenario are "declared as constituent systems of the flock and mediators are deduced according to the position of a UAV and its neighboring UAVs" (Oquendo 2019, p. 134). SoSADL supports the uncertainty in emergent behavior with concurrent constraints.

F. Oquendo wrote another paper called *Case Study on the Fuzzy Architecture Description of Cyber-Physical SoS under Uncertainty* and it expanded on Fuzzy Theory/Fuzzy SoSADL by using a Vehicle Platooning SoS to discuss fuzzy architecture descriptions under uncertainty. The Vehicle Platooning SoS deals with emergent behavior created by a platoon of self-driving vehicles. Platooning is defined as "the process of vehicles autonomously forming road convoys" (Oquendo 2021, p. 1). The

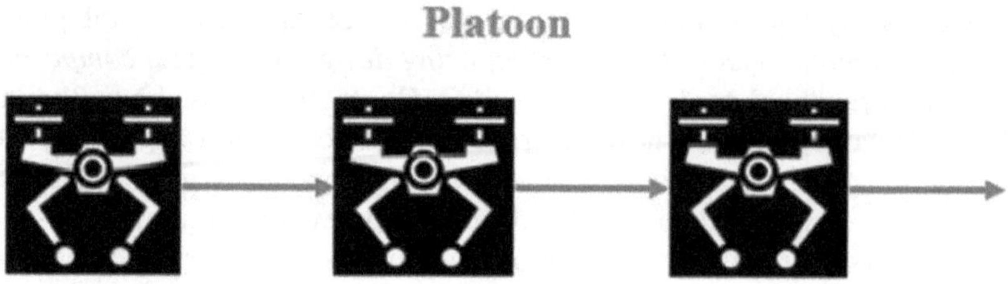

FIGURE 4.B11 Example of platooning. (Drone image from https://commons.wikimedia.org/wiki/File:198_Senpilota_aviadilo.svg, which was accessed May 26, 2022).

automated vehicle functions used for safety measures are radar/lidar sensors (used for positioning and identifying nearby vehicles), range perception (minimum safe distance from the vehicle in front and the maximum communication distance of the sensors), and controlled velocity (takes positioning and speed into consideration). The safety functions allow a group of automated vehicles to safely navigate and travel with each other to represent a platoon. Platooning and the three behaviors mentioned above also be used for the AUAV. Figure 4.B11 depicts a platoon of the AUAVs.

The Principle of Ambiguity can be applied to the AUAV system by reviewing a legacy diagram and identifying sources of ambiguity to improve on an updated version. Figure 4.B12 displays the legacy Use Case diagram. This diagram is not clear about who the stakeholders are or what their roles are. Also, there are no proper interactions identified and there is a confusion of who is responsible for what. Another source of ambiguity is how there are multiple activities within one bubble. Overall, this Use Case diagram is confusing and has many sources of ambiguity.

The legacy Use Case diagram can be improved by resolving those sources of ambiguity. Figure 4.B13 displays an updated Use Case diagram. The actors are clearly defined outside of the AUAV system boundary as Government Associations, Image Analyst, Flight Officer, Maintenance, and First Responders. The interactions between each actor and their activities are clearly identifiable. Each activity has its own bubble to reduce any confusion about what is occurring.

4.B.13 Principle of Reuse of Legacy Elements

The textbook *System Architecture, Strategy and Product Development for Complex Systems* by E.C. Crawley, B. Cameron, and D. Selva discussed the Principle of Reuse of Legacy Elements and supplied the following information on p. 228 in Box 10.2:

> All built systems involve reuse of legacy elements, either in the conceptual sense or in the sense of physical/informational elements. Understand the legacy system and its emergent properties thoroughly and include the necessary elements in the new architecture.
>
> - Legacy has attributes that one may not completely understand or appreciate, which might be lost in a complete redesign.
> - Legacy systems are built on an architecture that is often incrementally modified over time. The architect should assess whether the previous architecture is best positioned to meet new system goals, and whether that architecture reflects assumptions that have since been modified.
> - Legacy may involve substantial retesting to verify that the elements can be used in a system for which they were not originally intended.
> - Document design elements to facilitate their subsequent reuse where possible.
>
> *Crawley et al. (2016, p. 228)*

The Principle of Reuse of Legacy Elements is for the system architect to understand and think about the reuse of legacy elements of a system. The legacy element of a system can be either in a physical element sense or in a conceptual sense. The legacy system needs to be analyzed and understood

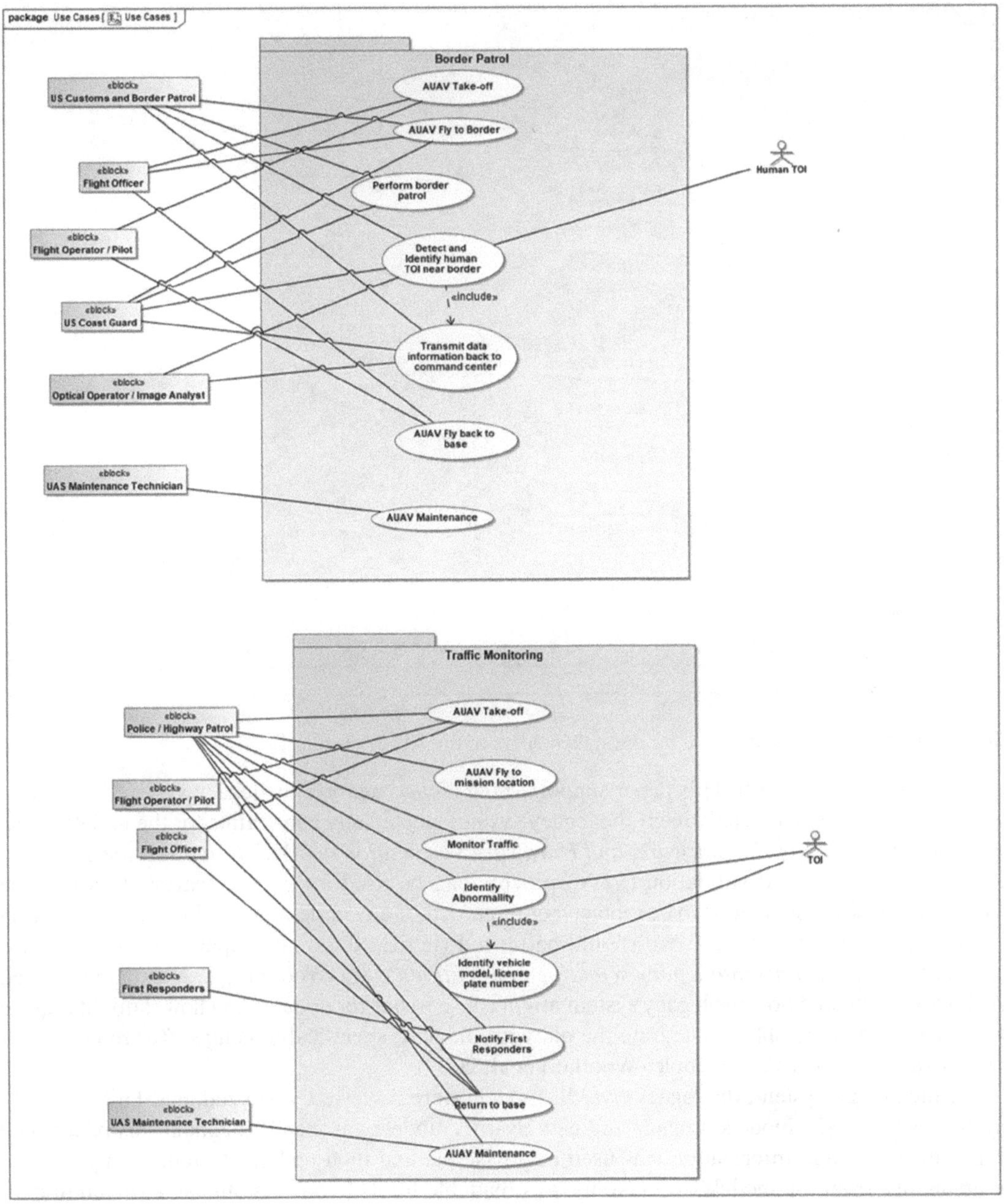

FIGURE 4.B12 Legacy use case diagram of the AUAV system.

prior to updates for optimizations. Handling legacy interfaces can lead to platform variant constraints on the open systems framework, as discussed in the paper *Optimizing an incremental Modular Open System Approach (MOSA) in avionics systems for balanced architecture decisions* by T. Gaska where the legacy system analysis and the resulting constraints are used for the approach inputs (Gaska 2012). The paper *Modeling intelligent transportation systems with multi-agent on SOA* by Dasheng Wang, Lei Ren, and Jing Li discussed how the legacy system must be analyzed to see if any host legacy functions in traditional transportation systems can be reused effectively or not and to see how many of the legacy applications can be reused for intelligent transportation

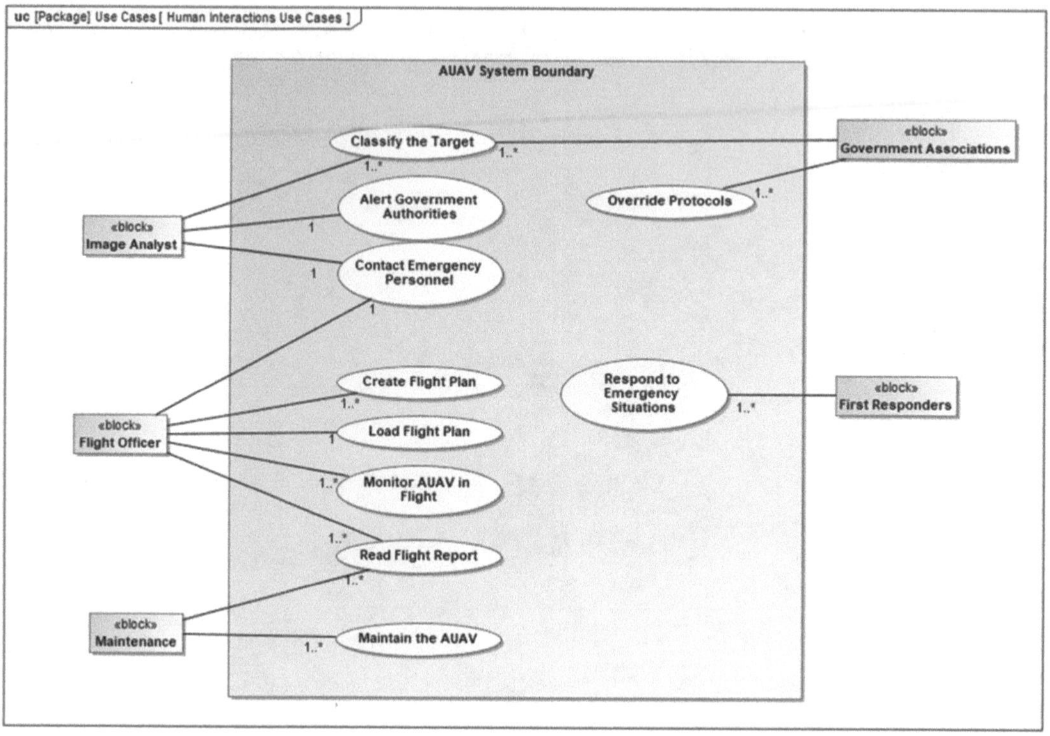

FIGURE 4.B13 Use case diagram for the stakeholders of the AUAV system.

systems (Wang et al. 2010). This paper supported the idea to analyze the legacy system, determine what can be reused, and determine if the legacy system created any constraints for the updates. The *IEEE Standard for System, Software, and Hardware Verification and Validation service* discussed how the verification and validation (V&V) process can be used for legacy analysis to see if the capabilities of the system need to be replaced or upgraded, if any updates to the installation location is needed, and if the legacy system should be retired (IEEE 2017). The paper *Injecting software architectural constraints into legacy scientific applications* by D. Woollard, C. Mattmann, and N. Medvidovic explored how the legacy system might cause issues for updates and how it might require additional work to be able to integrate the old with the new, specifically using software languages for software systems as an example (Woollard et al. 2009).

For the AUAV system, the legacy SysML models were reviewed, analyzed, and built off of to produce new SysML models. Legacy and new SysML models are seen throughout this paper. The legacy diagrams and information was used as a baseline and then updated to reduce any sources of ambiguity from the models, to make models readable for stakeholders, and add any additional information determined needed for the system.

4.B.14 Principle of Product Evolution

The textbook *System Architecture, Strategy and Product Development for Complex Systems* by E.C. Crawley, B. Cameron, and D. Selva discussed the Principle of Product Evolution and supplied the following information on p. 230 in Box 10.3:

> Systems will evolve or lose competitive advantage. When architecting, define the interfaces as the more stable parts of the system so that the elements can evolve. Plan for evolution with a master vision and sufficient reserve of resources.
>
> *Crawley et al. (2016, p. 230)*

The Principle of Product Evolution is for the system architect to think about how a product can grow, how much variety of the product can be demanded, the next step of product evolution, and to think about additional interfaces in the future of the system. There has been a shift in companies to adapt the product line approach to reuse legacy elements and increase the development productivity. The goal of product line engineering is "to leverage the reuse of core assets and processes, thereby reducing cost and time to market and increasing system reliability and quality" (Dhungana et al. 2008, p. 1). A constant issue for this approach is the evolution of technology (products, techniques, etc.), customer needs, and market circumstances (impacts profit). One crucial issue for product line evolution is that complex systems have multiple teams and stakeholders across the subsystems for development and maintenance, which can be difficult to manage and update overtime. A way to handle this issue is to use a multi-team modeling approach to distribute the development and upkeep of the product line models. Another crucial issue is the various speeds of evolution can differ between the components of the system. A way to handle this issue is to use variability model fragments to allow various teams to work on their specific subsystem separately and then use support tools to merge the fragmented parts.

4.B.15 Principle of Balance

The textbook *System Architecture, Strategy and Product Development for Complex Systems* by E.C. Crawley, B. Cameron, and D. Selva discussed the Principle of Balance and supplied the following information on p. 258 in Box 11.3:

> Many factors influence and act on the conception, design, implementation, and operation of a system. One must find a balance among the factors that satisfies the most important stakeholders.
>
> - In a built system, one may never be able to even enumerate, let alone quantify, all of the factors. One must make tradeoffs and compromise among the recognized factors to find solutions that come close to satisfying the identifiable influences, recognizing that formal optimality may never be achieved or may be meaningless in view of the un-enumerated factors.
> - The system must be balanced as a whole, and not just the sum of balanced elements.
> - In particular, an architect is always balancing optimality with flexibility. In general, the more optimized the architecture is for a given application, the less flexible it will be to respond to Use Cases and new technologies. In general, the more flexible the architecture, the more margin will be carried to enable flexibility, making it less optimized.
>
> *Crawley et al. (2016, p. 258)*

The Principle of Balance is for the system architect to find balance between the factors that satisfy the important stakeholders of a system. There must be a balance for interpreting stakeholder needs as goals and the implementation of those factors to produce system benefit. The system as a whole must be balanced and not just the summation of balanced elements. The system architect needs to think about the system as a whole and how it can be balanced for optimization and flexibility. The system architect should also balance and relate the functional and quality requirements "to define viable architectural abstractions that support them" (Hebisch et al. 2015, p. 1). Balance can be determined by decision-making, which is defined as "a problem-solving process related with selecting one satisfactory choice among several possible alternatives, applying a set of criteria" (Pedraza-García et al. 2014, p. 1). A system architect can use a MBSE approach to achieve an overall system balance and the satisfaction of stakeholder requirements. The article *Why Do We Need Model-Based Systems Engineering (MBSE)?* discussed the advantages of switching from a document-based approach to a model-based approach, which includes improved design coordination, satisfying requirements, and increasing the return value (Innoslate 2020). The MBSE approach provides a way to coordinate between the subsystem design teams to create an overall balanced system that will meet timeline, performance, and cost goals through the use of MBSE tools. The MSBE tools allow the various teams to share and collaborate, as well as track

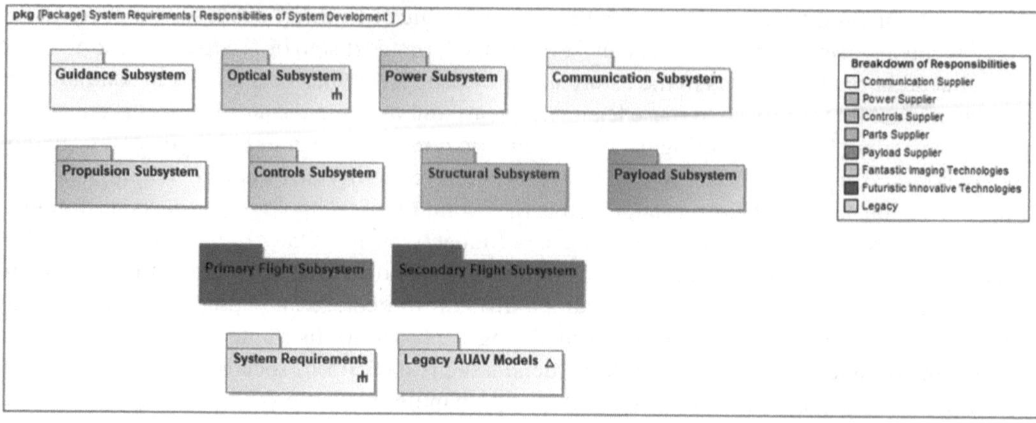

FIGURE 4.B14 Color-coded responsibilities of system development of the AUAV system.

traceability of information between the teams. Coordination and collaboration is important due to if one aspect of a system is dominating or underperforming, the overall system will not have optimal performance. The MBSE tools help analyze and manage the operational requirements the stakeholders need to reach their business objectives. The use of models and the descriptions of models provide a visual way to view the needs of the systems, as well as an explanation for stakeholders to understand. The MBSE tools can be used to run system simulations to verify models are working, reduce review time based on the active coordination during system development, and reduce the amount of documents due to the consolidation of documentation in modeling. This can reduce labor cost, optimize system cost, meet timeline milestones, and save time in system development, which overall leads to an increase in the return value.

Figure 4.B14 displays a color-coded subsystem Package Diagram identifying what supplier or team is responsible for each subsystem of the AUAV. The Communication Supplier is responsible for the Guidance Subsystem and the Communication Subsystem. The Power Supplier is responsible for the Power Subsystem. The Control Supplier is responsible for the Propulsion Subsystem and the Controls Subsystem. The Parts Supplier is responsible for the Structural Subsystem. The Payload Supplier is responsible for the Payload Subsystem. The Futuristic Innovative Technologies team is responsible for the Primary Flight Subsystem and the Secondary Flight Subsystem. The FIT team is responsible for the Optical Subsystem. System Requirements and Legacy AUAV Models are also identified in this diagram and marked as a Legacy responsibility. Identify who is responsible for what helps coordinate the leads and create balance of the system.

4.B.16 Principle of the System Problem Statement

The textbook *System Architecture, Strategy and Product Development for Complex Systems* by E.C. Crawley, B. Cameron, and D. Selva discussed the Principle of the System Problem Statement and supplied the following information on p. 261 in Box 11.5:

> The statement of the problem defines the high-level goal and establishes the boundaries of the system. It divides content from context. Challenge and refine the statement until you are satisfied that it is correct.
>
> - The statement of the problem has an enormous impact on the eventual design. It provides the origin for all other statements of goals, and it is a consistent and pervasive message to communicate to the team.
> - The statement of the problem is rarely stated properly upon its first formulation. Challenge the scope and statement of the problem early and often.
>
> *Crawley et al. (2016, p. 261)*

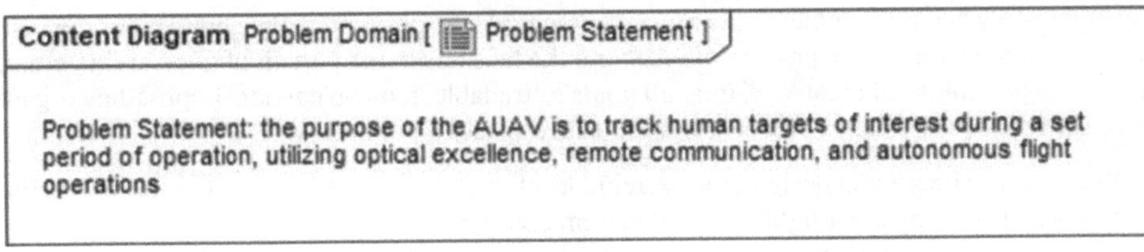

FIGURE 4.B15 Content diagram for the problem statement of the AUAV system.

The Principle of the System Problem Statement purpose is to create solvable goals around the idea of a system problem statement, which is defined as "the single assertion of what the system is intended to accomplish to deliver value" (Crawley et al. 2016, p. 261). The process of the problem statement is considered one of the key steps in solving problems due to it impacting the implementation of models and analytical tools (Gubarev and Chernov 2001, p. 1). A full system problem statement contains a solution-neutral function, a solution-specific function, and specific-system form for the system to be used. A good system problem statement should show the intent of the system, the function of the system, and the form of the system.

Figure 4.B15 is the content diagram for the AUAV Problem Statement, which is "the purpose of the AUAV is to track human targets of interest during a set period of operation, utilizing optical excellence, remote communication, and autonomous flight operations."

4.B.17 Principle of Ambiguity and Goals

The textbook *System Architecture, Strategy and Product Development for Complex Systems* by E.C. Crawley, B. Cameron, and D. Selva discussed the Principle of Ambiguity and Goals and supplied the following information on p. 266 in Box 11.6:

> The early phase of a system design is characterized by great ambiguity. The architect must resolve this ambiguity to produce (and continuously update) a small set of representative, complete and consistent, challenging yet attainable, and humanly solvable goals.
>
> - Ambiguity contains known unknowns and unknown unknowns, as well as conflicting and false assumptions.
> - In general, no one designs or rigorously controls the process upstream of architecting, so there should be no expectation of unambiguity. There will be uncoordinated, incomplete, and conflicting inputs.
> - The goals set must be representative of the needs of the beneficiaries. For a system to be successful, it must meet these real needs.
> - The goals must be set so as to be mindful of competitive pressures, consistent with enterprise strategy, challenging yet reachable with available resources and available technology, and respectful of regulatory and pseudo-regulatory influences.
> - Because the context and customer needs will evolve, there must be a process to continuously reexamine and update the goals.
>
> *Crawley et al. (2016, p. 266)*

This principle is not expanded on due to the closeness to the Principle of Ambiguity, which was expanded on in detail in that section of this research paper.

4.B.18 Principle of Creativity

The textbook *System Architecture, Strategy and Product Development for Complex Systems* by E.C. Crawley, B. Cameron, and D. Selva discussed the Principle of Creativity and supplied the following information on p. 281 in Box 12.1:

For any interesting and real problem, there will be essential tensions among the goals set for the system. Creativity in architecture is the process of resolving the tensions in the pursuit of good architecture. To maximize the possibility of creativity, treat all goals as tradable, remove barriers imposed by organizational or cultural environments, and consider the entire space of possibility.

- Because architecting is design at a strategic level, architects must resolve the tension in the high-level goals by creating the overall system concepts.
- The tensions are resolved by applying creativity to find a concept that allows all goals to be met. Alternatively, if no concept can be found within existing technology, the tension is resolved by revising the goals.
- Creativity is an iterative process of finding one's way when the destination is unclear. As E.L Doctorow put it, "Creativity is like driving a car at night. You never see further than your headlights, but you can make the whole trip that way."
- Generate creativity by provoking the mind to create new associations between existing pieces of information.
- Think outside the box and fully explore the space of alternatives to discover new and innovative solutions; otherwise, the solution set is confined to incremental improvements on existing designs.

Crawley et al. (2016, p. 281)

The Principle of Creativity is for the system architect to think outside of the box for system solutions and system improvement of pre-existing designs. Creativity can be applied during the development of the system's concept to meet high-level goals of the system. The system architect can creatively list and expand on the needs of the stakeholder in a structured approach (Crawley et al. 2016, p. 282) Three elements of creativity for system architecture are being innovative, being applicable, and being unexpected (Zou et al. 2017, p. 2). The paper *Art and architecture: Effectively communicating models of systems* by T. McDermott and A. Salado stressed the importance of art and engineering in the designing and communicating system architecture. McDermott and Salado defined a good systems architect as one that "combines systems engineering with systems thinking and creative arts to build an effective communication strategy for themselves" (McDermott and Salado 2018, p. 4).

The system architect can use a bdd (block definition diagram) for the optical subsystem hardware components, shown in Figure 4.B16, for the application of the Principle of Creativity. Creatively think about designs for the hardware components and what can be done better. For example, the camera can be thought about creatively when it comes to type of camera, type of lens, location on the AUAV, number of cameras on the AUAV, and the size of the camera. These variations impact

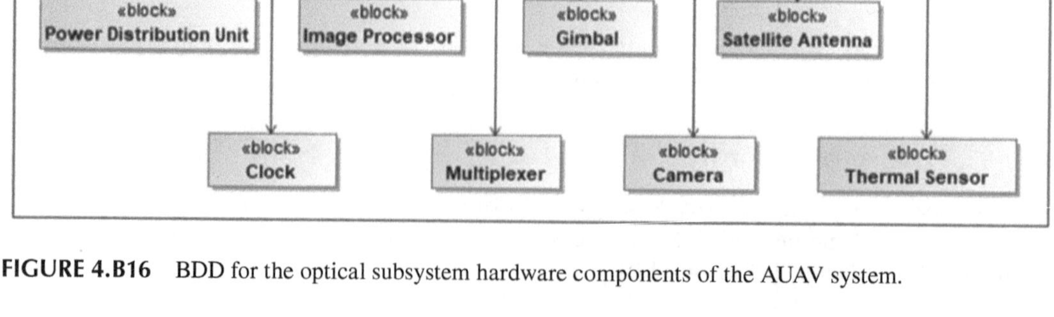

FIGURE 4.B16 BDD for the optical subsystem hardware components of the AUAV system.

design changes and support the idea of thinking outside of the box. The architect can decide on the creative options based on what's available and affordable.

4.B.19 Principle of Essential Complexity

The textbook *System Architecture, Strategy and Product Development for Complex Systems* by E.C. Crawley, B. Cameron, and D. Selva discussed the Principle of Essential Complexity and supplied the following information on p. 306 in Box 13.3:

> Functionality drives essential complexity. Describe the required functionality carefully, and then choose a concept that produces low complexity.
>
> - Essential complexity is the minimum complexity that must be incorporated in the system to deliver the robust functionality within a concept.
> - Gratuitous complexity is complexity added over and above the essential complexity for any given robust functionality within any concept. Gratuitous complexity should be avoided.
>
> *Crawley et al. (2016, p. 306)*

Complexity is defined "as the property of having many interrelated, interconnected, or interwoven elements and interfaces" (Crawley et al. 2016, p. 300). The Principle of Essential Complexity is for the system architect to define the minimum complexity for a system to function as specified. Minimum refers to the minimum data flow needed and the minimum set of functions needed to accomplish what the system needs to do. The system architect can take an attribute-driven approach to prioritize the drivers to determine the minimum complexity needed to accomplish the defined system concept (Sangwan and Neill 2009, p. 3). To define the essential complexity of the overall system, the complexity of integrating all the subsystems needs to be taken into account. The system integration complexity is defined as "an outcome of the interaction between degree of feasibility and level of effort required to understand, describe, implement, manage, and document the systems integration process for a given system development and operational environment" (Jain et al. 2008, p. 1).

An Entity-Relationship (E-R) diagram for the AUAV Optical Subsystem can be used to understand the essential complexity to use the camera and produce a processed image, shown in Figure 4.B17. The entities are identified by the rectangles, the relationships are specified in the diamonds, the attributes of the entities are shown in the circles, and lines connect the elements. Figure 4.B17 is showing the camera takes an image which is sent to the image processor to be processed and results in the processed image with the attribute of ATC/Raw (is the atmospherically turbulence

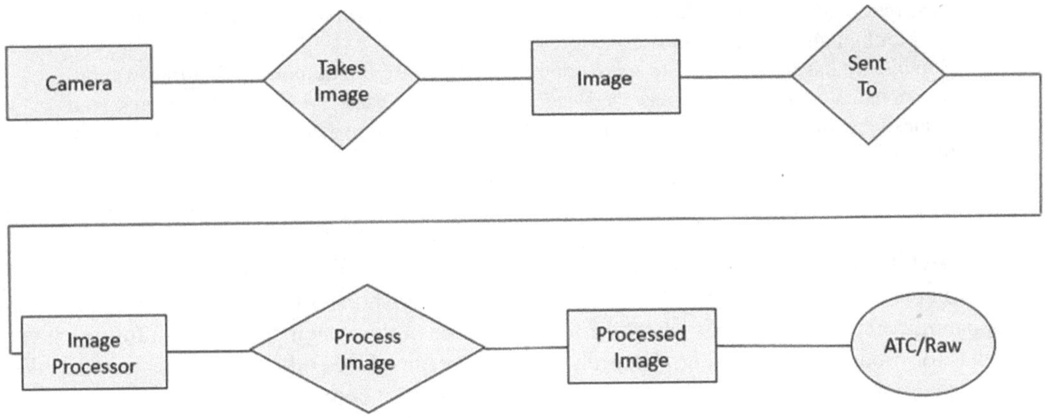

FIGURE 4.B17 E-R diagram for the camera and processed image of the AUAV optical subsystem - essential complexity.

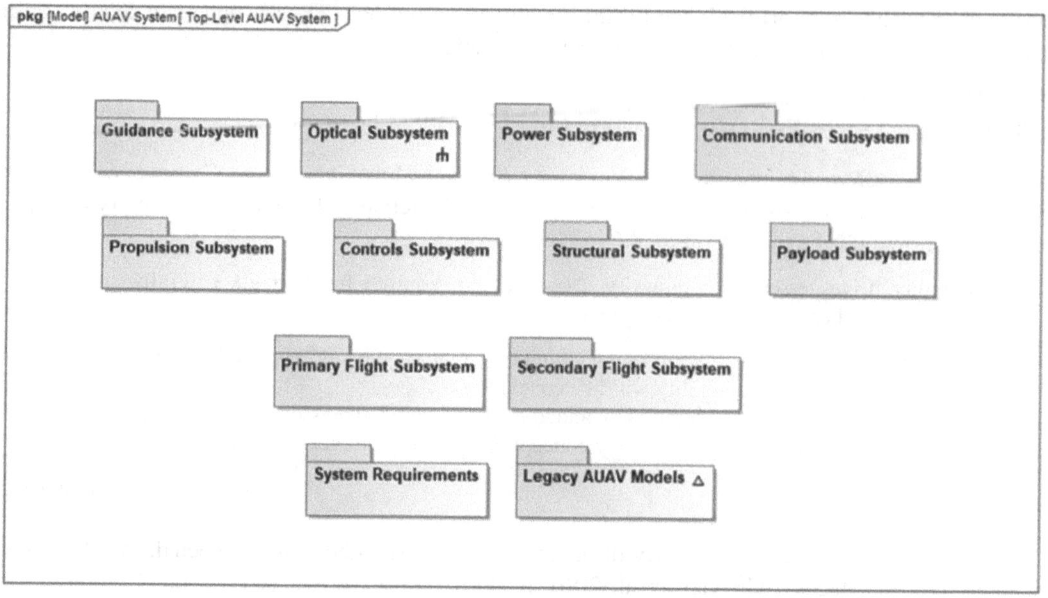

FIGURE 4.B18 Top-level package diagram for the AUAV system.

compensated?). This diagram assists the system architect in understanding the minimum functions needed to produce a final image from using the camera on the AUAV.

For the essential complexity of the overall AUAV system, the system architect must account for the complexity in each subsystem shown in Figure 4.B18 and for the complexity of integrating the subsystems together.

4.B.20 Principle of Apparent Complexity

The textbook *System Architecture, Strategy and Product Development for Complex Systems* by E.C. Crawley, B. Cameron, and D. Selva discussed the Principle of Apparent Complexity and supplied the following information on p. 303 in Box 13.2:

> The complexity of modern systems exceeds the human ability to comprehend. Create decomposition, abstraction, and hierarchy to keep the apparent complexity within the range of human understanding.
>
> - Apparent complexity is akin to how "complicated" the system is; complicated systems are hard for humans to understand.
> - Complexity has the potential to scale combinatorially with the number of features of robust functionality and/or the number of objects. One cannot scale up the processes and techniques used in simple systems to address complicated systems because of human cognitive limitations.
>
> *Crawley et al. (2016, p. 303)*

Complex architectures require the system architect to have the ability of assessing and understanding the complexity of the system. Apparent complexity is for how complicated a system appears. The apparent complexity captures any additional attributes of the system to deal with future aspects or models for product improvement. Apparent complexity introduces other design considerations or additional complexity that is more than the minimum (essential complexity).

The apparent complexity of the AUAV system can be represented in the Top-Level Package Diagram showing the various subsystems of the AUAV (Figure 4.1). There are ten subsystems listed, along with the system requirements and the legacy AUAV models. Having multiple subsystems and

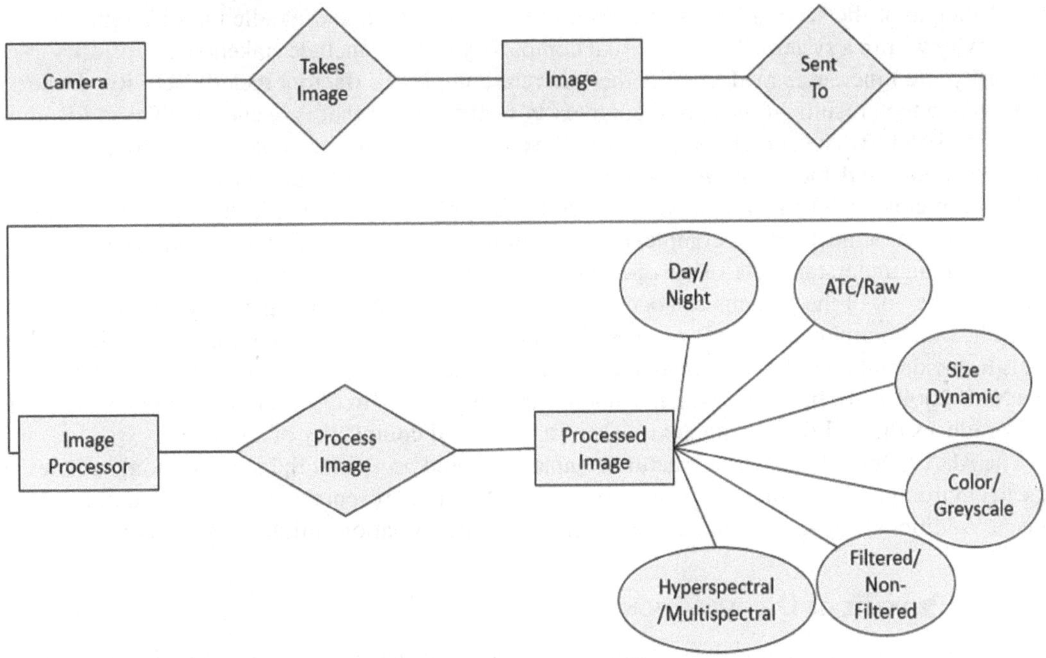

FIGURE 4.B19 E-R diagram for the camera and processed image of the AUAV optical subsystem - apparent complexity.

building off of legacy systems causes complexity of the system. The system architect must think about each subsystem, the impact of changes that occur for each subsystem, and the influences for each subsystem, which leads to understanding the complexity of the system.

Figure 4.B19 is building off of the E-R diagram for the essential complexity to use the camera and produce a processed image. This diagram has additional attributes for the system to deal with future needs depending on the AUAV missions. The additional attributes are Day/Night (daylight or Infrared), Size Dynamic (full image or a specific area), Color/Greyscale (color image or black and white), Filtered/Non-Filtered (any image or de-noise filtering needed?), and Hyperspectral/ Multispectral (determines the size and number of bands). This diagram assists the system architect in introducing other design considerations for the final image (fully processed image) produced by the camera of the AUAV optical subsystem.

4.B.21 Principle of the Second Law

The textbook *System Architecture, Strategy and Product Development for Complex Systems* by E.C. Crawley, B. Cameron, and D. Selva discussed the Principle of the second Law and supplied the following information on p. 309 in Box 13.4:

> The actual complexity of the system always exceeds the essential complexity. Try to keep the actual complexity close to the essential complexity, and the apparent complexity within the range of human understanding.
>
> - The actual complexity is the real measured complexity of the system. It exceeds the essential complexity because the extra complexity needed to create abstraction, decomposition, and hierarchy.
> - Actual complexity can also increase because of the creep of gratuitous complexity.
>
> *Crawley et al. (2016, p. 309)*

The Principle of the second Law serves as a way to understand and handle the different types of complexity within a system. The contextual complexity (factors include stakeholders, priority, risk, flexibility, etc.) increases or decreases the inherent complexity (factors include activity, duration, cost, etc.), which results in the actual complexity of the system that is occurring (Tie and Bolluijt 2014, pp. 4 and 5). The actual complexity will exceed the essential complexity due to abstraction, decomposition, and hierarchy. Actual complexity captures the complexity that was unexpected and not anticipated. The goal is to keep the actual complexity (the complexity the system is experiencing) close to the essential complexity (minimum complexity needed for system functionality) and within an understandable range of apparent complexity (how complicated a system appears). Various behavior of the systems can occur that were not originally anticipated by the system architect. The system architect focuses on specific aspects of the system and the aspects not focused on leads to incidental complexity, which adds to the complexity in the errors of the system (Sangwan and Neill 2009, p. 4). In some cases, it is not feasible or possible to test every aspect of a system and that produces the additional complexity seen in the actual complexity of a system.

The AUAV Optical Subsystem's actual complexity could be greater than the essential complexity due to environmental interactions. Rain, snow, or wind could prevent the camera from taking a clear image. Additional capabilities to handle different weather situations might be necessary.

4.B.22 Principle of Decomposition

The textbook *System Architecture, Strategy and Product Development for Complex Systems* by E.C. Crawley, B. Cameron, and D. Selva discussed the Principle of Decomposition and supplied the following information on p. 310 in Box 13.5:

> Decomposition is an active choice made by the architect. The decomposition affects how performance is measured, how the organization should be set up, the potential for supplier value capture, and how the product can evolve, among many other things. Choose the plane of decomposition to align as many of these factors as possible, in order to minimize the apparent complexity of the system.
>
> - Decomposition involves choices not only of how many and where to make internal boundaries, but also (and most important) in what plane to consider internal boundaries.
> - The choice of the plane of decomposition (form, function, operation, supplier, and so on) in which to consider internal coupling is a key decision.
>
> *Crawley et al. (2016, p. 310)*

The Principle of Decomposition is for the system architect to make a distinct choice on how the system will be decomposed and understand it is a high-leverage decision. Decomposition is an important part of system architecture because good decomposition will lead to highlighting the important couplings of the subsystems through interface testing, while wrong decomposition results in missing information (functions, systems, environmental factors) (Crawley et al. 2016, p. 310). There are various types of decomposition: issue, functional, etc., and all help reduce the complexity in the system and architecture design. Issue decomposition is defined as the intent to "treat the core task of architecture design as solving a collection of system-wide problems, so that the design can be conducted by means of decomposing the holistic design goals into the separated design issues, and then exploiting and making decisions on the solutions for these issues" (Cui et al. 2009, p. 2). Functional decomposition is defined as the intent to "determine which subsystems or entities will accomplish which functions, to determine which functions are necessarily emergent and cannot be decomposed, and to carry the intent of the architecture through the design" (Crawley et al. 2016, p. 310). The system architecture should be viewed from the perspective of the elements to decompose the system with the purpose of supporting the management of system complexity (Cui et al. 2009, p. 2).

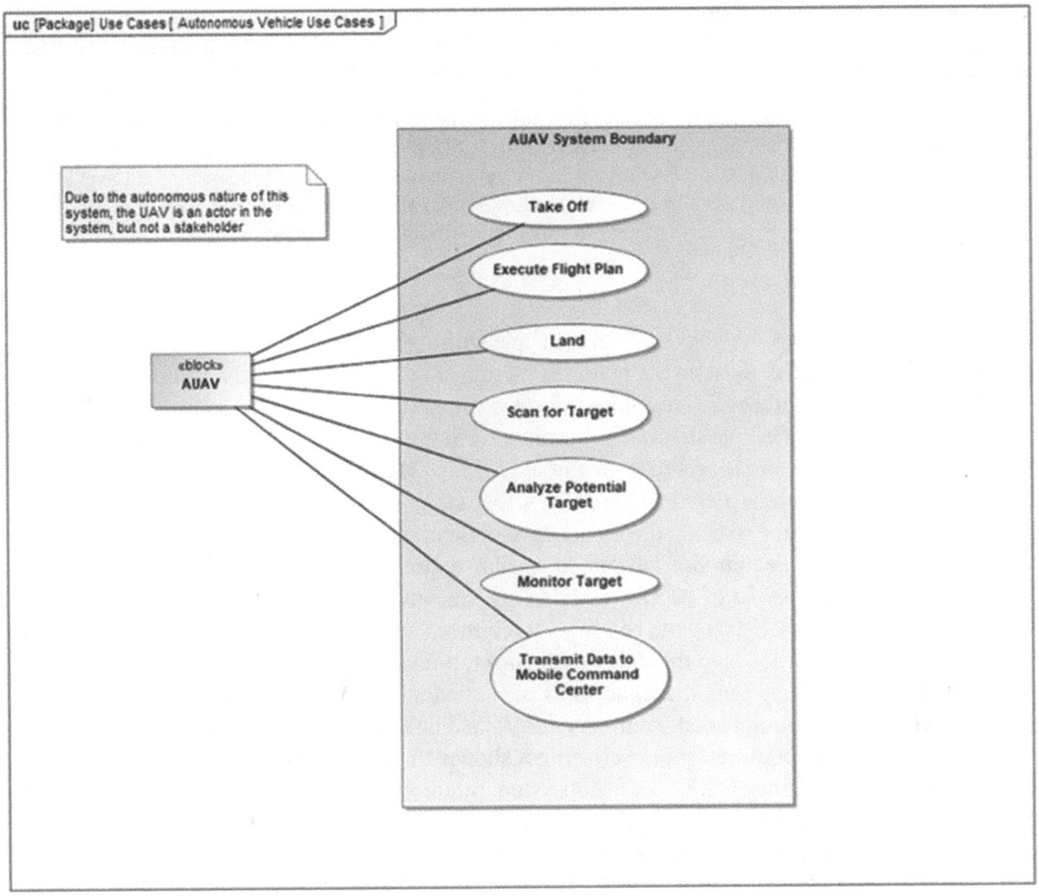

FIGURE 4.B20 Use case diagram for the AUAV system.

The paper *Ontology-Based Mission Modeling and Analysis for System of Systems* by W. Zhu, H. He and Z. Wang discussed mission modeling for a System of Systems (SoS) and how decomposition is used. The authors of this paper defined five decomposition principles: priority (sets "the scope of priority when modeling the mission model of SoS"), time (identifies the start and end times of a mission), resource (resources in the mission), event (events that perform the mission), and condition (condition for performing the mission) (Zhu et al. 2017, p. 3). The mission model is considered concluded once the mission model is decomposed into tasks.

The Principle of Decomposition can be seen for the AUAV system through a Use Case diagram for the system boundary, as seen in Figure 4.B20. This diagram displays the actor as the AUAV due to the autonomous nature of the system. The actions seen inside the AUAV System Boundary are identified as Take Off, Execute Flight Plan, Land, Scan for Target, Analyze Potential Target, Monitor Target, and Transmit Data to Mobile Command Center. The system architect can decompose the AUAV into a set of behaviors.

4.B.23 Principle of "2 Down, 1 Up"

The textbook *System Architecture, Strategy and Product Development for Complex Systems* by E.C. Crawley, B. Cameron, and D. Selva discussed the Principle of "2 Down, 1 Up" and supplied the following information on p. 312 in Box 13.6:

> The goodness of a decomposition at Level 1 cannot be evaluated until the next level down has been populated and the relationships identified (Level 2). Select the modularization at Level 1 that best reflects the clustering of relationships among the elements at Level 2.
>
> - The modularization should consider maximizing the internal couplings within a cluster and minimizing the couplings among clusters. This is often, but not always, advantageous.
> - The real information about clustering at Level 1 is contained in the interrelationships at Level 2.
> - Therefore, if decomposing from Level 0 to Level 1, go all the way to Level 2 (2 Down), identify the appropriate clustering, and use this for the final Level 1 decomposition (1 Up).
>
> *Crawley et al. (2016, p. 206)*

The Principle of "2 Down, 1 Up" is a useful principle within decomposition. The principle "embodies the idea that in a hierarchic system, the real information about how to cluster or group the system one level down from the reference is present in the structure and interaction two levels down" (Crawley et al. 2016, p. 311). The principle stresses that evaluating at the second level down is the best way to test the decomposition at the first-level (Crawley et al. 2016, p. 311). Model-Based Systems Engineering (MBSE), defined as "a formalized methodology that supports the requirements, design, analysis, verification, and validation associated with the development of complex systems" (Shevchenko 2021), can be used to record, decompose, and trace system requirements. The system architect can use MBSE to show the Principle of "2 Down, 1 Up" by decomposing the top-level requirements into low level requirements. Satisfying the below level requirements, satisfies the top level. Improving the lower-level requirements leads to the improvement of top-level requirements.

For the AUAV legacy system, the legacy optical subsystem level requirements (Level – 2)), shown in Figure 4.B21, were decomposed from the AUAV stakeholder requirements (level – 0, not shown) and the AUAV system requirements (level – 1, not shown). These legacy optical subsystem level – 2 requirements, show a representative set of subsystem requirements for the AUAV optical subsystem. However, these level – 2 subsystem requirements represent the system – level requirements (level – 1 requirements) for the optical sensor system developer's perspective. This is because the "system under consideration" from the optical systems developer is the optical system, not the AUAV. Said another way, the AUAV's optical subsystem requirements, are incorporated in the optical developer's system requirements. The stakeholders for the development of the optical system now are concerned only with the optical system, and not any other part of the AUAV. The AUAV becomes a host platform for the optical system developer and the optical systems developer just worries about interfacing the optical system with the AUAV. Consequently, the requirements levels change for the optical systems developer. The optical system stakeholder requirements are the level – 0 requirements for the optical system developer. The optical system requirements are the level – 1 requirements, and the optical subsystem requirements are the level – 2 requirements. In Figure 4.B22, we show the stakeholder requirements with respect to the optical systems development effort. Notice that we have left the "old" title "Optical Subsystem" to logically show where these requirements tie into the AUAV system. Often, there can be a separate development contractor for the optical system development effort, and they would refer to the top block in Figure 4.B22 as the "Optical System" and not the "Optical Subsystem" as shown. However, if the prime contractor for the AUAV development effort also has the responsibility for developing the optical sensors, then the top block would be as shown, "Optical Subsystem." However, in the latter case, the stakeholder requirements that affect the optical subsystem would be mixed in with the stakeholder requirements for the AUAV itself. In terms of the "2 – down, 1 – up" principle, the level – 2 (AUAV Optical Subsystem) requirements (2 – down), are the level – 1 system requirements for the optical system, and so the "1 – up" from the optical system requirements (with a separate development contractor) would be the optical system stakeholder requirements shown in Figure 4.B22. We wanted to point this out since the perspective of what the system is can change depending on the contract structure. This point is pretty clear in terms of a system-of-systems concept wherein multiple systems interact to produce something greater than the capabilities of the individual parts. From here on, we assume the AUAV is the "system" and the optical sensing functionality is a subsystem with respect to the AUAV.

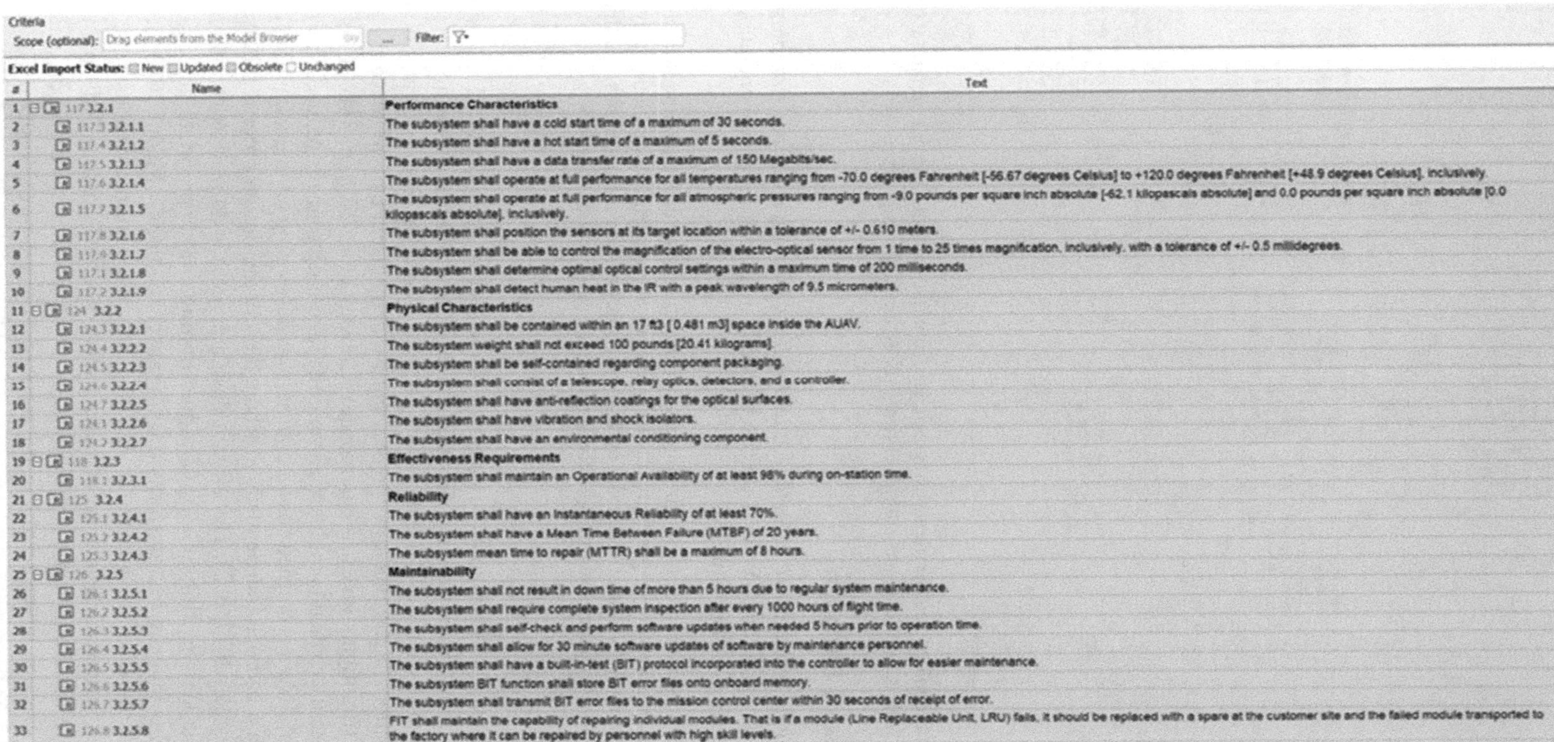

Criteria

Scope (optional): Drag elements from the Model Browser | Filter:

Excel Import Status: New Updated Obsolete Unchanged

#	Name	Text
1	117 3.2.1	**Performance Characteristics**
2	117.3 3.2.1.1	The subsystem shall have a cold start time of a maximum of 30 seconds.
3	117.4 3.2.1.2	The subsystem shall have a hot start time of a maximum of 5 seconds.
4	117.5 3.2.1.3	The subsystem shall have a data transfer rate of a maximum of 150 Megabits/sec.
5	117.6 3.2.1.4	The subsystem shall operate at full performance for all temperatures ranging from -70.0 degrees Fahrenheit [-56.67 degrees Celsius] to +120.0 degrees Fahrenheit [+48.9 degrees Celsius], inclusively.
6	117.7 3.2.1.5	The subsystem shall operate at full performance for all atmospheric pressures ranging from -9.0 pounds per square inch absolute [-62.1 kilopascals absolute] and 0.0 pounds per square inch absolute [0.0 kilopascals absolute], inclusively.
7	117.8 3.2.1.6	The subsystem shall position the sensors at its target location within a tolerance of +/- 0.610 meters.
8	117.9 3.2.1.7	The subsystem shall be able to control the magnification of the electro-optical sensor from 1 time to 25 times magnification, inclusively, with a tolerance of +/- 0.5 millidegrees.
9	117.1 3.2.1.8	The subsystem shall determine optimal optical control settings within a maximum time of 200 milliseconds.
10	117.2 3.2.1.9	The subsystem shall detect human heat in the IR with a peak wavelength of 9.5 micrometers.
11	124 3.2.2	**Physical Characteristics**
12	124.3 3.2.2.1	The subsystem shall be contained within an 17 ft3 [0.481 m3] space inside the AUAV.
13	124.4 3.2.2.2	The subsystem weight shall not exceed 100 pounds [20.41 kilograms].
14	124.5 3.2.2.3	The subsystem shall be self-contained regarding component packaging.
15	124.6 3.2.2.4	The subsystem shall consist of a telescope, relay optics, detectors, and a controller.
16	124.7 3.2.2.5	The subsystem shall have anti-reflection coatings for the optical surfaces.
17	124.1 3.2.2.6	The subsystem shall have vibration and shock isolators.
18	124.2 3.2.2.7	The subsystem shall have an environmental conditioning component.
19	118 3.2.3	**Effectiveness Requirements**
20	118.1 3.2.3.1	The subsystem shall maintain an Operational Availability of at least 98% during on-station time.
21	125 3.2.4	**Reliability**
22	125.1 3.2.4.1	The subsystem shall have an Instantaneous Reliability of at least 70%.
23	125.2 3.2.4.2	The subsystem shall have a Mean Time Between Failure (MTBF) of 20 years.
24	125.3 3.2.4.3	The subsystem mean time to repair (MTTR) shall be a maximum of 8 hours.
25	126 3.2.5	**Maintainability**
26	126.1 3.2.5.1	The subsystem shall not result in down time of more than 5 hours due to regular system maintenance.
27	126.2 3.2.5.2	The subsystem shall require complete system inspection after every 1000 hours of flight time.
28	126.3 3.2.5.3	The subsystem shall self-check and perform software updates when needed 5 hours prior to operation time.
29	126.4 3.2.5.4	The subsystem shall allow for 30 minute software updates of software by maintenance personnel.
30	126.5 3.2.5.5	The subsystem shall have a built-in-test (BIT) protocol incorporated into the controller to allow for easier maintenance.
31	126.6 3.2.5.6	The subsystem BIT function shall store BIT error files onto onboard memory.
32	126.7 3.2.5.7	The subsystem shall transmit BIT error files to the mission control center within 30 seconds of receipt of error.
33	126.8 3.2.5.8	FIT shall maintain the capability of repairing individual modules. That is if a module (Line Replaceable Unit, LRU) fails, it should be replaced with a spare at the customer site and the failed module transported to the factory where it can be repaired by personnel with high skill levels.

FIGURE 4.B21 Representative legacy optical subsystem requirements list for the AUAV system.

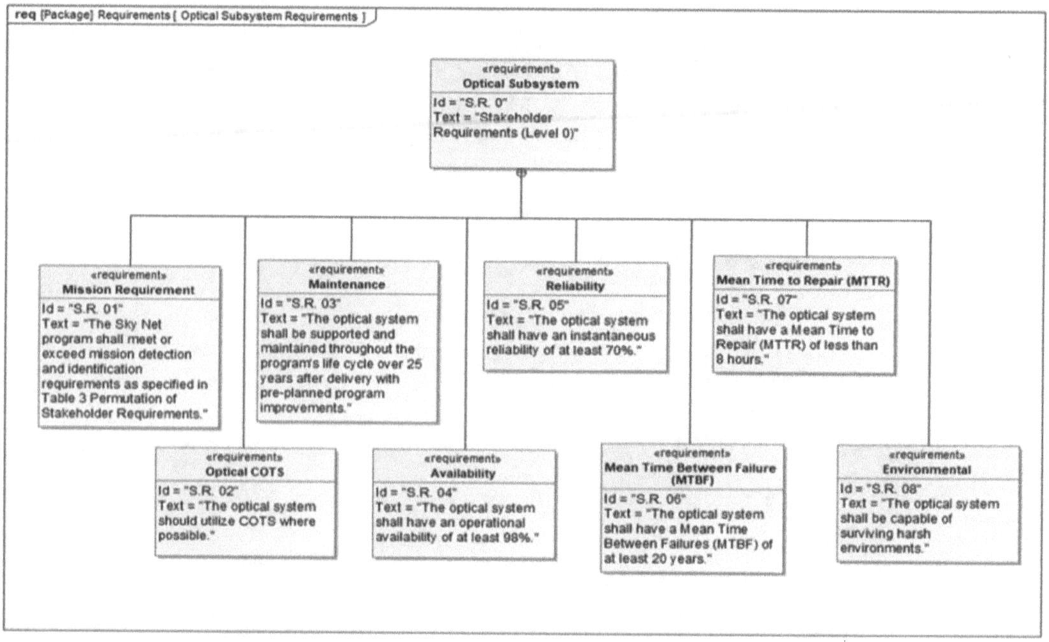

FIGURE 4.B22 Stakeholder requirements diagram for "optical system". (From separate optical system developer's perspective), or "optical subsystem". (From AUAV developer's perspective).

4.B.24 Principle of Elegance

The textbook *System Architecture, Strategy and Product Development for Complex Systems* by E.C. Crawley, B. Cameron, and D. Selva discussed the Principle of Elegance and supplied the following information on p. 314 in Box 13.7:

> Elegance is appreciated internally by the architect when a system has a concept with low essential complexity and decomposition that aligns many of the planes of decomposition simultaneously. Elegance is appreciated externally by a user when the system embodies a sense of aesthetics, quality, and low apparent complexity. Architect systems with these attributes that produce elegance
>
> *Crawley et al. (2016, p. 314)*

A good decomposition plan leads to more elegance in the system architecture. The system architect can use form decomposition or function decomposition. Form decomposition "allocates one or several elements of form to each Level 1 entity, with the end result that form is clearly divided", while function decomposition "highlights how the decomposition decision can influence the eventual system performance" (highlights types of emergence) (Crawley et al. 2016, p. 314). A higher elegance is appreciated by the user when complexity, ambiguity, and emergence is reduced from the system. The focus of elegant design is defined "as an approach to developing an understanding of how to tell a good design from a bad one independently of how closely the design process follows established SE practices" (Rousseau 2019, p. 3). The elegant design process is defined as "an iterative, creative process that exploits systems thinking, probing questioning, and appropriate analogies and metaphors to gain insights that can be transformed [into] novel solutions" (Madni 2012, p. 1). Qualities the system architect can take into account to add elegance to a system are robustness, productivity, usefulness, reducing incidental results, and balancing the system (Rousseau 2019, p. 3). The system architecture needs to determine what is important, how it should be decomposed, and then follow the 2 Down, 1 Up process, which results in the level of elegance in the system. There are various planes of decomposition the system architect can use to add elegance to the system, such as Delivered function and emergence, Form and structure, Design latitude and change propagation, Changeability

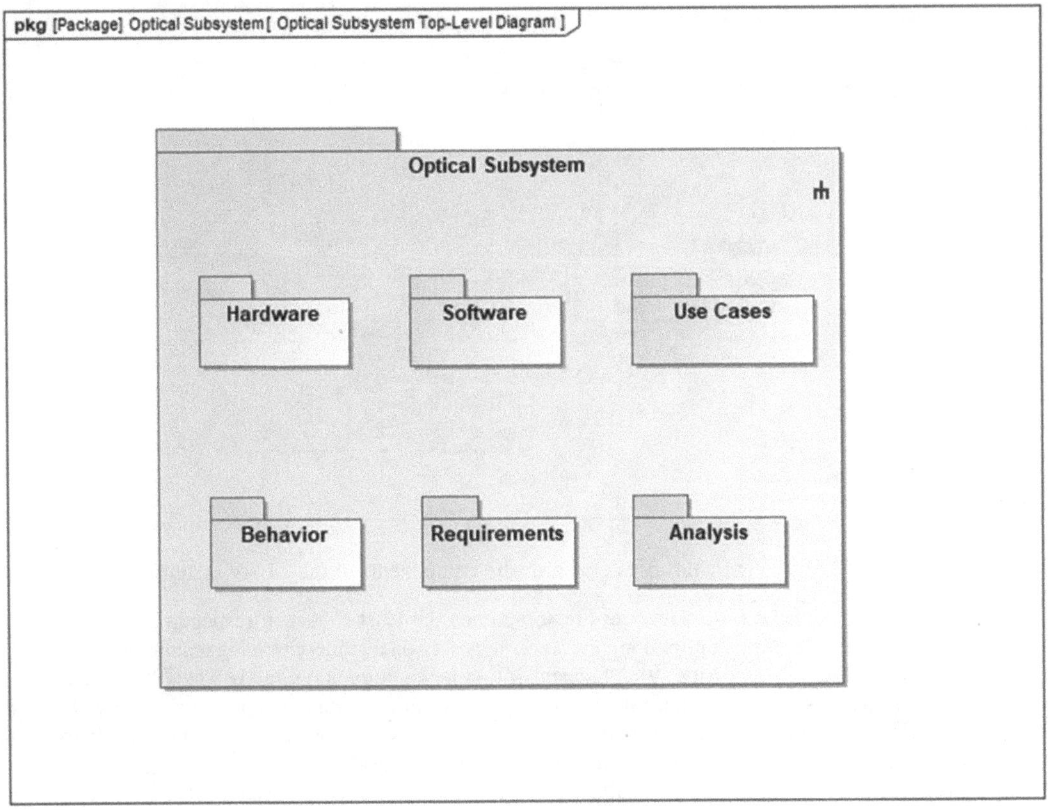

FIGURE 4.B23 Package diagram for the optical subsystem of the AUAV.

and evolution, Integration transparency, Suppliers, Openness, Legacy components, etc. to name a few (Crawley et al. 2016, p. 316). The collaboration of various planes can generate more elegant decompositions.

An example of the Principle of Elegance for the AUAV Optical Subsystem is using a top-level Package Diagram to show the breakdown of the optical subsystem. This is the decomposition of the Top-Level diagram shown for the Principle of Holism (Figure 4.B23) to specifically show the components that make up the optical subsystem. Figure 4.B23 is the Optical Subsystem Package Diagram, which identifies the components of the subsystem as hardware, software, Use Cases, behavior, requirements, and analysis.

4.B.25 Principle of Robustness of Architectures

The textbook *System Architecture, Strategy and Product Development for Complex Systems* by E.C. Crawley, B. Cameron, and D. Selva discussed the Principle of Robustness of Architectures and supplied the following information on p. 363 in Box 15.1:

> Good architectures need to respond to all manner of variations. They can respond to these variations by being robust (capable of dealing with variations in the environment) or by being adaptable (able to adapt to changes in the environment). Optimal architectures, in the Pareto sense, are often the least robust. Consider optimality, robustness, and adaptability in the choice of an architecture.
>
> - Variations can occur in the input assumptions to the architecting process, in the downstream execution of the design, and/or in the operating environment.

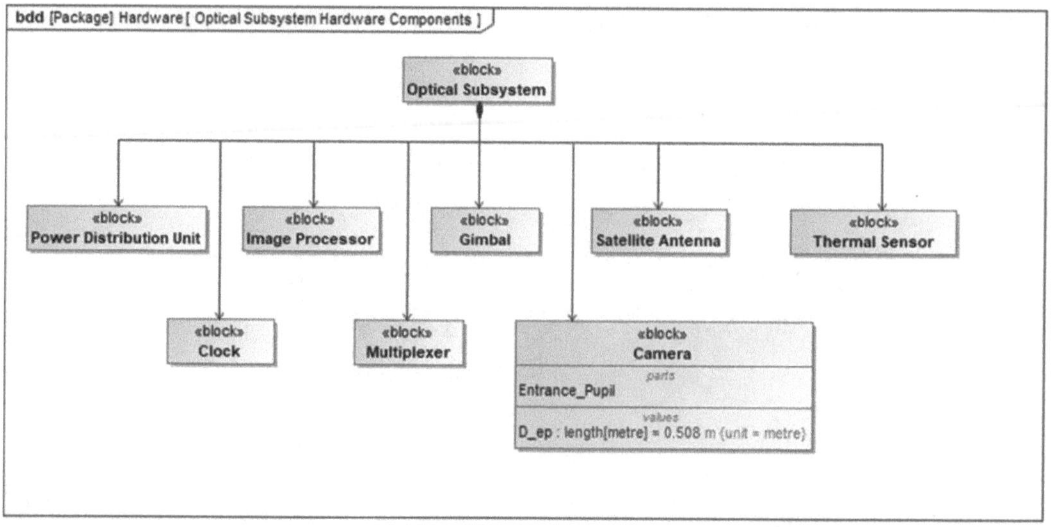

FIGURE 4.B24 BDD for the optical subsystem hardware components of the AUAV system.

> - Architectures on the Pareto frontier are sometimes the least robust in responding to these variations. For example, optimal architectures may rely on a game-changing technology that has low technology maturity. What happens if that technology is not ready when the system is developed? Is the architecture flexible enough to accommodate a different technology? Is there a strong tradeoff between optimality and robustness in your industry? The architect must decide whether a narrow peak in performance is achievable.
> - Robustness can be assessed by analyzing the architectures under a variety of technical and market scenarios and then picking the architectures that consistently perform well under a broad range of circumstances. Using fuzzy Pareto frontier instead of the Pareto frontier in the context of architecture optimization can also help identify the more robust architectures.
>
> *Crawley et al. (2016, p. 363)*

Robustness represents the capability of a system to handle a range of different scenarios. The level of robustness of a systems architecture is determined by the way the system responds to changes in the modeling assumptions. Robustness can be tested during sensitivity analysis which is "based on a reduced number of scenarios designed to cover most significant combinations of values for the relevant parameters" (Crawley et al. 2016, p. 363). Good system architecture should contain good robustness. A robust system is defined by "their ability to remain operational in the presence of errors, physical defects or environmental disruptions" (Krach 2017, p. 1).

Figure 4.B24 is the block definition diagram (bdd) for optical hardware components for the AUAV. This assists the system architect to think about how changes for one component could impact the other components, which helps develop capabilities to handle a range of scenarios. For instance, if there was a part shortage, the design should be adaptable and propagate the change through the entire system.

4.B.26 Principle of Coupling and Organization of Architectural Decisions

The textbook *System Architecture, Strategy and Product Development for Complex Systems* by E.C. Crawley, B. Cameron, and D. Selva discussed the Principle of Coupling and Organization of Architectural Decisions and supplied the following information on p. 368 in Box 15.2:

> Architecture can be thought of as the outcome of a relatively small number of important decisions. These decisions are highly coupled, and the sequence in which they are made is important. The sequence of

architectural decisions can be chosen by considering the sensitivity of the metrics to the decisions and the degree of connectivity of decisions.

- Architectural decisions are generally coupled. Couplings can be due to both constraints and metrics. The strongest couplings are due to hard constraints, when choosing a value for one decision precludes our choosing certain values for another decision.
- When a decision is strongly coupled with many other decisions through constraints, we say that this decision has a high degree of connectivity.
- Metrics have varying degrees of sensitivity to architectural decisions.
- Decisions that fall in the "first quadrant" (high sensitivity and high degrees of connectivity) should be made first, because they have the greatest impact on the remaining options.

Crawley et al. (2016, p. 368)

The system architect should identify the high-impact decisions of the system. A decision can be a high-impact decision if it has "a high sensitivity of the metrics, or a high degree of connectivity (or coupling) with other decisions" (Crawley et al. 2016, p. 364). All architectural decisions are coupled (linked or connected). The measurement of how one decision impacts another decision determines the degree of connectivity. Severity of impact depends on the degree of connectivity; higher the degree potentially means a higher effect on other decisions. The impact from connectivity spreads through an entire system. Weakly coupled decisions "should be treated in parallel because there is a high cost associated with waiting for one piece of analysis to begin before pursuing another" (Crawley et al. 2016, p. 368). Strong coupled decisions "should be combined in a single trade study, so that relevant interaction effects are considered and a thorough tradeoff can be performed" (Crawley et al. 2016, p. 368).

Architectural decisions can be organized based on the degree of sensitivity and the degree of connectivity. Decisions that are sensitive and have strong connectivity have the most impact on the system and determine the performance outcome. The second-highest priority are decisions that are sensitive and have weak connectivity, which can impact decisions the most if changed. Decisions that are insensitive and strongly connected are considered complementary in the system architecture. The lowest priority are decisions that are insensitive and weakly connected because they are not impactful to other decisions or connected to other decisions.

The Principle of Coupling and Organization of Architectural Decisions can be seen for the AUAV through the use of models. The CV-2 diagram for the AUAV Capability Taxonomy displays the entities that fulfill the capabilities, shown in Figure 4.B25. The capabilities are identified with a "C" in the top right corner of the element. The diagram identifies the capabilities of the AUAV Patrol as Rescue, Send Messages, Follow, Search, Transmit Images, ID Human, and Detect Human.

Figure 4.B26 is the operational view for the resource flow description of the AUAV. This displays the operation exchanges between the operational performers.

Figure 4.B27 is the operational view of the AUAV activities. The diagram displays the process of how the overall system works with the various operational performers. The AUAV flies the route and sends an image to the Image Specialist, which then determines what to do with the signal. Based on the Image Specialist determination, the signal goes to Border Patrol, Homeland Security, First Responders, or all. From there, the operational performer needs to determine what to send back to the AUAV: follow, abort mission, consider mission complete, and so forth. This view displays the coupled decisions and how the operations are organized and related to each other between the operation performers. In other words, Figure 4.B27 displays what the general purpose of the mission is, how the system is accomplishing the mission, and what capabilities are needed.

The SV-2 diagram for the AUAV Systems Resource Flow Description, Figure 4.B28, is to show how the Optics Package 1 (OP1) is improved by the Optics Package 2 (OP2), which contains better optics. This diagram also shows how the Optics Packets send the image and location to the Sat Comm link, which then sends an alert signal to the Patrol Comm link.

Figure 4.B29 is for the PV-1 diagram to showcase the AUAV project portfolio relationship. The left portfolio is for the current optics package (optics package 1) and the white in the pie chart is

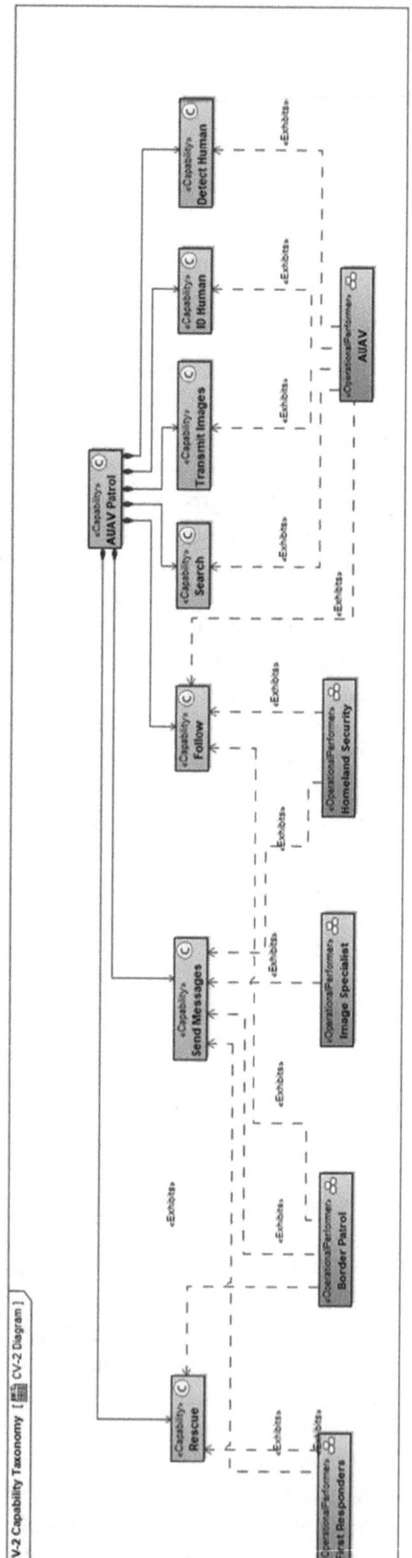

FIGURE 4.B25 CV-2 diagram for the AUAV patrol capabilities.

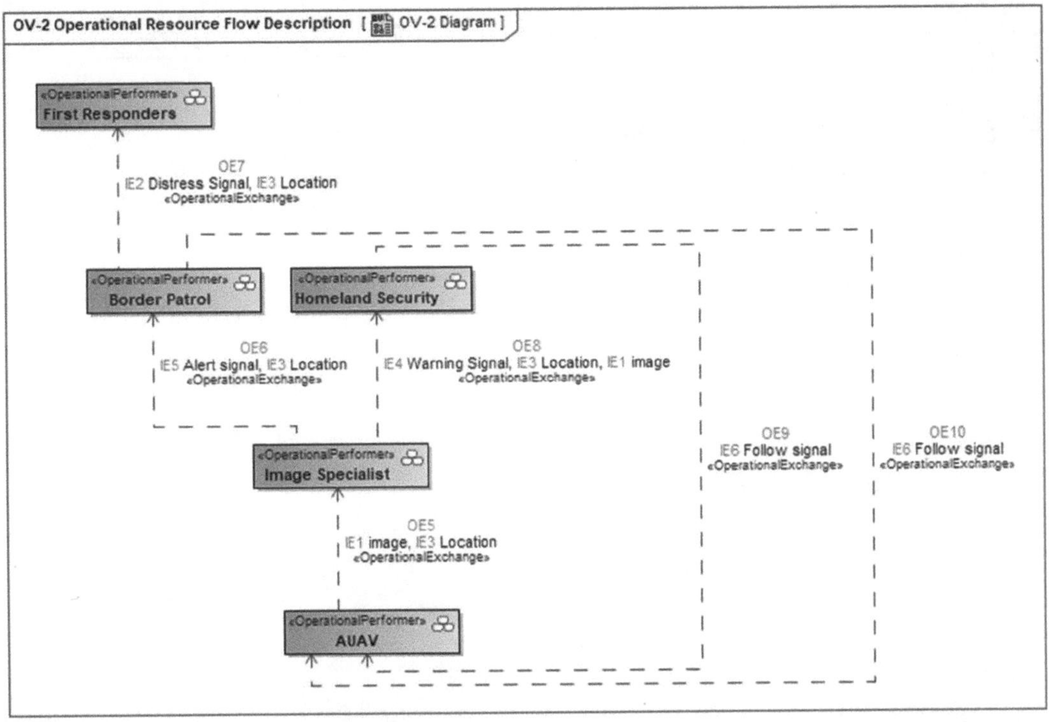

FIGURE 4.B26 OV-2 diagram for the AUAV operational resource flow description.

showing the optics package is in operation and maintenance. The right is for the technology refresh optics package (optic package 2) and the blue in the pie chart represents the optic package is in development.

The PV-2 chart, Chart 1 show in Figure 4.B30, shows the timeline for the two optic packages. Optics Package 1 is scheduled to be in the field for May 2022 to September 2022 and Optics Package 2 is scheduled to be in the field for October 2022 to December 2022.

These diagrams and the chart show how aspects of the AUAV system are coupled together. Changes to any aspect can lead to alterations in the other diagrams and an extension on optics package 2 in the timeline.

4.C APPENDIX: SYSTEMS ENGINEERING DEVELOPMENT MODELS

Three fundamental models of system engineering are the Vee Model (shown in Figure 4.C1), the Waterfall Model (shown in Figure 4.C2), and the Spiral Model (shown in Figure 4.C3). The Vee Model is the "graphical representation portraying the technical aspects of the project cycle [that] clarifies the role and responsibility of system engineering to a project" (Dickerson and Mavris 2016, p. 274).

The Vee model sequentially flows from the top left of the "V" down to the base and then up toward the top right of the "V". The left side of the model is for decomposition and definition, while the right side of the model is for integration and verification. The "time and project maturity flow from left to right, whereas the level of system decomposition increases from top to bottom, and then reverses as a level of integration from bottom to top" (Dickerson and Mavris 2016, p. 274).

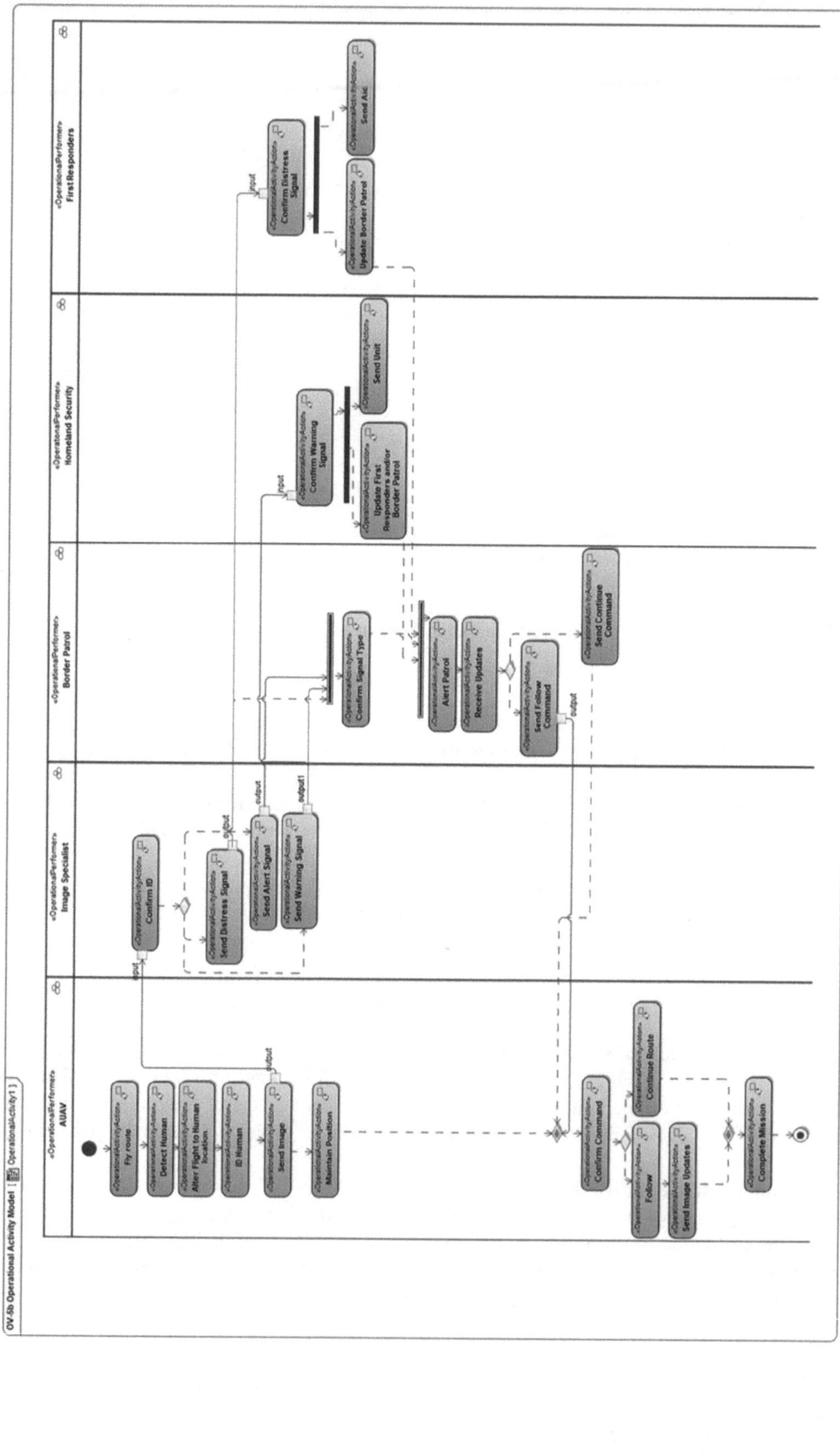

FIGURE 4.B27 OV-5 diagram for the AUAV operational activity model.

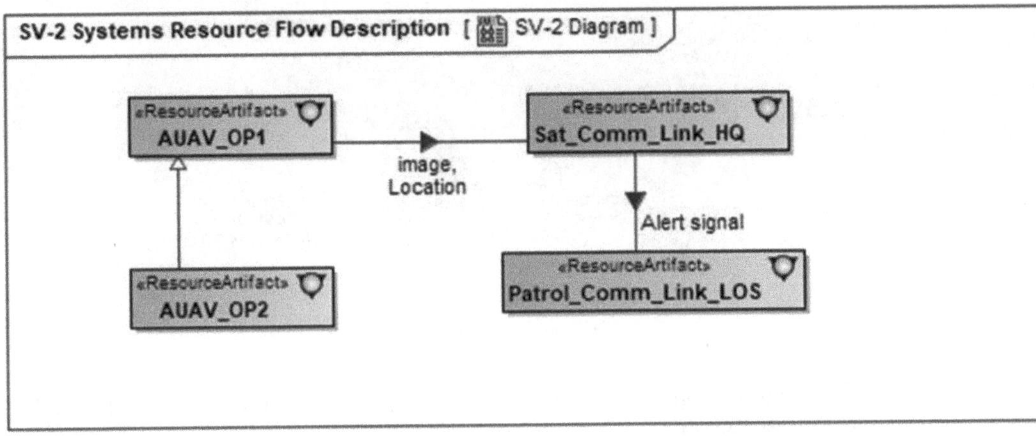

FIGURE 4.B28 SV-2 diagram for the AUAV system resource flow description.

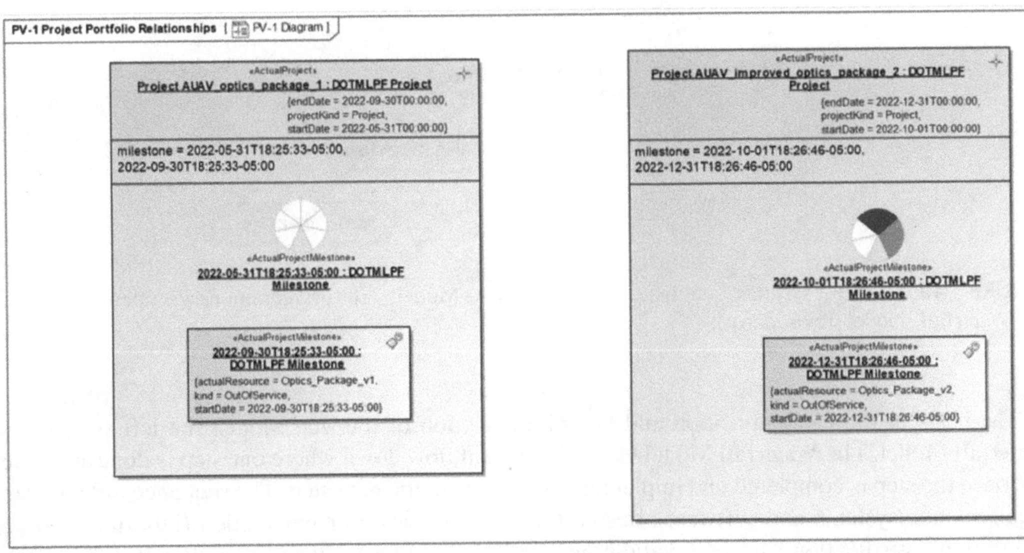

FIGURE 4.B29 PV-1 diagram for the AUAV project portfolio relationships.

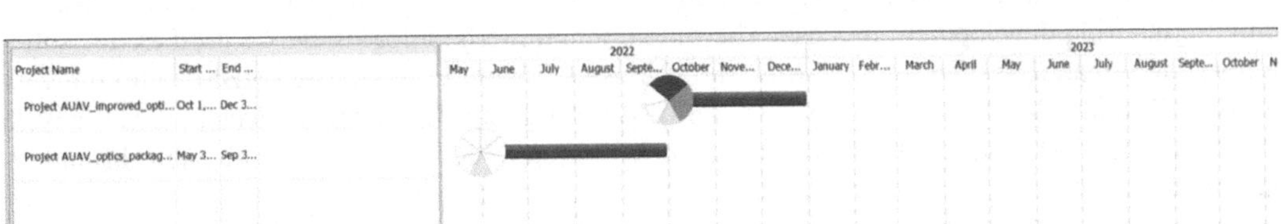

FIGURE 4.B30 PV-2 chart for operating in the field for the two optics packages of the AUAV.

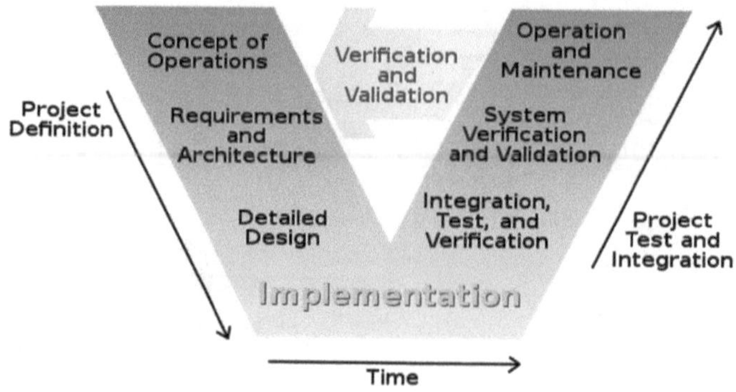

FIGURE 4.C1 The systems engineering Vee Model. (https://commons.wikimedia.org/wiki/File:Systems_Engineering_Process_II.svg).

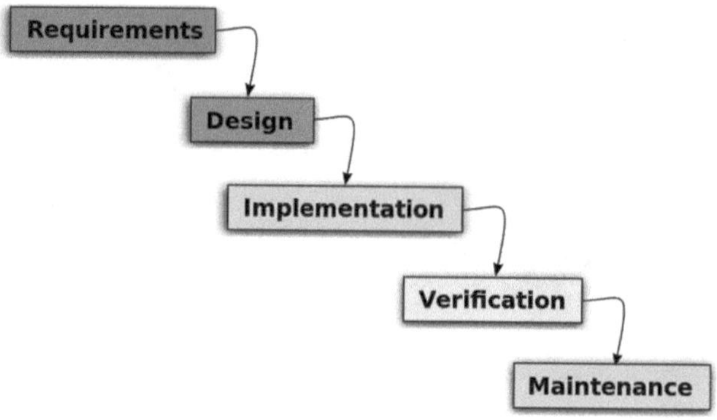

FIGURE 4.C2 The systems engineering Waterfall Model. (https://commons.wikimedia.org/wiki/File:Waterfall_model_revised.svg).

The basis of the Decomposition and Definition section of the Vee Model (the left side) is the Waterfall Model. The Waterfall Model has a subsequent flow down where one step is done at a time, and once the step is completed and implemented, it flows to the next step. The sequence order covers requirements (systems and software), design (of the program), implementation (building, coding, integrating), verification (includes validation), and maintenance (during operational and support phase; includes system reuse, retirement, and disposal).

Each loop of the Spiral Model represents an instance of the same progression of the following generic steps: "(1) determination of objectives, alternatives, and constraints; (2) evaluation of alternatives; (3) prototyping/simulation for risk analysis; (4) specification; (5) validation/verification; (6) planning for next cycle; and (7) review" (Dickerson and Mavris 2016, p. 286).

The fundamental terms, concepts, and models can be chosen and adapted as needed in the systems engineering development process. The fundamental methodologies can be used as a baseline or a way to enhance system architecture. Model-Based System Engineering (MBSE) and the various SysML and UML modeling artifacts can help throughout the systems development lifecycle. The various SysML and UML models can help with system development and be used throughout the system lifecycle.

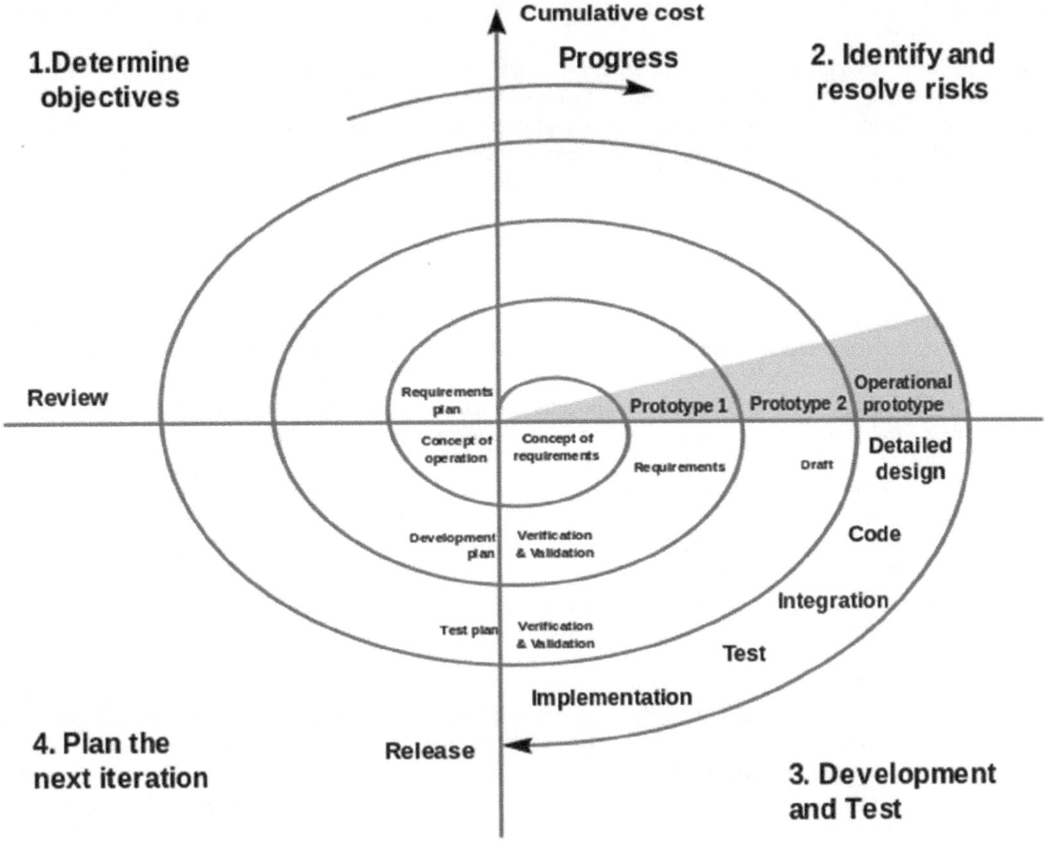

FIGURE 4.C3 The systems engineering Spiral Model. (https://commons.wikimedia.org/wiki/File:Spiral_model_(Boehm,_1988).svg).

REFERENCES

Anyanhun, A.I., A. Anzagira, and W.W. Edmonson. 2020. Intersatellite communication: an MBSE operational concept for a multiorbit disaggregated system. *IEEE Journal on Miniaturization for Air and Space Systems*, Vol. 1, pp. 56–65.

Asama, H., M. Yano, K. Tsuchiya, K. Ito, H. Yuasa, and J. Ota. 2003. System principle on emergence of Mobiligence and its engineering realization. *Proceedings 2003 IEEE/RSJ International Conference on Intelligent Robots and Systems (IROS 2003)* (Cat. No.03CH37453), Las Vegas, NV, pp. 1715–1720, Vol. 2. https://doi.org/10.1109/IROS.2003.1248891

Balta, D., V. Greger, P. Wolf, and H. Krcmar. 2015. E-government stakeholder analysis and management based on stakeholder interactions and resource dependencies. *2015 48th Hawaii International Conference on System Sciences*, Kauai, HI, pp. 2456–2465, https://doi.org/10.1109/HICSS.2015.294

Barroca, L., J. Hall, and P. Hall. 1999. *Software Architectures: Advances and Applications.* London: Springer-Verlag. https://doi.org/10.1007/978-1-4471-0367-7

Bass, L., P. Clements, and R. Kazman, 1998. *Software Architecture in Practice*, 1st edn. Boston, MA: Addison-Wesley.

Boella, G. and L. van der Torre. 2006. An architecture of a normative system. In: Goble, L. and J.-J.C. Meyer (eds), *DEON 2006. Lecture Notes in Computer Science.* Berlin: Springer, pp. 24–35.

Bouabana-Tebibel, T., S.H. Rubin, and M. Bennama. 2012. Formal modeling with SysML. *2012 IEEE 13th International Conference on Information Reuse & Integration (IRI)*, Las Vegas, NV. IEEE, pp. 340–347.

Buarqoub, I.A.S. 2019. Language barriers to effective communication. *Utopía Y Praxis Latinoamericana*, Vol. 24, No. 6, pp. 64–77.

Bushmann, F. n.d. The pragmatic architect. https://www.infoq.com/articles/pragmatic-architect/ (accessed March 9, 2024).

Cajori, F. 1899. *A History of Physics in Its Elementary Branches: Including the Evolution of the Physical Laboratories*. New York: The Macmillan Company.

Chami, M., C. Zoghbi, and J.-M. Bruel. 2019. A first step towards AI for MBSE: generating a part of SysML models from text using AI. *INCOSE Artificial Intelligence for Systems Engineering: 2019 Conference Proceedings*. INCOSE, Madrid, Spain. pp. 123–136.

Chen, B., S. Zhang, and B. Wang. 2018. A case study of MBSE method used in the EMU train design. *2018 International Conference on Intelligent Rail Transportation (ICIRT)*, Singapore. Singapore: IEEE, pp. 1–5.

Cheng, A.F., H.A. Weaver, S. Conard, F. Morgan, O.S. Barnouin, J. Boldt, K.A. Cooper, E.H. Darlington, M.P. Grey, J.R. Hayes, K.E. Kosakowski, T. Magee, E. Rossano, D. Sampath, C. Schlemm, and H.W. Taylor. 2007. Long-range reconnaissance imager on New Horizons. *Space Science Reviews,* Vol. 140, No. 1–4, pp. 189–215.

CIO. 2024. The Department of Defense Architectural Framework (DoDAF). Version 2.02. https://dodcio.defense.gov/library/dod-architecture-framework/ (accessed March 9, 2024).

Clinger Cohen Act. 1996. Title 40. 1996.

Compliance Testing Corporation. 2021. https://compliancetesting.com/what-are-fcc-class-a-emissions/ (accessed March 9, 2024).

Crawley, E.C., B. Cameron, and D. Selva. 2016. *System Architecture, Strategy and Product Development for Complex Systems*. Boston, MA: Pearson.

Crooke, J.A. and J.G. Hagiopian. 1998. Alignment and polarization sensitivity study for the Cassini composite infrared spectrometer (CIRS) far-infrared (FIR) interferometer. *SPIE's International Symposium on Optical Science, Engineering, and Instrumentation*, San Diego.

Cui, X., Y. Sun, S. Xiao, and H. Mei. 2009. Architecture design for the large-scale software-intensive systems: a decision-oriented approach and the experience. *2009 14th IEEE International Conference on Engineering of Complex Computer Systems*, Potsdam, pp. 30–39. https://doi.org/10.1109/ICECCS.2009.42

Dahmann, J. and K. Baldwin, 2011. Implications of systems of systems on system design and engineering. *2011 6th International Conference on System of Systems Engineering*, Albuquerque, NM, pp. 131–136. https://doi.org/10.1109/SYSOSE.2011.5966586

Dam, S.H. 2019. *Real MBSE: Model-Based Systems Engineering (MBSE) Using LML and Innoslate®*. Manassas, VA: SPEC Innovations.

Dauby, J.P. and C.H. Dagli. 2010. Complex systems architecting using design primitives, comparative analysis and fuzzy analytical feedback. *2010 IEEE International Systems Conference*, San Diego, CA, pp. 234–238. https://doi.org/10.1109/SYSTEMS.2010.5482496

Davila, P.S. 2006. Focus determination for the James Webb Space Telescope science instruments: a survey of methods. *Astronomical Telescopes and Instrumentation,* Vol. 6265, No. June, pp. 12–23.

Delligatti, L. 2013. *SysML Distilled: A Brief Guide to the Systems Modeling Language.* Crawfordsville, IN: Addison-Wesley.

Denford, M., J. Leaney, and T. O'Neill. 2004. Non-functional refinement of computer based systems architecture. *Proceedings. 11th IEEE International Conference and Workshop on the Engineering of Computer-Based Systems*, Brno, Czech Republic, pp. 168–177, https://doi.org/10.1109/ECBS.2004.1316696

Dhungana, D., T. Neumayer, P. Grunbacher, and R. Rabiser. 2008. Supporting the evolution of product line architectures with variability model fragments. *Seventh Working IEEE/IFIP Conference on Software Architecture (*WICSA 2008*)*, Vancouver, BC, pp. 327–330. https://doi.org/10.1109/WICSA.2008.23

Dickerson, C.E. and D.N. Mavris. 2016. *Architecture and Principles of Systems Engineering*. Boca Raton, FL: Auerbach Publications.

Dietrich, R. and Z.X. Zheng. 1993. The ordering of words in utterance production: an integrated view of the basic principles in German and Chinese. *Advances in Psychology*, Vol. 103, pp. 119–143, edited by Jeanette Altarriba. ScienceDirect. North–Holland. January 1, 1993.

Dilworth, D.C. 2008. New tools for the lens designer. *Optical Engineering and Applications*.

Douglass, B.P. 2014. *Real-Time UML Workshop for Embedded Systems.* Waltham, MA: Newnes.

EDS SB. 2000. Cat Herder Commercial. Electronic Data Systems (EDS) is part of Hewlett Packard. https://www.youtube.com/watch?v=eSotGdDKZDs (accessed March 10, 2024).

Ehrig, H., B. Braatz, M. Klein, F. Orejas, S. Perez, and E. Pino. 2005. Object-oriented connector-component. *Electronic Notes in Theoretical Computer Science,* Vol. 141, pp. 123–151. https://doi.org/10.1016/j.entcs.2005.02.048

Enoch, J.M. and V. Lakshminarayanan. 2000. Duplication of unique optical effects of ancient Egyptian lenses from the IV/V Dynasties: lenses fabricated ca 2620–2400 BC or roughly 4600 years ago. *Ophthalmic and Physiological Optics,* Vol. 20, No. 2, pp. 126–130.

Eppinger, S.D., D.E. Whitney, R.P. Smith, and D.A. Gebala. 1994. A model-based method for organizing tasks in product development. *Research in Engineering Design: Theory, Applications and Concurrent Engineering,* Vol. 6, pp. 1–13 (Springer-Verlag). https://doi.org/10.1007/bf01588087

Erickson, J. and K. Siau. 2013. Unified modeling language: the teen years and growing pains. In: *Human Interface and the Management of Information. Information and Interaction Design: 15th International Conference, HCI International 2013, Ed. Sakae Yamamoto*, Las Vegas, NV, USA, July 21–26, 2013, *Proceedings, Part* I 15. Berlin, Heidelberg: Springer, pp. 295–304.

Falessi, D., G. Cantone, and P. Kruchten. 2008. Value-based design decision rationale documentation: principles and empirical feasibility study. *Seventh Working IEEE/IFIP Conference on Software Architecture (WICSA 2008)*, Vancouver, BC, pp. 189–198. https://doi.org/10.1109/WICSA.2008.8

Ferrogalini, M. 2015. MBSE in rail transportation--product families and product lines. *OR Insight*, Vol. 18, pp. 19–21.

Firesmith, D.G., P. Capell, D. Falkentha, C.B. Hammons, D. Latimer, and T. Merendino. 2008. *The Method Framework for Engineering System Architectures.* Boca Raton, FL: CRC Press. https://doi.org/10.1201/9781420085761

Fixson, S. 2000. *Systems Architecture – A Knowledge Domain.* Cambridge, MA: Massachusetts Institute of Technology.

Franke, E. and D. Levis. 2000. The role of the architect in the design of modern military communications terminals. *MILCOM 2000 Proceedings. 21st Century Military Communications. Architectures and Technologies for Information Superiority (Cat. No.* 00CH37155*)*, Vol. 2, pp. 981–987. IEEE.

Friedenthal, S., A. Moore, and R. Steiner. 2015. *A Practical Guide to SysML: The Systems Modeling Language*, 3rd edn. Waltham, MA: Elsevier/Morgan Kaufmann.

Gubarev, V.V. and V.U. Chernov. 2001. The adaptive automated systems of statement and problem solving of a data analysis. *5th Korea-Russia International Symposium on Science and Technology. Proceedings. KORUS 2001 (Cat. No.01EX478)*, Tomsk, Russia, Vol. 1, pp. 140–143. https://doi.org/10.1109/KORUS.2001.975081

Gaska, T. 2012. Optimizing an incremental Modular Open System Approach (MOSA) in avionics systems for balanced architecture decisions. *2012 IEEE/AIAA 31st Digital Avionics Systems Conference (DASC)*, Williamsburg, VA, pp. 7D1-1–7D1-19, https://doi.org/10.1109/DASC.2012.6382420

Goldmakher, L. 2023. Propositional and predicate calculus. https://web.williams.edu/Mathematics/lg5/200/PropPred.pdf (accessed August 28, 2023).

Goodwin, M., D. Adams, R. Brown, V.N. Hartmann, C. Lacombe, C.H.N. Lahoz, J. Lawrence, D. May, E. O'Brien, and T. Zafar. 2022. Application of model-based systems engineering to MANIFEST conceptual design. *Modeling, Systems Engineering, and Project Management for Astronomy* X, Vol. 12187, pp. 34–53. SPIE.

Gore, G. 2003. Thoughts on the information system architect role. *Proceedings ITCC 2003. International Conference on Information Technology: Coding and Computing*, Las Vegas, NV, pp. 706–710. https://doi.org/10.1109/ITCC.2003.1197617

Hagopian, J.G., S.A. Getty, M. Quijada, J. Tveekrem, R. Shiri, P. Roman, J. Butler, G. Georgiev, J. Livas, C. Hunt, A. Maldonado, S. Talapatra, X. Zhang, S.J. Papadakis, A.H. Monica, and D. Deglau. 2010. Multiwalled carbon nanotubes for stray light suppression in space flight instruments. *SPIE Nanoscience and Engineering*, San Diego.

Haidrar, S., A. Anwar, and O. Roudies. 2017. A SYSML-based approach to manage stakeholder requirements traceability. *2017 IEEE/ACS 14th International Conference on Computer Systems and Applications (AICCSA)*, Hammamet. IEEE, pp. 202–207.

Haimes, Y.Y. 2004. *Risk Modeling, Assessment, and Management.* Hoboken, NJ: John Wiley & Sons.

Hampson, K. 2015. Technical evaluation of the systems modeling language (SysML). *Procedia Computer Science*, Vol. 44, pp. 403–412.

Hause, M. 2010. The unified profile for DoDAF/MODAF enabling systems of systems on many levels. *2010 IEEE International Systems Conference*, San Diego, CA (May 2010). IEEE Xplore.

Hause, M. and T. Francis. 2008. Building bridges between systems and software with SysML and UML. *INCOSE International Symposium*, Vol. 18, No. 1, pp. 797–811.

Hayat, F., M.W. Anwar, F. Azam, and A. Kiran. 2020. A SYSML-based approach for requirements risk management and change control. *Proceedings of the 2019 11th International Conference on Information Management and Engineering*, London. ICIME, pp. 20–24.

Hebisch, E., M. Book, and V. Gruhn. 2015. Scenario-based architecting with architecture trace diagrams. *2015 IEEE/ACM 5th International Workshop on the Twin Peaks of Requirements and Architecture*, Florence, Italy, pp. 16–19. https://doi.org/10.1109/TwinPeaks.2015.11

Hecht, E. 1998. Optics, 3rd edn. Reading, MA: Addison-Wesley.

Herrmann, C., A. Frad, I. Revnic, and T. Luger. 2005. Product architect, a new approach for transparency and controlling of the end-of-Llife performance. *2005 4th International Symposium on Environmentally Conscious Design and Inverse Manufacturing*, Tokyo, Japan, pp. 634–639. IEEE.

Herzig, S., R. Karban, G. Brack, S.B. Michaels, F. Dekens, and M. Troy. 2018. Verifying interfaces and generating interface control documents for the alignment and phasing subsystem of the Thirty Meter Telescope from a system model in SysML. *Modeling, Systems Engineering, and Project Management for Astronomy VIII*, vol. 10705, pp. 319–335.

Heylighen, F. 1989. Self-organization, emergence and the architecture of complexity. *Transdisciplinary Research Group, Free University of Brussels, Pleinlaan 2, B-1050*, Brussels, Belgium.

Holt, J. and S. Perry. 2018. *SysML for Systems Engineering: A Model-Based Approach*. Stevenage, Herts: The Institution of Engineering and Technology, p. 115.

Huygens, C. 1690. *Traité de la Lumière*. Leiden: Pieter van der Aa.

IEEE. 2017. IEEE standard for system, software, and hardware verification and validation. *IEEE Std 1012-2016 (Revision of IEEE Std 1012-2012/Incorporates IEEE Std 1012-2016/Cor1-2017)*, pp.1–260, September 29, 2017. https://doi.org/10.1109/IEEESTD.2017.8055462

Ielite, I., G. Olevsky, and T. Safiulins. 2015. Identification and prioritization of stakeholders in the planning process of sustainable development of the smart city. *2015 IEEE Seventh International Conference on Intelligent Computing and Information Systems (ICICIS)*, Cairo, Egypt, pp. 251–257. https://doi.org/10.1109/IntelCIS.2015.7397229

INCOSE Homepage. 2023. *International Council on Systems Engineering*. https://incose.org (accessed December 13, 2023).

Innoslate. 2020. Why do we need model-based systems engineering (MBSE)? Innoslate, 29 May 2020. https://www.innoslate.com/why_do_we_need_mbse/ (accessed May 17, 2022).

Irving, B. 2013. An introduction to optical design. Synopsys, Inc. [Online]. https://www.synopsys.com/optical-solutions/learn/gentle-intro-to-optical-design.html (accessed March 9, 2024).

ISO/IEC/IEEE. 2015. *Systems and Software Engineering—System Lifecycle Processes*. Geneva: International Organization for Standardization.

ISO/IEC. 2018. *ISO/IEC 29148, Systems and Software Engineering — Lifecycle Processes — Requirements Engineering*. Geneva: Standard, International Standards Organization/International Electrotechnical Commission.

Jaakola, H. and B. Thalheim. 2011. Architecture-driven modeling methods. *Frontiers in Artificial Intelligence and Applications*, Vol. 225, pp. 97–116. Finland.

Jain, R., A. Chandrasekaran, G. Elias, and R. Cloutier. 2008. Exploring the impact of systems architecture and systems requirements on systems integration complexity. *IEEE Systems Journal*, Vol. 2, No. 2, pp. 209–223. https://doi.org/10.1109/JSYST.2008.924130

Johnson, T.A., J.M. Jobe, C. Paredis, and R. Burkhart. 2007. Modeling continuous system dynamics in SysML. *ASME International Mechanical Engineering Congress and Exposition*, Vol. 42975, pp. 197–205.

Johnston, J.D., J.M. Howarda, G.E. Mosier, K.A. Parrisha, M.A. McGinnis, M. Bluth, K. Kim, and K.Q. Ha. 2004. Integrated modeling activities for the James Webb Space Telescope: structural-thermal-optical analysis. *Astronominal Telescopes and Instrumentation,* Vol. 5487, No. October, pp. 600–610.

Jones-Bey, H.A. 2005. Choosing software: different design paths lead to different solutions, Part I. *Laser Focus World,* Vol. 1, p. 9.

Kakola, T.K. and K.I. Koota. 1996. Redesigning work with dual information systems: the work process benchmarking service. *Proceedings of HICSS-29: 29th Hawaii International Conference on System Sciences*, Wailea, HI, Vol. 3, pp. 461–471. https://doi.org/10.1109/HICSS.1996.493242

Kancherla, A.K. and S. Kummamuru. 2011. Framework for Post Merger Integration using systems principles. *2011 Annual IEEE India Conference*, Hyderabad, India, pp. 1–4. https://doi.org/10.1109/INDCON.2011.6139602

Karcanias, N. and A.G. Hessami. 2011. System of systems and emergence part 1: principles and framework. *2011 Fourth International Conference on Emerging Trends in Engineering & Technology*, Port Louis, Mauritius, pp. 27–32. https://doi.org/10.1109/ICETET.2011.51

Karlik, A.E., I.A. Maksimtsev, and E.A. Iakovleva. 2016. Modern architecture of global value chains and value chain management of Russia. *2016 XIX IEEE International Conference on Soft Computing and Measurements (SCM)*, St. Petersburg, Russia, pp. 518–520. https://doi.org/10.1109/SCM.2016.7519832

Kennedy, R. 2023. Unleashing Data Storage: From Hardware to the Cloud. Spiceworks. August 8, 2023. https://www.spiceworks.com/tech/big-data/guest-article/the-history-and-future-of-data-storage/ (Accessed on March 5, 2025).

Kincade, K. and S. Anderson. 2005. Laser Marketplace 2005: consumer applications boost laser sales 10%. *Laser Focus World,* 1 January 2005. https://www.laserfocusworld.com/lasers-sources/article/16556215/consumer-applications-boost-laser-sales-10 (Accessed on March 5, 2025).

Krach, S.D. 2017. Model-based architecture robustness analysis for software-intensive autonomous systems. *2017 IEEE International Symposium on Software Reliability Engineering Workshops (ISSREW)*, Toulouse, France, pp. 104–107. https://doi.org/10.1109/ISSREW.2017.39

Krebs, R.E. 2004. *Groundbreaking Scientific Experiments, Inventions, and Discoveries of the Middle Ages and the Renaissance*. Westport: Greenwood Publishing Group.

Kulkarni, V. 2008. A conceptual model for capturing stakeholders' wish list. *2008 International Conference on Computer Science and Software Engineering*, Wuhan, China, pp. 275–278. https://doi.org/10.1109/CSSE.2008.878

Lamba Research Associates. 2024. OSLO leading lens design software. Lambda Research Associates, 2024. https://lambdares.com/oslo (accessed March 10, 2024).

Lambda Research Associates (Tracepro). 2024. TracePro suite of optical and illumination design software. https://lambdares.com/tracepro (accessed March 10, 2024).

Lang, J.T. 1987. *Creating Architectural Theory: The Role of the Behavioral Sciences in Environmental Design.* New York: Van Nostrand Reinhold Company.

Langheim, J., B. Guegan, L. Maillet-Contoz, K. Maaziz, G. Zeppa, F. Philippot, S. Boutin, H. Aboutaleb, and P. David. 2010. System architecture, tools and modelling for safety critical automotive applications - the R&D project SASHA.

Leonard, J. 1999. *Systems Engineering Fundamentals: Supplementary Text.* Fort Belvoir, VA: Defense Systems Management College.

Liu, R., Z.-Q. Ren, and Z.-Y. Wang. 2018. Dualism of knowledge creation for human-machine interactive processing. *2018 2nd IEEE Advanced Information Management, Communicates, Electronic and Automation Control Conference (IMCEC)*, Xi'an, China, pp. 1487–1490. https://doi.org/10.1109/IMCEC.2018.8469650

Long, D. 2020. MBSE: simple, complicated, or complex? Community.Vitechcorp.com, Vitech Corp, 9 April 2020, https://community.vitechcorp.com/mbse-simple-complicated-or-complex/ (accessed May 17, 2022).

Lucid Software Inc. 2022. System architecture diagram tool. https://lucidscale.com/create/system-architecture-diagram-tool (accessed August 31, 2022).

Luckham, D.C., J. Vera, and S. Meldal. 1995. *Three Concepts of System Architecture.* Technical Report CSL-TR-95-674, The Defense Advanced Research Projects Agency.

Madni, A.M. 2012. Elegant systems design: creative fusion of simplicity and power. *Systems Engineering*, Vol. 15, pp. 347–354. https://doi-org.portal.lib.fit.edu/10.1002/sys.21209

Madni, A.M., and M. Sievers. 2018. Model-based systems engineering: motivation, current status, and research opportunities. *Systems Engineering*, Vol. 21, No. 3, pp. 172–190.

Madni, A.M. and S. Purohit. 2019, Economic analysis of model-based systems engineering. *Systems*, Vol. 7, p. 12. https://www.researchgate.net/publication/331236880_Economic_Analysis_of_Model-Based_Systems_Engineering.

Maier, M.W. 1998. Architecting principles for systems-of-systems. *Systems Engineering*, Vol. 1, pp. 267–284.

Maier, M.W. and E. Rechtin. 2000. *The Art of Systems Architecting,* 2nd edn. Boca Raton, FL: CRC Press.

Mandel, C., M. Guenther, A. Martin, E. Windisch, N. Bursac, S. Rapp, H. Anacker, and A. Albers. 2022. Towards a system of systems engineering architecture framework. *17th Annual System of Systems Engineering Conference (SOSE)*, Rochester, NY. https://doi.org/10.1109/sose55472.2022.9812634

Mathew, P.G., S. Liscouet-Hanke, and Y. Le Masson. 2018. Model-based systems engineering methodology for implementing networked aircraft control system on integrated modular avionics – environmental control system case study. SAE Technical Paper 2018-01-1943, 2018. https://www.researchgate.net/publication/328638482_Model-Based_Systems_Engineering_Methodology_for_Implementing_Networked_Aircraft_Control_System_on_Integrated_Modular_Avionics_-_Environmental_Control_System_Case_Study

McDermott, T. and A. Salado. 2018. Art and architecture: effectively communicating models of systems. *2018 Annual IEEE International Systems Conference (SysCon)*, Vancouver, BC, pp. 1–7. https://doi.org/10.1109/SYSCON.2018.8369605

Min, I.A. and R.A. Noguchi. 2016. The architecture design and evaluation process: a decision support framework for conducting and evaluating architecture studies. *2016 IEEE Aerospace Conference*, Big Sky, MT, pp. 1–14. https://doi.org/10.1109/AERO.2016.7500565

Muller, G. 2011 *Systems Architecting: A Business Perspective*. Boca Raton, FL: CRC Press.

Muller, G. 2018. The challenge of increasing heterogeneity in systems of systems for architecting. *2018 13th Annual Conference on System of Systems Engineering (SoSE)*, Paris, France, pp. 521–526. https://doi.org/10.1109/SYSOSE.2018.8428772

Murphy, P.A. 2023. Stephen Hawking's model-dependent realism. Medium, April 14, 2023. https://medium.com/paul-austin-murphys-essays-on-philosophy/stephen-hawkings-model-dependent-realism-2b2973af41a6 (Accessed on March 5, 2025).

NASA. 2007. *Systems Engineering Handbook*. Washington, DC: National Aeronautics and Space Administration (NASA).

Newton, I. 1704. *Opticks or, a Treatise of the Reflexions, Refractions, Inflexions and Colours of Light: Also Two Treatises of the Species and Magnitude of Curvilinear Figures* (originally published in 1704). Palo Alto: Octavo, 1998.

Ojala, P. 2008. Developing value assessment for SW architecture. *Seventh Working IEEE/IFIP Conference on Software Architecture (*WICSA 2008), Vancouver, BC, pp. 245–248. https://doi.org/10.1109/WICSA.2008.7

Oleksiv, I., H. Lema, V. Kharchuk, T. Lisovych, O. Dluhopolskyi, and T. Dluhopolska. 2020. Identification of stakeholders importance for the company's social responsibility using the analytic hierarchy process. *2020 10th International Conference on Advanced Computer Information Technologies (ACIT)*, Deggendorf, Germany, pp. 1–4. https://doi.org/10.1109/ACIT49673.2020.9208897

Oquendo, F. 2018. On the emergent behavior oxymoron of system-of-systems architecture description. *2018 13th Annual Conference on System of Systems Engineering (SoSE)*, Paris, France, pp. 417–424. https://doi.org/10.1109/SYSOSE.2018.8428727

Oquendo, F. 2019. Coping with uncertainty in systems-of-systems architecture modeling on the IoT with SosADL. *2019 14th Annual Conference System of Systems Engineering (SoSE)*, Anchorage, AK, pp. 131–136. https://doi.org/10.1109/SYSOSE.2019.8753842

Oquendo, F. 2020. Fuzzy architecture description for handling uncertainty in IoT systems-of-systems. *2020 IEEE 15th International Conference of System of Systems Engineering (SoSE)*, Budapest, Hungary, pp. 000555–000562. https://doi.org/10.1109/SoSE50414.2020.9130553

Oquendo, F. 2021. Case study on the fuzzy architecture description of cyber-physical SoS under uncertainty. *2021 16th International Conference of System of Systems Engineering (SoSE)*, Västerås, Sweden, pp. 61–68. https://doi.org/10.1109/SOSE52739.2021.9497482

ORA. 2013. Optical Research Associates: CODE V: faster, easier, more flexible optical design. Synoptics. https://www.opticalres.com/cv/value.html (accessed March 9, 2024).

OSA. 2009. *Handbook of Optics, Volume* II, 3rd edn. New York: McGraw-Hill Professional. Optical Society of America.

Paech, B. 1999. On the role of activity diagrams in UML – a user task centered development process for UML. In: Bézivin, J. and P.A. Muller (eds), *The Unified Modeling Language. «UML» '98: Beyond the Notation. UML* 1998. Lecture Notes in Computer Science, Vol. 1618. Berlin, Heidelberg: Springer.

Pearl, J. 1984. *Heuristics: Intelligent Search Strategies for Computer Problem Solving*. Reading, MA: Addison-Wesley Pub. Co.

Pedraza-García, G., H. Astudillo, and D. Correal. 2014. Modeling software architecture process with a decision-making approach. *33rd International Conference of the Chilean Computer Science Society (SCCC)*, Talca, Chile, pp. 1–6. https://doi.org/10.1109/SCCC.2014.27

Photonics Media Online. 2024. Photonics buyers' guide. Photonics Media. https://www.photonics.com/BuyersGuide.aspx (accessed March 9, 2024).

Photonics Media Print. 2024. *Photonics Buyers' Guide*. Pittsfield, MA: Laurin Publishing Company.

Polya, G. 1945. *How to Solve It: A New Aspect of Mathematical Method*. Princeton, NJ: Princeton University Press.

Public Law 1007-347. 2002.

Ranganathan, P. and R. Aggarwal. 2017. Common pitfalls in statistical analysis: linear regression analysis. *Perspectives in Clinical Research*, Vol. 8, No. 2, p. 100.

Rashed, R. 1990. A pioneer in anaclastics: Ibn Sahl on burning mirrors and lenses. *Isis*, Vol. 81, pp. 464-491.

Rechtin, E. 1992. The art of systems architecting. *IEEE Spectrum*, Vol. 29, No. 10, pp. 66–69. https://doi.org/10.1109/6.158642

Rechtin, E. 2002. *The Art of Systems Architecting*. Boca Raton, FL: CRC Press, pp. 5,6, 158.

Rekhav, V.S. and H. Muccini. 2014. A study on group decision-making in software architecture. *2014 IEEE/IFIP Conference on Software Architecture*, Sydney, NSW, pp. 185–194. https://doi.org/10.1109/WICSA.2014.15

Rosenblatt, A. and G. Watson. 1991. Concurrent engineering. *Institute of Electrical and Electronics Engineers*, Vol. 28, pp. 22–37.

Rousseau, D. 2019. Advances in the prospects for realizing a scientific general theory underpinning systems engineering. *2019 International Symposium on Systems Engineering (ISSE)*, Edinburgh, pp. 1–5. https://doi.org/10.1109/ISSE46696.2019.8984520

Sanders, E. and P.J. Stappers. 2008. Co-creation and the new landscapes of design. *CoDesign*, Vol. 4, pp. 5–18. https://doi.org/10.1080/15710880701875068

Sangwan, R.S. and C.J. Neill. 2009. Characterizing essential and incidental complexity in software architectures. *2009 Joint Working IEEE/IFIP Conference on Software Architecture & European Conference on Software Architecture*, Cambridge, pp. 265–268. https://doi.org/10.1109/WICSA.2009.5290816

Saqui-Sannes, P., R.A. Vingerhoeds, C. Garion, and X. Thirioux. 2022. A taxonomy of MBSE approaches by languages, tools and methods. *IEEE Access,* Vol. 10, pp. 7–9.

Savage, N. 2007. Optical design software. *Nature Photonics,* Vol. 1, No. 10, pp. 598–599.

Sawhney, G.S. 2018. *Concurrent Engineering.* New Delhi, India: Laxmi Publications Pvt. Ltd.

Shevchenko, N. 2021. Requirements in model-based systems engineering (MBSE). *SEI Blog*, 22 February 2021. https://insights.sei.cmu.edu/blog/requirements-in-model-based-systems-engineering-mbse/ (Accessed March 5, 2025).

Sinclair, D. 2000. Whither optical design? *Optics and Photonics News*, Vol. 11, p. 34.

Stadter, P.A. and A.K. Garga. 1997. A neural architecture for fuzzy classification with application to complex system tracking. *Proceedings of International Conference on Neural Networks (ICNN'97)*, Vol. 4, pp. 2439–2444. https://doi.org/10.1109/ICNN.1997.614464

Tie, B.N. and J. Bolluijt. 2014. Measuring project complexity. *9th International Conference on System of Systems Engineering (SOSE)*, Glenelg, Australia. pp. 248–253. https://doi.org/10.1109/SYSOSE.2014.6892496

Townes, C.H. 2003. The first Laser. In: Garwin, L. and T. Lincoln (eds), *A Century of Nature: Twenty-One Discoveries that Changed Science and the World*. Chicago, IL: University of Chicago Press, pp. 107–112.

Trivedi, K.S. and A. Bobbio. 2017. *Reliability and Availability Engineering: Modeling, Analysis, and Applications*. Cambridge: Cambridge University Press.

Ulrich, K.T. and S.D. Eppinger. 2000. *Product Design and Development,* 2nd edn. Boston, MA: McGraw-Hill.

Vollmer, M., M. Schmitz, and J. Warschat. 2013. How function analysis of manufacturing processes can stimulate technology intelligence processes. *2013 Proceedings of PICMET '13: Technology Management in the IT-Driven Services (PICMET)*, San Jose, CA, pp. 2141–2146.

Wagner, D.A., M.B. Bennett, R. Karban, N. Rouquette, S. Jenkins, and M. Ingham. 2012. An ontology for state analysis: formalizing the mapping to SysML. *2012 IEEE Aerospace Conference*, Big Sky, MT. IEEE, pp. 1–16.

Walden, D.D., G.J. Roedler, K.J. Forsberg, R. Douglas Hamelin, and T.M. Shortell. 2015. *Systems Engineering Handbook: A Guide for System Lifecycle Processes and Activities*, 4th edn. San Diego, CA: Wiley.

Wang, D., L. Ren, and J. Li. 2010. Modeling intelligent transportation systems with multi-agent on SOA. 2010 *International Conference on Intelligent Computing and Integrated Systems*, Guilin, pp. 717–720. https://doi.org/10.1109/ICISS.2010.5657144

Weber, A. 2018. Smart manufacturing stakeholders and their requirements. *2018 e-Manufacturing & Design Collaboration Symposium (eMDC)*, Hsinchu, Taiwan, pp. 1–3.

Weilkiens, T. 2008. *Systems Engineering with SysML/UML: Modeling, Analysis, Design*. Burlington, VT: Elsevier, pp. 226–233.

Wilkinson, M., A. James, M. Emes, P. King, and P. Bryant. 2010. Belief systems in systems architecting: method and preliminary applications. *2010 5th International Conference on System of Systems Engineering,* Loughborough, UK. Institute of Electrical and Electronics Engineers, pp. 1–7. https://doi.org/10.1109/SYSOSE.2010.5544095

Winner, R.I., J.P. Pennell, H.E. Bertrand, and M.M.G. Slusarczuk. 1988. *The Role of Concurrent Engineering in Weapons System Acquisition.* Alexandria, VA: Institute for Defense Analyses.

Wolfram Homepage. 2023. Propositional calculus. From Wolfram MathWorld. https://mathworld.wolfram.com/PropositionalCalculus.html (accessed August 11, 2023).

Woods, E. 2016. Harnessing the power of architectural design principles. *IEEE Software*, Vol. 33, No. 4, pp. 15–17. https://doi.org/10.1109/MS.2016.88

Woollard, D., C. Mattmann, and N. Medvidovic. 2009. Injecting software architectural constraints into legacy scientific applications. *2009 ICSE Workshop on Software Engineering for Computational Science and Engineering*, Vancouver, BC, pp. 65–71. https://doi.org/10.1109/SECSE.2009.5069164

Wyrowski, F. and M. Kuhn. 2011. Introduction to field tracing. *Journal of Modern Optics,* Vol. 58, No. 3, pp. 449–466.

Zdanis, L. and R. Cloutier. 2007. The use of behavioral diagrams in SysML. *2007 IEEE Long Island Systems, Applications and Technology Conference,* Long Island. IEEE, pp. 1–1.

Zhong, S., A. Scarinci, and C. Alice. 2023. Natural language processing for systems engineering: automatic generation of systems modelling language diagrams. *Knowledge-Based Systems*, Vol. 259, p. 110071.

Zhu, W., H. He, and Z. Wang. 2017. Ontology-based mission modeling and analysis for system of systems. *2017 IEEE International Conference on Internet of Things (iThings) and IEEE Green Computing and Communications (GreenCom) and IEEE Cyber, Physical and Social Computing (CPSCom) and IEEE Smart Data (SmartData)*, Exeter, pp. 538–544. https://doi.org/10.1109/iThings-GreenCom-CPSCom-SmartData.2017.85

Zou, L., Q. Liu, H. Che, H. Yang, and H. Zhou. 2017. A creative approach to reusing data and generating knowledge in enterprise systems. *2017 IEEE 41st Annual Computer Software and Applications Conference (COMPSAC)*, Turin, Italy, pp. 425–430. https://doi.org/10.1109/COMPSAC.2017.13

Section II

Contemporary Systems Engineering Principles, Methods, and Techniques

5 Model-Based Systems Engineering

A good engineer thinks in reverse and asks himself about the stylistic consequences of the components and systems he proposes.

—**Helmut Jahn,**
Architect of the Sony Center (Helmut 2013)

5.1 OVERVIEW OF MBSE

Systems engineering (SE) was started to identify and analyze system requirements and determine if these requirements had been followed and properly implemented throughout the systems development lifecycle (SDLC). Hence, system requirements can be considered one of the most critical elements of the SE processes. At the core of the SE methodology, the requirements obtained from the system requirements generation process are used to drive the system's design and development processes. Among other things, to ensure that the system requirements are clearly communicated throughout the lifecycle, model-based systems engineering (MBSE) has been introduced to support the system requirements process (INCOSE 2007). Large, complex, systems development efforts are often conducted by a team of contractors who may (or may not) be managed by a prime contractor. For large teams, unambiguous requirements are essential for the successful integration of the constituent subsystems. In SE, a well-defined system is one that has its requirements clearly established and effectively communicated to the development team. In other words, the different parties responsible for designing and developing a system must clearly understand the requirements and ensure that the requirements are complete, measurable, and testable. A system that has good requirements will not only have fewer problems and surprises but will also save costs in the long term. MBSE tools are a tremendous aid in capturing, documenting, and effectively communicating requirements to stakeholders and the development team. They are also instrumental as analytical tools and are useful in capturing and communicating design and test artifacts.

Presently, a commercial or government entity that takes MBSE into consideration will usually use models to architect their enterprise and developmental systems. According to the Institute of Electrical and Electronics Engineers (IEEE 1990), "A model is an approximation, representation, or idealization of selected aspects of the structure, behavior, operation, or other characteristics of a real-world process, concept or system." Models are also applicable, developed, and used at multiple levels in a given systems development effort. Typically, there are three categories of models used in developing a system: operational, system, and component models.

According to the IEEE Standard 1471, which describes the architecture of a software-intensive system, "A model usually offers different views to serve different purposes. A view is a representation of a system from the perspective of related concerns or issues" (IEEE 2000). According to the International Council on Systems Engineering (INCOSE) article, "MBSE is the formalized application of modeling to support system requirements, [and] design[.] [T]hroughout history, the development and application of tools have enabled our ambitions and ideals to become a reality. Engineering has been responsible for some of humanity's greatest achievements, from landing a man on the Moon to developing semiconductors as small as a few nanometers. These complicated systems necessitate a considerable level of organization, planning, and documentation to be developed. Model-based systems engineering, or MBSE, is one strategy that can help us in our endeavor

DOI: 10.1201/9781003624097-7

to construct these complex systems, analysis, verification, and validation activities beginning in the conceptual design phase and continuing throughout development and later lifecycle phases (INCOSE 2007)."

In the next section, we provide an introduction and description of the MBSE process. In addition, an explanation of several MBSE methodologies, tools, and paradigms is presented. Lastly, we describe the importance of MBSE to the modern-day SE discipline.

5.1.1 Introduction to MBSE

Since the dawn of history, tools have allowed humanity to achieve our dreams and go further than we ever thought we could. Some of the greatest achievements of mankind have been realized through the application of tools in engineering and science, such as sending people into space, landing them on the Moon, and returning safely back to Earth. Humanity has made strides in understanding secret mysteries like the space-time continuum and discovering modern engineering marvels such as quantum computing, quantum entanglement, and the exploding accomplishments and promise of artificial intelligence (AI). We can make semiconductors that have structure on the order of nanometers, and technology is moving forward in leaps and bounds. As component sizes shrink, the systems made from these components become more complex with many parts, connections between these parts, and interconnections to and from other systems. To develop these complex systems, much organization, planning, and documentation is required. One way to help manage and understand the development of these complex systems is by using MBSE methods or MBSE. So, what is MBSE? According to INCOSE,

> Model-Based System Engineering (MBSE) is the formalized application of modeling to support system requirements, design, analysis, verification, and validation activities beginning in the conceptual design phase and continuing throughout development and later lifecycle phases
>
> *INCOSE (2007)*

Using traditional methods, system engineers employ a conventional document-based approach to manually generate artifacts that are essential in the system design process. Capabilities, stakeholder needs and requirements, specifications for components and systems, concepts of operations (ConOps) and Operational Concepts (OpsCons), System/Segment Specifications (SSS), Requirement Traceability and Verification Matrices (RTVM), Block Diagrams, and Interface Control Documents (ICDs), test case specifications, and specialty engineering analysis are among these essential systems development artifacts. The document-based artifacts lack any form of interconnection. An update to one document does not automatically manifest in the other document; manual updating and iterative review are required to prevent the artifacts from becoming riddled with errors, obsolete, or at best, inconsistent. The document-based approach is costly, labor-intensive, and prone to human error; it also requires substantial effort to preserve all the important system documents throughout the SDLC.

By employing model-based system engineering (MBSE), the shortcomings of the traditional document-based approach can be more effectively mitigated. With the MBSE approach, system engineers do comparable lifecycle tasks and provide analogous deliverables; yet the principal product of MBSE operations is a completely integrated model. The model is executed methodically and consistently, and this model provides cohesive and complementary views of the system in its environment. The model furthermore produces an additional artifact, the model itself, that is often mandated by stakeholders as a contractual deliverable (Walden 2015).

The use of models is growing in tandem with technological advancements, and many benefits result from effectively using models and modeling approaches, in particular MBSE, throughout the SDLC. For example, insights (views) into the system, or systems in general, relationships and interconnections to other systems, or components within the system, and connections to the environment in which the system operates become visible and understandable. Models help in establishing performance boundaries, become the lynch pins in feasibility and trade studies, and serve as the integrating elements between the stakeholder, design, development, operational, and support communities. The systems engineering community has made use of an MBSE methodology for many years, and this methodology is always being improved upon to meet the demands of engineers and the systems development community. A vast array of sectors and fields can benefit from MBSE's applications.

From a practical viewpoint, MBSE acts as an integrative vehicle that starts from the conceptual design phase of a project and carries the system's requirements, evolving design attributes, analytical capabilities/results, and verification and validation artifacts throughout all the SE lifecycle phases. In the system's lifecycle, all phases are unique, and the output of each phase acts as an input to the next phase, and together, all the phases act together to sequentially produce the system. The idea behind MBSE is to model key functional, behavioral, requirements, and parametric aspects of the system and to provide visualization tools, requirements management tools, and analytical tools for essential systems attributes throughout the system's lifecycle. When MBSE methods are applied and executed meticulously and diligently throughout the lifecycle of a systems development effort, the likelihood of program success is increased dramatically.

MBSE facilitates clear communication between stakeholders and the development team. MBSE provides organized and rational modeling tools to facilitate the supervision, analysis, evaluation, and observation of general systems, whether they are complex or not. As a result, MBSE ensures consistency, correctness, and completeness of data and facilitates information flow for decision-making throughout the system. Lastly, MBSE not only captures the requirements and aids in the performance analysis and verification and validation of a program (note we use program and project synonymously), but it also leverages and reuses these steps for other systems. Today, MBSE philosophies and methodologies are central to many top-performing SE organizations. MBSE has taken SE processes to a higher-level ensuring correctness, completeness, and aiding in integrating activities throughout an organization. MBSE, as the name implies, means engineering a system using MBSE methods and tools.

From a conceptual standpoint, the system requirements of a project/program/system service (PPSS) can be visualized as a tree structure. The prime node, or node from which all branches originate, is the "A" specification (A spec) or the system specification. From the prime node, any number of branches can stem depending on the system being modeled. Each subsequent node, or child node, of the prime node would represent another specification. For example, given a tree diagram used to model the system requirements of a weapons system, the "A" spec would outline the system requirements and present a clear operational view of the system in its environment. Two branches coming from the prime node could be two "B" specs or development specifications. One of the "B" specs would contain the hardware specifications of the system, while the other "B" spec would contain the software specifications. There could, of course, be more than two "B" specs stemming from the "A" spec. For instance, there could be multiple hardware subsystems and multiple software configuration items, each with its own "B" spec. It is important to note that regardless of the complexity of a system, or its modeled tree diagram, there is only ever one "A" spec, while there may be many other specifications at any given level subordinate to the "A" spec.

Traditionally, at the developer's facilities, system requirements were written first, and then models were developed based on the requirement documents. This traditional document-oriented process of developing the model using only requirements has created numerous problems in terms of originality, completeness, and efficiency. According to INCOSE, "MBSE is expected to replace the document-centric approach that has been practiced by systems engineers in the past and to influence the future practice of systems engineering by being fully integrated into the definition of systems engineering processes" (ECSS 2012). Therefore, we can model the system requirements from the beginning by applying new MBSE methodologies and techniques. Implementing these new MBSE techniques helps decrease requirement errors and subsequent analytical, design, or testing complications. MBSE is expected to provide good communications, reduce risk, and provide a better understanding of the elements of the system. For instance, in the *14th IEEE International Requirements Engineering Conference*, an article about the MBSE application to quantitative requirements analysis presented the idea of combining emergent and promising concepts and practices in the modern-day systems development environment (Cornford et al. 2006). We present a few of these emergent ideas to illustrate the impact that MBSE methods have on the development of modern systems.

Requirements engineering: results in clearly defined and characterized requirements (both functional and nonfunctional). These resulting requirements are framed in a systems architecture that produces constraints that help narrow design options and highlight the early assessment, investigation, and confirmation of developmental planning activities. MBSE methods and tools facilitate the requirements engineering processes by providing an integrative, actionable, instantiation of the requirements through enterprise and systems modeling, requirements management, and business process modeling tools. MBSE methods and tools can also directly link to design and test articles. For instance, rapid prototyping systems are made possible by MBSE methods and tools wherein changes in requirements directly and immediately affect a design or test element. For example, by making a change (or potential change) in top-level requirements, the effect is immediately observable in the design element, hardware-in-the-loop (HiTL), and/or software-in-the-loop (SiTL) testing element.

MBSE application to complex, vital, multi-entity systems: In a 2010 IEEE Systems Conference, one of the papers from A. Soyler on MBSE applications was to capture a disaster management system's (DMS) conduct and structure (Soyler 2010). The idea was to begin with the design framework required for identifying and constraining the DMS. The authors state that it is important to recognize that, "the management of disasters as complex adaptive systems (CAS), whose performance cannot be improved through the isolated optimization of their constituents" requires modern methods and practices to achieve success. MBSE methods could then be used to capture the system's behavior along with the causal structure for the system's activities. The Systems Modeling Language (SysML), which is a domain-independent language, can be used to put the MBSE into operation. The holistic application of MBSE methods is expected to bring a new standard into DMS and is intended to decrease their overall complexity. Application of a holistic management approach begins with initially addressing the DMS behavioral aspects and subsequently the use of the DMS. Afterward, the MBSE focuses on physical structural aspects that could lead to the observed behavior. The model can be applied to either the desired behavior (ideal DMS) or the actual behavior (existing DMS). Once instantiated, assessments and benchmarking studies can be conducted on the DMS. The authors conclude, "By enforcing traceability between the DMS subsystems, and by leveraging artifacts such as viewpoints to develop perspectives that are tailored to specific stakeholders and users, it is anticipated that an MBSE-based holistic approach will promote proper use of the scientific method for the design, verification, validation and improvement of the disaster management systems" (Soyler 2010).

At a *2011 IEEE Aerospace Conference*, one paper described the application of MBSE for future complex systems (Bajaj et al. 2011). According to this paper, building a complex system requires the combined effort of the collaborating team and disciplines including various modeling processes, software tools, and methodologies. The nature of MBSE is to facilitate a steady, logical, interoperable, and developed model during the lifecycle of the system. On the other hand, currently, there exists no comprehensive modeling language that shows all system characteristics (especially in the case of systems of systems [SoS]) throughout the SDLC (Bajaj et al. 2011). In the world of SE, it is likely that generally large and complex systems, with their subsystems of different scale, must not only coexist but need to work seamlessly together. MBSE is created to maximize the compatibility between subsystems. According to INCOSE, "Applying MBSE is expected to provide significant benefits over the document-centric approach by enhancing productivity and quality, reducing risk, and providing improved communications among the system development team" (ECSS 2012). Therefore, applying the MBSE method is becoming an essential aspect of modern-day SE practices.

According to an IEEE paper, the teams who are working together must "speak" a similar language and must work on a similar "matter" to build the more complex and interdisciplinary SoS. The system model is the "matter," so general, inclusive, adaptable, and friendly modeling languages are required for effective communications. The MBSE paradigm is advancing in the SE discipline, so it is likely to become a regular practice soon. "As an emerging paradigm for the systems of the 21st century, it seems useful to review the current state of the art concerning the developing standards, the embryonic formalisms, the available modeling languages, the methodologies, and the major applications" (Ramos 2012).

Conventionally, documents have been characterized as the principal medium to write down the requirements throughout the system's lifecycle, but after applying the new MBSE methodologies, we saw a dramatic reduction in the ambiguity of the requirements as the requirements became more clear, accurate, and complete. This notion is illustrated in an IEEE article that appeared in an Avionics Systems Conference in 1998. According to this paper, documents like system specifications and trade-off studies are expected to be the most important products for system engineers. Explaining the requirements to software and hardware designers by using these documents seems like adding unneeded complexity. By applying MBSE, no ambiguity exists in requirements specification documents. Therefore, the model built after applying MBSE methods completely describes the functional conduct, input/output (I/O), and necessary aspects of the system's physical architecture. In addition, the performance and resource requirements also give a combined, reliable, and traceable basis for the design. An example was provided for the development of an Automotive Personal Assistance System (APAS), using the Global Positioning System (GPS) capabilities, identical to General Motors OnStar System. The APAS model contains important artifacts from the SE development lifecycle, from originating requirements through system validation. Also, there exists no need to incur the high cost of prototype development because their model correctness can be verified by fully executing their behavioral model. Therefore, applying an MBSE methodology will support increasing the product quality while decreasing the overall development cost (Fisher 1998).

As a result, SE projects that use MBSE methodologies and paradigms, along with sound SE principles in their project, will likely enhance communications among team members, decrease errors and development time, and positively impact their chances of success.

In implementing MBSE methods, several factors such as the project scope, the functional decomposition of the system, and the allocation of requirements are important considerations in the SDLC. The new model-based approaches bring accurate and clear methods to define the structure, behavior, requirements, and parametrics of the system. Applications using the MBSE approach have been proven to increase communications between the development team, primary stakeholders, and other interested parties. MBSE methods facilitate integration, fluid workflow, organization, control, capitalization, and usability. Therefore, it becomes important to master MBSE concepts and apply them throughout the lifecycle of a project. In the next section, we provide some useful MBSE-related terminology and present some useful MBSE tools. Specifically, we focus on representative MBSE tools that are useful for optical SE projects. The following are examples:

- CATIA. *System-of-Systems Modeler*™ is an established Unified Modeling Language (UML)/SySML modeling tool with, enterprise modeling, systems modeling, and other useful plug-ins. There is also a requirements management and modeling tool called *CAMEO Systems Modeler.*
- *ENVI*™ (Environment for Visualizing Images) is a well-developed software tool used in image-processing applications and as an analytical tool for geospatial imagery.
- MATLAB®, *Simulink*®, and *Mathematica*™ are analytical, scientific, engineering, technical modeling, and rapid prototyping tools.

Note that these modeling tools are meant to provide a representative example and to provide a point of reference. There are many other excellent tools available on the commercial market.

These tools serve as illustrative examples here, and we present further introduction to modern MBSE and optical systems tools in their appropriate application-relative sections below.

5.1.2 Historical Aspects of MBSE

So, where does MBSE come from? The history of MBSE, as stated in the last two letters of its name, starts with system engineering. System engineering was fundamentally impacted by the Bell Telephone Laboratories in the 1940s. Bell Labs needed a methodical approach to systems development for their phone networks' business lines of development. They found that, even though the products that were under development at Bell Labs where good products, when inserted into a larger operational environment, their products did not satisfy all the needs and stakeholder requirements. To address these shortcomings, the early stages of the systems engineering discipline emerged, involving the integration of many disciplines to create well-functioning products and systems (Gao et al. 2019; INCOSE 2022).

In 1946, the United States Air Force started the Research and Development (RAND) project, which eventually became the RAND Corporation. The mission of RAND originally focused on research and development ideas but has evolved to a corporate non-profit think-tank covering a broad range of topics and policies. Systems engineering was recognized and emphasized in the 1940s by the Department of Defense in the development of weapon systems like missile systems (Goode and Machol 1957). In the 1940s, Paul Fitts made the connection between system *functions* and physical system components, which became a fundamental system engineering concept. Dr. Wayne Wymore, a pure mathematician, started contributing to the field of systems engineering and its evolution toward the development of MBSE. In 1967, he published his first book, *A Mathematical Theory of Systems Engineering,* while a professor at the University of Arizona (Wymore 1967). Among other books, Dr. Wymore wrote *Model-Based Systems Engineering* in 1993, and this was one of the early times that the term "Model-Based Systems Engineering" started to emerge and become recognized. Since the 1960s, there have been three eras of MBSE evolution.

There have been three distinct periods in the evolution of MBSE: the mainframe era, the mid-range era, and the contemporary age (Haskins 2011). In the 1960s, computers first gained widespread use, marking the beginning of the mainframe age. Coding was at the vanguard of every technological advancement as computers began to dominate the globe. Haskins is correct in his assertion that databases were essential to the creation of software. As a result of these databases, people started focusing less on how the computer worked and more on what code to use. During this period, new techniques emerged that enabled programmers to converse with one another regardless of the system they were developing. For MBSE, this was a watershed moment since effective communication is fundamental (Haskins 2011).

After that, the intermediate age arrived in the 1970s, a time of tremendous advancement in computer power. As computing power increased, some in the software industry saw the need for more effective tools, and so the Computer-Assisted Software Engineering (CASE) suite of programs was born. The original intent of this program was to provide a break from repetitive coding tasks and to implement a system for collecting requirements and associating them with other documentation. According to Haskins, coding was the preferred way to develop software at the time, but CASE served as the foundation for what would eventually become MBSE (Haskins 2011). Object-Oriented Programming (OOP) began to take shape around this time. An early model used to depict a system around this time was the E-R diagram. Using object-oriented methods, these diagrams would show how the system's components interacted with one another. This modeling notation brought together mathematical and modeling ideas (Dickerson et al. 2013). The languages used in MBSE would eventually make their way to these models, which were early useful depictions for modeling systems. Finally, peer reviews, which are essential to writing high-quality code, began at this time. All of this paved the way for MBSE because of the prevailing trends in greater communication among programmers, which would subsequently permeate other fields (Haskins 2011).

The present era started around 1980. According to Haskins, the current period saw the emergence of the UML, a language that was fundamental to systems and software engineering. Documents, protocols, and procedures were the foundation of systems engineering in its early days. With the advent of object-oriented methods (OOM), UML would provide a gateway for the emphasis on a model-based systems and software engineering foundation. Although it didn't really take off until the 1990s, the OOM had its roots in the 1980s. The two approaches that had the greatest impact on the development of UML were Object-Oriented Software Engineering (OOSE) by Booch and Jacobson, and Object Modeling Techniques (OMT) by Rumbaugh (Erickson and Siau 2013). It was from these sources that UML was born, and the Object Management Group quickly followed suit, making UML their de facto modeling language standard. The software development industry relied heavily on this language to facilitate communication among many, diverse teams (Erickson and Siau 2013). In connection with the previous CASE tools, a new range of capabilities manifested that helped with model organization and clearer, more understandable syntax (Haskins 2011). However, the Systems Modeling language (SysML) was also created as an adaptation of UML, with SE in mind, therefore UML wasn't the sole language to emerge.

The original intent of the SysML modeling language was to facilitate systems analysis, design, and verification. Directly relevant to MBSE, the SysML language expanded upon UML. There is an inherent hierarchy to the diagrams and models created with this language (Dickerson et al. 2013). SysML came about from a request for proposal change to UML which provided the impetus for the creation of SysML. The purpose of the change was to make the language more suitable for systems engineering. Development of the idea was authorized in 2003 and over the course of the following several years, the language would undergo numerous revisions and reviews until being completed and released in 2007 (Erickson and Siau 2013). In the same year, in January 2007, an MBSE initiative was announced by INCOSE, which provided traction for the MBSE approach and saw it rise in popularity. UML and SysML and other languages were created during this time, with contributions made toward MBSE by Tarski, Yourdon, Wymore, Klir, Lin, and Suh (Dickerson et al. 2013). Even so, there was a lack of standards to follow, despite the numerous languages and approaches that had emerged to help push the traditional document-centric systems engineering toward a more model-centered one.

To establish best practices for systems engineering, two distinct groups became prominent. As previously mentioned, the OMG group standardized on the UML language. In 1989, the consortium that made up the OMG group became fully established. The goal of the OMG is to facilitate the standard-setting process for businesses, particularly those wishing to embrace MBSE standards. According to the OMG, standards are required since there are so many different MBSE approaches, and the technology is advancing rapidly. But OMG wasn't the only organization at the time that sought to promote the interests of systems engineering and advance MBSE. Following the OMG's ascent, the International Council on Systems Engineering (INCOSE) emerged, and it has subsequently contributed significantly to the field of systems engineering. Examples include the definition of MBSE, standardization of tools, and the definition of processes (Honour 1998).

Back in 1989, Dr. Brian Mar initiated events that would eventually result in the establishment of the organization called INCOSE, and with Dr. Mar becoming known as the Father *of INCOSE*. The demand for engineers with a holistic view of systems, rather than narrow specializations, led to the founding of the organization (Honour 1998). In response to the growing demand for systems engineers and related standards, a small group of concerned citizens formed what was then known as the National Council on Systems Engineering (NCOSE). Over the course of the following year, the group responsible for systems engineering would host a plethora of events, including conferences, seminars, and workshops (Honour 1998). Numerous internal and external changes occurred within the organization. They went from establishing their own procedures to quickly expanding their membership, which included people from across the country.

In its early stages, NCOSE primarily served as a go-between for the US Department of Defense's MIL-STD-499 and other systems engineering-based publications. Though initially focused on government efforts, NCOSE was able to branch out into other areas over time and expanded to other efforts outside the government. The company's rapid expansion from 1992 to 1995 required a shift in emphasis to its internal processes (Honour 1998). The company's expansion and internal process changes caused it to look beyond its immediate geographic area for opportunities. The focus landed on the international arena and in 1995, the organization finally went global. The group has experienced tremendous growth since its inception, culminating in its transfer to an international organization. INCOSE has played an essential role in the growth and improvement of systems engineering since its inception in 1995. Beginning in 2007, there has been a continuous effort to integrate MBSE into many types of organizations (Dickerson et al. 2013).

A paradigm shift, from document-based to model-based systems, has occurred in systems engineering throughout the past two decades and this trend is continuing. Systems engineering has benefited from the development of new modeling languages, from the maturing of new methods like MBSE, and from organizations like OMG and INCOSE that promote the systems engineering discipline and assist others in systems optimization.

MBSE provides a level of visual organization and efficiency that can be applied throughout the system development effort, allowing engineers to gain valuable insight into the system, and use a consistent development methodology throughout the system development lifecycle. MBSE can also automate many tasks. Better documentation, more effective communication across teams, and lower costs are all possible outcomes of implementing an effective MBSE methodology throughout the systems development effort. In addition, MBSE aids in configuration management, provides for capturing and modeling requirements, and captures the system architecture and other important system elements that are critical to the system design and development process. Over time, new issues and challenges arise as the system evolves, and MBSE helps identify, capture, and mitigate these issues in a controlled, robust, and reliable way.

5.1.3 Theoretical Concepts of MBSE

In this section, we look at some of the important theoretical concepts relating to MBSE. We compare MBSE to the traditional document-centric systems engineering approach and highlight differences. We also present a short introduction to natural language processing (NLP) models and discuss their utility. As seen in Figure 5.1, MBSE integrates models from different domains.

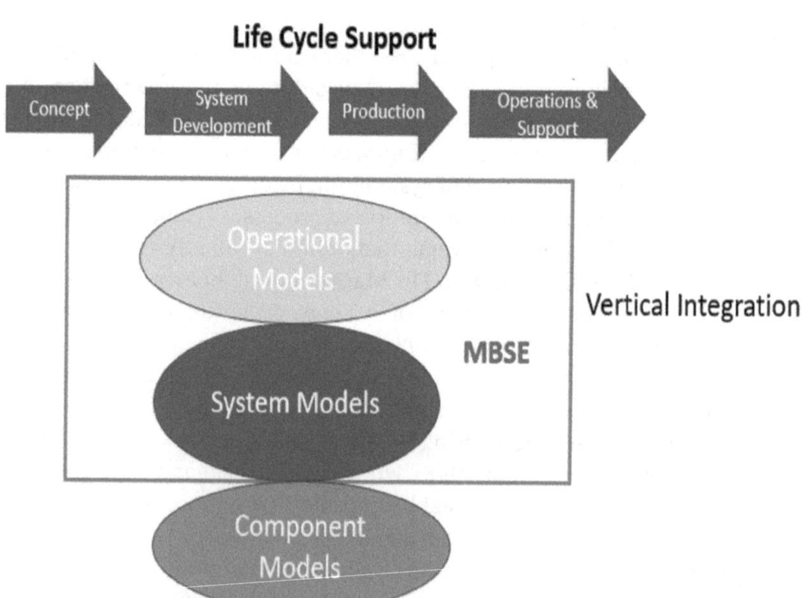

FIGURE 5.1 MBSE vertical integration of models.

TABLE 5.1
Document Based vs MBSE Method for Important SE Attributes

	Document-Based	MBSE-Based
Data	The data is primarily in text format and is not tightly linked across multiple documents, making it difficult to synchronize and navigate	The data is presented in both visual and text formats, which facilitates its reusability, sharing within and between domains, enhances consistency, and defines relations that may be used for data queries
View	Only by document. Tedious to generate and maintain	Filtering data to prevent clutter is possible, and there are standard and customized viewpoints. Different perspectives provided
Change impact	Making changes can be a tedious and error-prone process that may require updating many documents	Data in pertinent diagrams, perspectives, and related elements is immediately updated when changes are made in one location in the model due to the interconnected nature of its elements. This process can be automated with scripts
Integrity and quality	Manual inspection is laborious, error-prone, and time-consuming	Validation and scripting suites used to ensure data and model quality and integrity

The MBSE environment can integrate higher level abstract and conceptual models, such as those that are accomplished in operational and system-level modeling as shown, with lower-level models such as components-based models. The components-based models can capture important relations such as physics-based applications and constraints. From the top of Figure 5.1, the MBSE modeling environment spans the SDLC and may also cover aspects outside of the conceptual, design, and development phases, and include production, operations, support, and even disposal/retirement phases as necessary. MBSE has many advantages over traditional document-based methods. In Table 5.1 we see a comparison of the document-centric approach versus the MBSE approach for several important systems engineering attributes shown in the left column of the table (Data, View, Change Impact, Integrity, and Quality). Some of the advantages and disadvantages are pointed out under the header for each type of approach (Document-based, or MBSE). Some of the notable items in Table 5.1 are that data is relatively fixed in the document approach, whereas in the MBSE approach, connections can evolve through the modeling process. The data can also be hard-linked so that consistency between what is seen in the current baseline is ensured throughout the different views permitted by the model. In other words, the model contains configuration-controlled data, and guarantees can be made that everyone using the data is working with the appropriately approved set of data. Note also that security protocols can be easily implemented in the MBSE environment providing access to those with need to know, and restricting others as required. Another huge advantage is the ability to do scripting and generate validation suites to check the integrity of the model and the quality of the data.

MBSE provides *integrated* capabilities that, individually, are essential in successful systems development efforts, but through integration provide much more value added. That is, in MBSE, the sum is greater than its individual parts. Some of these integrated capabilities that the MBSE methodology and its digital environment provide are,

1. Digital repository for the system's requirements;
2. Robust configuration control;
3. Holistic systems, environmental, and operational views of structure, behavior, requirements, and system parameters for stakeholders including designers, developers, operators, servicers, and supporters;

4. Linking requirements to design and test artifacts;
5. Establishing and containing requirements and design baselines;
6. Linking models together;
7. Repository for system performance and benchmark data;
8. Capturing and evolving the system architecture (and other architectures);
9. Implementing the digital twin and digital thread;
10. Rapid prototyping, hardware-in-the-loop, software-in-the-loop;
11. Integrated performance, trade, and feasibility studies;
12. Natural Language Processing;
13. Parametric modeling;
14. Information assurance, cyber-security, and access control.

Each of these capabilities is essential when it comes to good systems development practices, but integrated, they provide so much more. For instance, requirements that have technical performance measures and metrics can use parametric modeling to ensure constraints are met. Say, that requirements dictate that the system weight cannot exceed 750 kg. If a remote design team changes the weight of one of the components in their design, a parametric model that is part of the MBSE environment automatically checks the new weight against the weight requirement constraint and flags the result if it is exceeded. The design team can then work on the trades to resolve the conflict.

What about the capability of NLP? What is NLP and why is it important? NLP is the use of a machine to process natural language. NLP is based on a variety of language models. To make text and voice more quantitative and machine-readable, researchers have developed natural language models for various purposes in industry. For example, some of the NLP models are used for industry for various apps, including Google, Excel, Microsoft Word, and many more. Consequently, language models are fundamental to NLP applications (Chaudhuri 2023). With the advancement of technology and the widespread use of electronic gadgets in daily life, there is an increasing demand for machines to interpret human natural language. Machines are replacing human tasks in many industries, making human-machine interaction a normal routine event. In one form, NLP uses a statistical model based on probability that predicts how many times a certain word will appear in a sentence based on past data (Chaudhuri 2023). It predicts which word in the sentence is more likely to appear. It is used in predictive text input systems by companies like Google, for machine translation, speech recognition like the iPhone, and spelling correction like Word and other applications. A set of phrases is fed into the model, and the model outputs the likelihood of the word sequence. It may anticipate the next word based on the last one word (unigram), last two words (bigram), last three words (trigram), or last n words (n-gram).

In another form, NLP can be used to introduce rigor and precision into the requirements generation process and the interpretation of requirements. Requirements statements can be methodically dissected and meaning defined. In the simplest sense, this can be illustrated with a program/system dictionary which defines the terms that are relevant to the program/system under development. For instance, the term *system engineering* should be well and consistently defined and understood by the system engineering community. However, looking up the definition of system engineering in different system engineering texts, and even different system engineering standards, there are marked and notable differences in definition. NLP can be used to introduce rigorous and precise meaning into words that are important to the system development effort, such as the words that are used in requirements statements. In addition to defining the precise meanings of the words within the requirements, the sentence structure is rigorously controlled and implemented. The intent is to implement requirements in digital form wherein the interpretation of the requirement statement is unambiguous and precise.

A good analogy is the precision and rigor that is evident in mathematics. The meaning behind mathematical statements is precise and rigorous enough to conduct logical reasoning and prove or disprove a hypothesis statement. To illustrate the ambiguity inherent in the English language, consider the words, "Mary had a little lamb." What do I mean? Am I quoting a nursery rhyme? Did Mary have a tasty lunch? Did Mary just eat a small portion of her lunch, or was the lamb itself small? Did Mary have a pet lamb and no longer has it? Did Mary take advantage of someone in the sense of being a con artist? These four words have so many different meanings and all of them could be "right" under the proper circumstances. The idea is to use NLP to eliminate the ambiguity in words, especially words used in requirements statements, by digitally implementing precise and rigorous meaning through definition and structure. The rigor and precision are applied to both the meaning of the words and how the words are combined into sentences, phrases, and paragraphs (structure). Combining the predictive capabilities of NLP above with the rigorous and precise aspects of NLP just described and mixing in AI-based machine learning leads to a whole new frontier of possibilities and challenges for the systems engineer, and the engineering community at large.

NLP evaluation and benchmarks: To assess the performance of models, benchmarks are necessary. In proper benchmarking, model results are compared to the best-of-the-best results. For NLP models, humans create the sample benchmarks typically from language-oriented tasks. Many such sample results have been created to evaluate language processing systems. For example, at Stanford, the Stanford Question Answering Dataset (SquAD) was developed (SquAD 2019). The Microsoft Research Paraphrase Corpus developed a data set relevant to extracting news sources on the internet (Microsoft 2019), and other developed data sets for language processing systems (NYU-MLL 2019; Glue 2019).

5.1.4 MBSE Principles, Methods, and Tools

MBSE borrows from SE and applies its own unique spin to the SE principles, methods, and techniques used to build models. Stakeholders, system developers, operators, and supporters should all be able to easily comprehend and navigate these models due to their clear visual orientation and organization. In addition to the structural elements of the system, these models must be capable of capturing the system's requirements, behaviors, and the interplay between its parts – the system flow. When it comes to systems modeling, a primary consideration is establishing the model's purpose, a consideration central to all models. The goal of building models in MBSE is to better explain the system's function and its constituent parts and interactions. The model should be straightforward and easy to grasp for everybody who uses it. It should also be easy to communicate model details and results. The interpretation of information from the model should be clear and help illustrate the points for which the model was intended. If this is not the case, ambiguity in model results may cause someone to do, or interpret results incorrectly leading to potentially severe problems. That is why it is critical to ensure that each model serves a purpose and is easy to understand (Hallqvist and Larsson 2016). In modeling, and developing a MBSE methodology within an organization, following a set of proven guiding principles can greatly benefit the overall effectiveness and success of the resulting organizational MBSE environment.

5.1.4.1 Principles Used for Model-Based Architecture, Systems Engineering, and MBSE

We now present a series of six principles that should be considered and applied when implementing a MBSE program in an organization. These principles have direct application in establishing a model-based architecture (MBA), and so adherence to these principles is fundamental in establishing a solid MBSE methodology. We provide these principles verbatim from the textbook *Architecture and Principles of Systems Engineering* by Charles E Dickerson and Dimitri N, Mavris so that we do not dilute, or alter the message in any way. We then apply the principles using SysML artifacts using examples from an autonomous unmanned aerial vehicle (AUAV) to illustrate the main points.

5.1.4.2 Reflection of Structure in Systems Design

The principle of Reflection *of Structure in Systems Design* is the observation that "the solution should reflect the inherent structure of the problem" (Dickerson et al. 2016, p. 54). This idea is based on structured design and requires models to properly understand. Once the system architecture has been validated and verified, the "design is inherently defined with a greater degree of detail and can be considered a baseline, which is the core output of the current design cycle and serves as the input of the next cycle" (Dickerson et al. 2016, pp. 303 and 304). Notice how aptly this principle fits into the framework of an Agile design methodology wherein the culmination of a *sprint* connects to the completion of a design cycle.

As an example, Figure 5.2 displays an AUAV patrol map structure tree for an optical subsystem. The structure tree displays the capabilities of the system and how the system architect needs those capabilities. The structure tree is to be read from left to right. The figure shows the AUAV Patrol is composed of capabilities (Detect Human, Follow, ID Human, Rescue, Search, Send Messages, and Transmit Images) which are displayed (AUAV, Border Patrol, Homeland Security, First Responders, Image Specialist, Optics Package 1, and Optics Package 2) and then implemented and is composed of resources. This figure displays the inherent structure of the optical subsystem in the proposed solution process.

5.1.4.3 Modular Structured Design

Modular Structured Design means that "systems should be comprised of modules, each of which is highly cohesive but collectively are loosely coupled" (Dickerson et al. 2016, p. 52). Cohesion is used between elements in different modules to relate certain aspects of the connectedness of a system. There are various levels of cohesion: Coincidental, Logical (sets rank in order), Procedural, Temporal, Sequential (sets sequential order), Communication, and Functional (Dickerson et al. 2016, p. 52).

Modular Structured Design can be seen through different SysML models to highlight certain aspects of the AUAV system. Figure 5.3 is the bdd (block definition diagram) for the analysis of the spatial resolution for the optical subsystem. The bdd is then used to develop the constraints and showcase the spatial resolution equations.

The parametric model for the detection capability of the AUAV optical subsystem, Figure 5.4, is built off the bdd (Figure 5.3). The parametric model (par) is the simulation of the bdd. The system designer can alter the parameters with the parametric model, such as various input parameters, try different entrance pupl plane diameter values, and so forth, to see what will pass and what will fail the required resolution given by ID_SR (for identification spatial resolution) and DT_SR (for detecting spatial resolution). Figure 5.5 represents the results of a simulation, and the highlighted green indicates the configuration passed. Figures 5.3–5.5 represent how the AUAV optical subsystem consists of modules, which are cohesive but still adjustable to account for design changes.

5.1.4.4 Conceptual Integrity and the Role of the Architect

Conceptual integrity and the role of the architect represent the idea that "conceptual integrity is the most important consideration in system design and the architect should be responsible for the conceptual integrity of all aspects of the product perceivable by the user" (Dickerson et al. 2016, p. 7). The use of architectural frameworks, set standards, and modeling languages and modeling are key components of conceptual integrity. The architectural framework helps develop the structure for the system architecture which leads to conceptual integrity down the line. Modeling languages and models help bring conceptual integrity by providing comprehensibility and precision to the systems development effort. An OV-1 (The High-Level Operational Concept Graphic) can be used to showcase details of the system that are of special interest and data for the integrated architecture, which results in preserving the conceptual integrity of the system (Dickerson et al. 2016, p. 200).

Figure 5.6 shows the OV-1 (Free Form Operational Concept Graphic) for the AUAV system. It displays that the AUAVs are fitted with daytime and nighttime optical subsystems that will transmit images and location from the field to the Image Specialist via a SATCOM link. Image Specialists will analyze images and determine if an Alert Message, Warning Message, or Distress Message should be sent to Homeland security and Border Patrol. Alert will simply alert Border Patrol of human life detected and proved location. Warning will provide image, location, and ID of person of

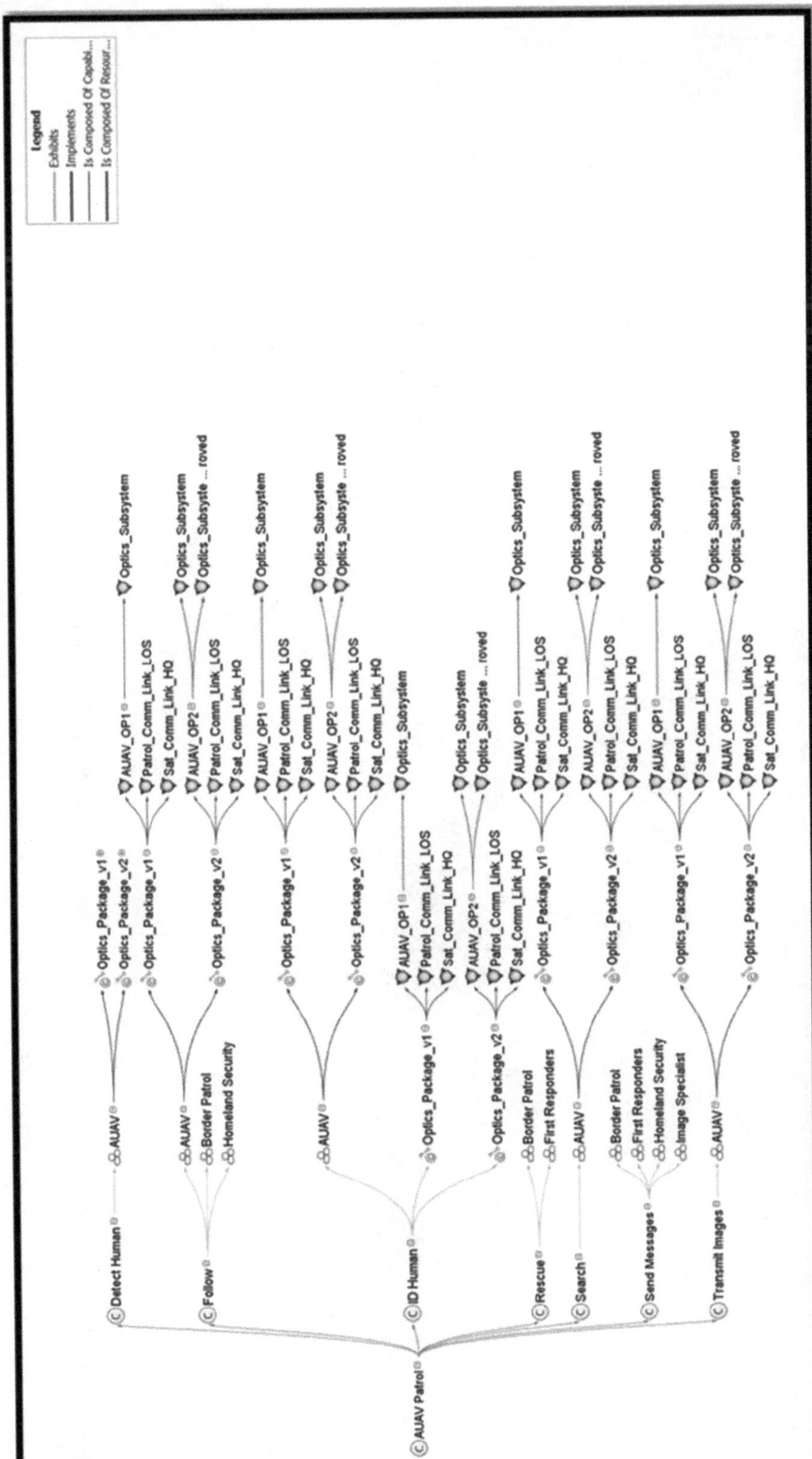

FIGURE 5.2 AUAV Patrol map structure.

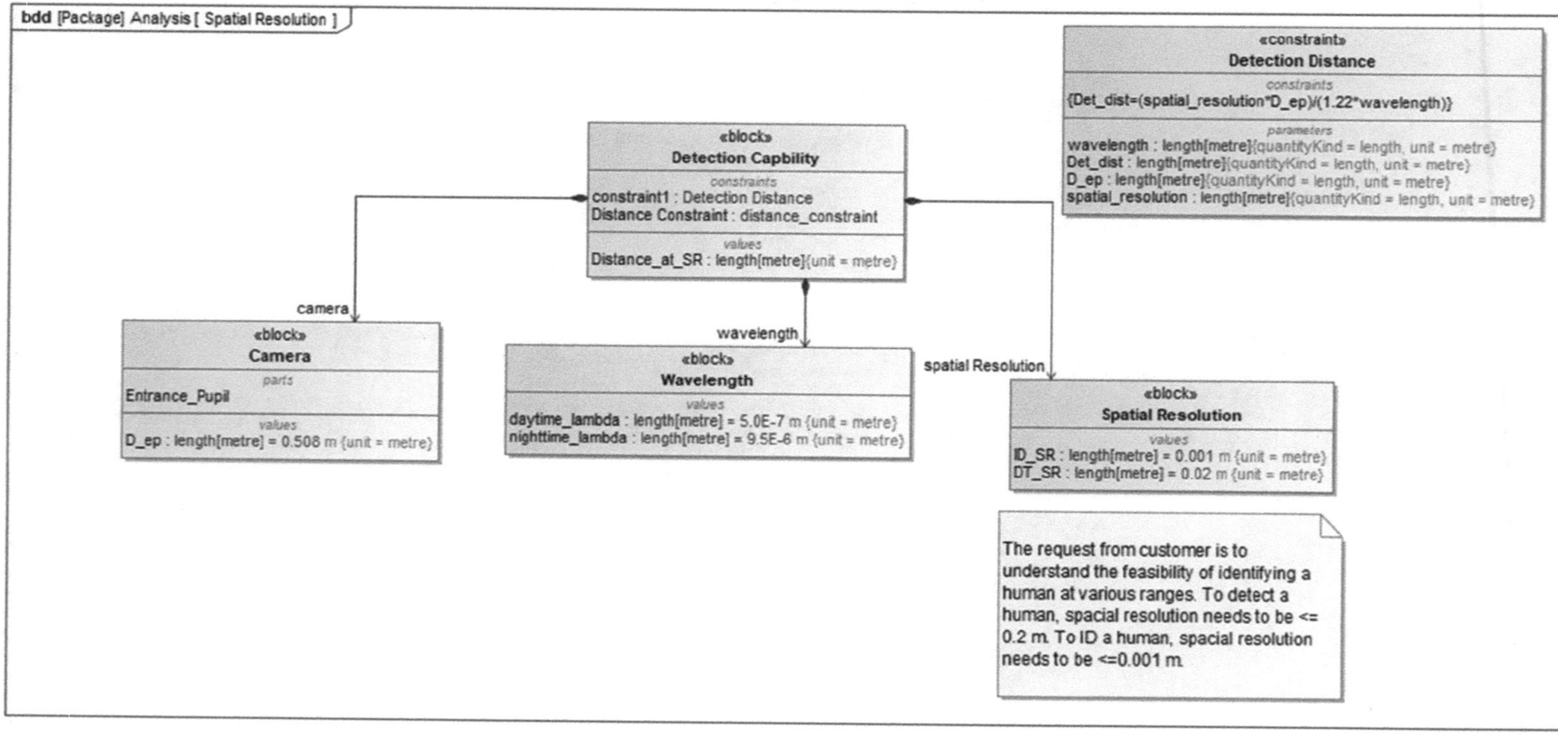

FIGURE 5.3 Block definition diagram (bdd) of the spatial resolution analysis of the AUAV optical subsystem.

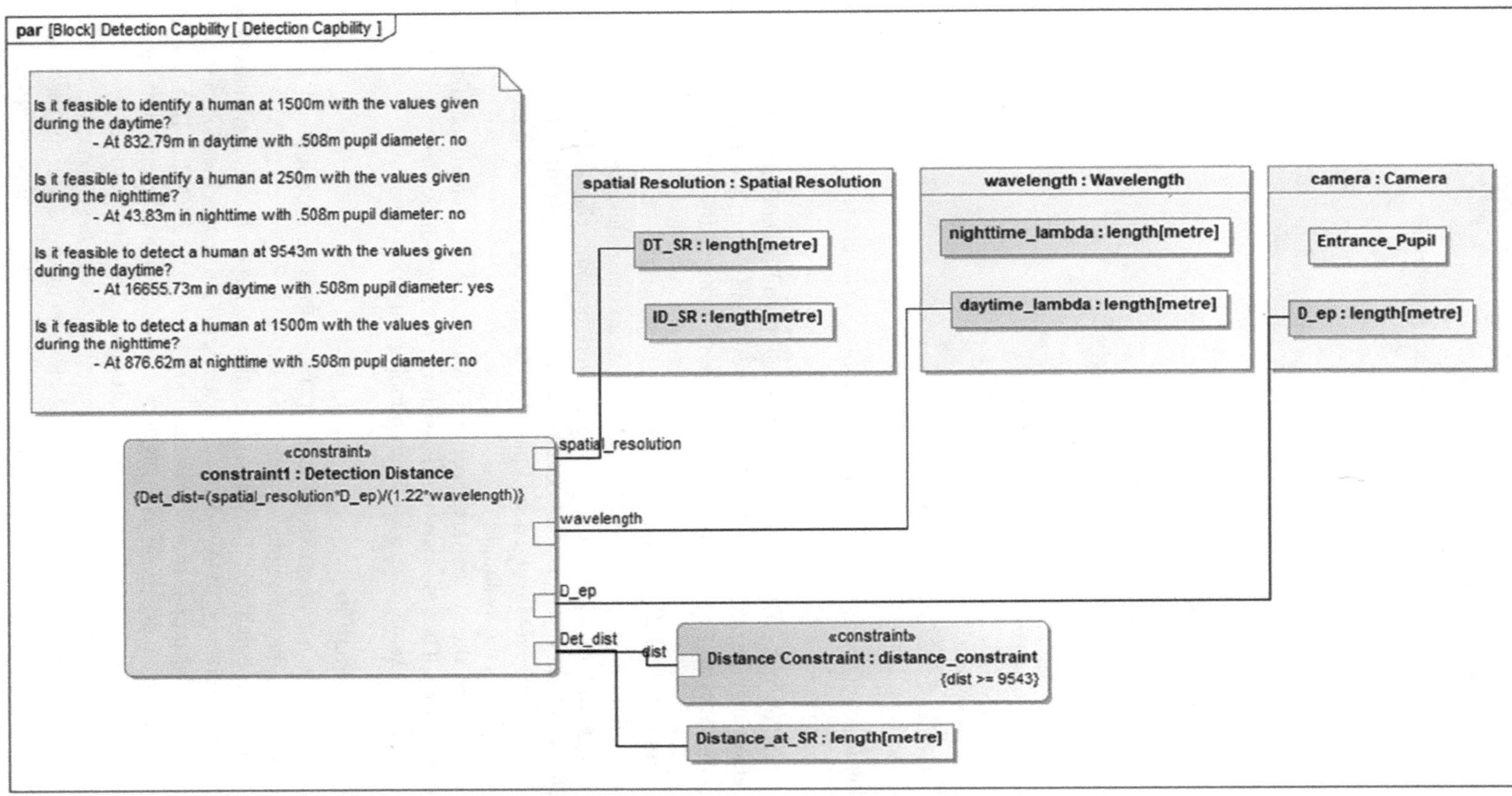

FIGURE 5.4 Parametric model for the detection capability of the AUAV optical subsystem.

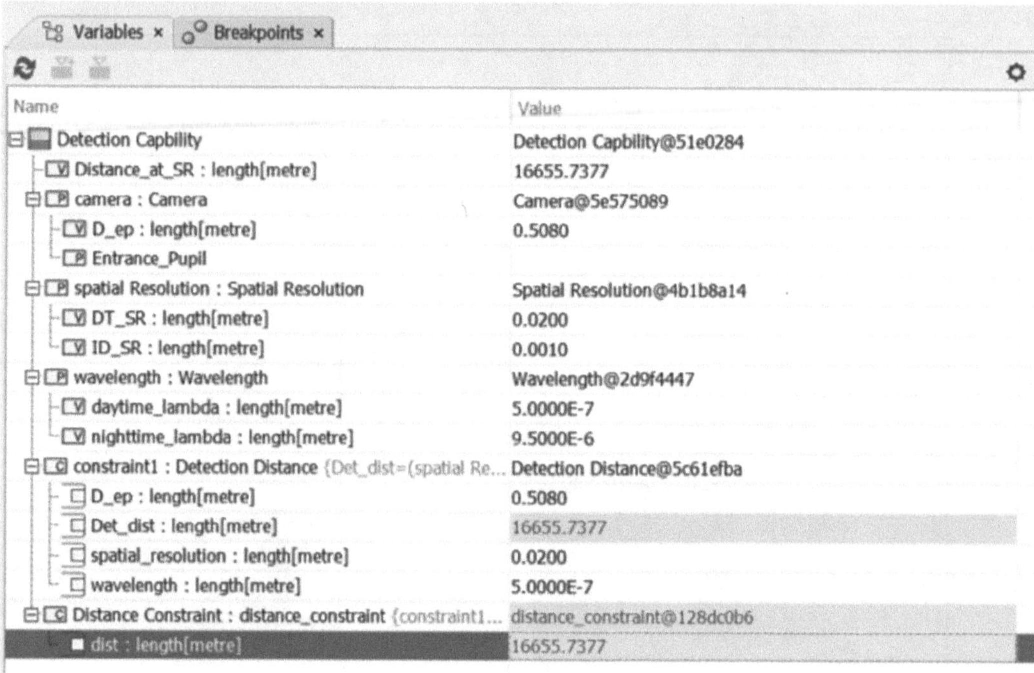

FIGURE 5.5 Results of the parametric model for the detection capability of the AUAV optical subsystem simulation.

interest to Homeland Security. Distress will provide location to Border Patrol and First Responders for immediate emergency response. Border Patrol or Homeland Security can send a message back to AUAV to follow a specific target, maintain a position, or continue the initial flight plan.

5.1.4.5 The Principle of Definition

The Principle of Definition is the notion that the system architect/designer "needs both a formal definition of a design, for precision, and a prose definition for comprehensibility" (Dickerson et al. 2016, p. 15). The definition of terms, components, and the overall system is important to reduce confusion among the engineers involved, and having clear definitions leads to a better understanding of what is needed from the system.

For the AUAV system, the system architect can create an AV-2 (integrated dictionary) to list terms and their definitions. The "glossary" will assist with making sure system engineers, and other stakeholders are all on the same page with the terminology of the system.

5.1.4.6 Model Transformation

The principle of Model Transformations is seen to "relate to system design and should preserve the relationships between the parameters being modeled" (Dickerson et al. 2016, p. xxi). Model transformation is used through Model Driven Architecture (MDA), which is a systems development approach. A model of a system in MDA is defined as "a description or specification of that system and its environment for some certain purpose" (Dickerson et al. 2016, p. 256). In MDA, the models are used to understand the system design, assist in system development, and track modifications through time. Model transformation, "is the process of converting one model to another model of the same system" and is considered part of the system development process (Dickerson et al. 2016, p. 257).

There are both internal and external model transformation methods. For the former, transforming the input variables of a model to output variables can be one example of *model transformation*. According to this principle, the mathematical relationships between input variables should

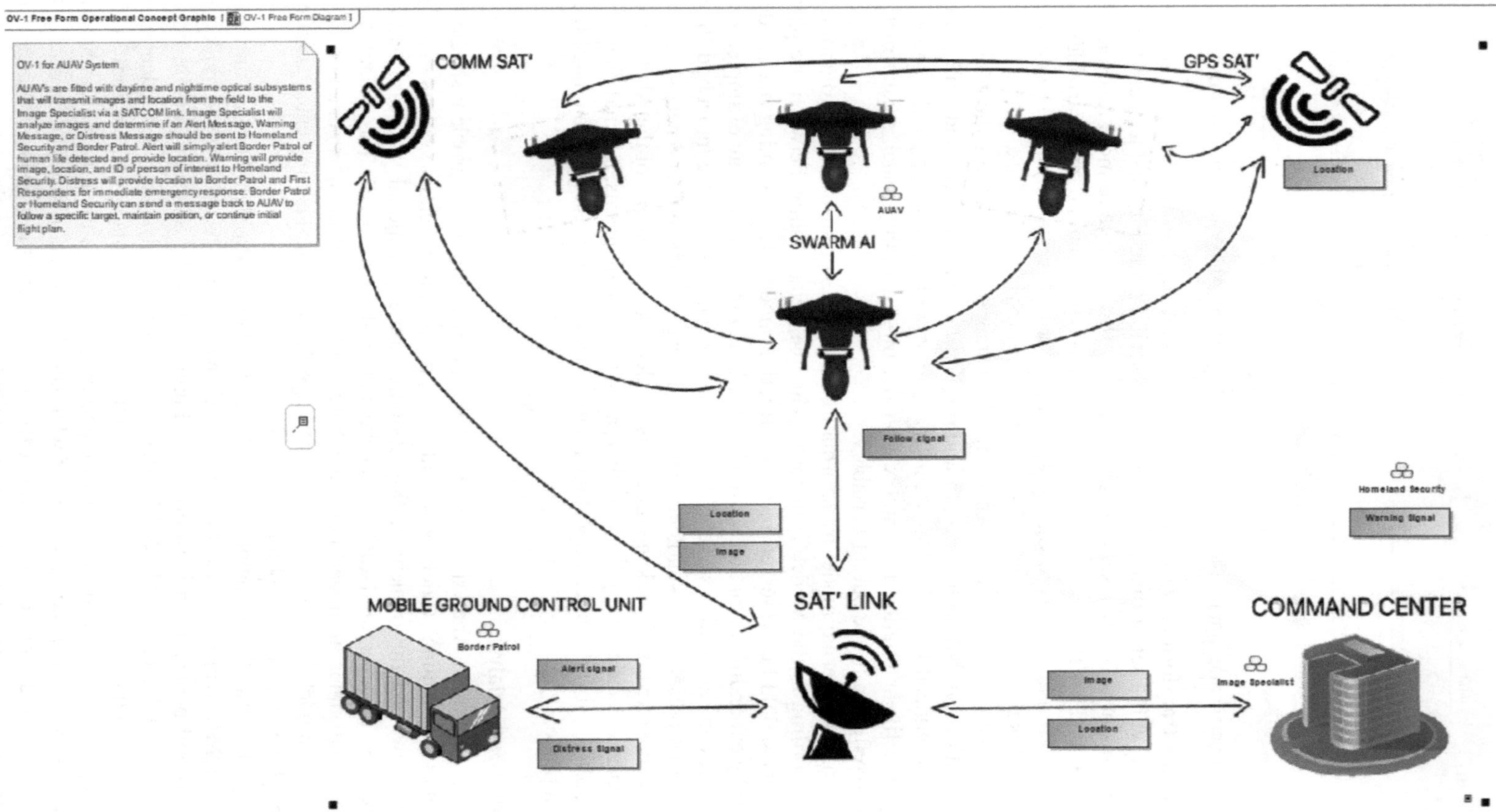

FIGURE 5.6 OV-1 for the AUAV system.

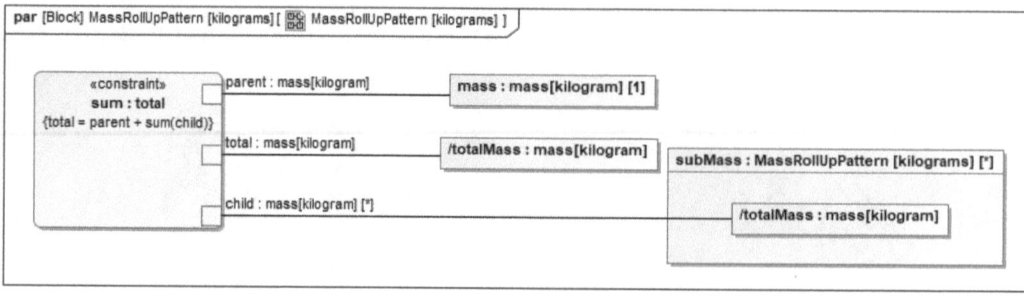

FIGURE 5.7 Parametric diagram for the detection capability of the AUAV optical subsystem.

be preserved in the transformed variables. As an example, using the AUAV, the system architect/designer can start with a set of input variables that describe some physical aspect of the system. The inputs are transformed into a set of output variables through one or more relations. Through the Principle of Transformation, any relationships that exist between the input variables should be reflected and preserved in the output variables. A linear, shift-invariant system is a good example of a system with this property. Figure 5.7 shows the parametric model for a "mass roll-up" basic cell that sums two masses together (parent and child masses) and outputs the total mass. The implemented relationship between the input and output variables is linear and shift-invariant. If the input mass is reduced by a fixed amount, the output will be reduced by the same amount. Consequently, any linear relationship imposed on the input variables, such as subtracting by a constant, is also reflected in the output variables.

For the case of the external model transformation, a model is transformed into another model (M2M) and, according to the Model Transform principle, relationships between the parameters in the first model should be preserved in the destination model. Transforming one model to the next can occur by expanding upon an existing model. Using the AUAV as an example, the system architect/designer can start with a stakeholder diagram which could evolve into a bdd (block definition diagram) that preserves stakeholder concerns. Using the stakeholders' needs, requirements diagram can be generated, and package diagrams to identify subsystems of the AUAV. Each subsystem can in turn be detailed using package diagrams, use case diagrams, internal block diagrams (ibd), and so forth. All the while, important relationships are preserved throughout the model transformation process.

5.1.4.7 Structured Analysis

The Principle of Structured Analysis states that, "the specification of the problem should be separated from that of the solution" (Dickerson et al. 2016, p. 53). The essential systems model (implementation-free view) and the implementation systems model (functional view) are two types of system models used for structured analysis (Dickerson et al. 2016, p. 53). The essential model is made up of the environmental model (boundary between the system environment and the system) and the behavioral model (how and what flows through the system). The implementation model is the fulfillment of the essential model for the system.

Another way of stating this principle is that there should be a Problem Domain and a separate Solution Domain. MBSE provides a variety of SysML-based artifacts to view and model essential system elements and behaviors in both domains. Figure 5.8 is the Solution Domain bdd for the data transmission function on the AUAV. The data transmission analysis is independent of the type of material parts and communication parts used. This bdd represents structured analysis because it shows only a specific part of the architecture and only considers the information necessary to determine the transmission speed. The idea of what should happen is disregarded and only what will happen is considered. The bdd objectively explains how transmit time is calculated in the first block. Here, *tt* represents the transmit time, in seconds, *ds* represent the data size in Megabits,

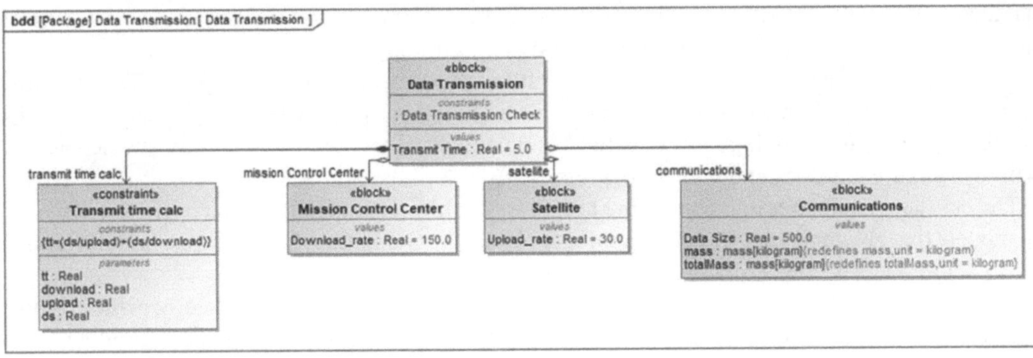

FIGURE 5.8 BDD for data transmission of an AUAV system.

upload represent the upload rate in *Megabits/s*, and *download* represents the download rate at Megabits per second.

Choosing the right model to represent a system is one of the most important considerations when implementing MBSE. Consequently, it is important to carefully analyze each system's unique characteristics when deciding which model to apply, as no two systems are identical. To accurately model complex systems, Ogren provides five principles in developing a model:

- A purpose must be defined for the model.
- There must be clarity in the model.
- It must be possible to show the mission of the system.
- The system structure must be showable by the model.
- Verification of the model must be possible.

Better team communication will result from a system that is clearly stated and easy for others to comprehend and interpret. This can be achieved by taking these model principles into account (Ogren 2000).

This section presented six useful model-based architecture and system engineering principles that are foundational elements of a good MBSE methodology. Where needed, we illustrated these principles using an AUAV application example and used SysML artifacts to aid in their understanding. The use of MBSE is a way to visually understand a system and create traceability for design decisions.

5.1.4.8 MBSE Methodologies and Methods

Sanford Friedenthal provides the following definition for MBSE methodologies, "a group of processes, methods and tools used to help system engineers in MBSE" (Friedenthal et al. 2014). In turn, processes are defined as, *the activities in which something comes out of*, which essentially communicates what is being done in the system. The method indicates *how* that process is being performed. Tools are used in concert in the development effort for increased efficiency. The combination of these things defines what is meant by methodology in a model-based systems context. Estefan says it this way, methodology is, "a way to complete specific tasks and outlines how we may look or solve something" (Estefan 2008). Estefan also conducted a survey that presented different types of methods and methodologies used in MBSE (Estefan 2008). Many of these methods are still in use in industry to date. We describe some of these notable methods and methodologies that are applicable to MBSE for effectively developing modern-day complex systems.

The embedded computer system analysis modeling methodology (ECSAM): The ECSAM technique has evolved from its original intent—to assist engineers in developing embedded systems using software applications—into a generalized approach applicable to a wide range of systems. The basic premise of this approach is to use a variety of analytical methods to define a system's characteristics in its environment. This occurs at the system and subsystem levels to determine which components are needed to meet the specified requirements.

Several models, meant to characterize the system, are generated by this process. The definition of the system is based on a conceptual model. The system's physical parts are defined by a design model. Part of the methodology is a structural and behavioral model, which specifies which aspects of the system are constant and which ones vary with time (Lavi and Kudish 2004).

The engineer can better understand the necessary requirements for each part of the system using the models that are generated from this approach. When applying this methodology, both the system and its subsystems follow a hierarchical structure (Battu et al. 2013).

Model-based architecture and system engineering (MBASE): This methodology primarily uses software systems that interface with a variety of models, including systems, process, product, property, and success models. Significantly less documentation is required for a system because of this approach (Battu et al. 2013).

IBM telelogic harmony SE: A key feature of the Harmony SE methodology is its neutrality toward tools and vendors. It closely follows the "Vee" model procedure for lifecycle management (Yang et al. 2021). The method's key features include its ability to aid in determining system requirements and functionality and in assigning that functionality to a specific physical component. Many MBSE applications have used SysML as their primary modeling language, and this one is no different.

Gathering requirements is the initial stage of this methodology. To do this, we combine SysML with a service request modeling strategy. Using SysML block diagrams, which reflect the system's functions, the service request is modeled. Any time one block needs to communicate with another block, a service request is initiated. Then, what is known as operational contracts alter the system's state. These operational contracts are deconstructed, and the process is carried out by the system's functionality. There are primarily three steps to Harmony SE: requirements analysis, system functional analysis, and architectural design (Estefan 2008).

In addition, as the methodology was designed to stand on its own, it may be applied using any tool. There may not be any dedicated tools for this approach, but the steps and methodology can be easily integrated into existing tool suites.

INCOSE object oriented system engineering method (OOSEM): The INCOSE-based OOSEM method is based on a top-down approach, and the main modeling language is SysML. This allows for system design, specification, analysis, and verification (Hoffmann 2008). Because OOSEM is more adaptable to system changes, it is excellent for systems with evolving requirements or technology. In addition, Estefan points out that this method ought to facilitate enhanced integration of technologies on the system's software, hardware, and testing fronts (Estefan 2008).

The primary steps of the OOSEM include requirements analysis to determine the system's requirements, developing an architecture, evaluating potential system alternatives, and verifying and validating the system.

The OOSEM starts with thoroughly examining stakeholder needs and requirements. The analytical approach used for this task involves determining the system's limitations and areas for improvement, and then translating the associated needs into requirements for the stakeholder and the mission.

After the development of stakeholder requirements, the following step is to determine the system requirements. First, the stakeholder requirements are modeled. Then, different scenarios are considered to determine the system's structure and behaviors. The system architecture is designed when the system requirements have been determined (Hoffmann 2008). To satisfy each requirement, the OOSEM determines which components are required. In doing so, the system architecture aids in determining the components of the system and therefore restricts the design modifications that can be made in the future. The system architecture contains both the logical architecture and the physical architecture. The OOSEM high-level activities are summarized as follows:

1. *Analyze stakeholder needs:* This activity involves identifying issues in current systems with the assistance of stakeholders. Following the stakeholder meeting, the system engineer evaluates the needs of the stakeholders. This activity is used for both implementing new systems and making changes to existing ones.
2. *Analyze system requirements:* The system engineer derives system-level requirements from stakeholder requirements after finishing the stakeholder requirement. A system engineer may return to stakeholders if further information is needed to finalize the system requirements.
3. *Define logical architecture:* The logical architecture is defined by many systems engineers as a best practice; this allows for design flexibility since it is not reliant on any particular physical implementation.
4. *Synthesize candidate physical architecture:* Once the logical architecture is defined, the systems engineer must proceed to define the structure for the physical components - the physical architecture. Each chosen component must be examined to verify it meets the component-level requirements that can be connected back to the system-level requirements.
5. *Optimize and evaluate alternatives:* This task is performed concurrently with other activities, as seen in Figure 5.9. Systems engineers perform a trade study to choose the most effective system design in this activity. The trade study evaluates many system attributes such as size, weight, cost, power, and cooling to determine a cost-effective solution and mitigate risks (Estefan 2008).
6 *Manage requirements traceability:* This activity manages the top-down, bottom-up, and cross-directional (from requirements to test for instance) traceability of the requirements at all levels.
7. *Verify and validate system:* Here, each requirement is verified, and collections of requirements are validated using appropriate tests of the system. When requirements are generated, testing considerations are simultaneously addressed such as the verification technique (demonstration, inspection, analysis, formal test, or where appropriate, simulation).

In the step that involves evaluating alternatives, typically various trade studies are used to ensure that the alternatives are thoroughly analyzed, and the optimum solution is selected. To minimize risk, the trade studies check that necessary components are being examined and that relevant technical and financial factors are being considered.

For the verification and validation step, several techniques are available. Some of these examples include evaluating specific use case scenarios, or through the formal testing process of the requirements.

IBM rational unified process for system engineering (RUP SE) for model driven systems development (MDSD): An offshoot of the Rational Unified Process (RUP), the RUP SE method was developed with systems engineering in mind. The goal of the RUP SE was to implement software development best practices to reduce development time, costs, and risks (Battu et al. 2013).

The system's roles, work products, and tasks that need to be accomplished, including the mission objective, are defined using a structure in RUP. RUP SE builds upon the RUP and applies it to the SE realm. Where RUP SE excels is in providing a framework for teams to methodically create, organize, and manage their requirements. The SE component of RUP SE provides a more encompassing view of the system. The four stages of plain RUP—initiating the system, elaborating the system, building the system, and transitioning the system, are all applicable to RUP SE.

However, RUP SE adds some additional structure on top of the basic RUP concepts and procedures. In creating new systems, new roles are required. These roles include things like system security, logistical support, and training for example. To ensure the system is in sync with the company's goals, objectives, and strategic initiatives, there is an extra emphasis on the system's business

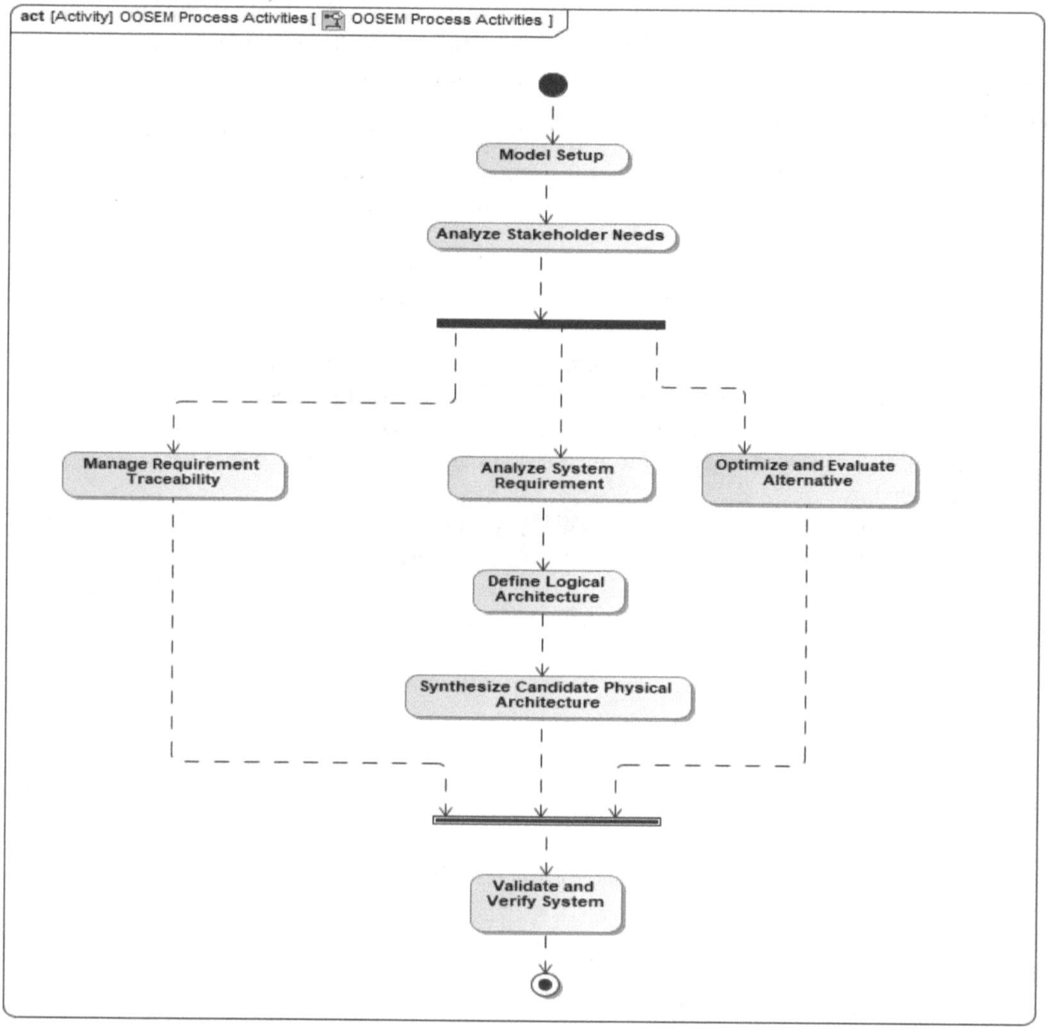

FIGURE 5.9 High-level OOSEM process activities.

drivers. Though it differs little from RUP, the RUP SE now includes a component to ensure that business requirements are also being documented and addressed.

The RUP SE architecture consists of three distinct levels. First, Estefan characterized the model level as the explicitly defined level; next, there is the model viewpoint, which characterizes the various perspectives on the system; and finally, the views, which characterize each perspective. At the model level (first level), we have the system definition, which specifies the overall model structure, subsystems, and components. To facilitate tracing, the system definition also specifies the relationship between model elements. The second level deals with different user viewpoints and so provides alternate user perspectives. The third level specifies the view for each viewpoint such that each view provides the kinds of information, and the level of detail, that each second-level viewpoint must have.

Throughout the system's lifetime, RUP SE additionally enhances the scalability and extensibility of projects. An advantage of RUP SE is that it allows for future scalability and adaptability when constructing subsystems and their subsequent features so that required changes and system upgrades can be accommodated.

System and subsystem requirements are captured by this method and satisfaction criteria for the lower-level requirements can be established such as whether or not a subsystem requirement satisfies the higher-level system requirement. The requirements can also be classified as allocated or derived wherein in allocation, the aggregate of distributed lower-level subsystem requirements satisfy the higher-level system requirement, and derivation where the lower-level requirement is deduced from a collection of higher-level requirements.

Vitech model-based system engineering methodology: The same company that developed the Compliance, Operations, Reliability, and Engineering (CORE) tool suite, developed the Vitech MBSE methodology. This methodology has four processes that are captured in a system design repository. The processes are requirements analysis, functional analysis, architecture, and verification and validation. Vitech MBSE uses the System Definition Language (SDL) in its methodology to manage the models that are produced. This language provides the benefit that it produces easy-to-understand models that provide enhanced communications between differing professional communities such as designers, analysts, and developers.

The Vitech methodology uses similar modeling principles as presented in the previous section. Recall that in model development, the models require explicit definitions with proper communication. The methodology is set up so that the system is developed in layers. The methodology uses a sequential process wherein a given layer must be completed before going on to the next layer.

The onion model process is the central procedure for this methodology. Like peeling an onion, this methodology starts at the top and works its way down. The system is fully developed when all layers and tasks are finished. Estefan suggests that this sequential, layered approach reduces the potential for cost and schedule overruns by allowing work on a single layer at a time. In the onion model, requirements are defined first, then system design occurs, then its physical characteristics are assigned, and lastly, testing is performed. One issue associated with this methodology and the onion model is that only one layer can be operated on at a time. This makes it hard to fix problems or address challenges as you go deeper into the system's development because only one layer can be modified at a time (Mazeika et al. 2016).

JPL state analysis: Models and state-based architecture are both used in the state analysis (SA) methodology. A system's state is its representation in its current configuration, and its model is the process by which that state changes. The system's behavior is the primary focus of this method. A system's behavior can be described by its state, its activities, and the interactions between them. When it is necessary to understand the system's behavior in relation to time, modeling the system state is an often-preferred modeling approach. Activities and interactions are used to depict the sequential movement of data through several stages. Each possible state that a system must encounter is specified by the systems engineers, along with defining the state variables and associating them with individual states. This methodology includes component integration, system upgrades, and future expansions. A fundamental advantage of this methodology is that it establishes a direct connection between requirements and the system that is being created.

A central model within this methodology is the "Control Diamond." This model connects the defined state variables of the system to the system states. A hardware adapter then establishes a connection between the control system and the system under control, allowing the system to be controlled through its defined states. With this methodology, there are some notable principles such as, the controlling system should have complete control over all parts of the controlled system's operations distinguishing between the controller and what is being controlled. Other appropriate models can be supplied to ensure that the right state is used.

Other methodologies can also be integrated thereby permitting the models to be enhanced and expanded over the life of the system. Given its iterative nature, this methodology is well-suited to complicated software-based systems and can yield excellent results when it comes to delivering autonomous and closed-loop commands (Estefan 2008).

5.1.4.8.1 Dori Object Process Methodology

To illustrate the system's structure, behavior, and functionality in a unified model, the OPM methodology uses object process diagrams and its proprietary language, Object Process Language (OPL). Everything is characterized as either a process or an object according to Object Process Methodology (OPM). An item might be a mental or physical representation of something else. An element's state describes its current condition and the range of possible states it can be in at any one moment, whereas its process describes the transformation it undergoes.

This methodology can borrow from other similar methodologies but describes the model using sentences. Dori observed that OPM can be used in a variety of systems including model systems, natural systems, and artificial systems.

This methodology has the advantage that while a model is created, excellent documentation is created at the same time by using the OPL during the model creation.

Furthermore, this methodology aids in dealing with the system's complexity by using three principles. First, the process of folding and unfolding links the system's hierarchical structure. Second, by adjusting the in and out zoom levels, the system's details can be seen or concealed. Third, facts that reveal or conceal the states of the system at any level can be expressed or suppressed. By adhering to these principles, this methodology can aid in defining and refining complex systems. Having a model and supporting language allows engineers to understand details at every level without losing sight of the system itself.

With their own OPM and OPL, each component of a system can be represented in multiple OPM models that cover requirements, design, analysis, and implementation.

5.1.4.8.2 Modeling Tools

Before we start with the description of some useful MBSE tools, it is important to recall that MBSE tools act as an aid to provide clear communication between stakeholders and the development team. To gain better understanding of the MBSE tools, and how they are applied in SE, it is important to understand the following definitions from the *Systems Engineering Guidebook* written by *James Martin*, which explain the concepts of methodology process, and tools (Martin 1996):

> A process (P) is a series of logical tasks carried out to accomplish a specific objective. A process describes *"what"* is to be done, without telling *"how"* each task is performed. The structure of a process provides several levels of aggregation to allow analysis and definition to be completed at various levels of detail, to support different decision-making needs. A method (M) has various techniques that specify *"how"* each task is performed, in this regard, the words "method," "technique," "practice," and "procedure," are frequently used interchangeably. At any level, process tasks are carried out using methods. Each method is also a process itself, with a series of tasks to be performed for that specific method. In other words, the *"how"* at one level of abstraction becomes the *"what"* at the next, lower level. A tool (T) is an instrument applied to a specific method to increase the effectiveness of the task, but only if it is applied in a correct manner, and by someone with the right skills and training. The use of a tool should be to assist the accomplishment of the "*how*" In a wider sense, a tool boosts the "*what*" and the "*how*." The majority of the tools used to support Systems Engineering are computer or software-based tools, which are known as Computer Aided Engineering (CAE) tools.
>
> *Martin (1996)*

As the MBSE paradigm is a relatively new advancement in the SE world, many tools related to MBSE use state-of-the-art and cutting-edge technology. For example, many MBSE tools use SysML, which is a very effective and popular modeling language. "The Modelling language is just the language and must be combined with a methodology to be useful" (INCOSE UK 2012). Today, SysML is regarded as one of the most powerful and widely used languages for modeling systems. SysML was built in partnership between the Object Management Group (OMG) and INCOSE and is based on the language known as UML, which has gone through many iterations and is now on version

UML 2.5.1 (OMG 2017). In a system model, SysML targets four major classes of diagrams. These four classes that SysML supports are structure diagrams, behavior diagrams, requirement diagrams, and parametric diagrams. The structure diagrams describe a systems structure and its parameters. The behavior diagrams describe how a system operates, functions, and behaves. The requirement diagrams describe what the system must do and accomplish from a functional and non-functional point of view. The parametric diagram allows relational aspects of the system to be modeled.

As an example of the utility of MBSE methods for space systems, an MBSE challenge project was launched to develop an imaginary FireSat satellite system to assess the appropriateness of the SysML for defining space systems. The hypothetical nature of the project prohibited the realization of the system and thus could not serve as an example to prove the effectiveness of the tool in that specific application. The study did, however, serve as a template for the standard CubeSat model, which was used in the Radio Aurora Explorer (RAX) mission, built by the Michigan Exploration Lab (MXL) and SRI International (Spangelo 2012).

If a SE project uses tools like CATIA System-of-Systems Modeler, various fundamental SE concepts and best practices are already built into the tool. CATIA System-of-Systems Modeler tool can use the basic components of the SysML modeling language. Various CATIA System-of-Systems Modeler elements such as blocks, properties, and constraints are used. Parametric models and diagrams can readily be implemented. One significant benefit of using the CATIA System-of-Systems Modeler is that by using constraint blocks, CATIA System-of-Systems Modeler can be linked to MATLAB® and Simulink® functions and scripts, thereby providing the ability to directly link requirements to executable code. Another benefit is the integration of the MATLAB and Simulink existing models into bigger CATIA System-of-Systems Modeler models, which can increase the speed of systems development. As an example, an integrated mechatronic system was achieved by building the links between SysML and MATLAB/Simulink models (Qamar 2009).

Another well-known and highly practical tool that can be used to build analytical models for complex SE problems is MATLAB. MATLAB is one of the most prevalent computational packages among engineers and scientists. It has the capability of providing complicated graphics in an easy manner. The simple plots are easy to model, but specialized plots are possible using "handles." Handles present the means to entirely control and customize graphical objects in MATLAB. After understanding the handle-based system, users can generate plots according to prescribed requirements and quality standards generally set by professionals. Various instances are developed to demonstrate this concept (Green 2007). MATLAB is a powerful prototyping tool that has come a long way since its inception. MATLAB comes embedded with numerous toolboxes that increase the power and usefulness of the tool while remaining relatively simple. For example, to solve parallel computing problems, MATLAB has a parallel computing toolbox add-on that makes it easy for users to write parallel computing native functions. Similarly for mathematics, statistics, and optimization problems, MATLAB offers a sequence of toolboxes such as partial differential equations, statistics, curve fitting, optimization, global optimization, neural networks, and model-based calibration to readily solve these complex problems. In control system design and analysis, MATLAB provides control system, system identification, fuzzy logic, robust control, model predictive control, and aerospace toolboxes. To deal with signal processing and communications problems, MATLAB provides signal processing, digital signal processing, communications systems, radio frequency phased-array systems, and wavelet toolboxes. Similarly, in image processing and computer vision applications, MATLAB provides a sequence of toolboxes including image processing, computer vision systems, and image acquisition and mapping toolboxes. MATLAB also contributes to the test and measurement area, as well, by providing data acquisition, instrument control, and vehicle network toolboxes. MATLAB has modeling capabilities in areas such as computational finance and computational biology by offering financial, econometrics, data feed, database, financial instruments, trading, and bioinformatics toolboxes.

When building complicated systems, using an MBSE methodology in developing systems is beneficial because of the adaptable nature of MBSE. Many studies have been carried out to identify the most suitable instruments for various MBSE tasks since it was first introduced to engineering practices (Rashid et al. 2015). The MBSE models are built with stakeholder needs for model transformation and verification in mind from the start. Figure 5.10 provides an example of how MBSE can be implemented to benefit an organization's developmental efforts. The systems engineering method still starts with a set of requirements. These requirements are implemented in the MBSE development environment by making use of modeling languages and approaches like SysML, UML, and MARTE (Modeling and Analysis of Real Time Embedded Systems) to specify system requirements (Kapos et al. 2014, pp. 531–538). Various tools and languages are used to model the system in its environment. Figure 5.10 also shows the benefits of using MBSE to provide model transformations to include model to model (M2M) transformations and model to text (M2T) transformations. The later can be very useful in automatically generating system documentation. The transformed models can be used in the verification process that is directed at the models themselves, and/or at the modeled requirements. The MBSE environment can also generate code that subsequently can be used in simulation. The idea here is that the MBSE environment is much more than just a repository of the system's requirements. It is an enabling technology that extends the functionality and utility of modeling so that it becomes a central, collaborative, and integral feature of the systems development effort.

Making the right tool selection for MBSE activities early in the program lifecycle can save time and effort later on. It can be tough to choose a suitable MBSE tool because embedded systems' behaviors vary from program to program. Before engineers can choose the right tool for the job, they must first identify the modeling needs, characteristics, and drivers that are relevant to the MBSE modeling activity. Does the Modeling allow XML formatting, for instance? This format is widely used in tools that require the import and export of data. Also, not all tools support the most popular profiles, such as UML, MARTE, and SysML. Finally, a tool's compatibility with its selected environment is dictated by its integration capacity. The tool's integration capacity can be further classified into separate levels: high, medium, and low. This means that the tool will provide a high level of integration, decent integration, or very little integration in the modeling environment (Rashid et al. 2015).

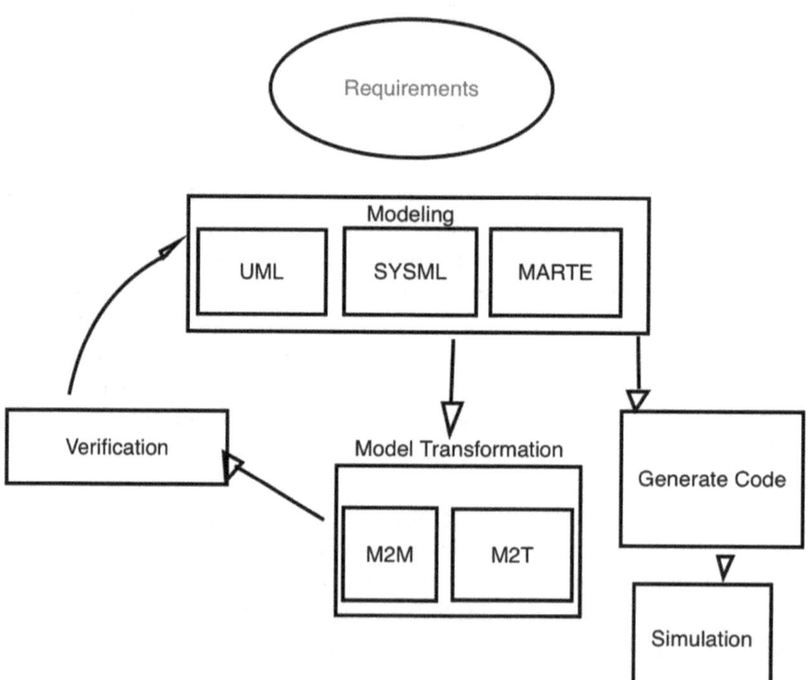

FIGURE 5.10 Primary MBSE for systems development. (Courtesy of N. Davis 2022).

In addition to serving as individual assets within their modeling environments, tools, and collections of tools can also serve as frameworks for a wide range of MBSE tasks. The framework is the complete environment that interacts with various MBSE tool sets and synchronizes their operations to accomplish various MBSE tasks (Rashid et al. 2015, pp. 150–163). A toolset is a collection of tools that can be employed to execute MBSE tasks within the specified environment. Various phases of the SDLC make use of popular MBSE modeling tools, which we will now summarize.

IBM rational rhapsody: One of the most popular tools in MBSE, IBM Rational Rhapsody provides assistance in a wide variety of settings (Rott et al. 2018, pp. 1856–1869). Rational Rhapsody and its companion products offer a solution for system modeling and design, allowing engineers to better handle the inherent complexity of any given system under development. Stakeholders and engineers alike will find these tasks presented in an easily digestible format. Systems engineers can use Rhapsody as a collaborative environment for design, development, and testing. It also allows importing and exporting from other popular tools that use UML and SysML.

The main advantage of using IBM Rational Rhapsody is its continuous validation capability, which allows for the early detection and correction of faults during rapid prototyping and simulation processes. To improve the efficiency of team cooperation, IBM Rational Rhapsody operates as a repository, lets engineers examine artifacts, and has a feature that enables automatic consistency checking.

Within IBM Rational Rhapsody, there is the Dynamic Object-Oriented Requirements System (DOORS) which is extensively used in government and industry such as the Aerospace and Defense sectors. A primary function of DOORS is to serve as a system requirements management repository. Requirements can be successfully managed throughout the SDLC and linked to other MBSE artifacts. Requirements can be developed through the classic systems engineering V-diagram and stored in different modules corresponding to different requirements levels (i.e., system-level requirements, subsystem-level requirements, and components/pieces/parts/assembly-level requirements). In turn, these requirements can be linked to each other, to test, and other useful UML and SysML generated artifacts for traceability (Egner et al. 2020). Information from different modules can be merged and presented to designers in convenient views. Figure 5.11 shows an example of some

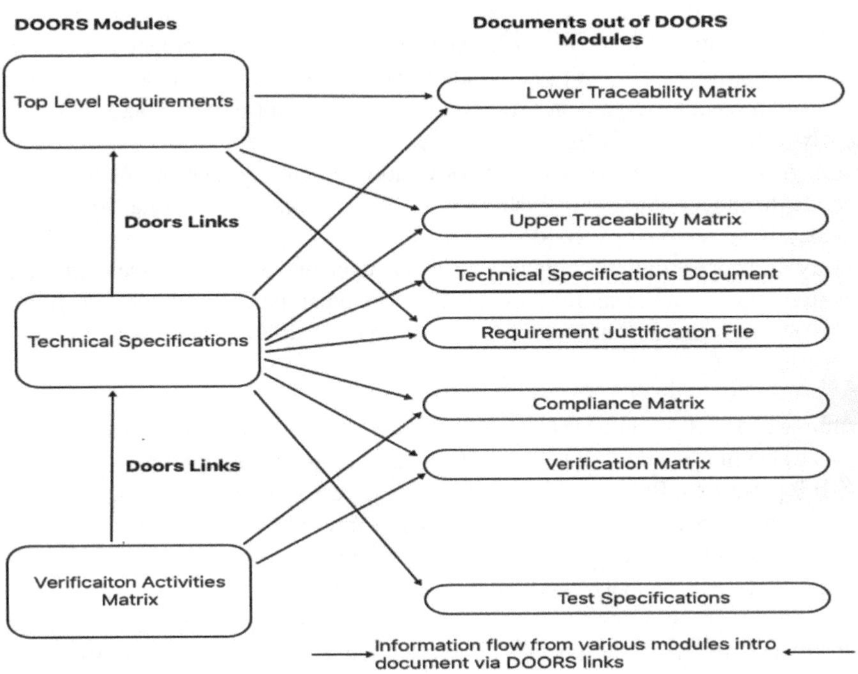

FIGURE 5.11 Representative information flow from various modules in DOORS (Davis 2022).

modules that can be implemented within the IBM Rhapsody modeling environment using DOORS. A requirements layer, such as the system- level requirements layer, is indicated by the *Top-Level Requirements* block at the top-left of the figure. Notice that links are used to connect technical specifications to the requirements and system verification activities. These technical specifications can also be used in the system modeling activities as constraints, and/or inputs to relationships between parameters that can be implemented using the parametric modeling process. On the right side of Figure 5.11 are several useful modules that create essential artifacts outside of DOORS that are used in systems analysis and documentation efforts. For example, the requirements justification file is important for developers to understand the purpose behind a given requirement, and the traceability matrices ensure that requirements link properly to parent requirements or child/children requirements. Figure 5.12 shows an example of a Requirements Verification Table wherein specific program verification activities are shown at the far left of the figure.

The ID for the verification activity is shown in the second column. The name of the verification activity is shown in the third column and the purpose and critical information about the activity is shown in the fourth column. This information is linked to the technical specifications which, in turn, are linked to the requirements themselves. IBM's products seamlessly interact with DOORS, providing increased flexibility throughout development. DOORS is compatible with Open Service for Lifecycle Collaboration (OSLC), providing convenient access for change, requirement, and quality management to engage with system changes during development.

Magicdraw: Another popular integrated system/software engineer toolset, MagicDraw is very model-oriented, offers support for MBSE methods, and is compliant with industry standards (3 Dynamic Systems 2024). Engineers spend less time learning how to use the controls of the modeling tool since MagicDraw, like IBM Rhapsody, offers a well-designed graphical user interface. While MagicDraw may function autonomously from other software development processes, it is also compatible with other systems modeling tools using plug-ins. For instance, MagicDraw can incorporate model-driven frameworks like MD Wizard as a plug-in to facilitate wizard-based modeling in UML tools. Users of MagicDraw can install the MD Wizard plug-in within the model and define an activity diagram with a sequence of activities for instance (Silingas et al. 2009). To facilitate the creation of the model, developers often employ wizards that lead them through several straightforward procedures. MagicDraw further facilitates team development and collaboration by allowing users to communicate with one another over a collaborative cloud. Also, time spent on configuration and project management can be reduced when multiple developers work on a project simultaneously (Silingas et al. 2009).

Model transformation tools: Model Transformation, which can take on the forms of a model-to-text or model-to-model approach (Rashid et al. 2015), is another essential MBSE activity that calls for a specialized technology. Model transformation is a powerful tool in MBSE that allows users to specify the process of producing target models from one or more source models. Developers can see how the elements of the source model can be linked and navigated to the destination model. We present two model transformation tools to illustrate this kind of functionality.

5.1.4.9 ATL Transformation Tool

A prime example of a MBSE model transformation language and toolkit is the Atlas Transformation Language (ATL) language and toolkit. The ATL toolkit provides the modeling transformation capability for MBSE processes. By including a developer's predetermined set of concepts, ATL lays out the steps necessary to generate a single target model or multiple target models from an established set of source models. When developing a model transformation, software engineers have found ATL to be extremely useful. Developers find that need for rework is reduced due to this tool's assurance that model transformations are done correctly and consistently, and that the model outcomes are the same (Boronat 2018). The features of ATL are unique in that they can work well with complex transformation algorithms, which often prove difficult in providing specific solutions. This includes Called Rules, which is a procedure within ATL that is invoked by name and takes arguments for

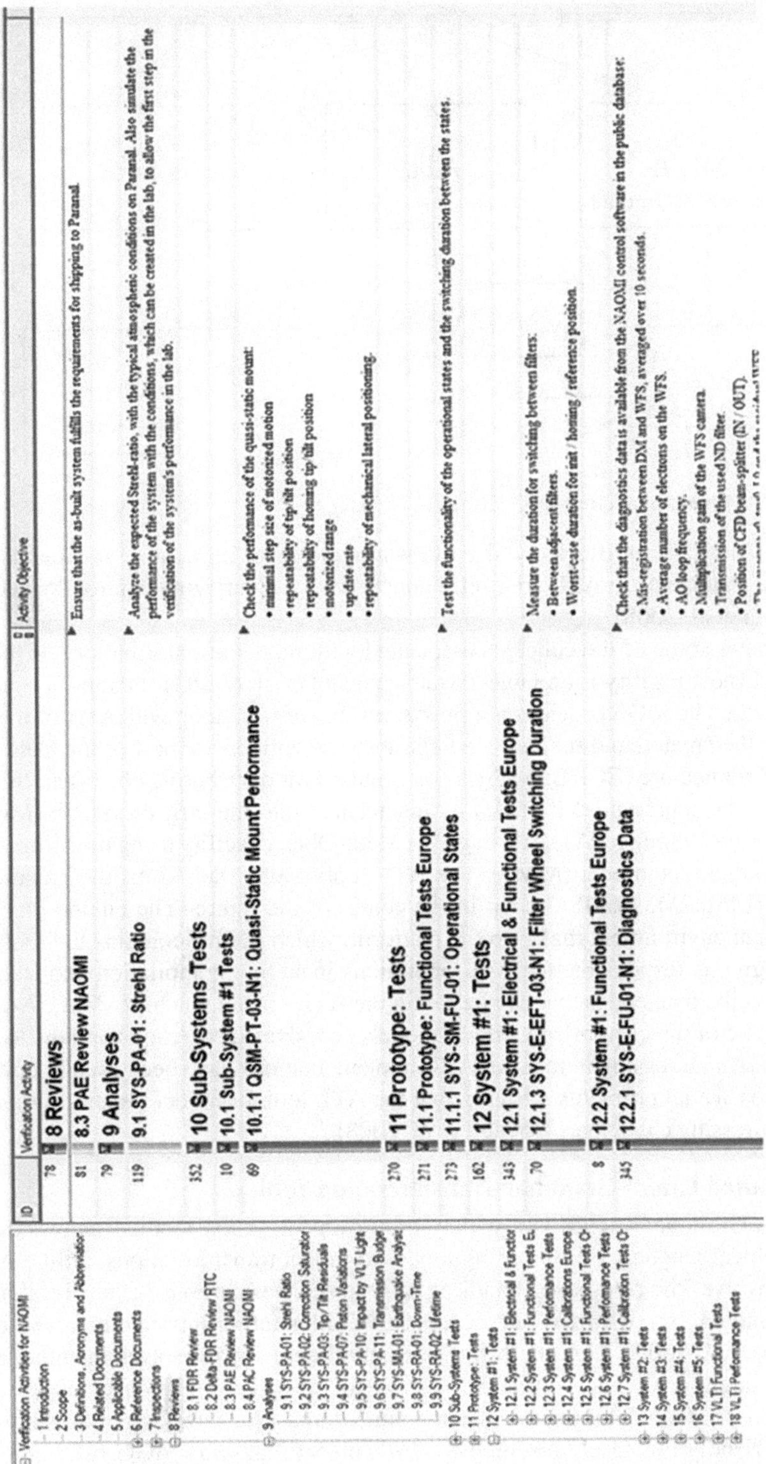

FIGURE 5.12 Requirements verification table for verification activities (Davis 2022).

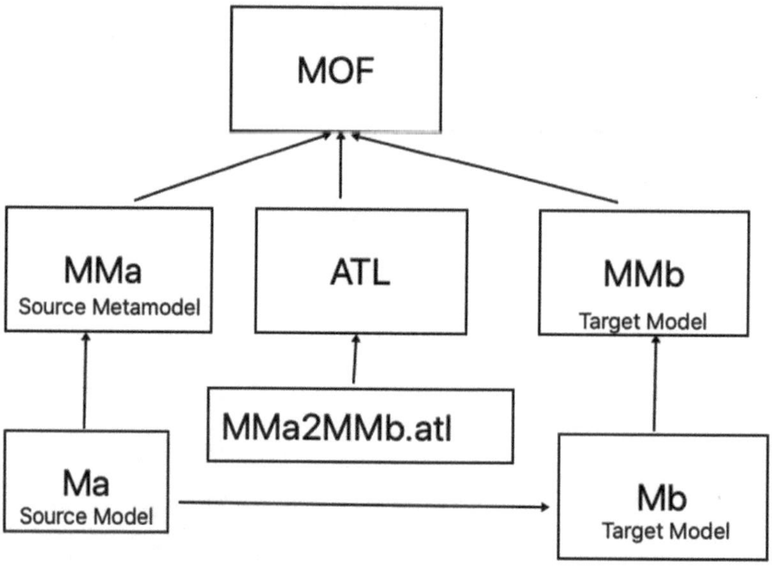

FIGURE 5.13 ATL operational context (Courtesy of N. Davis).

model checking. Also, Action Blocks, which allow a sequence of imperative statements to be used. These imperative statements in ATL are useful constructs for identifying control flow conditions of the model (Jouault et al. 2008).

Figure 5.13 shows some of the concepts associated with model transformation. At the top of the figure, the Meta-Object Facility is an Object Management Group (OMG) standard for model-driven engineering (MDE). The MOF provides a type system for entities along with a set of interfaces that can interact with the created and manipulated entities. The entities are part of the Common Object Request Broker Architecture (CORBA) which is a standard for distributing objects across a network.

In Figure 5.13, the source model a (Ma) is transformed into the target model b (Mb) using the Atlas Transformation Language (ATL) through the Meta-Object Facility using the source metamodel (MMa) and the target metamodel (MMb). The ATL toolkit also transforms the metamodels from source to target (MMa2MMb.atl) as seen in the center of the figure. The engine block and integrated development environment make up the platform, which in turn contains the tools that make up ATL. The engine is responsible for ATL operations including compilation and transformation execution into specific byte codes that can be run in the ATL virtual machine (VM). After that, this ATL VM takes care of the customized MBSE models and gives instructions on how to change the models if needed. An increasing number of ATL implementations have been used in developmental efforts to solve non-trivial problems, proving that the ATL tool for model transformation in MBSE is effective in addressing challenges (Jouault et al. 2008).

5.1.4.10 Attributed Graph Grammar Transformation Tool

Attributed Graph Grammar (AGG) is a widely used transformation method in system modeling. AGG offers graphical languages for defining model-to-model transformations, unlike ATL. Graph transformation involves the rule-based handling of graphs. These graphs have several applications in development and can serve varied purposes, such as illustrating intricate data and object hierarchies (Grønmo et al. 2009). The graph transformation described involves metamodel elements presented in a straightforward structure that allows for easy identification of nodes and arrows (Taentzer and Schürr 2012). The graphical environment and user interface of AGG are designed to be developer-friendly by closely resembling UML object diagrams. AGG offers model validation support capabilities in addition to implementing the editioning and interpretation functionality of graphs, which may be categorized into three approaches used by the tool. First, Graph

Parsing allows AAG to verify if a graph in the program conforms to a specific graph language. Developers may encounter a membership issue if something is out of place, potentially leading to delays in development. Second, Critical Pair Analysis is a feature that compares graphs and identifies conflicting circumstances. When identifying critical pairings, items and connections that lead to conflicts or dependencies are isolated and brought to the developer's attention for resolution. Third, Consistency Checking is a process that verifies fundamental aspects of a graph, ensuring that specific elements of the graph grammar stay consistent across different tools and applications (Grønmo et al. 2009). AGG applications through MBSE are highly versatile and can accommodate a wide range of needs. AGG is a prominent tool for model transformation in MBSE due to its easy interaction with UML models and validation support.

Model verification tools: An important aspect in implementing a robust MBSE environment is the ability to make sure the model in the system is accurate. Model verification tools are used for this purpose. The process of model checking involves creating a comprehensive representation of the system as it is in the real world, and then testing it at various points in its lifecycle (Ji et al. 2009). To verify the model, it is necessary to alter it using the previously mentioned model transformation tools. The result of this model transformation is to provide the correct format of the model to which the model verification tools can be applied, and the behavior and functionality of the system can be checked. If the model checker produces a negative result, the developer is provided with a detailed description of the system's behavior and the reasons why a step was not satisfied (Ji et al. 2009). A counterexample is also generated. When it comes to testing behavioral conditions on large and complex embedded system models, the tools used, however, tend to lose their validity (Rashid et al. 2015). We now provide some examples of useful model verification tools.

5.1.4.11 UPPAAL Model Verification Tool

UPPAAL is a real-time system verification, simulation, and modeling environment (UPPAAL 2023). A model checker, a description language, and a simulator are the three primary components of this product. Before sending the model to the model checker, the simulator can be used as a validation tool to look at potential dynamic executions for defect detection during the early stages of system design. At this point, the model checker verifies that the system model is consistent and has the desired functional features. Finally, the system behavior is described by the modeling/design description language. When it comes to MBSE, UPPAAL has a reputation for being both efficient and user-friendly. It has greatly improved the efficiency of model checking with features like on-the-fly searching approaches. By using a symbolic method that stores the potential complicated behavior of the system and gives a graphical representation, UPPAAL is able to reduce verification issues, which in turn increases efficiency during model verification. Statistical Model Checking (SMC) is another method UPPAAL uses to validate its models. SMC consists of statistical methods that track and assess several runs in relation to the model. A comprehensive estimate of the system design can be obtained by generating these statistics ensuring that the model meets its intended output properties. With UPPAAL SMC, the model's quantitative features can be tested using SMC model simulations (Basile et al. 2020). Because it is simple to use and understand, the SMC method is well-suited in general for engineering projects. Users can execute basic operations using an operational model of the system, which can be validated against properties, all within the UPPAAL modeling environment.

5.1.4.11.1 Prism Model Verification Tool

Prism is another interesting model verification tool used in systems development efforts. It functions as a probabilistic model checker that does modeling analysis for systems displaying problematic or erratic behavior. Prism has been used across various application fields for systems analysis and model verification (Prism Model Checker 2024). Prism uses cutting-edge algorithms and data structures that are based on user-friendly Binary Decision Diagrams. The software also facilitates approximation and statistical model checking with a discrete event simulation engine and is compatible with other widely used analysis methods. This tool functions by defining models using its proprietary PRISM language, which is a straightforward, state-based language suitable for automation. PRISM offers various quantitative properties of the model under analysis. These include determining the probability of system failure leading to shut down, identifying the worst-case probability of triggering a termination error during system test initialization, calculating the expected data exchange size after 30 minutes, and estimating the expected termination time for an algorithm in a worst-case scenario.

The most recent version of PRISM includes significant enhancements and new features. The new enhancements address the common issue of verifying larger or more complicated models, improving efficiency, and offering more precise error reporting options. Some additions aimed at enhancing usability for developers include a redesigned tool and model GUI interface, as well as better modeling language updates (Ji et al. 2009). Another nice feature is that PRISM is free and available to the public under a GNU General Public License (GPL).

Model simulation tools: Simulation tools in MBSE can be used as an alternative method for system validation. Validation through simulations involves using the created source code based on specified verified values (Girish et al. 2021). The simulation verification method has three primary methods: direct testing, constrained random testing, and random testing without limitations. The preferred simulation verification approach is the constrained random approach, which involves using random parameters that are restricted to certain values provided by the test scenario, while other functions can have random values. Simulation-based model verification offers engineers greater control over the verification environment and allows for precise direction of the verification process (Girish et al. 2021). Simulation verification, however, does have some drawbacks. One problem is choosing the right simulation tool for the MBSE task. Another issue is integrating simulation tools into a specific modeling environment. This can be challenging since the modeling tools are closely tied to their own modeling environment. Compatibility assessments are essential when choosing a tool for system development in a specific environment (Rashid et al. 2015). We present some hallmark simulation tools that are prevalent in industry and that can be integrated to great benefit in an MBSE environment.

5.1.4.12 Simulink Tool

MathWorks is a well-known software company that produces the highly popular scientific and computing tool known as MATLAB®. The MATLAB tool is a robust and well-supported mathematical analytical, modeling, and simulation platform for all sorts of scientific, technology, engineering, and mathematical (STEM) applications, and readily integrates into an MBSE environment. MATLAB has various toolboxes that expand its functionality such as in the areas of Math and Optimization, parallel computing, AI, Data Science and Statistics, Reporting and Database Access, Code Generation, Application Deployment, and Verification, Validation, and Test (Mathworks 2024). MathWorks also has a completely different product line known as Simulink, which as the name implies provides the capability to do first rate, flexible, and highly accurate simulations with drag-and-drop functionality. The Simulink tool provides Event-based Modeling, Physical Modeling, Real-Time Simulation and Testing, Reporting, Systems Engineering, Code Generation, Application Deployment, and Verification, Validation and Test capabilities and, like Matlab, can be readily integrated into an MBSE framework. MathWorks (2024) states that Simulink is a robust MATLAB add-on that helps users with system lifecycle activities such as design, analysis, and testing of software architectures and system development (Mathworks 2024).

The MathWorks Systems Engineering modeling functionality within Simulink includes the Requirements Toolbox and System Composer. The System Composer toolbox provides the ability to design and analyze systems and software architectures. In combination, these modeling tools allow users to build detailed models from descriptive architectural and conceptual models. When integrated into an MBSE environment, these tools seamlessly provide sophisticated modeling, analysis, and simulation capabilities to the conceptual and architectural modeling frameworks, such as those based on UML and SySML. Stakeholders and developers alike can benefit from the ability to create a digital twin, and generate digital threads that can inspect/evaluate critical program elements such as system requirements, implementation models, architectural models, and embedded software. When engineers use Simulink in conjunction with MATLAB and System composer, they can do things like manage system requirements, conduct trade studies by capturing architectural metadata and then using it in MATLAB analytics, customize model views to show engineering or stakeholder concerns, and use simulation-ready test components to verify and validate system requirements and architectural elements. Model-Based Design is a powerful tool in Simulink for translating and refining requirements into architectures (Mathworks 2024). Figure 5.14 shows a Simulink-based model checker flow example. Design models that are written in the Simulink environment are sent to a formal model translator shown at the top-left of the grey Box in the figure. Similarly in principle, the requirements and test specifications which also can be resident in the MBSE environment are sent to a property generator that generates formal properties that are used by the model checker, along with the formal model, to verify/check the model. There are both witness functions and counter-examples that are used in the model-checking process to validate the results, or supply information where the model did not perform adequately. The trace to test case translator function is used to isolate the model checker results. The suite of tools is directly integrated into the testing environment. Simulink can be used in a stand-alone mode, but its power and flexibility lie in integrating Simulink with other useful MBSE development tools. Also, a variety of useful third-party applications can readily be incorporated into a MBSE development environment by using Simulink as the interface to the MBSE development environment. For example, the Signature Tool is a third-party tool that operates with the Simulink applications to extract the signature of subsystems, representing their interface. Extracting signatures enhances the understanding of the subsystem and aids in identifying the data flow, permitting further analysis to evaluate the quality of the Simulink design. Signatures in Simulink Models can provide a simplified view of the data flow inputs and outputs to detect concealed issues within the model (Pantelic et al. 2015).

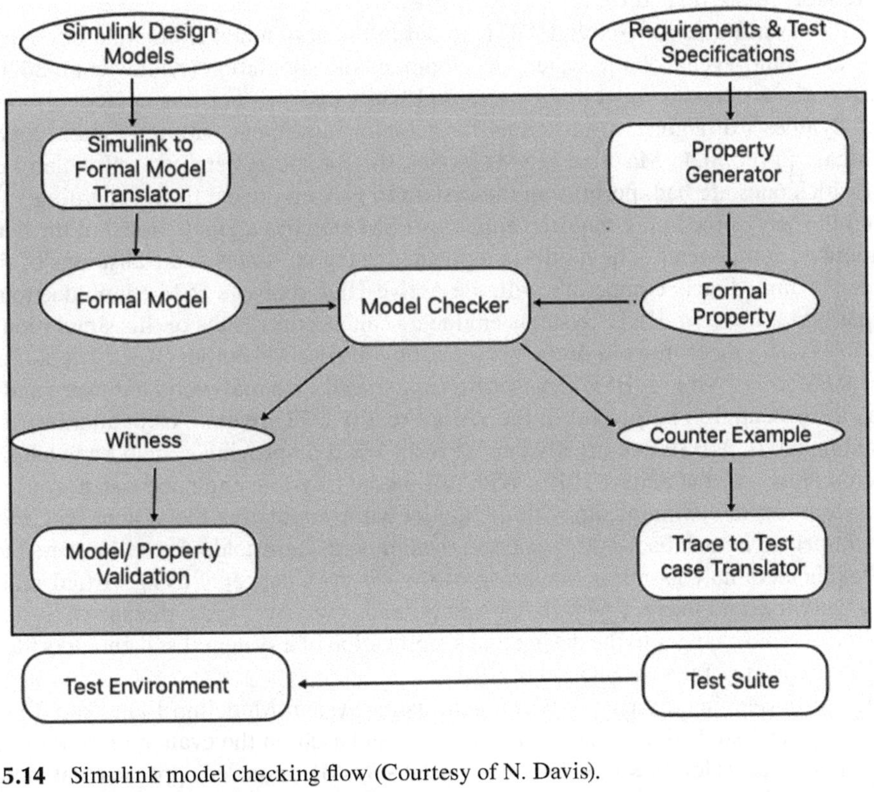

FIGURE 5.14 Simulink model checking flow (Courtesy of N. Davis).

Another third-party tool that works with Simulink is the Reach/Coreach Tool. Using blocks designated for reach and Coreach functionality, the Reach/Coreach Tool can zero in on particular model blocks. Model Slicing is the result of removing dependencies that the tool has detected. An easier and more understandable model is generated by the tool's extraction of Reach/Coreach relationships, which allows for change analysis during testing and verification. An excellent method for revealing the lifespan effects of a small change to the system architecture is impact analysis (Pantelic et al. 2015). Since changes to the system can be costly and time consuming, understanding the changes through this modeling activity can be beneficial.

5.1.4.13 ActiveHDL Simulation Tool

The Active-HDL (Hardware Descriptive Language) simulation environment has also proven helpful for MBSE projects. In terms of team-based operations, this tool is Windows-based and integrated with simulation solution environments for Field Programmable Gate Array (FPGA) technology (ALDEC 2024). FPGAs are programmable, sophisticated systems on a chip that systems engineers can use in conjunction with simulation tools to better understand and validate the model under development (Chochaev and Gavrilov 2021). In Active-HDL, you may find HDL—a full-featured graphical design toolkit—and an Integrated Design Environment (IDE) that helps with FPGA system design simulation, verification, and rapid deployment.

There are many useful features and benefits to using Active-HDL for simulation and debugging. The tool's adaptability is enhanced by its operation under a widely used language that offers support for numerous platforms. One important benefit is that metrics driven verification can help find untested sections of the system's architecture. Popular Active-HDL code coverage analysis tools perform this verification (ALDEC 2024). By being compatible with Unified Team-based design management, the technology simplifies project administration. Whether working remotely or in the same room, teams may stay consistent with each other using this tool. Teams may use a single modeling platform for the duration of development thanks to the Flow Managers' compatibility with so many vendors (more than 200).

One of the several uses for Active-HDL is to aid in the design and simulation of control device operations and multi-agent robotic system development and simulation (Kiktev et al. 2020). In this example, robotic systems for selecting berries and fruits make use of this approach to mimic the automated synthesis of algorithm programs for actuator movement within a system architecture. Using a Mealy Finite State Machine (FSM) model, the robotic system can tell which berries are good and which ones are bad, permitting the system to pick and reject them accordingly. To better understand the berry processing requirements, this FSM employs a graph model of the control system's automated components. The model is replicated using the states and transitions of the Mealy FSM, which is immediately compatible with the Active-HDL toolset's FSM editor, due to the broad tool compatibility of Active HDL. Systems engineers can use the results of this simulation to check if the robotic system meets the requirements set out by the desired output (Kiktev et al. 2020).

VR for MBSE tool: While MBSE is most often associated with real-world hardware and systems, it also has the potential to be pivotal in the virtual reality (VR) realm. As an alternative to document-based methods, MBSE can greatly benefit from VR in improving system knowledge through visualization (Taentzer and Schürr 2012). With this method, system engineers can now use a virtual reality environment to communicate with the model while simulating the system scenario according to the described model. Facilitating communication with stakeholders is still another advantage of VR. Stakeholders may gain a better grasp of the system's concepts using virtual reality (VR) since they can view and interact with the design perhaps even over great distances. Systems engineers can make rapid changes to the design and specification of a system if stakeholders identify any issues with the system (Taentzer and Schürr 2012).

The idea of developing a virtual environment using System Modeling Language (SysML) has been considered and studied. A systems lifecycle model based on the evaluation of the production stage served as the basis for the study. This study used a SysML-modeled approach that incorporated virtual reality scenario studies including the system, human actors, and the environment. The VR scenes presented components relating to the system's behavior and structure. Using SysML behavior diagrams, system developers could more easily make changes to the VR components. Over the course of the program's lifetime, this approach lessened the burden of scripting the full scenario with VR for individual updates, additions, or use cases. As a result, stakeholders were able to learn more about the system's behavior as the development progressed.

From the beginning of a system's creation all the way to its end of life, engineers and stakeholders can benefit from virtual reality's ability to aid in the MBSE method, enhance design, tell stories virtually, evaluate product performance, and verify requirements. Research into virtual reality's use in the MBSE framework has progressed, opening the technology to a wider audience (Taentzer and Schürr 2012).

We provide a summary of some additional useful MBSE tools in Appendix 5B. This list is not exhaustive but provides some insight into the powerful possibilities that are available with proper implementation of MBSE for systems development efforts. These tools, and others, undergo constant improvement and development and it is best to go to the developer's website and evaluate the current, ever evolving capabilities of these incredibly powerful MBSE tools.

5.1.5 The Shift: Document-Centric to Model-Centric

Historically, systems engineering has focused mostly on documents, resulting in lengthy development cycles for ever-changing systems. The INCOSE Systems Engineering Vision Working Group proposed a model-centric systems engineering strategy that would increase the interoperability of both the present and future products by creating a fully connected and dynamic system (INCOSE 2007).

The present MBSE emphasis promotes the use of digital models that are based on languages like SysML and UML. Teams can swiftly understand design modifications, their intentions, and their effects thanks to the transition from old document-centric techniques to a data-rich, dynamic platform environment. This change is also beneficial in the development of the systems architecture because it enables designers and developers to be informed of the evolving architecture's maturity level. According to Horowitz in his book, "The Importance of Architecture in DOD Software," systems engineers can more accurately assess the impact of a change and swiftly adjust to new requirements when they use models that accurately reflect the architecture (Horowitz 1991, p. 8).

In their 2012 research paper titled, "Document the Model, Don't Model the Document" (Harvey et al. 2012), David Harvey and his colleagues highlight the benefits of an MBSE strategy as opposed to a method that is solely focused on documents. The authors of this paper argue that systems engineering done with a singular focus on documents is simplistic and uninformed. This style of system design fails to take stakeholders and the surrounding environment into account in a comprehensive manner since it lacks system models, context, and definitions (Harvey et al. 2012, p. 3). It is likewise inadequate to have a strategy that is model-centric but document-based. In this setup, a model's focus on the document population obscures its ability to do thorough analysis (Harvey et al. 2012, p. 3).

What about using an MBSE tool to select chunks of documents to model? Additionally, this setup is inadequate since it does not adequately depict the interdependencies among the parts. According to Harvey and colleagues, a document-centric, model-based approach ignores system interconnections and interrelationships in favor of extracting parts from stakeholder material and modeling them in an MBSE tool (Harvey et al. 2012, p. 3).

The method that works best is the model-centric, model-based approach. In this method, the model describes the system and can show all the parts, interfaces, and linkages between them in one location (Harvey et al. 2012, pp. 3–4). Improved system definition clarity, consistency, and traceability result from this setup.

In comparison to the model-centric approach, the document-centric method is not as preferred. As the authors put it, "if the focus remains on documents, then only information needed for those documents will be captured" (Harvey et al. 2012, p. 5), hence it doesn't matter if MBSE techniques are used. Modern challenges require addressing dynamic characteristics and interoperability; the document-centric approach fails to do so when scaling systems or taking system complexity into account. A document-centric strategy, according to Harvey and his team, may yield faster results for simpler systems, but it doesn't offer any technical benefits when it comes to dealing with complicated, or complex systems in real-world scenarios.

The MBSE model repository receives input from requirements specifications, use cases, and the Concept of Operations (CONOPS). The MBSE model repository is updated as the system undergoes transformation with revised requirements, subsystem decomposition, and defined functionality. The model can maintain its integration with the system description through this iterative approach and constant cycles of updates. Reports can be generated from the MBSE model by running it through a report generator. According to Harvey, system engineers and architects may collect, analyze, and record data as it becomes available, which is one of the reasons this model-centric process is beneficial. It is also possible to assess the model's robustness and quickly trace information. As a result, systems engineers are shifting their focus from documents to models and adopting a model-centric, model-based approach.

5.1.6 MBSE Use in the System Development Process

More and more, during system development, Model-Based System Engineering methodologies are being used. Every step of the V-diagram, from requirements gathering and functional analysis to design, development, production, and qualification (testing including verification and validation), is backed by the MBSE approach. All the way from the system's conception to its retirement and disposal, MBSE makes modeling, organizing, and developing an achievable, practical, and beneficial activity. Applying this model-based approach consistently throughout system development fosters a diverse environment that facilitates the establishment of an inclusive system-wide standard (Shevchenko 2021).

Formulating, organizing, and ultimately completely embracing an idea is the first step in building a model-based system. In such a system, there are three fundamental properties (Long and Scott 2011):

1. Entities
2. Attributes
3. Relationships

The real parts and components that make up the system are called *entities*. Dimensions, weight, maintainability, repairability, cost, and system reliability are all examples of *attributes*, which are quantifiable features of entities. The connections between entities and attributes occur through their *relationships* which describe their interactions. According to Long and Scott, the MBSE approach relies on these three properties as foundational building blocks.

The primary deliverables of a model-based system design are the models themselves. Combined with the many other useful UML-based, or SysML-based views of the system, these model artifacts can provide a clear picture of the system's management and control structure and mechanisms at work within the system. In MBSE, the system development effort can benefit from the use of a standard approach, standard values, and standard language when developing the models and artifacts. We now present some example artifacts that are generated using UML and SysML as applied to a representative operational scenario.

Example MBSE artifacts: As we have stated earlier, one of the primary advantages of an MBSE approach is obtaining different views of the system in its environment with complimentary perspectives that provide critical insights into the nature of the system, to include the system's structural, behavioral, requirements, and relational elements. The provided perspectives cover the system's, technical, operational, and program interests for stakeholders, developers, operators, and supporters alike. The different UML and SysML artifacts generated by the MBSE help capture essential aspects of the system required for its successful development and evolution. By methodically determining the MBSE artifacts, critical system features such as the system architecture emerge. The MBSE artifacts span the gamut from the Problem Domain through the Solution Domain and are essential in the understanding, capture, preservation, and configuration control of the system's requirements, and for establishing and maturing the system design. We use an Autonomous, Artificial Intelligence-based Unmanned Aerial Vehicle (A/AI-UAV) for a Department of Homeland Security (DHS), US Customs and Border Patrol mission (US CBP) project as an operational example for illustration. We show some of the different artifacts that can be generated using MBSE and explain how, together, they are useful in understanding essential system details necessary to the design and development effort.

One of the initial artifacts that is essential in understanding the big picture of the system, the expected operations, and the system in its environment is the CONOPS and/or OpsCon. The CONOPS shows how the stakeholders view the mission, and what will be done from their perspective in terms of required capabilities and the system's operations in its greater organizational environment. In contrast, the OpsCon deals with the actual use of the technology in a real-world scenario from the perspective of the system user. Below we show a CONOPS for a notional US Department of Homeland Security (DHS), Customs and Border Protection (CBP) project where the stakeholders at the DHS CBP have a need for a border surveillance system that is capable of detecting, identifying, and/or tracking targets-of-interest (TOIs) that are attempting to illegally cross into our country. Additionally, DHS CBP wants to be able to detect concealed TOIs such as camouflaged TOIs, which have a low-signal detection capability, material characterization capability, and an anti-glare capability. The detection system also must be able to act independently, or as part of an integrated, cooperative swarm of A/AI-UAVs. Figure 5.15 shows the CONOPS for these capability needs expressed from the user's perspective and shows the technologies that

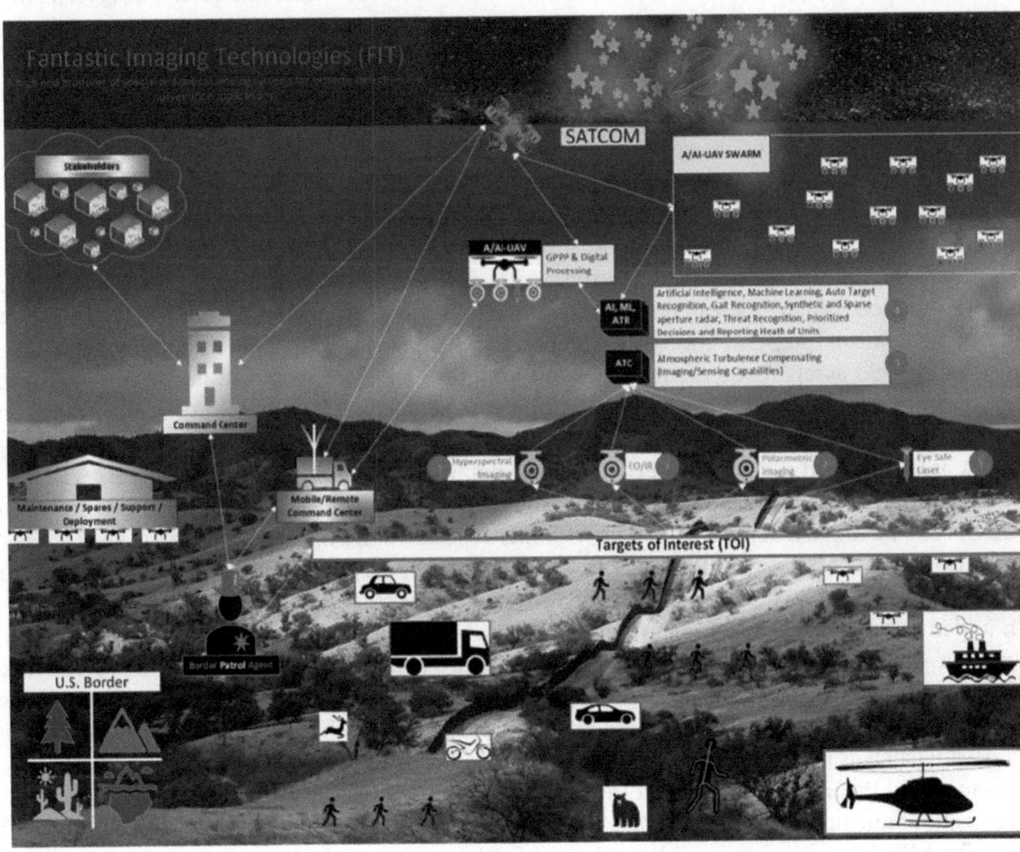

FIGURE 5.15 Concept of operations (CONOPS) for A/AI-UAV Border Patrol scenario (Schwartz 2023, with permission).

satisfy the needed capabilities and features. Notice the wealth of information in the CONOPS. The CONOPS is determined through interaction with the stakeholders to determine their needs/wants and required capabilities and features. There is usually a lot of give and take, with trade/feasibility studies involved before the CONOPS is sufficiently refined. So, what is our system? The answer is it depends (useful right)! The *system* can be different depending on the stakeholder viewing the CONOPS. For instance, from the point of view of DHS CBP, everything working together could be the system-of-systems that they are interested in. From the point of view of the UAV manufacturer, the host UAV platform is the system from their perspective. From the optical and sensor suite development company, the system could be the illumination and sensing systems suite (hardware, software, and laser illuminating system) that fits on the host UAVs. Usually, different contracts are awarded to different suppliers of different elements shown in the CONOPs and each will have their system documentation that is relative to their effort/system. Note, however, that all these separate systems, still interact according to the vision that is manifested in the CONOPS (e.g., the sensors are still on the UAV which communicates with the satellite and performs the mission according to DHS CBP directives). The MBSE architecture can link all these systems together and provide the integrated system-of-systems overarching model. As the understanding of the capabilities and features of the system emerges, the CONOPS evolves until it represents a good enterprise operational picture from the stakeholder's perspective, and that captures and reflects the essential stakeholder wants and needs. The system itself (A/AI-UAV) is viewable in its environment (surrounding connected parts) and illustrates its role in the enterprise. From the CONOPS, we can see that the A/AI-UAV must be able to operate autonomously, or as part of a swarm of similar A/AI-UAVs. We see the types of capabilities used to satisfy the stakeholder needs/wants (material characterization, low-light imaging conditions, atmospheric turbulence compensation (ATC) for best possible spatial resolution in imagery for detection, identification, and/or tracking, and anti-glare). Special AI and signal processing capabilities features such as machine learning mechanisms, automatic target recognition, synthetic aperture, sparse aperture, radar, and gait recognition features are called out. Different types of targets of interest are shown to include vehicles from motorcycles to trucks to cars, people, animals, slow flyers, and boats/ships. We can also see that there is a remote, control center, as well as a field control unit, so the A/AI-UAV must be capable of manual local or remote control/override in addition to its autonomous solo/swarm mode. At the bottom-left, different types of terrain indicate the different environmental operating conditions of the A/AI-UAV Border Control Detection System. Communication/control links are also shown through the satellite communications/control link, and direct communication with leadership and responders, including border patrol field assets. Even maintenance and support facilities are shown indicating long-term program support requirements. With a slightly different perspective, where the point of view is from the end user and not the aggregate stakeholders, and where mission operations and specific technologies are concerned (i.e., hyperspectral sensor, polarization sensor, laser illuminator, ATC EO/IR sensors), we then have the Operational Concept representation (OPSCON). The CONOPS and/or OPSCON can be imported into a MBSE environment as shown in Figure 5.16. Here we used CATIA System-of-Systems Modeler to generate the MBSE model of the CONOPS. In Figure 5.16, you can see actors represented by little stick figures, entities by the blocks, explanatory information in boxes with folded corners on the right, and relations by the connections. Note that in Figure 5.16 all the essential aspects necessary to the operational mission are contained in the MBSE model. The mission support facility is not shown in this model because it may not be relevant in direct operations. However, it could be very easily added, as needed for Use Cases wherein the support requirements are relevant (such as in contingency planning scenarios for instance). Note also that there is a person with a man pad that represents a direct threat to the A/AI-UAV and mission operations that wasn't in the original document-based CONOPS but could be relevant in a set of possible encounters with TOIs. The MBSE version is very adaptable and further can be linked to other models to provide added visibility into the systems structure, behavior, requirements, and/or system performance and parametric relationships. Notice also, that the system architecture is already beginning to emerge,

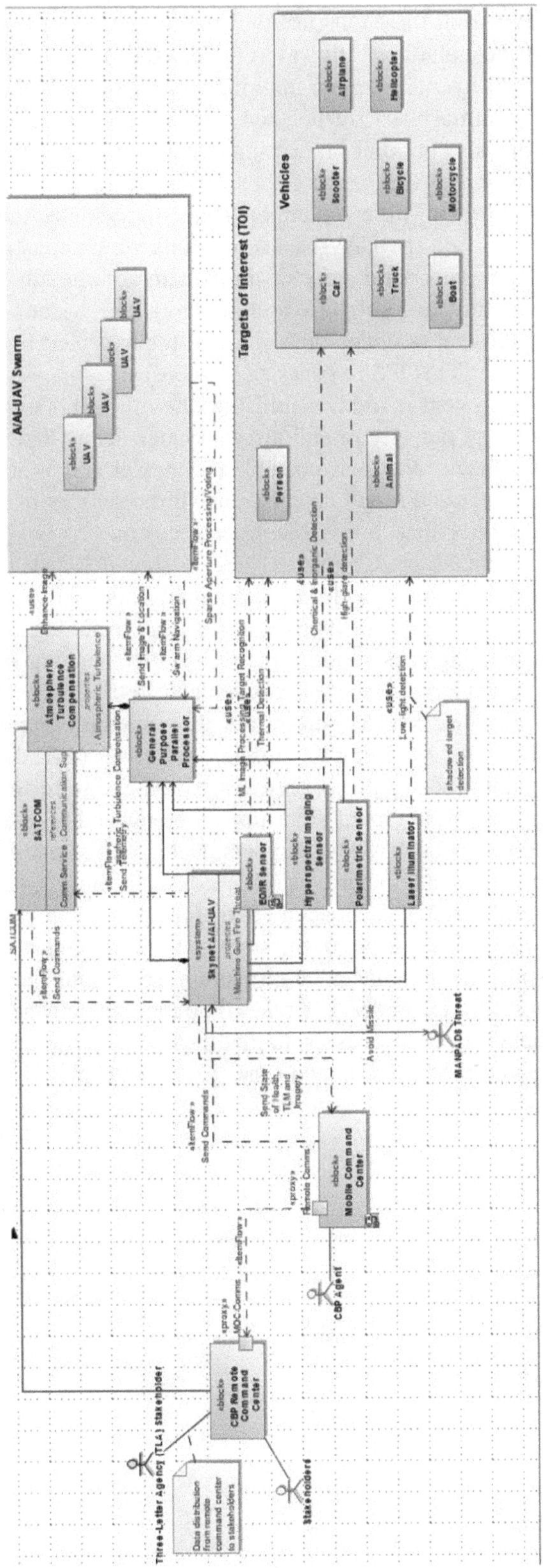

FIGURE 5.16 CONOPS of A/AI-UAV US Customs and Border Patrol Detection System (MacDonald 2023, with permission).

even at this early stage. Let's say that we are the optical sensor suite developer, Fantastic Imaging Technologies (FIT). We know from the CONOPS that the border surveillance solution is an airborne one, that it uses a UAV host platform, that it has optical sensors to include an EO/IR imaging capability, hyperspectral capability, polarization imaging capability, with all sorts of advanced signal/image processing features (atmospheric turbulence compensation, machine learning, automatic target recognition, general purpose parallel processing, advanced digital filtering, and more). We have insight into the platform, sensor types, and even software features that enable us to eliminate a variety of non-solutions (like ground-based detection, space-based detection, infrasound queuing, and more). Instead, we can focus on the UAV-based solutions which constrains our design space accordingly and consequently fleshes out a piece of the system architecture (as well as the logical architecture, hardware architecture, and software architecture). The system architecture is already *beginning* to emerge. Next, we need to understand some of the expected behaviors of our system. We can do this by analyzing our CONOPS to generate Use Cases.

Use Cases capture how the system is used in fulfilling the mission. This provides insights into some of the behavioral aspects of the system and also indicates actors and their roles. The actors participate by acting upon a behavior that is executable (Schneider and Winters 2001). Figure 5.17 below shows a Use Case for concealed target detection using the sensors in our A/AI-UAV Border Patrol surveillance problem. Notice that in this Use Case the drone carrying the illumination and sensing suite has the AI subsystem engaged. The A/AI-UAV is shown as an actor at the left. Once it is in a location of interest (operational area), the drone autonomously engages in scanning the scene for items on its target set. The drone has several sensors at its disposal including a hyperspectral imaging sensor (HSI), a polarization imaging sensor, and a regular EO/IR imaging system. The DHS CBP mission in this case is a concealed target detection mission and so the drones AI is engaged to carry out the concealed detection mission. The AI determines the imaging parameters such as standoff range, required spatial resolution for satisfying the mission objectives (TOI detection, identification, tracking, material characterization, scan pattern, optical settings, and so forth) and goes through the data/image/signal collection process. Onboard algorithms are engaged to extract meaningful information from the data set and report back to stakeholders. Data is also stored onboard the drone. A variety of Use Cases are generated, and this provides a great deal of insight into the behaviors of the system and the interactions of the system with actors and entities in the environment of the system. Use Cases can be generated for different types of missions such as tracking a human target, identifying contents of a material spill, finding concealed targets under camouflage netting, and engaging in search and rescue operations, for example. Any type of operational engagement that is of interest to DHS CBP can be represented by Use Cases. Note that these Use Cases are also in effect models that can be linked to the CONOPS or OpsCon and other parts of the model. As such attributes can

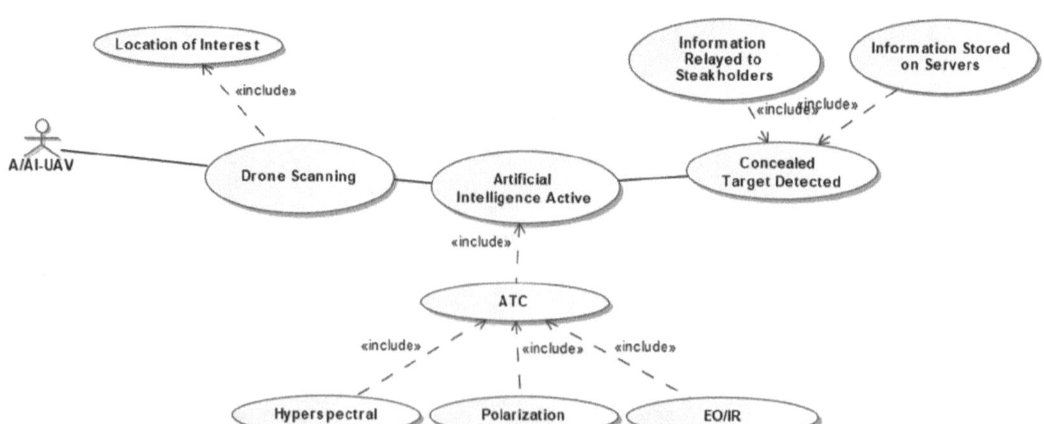

FIGURE 5.17 Use case diagram of A/AI-UAV Border Patrol application (McNabb 2023, with permission).

be assigned to each of the actors and entities such as weight, height, speed, test criteria and more, and relationships established in the model through the connection lines between them. As many Use Cases as are necessary can be generated. For each Use Case, the next step is to generate a set of *Scenarios* using, for example, Activity Diagrams, that provide the sequential steps that need to be carried out for the Use Case in question. Usually, a sunny-day scenario, where everything goes right, and a rainy-day scenario, where everything goes wrong are generated. Also, a nominal scenario which represents a normal mission with expected perturbations is generated. For instance, accounting for the observation area having partial cloud cover would be something that would be relatively commonplace depending on the location of the operations. Mission planning changes such as altitude changes, flight pattern changes, and such could be used to adapt to these circumstances.

The Activity Diagrams can be used to illustrate the operational or flow details of the system. Specifically, for our A/AI-UAV for the DHS CBP border surveillance problem for automatic sensor selection, we have the following activity diagram shown in Figure 5.18. Activity diagrams may be broad in scope or narrowly tailored to fit the needs of the system that is under development. The stakeholders, the entire system, or only a subsystem can be the focus in these diagrams (Eriksson et al. 2003, Chapters 5 and 6). Notice the AI Automatic Sensor Selection Scenario in Figure 5.18 has rounded rectangles that show the actions that are being taken. The diamond represents a decision point and, as before, the rectangles with the fold at the top-right are information blocks. The circles are entry or exit nodes to the Activity Diagram. Notice that this Activity Diagram shows which sensors are engaged and in what order. It also shows what happens if a sensor has a positive, or negative result. A self-diagnostic routine is engaged if none of the sensors provide a result, and reporting actions, and actions taken for a successful, or unsuccessful detection are provided. When all the steps required for the Scenario are captured by the Activity Diagram, along with anticipated problems and workarounds, the scenario is complete, and the next scenario is generated. Once all the scenarios have been generated for all the relevant Use Cases, these artifacts and other *problem definition* artifacts are used to generate requirements. For example, in the above scenario, there is a step to lower the A/AI-UAV. How fast must this be initiated? Immediately? Wait for a more opportune time? How fast will the A/AI-UAV descend? What route will it take? What information needs to be reported back to the command center, and how fast does this information take to get there? Several requirements need to be generated to nail these things down. Once this is determined, a requirement can be written such as, *The A/AI-UAV shall send confirmation of a positive detection to the Mission Control Center (MCC) within 5 seconds of making the detection.* If the image data is also required at the MCC, then another similar requirement could be generated for the image and/or video. Further, the 5 second response constraint can be attached to a separate communications block activity diagram model and used as a constraint on the evolving system design. The model can automatically check to ensure design changes/implementations do not violate this constraint. At this point, we can once again refine the system architecture from what we know. Figure 5.19 shows the evolving architecture for the A/AI-UAV which consists of all the functionality for the host UAV platform, including the avionics subsystem, communications subsystem, propulsion subsystem, structure and mechanics subsystem, autonomy, and AI subsystem, as well as the illumination and sensing system. Note that at the more detailed level, we see from the Optical AI Signal Processing element, that there is an Aspheric Turbulence Compensation element that connects to the atmospheric turbulence compensation functionality. This indicates that there is a required fixed spherical aberration correction associated with the legacy hardware that must be properly dealt with in the atmospheric turbulence compensation functionality. From the block definition diagram, we can create an internal block definition diagram to further detail these blocks. For example, the Hyperspectral Imaging Sensor block is shown in Figure 5.20.

The project name, Skynet is shown in the header of the IBD. The functionality of the internal components of the HSI sensor is listed in the rectangular boxes. The items, in turn, can be further decomposed with IBDs. Note that the *ImSpector Optics* block refers to the specialized optics that

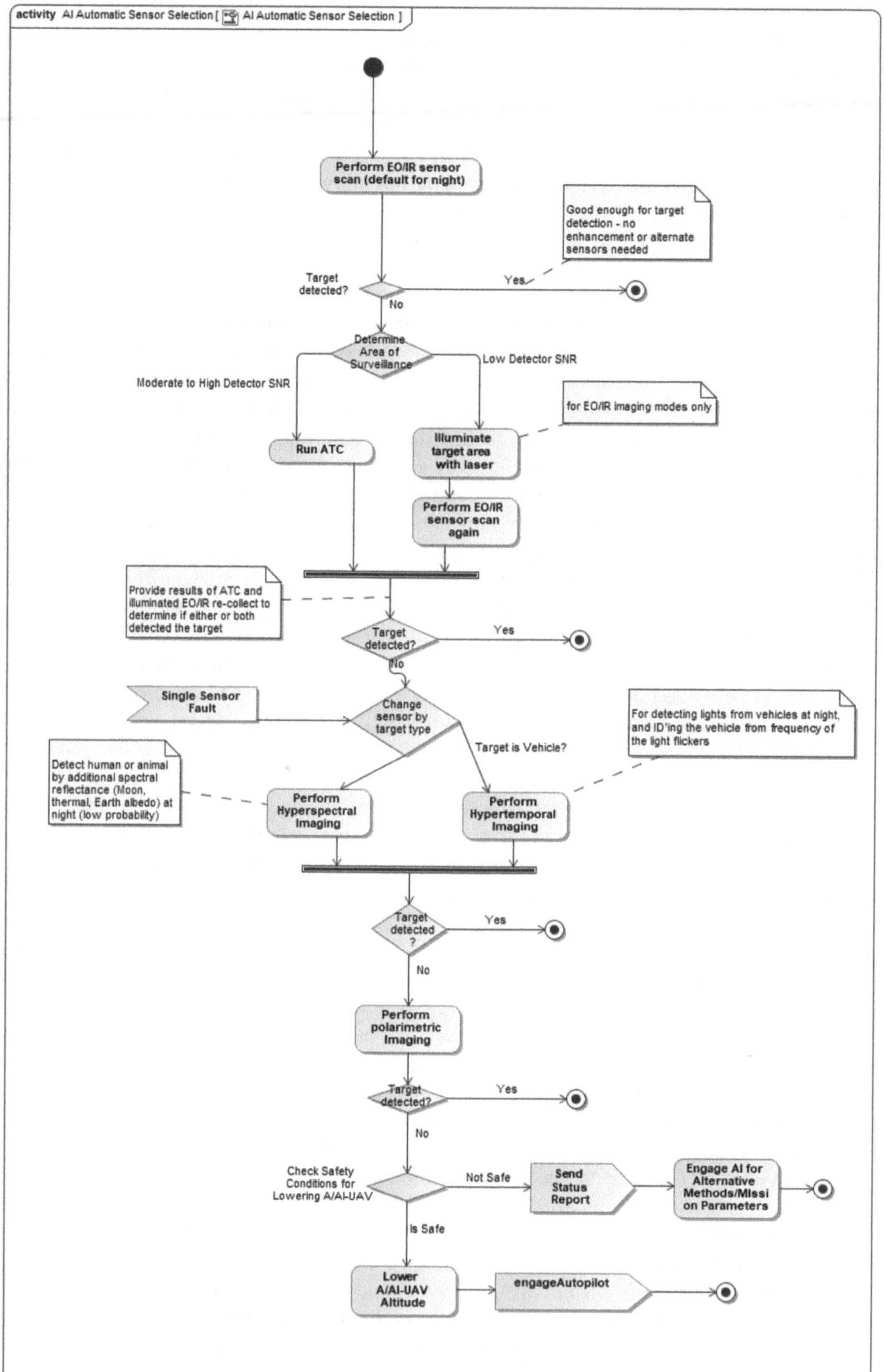

FIGURE 5.18 Activity Diagram for A/AI-UAV AI automatic sensor selection (McDonald 2023, with permission).

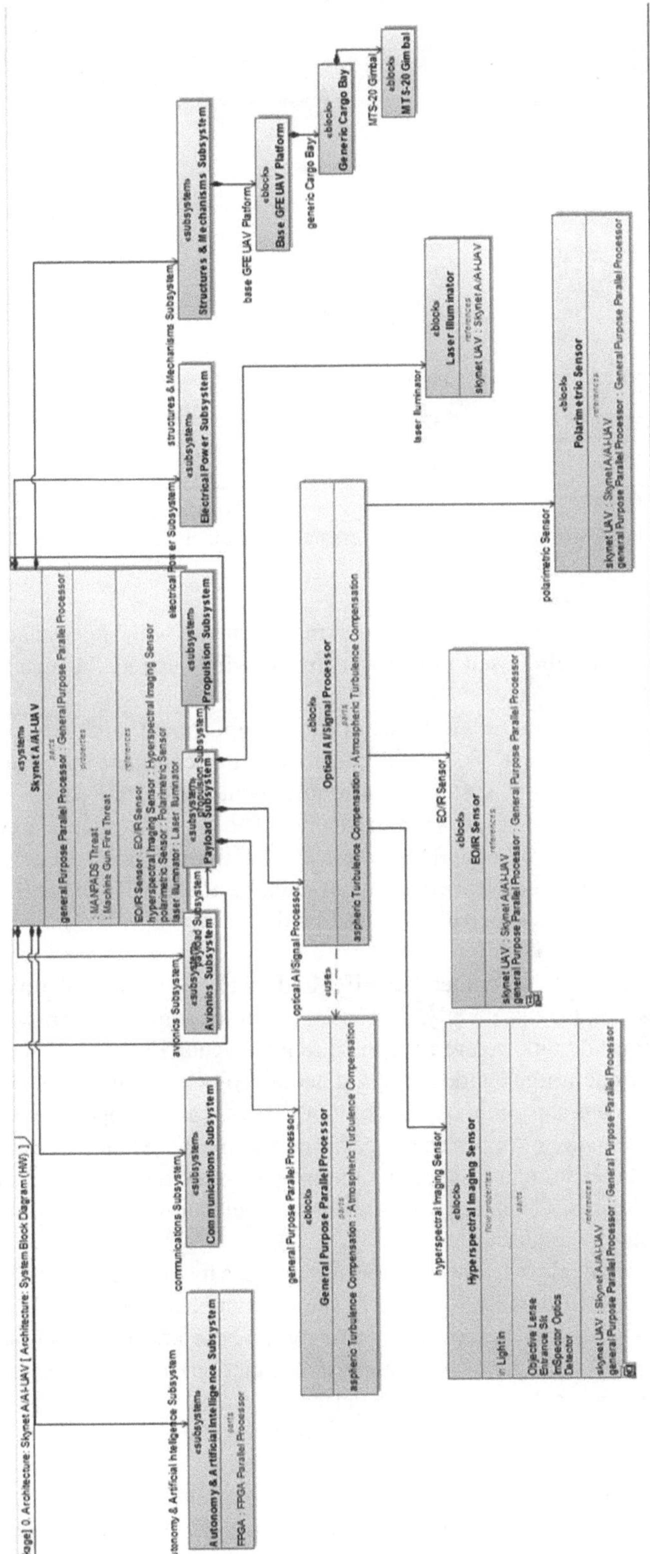

FIGURE 5.19 Block definition diagram for an A/AI-UAV (Macdonald 2023, with permission).

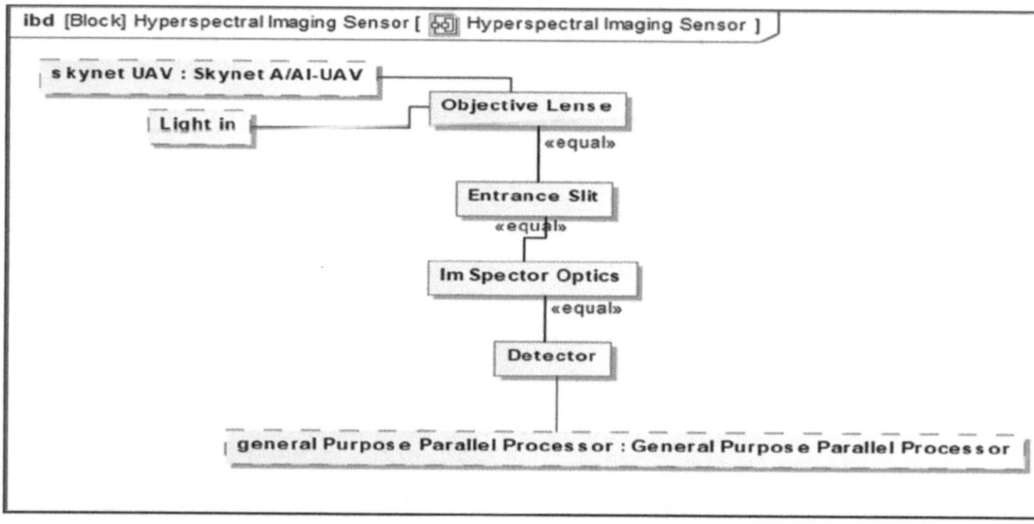

FIGURE 5.20 Internal block definition diagram hyperspectral scanner (Courtesy of McNabb 2023).

are needed for the hyperspectral sensor in either the imaging mode with spectral analysis (*ImSpect), or,* the hyperspectral sensor being used just for spectral sensing such as for material characterization, or material identification.

Requirements diagrams are also a vital part of the system model, and these can also be integrated into the MBSE environment. The requirements can be separated into their respective levels (stakeholder requirements level, system level requirement, subsystem level requirements, and components/pieces/parts/assemblies level requirements). These requirements are captured in the system model and linked to each other and elsewhere needed in the model. Figure 5.21 shows some *functional* requirements for the SkyNet program that must be accomplished for the mission to be a success. Functional requirements specify *what* must be done whereas non-functional requirements specify *how well* the system works.

For example, the capabilities to detect, identify, and track a TOI are of fundamental importance to mission success, and consequently, detection, identification, and/or tracking requirements cover what the system must do, and so are functional requirements. Figure 5.21 shows the detection requirement for various operational modes (e.g., detection mode at nighttime or daytime, using atmospheric turbulence compensation (ATC), or not). The separate operational modes can be parameterized, and links made to the technical performance measures and metrics for the individual modes. These linked metrics can be used in other places in the model as constraints or as input parameters to model relations. Changing a technical parameter in the requirements, immediately affects results throughout the model.

Reliability requirements deal with how well the system does its job and so these types of requirements are *non-functional* requirements. Figure 5.22 shows examples of non-functional requirements that are related to the Availability requirements category. These are non-functional requirements because they deal with how long and how well the system functions and not specifically what it must do. For example, the Operational Availability requirement states what percentage of the time the system must be operational. Here, the system is operational and doing its thing (what it must do) and this requirement deals with how often percentage-wise it must do it.

A major advantage of using MBSE is that the models are often stored in a centralized database so that all authorized personnel have access and can see the same thing (as opposed to distributed documents that are maintained at separate locations and can have errors). If anyone looking at the database finds an error, it can be brought to the attention of the development team and corrected

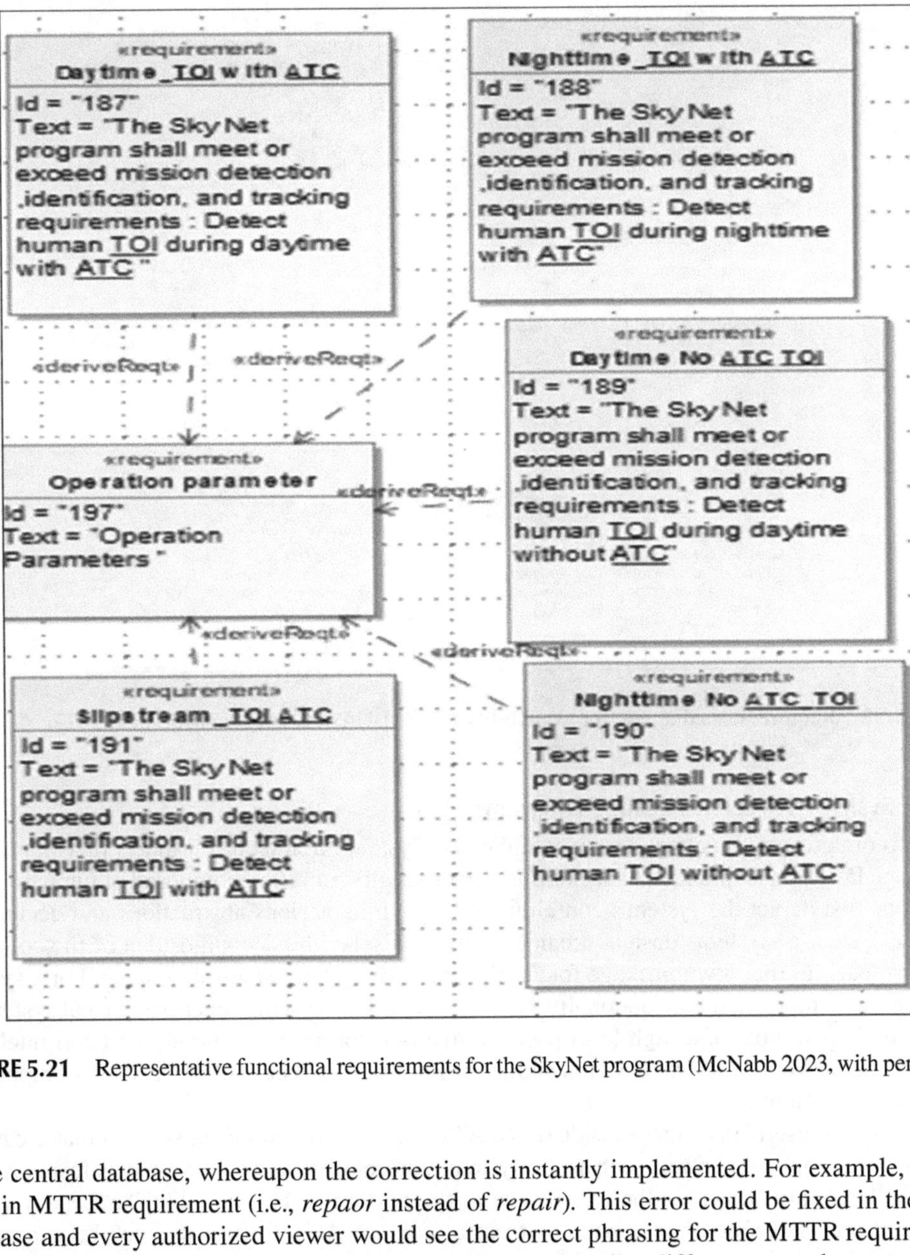

FIGURE 5.21 Representative functional requirements for the SkyNet program (McNabb 2023, with permission).

in the central database, whereupon the correction is instantly implemented. For example, note the error in MTTR requirement (i.e., *repaor* instead of *repair*). This error could be fixed in the central database and every authorized viewer would see the correct phrasing for the MTTR requirement.

Also, note how these artifacts are used together to describe different, complementary views of the system. The MBSE modeling effort continues in this fashion, with each artifact providing a different glimpse into the nature of the system, its behavior, and the system's interactions with the environment. We end up with new insights that help us define the system architecture and the system's requirements. These artifacts ultimately help us articulate and evolve the system's design. We now want to show examples of the different types of artifacts that are available in SysML along with a short description of how they are used. We also provide additional examples using the A/AI-UAV per above to illustrate the points. There are 9 SysML diagrams that are primarily used to model systems. We address each of these diagrams and also present a few additional examples from a Department of Defense Architectural Framework (DODAF) plugin that are often used. We use the CATIA System-of-Systems Modeler to produce these artifacts. There are more artifacts that can be generated with this excellent MBSE tool, but we focus on these primary ones for now.

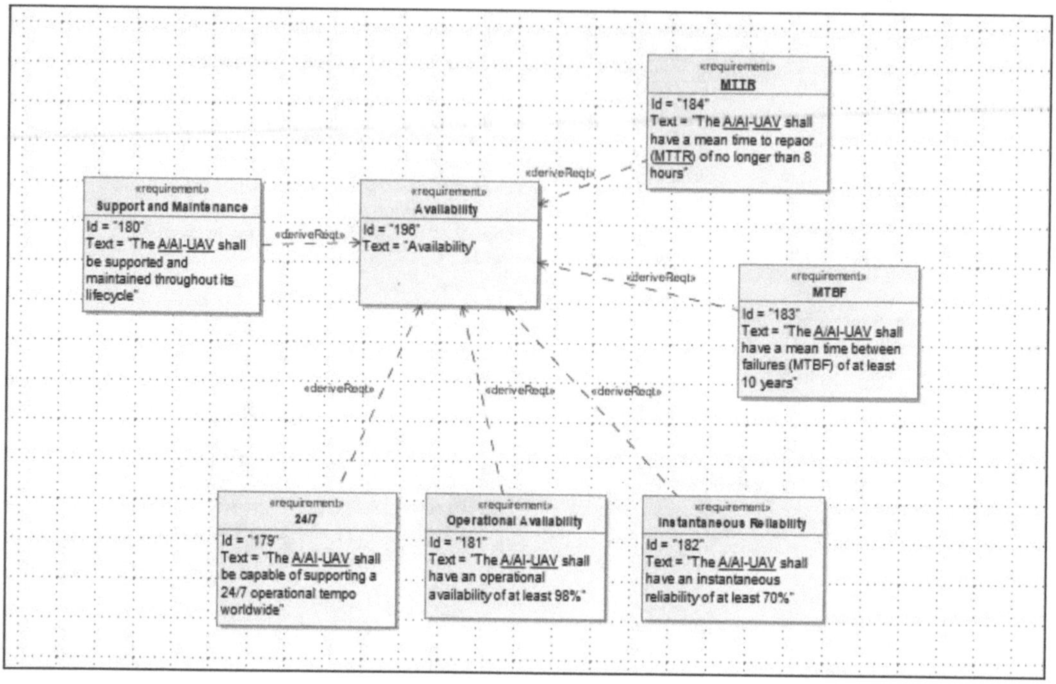

FIGURE 5.22 Requirements diagram for availability of A/AI-UAV (Courtesy of McNabb 2023).

5.1.6.1 SysML Artifact 1: Package Diagrams

Helping to organize and manage large, complicated System Architecture Models (SAMs) is what the Package Diagram is intended to accomplish. Package diagrams are composed of interconnected folder icons that depict the system's containment through its various abstractions and decompositions. In a system-of-systems design, it can effectively display the decomposition of the collection of systems, each in their own package folder, that comprises the system-of-systems. For a systems design, the decomposition can optionally be with respect to system functions, or subsystems of the system. An example of a high-level package diagram for an autonomous, artificial intelligent unmanned aerial vehicle (A/AI-UAV) is shown in Figure 5.23. The A/AI-UAV is decomposed in terms of its subsystems.

There are also useful packages outside of the A/AI-UAV decomposed subsystems that are necessary for the developmental effort such as the system requirements and legacy models from previous programs. These are shown as the two blocks at the bottom of Figure 5.23. The Legacy AUAV models house reusable/modifiable artifacts from a previous AUAV effort that can be adapted for the new program. Notice that the system that is represented in the package diagram in Figure 5.23 represents the entire A/AI-UAV and the optics, illumination, and sensing elements are viewed as subsystems with respect to the A/AI-UAV system. As will be seen in our Integrated Case Study in Section 5.3, the optics, illumination and sensing elements can be implemented by a separate contractor that views the illumination and sensing suite developmental effort as their *system*. So, whether or not a particular block in a SysML artifact, like the package diagram in Figure 5.23, is viewed as a system or subsystem element, depends on the point of view of the stakeholders, and the relative decomposition that is being considered. For example, from the perspective of a governmental agency that wants to use an A/AI-UAV in part to satisfy a mission requirement, their perspective could be the entire system-of-systems in its operational environment(s) and the A/AI-UAV could just be one of many entities making up their system-of-systems. For the A/AI-UAV developer, their perspective and system is the A/AI-UAV itself. On the other hand, for the optical systems developer,

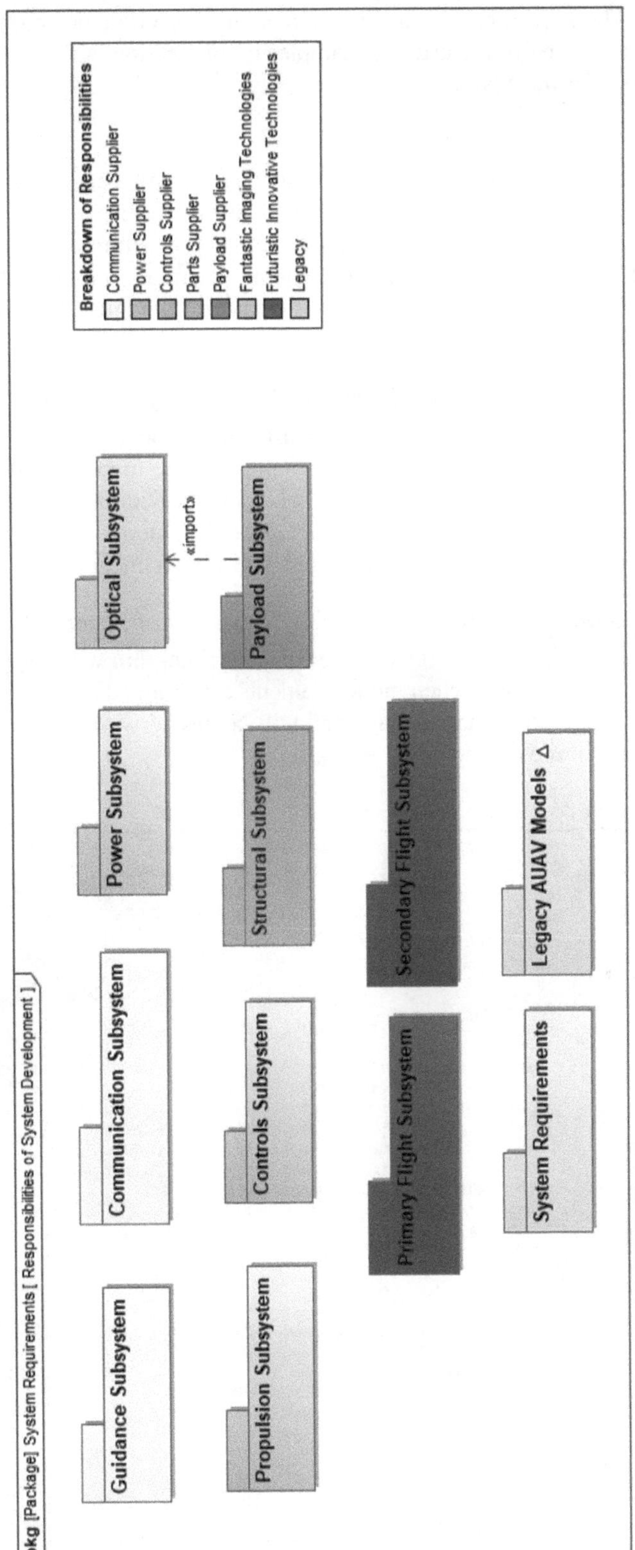

FIGURE 5.23 A/AI-UAV system-level package diagram.

their perspective and system is the optical system itself, which is considered a subsystem to the A/AI-UAV developer. The MBSE environment can keep all these linkages and different perspectives straight. For the remainder of the artifact examples in this section, we adopt the perspective of the A/AI-UAV developer for the examples.

5.1.6.1.1 Artifact Methodology and Technique

With the use of package diagrams, a modeler can arrange model pieces into folder-like "packages" that are visually separated. These packages provide for the representation of model elements arranged into groups as can be seen in Figure 5.24. The folder structure of the packages shown in Figure 5.23, is generated by the SysML modeling tool and essentially matches the CATIA System of Systems Modeler "Containment Tree" structure, as shown in Figure 5.24 in diagram format (Batarseh and McGinnis 2012).

5.1.6.1.2 Application: How the Systems Engineer Uses Package Diagrams

Package diagrams are used by systems engineers to illustrate model hierarchy and containment. Containment is essential since it establishes the components of the model and their relative hierarchy. Containment aids in communicating with stakeholders by providing a simplified model for the complex system that includes ordered groupings of the model elements. These model elements are available to other system artifacts and models in the MBSE environment.

5.1.6.1.3 Implementation Description/Steps in CATIA System of Systems Modeler

Dragging a package from the containment tree onto a package diagram will display all other model items contained within that package. When model elements are dragged onto a package in the package diagram, they are contained in that package and will be nested within the containment of that package in the containment tree (Lempia and Ray 2011).

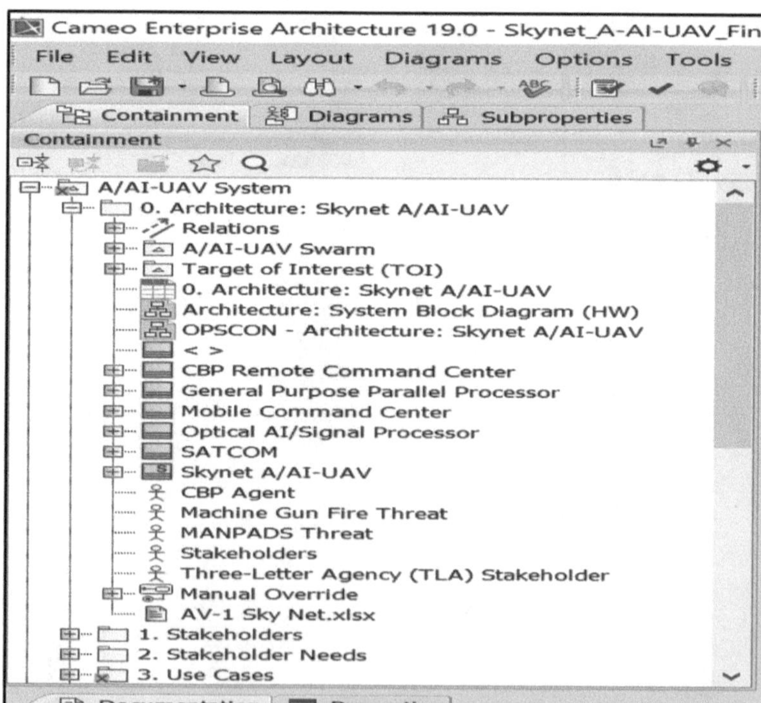

FIGURE 5.24 A/AI-UAV containment tree.

By using the import relationship connection arrow from one package to another, as shown by the arrow direction in Figure 5.23, the optical subsystem is imported into the payload subsystem package. Consequently, the optical subsystem becomes a part of the total payload subsystem in this A/AI-UAV illustration.

5.1.6.2 SysML Artifact 2: Requirement Diagrams

According to Tsadimas, Nikolaidou, and Anagnostopoulos, requirement diagrams are used to outline functional and non-functional requirements within the model. This allows for their verification through test cases and their satisfaction by other model elements (Tsadimas et al. 2012).

The following is a preliminary system requirements diagram for the A/AI-UAV example:

Note that some of these requirements will require further definition, analysis, and refinement and can be improved (they are preliminary after all). Once in the MBSE environment, these requirements can be linked to each other, to test artifacts, to parametric models, and more, as can their technical performance metrics.

5.1.6.2.1 Artifact Methodology and Technique

As an example of a typical requirements decomposition in system design, a requirement diagram can represent a hierarchy of requirements, their attributes, relationships to other requirements (e.g., derived requirements), links to other artifacts and modeling elements (e.g., Use Cases, OpsCon, and Determine Gait Profile model), provide linkages for trace functionality and satisfaction arguments, and provide supporting information (as indicated by the blocks with the folded corners at the top-right side of the rectangle in Figure 5.25). The stereotypes <<derive>> and <<refines>> provide derivation and refinement relationships. The arrowhead on the <<derive>> indicates the source to which the derived requirement at the tail of the arrow originated from. The <<refines>> feature can provide additional details through other modeling elements (e.g., artifacts like use cases or other MBSE models). The refining model or artifact is shown at the tail of the connecting arrow, and the refined requirement is shown at the arrowhead.

5.1.6.2.2 Application: How the Systems Engineer Uses Requirements Diagrams Implementation Description/Steps in CATIA System of Systems Modeler

Another approach to creating a requirements diagram is to use satisfy relationship arrows to connect requirements to model objects or elements. You can also use verify relationship arrows to connect test case elements to model elements, showing that a specific test case will test a model component and verify a requirement. In essence, the modeler would add requirement diagrams onto block definition diagrams (BDD), packages, and use cases, and then connect them with the relevant <<satisfy>> and/or <<verify>> relationships.

The Requirements Table is an additional SysML item that can be used to convey requirements in the context of the CATIA System of Systems Modeler application. By selecting a containment level in the containment tree, a requirements table—a specific kind of allocation table—can be configured to list requirements that are scoped in the table settings. Figure 5.26 shows the addition of user-selectable "Satisfied By" and "Derived From" allocation views as new columns. You can then list columns such as requirement ID, requirement text, and so forth.

5.1.6.2.3 SysML Artifact 3: Activity Diagrams

As a kind of behavioral diagram, activity diagrams (SysML and UML) provide information about the execution of tasks rather than the states of objects within a system. A state machine diagram executes transitions and actions in response to predefined events or conditions, while an activity diagram depicts a system carrying out a specific task, like a flowchart.

Figure 5.27 shows the previous SysML activity diagram presented in Figure 5.18 above. This figure was developed in CATIA System of Systems Modeler software for an automatic sensor switching function for an A/AI-UAV.

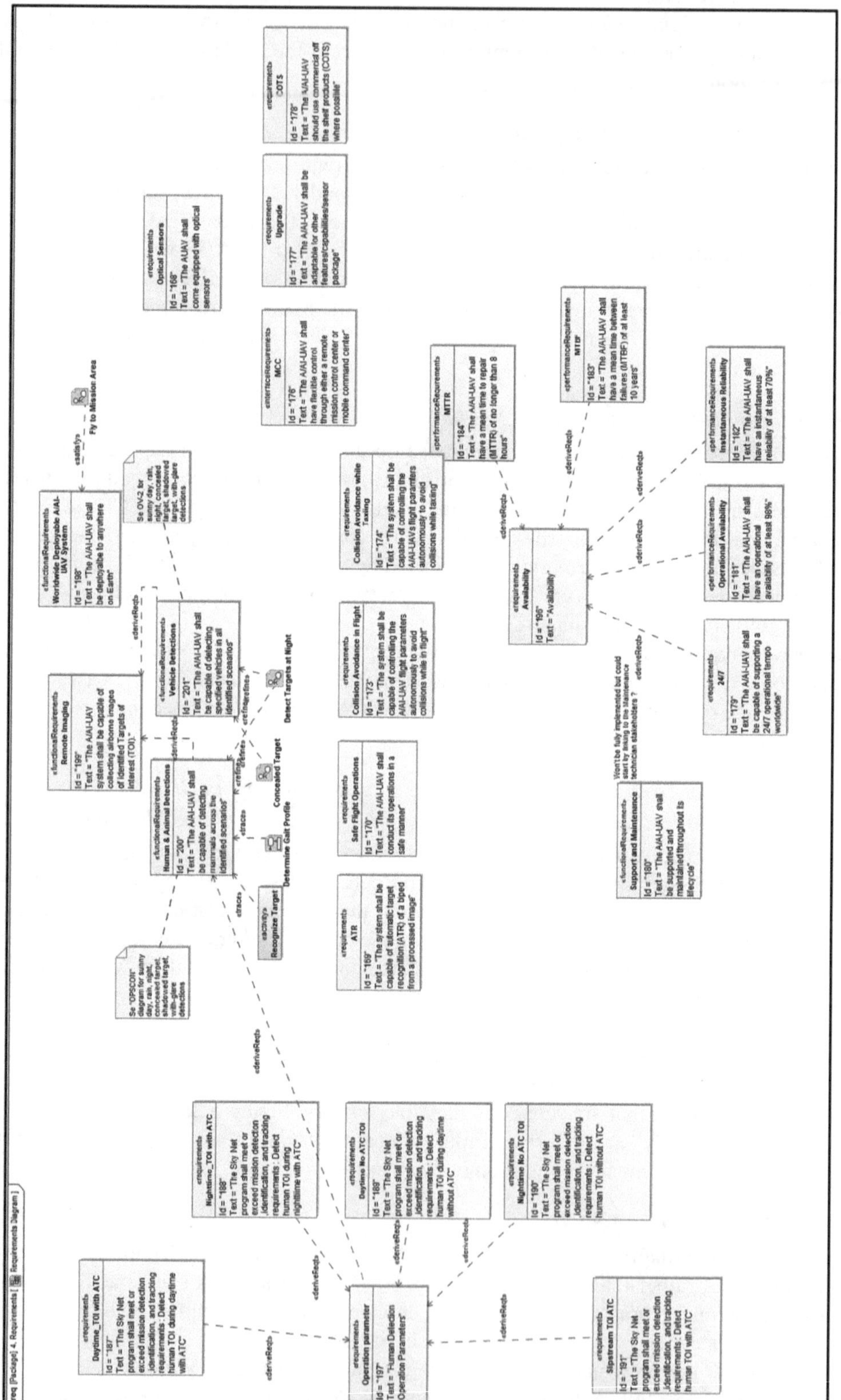

FIGURE 5.25 A/AI-UAV representative preliminary requirements diagram.

#	Name	Text	▽ Satisfied By	Derived From
1	195 Resolution	The minimum spatial resolution of daytime imaging shall be seven resolution cells across the linear extent of the target	Telescope	
2	198 Worldwide Deployable A/AI-UA	The A/AI-UAV shall be deployalbe to anywhere on Earth	Fly to Mission Ar	
3	185 Data Transfer	The AUAV shall be capable of sending critical information to the Mission Command Center within a configurable time initially set to 5 seconds.	Communications	
4	167 UAV MCC Interface	The A/AI-UAV system shall interface with the existing AUAV mission control center to provide the best possible imaging technology and data processing for detection, identification, and tracking capabilities.		
5	187 Daytime_TOI with ATC	The Sky Net program shall meet or exceed mission detection ,identification, and tracking requirements : Detect human TOI during daytime with ATC		197 Operation pa
6	188 Nighttime_TOI with ATC	The Sky Net program shall meet or exceed mission detection ,identification, and tracking requirements : Detect human TOI during nighttiime with ATC		197 Operation pa
7	189 Daytime No ATC TOI	The Sky Net program shall meet or exceed mission detection ,identification, and tracking requirements : Detect human TOI during daytime without ATC		197 Operation pa
8	190 Nighttime No ATC TOI	The Sky Net program shall meet or exceed mission detection ,identification, and tracking requirements : Detect human TOI without ATC		197 Operation pa
9	191 Slipstream TOI ATC	The Sky Net program shall meet or exceed mission detection ,identification, and tracking requirements : Detect human TOI with ATC		197 Operation pa
10	168 Optical Sensors	The AUAV shall come equipped with optical sensors		
11	169 ATR	The system shall be capable of automatic target recognition (ATR) of a biped from a processed image		
12	170 Safe Flight Operations	The A/AI-UAV shall conduct its operations in a safe manner		
13	192 Pitch	The system shall be capable of sensing the AUAVs pitch		
14	193 Roll	The system shall be capable of sensing the AUAVs roll		
15	194 Heading	The system shall be capable of sensing the AUAVs heading		
16	171 Altitude	The system shall be capable of sensing the AUAVs position relative to the ground		
17	172 US Airspace	The system shall prevent the AUAV from leaving the US airspace		
18	173 Collision Avoidance in Flight	The system shall be capable of controlling the A/AI-UAV flight parameters autonomously to avoid collisions while in flight		
19	174 Collision Avoidance while Taxiin	The system shall be capable of controlling the A/AI-UAVs flight paramters autonomously to avoid collisions while taxiing		
20	175 Power	The system shall transform the incoming power voltage to suit components		
21	176 MCC	The A/AI-UAV shall have flexible control through either a remote mission control center or mobile command center		
22	177 Upgrade	The A/AI-UAV shall be adaptable for other features/capabilities/sensor package		
23	178 COTS	The A/AI-UAV should use commercial off the shelf products (COTS) where possible		
24	179 24/7	The A/AI-UAV shall be capable of supporting a 24/7 operational tempo worldwide		196 Availability
25	180 Support and Maintenance	The A/AI-UAV shall be supported and maintained throughout its lifecycle		
26	181 Operational Availability	The A/AI-UAV shall have an operational availability of at least 98%		196 Availability
27	182 Instantaneous Reliability	The A/AI-UAV shall have an instantaneous reliability of at least 70%		196 Availability
28	183 MTBF	The A/AI-UAV shall have a mean time between failures (MTBF) of at least 10 years		196 Availability
29	184 MTTR	The A/AI-UAV shall have a mean time to repair (MTTR) of no longer than 8 hours		196 Availability
30	186 Auto-Pilot	The AUAV shall have an artificial system with autonomous decision-making capabilities		
31	196 Availability	Availability		
32	197 Operation parameter	Human Detection Operation Parameters		200 Human & Ar
33	199 Remote Imaging	The A/AI-UAV system shall be capable of collecting airborne images of identified Targets of Interest (TOI).		
34	200 Human & Animal Detections	The A/AI-UAV shall be capable of detecting mammals across the identified scenarios		199 Remote Ima
35	201 Vehicle Detections	The A/AI-UAV shall be capable of detecting specified vehicles in all identified scenarios		199 Remote Ima

FIGURE 5.26 A/AI-UAV representative requirements table.

5.1.6.2.4 Artifact Methodology and Technique

CONOPS and OpsCons can be good sources for generating Use Cases since they capture important operational aspects of the mission and illustrate the system in its environment. Activity diagrams can be used to determine various scenarios from the CONOPS and OpsCons such as the rainy-day scenario (in which problems are anticipated), the sunny day scenario (in which all goes well and as planned), and the nominal scenario (wherein things go as expected and some expected issues can be addressed). The activity diagrams are useful in capturing these behavioral aspects of the system as it performs its required operations.

5.1.6.2.5 Application: How the Systems Engineer Uses Activity Diagrams

Stakeholders can understand the system's intended purpose and implementation with the use of activity diagrams, which show the flow of functions and activities. The SE can use these diagrams, together with state machine diagrams and sequence diagrams, to illustrate logic flow and incorporate interface requirements into the model.

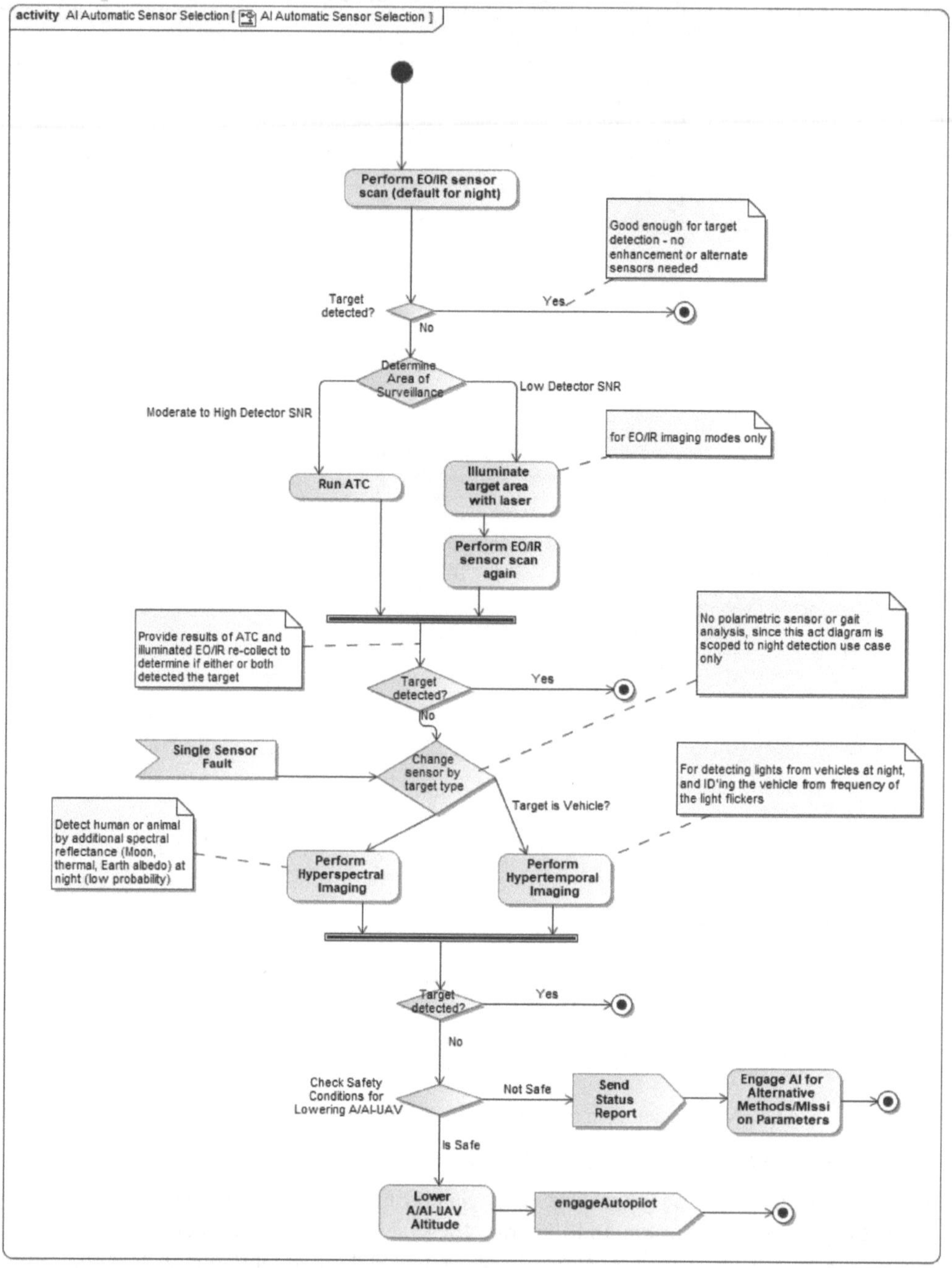

FIGURE 5.27 A/AI-UAV automated sensor switching activity diagram.

5.1.6.2.6 Implementation Description/Steps in CATIA System of Systems Modeler

A SysML activity diagram makes use of the following symbols:

- *Rounded rectangles:* Represent activities.
- *Arrows:* Connect tasks and represent control flows.
- *Diamonds:* Represent decision nodes.

- *T-junctions:* Illustrate forks with corresponding flow labels indicating which flows must be satisfied.
- *Y-junction:* Labels the incoming flows with the conditions that must be satisfied for the join to take place.

In CATIA System of Systems Modeler, vertical or horizontal swim lanes can be created. Activities can then cross these lanes and connect to other swim lanes to illustrate how processes and their intra-system flows are connected to each other. This diagram can be useful for generating requirements at the interfaces.

5.1.6.3 SysML Artifact 4: Sequence Diagrams

One kind of interaction diagram that shows how items interact with one another over time is a sequence diagram. These diagrams are used in both UML and SysML. Sequence diagrams can illustrate messages or data transfers that are required to carry out operations that must be exchanged in Use Cases, or certain tasks. As an example, a general flow diagram for the A/AI-UAV system's flying operations is shown in Figure 5.28.

As another example, the Scan for Target sequence can be represented using a sequence diagram with units and model elements, as illustrated in Figure 5.29. Here we illustrate the scanning process of the EO/IR camera on the A/AI-UAV when attempting to find motion in the scene. In this mode, the camera continuously scans the scene as indicated in the block that is labeled *Loop* in Figure 5.29. The camera undergoes a scan pattern chosen by the Illumination and Sensing System AI and continues this pattern until either motion is detected, or the AI changes the scan pattern. If motion is detected, the image is sent to the image processor for processing.

5.1.6.3.1 Artifact Methodology and Technique

When working with a SysML model for more complex parametric simulations, it is important to make use of several sequence diagram symbols found in the SysML ontology. These sequence diagram symbols provide enriched information that is required for parametric simulation such as defining constraints and parameters.

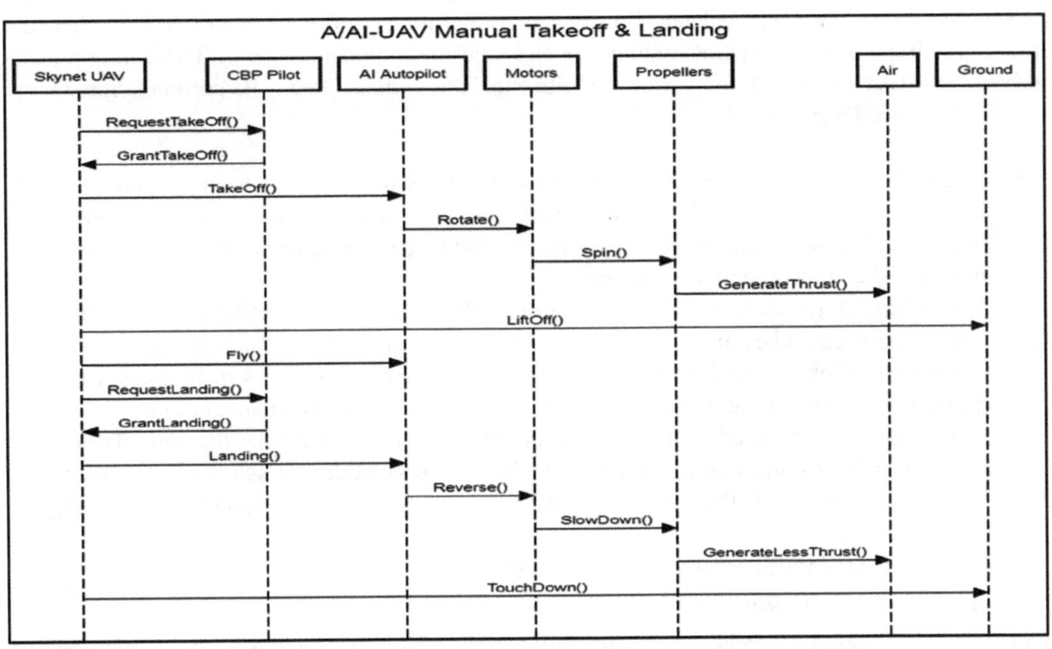

FIGURE 5.28 A/AI-UAV takeoff & landing sequence diagram.

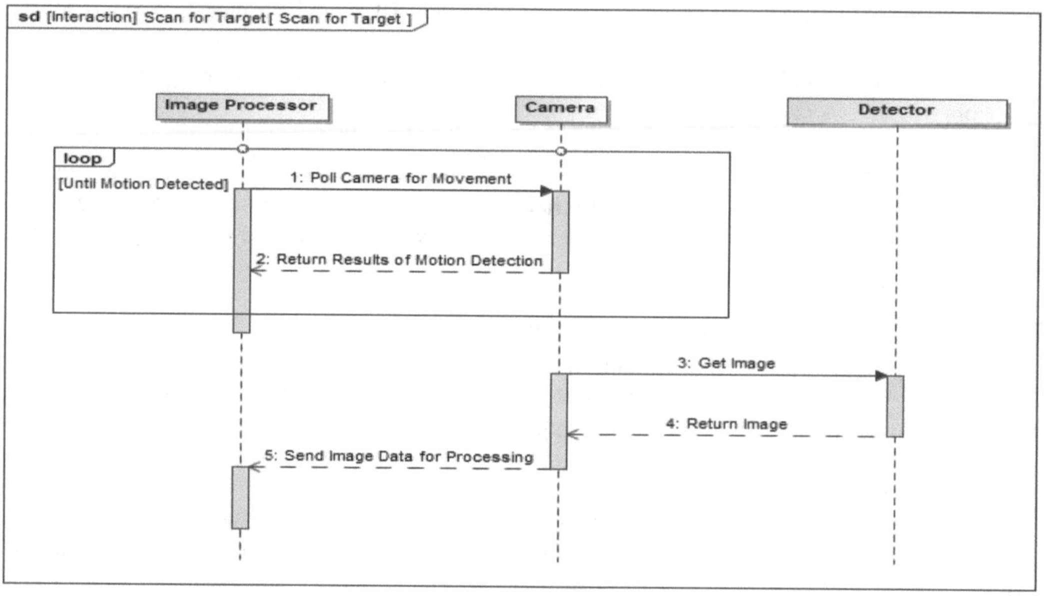

FIGURE 5.29 A/AI-UAV scan for target sequence diagram.

5.1.6.3.2 Application: How the Systems Engineer Uses Sequence Diagrams

Systems engineers rely on sequence diagrams for a variety of reasons, including understanding the flow of operations inside their system, seeing how the system is behaving dynamically, finding logic or dependency issues, and communicating with end users. The verification and validation processes involved in developing and deploying a system can also make use of sequence diagrams, since satisfaction of the requirement may depend on following the sequence of steps in the sequence diagram.

5.1.6.3.3 Implementation Description/Steps in CATIA System of Systems Modeler

CATIA System of Systems Modeler allows users to add images called "swim lanes" to their diagrams, which are used to illustrate changes or flows between model elements. These diagrams are derived from Use Cases and are associated with the Containment Tree. The correct symbols for a SysML Sequence Diagram are:

- *Object lifelines*: Object lifelines represent the objects that participate in the interaction. They are shown as vertical lines that extend from the top of the diagram to the bottom.
- *Messages*: Messages represent the communication between objects. They are shown as arrows that connect the object lifelines.
- *Activations*: Activations represent the period when an object is active and can send or receive messages. They are shown as a horizontal bar on the object lifeline.
- *Constraints*: Constraints can be used to show the order of messages or to specify other restrictions on the interaction. They are shown as dashed lines with annotations.
- *Notes*: Notes can be used to provide additional information about the diagram. They are shown as balloons attached to the diagram. Notes can be added by selecting the "anchor" context in the tool to add them to the sequence diagram model elements and flows for context.

5.1.6.4 SysML Artifact 5: State Machine Diagrams

5.1.6.4.1 Artifact Methodology and Technique

An object's (i.e., a model element's) "lifetime" in reaction to events or triggers can be visually represented by a state machine diagram, which is a type of dynamic behavioral diagram in SysML or

UML. When using SysML in simulation mode, this diagram is second only to the SysML parametric diagram for built-in simulation in terms of importance for describing parametric modeling and discrete event simulation.

As an example, Figure 5.30 shows an example of a state machine diagram for a hyper-temporal imaging (HTI) mode for an EO/IR sensor that can be used to identify vehicle types by differences in their patterns of how the headlights flash. Hyper-temporal imaging accomplishes this by taking very high frame rate images and looking at pattern changes between the image frames. The EO/IR sensor receives a rate change command and then switches to either a normal EO/IR frame rate of 60 frames per second (FPS) or a change to the HTI mode with a frame rate of 240 FPS. The EO/IR sensor is commanded to collect data until sufficient data is captured, or the collection is aborted. For either circumstance, a stop collection request is sent to the EO/IR sensor.

5.1.6.4.2 Application: How the Systems Engineer Uses State Diagrams

A state machine diagram is used by systems engineers in a manner like an activity diagram; both allow for the general representation of model elements or systems and their state transitions in response to events or triggers. It differs from general activity flows in that state transition diagrams focus on displaying system states and transitions.

5.1.6.4.3 Implementation Description/Steps in CATIA System of Systems Modeler

There are several key attributes of a state diagram (Friedenthal 2014):

- *Initial and final states:* These states are shown as a black dot for the input, and black dot(s) within circle(s) for outputs.
    - This means that multiple output states can be represented.
- *State:* Depicted as a rectangle with rounded corners.
- *Actions*: Can be "guarded" meaning the actions can be automatic or require conditions. Actions are considered to be executed instantly not like in a discrete event simulation where there can be temporal aspects to actions.
- *Decision blocks*: Decision blocks are modeled by diamonds and are also known as decision nodes.
- *Transitions*: Transitions are illustrated with arrows and indicate changes between states.

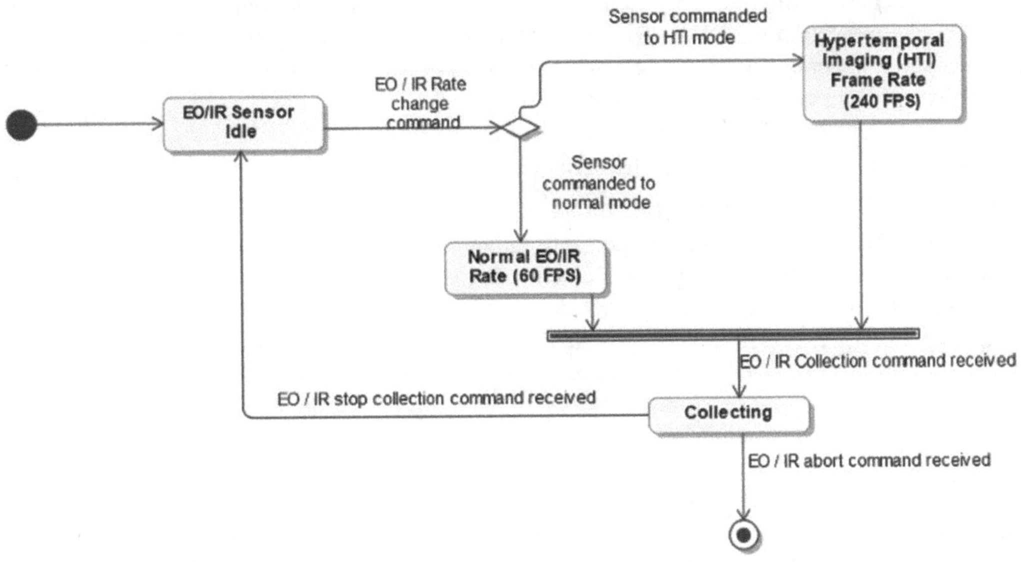

FIGURE 5.30 A/AI-UAV state machine diagram for transitions within the EO/IR sensor.

5.1.6.5 SysML Artifact 6: Use Case Diagrams

5.1.6.5.1 Artifact Methodology and Technique

Use case diagrams are tools for communicating the system's high-level requirements to stakeholders in a way that is both visually appealing and easy to understand, without resorting to technical jargon. According to Weilkiens, Use Cases establish connections between the various functions, and activities of a system, as well as their interactions with users and actors in the system's external environment (Weilkiens 2020).

As a representative example, we see a Use Case in Figure 5.31 for an A/AI-UAV flying to the area where the mission is to occur. The stick figures are actors and the ovals represent operational actions that need to be accomplished for the A/AI-UAV to get to its destination. Notice that there is also an ordering to the actions (e.g., first communicate with the operations center, then taxi, then take-off, and so forth). Notice also that in this Use Case, the AI is controlling the navigational and flight details to get the A/AI-UAV to its destination.

5.1.6.5.2 Application: How the Systems Engineer Uses Use Case Diagrams

The Use Case diagram can be created from high-level system information like the OV-1 and OV-2 artifacts in SysML to help in early architecture view creation and understanding of system functions involving important actors. This process aids in decomposing the system into a logical structure and establishing relations between objects, flows, and early requirements.

5.1.6.5.3 Implementation Description/Steps in CATIA System of Systems Modeler

The Use Case diagram can be created in MBSE modeling software such as CATIA System of Systems Modeler by dragging actors into the diagram and connecting them to distinct oval Use Case parts. These Use Cases should be linked to each other using the <<include>> and <<extend>> relationships based on the functions and activities the system will execute with its users. An example is shown in Figure 5.32 where the illumination and sensing system (IaSS) is being used to detect targets of interest at nighttime. Relevant actors are shown as stick figures outside the central part of the Use Case. Actions are shown in the ovals and some order of precedence can be inferred. For instance, note the oval that says "Fly to Mission Area" at the near the top-left of the figure. Recall that this oval has a Use Case assigned to it already that we just previously discussed. There are many actions and interactions that are shown on this representative Use Case and even some links to other SySML artifacts (like the Fly to Mission Area Use Case, and the hyper-temporal imaging processing state machine diagram, and the AI Sensor Switching Activity Diagram). In general, Use Cases can be linked to other useful MBSE artifacts to provide better insight into how the system functions and behaves in its operational environment.

5.1.6.6 SysML Artifact 7: Parametric Diagrams

5.1.6.6.1 Artifact Methodology and Technique

A parametric diagram can capture important mathematical relationships, system parameters, and connections between blocks, or system elements. Constraint blocks are used to establish the rules and parameters that restrict the value attributes of the blocks (Hart 2015).

Figure 5.33 displays a parametric diagram that assesses the feasibility of dynamic picture collection by comparing it to a detection distance equation. The inputs include the mean, narrow-band optical wavelength, the required spatial resolution, the focal length of the optical system, and whether it is day or nighttime. The note shows some of the details of the optical system parameters (not all details are shown in the note, such as atmospheric turbulence compensation on or off, or the required spatial resolution, or special collection considerations due to space). The representative parametric diagram can determine the maximum stand-off distance as a function of the mean, narrow-band wavelength, entrance pupil diameter $D_{\_ep}$, and required spatial resolution *spatial_resolution*. The power of parametric diagrams is that they can be coupled to other parametric diagrams to build

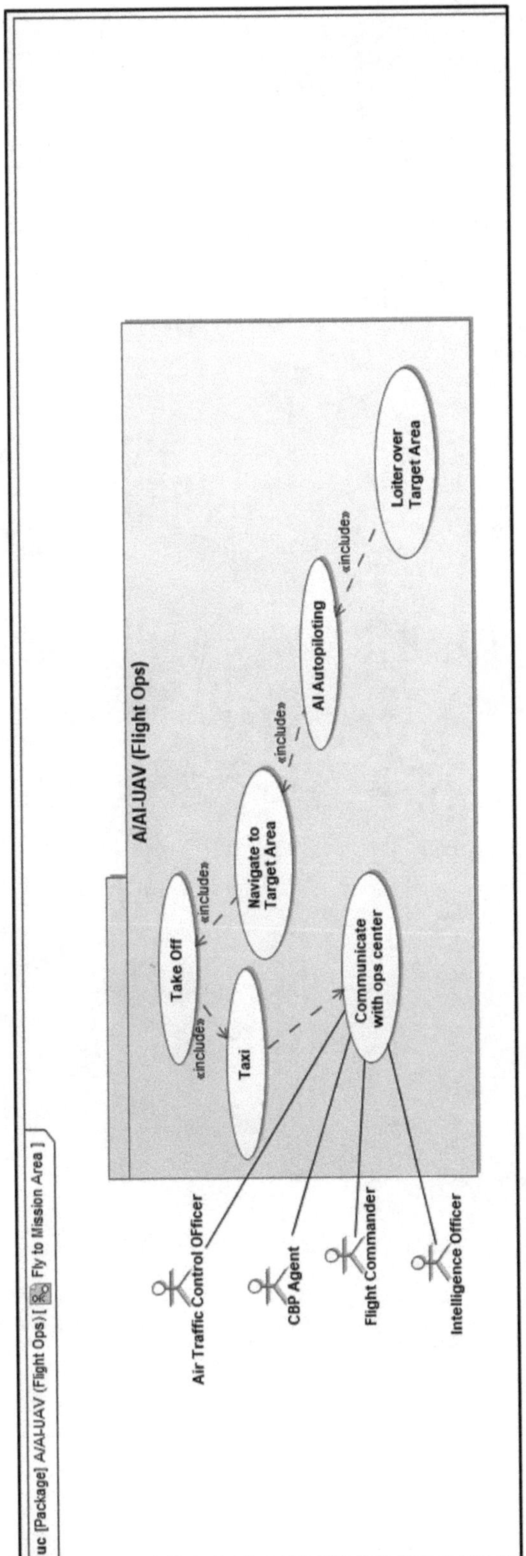

FIGURE 5.31 A/AI-UAV fly to mission area use case diagram.

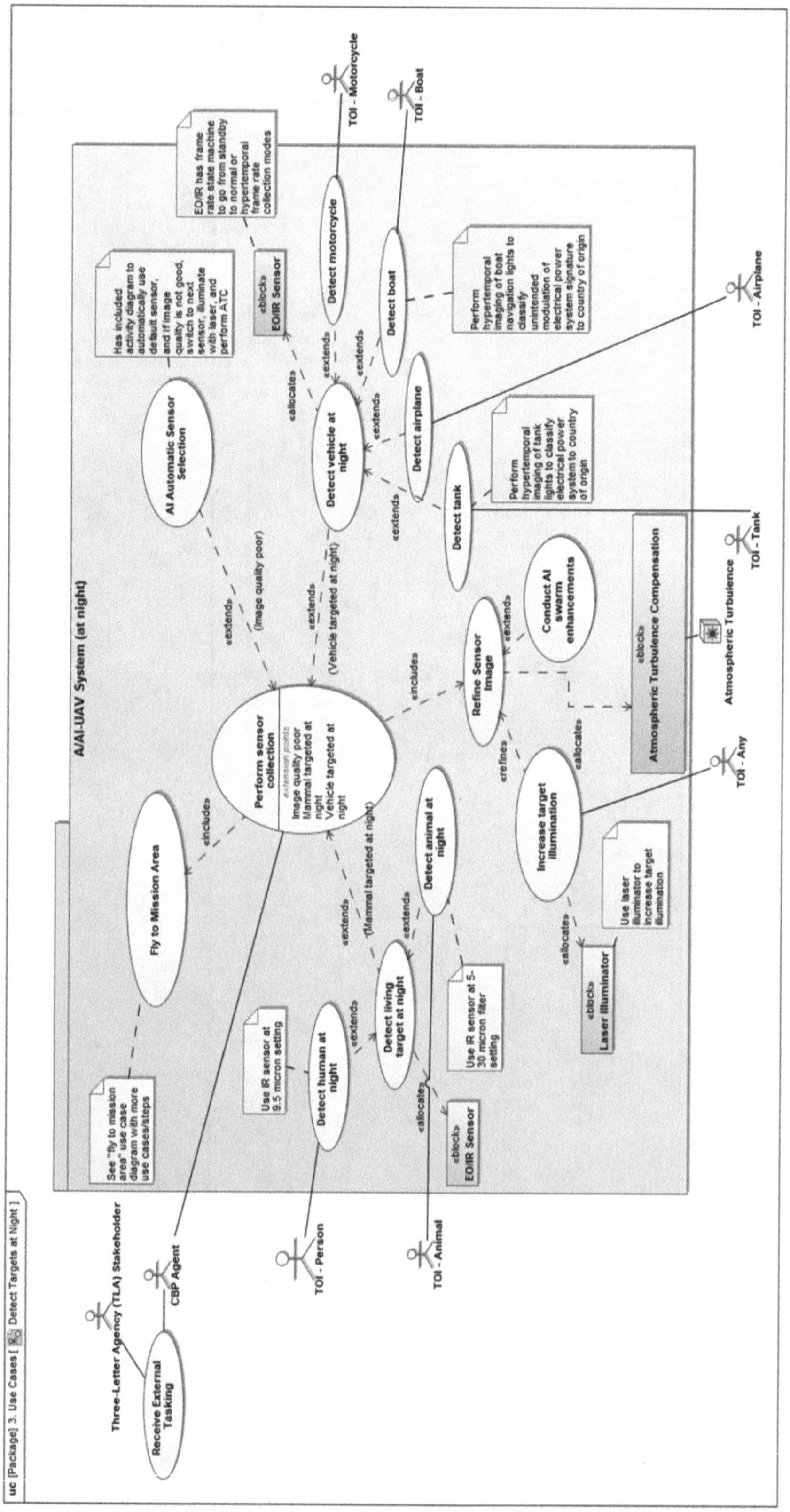

FIGURE 5.32 A/AI-UAV detect targets at night use case diagram.

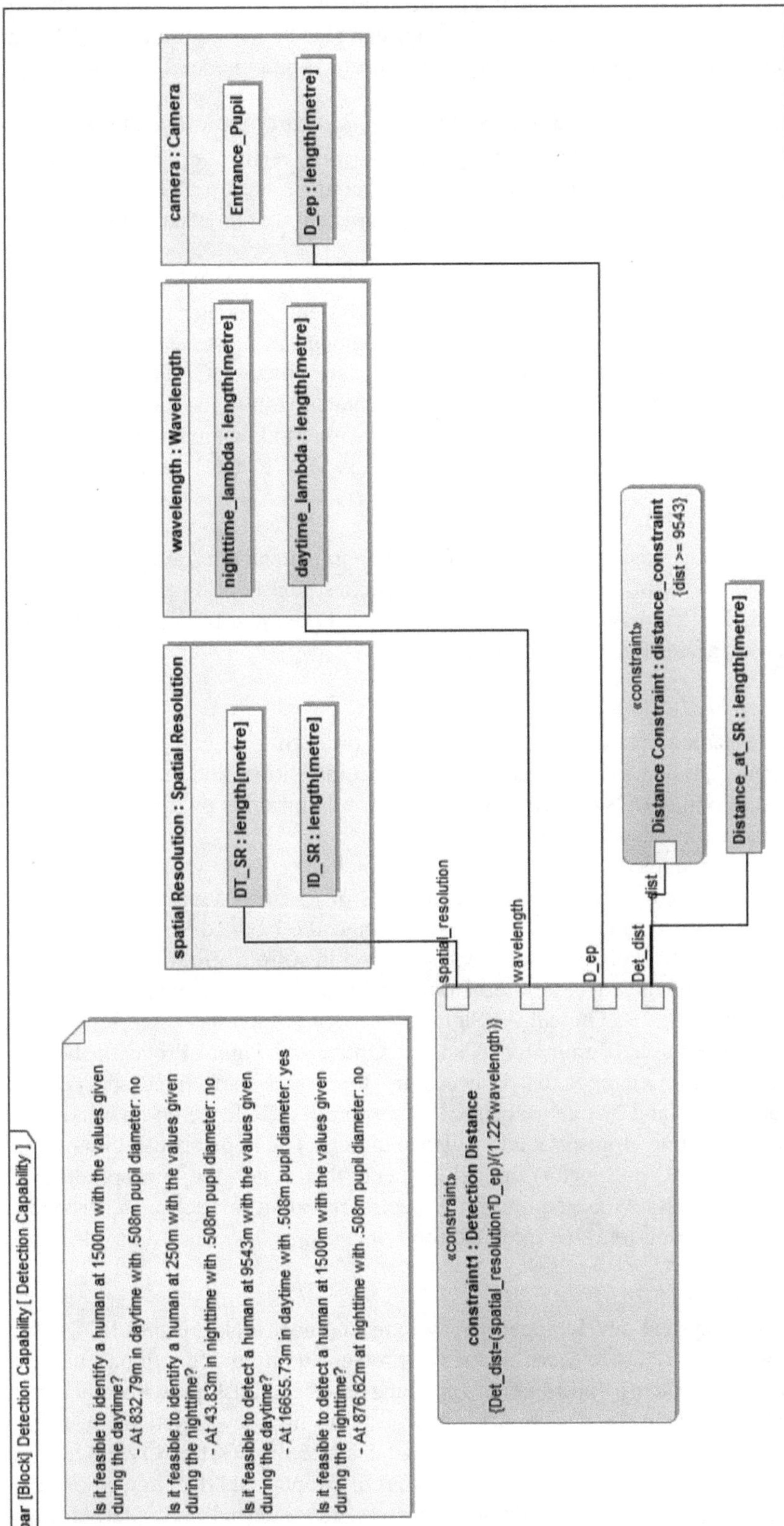

FIGURE 5.33 Parametric diagram for detection feasibility (adopted from legacy).

larger connected, executable models that capture scientific, engineering, and mathematical relationships important to the development effort. They can also be linked to other SysML artifacts to illustrate connections and apply constraints where required and beneficial.

5.1.6.6.2 Application: How the Systems Engineer Uses Parametric Diagrams

Engineers can use parametric diagrams for systems analysis, verification, and validation. To gain a better grasp of system performance or to analyze and confirm requirements, parametric diagrams can be run in SysML software tools to provide a dynamic simulation capability.

5.1.6.6.3 Implementation Description/Steps in CATIA System of Systems Modeler

The wizard tool in software like CATIA System of Systems Modeler can be used to construct parametric diagrams from mathematical equations, although in most cases, constraints and value properties (parameter values) are defined before they are applied. CATIA simulations can verify, validate, and debug the model by providing executable results, or they can dynamically run functions (called "constraints") in the model, quantitatively modify attribute values, and introduce logic-based relations such as IF/ELSE, TRUE/FALSE, AND/OR, NOT.

5.1.6.7 SysML Artifact 8: BDD

BDD in SysML are the logical extension of UML component diagrams. Inherently modeling object-oriented blocks, defining their parts and properties, and relating them amongst each other with various SysML relationships and containment structures, it is the most fundamental type of SysML structure diagrams (Mori et al. 2018).

5.1.6.7.1 Artifact Methodology and Technique

Using the object-oriented breakdown of the system in the form of rectangular blocks, a BDD can provide valuable insight into the systems functional decomposition, and the evolution of the system architecture. In addition, BDDs can display inter-block relationships including dependencies, linkages, and generalizations (Bruel and Roques 2013).

A representative example of the BDD is shown in Figure 5.34 where an A/AI-UAV has its top-level functional decomposition illustrated in terms of its subsystems. Subsystems such as the avionics subsystem, the communications subsystem, and the payload system (among others) are defined, and some structural elements of the system architecture begin to emerge. Notice that the Payload Subsystem is further decomposed into three functional elements, the General Purpose Parallel Processor (GPPP), the Optical AI Signal Processor (which has three sensors functionally attached to it), and the Laser Illuminator. Also, the Optical AI Signal Processor has an Aspheric Turbulence Compensation element that is necessary to correct some fixed spherical aberrations on existing legacy optics and has an associated atmospheric turbulence compensation element for correcting general dynamic atmospheric turbulence effects. The hyperspectral imaging sensor has both imaging, or spectral (imSpector) optical elements that match the appropriate hyperspectral detection mode (i.e., a mission that requires only spectral information such as material identification and/or characterization (non-imaging mode), or both imaging and spectral data (HSI mode).

5.1.6.7.2 Application: How the Systems Engineer Uses BDDs

At every stage of the system development, the systems engineer makes use of BDDs. In the initial stages, the SE will use BDDs to learn about the problem domain's functional and architectural system elements, and system's intrarelationships, and interrelationships with other entities in its environment. This will help the development team capture high-level system functions, establish the conceptual architecture, and produce useful artifacts like the OV-1 or OV-2. Once in the solution domain, the SE will use BDDs to establish and refine functional decompositions, develop and evolve the system architecture, and understand the system's internal and external interfaces and hierarchies, as well as to convey the system design to stakeholders.

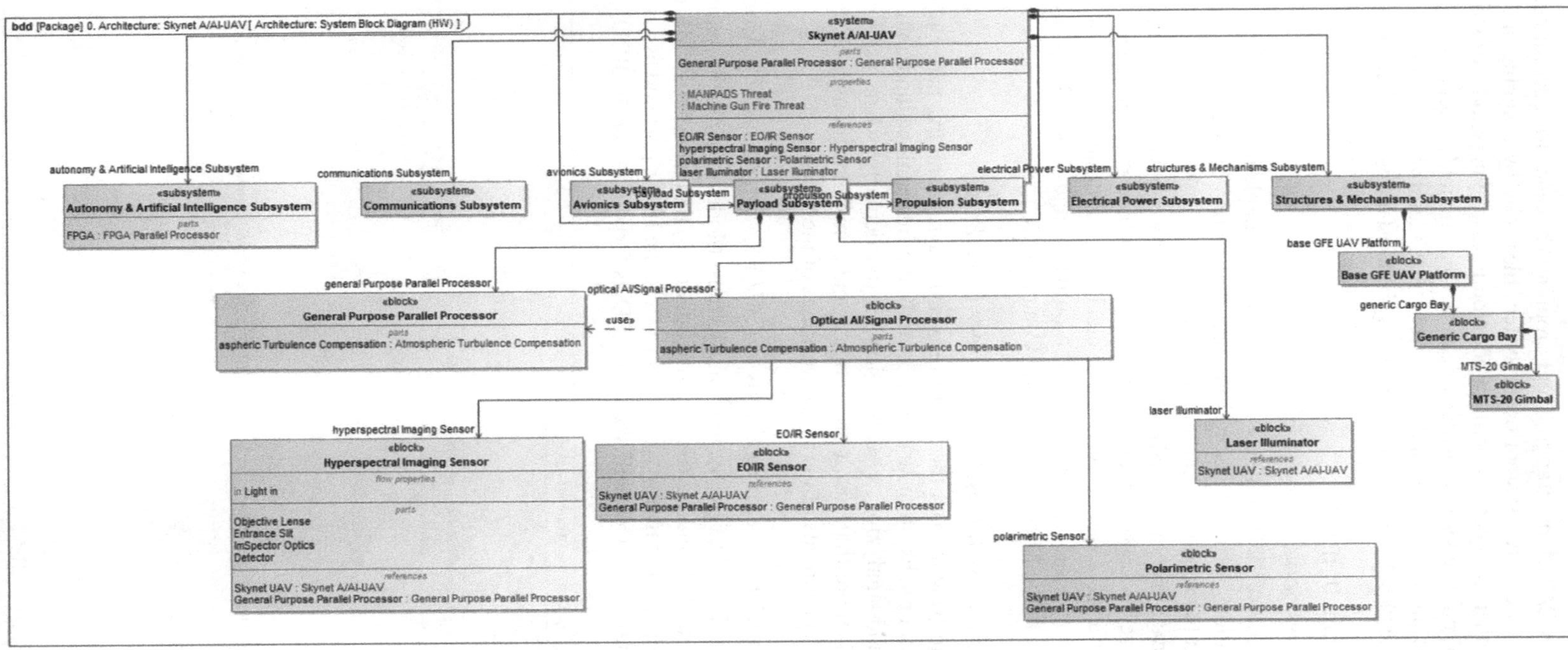

FIGURE 5.34 A/AI-UAV system-level hierarchy block definition diagram.

5.1.6.7.3 Implementation Description/Steps in CATIA System of Systems Modeler

A BDD can be built in CATIA by dragging and dropping packages for containment into its canvas area. The next step is to create a system or subsystem hierarchy by dragging and dropping blocks onto the diagram and then connecting them. In addition to displaying model or structural hierarchy, BDDs may be used to generate various artifacts such as Operational Concepts (OpsCons), Concept of Operations (CONOPS), and other types of free-form context diagrams. Many model elements and connections are permissible in BDDs since they are among the most unrestricted diagram types in SysML.

5.1.6.8 SysML Artifact 9: IBD

5.1.6.8.1 Artifact Methodology and Technique

An IBD in SysML is like a BDD, only it is included within an element block and shown as part of the BDD. The associations, generalizations, and proxy ports can be the same as in BDDs, but it is nested inside a scoped block of the containment tree.

A representative example of an IBD for the A/AI-UAV optical subsystem is illustrated in Figure 5.35. The Image Processor receives movement, heat, and time information, from the connected camera, temperature sensor, satellite, and clock ports. Adding a contextual comment helps illustrate the flow of input and output data as the output image leaves the processor and is sent through a satellite link to the Mission Command Center (MCC).

5.1.6.8.2 Application: How the Systems Engineer Uses IBDs

Systems engineers can benefit from the IBD in several ways, including better understanding and communicating system or subsystem structure, visualizing interactions, and containment between block components, and analyzing system behavior (particularly when used with Parametric Diagrams). The IBD is more than just a presentation and communications tool. It provides insight into important technical elements of the inner workings of the system.

5.1.6.8.3 Implementation Description/Steps in CATIA System of Systems Modeler

To implement IBDs using CATIA System of Systems Modeler, pick the parent BDD in the Containment Tree, right-click and choose construct diagram, then choose SysML IBD to build the IBD. After this, a new pop-up window will appear where the IBD's port and assets can be selected. At this point, the Containment Tree will show the IBD as formed under the BDD.

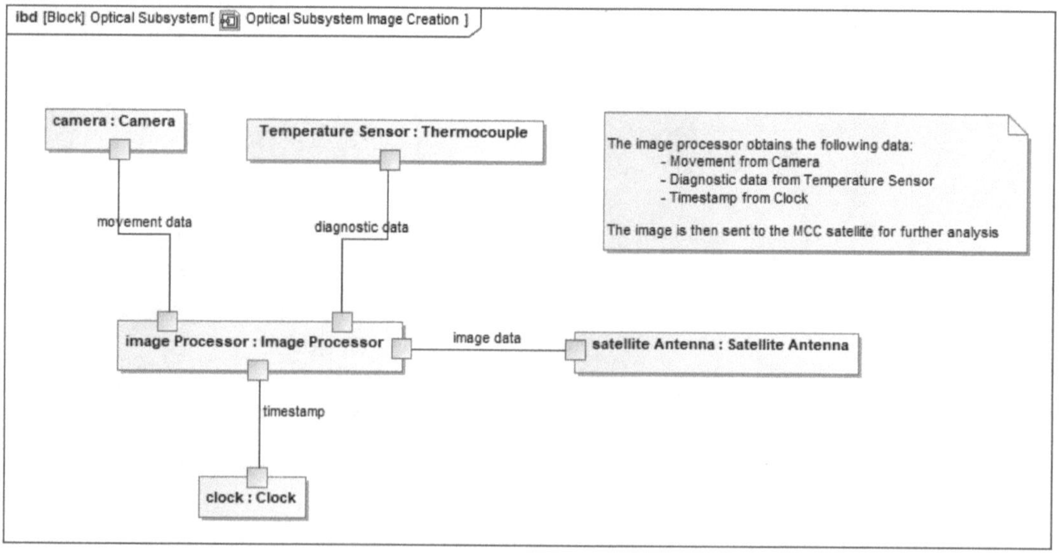

FIGURE 5.35 A/AI-UAV optical subsystem internal block diagram.

5.1.6.9 SysML Artifact 10: Allocation Matrix

5.1.6.9.1 Artifact Methodology and Technique

In SysML, there is a general row-versus-column table called the allocation matrix. This table compares two model elements using filterable conditions; it is also known as a relationship or dependency matrix. For example, when representing orphaned vs. met requirements, or any other relationship with directionality between two classes of model elements, this table is helpful in showing which elements are linked and which ones are not.

5.1.6.9.2 Application: How the Systems Engineer Uses Allocation Matrices

If a systems engineer uses SysML-based MBSE software to build a model of the system or one of the systems essential elements (e.g., the system's architectural descriptions, Use Cases, functional decompositions, and so forth), they can use a relationship matrix to display the model's elements in a tabular format and illustrate key associations. This can be beneficial when debugging or illustrating which model elements connect with each other in some meaningful way. A representative example for the A/AI-UAV is to demonstrate which actors from the set of all Use Cases are associated with which Use Cases as shown in Figure 5.36 (excerpts of the allocation matrix shown). Another useful example could be to allocate a list of requirements to functional blocks of the decomposed system where they will be implemented or link them to testing artifacts where they will be verified, and/or validated.

5.1.6.9.3 Implementation Description/Steps in CATIA System of Systems Modeler

The systems engineer must take care in selecting the criteria used in implementing the allocation matrix, as shown in Figure 5.36. Here, the A/AI-UAV model containment tree's "*3. Use Cases*" package was the only package that the matrix was restricted to, meaning that only the Actors (rows) and Use Cases (columns) in this package are contained in this allocation matrix. Additionally, the automatic tally of the associations for each actor (shown in the gray columns to the left of the figure) makes it immediately evident in table format that all actors in the use case package were related with a use case. Note that no more than one actor was associated with any of the current Use Cases, so additional Use Cases may need to be generated and/or some of the existing Use Cases may need to be revisited.

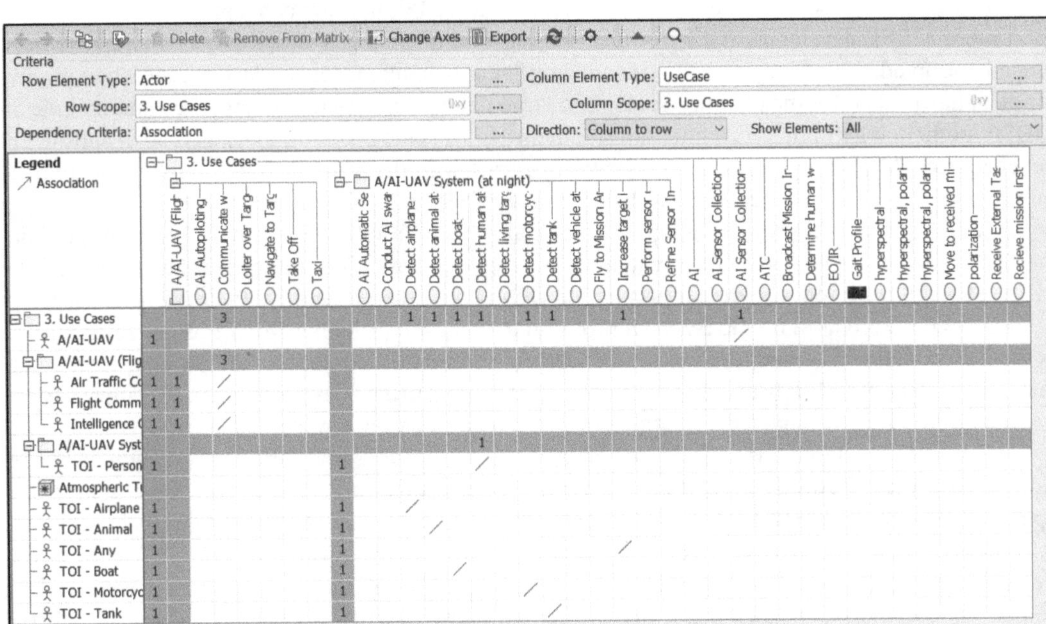

FIGURE 5.36 A/AI-UAV-Model allocation matrix relating actors to use cases.

In the next section, we consider some of the economic benefits associated with implementing an MBSE methodology. We illustrate some of these benefits using a Navy submarine program example wherein MBSE methods were directly compared to document-centric program development practices. We also look at a more general study that assesses the benefits of MBSE across a variety of sectors such as government, industry, and academe for varying organizational sizes. Additionally, we present a set of representative MBSE assessment metrics that are conceptually logical and may be a good place to start in assessing the pros and cons of implementing an MBSE.

5.1.7 Economics of MBSE

There will undoubtedly be a rise in cost to support the development of contemporary systems as their complexity increases. Designing, developing, manufacturing, and supporting complex systems has become increasingly expensive over the years across a variety of programs. Efforts to reduce costs, improve efficiency, and discover new time-saving methods to develop systems seem to be common threads in today's development environment. It is possible to compare traditional methods such as a document-based approach to systems engineering, to modern alternatives that perhaps improve development efficiency and effectiveness. For example, MBSE approaches are becoming more prevalent and are providing boosts in productivity, and improvements in efficiency throughout the SDLC.

As technology and the world progress, the methods for developing systems must also advance. In the current digital era, traditional development practices and procedures may no longer be the best fit for developing contemporary systems, and may not provide the most cost-effective, efficient, or even the most appropriate development approach in all circumstances. Even so, transitioning from traditional systems engineering methods, tools, and procedures such as in a document-centric systems development environment, to a model-driven engineering (MDE) systems development environment is a complex task. Transitioning from the traditional document-centric method to fully adopt a new methodology like MBSE may be costly and time-consuming, despite MBSE being more widely recognized as being beneficial. Companies and teams need to perform research to thoroughly understand the implications of transitioning to new methodologies, procedures, and a new development culture. Performing a cost-benefit analysis would assist in determining whether to switch approaches or not (Allen et al. 2016, pp. 1–7). Let's look at some of the benefits of adopting an MBSE development approach.

A more MBSE development environment has several advantages, and many of those who have recently adopted a MBSE approach noticed significant benefits. Over the last decade, the US government has pushed for the military sector and large defense contractors to embrace and include an MBSE methodology in their program development efforts. One program that has reaped significant benefits from MBSE and system-of-systems integration is the United States Submarine Fleet's, Submarine Warfare Federated Tactical System (SWFTS).

In 2011, the SWFTS program adopted an MBSE-style methodology. It took 3 years to go from the document-based method to the present MBSE methodology. The changeover has improved the SDLC at every level, led to a greater production rate, and increased quality. Table 5.2 shows a series of metrics that were used on the SWFTS program to assess the value of MBSE on the program.

Benefits of MBSE: By moving away from document-centric methods and toward a model-centric one, many advantages of MBSE become apparent. When compared to the more conventional document-based systems engineering method, the advantages of MBSE can be cast in terms of good outcomes.

"How is Model-Based Systems Engineering Justified?" was a literature analysis conducted by Robert Malins and Edward Carroll that examined the advantages of an MBSE-based systems development approach. The MBSE method is effective because it provides for thoroughness in all areas, improved consistency, and excellent communication (Carroll and Malins 2016, p. 3). Moreover, Carroll and Malins detail several MBSE cost-cutting features, including methods for avoiding defects. The authors conclude that a MBSE methodology has five primary advantages, with the most essential benefit of MBSE being the knowledge it provides for teaching and passing on the principles of systems engineering. The other four primary advantages of MBSE are also important

TABLE 5.2
SWFTS Program Benefit Analysis after Switching to MBSE

Metrics	Percentage Increase (%)
Number of baselines produced	30
Subsystem integration	60
Number of requirements	7.5
Number of discrete SE products	60

and include, helping with better communication, managing system complexity, increasing product quality, and learning about the system (Carroll and Malins 2016, p. 11).

According to their research, the MBSE methodology enhanced communications. According to Carroll and Malins, it was found that members of disciplinary teams had a superior grasp of the requirements and their interdependencies. In addition, programs observed that design reuse resulted in substantial cost reductions for companies that used MBSE models. Also, better planning and more detailed traceability occurred, allowing for more effective verification and validation of the models. All things considered, the models served as a unified repository for comprehensive and consistent data (Carroll and Malins 2016, p. 16).

A well-designed, fully integrated, and adaptable system architecture is the result of following a correct MBSE methodology. As an example, the positive effects of MBSE on the Orion F-1 architecture are discussed by Thomas McVittie, Kimberly Simpson, and Oleg Sindiy (McVittie et al. 2012, p. 5). The authors discuss how MBSE offered various levels of design definitions, ranging from high-level diagrams for managers, to comprehensive and detailed diagrams for engineers. They also note that the architecture's configuration management contributed to the system's robustness, which in turn allowed them to be used in subsequent endeavors.

A strong systems architecture can be built with the help of MBSE. The ability to collaborate with other fields and think of new and innovative development approaches is a huge boon to systems engineers and architects. MBSE enables systems to be adaptive, meeting the needs of a complex and ever-changing world. With the correct knowledge and resources, MBSE can provide time-saving, and money-saving solutions.

Measuring the benefits of MBSE: We have presented some of the advantages of MBSE. However, the fundamental question of, "How can the benefits of MBSE be quantified?" is still debatable. Is MBSE genuinely advantageous? A group of researchers from the Systems Engineering Research Center undertook a thorough investigation of the present state of MBSE. Thomas McDermott and his team conducted a survey of 168 participants from industry, the government, and academe to assess the effectiveness of MBSE in organizations, and its impact on business outcomes. The poll was divided into six subsections to guarantee a comprehensive analysis in the report. The six subsections of the survey included, *survey time and responses*, *survey demographic information*, *maturity analysis and participant ratings*, *analysis of MBSE benefits and metrics*, *analysis of enterprise adoption*, and *analysis of workforce development* (McDermott et al. 2020, p. E-2).

There were 109 participants from business, 48 from government facilities, and 11 from academic institutions. Their enterprise sizes varied from fewer than 500 employees to over 10,000. Furthermore, the poll provided details about the type of establishment where the participants were employed, including their precise functional area and their role in the organization. Figure 5.39 provides the breakdown of the responses. Most participants in the study are employed in the aerospace and defense sector, with a significant number working as systems engineers, as depicted in Figure 5.37. The authors acknowledge that MBSE is predominantly used in the aerospace and defense sectors (McDermott et al. 2020, p. E-3).

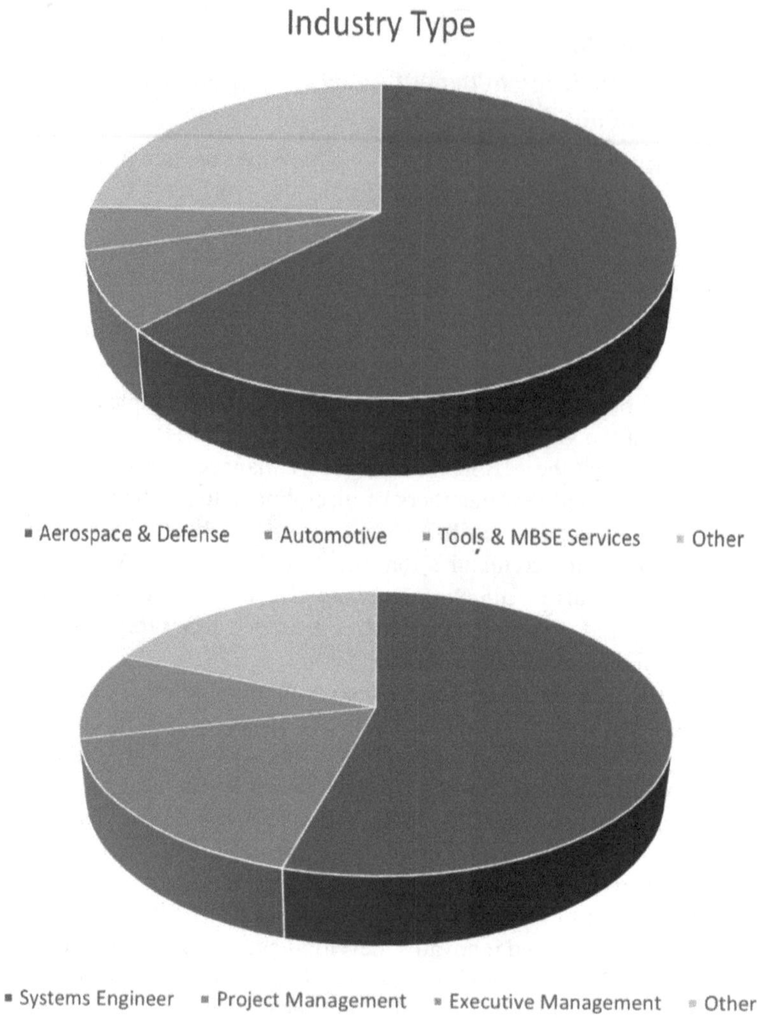

FIGURE 5.37 Industry type (a) and roles (b). (Adapted from McDermott et al. 2020. Created by Anagha Indic on May 30, 2022 and presented here with permission.)

The researchers used a Likert-type agreement scale in the maturity analysis to assess the level of maturity of MBSE practices in the participants' field or business. Participants were asked to select one of the provided options for each question. Strongly agree, Agree, Disagree, Strongly Disagree. Notice the absence of a neutral choice. The researchers aimed to confirm that participants held a specific viewpoint (McDermott et al. 2020, p. E-4). This scaling method is important for evaluating the benefits of MBSE as it prevents any individual from remaining neutral. Participants needed to determine whether their business aligns with the stated criteria in the survey, or not. Interestingly, the authors discovered that most of the comments were *unfavorable*. Government establishments lagged business and academe in the scores. Participants in aerospace/defense businesses may have perceived that MBSE techniques were not fully developed. Also, according to the survey, smaller firms were more likely to embrace MBSE methods than larger organizations.

The results of their study offer intriguing insights into present MBSE benchmarks; however, the research and survey findings are not comprehensive. Initially, a total of 168 participants were polled, with a strong emphasis on the industry sector. The survey report lacks graphical differentiation between government and industry participants to accurately indicate the sources of the unfavorable answers. Furthermore, the report fails to identify the specific demographic that exhibits a negative reaction to MBSE methods. Do systems engineers prefer using MBSE while other groups tend to oppose it? Do systems engineers believe that MBSE is immature? The survey doesn't provide any answers to these questions.

The research indicates that participants are hesitant toward MBSE, perhaps due to its limited prevalence and lack of professionals to promote effective MBSE practices. People acknowledge that modeling leads to tangible enhancements in systems development, but they also point out the lack of uniform MBSE assessment metrics across the organization. Participants recognize that model libraries facilitate reuse but do not believe that they offer any identifiable new knowledge or consistent use of the shared models. Also, participants can discuss the advantages of MBSE, but they are less confident about implementing and adopting its methods.

McDermott's team acknowledged that there were obstacles and challenges in implementing MBSE. The team requested participants to recognize the advantages or obstacles in the implementation of MBSE. The participants identified various obstacles related to MBSE methods/processes, organizational culture, communication practices, customer/stakeholder engagement, MBSE terminology and libraries, legacy versus current processes, willingness to use MBSE tools, training on MBSE tools, the tools themselves, cost, demonstrating benefits, stakeholder involvement, and leadership support (McDermott et al. 2020, p. E-5).

These results are intriguing! Based on the survey outcome, many participants concluded that MBSE methods and processes were an obstacle. However, only a small number of individuals saw training, MBSE equipment, and the associated costs as impediments. Why is MBSE difficult to embrace if cost and training are not major obstacles? Participants may have blamed MBSE methods/processes in general, when the actual culprit that impedes the effective implementation of MBSE may have been lack of a proper MBSE-oriented culture and/or organizational support.

On the other hand, participants acknowledged numerous advantages of MBSE. The benefits mentioned are enhanced traceability, multiple perspectives visibility, improved communication and information sharing, enhanced reusability, better system comprehension, risk mitigation, schedule and error reduction, decreased workload for systems engineering tasks, improved uniformity, and better demonstration of benefits/results (McDermott et al. 2020, p. E-5).

The research indicates that MBSE offers advantages such as decreased risk, time, errors, and SE task burden, as well as improved consistency and clear demonstration of benefits/results. The findings are somewhat perplexing as participants acknowledge the benefits of MBSE, but its lack of widespread adoption raises questions. This relates to a previous statement suggesting that participants may understand the advantages of MBSE but struggle to put it into practice or fully grasp best practices.

McDermott and his team specifically inquire about three questions regarding the advantages of MBSE. Question Q3 prompts participants to identify the primary rationale for incorporating MBSE techniques. Question 7 encourages participants to explain the business value it offers. The third question, labeled Q26, requires participants to recognize advantages of interdisciplinary model sharing (McDermott et al. 2020, p. E-7). Table 5.3 below displays the top eight benefits along with the corresponding number of participants who provided each response.

The participants acknowledge the advantages of MBSE, which aims to save costs, save time, enhance efficiency, facilitate better communication, and enhance system comprehension. Still, there is a noticeable difference between the theoretical advantages and the actual perceived benefits of implementing MBSE.

Debate continues regarding the most effective methods to quantify the advantages of MBSE, with most of the discussion stemming from the early adoption phase of MBSE. The sample size of McDermott and his team's research is limited, possibly due to a lack of participants having MBSE experience. Many believe that MBSE can offer advantages to the development process over the course of the SDLC. However, there is still much progress to be achieved in terms of effective MBSE implementation.

TABLE 5.3
Top Eight Stated Benefits of MBSE

Question 3	Question 7	Question 26
Reduced cost (17)	Increased capacity for reuse (18)	Better communication/information sharing (13)
Reduced time (17)	Improved consistency (16)	Improved system understanding (10)
Better accessibility of info (16)	Improved system understanding (9)	Better accessibility of information (6)
Increased efficiency (14)	Reduced time (9)	Improved consistency (5)
Improved consistency (13)	Better communication/information sharing (7)	Reduced errors (5)
Increased traceability (11)	Better accessibility of information (7)	Reduced time (5)
Improved system understanding (10)	Reduced cost (7)	Increased capacity for reuse (5)

Source: Data courtesy of McDermott et al. (2020). Courtesy of Anagha Indic, created on May 19, 2022.

MBSE in requirements engineering: So, what about the benefits of MBSE when it comes to requirements engineering? Requirements engineering is a subset within systems engineering that deals with the generation, configuration control, qualification (test, verification, validation), documentation, and management of the evolution of good requirements for the system. Herein, the requirements generation and requirements use processes are holistically engineered to successfully drive the system design and development effort to meet stakeholder needs/requirements throughout the system development lifecycle (SDLC). As discussed extensively above, the two primary methodologies used by system engineers to define a system are,

1. Document-centric methodology,
2. MBSE methodology.

The classic requirements engineering method is document-centric and has been widely used in system engineering and related domains. An effective document-based methodology ensures that all pertinent information, such as nomenclature, processes, specifications, architectural design, and related essential system artifacts are precisely identified, regulated, and supervised. A comprehensive document-based engineering methodology can establish a clear paper trail for verifying, validating, and tracing essential system requirements. Although this methodology is proven, it does have certain disadvantages. Information is frequently not centralized but rather spread out across various platforms. Moreover, when this methodology is applied in extensive and complex systems, multiple documents may have inconsistencies and discrepancies, leading to redundant and/or erroneous information, and making it challenging to precisely and consistently mitigate design and development errors. This can have a detrimental impact on the length of the development cycle, leading to higher expenses and schedule slips.

Nevertheless, the transition to MBSE methods is becoming increasingly popular, with MBSE being used to support requirements engineering. The MBSE methods mainly depend on using models instead of the usual text-based document strategy. Using an MBSE-based requirements engineering method appropriately can significantly improve the efficiency, clarity, and quality of the requirements engineering process. MBSE methods promote the mentality of reducing, reusing, and recycling. Models can streamline the system development lifecycle. A recent study found that employing a model-based method for requirements engineering can decrease the time spent by as much as 50% compared to a document-based method (Subarna et al. 2020). This study was centered on the development of a fire detection system that revealed that the MBSE method resulted in saving around 128 hours of effort in comparison to the traditional document-based approach. The decrease in time was especially noticeable during the validation stages of the A-spec system level requirements, B-spec subsystem level requirements, and C-spec component level requirements. MBSE allows for the linking of model objects and elements, unlike document-based methods that rely solely on textual descriptions to indicate connections. Table 5.4 below demonstrates comparative performance of an MBSE approach for this fire detection sysem compared to a document-based strategy (Subarna et al. 2020).

It is interesting to note the increased time of MBSE over the document-based approach when it comes to stakeholder requirements capture, and requirements verification for the A-Spec and B-Spec level requirements. This might be due to the nature of the verification process itself, and the better visibility/readability/co-located nature of the requirements in the completed document, versus the distributed nature of the requirements and screen limitations in the MBSE. Even so, the overall time savings for this project are notable. Many studies support the notion that MBSE has observable benefits over the document-based traditional systems engineering approach. So, how can we tell? What metrics and observables are needed to assess the pros and cons of adopting an MBSE methodology?

MBSE assessment metrics: While stakeholders and system designers are likely to establish measurements of effectiveness and performance for systems, services, products, and/or processes, assessment metrics for MBSE methods, techniques, languages, and tools are less commonplace. Despite this, the success of systems engineering processes and procedures using MBSE methods can be evaluated. Henderson et al. established digital engineering assessment metrics with metric sub-factors by defining and measuring user experiences and productivity metrics, as presented in Table 5.5 (Henderson et al. 2023).

Since subject matter experts (SMEs) regularly use these metrics as part of the regular review process, they make a certain intuitive sense. Notice that several of these metrics (e.g., reduced defects/

TABLE 5.4
A Comparison between a Document-Based Requirement Approach vs. a Model Based Systems Engineering Requirement Approach

Requirement Activity	Document-Based Approach Time (Hours)	MBSE Approach Time (Hours)
	System Level Requirements (A Spec)	
Business requirements capture	1	1
Stakeholder requirement capture	<1 (nominal)	10
System level requirements	1	1
Subsystem requirement decomposition	3	2
Validation	60	30
Verification	10	20
	Subsystem Level Requirements (B Spec)	
Component requirement decomposition	3	2
Validation	60	10
Verification	15	20
	Component Level Requirements (C Spec)	
Validation	80	10
Verification	20	20
Total hours	254	126
Difference in hours	+128	−128
Percentage saved		50%

TABLE 5.5
Assessment Metrics for MBSE Frameworks and Tools like SysML

MBSE/Digital Engineering Assessment Metrics	Metric Sub-Factors
Quality	Increased traceability
	Reduced defects/errors
	Reduced cost
Velocity/agility	Reduced time
	Improved consistency
	Better support for integration, early V&V
	Increased capacity for reuse
User experience	Multiple viewpoints of model
	Higher level support for automation
Knowledge transfer	Better accessibility of information
	Better communication/info sharing
Adoption	Decision engineering (DE)/MBSE methods and processes
	Training
	People willing to use DE/MBSE tools

errors, early verification, and validation) have no subjectivity. Both of these metrics were used, and showed benefits, in the case study presented in Section 5.1.9 that evaluated the utility of adopting a system-of-systems MBSE approach for the development of the Virginia Class Submarine Warfare Tactical System (SWFTS).

5.1.8 MBSE in AGILE Systems Engineering

The original concept of Agile development was focused on software development efforts, but it has now transformed the way teams plan and carry out the development of some very large and complex, mixed systems (containing both hardware and software elements). Fowler and Highsmith describe how 17 software developers met in the Utah highlands to formally write the Agile Manifesto, which includes the following four principles of Agile development (Fowler and Highsmith 2001):

1. *Individuals and interactions over processes and tools*
2. *Working software over comprehensive documentation*
3. *Customer collaboration over contract negotiation*
4. *Responding to change over following a plan*

Despite Agile's roots in software development, the previous four principles are now integral to systems engineering and, in the correct contexts, can be applied to systems that consist of both software and hardware. Initially, the Agile Manifesto appears to directly challenge conventional wisdom in systems engineering. However, according to James A. Highsmith III, a co-founder of the Agile Manifesto, this is a prevalent misunderstanding:

> The Agile movement is not anti-methodology, in fact many of us want to restore credibility to the word methodology. We want to restore a balance. We embrace modeling, but not in order to file some diagram in a dusty corporate repository. We embrace documentation, but not hundreds of pages of never-maintained and rarely-used tomes. We plan, but recognize the limits of planning in a turbulent environment.
>
> ***(Highsmith 2001)***

The systems engineering community has long adhered to the waterfall methodology, which organizes the process of developing a system into a strict sequential order. The methodology places a strong emphasis on linear development and documentation. According to Darrin and Devereux, a MBSE approach can be used to modify the Agile methodology so that it better fits this waterfall design, which follows the systems engineering V Diagram (Darrin and Devereux 2017, pp. 1–5). All through the system development lifecycle, teams can be encouraged to be agile and nimble by using an MBSE development strategy. When requirements, designs, testing, and production are subject to change or updates, using a model allows for easier modification and less restriction.

MBSE in an Agile world – a cautionary tale: Even though it is recognized that MBSE and Agile can be individually beneficial, caution needs to be exercised when combining these approaches. The following example illustrates that two good things are not always better than one. Care needs to be taken in combining beneficial methodologies to avoid counterproductive, or even opposing effects. The following example is provided by a practicing software – systems engineer, and training systems architect, in the U.S. defense industry and is repeated here from this professional's perspective, exactly how it was presented.

> I had the honor of having a philosophical conversation with one of my mentors, Michael "Mic" H. at [Blanked Out] Technologies, Inc., a space and defense industry leader, about implementing model-based systems engineering in an Agile environment. He talked about how his program was struggling to utilize MBSE practices because the customer really wanted Agile development. The customer wanted those incremental builds over models or any theoretical artifacts. The conversation made me realize that sometimes, model-based systems engineering and the agile development process are at odds with one another. I see it with my own team. As a software-systems engineer at work, training to be a systems architect, I see my two teams struggling to find common ground. The software team loves to build. For them, it is all about producing tangible results. Every two week[s] at the end of the sprint, they demonstrate a capability for the program using simulations and real hardware. These products are created from a document over here or a conversation from over there. They do not have a central location in which they maintain all their artifacts - and no, you cannot consider GIT as a repository in this scenario. They barely do that right. On the other side, the MBSE team is trying to play catch up by creating models that capture the architecture. They often find themselves looking at the software team on how the product was implemented, and then backtracking their models accordingly to see if the architecture works. And of course, management is thrilled that software has something to show for when systems is running around clueless. While the program is still in its early stages, actively pursuing the bid, it is critical that the software team, determined to produce physical results with incremental products, learns to cooperate with the MBSE team who wants to ensure that the product will meet all the standards of the customer. *MBSE can only work in an agile environment when both forms of development are executed correctly and in sync.*

From this example, the importance and necessity of having good communication between various elements of the development team is observable. Also, having the systems architecture and MBSE artifact development lag the software/system build instead of driving the design evolution through the system architecture and modeling environment is fraught with peril and may result in expensive downstream problems to fix, or worse.

5.1.9 MBSE Case Studies and Applications

In this section, we provide some case studies and application examples that show how MBSE has been used in a variety of different sectors to great effect. These case studies are broken out into different application areas and illustrate the depth of beneficial possibilities that are realizable by effectively implementing MBSE methods.

5.1.9.1 United States Navy Submarine Case Study

Under the Submarine Warfare Federated Tactical System (SWFTS) program, the United States Navy and its contractors, mostly Lockheed Martin, have been involved in the modeling, production, and testing of submarine fighting systems for many years (Rogers and Mitchell 2021). Following a feasibility study conducted in 2010, the SWFTS systems engineering, and integration program office began the process of shifting from a document-centric systems engineering (DCSE) to a MBSE methodology. At the time, there was not much research supporting the benefits of switching from a DCSE-centric, to a MBSE approach, so the decision to move away from their traditional requirements database (IBM DOORS 2024) and DCSE methodology was quite interesting.

A government and contractor team developed an MBSE business framework for managing sophisticated combat system-of-systems for several submarine variants and conducted a feasibility study finding that the MBSE methodology had several important advantages over the conventional, document-centric methods. One finding was relative to efficiency where the conventional, document-centric processes were found to require an average of 12.1 hours to finish each requirement. For the MBSE approach, establishing a baseline requirement took only 9.9 hours (Carroll and Malins 2016). This led to MBSE methods outperforming DCSE methods by 18%. The MBSE methodology was also 5% better in making program predictions prior to the program's kickoff (Mitchell 2014). The MBSE approach made use of MagicDraw which is an MBSE tool originally developed by NoMagic, and now is one of the offerings of CATIA which also has the CATIA System-of-Systems Modeler tool.

Another benefit was found in finding interface errors earlier in the lifecycle than when using DCSE methods. SysML-based tools used in a laboratory setting found 32% more interface defects than document-based methods. At first blush, this would appear to be an issue with the MBSE approach since the MBSE methodology should have fewer errors than the DCSE methodology if it was the better approach, but the findings were in line with a 37% decrease in found flaws in subsequent hardware integration testing for the submarine platforms. It is important to note that defects discovered during system-level acceptance testing cost 4–50 times more than those discovered during early integration and lab testing. Using MBSE, 18% of defects were discovered earlier in the lifecycle compared to traditional, document-centric systems engineering methodologies (Feiler et al. 2013). In addition to saving time and money, the system (a fast-attack submarine) is of higher quality due to a 9% reduction in total interface problems found using MBSE methods and SysML tools.

A data-centric strategy was implemented with the conversion to MBSE tools with SysML, which led to improved efficiency and quality metrics for the SWFTS. This approach also enabled the automatic development of interface control documents (ICDs) and verification and validation (V&V) artifacts. In addition, the SysML model became a repository for all information pertaining to verified capability/tests, designs, and requirements, as well as a central source for design references and a container for more extensive information about system functions. One major benefit of a model-based approach, which SysML software offers for managing the system architecture and associated artifacts, is having a single baseline for managing inter-system interface and service needs. The study conducted by the SWFTS MBSE team demonstrated that the common baseline MBSE approach offers a more cost-effective means of automating the requirements artifact creation for the various submarine combat system variants (Cameron and Crawley 2012).

Using SysML artifacts to maintain various baselines of system-of-system interfaces had three major benefits. First, a customized user interface was developed in MagicDraw starting with the 2012 migration to MBSE. This UI served as a common ground for both experts and end users to contemplate and understand the interfaces they would be testing and to demonstrate consistency with the documents they were already familiar with from the previous system. Second, scripts were developed in SysML tools to automate SE processes. As a result, efficiency was increased by the auto-generation of tailored SE artifacts. By gradually adopting MBSE methods, they were able to focus on the interface management and sustainment phase of the baselined submarines' lifecycles. This freed the development team up to explore other areas, like sophisticated modeling methods, and the design of follow-on combat systems (Mitchell 2014). One notable distinction in implementation approaches to MBSE is illustrated by comparing the SWFTS MBSE-based development effort to a NASA Jet Propulsion Laboratory (JPL) MBSE project. In contrast to the more ambitious NASA JPL Integrated Model-Centric Engineering Initiative, which aimed to completely convert their entire Flight Project Development Enterprise to become MBSE-centric, the SWFTS case study's MBSE transition primarily concentrated on the evolution and sustainment phases of a single Virginia-class submarine combat system. With an investment savings of $900K+, this case study proved that a gradual, rather than a sudden shift to MBSE — in this instance, with the help of SysML artifacts—is necessary for success (Stepanchick 2016).

Additionally, there were further, less quantifiable benefits to mission readiness and future process enhancements, and the ROI for switching to MBSE for the US Navy's Program Executive Office (PEO) Subs was at least $900K (Rogers and Mitchell 2021). With an acquisition cost of $3.2 billion in FY20 for a single Virginia class submarine, it is clearly advantageous to automate processes whenever feasible. SysML tools for system architecture development have proven to be effective in this regard, particularly when it comes to managing and verifying interfaces across complex system-of-systems and variants of submarines.

5.1.9.2 Civil Aircraft Application

The commercial airline industry has many examples of complex development efforts that can benefit greatly from the application of MBSE methods. Important developmental activities such as understanding stakeholder needs, establishing the various architectures for the system (e.g., the functional architecture and physical architecture), generating and capturing the system's requirements that drive the design activities, and the system's design, evolution, and support are all made easier with MBSE methods.

Developing aircraft systems in the civil aviation sector is multifaceted and involves diverse elements of engineering and engineering management, including a robust systems engineering component that is necessary for successful systems developments. The combined efforts of this diverse team include conducting and managing the system design and development activities throughout the system development lifecycle (Wenhao et al. 2021).

The modern-day challenges of developing complex systems in the civil aviation sector are straining the traditional document-centric systems engineering methodology's ability to meet program demands, requirements, and satisfy expectations. In the process of designing and developing large-scale civil aircraft, traditional systems engineering and MBSE methodologies diverge (Wenhao et al. 2021). MBSE methods have the advantage of generating models that can be directly linked to requirements that provide different insights to the development team and stakeholders through separate viewpoints and MBSE artifacts. These artifacts can capture structural and behavioral aspects of the system, as well as capture relational aspects and system constraints through the models themselves. Artifacts such as Use Cases, Activity diagrams, Sequence diagrams, and State Transition diagrams can provide critical operational details and information about system flow, whereas Parametric diagrams can provide relational information, capture system constraints, and actively link the models to the system's requirements. This integrated modeling capability provides developmental features that simply cannot be accomplished with the traditional document-centric methodology. Architectural elements of the system naturally emerge from the MBSE activities. The system architecture, conceptual/logical architecture, functional architecture, and physical architecture (both hardware and software) naturally emerge through creation of the distinct artifacts supporting the various viewpoints called out in the systems architectural framework. Since models can be linked to each other and to the system's requirements, traceability is enhanced and top-level requirements errors and their impact on design can better be mitigated (Li et al. 2019). The models can also be used to verify the consistency between requirements and design (Cloutier et al. 2015).

The Flight Control System (FCS), Environmental Control System (ECS), Landing Control System (LCS), Fuel System (FS), and Fire Protection System (FPS) are all critical systems in the civil aircraft system-of-systems architecture. Of these, the FCS is the most important and complex system in keeping the aircraft in the air. Operational scenarios can be developed for the FCS by using MBSE methods. In this approach, the expected structural and behavioral aspects of the system or complex system-of-systems are modeled. As an example, an automatic FCS can be broken down into two scenarios, one for the Automatic Flight functions, and another for the Stabilization

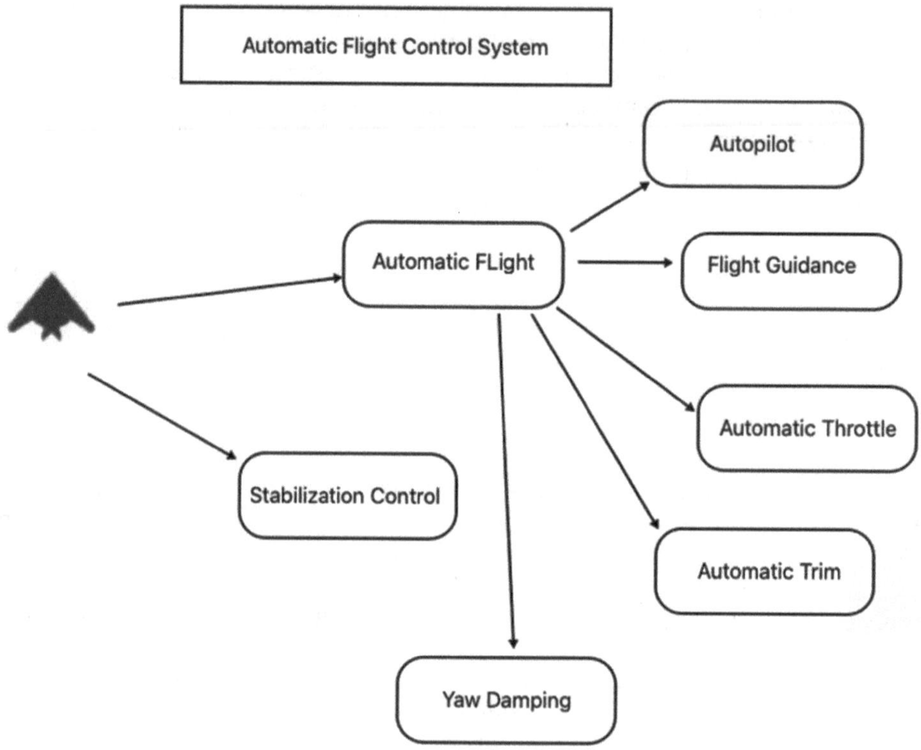

FIGURE 5.38 Automatic flight control system functions. (Courtesy of Davis 2022.)

Control functions. Each of these scenarios consists of many functions to provide the respective capabilities (i.e., flight control capability and stabilization capability). Figure 5.40 shows the functions for the automatic flight control system. The Automatic Flight function includes the Autopilot and other functions that are necessary to control the essential flight parameters of the aircraft, such as the flight guidance, the automatic throttle and trim, and the yaw damping functions. Use Cases can be generated for a variety of expected and unexpected scenarios to describe the operational details relevant to the automatic FCS. The steps in the scenario along with other MBSE artifacts can be used to generate requirements for the system (Figure 5.38).

Since the civil aircraft development process involves complex systems, emergent behavior from the system should be expected and anticipated. According to Cloutier, it is feasible to efficiently build programs for complex civil aviation systems by using system-based approaches such as functional modeling, requirements modeling, and process management (Cloutier et al. 2015). MBSE methods can serve as a central repository for important system development artifacts such as a hub for the system's requirements, its architectures, design artifacts, and integrated models. This, and the integration of the system's conceptual and engineering models used in the system's development and evolution provides new features and capabilities for modeling complex systems and successfully meeting the design and developmental challenges of the future.

5.1.10 Some Challenges of MBSE

From everything we've covered so far, it appears that MBSE is here to stay, and there are many advantages to using it. However, there are many who are still skeptical about the ROI of adopting an MBSE approach, and there are opposing viewpoints, who raise issues with using MBSE. Even some scholars cast doubt on the validity of MBSE's purported benefits. For instance, to determine the full scope of the advantages of MBSE methods, Kaitlin Henderson and Alejandro Salado compiled information from a number of MBSE case studies and literature reviews. According to their results, only two of the 260 studies evaluated were able to provide evidence that MBSE improved system performance using systematic metrics (Henderson and Salado 2020, p. 59). Henderson and Salado contend that one should seriously consider the claims of any author asserting the benefits of MBSE. Put simply, these researchers fail to see sufficient data to support the claim that MBSE significantly enhances the systems engineering process. While they acknowledge that their study did not find any concrete evidence of MBSE's benefits, they stress that this in no way implies that MBSE is useless in the context of system development (Henderson and Salado 2020, p. 61). Rather, rigorous assessment is required to quantify and study the advantages of MBSE; this rigorous assessment was found lacking at the time when Henderson and Salado published their findings.

Although the authors claim that there *is no real evidence for the benefits of MBSE* is debatable, the conclusion that *the benefits of MBSE need to be quantified and studied through formal measurement* is spot on. For some published results, the matter of generalizability of results may surface. For instance, for our case study involving the Navy SWFTS program, the argument could be made that this was data based on one program and the program was in the defense sector. How do these results generalize to other industries or even over other defense systems? Other studies may have issues with having a large enough sample size or perhaps issues in how the sampling method was implemented. What is the best choice of metrics to assess the MBSE so that there are generalizable results? There seems to be plenty of anecdotal information that *indicates* the benefits of MBSE in certain situations, however, continued formal study with the mindset of having scalable and generalizable results is still of practical value.

Furthermore, there can be issues with personnel when implementing MBSE methods. In some organizations, proper training in sound MBSE principles and methodologies is lacking. MBSE can't be successfully implemented if the development team and leadership are not properly grounded in the MBSE methodology. Even with training the workforce, there must also be commensurate support from leadership in terms of supporting MBSE migration activities both verbally and with institutional resources. With lack of training, commitment, and resources, attempts to dabble in MBSE can lead to counterproductive results. The adage "garbage in, garbage out" applies well in this situation. Without proper grounding in the MBSE methodology, bad development practices can be passed on and degrade the potential benefit of MBSE. For example, using a MBSE environment just for an electronic repository of the requirements may not justify the switch to MBSE. Even worse, false expectations coupled with poor MBSE implementation practices can lead to managerial disappointments and wrong conclusions being drawn from the misapplication of MBSE. For example, when MBSE is not applied correctly or is applied in a hasty fashion, it can lead to negative outcomes. In turn, when program leadership sees the negative outcomes and erroneously interprets this as a failure of MBSE, rather than a misapplication of MBSE, they may be discouraged from using MBSE in the future. Program management and/or the development team may react negatively to the time and money needed to implement MBSE under the *perception* that there isn't enough time to do it now. After all, who has time to do anything except meet current commitments in today's hectic development environment. Consequently, program leadership and/or the development team may also be resistant to the change to model-based systems. This stresses the importance of having a positive MBSE culture with informed, realistic expectations throughout the stakeholder, leadership, and development team. Qualified and trained systems engineers with expertise in MBSE are necessary for the time-consuming and complex process of correctly implementing MBSE.

Possibly, there may also be egos to contend with. Egos can also adversely affect the good implementation of MBSE and good systems engineering practices in general. Whether from *old school greybeards* that say, we have always done it this way, so why change; to non-trained engineers *wearing the systems engineering/MBSE hat* that make key MBSE and SE decisions; to inexperienced leaders that don't understand the strategic and tactical advantages of a solid MBSE program and culture, the profitable, efficient and effective shift from a document-centric to model-centric development environment can be hindered if not down-right checked. Consequently, the myth that

MBSE is ineffective continues to spread and gives MBSE a bad reputation. Implementing a bad MBSE environment will not provide the expected benefits and will not yield advantages greater than the effort and expense required to execute the MBSE. The good news, however, is that implementing a good, holistic MBSE environment will produce benefits that outweigh the cost of implementing the MBSE environment in the long run.

5.1.11 Summary

Systems engineering has always been open to new effective ideas and technologies, and it is now an essential element in developing large, complex systems. Engineers need to discover better methods to help systems adapt and evolve as they go through the system development lifecycle, especially as systems become more complex. A growing number of systems engineers are turning to MBSE as a primary tool in their systems development toolbox.

The potential benefits to efficiency and ROI that MBSE methods provide, and their ease of use have led to their adoption across several industries. Where appropriate, newer, more agile development methods are gradually replacing older, traditional ones. Adopting an MBSE approach does not have to be done all at once and can instead be done incrementally. Benefits of MBSE can manifest in collecting and organizing requirements, creating a visual representation of the design and important systems development artifacts, and optimizing the developmental effort across the entire systems lifecycle.

Although the industry has an increasing familiarity with MBSE methods needed to build complex systems, MBSE methodologies haven't yet completely supplanted the more traditional, document-centric systems engineering frameworks and acquisition methodologies. This is likely due to MBSE concepts still being relatively new and still evolving. There are also many choices in software tools and languages available. Also, determining the ROI of adopting an MBSE strategy is not always straightforward or readily generalizable. The combination of these factors likely affects MBSE's widespread adoption (Madni and Sievers 2018). Nevertheless, according to evaluation criteria and real-world case studies, MBSE has quantifiable benefits and a better return on investment than DCSE procedures (Goodwin et al. 2022). SysML-based MBSE artifacts facilitate better program communications, aid in representing different useful system views, and provide a better analytical environment for complex systems developments. MBSE methods also help avoid expensive design errors. Combining the standardization of languages such as SysML, the development of reusable model building blocks, the use of automation, and advances in AI, with stakeholder model-centric frameworks will result in more efficient and cost-effective complex systems development efforts. Upcoming systems development efforts that use an MBSE approach will undoubtedly benefit from continued MBSE research. In a nutshell, the MBSE methodology has transformed engineering practices and will be a key element of successful complex systems developmental efforts of the future.

In the next section, we will discuss the building blocks of optical systems, in which various elements of optical systems are analyzed. Additionally, the next section also discusses how the various MBSE tools are used throughout the optical system's lifecycle. These tools are primarily responsible for relating requirements of the various building blocks of the optical system and for analytical and simulation purposes.

5.1.12 Transition to Optical Systems Building Blocks

Modeling the components of an optical system can be difficult due to its complex nature. Understanding the optical components often requires a multidisciplinary approach and expertise in various fields. The visible sensor is operated in a different band of wavelengths than the infrared (IR) sensor. With visible light, we deal primarily with reflected light, whereas IR light may be both reflected and emitted (in the mid-wave IR) or emitted radiation (long-wave IR). The visible light interacts differently with the atmosphere than radiation in the IR part of the electromagnetic spectrum. Dominating noise terms, spatial resolution capabilities, thermal shielding requirements, optical coatings, and even signal processing requirements can differ substantially between the visible and IR detection regimes. Thus, to build the complete system, all of these optical system building blocks must work together to generate a common output.

Consequently, to connect the optical system's requirements from the architectural level down to a component level, it becomes essential to employ MBSE tools. Different optical system components such as the IR sensors, visible sensors, data storage, image processors, or optical modulators interact with each other. The output obtained from one component is used as the input to other components. These different components perform different and distinctive tasks but work in concert to generate the final system's output. This stresses the significance of using MBSE tools to connect the requirements of the different optical system components and build integrated system models used to analyze and simulate the optical system's performance. As an example, all optical system requirements can be divided into functional and nonfunctional requirements. The interfaces definitions between the functional decomposition of the requirements are also critical and can be captured by MBSE tools. For complex systems like optical systems, it is essential to manage requirements and model systems by using MBSE tools such as CATIA system-of-systems modeler (or some of the earlier/related MBSE tools such as MagicDraw, Cameo Systems Modeler and Requirements Management Software, and/or other similar contemporary MBSE products). These products can also interface directly with the prevalent Dynamic Object-Oriented Requirements System (DOORS), a well-known and widely used requirements management software suite from IBM. These tools provide MBSE support for general-purpose traditional requirements management as well as provide a variety of modeling functionality with different plug-ins. The following sections detail optical building blocks, which introduce the fundamentals of optical SE, and a case study, which demonstrates these optical methods and MBSE principles.

5.2 OPTICAL SYSTEMS BUILDING BLOCK: THE BASIC OPTICAL SYSTEM

We have looked at electromagnetic light and its propagation characteristics in previous chapters and now want to characterize how light interacts with some basic optical devices such as lens mirrors, beam splitters, turning mirrors, collimators, corner cubes, filters, and telescopes. We will treat more specialized optical elements such as acousto-optical devices, liquid crystals, and special devices elsewhere in the text and, instead, focus on basic optical elements here in this chapter. According to Saleh and Teich, different approaches to characterize electromagnetic phenomena in decreasing amounts of rigor are (1) quantum optics, (2) electromagnetic optics, (3) wave optics, and (4) ray optics also known as geometrical optics (Saleh and Teich 1991). Quantum optics describes virtually all interactions of light with matter and presents the most complete and rigorous theory. Electromagnetic optics describes the interaction of light with matter in a classical sense, and wave optics, also known as scalar diffraction theory, is a theoretical simplification that is useful in Fourier optics or studying the propagation of optical light through atmospheric turbulence. In this section, we focus on the simplest form of describing the interaction of light with materials and devices—the geometrical optics approach, also known as ray optics.

This simple description of the propagation of light and its interaction with matter does not describe diffraction effects and also assumes that light is monochromatic. Nevertheless, the geometrical optics approach is useful in many applications including describing the propagation of limiting "rays" of light such as paraxial rays that indicate the relative size of the image on a detector element. As an example, a typical purpose of the optics between the telescope entrance pupil and the optical detector element is often to place the spatial Fourier transform of the electromagnetic wave in the optical system's entrance pupil completely on the imaging system's detector element in as efficient way as possible. As will be seen in this chapter, lenses, turning mirrors, beam expanders, collimators, beam splitters, neutral density filters, and other basic optical elements are used to accomplish this task.

Geometrical optics methods are useful in such an application as well as other applications where diffraction theory does not play a significant role. The advantage of the geometrical optics approach is that it provides relatively simple mathematics to describe the propagation of light and the interaction of light with optical devices. An excellent text on a concise overview of geometrical optics is found in Saleh and Teich's book on photonics (Saleh and Teich 1991). More comprehensive information can be found in standard modern optics texts such as Guenther's book on *Modern Optics* (Guenther 1990). In this section, we present some fundamental geometrical optics principles that need to be understood by the optical systems engineer and working professional working on EO/IR systems.

5.2.1 Basic Optical Principles

Most objects are observable to us because they reflect, refract, transmit, or radiate light in the visible part of the electromagnetic spectrum. Our eyes are keenly tuned to receive light in the visible part of the electromagnetic spectrum having a peak transmission at the yellow color of our sun. What we visually perceive around us is the superposition of the electromagnetic fields in the visible part of the electromagnetic spectrum that are reflected, transmitted, refracted, or radiated into our eyes field of view. When dealing with the propagation of light through a chain of optical elements, frequently some of the more elegant and mathematically rigorous methods are not required, and a simpler and a mathematically more tractable approach such as geometrical optics suffices. An example would be in determining the size of an image on an imaging system's detector element. Another example would be in estimating how light refracts through a series of optical elements. If accounting for diffraction effects is not important to a particular application, and we are dealing with larger objects and distances as compared to the wavelength of light under consideration, then a geometrical optics approach provides a reasonably accurate and tractable solution for our needs. In this section, we will develop a geometrical optics-based interpretation of refraction, reflection, and the interaction of light with mirrors, lenses, optical coatings, prisms, and filters. We present a matrix-based ray tracing method and provide insight into some commercial software that is available for this purpose. We end with a discussion on telescopes and provide a useful model in describing their functionality. By understanding these methods, the reader should be able to model basic optical elements found in a collection of basic optical components.

5.2.1.1 Refraction

An incident wave on a surface may also be refracted. Refraction occurs when light propagating in one medium encounters another medium that permits transmission of the light. At the interface boundary, the transmitted light leaves the boundary at a different angle. For example, a train of plane waves of light incident on a plane surface separated by two transparent substances such as air and glass is split: part of the incident light is reflected at the surface, and part passes into the medium and is refracted (e.g., changes direction).

The directions of the incident, reflected, and refracted light are specified in terms of the angle they make with the surface normal at the point where the incident light meets the substance boundary. The following observations can be made by looking at the interaction of the incident, reflected, and refracted light at the point of incidence on the boundary surface:

1. There is one plane that contains the rays of light from the incident, reflected, refracted, and surface normal vector.
2. The angle of incidence is equal to the angle of reflection at all wavelengths.
3. For monochromatic light, the ratio of the sine of the angle of incidence to the sine of the angle of refraction is a constant.

These three observations are quite useful and form the basis for ray tracing methods that allow us to describe the path of an optical ray through a collection of optical elements. The mathematical articulation of item (3) is given by the law of refraction:

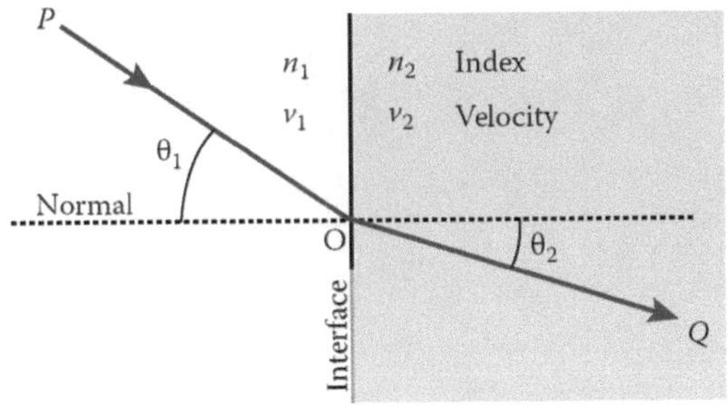

FIGURE 5.39 Graphical rendering of Snell's law. (Image courtesy of Cristan 2006. Wikimedia Commons. http://commons.wikimedia.org/wiki/File: Snells_law.svg, accessed December 30, 2014.)

$$C = \frac{\sin(\theta_1)}{\sin(\theta_2)}, \tag{5.1}$$

where
C is a constant
the angles θ_1 and θ_2 are the angles of incidence and refraction, respectively, as shown in Figure 5.39.

The left side of Figure 5.39 shows the incident ray of light propagating in Medium 1 with a refractive index of n. The refractive index is the ratio of the speed of light in a vacuum to the speed of light in a particular material:

$$n = \frac{c}{v}. \tag{5.2}$$

The index of refraction is a function of temperature and, as will be seen in a later chapter, the variations in the index of refraction are the primary contributing factor in atmospheric turbulence effects. For now, we note that as propagating rays encounter other materials, the velocity of light changes in that material and Equation 5.2 defines the index of refraction of the material. Notice that the numerator and denominator in Equation 4.2 have the same units, and so the index of refraction is unitless.

When light passes from one material with one index of refraction to a different material with a different index of refraction, the light is refracted, or bent. Snell's law determines the amount that the light is bent:

$$n_1 \sin(\theta_1) = n_2 \sin(\theta_2). \tag{5.3}$$

Here
n_1 is the index of refraction in Material (or Medium) 1
n_2 is the index of refraction in Material 2
the angles θ_1 and θ_2 are the angles that the incident/refracted light make with the material boundary normal vector, respectively.

This phenomenon is well known by spear fishermen or birds hunting fish. If Material/Medium 1 is air and Material/Medium 2 is water, and the point of observation makes an angle θ_1 with the normal to the water line (shown vertically in Figure 5.1), then the "apparent" position of the fish would be on the line obtained by extending the incident ray through the boundary at the same angle. However, the actual position of the fish would be along the trajectory of the refracted ray. Not accounting for this difference in angles would likely send the fisherman or bird home empty-handed and hungry.

Another interesting phenomenon is the case of total internal reflection. Total internal reflection occurs when the medium that contains the incident light has an index of refraction greater than the medium that has the refracted light. Additionally, the angle of incidence of the incoming ray must be equal to or greater than a critical angle determined by

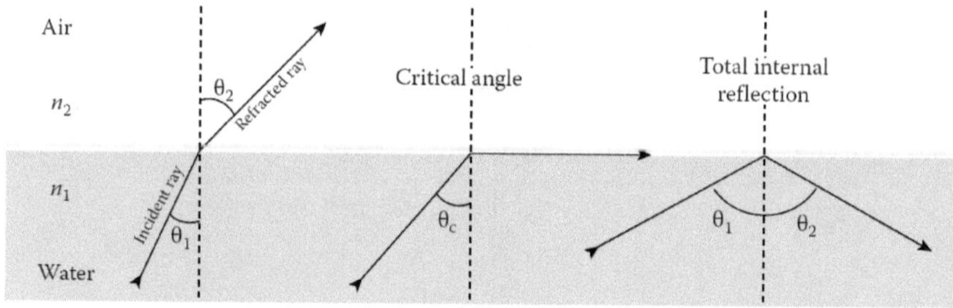

FIGURE 5.40 Total internal reflection. (Image courtesy of Josell7 2012. Wikipedia. http://en.wikipedia.org/wiki/File:RefractionReflextion.svg, accessed December 30, 2014.)

$$\theta_C = \sin^{-1}\left(\frac{n_2}{n_1}\right), \tag{5.4}$$

where

θ_C is the critical angle
n_1 and n_2 are the indices of refraction of the first medium (medium with incident ray) and second medium (medium with refracted ray), respectively, as depicted in Figure 5.40

The case for total internal reflection is shown at the right side of Figure 5.40: where the angle of the refracted ray θ_2 is 90°. Equation 5.4 becomes apparent when substituting θ_2 equal to 90° into Snell's law. Physically, if an observer were to look along the incident ray under conditions for total internal reflection, the observer would not see anything in Medium 2 – Medium 2 would appear opaque to the observer. It is interesting to note that the critical angle condition given in Equation 5.4 is only possible for boundaries where the index of refraction of the material containing the incident ray is greater than that of the refracted ray. If this is not the case, as is depicted on the left of Figure 5.40, there is no possibility of total internal reflection, and rays for any valid angle θ_1 are refracted into Medium 2. Also, if the incident ray has an angle θ_1 that is less than the critical angle, the refracted ray is transmitted into Medium 2 with an angle $\theta_2 < 90°$. We now look at some properties of reflected light.

5.2.1.2 Reflection

During daytime, the majority of what we see is due to reflected light. Commonly, the reflected light is a diffuse reflection having the property that it is scattered in all directions. Diffuse reflection occurs when the reflecting surface is rough when compared to the optical wavelength undergoing the reflection.

Another type of reflection is specular reflection. In this type of reflection, incoming light rays are reflected in a given direction. Specular reflection occurs when the reflecting surface is smooth compared to the size of the wavelength of the incoming light. The following are examples:

Reflection from a blotter is diffuse.
Reflection from a mirror is specular.

There are also cases in which there are both diffuse and specular reflections. Part of a material may be "optically rough" and cause the incident light to undergo diffuse reflection, whereas parts of the material might be highly polished and optically smooth thereby producing a specular reflection component to the reflected light. These reflecting components are additive at the electromagnetic field level. In other words, given an observation point P some distance away from a reflecting surface, the observed electromagnetic field would be the sum of all the reflected components (due to both diffuse and specular reflections) that are incident at point P.

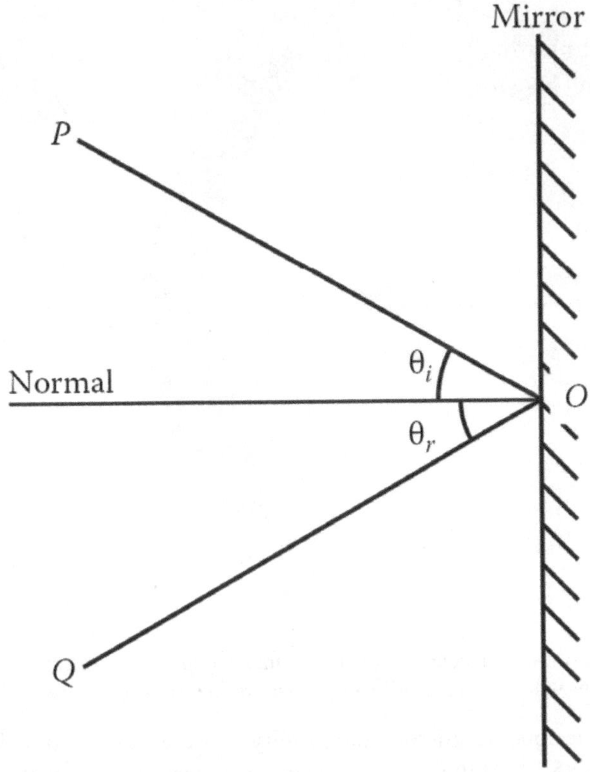

FIGURE 5.41 Illustration of a reflected light ray from a flat surface (e.g., mirror). (Image courtesy of Johan Arvelius 2005. Wikipedia. http://en.wikipedia.org/wiki/File:Reflection_angles.svg, accessed December 30, 2014.)

Figure 5.41 represents light reflected from a boundary such as a mirror. When light is reflected, there is no change in medium; thus, there is no change in index of refraction. By Snell's law, the incident angle is equal to the reflected angle.

In Figure 5.41, the incident light is shown at the top of the figure making angle θ_i with the normal vector to the surface boundary of the mirror. It is reflected back as illustrated by the departing ray at the bottom of the figure making angle θ_r with the surface normal vector. Snell's law still holds with both the medium for the incident and reflected ray being the same and having the same index of refraction. Looking at Equation 5.3 with equal indices of refraction, the incident ray angle θ_1 and the exiting ray angle θ_2 must be the same.

5.2.1.3 Mirrors

Mirrors are highly reflective surfaces that use the principle of reflection. The geometrical optics approach to understanding the interaction of light rays with the optical surface of the mirror has already been presented in Snell's law. The task in modeling the interaction of light with a mirror surface is simply applying Snell's law at each surface. Mirrors come in three types, convex, planar, and concave, according to the radius of curvature of the optical surface. Assuming that light is traveling from left to right, a concave mirror has its focal point to the left of the mirror surface at some distance *F*. A planar mirror has its focal point at an infinite distance from the optical surface, and a convex mirror has a virtual focal point to the right of the optical surface. An example of the focal point for a concave mirror is shown in Figure 5.4.

The focal length *F* is given as half the radius of curvature *R* of the concave mirror:

$$F = \frac{R}{2}. \tag{5.5}$$

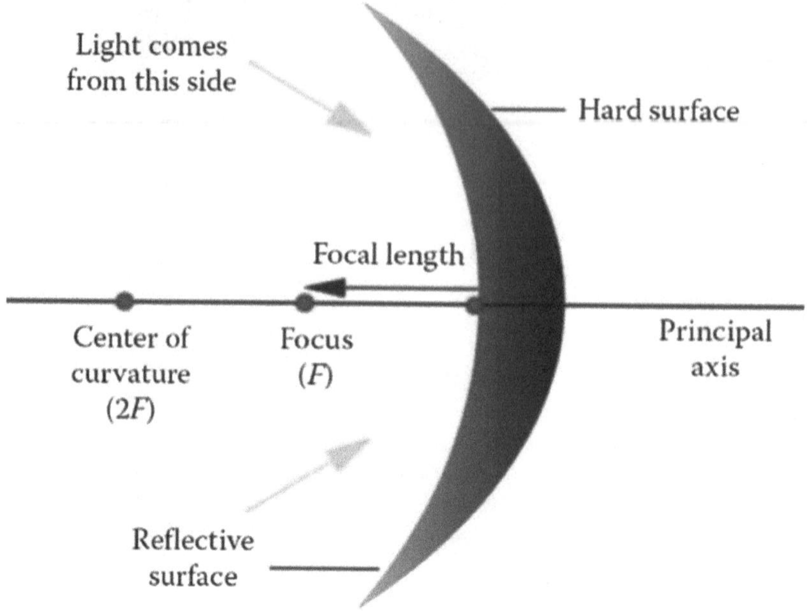

FIGURE 5.42 Reflection from a concave mirror. (Image courtesy of Cronholml44 2007. Wikimedia Commons. http://commons.wikimedia.org/wiki/File:Concave_mirror.svg, accessed December 30, 2014.)

Light rays arriving at the concave mirror from infinity (parallel rays) will be focused onto the focal point as shown in Figure 5.42. At this point, we need to choose a sign convention used when dealing with distances in relationship to mirrors and lenses. There are various conventions, but the one we will use is the same for spherical mirrors and for lenses. In building our optical model, we assume that light always comes in from the object side and is called the incoming light if it is on the object side of a particular surface (mirror or lens). In the figure, we show the incoming light coming in from left to right corresponding to a source object on the left. For a mirror (or lens) surface, the radius of curvature is the distance from the vertex V to the center of curvature of the mirror (or lens) surface. If the center of curvature lies to the left of the mirror, as is the case for a concave mirror, then the radius of curvature R is positive. Also, according to Equation 5.5, the corresponding focal length is positive. If the radius of curvature of a mirror is to the right of the mirror, as in a convex mirror, then the radius of curvature is negative as is the focal length. If is the distance from the vertex V of the mirror to the object, and s_2 is the distance from the mirror vertex to the image, then for both convex and concave mirrors, a useful sign convention is that the object and image distances are positive if located to the left of the vertex V and negative if located to the right of vertex V. Another way to say this is the object and image distances are positive if the spherical mirror makes a real image (in front of the mirror) and the object and image distances are negative if they make virtual images (behind the mirror) for either concave or convex mirrors. With these assumptions, a mirror equation that is valid for all spherical mirror types can be given that relates the radius of curvature of the mirror surface to the object and image distances:

$$\frac{1}{s_1}+\frac{1}{s_2}=\frac{2}{R}. \tag{5.6}$$

With regard to the sign convention of vertical dimensions such as the object and image height, the adopted convention is that distances are positive if they are above the optical axis and negative if they are below the optical axis. With this convention, we can define a magnification for the mirror given by

$$m_m=-\frac{h_i}{h_o}=\frac{s_2}{s_1}, \tag{5.7}$$

where h_i and h_o are the image height and object height, respectively, in meters. The subscript m is used to indicate the magnification due to a mirror element. Notice that for a planar mirror, both s_1 and s_2 are positive and infinite, so the mirror magnification is one. The magnification and mirror equations are very useful in understanding the interaction of light rays with mirrors. We will discuss telescope systems that consist of mirrors and lenses later on in this section.

5.2.1.3.1 Types of Mirrors

In addition to characterizing mirrors as planar or spherical, there are characteristics of mirrors that need to be understood. One such characteristic is the mirror type. Predominantly, there are two types of mirrors, *front-side mirrors* and *back-side mirrors*. Front-side mirrors are often used in optical systems applications since they reflect light from the front surface. This type of mirror requires special handling since the optical reflective coating is on the front surface of the mirror element and can easily be damaged. In field experiments, or operational environments, these types of mirrors need to be protected against the elements and incidental handling and are usually contained within an environmental conditioning chamber. The second-type mirror is the *back-side mirror*. This type of mirror is often used in day-to-day applications such as in bathroom mirrors. For back-side mirrors, the optical coating is put on the rear side of a mounting surface such as glass. The glass itself acts as a protective cover, and so these mirrors are quite sturdy and practical for day-to-day use. Notice that although the back-side mirror is very rugged, it introduces an extra material interface by the addition of the glass mounting surface. An optical ray first undergoes a reflection at the air-glass interface that is on the order of 4% of the incident optical power. Consequently 96% of the light is transmitted into the glass and then propagates to the reflective surface at the back of the mirror. At the back end of the mirror, the light is reflected back through the glass where it again encounters the glass/air interface. At this interface, some light is again reflected and the remaining light is transmitted into the air. Notice that in this case, Snell's law must be applied to three surfaces (the original air/glass interface, the glass/polished surface interface, and then the glass/air interface), and so the angle of incidence and the angle of reflection are not exactly the same. Also, the phase of the electromagnetic wave undergoes a phase shift due to the extra optical path difference between the incident ray at the air/glass interface and the emerging ray. Because of these differences, front-side mirrors are typically preferred in optical applications that require high accuracy.

5.2.1.4 Lenses

Lenses are optical elements that are of primary importance in building an optical system. As we learned earlier, lenses take the optical spatial Fourier transform of the incident irradiance thereby producing the spectrum of the object in the lens's focal plane. Another lens placed in the focal plane of the first lens would produce an inverted image of the object in its focal plane. In this section, we present an overview of basic optical principles needed to understand how lenses work. Before we do this, we first want to develop a useful model for analyzing a general-purpose optical lens system and also define some terms that are used to understand the propagation of light through these optical elements using the geometrical optics approximation. We start with Figure 5.43, which presents an example of how to model a general lens system. More optical elements could be present, but the approach for determining important optical parameters that are essential in describing the optical system is the same as presented in this example.

In Figure 5.43, consider one dimension of an object that is shown at the left of the figure. For any series of optical elements, there is a limiting aperture that acts as the *aperture stop* of the lens system (shown at the center of the figure). The aperture stop is the object in the chain of optical elements that limits the size of the angle U_2 through the optical system (e.g., it determines the maximum value of U_2). The *marginal ray* is the ray from the object that skims the edge of the aperture stop and just barely makes it through the optical system. Conversely, any ray that goes through the center of the aperture stop is called a *principal ray* (or *chief ray*). One important point to consider is that for any ray in a given medium, light will continue linearly in that medium if there is no change in the index of refraction. Consequently, in a constant index of refraction medium, a ray of light will continue

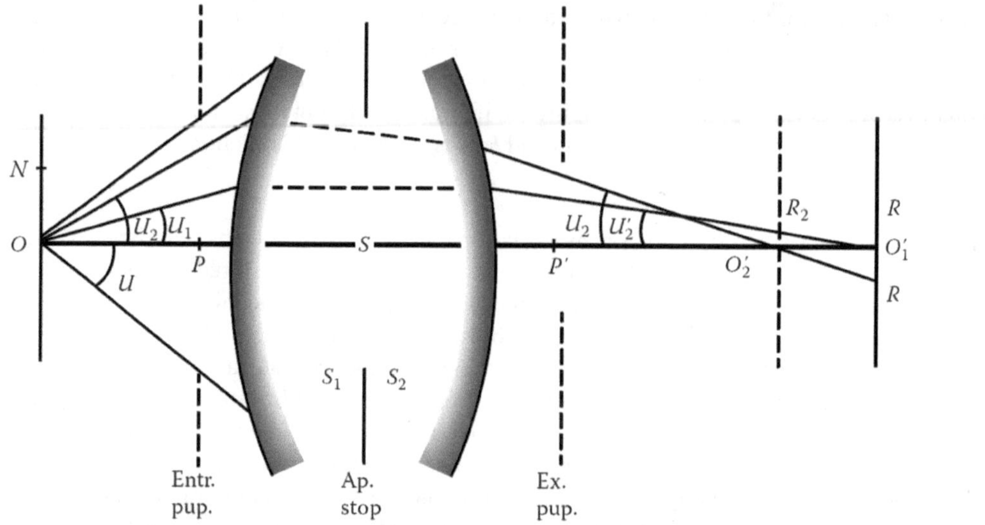

FIGURE 5.43 Representative example of a general lens system. (Image courtesy of Blleininger 2010. Wikimedia Commons. http://commons.wikimedia.org/wiki/File:ABERR1.svg, accessed December 30, 2014.)

in a straight path until it encounters an interface. At the interface, Snell's law can be used to determine the change in direction of the ray of light. As will be seen later, this simple fact can be used to trace a collection of rays emanating from an object through an optical system to determine the image. Snell's law is a fundamental principle for many ray tracing algorithms used in geometrical optics modeling software. The *entrance pupil* is the image of the aperture stop when viewed from the object side, whereas the *exit pupil* is the image of the aperture stop when viewed from the image side. It is interesting to note that the principal ray will also appear to pass through the center of the entrance pupil and the exit pupil. For well-designed telescope systems, the entrance pupil is also the location of the primary telescope mirror. Hence, the diameter of the telescope mirror (e.g., an 8.5 in. telescope) is also the diameter of the optical system's entrance pupil. In general, light interacting with an optical system that consists of an arbitrary number of optical elements can be reduced to a simple model of refraction at so-called principal planes. Figure 5.43 represents a simple two-lens example model showing only the exit pupil (Ex. Pup.) on the right side of the figure.

Any combination of simple optical elements like lenses and mirrors can be modeled by introducing two fictitious planes known as principal planes. Two of these planes are shown as the entrance pupil (called H_1 for brevity's sake) and the exit pupil (called H_2 hereafter) in Figure 5.43. As will be seen in the following ray tracing section, a single transfer matrix can model an arbitrary number of optical elements between these principal planes. A property of the principal planes is that a ray leaving the rear principal plane (H_2) appears to be at the same height from the optical axis as a ray viewed from the front of the first principal plane (H_1). As a consequence, all the optical elements between the principal planes can be reduced to a set of single refractions that occur at points on the principal planes. The intersections of the principal planes with the optical axis are the principal points P and P'. If the medium surrounding the lens system is the same (e.g., air or water), then the location of the principal points is also the location of the nodal points. Nodal points have the property that if a ray is aimed at a nodal point at a particular angle, that ray appears to leave the other nodal point at the same angle. The collection of the three types of points—principal points, nodal points, and focal points—makes up the set of *cardinal points* of an optical system. For an ideal, rotationally symmetric optical system, knowing the location of these points on the optical axis is sufficient to determine basic imaging properties like the image size, the location of the image plane, and the orientation of the image. There are also other important stops in an optical system such as field stops that limit the field of view of the optical system. Other stops are used to control aberrations, for mounting purposes, and/or to reduce glare. We now look at some useful properties of a thin lens.

A lens is an optical device that interacts with the incident light, converging, diverging, or collimating the incident beam. Figure 5.44 shows a positive (converging) lens.

There are two spherical surfaces associated with the lens, and we refer to the radius of curvature of the first surface encountered by the light as R_1 and the second surface encountered by the light as R_2. We illustrate in the previous model that incoming light is to the left of the lens and outgoing light from the lens is to the right of the lens. Our sign convention for lenses is similar to that of the mirror in all respects except for the radius of curvature. If the center of curvature for the surface is to the left of the surface vertex, as in the case of a diverging lens, then the radius of curvature is negative. This occurs if the first surface is concave. Figure 5.45 shows a negative (diverging) lens. On the other hand, if the center of curvature is to the right of the surface vertex, then the radius of curvature and the focal length are positive.

In Figure 5.45, the radius of curvature R_1 for the first surface (left-most surface) is negative, whereas the radius of curvature R_2 for the second surface is positive. The remaining sign conventions are the same as for the case of the spherical mirror. Namely, if the object is to the left of the lens, it is a real object and the distance to the object is positive. If the object lies to the right of the lens, it is a virtual object and the distance from the lens to the object is negative. If the image is a real image (e.g., formed by a real object and to the right of the lens as in a converging lens), then the image distance is positive. If the image is a virtual image (formed to the left of the lens from a real object as would be the case for a diverging lens), then the image distance is negative. The vertical distances, just as in the case for the spherical mirrors, are positive if they are above the optical axis and negative if they are below the optical axis.

5.2.1.4.1 *Thin Lenses*

The simplest kind of lens is a thin lens. A lens can be classified as a thin lens if its thickness along the optical axis is small compared with the focal length of the lens. To determine the focal length, f', the previous sign convention is used along with the so-called lens maker's equation:

$$\frac{1}{f'} = \left(\frac{n_1}{n_2} - 1\right)\left(\frac{1}{R_1} - \frac{1}{R_2}\right), \tag{5.8}$$

where,

n_1 is the index of refraction of the lens material
n_2 is the index of refraction of the surrounding medium

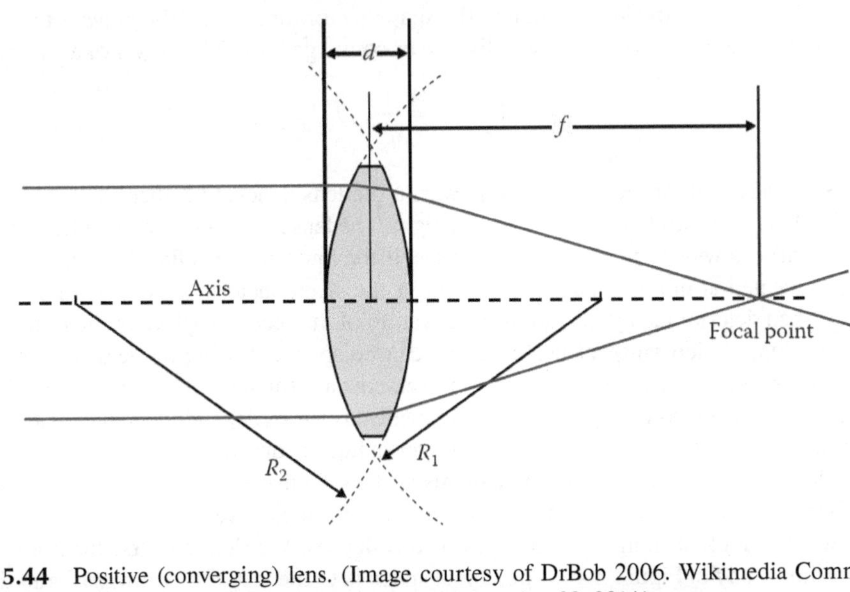

FIGURE 5.44 Positive (converging) lens. (Image courtesy of DrBob 2006. Wikimedia Commons. http://commons.wikimedia.org/wiki/File:Lensl.svg, accessed December 30, 2014.)

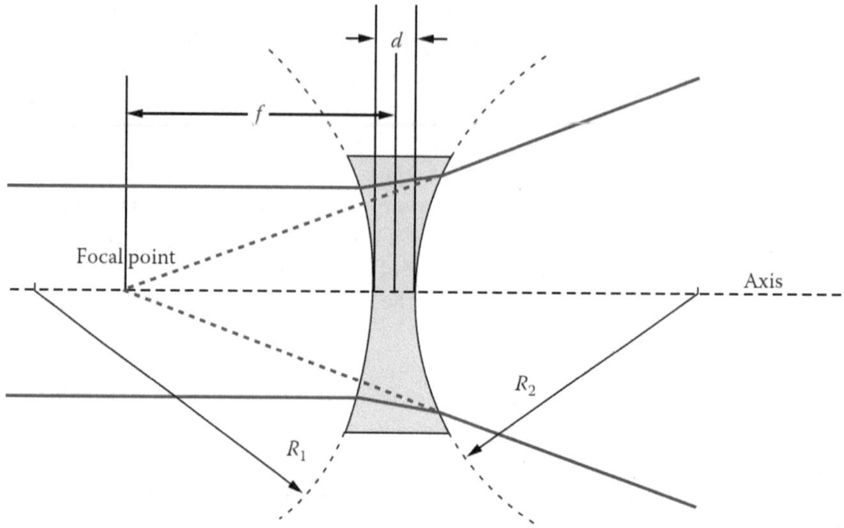

FIGURE 5.45 Negative (diverging) lens. (Image courtesy of DrBob 2006. Wikimedia Commons. http://commons.wikimedia.org/wiki/File:Lenslb.svg, accessed December 30, 2014.)

Notice that for a converging lens like the one shown in Figure 5.6, the radius of curvature R_1 of the first surface is positive and the radius of curvature of the second surface R_2 is negative. For this case, the shape factor K_s is positive and is given by

$$K_s = \left(\frac{1}{R_1} - \frac{1}{R_2}\right). \tag{5.9}$$

The shape factor is a useful quantity. If the lens is in air, then the index of refraction of the surrounding medium n_2 will be 1 and the index of refraction of the lens itself (n_1) will be >1, and so the shape factor then determines whether the lens is a converging lens or a diverging lens. In air, if the shape factor K_s is positive, the lens will be a converging lens. If the shape factor K_s is negative when the lens is in air, then the lens will be a diverging lens. If the surrounding medium is not air, then whether or not the lens is a converging or diverging lens depends on the relative nature of the material surrounding the lens, the material of the lens itself, and the shape factor. In general, the *power* of a lens can be used to determine whether or not a lens will diverge or converge light. The power of a lens is given by

$$\text{Power} = \frac{1}{f'}, \tag{5.10}$$

where f' is the focal length of the lens. If the power of the lens is negative, then the lens will diverge light; if the focal length is positive, it will converge light. The lens maker's equation in Equation 5.8 can be used to determine a wide range of practical lenses. If the first surface is flat (Plano), then the radius of curvature is located at infinity and the first term of the shape factor goes to zero. The sign of the shape factor is then determined by the radius of curvature of the second surface of the thin lens. Using our sign convention, a wide range of useful lenses can be specified using the lens maker's equation such as biconvex, plano-convex, plano-concave, convex-concave (meniscus), and biconcave. Regarding meniscus lenses, they come as positive and negative meniscus lenses depending on the relative radius of curvature of the surfaces. If the first surface has a larger radius of curvature than the second surface, the shape factor shows that this will be a negative meniscus lens. Conversely, if the radius of curvature for the first surface is smaller than the second surface, the result is a positive meniscus lens.

We now want to look at imaging properties of thin lenses. We define a first focal length (front focal length) shown as f in Figure 5.46. The object space is to the left of the thin lens, and the image

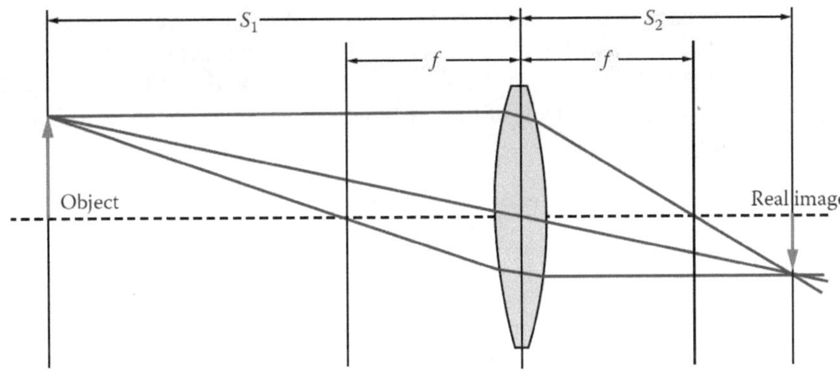

FIGURE 5.46 Imaging model for a simple thin lens. (Image courtesy of DrBob 2006. Wikipedia. http://en.wikipedia.org/wiki/File:Lens3.svg, accessed December 30, 2014.)

space is to the right of the lens. We see that for a thin lens, there is one principal plane that vertically bisects the lens and separates the object space from the image space.

The entrance pupil, exit pupil, aperture stop, and field stop, as well as the nodal points and principal plane points, all lie in this plane. In the image space, f' is the focal length given by the lens maker's equation in Equation 5.8. For imaging applications, we define the distance from the thin lens to the object as S_1 and the distance from the thin lens to its corresponding image as S_2. If a real object is in the object space as shown, then the biconvex lens will form a real image in the image space as shown. The image will be inverted and will appear in the image plane. The image plane can be found by projecting a parallel ray from the top of the object through the lens. The lens will focus the ray through the focal point/By projecting another ray leaving the top of the object through the nodal point at the center of the lens on the optical axis, the intersection of the two projected rays will then determine the image plane. Besides helping determine the image plane (e.g., the plane where the optical image is formed), the focal length is useful in a variety of applications. If optical rays are projected through the first focal point as shown in Figure 5.47, the rays leave the thin converging lens in parallel lines and are *collimated.* This principle can be used quite effectively in expanding or contracting the dimensions of a beam of light.

An interesting application is when two lenses are combined as follows. Beams of parallel rays arrive at a lens L_1 and are focused through its focal point. A second lens L_2 has its front focal point at

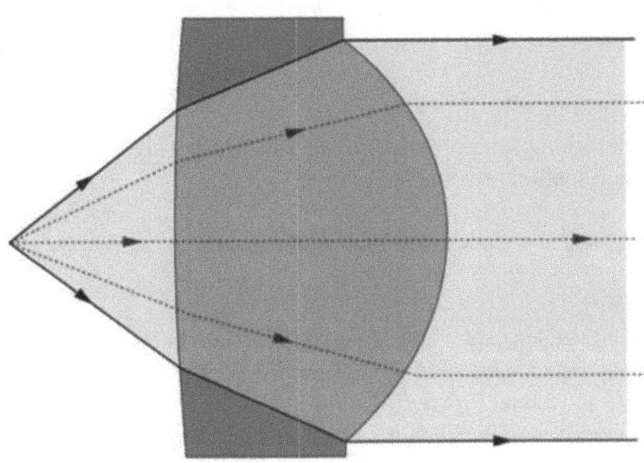

FIGURE 5.47 Generating parallel rays from a thin lens. (Image courtesy of RCarboni 2009. Wikipedia. http://en.wikipedia.org/wiki/File:CollimatingLensSVG.svg, accessed December 30, 2014.)

the same position as the back focal point for the first lens L_1. When the beam emerges from L_2, it is once again collimated but with a different diameter that is a simple function of the ratio of the focal lengths of the lenses. If f_1 is the back focal length of the first lens and f_2 is the front focal length of the second lens, then the output diameter of the ray bundle leaving the second lens is related to the diameter of the incoming ray bundle by

$$D_{\text{out}} = D_{\text{in}} \cdot \frac{f_2}{f_1}. \tag{5.11}$$

If the focal length of the second lens is larger than that of the first lens, then the two-lens system is a *beam expander.* Another important parameter related to the optical throughput of a lens (or imaging system) is its *f*-number. The *f*-number is given by,

$$N = \frac{f}{\#} = \frac{f}{D}, \tag{5.12}$$

where,

D is the diameter of the lens (e.g., the diameter of the entrance pupil of the imaging system)
f is the focal length of the lens

For the case of a wide angle, wherein the focal length is small and the diameter of the lens is large, the *f*-number is very small. Conversely, if the diameter is very small and the focal length is very large, then the *f*-number is very large. This metric is counterintuitive, and so another metric called the numeric aperture is often stated. The numeric aperture is given by

$$\text{NA}_i = n \sin\left(\arctan\left(\frac{D}{2f}\right)\right) \cong n\frac{D}{2f}, \tag{5.13}$$

where n is the index of refraction of the medium surrounding the lens. The subscript i indicates that the numeric aperture is viewed from the image side. By accounting for the magnification m of the optical system, the numeric aperture as viewed from the object side is given by

$$\text{NA}_o = \left(\frac{2(m-1)}{nm}N\right)^{-1}, \tag{5.14}$$

where

N is the *f*-number given by Equation 5.12
n is the index of refraction of the surrounding medium (1 for air)
m is the lateral magnification (assumed negative)

We now want to extend some of these useful results to thick lenses.

5.2.1.4.2 Thick Lenses

Other than thin lenses, all real lenses are considered thick lenses and have additional complexity. Recall that for a thin lens, there was only one principal plane that served as the entrance pupil, exit pupil, and aperture stop of the optical system. Figure 5.48 shows a model for a thick lens. Thick lenses have two principal planes whose locations depend on the radius of curvature of the front and back surfaces. The focal points are illustrated by f and f'. The remaining cardinal points are shown as C_1 and C_2.

Note that the nodal points and principal points coincide at the principal plane for a general optical system. The lens maker's formula determines the optical power of the thick lens as

$$\phi = \frac{1}{f'} = \frac{(n_L - n_1)}{n_1 R_1} - \frac{(n_L - n_2)}{n_1 R_2} + \frac{(n_L - n_2)(n_L - n_1)}{n_1 n_L}\frac{t}{R_1 R_2}, \quad (5.15)$$

where

n_L is the index of refraction of the thick lens
n_1 is the index of refraction of the medium that interfaces on the first surface (assumed to the left of the thick lens in Figure 5.48)
n_2 is the index of refraction of the medium on the object side (right) of the thick lens
t is the thickness of the lens along the optical axis
R_1 and R_2 are the respective radii of curvature as before

If the thick lens is in air, as is often the case, Equation 5.15 reduces to

$$\phi_{\text{air}} = \frac{1}{f'} = (n_L - 1)\left[\frac{1}{R_1} - \frac{1}{R_2} + \frac{(n_L - 1)}{n_L}\frac{t}{R_1 R_2}\right]. \quad (5.16)$$

Notice that if the thickness t is reduced to zero, we have the thin lens result. There is a useful relationship to relate the image and object distances to the front and back focal lengths. If the distance from the object to the first principal plane of the lens is given by S and the distance from the lens' second principal plane to the image is given by S', then the relationship between the front and back focal lengths and the object and image distances is given by

$$\frac{n_3}{f'} = \frac{n_1}{f} = \frac{n_1}{S} + \frac{n_3}{S'}, \quad (5.17)$$

where

n_1 is the index of refraction in the object space
n_3 is the index of refraction in the image space

Not only is the general shape and thickness of a lens important but also the amount of light that makes it through our optical system. We provide a summary of some of the notable results relating to mirrors and lenses in Table 5.6.

In the next section, we look at antireflection coatings to maximize the amount of light that reaches the detector.

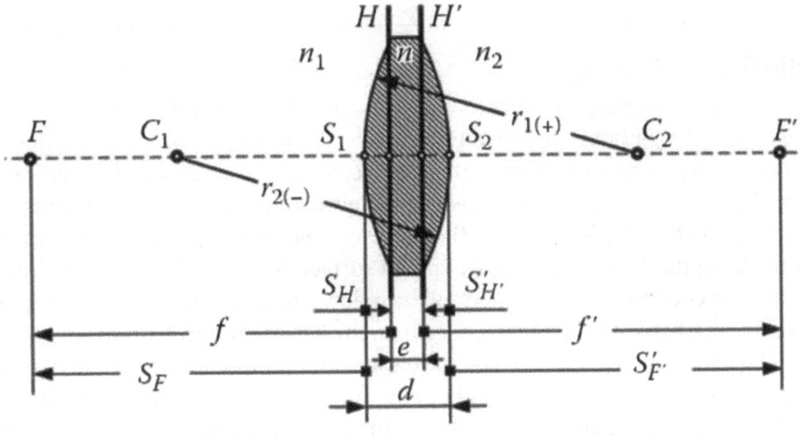

FIGURE 5.48 Model of a thick lens. (Image courtesy of Tamasfiex. 2011. Wikipedia, http://en.wikipedia.org/wiki/File:ThickLens.png, accessed December 30, 2014.)

TABLE 5.6
Summary of Notable Equations

Image formation	S=object distance from principal plane (always positive for real object) S'=image distance from principal plane (positive, real image; negative, virtual image) f=front focal length f'=back focal length R=radius of curvature h_i=image height h_o=object height D=diameter of entrance pupil t=thick lens thickness at optical axis Spherical mirror (in air) $\frac{1}{S}+\frac{1}{S'}=\frac{2}{R}$ Plane mirror (spherical mirror for $R\to\infty$) $S=-S'$ Spherical refracting surface $\frac{n_1}{S}+\frac{n_2}{S'}=\frac{(n_2-n_1)}{R}$ Thick lens $\frac{n_3}{f'}=\frac{n_1}{f}=\frac{n_1}{S}+\frac{n_3}{S'}$
Thin lens	$\frac{1}{f'}=\left(\frac{n_1}{n_2}-1\right)\left(\frac{1}{R_1}-\frac{1}{R_2}\right)$
Thick lens optical power	$\phi=\frac{1}{f'}=\frac{(n_L-n_1)}{n_1R_1}-\frac{(n_L-n_2)}{n_1R_2}+\frac{(n_L-n_2)(n_L-n_1)}{n_1n_L}\frac{t}{R_1R_2}$
Lateral magnification m	$m=-S'/S$
Magnitude of lateral magnification $\lvert m\rvert$	$\lvert m\rvert=h_i/h_o$
Angular magnification	$m_\gamma=\frac{\gamma_i}{\gamma_0}$
F/# (f-number)	$N=\frac{f}{\#}=\frac{f}{D}$
Numerical aperture (image side) NA_i	$NA_i=n\sin\left(\arctan\left(\frac{D}{2f}\right)\right)\cong n\frac{D}{2f}$
Optical invariant	$m\cdot m_\gamma=\frac{n_o}{n_i}$

5.2.1.5 Antireflective Coating

When setting up optical imaging systems that are designed to work with low-light levels, the treatment of the surfaces of the optical components becomes important. As we stated earlier, if we leave a surface untreated, a certain portion of the light is reflected from the surface. In glass, this is roughly 4% of the optical power at each glass interface. The more untreated optical elements we introduce in the optical chain of the imaging system, the more loss of optical power there is on the detector. A method for reducing the reflection from an optical surface is to coat it with an *antireflection coating*. An antireflection coating, as the name implies, reduces the amount of reflected light by causing destructive interference between the reflected light on the surface of the coating, and the rays of light that are reflected back from the glass interface are transmitted through the air/film interface.

The film thickness is designed to be equal to one-quarter of the wavelength of the incident light, so a down and back round trip path of the ray (from the air/film boundary to the film/glass boundary and back) is $\lambda/2$ m long. The two reflected rays are roughly of equal magnitude and 180° out of phase and so destructively interfere with each other—no power is carried away in the reflected direction. Consequently, a high percentage of the optical power in the incident ray is transmitted into the glass.

5.2.2 Achromatic Principle

The index of refraction is not constant with changes in wavelength. The use of lenses to focus a broadband image may result in a poor image due to differences in focal length for the various wavelengths. Variations due to color (wavelength) in the index of refraction must be corrected. Figure 5.49 shows the differences between a normal lens (shown at the right of the figure) and an achromatic lens (shown at the left of the figure).

Achromatic lenses are designed to focus the different colors of light onto the same focal point. This is done by adding a second material of a different index of refraction and designing the thickness and curvature of the interface, to focus the wavelengths onto the same focal point.

5.2.3 Ray Tracing

Ray tracing is a tool used to describe optical systems, where each optical component is expressed in a matrix form. This is a powerful methodology that traces rays through a host of optical components using the geometrical optics principles discussed earlier. The input ray to the optical system is defined by its height and also by the angle that it makes to the horizontal plane at the point of interest. This ray can be propagated through the chain of optical elements by repeated application of Snell's law when the ray encounters an optical surface. In the paraxial limit, the propagation of light rays can be described by a series of matrices that characterize the interaction of the light ray when it encounters a particular surface or propagates through a particular medium. The new height and angle of the ray can be expressed as the matrix multiplication of the old height and angle of the ray and the matrix representing the particular surface or propagation distance. The following equation defines the relationship between the height of the original ray from the optical axis and the angle it makes to the horizontal surface and a particular propagation or surface encounter represented by the matrix element:

$$\begin{bmatrix} h_{\text{out}} \\ \theta_{\text{out}} \end{bmatrix} = \begin{bmatrix} A & B \\ C & D \end{bmatrix} \begin{bmatrix} h_{\text{in}} \\ \theta_{\text{in}} \end{bmatrix}, \tag{5.18}$$

where the A through D elements of the matrix are uniquely defined depending on the type of interaction the ray encounters. For instance, for a thin lens with unit magnification, the A, B, C, and D parameters are given by

$$\begin{pmatrix} 1 & 0 \\ -\frac{1}{f} & 1 \end{pmatrix}. \tag{5.19}$$

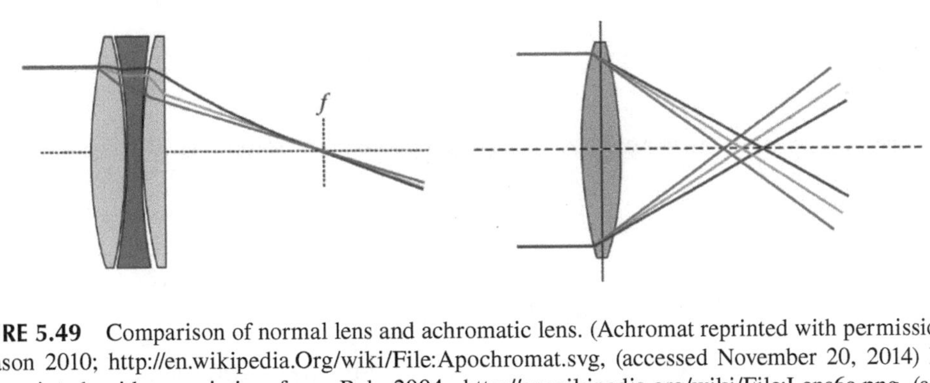

FIGURE 5.49 Comparison of normal lens and achromatic lens. (Achromat reprinted with permission from Egmason 2010; http://en.wikipedia.Org/wiki/File:Apochromat.svg, (accessed November 20, 2014) Normal lens reprinted with permission from Bob 2004, http://en.wikipedia.org/wiki/File:Lens6a.png (accessed November 20, 2014.)

By substituting Equation 5.19 into Equation 5.18, we see that, at the point where the light ray intersects the thin lens, the output height is equal to the input height and the output angle is equal to the input angle minus the input height divided by the focal length of the lens. Table 5.7 shows some common matrices for the ray propagation principles discussed previously.

The utility of the matrix method becomes apparent when tracing a ray of light through a collection of optical elements using simple transfer matrices. By multiplying all the matrices of our optical system, we end up with one matrix that describes the whole optical system and relates the input ray and input angle to the output ray and output angle. We can illustrate this concept by deriving the *ABCD* matrix for a thick lens. For a thick lens, the optical ray in a medium with refractive index first encounters a refractive surface. The optical ray then propagates through the thick lens of width t in a medium that has an index of refraction n_L. The optical ray then encounters another refractive surface that separates the thick lens from a medium with refractive index n_3. The resulting *ABCD* matrix M_{tl} is given by

$$M_{tl} = \begin{pmatrix} 1 & 0 \\ \dfrac{n_L - n_3}{n_3 R_2} & \dfrac{n_L}{n_3} \end{pmatrix} \begin{pmatrix} 1 & t \\ 0 & 1 \end{pmatrix} \begin{pmatrix} 1 & 0 \\ \dfrac{n_1 - n_L}{n_L R_1} & \dfrac{n_1}{n_L} \end{pmatrix} = \begin{pmatrix} A & B \\ C & D \end{pmatrix}, \tag{5.20}$$

where R_1 and R_2 are the respective radii of curvature for the first and second surfaces of the thick lens. Notice the way the matrices are "stacked" is from right to left. The first operation on the incident ray is shown as the right-most matrix. Subsequent operations are ordered from right to left. In general, an *ABCD* matrix M that consists of a sequence of N operations, each having a 2×2 matrix formulation, is given by

$$M = M_N \cdot M_{N-1} \rightleftharpoons M_2 \cdot M_1. \tag{5.21}$$

TABLE 5.7
ABCD Matrices for Simple Optical Surfaces and Components

Optical Element (or Operation)	ABCD Matrix	Remarks
Free space propagation or index of refraction is constant	$\begin{pmatrix} 1 & d \\ 0 & 1 \end{pmatrix}$	d=distance
Refraction at a flat interface	$\begin{pmatrix} 1 & 0 \\ 0 & \dfrac{n_1}{n_2} \end{pmatrix}$	n_1=first refractive index, n_2=second refractive index
Refraction from a curved boundary	$\begin{pmatrix} 1 & 0 \\ \dfrac{n_1 - n_2}{R \cdot n_2} & \dfrac{n_1}{n_2} \end{pmatrix}$	R=radius of curvature $R>0$ for convex surface $R<0$ for concave surface n_1=first refractive index n_2=second refractive index
Reflection from flat surface (e.g., flat mirror)	$\begin{pmatrix} 1 & 0 \\ 0 & 1 \end{pmatrix}$	Identity matrix
Reflection from a curved surface (e.g., curved mirror)	$\begin{pmatrix} 1 & 0 \\ -\dfrac{2}{R} & 1 \end{pmatrix}$	R=radius of curvature $R>0$ (convex surface) $R<0$ (concave surface)
Thin lens	$\begin{pmatrix} m & 0 \\ -\dfrac{2}{f} & \dfrac{1}{m} \end{pmatrix}$	f=back focal length of lens $f>0$, for convex lens $f<0$, for concave lens Valid for thin lens model only. The factor m is the lateral magnification of the optical system

In this manner, rays can be propagated through a complicated set of optical elements, and the resulting output ray position and angle can be determined by a single matrix multiplication consisting of the reduced *ABCD* matrix from the sequence of 2×2 matrices. The *ABCD* matrix method applies to a wide variety of optical operations and optical devices such as turning mirrors, corner cubes, beam splitters, and prisms. *Turning mirrors* are used to change the direction of propagation of the light. A common use is to extend the focal length of an imaging system by folding the optical path length so that a larger focal length fits into a given small space. *Corner cubes* are simply two, 90° reflections from mirrored surfaces (canted at 45° to the incident light) that reflect the light back in the direction that it came. A corner cube was placed on the Moon so that lasers fired from the Earth to the Moon would have their concentrated laser pulses reflected back toward the Earth. In that manner, lasers can be used to measure the time of flight of an optical pulse from the Earth to the Moon and back.

5.2.4 Optical Filters

Optical filters are useful to select certain wavelength to be processed in an optical system. A *narrow-band filter* passes light in a certain frequency band (e.g., 50 nm wide and centered at a particular wavelength). A "red" filter passes red light but rejects other wavelengths. Looking through a red filter, the transmitted light all appears red. Filters also come in *high-pass, low-pass,* or *notch* varieties based on whether they pass high-frequency light, low-frequency light, or light in a particular frequency range. *Polarizers/polarization filters* are designed to select/remove polarization components from the transmitted imagery. *Beam splitters* split the optical power in different directions. For instance, a 50/50 beam splitter reflects half the optical power in one direction and transmits half the optical power through a partially reflecting mirror in another direction. Different ratios of reflected to transmitted power can be specified. *Dichroic beam splitters* can pass light in a certain wavelength range and reflect light in another wavelength range. *Neutral density filters,* polarization components, and *optical gratings* can all be used to adjust the optical power through a chain of optical elements. *Optical gratings* and *prisms* can be used to separate different colors or reflect a particular color at a determinable angle. All these optical components can be used to control light, as it propagates through an optical system. We will look at some of these components in closer detail in later sections. We will also take a closer look at optical aberrations that arise in our optical system due to imperfections in the optical components, effects of the turbulent atmosphere, and even the random nature of the light itself.

5.2.5 Optical Design Software

In the previous sections, we discussed the ray tracing technique used for modeling the optical systems by developing the matrices and where understanding of the matrices calculation is clear-cut and simple. But using the ray tracing method for 3D structures of an optical system to mark out bundles of rays can be a complex and tedious task. There exist software packages that are useful in designing and analyzing optical systems. Several of these software packages are licensed, and some packages are free. What follows is a list of optical design and analysis software that can be helpful for ray tracing, designing, and analysis of an optical system:

- Advanced Systems Analysis Program (ASAP)
- Code V
- dbOptic
- LightTrans Virtual Lab 5
- Optics Software for Layout and Optimization (OSLO)
- HEXAGON
- Optix
- FRED Optical Engineering Software
- Zemax.

In general, commercially available software is better documented and maintained and less error-prone than freeware. For example, "Zemax optical design and analysis software" has been available on the commercial market for a long time, and they have a refined and excellent product (Radiant Zemax 2014). These optical design and analysis software suites simplify the process of designing an optical system and have become fundamental tools for the optical designer. The specific tool used, however, is determined by the needs of the organization and hopefully determined by a requirement-driven SE process.

In the next section, we will use our integrated case study to apply some of the methods learned so far. We will demonstrate how the MBSE tools are applied to the optical system. As part of the case study, the SE concept of the "analytical hierarchy process" (AHP) is explained.

5.3 INTEGRATED CASE STUDY: MBSE, MATLAB®, AND RAPID PROTOTYPING AT FIT (BASIC OPTICAL COMPONENTS)

The objective of this case study is to present how MBSE tools may be used in a working environment and show how they are an integral aspect of the modern SE-oriented development environment. We also use this section to provide an example of how the optical building blocks material is used in the development of the optical system. The AHP is presented as a tool for conducting a trade-off study for selecting evaluation software for a critical component for our scenarios unmanned aerial vehicle (UAV) optical system, an optical modulator.

As part of the Department of Homeland Security UAV contract that fantastic imaging technologies (FITs) have recently won, FIT has been selected to design and develop the basic optical components of the optical system mounted on a UAV. In this application, the optical modulator is used to modulate a beam of light before it reaches the optical detectors to help reduce background noise and help in signal detection. As part of these discussions, the technical aspects of the basic optics systems needed to be understood before embarking on the optical modulator design. The company is also applying MBSE methods to help manage the requirements, design, and testing aspects of the system. For this project, FIT uses the CATIA System of Systems modeling tool with its No Magic Cameo requirements management plug-in, to manage the requirements of the system, and Zemax™ as the primary optical design and analysis software tool. For their high-speed, atmospheric turbulence compensation (ATC) applications, FIT has developed a rapid prototyping system that implements their patented MATLAB-based ATC algorithms.

A number of brainstorming meetings occurred between the systems engineers at FIT to discuss the basic optical system and the optical modulator. An important topic of discussion between the systems engineers and the optical designers was the technical requirements for the basic optical system for the imaging sensors for the DHS UAV project. Karl Ben, the senior systems engineer, called a meeting with one of his junior systems engineers, Jennifer O. (Jen) to discuss the basic optical system. We join the two as they begin their discussions.

(Tuesday morning, bright and early.)

Karl: "Hi Jen, how's life been treating you?"

Jen: "I am doing good and how are you?"

Karl: "Great! Thanks for asking! So, how is that radiometric study coming along?"

Jen: "Pretty good! We are wrapping up some of the final details, but we are practically certain that we can get away with a passive imaging system."

Karl: "That is good news. When do you think the study will be finished?"

Jen: "Fairly soon. We need to have a better understanding of the basic optical system so that we can put that into our models. We are coming along pretty well on that front also."

Karl: "Great, tell me about it."

Jen: "We were looking at the Raytheon MTS-B-20 gimbal pod, and we now know that it has a 20 inch (0.508 m) primary mirror for both the visible and infrared parts of the spectrum.

We have identified representative sensors for both the visible and infrared parts of the spectrum that we can use as placeholders to give us a feel for the imaging scenario."

Karl: "That's good. We can use these in our current MBSE models and then use the actual sensor parameters once the trade study completes. That way, we can get started with a soft design and see what issues we run into."

Jen: "Right. That brings up an issue though. We haven't chosen our component-level MBSE tool yet. I was wanting to do a trade-off study to select the right tool but I have been tied up with the radiometric model, and basic optical systems analysis and so haven't gotten a chance to do that yet."

Karl: "Well, you still have the weekend, right?! Just kidding! The radiometric study has priority, followed by understanding the optical system. Let's continue with that."

Jen: "Hah, good one! Ok, the sensors we are using for both the visible and imaging sensors both have a 1 cm^2 active area with 100 percent fill factor. The read-out electronics are on the back of the sensor elements. They each have 2,048×2,048 pixels. With some relay optics, we can get an effective focal length that ranges between (0.2 and 2 m). We are using a nominal 1 m effective focal length to start.

Karl: "What does the Raytheon gimbal provide?"

Jen: "It provides the primary telescope mirrors, the pointing and tracking mechanisms, servo-control mechanisms, drivers, gimbal command software, and a PC interface. The rest is up to us."

Karl: "OK, what about the transmittance of the optics?"

Jen: "We are going to minimize the number of optical surfaces in our relay optics and want to keep the transmittance through all the optical components greater than 95%. The optical engineers are running ZeMax to work on some preliminary designs."

Karl: "Ok, what are we using for the Fried parameter in our analysis?"

Jen: "Well, at visible wavelengths, the atmospheric coherence length often ranges between 8 and 12 cm for air-to-ground applications. We decided to use 10 cm for a nominal value of r_o. For the infrared spectrum, ro scales as the ratio of the center wavelength at the IR wavelength to that of the visible wavelength with the result raised to the 6/5th. So, the atmospheric coherence length can be found by multiplying our 10 cm nominal ro by $(\lambda_{ir}/\lambda_{vis})^{6/5}$."

Karl: "Sounds reasonable. We may want to do a worst-case-best-case analysis, along with the nominal case once we have all the necessary parameters. What are we using for the separation distance between the sensor and our intended human object."

Jen: "The nominal UAV altitude above ground-level is 15,000 ft (4,572 m). Maximum altitude is 50,000 ft (15, 240 m), and minimum altitude is 13,123 ft (4,000 m). With the human object standing at the fence line, that gives a nominal sensor-to-human standoff distance of 9,543 m."

Karl: "OK, again, we should probably look at the best and worst cases to go along with the nominal case."

Jen: "Sure, no problem."

Karl: "What are the transverse and lateral magnifications of the basic optical system?"

Jen: "We still must determine that during the actual design process. For now, we are going with the variable ranges of 1 to 25 for both longitudinal and transverse magnifications for both the visible and IR optics."

Karl: "What mean wavelengths are you using for the visible and IR sensors?"

Jen: "For the visible, we are using 500 nm. We are using 9.5 μm for the infrared mean wavelength based on the peak emission wavelength for our human object."

Karl: "Good. We will have to talk about sources soon. Sounds like you have a good handle on the basic optical properties. I should be able to determine a good number of the attributes and capabilities from the information you just provided. I think that is enough for now. Maybe you can get started on trade-study that you mentioned earlier."

Jen: "I was hoping you'd ask!"

Karl: "I learned that you are studying Systems Engineering at that great local university. How is the class and coursework going?"

Jen: "I like every bit of the class. I am studying Systems Engineering and its principles and gaining a lot of knowledge. Most importantly the knowledge I am getting can be directly applied to several projects that I have been working on at my job."

Karl: "Make sure that when you do that trade study, we can use the resulting tool to help model the optical modulator that we have been discussing for background suppression both for this and future applications."

Jen: "Sure, I will make that a goal of the study."

Karl: "Oh, and make sure that you use a formal, repeatable, and justifiable analytical methodology so that we can defend the results."

Jen: "Oh, what a coincidence! In the system engineering class, we just finished studying the Analytical Hierarchy Process or AHP for short. AHP is an analytic method used to analyze alternatives and aid in decision-making. I think it's perfect for these types of problems."

Karl: "That's great! I would like you to consider at least three MBSE tools and determine which would be best in analyzing the optical modulator, integrating with the rest of our MBSE tools, and providing the best features for our future work."

Jen: "I understand. I would like to assess MATLAB, ENVI, and Mathematica, as they are all prominent analytical tools, and some may even be useful in rapid prototyping applications"

Karl: "I want you to specifically focus on development time, cost, maintenance and support, and also ease of use."

Jen: "Alright. I'd like to get started right away."

Karl: "Let me know as soon as you're done."

As Jen started the trade-off study, her mind drifted back to what she learned about the AHP method in her SE class. She recalled that the AHP method was well suited for choosing the best alternative given a set of (possibly qualitative) criteria. Dr. Thomas Saaty developed the AHP methodology in the 1970s. It has been applied in many disciplines, such as social sciences, engineering/science, national-level models, and private- and public-sector models. In implementing the AHP methodology, the first step is to choose a particular goal for the analysis. In this case, the goal is to select the best long-term analytical tool used for analyzing key optical components such as an optical modulator. To use AHP, one must understand matrix manipulation and be familiar with the calculation of eigenvalues and eigenvectors. Two matrices are created with dimensions equal to the number of decision criteria that are to be evaluated. The first matrix is known as the criteria versus criteria (CvC) matrix, which represents the previously mentioned decision criteria. The second type of matrix is called the alternative versus alternative (AvA) matrix, of which, one is created for every criterion. Four simple steps are followed in using the AHP method to select alternatives:

1. Establish decision criteria (e.g., price, development effort, maintenance, quality) and alternatives (e.g., Job A, Job B).
2. Collect pair-wise importance/preference data for paired criteria and for each alternative.
3. Synthesize and determine the relative priority of each criterion using the AHP method.
4. Aggregate all priorities and compute weights of all alternatives; select the best alternative.

In these steps, CvC matrices are generated for each alternative based on a pair-wise comparison and the *decision consistency* or perhaps more accurately the *decision inconsistency* is determined. Eigenvalues and eigenvectors help to quantify this aspect.

Figure 5.50 shows a block diagram for the decision-making process in this scenario. The top level of the block diagram shows the goal of the analysis. The middle row shows the criteria that will be used in the analysis. The third level shows the alternatives available for comparison. Jen noted that she was given the criteria for evaluating the MBSE tools:

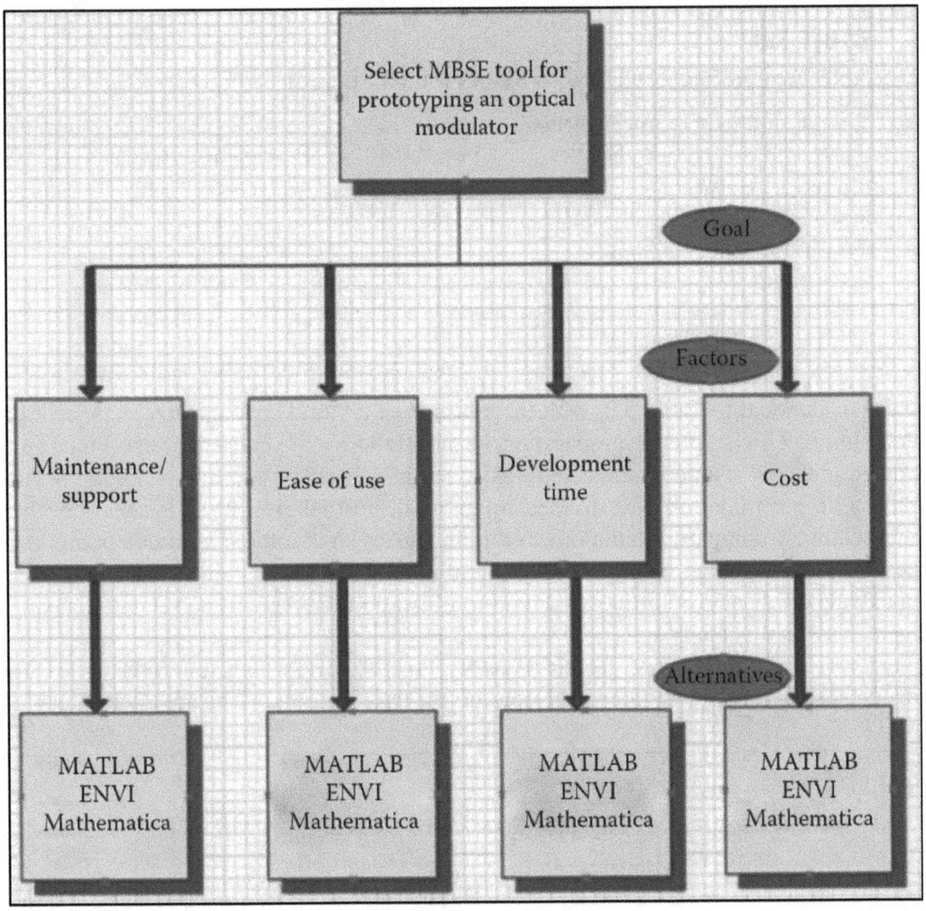

FIGURE 5.50 Analytical hierarchy process criteria and alternatives for selecting an analytical MBSE tool. (Reprinted with permission from Dr Bob 2004, http://en.wikipedia.org/wiki/File:Lens6a.png, accessed November 20, 2014.).

- Maintenance support
- Ease of use
- Development time
- Cost

She was also given a list of alternatives:

- MATLAB
- ENVI
- Mathematica.

Jen then proceeded with her analysis. Table 5.8 illustrates the CvC comparison matrix to pick from the different MBSE tools. Table 5.8 has four rows and four columns, where every row and every column correspond to a chosen criterion. When the comparison of the same criteria is performed, then its matching cell is allocated a weight factor of 1 as shown. An arbitrary, with appropriate granularity, is chosen in making the pair-wise comparisons. The results of the preferences in Table 5.8 can be understood as follows: Cost is preferred five times more than maintenance/support and three times more than development time but only half as much as ease of use. Notice that the comparisons are made on a pair-wise basis, from row to column. The diagonals are all one because any criteria must be preferred the same to itself.

TABLE 5.8
AHP Weight Factors Used to Select an MBSE Tool

	Maintenance/ Support	Ease of Use	Development Time	Cost
Maintenance/support	1	2	1/3	1/5
Ease of use	1/2	1	3	2
Development time	3	1/3	1	1/3
Cost	5	1/2	3	1

Source: Data from Barteotti (2014).

Table 5.9 is identical to Table 5.8, with the addition of a "total" row, which includes the sum of the given column. This step is in preparation for normalizing the cell entries so that each matrix entry is listed in terms of a percentage and not its actual value.

In Table 5.10, each column was divided by its total, normalizing it. This allows each column entry to be relatively compared to the other column entries so that the magnitude of any individual entry does not dominate over equally valid but smaller magnitude cell values. The mean of each row was then calculated. A new column was added to illustrate this. As will be seen, the mean column is the eigenvector in this analysis.

The next step in the AHP is to create one AvA matrix for every decision criterion that indicates how each alternative meets the stated criterion. In this case, four matrices are created, representing

TABLE 5.9
AHP Weight Factors with a Total Row Used to Select an MBSE Tool

	Maintenance/ Support	Ease of Use	Development Time	Cost
Maintenance/support	1	2	1/3	1/5
Ease of use	1/2	1	3	2
Development time	3	1/3	1	1/3
Cost	5	1/2	3	1
Total	9.5	3.83	7.33	3.53

Source: Data from Barteotti (2014).

TABLE 5.10
AHP Weight Factors with a Total Row Used and Row Mean to Select an MBSE Tool

	Maintenance/ Support	Ease of Use	Development Time	Cost	Mean
Maintenance/support	0.105	0.522	0.045	0.057	0.182
Ease of use	0.053	0.261	0.409	0.566	0.322
Development time	0.316	0.087	0.136	0.094	0.158
Cost	0.526	0.130	0.409	0.283	0.337
Total	9.5	3.83	7.33	3.53	

Source: Data from Barteotti (2014).

TABLE 5.11
AHP Weight Factors of the Alternatives with Respect to Maintenance/Support

Maintenance/Support	MATLAB®	ENVI	Mathematica	Mean
MATLAB	1	4	3	0.623
ENVI	1/4	1	1/2	0.137
Mathematica	1/3	2	1	0.239
Total	1.583	7	4.5	

Source: Data from Barteotti (2014).

maintenance/support, ease of use, development time, and cost. Table 5.11 shows the matrix created for maintenance/support. These matrices are also normalized in the same fashion as the CvC matrix was normalized. Additional rows and columns were added to show the means and totals, respectively.

Table 5.12 shows the repeated process for the ease-of-use criterion.

Next, development time was considered, and a corresponding AvA matrix was generated. Table 5.13 shows the result.

Finally, the AvA matrix for cost was generated and normalized as for the previous matrices. The results are shown in Table 5.14.

The next step in the AHP is to construct a matrix that summarizes the row means of each AvA with its decision criteria. Table 5.15 shows the summary. Each row represents a different alternative. Each column of the matrix represents the row mean of the AvA for each respective decision criterion.

The final step in obtaining a ranked preference order for each alternative is to multiply, by matrix multiplication, the alternative versus criteria matrix shown in Table 5.15 by the mean column (eigenvector) obtained in Table 5.10. The results are shown in Table 5.16.

TABLE 5.12
AHP Weight Factors of the Alternatives with Respect to Ease of Use

Ease of Use	MATLAB	ENVI	Mathematica	Mean
MATLAB	1	5	3	0.671
ENVI	1/5	1	1/3	0.114
Mathematica	1/3	3	1	0.287
Total	1.533	9	4.333	

Source: Data from Barteotti (2014).

TABLE 5.13
AHP Weight Factors of the Alternatives with Respect to Development Time

Dev. Time	MATLAB	ENVI	Mathematica	Mean
MATLAB	1	3	7	0.872
ENVI	1/3	1	3	0.340
Mathematica	1/7	1/3	1	0.120

Source: Data from Barteotti (2014).

TABLE 5.14
AHP Weight Factors of the Alternatives with Respect to Cost

Cost	MATLAB	ENVI	Mathematica	Mean
MATLAB	1	6	3	0.718
ENVI	1/6	1	3	0.305
Mathematica	1/3	1/3	1	0.160
Total	1.500	7.333	7	

Source: Data from Barteotti (2014).

TABLE 5.15
AHP Alternatives versus Criteria

	Maintenance/Support	Ease of Use	Dev. Time	Cost
MATLAB	0.623	0.671	0.872	0.718
ENVI	0.137	0.114	0.340	0.305
Mathematica	0.239	0.287	0.120	0.160

Source: Data from Barteotti (2014).

TABLE 5.16
AHP Alternatives versus Criteria into Row Mean Vector

	Maintenance/Support	Ease of Use	Dev. Time	Cost		Mean		
MATLAB	0.623	0.671	0.872	0.718		0.182		0.710
ENVI	0.137	0.114	0.340	0.305	×	0.322	=	0.219
Mathematica	0.239	0.287	0.120	0.160		0.158		0.209
						0.337		

Source: Data from Barteotti (2014).

The resultant vector provides the ranked order of alternatives based on the relative ranking of the specified criteria. The highest number is the best choice. In this case, the MATLAB MBSE tool clearly scores higher than the other MBSE tools. Table 5.17 presents the resultant rankings of the three MBSE alternative tools.

Although the AHP analysis is complete, there is an additional step that will reveal the consistency of the rankings given to the criteria. The idea is to determine a consistency index that

TABLE 5.17
Alternatives Ranking

MATLAB	1
ENVI	2
Mathematica	3

Source: Data from Barteotti (2014).

TABLE 5.18
Decision Consistency

	Maintenance/Support	Ease of Use	Development Time	Cost	Mean	Lambda *P*	Lambda
Maintenance/support	1	2	1/3	1/5	0.182	0.947	5.195
Ease of use	1/2	1	3	2	0.322	1.563	4.851
Development time	3	1/3	1	1/3	0.158	0.925	5.841
Cost	5	1/2	3	1	0.337	1.885	5.589
Total	9.5	3.83	7.33	3.53			

Source: Data from Barteotti (2014).

provides an indication for how consistent (or inconsistent) the weighting factors were determined. An example is as follows: say that you prefer ease of use three times to cost and prefer cost three times as much to development time. This implies that you prefer ease of use to development nine times as much (3×3). If, on the other hand, during the pair-wise comparison step, you assigned a value of 7 to the ease of use to development time matrix entry, then you would have an inherent "inconsistency" in assigning the weights. The consistency index captures this aspect of the analysis. If all the pairwise comparisons were perfectly consistent, when the relative weights were assigned, then Saaty showed that the eigenvalue estimates of the CvC matrix would exactly equal the number of criteria used in the AHP analysis (in this case 4). If any decision inconsistency were present, the eigenvalue estimates would deviate from 4. Determining the eigenvalue estimates can be done for any of the matrices using their row means, but we will illustrate using the original CvC matrix in Table 5.8 and its eigenvector shown as the mean in Table 5.10. Multiplying the original CvC matrix by the eigenvector produces a 4×1 vector. This vector can be divided elementwise by the original eigenvector to estimate the eigenvalues. The result is shown in the last row of Table 5.18.

Subtracting the number of criteria (4 in this case) from each element of the result and dividing each element by the number of criteria minus 1 (3 in this case) produce the consistency index. Dividing the consistency index by a statistical correction factor known as the ACI that is associated with the table of randomly generated pair-wise comparisons (equals 0.9 for 4 criteria) produces the consistency ratio. If the consistency ratio is larger than 10% (which it is in this case), then the decision maker should revisit the preference weights because there is a large degree of inconsistency present in assigning the preference weights to the pair-wise comparisons. (Jen wonders what could have gone wrong.) (The next day….)

Karl: "Hi Jen, it's good to see you again. Were you able to finish the analysis on which MBSE tool we should use to evaluate the optical modulator?"

Jen: "Hi Karl, yes I was able to apply the Analytical Hierarchy Process to our problem, but I ran into a problem."

Karl: "What's wrong?"

Jen: "Well, I was able to conclude that MATLAB is the best MBSE analytical tool from applying the AHP. However, when I calculated the consistency ratio, I found that I had a large degree of inconsistencies present when the weights were assigned. This means we can't really trust the results until we straighten out the inconsistent rankings."

Karl: "Well, at least we have a metric that lets us know when we are off base. How did you determine the weights?"

Jen: "I asked different people in the organization who had a stake in a criterion to make an estimate. For instance, I asked those that were going to do the analysis to rate the importance of 'ease of use' to the other criteria. I asked Joe in finance, the importance of cost, and Roger in software to evaluate the development time, and the lead service tech on the maintenance aspects."

Karl: "Did you ask them to be consistent in their relative rankings and with each other?"

Jen: "Oh, no! I didn't!"

Karl: "Well, there is your problem. None of them were enforcing consistency and so these inconsistent weightings crept into your analysis. I would get them all together, explain the AHP method, and ask them to revise their rankings while considering the overall consistency. I think you will get better results next time."

Jen: "Thanks Karl, I'll get right on it!"

Karl: "Ok, good luck! Let me know when you finish."

Jen conducted a second AHP; this time gathering all persons giving input together to discuss the criteria ranking. She was able to achieve a consistency ratio of 0.0023. MATLAB was still the MBSE analytical tool of choice.

In this scenario, Jen used AHP to analyze which MBSE analytical tool would best meet the organizational needs for an analytical tool for analyzing optical components like an optical modulator for the UAV project and future applications. This scenario illustrates the usefulness of AHP as an alternative and decision-making tool. One unique feature of the AHP method is its ability to accept "fuzzy" inputs. Simply put, AHP can be used to mathematically analyze attributes such as "ease of use," which was used in this example. Another example of "fuzzy" input would be "works well." Another unique feature is its ability to provide a consistency check. Due to its unique features and its relative ease of use, AHP is an essential and robust SE method for considering alternatives and making decisions.

5.A APPENDIX: ACRONYMS

AHP	Analytical hierarchy process
APAS	Automotive Personal Assistance System
ASAP	Advanced Systems Analysis Program
CAE	Computer-aided engineering
CAS	Complex adaptive systems
DMS	Disaster management systems
DOORS	Dynamic Object-Oriented Requirements System
ENVI	Environment for Visualizing Images
FIT	Fantastic imaging technologies
GUI	Graphical user interface
IEEE	Institute of Electrical and Electronics Engineers
INCOSE	International Council on Systems Engineering
MBSE	Model-based systems engineering
MXL	Michigan Exploration Lab
OMG	Object Management Group
OSLO	Optics Software for Layout and Optimization
RAX	Radio Aurora Explorer
SDLC	Systems development lifecycle
SE	Systems engineering
SoS	Systems of systems
SysML	Systems Modeling Language
UAV	Unmanned aerial vehicle
UML	Unified Modeling Language

5.B APPENDIX: DESCRIPTION OF USEFUL MBSE TOOLS

5.B.1 Cameo Systems Modeler

Introduction: Cameo Systems Modeler is a cross-platform MBSE environment tool that can define, track, and visualize all aspects of systems using SysML. System models can be managed in a remote repository like Teamwork Cloud (TWC) (Dassault, DS CATIA - Cameo Enterprise Architecture 2002–2022). This tool is an evolution of No Magic's Magic Draw tool.

Features: Cameo System Modeler supports features such as,

Open standards: This tool supports SysML and Business Process Modeling Notation (BPMN).

Requirements management: This tool supports requirements management by creating SysML requirements models, SysML requirements diagrams, requirements traceability, predefined *satisfy* requirement and *verify* requirement matrices, and requirements tabular representation. This tool supports assigning unique numbers to requirements, wherein custom requirements identification can be used depending on project need. To keep track of changes in requirements it is important to have version control which is also supported in this tool.

Traceability: This tool supports model elements traceability at different levels of abstraction. It has a standard dependency matrix which can be customized based on project need. The dependency matrix provided by this tool plays a critical role in traceability. Standard matrices are available which can be customized based on project requirements. These matrices provide information like how operational activity is mapped to capability, descriptions of available organizational resources and associated organizational responsibilities using a project matrix, and how operations are fulfilled by allocated functionality, and how system components relate to operational activity. This tool comes with a predefined SysML matrix similar to an *allocation* matrix that shows the allocated relationships between model elements. The tool also has a *satisfy* matrix which shows "satisfy" relationships to display which component of a system satisfies which requirement. It also has a *verified* matrix which shows verification relations that display which test case *verifies* which requirement. All these predefined matrices can be customized as needed.

Reports: This tool uses velocity engineer to build reports. Even though the model represents a single source of truth, generating and exporting data in different formats is still required as part of contractual submittal requirements, or data distribution within different divisions of an organization. Report wizards can be used to generate data in traditional formats like MSWord, MSExcel and HTML formats.

Solving parametric models and system Measure of Effectiveness (MoEs): This tool supports constraint blocks which are used mainly in parametric diagrams. Constraint blocks represent mathematical expressions wherein the constraint block takes input parameters and evaluates them based on the mathematical expression. The constraint can be applied to *value properties* wherein the value properties are evaluated within a valid range. If the value properties fall outside of the valid range, the tool flags it during simulation. Constraint expressions are evaluated during the initialization of a given simulation before other behaviors are executed. Parametric diagrams display the dependencies among different parameters. Parametric trade studies can be done using a parametric plugin to the model which simulates a set of instance values to determine optimality.

Model repository: Many organizations purchase a teamwork cloud server along with this tool for the model repository. The teamwork cloud server allows multiple users to work on the same project at the same time. The Model Manager provides different levels of access to each user depending on the user's role. Users must lock the model part for editing, users can save work locally or save it to the teamwork cloud as needed. It is a good practice to store changes on the cloud server every day to keep models up to date. Users need to update their local models to the latest version after other users save their changes to the cloud server.

Configuration management: MagicDraw Teamwork Cloud server supports model version control. Every user can have the latest version of the model by updating their local model from the cloud server, and every user is allowed to work in parallel on their own subsystem or component by locking out their respective part.

Security: Administrators provide different levels of permissions of the data stored in the TWC server to every user. Every user requires a login and password to access the projects stored in the MagicDraw Teamwork Cloud repository.

Interoperability: It is important to have flexible tools that can work on models developed using other tools. Cameo Systems Modeler is designed to work with other tools as long these other tools are based on UML and SysML. The Cameo Datahub plugin enables the import and export and automatic synchronization of external data to internal model data.

Adjustments: Domain-specific profiles and modeling domains can be adapted using the Domain Specific Language customization engine. Model elements can have Object Constraint Language (OCL) expressions for execution to validate models. Cameo Systems Modeler supports the open application programming interface (API). Additionally, Cameo Systems Modeler supports various scripting languages such as BeanShell, JRuby, Jython, Groovy, and Javascript.

5.B.2 Cameo Enterprise Architecture

Introduction: Cameo Enterprise Architecture is based on MagicDraw and supports all architectural frameworks. (Dassault, DS CATIA - Cameo Enterprise Architecture 2002–2022)

Features: Cameo Enterprise Architecture supports the following features,

Open standards: This tool supports open standards like UAF, UPDM, DODAF, NAF, UML, SysML, SoaML, BPMN, TOGAF, and Zachman.

User interface: This tool has a standard and customizable user interface based on user roles. It supports *cut*, *copy*, and *paste* of model elements. Up to 500 steps of undo and redo are allowed. This tool also has a spell checker. Users can generate customizable views of UAF, SysML, and UML. Containment trees provide tree-like interfaces to manage model structure. Hyperlinks can be inserted for model elements for easy navigation. An advanced search engine is available to search model elements and diagrams.

Adjustments/Tailoring: Domain-specific profiles and modeling domains can be adapted using the Domain Specific Language customization engine. Model elements can have OCL expressions for executing and validating models. This tool supports the open application programming interface. Cameo Enterprise Architecture supports various scripting languages like BeanShell, Groovy, JRuby, Jython, and Javascript.

Distributed use/Parallel development: Teamwork cloud needs to be purchased from Dassault Systems for a team-oriented, cloud-based model repository. Teamwork cloud provides a collaborative development environment for multiple users to work on models in parallel. It also provides a model versioning capability.

Configuration management: The entire model can be saved on the Teamwork cloud, and users can obtain the latest version from the Teamwork cloud to work on. Multiple users can work in parallel on their own parts of model. Due to the single common cloud-based repository, it is easy to manage model configurations.

Security: Teamwork cloud allows different levels of access to different users providing for secure access, and helping prevent unauthorized user accessing to repository data. The level of access can be different for each project and individual.

Interoperability: It is important to have tools which are capable of working with other models that have been developed using other tools. Cameo Enterprise Architecture is designed to work with other tools as long these tools are based on UML and SysML. The Cameo Datahub plugin enables the import/export and automatic synchronization of external data to the model data.

Requirement management: Cameo Enterprise Architecture allows for management of requirements at different levels, like stakeholder requirements, system requirements, subsystem requirements, and component/pieces/parts-level requirements. The requirements model element is standard and can be customized depending on project needs. Requirements management is based on SysML and supports SysML requirements diagrams. Predefined matrices based on the *satisfy requirement* and *verify requirement* stereotypes are available for traceability. A customized requirements verification matrix can be created to manage traceability between the *satisfying* element and requirements, *test cases* and requirements, and the *refined* elements and requirements. Cameo Enterprise Architecture facilitates automatic requirements numbering as well as customized requirements numbering schemes.

Traceability: Cameo Enterprise Architecture supports model elements traceability at different levels of abstraction. It has a standard dependency matrix which can be customized based on project needs. This tool provides predefined matrices for standard frameworks like UAF/DODAF/MODAF/NAF. The matrices can be customized to implement operational views and can provide system view gap analysis. Cameo Enterprise Architecture has some other predefined matrices to display how operational activities are related to system capabilities; how organizational resources connect to project responsibilities; how system capabilities relate to service capabilities; the interaction of resources; the allocation of functions to operational activities; and the relationships between systems and operational activities. This tool also provides other predefined matrices for SysML which includes an allocation matrix to show how model elements are allocated, and a matrix that shows which model element satisfies which requirement. The verification matrix is used to show relationships between test cases and requirements, and can be used to ensure each requirement has an assigned verification method. A capability behavior map shows the relationship between a capability and related operations and functions. Implementation maps show the relationship between operational performers and operational activities they are performing. Operational activity maps can be used to capture the decomposition of behavior. Organizational structure maps show the relationship between the organization and its people, such as the position people hold in an organization. System resource maps show relationships between different system resources.

Reports: Cameo Enterprise Architecture uses velocity engineer to build reports. Just as in Cameo Systems Modeler, generating and exporting data in different formats is still an essential requirement. The report wizard can be used to generate data in tradition formats like MSWord, MSExcel, and in HTML format.

Quality check: Cameo Enterprise Architecture has predefined validation suites which can be customized to validate models. Validation is used to check model accuracy and completeness based on preset modeling criteria set by the model manager and development team. Errors are captured and displayed in the console window, in table format, and recorded in a log file.

Multilingual models: Cameo Enterprise Architecture supports predefined and custom languages including English, French, German, Japanese, Russian and Thai and has different interfaces based on different regions of the world.

5.B.3 Enterprise Architect

Introduction: Four different editions are available from Sparx Systems. Since each organization has its own needs, one standard tool can't isn't suitable for all organizations. Enterprise Architect allows customization to support different applications for different organizations (SPARX 2000–2022).

Features: Enterprise Architect supports the following features,

Open standards: This tool supports open standards like UML for software engineering, Business Process Modeling Notation (BPMN) for business development and SysML for system engineering. Since enterprises need to use different frameworks to meet enterprise-level requirements, this tool supports the Open Group Architecture Framework (TOGAF) and Unified Profile for DODAF/MODAF (UPDM).

Track tasks: In general, project managers need to define tasks, set task priorities, assign tasks, and subsequently track the assigned tasks. Enterprise Architect has features to make these needs easy to accomplish. For instance, Gantt charts can be used to view task allocation.

Test and debug: This feature provides the ability to test and verify the correct behavior of a system under development, based on criteria set up by the model manager. This tool sets up the test point, manages it, and provides debugging software for inspection.

Simulation: A capability for simulating system behavior is a necessary feature that many organizations require of their modeling software. System Architect provides this capability.

5.B.4 IBM Engineering Rhapsody

Introduction: IBM has been at the forefront of systems modeling development software and has developed many useful products. IBM's Engineering Rhapsody is an example of one of these products. Rhapsody supports systems modeling and can also manage other complex activities for an organization. This tool provides a collaborative environment for design and development and provides a means for interfacing with other with other high-demand modeling tools. Rhapsody is capable of importing and exporting data in a variety of useful formats and provides a test environment to support systems testing.

Features: Rhapsody supports the following features,

Open standards: This tool supports open standards like UML, SysML, UAF, DODAF, MODAF and UPDM.

Requirements engineering: This tool can be used to generate, capture, link, and manage the requirements for a system under development. It can also help in capturing system interfaces, and link test cases using standard languages and frameworks like SysML, UML, and UAF. Engineering Rhapsody supports importing and exporting data from/to external sources.

Rapid design implementation: Engineering Rhapsody provides an integrated environment for software engineering to design and develop systems rapidly. The systems developer uses UML to provide visualization for C, C++, and java languages.

Automate design reviews: Engineering Rhapsody provides a collaborative and data sharing environment for diverse teams composed of members with different disciplines. Team members can share documents for review using a common repository. This common repository can also be used manage the evolving system design and system models. Engineering Rhapsody supports web-based access for external stakeholders like customers and vendors. Consequently, this tool facilitates an automatic review process wherein team members and stakeholders can review, approve, and make decisions quickly.

Generate documentation: Engineering Rhapsody supports document generation requirements such as automatically generating a systems specification, or other useful reports. The format of these documents can be customized based on project needs.

Prototype, simulation and execution: Engineering Rhapsody can provide a prototyping capability that allows the system to be simulated and critical parts executed before the final system design is established. In this manner the system design can be verified and validated in the early stages of systems development. Simulation and execution of critical aspects of the design can be used to identify and mitigate program risks and help ensure that the system's requirements are being met.

Agile engineering environment: Engineering Rhapsody can also be used in an Agile engineering development environment and help manage complexity in concurrent engineering activities.

5.B.4.1 SolidWorks

SolidWorks is a computer-aided design (CAD) and computer-aided engineering (CAE) application for solid modeling published by Dassault Systems.

License: Proprietary, term
Initial release: November 1, 1995
Developer: Dassault Systems

Operating system: Microsoft Windows
Stable release: SolidWorks 2021 SP2.0/January 11, 2021

5.B.4.2 CATIA

CATIA (Computer-Aided Three-dimensional Interactive Application) is a software tool that supports computer-aided design primarily for mechanical engineering applications in designing physical parts. This tool also supports computer-aided manufacturing (CAM) facilitating production design. CATIA can also be used in computer-aided engineering (CAE) in validating the design through simulation.

Developer: Dassault Systems
License: Proprietary software
Initial release: 1982
Operating system: Windows, Unix (server)
Stable release: P3 V5 6R-2020

5.B.4.3 FreeCAD

FreeCAD is an open-source parametric 3D computer-aided design (CAD) modeler primarily used to design real-life objects.

License: LGPL-2.0 or later
Original author(s): Jurgen Riegel, Werner Mayer, Yorik Van Havre
Initial release: 29 October 2002
Operating system: Linux; MAC OS; Unix; Windows; FreeBSD
Stable release: 0.20/14 June 2022
Written in C++, Python

5.B.4.4 Creo Parametric

Creo Parametric, formerly known, together with Creo Elements/Pro, as Pro/Engineer and Wildfire. This application is installed on platforms which use Microsoft Windows. This tool provides 3D modeling and is often used in the mechanical engineering discipline.

Developer: PTC
Initial release: 1987
Stable release: 7.0
Operating system: Windows
License: Trialware

5.B.4.5 Siemens NX

Siemens NX was previously known as "Unigraphics." It supports 3D designing and can be flexibly adapted in organization for various engineering applications. The tool can be readily integrated with other tools in the organization. Siemens took ownership of this tool in 2007 (SIEMENS 2022).

Developer: Siemens Digital Industries Software
License: Proprietary
Initial release: October 1973
Stable release: June 2019, NX 2206
Operating system: MAC OS up to NX12

5.B.4.6 Rhino with Grasshopper

Grasshopper is now included with Rhino. This tool applies to solid and surface modeling. Grasshopper is a visual scripting tool that turns Rhino into a parametric modeling tool.

Developer: Robert McNeel and associates
Initial release: September 2007
Stable release: 1.0, April 4,2014
License: Proprietary
Operating system: Windows 2000 and later, MAC OS

5.B.4.7 Fusion 360

Fusion 360 is a developed by Autodesk which can be used for design engineering, manufacturing and detail component design like for printed circuit board (PCB) production.

Developer: Autodesk
Operating system: Microsoft Window 8.1, MAC OS
Initial release: September 24, 2013
Latest version: 4.4
License: Trialware

5.B.4.8 Inventor

Autodesk Inventor is a computer-aided design (CAD) application for mechanical design that is produced by Autodesk. It supports 2D and 3D design, harness design, and sheet metal design. Autodesk Inventor also support simulation for design verification.

Developer: Autodesk
License: Proprietary
Initial release: 1999
Stable release: January 19, 2021
Operating system: Windows

REFERENCES

3 Dynamic Systems. 2024. MAGICDRAW. https://www.3ds.com/products-services/CATIA/products/no-magic/magicdraw/ (accessed March 21, 2024).

ALDEC. 2024. ACTIVE-HDL and FPGA design and simulation. https://www.aldec.com/en/products/fpga_simulation/active-hdl (accessed March 21, 2024).

Allen, C., M. Di Maio, G. Kapos, and N. Klusmann. 2016. Reducing the cost and complexity of variant exploration with MBSE and LSP. *IEEE International Symposium on Systems Engineering (ISSE),* Edinburgh, October 3–5, pp. 1–7.

Bajaj, M., D. Zwemer, R. Peak, A. Phung, A.G. Scott, and M. Wilson. 2011. SLIM: collaborative model-based systems engineering workspace for next-generation complex systems. Paper presented at *Aerospace Conference, 2011 IEEE,* InterCAX LLC, Atlanta, GA.

Barteotti, B. 2014. *Technical Report on AHP.* Submitted to Florida Institute of Technology, Melbourne, FL.

Basile, D., M.H. Beek, F. Di Giandomenico, A. Fantechi, S Gnesi, and G.O. Spagnolo. 2020. 30 years of simulation-based quantitative analysis tools: a comparison experiment between Möbius and Uppaal SMC. In: Margaria, T. and B. Steffen (eds), *Leveraging Applications of Formal Methods, Verification and Validation: Verification Principles.* Cham: Springer, pp. 368–384.

Batarseh, O., and L.F. McGinnis. 2012. SysML to discrete-event simulation to analyze electronic assembly systems. *Proceedings of the 2012 Symposium on Theory of Modeling and Simulation-DEVS Integrative M&S Symposium,* Orlando, FL, pp. 1–8.

Battu, H.R., K. Padmaja, and P. Gurulingam. 2013. A brief view of model based systems engineering methodologies. *International Journal of Engineering Trends and Technology (IJETT),* Vol. 4, No. 8, pp. 3266–3271.

Blleininger. 2010. Wikimedia Commons. https://commons.wikimedia.org/wiki/File:ABERR1.svg (accessed April 10, 2025).

DrBob. 2004, October 9. http://en.wikipedia.org/wiki/File:Lens6a.png (accessed November 20, 2014).

DrBob. 2006a. Wikimedia Commons. https://commons.wikimedia.org/wiki/File:Lens1.svg (accessed April 10, 2025).

DrBob. 2006b. Wikimedia Commons. https://commons.wikimedia.org/wiki/File:Lens1b.svg (accessed April 10, 2025).

DrBob. 2006c. Wikipedia. http://en.wikipedia.org/wiki/File:Lens3.svg (accessed December 30, 2014).

Boronat, A. 2018. A formal framework for prototyping executable semantics in ATL. In: Rensink A. and J. Sánchez Cuadrado (eds), *Theory and Practice of Model Transformation*. Lecture Notes in Computer Science, Vol. 10888. Cham: Springer. Pp 157-172.

Bruel, J.-M. and P. Roques. 2013. Presentation of SysML concepts. *Embedded Systems*. Eds. Fabrice Kordon, Jérôme Hugues, Agusti Canals, Alain Dohet. Wiley,Hoboken, NJ. pp. 45–69.

Cameron, B.G. and E.F. Crawley. 2012. Costing commonality: investigating the impact of platform divergence. *IEEE Aerospace Conference,* Big Sky, MT, March 3–10. IEEE.

Carroll, E.R. and R.J. Malins. 2016. Systematic Literature Review: How Is Model-Based Systems Engineering Justified? *Sandia National Laboratories*. Report: SAND2016–2607.

Chaudhuri, K.D. 2023. Building language model in NLP. *Data Science Blogathon*. https://www.analyticsvidhya.com/blog/2022/01/building-language-models-in-nlp/ (accessed March 21, 2024).

Chochaev, R.Z. and S.V. Gavrilov. 2021. Evaluating FPGA routing architectures with complex grid layouts. *IEEE Conference of Russian Young Researchers in Electrical and Electronic Engineering*, St. Petersburg, Moscow, January 26–29, pp. 2604–2635.

Cloutier, R., B. Sauser, M. Bone, and A. Taylor. 2015. Transitioning systems thinking to model-based systems engineering: systemigrams to SysML models. *IEEE Transactions on Systems,* Man, and Cybernetics*:* Systems, Vol. 45, no. 5, pp. 662–674.

Cornford, S.L., M.S. Feather, V.A. Heron, and J. Steven Jenkins. 2006. Fusing quantitative requirements analysis with model-based systems engineering requirements engineering. Paper presented at *14th IEEE International Conference,* Jet Propulsion Lab, California Institute of Technology, Pasadena, CA, September 11–15, 2006.

Cristan. 2006. Wikimedia Commons. https://commons.wikimedia.org/wiki/File:Snells_law.svg (accessed April 10, 2025).

Cronholml44. 2007. Wikimedia Commons. https://commons.wikimedia.org/wiki/File:Concave_mirror.svg (accessed April 10, 2025).

Darrin, M.A.G. and W.S. Devereux. 2017. The Agile Manifesto, design thinking and systems engineering. *Annual IEEE International Systems Conference (SysCon),* Montreal, QC, pp. 1–5.

Dickerson, C.M. and N. Dimitri. 2013. A brief history of models and model based systems engineering and the case for relational orientation. *IEEE Systems Journal*, Vol. 7, no. 4, pp. 581–592. https://hdl.handle.net/2134/13085.

Dickerson, C.E. and D.N. Mavris. 2016. *Architecture and Principles of Systems Engineering*. Auerbach Publications, Boca Raton, FL.

Egmason. 2010, May 8. http://en.wikipedia.org/wiki/File:Apochromat.svg (accessed November 20, 2014).

Egner, S.E., J.C. Gonzalez-Herrera, O. Bellido-Tirado, and S. Frey. 2020. Verification control using DOORS. In: Angeli G.Z. and P. Dierickx (eds), *Modeling, Systems Engineering, and Project Management for* Astronomy, IX. Proc. SPIE 11450, https://doi.org/10.1117/12.2561976 (online only).

Erickson, J. and K. Siau. 2013. *Unified Modeling Language: The Teen Years and Growing Pains*. Berlin: Springer, pp. 295–304.

Eriksson, H.-E., M. Penker, B. Lyons, and D. Fado. 2003. *UML 2 Toolkit*. Indianapolis, IN: John Wiley & Sons.

Estefan, J.A. 2008. Survey of model-based systems engineering (MBSE) methodologies. INCOSE. https://www.omgsysml.org/MBSE_Methodology_Survey_RevB.pdf (accessed March 21, 2024).

European Cooperation for Space Standardization (ECSS). October 1, 2012. *Space Engineering—Software Engineering Handbook*. Noordwijk, the Netherlands: European Space Agency (ASA) Requirements and Standards Division. ECSS-E-HB-40A PR-Draft.

Feiler, P.H., J.B. Goodenough, A. Gurfinkel, C.B. Weinstock, and L. Wrage. 2013. *Pillars for Improving the Quality of Safety-Critical Software-Reliant Systems*. White Paper, Pittsburgh: Software Engineering Institute, Carnegie Mellon University.

Fisher, G.H. 1998. Model-based systems engineering of automotive systems. Paper presented at *Proceedings of the AIAA/IEEE/SAE Digital Avionics Systems Conference, 17th DASC,* Vitech Corporation, Vienna, VA.

Fowler, M. and J. Highsmith. 2001. The Agile Manifesto. https://www.usna.edu/Users/cs/needham/courses/ic470/fallAY11/TheAgileManifesto.pdf (accessed March 21, 2024).

Friedenthal, S., M.A. Moore, and R. Steiner. 2014. *A Practical Guide to SysML: The Systems Modeling Language*. San Francisco: Elsevier Science & Technology. https://www.academia.edu/62142117/A_Practical_Guide_to_SysML_The_Systems_Modeling_Language (accessed March 6, 2025).

Gao, Y., X. Gao, X. Yang, J. Liu, and G. Chen. 2019. An efficient ring-based metadata management policy for large-scale distributed file systems. *IEEE Transactions on Parallel and Distributed Systems*, Vol. 30, no. 9, pp. 1962–1974.

Girish, M., G. Gopakumar, and D.S. Divya. 2021. Formal and simulation verification: comparing and contrasting the two verification approaches. *2nd International Conference on Advances in Computing, Communication, Embedded and Secure Systems*, Ernakulam, India, pp. 41–44.

GLUE. 2019. Glue Benchmark. https://gluebenchmark.com/ (accessed March 21, 2024).

Goode, H.H. and R.E. Machol. 1957. *System Engineering; an Introduction to the Design of Large-Scale Systems.* New York: McGraw-Hill.

Goodwin, M., D. Adams, R. Brown, V.N. Hartmann, C. Lacombe, C.H.N. Lahoz, J. Lawrence, D. May, E. O'Brien, and T. Zafar. 2022. Application of model-based systems engineering to MANIFEST conceptual design. *Modeling, Systems Engineering, and Project Management for Astronomy* X, Vol. 12187, pp. 34–53. SPIE.

Green, R.A. July 2007. Getting a handle on MATLAB graphics. *Potentials IEEE*, Vol. 26, no. 4, pp. 31–37.

Grønmo, R., B. Møller-Pedersen, and G.K. Olsen. 2009. Comparison of three model transformation languages. In: Paige R.F., A. Hartman, and A. Rensink (eds), *Model Driven Architecture - Foundations and Applications.* Berlin, Heidelberg: Springer.

Guenther, R. 1990. *Modern Optics.* New York: John Wiley & Sons.

Hallqvist, J.L. 2016. Introducing MBSE by using systems engineering principles. *26th Annual INCOSE International Symposium*, Vol. 26, Edinburgh, Scottland, pp. 512–525.

Hart, L.E. 2015. Introduction to model-based system engineering (MBSE) and SysML. *Delaware Valley INCOSE Chapter Meeting*, Vol. 30. Mount Laurel (New Jersey): Lockheed Martin, Information Systems and Global Solutions (IS&GS).

Harvey, D., P. Logan, M. Waite, and T. Liddy. 2012. Document the model, don't model the document. Proceedings in Systems Engineering and Test Evaluation (SETE) Conference. 6th Asia Pacific Conference on Systems Engineering. Brisbane, Australia (May 2012). https://www.researchgate.net/publication/267204157_Document_the_Model_Don't_Model_the_Document (accessed Mar 6, 2025).

Haskins, C. 2011. A historical perspective of MBSE with a view to the future. *INCOSE International Symposium*, Vol. 21, pp. 493–509.

Helmut, J. 2013. BrainyQuote.com, Xplore Inc. Hemult Jahn at BrainyQuote.com. https://www.brainyquote.com/quotes/quotes/h/helmutjahn325250.html (accessed April 2, 2014).

Henderson, K. and A. Salado. 2020. Value and benefits of model-based systems engineering (MBSE): evidence from the literature. *Systems Engineering*, Vol. 24, pp. 51–66.

Henderson, K., T. McDermott, A. Salado, and E. Van Aken. 2023. Measurement framework for model-based systems engineering (MBSE). *Systems Engineering*, Vol. 26, no. 1, pp. 3–31.

Highsmith, J. 2001. History: the Agile Manifesto. Agilemanifesto.org. https://agilemanifesto.org/history.html (accessed March 22, 2024).

Hoffmann, H.-P. 2008. SysML-based systems engineering using a model-driven development approach. White Paper, Telelogic. https://rose-hulman.edu/Class/cs/OldFiles/csse490-mbse/Readings/SysML_based_systems_engineering-08.pdf (accessed on March 6, 2025).

Honour, E.C. 1998. INCOSE: history of the international council on systems engineering. *System Engineering*, Vol. 1, pp. 4–13.

Horowitz, B.M. 1991. The Importance of Architecture in DOD Software. DTIC Report: ADA244848.

IBM DOORS. 2024. Overview of DOORS. https://www.ibm.com/docs/en/engineering-lifecycle-management-suite/doors/9.7.2?topic=engineering-requirements-management-doors-overview (accessed March 22, 2024).

IEEE (Std 610.12–1990). September 28, 1990. *IEEE Standard Glossary of Software Engineering Terms.* New York: The Institute of Electrical and Electronic Engineers.

IEEE (Std 1471–2000). October 9, 2000. *IEEE Recommended Practice for Architectural Description of Software-Intensive Systems.* New York: The Institute of Electrical and Electronic Engineers.

INCOSE. 2007. *Systems Engineering Vision 2020,* version 2.03. Seattle, WA: International Council on Systems Engineering. INCOSE-TP-2004-004-02.

INCOSE. 2022. History of systems engineering. https://www.incose.org/about-systems-engineering/history-of-systems-engineering (accessed March 21, 2024).

INCOSE UK. 2012. Chapter Z9—model based systems engineering. https://incoseuk.org/Documents/zGuides/Z9_version2.1_2020_web.pdf (accessed April 10, 2025).

Ji, M., S. Huang, H. Huang, and T. Zhang. 2009. The new development of PRISM for probabilistic model checking. *Fourth International Conference on Internet Computing for Science and Engineering*, Harbin, China, December 21–22, pp. 64–66.

Johan Arvelius. 2005. Wikipedia. http://en.wikipedia.org/wiki/File:Reflection_angles.svg (accessed December 30, 2014).

Josell7. 2012. Wikipedia. http://en.wikipedia.org/wiki/File:RefractionReflextion.svg (accessed December 30, 2014).

Jouault, F., F. Allilaire, J. Bézivin, and I. Kurtev. 2008. ATL: a model transformation tool. *Science of Computer Programming*, Vol. 72, pp. 31–39.

Kapos, G., V. Dalakas, A. Tsadimas, M. Nikolaidou, and D. Anagnostopoulos. 2014. Model-based system engineering using SysML: deriving executable simulation models with QVT. *IEEE International Systems Conference Proceedings*, Ottawa, ON, March 31 to April 3, pp. 531–538.

Kiktev, N., A. Didyk, and M. Antonevych. 2020. Simulation of multi-agent architectures for fruit and berry picking robot in active-HDL. *IEEE International Conference on Problems of Infocommunications. Science and Technology, Kharkiv*, Ukraine, October 6–9, pp. 635–640.

Lavi, J.Z. and J. Kudish. 2004. Systems modeling and requirements specification using ECSAM: a method for embedded computer-based systems analysis. *Proceedings 11th IEEE International Conference and Workshop on the Engineering of Computer-Based Systems*, Brno, Czech Republic, May 27.

Lempia, D. and J. Ray. 2011. Practical SysML applications: a method to describe the problem space. *INCOSE International Symposium*, Vol. 21, no. 1, pp. 85–98. INCOSE.

Li, H., M. Zou, D. Weidmann, S.A. Cheaib, M. Mörtl, and B. Vogel-Heuser. 2019. Model-based systems engineering process for supporting variant selection in the early product development phase. *IEEE International Conference on Industrial Engineering and Engineering Management*, Macao, China, December 15–18, pp. 637–643.

Long, D. and Z. Scott. 2011. *A Primer for Model-Based Systems Engineering*, 2nd edn. Blacksburg, VA: Vitech Co. https://www.amazon.com/Primer-Model-Based-Systems-Engineering/dp/1105588106 (accessed March 21, 2024).

Ogren, I. 2000. On principles for model-based systems engineering. *Systems Engineering*, Vol. 3, no. 1, pp. 38–49.

Madni, A.M. and M. Sievers. 2018. Model-based systems engineering: motivation, current status, and research opportunities. *Systems Engineering*, Vol. 21, no. 3, pp. 172–190.

Martin, J.N. 1996. *Systems Engineering Guidebook: A Process for Developing Systems and Products.* Boca Raton, FL: CRC Press.

Mathworks. 2024. MATLAB and simulink for model-based system engineering. https://www.mathworks.com/solutions/model-based-systems-engineering.html (accessed March 21, 2024).

Mazeika, D., A. Morkevicius, and A. Aleksandraviciene. 2016. MBSE driven approach for defining problem domain. *11th System of Systems Engineering Conference*, Kongsberg, Norway, June 12–16, pp. 1–6.

McDermott, T.A., Hutchinson, M. Clifford, N.E. Van Aken, A. Salado, and K. Henderson. 2020. Systems Engineering Research Center (SERC). https://sercuarc.org/wp-content/uploads/2020/03/SERC-SR-2020-001-Benchmarking-the-Benefits-and-Current-Maturity-of-MBSE-3-2020.pdf?force_isolation=true (accessed March 22, 2024).

McVittie, T.I., O.V. Sindiy, and K.A. Simpson. 2012. Model-based system engineering of the Orion flight test 1 end-to-end information system. *2012 IEEE Aerospace Conference*, Big Sky, MT, pp. 1–11. https://doi.org/10.1109/AERO.2012.6187440

Microsoft. 2019. Microsoft reserach paraphrase corpus. https://www.microsoft.com/en-us/download/details.aspx?id=52398 (accessed March 21, 2024).

Mitchell, S.W. 2014. Transitioning the SWFTS program combat system product family from traditional document-centric to model-based systems engineering. *Systems Engineering*, Vol. 17, no. 3, pp. 313–329.

Mori, M., A. Ceccarelli, P. Lollini, B. Frömel, F. Brancati, and A. Bondavalli. 2018. Systems-of-systems modeling using a comprehensive viewpoint-based SysML profile. *Journal of Software: Evolution and Process*, Vol. 30, no. 3, pp. 1–20.

NYU-MLL. 2019. The corpus of linguistic acceptability. https://nyu-mll.github.io/CoLA/ (accessed March 21, 2024).

Object Management Group (OMG). 2017. OMG unified modeling language specification. Version 2.5.1. https://www.omg.org/spec/UML/2.5.1/About-UML (accessed September 3, 2024).

Pantelic, V., et al. 2015. A toolset for Simulink: improving software engineering practices in development with Simulink. *3rd International Conference on Model-Driven Engineering and Software Development*, Angers, France, February 9–11, pp. 1–12.

Prism Model Checker (PRISM). 2024. Probabilistic Model Checker. https://www.prismmodelchecker.org/ (accessed March 21, 2024).

Qamar, A. 2009. Designing mechatronic systems, a model-based perspective, an attempt to achieve SysML-MATLAB/Simulink model integration. Paper presented at *2009 IEEE/ASME International Conference on Advanced Intelligent Mechatronics,* Singapore, July 14–17, 2009.

Radiant Zemax 13. 2014. Release 2—The industry standard, powerful optical and illumination design software. https://www.radiantzemax.com/rz/news (accessed July 21, 2014).

Ramos, A.L. 2012. Model-based systems engineering: an emerging approach for modern systems. *IEEE Transactions on Systems, Man, and Cybernetics, Part C: Applications and Reviews,* Vol. 42, no. 1, pp. 101–111.

Rashid, M., M.W. Anwar, and A.M. Khan. 2015. Toward the tools selection in model based system engineering for embedded systems – A systematic literature review. *Journal of Systems and Software*, Vol. 106, pp. 150–163.

RCarboni. 2009. Wikipedia. http://en.wikipedia.org/wiki/File:CollimatingLensSVG.svg (accessed December 30, 2014).

Rogers, E.B. and S.W. Mitchell. 2021. MBSE delivers significant return on investment in evolutionary development of complex SoS. *Systems Engineering*, Vol. 24, No. 6, pp. 385–408.

Rott, J., J. Weixler, A. Rabl, P. Sandl, M. Wei and B. Vogel-Heuser. 2018. Integrating hierarchical task analysis into model-based system design using airbus XHTA and IBM rational rhapsody. *IEEE International Conference on Industrial Engineering and Engineering Management*, Bangkok, Thailand, December 16–19, pp. 1856–1860.

Saleh, B.E.H. and M.C. Teich. 1991. *Fundamentals of Photonics.* New York: John Wiley & Sons.

Schneider, G. and J.P. Winters. 2001. *Applying Use Cases.* India: Pearson Education.

Shevchenko, N. 2021. Requirements in model-based systems engineering (MBSE). *SEI Blog*, 22 February 2021. https://insights.sei.cmu.edu/blog/requirements-in-model-based-systems-engineering-mbse/ (Accessed March 6, 2025).

SIEMENS. 2022. NX. https://www.plm.automation.siemens.com/global/en/products/nx/ (accessed March 23, 2024).

Silingas, D., S. Pavalkis, and A. Morkevicius. 2009. MD wizard - a model-driven framework for wizard-based modeling guidance in UML tools. International Multiconference on Computer Science and Information Technology, Mragowo, Poland, October 12–14, pp. 609–615.

Soyler, A. 2010. A model-based systems engineering approach to capturing disaster management systems. Paper presented at *Fourth Annual IEEE Systems Conference,* Department of Industrial Engineering and Management Systems, University of Central Florida, Orlando, FL.

Spangelo, S.C. 2012. Applying model based systems engineering (MBSE) to a standard CubeSat. Paper presented at *2012 IEEE Aerospace Conference,* University of Michigan, Ann Arbor, MI.

SQu AD. 2019. The Stanford question answering dataset. https://rajpurkar.github.io/SQuAD-explorer/ (accessed March 21, 2024).

Stepanchick, J. 2016. Integrating Model Based Engineering and Trade Space Exploration into Naval Acquisitions. Masters Thesis, Cambridge: Massachusetts Institute of Technology.

Subarna, S., A.K. Jawale, A.S. Vidap, S.D. Sadachar, S. Fliginger, and S. Myla. 2020. Using a model based systems engineering approach for aerospace system requirements management. *2020 AIAA/IEEE 39th Digital Avionics Systems Conference (DASC)*, San Antonio, TX, October 11–15.

Taentzer, G. and A. Schürr. 2012. AGG: a graph transformation environment for modeling and validation of software. *Applications of Graph Transformations with Industrial Relevance: 4th International Symposium (Revised Selected and Invited Papers)*, Budapest, Hungary (October 7–11, 2011).

Tamasfiex. 2011. Wikipedia. http://en.wikipedia.org/wiki/File:ThickLens.png (accessed December 30, 2014).

Tsadimas, A., M. Nikolaidou, and D. Anagnostopoulos. 2012. Extending SysML to explore non-functional requirements: the case of information system design. *Proceedings of the 27th Annual ACM Symposium on Applied Computing.* Association for Computing Machinery, Trento, Italy (March 26–30, 2012), pp. 1057–1062.

UPPAAL. 2023. UPPAL features. https://uppaal.org/ (accessed December 27, 2023).

Walden, D.D. 2015. *INCOSE, Systems Engineering Handbook, a Guide for System Lifecycle Processes and Activities*, 4th edn. San Diego, CA: INCOSE.

Weilkiens, T. 2020. SYSMOD-The systems modeling toolbox-pragmatic MBSE with SysML. MBSE4U. ISBN-13: 978-3981852981.

Wenhao, I., W. Wang, A. Zhang, and Y. Wang. 2021. Extending model-based systems engineering into integrated platform designed for civil aircraft developments. *Journal of Aerospace Technology and Management*, Vol. 13, p. e2321.

Wymore, W.A. 1967. *A Mathematical Theory of Systems Engineering.* New York: John Wiley & Sons.

Yang, Z., H. Du, Y. Liu, R. Liu, and Y. Liu. 2021. Use the harmony-SE approach to extend the advantages of MBSE. *2021 IEEE 16th Conference on Industrial Electronics and Applications (ICIEA)*, Chengdu, China, August 1–4, pp. 223–227. https://doi.org/10.1109/ICIEA51954.2021.9516269

6 Research Methods and Experimental Design in Systems Engineering

All life is an experiment. The more experiments you make, the better.

—**Ralph Waldo Emerson**

6.1 OVERVIEW

Some research methods in systems engineering are very mathematically intensive and borrow heavily from the fields of calculus, linear algebra, and probability and statistics, among others. A firm foundation in the mathematics of probability, statistics, and random processes is fundamental in understanding systems engineering methods and techniques in systems optimization, statistical process control (SPC), design of experiments (DoE), decision and risk analysis, and reliability engineering.

It is common practice to gather and analyze data throughout the systems development lifecycle to gain insight or predict fundamental aspects of system performance. For example, it is commonplace to compare data sets from experiments through statistical analysis to verify or validate that requirements, at all levels, are being met. In this chapter, we focus on statistically based research methods that are often used throughout the system lifecycle. We especially focus on DoE methods since these methods are required to successfully generate believable experimental results. By *experiment* we mean *...a test or series of runs in which purposeful changes are made to the input variable of a process or system* [or product or service] *so that we may observe and identify the reasons for changes that may be observed in the output response* (Montgomery 2020). The quantity in brackets was added by us for consistency with the general class categorization/decomposition scheme of Blanchard and Fabrycky's where, in general, items under consideration can be decomposed and categorized into systems, services, products, and/or processes (Blanchard and Fabrycky 2011). With this definition for experiment, we can see that we will encounter *experiments* regularly throughout the systems lifecycle, especially in the developmental phases but also in the support and operational phases, and possibly even in the system retirement and disposal phase. Our focus will be research methods for DoE but will also briefly address SPC methods and optimization methods. The methods for SPC and optimization are covered in much more detail in other chapters where they are directly relevant. For example, SPC methods fit very well into production and manufacturing applications, and so we cover this topic in our chapter on Quality Production and Manufacturing. The optimization topic is more in line with our Decision and Risk Analysis in Systems Engineering chapter and so optimization methods are covered there. With regard to notation for the DoE content of this chapter, we use consistent notation with the excellent text, *Design and Analysis of Experiments* by Douglas C. Montgomery (2020). The notation used therein is seen elsewhere but is consistent, logical, and practical. Two other good references for this topic are, *Statistics for Research, 3rd Ed.,* by Shirley Dowdy, Stanley Wearden, and Daniel Chilko (2004), and Andy Field's *Discovering Statistics Using IBM SPSS Statistics, North American Edition* (2017). The later book is especially colorful, full of useful information, and practical. We cannot cover all the topics in the DoE category since there are volumes on this subject alone. If we attempted to do this, we would be writing a DoE book and not introducing this important topic as part of one of this

DOI: 10.1201/9781003624097-8

chapter. Our intent is to useful information about DoE methods and provide the fundamentals to understand and apply these concepts. It is expected that the reader refers to some of these excellent texts or similar texts for further information, as needed. For the optics material in Section 6.2, we focus on the statistical nature and propagation characteristics of light. There is a logical connection to the statistical methods presented in the research methods section, and these sections are complementary to each other. For the optical material in Section 6.2, we use the notation of Joseph W. Goodman in his classic text, *Statistical Optics*, which surely is one of the fundamental and essential texts in understanding the nature of light for coherent and incoherent imaging systems applications (Goodman 1985). So what impacts can these research methods have on systems in general?

There are both good and bad examples of research method applications ranging over a wide range of systems, services, products, and processes extending from Department of Defense UAV communication systems experiments, to ships, planes, and more. It can be argued that every systems development effort requires experimentation in some form or another. For example, in World War II, there was the Liberty Ship Project wherein many Liberty ship hulls ended up splitting during the war effort. The flaws in the hulls were eventually linked to a lack of research prior to going into production. On a more positive note, DoE methods were used in India to significantly reduce emissions in internal combustion engines. On the national front in the United States, semiconductor companies are using research methods to improve production capacity in their factories to gain a competitive edge in international markets.

In this chapter, we start with an introduction to research methods and then provide the historical background of some of the important contributions made in the areas of research methods and DoE. We then present some of the important concepts and principles of research methods and follow this with the development of a general theoretical and mathematical framework for a necessary foundational baseline for the later material. In this section, we introduce basic statistical and probability concepts which underly the majority of the mathematical descriptions in this chapter. We provide a section on research methods, tools, and techniques and follow this with research methods applications. In the application section, we address hypothesis testing, ANOVA, SPC, optimization methods, and regression analysis. We also develop assessment metrics like the P – statistic, R^2 some variants, the coefficient of variation (CV), and the Pearson coefficient r. We then talk about implementation methodologies and end this section with some national and international case studies. We analyze these projects in terms of their use of research methods. In Section 6.2, we look at the propagation and characterization of light using statistical methods as the optical analog to the research methods presented in Section 6.1. We cover both coherent and incoherent imaging systems and look at correlation and polarimetric imaging effects. In the last section of this chapter, we apply what we have learned in Sections 6.1 and 6.2 to a sensor characterization effort where FIT is using DoE methods to compare experimental field data captured for their hyperspectral, EO, and IR sensor suite to theoretical, and verified and validated laboratory simulation data. Both ANOVA and hypothesis testing methods are employed and checked for accuracy using assessment metrics.

6.1.1 Introduction

According to the writings of Marcus Aurelius, "Nothing has such power to broaden the mind as the ability to investigate systematically and truly all that comes under thy observation in life" (Holiday and Hanselman 2016). Evidently, the benefits of experimentation were well understood by this ancient Roman philosopher who also happened to be an emperor. As a species, we have come a long way from those ancient days, but it is interesting to note that, even then, the ancient Roman philosopher and emperor recognized the value of research. Our species has advanced from hurling sticks and stones to controlling precision strike drones from halfway around the world thanks in large part to research, which like the unquenchable thirst for knowledge, is a critical component of innovation. But how did research get us here? Research methods include the equipment used to collect data as well as the procedures used to interpret and draw conclusions from that data (Dawson 2019). Like selecting the best questions for an interview, picking the right research methods is essential in producing good, defensible research products. There are many common kinds of research methods, including surveys, quantitative research and experiments, and qualitative research and experiments, however, we will focus on quantitative research and experiments since they are most often used in engineering and science circles.

6.1.2 Historical Background

Research methods in systems engineering are heavily reliant on statistical methods as are optical imaging methods, and most, if not all, scientific and engineering disciplines. Successful civilizations have used probability and statistics for thousands of years. Without realizing it, the Romans of the classical era, for example, employed statistics to conduct their censuses. How the empire's resources needed to be distributed was directly influenced by the census results. The ideas from the Roman Empire and the early Greeks, from whom they borrowed many of their ideas and concepts, continue to influence current Western thought. As a result, these ideas largely impact our foundational understanding of mathematics and engineering. Some highly interesting historical applications of statistics have come from other well-known sources, such as Galileo in astronomy, and Florence Nightingale in nursing. Various leaders throughout history have also used statistics to understand their populations and manage their forces.

During his time, Galileo made a statistical discovery in the realm of astronomy. He discovered that inaccuracies in the then-present measuring tools were causing him and his peers to have trouble precisely gauging the distance and motion of stars and planets. As a result, they wrote, "Those observations must be termed the more exact, or the less in error, which by adding or removing the fewest minutes, restore the star to a probable location. The real location must be thought to be that among the alternative locations where the greatest number of distances, measured using the most precise observations, agree" (Hald 2003). Galileo and his contemporaries shared data so that they might build on each other's models and get a better understanding of the structure of our universe.

During the 19th century, Florence Nightingale used statistics to promote change and reform. Many members of the English nobility were persuaded that changes needed to occur in the way patients were cared for, by the data she helped to gather and diagram. The "Polar Area Diagram," shown in Figure 6.1, was one of the most important instruments in her research (Science Museum 2018).

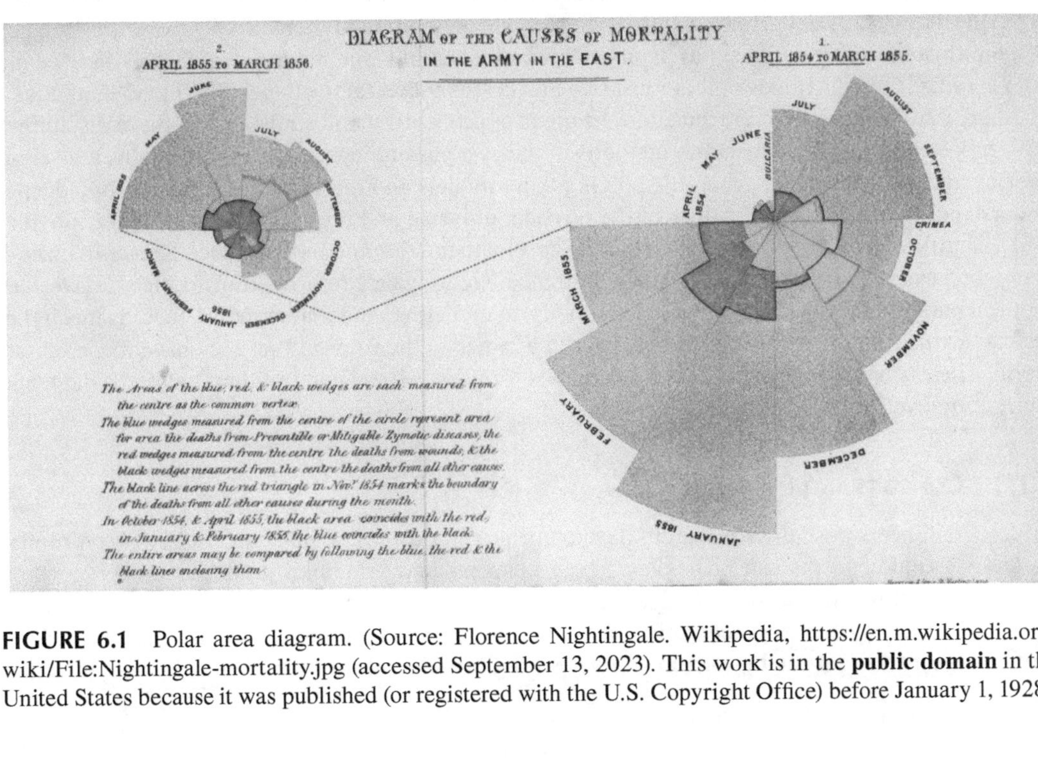

FIGURE 6.1 Polar area diagram. (Source: Florence Nightingale. Wikipedia, https://en.m.wikipedia.org/wiki/File:Nightingale-mortality.jpg (accessed September 13, 2023). This work is in the **public domain** in the United States because it was published (or registered with the U.S. Copyright Office) before January 1, 1928.)

There is no denying of the positive impacts she had on nursing processes and procedures, and the ultimate affect on the sick and dying during the Russian Crimean War of the 1850s, as shown in this graphic. Because of her many contributions, she became the first female to be admitted to the Royal Statistical Society. "The first year (shown on the right of the diagram) was 1854–1855, following her arrival in the region. The second (on the left) was 1855–1856, after she had implemented a series of reforms to the hospital and nursing practices (Science Museum 2018)."

Because of her efforts, there is a marked decrease in deaths due to preventable diseases illustrated in the left figure (shown by the relative area of the blue regions in Figure 6.1).

Research methods in systems engineering use a variety of tools, methods, and techniques, as will be seen below, however, DoE is of fundamental importance. DoE can be thought of as a method of conducting research in which there are multiple variables. The approach is to keep one or more variables constant while altering others, with the results being examined to identify the set of variables that produce the best intended result (Antony 2014). DoE was first created when Sir Ronald Fisher was conducting research on the impacts of various fertilizers on different patches of land in the early 1920s (Deboeck 2010). The history of DoE refers to this as the Era of Agricultural Origins. DoE was immediately adapted for use by the military, as is the case with most useful concepts. During World War II, renowned statistician Besse Day started using DoE at the Naval Experimentation Laboratory to assist the Navy in identifying ship flaws (Telford 2007). George Box, a statistician, and K.B. Wilson, a chemist, launched DoE's first Industrial Era in 1951 (Kotz and Johnson 1992). While working for Imperial Chemical Industries in England, Box and Wilson employed DoE and developed the Response Surface Methodology (RSM). Their work aided in demonstrating the value of statistics in a variety of industries, including the chemical sector (Hunter 2013).

Japan experienced the DoE's Second Industrial Era just slightly after the first Industrial Era (Telford 2007). Early in the 1950s, an American engineer and statistician by the name of W. Edwards Deming traveled to Japan and took DoE with him. He imparted research techniques to Japanese scientists and engineers, including Genichi Taguchi, after whom the Taguchi method was named. To help numerous industries, including the automotive industry, improve, a group of Japanese scientists used DoE to develop Total Quality Management (TQM) and Continuous Quality Improvement (CQI). Japanese goods are now renowned for their reliability, particularly their automobiles.

The Modern Era is the current era of DoE (Telford 2007). Leaders in industry in the US recognized the importance of CQI and wanted to implement these methods here. Six Sigma was created by Bill Smith, a Motorola engineer, as a quality-control technique that made use of DoE (Antony et al. 2017). Today, GE and Honeywell are just two businesses that effectively use Six Sigma methods.

Many individuals have contributed to the areas of probability and statistics and these disciplines have profoundly impacted research methods in the systems engineering discipline as well as engineering and the sciences in general. Applying these models and mathematics provides for a deeper understanding of the random nature of the physical universe and provides useful insights into general scientific and engineering problems. Galileo's mathematics and models aided Kepler in discovering the laws of planetary motion, while Florence Nightingale's reforms transformed the western medical paradigm. From data collection methods, to modeling, and mathematics, there is much that we can learn from these titans. Further, we can appreciate the imprint that they have made on our world. Their achievements can serve as examples for us, while we can use their methods, data, and spirit of innovation to develop better systems.

6.1.3 Concepts and Principles

Systems engineers base their research, development, and design methods and decisions on fundamentally sound concepts and principles. These concepts and principles guide researchers as they seek out the optimal means of making their system a reality. Systems thinking is an important starting point for any system engineer doing research. Engineers who practice systems thinking, as the name suggests, look at the big picture before getting into the nitty-gritty (Bui and Galanou 2022).

According to Lavi and associates, engineers need to think about how the system functions, how its elements are sized and shaped, and how the physical and electrical components interact with each other (Lavi et al. 2020). Systems thinking addresses the interdisciplinary as well as multidisciplinary aspects of the system under development (Ford 2012). The engineering discipline now encompasses a wide range of specializations, from electrical and mechanical to software and beyond (Camelia et al. 2020). Ensuring that the developed system works as intended and making sure all the parts, and the whole work together efficiently is central to the systems engineering discipline.

Research methods in systems engineering encompass several important principles that apply to the complete systems development lifecycle. Having a *lifecycle* perspective implies that the system engineer is accountable for the complete lifecycle of the system, spanning from problem definition to decommissioning the system (Muda et al. 2020). Consequently, research methods must be applicable to all stages of the systems development lifecycle to aid in fulfilling this task. Also, if a system engineer focuses only on the functional aspects of the system and neglects non-functional details such as reliability, they are setting up the systems development effort for failure.

The systems engineer needs to carefully plan when selecting research methods. Important considerations like ethics, sustainability, and continual assessment throughout the systems lifecycle must be carefully addressed and implemented throughout the systems development effort. Ethical decisions must be made in research methods to avoid negative consequences and major ramifications for all parties involved, such as consumer backlash if unethical practices come to light and the public is made aware (Zhu and Jesiek 2020). Seriously considering and addressing sustainability and the system's lifecycle support requirements is fundamentally important for program success. Sustainability encompasses both environmentally favorable and production-friendly practices (Vankatesan et al. 2022). The systems engineer should research and find solutions that are environmentally friendly and lead to a sustainable ecosystem. With regard to supportability, the systems engineer must examine the future availability of parts and support items used in the system (Randrianariosoa 2019). The unexpected and unplanned for discontinuation of a component can significantly impede upcoming manufacturing processes, the operational tempo, and the supply chain. Thorough research needs to consider the availability of parts, the reliability of the supply chain including the feasibility of reproducing the item by another manufacturer (Randrianariosoa 2019).

According to Montgomery and Runger, "The field of statistics deals with the collection, presentation, analysis, and use of data to make decisions, solve problems, and design products and processes. Because many aspects of engineering practice involve working with data, obviously some knowledge of statistics is important to any engineer (Montgomery and Runger 2019)".

We can see the "Engineering Method" shown in Figure 6.2. The dashed box around the second two blocks shows the relationship between generating or refining a model and experimentation. The engineering method starts with developing a clear description of the problem. Notice that before arriving at conclusions and making recommendations in the last block of Figure 6.2, the solutions from the model need to be confirmed.

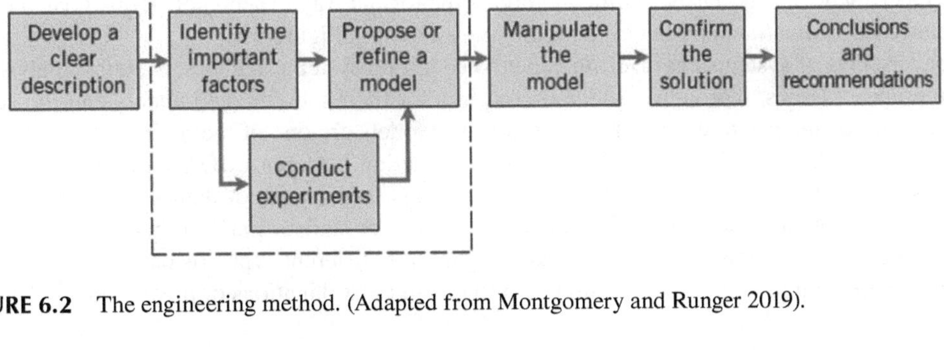

FIGURE 6.2 The engineering method. (Adapted from Montgomery and Runger 2019).

Systems engineers generally rely heavily on statistical modeling. Modeling is often a preferred approach over testing because engineers can gain insights into the inner workings of the system and spend less time and money than what it would take to attain this insight through real world testing. In statistical modeling, often there is a term of variability that is added. For example, if we look at the very familiar Ohm's law from electrical engineering and solve for electrical current, we have

$$I = \frac{V}{R}, \tag{6.1}$$

where I is the current in amps, V is the voltage in volts, and R is the resistance in ohms.

Repeated tests of Ohm's law in the real world will have variations of the output current due to system noise effects like thermal noise. This noise effect can readily be modeled by adding a variable term ε to Equation 6.1 as in,

$$I = \frac{V}{R} + \epsilon. \tag{6.2}$$

Here, ϵ represents an additive noise term that accounts for the variation in the current I. This additive noise term can consist of the sum of different noise terms such as shot noise, $1/f$ noise, generation and recombination noise, or in the case of low-light optical applications, noise that is represented as a doubly stochastic random process that is due to the random photon locations on the optical detector, and their random arrival rate. These noise terms can be theoretical and mathematically described, and systematically modeled and/or simulated, through the application of probabilistic methods as seen throughout this and other books.

To conduct research, it is often beneficial to have some foundational methods and processes to help guide the effort (Du Plooy 2022). Trade-Off Analysis, Data-Driven Decision Making, Iterative and Incremental Development, and Continuous Improvement are some useful concepts/methods that all provide great benefits. Researchers use trade-off analysis to choose between multiple options when faced with competing demands on performance, cost, and schedule. According to Alijosiene and Gudonaviciene, there are several useful trade-off methods, such as MaxDiff analysis, Choice-based conjoint (CBC), and Complete-profile conjoint (Alijosiene and Gudonaviciene 2010). While Iterative and Incremental Development is mostly associated with software engineering, its principles have wide-ranging applicability (Oakley 2023). The core idea in iterative and incremental development is to iteratively repeat a procedure in cycles by dividing it into smaller parts. This type of approach works well when there are changes in stakeholder needs or volatility in requirements. According to the idea of data-driven decision-making, decisions should be made, and conclusions drawn from facts and data obtained via thorough research (Jiang et al. 2023). This is important to help minimize biases and remain objective. According to Quick, researchers often use the principle of continuous improvement when they intend to construct their system by building upon an existing system (Quick 2019). Systems engineers may save a lot of time and energy by studying current systems and finding methods to make them better. A primary goal of Systems Engineering's application of economics, statistics, and mathematics is to generate evidence to back up engineering decisions. The scientific method is the formal way to accomplish this goal. The scientific method is illustrated by the cyclical diagram shown in Figure 6.3.

Many aspects of systems engineering depend heavily on systems analysis. Systems analysis is a way to pinpoint issues, thoroughly study alternatives for their solution, and understand the advantages and disadvantages of each of the alternatives. Models are one of the main instruments used to do systems analysis. Figure 6.4 shows some interaction between different types of modeling elements and output types after systems analysis. Inputs in the form of data and information are used by modeling elements like mathematical equations, or statistics and probability applications for instance, in the systems analysis process to generate different types of outputs (Descriptive, Qualitative, Quantitative) to produce outputs in the form of explanations or predictions.

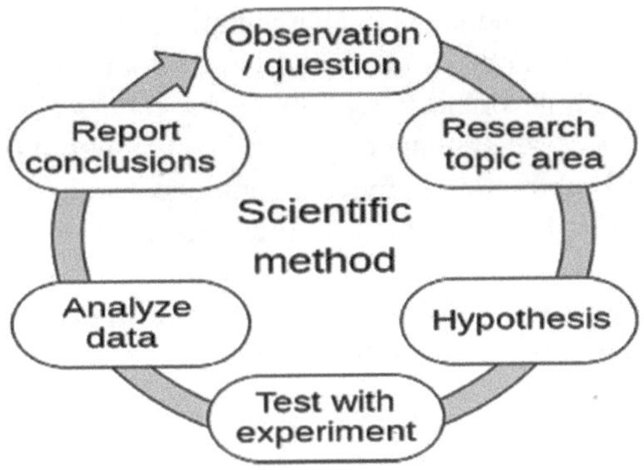

FIGURE 6.3 The scientific method. (Source: EFbrazil. Wikipedia, https://commons.wikimedia.org/wiki/File:The_Scientific_Method.svg (accessed September 13, 2023). Creative Commons Attribution – Share Alike 4.0 International License.)

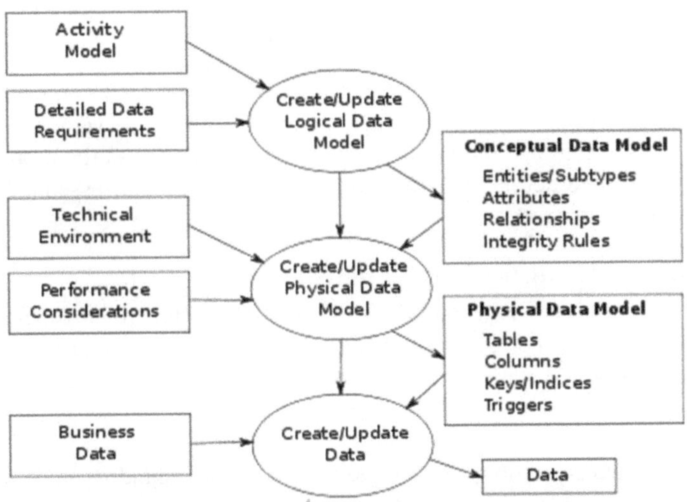

FIGURE 6.4 Data modeling process. (By 4-3_Data_Modelling_Today.jpg: Matthew West and Julian Fowler, derivative work: Razorbliss (talk) - Developing High Quality Data Models. The European Process Industries STEP Technical Liaison Executive (EPISTLE).4-3_Data_Modelling_Today.jpg, CC BY-SA 3.0, https://commons.wikimedia.org/w/index.php?curid=15895987 (accessed October 13, 2023).)

These ideas and concepts serve as important guidelines for systems engineers who find themselves with the need to conduct research. The research methods detailed below will provide an effective statistically-based foundation for helping the contemporary practicing systems engineer deliver high-quality, and resilient systems to their stakeholders.

6.1.4 General Theories and Mathematics

Systems engineering research methods rely heavily on mathematics (Bali and Goyal 2020). Accordingly, there are many applications of mathematics throughout the systems engineering discipline ranging from common basic quantitative analysis methods to more advanced methods involving higher-level math and mathematical concepts. As will be seen later, probability, statistics, and stochastic methods are in widespread use such as in DoE methods, SPC methods, quality engineering, systems reliability, maintainability, and availability (SRMA) methods, system optimization, and more.

Even though it is one of the foundations of mathematics, basic arithmetic is widely employed in systems engineering research methods (Bali and Goyal 2020). Calculations as simple as adding and subtracting are used quite often. For example, the total weight of system components is often tracked and used as an overall system constraint. To stay within the weight limit, some basic math is needed in adding and subtracting individual component weights. As another example of the application of basic math skills, knowing the overall circuit resistance is important for any electrical engineer (Hacker and Sumereder 2019). To calculate the total resistance of a group of resistors connected in parallel, one must use a different formula than that used for resistors connected in series. When resistors are connected in series, their total resistance is just the sum of their individual resistances as seen in Equation 6.3,

$$R_{\text{Total}} = R_1 + R_2 + \cdots + R_{n-1} + R_n. \tag{6.3}$$

However, when the resistances are connected in parallel, their total resistance is the inverse of the sum of their individual inverse resistances as seen in Equation 6.4,

$$\frac{1}{R_{\text{Total}}} = \left(\frac{1}{R_1} + \frac{1}{R_2} + \cdots + \frac{1}{R_{n-1}} + \frac{1}{R_n}\right). \tag{6.4}$$

Calculus is a somewhat more advanced branch of mathematics that finds common application in the systems engineering discipline. Calculating the area under a curve, the velocity and acceleration of bodies in motion, and describing complex physical phenomena such as the propagation of electromagnetic waves and the interaction of light with matter can all be addressed (albeit often not very simply) with the Calculus.

6.1.4.1 Probability and Statistics in Research Methods

The field of mathematics known as probability and statistics is often employed in systems engineering's research methods (Nelson et al. 2003). Additionally, data collection, and analysis rely heavily on the tools of probability and statistics. In the context of research methods, data collecting is the process of amassing useful information from which important design decisions can be made for the system (Ren et al. 2023).

Sampling method: Data collection isn't complete without understanding some of the fundamental concepts of sampling. For a sample set of data to be representative of a whole, large population, the number of samples (sample size) drawn from this population must be sufficiently large (Hassan et al. 2022). If the sample size is too small, the results will be unreliable, but if the sample size is too large, the experiment will take too long or cost too much money. To get a reliable result, a representative sample from the whole population needs to be determined using the correct sampling method. Probability sampling and non-probability sampling are the two main categories of sampling methods. The method of probability sampling is also known as the method of random selection. With this approach, every member of the population has an equal chance of being included in the sample. Probability sampling methods are used if it is necessary that the true population at large is reflected in the analysis. There are four primary categories of probability sampling, each of which can be applied in different contexts,

A. Standard random sampling
B. Stratified sampling,
C. Clustered sampling, and
D. Systematic sampling.

As an example, use a standard random sample of all Ford Focus owners to generate an average mile per gallon (MPG) to test the gas mileage of the Ford Focus across all Ford Focus variants. Ford offers several different Focus models; thus a random sampling of owners of a particular Ford Focus variant wouldn't be accurate to look at gas mileage of all Ford Focus versions. Using this approach, the Ford Focus vehicles can be classified into subsets and appropriately sampled to guarantee that all Ford Focus models are represented (Figure 6.5).

Non-probability sampling is used when evaluating a smaller sample size from a population in which no two individuals have an equal chance of being selected. Rapid results are possible with this method of sampling, but so is the possibility of bias. Given that not everyone will be selected, the sample size may be smaller than with a probability sample. Non-probability samples are used primarily to gain an early understanding of the small target group, rather than to test the hypothesis of the larger population. Non-probabilistic methods also have four categories which are illustrated in Figure 6.6.

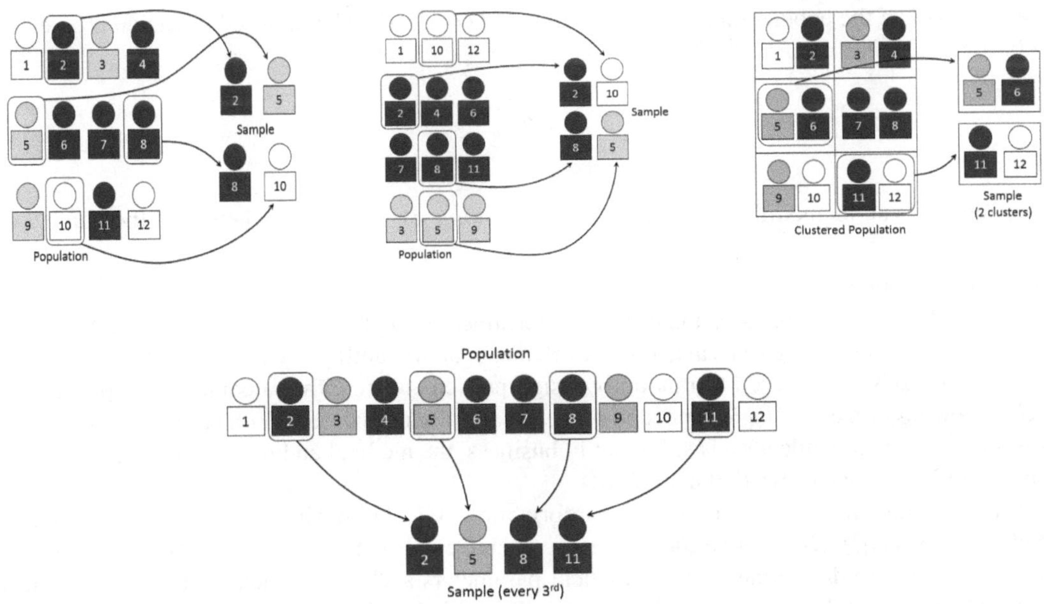

FIGURE 6.5 Types of probability sampling methods. (a) Simple random sampling, (b) stratified sampling, (c) clustered sampling, and (d) systematic sampling. (By Dan Kernler. Creative Commons Attribution – Share Alike 4.0 International License. https://commons.wikimedia.org/wiki/File:Simple_random_sampling.PNG (accessed October 13, 2023).)

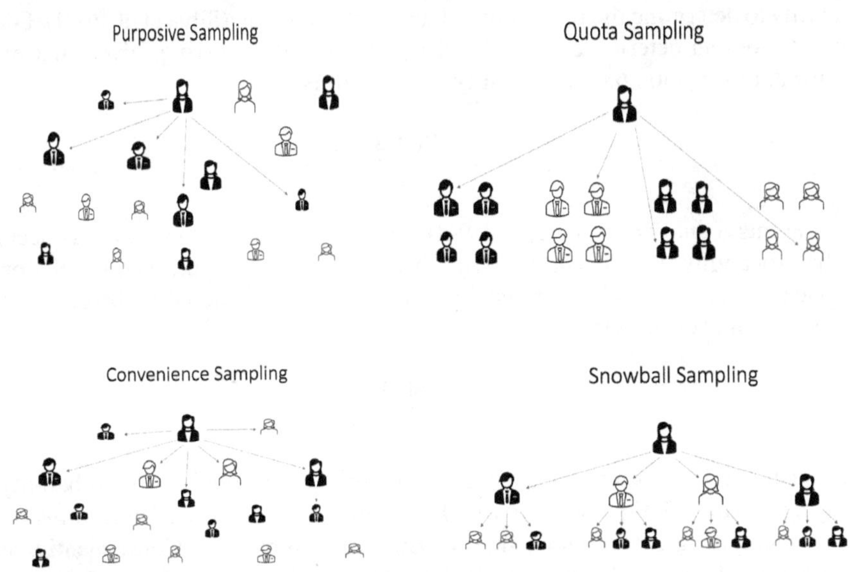

FIGURE 6.6 Example of non-probabilistic sampling methods.

Let's look at an example. Consider the case of a corporation that wants to survey its employees about the pros and cons of switching to a 10-hour, 4-day work week schedule, but only distributes the survey to its first shift workers. Even if the corporation can quickly and cheaply conduct a survey, the results will not be representative of the entire population.

Cochran's sample size formula (Cochran 1977) can be used by researchers to determine the appropriate sample size n_o for very large unknown population with a certain level of precision, as seen in Equation 6.5 (Bartlett et al. 2001). The parameter Z is a draw from the Z-distribution found from the normal distribution lookup table and is based on the required confidence level that the researchers need in the study. The value P is the fraction of the population expressed as a percentage (e.g., 50% of the population means $P = 0.50$) and $Q = (1 - P)$. The required precision is given by e (Bartlett et al. 2001),

$$n_0 = \frac{Z^2 PQ}{e^2}. \tag{6.5}$$

This equation can be modified for finite populations to, $n = \dfrac{n_0}{1 + \dfrac{(n_0 - 1)}{N}}$, where N is the number of the total population.

A cost-effective and precise sample size is determined by applying probability theory. Assigning people to their respective groups also involves elements of probability and statistics. Probability and statistical models can be used to randomly assign participants to either a control or experimental group. How precise of a sample is needed must be carefully considered. With regards to how much precision is enough, while 95% is sufficient in business, the medical and pharmaceutical industries require 99.9% precision (Bartlett et al. 2001).

Data analysis involves interpreting the information collected during the data collection phase (Nelson et al. 2003). *Descriptive Statistics* is a fundamental method of data analysis wherein the data is analyzed to determine useful statistical parameters such as the mean, median, mode, and standard deviation (Hou and Gao 2022). Descriptive statistics aid in making rapid data interpretations without getting too deep into statistical methods. *Inferential statistics* use probability theory to study sample populations to gain insight into the entire population (Li and Sonnier 2023).

The systems engineering discipline is rife with interesting applications of probability and statistics. One useful example is conditional probability. Historical information can be used with conditional probability to determine the probability of an event occurring (Silva et al. 2021). For example, in Equation 6.6, we can determine the probability of event B occurring, given that event A has already occurred. In Equation 6.6, the quantity $P(A \text{ and } B)$ is the

$$P(B|A) = \frac{P(A \text{ and } B)}{P(A)}, \tag{6.6}$$

probability of events A and B occurring, and $P(A)$ is the probability that event A has occurred.

To find the probability of an event occurring based on its past occurrence in relation to previous events, one can use Bayes' Theorem, which is a variant of conditional probability (Chuan et al. 2023). Equation 6.7 makes this clear,

$$P(A|B) = \frac{P(B|A) \times P(A)}{P(B)}. \tag{6.7}$$

Notice that Equation 6.7 can be obtained directly from the general conditional probability equation by first changing variables in Equation 6.6 from $P(B|A)$ to $P(A|B)$, and then determining the quantity $P(A \text{ and } B)$ by multiplying through both sides of Equation 6.6 by $P(A)$. Consequently, researchers can calculate the conditional probability of event A happening if event B occurs using the individual event probabilities $P(A)$ and $P(B)$, and the formula $P(B|A)$ (Chuan et al. 2023).

Researchers can use distribution functions to determine the likelihood that a given value will be attained by a given variable (Britannica 2023). For example, the failure analysis field makes extensive use of this fact. Researchers, however, can use the distribution function for many other applications in systems engineering as well (Blanchard and Fabrycky 2011). Depending on the specifics of an experiment, researchers can choose from a wide variety of distribution functions. Some of the more notable distribution functions include the Poisson Distribution Function, the Binomial Distribution Function, and the Gaussian Distribution Function.

The Poisson Distribution Function [probability density function (PDF) for the continuous case and the probability mass function for the discrete case] is used when there is a counting process such as counting the number of failures in the failure rate in reliability engineering or counting the number of photons on a detector in low-light imaging conditions. Equation 6.8 shows the Poisson distribution function which provides the probability of the event x occurring (Blanchard and Fabrycky 2011).

$$P(x) = \frac{\mu^x e^{-\mu}}{x!}. \tag{6.8}$$

In Equation 6.8, μ represents the mean value of the random variable $\boldsymbol{x}$, and x represents a realization of that random variable. For the Poisson distribution, the mean μ is equal to the variance σ^2 and can be calculated based on the sample size n and probability p. The mean and variance for the Poisson distribution are given by

$$\mu = \sigma^2 = np. \tag{6.9}$$

Another useful probability distribution function is the Binomial Distribution Function. The Binomial Distribution Function is useful when looking for x successes in n trials where there are only two possible states (heads/tails, on/off, true/false, working/not working, …). The single trial probability of success is given by p. The Binomial Distribution Function for n trials, and for x occurrences, is given by

$$P(x) = \frac{n!}{x!(n-x)!} p^x q^{n-x}, \tag{6.10}$$

with $q=(1-p)$. For example, the probability that there are three occurrences ($x=3$) of a six-sided die showing the number 6, out of five throws of the die ($n=5$) can be directly determined by Equation 6.10. At first glance, this may not seem like a problem involving the binomial distribution since there are six different faces on the die. However, the way the problem was stated was how many times a single face occurs. So, we do have a binary state in that either the face occurs (six in this case) or it doesn't (any of the other faces of the die. The probability of seeing the number 6 on a single trial (throw) of the die is one chance out of six ($p=1/6$). For this example, the probability of seeing three sixes in five throws of the die using Equation 6.10 is about 3.215%. The mean is given by

$$\mu = np, \tag{6.11}$$

and the variance is given by

$$\sigma^2 = npq. \tag{6.12}$$

Another very useful probability distribution function is the Gaussian or Normal distribution function. The Gaussian distribution function typically occurs when describing lots of independent events. This represents the case for normally distributed random variables. The Gaussian distribution is well known and has the characteristic bell curve on a graph (Blanchard and Fabrycky 2011). Because of the Central Limit Theorem, the distribution of sample means approaches a Gaussian as the number of samples gets larger, regardless of the distribution of the underlying population. An example of this is for the case of optical imaging, as the number of photons increases, the underlying population approaches a Gaussian distribution. For the normal distribution, only the mean μ and variance σ^2 are required to determine the probability of event x occurring. The Gaussian distribution function is given by

$$P(x)=\frac{1}{\sigma\sqrt{2\pi}}e^{\frac{-(x-\mu)^2}{2\sigma^2}}. \tag{6.13}$$

The probability that a random variable X has a specific realization x that falls within the range $a \le x \le b$, can be determined if the PDF $f(x)$ is known. This is accomplished by calculating the integral of the PDF, with the bounds of integration corresponding to the desired range

$$P(a \le x \le b)=\int_a^b f(x)dx. \tag{6.14}$$

In general, the mean and variance for a given PDF can be found by

$$\mu=\int_{-\infty}^{\infty} xf(x)dx, \tag{6.15}$$

and

$$\sigma^2=\int_{-\infty}^{\infty}(x-\mu)^2 f(x)dx. \tag{6.16}$$

We can generalize this expression to an arbitrary integer value of n on the standard deviation shown on the left side of Equation 6.16, and so obtain an expression for the nth central moment as follows:

$$\sigma^n=\int_{-\infty}^{\infty}(x-\mu)^n f(x)dx. \tag{6.17}$$

If the nth moment is desired (as opposed to the nth central moment), then the mean μ in Equation 6.17 is set to zero. Equation 6.17 is very useful in statistics because different moments provide different insights into the characteristics of the data set. Some notable moments are the first moment as shown in Equation 6.15 (the mean of the data set). The second moment is provided by Equation 6.16 with the mean μ set to zero. The second central moment shown by Equation 6.16 is the variance of the data set and, as the name implies, gives some indication as to how the data set varies about its mean. The third central moment,

$$\gamma=\sigma^3=\int_{-\infty}^{\infty}(x-\mu)^3 f(x)dx, \tag{6.18}$$

is the Skewness and indicates the amount of lopsidedness in the distribution. If the distribution is symmetric about its mean such as in the Gaussian distribution, then the skewness is zero. The fourth central moment is Kurtosis given by

$$K=\int_{-\infty}^{\infty}(x-\mu)^4 f(x)dx, \tag{6.19}$$

which provides some measure of how much of the data set is in the tails of the distribution as opposed to the center part of the distribution. As such, the Kurtosis can provide insight into the outliers of the data set. The fifth central moment is determined by letting *n* in Equation 6.17 be equal to 5. This moment is like the skewness but provides insight into the lopsidedness of the tails of the distribution as opposed to its center. Higher order moments can be determined according to Equation 6.17 but tend to be impractical. There are also a variety of useful metrics that provide insights into data sets like the Pearson correlation coefficient *r*, the coefficient of determination R^2, the *F*-Statistic and the *P*-Statistic that will be presented and discussed in Section 6.1.8.

To determine many statistical parameters and assessment metrics, the underlying PDF needs to be known. Besides the often used Poisson, Gaussian, and Binomial PDFs described above, some other useful distribution functions include: (1) the Chi-square Distribution (χ^2), and (2) Student's *t*-Distribution. In the Chi-square distribution, an example used would be for determining how well data fits a theoretical distribution. The equation for the Chi-square PDF is given by

$$f(x|\upsilon) = \frac{x^{(v-2)/2} e^{-x/2}}{2^{\frac{\upsilon}{2}} \Gamma(v/2)}, \tag{6.20}$$

where the parameter v is the *degree of freedom*, and Γ is the Gamma function. The degree of freedom represents the maximum number of independent variables that are free to change and are allowed to vary. It indicates the number of independent information pieces in the estimate. The *t*-distribution is useful when the population variance is unknown, and there are relatively few samples to estimate the population statistics. It is sometimes called the Student's *t*-distribution because the originating paper was from William Sealy Gousset who at the time (1908) was a student of the mathematician Karl Pearson and couldn't use his name because of employee publication restrictions of where Gousset worked (the Guinness brewery in Dublin, Ireland). Gousset found in his paper that as the number of samples increases, the *t*-distribution approaches the normal distribution but varies for low sample numbers. The equation for the PDF of the *t*-distribution is given by

$$f(x) = \frac{\left(1 + \frac{x^2}{v}\right)^{\frac{-(v+1)}{2}}}{B(0.5, 0.5v)\sqrt{v}}, \tag{6.21}$$

where v is a fitting parameter related to the degree of freedom, and $B(\alpha, \beta)$ is the Beta function given by

$$B(\alpha, \beta) = \int_0^1 t^{\alpha-1} (1-t)^{\beta-1} \, dt. \tag{6.22}$$

A probability tree diagram is an excellent tool for calculating the probabilities of conditional events as seen in Figure 6.7. For events *A* and *B*, the probability tree provides probabilities for the combination of events *A* and *B* occurring (or not occurring). To arrive at these probabilities using the Probability Tree, let $\bar{A}$ and $\bar{B}$ represent that the events *A* and *B* did not occur. Then for two events *A* and *B*, there are four possible combinations of occurrences as seen on the right side of the Probability Tree. To get to each of the joint probabilities on the right, we must first multiply the probability values of the connected branches as shown. For example, notice that the top joint probability $P(A \cap B)$ corresponds to both events *A* and *B* occurring, and $\cap$ is the *intersection* symbol.

The second (from the top-right) joint probability $P(A \cap \bar{B})$ represents the probability that event *A* happened but event *B* did not occur. Note that the joint probabilities on the right side of Figure 6.6 are directly attainable from Bayes' Theorem in Equation 6.6. Let's look at an example from the airline industry.

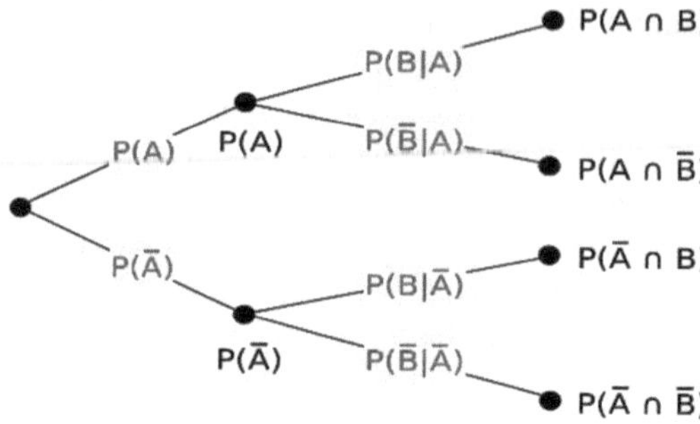

FIGURE 6.7 Probability tree diagram (probability theory). (Obtained from, Wikipedia, https://en.wikipedia.org/wiki/Tree_diagram_(probability_theory), (accessed May 27, 2021).)

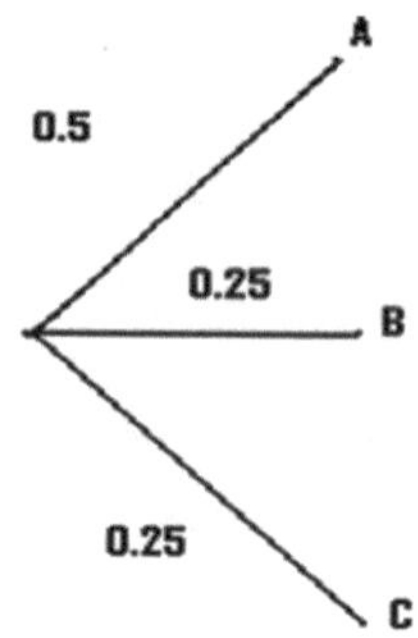

FIGURE 6.8 Initial step in probability tree for Boeing example.

Assume that the Boeing Aircraft company has three plants across the United States designated as *A*, *B*, and *C*. These plants produce 50%, 25%, and 25% of the company's airplanes respectively. Factory *A* produces passenger airplanes accounting for 70% of its overall output (i.e., factory *A* produces 30% non-passenger aircraft), whereas Factory *B* produces passenger airplanes accounting for 25% of its total output, and Factory *C* produces passenger airplanes accounting for 25% of its total output. If an airplane was selected at random by the manufacturer, what is the likelihood that it would be a passenger airplane?

Step 1: We know that there are three factories so we can start by drawing three lines from the starting point that represent each of the three factories. On each of the lines, we place the percentage of airplanes that are produced by each factory in decimal form. As an example, we know that factory *A* produces 50% of all airplanes (passenger and other) and so we represent 50% as 0.50 on the first line segment as shown in Figure 6.8. The other two line segments show the 25% overall airplane production percentages for factories *B* and *C* respectively.

Step 2: From each line segment endpoint, draw two, line segments that represent the percentages of passenger airplanes produced and the number of non-passenger airplanes produced, as shown in Figure 6.9. Note that for factory *A*, 70% of all the airplanes produced are passenger airplanes. This means that there are 30% of the aircraft produced at this facility that are not passenger aircraft. The symbol *P* represents passenger aircraft, and NP represents non-passenger airplane.

The process in step 2 is now repeated for the remaining two-line segments at factories *B* and *C*. At factory *B*, there are 25% of the airplanes produced as passenger airplanes, leaving 75% of the aircraft produced at *B* being non-passenger airplanes. Similarly, at factory *C*, there are the same

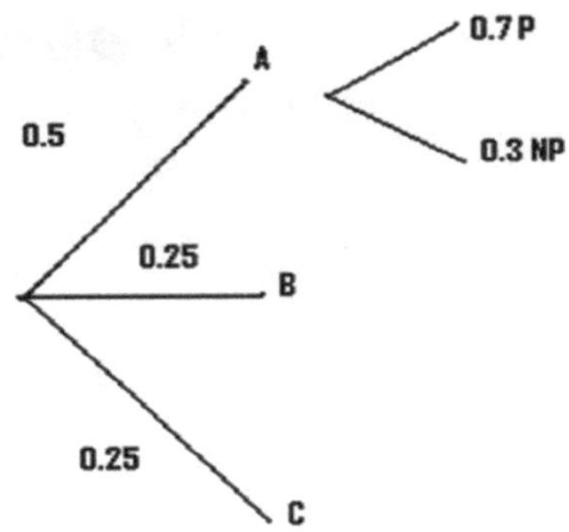

FIGURE 6.9 Expanding node A in probability tree for Boeing example.

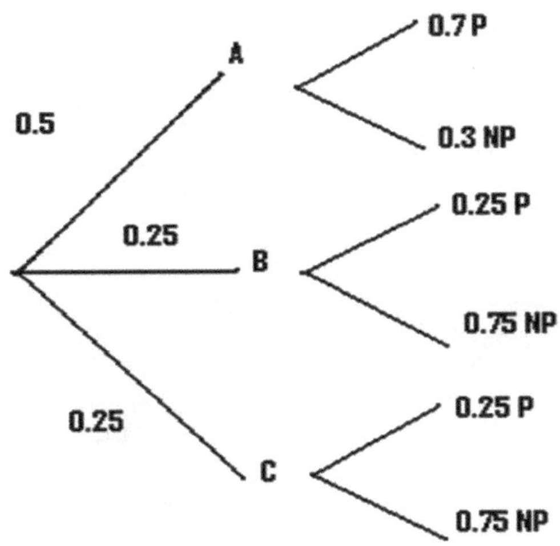

FIGURE 6.10 Expanding nodes B and C in probability tree for Boeing example.

percentages in effect as for factory *B* and the branches leaving points *B* and *C* are identical. The completed probability so far is shown in Figure 6.10.

Step 3: For each of the six numeric results at the end of the second set of branches, multiply these values by the corresponding probabilities of the attached preceding branches (i.e., the probability values for the attached first set of branches). For example, for factory *A*, multiply 0.7 by the probability value of the preceding branch, 0.5 as shown in Figure 6.11. We see that the combined value is, $0.5 \times 0.7 = 0.35$. Note that since we are only interested in the likelihood of producing a passenger aircraft across all three factories, we don't have to evaluate the probability tree for the non-passenger cases shown in Figure 6.10.

Step 4: We can repeat the process carried out in Step 3 for the remaining entries for factories *B* and *C*. There are only two others (three in total) of the four branches that deal with passenger airplanes, so only these two additional branch paths need to be calculated. The result is shown in Figure 6.12.

Step 5: We can now determine the answer to our question, "If an airplane was selected at random by the manufacturer, what is the probability that it would be a passenger plane?" by summing the resulting probabilities ($0.35 + 0.0625 + 0.0625 = 0.475$). So, 47.5% of the airplanes picked at random would end up being passenger airplanes (in a statistical sense).

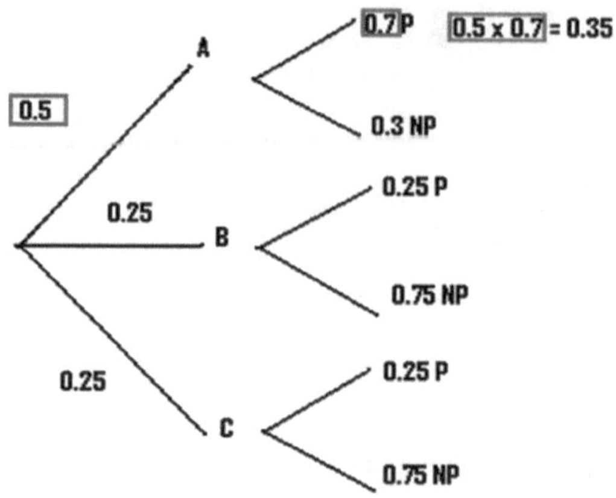

FIGURE 6.11 Calculating combined probabilities in top branch in probability tree for Boeing example.

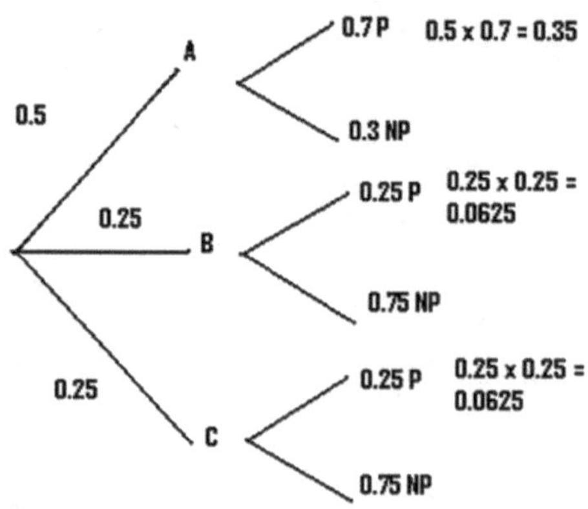

FIGURE 6.12 Calculating combined probabilities in remaining branches in probability tree for Boeing example.

6.1.5 Methods, Tools, and Techniques

Systems Engineering optimizes the design and development of a system by using a variety of methods, tools, and techniques like "probability, mathematical statistics, computing, system logic, queuing theory, game theory, linear programming, cybernetics, group dynamics, simulation, information theory, servomechanism theory, and human engineering" as well as "decision theory, nonlinear programming, some elements of econometrics, and communications theory" (Encyclopedia Britannica, Systems Engineering Techniques, Tools, and Procedures, 2023). Systems engineering is seen to be multi-disciplinary and diverse, and many of these methods, tools, and techniques are essential in modeling and understanding contemporary systems

In many areas of systems engineering, modeling the system is an essential tool in the systems development process, and works together with theoretical and experimental methods to gain proper insights into the nature, functionality, and behavior of the system in its environment. Modeling systems also helps in conducting feasibility and trade studies and helps evolve the system through its development cycle. Figure 6.13 shows the application of modeling and illustrates the practical relationship between modeling, theory, and experimentation and how these synergistic elements help improve understanding of the system in its environment.

Models can vary widely in terms of accuracy, precision, and detail. All models are approximations of reality and so have some degree of error. As George Box so famously said, "All models are

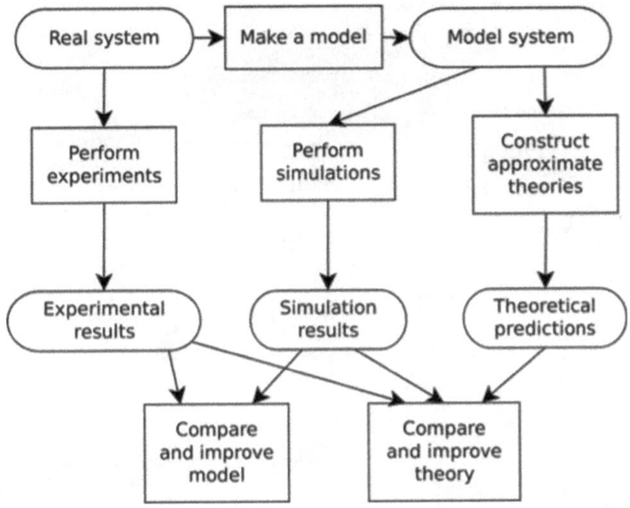

FIGURE 6.13 Application of modeling. (Obtained from Danski14, Wikipedia, https://commons.wikimedia.org/wiki/File:Molecular_simulation_process.svg (accessed October 13, 2023).)

wrong, some are useful." This statement needs to be always kept in mind when engaged in modeling. The modeling process should be kept as simple as possible yet must provide something useful. In addition to being simple, and useful, a good model according to Stephen Hawking should also be verifiable. The precision and accuracy of a model should be set such that there are useful and repeatable results while still retaining computational efficiency and ease of use. Modeling is useful in understanding system attributes and for making predictions.

Many mathematical models are useful in some aspect of prediction or control. For example, with physical phenomena, laws such as Boyle's law, Ohm's law, and Newton's laws of motion are well established and can be used in making predictions in terms of the states of their variables. In engineering, linear programming methods may be used in all kinds of engineering applications such as computational fluid dynamics (CFD) or finite element analysis (FEA). DoE, SPC, systems reliability and maintainability analysis, queuing methods, game theory applications, and many quality engineering techniques are all relevant to, and in the domain of, the systems engineering discipline and require statistical modeling methods. In a nutshell, modeling is pervasive and central in the systems engineering community (and many others).

Figure 6.14 shows the advantages of modeling throughout the systems development lifecycle. Initially, models can be used to understand the operational situation as is shown by the top block in the figure.

Models can be used to understand and help resolve schedule issues. Modeling can also be used to test concepts by generating simulations to gain insights into system functionality and behavior. Refining requirements, reducing program risk, and easing the systems transition into operations are all advantages that modeling brings to the systems development process. An interesting use of models is that sometimes, modeling is the only way that verification and validation can be accomplished. In the end, modeling saves time and money and is an integral aspect of systems engineering.

With regards to testing, it is important to distinguish between the term's *verification, validation*, and *accreditation*. As seen in Figure 6.15, verification is typically used when the developer wants to determine if a requirement, at any level, has been properly implemented. Here the verification agent makes sure that the system elements under consideration satisfy the driving requirements, usually through testing. Validation then demonstrates that the system can work in the real world. Usually, the validation agent is a functional expert who understands how the system or article under validation is supposed to function in the real world. These concepts carry over to modeling such that

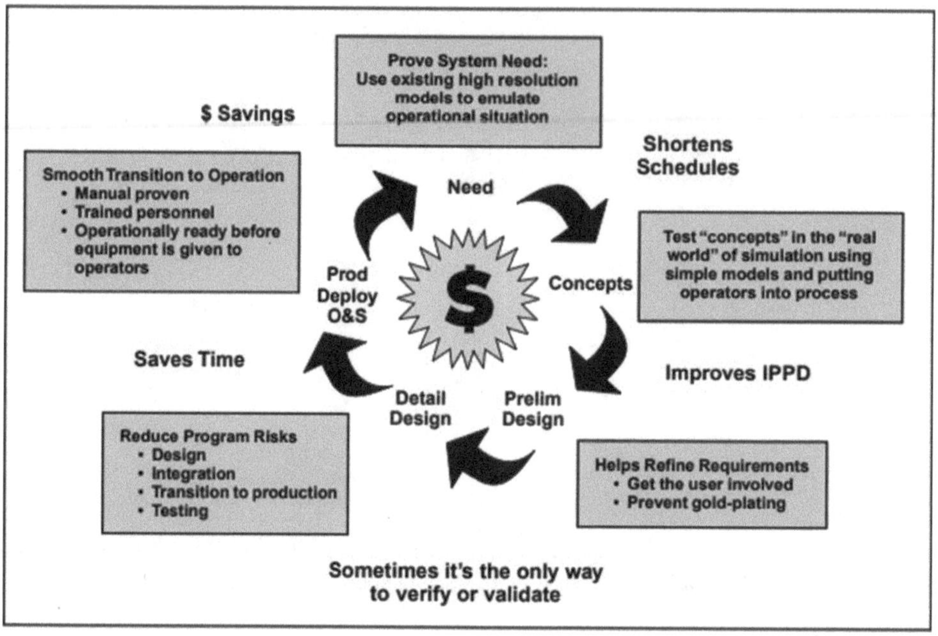

FIGURE 6.14 Advantages of modeling and simulation (Wikimedia Commons 2007).

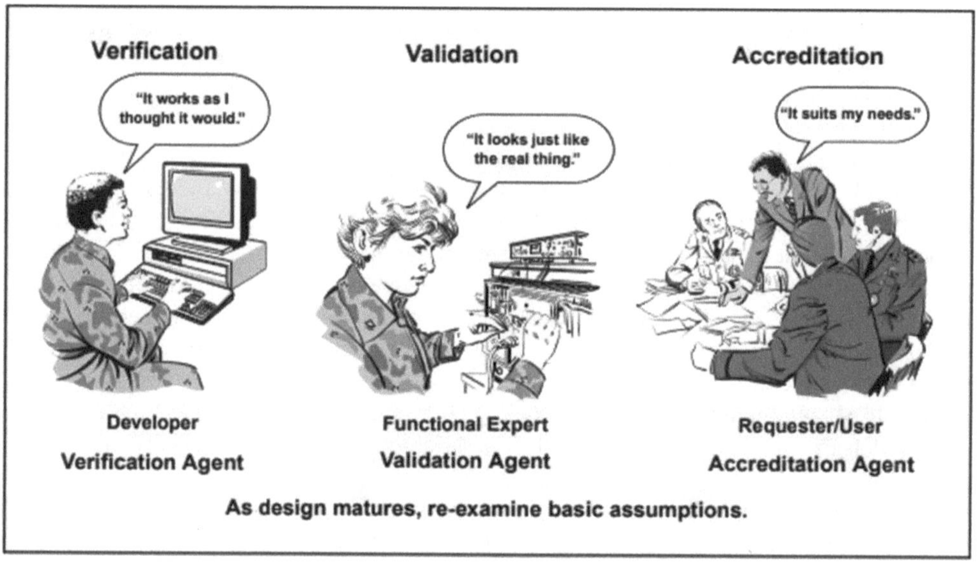

FIGURE 6.15 Verification, validation and accreditation (Wikimedia Commons 2007).

verifying a model means to ensure the model works as it was intended by satisfying the modeling requirements (what the model was meant to accomplish). Model validation ensures that modeling results are valid in the real world. Model accreditation then makes sure the model provides what is needed by the user community through the accreditation process (e.g., provides useful results in general to a set of stakeholders, or to the community at large). In other words, accredited models signify that they are suitable for addressing the stakeholders' needs. The accreditation agent is typically from the user community, or stakeholder that requested the model. Verified, validated, and accredited models can be used in lieu of formal testing. This is of great benefit since modeling is often much cheaper than formal testing.

6.1.6 Applications

Systems engineering uses research methods in various theoretical and mathematical applications. Some typically useful research methods applications include Hypothesis Testing, Analysis of Variance (ANOVA), Analysis of Covariance (ANCOVA), Latin Squares, and DoE. We provide an introductory overview of these methods along with citations for more in-depth supporting information. Some methods, like SPC, are covered in more detail in other chapters where these methods are more commonly used, and where they are directly relevant in practice (i.e., in Production and Manufacturing for SPC).

6.1.6.1 Design of Experiments

Throughout the development cycle, there is a need for understanding how design parameters affect critical system, subsystem, and/or components/pieces/parts. Often, we conduct experiments to investigate the behavior of one or more dependent variables (outputs) as a function of one or more independent variables (inputs) to gain insight as to how the independent variables interact and affect the dependent variables. We look for *Cause and Effect* (C&E) relationships through controlled experimentation to aid in making design, production, and process decisions. Instead of relying on luck in our experimentation process, we need to design the experiment, using a "systematic plan of investigation, based on established statistical principles, so that the interpretation of the observations can be defended" (Kolarik 1995).

When conducting an experimental study, it is important to properly identify the variables of interest. These variables can be referred to as *factors* in the study, and these factors are carefully controlled throughout the experiment. The intent is to not only gain insight into the role that these factors play in the experiment but also into the effect that they have on an output variable called a *response variable* (also known as *response*). Some key ideas and definitions are:

Factors: Independent variables that may be qualitative or quantitative. Factors are the variables investigated when interested in determining their effect on one or more output variables.

Factor levels: The physical measure of the factor. For example, if temperature is a factor, then 400°K would be a factor level.

Treatment level: Specific combination of factor levels under consideration. This is sometimes called *treatment combinations.*

Response: The measured/observed result for a given factor setting or treatment level.

Balance: Experimental balance means that there are the same number of observations made for every treatment or treatment level.

Degree of freedom: Provides the available number of independent pieces of information.

Experimental error: The measured variation in the responses of experimental units that have undergone the same treatment.

P-value: The smallest level of statistical significance (observable significance level) for which the null hypothesis in a hypothesis test can be rejected.

In DoE, the objectives of an experiment are usually phrased in terms of a hypothesis statement, and hypothesis tests are used to either accept or reject the hypothesis using statistical analysis. The *null hypothesis Ho* is a hypothesis that there is no difference between the statistical parameters of the variables under consideration. For instance, if we are considering the diameters of soccer balls manufactured on several different production lines in a factory, then the null hypothesis would be that there is no statistically significant difference between the mean diameters of the soccer balls produced from each of the production lines. The alternate hypothesis H_a is that there is a statistically significant difference between the means of the soccer ball diameters produced in one or more of the production lines. Typically, it is assumed that the null hypothesis holds true until there is enough statistical evidence gathered to show that this is not the case. The model framework used in DoE is shown in Figure 6.16. The entire experiment is set up under controlled conditions and often framed

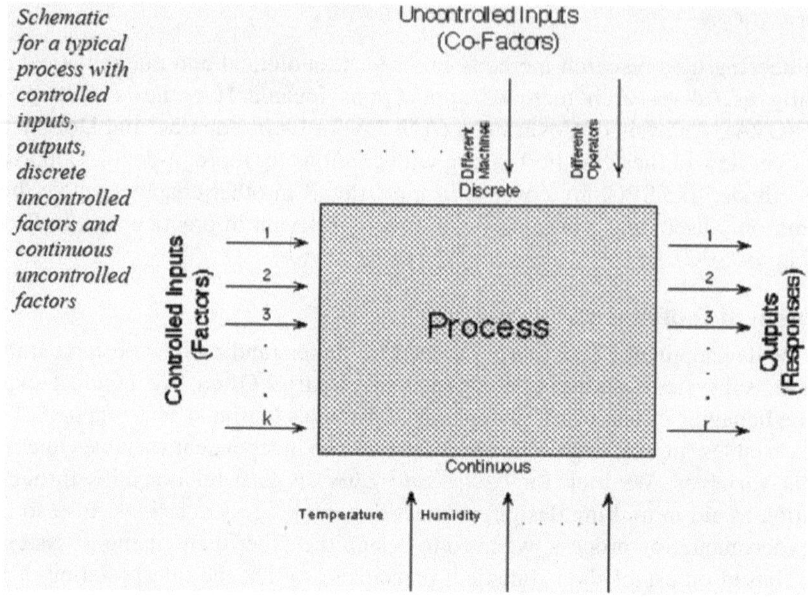

FIGURE 6.16 Model for design of experiments framework (NIST 2025).

in terms of hypothesis statements. The inputs (factors) shown on the left side of the figure are controlled. These are the independent variables such as temperature, velocity, and pressure that are under investigation. At the top of the figure, discrete uncontrolled cofactors are shown. These can consist of uncontrolled discrete noises (unwanted signals) such as different machines being used or different operators. Continuous uncontrolled signals such as uncontrolled temperature or humidity effects are shown at the bottom of the figure. When conducting these experiments, statistical thinking is necessary. The act of sampling introduces random variations in the process under investigation, and sampling effects need to be carefully considered and included in the experimental design. When designing the experiment, there are six fundamental steps needed in planning:

1. *Define the purpose of the experiment:* Set goals, objectives, and targets for each experiment and understand what is expected.
2. *Determine the variables:* Identify controllable and uncontrollable factors, and responses, and choose factors and factor levels.
3. *Design the experiment:* Organize the experiment to account for randomness in the experiment. How the experiment is replicated and randomized determines the appropriate model and type of analysis that is used.
4. *Run the experiment:* Conduct the experiment and carefully record the results. Consider and avoid potential biases or errors in the experimental setup.
5. *Conduct analysis on the experimental results:* Use the structure of the model to conduct the analysis of the experiment and develop the inferential, descriptive, and predictive information as required.
6. *Interpret and inform:* Determine and prepare the findings and communicate results to stakeholders. Be prepared to defend the results.

Experiments can be single-factor experiments or consist of multiple factors. In single-factor experiments, only one controllable factor is manipulated. In multi-factor experiments, multiple controllable factors are allowed to vary but typically only one at a time. The other factors are kept constant as one of the factors is varied. The treatment levels are changed in a methodical fashion and all sampling and population statistics are properly represented and evaluated. A common and effective way to introduce a defendable statistical analysis methodology into experimental design is through hypothesis testing which we now present.

6.1.6.2 Hypothesis Testing

In hypothesis testing, statistical analysis can be used to compare different data sets or variables based on limited amounts of data and infer whether or not statistically meaningful results have been obtained. For instance, relationships between several data sets such as the differences in their means can be evaluated as a function of confidence that there is an observable difference in the result. Included are metrics that determine if the results are statistically significant or not and that provide insight into the degree of confidence that the results can be believed. In hypothesis testing, there is a null hypothesis that is stated that frames the data comparison in terms of a binary comparison. For instance, in one implementation, the null hypothesis H_o asserts that there is no difference in the means between two data sets. The binary condition is then expressed in the alternate Hypothesis H_A that there is a difference in the means of the two data sets. It is a binary condition since there are only two possible states for this hypothesis, the means are either the same or not. For these binary conditions, some important statistical results and parameters relative to hypothesis testing are illustrated in Figure 6.17. We can see in the top-left and bottom-right of the figure that the correct results are obtained in that the rejection or acceptance of the null hypothesis of the experiment/test (shown as row elements in the figure) matches what is expected based on the reality of the experiment/test (shown as columns in the figure). Type I error shown at the top right of the figure occurs when we reject the null hypothesis when in fact the null hypothesis is true. If for instance, two means were the same in a statistical sense, but our test declared them to be different, we would have a Type I error. This error is a "false alarm" type of error in that even though the signals are the same, we incorrectly claim that they are different. The bottom left error (Type II) error occurs when the null hypothesis is false, but our test says our null hypothesis is true. Sometimes it is easier to with the *power of the test* given by $1 - \beta$ which is the probability of rejecting the null hypothesis H_o given that H_o is false. Our test indicates that the means are the same, when in fact they are different. This type of test falls into the category of a "failure to detect" since there was an actual difference in the means that was not flagged by the test. The value α for Type I error is the probability of having a Type I error, and β is the probability of having a Type II error. The typical process in hypothesis testing is to set/choose the probability of having a Type I error and then design the test so that a reasonably small probability β of a Type II error is obtained. The value α is referred to as the *significance level of the test.* Figure 6.18 shows some important statistical concepts with regard to two sampling distributions associated with two random variables. The α value discussed above, which is related to Type I error, is directly relevant to hypothesis testing of the differences of the means of the two random variables, since it relates to the transition point between where the null hypothesis is accepted, and where it is rejected. The transition point is readily seen in Figure 6.18 where we see the critical value associated with our soon-to-be-defined rejection rule for the null hypothesis (that the means of the two distributions are the same). Note the separation of the means in the two sampling distributions in the figure. If these two distributions are associated with two independent variables in our Designed

		Reality	
		H_0 False	H_0 True
Test	Reject H_0	Correct rejection H_0 = Power = 1 - β	Type I error = α
	Accept H_0	Type II error	Correct acceptance of H_0

FIGURE 6.17 Type I and Type II error table in hypothesis testing. (By Jtneill - Own work, CC BY-SA 3.0, https://commons.wikimedia.org/w/index.php?curid=255971 (accessed October 16, 2023).)

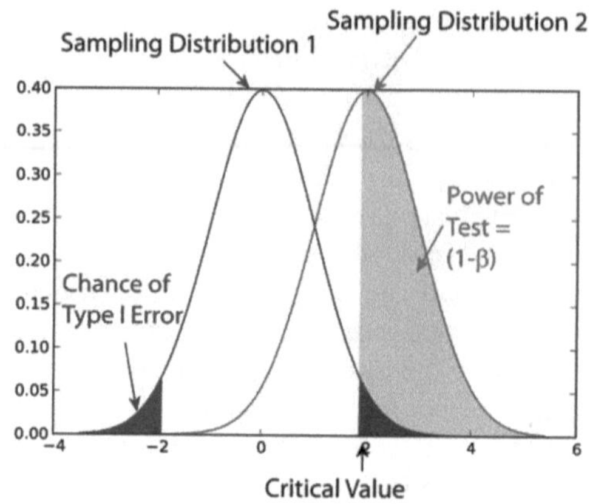

FIGURE 6.18 Probability density functions for two random variables in hypothesis testing for fixed level of significance. (Obtained and licensed as CC BY-SA 4.0. https://commons.wikimedia.org/w/index.php?curid=113194144 (accessed April 10, 2025).

Experiment, and if we are interested in whether these two means are the same or not in a statistical sense, then we need to have a rejection rule that tells us whether or not the means are the same. One way to do this is to look at the differences in the means of the two distributions. As shown in the figure, we can set a critical value based on the particulars of the experiment that define the boundary between the null hypothesis being true, and it being false. If the difference in means between the two sampling distributions/variables/data sets is less than the critical value, then the two means are considered the same in a statistical sense and the null hypothesis is not rejected. If on the other hand the difference in means is greater than the critical value, then there is a significant difference between the means and the null hypothesis must be rejected. From this Type I error is the integral of the PDF of the distribution outside and up to the critical value as shown by the dark (blue) regions in the figure. This value is the α value mentioned above. Typically, the α value is chosen based on a decision maker's tolerance for risk. Common values of α are 0.05 and 0.01 (for the more conservative decision makers). One issue with using the α value is that, once α has been chosen, there is no good way to determine how far a computed test statistic is in the null hypothesis rejection region. Is it just barely in the rejection region, or is the test statistic quite far into the region? One way around this is to use the *P*-value instead of the fixed significance value of α. Most statistical software packages like Minitab, or IBM's Statistical Package for the Social Sciences (SPSS) software, or some sophisticated calculators like some higher-end HP models can calculate the *P*-value along with a host of other useful statistical quantities.

In Designed Experiments, we first generate a model of the output of the experiment – the data. We assume that we are using a *completely randomized experimental design* (CRD) in which the data is a random sample from a normal distribution. With large number of independent data elements, this is naturally the case because of the central limit theorem. A simple statistical model for two levels of a factor (e.g., two different temperature levels with the factor being temperature) is given by Montgomery (2020) as

$$y_{ij} = \mu_i + \epsilon_{ij} \left\{ \begin{array}{l} i = 1,2 \\ j = 1,2,\ldots,n_i \end{array} \right. , \tag{6.23}$$

where the index i refers to the first or second factor level, and j refers to the jth observation in the factor level i, data set in a two-factor experiment, the index j refers to the jth data element in the ith data set, μ_i is the mean of the ith factor level, and ϵ_{ij} is the random error component of the model that is normally distributed with mean μ_i and variance σ_i^2. Consequently, y_{ij} are normally distributed with means μ_i and variances σ_i^2, which is written as NID(μ_i,σ_i^2). Having this model for the data helps in understanding and quantifying how the factors relate to each other and what effects the interactions have on the outputs.

In hypothesis testing, there needs to be a rejection rule for knowing when to reject (or not reject) the null hypothesis. This rejection rule is based on an appropriate test statistic, and the test statistic in turn is based on the statistical formulation of the problem (e.g., are the means of the random variables known or not, are the variances of the random variables known or not, are they normally distributed or not, are there a large or small number of samples). For instance, if the problem statement for the completely randomized experiment is expressed as the difference in means of two distributions/random variables/data sets, that have unknown variances and small sample sizes, then the appropriate test statistic is the student's t-distribution and not the normal distribution. Consequently, for large numbers of independent data items, the normal distribution is used as the underlying PDF for the sampled random variables, whereas if relatively few data items are present (say less than 30 for many practical applications), then the student's t-distribution is more appropriate than the normal distribution. As the number of data items increases, the student's t-distribution approaches the normal distribution.

Small number of samples and equal unknown variances: Suppose we have a small number of samples from our data that are sampled under CRD conditions, and we know that the variances are the same but not are not known explicitly. We can use the two-sample t-test with the following test statistic:

$$t_0 = \frac{\overline{y}_1 - \overline{y}_2}{S_p\sqrt{\frac{1}{n_1}+\frac{1}{n_2}}}, \tag{6.24}$$

where $\overline{y}_1$ and $\overline{y}_2$ are the sample mean of the first and second sample respectively, n_1 and n_2 are the total number of samples in the first and second sample, and S_p is the estimate of the common variance. The sample means are the estimates from the data of the true means given by

$$\overline{y}_i = \frac{\sum_j y_{ij}}{n_i}, \tag{6.25}$$

where the subscript i refers to the data sample set, the subscript j is the jth element in the ith data set, and n_i is the total number of elements in the ith data set. The sample variance is given by

$$S_i^2 = \frac{\sum_j \left(y_{ij} - \overline{y}_i\right)^2}{n_i - 1}. \tag{6.26}$$

The quantity in the numerator is the sum of squares (SS) and will make a common appearance in our estimation process. The term S_i^2 is a biased estimate and so requires a correction term to produce the *corrected sum of* squares that is shown in the numerator of Equation 6.26. Namely, by taking the expectation of Equation 6.26, the result of the expectation in the numerator produces the statistical variance σ_i^2 divided by the degree of freedom (n_i-1). Multiplying Equation 6.26 by the correct degree of freedom $n_i - 1$ for the ith data set produces the corrected sum of squares. The estimate of the common variance is then given by

$$S_p^2 = \frac{(n_1-1)S_1^2+(n_2-1)S_2^2}{(n_1+n_1-2)}, \tag{6.27}$$

where Equation 6.26 was used for two sample data sets (i=1, 2) to produce S_1^2 and S_2^2 in Equation 6.27. The term in the denominator of the test statistic in Equation 6.24 is typically called the *standard error.*

To conduct the statistical analysis, we would use the following rejection rule to decide whether the null hypothesis is true or not

$$|t_0| > t_{\frac{\alpha}{2}, n_1+n_1-2}. \tag{6.28}$$

In Equation 6.28, we are comparing the value of the magnitude of the test statistic for our problem using Equation 6.24 with the entry from the t-distribution table evaluated at our accepted probability of a Type I error α, and our degree of freedom $(n_1 + n_2 - 2)$. Note that we are using the value of $\frac{\alpha}{2}$ since we are using only one tail of the t-distribution in the statistical evaluation. If the left side of Equation 6.28 is larger than the right side, then we reject the null hypothesis that the means are the same. However, if the left side is smaller than the right side of Equation 6.28, we accept the hypothesis that the means are the same in a statistical sense.

The decision maker can also establish a confidence level in the statistical result by calculating the confidence interval. The confidence interval provides an interval in which the parameters under observation are expected to fall with some degree of confidence. A good application of this is to quantify the differences in the means with some certainty. If we again look at the differences between two sample means and this time expect a difference, we can use confidence intervals to give us a probability of the difference in the means falling in the stated interval. Where possible, confidence intervals should complement hypothesis testing results. Confidence intervals are set up in the form:

$$P(L \le \theta \le U) = 1 - \alpha, \tag{6.29}$$

where θ is an unknown parameter, L is the lower confidence limit, U is the upper confidence limit, and α is the probability of having a Type I error as before. Note that $1-\alpha$ is the probability of not having a Type I error. The interval inside the probability in Equation 6.29 is called the $(100-\alpha)$ confidence interval. So, for the random variable θ with a value of α given by 0.05, we would have a 95% confidence interval. If we want to consider the difference between the sample mean differences and the true difference of the means, the test statistic in Equation 6.24 is modified correspondingly to

$$t_0 = \frac{\bar{y}_1 - \bar{y}_2 - (\mu_1 - \mu_2)}{S_p\sqrt{\frac{1}{n_1} + \frac{1}{n_2}}}, \tag{6.30}$$

where all the terms have been previously defined. Letting θ be this test statistic, the lower confidence level corresponds to the value of the t-distribution on the left side of the distribution evaluated at $\frac{\alpha}{2}$ (half the critical value with degrees of freedom $(n_1 + n_1 - 2)$. Substituting these values in the description of the confidence interval in Equation 6.29, we get

$$P\left(-t_{\frac{\alpha}{2}, n_1+n_1-2} \le \frac{\bar{y}_1 - \bar{y}_2 - (\mu_1 - \mu_2)}{S_p\sqrt{\frac{1}{n_1} + \frac{1}{n_2}}} \le t_{\frac{\alpha}{2}, n_1+n_1-2}\right) = 1 - \alpha, \tag{6.29}$$

and manipulating the argument of the interval in Equation 6.29 to isolate the difference of the true means in the center, we get

$$P\left(\bar{y}_1 - \bar{y}_2 - t_{\frac{\alpha}{2}, n_1+n_1-2} S_p\sqrt{\frac{1}{n_1} + \frac{1}{n_2}} \le \mu_1 - \mu_2 \le \bar{y}_1 - \bar{y}_2 + t_{\frac{\alpha}{2}, n_1+n_1-2} S_p\sqrt{\frac{1}{n_1} + \frac{1}{n_2}}\right) = 1 - \alpha. \tag{6.30}$$

Equation 6.30 is the probability of not making a Type I error. The quantity $100(1-\alpha)$ can therefore be interpreted as the percent confidence of not making this type of error. The confidence interval for the difference in means is given by the argument of Equation 6.30:

$$\bar{y}_1-\bar{y}_2-t_{\frac{\alpha}{2},n_1+n_1-2}S_p\sqrt{\frac{1}{n_1}+\frac{1}{n_2}}\le\mu_1-\mu_2\le\bar{y}_1-\bar{y}_2+t_{\frac{\alpha}{2},n_1+n_1-2}S_p\sqrt{\frac{1}{n_1}+\frac{1}{n_2}}. \tag{6.31}$$

The $100(1-\alpha)$ confidence interval provides useful and complementary information to the hypothesis testing process and should be used when possible. At this point, an example may be appropriate and beneficial.

Let's say that we have two factory production lines that are producing rulers. We want to find out whether the mean ruler length μ_1 in production line one is the same as the mean ruler length μ_2 in production line two. We assume that we have a completely randomized design with two sample sets each with 15 samples ($n_1 = n_2 = n = 15$). We also assume that the variances of both production lines are the same. Since we have relatively few samples that are randomly drawn from each of the two production lines, we use the *t*-statistic as our statistical metric. We can set up a hypothesis test that statistically compares the means of the two production line data sets. The null hypothesis H_o assumes that the means of both production lines are statistically the same and is given by

$$H_0:\mu_1=\mu_2. \tag{6.32}$$

The alternate hypothesis H_A is then given by

$$H_A:\ \mu_1\ne\mu_2. \tag{6.33}$$

These are the only two possibilities that can occur in the comparison of the means of these two production line data sets. For Equation 6.32, the assumption is that we are using a two-tailed test where the variation of the means can shift either to the left or right of the expected 12 in. If the decision makers accept a 5% probability of making a Type I error (the means are the same, but our test says they are not), then α is set to 0.05. If we take 15 random samples from each production line, we can determine the sample means and variances for each production line using Equations 6.25 and 6.26. Say that we determine the values $\bar{y}_1 = 12.07$ in., $S_1^2 = 0.0009$ in.^{2}, $\bar{y}_1 = 11.95$ in., $S_1^2 = 0.0025$ in.^{2} from the sample data. We can use Equation 6.27 to determine the common sample variance of 0.0017 in.^{2}. The test statistic for the hypothesis that the means are the same is given by Equation 6.28 and, with an α value of 0.05 and degree of freedom of 28, we get a value of 2.048 for $t_{\frac{0.05}{2},28}$. The value of our test statistic for the null hypothesis is given by Equation 6.24 and, after substitution of the above parameters yields a value of t_o=7.971. According to Equation 6.28, the magnitude of t_o is greater than the percentage points of the *t*-distribution evaluated at $\frac{\alpha}{2}$, and degree of freedom $n_1+n_2-2=28$ given by $t_{\frac{0.05}{2},28}=2.048$ so we must reject the null hypothesis that the means are the same. By using Equation 6.31, and substituting the values above, we can get a 95% confidence interval of, $0.08917 \le \mu_1-\mu_2 \le 0.15083$ in. This means that we are 95% certain that the difference in means between the first and second production lines falls between 0.08917 and 0.15083 in. If a difference in means of less than 0.05 in. is desired by the production team (to conform to less than half of the smallest resolution on the ruler (assumed one-tenth of an inch)), then there is an issue that requires further study and improvement of the production process. For this example, we assumed that we had relatively few samples from an experiment and equal variances. What if this was not the case and we had a lot of samples (i.e., typically more than 30 samples for each sample set)? What if the variances weren't equal? Dealing with these experimental variations are all relatively simple modifications to the approach just presented. For the following variations to the hypothesis testing of the equality of means, we assume as before that the samples are randomly selected as part of a completely randomized experimental design process. Correspondingly, we describe the hypothesis testing process impacts for some of the common hypothesis testing variations.

Large number of samples and equal known variances ($\sigma_1^2 = \sigma_2^2 = \sigma^2$): If instead of having relatively few numbers of samples, we had many samples from our independent sample sets, or we are sampling under CRD conditions, then the only difference in the "hypothesis testing of the means" process from before is in the underlying statistical distributions. Because of the central limit theorem, the underlying sampled data in our two independent data sets each have a normal distribution instead of a student's *t*-distribution as above. The null hypothesis and alternate hypothesis are the same as shown in Equations 6.32 and 6.33, but the test statistic with assumed known and equal variances becomes

$$Z_0 = \frac{\overline{y}_1 - \overline{y}_2}{\sigma\sqrt{\frac{1}{n_1} + \frac{1}{n_2}}}, \tag{6.34}$$

where σ is the standard deviation from the normal distribution. The rejection rule is also like the rejection rule shown in Equation 6.28 with the only change being the distribution to a normal distribution. The rejection rule for rejecting the null hypothesis takes the form:

$$|Z_0| > Z_{\frac{\alpha}{2},}. \tag{6.35}$$

If the magnitude of the test statistic in Equation 6.34 is greater than the critical value of the normal distribution given by the right-hand side of Equation 6.35, then the null hypothesis must be rejected in favor of the alternate hypothesis. The confidence interval with the normal distribution becomes

$$\overline{y}_1 - \overline{y}_2 - Z_{\frac{\alpha}{2},}\,\sigma\sqrt{\frac{1}{n_1} + \frac{1}{n_2}} \le \mu_1 - \mu_2 \le \overline{y}_1 - \overline{y}_2 + Z_{\frac{\alpha}{2},}\,\sigma\sqrt{\frac{1}{n_1} + \frac{1}{n_2}}, \tag{6.36}$$

The process for conducting the hypothesis test and establishing the confidence interval is the same as before, only the underlying distribution has changed from the *t*-distribution to the normal distribution.

Large number of samples and different known variances ($\sigma_1^2 \neq \sigma_2^2$): If we have a large number of samples from our independent sample sets with different known variances, or we have CRD conditions and sample sets with different means, then the data is normally distributed with different known variances. The null hypothesis and alternate hypothesis are the same as shown in Equations 6.32 and 6.33. The test statistic becomes

$$Z_0 = \frac{\overline{y}_1 - \overline{y}_2}{\sqrt{\frac{\sigma_1^2}{n_1} + \frac{\sigma_2^2}{n_2}}}, \tag{6.34}$$

where σ_1^2 and σ_2^2 are the known variances from our two independent data sets. The rejection rule for rejecting the null hypothesis is now related to the standard normal distribution and takes the following form:

$$|Z_0| > Z_{\frac{\alpha}{2},}, \tag{6.35}$$

where $Z_{\frac{\alpha}{2}}$, is the $\frac{\alpha}{2}$ entry of the standard normal distribution table. If the magnitude of the test statistic in Equation 6.34 is greater than the normal distribution table value given by the right-hand side of Equation 6.35, then the null hypothesis must be rejected in favor of the alternate hypothesis. The confidence interval with the normal distribution and different variances becomes

$$\overline{y}_1 - \overline{y}_2 - Z_{\frac{\alpha}{2}}\sqrt{\frac{\sigma_1^2}{n_1} + \frac{\sigma_2^2}{n_2}} \leq \mu_1 - \mu_2 \leq \overline{y}_1 - \overline{y}_2 + Z_{\frac{\alpha}{2}}\sqrt{\frac{\sigma_1^2}{n_1} + \frac{\sigma_2^2}{n_2}}. \tag{6.36}$$

The process for conducting the hypothesis test and establishing the confidence interval is the same as before in the *large number of samples and equal variances* case, only the variances are not combined and factored as was done before.

Small number of samples with different and unknown variances: For the case of a completely randomized design with few samples, we once again use the t-distribution in conducting our hypothesis test of comparing the means of the two data sets. However, because of the differences in the variances between the two sample sets, we don't quite have a t-distribution. This can be fixed by modifying the expression for the degrees of freedom. The null hypothesis is still $H_0 : \mu_1 = \mu_2$, and the alternate hypothesis is also $H_A : \mu_1 \neq \mu_2$. Because the variances are different and need to be estimated, the test statistic becomes

$$t_0 = \frac{\overline{y}_1 - \overline{y}_2}{\sqrt{\frac{S_1^2}{n_1} + \frac{S_2^2}{n_2}}}, \tag{6.37}$$

which is intuitive when compared to the previous test statistic for the t-distribution in Equation 6.24 with equal variances where the variances were factored out of the square root in the denominator. We get good agreement between the distribution of the test statistic t_o and the t-distribution if the number of degrees of freedom is given by

$$v = \frac{\left(\frac{S_1^2}{n_1} + \frac{S_2^2}{n_2}\right)^2}{\frac{\left(S_1^2/n_1\right)^2}{n_1 - 1} + \frac{\left(S_2^2/n_2\right)^2}{n_2 - 1}}. \tag{6.38}$$

The rejection rule is given by

$$|t_0| > t_{\frac{\alpha}{2},v}. \tag{6.39}$$

The only difference from the confidence interval shown in Equation 6.31 is that the individual variances are not known and are estimated and shown explicitly in the square root terms of the $100(1-\alpha)$ confidence interval, and the degree of freedom v is calculated differently. The $100(1-\alpha)$ percent confidence interval is then given by

$$\overline{y}_1 - \overline{y}_2 - t_{\frac{\alpha}{2},v}\sqrt{\frac{S_1^2}{n_1} + \frac{S_2^2}{n_2}} \leq \mu_1 - \mu_2 \leq \overline{y}_1 - \overline{y}_2 + t_{\frac{\alpha}{2},v}\sqrt{\frac{1}{n_1} + \frac{1}{n_2}}. \tag{6.40}$$

Comparison of single-factor data set to constant mean: In many experiments, there may only be one factor that needs to be compared to a constant mean. For instance, in our ruler factory example from before, we may want to check one production line to see whether the mean length of our rulers is exactly 12 in. within our statistical limits. The hypothesis testing formulation of this type of problem has a null hypothesis of,

$$H_0 : \mu = \mu_0. \tag{6.41}$$

The alternate hypothesis H_A is then given by

$$H_A : \mu \neq \mu_0. \tag{6.42}$$

If we have an underlying distribution that is normal with known variance, or if we have a distribution that is not normal, but we have a large number of independent samples, then we can use a one-sided normal distribution and the one-sided Z-test with a test statistic given by

$$Z_0 = \frac{\overline{y} - \mu_0}{\sigma / \sqrt{n}}, \tag{6.43}$$

where σ is the known standard deviation of the normal distribution, and n is the total number of data samples. The test statistic for rejecting the null hypothesis H_0, that the mean of the sampled data is the same as μ_0 is,

$$|Z_0| > Z_{\frac{\alpha}{2}}. \tag{6.44}$$

The corresponding $100(1-\alpha)$ confidence interval is,

$$\overline{y} - Z_{\frac{\alpha}{2},} \sigma / n \leq \mu \leq \overline{y} + Z_{\frac{\alpha}{2},} \sigma / n. \tag{6.45}$$

Note that the comparison between the single random factor data sets does not have to be the comparison of the means, rather other statistical parameters such as the variance, or covariance of the data sets can be used in a hypothesis test. For instance, in our ruler factory example, if after establishing that rulers were being built with an average mean of 12 in., we wanted to ensure that the variations about that mean were less than a constant value (say a standard deviation of 0.05 in.), then we would have a null hypothesis based on the variance and not the mean as follows:

$$H_0 : \sigma^2 = \sigma_0^2. \tag{6.46}$$

The alternate Hypothesis H_A is then given by

$$H_A : \sigma^2 \neq \sigma_0^2. \tag{6.47}$$

The distribution for this case is the chi-square distribution with a test statistic of

$$\chi_0^2 = \frac{(n-1)S^2}{\sigma_0^2}, \tag{6.48}$$

and degrees of freedom $(n-1)$. Since the chi-distribution is non-symmetrical, we need to specify two rejection rules for the boundary between the upper and lower critical regions. The rejection rules for rejecting the null hypothesis, that the variance is equal to σ_0^2 is given by

$$\chi_0^2 > \chi_{\frac{\alpha}{2},n-1}^2, \tag{6.49}$$

or,

$$\chi_0^2 < \chi_{1-\frac{\alpha}{2},n-1}^2. \tag{6.50}$$

Here, $\frac{\alpha}{2}$ and $1-\frac{\alpha}{2}$ are the respective upper and lower percentage points from the chi-square distribution. If Equation 6.49 or 6.50 are true, then the null hypothesis is rejected, and the variance of the data set is not equal to σ_0^2. The $100(1-\alpha)$ confidence interval is given by

$$\frac{(n-1)S^2}{\chi^2_{\frac{\alpha}{2},n-1}} \leq \sigma^2 \leq \frac{(n-1)S^2}{\chi^2_{1-\frac{\alpha}{2},n-1}}. \tag{6.51}$$

The quantity in the numerators in Equation 6.51, namely $(n-1)S^2$ is of interest. As can be seen from Equation 6.26 with $S_i^2 = S^2$, and $n_i = n$ for one data set, by multiplying Equation 6.26 by $(n-1)$, we get the corrected sum of squares (SS) term which makes a common appearance throughout hypothesis testing. We can then write Equation 6.51 as,

$$\frac{\text{SS}}{\chi^2_{\frac{\alpha}{2},n-1}} \leq \sigma^2 \leq \frac{\text{SS}}{\chi^2_{1-\frac{\alpha}{2},n-1}}. \tag{6.52}$$

If we are interested in comparing the variance of two data sets as before when we compared the means of two CRD data sets, such as if we wanted to compare the variances of two production lines in our ruler factory to make sure they were the same, then the null hypothesis and alternate hypothesis take on the following form:

$$H_0 : \sigma_1^2 = \sigma_2^2, \tag{6.53}$$

and

$$H_A : \sigma_1^2 \neq \sigma_2^2, \tag{6.54}$$

where the subscripts refer to the respective production line sample data sets. The test statistic is the ratio of the sample variances which has an F-distribution given as

$$F_0 = \frac{S_1^2}{S_2^2}. \tag{6.55}$$

For the numerator, there are $n_1 - 1$ degrees of freedom, and for the denominator, there are $n_2 - 1$ degrees of freedom. The rejection criteria for the null hypothesis that the two variances are the same is given by

$$F_0 > F_{\frac{\alpha}{2},n_2-1,n_1-1}, \tag{6.56}$$

and

$$F_0 < F_{1-\frac{\alpha}{2},n_2-1,n_1-1}. \tag{6.57}$$

Of course, for different problem types wherein the two-tailed hypothesis test is not required, such as in cases where the variation of one sample data set needs only be less than (or greater than) the other data set, then the appropriate rejection criteria in Equation 6.56, or Equation 6.57 needs to be selected. The $100(1-\alpha)$ percent confidence interval for the hypothesis test of equal variances is given by

$$\frac{S_1^2}{S_2^2} F_{1-\frac{\alpha}{2},n_2-1,n_1-1} \leq \frac{\sigma_1^2}{\sigma_2^2} \leq \frac{S_1^2}{S_2^2} F_{\frac{\alpha}{2},n_2-1,n_1-1}. \tag{6.58}$$

In the next section, we generalize our results for hypothesis testing cases with single random factors but at multiple factor levels (two or more). The method we will use is the ANOVA method.

6.1.6.3 Analysis of Variance

The ANOVA is a statistical method that compares two or more groups to determine whether there is a significant difference between them. One-Way and Two-Way ANOVAs are the most fundamental forms of the ANOVA statistical analysis. The analysis of variance known as the one-way ANOVA uses just one independent variable, whereas the analysis of variance known as the two-way ANOVA employs two independent variables (Wilcox 1989). Both the intra group variance and the inter group variance are discovered in both scenarios. The terms "degrees of freedom inside groups (DFW)" and "degrees of freedom between groups (DFB)" are used to refer to these two categories, respectively (Rutherford 2001). In an ideal situation, the variance between groups should be high while the variance within groups should be as small as possible. After conducting an analysis on the information gathered throughout the course of the experiment, a probability value, known as a p-value, will be determined (Grimmett and Welsh 2014). It is possible to draw a conclusion about whether there is a connection between the dependent variable and the independent variable by determining whether the variance is significant enough. If this is the case, then we can conclude that the Null Hypothesis, which predicts that there would be no substantial differences between the groups, is incorrect, and we can proceed to test the Alternate Hypothesis (Tan 2023).

In this section, we generalize the hypothesis testing criteria of only one random factor with up to two treatment levels from the preceding section, to multiple treatment levels (say a different factor-levels/treatment levels). We are interested in comparing sampled experimental data over many treatment levels (or factor-levels) with each other. For example, we may have a single random factor being the temperature of an environmental conditioning unit of a sensor package with five different temperature levels. We assume a completely randomized design and want to conduct hypothesis testing on the five different treatment levels (temperature settings, or factor levels). Since we are comparing the variances over multiple factor levels, we use a slightly different data model given by Montgomery (2020),

$$y_{ij} = \mu + \tau_i + \epsilon_{ij} \begin{cases} i = 1, 2, \ldots, a \\ j = 1, 2, \ldots, n \end{cases}, \tag{6.59}$$

where μ is a constant mean, τ_i is the variation about the mean of the ith treatment level (e.g., ith temperature level), and j represents the sample number in the ith treatment level. The parameter ϵ_{ij} is the random error term that captures all of measurement, system, and data errors that are present in the experiment (say, system noises, sampling errors, uncontrolled random variables, or any unwanted signal or random effect present in the data). There are an equal number of n samples in each treatment level data set (e.g., we have n temperature samples (measurements) at every i temperature treatment level). Another way of saying this, we don't have different numbers of samples at the different i treatment levels. For the data model in Equation 6.59, we can consider this either as a *fixed effects model* (where the parameters, μ, τ_i, and σ^2 are estimated), or a *random effects model* where the a treatments could themselves be samples of a larger population. We first focus on the fixed effects model and define some commonly used terms and expressions in ANOVA:

Total sum of samples in ith treatment: Here we sum all the individual n terms in the ith treatment level

$$y_{i.} = \sum_{j=1}^{n} y_{ij}. \tag{6.60}$$

Note that the dot after the subscript i represents the summation over that parameter (in this case j). This notation obtained from Montgomery (2020) is quite convenient and is used throughout this section. The mean of the total sum of samples in the ith treatment is then simply

$$\bar{y}_{i.} = \sum_{j=1}^{n} \frac{y_{ij}}{n}. \tag{6.61}$$

Total sum of samples across all observations: To determine this value, we also must sum over all a treatment levels by

$$y_{..} = \sum_{i=1}^{a} \sum_{j=1}^{n} y_{ij}. \tag{6.62}$$

The mean value of all observations is then

$$\bar{y}_{..} = \sum_{i=1}^{a} \sum_{j=1}^{n} \frac{y_{ij}}{an} = \sum_{i=1}^{a} \sum_{j=1}^{n} \frac{y_{ij}}{N}, \tag{6.63}$$

where N is the total number of observation points given by $N = an$. The total corrected sum of squares is given by

$$SS_T = \sum_{i=1}^{a} \sum_{j=1}^{n} \left(y_{ij} - \bar{y}\right)^2. \tag{6.62}$$

In ANOVA, the sum of squares term is broken into more fundamental components that are quite useful in statistical analysis. By adding and simultaneously subtracting the term $\bar{y}_{i.}$ inside the parenthesis of Equation 6.62 (a net effect of adding zero) but rearranging terms and simplifying, we obtain the result

$$SS_T = \sum_{i=1}^{a} \sum_{j=1}^{n} \left(y_{ij} - \bar{y}\right)^2 = n \sum_{i=1}^{a} \left(\bar{y}_{i.} - \bar{y}_{..}\right)^2 + \sum_{i=1}^{a} \sum_{j=1}^{n} \left(y_{ij} - \bar{y}_{i.}\right)^2, \tag{6.63}$$

which is the fundamental expression in ANOVA. The first term to the right of the equal sign in Equation 6.63 is the sum of squares of the difference ***between*** the averages of the treatments/factor-levels and the average over all observations. The second term after the equal sign is the sum of squares of the differences ***within*** treatments/factor-levels and the average over the treatments/factor-levels. Consequently, we can express Equation 6.63 as

$$SS_T = SS_{\text{Treatment}} + SS_E, \tag{6.64}$$

where the terms on the right represent the first and second terms after the equal sign in Equation 6.63 respectively. The term SS_E is the sum of squares due to the errors in the experiment. Since there are n samples in every treatment/factor-level, there are $n-1$ degrees of freedom for each treatment/factor-level. Similarly, across the treatments/factor-levels, there are a treatment levels and so there are $a-1$ degrees of freedom across all treatment levels. Across all the observations, we have $a\ (n-1) = (an-a) = (N-a)$ degrees of freedom for the error. If we take a closer look at the inner summation in SS_E, we see that the summation over j produces the corrected sum of squares over the samples of the ith treatment/factor-level. In a similar manner to Equation 6.26, we can get the

ith treatment/factor-level sample variance by dividing the sum of squares over the samples in the ith treatment/factor-level by the degree of freedom over the n samples within the ith treatment/factor-level $(n-1)$,

$$S_i^2 = \frac{\sum_{j=1}^{n}\left(y_{ij} - \bar{y}_{i.}\right)^2}{n-1}. \tag{6.65}$$

If we now sum the ith treatment/factor-level sample variances over the a treatments/factor-levels, we get

$$\frac{\left[\sum_{i=1}^{a}\sum_{j=1}^{n}\left(y_{ij} - \bar{y}_{i.}\right)^2\right]}{\sum_{i=1}^{a}(n-1)} = \frac{(n-1)S_1^2 + (n-1)S_2^2 + \cdots + (n-1)S_a^2}{(n-1)+(n-1)+\cdots+(n-1)} = \frac{SS_E}{(N-a)}, \tag{6.66}$$

where Equation 6.66 gives an estimate of the common variance in all of the a treatments/factor-levels. The result is the corrected sum of squares of the error divided by the degree of freedom over all observations $(N-a)$. Equation 6.66 also provides a connection between the individual treatment/factor-level sample variances and the sum of squares of the errors SS_E. If there are no differences between the treatment/factor-level means, the variance σ^2 can be estimated from the sum of squares between treatments divided by the degree of freedom for the a treatments/factor-levels $(a-1)$,

$$\frac{SS_{\text{Treatment}}}{a-1} = \frac{n\sum_{i=1}^{a}\left(\bar{y}_{i.} - \bar{y}_{..}\right)^2}{a-1}, \tag{6.67}$$

which is equal to σ^2 (only if there are equal treatment/factor-level means). The expressions on the far right of Equation 6.66 and far left of Equation 6.67 are called the mean squares as follows:

$$MS_{\text{Treatment}} = \frac{SS_{\text{Treatment}}}{a-1}, \tag{6.68}$$

and

$$MS_E = \frac{SS_T}{N-a}, \tag{6.69}$$

where the respective treatment and error-corrected sum of squares are divided by their respective degree of freedom. If we take the expected value of the respective mean squares, simplify the terms, and substitute in the data model parameters in Equation 6.59, we get

$$E(MS_E) = \frac{1}{N-a}\left[N\mu^2 + n\sum_{i=1}^{a}\tau_i^2 + N\sigma^2 - N\mu^2 - n\sum_{i=1}^{a}\tau_i^2 - a\sigma^2\right] = \sigma^2, \tag{6.70}$$

and

$$E(MS_{\text{Treatments}}) = \sigma^2 + \frac{n\sum_{i=1}^{a}\tau_i^2}{a-1}. \tag{6.71}$$

We next show how to apply these expressions in a hypothesis test where we have multiple treatments of a single random factor. Let's look at the equality of means in our problem such that we have for the hull hypothesis

$$H_0: \ \mu_1 = \mu_2 = \cdots = \mu_a, \tag{6.72}$$

which has a corresponding alternative hypothesis given by

$$H_0: \ \mu_1 \neq \mu_2 \neq \cdots \neq \mu_a, \quad \text{for at least one pair } (i, j), \tag{6.73}$$

The variations τ_i about the mean μ themselves are zero mean random variables so that they don't contribute to the mean μ (they are variations about the mean μ after all). Consequently, we can write the null and alternative hypotheses in terms of the variations, that we have for the null hypothesis

$$H_0: \ \tau_1 = \tau_2 = \cdots = \tau_a = 0, \tag{6.74}$$

which has a corresponding alternative hypothesis given by

$$H_0: \ \tau_1 \neq \tau_2 \neq \cdots \neq \tau_a \neq 0, \quad \text{for at least one } i. \tag{6.75}$$

The corrected sum of squares for the ith treatment is normally distributed with mean $\mu_i = \mu + \tau_i$ and variance σ^2. The errors are assumed normally distributed and independent. The test statistic is

$$F_0 = \frac{\text{MS}_{\text{Treatments}}}{\text{MS}_E}, \tag{6.76}$$

And has an underlying F-distribution. Note that the mean square terms at the right of Equation 6.76 each have an underlying chi-square distribution. The rejection rule for the test statistic shown in Equation 6.76 is

$$F_0 > F_{\alpha, a-1, N-a}, \tag{6.77}$$

where the degrees of freedom shown to the right of the α-term are associated with the $\text{MS}_{\text{Treatments}}$ and MS_E terms respectively. If we use the MS_E as an estimator of the variance σ^2, then we would get a confidence interval for the ith treatment of mean μ_i

$$\bar{y}_{i.} - \bar{y}_{j.} - t_{\frac{\alpha}{2}, N-a}\sqrt{\frac{\text{MS}_E}{n}} \leq \mu_i - \mu_j \leq \bar{y}_{i.} - \bar{y}_{j.} + t_{\frac{\alpha}{2}, N-a}\sqrt{\frac{\text{MS}_E}{n}}, \tag{6.78}$$

and for the difference between the ith and jth treatment means, the confidence interval is

$$\bar{y}_{i.} - \bar{y}_{j.} - t_{\frac{\alpha}{2}, N-a}\sqrt{\frac{\text{MS}_E}{n}} \leq \mu_i - \mu_j \leq \bar{y}_{i.} - \bar{y}_{j.} + t_{\frac{\alpha}{2}, N-a}\sqrt{\frac{\text{MS}_E}{n}}. \tag{6.79}$$

To estimate the sum of squares terms (or mean square terms), this can be done by using the direct equations from the fundamental form of the ANOVA process shown in Equation 6.63 according to Equation 6.64 to get the expressions for the corrected sum of squares for the Total, Treatment, Error as follows:

$$\text{SS}_T = \sum_{i=1}^{a}\sum_{j=1}^{n}\left(y_{ij} - \bar{y}\right)^2, \tag{6.80}$$

and for the treatments

$$\text{SS}_{\text{Treatment}} = n\sum_{i=1}^{a}\left(\bar{y}_{i.} - \bar{y}_{..}\right)^2, \tag{6.81}$$

TABLE 6.1
Analysis of Variance Table (Single Random Factor, Fixed Effects Model)

Variation	Corrected Sum of Squares	Degrees of Freedom	Mean Square	F_o
Between treatments	$SS_{\text{Treatment}} = n\sum_{i=1}^{a}(\bar{y}_{i.} - \bar{y}_{..})^2$	$a-1$	$MS_{\text{Treatment}}$	$\frac{MS_{\text{Treatments}}}{MS_E}$
Error	$SS_E = SS_T - SS_{\text{Treatment}}$	$N-a$	MS_E	
Total	$SS_T = \sum_{i=1}^{a}\sum_{j=1}^{n}(y_{ij} - \bar{y})^2$	$N-1$		

and for the error

$$SS_E = SS_T - SS_{\text{Treatment}}. \tag{6.82}$$

The first two of these expressions can be directly determined by the data set over all observations, and the last expression in Equation 6.82 determines the corrected sum of squares for the error in terms of the other two results. Table 6.1 shows a useful way to order the ANOVA results for a single-factor experiment with multiple (two or more) treatment levels for a fixed random factor with a fixed effects model.

We provide an ANOVA example that compares three production lines (A, B, and C) from an adhesive manufacturer for a given adhesive that is supposed to be able to withstand up to 5 lbs of weight before detaching from a standard painted wall. Seven random samples were taken from each production line and tested. Weight was added until the adhesive detached and the "pull-off" weight was noted and tabulated. The initial assumption was that all the production lines were producing equally good adhesives with the same average pull off weight. The null hypothesis is then that all three means are the same as follows:

$$H_0 : \mu_A = \mu_B = \mu_C. \tag{6.83}$$

It follows that the alternative hypothesis is then,

$$H_0 : \mu_A \neq \mu_B \neq \mu_C, \tag{6.84}$$

for at least one comparison. The individual data samples are shown in Table 6.2.

In this example, we have three production lines A, B, and C that have seven production samples in each line. We can let a be the number of production line, $a=3$. The total number of production line samples for three lines with seven line samples each is, $N=21$. The values for a and N can be used to determine the various degrees of freedom needed, such as the degrees of freedom between groups, df_{between}, within groups, df_{within}, and the total degrees of freedom, df_{total}, as follows:

$$df_{\text{between}} = a - 1 = 3 - 1 = 2, \tag{6.85}$$

and for within,

$$df_{\text{within}} = N - a = 21 - 3 = 18, \tag{6.86}$$

and a total degree of freedom of,

$$df_{\text{total}} = N - 1 = 21 - 1 = 20. \tag{6.87}$$

TABLE 6.2
ANOVA Example with List of Number of Errors

A	B	C
8	7	2
8	6	3
9	8	4
8	7	2
9	6	4
8	6	3
7	7	3

These values for the various degrees of freedom can be used to determine the Sum of Squares for similar quantities (i.e., Sum of Squares between, Sum of Squares within, and Sum of Squares total) from

$$\mathrm{SS}_{\text{between}} = \frac{\sum\left(\sum a_i\right)^2}{n} - \frac{T^2}{N} = \frac{57^2 + 47^2 + 21^2}{7} - \frac{125^2}{21} = 98.67, \tag{6.88}$$

and

$$\mathrm{SS}_{\text{within}} = \sum Y^2 - \frac{\sum\left(\sum a_i\right)^2}{n} = 853 - \frac{57^2 + 47^2 + 21^2}{7} = 10.29, \tag{6.89}$$

and

$$\mathrm{SS}_{\text{total}} = \sum Y^2 - \frac{T^2}{N} = 853 - \frac{125^2}{21} = 108.95. \tag{6.90}$$

Once Equations 6.85–6.90 have been determined, we can calculate the Mean of Squares between the groups by

$$\mathrm{MS}_{\text{between}} = \frac{\mathrm{SS}_{\text{between}}}{\mathrm{df}_{\text{between}}} = \frac{98.67}{2} = 49.34, \tag{6.91}$$

and the Mean of Squares within the groups by,

$$\mathrm{MS}_{\text{within}} = \frac{\mathrm{SS}_{\text{within}}}{\mathrm{df}_{\text{within}}} = \frac{10.29}{18} = 0.57. \tag{6.92}$$

Having the Mean of Squares between groups and within groups, we can finally determine the F-value by,

$$F = \frac{\mathrm{MS}_{\text{between}}}{\mathrm{MS}_{\text{within}}} = \frac{49.34}{0.57} = 86.56. \tag{6.93}$$

Table 6.3 shows these results organized in tabular form.

In the last column, we can see the critical value which is found using the F distribution table with the degrees of freedom between, and the degrees of freedom within the results shown above.

TABLE 6.3
Table of Values Found from ANOVA Calculations to Determine the *F*-Value

	SS	df	MS	*F*-value	Critical Value
Between	98.67	2	49.34	**86.56**	3.5546
Within	10.29	18	0.57		
Total	108.95	20			

Inspecting the F-table at $F(2,18)$, the table value produces a critical value of 3.5546. Consequently, we must reject the null hypothesis that all three means are the same, in a statistical sense, if the F-value in the table is greater than the critical value. Since the F-value is indeed larger than the critical value, we reject the null hypothesis that means are all the same. We now know that there are statistically significant differences present between the production lines and need to investigate further to determine the root cause(s).

Estimating parameters of the model: For the data model shown in Equation 6.59, it is often necessary to estimate the individual parameters in the model such as the mean μ and the ith treatment variation τ_i. This can be accomplished by

$$\hat{\mu} = \bar{y}_{..}, \tag{6.83}$$

and

$$\hat{\tau}_i = \bar{y}_{i.} - \bar{y}_{..}, \tag{6.84}$$

where the ^ symbol means estimate.

ANOVA with different number of observations in treatments: Sometimes the number of samples changes between treatment levels. This is called *unbalanced data* and can occur quite often if deliberate care is not taken to ensure sample sizes are the same between treatments. For instance, if a sampling plan is not formally developed, there may be differences in the number of samples taken between treatments. Fortunately, the ANOVA approach above can still be used but there are slight changes to the corrected sum of square terms SS_T and $SS_{\text{Treatments}}$ as follows:

$$SS_T = \sum_{i=1}^{a}\sum_{j=1}^{n_i} y_{ij}^2 - \frac{y_{..}^2}{N}, \tag{6.85}$$

and

$$SS_{\text{Treatment}} = \sum_{i=1}^{a} \frac{y_{i.}^2}{n_i} - \frac{y_{..}^2}{N}. \tag{6.86}$$

These are the only changes required, and the rest of the ANOVA analysis can proceed as before. Table 6.2 provides a summary of the necessary formula for conducting ANOVA for a single random factor with a fixed effects model.

Checking the adequacy of our data model: One question that needs to be addressed is how adequate our data model, shown in Equation 6.59, is in describing the experimental observations. We have assumed that we can represent the error with an independent Gaussian distributed random variable with zero mean and unknown constant variance. If these assumptions are correct, then our ANOVA approach gives an exact result for the hypothesis test of no difference in the means of the treatments. These assumptions are critical in the validity of our results and so need to be investigated. The approach taken in checking our assumptions is to evaluate the *residuals* from our experimental data. The residual for the jth observation in the ith treatment is given by

$$e_{ij} = y_{ij} - \hat{y}_{ij}, \tag{6.87}$$

where the expression is basically the difference between the actual value of y_{ij} and the estimate value $\hat{y}_{ij}$. We can determine the estimated value of $\hat{y}_{ij}$ as follows:

$$\hat{y}_{ij} = \hat{\mu} + \hat{\tau}_i, \tag{6.88}$$

where the model parameters on the right-hand side are determinable through Equations 6.83 and 6.84. Substituting these model parameters into Equation 6.88 and simplifying we get,

$$\hat{y}_{ij} = \overline{y}_{i.}, \tag{6.89}$$

which is just the mean of the ith treatment observations. Substituting Equation 6.89 into Equation 6.87 provides a way to estimate the residuals for any observation in the collected data set using only the data itself.

Checking for normality of the residuals: We know that the residuals should be normally distributed with zero mean and constant (unknown) variance, so we can plot the histogram of the residuals to see if this is true or not. For this approach, we require a lot of residuals to properly fill out the histogram. Another useful approach is to construct a *normal probability plot*. This is accomplished by first ordering the residuals e_{ij} from the smallest residual to the largest residual to provide an ordered residual data set $e_{(l)}$. The l index runs from 1 to the total number of observations N such that $e_{(1)}$ is the smallest residual in the data set, $e_{(2)}$ the next smallest, all the way to $e_{(N)}$ which is the largest residual in the data set. If this is plotted against the *observed cumulative frequency* $100(l-0.5)/N$, this will produce a plot that varies around a straight line fit to the ordered residuals. In other words, if we use linear regression on the data set, we expect a good fit of the data if the residuals are normally distributed.

Checking for outliers in the data set: In addition to the normality check, we need to check for some other errors that can appear in the dataset. For instance, sometimes the *standardized residuals* given by

$$d_{ij} = \frac{e_{ij}}{\sqrt{\text{MS}_E}}, \tag{6.90}$$

are used in determining *outliers* in the data set. An outlier is a residual that is much larger than the other residuals and, if present in the data, can distort the ANOVA results. If one or more outliers are discovered in the data set, then the root cause of the outlier must be investigated and understood, and corrections made to the designed experiment as required.

Check for equivalency in variance between treatment levels: For the ANOVA test, in addition to checking for normality of the residuals and ensuring that the data are independent, the variance between treatments/factor-levels must be equal. This can be checked by visually examining box-plots of the data. Another way is to use a statistical package to run a Bartlett test. The Bartlett test checks that the variances are equal over a set of samples.

Ordering residuals in time: If the residuals are ordered sequentially according to when they occurred in the experiment, some insight into the correlations between the residuals is obtained. This indicates if the *independence* assumption of the data has been violated. This type of error can be hard to correct and can occur, for instance, if the data hasn't been properly randomized. Another thing to watch out for is drift in the variance over time. This can be seen by the spreading of the residuals as time progresses like a funnel that progresses to the right as time increases.

The approach for dealing with a changing variance during the experiment is to apply a *variance stabilizing transformation* and repeat the experiment on the transformed data (Montgomery 2020). The residuals should be investigated for any application of ANOVA to ensure that they contain no observable pattern (i.e., that residuals are *structureless*).

Random effects model for single random factor: This type of model occurs if there are a large number of treatment levels. In this case, the a treatment levels are themselves samples of a larger population and the appropriate model is the *random effects model*. For this case, we can use the same data model as is shown in Equation 6.59 and repeated here for convenience,

$$y_{ij} = \mu + \tau_i + \epsilon_{ij} \begin{cases} i = 1,2,\ldots,a \\ j = 1,2,\ldots,n \end{cases}. \tag{6.91}$$

As before, the index i refers to the treatment level and the index j refers to the observation within a treatment level. The assumption is that the treatment random variable τ_i is normally distributed with zero mean and variance σ_τ^2, and the errors are independent, normally distributed random variables with zero mean and variance σ^2. Because of the independence assumption, the variance in our data model is given by

$$V(y_{ij}) = \sigma_\tau^2 + \sigma^2, \tag{6.92}$$

and are defined as the *variance components* in the random effects model. In the random effects model, observations are only independent if they come from different treatment levels, and so we can use the covariances between observations to capture this fact,

$$\begin{aligned} &\mathrm{Cov}\left(y_{ij}, y_{i,j'}\right), \quad j \neq j' \\ &\mathrm{Cov}\left(y_{ij}, y_{i',j'}\right), \quad i \neq i' \end{aligned}. \tag{6.93}$$

Our fundamental expression in the ANOVA process shown in Equation 6.64 is still the same, and we can proceed as usual with the ANOVA with the following changes. We want to conduct the hypothesis test about the variance component since we are interested in the population of the treatments and not so much in the individual treatment effects. Consequently, we test σ_τ^2 according to,

$$H_0 : \sigma_\tau^2 = 0, \tag{6.94}$$

and

$$H_A : \sigma_\tau^2 > 0. \tag{6.95}$$

The characteristics for the sum of squares for the treatments and error are the same as before, with SS_E/σ^2 having a chi-square distribution with a degree of freedom of $N - a$, and $SS_{\text{Treatment}}/\sigma^2$ also having a chi-distribution with degree of freedom $a - 1$. The test statistic is the ratio of the mean squares

$$F_0 = \frac{MS_{\text{Treatments}}}{MS_E} = \frac{SS_{\text{Treatment}}/a-1}{SS_E/N-a}, \tag{6.96}$$

and follows the F-distribution. Taking the expected value of the corresponding corrected mean squares, we get after some simplification

$$MS_{\text{Treatment}} = \sigma^2 + n\sigma_\tau^2, \tag{6.97}$$

and

$$\mathrm{MS}_E = \sigma^2. \tag{6.98}$$

We can then estimate the model parameters from the mean squares by

$$\hat{\sigma}^2 = \mathrm{MS}_E, \tag{6.99}$$

and

$$\hat{\sigma}_\tau^2 = \frac{\mathrm{MS}_{\text{Treatment}} - \mathrm{MS}_E}{n}. \tag{6.100}$$

If the number of samples is different between treatment levels, then the n in the denominator of Equation 6.100 is replaced with n_o where

$$n_0 = \sum_{i=1}^{a} n_i - \frac{\sum_{i=1}^{a} n_i^2}{\sum_{i=1}^{a} n_i}. \tag{6.101}$$

Finally, the $100(\alpha - 1)$ percent confidence interval for the variance σ^2 is given by

$$\frac{(N-a)\mathrm{MS}_E}{\chi^2_{\alpha/2, N-a}} \le \sigma^2 \le \frac{(N-a)\mathrm{MS}_E}{\chi^2_{1-\alpha/2, N-a}}. \tag{6.102}$$

Maximum likelihood estimate of the variance components: In Equation 6.102, the confidence interval was found for the variance of the error. However, the confidence interval for the treatment variance σ_τ^2 is not so readily found. One method for finding this confidence interval is based on the maximum likelihood method wherein a *likelihood function* is developed that contains a parameter θ to be estimated, and x_i are n random samples from a joint probability distribution function $f(x_i;\theta)$. The likelihood function considers the random samples fixed, and θ is estimated such that the likelihood function,

$$L(x_1, x_2, \ldots, x_n; \theta) = \prod_{i=1}^{n} f(x_i;\theta), \tag{6.103}$$

is maximized. In Equation 6.103, the likelihood function is just the product of the joint probability distribution functions as shown. We can apply this to estimate our variance components as follows. If we let $\boldsymbol{y}$ represent the $an \times 1$ vector of observations where we have a treatments/factor-levels and n replications, and where Σ is the covariance matrix with dimensions $an \times an$, then the likelihood function is given by

$$L\left(x_{11}, x_{12}, \ldots, x_{a,n};\ \mu,\ \sigma_\tau^2, \sigma^2\right) = \frac{1}{2\pi^{N/2}[\Sigma]^{1/2}} \exp\left[-\frac{1}{2}(y - j_N\mu)' \, \Sigma^{-1} (y - j_N\mu)\right], \tag{6.104}$$

where $N = an$ as before, j_N is a vector of $N \times 1$ ones, and the apostrophe represents conjugation. In Equation 6.104, the random a draws from the larger population are *fixed* and covariance matrix has the sum $\sigma_\tau^2 + \sigma^2$ on the main diagonal, and σ_τ^2 is on the appropriate off-diagonal elements in accordance with Equations 6.92 and 6.93. The combination of μ, σ_τ^2, and σ^2 that maximizes Equation 6.104 provides the best maximum likelihood estimate of μ, σ_τ^2, and σ^2. An often-used variant of this approach is the *residual maximum likelihood (REML)* which uses the following estimates that are based on the sampled data

$$\hat{\mu} = \frac{\sum_{i=1}^{n} y_i}{n}, \tag{6.105}$$

and

$$\hat{\sigma}^2 = \frac{\sum_{i=1}^{n} (y_i - \bar{y})^2}{n}. \tag{6.106}$$

The sample variance is an unbiased estimate and is given by

$$S^2 = \frac{\sum_{i=1}^{n} (y_i - \bar{y})^2}{n-1}. \tag{6.107}$$

If we have $\hat{\theta}$ be our maximum likelihood estimate of our unknown parameter θ, then the $100(1-\alpha)$ percent confidence interval is approximately

$$\hat{\theta} - Z_{\frac{\alpha}{2}}\hat{\sigma}\left(\hat{\theta}\right) \le \theta \le \hat{\theta} + Z_{\frac{\alpha}{2}}\hat{\sigma}\left(\hat{\theta}\right), \tag{6.108}$$

where $\hat{\sigma}(\hat{\theta})$ is the estimated standard error which is the square root of the main diagonal elements of the covariance matrix.

Generalization to more than one random factor (factors ≥ 2): We have seen various methods for hypothesis testing in working with a single factor in a CRD experiment. Now we generalize our approach to multiple factors, at multiple levels. For instance, if we have three factors, such that factor A has a treatments/factor levels, factor B has b treatments/factor levels, and factor C has c treatments/factor levels, with n replicates of the experiment, we have a total of $abcn$ observations. Notice that this pattern continues for a higher number of factors wherein the total number of observations is just the product of all the treatments/factor levels with each other and then multiplied by the number of replicates n. If all the factors are fixed, then we can conduct hypothesis testing of the main factors and their interactions using ANOVA. For the fixed effects model, we can determine the required test statistic by dividing the mean square term for the desired term or interaction by the mean square error. As before, dividing the mean square terms by the mean square error leads to an F-statistic for all the terms, with upper, one-tail tests. The degrees of freedom are one less than the total number of observations in the level. For example, for the random factor B with b levels, the degree of freedom is $b-1$. We use a three-factor model as an example but comment on the generalizations. For a CRD experiment with three factors, the data model becomes

$$y_{ijkl} = \mu + \tau_i + \beta_j + \gamma_k + (\tau\beta)_{ij} + (\tau\gamma)_{ik} + (\beta\gamma)_{jk} + (\tau\beta\gamma)_{ijk} + \epsilon_{ijkl} \left\{ \begin{array}{l} i = 1,2,\ldots,a \\ j = 1,2,\ldots,b \\ l = 1,2,\ldots,c \\ l = 1,2,\ldots,n \end{array} \right. . \tag{6.109}$$

We notice that the data model consists of the sum of the main effects $(\mu + \tau_i + \beta_j + \gamma_k)$ as in the random single factor case but with two additional random factor main effects included $(\beta_j + \gamma_k)$, Additionally, there is the sum of all permutations of the joint effects $((\tau\beta)_{ij} + (\tau\gamma)_{ik} + (\beta\gamma)_{jk})$ and the effect between all three random factors $((\tau\beta\gamma)_{ijk})$ and the error term for all observations including the replicates (that include the n replicates (ϵ_{ijkl})). Extensions to higher numbers of factors, sum all the main effects, sum all permutations of the factors to include all permutations of joint interactions, all permutations of triple interactions, all permutations of quadruple interactions and so forth until the combined interaction of all factors is reached. The random, normally distributed error is then added to every *ijk*th observation and for each replication (*n* replicates assumed here). Mathematically, we can write the total sum of squares term as

$$SS_T = \sum_{i=1}^{a}\sum_{j=1}^{b}\sum_{k=1}^{c}\sum_{l=1}^{n} y_{ijkl}^2 - \frac{y_{....}^2}{abcn}, \quad (6.110)$$

which can be determined directly from the data. Generalizing Equation 6.110 to more than three factors would require an additional sum and index for each new factor above the three shown in Equation 6.110, with the last sum remaining as a summation over the replicates. The numerator of the last term on the right side of Equation 6.110 would be summed over all observations (abcn of them in the above example), then squared, and then divided by the total number of observations. We now show the equations for the sum of squares for the main effects:

$$SS_A = \frac{1}{bcn}\sum_{i=1}^{a} y_{i...}^2 - \frac{y_{....}^2}{abcn}, \quad (6.111)$$

and

$$SS_B = \frac{1}{acn}\sum_{j=1}^{b} y_{.j..}^2 - \frac{y_{....}^2}{abcn}, \quad (6.112)$$

and

$$SS_C = \frac{1}{abn}\sum_{k=1}^{c} y_{..k.}^2 - \frac{y_{....}^2}{abcn}. \quad (6.113)$$

The pattern for generalizing to more than three factors is easily seen by observing these equations. The product in the denominator at the left of the summation signs above includes the products of all factor levels except the one being summed over in the summation. The first *y*-term to the right of the summation signs in Equations 6.111 through 6.113 sum over the index associated with the factor, with all the other indices summed already, and the last term on the right side of the summation being the square of all observations in all levels and replicates, squared, then divided by the total number of observations in all levels and replicates. The equations for all permutations of the joint sum of squares are

$$SS_{AB} = \frac{1}{cn}\sum_{i=1}^{a}\sum_{j=1}^{b} y_{ij..}^2 - \frac{y_{....}^2}{abcn} - SS_A - SS_B, \quad (6.114)$$

and

$$SS_{AC} = \frac{1}{bn}\sum_{i=1}^{a}\sum_{k=1}^{c} y_{i.k.}^2 - \frac{y_{....}^2}{abcn} - SS_A - SS_C, \quad (6.115)$$

and

$$SS_{BC} = \frac{1}{an}\sum_{j=1}^{b}\sum_{k=1}^{c} y_{.jk.}^2 - \frac{y_{....}^2}{abcn} - SS_B - SS_C. \quad (6.116)$$

The generalization of Equations 6.114–6.116 follows the same rules as those for the sum of squares for the main effects, with the addition of subtracting out the sum of squares for the main effects determined in Equations 6.111–6.113. The sum of squares for the triple factor interaction is given by

$$SS_{ABC} = \frac{1}{n}\sum_{i=1}^{a}\sum_{j=1}^{b}\sum_{k=1}^{c} y_{ijk.}^2 - \frac{y_{....}^2}{abcn} - SS_A - SS_B - SS_C - SS_{AB} - SS_{AC} - SS_{BC}, \tag{6.117}$$

where all the previous sum of square terms are subtracted from the triple summation along with the constant term $\frac{y_{....}^2}{abcn}$. The generalization to more than three factors is like the jointly interacting terms and is readily apparent from the structure of Equation 6.117. By naming the first two terms in Equation 6.117, the ABC subtotal sum of squares, we get a compact form of Equation 6.117 in

$$SS_{ABC} = SS_{\text{Subtotals }(ABC)} - SS_A - SS_B - SS_C - SS_{AB} - SS_{AC} - SS_{BC}. \tag{6.118}$$

We summarize the ANOVA process for three random factors in a fixed effects model in Table 6.4.

The ubiquitous 2^k-factorial design: In the last section, we presented a general solution for an arbitrary number of factors, with an arbitrary number of levels per factor. Here we present the case for an arbitrary number of factors with only two levels per factor. An example would be k random

TABLE 6.4
ANOVA for Three Factors with Fixed Effects, Completely Randomized Design, and Normality Assumptions

Variation Source	Sum of Squares	DF	MS	Expected Mean Square	Statistic F_0
A	SS_A	v_a $(a-1)$	MS_A	$\sigma^2 + \frac{bcn\sum_{i=1}^{a}\tau_i^2}{v_a}$	$\frac{MS_A}{MS_E}$
B	SS_B	v_b $(b-1)$	MS_B	$\sigma^2 + \frac{acn\sum_{j=1}^{b}\beta_j^2}{v_b}$	$\frac{MS_B}{MS_E}$
C	SS_C	v_c $(c-1)$	MS_C	$\sigma^2 + \frac{abn\sum_{k=1}^{c}\gamma_k^2}{v_c}$	$\frac{MS_C}{MS_E}$
AB	SS_{AB}	v_{ab} $(a-1)(b-1)$	MS_{AB}	$\sigma^2 + \frac{cn\sum_{i=1}^{a}\sum_{j=1}^{b}(\tau\beta)_{ij}^2}{v_{ab}}$	$\frac{MS_{AB}}{MS_E}$
AC	SS_{AC}	v_{ac} $(a-1)(c-1)$	MS_{AC}	$\sigma^2 + \frac{bn\sum_{i=1}^{a}\sum_{k=1}^{c}(\tau\gamma)_{ik}^2}{v_{ac}}$	$\frac{MS_{AC}}{MS_E}$
BC	SS_{BC}	v_{bc} $(b-1)(c-1)$	MS_{BC}	$\sigma^2 + \frac{an\sum_{j=1}^{b}\sum_{k=1}^{c}(\beta\gamma)_{jk}^2}{v_{bc}}$	$\frac{MS_{BC}}{MS_E}$
ABC	SS_{ABC}	v_{abc} $(a-1)(b-1)(c-1)$	MS_{ABC}	$\sigma^2 + \frac{n\sum_{i=1}^{a}\sum_{j=1}^{b}\sum_{k=1}^{c}(\tau\beta\gamma)_{ijk}^2}{v_{abc}}$	$\frac{MS_{ABC}}{MS_E}$
Error	SS_E	$abcv_n$ $abc(n-1)$	MS_E	σ^2	
Total	SS_T	$abcn-1$			

factors in a fixed effects model; under CRD conditions each has binary levels such as true/false, yes/no, high/low, and so forth. The binary levels for each factor can be quantitative, or qualitative. This of course is a degenerate case of the general solution in the previous section, but this type of designed experiment is quite common. The 2^k-factorial design also provides the smallest number of runs that are needed to have a complete factorial design, so this method is good early in the design process in such things, for example, choosing important factors from a larger group. We present the method for a general 2^k-factorial design which will have $(2^k - 1)$ effects. We use the following common notational convention, lower case letters, if present, represent the high level of a given factor. The factor itself is represented by a capital letter. For example, factor B out of k factors, has a high state if the letter b is present. As another example, in a four-factor experiment ($k=4$), the letters *abd* mean that factors A, B, and D are at the high state, and factor C is at the low state. There is a *standard order* for the combination of factor levels that starts with the main effects and then brings in higher order combinations by combining them with earlier factor-level combinations. For example, say we have a three-factor experiment ($k=3$). The factors would be represented by A, B, and C with corresponding factor levels a, b, and c. We would have $2^3=8$ possible combinations of factor levels with a standard order of (1), a, b, ab, c, ac, bc, abc. The quantity (1) represents all factor levels at the low state (i.e., not seen with our convention of seen letters representing the high states of those letters). As another example in a seven-factor experiment, we would have random factors A, B, C, D, E, F, and G. If we show the following, *acfg*, this means that factors A, C, F, and G are at their high state, and factors B, D, and E are at their low state. In a 2^k-factor analysis, as before we determine the corrected sum of squares for the main effects and various combinations of interactions as follows:

1. Determine the *contrast* for the effects,
2. Use the contrast to determine the effects themselves,
3. Use the effects to determine the appropriate corrected sum of squares.

The contrast can be calculated by

$$\text{Contrast}_{AB\cdots K} = (a \pm 1)(b \pm 1)\cdots(k \pm 1), \tag{6.119}$$

where the "+" or "–" is determined whether-or-not the factor level is at the high state (–), or at the low state (+). So, for example, the seven-factor experiment in the preceding paragraph, a contrast of ABCDEF would be

$$\text{Contrast}_{ABCDEF} = (a-1)(b-1)(c-1)(d-1)(e-1)(f-1)(g+1). \tag{6.120}$$

These terms are multiplied out and the last term of one is replaced with all factors at their low state (1). The effects themselves in step (2) above can be determined by

$$AB\cdots K = \frac{2}{n2^k}\left(\text{Contrast}_{AB\cdots K}\right), \tag{6.121}$$

and the corrected sum of squares is given by

$$\text{SS}_{AB\cdots K} = \frac{2}{n2^k}\left(\text{Contrast}_{AB\cdots K}\right)^2. \tag{6.122}$$

Table 6.4 shows the ANOVA for the 2^k design. The sum of squares terms is determined by Equations 6.119 through 6.122. As seen in Table 6.5, we have k main effects, $\binom{k}{2}$ two-factor

TABLE 6.5
ANOVA for 2^k Factorial Design Assuming Completely Randomized Design, Fixed Effects, and Normality

Variation Source	Number of Terms	Sum of Squares	Degrees of Freedom
Main effects	K		
A	1	SS_A	1
B	1	SS_B	1
$\vdots$	$\vdots$	$\vdots$	$\vdots$
K	1	SS_K	1
Two-factor interactions	$\binom{k}{2}$		
AB	1	SS_{AB}	1
AC	1	SS_{AC}	1
$\vdots$	$\vdots$	$\vdots$	$\vdots$
JK	1	SS_{JK}	1
Three-factor interactions	$\binom{k}{3}$		
ABC	1	SS_{ABC}	1
ABD	1	SS_{ABD}	1
$\vdots$	$\vdots$	$\vdots$	$\vdots$
IJK	1	SS_{IJK}	1
$\vdots$	$\vdots$	$\vdots$	$\vdots$
k –Factor interactions	$\binom{k}{k}$		
$\boldsymbol{ABC \cdots K}$	1	$SS_{ABC \cdots K}$	1
Error		SS_E	$2^k(n-1)$
Total		SS_T	$n2^k - 1$

effects, $\binom{k}{3}$ three-factor effects, and $\binom{k}{x}$ x-factor effects, where $\binom{k}{x}$ is the binomial coefficient given by

$$\binom{k}{x} = \frac{k!}{x!(k-x)!}. \tag{6.123}$$

Factorial designs and blocking: In all our previous examples, we have assumed that it is possible to completely randomize a design. What if this is not possible? For instance, what if because of experimental time constraints, the experiment cannot be run at once and must be continued on a separate day (or days)? This introduces a non-random aspect into the design which impacts our "completely randomized design" assumption. Another example is the presence of a nuisance factor in our data, where a nuisance factor is a controllable or uncontrollable unwanted signal. These situations and others may require that we run our data in *blocks* to either account for the restriction on the randomness of the experiment or mitigate the effects of the nuisance parameter. As an example of blocking, let's assume we have two random factors with respective levels a and b, with n replicates. Our general equation for the data model in Equation 6.109 becomes

$$y_{ijk} = \mu + \tau_i + \beta_j + (\tau\beta)_{ij} + \epsilon_{ijk} \begin{cases} i = 1,2, \ldots, a \\ j = 1,2,\ldots,b \, . \\ k = 1,2,\ldots,n \end{cases} \tag{6.124}$$

Let's say that due to resource availability issues, we cannot run a fully randomized design and must instead run the replicates on different days. Consequently, we have introduced a restriction on the randomization of the n replicates and need to run these in n separate, sequential blocks. The effect on the data model is to account for the effect of the kth block by adding a term to our data model as follows:

$$y_{ijk} = \mu + \tau_i + \beta_j + (\tau\beta)_{ij} + \delta_k + \epsilon_{ijk} \begin{cases} i = 1,2, \ldots, a \\ j = 1,2,\ldots,b \, , \\ k = 1,2,\ldots,n \end{cases} \tag{6.125}$$

where the term δ_k represents the effect of the kth block. The interaction between the effect of the block and the two-factor effects is assumed negligible and so these joint effects are zero. If this is not the case, then Equation 6.125 needs to include the interaction terms between δ_k and the other factors (namely, $(\tau\delta)_{ik}$, $(\beta\delta)_{jk}$, and $(\tau\beta\delta)_{ijk}$) which would be indistinguishable from the error. The corrected sum of squares for the blocks is given by

$$SS_{\text{Blocks}} = \frac{1}{ab}\sum_{k=1}^{n} y_{..k}^2 - \frac{y_{...}^2}{abn}, \tag{6.126}$$

which has $n-1$ degrees of freedom. The mean square for the blocks is the SS_{Blocks} divided by the degree of freedom,

$$MS_{\text{Blocks}} = \frac{SS_{\text{Blocks}}}{n-1}. \tag{6.127}$$

The block variance component is given by

$$\sigma_\delta^2 = \frac{MS_{\text{Blocks}} - MS_E}{ab}. \tag{6.128}$$

Table 6.6 shows the inclusion of blocking in an ANOVA with two factors at multiple levels for each factor, with n replicates and fixed effects.

6.1.6.4 Latin Squares

The concept of the Latin Square design can be used by systems engineers to reduce the risks associated with unknown factors (Reese 1997). The Latin Square method uses n treatments arranged in an $n \times n$ grid (Cavenagh 2018). In the Latin Square experimental layout, the rows and columns represent blocks on the experiment due to some constraint(s) on the randomness used in the experimental methodology. For example, the rows may represent different batches of a material that were necessitated by the lack of enough materials for a truly random sample (hence the introduction of a material block). The columns could also reflect a constraint on the true randomness of the experiment in that different shifts (or operators) could have been used imposing another block on the

TABLE 6.6
ANOVA for Two Factors with Randomized Complete Block

Variation Source	Degrees of Freedom	Sum of Squares	Expected Mean Square	Test Statistic F_0
Blocks	ν_n $(n-1)$	$\frac{1}{ab}\sum_{k=1}^{n} y_{..k}^2 - \frac{y_{...}^2}{abn}$	$\sigma^2 + ab\sigma_\delta^2$	
A	ν_a $(a-1)$	$\frac{1}{bn}\sum_{i=1}^{a} y_{i..}^2 - \frac{y_{...}^2}{abn}$	$\sigma^2 + \frac{bn\sum_{i=1}^{a}\tau_i^2}{\nu_a}$	$\frac{MS_A}{MS_E}$
B	ν_b $(b-1)$	$\frac{1}{an}\sum_{j=1}^{b} y_{.j.}^2 - \frac{y_{...}^2}{abn}$	$\sigma^2 + \frac{an\sum_{j=1}^{b}\beta_j^2}{\nu_b}$	$\frac{MS_B}{MS_E}$
AB	ν_{ab} $(a-1)(b-1)$	$SS_{AB} = \frac{1}{n}\sum_{i=1}^{a}\sum_{j=1}^{b} y_{ij.}^2 - \frac{y_{...}^2}{abn} - SS_A - SS_B$	$\sigma^2 + \frac{n\sum_{i=1}^{a}\sum_{j=1}^{b}(\tau\beta)_{ij}^2}{\nu_{ab}}$	$\frac{MS_{AB}}{MS_E}$
Error	ν_E $(ab-1)(n-1)$	$SS_E = SS_T - SS_{AB} - SS_A - SS_B$	σ^2	
Total	ν_T $(abn-1)$	$SS_{AB} = \frac{1}{n}\sum_{i=1}^{a}\sum_{j=1}^{b}\sum_{k=1}^{n} y_{ijk}^2 - \frac{y_{...}^2}{abn}$		

randomness of the experiment. In the Latin Square methodology, the treatments are laid out such that any treatment occurs once and only once for any row and column combination. This is very similar to a Japanese Sudoku puzzle. The Latin Square method allows the blocking of nuisance factors in two directions simultaneously (rows and columns). Latin Squares are used to minimize bias in experimental results by guaranteeing that each treatment is evaluated in every permittable combination of testing criteria (Reese 1997).

Latin Squares have been applied to various industries, such as in the electronics industry in chip implantation scenarios. A machine vision method was employed to confirm the accurate positioning of capacitors on microchips (Shi et al. 2016). The manufacturer aimed to assess chip position, chip variability, and the angles of chip placement relative to each other. According to Shi, analysis of the visual system required 1,536 trials. The research team then recommended using a Latin Square design for the experiment. The outcome was a reduction in the required number of trials to only 96. An additional step including ANOVA was necessary to assess the variance; however, the overall duration of the experiment was shortened as the chip analysis was the most time-consuming element of the process.

In Table 6.7, a simple example of the Latin Square concept is shown. Three lab stations are shown as rows on the left side of the table. A set of three treatments, *A*, *B*, and *C* are planned for the designed experiment. To reduce the potential variations that these factors may introduce, each treatment is carried out at a different lab station and at a different time (morning, afternoon, and evening). Note that the treatments occur only once across the space of lab stations and time slots (i.e., only once in any of the rows or columns). One technique for generating this type of layout is to label each treatment in a sequence of Latin letters (e.g., *A*, *B*, *C*, …) and arrange them in sequence from the first row down and across the columns. For example, for an experiment with five treatments labeled *A*, *B*, *C*, *D*, *E*, and *F*, we would have the first column (front the top-left downwards) given by *A*, *B*, *C*, *D*, *E*, and *F*. The first row from the top left would also be *A*, *B*, *C*, *D*, *E*, and *F*. The second column would start with *B* and continue the sequence rolling over the first letter as follows: *B*, *C*, *D*, *E*, *F*, and *A*. The third column from the top-left would be *C*, *D*, *E*, *F*, *A*, *B*, and this pattern would continue until the end of the columns. In this fashion, there would be a square matrix

TABLE 6.7
Example of a 3 by 3 Latin Square Experimental Arrangement

	Time Slot		
Lab Station	**Morning**	**Afternoon**	**Evening**
1	A	B	C
2	B	C	A
3	C	A	B

that for every column and row has (1) all the treatments and (2) each treatment listed only once in any intersecting row and/or column. The method received its name from using Latin letters to describe the treatments and arranging them in this fashion resulting in a square layout of treatment levels.

The data model for the Latin Square experimental methodology is given by

$$y_{ijk} = \mu + \alpha_i + \tau_j + \beta_k + \epsilon_{ijk} \begin{cases} i = 1,2,\ldots,p \\ j = 1,2,\ldots,p\,, \\ k = 1,2,\ldots,p \end{cases} \tag{6.129}$$

where we have p-treatments that are arranged in a square fashion as just presented. The rows and columns each represent a type of blocking that is in effect. The Latin Square matrix has dimensions p^2. The i-index runs over the rows, the k-index runs over the columns, and the j-index runs over the treatments. The parameter μ is the overall mean, α_i is the ith row effect, τ_j is the jth treatment effect, β_k is the kth column effect, and ϵ_{ijk} is the experimental random error that is assumed normally and independently distributed, with zero mean and variance σ^2. Note that there are no interaction terms. Only two of the three indices are needed in the Latin Square methodology since there is only one treatment in any cell location of the Latin Square. The ANOVA is decomposed into components of treatments, column effects, row effects, and random error as follows:

$$\mathrm{SS}_T = \mathrm{SS}_{\mathrm{rows}} + \mathrm{SS}_{\mathrm{columns}} + \mathrm{SS}_{\mathrm{Treatments}} + \mathrm{SS}_E. \tag{6.130}$$

If we test for the hypothesis of the equivalency of the means, we have as before each sum of square terms divided by σ^2 follows a chi-square distribution and so the appropriate test statistic is the F-statistic,

$$F_0 = \frac{\mathrm{MS}_{\mathrm{Treatment}}}{\mathrm{MS}_E}, \tag{6.131}$$

which has a distribution given by $F_{p-1,(p-2)(p-1)}$. The ANOVA is shown in Table 6.8.

6.1.6.5 Statistical Process Control

A useful methodology for determining if a system in operation is in control (or not) is statistical process control. Many important manufacturing and operational parameters and variables such as temperature, pressure, assembly line speed, torque, and more can be characterized and understood through SPC methods. We will treat SPC in detail in our chapter about Quality Production and Manufacturing where this topic is more directly relevant. We provide a short example here of SPC since many of the statistical principles presented above are also applicable to SPC methods. Figure 6.19 shows a useful SPC method to tell whether a process such as a production line process is in control. Figure 6.19 shows control limits above and below the mean (Center Line) of the initial

TABLE 6.8
ANOVA for Two Factors with Randomized Complete Block

Variation Source	Degrees of Freedom	Sum of Squares	Mean Square	Test Statistic F_0
Treatments	$v_p(p-1)$	$SS_{\text{Treatments}} = \frac{1}{p}\sum_{j=1}^{p} y_{.j.}^2 - \frac{y_{...}^2}{p^2}$	$\frac{SS_{\text{Treatments}}}{v_p}$	$\frac{MS_{\text{Treatments}}}{MS_E}$
Rows	$v_p(p-1)$	$SS_{\text{Rows}} = \frac{1}{p}\sum_{i=1}^{p} y_{i..}^2 - \frac{y_{...}^2}{p^2}$	$\frac{SS_{\text{Rows}}}{v_p}$	
Columns	$v_p(p-1)$	$SS_{\text{Columns}} = \frac{1}{p}\sum_{k=1}^{p} y_{..k}^2 - \frac{y_{...}^2}{p^2}$	$\frac{SS_{\text{Columns}}}{v_p}$	
Error	v_E $(p-2)(p-1)$	$SS_E = SS_T - SS_{\text{Columns}} - SS_{\text{Rows}} - SS_{\text{Treatments}}$	$\frac{SS_E}{v_E}$	
Total	$v_T(p^2-1)$	$SS_T = \sum_{i=1}^{p}\sum_{j=1}^{p}\sum_{k=1}^{p} y_{ijk}^2 - \frac{y_{...}^2}{p^2}$		

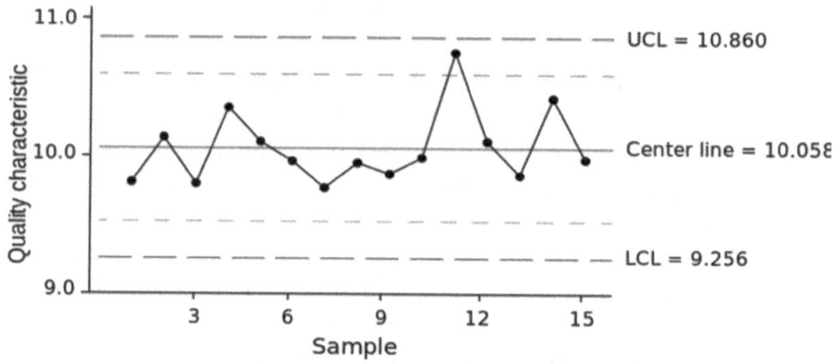

FIGURE 6.19 *P*–Chart (in control) in statistical process control. (By Daniel Penfield (no restrictions), from https://commons.wikimedia.org/wiki/File:ControlChart.svg (accessed October 3, 2023).)

stable control pattern shown by the connected, sampled dots. The method and motivation for generating a *P*-chart is discussed in our chapter about Quality Production and Manufacturing and if the points generated from this method fall between the upper control limit (UCL) and lower control limit (LCL), the process is in statistical control. If a point falls outside of the UCL (greater than 10.860 in Figure 6.19) or the LCL (lower than 9.256 in Figure 6.19), then the system is not in statistical control. If a sample point falls outside the region between the upper and lower control limits but should not have done so, then we have a Type I error. Alternatively, if a sample point falls within the region between the UCL and LCL but should have fallen outside of it, we have a Type II error.

Correctly applied SPC methods result in processes that are stable and in control. These stable and controlled processes improve overall product quality and minimize scrap and waste. SPC methods are very beneficial to manufacturing and production operations but are also applicable in other systems development lifecycle phases where control systems and feedback are important, and stability methods can be used to improve the system design. Common SPC tools are variable control charts and attribute control charts and process capability methods, and these are presented and described in our later Quality Production and Manufacturing chapter. These tools are fundamental to SPC and are regularly used to improve the performance of production lines, and more.

6.1.6.6 Regression Analysis

According to the Encyclopedia Britannica's article on Experimental Design 2025, "Regression analysis involves identifying the relationship between a dependent variable and one or more independent variables. A model of the relationship is hypothesized, and estimates of the parameter values are used to develop an estimated regression equation (Encyclopedia Britannica, Regression Analysis (statistics) 2025)." As an example, a simple linear regression equation can be given by $y = \beta_0 + \beta_1 x + \varepsilon$, where y is the dependent variable, x is an independent variable, and weights B_0 and β_1 are the model parameters. The term ε represents the variation, or error that is present. This simple linear regression equation can be used to show trends in data such as does the data match/fit to a line with slope β_1 and y-intercept β_0. Various methods can be used to determine the best/optimal fit. For instance, the best fit line could be drawn so that the standard deviation between the fitted points and the chosen line with parameters (β_1, β_0) is minimized. In this case, under the assumption of independent data points, the standard deviation between the fitted line and data points is determined as the square root of the sum of the square (RSS) of the differences between each data point and the line generated with the parameters (β_1, β_0). The parameters (β_1, β_0) are varied and the root of the sum of squares of line – sample point difference is determined, and the best fit line parameters are chosen based on the smallest RSS difference value. Figure 6.20 shows a simple linear regression to a set of data points. By observation, the y-intercept β_0 is 5, and the slope of the line β_1 is approximately 10/30=0.3. In Figure 6.20, we can see a clear positive trend in the data. This is not always the case, however. Figure 6.21 shows an example of data that doesn't necessarily fit a single straight line. Notice the green clusters of data at the edges of the one solution is to fit two lines through the blue center region through the green clusters as shown. This is a bilinear (multilinear) regression model and has four fitting parameters (two slopes and two intercepts for both lines). This idea can be expanded to multiple lines in the multilinear regression model. The linear model can be extended to a variety of multiple linear regression models in the general linear model, which can be used for many variables and works well with statistical models such as ANOVA, ANCOVA, and hypothesis testing methods. It is possible to have a compact form of multiple linear regression models using the general linear model. This model can have many dependent variables that map to the same independent variables. The response variable is assumed normally distributed and has a mean that is a function of a linear combination of independent variables (predictors) and known variance. If the output follows a distribution of the exponential family, then this model belongs to the generalized linear model (GLM) class of models, and if the output has randomness, then the generalized linear mixed model is appropriate. Data can also be fit by polynomial regression methods. Polynomial regression analysis is a technique

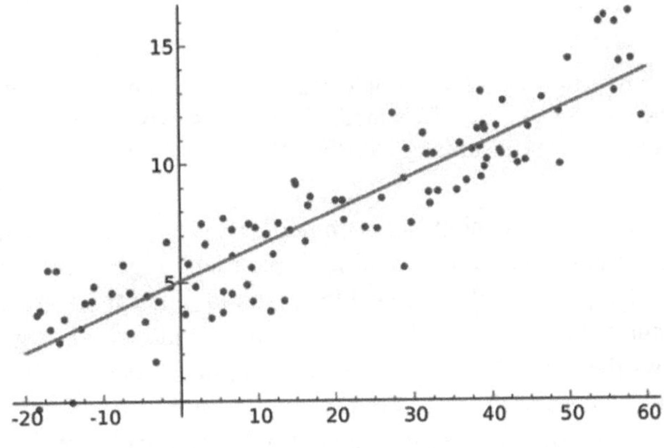

FIGURE 6.20 Simple linear regression. (By Sewaqu (no restrictions), from https://commons.wikimedia.org/wiki/File:Linear_regression.svg (accessed October 13, 2023).)

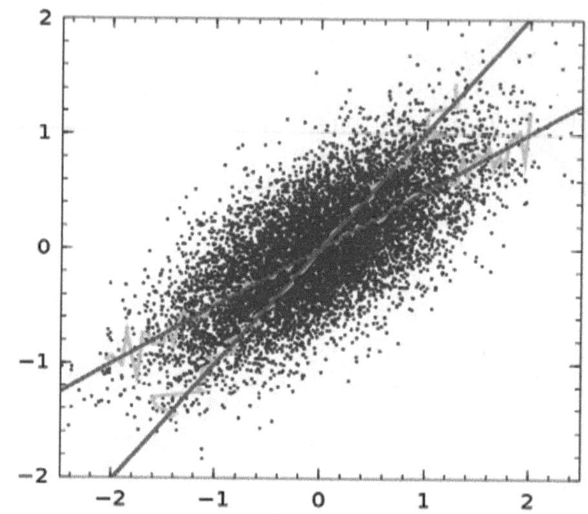

FIGURE 6.21 Multi-linear regression with larger data spread (poorer fit and lower Pearson coefficient). (By *Rickedanielson*, Creative Commons Attribution-Share Alike 4.0 International, from https://commons.wikimedia.org/wiki/File:Visualization_of_errors-in-variables_linear_regression.png (accessed October 13, 2023).)

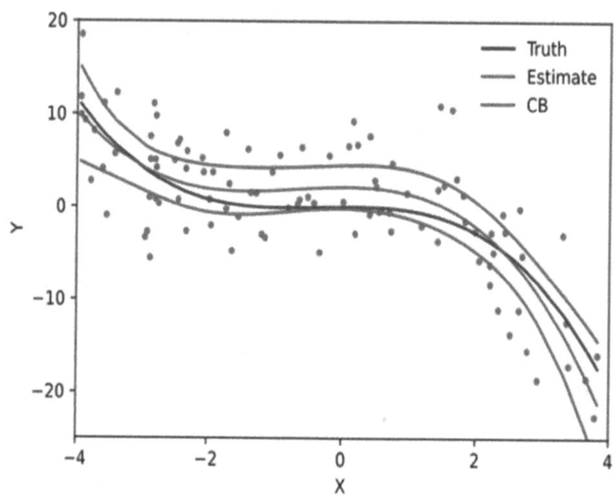

FIGURE 6.22 Cubic polynomial fit with multiple least squares with point estimate (green), truth plot (blue) and 95% confidence band using Scheffe's method. (By *Skbkekas*, Creative Commons Attribution 3.0 Unported, from https://commons.wikimedia.org/wiki/File:Polyreg_scheffe.svg (accessed October 13, 2023).)

that can be used to show the relationship between dependent and independent variables that is not linear in nature. It is a variation on the model of multiple linear regression, with the primary difference being that the line that provides the greatest fit is curved rather than straight. Figure 6.22 shows a polynomial fit using cubic polynomials. The nonlinear fit is evident in the figure. The blue line (second from top-left) shows the true data. The estimated data of the cubic polynomial regression fit is shown in green (third top-left), and the red line (first and last top-left) represents the 95% confidence band for the data. Scheffé's method is a single-step multiple comparison procedure that applies to the set of estimates of all possible contrasts among the factor level means, not just the pairwise differences considered by the Tukey–Kramer method. The Scheffé method can do multiple comparisons in a single step. It works with the contrast among the means of the factor levels as opposed to using only pairwise differences like in the Tukey–Kramer method.

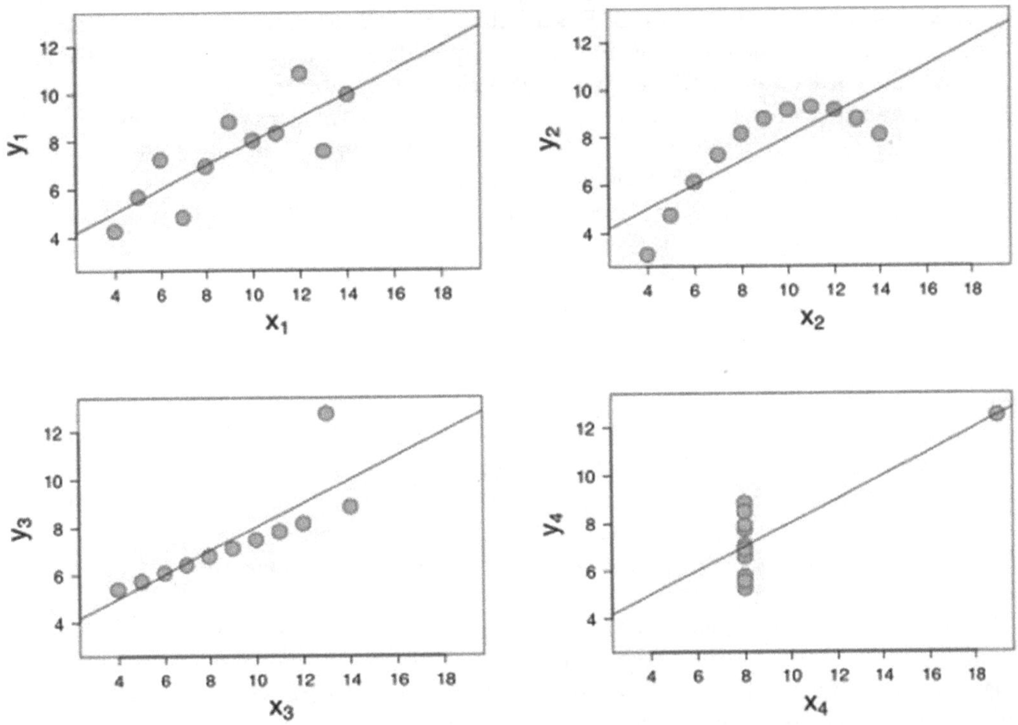

FIGURE 6.23 Pitfalls of model fitting. (By Anscomb.svg: Shutz, *Creative Commons Attribution-Share Alike 3.0 Unported*, from https://en.wikipedia.org/wiki/File:Anscombe%27s_quartet_3.svg (accessed October 13, 2023).)

Some care must be taken when interpreting data using statistical parameters alone. The following example in Figure 6.23 shows a data set by Anscomb that has the same statistical values for the mean, variance, correlation coefficient, and straight–line fit of the data. The human eye can easily tell the difference between these patterns, however these statistical parameters, in this instance cannot distinguish any difference. Other methods and models are prevalent for selecting the best data fit. Here is a short description of these regression methods and when they are used.

Logistic regression: Both categorization and predictive analytics frequently make use of a statistical model referred to as the Logit model which deals with logistic regression. Logistic regression is a statistical method that determines the likelihood of a binary process, such as voting or not voting, true or false, yes or no, that is based on an interpolation of a sigmoid function between the binary states. The technique involves the computation of the probability of these mutually exclusive states that are connected to the independent variable through the sigmoid function.

Ridge regression: This method is used when data exhibits multicollinearity. For example, if the independent variables in an experiment are highly correlated, then the ridge regression technique is applicable. Even though least squares methods are unbiased with respect to multicollinearity, differences in the observed value can occur due to the significance of the variances in the data. For Ridge Regression, the regression estimates are biased which can reduce the standard errors.

Lasso regression (LR): *Lasso* stands for Least Absolute Shrinkage and Selection Operator. This approach introduces a penalty to the absolute magnitude of the regression coefficient just as in the ridge regression method. *Lasso* uses chosen variables that result in coefficient values shrinking to absolute zero.

Quantile regression (QR): This method belongs to the set of linear regression techniques. Quantile regression is effective if outliers are present in the data or if the requirements for linear regression have not been satisfied. This method is often used in statistics and econometrics.

Bayesian linear regression (BLE): BLE is a technique that is often used in machine learning. It is a form of linear regression that uses Bayes' theorem to determine the coefficients of the regression. This approach determines the posterior distribution of the features, instead of the least squares solution. This approach has better stability performance than the standard linear regression method.

Principal components regression (PCR): PCR is often used to evaluate multicollinear data. Similar to ridge regression, principal components regression biases the regression estimates leading to reduced standard errors. In PCR, principal component analysis (PCA) is accomplished first to modify training data. These modified data are then used to train the regressors.

Partial least squares regression (PLSR): The covariance-based regression analysis technique known as partial least squares regression is a technique that is both quick and effective. It is helpful for regression issues that involve many independent variables and a high possibility of multicollinearity among the variables. This technique first reduces the number of variables to a more manageable number of predictors and then applies that smaller set of variables to a regression.

Elastic net regression: The Ridge and Lasso regression methods can be effectively combined into the elastic net regression model. This model is useful if the data is strongly correlated. Penalty functions are used to regularize the models associated with the Ridge and Lasso regression methods.

Stepwise linear regression: If there are large data sets that cannot be practically fit by the individual regression models, then the solution space is broken into different regions of data that can be fit and connected.

If more than one factor in an experimental design is expressible as a quantitative relationship with the observable output, then these factors can be used as predictors of the output. Regression models can be used to predict and visualize the output based on variations of the factors. Instead of plotting single lines as in the common case of a single factor (independent variable) and output response (dependent variable), two or more quantitative factors can produce *response surfaces* that can predict and visualize the output as a function of the two (or more) factors.

Optimization methods: Optimization problems abound in the systems engineering discipline. Finding an optimal solution to a design problem can take on many forms ranging from as simple as basic calculus, in finding the minimum, maximum, or saddle point of functions of interest, to statistical methods to optimally design an experiment as described above, or optimally control processes as in using SPC methods. The optimal solution is the best obtainable solution to a problem. For instance, if there is an objective function, this function is maximized (or minimized) when the optimal solution is reached. If there are a set of corresponding constraints, then we have constrained optimization whereas not including constraints leads to unconstrained optimization methods. There are a variety of methods to solve optimization problems such as linear and nonlinear optimization methods, genetic algorithms, linear programming methods, convex optimization, point optimization, global optimization, and particle swarm optimization. We will look at some of these methods in more detail in our Decision and Risk Analysis in Systems Engineering Chapter 7 later on.

6.1.7 Implementation Methodology and Procedures

The approach taken to implement systems engineering research methods closely resembles that of the scientific method. Although research methods can be interpreted differently across engineering disciplines, they all consist of the following: a problem statement, a hypothesis, an experiment, data analysis, and interpretation (Nunamaker et al. 1991). In systems engineering, developing a good problem statement is an essential initial step when conducting research. As previously mentioned, the systems engineer must consider the entire lifecycle. Who, what, when, where, and why are the five "Ws" that the problem statement must address. Stakeholder requirements must completely and accurately address the questions raised by the five "Ws" (Grady 2006). As an illustration, suppose a stakeholder specifies a need for an unmanned aerial vehicle (UAV) with a longer duration of flight than the current model that the system engineer had planned to implement. In this case, the problem statement derived from the requirement would be devising a method to extend the UAV's flight time.

A way to think about the meaning of a hypothesis is to make an educated guess as to the likely outcomes of the research activity. According to Wilcox, in Introduction to Robust Estimation and Hypothesis Testing, the hypothesis statement motivates the systems engineer to design an experiment to confirm or refute the hypothesis (Wilcox 2005). Remaining with the UAV as an example, a researcher may be presented with multiple choices on extending the UAV's flight duration. Possible solutions could include developing a new UAV design, increasing the UAV's fuel efficiency, or incorporating more fuel sources. Based on the systems engineer's best judgment, a hypothesis could be established on which approach would achieve the requirement in the most cost-effective manner.

Gathering data through research and experiments is the next step for the systems engineer. Finding previous experimental results or existing data that can answer the questions raised by the problem statement is a good way to start. Collecting information and data for use in modeling and/or simulation efforts is another kind of research (Panchal and Szajnfarber 2017). For instance, if the information sought is basic enough and readily available (such as looking up engine weights or comparing fuel efficiency), this and similar information can be found on a manufacturer's data sheets. However, this effective method of data collection, in most cases, is not always an option. Sometimes, it is necessary to conduct an experiment when the required data is unavailable (Panchal and Szajnfarber 2017). Though it is advisable and typical for systems engineers to enlist the support of domain experts, it is the systems engineer's task and responsibility to ultimately design the experiment (Antony 2014).

After collecting data from an experiment or some research activity, the data must then be analyzed. There are many choices available to the systems engineer when it comes to data analysis methods (Govindan et al. 2020). Various software tools are available for data analysis each with their pros and cons. When dealing with large datasets, Python models can be quite effective, especially over manual data analysis methods (Layton 2017). However, there are certain cases where hand analysis may be more efficient than software implementation, such as when the amount of data collected is little or when the input parameters are few and calculations are straightforward. Using the analytical hierarchy process method for comparing choices is also very effective. This method just requires a set of evaluation criteria and a set of alternatives and with a little elbow grease and some matrix manipulation capability like a spreadsheet, the optimal choice, based on pairwise rankings of the decision criteria, can be made.

Systems engineers draw conclusions from the experimental results and research by interpreting the data (Villacampa et al. 2018). Since the highest-ranked result may not always be the best overall solution when using the AHP method (such as in the presence of decision inconsistency), it is necessary to take an additional step before accepting the AHP results. Before accepting the highest ranked solution from the AHP analysis, the consistency ratio needs to be determined (Raharjo and Endah 2006). This is to make sure that the relative preferences between the evaluation criteria choices have been consistently made throughout the evaluation process. The results of the AHP analysis may prove to be erroneous and point to an undesirable alternative if the consistency ratio is more than 0.10.

6.1.8 Assessment Metrics

For establishing the validity of conducted research and experiments, assessment metrics are often the tools that are used (Abreu et al. 2022). Some of the assessment metrics such as various test statistics were presented earlier in the Design of Experiments section and are not repeated here. Some statistical assessment metrics that are worthy of some discussion are the *P*-value and the *R*-Squared metric (R^2).

P-value: For a hypothesis test, to determine whether or not the results of the test are significant, the *P-value* metric is commonly used. It can be used as an alternative to the fixed significance level method set by the α-level in a standard hypothesis test. The *P*-value is the probability that the test statistic has a value that is as extreme or greater than the observed value of the test statistic under null hypothesis assumptions. Since the α-parameter is typically used to set the significance level in a fixed significance level hypothesis test (the boundary between accepting or rejecting a null hypothesis), then the P-value in relation to the α-value represents the probability of the test statistic at the boundary of the null hypothesis having an equal to or more extreme result than that set by the observed point where the P-value is determined. Said another way, the α-value is the probability of having a Type I Error and, in the upper one-sided case for example, is the integral of the distribution function from the critical value to infinity. If the P-value is associated with a more extreme point, this point will lie to the right of the critical value and have a smaller integrated value (from the point of observation associated with the P-value to infinity). Consequently, if the P-value is smaller than the α-value, then there is significance in the result and the null hypothesis is rejected. On the other hand, if the P-value is greater than the α-value, then there is no statistical significance, meaning the observation point p associated with the P-value lies within the boundary established by the null hypothesis, and the null hypothesis cannot be rejected. The P-value for a normal distribution case can be determined by

$$P = \begin{cases} 2\left(1-\phi\left(|Z_0|\right)\right), & |Z_0| > Z_{\frac{\alpha}{2}} \\ \phi(Z_0), & Z_0 < -Z_\alpha \\ 1-\phi(Z_0), & Z_0 > Z_\alpha \end{cases}, \tag{6.132}$$

where $\phi\left(|Z_0|\right)$ is the cumulative distribution function evaluated at Z_0, and α is the α-value. For other distributions, Equation 6.132 can be used with the change that Z_0 is replaced by the appropriate distribution (e.g., $F_0, t_0, \ldots$).

Coefficient of determination (R^2): Another useful assessment metric is the coefficient of determination R^2, or R-squared, which roughly indicates the proportion of variation in the data that is explained by the model. For our ANOVA results, this can be given by

$$R^2 = \frac{SS_{\text{model}}}{SS_E}, \tag{6.133}$$

where SS_{model} in the numerator typically relates to $SS_{\text{Treatments}}$ in the ANOVA. This assessment metric falls in the range $0 \le R^2 \le 1$ with higher values of R^2 being better. Often, R^2 values greater than 0.5 are required to *believe* the results of the model. This value is chosen by the development team and is based on the problem at hand. For some applications, higher values of R^2 are required to trust the results of the experiment (e.g., $R^2 > 0.7$ can be common). Lower values of R^2 indicate that the model does not adequately explain the observed variations in the designed experiment, and that the experimental results are suspect. There are also different versions of R^2 that are used in other instances. For example, this assessment metric always increases when additional factors are added to an experiment even if these factors do not contribute much. To compensate for this, an adjusted R^2 value is used. The adjusted R^2 statistic applies if there are many factors in the experiment and is given by

$$R^2_{\text{adj}} = \frac{SS_E/\upsilon_E}{SS_T/\upsilon_T}, \tag{6.134}$$

where υ_E is the degree of freedom for the error, and υ_T is the total degree of freedom for the designed experiment. The adjusted R-squared value is good for experiments with a lot of factors. However, if the added factors don't contribute (are not significant), they can reduce the adjusted R-squared value. An R^2 assessment metric that looks at how well the model predicts new data is the Prediction Error Sum of Squares (PRESS) statistic used in the predicted R-squared value given by

$$R^2_{\text{Pred}} = 1 - \frac{\text{PRESS}}{SS_T}. \tag{6.135}$$

The PRESS statistic in Equation 6.135 is determined by calculating the sum of squares prediction errors by predicting the ith data point with a model that uses all the observations except the ith observation. The PRESS statistic is an indicator of how well the model can predict the results of a new experiment. If the PRESS statistic is small, then the indication is that the model is good at prediction. The PRESS statistic is an output in many modern statistical packages.

Coefficient of variation (CV): The standard deviation in an experiment is the square root of the error mean square. It can be related to the CV which is the standard deviation divided by the output mean $\overline{y}$,

$$\text{CV} = 100\left(\sqrt{\text{MS}_E}\,/\overline{y}\right), \tag{6.136}$$

which is given as percentage and reflects the residual variation with respect to the mean response variable $\overline{y}$.

Correlation coefficient (R): It is often of interest to see how two variables are correlated with each such as if one variable increases, does the other one increase, decrease, or have any relationship to the first variable. The Pearson product-moment correlation coefficient, or Pearson correlation coefficient r can provide this information. The Pearson correlation coefficient was invented by Karl Pearson who worked with Florence Nightengale David and is given by Field (2017),

$$r = \frac{\text{COV}_{xy}}{s_x s_y} = \frac{\sum_{i=1}^{n}(x_i - \overline{x})(y_i - \overline{y})}{(N-1)s_x s_y}, \tag{6.137}$$

where COV_{xy} is the covariance function of variables x and y, the parameters s_x and s_y are the standard deviations for the x and y variables respectively, $\overline{x}$ and $\overline{y}$ are the respective means for the x and y variables, n is the total number of paired sampled data points in the x and y variables (each variable has n sample points), and N is the total number of observations. Note that for the single sum paired case with a single subscript i in Equation 6.137, the number of observations N equals the number of paired data points n. The correlation coefficient r can take on values that range from $-1 \le r \le 1$. If r is -1, then the variables are negatively correlated and an increase in one variable causes a decrease in the other. Conversely, if r is 1, then the variables are positively correlated and an increase in one variable causes an increase in the other. A deterministic example would be if x is the periodic amount of money invested in a fixed interest investment, and y is the overall return on the investment. Say that the fixed interest rate is 7%, then if the periodic investment amount is increased, then the overall return on the investment is increased (assuming no fees or expenses). If the periodic amount invested is decreased, then the overall return is decreased illustrating a positive correlation between these two variables. An interesting point is that the square of the Pearson correlation coefficient is the coefficient of determination R^2 discussed above. By convention, the lower-case r is replaced by a capital R in the R^2 term.

6.1.9 Applications

Research methods in system engineering have been used in all sorts of different applications. When applied correctly, research methods generally provide a benefit. However, when misapplied, some dire consequences can result. We provide two short examples of the application of research methods to illustrate.

The Department of Defense's (DoDs) dealings with autonomous unmanned aerial vehicles (AUAVs) provide a prime illustration of how the department uses research methods to tackle tough problems. Jones, Djahel, and Welsh note that the development of AUAVs for DoD applications has received substantial research funding (Jones, Djahel, and Welsh 2023). The DoD noted that there were path-planning issues in enabling an AUAV to process data about its surroundings and find its way around potential dangers. According to Li and Shang, AUAVs have experienced persistent communication loss (Li and Shang 2022). Through the application of good research methods, researchers found that the communication loss was due to a communications delay resulting from environmental and meteorological factors. The study's findings mandated the inclusion of emergency communication software options in the program. According to Li and Shang, this would enable the AUAV to restore communications while preserving sensitive information if the emergency communications are turned off.

On the negative side, the Department of Defense's failure to properly investigate the materials and welding techniques employed by the US Liberty Ship Building Program during WWII is an example of a poor use of research methods (Zhang 2016). The United States entered WWII in the aftermath of the Pearl Harbor attacks, and the Liberty Ships program was launched to aid in the war effort. In less than 5 years after construction of more than 2,700 Liberty Ships began, more than a thousand damage reports were filed owing to brittle hull fractures. The hull cracks caused the loss of several ships, where some even split in half.

The hull damage in these ships was determined to be caused by two main factors: first, the steel used in their construction became more brittle due to the freezing temperatures, and second, the ships were welded instead of rivetted (Kobayahi and Onoue 1943). The steel used in the Liberty Ships became brittle at the critical ductile-brittle transition point, according to research on the material. Due to the steel's transformation from ductile to brittle in the frigid North Atlantic seas, the ships could crack with sufficient stress. This, however, was not the only factor.

While other ships did use the same steel as the Liberty Ships, they were not subject to the same cracking in frigid waters (Thompson 2001). The Liberty ships were welded into one massive metal structure, in contrast to the riveted ships, which were constructed from numerous smaller metal sections joined by rivets. As a result, a stress fracture in one area of the hull of the Liberty Ships might potentially extend to other parts of the vessel, unlike the riveted ships where such a crack would find it considerably harder to propagate. Early in the Liberty Ship program, the use of good engineering research methods likely took a back seat to meeting the war effort's strong demand for rapid output. However, the Liberty Ships Program was successful in the end because it was able to address the issues promptly; nevertheless, with better research methods, they likely could have prevented the loss of numerous ships and lives.

6.1.10 Case Studies

Research methods are used in a variety of ways in systems engineering and are important across all disciplines. There are many useful examples of good and bad research methods and engineering practices that can be instructive and provide useful lessons learned. In this section, we provide a national and international case study relating to the impact of research methods in systems engineering. For each of the case studies, we provide a short introduction to the case study scenario. This is followed by an overview of the case study that provides some top-level details. Next, we describe how the case study relates to systems engineering in general and provide some lessons learned.

6.1.10.1 National Case Study

Introduction: Electricity can be controlled by using electronic devices known as semiconductors. Semiconductors are fundamental to the operation of many contemporary electronic gadgets, including computers and electronic storage devices. The US semiconductor sector is far weaker now than it was in the past. Japanese companies supplanted American giants like IBM and Bell Labs in the semiconductor manufacturing industry because of foreign competition and a lack of domestic talent (Logar et al. 2014). A case study was carried out by the Semiconductor Research Corporation (SRC) to investigate potential solutions to this issue.

Many problems arose for the United States due to the loss of leadership in the semiconductor industry. One issue is the availability of chips during times of crisis where demand exceeds supply, the United States may not be able to obtain the chips that are manufactured elsewhere. The effects on large American corporations in the tech, telecom, and automotive sectors might be devastating. As an example, the pandemic's impact on the automotive industry's supply chain was demonstrated during the COVID-19 pandemic (Banerjee and Moore 2022). Security is another issue. The potential for tampering with semiconductors sent to us by foreign suppliers raises security concerns and may leave the United States and the Department of Defense vulnerable to cyberattacks (Rosenberg et al. 2022).

Overview: There have been many advancements in semiconductor technology due to the establishment of the SRC. Consequently, the semiconductor industry has kept up with Moore's Law, which was developed by Gordon Moore (Van Schoot 2023). Moore's law states that the density of transistors on a given surface will double every 2 years). In addition to pressuring leaders in the semiconductor industry to maintain pace with Moore's Law, the SRC also financed innovative research to advance semiconductor technology resulting in over 300 patents issued in the field of semiconductor research (Logar et al. 2014).

As an example, the Fin Field-Effect Transistor (FINFET) is one of the most cutting-edge inventions that will be supported by SRC research funds. The FINFET gets its name from the fact that the oxide forms a channel resembling a shark's fin. Reduced power consumption and increased switching speeds are both made possible by the FINFET. Since the transistor can accommodate multiple channels, or fins, it is also smaller.

Relation to research methods in systems engineering: SRC is committed to investing in both technology and the people that will drive the semiconductor industry sector forward. Those who are interested in working in semiconductors may also be able to get financial support from SRC to further their study (Logar et al. 2014). The SRC also provides advisory and leadership support. The Scientific Research Committee identified a deficiency in research conducted in the United States in a key technology area that could significantly impact both the U.S. business and government sectors (Logar et al. 2014). Research was used by the SRC to help drive innovation in semiconductor research and foster competition in the semiconductor industry. These research efforts were aimed at finding new methods for keeping up with Moore's law, and for advancing the state-of-the art of the semiconductor industry here in the US. Engineers can use research methods such as DoE, modeling, and simulation methods to optimize the testing process and to determine the placement of semiconductors without having to produce physical wafers or go through the manufacturing process of the chip (Greengard 2017). This can help chipmakers save significant time and costs while maintaining innovation.

Lessons learned: The nation faced troubling times in the semiconductor industry in the past due to neglecting research and allowing foreign competitors to gain the upper hand. This technology holds significance for both national security and the global market. The emphasis on research methods by the SRC enabled the US to establish a stronger position in the global semiconductor market, realizing an uptick in the US semiconductor market of 23.5%. Although the US sector is not as large as Japan's or China's sector, it is steadily growing and capturing a larger portion of the market.

6.1.10.2 International Case Study

Introduction: Worldwide, nonrenewable energy sources constitute the bulk of the energy market, leading some experts to predict an impending energy crisis (Feroskhan and Ismail 2017). Further, there is increased pressure on the United Nations and also the manufacturing sector regarding the effect of emissions on the environment and global warming in general. Other concerns about the environment include acid rain, water pollution, and too much carbon dioxide being released into the atmosphere. Responding to these environmental concerns, an Indian research team optimized engines and decreased emissions by employing DoE (Mahla et al. 2022). The DoE method was used because the researchers wanted to find which combination of input parameters would provide the best results.

Overview: Researchers in India aimed to find solutions to lessen the dangerous pollution in their country's air and water by using DoE methods on biogas. Since biogas may be used for both fueling renewable energy and disposing of biowaste, its potential applications piqued the researchers' interest. The Indian researchers received the following input parameters: Compression Ratio (CR), Biogas Flow Rate (BFR), and Engine Load (EL). Researchers were able to adjust the CR from 16:1 to 17:1, and 18:1 using modifications made to the engine, which had a nominal CR of 17.5:1. The rate at which researchers introduce biogas into the engine is called the biogas flow rate. A valve was inserted by the researchers to enable the injection of biogas, which is produced from cow manure and kitchen waste, into the engine. Normally, the engine works on diesel fuel. Three different biogas flow rates were introduced: 1.22, 2.2, and 3.22 kg/h. Engine load is the third input parameter investigated. To raise or lower the engine's output, a restrictive load bank was employed to increase the engine load. Different values for the engine load were tested, including 20%, 60%, and full throttle. The output parameters (responses) had five factors: Brake Thermal Efficiency (BTE) in percentage, Brake-specific Energy Consumption (BSEC) in megajoules per kilowatt-hour, hydrocarbons (HC) in grams per kilowatt-hour, carbon monoxide (CO) in volume percent, nitrogen oxides (NO_x) in grams per kilowatt-hour, and smoke in percentage. To measure the response, the researchers aimed to change the three input parameters (CR, BFR, and EL).

Relation to research methods in systems engineering: The Taguchi method, which is a well-known quality engineering technique and often used in systems engineering, was implemented by Indian researchers to analyze their experiment. The Taguchi method can be used to determine the optimal combination of input parameters by varying them against controlled parameters. With the Taguchi method, a "the larger the better" or "the lower the better" signal-to-noise ratio is determined by the experimental conditions. The software used to analyze their experiments was MINITAB.

Lessons learned: Experimental results showed that at low biogas flow rates and maximum CR and engine loads, BTE produced the best production (Mahla et al. 2022). Engine load, followed by biogas flow rate, and CR, had the least impact on emission responses, according to the results. The researchers found that the use of the Taguchi Method was a good way to analyze data and derive conclusions, which could help develop future engines with lower emissions (Mahla et al. 2022).

6.1.11 Transition to Optical Systems Building Blocks: The Statistical Nature and Propagation Characteristics of Light

In this section, we developed some insights into research methods that are useful in the systems engineering discipline. Several of the methods presented in this section used probabilistic and statistical underpinnings to provide the foundational framework to make them work. In the next section, we look at the statistical nature of light and gain some theoretical and practical insights into its propagation and detection characteristics.

6.2 THE STATISTICAL NATURE AND PROPAGATION CHARACTERISTICS OF LIGHT

In this section, we present an overview of the statistical properties of light that are useful in understanding the theoretical and mathematical aspects of light, how it propagates, and how it is characterized and detected. We present various useful mathematical and theoretical representations of light such as monochromatic, narrow-band, thermal, and quasi-monochromatic light that will permit the characterization and analysis of a wide range of useful optical systems engineering imaging and detection scenarios and applications. We adapt materials from the excellent texts, *Statistical Optics*, by Joseph W. Goodman for theoretical and mathematical underpinnings (Goodman 1985); Random Signals: Detection, Estimation, and Data Analysis by K. Sam Shanmugan and A. M Breipohl for characteristics of random processes (Shanmugan and Breipohl 1988); and Athanasios Papoulis for general probability and statistics where needed (Papoulis 1991). We use the logically straightforward notation of Goodman here and frame everything from a systems engineering point of view. For additional insights and further reading, please see these excellent texts. We assume the reader has a basic knowledge of statistics, probability, and random processes as presented earlier in Chapter 2, and in the mathematical appendix herein. We start with the basic case such as for a laser, wherein we describe light propagation at a single wavelength – monochromatic light. For monochromatic light, let $u(P,t)$ represent the scalar field of one polarization state of the electric or magnetic field. Mathematically and conceptually, the monochromatic optical wave can be represented by

$$\boldsymbol{u}(P,t)=\boldsymbol{U}(P,\nu)\exp(-j2\pi\nu t), \tag{6.138}$$

where $\boldsymbol{U}(P,\nu)$ is the phasor amplitude at point P and optical frequency ν. The optical electric or magnetic field in this phasor form has the phase given by the argument of the complex exponential, $-2\pi\nu t$. The actual value of the *analytic signal* representation of the electric or magnetic field at point P and time t shown in Equation 6.1 is obtained by taking the real value of this expression. Figure 6.24 shows the spatial orientation for source point P_1 and observation point P_o where the angle θ is between the normal to the scattering surface at point P_1, and the line connecting the scattering point P_1 to the observation point P_o. The distance between P_1 and P_o is r. The scattering angle θ is assumed small to moderate which is in line with many long-range imaging systems since the detector's clear aperture is typically far away from the scattering surface. At the observation point P_o, the observed phasor amplitude can be determined from the superposition of the many individual contributions from each of the scattering elements on surface Σ according to the Huygens-Fresnel principle. The Huygens-Fresnel principle states that provided that the distance r is much greater than the length of the wavelength of the optical radiation λ, the phasor amplitude at point P_o is determinable based on the superposition of the spherical waves that emanate from each of the source points along surface Σ. Every point that is on the surface Σ radiates with a strength proportional to $(j\lambda)^{-1}\boldsymbol{U}(P,\nu)$ at that point. At the point P_1, the strength of the optical radiation is $(j\lambda)^{-1}\boldsymbol{U}(P_1,\nu)$. The points other than at P_1 can be considered "secondary points" that radiate as point sources with the proportional strength shown above. At the observation point P_o, the contributions of all the spherically radiating points from the surface Σ are superimposed. These ideas are mathematically presented by the Huygens-Fresnel principle given by

$$\boldsymbol{U}(P_0,\nu)=\frac{1}{j\lambda}\iint_{\Sigma}\boldsymbol{U}(P_1,\nu)\frac{e^{j2\pi\left(\frac{r}{\lambda}\right)}}{r}\chi(\theta)\,dS, \tag{6.139}$$

where

$\lambda=\dfrac{c}{\nu}$ is the mean wavelength of the light,
c is the speed of light in the medium between the object surface Σ and observation point P_0,

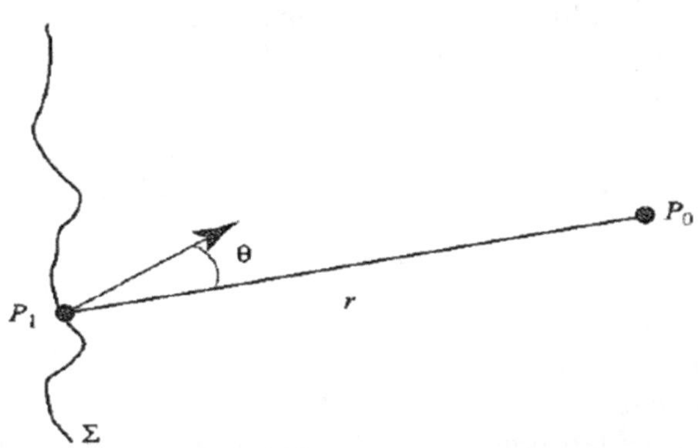

FIGURE 6.24 Spatial geometry relating scattering point P_1 to observational point P_0.

ν is the spatial frequency of the light,
r is the distance between P_1 to P_0,
θ is the angle between the line joining P_0 to P_1 and the normal to the surface Σ at P_1,
$\chi(\theta)$ is an off-axis "obliquity factor" with the properties $\chi(0)=1$ and $0 \le \chi(\theta) \le 1$.

In Equation 6.139, the integration is over the surface Σ which is in the scattering plane typically associated with the object. We refer to this as the object plane for conciseness. The obliquity factor is often close to one for long-range imaging scenarios since the collecting aperture of the imaging system is usually small compared to the separation distance between the object and the imaging system, and so only smaller angles of the θ can make it through the collection aperture (entrance pupil) of the imaging system.

We can extend the analytic signal representation to also represent non-monochromatic light. Even though in general, non-monochromatic optical radiation cannot be Fourier transformed, by truncating the optical radiation over the interval ($-T/2$, $T/2$) we can get an analytical signal $\boldsymbol{u}_T(P,t)$ that represents the optical radiation over that interval, and is Fourier transformable. This needful truncation is practical since many objects of interest in imaging scenarios have finite spatial extent and are well represented by a spatially truncated reflected optical field. Note that the truncation is required on the real part of the analytic signal and the imaginary component may stay unbounded. For the non-monochromatic case, a general expression for the truncated analytic signal for the optical wave can be obtained from analytic signal properties (Goodman 1985),

$$\boldsymbol{u}_T(P,t) = \int_0^\infty 2\mho_T(P,\nu)e^{-j2\pi\nu t}d\nu, \tag{6.140}$$

where $\mho_T(P,\nu)$ is the Fourier Transform of the real truncated signal $u_T(P,t)$. By using this analytic signal representation and the Huygens–Fresnel principle, along with some substitution and mathematical simplifications, we can get an expression of the truncated analytic signal at the observation point, at time t as

$$\boldsymbol{u}_T(P_0,t) = \iint_\Sigma \frac{(d/dt)\boldsymbol{u}\left(P_1,t-(r/c)\right)}{2\pi cr}\chi(\theta)dS. \tag{6.141}$$

This interesting expression for non-monochromatic light can be used to provide insight into a variety of useful light propagation conditions such as the propagation of narrow-band light, polarized thermal light, unpolarized thermal light, partially polarized thermal light, single-mode laser light, and multi-mode laser light which we now present.

Narrowband Light: Given that we have non-monochromatic light we know that we have the condition where the propagating light is not propagating at one wavelength but rather over a range of frequencies $\Delta\nu$. If this range of frequencies is small compared to the center frequency of the optical radiation $\bar{\nu}$, then the light can be classified as *narrow-band* light. If we let the truncation interval grow infinitely large ($T \to \infty$), then Equation 6.141 reduces to

$$\boldsymbol{u}_T(P_0,t) \approx \iint_\Sigma \frac{1}{j\lambda r}\boldsymbol{u}\left(P_1,t-\frac{r}{c}\right)\chi(\theta)dS. \tag{6.142}$$

Equation 6.142 provides the fundamental law of propagation for narrow-band signals and is readily implementable in computational and in simulation environments. We note that the above expressions are proportional to the electromagnetic field, whereas the eye and classical detectors respond to optical power and not directly to the field itself. To relate the electromagnetic field quantities to brightness quantities that are proportional to optical power, we first start with the definition of the instantaneous intensity (for point sources), or instantaneous irradiance (for extended objects). The instantaneous intensity (or irradiance) $I_x(P,t)$ is given by

$$I_x(P,t) = |\boldsymbol{u}_x(P,t)|^2, \tag{6.143}$$

where the instantaneous intensity (irradiance) at point P and time t is just the magnitude squared of the analytic signal representation for the instantaneous optical radiation (field) $\boldsymbol{u}_x(P,t)$. Since detectors (including human or other eyes) integrate the signal over time until there is enough signal present to elicit a detector response, we need to take the time average of Equation 6.143 to represent the time-averaged intensity or irradiance,

$$I_x(P,t) = \langle I_x(P,t)\rangle = \left\langle |\boldsymbol{u}_x(P,t)|^2 \right\rangle, \tag{6.144}$$

wherein the outer brackets on $\left\langle |\boldsymbol{u}_x(P,t)|^2 \right\rangle$ denote an infinite time average. Under ergodic conditions, the time average can be interchanged with a statistical average over random states of the analytic signal and so we can use the time average to estimate statistical quantities (Shanmugan and Breipohl 1988).

Thermal light: A very large number of optical sensors are thermal sensors. The physical principle behind these sensors is that they detect spontaneously emitted light resulting from many independent radiators. Consider for instance atoms or molecules that have been illuminated by a thermal energy source like the Sun and have their electrons raised to higher energy levels. When these electrons decay to a lower energy state and emit a photon in the process, this optical radiation is classified as *thermal light.* The large number of atoms or molecules radiate independently from each other and so because of the Central Limit Theorem, we know that thermal light has a Gaussian distribution. Consequently, we can use Gaussian statistics to determine statistical quantities of interest such as the temporal or statistical mean, variance, correlation, and propagation properties of the thermal light. Besides the Sun, incandescent bulbs, and gas discharge lamps are examples of thermal light generators. Also, as will be seen later, laser light that has gone through a diffusive element such as moving ground glass or reflected off a moving optically rough surface are examples of pseudo-thermal light generators and have similar, but not equivalent properties to thermal radiation sources. Laser light itself of course is not a thermal light source, and we will discuss laser light further on in this section.

Polarized thermal light: If we take a thermal radiation source and run it through an optical element capable of selecting a polarization state of the thermal optical radiation, then we have the case of *polarized thermal light.* An example of such a device is a polarization analyzer. Let's assume that thermal optical radiation passes through the polarization in the z direction and that the polarization analyzer passes the X component of the transmitted light while blocking the Y component completely. The passed linearly polarized thermal optical radiation can be represented by a real-valued function $u_X(P,t)$, whereas the blocked optical radiation in the Y direction is given by $u_Y(P,t) = 0$. Because thermal radiation sources radiate independently from each other, the passed linearly polarized light can be well represented by the sum of the X-components of each of the independently radiating elements

$$u_X(P,t) = \sum_{\substack{\text{all elementary} \\ \text{sources}}} u_i(P,t), \tag{6.145}$$

where the subscript i on the right side of Equation 6.145 denotes the X-component of the electric or magnetic field from the ith atom or molecule. Instead of directly working with the analytic signal itself, it can be useful to work with the complex envelope of the signal as follows:

$$A_X(P,t) = u_X(P,t)e^{j2\pi\bar{\nu}t}. \tag{6.146}$$

Here, $\bar{\nu}$ is the center frequency of the optical wave and the complex envelop can be equivalently represented by the sum of random, independent elementary sources

$$A_X(P,t) = \sum_{\substack{\text{all elementary} \\ \text{sources}}} A_i(P,t). \tag{6.147}$$

As in the case of the analytic signal $u_i(P,t)$ for the ith radiating element, with the assumptions that each radiating element is independent of each other and randomly phased, then both the X-component of the complex envelope $A_X(P,t)$ and the X-component of the electric (or magnetic) field component observed at point P are both circularly complex Gaussian random processes. For *circularly complex Gaussian random processes*, the real and imaginary parts of the random variables are independent, identically distributed, Gaussian random variables, with zero means and equal variances. We see from Equations 6.145 and 6.146 that the instantaneous intensity of the polarized wave (we drop the irradiance term with the understanding that if the context of the radiating element is a point source, then intensity is correct, and if the radiating element has finite extent, then irradiance is the appropriate term) can be obtained from either the analytic signal or complex envelope representations as follows:

$$I_X(P,t) = |u_X(P,t)|^2 = |A_X(P,t)|^2. \tag{6.148}$$

We can simplify this notation considerably and just remember that these quantities are linearly polarized functions of position and time

$$A \triangleq |A_X(P,t)| \quad I \triangleq I_X(P,t). \tag{6.149}$$

Using these abbreviations, we can write the PDF for the magnitude of the linearly polarized complex envelope which obeys a Rayleigh PDF

$$p_A(A) = \begin{cases} \left(\frac{A}{\sigma^2}\right)\exp\left(-\frac{A^2}{2\sigma^2}\right) & A \geq 0 \\ 0 & \text{otherwise} \end{cases}. \tag{6.150}$$

Here, σ^2 is the identical variance for the real and imaginary parts of the linearly polarized complex envelope $A_X(P,t)$. If we transform the linearly polarized complex envelope and the linearly polarized instantaneous intensity as $I = A^2$ and $A = \sqrt{I}$ we can then get the probability of the polarized instantaneous intensity as,

$$p_I(I) = p_A\left(\sqrt{I}\right)\left|\frac{dA}{dI}\right| = \begin{cases} \left(\frac{1}{2\sigma^2}\right)\exp\left(-\frac{I}{2\sigma^2}\right) & I \geq 0 \\ 0 & \text{otherwise} \end{cases}. \tag{6.151}$$

We know that is a negative exponential PDF and consequently the standard deviation of the polarized instantaneous intensity is equal to the mean of the random variable I, so the standard deviation for the polarized instantaneous intensity is,

$$\sigma_I = \bar{I} = 2\sigma^2. \tag{6.152}$$

Substituting these values into Equation 6.151 we get the following simplified expression:

$$p_I(I) = \begin{cases} \left(\frac{1}{\overline{I}}\right)\exp\left(\frac{-I}{\overline{I}}\right) & I \geq 0 \\ 0 & \text{otherwise} \end{cases}, \tag{6.153}$$

which is expressed solely in terms of the mean of the linearly polarized instantaneous intensity. Once we have the PDF, a host of useful statistical quantities are at our fingertips! For instance, the qth moment of the linearly polarized instantaneous intensity can be obtained from the basic definition of the statistical moment,

$$\overline{I^q} = \int_0^\infty I^q p_I(I)\,dI, \tag{6.154}$$

which can be reduced to the following expression:

$$\overline{I^q} = \left(\overline{I}\right)^2 q!. \tag{6.155}$$

From Equation 6.18, some useful statistical quantities include the mean (first moment), variance (second central moment), skewness (third moment), and Kurtosis (fourth moment). The central moments can be readily determined by first subtracting the mean $\overline{I}$ from I in Equation 6.154. For example, in general, the central moments are given by

$$\overline{I_c^q} = \int_0^\infty \left(I - \overline{I}\right)^q p_I(I)\,dI. \tag{6.156}$$

As an example, the variance (second central moment) is given by

$$\overline{I^2} = \int_0^\infty \left(I - \overline{I}\right)^2 p_I(I)\,dI, \tag{6.157}$$

where after the term in parentheses is multiplied out, Equations 6.154 and 6.155 can be used to reduce the expressions to terms just involving the mean of the linearly polarized instantaneous intensity.

Unpolarized thermal light: For the case of unpolarized thermal light, we still have all of the conditions for thermal light in the previous section holding true. Consequently, we still have the case where the orthogonal components of the analytic signal representation of the polarized wave, $\boldsymbol{u}_X(P,t)$ and $\boldsymbol{u}_Y(P,t)$ are circularly complex Gaussian random processes. For unpolarized light, we also have the following two conditions:

1. The transmitted light that is in a plane perpendicular to the direction of propagation of the light, must be independent of the rotational positioning of the polarization analyzer.
2. For orthogonal field components $\boldsymbol{u}_X(P,t)$ and $\boldsymbol{u}_Y(P,t)$, the expectation $\left\langle \boldsymbol{u}_X(P,t+\tau)\boldsymbol{u}_Y^*(P,t)\right\rangle$ must be zero for any X–Y positioning/orientation of the polarization analyzer and for any value of the delay τ.

In addition to being considered circular complex Gaussian random processes, the field components $\boldsymbol{u}_X(P,t)$ and $\boldsymbol{u}_Y(P,t)$ are also statistically independent, and consequently statistically uncorrelated. The instantaneous intensity of the optical wave therefore does not have a contribution from correlated cross terms and can be written solely based on the individual orthogonal field components as in

$$I(P,t)=|\boldsymbol{u}_X(P,t)|^2+|\boldsymbol{u}_Y(P,t)|^2=I_X(P,t)+I_Y(P,t). \tag{6.158}$$

Equation 6.158 can also be written in terms of the complex envelope wherein,

$$I(P,t)=|\boldsymbol{A}_X(P,t)|^2+|\boldsymbol{A}_Y(P,t)|^2=I_X(P,t)+I_Y(P,t). \tag{6.159}$$

Since we have circularly complex Gaussian random processes in effect, both orthogonal field components $I_X(P,t)$ and $I_Y(P,t)$ have the same means and so we can simplify the notation for the means as follows:

$$\bar{I}_X(P)=\bar{I}_Y(P)=\frac{1}{2}\bar{I}(P), \tag{6.160}$$

wherein the last term is the total instantaneous intensity, and each of these orthogonal field components are statistically independent random processes. If we now use a negative exponential PDF like Equation 6.16 for each of the independent orthogonal components of the instantaneous intensity and substitute the terms for the means according to Equation 6.160, we get,

$$p_{I_X}(I_X)=\left(\frac{2}{\bar{I}}\right)\exp\left(\frac{-2I_X}{\bar{I}}\right),\quad p_{I_Y}(I_Y)=\left(\frac{2}{\bar{I}}\right)\exp\left(\frac{-2I_Y}{\bar{I}}\right), \tag{6.161}$$

where $I_X \geq 0$ and $I_Y \geq 0$, and the component PDFs are zero otherwise. Since the component PDFs are independent, the joint PDF is just the product of the individual component PDFs and we can write the PDF for the total instantaneous intensity for unpolarized thermal light as

$$p_I(I)=\begin{cases}\left(\frac{2}{\bar{I}}\right)^2 I\exp\left(\frac{-2I}{\bar{I}}\right) & I\geq 0\\ 0 & \text{otherwise}\end{cases}. \tag{6.162}$$

If we use Equation 6.162 to determine the mean and standard deviation for the total instantaneous intensity, we see from Equation 6.152, that the ratio of the mean to the standard deviation for polarized light is one, whereas for unpolarized light it is reduced to $\sqrt{1/2}$ indicating that the likelihood of finding extremely small values of the total instantaneous intensity is much smaller for unpolarized light than it is for polarized light. Also, the ratio of the signal mean to its standard deviation is often interpreted as some form of a signal-to-noise ratio (SNR) and so we see that the polarized light has a slightly higher SNR over the unpolarized light using this metric. We now look at the case for partially polarized light.

Partially polarized light: In previous sections, we looked at the extreme cases of fully polarized light and completely unpolarized light. So, what happens if we have partial polarization? To understand this case, we find a matrix representation useful wherein we treat the instantaneous, orthogonal, polarized, field components as a vector. We also assume that we have narrow-band light assumptions prevalent wherein the spread of frequencies of the light $\Delta\nu$ is small compared to the center frequency of the optical radiation $\bar{\nu}$. If the electric or magnetic field components at time t and position P are represented by $\boldsymbol{u}_X(t)$ and $\boldsymbol{u}_Y(t)$, then we can generate a column vector from these field components as follows:

$$\underline{U}=\begin{bmatrix}\boldsymbol{u}_X(t)\\ \boldsymbol{u}_Y(t)\end{bmatrix}. \tag{6.163}$$

If these field components are passed through optical elements that can affect the polarization states of the field, then the field components at P' (the geometrical projection of P through the optical elements) are described by the following Jones matrix representation (Jones 1948),

$$\underline{U}' = \begin{bmatrix} u'_X(t) \\ u'_Y(t) \end{bmatrix} = \begin{bmatrix} l_{11} & l_{12} \\ l_{21} & l_{22} \end{bmatrix} \begin{bmatrix} u_X(t) \\ u_Y(t) \end{bmatrix} = \underline{L}\underline{U}. \tag{6.164}$$

Here, the output field in terms of its components is represented by $\underline{U}'$, the input field is given by $\underline{U}$, and $\underline{L}$ is a 2×2 *polarization matrix* or Jones matrix that represents the effects that the polarization-sensitive optical elements have on the input field components. This representation is quite useful in that many optical devices and operations on the optical field components can be characterized in terms of their Jones matrix elements. For instance, the rotation operation for polarized light, which rotates the orthogonal input field components in the X–Y plane, can be accomplished by a simple Jones matrix multiplication in Equation 6.164. Figure 6.25 shows a graphical representation of the rotation operation. Here both X and Y components of the field are rotated by the angle θ in the counterclockwise direction. The original field components $u_X(t)$ and $u_Y(t)$ are transformed into the outputs $u'_X(t)$ and $u'_Y(t)$ according to the rotation operator,

$$\underline{L} = \begin{bmatrix} \cos\theta & \sin\theta \\ -\sin\theta & \cos\theta \end{bmatrix}. \tag{6.165}$$

The quantity θ is the rotation angle. Another useful operation is the retardation plate which serves to delay one of the polarization elements with respect to the other. This can be used by employing birefringent materials. The effect is to cause a relative delay between the polarization states given by

$$\tau_d = d\left(\frac{1}{v_X} - \frac{1}{v_Y}\right), \tag{6.166}$$

where v_X is the velocity of propagation of the X-polarization component, v_Y is the velocity of propagation of the Y-polarization component, and d is the thickness of the birefringent material. The delay is between the polarization states of X and Y components. For the narrow-band assumption, the delay τ_d must be small compared to $1/\Delta\nu$ (the reciprocal of the optical bandwidth). Given this condition, the Jones matrix for the effect of a retardation plate is given by

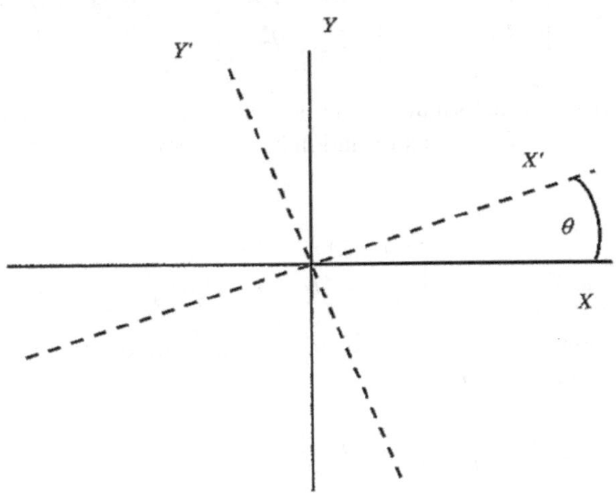

FIGURE 6.25 Rotation operation for orthogonal components of the optical field.

$$\underline{L} = \begin{bmatrix} e^{j\delta/2} & 0 \\ 0 & e^{-j\delta/2} \end{bmatrix}, \tag{6.167}$$

where the parameter δ is further defined as

$$\delta = \frac{2\pi dc}{\bar{\lambda}}\left(\frac{1}{v_X} - \frac{1}{v_Y}\right) = \frac{2\pi d}{\bar{\lambda}}(\eta_X - \eta_Y). \tag{6.168}$$

In Equation 6.31, $\bar{\lambda} = \frac{c}{\bar{v}}$ is the center wavelength of the optical radiation, c is the speed of light in vacuum, and η_X and η_Y are the refractive indices for the respective polarization components. From Equation 6.168, the respective refractive indices are the inverse of the propagation velocities of the individual polarization components.

Before proceeding on to some additional useful Jones matrix representations, we first make some comments and observations about lossless transformations. The two matrix operations that we just looked at, namely the rotation operation and the retardation operation have the property that they are both unitary matrices. This means that if you multiply either of these matrices by their Hermitian conjugate, you end up with the identity matrix $\underline{\mathfrak{I}}$. Mathematically this is expressed as $\underline{L}\underline{L}^{\dagger} = \underline{\mathfrak{I}}$, given that $\underline{L}^{\dagger}$ is the Hermitian conjugate of $\underline{L}$. The Hermitian conjugate of a matrix can be obtained by first taking the complex conjugate of each element of the matrix followed by taking the transpose of the resulting matrix. For a matrix

$$\underline{L} = \begin{bmatrix} l_{11} & l_{12} \\ l_{21} & l_{22} \end{bmatrix}, \tag{6.169}$$

the Hermitian conjugate is given by

$$\underline{L}^{\dagger} = \begin{bmatrix} l_{11}^* & l_{21}^* \\ l_{12}^* & l_{22}^* \end{bmatrix}, \tag{6.170}$$

Where the asterisk represents complex conjugation. Multiplying the matrix $\underline{L}$ by its Hermitian conjugate $\underline{L}^{\dagger}$ we get

$$\underline{L}\,\underline{L}^{\dagger} = \begin{bmatrix} l_{11} & l_{12} \\ l_{21} & l_{22} \end{bmatrix}\begin{bmatrix} l_{11}^* & l_{21}^* \\ l_{12}^* & l_{22}^* \end{bmatrix} = \underline{\mathfrak{I}} = \begin{bmatrix} 1 & 0 \\ 0 & 1 \end{bmatrix}. \tag{6.171}$$

This property applies to any transformation that is lossless, and so there is an easy and convenient test of any Jones matrix to tell if the transformation has loss or not. As an example, let's say we have a matrix

$$\underline{L} = \begin{bmatrix} 1+2*i & 2+3*i \\ 3+2*i & 3+3*i \end{bmatrix}, \tag{6.172}$$

That represents a Jones matrix in our optical system. To determine if this matrix is lossless, we first calculate the complex conjugate of $\underline{L}$ as

$$\underline{L}^* = \begin{bmatrix} 1-2*i & 2-3*i \\ 3-2*i & 3-3*i \end{bmatrix}. \tag{6.173}$$

Then, we calculate the transpose of $\underline{L}^*$ to get $\underline{L}^\dagger$

$$\underline{L}^\dagger = \begin{bmatrix} 1-2*i & 2-3*i \\ 3-2*i & 3-3*i \end{bmatrix}. \quad (6.174)$$

Notice that the off-diagonal elements are not zero, so this is not a lossless transformation matrix. An interesting property is if some of these operations need to be carried out in a sequence (such as if first the polarization components need to be rotated, and the one polarization state needs to be delayed with respect to the other one, then the combination of these Jones matrix operations is simply the product of the rotation matrices. In general, if the input field $\underline{U}$ passes through a sequence of instruments in the order $1 \rightarrow N$, that have Jones matrices $\underline{L}_1, \underline{L}_2, \ldots, \underline{L}_N$, then the effect of the combined instruments can be represented by a single Jones matrix,

$$\underline{L} = \underline{L}_N \cdots \underline{L}_2 \underline{L}_1, \quad (6.175)$$

The output field is given by

$$\underline{U}' = \underline{L}_N \cdots \underline{L}_2 \underline{L}_1 \underline{U} = \underline{L}\ \underline{U}. \quad (6.176)$$

Notice the reversal of order of the individual Jones matrices from the direction in which they were first encountered by the propagating field, and that the multiplication of the Jones matrices occurs to the left side of the field vector. As will be seen later, Equation 6.176 can be used to describe a range of useful polarization effects that an optical system can impart on an optical beam that propagates through it. We now look at some useful statistic optics metrics.

The coherency matrix: One useful metric for polarized light is the coherency matrix which provides insight into the polarization states of the optical radiation. The optical radiation can have deterministic properties and random fluctuations and the coherency matrix provides a useful method to capture these effects. The coherency matrix is defined as the time average of the field multiplied by its Hermitian conjugate

$$\underline{J} \triangleq \underline{U}\underline{U}^\dagger. \quad (6.177)$$

As before, the time average under ergodicity also provides the statistical average over the ensemble of statistical realizations. If the coherency matrix $\underline{J}$ can be expressed in terms of the individual polarized field components as

$$\underline{J} = \begin{bmatrix} J_{xx} & J_{xy} \\ J_{yx} & J_{yy} \end{bmatrix}, \quad (6.178)$$

then the infinite time average can be distributed to each of the individual elements of the coherency matrix as follows:

$$\begin{matrix} J_{xx} \triangleq \langle u_X(t)u_X^*(t) \rangle & J_{xy} \triangleq \langle u_X(t)u_Y^*(t) \rangle \\ J_{yx} \triangleq \langle u_Y(t)u_X^*(t) \rangle & J_{yy} \triangleq \langle u_Y(t)u_Y^*(t) \rangle \end{matrix}. \quad (6.179)$$

By inspecting the elements in Equation 6.179 and recalling the Schwarz inequality, we note that the determinant of Equation 6.178 is nonnegative (must be positive or zero). Also, the trace of the coherency matrix $\underline{J}$ gives the average time-averaged intensity

$$\mathrm{tr}\left[\underline{\boldsymbol{J}}\right] = \boldsymbol{J}_{xx} + \boldsymbol{J}_{yy} = \overline{\boldsymbol{I}}. \tag{6.180}$$

Let's now consider how the coherency matrix $\underline{\boldsymbol{J}}$ propagates through an optical system. If we start with Equation 6.176 and substitute this into Equation 6.177 while recalling that for Hermitian conjugation, $\left(\underline{\boldsymbol{LU}}\right)^{\dagger} = \underline{\boldsymbol{U}}^{\dagger}\underline{\boldsymbol{L}}^{\dagger}$, we find that after substitution and re-orientation, the center part of the result is just the coherency matrix and we have the following input/output transformation relationship for the coherency matrix

$$\underline{\boldsymbol{J}}' = \underline{\boldsymbol{LJL}}^{\dagger}. \tag{6.181}$$

This type of transformation occurs quite often in optics and deserves special note. We now can look at a variety of special coherency matrix forms that represent common polarization states. Equation 6.181 can be used to specify the output coherence matrix as a function of the Jones matrix operations given by Equation 6.175 for a particular optical system in question. The coherency matrix describing the different polarization states of the input optical field are given as

1. Linear polarization in the X direction,

$$\underline{\boldsymbol{J}} = \begin{bmatrix} 1 & 0 \\ 0 & 0 \end{bmatrix}, \tag{6.182}$$

2. Linear polarization in the Y direction,

$$\underline{\boldsymbol{J}} = \begin{bmatrix} 0 & 0 \\ 0 & 1 \end{bmatrix}, \tag{6.183}$$

3. Linear polarization at +45° to the X axis,

$$\underline{\boldsymbol{J}} = \frac{\overline{I}}{2}\begin{bmatrix} 1 & 1 \\ 1 & 1 \end{bmatrix}, \tag{6.184}$$

4. Right-hand circular polarization, for input polarization states,

$$\boldsymbol{u}_X(t) = \mathbf{A}(t)e^{-j2\pi\bar{\nu}t}, \qquad \boldsymbol{u}_Y(t) = \mathbf{A}(t)e^{-j[2\pi\bar{\nu}t+(\pi/2)]},$$

the coherency matrix is given by

$$\underline{\boldsymbol{J}} = \frac{\overline{I}}{2}\begin{bmatrix} 1 & j \\ -j & 1 \end{bmatrix}, \tag{6.185}$$

5. Left-hand circular polarization, for input polarization states,

$$\boldsymbol{u}_X(t) = \mathbf{A}(t)e^{-j2\pi\bar{\nu}t} \qquad \boldsymbol{u}_Y(t) = \mathbf{A}(t)e^{-j[2\pi\bar{\nu}t-(\pi/2)]},$$

the coherency matrix is given by

$$\underline{\boldsymbol{J}} = \frac{\overline{I}}{2}\begin{bmatrix} 1 & -j \\ j & 1 \end{bmatrix}. \tag{6.186}$$

As before, for the two cases of right and left circular polarization in items (4) and (5) above, the polarization components in the X and Y directions have a slowly varying over time envelope given by $\mathbf{A}(t)$. The condition for circular polarization applies to when the wave is observed in line with the direction of propagation. For the circular polarization condition, the vector field rotates at a steady rate as a function of time around the propagation axis (i.e., Z-direction for the X and Y polarized fields presented in our examples). An excellent metric to characterize the state of polarization of an optical wave is the *correlation coefficient* given by

$$\left|\mu_{xy}\right| = \frac{\left|J_{xy}\right|}{\left[J_{xx}J_{yy}\right]^{\frac{1}{2}}}, \tag{6.187}$$

which has a maximum value of one for perfectly correlated polarization components, and a value of zero for uncorrelated polarization components of the optical field. For both left and right circular polarized optical waves, the correlation coefficient is also fully correlated and so the correlation coefficient has a value of one. For *natural light*, the average intensities are equal in all directions just like circularly polarized light, however, the direction of polarization fluctuates randomly in all directions. For natural light, the analytic signal representation of the polarization components of the optical field, and the associated Jones matrix for natural light are given by,

6. Natural light, for input polarization states,

$$u_X(t) = \mathbf{A}(t)\cos\theta(t)e^{-j2\pi\bar{v}t} \qquad u_Y(t) = \mathbf{A}(t)\sin\theta(t)e^{-j[2\pi\bar{v}t-(\pi/2)]},$$

the coherency matrix is given by

$$\underline{J} = \frac{\bar{I}}{2}\begin{bmatrix} 1 & 0 \\ 0 & 1 \end{bmatrix} = \frac{\bar{I}}{2}\underline{\mathfrak{I}}, \tag{6.188}$$

where just as before, the symbol $\underline{\mathfrak{I}}$ refers to the identity matrix. For natural light, there is no cross-correlation of the polarization components as evidenced by the zero of the off-diagonal elements of Equation 6.188. The analytic signal representation of the input polarization components has the slowly varying amplitude term just as for circularly polarized light but there is also a slowly varying phase term illustrated by $\theta(t)$. Recall that for narrow-band light $\Delta t \ll 1/\Delta v$ so the envelope $A(t)$ and phase $\theta(t)$ vary slowly compared to the complex exponential term that has the center frequency of the optical radiation $\bar{v}$. For the coherency matrix in Equation 6.188, the phase $\theta(t)$ is assumed uniformly distributed over the interval $(-\pi, \pi)$. Another useful metric for partially polarized light is the degree of polarization which is intended to define to what degree an optical wave is polarized. The definition for the *degree of polarization* is obtained by the operation of dividing the intensity of the polarized component by the total intensity. Before, presenting this concept mathematically, we need to draw attention to another fundamental principle of Hermitian matrices. For Hermitian matrices, there exists a unitary matrix transformation that produces the real-valued eigenvalues of the Hermitian matrix. A unitary matrix transformation is a transformation in which the Hermitian properties of the matrix, multiplication and addition relationships between operators, and eigenvalues of the matrix are preserved. This unitary matrix transformation diagonalizes the matrix and produces the Eigenvalues of the matrix. If we use the Hermitian coherency matrix $\underline{J}$ for example, then there exists a unitary matrix transformation $\underline{P}$ so that the coherency matrix is diagonalized (has no off-diagonal components) and so produces the Eigenvalues of the coherency matrix

$$\underline{P}\underline{J}\underline{P}^{\dagger} = \begin{bmatrix} \lambda_1 & 0 \\ 0 & \lambda_2 \end{bmatrix}, \tag{6.189}$$

where λ_1 and λ_2 are the real-valued Eigenvalues of the coherency matrix $\underline{\boldsymbol{J}}$. Since the coherency matrix is nonnegative definite, so are its Eigenvalues. Recall that the lossy elements of an instrument show up as an off-diagonal component in the coherency matrix, so Equation 6.189 is saying that there is an instrument that must be possible that can remove the lossy effects from any optical wave. That instrument would be represented by the unitary matrix transformation $\underline{\boldsymbol{P}}$ in Equation 6.189 which has the effect of diagonalizing the coherency matrix $\underline{\boldsymbol{J}}$. With this preparation, we can define the *degree of polarization* as

$$\mathcal{P} \triangleq \frac{\lambda_1 - \lambda_2}{\lambda_1 + \lambda_2}, \tag{6.190}$$

which requires the determination of the Eigenvalues of the coherency matrix $\underline{\boldsymbol{J}}$. However, with a little bit of matrix algebra, we can write the degree of polarization in the original form of the coherency matrix as (Goodman 1985)

$$\mathcal{P} = \sqrt{1 - 4\frac{\det[\underline{\boldsymbol{J}}]}{\left(\mathrm{tr}[\underline{\boldsymbol{J}}]\right)^2}}, \tag{6.191}$$

where det and tr are the basic determinant and trace operations from matrix algebra. We can also express the average intensities for the polarized components in terms of the degree of polarization

$$\lambda_1 = \frac{1}{2}(1+\mathcal{P})\bar{I} \quad \lambda_2 = \frac{1}{2}(1-\mathcal{P})\bar{I}, \tag{6.192}$$

where the intensity of a partially polarized optical wave can be obtained from the sum of the uncorrelated individual components λ_1 and λ_2, and if the optical wave comes from a thermal source, then the average intensities of the two thermal field components with arbitrary polarization are given by

$$I_1 = \frac{1}{2}(1+\mathcal{P})\bar{I} \quad I_2 = \frac{1}{2}(1-\mathcal{P})\bar{I}. \tag{6.193}$$

For partially polarized light, since the underlying statistics of the optical field obey circularly complex Gaussian statistics, the component intensities I_1 and I_2 have negative exponential statistics wherein their PDFs are given by

$$p_{I_1}(I_1) = \frac{2}{(1+\mathcal{P})\bar{I}}\exp\left(\frac{-2I_1}{(1+\mathcal{P})\bar{I}}\right) \quad p_{I_2}(I_2) = \frac{2}{(1-\mathcal{P})\bar{I}}\exp\left(\frac{-2I_1}{(1-\mathcal{P})\bar{I}}\right), \tag{6.194}$$

wherein $I_1 \geq 0$ and $I_2 \geq 0$. Since I_1 and I_2 are independent of each other for thermal light

$$p_I(I) = \begin{cases} \dfrac{1}{\mathcal{P}\bar{I}}\left(\exp\left[-\dfrac{2I}{(1+\mathcal{P})\bar{I}}\right] - \exp\left[-\dfrac{2I}{(1-\mathcal{P})\bar{I}}\right]\right) & I \geq 0 \\ 0 & I < 0 \end{cases}, \tag{6.195}$$

where I is the instantaneous intensity of a thermal source, and the standard deviation for the partially polarized thermal source is determined to be

$$\sigma_I = \sqrt{\frac{1+\mathcal{P}^2}{2}}\bar{I}, \tag{6.196}$$

where $\mathcal{P}$ is the degree of polarization, and $\overline{I}$ is the average instantaneous intensity of the partially polarized thermal optical wave. We now look at the case of coherent light like in a laser. In this section, we treat the statistical nature of coherent light. Later in another chapter, we provide some fundamental deterministic aspects of lasers. We start this section with an idealized version of single-mode laser light, that of pure monochromatic light. With purely monochromatic light, the fundamental characteristics of the wave, amplitude S, frequency $\overline{v}$, and phase ϕ are represented as

$$u(t) = S\cos\left(2\pi v_0 t - \phi\right), \tag{6.197}$$

where the amplitude S and frequency v_0 are assumed known, and the phase ϕ is assumed fixed but not known. Since we are modeling a laser, the frequency v_0 is assumed to be well-defined and deterministic. The phase ϕ is regarded as a random variable that is uniformly distributed between $(-\pi, \pi)$. The random process associated with Equation 6.197 is both stationary and ergodic. To determine the PDF of the instantaneous amplitude U, the easiest way to do this is to first determine the characteristic function for the instantaneous amplitude U and then take its inverse Fourier transform (Papoulis 1991). We first set $t=0$ since we have a stationary process and so we get the characteristic function

$$M_U(\omega) = E\left[\exp\left(j\omega S\cos\phi\right)\right] = \frac{1}{2\pi}\int_{-\pi}^{\pi}\exp\left(j\omega S\cos\phi\right)d\phi = J_0(\omega S), \tag{6.198}$$

which is just the Bessel function of the first kind of order zero. The inverse Fourier transform of this function is (Goodman 1985),

$$p_U(u) = \begin{cases} \left(\pi\sqrt{S^2 - u^2}\right)^{-1} & |u| \le S \\ 0 & \text{otherwise} \end{cases}. \tag{6.199}$$

Note that this PDF tends toward infinity as u^2 tends toward S. The instantaneous intensity can be obtained by taking the magnitude squared of the analytic signal of Equation 6.197

$$I = \left|S\exp\left[-j\left(2\pi v_0 t - \phi\right)\right]\right|^2 = S^2, \tag{6.200}$$

which is a constant equivalent to the amplitude squared. The PDF of the instantaneous intensity can then be represented by a Kronecker delta function centered S^2

$$p_I(I) = \delta\left(I - S^2\right). \tag{6.201}$$

These results for the purely monochromatic form of light are theoretically and mathematically convenient, however, light encountered in nature doesn't follow this idealistic representation. We already looked at the narrow-band assumption wherein the center frequency is allowed to vary over a small interval around its center. We can also have time variations in the amplitude, and/or the phase, and we also often need to deal with additive noise. We will discuss how to properly model these things. We start with looking at the phase term $\theta(t)$ which is allowed to slowly vary in much the same fashion as the amplitude earlier (namely $\Delta t \ll 1/\Delta \upsilon$). We relax the conditions for the constant phase in Equation 6.197 and allow the phase to have time fluctuations

$$u(t) = S\cos\left(2\pi v_0 t - \theta(t)\right), \tag{6.202}$$

where the time variation is explicitly shown as a function of time in the phase term at the right. We can represent the oscillating mode of the laser as the argument of the cosine term in Equation 6.202 as follows:

$$\Psi(t) = 2\pi\nu_0 t - \theta(t), \tag{6.203}$$

where $\Psi(t)$ is the total phase of the oscillating mode. We can define the instantaneous frequency of the oscillating mode as the time derivative of the total phase, normalized by dividing out the 2π radial component,

$$\nu_i(t) \triangleq \frac{1}{2\pi}\frac{d\Psi(t)}{dt} = \nu_0 - \frac{1}{2\pi}\frac{d\theta(t)}{dt}. \tag{6.204}$$

By looking at the right side of Equation 6.204, we see that the instantaneous frequency is made up of a constant term ν_0, and a random term given by

$$\upsilon_R(t) = \frac{1}{2\pi}\frac{d\theta(t)}{dt}. \tag{6.205}$$

By integrating Equation 6.205 from $-\infty$ to t, and multiplying by 2π, we can get an integral expression for the phase term $\theta(t)$ as follows:

$$\theta(t) = 2\pi\int_{-\infty}^{t} \nu_R(\xi)\,d\xi. \tag{6.206}$$

For most applications dealing with single-mode laser light, the underlying physical processes of the laser cause the random term of the instantaneous frequency $\upsilon_R(t)$ to produce stationary, zero mean fluctuations. The phase fluctuations in $\theta(t)$ are generally non-stationary, although arguments can be made using the structure function to show $\theta(t)$ is stationary in first increments. The PDFs for the amplitude and intensity are the same as in Equations 6.199 and 6.201 since the phase is still uniformly distributed across $(-\pi, \pi)$. We now generalize Equation 6.202 once more by adding varying noise term $u_n(t)$ to Equation 6.202,

$$u(t) = S\cos\left(2\pi\nu_0 t - \theta(t)\right) + u_n(t). \tag{6.207}$$

The noise term $u_n(t)$ is assumed narrow band with $\Delta\upsilon \ll \nu_0$. If the noise term is considered small compared to the laser signal itself such as arising from residual spontaneous emission, then it can be modeled as independent from the stimulated emission term to its left and treated as having Gaussian statistics. We already know the PDFs for the amplitude and instantaneous intensity for the first term to the right of the equal sign in Equation 6.207. The effect of the noise term on the amplitude PDF is to slightly smooth the PDF in Equation 6.199 if the signal strength of the laser is much higher than the noise. For the intensity statistics, we have the magnitude squared of a strong constant amplitude, random phase phasor summed with a weak circularly complex Gaussian random variable

$$\boldsymbol{I} = |\boldsymbol{S} + \boldsymbol{A}_n|^2, \tag{6.208}$$

where S and A_n are the analytical signal representation of the laser signal and the Gaussian noise term given by

$$\boldsymbol{S} = S\exp(j\theta), \quad \text{and} \quad \boldsymbol{A}_n = A_n\exp(j\varphi), \tag{6.209}$$

and the amplitude S is a constant value, and the phases θ and $\varphi$and the noise amplitude A_n are independent of each other. The phases each are uniformly distributed across $(-\pi,\pi)$. The Gaussian noise term has zero mean so the variance of the instantaneous intensity for a signal represented by Equation 6.207 is given by

$$\sigma_I^2 = 2I_S\bar{I}_N, \tag{6.210}$$

where I_S is the instantaneous intensity for the signal, and $\bar{I}_N$ is the average noise intensity due to the narrow-band additive Gaussian noise. The PDF for the instantaneous intensity for a signal that is represented by Equation 6.207 is given by

$$p_I(I) \approx \frac{1}{\sqrt{4\pi I_S\bar{I}_N}}\exp\left[\frac{(I-I_S)^2}{4I_S\bar{I}_N}\right]. \tag{6.211}$$

Knowing this and the previous PDF provides significant insight into the statistical properties of the PDF (e.g., the determination of all the moments of the instantaneous intensity for instance). We now make one more generalization in relaxing the requirement for a single mode.

Multimode laser light: Many lasers can operate in single mode given sufficient design considerations. However, there are many lasers that operate over multiple transverse and longitudinal modes. If the laser is operating at steady state in a mode far above its threshold, then we can model the multimode output signal at point P and time t as a real-valued signal given by

$$u(P,t) = \sum_{n=1}^{N} S_n(P)\cos\left[2\pi v_n t - \theta_n(P,t)\right], \tag{6.212}$$

where N is the number of modes, $S_n(P)$ is the amplitude of the nth mode at point P, v_n is the center frequency of the nth mode, and $\theta_n(P,t)$ is the slowly varying phase relative to the nth mode at point P and time t. The easiest way to get to the PDF expression is by first determining the characteristic function. For a single-mode laser, the characteristic function is,

$$M_n(\omega) = J_0(\omega S_n). \tag{6.213}$$

If the modes are independent, then the characteristic function for N multiple modes is just the product of the individual characteristic functions

$$M_U(\omega) = \prod_{n=1}^{N} J_0(\omega S_n). \tag{6.214}$$

Further, if the modes are all equal strength, then the characteristic function for the multi-mode, equal strength laser is

$$M_U(\omega) = J_0^N\left(\omega\sqrt{\bar{I}/N}\right), \tag{6.215}$$

where each mode has strength $\sqrt{\bar{I}/N}$. To obtain the PDF, the amplitude characteristic function in either Equation 6.214 or 6.215 (for equal amplitudes) must be inverse Fourier transformed. Analytically, this can be quite cumbersome, and a better approach is to computationally solve the inverse Fourier transform of Equation 6.214 or 6.215. If there are as few as five independent modes, with no phase locking of the laser light, then thanks to the Central Limit Theorem (CLT), the laser light looks like thermal light with a Gaussian distribution. The probability function for the intensity statistics is more difficult than for the amplitude. Goodman has worked out the details for the characteristic function of the instantaneous intensity for two modes (Goodman 1985; pp. 148–149). Luckily for more than five modes, a thermal light approximation works quite well. Analytically, the characteristic function for two modes is given by

$$M_I(\omega) = \exp\left(j\omega\bar{I}\right) J_0\left(2\omega\bar{I}\sqrt{k(1-k)}\right), \tag{6.216}$$

where k is the fraction of the total intensity in one mode. The inverse Fourier transform produces the probability transfer function for the instantaneous intensity for two modes as

$$p_I(I) = \begin{cases} \left[\pi\sqrt{\left(2\bar{I}\sqrt{k(1-k)}\right)^2 - \left(I-\bar{I}\right)^2}\right]^{-1}, & \text{for} \quad \bar{I}\left[1-2\sqrt{k(1-k)}\right] < I < \bar{I}\left[1+2\sqrt{k(1-k)}\right] \\ 0 & \text{elsewhere} \end{cases}. \tag{6.217}$$

The PDF for more than two modes is difficult to find. However, it is possible to determine the standard deviation for the instantaneous intensity as a function of the number of modes N, for equal strength modes that are independent from each other. This standard deviation is given by

$$\sigma_I = \bar{I}\sqrt{1-\frac{1}{N}}, \tag{6.218}$$

where $\bar{I}$ is the mean intensity. For five or more modes, the thermal light approximation leads to more tractable solutions for the PDF of the instantaneous intensity. One other consideration is that of pseudo-thermal light which we now discuss.

Pseudo-thermal light: Pseudo-thermal light is obtainable when a laser is transmitted through a moving diffusive element such as rotating glass. The laser, whether single-mode or multi-mode approximates polarized thermal light but has a much narrower bandwidth than thermal light. If the diffusive element is rough on the scale of the optical wavelength, the spatial coherence of the laser is destroyed and the phase of the laser amplitude after the diffusive element can undergo large phase shifts. In an observation plane far away at an observation point P_o, the observed optical field consists of the summation of randomly phased contributions that obey Gaussian statistics. The intensity obeys negative exponential statistics. If the diffuser rotates, then the fluctuates vary randomly in time and polarized thermal light is approximated. In the next section, we present an application of the statistical methods developed so far, to the optical sensing system being developed by Futuristic Imaging Technologies. Here some of the statistical methods learned in this section are applied to characterize and analyze the expected light at the optical sensing element, and systems engineering research methods are used to analyze the results.

6.3 INTEGRATED CASE STUDY: STATISTICAL ELEMENTS OF SENSOR DESIGN

In this section, we find the FIT design team at an internal Technical Interchange Meeting (TIM) to discuss how to ensure that they are getting good test data to show that their prototype electro-optical (EO) sensor is performing as required. The discussion centers around a set of EO sensor lab data that was generated to mimic operational conditions under a controlled environment and field data that was collected. At the meeting are Tom Phelps, FIT's Chief Technical Officer, Karl Ben, the Senior Systems Engineer, Jennifer O., systems engineer, Amanda R., Optical Engineering, George B., Hardware Specialist, Phil K., Software Engineering

Tom: "I am glad everyone is here and that the EO sensor prototype appears to be working well. Can you give me an update on the overall sensor tests? Amanda?"

Amanda: "Sure Tom. George finished the deployable prototype EO sensor and Phil has the atmospheric turbulence compensation (ATC) module going. We ran our simulation models for the prototype at nominal conditions, noisy conditions and for all the different Use Case scenarios and checked that the simulation results matched our theoretical predicted results and satisfied all our baselined requirements. We then took the EO sensor with the ATC module into the test room and simulated the sensor acquisition profiles for the various Use Case scenarios under simulated environmental conditions. We could control the ambient temperature, pressure, humidity, and could introduce various types and levels of sensor and system noises including sensor jitter and shock. We then compared the theoretical, simulation, laboratory results to make sure the data was similar."

Tom: "That sounds like we are off to a great start! I noticed we had theoretical predictions, simulated performance predictions, and controlled laboratory results, but what about real data results like experimental field data? Have we collected any of that and compared it to what we have?"

Karl: "Tom, I am glad you asked! The short answer is, yes, we have. And we made darn sure our results are right, believable, and justifiable. I will let Jennifer explain the process."

Tom: "Can't wait to hear it."

Jennifer: "Thanks Tom and Karl. The systems engineering, design, and test teams together developed designed experiments to make sure that we would conduct a robust, valid, and justifiable EO sensor data collection experiment that we could use demonstrate that our EO sensor and ATC module are working and that the results are believable."

Tom: "I am liking what I am hearing. Go on!"

Jennifer: "We used several design of experiments statistical methods to get objective insight as to how well our EO sensor was performing. The laboratory EO sensor data we collected served as our truth data since it modeled the sensor performance well enough to satisfy stated requirements for all Use Cases. The experimental data we collected was also collected under similar environmental conditions to our truth data. Since all the Use Case scenarios and requirements were met with our truth data, if our field results also matched our laboratory data, we would have consistency between theoretical predictions, simulated predictions, laboratory results, and actual field data. So, we devised a series of comparison tests using statistical methods to see if our laboratory and field results were equivalent for equivalent sensor environmental conditions and see if our results were believable in a statistical sense."

Tom: "What kind of tests did you do?"

Jennifer: "Well, we looked at all the Use Case scenarios for all of the different mission profiles such as concealed target detection, target tracking, human target identification, and so forth and developed the theoretical and simulated sensor parameters for the corresponding Use Case scenario. We included nominal, optimal, and "rainy day (degraded)" simulated test profiles and generated simulated test data that serves as our "truth" data."

Tom: "So we covered all the expected types of missions according to the stakeholders ConOps? That sounds like a lot of data."

Jennifer: "Yes, that is correct. We then developed the truth data for our sensor package that aligns with each of the different types of missions that need to be supported. For example, for a material identification type of mission, we would use the hyperspectral sensor to determine what type of materials were present in the scene. We then generated truth data for our hyperspectral sensor using our theoretical and simulated results under nominal, optimal, and rainy-day conditions."

Tom: "What do you mean by rainy day conditions?"

Jennifer: "That is an expression used for sub-optimal conditions such as bad seeing where we have a small Fried parameter in the atmosphere. Or, unanticipated jitter due to rough turbulence for instance. We can even conduct sensor performance assessments and predictions based on partial system failures to check out our redundant systems. The idea is to not just test our sensor when we assume everything is going as well as possible, but also look at some of the limiting cases."

Tom: "Well, that is excellent. How do we know we can believe our results?"

Jennifer: "Excellent question Tom. When we do our sensor performance predictions, we start with a theoretical analysis for what we expect as a sensor output for the various Use Case scenarios. We check these theoretical predictions, among other things, in our MBSE implementation. We have a digital twin for all the different sensors in our sensor package for this project that allows us to conduct assessments and predictions for how these sensors will operate across a wide range of scenarios and operational conditions. Many theoretical and operational constraints are built into our MBSE environment with consistency and constraint checks automatically enforced and linked to requirements at all levels. We have verified and validated our models and simulation tools and have high confidence in their predictions. That said, as George Box has said, "All models are wrong, some are useful." Ours are useful in providing believable truth data. Even so, the sensor package needs to be experimentally verified and validated, and so the sensor data from our laboratory serves as the truth data for actual sensor experimental results."

Tom: "Ah, then we will have experimental evidence to support our theoretical, and simulated results?"

Jennifer: "Naturally! There is one other thing. We need to design the experiment and gather statistical evidence that supports our sensor performance conclusions very carefully. We have to make sure that we don't introduce any biases into our results and that we use correct sampling techniques. We will use design of experiments methods and set up a completely randomized design for the collected experimental field data. We will then use statistical analysis to infer how well our sensor package performed under each of the different Use Case scenarios. We will use hypothesis testing and compare the collected field data to the laboratory truth data and conduct the appropriate statistical checks to make sure our data is valid in a statistical sense."

Tom: "Excellent! Let's go over a few of these tests to make sure we are on the right track."

Jennifer: "Absolutely! Let's start with the concealed target detection Use Case. One of the Use Case scenarios for this is to use our hyperspectral line sensor to check for abrupt changes in the material, and then compare it to some material types that we are likely to encounter. For instance, a brown camouflage tarp that matches the surrounding landscape that is stretched over a hole that conceals targets or items of interest. I will let George tell you about the details of the hyperspectral sensor and what we are comparing."

George: "Thanks Jennifer. We can run the hyperspectral imaging sensor in a push broom mode using only one line of pixels. This substantially reduces the processing time since we don't have to actually image the object of interest in this Use Case scenario. All we need to do is to check for abrupt changes in material over a consistent area and, when found, analyze the changes to see if it matches some class or type of material. Ultimately, what comes out of the sensor at the back end is detector current. We want to compare the sensor output current of the experimental data to that which we generated in the lab for a given target set like camouflage tarp. For the hyperspectral line-scan application, we have a 256-pixel linear array. This corresponds to 127 spatial samples in the entrance pupil of the optical system. We are collecting data at 127 distinct wavelengths with the hyperspectral sensor. Some summary data that we have from the lab is that the total expected output mean current is 76.8 nA per sample under uniform illumination conditions. Under constant uniform illumination conditions, we expect 4.76 pA per pixel for an initial baseline wavelength. We generated spatial output current data per pixel as a function of wavelength for the constant illumination across all the pixels to determine the uniformity of the pixel response as a function of wavelength. For uniform illumination, we expect the output current to be the same within statistical limits for each pixel. The response of course will be different for different wavelengths but if all the pixels for a given wavelength are seeing the same irradiance levels, then the output current should be the same to within statistical limits. Jenn can tell you more about the actual data collect and the subsequent statistical analysis."

Jennifer: "Thanks George. As George said, for the pixel uniformity test we collected a set of simulated truth data. We then set up a field experiment where we placed our prototype sensor onto a test UAV and collected sensor data according to our completely randomized design. We set the testing conditions for the field data to match those in the Use Case scenario that we ran in the laboratory and collected the data. We then did an Analysis of Variance (ANOVA) analysis using an *F*-test statistic to compare our field data with the laboratory data."

Tom: "Why did you use an *F*-statistic for the test?"

Jennifer: "Because we are conducting an equality of means ANOVA on a single factor, with a fixed effects model, using a completely randomized design."

Tom: "That's correct, go on."

Jennifer: "As I said, we used a single factor, fixed effects model for the ANOVA analysis comparing the output current for each pixel as our single factor. The wavelengths were our different treatment levels. For a single factor multiple treatment levels under the equality of means null hypothesis assumption with completely randomized design conditions, the appropriate test statistic is the *F*-statistic. Consequently, we compared the field data to our truth data using these assumptions. We assumed that the mean of our laboratory data was equal to the mean of our field data for all treatment levels (all different wavelengths). Our null hypothesis was then,

$$H_0 : \mu_1 = \mu_2, \tag{6.219}$$

where μ_1 is the mean of the laboratory truth data, and μ_2 is the means of the sampled experimental field data. The alternate hypothesis is then that the means are not the same, or,

$$H_o : \mu_1 \neq \mu_2. \tag{6.220}$$

We then ran the ANOVA with a significance level of 0.05, meaning that if the computed *p*-value exceeds 0.05 there is no significant difference between the means of the datasets. The output of the ANOVA is shown below (Table 6.9):

Given that the test resulted in a *p*-value of 0.066089, the null hypothesis can be accepted. This means that there is no statistically significant difference between the data collected by the sensor in the lab setting and in the field setting. Therefore, our hyperspectral field data matches our laboratory truth data for this uniform illumination test case. We also tested and compared many other Use Case scenarios as I said but the test all came back as expected."

Tom: "Well that is great news! How about EO/IR sensors with the atmospheric turbulence compensation module? Did we test the Use Case scenarios that require these sensor modalities?"

Jennifer: "Yes, we tested these too. George, do you want to run down the sensor descriptions for EO/IR Use Case scenarios?"

George: "Happy to Jenn!" For the electro-optic (EO) and infrared (IR) Use Case scenarios, most of the Use Cases fell in line with the high spatial resolution imaging application and so the atmospheric turbulence compensation module is a critical element for these examples.

TABLE 6.9
ANOVA Results

Anova: Single Factor						
		SUMMARY				
Groups	Count	Sum	Average	Variance		
Column 1	16129	7.68E-08	4.76E-12	3.39E-27		
Column 2	16129	7.68E.08	4.76E-12	3.34E-27		
			ANOVA			
Ce of Variable	SS	Df	MS	*F*	*P*-Value	*F* Crit
Between G	1.14E-26	1	1.14E-26	3.377801	0.066089	3.841747
Within Grc	1.08E-22	32,256	3.36E-27			
Total	1.08E-22	32,257				

We followed the previous process in generating laboratory truth data that was anchored to different Use Case scenarios. In this case, this is the high spatial resolution daytime, or nighttime imaging Use Cases for target detection, target classification and/or identification, with possible tracking. I will go over the basic human target detection Use Case for the daytime sensor. For daytime imaging, we use the visible EO imaging system with the optional atmospheric turbulence compensation module, so this is what is being tested and compared. For this Use Case scenario, the laboratory average detector per pixel output current was determined to be 92 pA. We also have a noise floor in the operational sensor of about 20 pA per pixel. For this imaging application, we are looking at one wavelength, grayscale imagery and we use a 256 by 256 image segment. This corresponds to 127 by 127 entrance pupil plane samples in the vertical and horizontal directions of the imaging systems entrance pupil. Jenn, I will let you talk about comparing the laboratory and field data in the presence of this constant 20 pA noise term."

Jennifer: "Sure thing George! In this case, if we perform a paired hypothesis test that looks at the difference in means but sets a constant threshold of 20 pA, we have a null hypothesis of the difference in means set to less than or equal to the constant value of 20 pA,

$$H_o : \mu_1 - \mu_2 \leq 20 \text{ pA}, \tag{6.221}$$

with an alternative hypothesis of,

$$H_o : \mu_1 - \mu_2 \nleq 20 \text{ pA}. \tag{6.222}$$

If the laboratory and field experimental data are the same, then the statistical difference of an acceptable solution of the difference in means is less than the 20 pA noise floor. We again carefully set up our sampling plan so that we had a completely randomized design with fixed effects. For this pixel uniformity test, constant illumination at a given wavelength was provided to both the laboratory and field EO sensors and the appropriate output pixels measured. The appropriate test statistic for the difference in means hypothesis test is the *t*-test statistic. The results of the hypothesis test analysis are shown in Table 6.10 for the comparison of the IR lab results and field data. For this test, we had a critical *t*-value of 1.65, which relates to a *p*-value of 0.051, the null hypothesis can

TABLE 6.10
Hypothesis Test Comparison of Lab (Truth) Data, and Experimental Flight Data for IR Sensor Uniformity Test

***t*-Test: Paired Two Sample for Means**

	IR Lab Data	IR Field Data
Mean	8.66076E-11	9.89202E-11
Variance	1.45232E-23	2.72418E-23
Observations	16,129	16,129
Pearson correlation	−0.001000828	
Hypothesized mean difference	20	
df	16,128	
t Stat	−3.92845E+14	
$P(T \leq t)$ one-tail	0	
t critical one-tail	1.644948112	
$P(T \leq t)$ two-tail	0	
t critical two-tail	1.960111086	

TABLE 6.11
Hypothesis Test Comparison of Lab (Truth) Data, and Experimental Flight Data for EO Sensor Uniformity Test

***t*-Test: Paired Two Sample for Means**

	Optical Lab Data	Optical Field Data
Mean	1.48741E-12	4.30041E-12
Variance	2.54498E-27	2.97797E-26
Observations	16,129	16,129
Pearson correlation	0.005883439	
Hypothesized mean difference	20	
df	16,128	
t Stat	−1.415E+16	
$P(T \leq t)$ one-tail	0	
t critical one-tail	1.644948112	
$P(T \leq t)$ two-tail	0	
t critical two-tail	1.960111086	

be accepted. This means there is no statistically significant difference between the data collected by the infrared sensor in a lab setting and a field setting. We also conducted the same uniformity experiment for the EO sensor in the visible part of the electromagnetic spectrum and the results are shown in Table 6.11.

The EO sensor test also resulted in a critical *t*-value of 1.65, which again relates to a *p*-value of 0.051, and the null hypothesis can be accepted. This means there is no statistically significant difference between the data collected by the optical sensor in a lab setting and a field setting. We didn't show it in the chart above, but we have an adjusted R^2 value that is greater than 0.7 and checked our normality assumption of the data by plotting a normal probability plot. This is standard operating procedure to make sure our results are valid."

Tom: "Excellent! What about the ANOVA analysis, did you plot a normal probability plot there too?"

Jenn: "Yes, but for the ANOVA, we plotted the normal probability plot of the residuals as well and ensured that the variances were the same in all the treatment levels, whereas for the *t*-test, we plotted the normal probability chart for the data itself to check the normality assumption. We also plotted the residuals in time sequence and didn't see any runs of positive or negative residuals. Also, the residuals for our ANOVA were structureless indicating that our assumption of independence of our data is valid."

Tom: "Very nice! Anything else?"

Jenn: "Yes, we also did some regression analysis on the data to understand some of the sensitivities of our sensors."

Tom: "Great! Tell me about it."

Jennifer: "We conducted a set of regression analysis to see if our experimental data was matching what we obtained in the lab, and to see how much variation we were getting in our experimental data. Since spatial resolution was so important for most of our imaging Use Case scenarios, we investigated factors that would affect the spatial resolution such as wavelength variations, distance differences between the target and optical system, and changing atmospheric conditions as characterized by the atmospheric coherence length (Fried parameter). We mounted our EO/IR ATC sensing system on a Reaper that we managed to borrow from Homeland Security and set up a designed experiment to compare both lab and field data sets. Let me bring up the regression analysis results. You will see something interesting."

[Jenn clicks the remote and the first chart (Figure 6.26) pops onto the screen.]

Jenn: "Here we see our laboratory predictive results showing the mean of the expected spatial resolution of our lab data in orange. As expected, the spatial resolution gets progressively worse as the wavelength increases. Our field data is shown in blue with the dashed blue line giving the best linear fit to our field data."

Tom: "I see the variation in the field data, but why does the laboratory truth data not have any variations?"

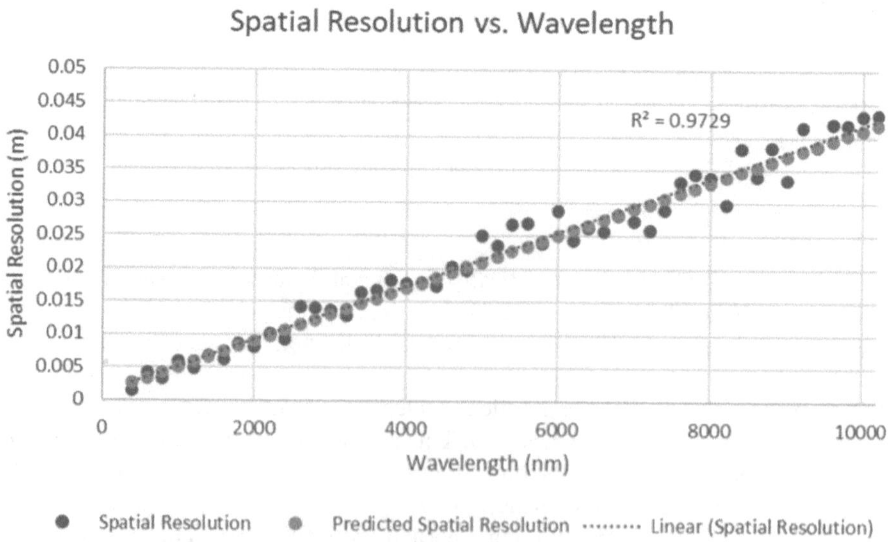

FIGURE 6.26 Regression analysis plot for wavelength variations.

Jenn: "That is a good point, Tom. We are only plotting the mean in the lab data since we are not flying the development article, only a fieldable prototype, and so we don't really have all the noise cancellation, and vibration isolation infrastructure in place. Consequently, our fielded system is expected to have higher system noise and vibrations than our developed system will have. We just wanted to see how the system scales with respect to the three design parameters (wavelength, sensor-target separation distance, and Fried parameter). That said, you can see that there is an extremely good fit between our laboratory "truth" data and the experimental results as indicated by the high R^2 value in the regression results for the above chart. The means of the experimental data fits the means of the lab data pretty well. Also, the data variations were in line with what we expected to see, and overall, we were pretty happy with these results. Next, we looked at the variation of sensor-to-target distance. You can see this in the next regression plot."

[Jenn brings up the next plot (Figure 6.27) on the monitor.]

Jenn: "Here the Reaper was flown at a slant distance to a fixed target of interest on the ground. The flight path was chosen so that the sensor could look at the target at the same angle but at progressively further separation distances. Again, the sensor data scales as expected in that as the sensor-to-target separation distance increases, the spatial resolution gets worse. Based on the high R^2 value, this also is a good fit for the regression analysis indicating a good match between laboratory and field results. Finally, let me bring up our last result."

[Jenn, once again brings up a regression analysis plot (Figure 6.28).]

Jenn: "Here we found something interesting. The atmospheric coherence length changes as a function of altitude, and can change as a function of height, time of day, and change from day to day. We decided to instrument the Reaper to gather data to estimate the atmospheric coherence length. We then wanted to gather data over a wide range of atmospheric conditions. Well, this test ended up being a bit harder than we anticipated since for some of the atmospheric conditions we needed to measure, we experienced unanticipated noise in terms of shock and vibrations that the fieldable prototype sensor couldn't really deal

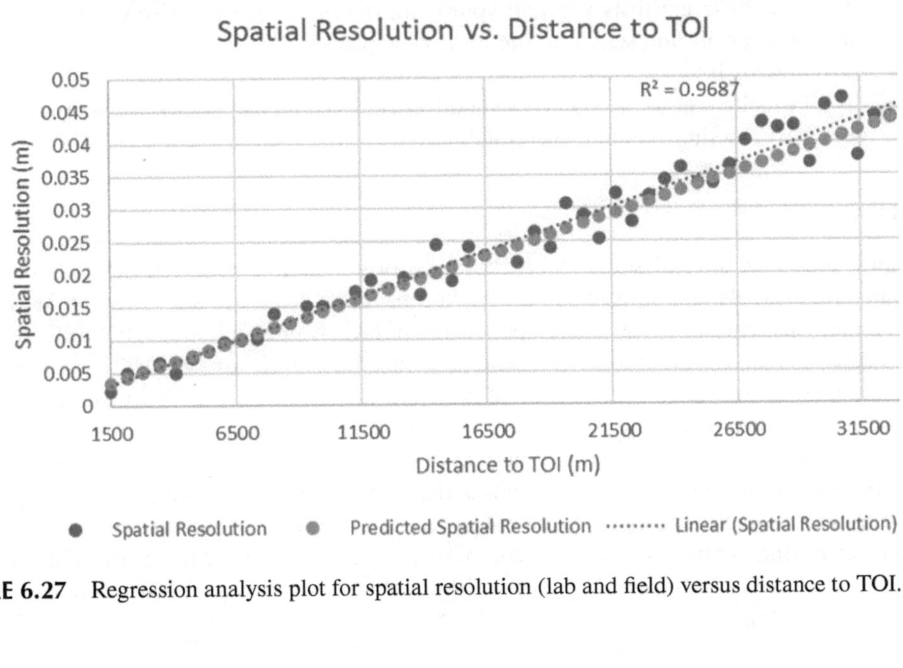

FIGURE 6.27 Regression analysis plot for spatial resolution (lab and field) versus distance to TOI.

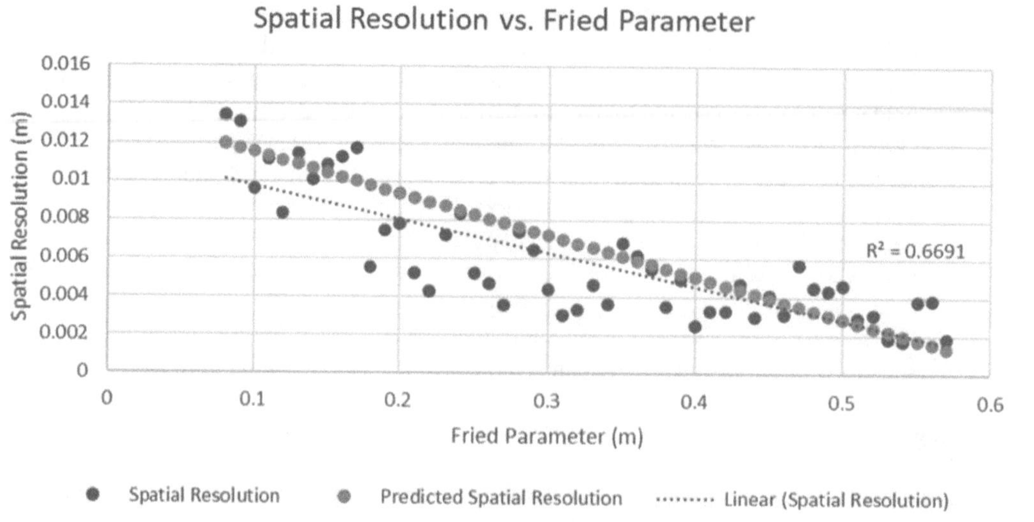

FIGURE 6.28 Regression analysis plot for spatial resolution versus fried parameter (both lab and field data).

with. We had a lot more variations in the sensor data for this test than the other two, as indicated by the lower R^2 value for this result. We need to work on some stabilization and vibration isolation mechanisms before we try this again. The good news is that the data still generally appears to scale as expected in that the smaller the atmospheric coherence length (Fried parameter), the worse the "seeing" prior to atmospheric compensation, and consequently the worse the spatial resolution gets. Also, for the first two regression plots that had the high R^2 values, we went back and checked the upper and lower variations of these parameters against our spatial resolution requirements. Well, our MBSE architecture did a lot of that automatically. We used our digital twin to check and see if the potential variations we saw in the field data still met the spatial resolution requirements with a sufficient margin of error, and they did!"

Tom: "It looks like you did your homework. Great job team! Let's go over the test results for some of the other tests and Use Cases later. Jenn, can you and your team write up a summary for all the different tests you and your team ran organized per Use Case scenario with summary results and send it to me for review please?"

Jennifer: "No problem, Tom."

Tom: "This is all excellent news. Keep up the good work. I need a cup of coffee! See you all in the next staff meeting. Thanks everyone!"

6.4 SUMMARY

This chapter presented several statistical techniques for conducting research methods on experimental data and for understanding the nature and propagation characteristics of light. With regard to research methods, emphasis was placed on Design of Experiments methods, but SPC, optimization, and regression analysis methods were also introduced. In DoE, we stressed the importance of a completely randomized design and looked at the importance of carefully setting up the sampling plan for a designed experiment to ensure as much as possible that we have a completely randomized design of the experiment. We covered a variety of hypothesis testing methods including testing for the equality of means and equality of variances under various conditions. We presented appropriate test statistics such as the t-test statistic, the F-test statistic, and Z_0-test statistic and others for useful hypothesis testing applications such as comparing the equality of means of two data sets for a small (or large) number of samples, comparing the equality of variances for various conditions and developed useful tests for single output single factor experiment. We developed analysis of variance methods for single factors with one or more treatment levels and extended this to the general case of multiple factors at multiple factor levels. We looked at requirements for ensuring that results are believable such as normality of the data, independence of the data, and equality of the variances between different treatments/factor-levels. We also presented useful assessment metrics such as the Coefficient of Determination R^2 value (along with several variants), the CV, the Pearson correlation coefficient, and the P-value to help us assess the quality of our data. We presented different types of regression analysis and introduced SPC and optimization as useful research methods that would be expanded upon in other chapters. We extended the statistical analytical approach to the optical imaging discipline where we looked at electromagnetic propagation effects and presented the statistical distributions for describing light under lowlight illumination conditions and highlight illumination conditions. We generated expressions for the moments and central moments of the instantaneous intensity and evaluated the statistical implications of coherent, incoherent, and narrow-band light. We looked at the effects of polarization on the instantaneous intensity and generated the coherency matrix for describing various polarization states and presented the useful degree of polarization metric. We presented the statistical description of thermal light, polarized light, and laser light (both single mode and multi-mode) and developed relevant characteristic functions. In the last section of this chapter, we provided an application example of the statistical methods that were developed in this chapter. We used our running Fantastic Imaging Technologies example to evaluate laboratory truth data for a sensor suite of hyperspectral, EO and IR imaging sensors, to experimental field data. FIT used hypothesis testing methods to conduct a comparison of the means of the laboratory and field data for the EO and IR sensors. The statistical comparison was for comparing the equality of means for a single-factor experiment. An ANOVA was run for a hyperspectral sensor that had a single factor at multiple treatments (wavelengths) with an equality of means experiment. The validity of the data was verified by checking for the normality of the residuals through a normal probability plot. The independence assumption of the ANOVA data was checked by plotting the residuals in time sequence. The residuals were evaluated to see that there were no noticeable patterns such as runs of positive and negative residuals. The lack of positive or negative residuals indicated that there were no noticeable correlations in the data and that the data was structureless and independent. A check was also made to ensure that variances between the different treatment levels were the same and that there were no drifts in the treatment level variances over time. A Bartlett test was used to demonstrate the equality of variances over all the treatment levels. Linear regression was performed to evaluate the effect of how close the predicted values of key performance parameters were to those that were gathered in the field. The coefficient of determination R^2 was used to indicate the goodness of fit of the data.

6.A APPENDIX: ACRONYMS

AHP	Analytic hierarchy process
ANCOVA	Analysis of Covariance
ANOVA	Analysis of Variance
ATC	Atmospheric turbulence compensation
C&E	Cause and Effect
CBC	Choice Based Conjoint
CFD	Computational Fluid Dynamics
CI	Consistency index
CL	Center Line
CLT	Central Limit Theorem
CONOPS	Concept of operations
CR	Consistency ratio
CRD	Completely Randomized Experimental Design

CV	Coefficient of Variation
CTO	Chief technology officer
CQI	Continuous Quality Improvement
DHS	Department of Homeland Security
DoE	Design of Experiments
DoF	Degree of Freedom
EO	Electro-Optical
EV	Eigenvector
FEA	Finite Element Analysis
FIT	Fantastic Imaging Technologies
GLM	Generalized Linear Model
Ho	Null Hypothesis
INCOSE	International Council on Systems Engineering
IPT	Integrated product team
LCL	Lower Control Limit
MS	Mean Sum
MSS	Mean Sum of the Squares
NID	Normal (Gaussian), Identically Distributed
NASA	National Aeronautics and Space Administration
PDF	Probability Density Function
RSM	Response Surface Methodology
RSS	Root Sum of the Squares
SE	Systems engineering
SNR	Signal-to-Noise-Ratio
SPC	Statistical Process Control
SPSS	Statistical Package for the Social Sciences
SRMA	System Reliability, Maintainability and Availability
TIM	Technical Interchange Meeting
TQM	Total Quality Management
UAV	Unmanned aerial vehicle
UCL	Upper Control Limit
U.S.	United States

REFERENCES

Abreu, A., Y. Morere, and E. Naves. 2022. Skills assessment metrics of electric powered wheelchair driving in a virtual environment: a survey. *Medical & Biological Engineering & Computing*, Vol. 60, pp. 323–335.

Alijosiene, S. and R. Gudonaviciene. 2010. Analyzing price-quality relationship using conjoint analysis. *Economics & Management*, Vol. 15, pp. 350–358.

Antony, J. 2014. *Design of Experiments for Engineers and Scientists.* London: Elsevier Ltd.

Antony, J., R. Snee, and R. Hoerl. 2017. Lean Six Sigma: yesterday, today, and tomorrow. *International Journal of Quality & Reliabilty Management*, Vol. 34, pp. 1073–1093.

Bali, N. and M. Goyal. 2020. *Concise Engineering Mathematics.* New Delhi: Laxmi Publications.

Banerjee, A. and J. Moore. 2022. The automotive semiconductor crisis and the way forward. *ISE: Industrial & Systems Engineering at Work*, pp. 34–39.

Bartlett, J., J. Kotrlik, and C. Higgins. 2001. Organizational research: determining appropriate sample size in survey research. *Information Technology, Learning, and Performance Journal*, Vol. 19, pp. 43–50.

Blanchard, B.S. and W.J. Fabrycky. 2011. *Systems Engineering and Analysis*, 5th edn. Upper Saddle River, NJ: Pearson Education, Inc.

Britannica, T. 2023. Editors of Encyclodpedia. *Encyclopedia Britannica.* August 28. https://www.britannica.com/science/distribution-function (accessed September 17, 2023).

Bui, H. and E. Galanou. 2022. Translation of systems thinking to organizational goals: a systematic review. *Journal of General Management*, Vol. 47, pp. 233–245.

Camelia, F., T.L. Ferris, and M.B. Behrend. 2020. The effectiveness of a systems engineering course in developing systems thinking. *IEEE Transactions on Education*, Vol. 63, pp. 10–16.

Cavenagh, N.J. 2018. Lower bounds on the sizes of defining sets in full n-latin squares and full designs. *Graphs & Combinatorics*, Vol. 34, pp. 571–577.

Chuan, Y., L. Hee, H. Hoou, and E. Cheng. 2023. Bayes' theorem for multi-bearing faults diagnosis. *International Journal of Automotive and Mechanical Engineering*, Vol. 20, pp. 10371–10385.

Cochran, W.G. 1977. *Sampling Techniques*, 3rd edn. New York: John Wiley & Sons.

Dawson, C. 2019. *Introduction to Research Methods: A Practical Guide for Anyone Undertaking a Research Project.* London: Little, Brown Book Group.

Deboeck, P. 2010. Estimating dynamical systems: derivative estimation hints form Sir Ronald A. Fisher. *Multivariate Behavioral Research*, Vol. 45, pp. 725–745.

Dowdy, S., S. Weardon, and D. Chilko. 2004. *Statistics for Research*, 3rd edn. Hoboken, NJ: Wiley Series in Probability and Statistics.

Du Plooy, D. 2022. Business intelligence through the application of data analytics and systems engineering principles with focus on project success. *PM World Journal*, Vol. XI, pp. 1–5.

Encyclopedia Brittanica. 2025. *Regression Analysis (Statistics). https://www.britannica.com/search?query=regression%20analysis* (Accessed March 7, 2025).

Feroskhan, M. and S. Ismail. 2017. A review on the purification and use of biogas in compression ignition engines. *International Journal of Automotive & Mechanical Engineering*, Vol. 14, pp. 4383–4400.

Field, A. 2017. *Discovering Statistics Using IBM SPSS Statistics North American Ed*, 5th edn. Los Angeles, CA: Sage Edge.

Ford, J.D. 2012. Integrating communication into engineering curricula: an interdisciplinary approach to facilitating transfer at New Mexico Institute of Mining and Technology. *Composition Fourm*, Vol. 26, pp. 1–16.

Goodman, J.W. 1985. *Statistical Optics*. New York: Wiley-Interscience.

Govindan, G., S. Hora, and K. Palagachev. 2020. *The Data Analysis Workshop: Solve Business Problems with State-of-the-Art Data Analysis Models, Developing Expert Data Analysis Skills Along the Way.* Birmingham: Packt Publishing.

Grady, J. 2006. *System Requirements Analysis.* Amsterdam: Academic Press.

Greengard, S. 2017. The Future of Semiconductors. *Communications of the ACM*, Vol. 60, pp. 18–20.

Grimmett, G. and D. Welsh. 2014. *Probability: An Introduction.* Oxford: Oxford University Publications.

Hacker, V. and C. Sumereder. 2019. *Electrical Engineering: Fundamentals.* Berlin: De Gruyter.

Hald, A. 2003. *A History of Probability and Statistics and Their Applications before 1750.* Hoboken, NJ: John Wiley & Sons.

Hassan, A., M. Lone, I. Dar, and P. Ahmad. 2022. A new continuous probability model based on a trigonometric function: theory and applications. *Reliability: Theory & Applications*, Vol. 17, pp. 261–272.

Holiday, R. and S. Hanselman. 2016. *The Daily Stoic.* New York: Penguin.

Hou, J. and T. Gao. 2022. A method of shear line detection in vector fields based on descriptive statistics of circular data. *Multimedia Tools & Applications*, Vol. 81, pp. 20853–20870.

Hunter, J.S. 2013. In Memoriam: George E. P. Box. *Technometrics*, Vol. 55, pp. 119–120.

Jiang, S., T. Wang, and K.-H. Zhang. 2023. Data-driven decision-making for precision diagnosis of digestive diseases. *BioMedical Engineering OnLine*, Vol. 22, pp. 1–30.

Jones, M. R., S. Djahel, and K. Welsh. 2023. Path-planning for Unmanned Aerial Vehicles with Environment Complexity Considerations: A Survey. *ACM Computing Surveys*, *55*(11), [234]. https://doi.org/10.1145/3570723 (Accessed March 7, 2025).

Jones, R.C. 1948. A new calculus for the treatment of optical systems. IV. *Journal of the Optical Society of America*, Vol. 31, p. 488.

Kobayashi, H., and H. Onoue. 1943. Brittle fracture of Liberty Ships. Failure Knowledge Database/100 Selected Cases. https://www.shippai.org/fkd/en/hfen/HB1011020.pdf (Accessed March 7, 2025).

Kolarik, W.J. 1995. *Creating Quality, Concepts, Systems, Strategies, and Tools.* New York: McGraw-Hill.

Kotz, S. and N. Johnson. 1992. *Breakthroughs in Statistics.* New York: Springer.

Lavi, R., Y.J. Dori, N. Wengrowicz, and D. Dori. 2020. Model-based systems thinking: assessing engineering student teams. *IEEE Transactions on Education*, Vol. 63, pp. 39–47.

Layton, R. 2017. *Learning Data Mining with Python – Second Edition: Harness the Power of Python to Develop Data Mining Applications, Analyze Data, Delve into Machine Learning, Explore Object Detection Using Deep Neural Networks, and Create Insightful Predictive Models.* Birmingham: Packt Publishing.

Li, J., and, F. Shang. 2022. Research on Performance Optimization Method of UAV Communication Link in Emergency Communication Environment. Se3curity and Communication Networks (May 2022), 2022(7):1-7. DOI: 10.1155/2022/5098730. https://www.researchgate.net/publication/360739545_Research_on_Performance_Optimization_Method_of_UAV_Communication_Link_in_Emergency_Communication_Environment (Accessed March 7, 2025).

Li, K. and G. Sonnier. 2023. Statistical inference for the factor model approach to estimate causal effects in quasi-experimental settings. *Journal of Marketing Research*, Vol. 60, pp. 449–472.

Logar, N., L. Anadon, and V. Narayanamurti. 2014. Semiconductor research corporation: a case study in cooporative innovation partnerships. *Minerva*, Vol. 52, pp. 237–261.

Mahla, S., T. Goyal, D. Goyal, H. Sharm, A. Dhir, and G. Goga. 2022. Optimization of engine operating variables on performance and emissions characteristics of biogas fuelled CI engine by the design of experiments: Taguchi approach. *Enviornmental Progress & Sustainable Energy*, Vol. 41, pp. 1–16.

Montgomery, D.C. 2020. *Design and Analysis of Experiments,* 10th edn. Hoboken, NJ: John Wiley & Sons.

Montgomery, D.C. and G.C. Runger. 2019. *Applied Statistics and Probability for Engineers*. Milton, QLD: John Wiley & Sons.

Muda, S., M. Abdul, N. Hamzah, and N. Saleh. 2020. Intellectual capital and SMEs' business performance from an organisational lifecycle perspective. *South East Asian Journal of Management*, Vol. 14, pp. 78–105.

Nelson, P., K. Copeland, and M. Coffin. 2003. *Introductory Statistics for Engineering Experimentation.* Amsterdam: Academic Press.

NIST. 2025. Model for Design of Experiments Framework. https://www.itl.nist.gov/div898/handbook/pri/section1/pri11.htm (Accessed March 7, 2025).

Nunamaker, J., M. Chen, and T. Purdin. 1991. Systems development in information systems research. *Journal of Management Information Systems*, Vol. 7, pp. 89–106.

Oakley, S. 2023. Iterative cycles enable rapid delivery of complex, innovative products. *GAO Reports*, pp. 1–53.

Panchal, J. and Z. Szajnfarber. 2017. Experiments in systems engineering and design research. *Systems Engineering*, Vol. 20, pp. 529–541.

Papoulis, A. 1991. *Probability, Random Variables, and Stochastic* Processes, 3rd edn. New York: McGraw-Hill.

Quick, T. 2019. *Splitting the DMAIC: Unleashing the Power of Continuous Improvement.* Milwaukee, WI: ASQ Quality Press.

Raharjo, H. and D. Endah. 2006. Evaluating relationship of consistency ratio and number of alternatives on rank reversal in the AHP. *Quality Engineering*, Vol. 18, pp. 39–46.

Randrianariosoa, H. 2019. *Sustainable Manufacturing.* Ontario: Archer Press.

Reese, H. 1997. Counterbalancing and other uses of repeated-measures latin-square designs: analyses and interpretation. *Journal of Experimental Child Psychology*, Vol. 64, pp. 137–158.

Ren, Y., T. Wang, S. Zhang, and J. Zhang. 2023. An intelligent big data collection technology based on micro mobile data centers for crowdsensing vehicular sensor network. *Personal & Ubiquitous Computing*, Vol. 27, pp. 563–579.

Rosenberg, I., A. Shabtai, Y. Elovici, and L. Rokach. 2022. Adversarial machine learning attacks and defense methods. *ACM Computing Surveys*, Vol. 54, pp. 1–36.

Rutherford, A. 2001. *Introducing ANOVA and ANCOVA: A GLM Approach.* London: Sage Publications Ltd.

Science Museum. 2018. Florence nightingale: the pioneer statistician. Science Museum. https://www.sciencemuseum.org.uk/objects-and-stories/florence-nightingale-pioneer-statistician (Accessed March 7, 2025).

Shanmugan, K.S. and A.M. Breipohl. 1988. *Random Signals Detection, Estimation and Data Analysis*. New York: Jhon Wiley & Sons.

Shi, L., H. Ren, J. Wang, and Z. He. 2016. Latin square design for chip length machine vision measurement system analysis. *Quality Engineering*, Vol. 28, pp. 381–387.

Silva, L., A. Oliviera, M. Leite, and F. Marins. 2021. Risk assessment model using conditional probability and simulation: case study in a piped gas supply chain in Brazil. *International Journal of Production Research*, Vol. 59, pp. 2960–2976.

Tan, T. 2023. Informative hypothesis for group means comparison. Practical Assessment, Research, and Evaluation, Vol. 28, Article 1, pp. 1-23.

Telford, J. 2007. A brief introduction to design of experiments. *John Hopkins APL Technical Digest*, Vol. 27, pp. 224–232.

Thompson, P. 2001. How Much did the Liberty Shipbuilders Learn? New Evidence for and Old Case Study. Journal of Political Economy, Vol. 109, Number 1, February 2001, Pp. 103-137.

Vankatesan, G., S. Prabu, and M. Rengasamy. 2022. *Sustainability Studies: Environmental and Energy Management.* Hauppauge: Bentham Science Publishers.

Van Schoot, J. 2023. The Moore's Law Machine: the next trick to tinier transistors is high-numerical-aperture EUV lithography. *IEEE Spectrum*, Vol. 60, pp. 44–48.

Villacampa, Y., G. Carlomagno, S. Ivorra, and C. Brebbia. 2018. *Computational and Experimental Studies.* Southampton: WIT Press.

Wikimedia Commons. 2007. Advantages of Modeling and Simulation. https://commons.wikimedia.org/wiki/File:Modeling_and_Simulation_Advantages.jpg (Accessed March 7, 2025).

Wikimedia Commons. 2007. Verification, validation and accreditation. https://commons.wikimedia.org/wiki/File:Verification_Validation_Accreditation.jpg (Accessed March 7, 2025).

Wilcox, R. 1989. Adjusting for unequal variances when comparing means in one-way and two-way fixed effects ANOVA models. *Journal of Educational Statistics*, Vol. 14, pp. 269–278.

Wilcox, R. 2005. *Introduction to Robust Estimation and Hypothesis Testing*. Amsterdam: Academic Press.

Zhang, W. 2016. "Technical Problem Identification for the Failures of the Liberty Ships," Challenges, MDPI, vol. 7(2), (November 2016). Pp. 1-8.

Zhu, Q. and B. Jesiek. 2020. Practicing engineering ethics in global context: a comparative study of expert and novice approaches to cross-cultural ethical situations. *Science & Engineering Ethics*, Vol. 26, pp. 2097–2120.

7 Decision-Making and Risk Analysis in Systems Engineering

Good decisions come from experience. Experience comes from bad decisions.

—**Mark Twain**

7.1 MAKING GOOD DECISIONS AND UNDERSTANDING RISK

7.1.1 Introduction to Risk Analysis, Decision-Making, and Game Theory

In Section 7.1, we present analytical methods to help understand the decision-making process, and how to make good, and sometimes tough decisions. We address decision-making when faced with incomplete or conflicting data, unclear outcomes, and/or competing priorities. We also introduce risk analysis and show how to assess, quantify, and manage risk that is associated with a development effort across the systems development lifecycle (SDLC). Additionally, we introduce game theory (GT) and show how useful GT can be in making decisions in competitive environments that can be modeled as a game. We start with a historical background on risk and decision analysis and include historical elements of GT. We then provide an understanding of risk methods and tools, and how these relate to systems engineering. Afterward, we provide a series of national and international case studies that illustrate systems engineering-related risk experiences on the national and international level. We follow this with an introduction to decision-making and GT. We start first with decision-making and present decision analysis methods, tools, and techniques and illustrate these topics with some decision analysis examples. Afterward, we present some interesting and useful GT methods, show how GT relates to decision-making, and provide some application examples. We end Section 7.1 by going over some special considerations in modern-day decision-making. In Section 7.2, we provide the optical building blocks for effectively modeling atmospheric effects and determining important attributes of an optical system like the optical transfer function (OTF) and the detector signal-to-noise ratio (SNR). We use statistical methods to describe these attributes and provide analytical expressions that aid the decision-maker in understanding and predicting the performance of an imaging system. In Section 7.3, we bring the concepts from Sections 7.1 and 7.2 together in an integrated case study that uses a fictitious optical systems development company, Fantastic Imaging Technologies (FIT) that is working on an optical systems development program for their customer from the Department of Homeland Security, Customs and Border Protection (DHS CBP). FIT is getting ready for a technical interchange meeting (TIM) with their sponsor to go over the expected performance of the imaging system. The plan is to cover important imaging systems metrics such as the detector SNR and present some ideas on the use of GT methods to aid in the decision-making process of an artificial intelligence capability that is resident on the imaging system's host platform, a predator class Autonomous/Artificial Intelligent Unmanned Aerial Vehicle (A/AI–UAV).

7.1.2 Historical Background on Risk and Decision-Making, and GT

Principles of risk analysis have been cited in academic works since the early 1900s. Standardized risk assessment templates appeared in the insurance and finance industries in the 1950s (Dionne 2013). The late 1900s saw the formation and shaping of standard risk guidelines in a variety of

DOI: 10.1201/9781003624097-9

disciplines, and the late 2000s saw the development of a flexible risk management standard with the release of ISO 31000 Risk Management in 2009 (International Organization for Standardization 2018). When it came time to manage security, privacy, and supply chain risk within a system development lifecycle, the National Institute of Standards and Technology (NIST) developed the Risk Management Framework in 2018 with the release of NIST Special Publication 800–37 Rev 2 (National Institute of Standards and Technology 2018). The U.S. Department of Defense (DoD) quickly adopted this and now uses it across the board for all defense acquisitions across the entire systems lifecycle.

There are risks at every point in a system's lifecycle, from the initial need to bring an idea to life, to the eventual shutdown of a program (West and Trainor 2006). Every area of our lives requires us to make decisions in the face of uncertainty (Sage and Rouse 2014, p. 155). Therefore, we need to expect and effectively deal with risk throughout the SDLC. Evidence suggests that the effectiveness of a project or program is directly related to the quality of its risk management (Kishk and Ukaga 2008). To lead such a risk management activity, there must be a complete understanding of the program's or project's financials, timelines, and technical requirements (Kockler et al. 1990). Consequently, the risk management team must have individuals with this type of experience to be effective. It is also important to establish a framework for managing risks early in the program's development and maintain that framework throughout the system's lifecycle (Kockler et al. 1990).

With regard to the historical aspects of decision analysis, J. E. Smith and Detlof von Winterfeldt looked to Bernoulli in 1738 as the point of origin for Decision Analysis. Bernoulli was looking into why people don't use the expected value strategy when buying insurance. To provide an explanation to this observation, he developed the expected utility (EU) model using a logarithmic utility function. Utility functions are now a quite commonplace in contemporary risk and decision analysis methods. Thomas Bayes, another 18th-century mathematician, was investigating probability-based decision-making. He was especially interested in how to revise probability theory when including prior information/knowledge. The outcome was the well-known Bayes Theorem (Smith and Detlof 2004). In 1931, Ramsey investigated the link between probability and usefulness or utility. He asserted that subjective probabilities can be used to represent beliefs or preferences. DeFinetti conducted a similar study in 1937 to derive subjective probability. This created the framework for Morgenstern and Neumann to establish what is now known as the EU model in 1944, with inputs from Savage later. Their work, *Theory of Games and Economic Behavior*, concentrated on the study of games; their contribution to EU was supplementary. They offered the model as an axiom in an appendix in their work (Smith and Detlof). Nonetheless, their work significantly impacted decision analysis.

In his book Game Theory, a Multilevel Approach, Dr. Hans Peters discusses what he calls "'game-theoretic' scenarios" that can be found in ancient literature. He draws on the Bible, the Talmud, and the works of Sun Tzu, a Chinese warrior-philosopher who lived more than 2000 years ago (Peters 2008; Brams 1980; Aumann and Maschler 1985). Gambling among French nobility in the 17th century paved the way for the current GT. It arose from conversations about the gambling problem among contemporary mathematicians, particularly Pascal and Fermat. Early applications of GT focused on actual games, such as chess. Early theories were primarily concerned with what is known in GT as two-person, zero-sum games, which will be discussed in more detail further below (Kelly 2003). Ernst Zermelo, a German mathematician, made many contributions to GT in the early 20th century. The minimax theorem is one of the most well-known, having become "the core theorem of Game Theory" (Kelly 2003). The theorem related to competitive (zero-sum) games, but the Hungarian mathematician John Von Neumann eventually proved it to be true in general. Von Neumann's contributions were not limited to just this. In 1941, he collaborated with Austrian economist Oskar Morgenstern at Princeton University. They worked on the well-known book, Theory of Games and Economic Behavior, which was published 3 years later. This research brought GT to the attention of economists (Kelly 2003). Previously, it was exclusively the domain of mathematicians.

It has been widely employed in economics since then. With John Harsanyi's thesis in 1967 and John Nash's discovery that every competitive game has at least one equilibrium point, in both mixed and pure strategies, GT evolved to incorporate games with incomplete information. Since then, GT has expanded to cover increasingly complicated gaming scenarios, as will be seen later in this chapter.

7.1.3 Understanding Risk and Its Connection to Systems Engineering

Effectively managing systems, services, products, and/or processes (SSPPs) often comes down to effectively managing risks and making good decisions. The familiar concept of managing cost, schedule, and technical effort/performance is well known in systems engineering and program management circles and has been understood to be of fundamental importance for decades if not generations or even longer. However, it can be argued that instead of the pillars of cost, schedule, and technical effort/performance, only one thing needs to be managed – risk. Of course, with a little thought, we realize that risk can be broken down into cost risk, schedule risk, and technical effort/performance risk, and so we really don't lose anything from this simplified alternate interpretation and perspective. The risk management interpretation does not necessarily limit us to these three pillars and so begs the question, what else should be considered and managed – a healthy question indeed! Let's consider risk in the context of systems engineering and program management. Figure 7.1 illustrates the many parallels between the systems engineering and program management fields. It's possible to think of systems engineering as a subset of the larger program management domain (Iyer and Mazzara 2009).

Because of this substantial overlap in responsibilities, developing an integrated and systematic approach to risk management requires that system engineers and PMs, at the very least, have effective communication skills and communicate well with each other throughout the systems development effort. Additionally, PMs should be knowledgeable about systems engineering principles, techniques, and methods.

Definition of risk: There are several ways to define risk and an almost infinite number of situations in which the definitions can be applied. We will provide a few common and some useful definitions of risk in this section, along with a Systems Engineering (SE) viewpoint of its underlying theory.

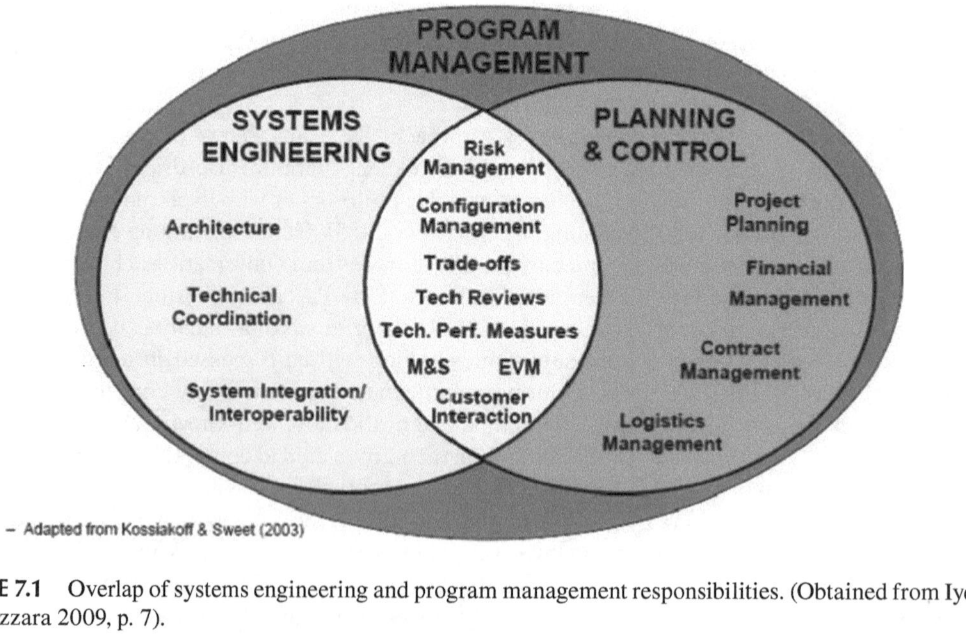

FIGURE 7.1 Overlap of systems engineering and program management responsibilities. (Obtained from Iyer and Mazzara 2009, p. 7).

A common example of the definition of risk is the "possibility of loss or injury" (Merriam-Webster 2022). Although this qualitative definition of risk is correct, it is not particularly helpful for the SE practice. A more practical definition for systems engineers from the Society of Risk Analysis (SRA) is "the combination of probability and magnitude/severity of effects" or "the combination of the likelihood of a hazard occurring and a vulnerability meter given the occurrence of the hazard" (Society for Risk Analysis 2018). These more quantitative definitions are used in the context of systems engineering to establish mathematically defined risk metrics and develop risk management tools for identifying, analyzing, and treating risk associated with a system under construction (ISO/IEC/IEEE 2015).

Risk management: The purpose of risk management is defined as follows: "the creation and protection of value… [to] improve performance, encourages innovation, and supports the achievement of objectives, according to ISO 31000:2018 Risk Management – Guidelines" (International Organization for Standardization 2018). There are eight ISO guiding principles that shed light on characteristics of effective and efficient risk management. These principles provide the foundation for risk management and must be considered when establishing an organizational risk management program. According to ISO 31000: 2018, these guiding principles must be part of an organization's risk management foundation:

- *Best available information*: information used as inputs to risk management includes accurate historical, current, and future expectations;
- *Human and cultural factors:* must account for human behavior and culture since these significantly influence risk management;
- *Integrated:* risk management must be an integral part of an organization's activities;
- *Structured and comprehensive:* the risk management approach must be structured and comprehensive for consistent and comparable results;
- *Customized:* risk management framework and process are customized and attuned to the organization's objectives;
- *Inclusive:* include stakeholders in an appropriate and timely manner;
- *Dynamic:* risks are dynamic and can disappear, change, or enter the picture based on changes in the organization;
- *Continual improvement:* risk management is continually improved through learning and experience.

Although these principles are directly applicable to corporate risks, they may be tailored to any kind of organization. For example, Army operations can employ customized methodologies to estimate risk during large-scale combat operations (Taylor 2019). On the other hand, as seen in Figure 7.2, the U.S. Government Accountability Office (GAO) defines risk management in a more unified manner that makes adoption simpler (U.S. Government Accountability Office 2013).

Systems engineers play a significant role in risk management, especially when it comes to identifying technical risks, creating plans for risk mitigation, and keeping track of developments (Dunbar 2019). The four primary activities of risk management are assessment, handling, planning, and monitoring/reporting (Defense Acquisition University Press 2001). While the remaining three categories represent the tactical process stages for managing risks, planning for risk incorporates the formation of a proactive mindset to support risk management processes (e.g., building a risk management plan). Every step in the risk management cycle, shown in Figure 7.3, is essential. Following the establishment of a risk management plan, the assessment process consists of two main steps: (1) examining and identifying potential risks and (2) analyzing those risks.

Risks can be assessed and identified using various methods, including trade studies, stakeholder surveys, and user inquiries about issues (Kockler et al. 1990). Focusing on the review and identification of risk requires attention to detail because these activities directly affect how well the project's risk management program is implemented. Risks can be considered as possible negative effects on

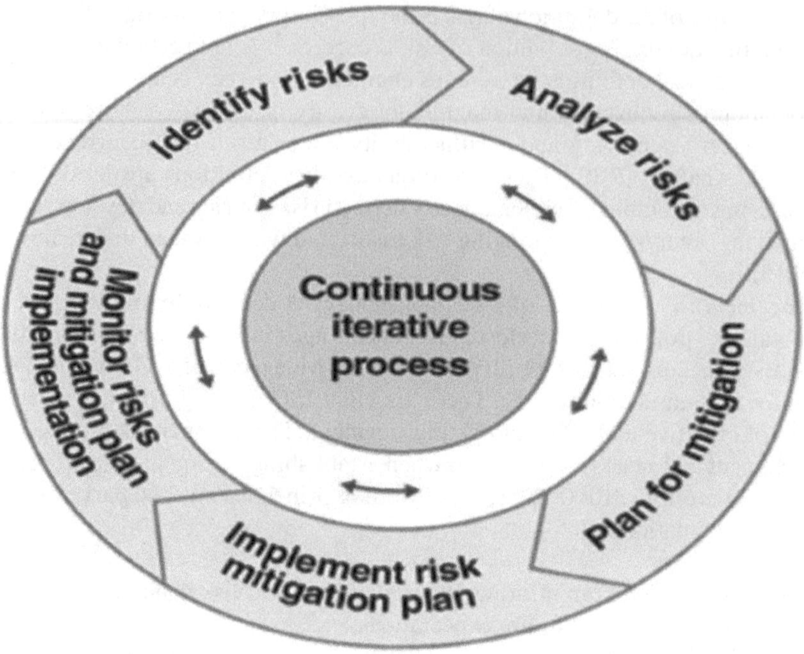

FIGURE 7.2 Principles of an effective risk management system. (Obtained from U.S. Government Accountability Office 2013, p. 9).

FIGURE 7.3 Primary activities for risk management. (Obtained from Defense Acquisition University Press 2001, p. 134).

budget, schedule, or technical performance. Since the systems development effort "covers the full lifecycle of the product [also process, system, and/or service] and directly impacts cost, schedule, and technical performance of the system," risk management is a fundamentally important activity throughout the SDLC (Ortiz 1992, p. 63). Using the comprehensive risk framework provided by DoD manual 4245.7-M, which is specifically designed for identifying risk areas when transitioning from development to production, helps ensure that risk areas are confidently addressed (Under Secretary of Defense for Acquisition 1985). A complete transition template is depicted in Figure 7.4, which also contains a thorough description of the key domains and important systems engineering tasks.

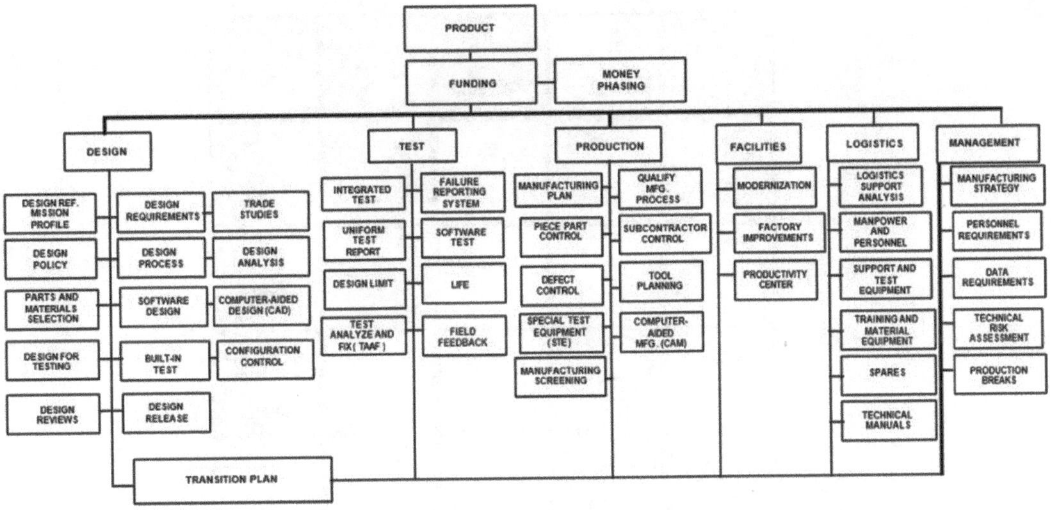

FIGURE 7.4 DoD 4245.7-M risk management template. (Adapted from Defense Acquisition University Press 1985, pp. 3–2).

To ensure completeness, the DoD manual also offers templates for the blocks shown in Figure 7.4. A PM can examine the full collection of these finished templates and generate a comprehensive list of risk areas before moving forward into the production phase. These risks can then be prioritized and handled as part of the risk management program to guarantee a smooth transition.

The analysis of risks, which determines each risk's probability of occurrence and potential severity, constitutes the second phase of the risk assessment process. There are many models available to assess risks including scheduling network models and lifecycle cost models. During risk assessment, the objectives are to determine the origin and consequences of the risks as well as to create and evaluate backup plans and risk-reduction methods. One common output from the risk assessment process is a watch list that lists the probable consequences of any potential risks that may be present. This is typically categorized to display the probable impact of the risk, the event that starts it, and the mitigation strategies. For instance, the first theoretical unit cost may be greater than anticipated if a system is produced before the technical data package has been sufficiently developed. The reason for this is the likelihood of the need for extensive rework due to flaws found later in the development cycle. Mitigation strategies could include producibility reviews, modeling, and process checks to ensure a quality technical package so that management is aware of the issue and supports the mitigation strategy, and the watch list must be evaluated frequently.

In risk assessment, quantifying the risk is required so that the risk events can be prioritized. Consequently, to guarantee proper risk prioritization, criteria for the scale of the consequence and the associated probability of occurrence are created for each risk (Defense Acquisition University 2003). A straightforward risk matrix can be used as a method for graphically representing risk as shown in Figure 7.5 (Iyer and Mazzara 2009). In the figure, an increase in the vertical direction represents an increase in the consequence of the risk event, whereas an increase in the horizontal direction represents an increase in the probability of occurrence of the risk. To create distinct priorities, it is customary to assign a single number, known as a Risk Priority Number (RPN), to each risk event. Often, the scale of the consequence is mapped into the range of (0, 1) to match the range of the probability of occurrence. This results in the RPN also falling into the range of (0, 10).

The next step in the risk management process is called "Handling," during which time efforts are made to try and prevent or control the risk, assume the risk, transfer the risk, or address the risk through development and research (Kockler et al. 1990). When choices are made to totally stop an action from occurring, such as choosing certain sources or design elements to reduce possible risks,

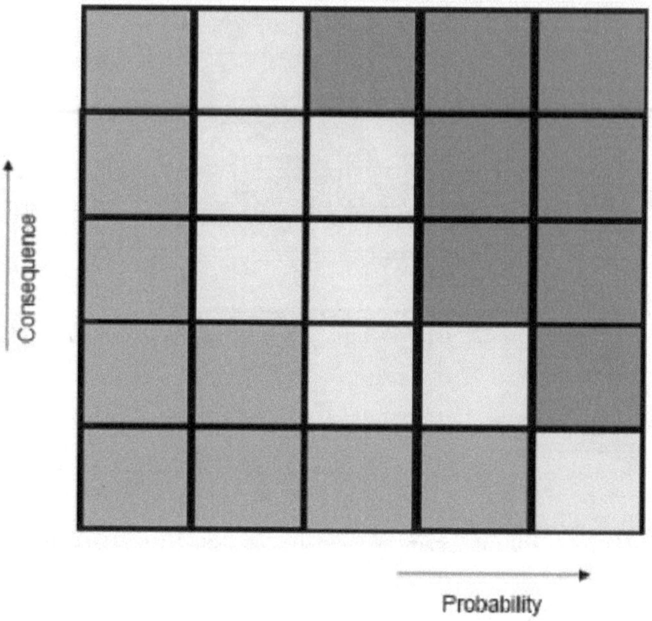

FIGURE 7.5 Simple risk matrix. (Obtained from Iyer and Mazzara 2009, p. 18).

risk avoidance can be achieved. Preventing risk means being able to create other ways to deal with the risk, such as finding different suppliers for deliverables that have short deadlines. When a deliberate choice is made to embrace the risk's consequences, and particular actions have been taken to minimize those consequences, the risk has been assumed. When a possible failure mode can be shifted to an entity outside of an organization, this is known as risk transfer. Lastly, early development efforts, thorough testing to guarantee technical sufficiency, and/or performance prediction using simulations are all ways that research and development efforts can be used to manage risk. Another way to assess and mitigate risk is to use experimental prototyping. Monitoring & Reporting is the last step in the risk management process. Since risks arise at every stage of a system's lifecycle, it is essential to monitor and assess risks on an ongoing basis. Regular organizational input on potentially increasing risks, watch list item feedback, and metric reporting are examples of typical risk monitoring and reporting processes (Defense Acquisition University Press 2001). The next step in the risk monitoring and reporting process is to ensure that everyone in the organization is aware of the risks that have been found within their areas of responsibility and to highlight the significance of these risks. In this step, an update of open risks is generated and distributed to everyone.

Risk assessment enhancements: The technology sector is changing so quickly that adopting new technologies is now seen as a need rather than a choice (Valerdi and Kohl 2004). Technology readiness levels (TRL) are a concept that the National Aeronautics and Space Administration (NASA) developed in the 1970s to handle the rapid speed of technological advancement. In the late 1980s and early 1990s, TRL was further developed and published externally (Olechowski et al. 2015). TRL is intended to provide "A method for assessing the maturity of technologies throughout complex system development" (Olechowski et al. 2015, p. 1). This idea, which the Air Force Research Laboratory (AFRL) has embraced, supports the use of TRL to assess a technology's suitability for integration into a weapons system (or any other type of system for that matter). Since most DoD weapons system developments involve cutting-edge technologies, new DoD regulations mandate that a TRL assessment be performed for any key technologies before the start of the Engineering and Manufacturing Development (EMD) phase of the system lifecycle (Valerdi and Kohl 2004). TRLs are marketed as an evaluation of the gap between the technology's current maturity and what

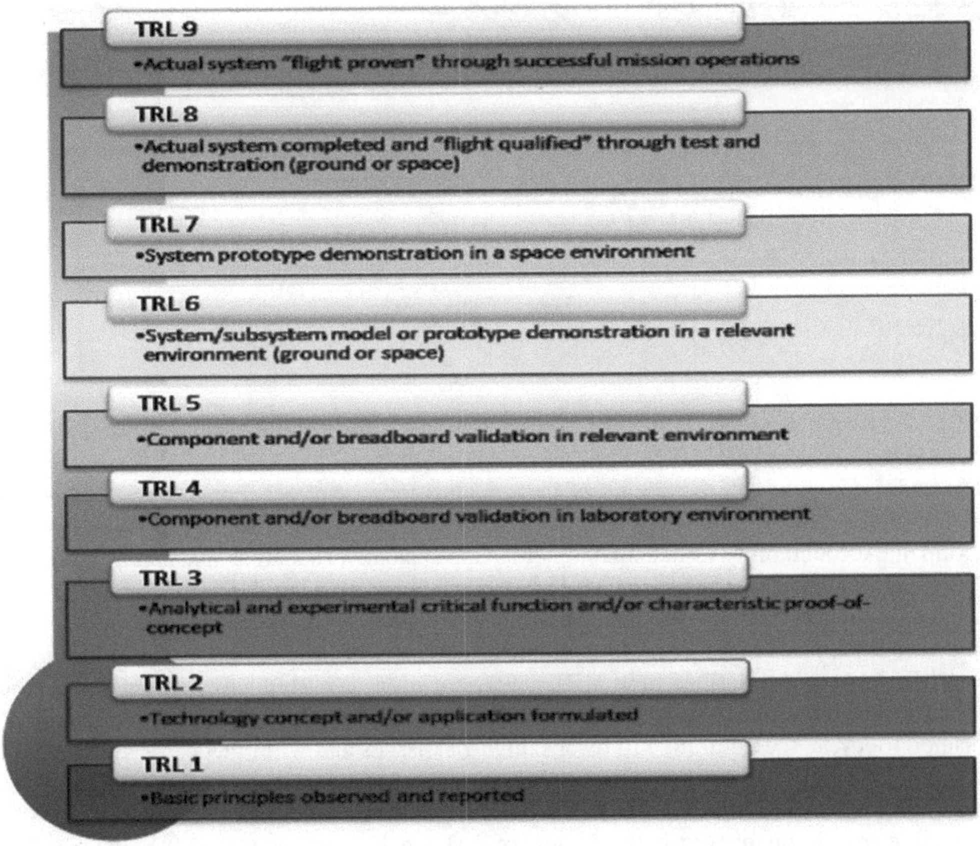

FIGURE 7.6 NASA technology readiness levels and descriptions. (Obtained from National Aeronautics and Space Administration 2021).

is needed for successful inclusion and use. Critical technologies usually need a TRL of 6 or above to move on to Milestone B of the Defense Acquisition Process (ACQNotes 2021). TRLs are rated from one to nine, where one represents a conceptualized idea and nine represents technology that has been proven on a mission. A complete explanation of the nine TRL levels developed and implemented by NASA is found in Figure 7.6 (National Aeronautics and Space Administration 2021). The DoD usually invests in TRL 4 technologies, pushing them toward a TRL 6 level of readiness, which is the general target of government acquisitions.

It is noteworthy that the degree of risk associated with completing budget, schedule, and performance goals in a systems development effort is closely correlated with the TRL. Program developments with higher TRLs point to a relatively more mature technology with less implementation risk, whereas lower TRLs point to a higher risk of budget overruns, missed deadlines, and/or risk of failure to achieve minimal technical performance criteria. In attempting to examine possible risks within a systems development effort, TRLs are an important concept that can be included as a component of the risk assessment process.

The DoD has embraced this concept and developed manufacturing readiness levels (MRL) and descriptions to assess the level of manufacturing readiness maturity (Stanley 2010). The descriptions and ranking scale are comparable. According to the Government Accountability Office (GAO), when used in conjunction with TRLs, MRLs are critical measures that identify risk when a technology or process is mature and migrated to a system (U.S. Government Accountability Office 2020, p. 122; OSD Manufacturing Technology Program 2020). As a result, the defense sector uses risk management concepts that are supported by the evaluation of both TRLs and MRLs. It is also important

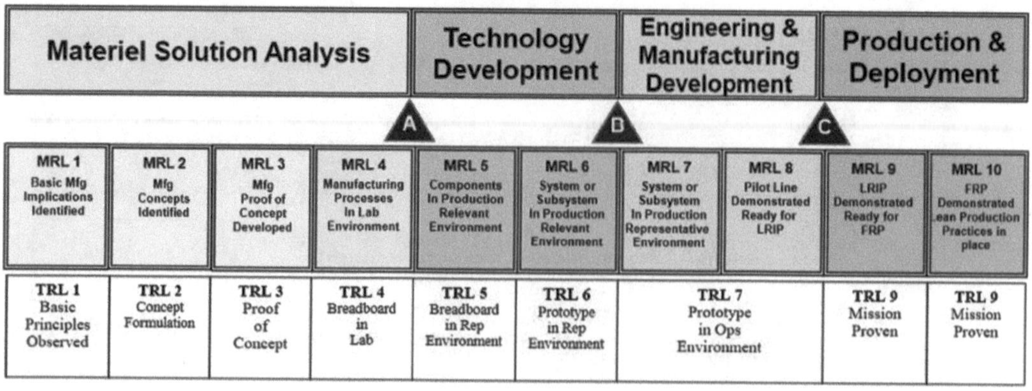

FIGURE 7.7 Acquisition lifecycle and TRL & MRL maturity. (Obtained from Stanley 2010, p. 14).

to take under consideration the timing of these maturity assessments. For example, according to the GAO, 'A component's low MRL at an early stage of acquisition is not always a bad thing (U.S. Government Accountability Office 2020, p. 45).' Consequently, risk areas that can be addressed to raise the MRL in the next acquisition phase are recognized when a low MRL is assigned, indicating a natural evolution in MRL over the course of the acquisition as shown in Figure 7.7.

This idea is captured in the program monitoring activities of the Defense Contract Management Agency (DCMA) and describes how MRL levels are used as a standard instrument to conduct consistent evaluations in assessing manufacturing hazards (Agosto 2016). Over the course of the acquisition lifecycle, the program will resolve high-risk issues and will practically and sequentially mature the technology, provided that the system is developed under a robust risk management and decision-making framework.

Assessing risk: To quantify risk, the Society of Risk Analysis provides the following mathematical description for the risk value,

$$R(X) = xP(x), \tag{7.1}$$

where x is a realization of the random variable X, and $P(x)$ is the probability of the event occurring. The range of $P(x)$ is [0, 1]. If these conditions are true, then Equation 7.1 can be restated as follows: $P(x)$ is the probability that an event will occur, while x signifies the aggregate risk resulting from its occurrence (often a monetary value or criticality value). This formula can be rewritten as follows:

$$R = S \times P, \tag{7.2}$$

when R represents the risk value, S represents the risk's severity, and P represents the risk's probability of occurrence. The following example illustrates the concepts behind Equation 7.2. Let's say that the consequence of the risk is given by a criticality number S. The criticality number relates to the severity of the risk and is mapped into a range [0, 1], where the higher number means a more severe consequence. If the determination is made that a risk on the project has a criticality value of 0.9, and that the risk has a 40% probability of occurring, then the risk value given by Equation 7.2 is

$$0.9 \times 0.4 = 0.36. \tag{7.3}$$

Equation 7.3 shows a risk value of 0.36 which is a metric that accounts for both the consequence of the risk and the probability of the risk occurring in a single number that has a range of [0, 1]. In an actual situation, each risk that has been identified would be assessed, and its risk metric value would be compared to a pre-established risk threshold to determine whether any action is required for the risk event under consideration. Equation 7.2 may be extended if necessary to also incorporate a metric D related to the probability of detection as follows:

$$R = S \times P \times D. \tag{7.4}$$

While not always required, the inclusion of D can serve as a means of additionally accounting for the ease or difficulty in detecting the risk. According to Dhillon (1999), the probability of detection is a quantity that is "better" the higher that it gets. After all, it is better to have a higher probability of detecting something dangerous than not. However, the severity in Equation 7.4 means that the more severe something is (i.e., S increase), the more dramatic and hurtful the impact of the risk event gets. In this case, as S increases, the impact is worse. Consequently, the D metric should be cast in the form that, as it increases, the worse the impact becomes. We can do this by changing D into the probability of not detecting the risk event instead of the probability of detecting it. An example would be an odorless, hazardous gas leak which would be hard to detect and so have a low probability of detection. The value D can then be considered the probability of not detecting the risk event (gas leak), where $D = 1 - P_d(x)$, and $P_d(x)$ is the probability of detecting the risk event x. With this definition of D, a low D means the probability of not detecting the risk event is low, and conversely, a high D means the probability of not detecting the risk event is high (low is good, high is bad), and we have a consistent scale with S. Sometimes, the risk value R is called the risk priority number (RPN), and if D is relevant, is given by

$$\text{RPN} = S \times P \times D. \tag{7.5}$$

If D is not relevant to the problem, it is just dropped from the equation such as in Equation 7.2. According to Dhillon (1999), the RPN is also the foundation for the popular risk assessment method, Failure Modes Effect and Criticality Analysis (FMECA). As presented throughout this book, FMECAs are often used in the systems engineering discipline to assess, quantify, and articulate risk with regard to a system, service, product, and/or process. Figure 7.8 shows an example of an FMECA for the functionality of the optical system using Equation 7.2. Note that in the FMECA, the probability of the risk occurrence is termed "Liklihood."

If there was a need to include the information about the detectability of the risk, then the two-variable RPN shown in the FMECA of Figure 7.8 would be replaced with the three-variable expression in Equation 7.5. Also, while there are certainly other expressions in the literature for mathematically describing risk, the premise presented above is a popular one. These other expressions are equally valid and can be used, if necessary, by the engineering team. For instance, a linear combination of weighted risk factors such as hardware risk, software risk, financial risk, schedule risk, and a host of other factors may be generated as consequences, along with their associated probability of occurrence, and the linear combination of these factors could be used as an equally valid RPN.

Feasibility analysis: One popular systems engineering technique for assessing a system's risk is feasibility analysis. Feasibility is complementary to risk in the sense that as one increases, the other decreases. The more feasible something becomes the less risky it is in general. The identification, recognition, and acceptance of a risk-related problem is followed by the performance of a feasibility investigation. Feasibility analyses formally appear after the completion of the Problem Definition stage of the waterfall systems engineering process in the system lifecycle. Feasibility analyses are applied in a holistic sense and include feasibility criteria such as Technical, Lifecycle Cost, Legal, Operational, and Schedule feasibility for example (Heathcote 2005). The acceptable degrees of risk associated with each of these feasibility criteria differ from program to program. For new development initiatives, risk is a major concern. Feasibility analyses might be explicitly tied to the system deliverables or to the system development process itself. Identified risks must have acceptable procedures for mitigating them, bringing RPNs down to levels that the program accepts for certification to proceed.

Failure Mode, Effect and Criticality Analysis (FMECA)
System: Optical Function
Function Block: 7.0

RPN Threshold Values
RPN < 0.25
0.25 = RPN < 0.5
0.5 = RPN < 1

Requirement	Reference Number	Potential Failure Mode	Potential Cause of Failure	Potential Effects of Failure	Before Mitigation Likelihood	Before Mitigation Criticality	Before Mitigation RPN	Mitigation	After Mitigation Likelihooc	After Mitigation Criticality	After Mitigation RPN
3.2.1.30 The system shall have an optical system that can detect human heat in the IR with a peak wavelength of 9.5 micrometers.	3.2.1.30. a	a. Defective Components b. Loss of Data c. Loss of human tracking	Hardware Related Issues	Requirement not met to detect human heat.	0.4	0.7	0.28	System shall reliable detection hardware.	0.3	0.7	0.21
				Cost and Schedule: Potential damage to equipment and design re-work necessary > 8 weeks.	0.3	0.5	0.15	Equipment Maintenance and Support. Increase Inspection and installation.	0.2	0.5	0.1
				Operational: System may fail in service resulting in ability to detect human heat.	0.4	0.6	0.24	AUAV Optical System shall transmit an alert and engage safe mode.	0.3	0.6	0.18
	3.2.1.30. b	A. Inability to detect human heat using IR sensor application B. Loss of human/object tracking C. IR Software errors	Software Related Issues	Requirement not me to detect human heat.	0.1	0.7	0.07	Increase in testing and verification of software	0.1	0.7	0.07
				Cost and Schedule: Potential damage to equipment and design re-work necessary > 8 weeks.	0.4	0.5	0.2	Development of Software Test Plan	0.2	0.5	0.1
				Operational: System may fail in service resulting inability to detect human heat.	0.6	0.6	0.36	Review procedures to incorporate software application failure alert and mode warnings.	0.2	0.6	0.12
3.2.2.2 The system weight shall not exceed 4,901 pounds [2,223 kilograms].	3.2.2.2. a	A. Inability to flight at it's capacity. B. Inability to shift and move during tracking of objects. C. Inability to maneuver through objects.	Hardware Related Issues	Technical: AUAV weight will cause technical problems with flying, maneuvering, and track objects / humans.	0.2	0.7	0.14	Hardware Upgrades, weight testing, and additional operational testing of all equipment and missions.	0.1	0.7	0.07
				Cost and Schedule: Potential damage to other equipment and design re-work necessary > 8 weeks.	0.3	0.5	0.15	Equipment Maintenance and Support. Increase Inspection and installation.	0.2	0.5	0.1
				Operational: System may fail in service resulting inability to fly at its potential height and maneuver around objects/humans.	0.3	0.6	0.18	Testing of all equipment and mission capacity.	0.2	0.6	0.12
3.2.2.5 The system shall have vibration and shock isolators.	3.2.2.5.a	A. Inability to provide clear images and videos due to equipment failure.	Hardware Related issues	Technical: Spring soften vibrations causing images and videos from capturing objects/human correctly.	0.4	0.5	0.2	Installation inspection required prior to initial energization. Good Engineering Practices	0.2	0.5	0.1
				Cost and Schedule: Potential damage to equipment and design re-work necessary > 2 weeks.	0.3	0.3	0.09	Installation inspection required prior to initial energization. Procure / Produce robust and reliable components	0.1	0.3	0.03
				Operational: AUAV unable to provide vibration and shock isolators.	0.4	0.7	0.28	AUAV Optical System shall transmit an alert if vibration an shock isolators are failing.	0.2	0.7	0.14
3.2.10.3 The system shall have an optical system with environmental conditioning protection features(temperature regulators, moisture and humidity control, dust, pressure, and light shields).	3.2.10.3. a	A. Failure of equipment B. Failure of sensors C. Failure of lighting shields	Hardware Related Issues	Technical: Environmental Issues can cause equipment failure, mission to fail, and AUAV to crash.	0.2	0.7	0.14	Hardware Upgrades, weight testing, and additional operational testing of all equipment and missions during multiple environmental occasions. .	0.1	0.7	0.07
				Cost and Schedule: Potential damage to other equipment and design re-work necessary > 8 weeks.	0.3	0.5	0.15	Equipment Maintenance and Support. Increase Inspection and installation.	0.2	0.5	0.1
				Operational: Environmental issues can cause mission failure.	0.3	0.6	0.18	Testing of all equipment and mission capacity.	0.2	0.6	0.12

FIGURE 7.8 Example of a FMECA for the optical system.

As an example, to demonstrate that different engineering systems are structurally satisfactory, there are typically three approaches used to meet certification criteria: testing, analysis, and comparison. Acceptance through comparison is predicated on resemblance to previously completed and accepted work. Large finite element models combined with structural simulation software are usually required for analysis acceptability. For certification by analysis, one approach is through product certification testing programs that can provide lower expenses while retaining equal standards of safety (NASA 2021). Program testing can be costly and may be restricted to the most important components, so identifying lifecycle risk issues early is important.

Risk assessment using Weibull analysis: A key concept in aircraft certification is lifecycle risk, which can typically be determined analytically using Waloddi's Weibull analysis methods to determine risk likelihood. Weibull analysis, also known as life data analysis (Life Data Analysis 2022), is a process that takes a parts installation date, and the current date to statistically analyze failure, including the failure pattern and frequency (Pham and Lai 2007). A primary objective of Weibull analysis is to determine the time to failure. In Weibull analysis, the Beta value is one of the main analytical results (shape factor). According to Murthy et al. (2004), the Beta value can show a generated probability plot's slope as well as the failure pattern (Mudholkar and Kollia 1994). The location of the failure rates can be obtained by plotting these variables on the reliability bathtub curve. Some significant Beta value results are

- $\beta < 1$ Early failure
- $\beta = 1$ Useful Life (random failure occurrences)

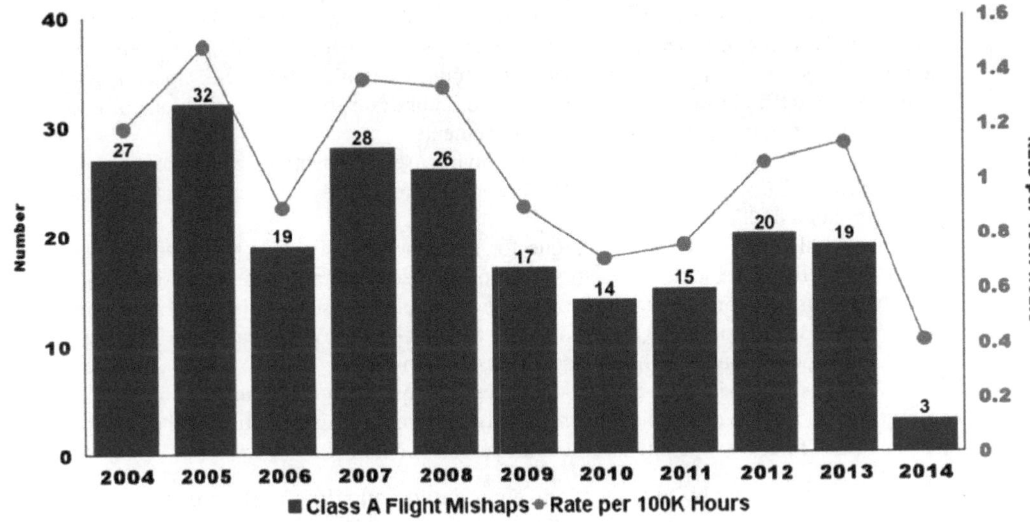

FIGURE 7.9 Class A aviation mishaps 2004 through 2014. (HQ AFSEC/SEH 2014).

- $1 < \beta < 4$ Wear-out Failure
- $\beta > 4$ Indicates rapid wear-out failure.

Another important Weibull analysis output metric is the Eta value, sometimes referred to as the characteristic life. This metric represents the parts expected life, or the moment at which, statistically speaking, 63.2% of the population has failed. The Weibull analysis can represent the true expected life more accurately, and is a better indicator, than the mean time between failures (MTBF) metric, which represents a straight average. The Weibull distribution, which has seen usage in survival data modeling, is part of a broader family of distributions that is produced by adding a second shape parameter. This generalized family has a wider class of monotonic hazard rates in addition to distributions with unimodal and bathtub hazard shapes (Mudholkar et al. 1996). Weibull analysis, which can forecast future failure probabilities based on past performance, is typically plotted as a cumulative distribution function (though it can also be occasionally represented as "probability density functions, hazard rates, and parameter estimates") (Hallinan 1993).

Risk mitigation example: Generally, the success of a program is intrinsically dependent on risk mitigation. As an illustration, a thesis (Vasquez 2015) was produced about an analysis of data from the Accident Investigation Boards regarding Class A mishaps for the F-15, F-16, and F-22 aircraft that occurred since 2004. Figure 7.9 shows the number of mishaps for these aircraft over this 10-year time from 2004 through 2014. It is noteworthy – and regrettably so – that, according to the DoD, human error – rather than technical problems – accounted for the great majority (80%) of accidents. Secretary of Defense Gates stated back in 2002, that by 2008, a force-wide aim of lowering this number by 75% had been established. The Air Force extended this deadline to 2012. In addition to providing a detailed overview of typical tools used for locating defects and root causes within problematic systems, Vasquez's thesis (2015) evaluated the efficacy of the risk management program for the F-15E Strike Eagle Community (although it also covered the F-16 and F-22 models). Among them are,

> An inductive, bottom-up analytical technique called Failure Mode and Effects Analysis (FMEA) which, "is focused on understanding the effects of a single component of function failures on equipment or subsystems" (Long 2010). FMEAs save costs and shorten the time needed for process development while also increasing the process's quality and dependability.

A Reliability Block Diagram (RBD) is a diagrammatic technique that illustrates how the dependability of individual components affects a complex system's success or failure. Dependence Diagrams (DDs) are another name for RBDs (Modarres et al. 1999; U.S. Department of Defense 2025). RBDs have the potential to reduce failure rates by detecting regions of poor reliability and helping to provide focused improvements.

Root Cause Analysis (RCA) is "A method of issue solving used for identifying the root causes of flaws or problems" (Wilson et al. 1993). RCAs concentrate on fully understanding the fundamental cause of a fault's occurrence.

Why-Because Analysis (WBA) is a technique for accident analysis. A Why-Because Graph (WBG), which lists the incidental factors that contributed to an accident, is the output of a WBA study. It is an unbiased examination of the underlying causes of mishaps (Ladkin and Loer 1998). WBAs are mostly employed in the analysis of behavioral causality in complicated technological systems during accident investigations.

Fault Tree Analysis (FTA) is a top-down, deductive technique designed to analyze how well a complex system may be made to experience faults and events. This analytical tool allows the investigation to filter down to any level that the researcher chooses.

A fault tree analysis can be used by those who have good understanding of the specific system issue to help account for possible circumstances leading to the fault. Diagrams containing gates, transfer lines, and events make up fault transfer architectures (FTAs), which are designed to illustrate the connections between the many elements that contribute to a fault. In the end, the FTA (reference Figure 7.10) was employed in the analysis for the risk mitigation program effectiveness assessment. The results were then compared with several formal spreadsheets that are used by different Air Force fighter units to evaluate system defects.

Finding the source of a specific defect might be difficult, as Figure 7.10 illustrates. Several "underlying factors that weaken human components" are depicted in the FTA. Stressors, weather (operating at near minimum allowable conditions), experience level, mission complexity, late changes, sleep problems, ops tempo, and proficiency/currency levels are a few of these underlying factors (Long 2010). The likelihood of failure and the underlying cause of the failure that arises can be determined and then addressed to create considerably more dependable systems with a higher degree of confidence in the systems.

There are various add-on tools for Excel spreadsheet computations that have recently attracted significant interest that are available to help with risk assessments with challenging or unknown parameters. Some representative addons are:

Decision trees and influence diagrams are produced by Precision Tree.
What-If Analysis, for data table creation as well as goal-seeking applications.
@RISK, which produces simulations, fitting distributions, and probability distributions
The program RISK Optimizer produces stochastic optimizations.
Analysis of sensitivity.

"While making tough decisions is a typical occurrence in life, it's not as difficult as it seems to grasp how to study them thoroughly" (Clemen and Reilly 2014). Several organizations, including those in the gas and oil, aviation, and automotive industries, use these tools to speed up decision-making and develop projects while adhering to systems engineering procedures. In the oil industry, "Risk models have been used as a regular industry practice for the past 20 years to rank pipeline segments for repair, identify serious failure hazards, and guide operational and integrity management choices as part of pipeline maintenance programs" (Koduru et al. 2017). A huge benefit to systems development efforts in general is providing the ability to make logical decisions in terms of risk and reliability from the probabilistic data produced by these tools (Kossiakoff et al. 2011; Sage and Rouse 2014).

A communications issue between development program Systems Engineers (SE) and DoD PMs has been identified. Systems developers must generate the systems engineering documentation

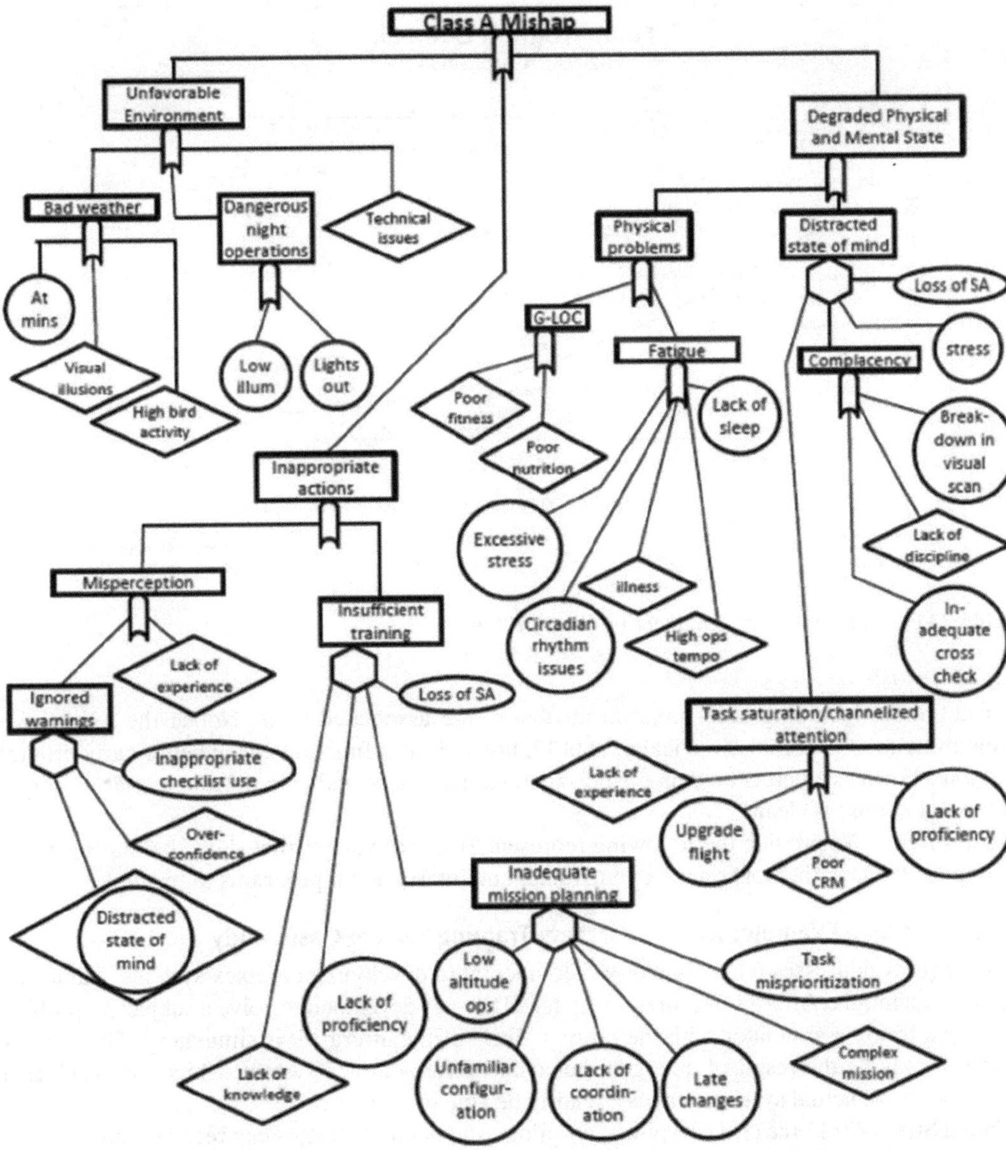

FIGURE 7.10 Fault tree analysis of Class A mishaps (Long 2010).

needed for the PM to operate in a political environment with limited resources (Irwin 2009). According to Pernin, milestone gates require the generation of essential documents that need a great deal of engineering work and approval from an often-outdated stakeholder review process before they can be formally submitted (Pernin et al. 2012). Ideally, the critical artifacts from a quantified risk assessment technique during program planning can be provided to key decision-makers in both Program Management and Systems Engineering groups by incorporating them into the initial program risk assessment activities.

Throughout the course of the SDLC, proper risk assessment planning included in the early phases makes a big difference. Figure 7.11 (QRA Corp 2025; ISO/IEC/IEEE 2011) helps show how important it is to introduce systems engineering principles early on, how much of an influence suitable processes have on the various programs, and how overruns might potentially be minimized. Programs that have early on systems engineering processes better defined, lead to programs with

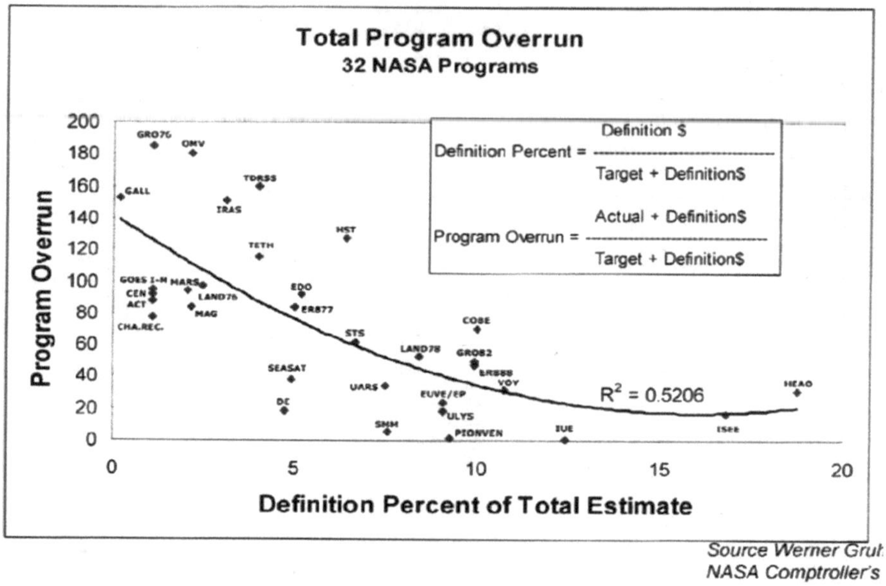

FIGURE 7.11 Value of systems engineering (Irwin 2009).

better definition including core program attributes and associated risks. Notice the reduction in percent overrun of programs (vertical axis of Figure 7.11) as a function of the better early program definition (shown as a percentage on the horizontal axis). The R^2 value in the above chart is not that stellar but the trend is clear.

Case studies: We provide the following representative case studies that show the impacts of risk planning (or the lack thereof) on notable national and international programs as examples.

7.1.3.1 National Example: KC-135 Aircrew Training System Case Study

This case study addresses the technology risk inherent in developing complex systems. Namely, an objective system, the Aircrew Training System (ATS), was developed to solve a subjective problem, providing a training simulator with the correct "feel" of the aircraft it is simulating. The "feel" of an aircraft may be the result of emergent behavior of the complex system, which the ATC, as an abstraction of the actual system, may not readily be able to replicate.

United States Air Force (USAF) pilots, co-pilots, and boom operators can receive training on the procedures and techniques necessary to operate the KC-135 safely from the ATS, a ground-based simulator (Macaulay-Brown, Inc., Chrislaghi et al. 2009). Boeing created the KC-135 Stratotanker, an aerial refueling aircraft, based on its 367–80 prototype. It was initially put into service in 1957 (Boeing 2022). The ATS is the most recent iteration of training systems, having begun with the Cockpit Procedures Trainer (CPT) in the 1950s and going into service in 1992.

The KC-135 ATS team's use of the System Engineering (SE) approach represented a collection of integrated SE procedures thought to be especially suited to the creation, execution, and upkeep of ATSs (Macaulay-Brown, Inc., Chrislaghi et al. 2009). During this time, the DoD leadership was promoting a wave of simplifying activities, which resulted in the cancellation of many MIL-SPECS and Standards that were formerly applicable to this program. Consequently, FlightSafety International (FI), the main contractor for the simulator's development, was exempt from government and military SE criteria at the time of contract award in 1992. Instead of complying with these SE standards, the then-current commercial best practices for SE applied to contracts awarded during this period. These best practices for FI were their own corporate-level SE procedures, which they had initially taken from the American National Standards Institute Standard/Electronic Industries Association

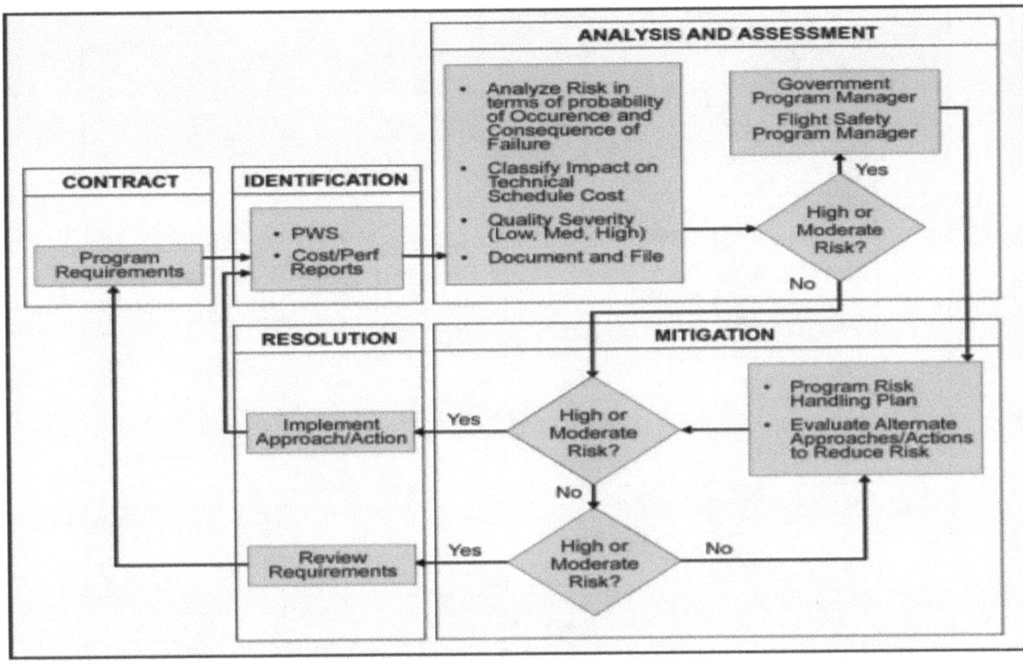

FIGURE 7.12 Risk management process (Macaulay-Brown Inc. 2009).

632 (ANSI/EIA-632). The risk management procedure used by FI seemed to be rather typical. Any member of the product team, including the government, contractor, and financial analyst, may identify risks through the process, which was based on the standard risk management categories of risk identification, evaluation, handling, and monitoring. The processes used by FI for risk management and risk assessment are shown in Figures 7.12 and 7.13, respectively.

A primary drawback of the KC-135 ATS was in having an objective system (the ATS) that is trying to simulate a subjective mission. It is intrinsically hard to duplicate the "feel" of an aircraft in flight, regardless of the volume of data gathered, analyzed, and extrapolated (never mind that each aircraft has its own "feel"). In this instance, a well-defined risk management process resulted in the establishment of several requirements deemed to improve the simulator's quality. Unfortunately, the determination of whether the system was deemed "excellent" or "ready" ended up being contingent upon the subjective opinions of test crews and Air Force instructors. However, the risk management and SE procedures produced a system that was rated "good enough" in the absence of quantitative proof that the system adequately accomplished the intended task. The risk management process here was good and resulted in a "good enough" rating for the developed system, but satisfying some requirements such as providing the actual "feel" of the aircraft, remained elusive, and required something more.

7.1.3.1.1 National Case Study: Global Hawk Systems Engineering

Here we have an example of the potential negative impacts of cost risk. An excellent national decision and risk case study example is that of the Northrop Grumman RQ-4 Global Hawk program, which developed a high-altitude remotely piloted surveillance aircraft (Northrop Grumman RQ-4 Global Hawk 2024). Included within the Global Hawk System was the unmanned aerial vehicle (UAV) and supporting ground stations, associated links (data, control, and communications), and logistical support elements. The common ground segment (CGS) configuration of the Global Hawk consisted of the command, control, and data collection activities. Up to three vehicles can have their command, control and data requirements serviced by only one CGS along with one payload. Besides the UAV and CGS, a third formal system element is the support system. The motivation for

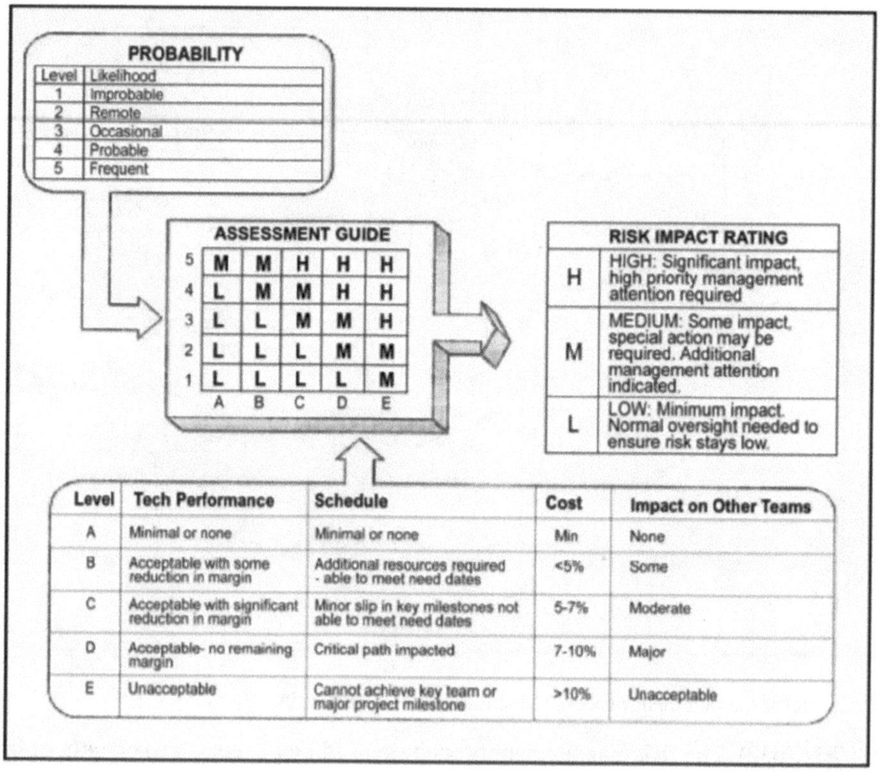

FIGURE 7.13 Risk Management Process (Macaulay-Brown Inc. 2009).

developing system like the Global Hawk came from Operation Desert Storm where the Air Force was looking for mobile SCUD missile launchers (Air Force Institute of Technology 2010).

The Defense Advanced Research Projects Agency (DARPA) was assigned to oversee the initial stages of the Advanced Concept Technology Demonstration (ACTD) leading to the Global Hawk, but the program that gave rise to the Global Hawk system began in the mid-1990s and was funded by the Defense Airborne Reconnaissance Office (DARO). Given that the general program included deliverables for both the Air Force and the Navy, a Joint Program Office (JPO) was created. Four phases, each with specific durations and milestone objectives, were to be used to administer the ACTD.

Addressing the primary cause of previous UAV program failures – unit prices that were higher than what the end user or customer was willing to pay – was the main objective of the ACTD. Dr. Kaminski, the Under Secretary of Defense for Acquisition and Technology at the time, stated:

> There are three characteristics… which are the hallmark of the program. The first is that there is usually joint service involvement in an ACTD. Second, ACTDs allow our warfighter to perform a very early operational assessment of a system concept before we've invested a lot of money in this concept. And third, there is usually some residual operational capability left in the field at the completion of an ACTD, even if we haven't decided to put the program into a full development phase
>
> **Perry (1995)**

The ACTD aimed to differentiate itself from other reform initiatives for the defense acquisition process by virtue of its high level of user involvement. It was decided to use an iterative method for both the hardware deliverable and the operational needs that it was intended to fill. DARPA is known for creating programs with extremely high-performance goals. The issue with this strategy is that failure risk typically accompanies high performance risk. Cost is also positively correlated with a high chance of failure; in addition, high technical risk often means that program costs go over budget, which is a surefire way of getting a program on the chopping block, or even terminated. The JPO made decisions to force the contractors to choose designs with fewer failure risks because they wanted the Global Hawk program to succeed (and have lower costs). Limiting the amount of development funds that contractors may access was a particularly effective approach to enforce this risk mitigation strategy.

As it happened, a few choices made for Global Hawk with the intention of reducing risk (and consequently expense) resulted in unintended consequences. Several problems surfaced in the process of moving from concept to engineering development. The RAND Corporation was hired to conduct a post-mortem to determine the impact of the "innovative" strategies used during the concept phase on the program's outcome. The following summarizes the key findings from that examination of the program's Phase II development efforts:

"The competitive environment in Phase II was lost due to budget reductions."

Although the budget cuts occurred in Phase I, the Phase II execution was significantly impacted by it. The contractual and management environment was drastically altered from one that relied on competition to a sole-source contractual arrangement that just required best-effort, had poor incentives, and did not provide the means for the government to intervene. The risks associated with systems integration and software development were understated.

The lead contractor, which was originally Teledyne Ryan Aeronautical (TRA) before being acquired by Northrop Grumman, was a tiny company with expertise in small UAV platforms but not much in huge, complex programs like Global Hawk. TRA was especially deficient in important software and integration areas (Drezner et al. 2007). Many of the choices made to save costs by lowering risk turned out to be poorly thought out when it came time to develop hardware. For instance, when engineers believed that the interfaces would function like purpose-built systems, they placed a greater emphasis on using less expensive, commercial off-the-shelf (COTS) equipment to cut expenses. However, the systems engineers in charge of establishing the requirements and managing risk underestimated the risks and issues related to integrating COTS. Moreover, the drive toward essentially design-to-cost (DTC) had been shown to be, at most, cost neutral and generally more expensive than alternative program design structures (Nilchiani and Pugliese 2016).

The ACTD's legacy is mixed, despite being officially declared a success. It proved that there was technology available to build a UAV whose main purpose complemented the mission's overarching objectives (continuous, long-endurance reconnaissance and surveillance in all-weather conditions). But, because of the JPO's clear cost-related limitations on the contractor, the program's scope and content were so compromised that the final physical product was not prepared for manufacturing.

7.1.3.2 National Example: Lockheed Martin C-5A Galaxy Case Study

This case study provides an example of risk associated with setting unrealistic program/goals and requirements (empty weight guarantee) and cost risk in choosing the wrong contract vehicle (Total Package Procurement Contract). Being a large military transport aircraft, the Lockheed C-5 Galaxy was created and manufactured by Lockheed (now Lockheed Martin). Its main function for the US Air Force (USAF) is as an intercontinental strategic airlift vehicle, able to transport all air-certifiable goods and large cargoes (Lockheed 2022). The late 1950s saw the start of the C-5A Galaxy transport aircraft's development. When the C-5 was originally deployed, it was the largest airplane in the world and a revolutionary machine. It is still the largest transport and cargo aircraft in the US military's current inventory. This kind of vehicle's development was a major engineering program, and it offers useful material for analyzing the state of systems engineering in the middle of the 20th century.

Lockheed developed the C-5A at their Marietta, Georgia, factory. The technical and managerial team on the C-5A program considered systems engineering and systems procedures and processes to be quite effectively incorporated into their program culture. This was mostly because of the robust development methods at the Marietta location and the persistence of the core design team across multiple programs. The senior subsystem engineers had extensive working knowledge of the skills of other team members and were well-versed in Lockheed procedures. The Lockheed method was a well-documented set of formal procedures, processes, and tools that were fully integrated into the regular execution rhythm for a new program, while being relatively antiquated and incomplete by today's standards of systems engineering. PERT-TIME, the leading scheduling technology at the time, had to be used according to the terms of the Air Force contract. Lockheed's systems engineering methodology required extra flexibility to accommodate schedule changes even if PERT-TIME was the industry norm.

A motivated design team was essential to Lockheed's risk assessment methodology. Each design group was tasked with identifying risks, creating plans for mitigating those risks, and making sure that technical disciplines that could be impacted by the risk were involved in the risk solutions. Operations, manufacturing, and logistics were all included in this duty. Every day, risks were examined by the head engineer and everyone who directly reported to him (Griffin 2005). These were very significant meetings that the principals almost never missed.

Through the System Program Office (SPO), the government additionally took an active part in the process of identifying and mitigating risks. Additionally, the SPO assigned its engineers, logisticians, and other professional disciplines to handle the risk assessment activities. The purpose of these SPO internal processes was to inform SPO leadership of their own risk assessment findings.

Some of the major achievements and setbacks of the C-5A program can be linked to the application of systems engineering principles and procedures. Four important "learning principles" were identified in a post-mortem of the airplane's development (Griffin 2005):

1. *Requirements*: The defined processes and procedures produced a set of requirements that were balanced, easily understood by those involved, and mostly unchanged during the development process.
2. *Total package procurement concept (TPPC)*: A fixed-price, incentive-oriented contract was the outcome of the TPPC. Although the idea was to reduce the government's exposure to costs, it led to cost overruns and the delivery of fewer aircraft than anticipated under the original contract.
3. *Empty weight guarantee*: The initial contract's use of weight as a cost penalty (for overweight circumstances) and performance requirement had a major detrimental impact on the creation of the C-5A.
4. *Independent review teams (IRTs)*: The Air Force C-5 Systems Program Office applied the IRT concept to overcome technical obstacles in the development stage.

During the development of the C-5A, both Lockheed and the government stakeholder appeared to have implemented cutting-edge systems engineering processes and procedures. It might be claimed that while the methods had some worth and are accountable for some of the program's acknowledged results, they were deficient, especially in risk analysis tools. More specifically, the effects of early programmatic decisions were not adequately evaluated. For instance, choices were made to satisfy an excessively optimistic Weight Empty Guarantee, which ultimately led to fatigue problems with the pylons and wings, seriously impairing operational readiness and increasing costs. The processes and procedures gave Lockheed and the Air Force a way to communicate and work together to reduce problems during the design and development phases, but they did not give the necessary information to prevent choices that led to the problems in the first place.

The considerable cost overruns were one of the most prominent failures of the C-5 development. The USAF took action to address the lessons learned (learning principles) in the following two major procurement programs after the C-5A, the F-15 and the B-1, in response to the unfavorable attention this received inside the Government. The following programs seemed to perform a far better job of avoiding the C-5's faults, despite their imperfections.

7.1.3.3 International Example: International Space Station Case Study

This is an example of organizational, cost, and schedule risk. In many respects, the Shuttle-Mir Program in the early 1990s and the Space Station Freedom (SSF) concept served as a foundation for the creation of the International Space Station (ISS). The SSF program evolved into the ISS program in 1993 because of acting on a goal to bring down project costs and building on lessons learnt, in part, from the US-Russian partnership. The National Aeronautics and Space Administration (NASA), the Canadian Space Agency (CSA), the European Space Agency (ESA), the Japan Aerospace Exploration Agency (JAXA), and the State Space Corporation (Roscosmos) in Russia were among the international participants in the collaboration (ISS 2022).

On the surface, NASA was entrusted with the goal of providing its ten main operations centers with robust systems engineering process supervision. A Systems Engineering (SE) Handbook was kept up to date by the agency to outline procedures and best practices. Nonetheless, present and former engineers observed that established SE leads at NASA locations frequently relied on legacy programs with which they were acquainted or at ease (Stockman et al. 2010). Depending on the kinds of systems being designed or run, several historical methods to systems engineering were taken. The ISS program also introduced the element of multinational partners with differing ways of doing things and whose approaches to SE could differ greatly. Ultimately, the prolonged phases of research and implementation led to a turnover of skilled systems engineers who pursued alternative avenues for advancing their careers.

While NASA employed a variety of strategies and tactics to evaluate and control risk on the ISS, the primary emphasis fell on the application of Probabilistic Risk Assessment (PRA). Probabilistic Risk Assessment is defined as:

> … a comprehensive, structured, and logical analysis methodology aimed at identifying and assessing risks in complex technological systems. PRA is generally used for low-probability, high-consequence events for which limited statistical data exist. Its application is started at risk environments common within NASA that may involve the compromise of safety, inclusive of the potential loss of life, personal injury and loss or degradation of high-value property that may be found in NASA mission-related programs
>
> **National Research Council (U.S.) (2003)**

Simply said, PRA is an evaluation framework that assists engineers and management in identifying vulnerabilities in complex systems (in this case complex space systems). There are many successes and typically a good number of failures to learn from in a program with the size, complexity, reach (in terms of participants), and duration of the ISS program. It's no little thing that the International Space Station has run accident-free for the past 20 years, considering the harsh climate and constant danger of space travel. Therefore, it is important to consider the following specific risk management process failures in relation to the program's engineering successes:

Major schedule and cost overruns: NASA programs have a history of major schedule and expense overruns (Bearden 2008). It seems that the NASA model generates success-oriented timetables that are insufficient in their evaluation, quantification, and mitigation of development risk. The ISS Management and Cost Evaluation Task Force's conclusion that the program plans and budget were not credible is especially noteworthy. NASA was unable to create and carry out a thorough cost estimate for the International Space Station. NASA lacked the finance staff and basic tools necessary to develop and monitor a comprehensive lifecycle cost estimate.

Logistical support: Maintenance appears to be a recurring problem for space stations. NASA could not escape having maintenance be the ISS residents' principal responsibility, even though they were aware of the problem throughout the Shuttle-MIR cooperation. Counting the time it took to assemble the station, astronauts worked on preventative and corrective

> maintenance of the International Space Station (ISS) for more than 4,000 hours in the 5 years following the station's inaugural occupation from 2000 to 2005 (Bertels 2006). This resulted in part from the systems engineers' difficulties in creating actual Mean Time Between Failure (MTBF) criteria. Because the engineers tended to base their evaluations on "earth values," space station storage was squandered holding spare parts for systems that proved to be more reliable than anticipated, while other vital systems experienced far more frequent failures than anticipated. Major interruption to the computer occurred during STS-117. Even though no astronaut was lost in the incident, the main cause – which was ultimately identified as a single point failure built into the system – was a humiliating oversight that should have been prevented.

The worldwide consortium in charge of the mission added to the developmental difficulties of establishing the ISS by introducing uncertainty. In addition to the basic problems with communication and conflicting systems engineering philosophies, the systems engineers were continuously faced with schedule and financial uncertainties, which made it difficult for them to come up with solutions to achieve their goals. One of the most obvious lessons from the ISS example is that NASA's systems engineering techniques have a long history of failing to anticipate and reduce schedule and expense overruns (schedule and cost risk). This implies that if a more thorough procedure could be undertaken to address the root cause(s) for these overruns, there would be a large chance for improvement, and reducing risks in these areas.

7.1.4 Understanding Decision-Making and Its Connection to Systems Engineering

Mathematicians, scientists, and engineers have long explored decision-making and made useful contributions that can be exploited by systems engineers designing and developing all kinds of systems, including optical and imaging systems. In this section, we present an overview of some useful decision-making methods including an introduction to the powerful methods of GT. GT and decision analysis are two complimentary fields of decision-making with decision analysis methods emphasized in this section, and GT emphasized in Section 7.1.5. In this section, we focus on helping the reader understand how to effectively analyze decisions, and so introduce decision analysis along with relevant examples. Some instances of contemporary uses of the decision analysis round out this section.

7.1.4.1 Decision Analysis and GT in Systems Engineering

An inherent feature of SE is in providing the tools and methods to make good decisions. To that end, a few useful systems engineering decision-making tools will be presented in this section, covering both the method and its utility. As we have done in the prior section on risk, we start with a practical definition of decision analysis to help focus our thoughts.

In decision-making, a choice is described as an "irrevocable allocation of resources" in the context of decision analysis, which means that it is very difficult to reverse the actions or expenditures made (Howard 1966). Howard states that a decision is an action made toward a goal rather than a mental choice to do something. Additionally, he distinguishes between a good choice and a good result. We desire a good result, and good results likely stem from logical decisions that we make that we believe have a high probability of leading to that good result (Howard 1966).

According to Buede and Miller, "decision making is a process that is conducted by an individual or organization," and the goal of the process is to use some factor or set of measures to positively impact the future of the entity undertaking it (Buede and Miller 2016). This decision-making process in SE can be facilitated through trade studies, modeling, and/or analytical activities that provide insights into favorable program schedule, cost, and/or technical elements. For example, one of the most common tasks for systems engineers is to select alternatives. Generating (and selecting) the best alternatives is an important aspect of SE when it comes to evaluating a set of options.

Many of the alternatives may not even be attainable at first, nor are they necessarily well-defined or easily accessible. An underlying idea is that it is preferable to think about some initially thought to be impractical options, then to overlook one that might be more efficacious. An option cannot be chosen if it is not considered (Blanchard and Fabrycky 2010).

Various works talk about what they consider to be the three elements of a decision. Summarizing these thoughts, we have: (1) What is achievable, or what are the available options? (2) What is the current state of knowledge, or what kind of data is available for use and how much of it is there? (3) What is wanted, or preferred (Howard 2007)? In terms of systems engineering, these three elements can be phrased as (Buede and Miller 2016):

1. What needs or requirement do the stakeholders have?
2. What alternatives are available?
3. What do we know (or not know)?

The combination of these elements, taken together, meet the requirements for a choice, but one essential piece is still missing. According to Howard's previous research, deciding what decision must be made should be the first priority in the decision-making process. He goes on to say that this is a question that has frequently not been asked, before considering potential solutions. Additionally, we need to distinguish between evaluating likely outcomes in situations where there is no inherent control, and evaluating decisions where resources need to be allocated (there is inherent control). Decision analysis is only useful in the later instance (Howard 1966).

Decision analysis is defined as the "logical technique for the weighing of the factors that influence a decision" (Howard 1966). Decision analysis can be done with many techniques available to systems engineers, many of which are discussed throughout this book. For example, systems engineers frequently view decision analysis within the framework of trade studies, as was previously indicated. This makes logical sense since trade studies are essentially exercises in making decisions (Ullman and Spiegel 2006). Formally, a trade study is an analysis carried out to rationally assess options and suggest the most desirable workable option that enhances the system or makes it possible for it to satisfy functional criteria (FAA 2014; NASA 2016). Typically, a predetermined set of cost, performance and schedule-based metrics are used to analyze the options. Another technique for decision-making that enables the analytical comparison of quantitative or qualitative criteria for a set of alternatives is the analytical hierarchy process (AHP). AHP can compare and preferentially rank system features using a pairwise preference ranking method. The AHP is a useful tool that also includes a feature for verifying the consistency of analytics used in determining the preferences (Saaty 1987).

Decision criteria: Decision criteria are used in many disciplines, including systems engineering and operations research, to great effect, and are generally useful in characterizing different decision-making methods and approaches. The decision criteria can consist of a strategy, set of rules, or a set of guidelines that are used when decisions are made. We provide some examples of typical decision criteria that are often used in making decisions.

1. *Maximin criterion*: Find the worst potential outcome for each of the provided actions. Choose the "best" worst-case scenario. Without considering the best-case situation, the maximin criterion's main goal is to minimize bad results in the worst-case scenario (Winston 1994). This approach to decision-making is sometimes described as being highly pessimistic (Denœux 2019; Maschler et al. 2013).
2. *Maximax criterion*: Find the optimal result for each of a given set of actions. Choose the course of action that will yield the "best" result. This criterion does not consider the worst-case situation to determine the maximum benefit given all feasible actions (Winston 1994). This is associated with a very positive mindset on the decision maker's (DM) behalf (Denœux 2019).

3. *Minimax regret*: Determine the maximum regret for each of a set of specified actions. Decide which action you regret the least. The gap between an action's reward and the best potential reward for a particular world state is known as regret. This is also known as the loss of opportunity (Winston 1994). Another name for this strategy is pessimistic (Savage 1951).
4. *Expected value criterion*: Find the expected value of each of the provided actions. The action with the highest expected value should be chosen (Winston 1994). To use this approach, the probability for each world state needs to be known (Li 2003).

An aside on bias: Addressing potential biases in decision-making processes is one goal of risk and decision analysis. A basic overview of bias and its function in risk and decision analysis within the framework of SE is provided in this section. A report from the Committee on Risk and Decision-Making of the U.S. National Research Council states that a person's beliefs and impressions affect how they perceive risk. These impressions are usually created, "using faulty processes of inference, which may result in systematic distortions and misperceptions of reality," based on incomplete and frequently biased information (National Research Council 1982). We call these false impressions of reality cognitive biases. Another definition of cognitive biases is, "Cases in which human cognition reliably produces representations that are systematically skewed compared to some feature of objective reality" (Haselton et al. 2015, p. 968). Cognitive bias can appear in a variety of ways. A summary of typical cognitive biases is shown in Table 7.1, along with an example given in the context of a systems engineering-oriented circumstance (Ehrlinger et al. 2016).

TABLE 7.1
Cognitive Biases and Their Applications to Systems Engineering

Bias	Description	SE Example
Anchoring bias	The tendency for a reference point, or "anchor," to have an impact on a decision or view (Tversky and Kahneman 1974). The anchor is frequently the first item of information that is found	In a trade study, leadership is given two options (a $2M option mentioned first, and a $500K option mentioned later). The second choice seems more affordable than the first one with respect to the anchored $2 million option
Confirmation bias	The tendency to choose, look for, or interpret data in a way that supports beliefs or viewpoints that are currently held (Nickerson 1998)	Discounting outlier test data points that potentially skew results unfavorably, based on previous tests that had gone well, and schedule pressures to move things along. Our tests were good before, so these test results must be wrong
Focusing illusion (focalism)	The tendency to judge or make decisions based too much on a specific portion of the available data (Schkade and Kahneman 1998)	Focusing too much on the price of a particular alternative while undervaluing other facets of the design process
Framing	The tendency for preferences or beliefs to change according to how information is "framed" or presented (Tversky and Kahneman 1974).	Describing the anticipated outcomes of a flying test as an 85% success rate compared to a 15% failure rate
Overconfidence bias	The tendency for a person to have greater faith in their skills and expertise than is justified by data (Moore and Healy 2008)	The decision to develop a complicated system component all in-house instead of partnering or subcontracting, when critical subject matter expertise is lacking
Planning fallacy	The occurrence of people underestimating how long a task will take, even when past performance indicates otherwise (Buehler, Ross, and Griffin 1994)	Project managers and engineers are seldom quite on schedule and are usually overly optimistic about the amount of time, money, and risk involved in finishing a given set of activities

Because these biases exist, this may be sufficient justification to enlist the support of analysis tools that can mitigate these biases and relative subjectivity in decision-making. While these analytical tools certainly can help, care must be taken so that the inputs given to these models aren't also subjective or biased (Montibeller and von Winterfeldt 2015). Depending on how input is requested (framed) of them, or how they observed someone else providing the information initially, engineers who are asked to contribute to various tools and analyses may give information in different ways (anchoring). Systems engineers need to take these input biases into consideration. For example, when presenting the outcomes of risk and decision analysis, they need to take bias and the presence of subjectivity into consideration. For instance, engineers contributing to an analysis may have subconscious schedule pressures that may lead to overly optimistic analysis of the risks (suffering from the planning fallacy). As a result, an analysis may yield conclusions that are unduly optimistic. In the context of planning fallacy, on occasion, there are heuristics or guidelines to help combat biases. However, Hofstadter's Law states that "it always takes longer than you think, even when you take Hofstadter's Law into account" (Hofstadter 1979). The practicing systems engineer should be practically aware of these biases and vigilant for methods to mitigate them.

In addition, systems engineers should avoid motivational biases, which typically involve an individual's conscious prejudices resulting from "self-interest, societal pressures, or organizational context" (Montibeller and Von Winterfeldt 2015). It is relatively easy to envision how motivational bias can occur in the context of systems engineering. Both engineers and PMs may experience pressure to deliberately make expedient and suboptimal decisions to have short-term success, particularly when faced with financial incentives or, on negative pressures such as threats of losing their job or contracts. Systems engineers need to be mindful of these potential biases while doing risk and decision analysis, even if they may not have much control over them.

In Systems Engineering, decision analysis and GT have been widely applied. Simple decisions are rare in today's complicated or complex system engineering world. Most of the time, decisions must be made based on a trade-off between several criteria. For this reason, developing and applying Multiple Criteria Decision Making (MCDM) methods are essential in making and defending good decisions.

It is also true that, while making decisions about alternatives, we frequently don't know with certainty what will happen. Said another way, many decisions entail risk, and the probability of the risk needs to be part of the consideration. Maybe now more than ever, having the ability to defend a decision is required. This is due to the painful reality that, sometimes, there are no good choices available, yet still an optimal choice needs to be made. Understanding and being able to apply techniques to support risk-related decision-making must be a core part of a systems engineer's tool kit and professional makeup.

Finally, there are situations where several parties must make decisions based on the same parameters, either simultaneously or sequentially. This decision-making scenario is such that the decisions made by each party can affect those of the other parties. In this situation, a strategy must be developed for making competitive decisions and the field of GT is highly beneficial. GT is based on strategies for different kinds of competitive scenarios, or "games." GT and decision analysis are related and complementary fields. The focus of each is on making better decisions. On many occasions, the same people made significant contributions to both fields. The most significant contributors were arguably John Von Neumann and Oskar Morgenstern, who established the foundation for the current EU model in decision analysis through their groundbreaking work on GT, "Theory of Games, and Economic Behavior." GT focuses on the interactions between several DMs, whereas Decision Analysis concentrates on a single DM making decisions to optimize an outcome. This section continues with focusing on decision analysis, and Section 7.1.5 will focus on GT.

Ronald Howard once said, "Making decisions is what you do when you don't know what to do" (Howard 1980). According to Howard, decision analysis is a process for creating effective decisions. Howard believed that effective decision-making includes an artistic component and did not restrict decision analysis to only the realm of the rational. From a professional perspective, decision analysis provides support to individuals who make decisions. Hence, herein we focus on logical strategies for making decisions in the context of systems engineering.

There are a variety of models and methods for decision analysis that help with decision-making, particularly in complex situations. For example, there are methods that can be used for quantifying goals and objectives, choosing objective metrics for evaluating choices and options, and for determining weighting factors for comparing prospective alternatives/criteria in the evaluation process. There is much information available in the literature concerning these topics. To understand decision analysis, it is helpful to first start with a definition for decision-making. The fundamental elements of making a decision, according to Bierman, Fouraker, and Jaedicke, can be broken down into abstraction, modeling, and solutions, and each of these is defined as follows (Bierman et al. 1961).

Abstraction: There are practically an infinite number of surrounding facts to a real-world problem. There is no way the DM can consider them all. He or she must pick the most important data and disregard the rest. Abstraction is the definition of this process (Bierman et al. 1961).

7.1.4.1.1 Model Building

To create a model of a problem, the decision-maker must identify and integrate essential and representative elements of the problem. The best model is a basic and simple one. According to Bierman, Fouraker, and Jaedicke, a simple model must be efficient in time and well thought out. The DM must also be able to understand and modify the model. The model must also be able to determine the anticipated outcomes associated with the planned actions of the DM (Bierman et al. 1961).

7.1.4.1.2 Solutions

A logical analysis of the results of the models can be used by the DM to obtain an effective decision. If the model that was used was effectively abstracted and contains the essential elements for the decision-making problem so that useful results are provided, then the model will aid in real-world decision-making (Bierman et al. 1961).

The modeling aspect of decision-making is the focus of the remaining discussion on decision analysis. The types of decisions that need to be made can be divided into two categories: those made in the face of certainty, and those made in the face of uncertainty. The same rules apply when certainty is approximated.

In a quantitative analysis, decision-making under uncertainty is typically described by a goal, like profit or loss, that must be maximized or minimized under certain restrictions, such as costs. The problem can usually be quantitatively modeled if the objective(s) and constraint(s) are quantifiable (Bierman et al. 1961). There are numerous models available that accomplish this. These kinds of models are the focus of the field of operations research. Queuing models, transportation algorithms, and linear programming are a few illustrations (Philips et al. 1976).

Making decisions for a single criterion is relatively straight forward. Depending on the goal, the option that maximizes the goal with a minimum, targeted, or maximum value wins. A rank order of preferences can produce a "solution" to qualitative problems like which color BMW to purchase. Decisions sometimes are also made from the perspective of a given site. This is the so-called site-centric decision making (SCDM).

Decision-making becomes more complicated when there are several criteria involved. A decision-maker now needs to weigh the relative benefits of each criterion in addition to the many options that need to be considered. As a result, Multi Criteria Decision Making (MCDM) has received much favorable attention in the literature. In his book *Fuzzy Multi-Criteria Decision Making: Theory and Applications with Recent Developments*, Kahraman Cengiz presents the two MCDM methodologies of Multiple Objective Decision Making (MODM) and Multiple Attribute Decision Making (MADM) (Kahraman 2008). In MODM, there are multiple objectives that need to be considered and addressed in the modeling process. MODM uses several continuous objective functions to characterize the trade-offs between evaluation criteria. Conversely, MADM problems have multiple attributes instead of multiple objectives and have discrete decision spaces. We focus on MADM methods below because of their common usage.

Sorting and ranking techniques are an important aspect of MADM models. In most MADM problems, the DM is actively involved in the decision-making process and contributes in some way to the ultimate decision. MADM techniques fall into one of two categories: compensatory or non-compensatory. The DM's belief that strong success in one area can offset poor performance in another is reflected in compensatory MADM methods, which entail trade-offs. Such trade-offs are absent from non-compensatory methods. There are two types of MADM methods: one is for reaching a single decision, and the other is for screening, which involves eliminating certain alternatives (Kahraman 2008). We describe some of the MADM methods below.

7.1.4.1.3 Dominance

Alternatives are compared using this screening process. The dominance of one alternative over another is defined as the degree to which one option meets or exceeds all requirements. The dominant solution can either be accepted or rejected depending on the application (Kahraman 2008).

7.1.4.1.4 Maximin

As previously summarized, according to the maximin method, the saying that, "a chain is only as strong as its weakest link" applies and is considered in the decision-making process. The alternative in this model is assigned the score of the criterion that performs the lowest. This is predicated on the idea that an attribute's performance metrics are normalized and comparable (Kahraman 2008).

7.1.4.1.5 Maximax

The opposite of maximin is called Maximax. Here, the alternative takes the highest performing criterion's score as its default. As for Minimax, performance metrics must be standardized or comparable (Kahraman 2008).

7.1.4.1.6 Conjunctive

The conjunctive approach is a screening technique that is also known as *satisficing*. Every criterion in this method has a minimal bar for performance. It is possible to exclude any option that does not satisfy *any one* of the minimal requirements (Kahraman 2008).

7.1.4.1.7 Disjunctive

The disjunctive approach, another screening technique, is connected to the conjunctive approach. For every criterion, it uses a minimal standard, just like in the conjunctive method. However, if an option doesn't achieve the minimum requirement for *all* criterion, it is eliminated, as opposed to if it just fails to reach the minimum level for one criterion as in the Conjunctive case (Kahraman 2008).

7.1.4.1.8 Lexicographic

The significance of order is what gives this approach its name. The attributes in this strategy are ranked according to how important they are to the choice. The selection factor is the most significant attribute. Whichever option performs the best for that given attribute wins. The next most significant characteristic is employed as the discriminator in the event of a tie, and so on (Kahraman 2008).

7.1.4.1.9 Elimination by Aspects

This approach, which is more widely referred to as "the process of elimination," can be conceptualized as a synthesis of the lexicographic and conjunctive procedures. Just like the lexicographic, a ranking of the attributes is needed. It uses minimum requirements, just like the conjunctive technique. This method begins with the most preferable characteristic and eliminates any options that fall short of the required minimum. The procedure is then carried out again for the next most important attribute, and so on, until just one option is left. This method can be a useful screening technique (Kahraman 2008).

7.1.4.1.10 Linear Assignment

Using this method, each attribute is given a cardinal number as a weight. Ordinal rankings are assigned to the alternatives for each attribute based on how well they perform in that attribute. Keep in mind that compensatory analysis can be done with the ordinal ranking. However, this approach might only be regarded as semi-compensatory because attribute value preference is often nonlinear (Kahraman 2008).

7.1.4.1.11 Additive Weighting

An alternative's preference ratings for an attribute are cardinal in additive weighing systems. Each attribute has a cardinal weight given to it, just like in a linear assignment. Summing the weighted ratings gives the alternatives a score. One popular application of additive weighing is the Analytical Hierarchy Process (AHP), which will be covered in more detail below (Kahraman 2008).

7.1.4.1.12 Technique for Order Preference by Similarity to Ideal Solution

The objective behind this approach is to identify the option that comes the closest to the perfect answer. Like a fantasy sports team, the ideal answer is a combination of the best performance metrics attained by any of the options for each attribute. Next, each option is assessed based on how close it is, in an Euclidean sense, to the optimal answer. Put otherwise, a sum of the squared deviations from the ideal for each attribute in the "attribute space" is calculated for each choice. The score is the square root of the total. An ideal that is a composite of the best conceivable performance measures rather than a combination of the best performance measures that were demonstrated is the distinguishing feature of Technique for Order Preference by Similarity to Ideal Solution (TOPSIS) (Kahraman 2008).

Let's look at a few of these techniques using an example. Let's say a lens material is being assessed by an optical engineer. Several criteria that were determined as important are shown in first column in Table 7.2. Additionally, the minimum standard for these attributes is shown in the last column of this table. These minimum standards were determined by the stakeholder. Materials A through D consist of four possible options for constructing the lens material.

Higher is considered better when it comes to the Abbe number; in this case, 35 is the lowest acceptable value. Scratch resistance is rated from poor to exceptional in ascending order of preference, with poor being ranked lowest (1), fair coming in next (2), good coming in higher (3), and excellent coming in highest (4). Two is the bare minimum of an acceptable rating. The lower the numerical value that represents impurities, the better. The highest number that can be used with regard to impurities is 0.03. It is possible to exclude one of the options by applying the Conjunctive technique. In other words, material D can be eliminated from the decision since it does not satisfy the minimal criteria of 35 for the Abbe number. Dominance allows for the elimination of a different

TABLE 7.2
Lens Material Comparisons

Criterion/	Lens Material				Minimum
Attribute	A	B	C	D	Std.
Index of refraction	1.50–1.54	1.60–1.62	1.490–1.492	1.584–1.586	$1.4 < n < 1.65$
Abbe number	59	58	43	31	35
Scratch resistance	3	3	2	2	2
Impurities	0.0034	0.0025	0.013	0.015	0.03

option. It is evident from a comparative analysis that options A and B outperform C in every way. Consequently, option C can be eliminated because the other two options dominate.

Unfortunately, using elimination to reduce the options to just one is rarely feasible. Instead, we can use Linear Assignment to determine which of the remaining options to choose. Assume that the following weights have been applied by our engineer to the criteria. Each weight for the criterion can be a percentage of its overall importance. Let's say that the engineer places a great deal of importance on the Abbe number, which is assigned a relative weight of importance of 0.45. The engineer also values the relative importance of the impurities, which are assigned a weight of 0.4. With an assigned weight of 0.1, scratch resistance is not as significant as the other criteria. Finally, the refractive index is given an assigned weight of 0.05.

Next, each of the criteria is evaluated to ensure they fall into a "more is better" category since weighted scores of the criteria are summed for each option. Note that the impurities metric follows a "less is more" scaling and so needs revision. One can rewrite the impurity percentage metric as a "purity" measure, which is equivalent to 1 minus the impurity. To ensure that the index of refraction falls into the required range, a Boolean value of 1 is assigned to the refractive index when it falls inside the range, and 0 otherwise. Finally, adding up the weighted ranks of the various criteria would punish the smaller scale measures because they do not contribute as much as their larger counterparts. Consequently, our engineer has normalized the range of each measurement to a scale between 1 and 100 so that each attribute has the same influence under summation. Remember that options C and D were ruled out earlier and only options A and B need to be compared. The adjusted values are shown in Table 7.3.

The weighted ratings for Material A and Material B are now summed and seen to be 96.61 and 95.9, respectively. In this case, Material A ends up being the best option – but not by much.

Assigning weights to criteria can be challenging since it involves a lot of subjectivity. One variation of linear assignment is the analytic hierarchy process (AHP). This approach promotes uniformity to enhance the process of assigning subjective weights. The creator of this method is Thomas Saaty. Saaty generated a method of choosing the best alternative from a set of criteria by casting the decision problem into a linear algebra framework using Eigen vectors and Eigenvalues. The decision problem takes on the form

$$Av = \lambda v, \tag{7.6}$$

where the preference matrix A multiples the Eigen vector $\boldsymbol{v}$ on the left side of Equation 7.6, and λ is the Eigen value that multiples the Eigen vector on the right side of the equation (Kandiller 2007).

During this process, each criterion is compared pairwise with all other criteria. Let's say the engineer uses AHP rather than giving each criterion a set weight. By doing this, he creates the pairwise comparison matrix that is shown below in Table 7.4.

The engineer is now employing ratios, which indicate pair-wise comparisons for each row and column, in place of weights. The row/column is used to express the ratios. Each combination simply

TABLE 7.3
Weighted Method

		Material A		Material B	
Criterion	**Weight**	**Rating**	**$W \times R$**	**Rating**	**$W \times R$**
Refraction index	0.05	100	5	100	5
Abbe number	0.45	98.33	44.25	96.67	43.5
Scratch resistance	0.1	75	7.5	75	7.5
Impurities	0.4	99.66	39.86	99.75	39.9
		Total	**96.61**	Total	**95.9**

TABLE 7.4
Criterion Comparison Matrix

	Refraction	Abbe	Scratch	Purity
Refraction	1	1/9	1/2	1/8
Abbe	9	1	4.5	4.5/4
Scratch	2	1/4.5	1	1/4
Purity	8	4/4.5	4	1
	20	2.12	10	2.51

TABLE 7.5
AHP Criteria Preference Matrix

	Refraction	Abbe	Scratch	Purity
Refraction	1.00	0.01	0.50	0.13
Abbe	9.00	1.00	4.50	1.13
Scratch	2.00	0.22	1.00	0.25
Purity	8.00	0.89	4.00	1.00
	20.00	2.12	10.00	2.51

TABLE 7.6
Normalized Comparison of the Criterion versus Criterion Matrix

	Refraction	Abbe	Scratch	Purity	
Refraction	1/20	01/2.12	0.5/10	0.13/2.51	0.039
Abbe	9/20	1/2.12	4.5/10	1.13/2.51	0.455
Scratch	2/20	0.22/2.12	1/10	0.25/2.51	0.101
Purity	8/20	0.89/0.212	4/10	1/2.51	0.405
	20.00	2.12	10.00	2.51	

needs to be compared once. For instance, the engineer enters nine for the Abbe to Refraction ratio and the inverse of that, or one-ninth, for the Refraction to Abbe ratio after figuring out that he values the Abbe number nine times more than the Refraction index. Observe that since a factor cannot be preferred greater or lower than itself, the diagonals will always equal one. After adding up the columns, our engineer displays the results in Table 7.5. For the sake of the next step, Table 7.5 ratios have been transformed from fractions to decimals for clarity.

As seen in Table 7.6, the following step is to normalize each column by dividing each cell by the sum of the column entries. The row means, which is displayed on the right, is calculated by averaging the rows.

Our final calculations will use the row means as the Eigen vector. After generating an Eigen vector based on our criterion versus criterion matrix, we must repeat the procedure for every alternative, comparing all alternatives for each criterion as shown in Table 7.7. Our optical systems engineer rates the performance of materials A and B in comparison to one another for each criterion. These ratings are pairwise comparisons same as before. A table for each option is produced after comparing, summing, normalizing, and determining the row means as seen in Table 7.7.

TABLE 7.7
Alternative Comparison Matrices

	Refraction				Abbe Number		
	A	B			A	B	
A	0.500	0.500	0.500	A	0.468	0.468	0.468
B	0.500	0.500	0.500	B	0.532	0.532	0.532
	Scratch Resistance				**Purity**		
	A	B			A	B	
A	0.500	0.500	0.500	A	0.468	0.468	0.468
B	0.500	0.500	0.500	B	0.532	0.532	0.532

TABLE 7.8
Solution Matrix

	Refraction	Abbe	Scratch	Purity		Means			
A	0.500	0.468	0.500	0.468		0.039			
B	0.500	0.532	0.500	0.532	×	0.455	=	0.473	A
						0.101		0.527	B
						0.405			

Notice that the refraction and scratch resistance were equally preferred in both potential solutions. In the last stage, the normalized means of the alternatives create a new matrix that is used in the Eigen vector computations. Table 7.8 shows this matrix on the left side.

Notice that the new preference matrix shown on the left is just the row means determined from the alternative versus alternative matrices for each criterion. The vector in the center of Table 7.8 is the Eigen vector that was determined from the criterion versus criterion matrix.

The solution to the equation $Av = \lambda v$, is shown in Table 7.8 where the preference matrix A is given by the 2 row by 4 column matrix on the left, and the Eigen vector is the 4 row by 1 column matrix at the center. The vector on the right of the equal sign, which shows the value of options A and B, is the solution. Option B is the better choice in this instance since it has a higher value.

As a separate example, probability needs to be considered while making decisions in the face of uncertainty. Often when making a decision, the likelihood of an event happening or not happening can be determined. For independent events, a decision tree can be constructed to determine the expected value of a series of decisions. The expected value is determined by multiplying a set of outcomes in the decision-making process by their probability of occurrence. For the "bigger is better" case, the expected value with the largest value is the best option. Let's look at an example.

We use a scenario where an optical engineer (OE) is building a lens for a military system that will be deployed for 5 years. Anti-reflective coatings are required for the lens. The OE examines the criteria for supportability as part of the systems engineering process and attempts to choose between two coating vendors. Each provider has the capability of producing a product of comparable quality.

A given provider (A) has a cost of $1,200, which is less than the costs for the other providers, but there is a reduced chance that they will fulfill all of the scheduled deliveries. The other provider (B) has an increased cost of $1,700, but they are more likely to meet the delivery demands. The OE calculates that the financial impact of missing deliveries during the system's lifecycle will be $5,000 as part of the systems engineering process. The decision tree for this probabilistic decision is initially shown in Figure 7.14.

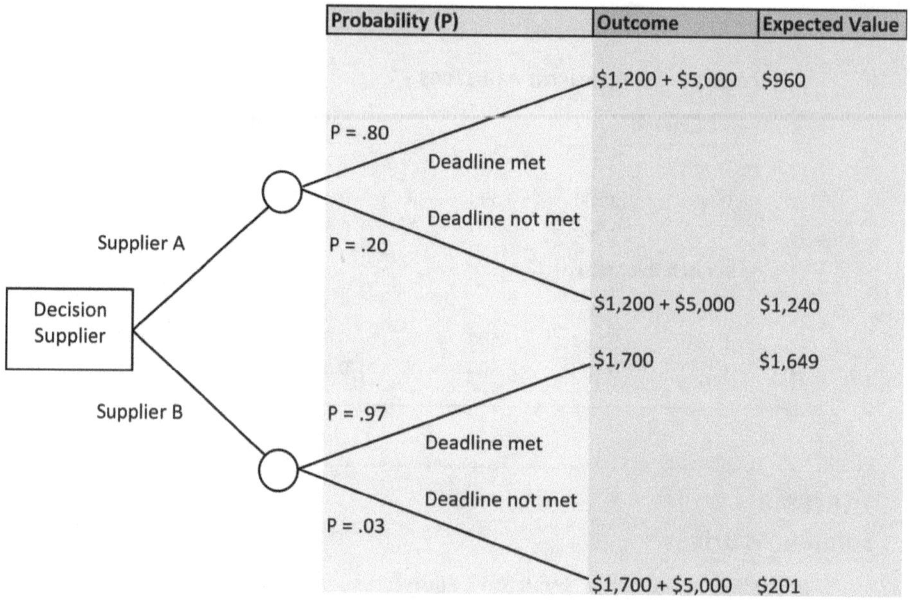

FIGURE 7.14 Decision tree.

A branch connecting the two supplier options – Supplier A and Supplier B – represents the fundamental choice. The circle stands for the decision's uncertainty. There are two possible outcomes for each supplier, which are represented as branches, and each eventuality has a probability *P* that is shown along the branch lines. The Outcome column indicates how much the outcome will cost. Keep in mind that the cost of every scenario where the scheduled delivery is not made is increased by $5,000.

We first start with modeling the impact of not meeting the schedule. If the probability that vendor A meets the deadline is 80%, then the probability that vendor A does not meet the deadline is 100−80=20%. Similarly, if vendor B's probability of meeting the deadline is 97%, then vendor B's probability of not meeting the event is 100−97=3%. The decision node can then be split into two cases as shown that illustrate all the possible outcomes. The entire cost of an occurrence multiplied by its likelihood/probability yields the expected value for each outcome. Every decision's expected value is determined by adding up all the expected values on the connected branches that result from that decision. Notice that for the top outcome, it might be tempting to assess the $5,000 scheduled delivery charge to this cost since there isn't a 100% chance of meeting the delivery. However, that $5,000 charge for not meeting the deadline does not apply on this line since the assumption is that the "deliveries are met." Consequently, this charge should be removed from the calculation for any of the decision nodes that assume deliveries have been met. In this case (eliminating the $5,000 in the top line) the expected value for the top line of branches results in $1,200×0.80=$960 as shown. From the top two sets of branches in the decision tree, the expected value of choosing Supplier A in this case is $1,240+$960, or $2,200. If Supplier B is chosen, the expected value will be $1,649+$201, or $1,850. For this example, Supplier B is the recommended option because of the lower comparative expected value (lower is better in this case). When using expected value to guide decision-making, a choice with a higher cost but a lower likelihood can very well be the best one. On the other hand, humans tend to focus more naturally on the most likely outcome than on the best possible expected value.

7.1.4.2 Decision Analysis Applications

There are many different types of applications in decision analysis. Whether or not the decision made has subjective or objective decision criteria, or whether the decision is made under certainty or uncertainty, there are useful models that can be used. For example, in industry, Linear Analysis methods can be used when it comes to vendor selection for software purchases. These methods also apply to personal decisions such as in helping decide which TV to buy. A list of desired features can be generated, ranks and relative weights assigned, and the decision can be analytically formulated. This can be a good way to eliminate buyer's remorse.

Systems engineering and organizations that use decision management have long employed decision analysis. The medical industry has applied MCDM approaches extensively in the recent past. Medical personnel often must make decisions within a framework of uncertainty. More specifically, choices on various courses of action include associated probabilities. A decision tree can be used to model these circumstances to great effect. The use of expected value decision trees for treatment selection was published in *Anesthesia* by Dr. S. M. Yentis. This kind of analysis was done on a fictional tumor case and a caesarian case in the publication (Yentis 2006). An article detailing the application of decision analysis in the field of endocrinology was published by Evelyn Mentari, MD, and David Aron, MD, MS (Mentari and Aron 2006). Even the agreement to participate in clinical experiments has been explored in relation to decision analysis and predicted utility (Baron 2006).

Environmental science is another area where decision analysis has been used. In their discussion of multiple decision analysis tools, Martin Drechsler and Mark A. Burgman address how population viability evaluations are carried out. The models covered in this article include statistical models as well as decision trees (Drechsler and Burgman 2004).

In traditional decision analysis, often a single person or entity chooses between several available options. In this case, decision analysis considers only one decision-maker who is attempting to optimize a result to a potential decision, whether that is to maximize, minimize, or hit a certain target. On the other hand, choices are usually not made independently and in a vacuum in the real world. Decisions made by one person or organization regarding specific factors may also be made by other individuals or organizations regarding the same parameters, perhaps in a competitive environment. Specialized modeling techniques are needed to account for the complicated or complex interactions that are possible with these types of decisions. GT provides an excellent approach to deal with these kinds of problems and decision-making in a competitive environment.

7.1.5 Game Theory

Although it seems like a lot of fun, GT is mostly about conflict. The concept comes from board games like chess, checkers, mancala, and other games where two competitors must battle it out to win. This antagonistic dynamic is taken up by GT, which aims to mathematically model it. The 1983 film War Games, which uses a then cutting-edge computer to simulate nuclear conflict between NATO and the former Soviet Union, serves as an excellent illustration of this. All those mathematical models led to "complete destruction," of course. Keeping this image in mind, every opposing force tries to "win the game" by taking actions that will advance their own interests and impede those of their opponents (Myerson 1997).

A few definitions must first be understood to have a meaningful conversation about GT. Players are "participating, independent decision makers" who are essential to a game (Kelly 2003). Participants can symbolize groups of people, entities, creatures, or the natural world. Every player has at least two options or courses of action. When every player has made their strategic choice, there is an outcome. Payoffs are values that are connected to desired results.

A pure strategy is the entirety of what a player would do to address every scenario. A player's decisions and their pure strategy are the same if they are made without being aware of the choices made by other players. There will be more pure strategies than options if the player chooses a strategy after learning about the strategies of the other players (Kelly 2003).

Games fall into one of three categories: strategy, chance, or skill. One way to think of games of skill is as one-player games where the player controls every possible outcome. Put otherwise, every strategic decision has a predetermined result. GT typically does not account for skill-based games (Kelly 2003).

Games of chance are "two-person" games in which nature, the symbol for chance, is the other participant. Games of chance can also be divided into subcategories such as risk and uncertainty. A game of risk is played if the first player is aware of the probability involved in Nature's choices. It is an uncertain game, however, if the probabilities are unknown.

Two or more players – none of whom are Nature – play strategy games. Each participant can influence some of the results. Cooperative, non-cooperative, and mixed-motive strategy games are further subdivided based on the motivations of the players (Kelly 2003).

Zero-sum games are those in which the victory of one player equals the loss of the other in non-cooperative games. According to Bierman, Fouraker, and Jaedicke, "If (1) the rankings of the payoffs by one party is the exact opposite of the rankings by the other, and (2) the utility functions of the two parties over these results add to zero for any given payoff," then the game is considered zero-sum Jaedicke (Bierman et al. 1961).

One famous instance of a zero-sum, non-cooperative game is the Battle of the Bismarck Sea. During World War II, the South Pacific is the setting for this example. American Admiral Kenney is preparing to destroy Japanese Admiral Imamura's transport while Imamura is moving soldiers across the Bismarck Sea. The two admirals can choose between a three-day Southern route and a two-day Northern route. Without knowing one another's route, the admirals must each select a course of action. The amount of bombing days, which is a gain for Kenney and a loss for Imamura, is the payout. Kenney can redirect his jets at the price of 1 day of bombing if he makes the incorrect decision.

Table 7.9 illustrates how a two-row, two-column payout matrix might be used to model this problem. Imamura is the column player, and Kenney is the row player. In bombing days, every reward signifies a win for Kenney and a defeat for Imamura. Since Imamura's goal is to minimize his loss, it is obvious that he will select option north because his position is either superior to or equal to that of option south. This decision is referred to as "dominant." Assuming Imamura will make the dominant (logical) decision, Kenny also goes with the north. This would optimize his earnings considering Imamura's choice of North (Peters 2008; Rasmusen 1989).

One famous illustration of a non-zero-sum game is the Prisoner's Dilemma. Melvin Dresher and Al Tucker at the Rand Corporation conducted the first analysis of this game in 1953 (Dutta 1999). In this instance, two inmates were co-conspirators. They each get a separate interrogation. Every prisoner has the option to either "defect" (D), which means to betray the other prisoner, or "cooperate" (C) with his partner. Ten years in jail are the penalty for the offense. A year is taken off a traitor's sentence for betrayal. The sentence of a prisoner is reduced to 1 year for a small offense if they are not betrayed.

This table can be used to model this problem. Two payouts are displayed for each scenario because this is a non-zero-sum game: the row player's payoff and the column player's payoff. Given that the reward is expressed in terms of years spent behind bars, it is seen as negative (loss).

Since defecting (D) always results in the biggest payout, regardless of the other player's actions, it is the plainly dominant alternative for each player. Cooperative play (C) yields the maximum total payment (–1 for each prisoner), which is what makes the game intriguing (Peters 2008; Dutta 1999). However, this will never occur if the prisoners each make a rational choice.

TABLE 7.9
Battle of the Bismarck Sea

	North	South
North	2	2
South	1	3

TABLE 7.10
Prisoner's Dilemma

	C	D
C	−1, −1	−10, 0
D	0, −10	−9, −9

TABLE 7.11
Prisoners Dilemma with Example Logic

	Prisoner B Keeps Silent (C)	Prisoner B Defects (D)
Prisoner A keeps silent (C)	(Silent, Silent) = jail time (−1, −1)	(Silent, Defect) = jail time (−10, 0)
Prisoner A defects (D)	(Defect, Silent) = jail time (0, −10)	(Defect, Defect) = jail time (−9, −9)

Let's look at Table 7.10 in more detail showing the important decisions required for each cell. The logic and various outcomes for both prisoners are shown in Table 7.11. Using the numbers for the example above, we'll say the optimal jail time from the prisoner's perspective should of course be 0. Moreover, the cell entries in the table are arranged as (P_A, P_B) where A and B represent which player is making the decision and P_A and P_B corresponding to the respective payoffs in jail time (shown as a non-positive number since jail time is a negative event from both players point of view). The payoff for each prisoner is dependent on the choices of the other prisoner.

Defecting (D) is the plainly dominant option available to each player, as it offers the biggest payout *independent* of the other player's actions. The reason this game is so intriguing is that if both cooperate (C), the maximum total payoff (minus one for each) would be reached; but, if both prisoners choose to defect, the payoff for both is bad (Peters 2008; Dutta 1999).

The goals of the players in a cooperative game line up with each other. For example, two bikes on a collision path approaching one another from opposite directions is representative of a completely cooperative game. It is in both bikers' best interests to avoid the accident. Every rider has the option to stay on route, go to the left, or go to the right (Kelly 2003). A more engineering-related example can be given as follows. A mechanical engineer designs a telescope's mount, while an optical engineer designs the telescope itself. The optical engineer needs a diameter of 36 cm to meet standards. The mechanical engineer creates an inadequate amount to meet budgetary constraints. The optical engineer has the option to maintain course or decrease the diameter and weaken the optical system. The mechanical engineer has the option to go over budget or stick to it. To achieve the best option, they both cooperate.

The interests of the partners in a mixed-motive game are partially aligned and partially at odds with one another. An illustration of a mixed motive game is provided by Anthony Kelly in Decision-Making with GT: An overview intended for managers. In his scenario, school administration and the teacher's union are at odds. If their requests are not honored, the union threatens to stop taking part in parents' evenings and the government inspection preparations. The management can either give in or resist. Resuming parents' evenings, starting inspection preparations, starting both, or starting neither is the union's option. Both parties want the return of parents' evenings, while management wants to deny the requests, and the union prefers management concession. This game has conflicting motives (Kelly 2003).

It is possible to characterize strategy games further. Complete or partial information can be found in games. Players in games with full knowledge are aware of both their own and other players' plans and payoffs. Players in games with imperfect knowledge are aware of their own payoffs and strategies, but they are unaware of their opponents' payoffs and strategies. Strategies are chosen sequentially in games of perfect knowledge so that players know what other players have selected. Players in games with imperfect knowledge must make decisions without knowing what the other players have selected (Kelly 2003).

7.1.5.1 GT Applications

The origins of GT are rooted in mathematics. Most of its early applications were in the subject of economics. Since then, GT has been applied in several domains. These include, among others, physics (Hauert and Szabo 2005), artificial intelligence (Chalkiadakis 2012), education, and sociology (Swedberg 2001). GT can be used to describe almost any situation in which there are several decision-makers who influence one another's decisions. More and more intricate game scenarios have been simulated and resolved using the GT approach. It has even penetrated the jewelry industry (Whitney 2010)! Figure 7.15 shows a GT classification tree that summarizes many of the points that we have made, and shows how GT can be effectively decomposed.

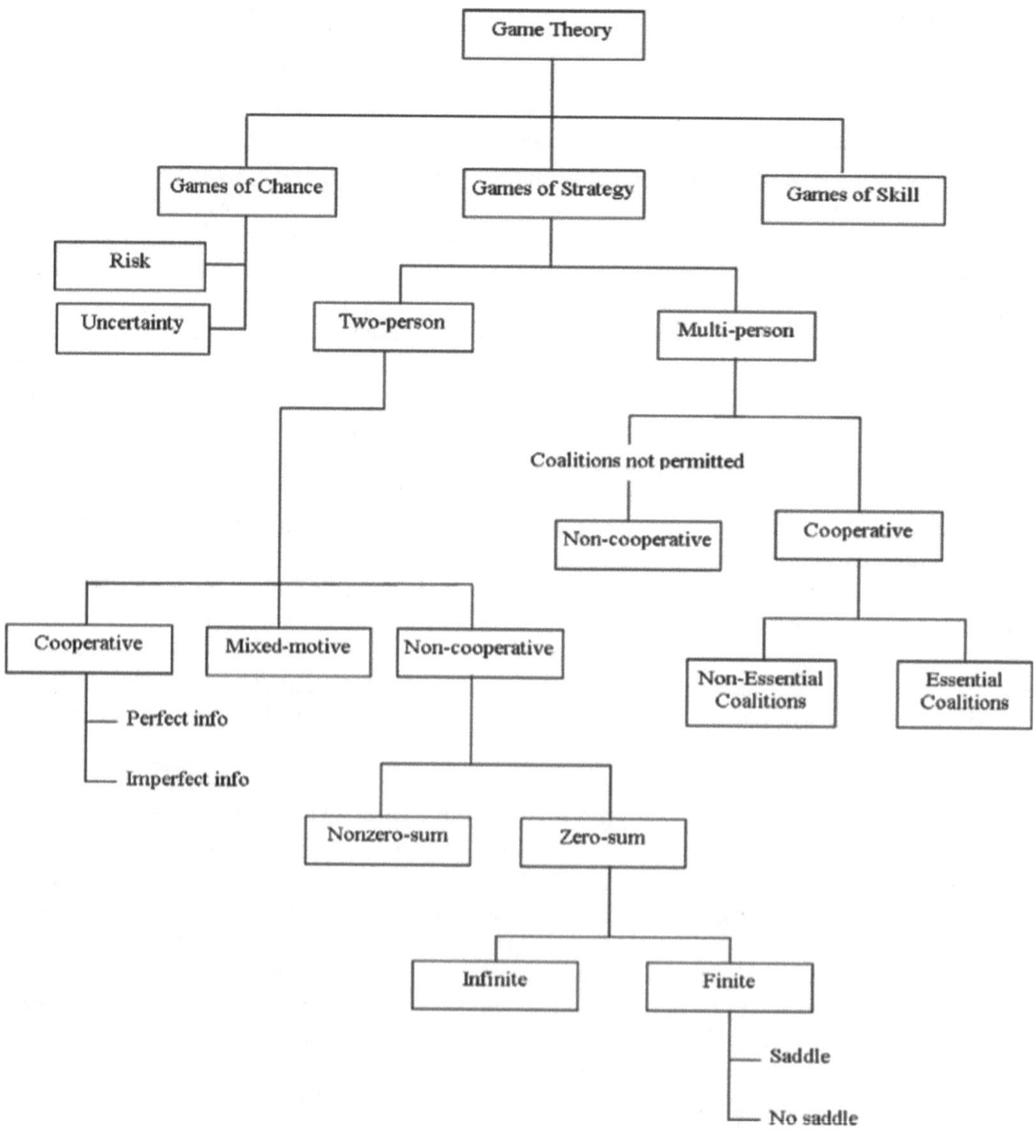

FIGURE 7.15 Game theory classification tree.

GT has seen several intriguing applications in optical recognition technology in recent years. In two related studies, Kaushik, Bhattacharya, and Suen used GT to extract iris information from eye photographs for the purpose of biometric scanning. The topics of "Iris segmentation using game theory" and "Iris recognition using shape-guided technique and game theory" investigated the challenges in separating the details of an eye's iris from its surrounding image. The authors show how to combine the complementing advantages of gradient-based and region-based iris boundary identification techniques using a game-theoretic approach. The two segmentation methods are the "players" in their game model, which is a cooperative two-player game with iteration and information sharing (Roy et al. 2011, 2012).

Iris feature selection is another area of biometric optical recognition where GT has been successfully employed. The same authors as for the game-theoretic segmentation methods, along with Jane You ranked the iris features using the Shapley value from GT. Here, the players represent characteristics of a data set, and the game is cooperative. The four metrics that represented the utility function are the classifier's Recognition Error (RE), False Accept Rate (FAR), False Reject Rate (FRR), and Feature Subset Size (FSS) (Roy et al. 2010).

Their approach was partially inspired by the work of Amit Chakraborty and James S. Duncan, who integrated boundary-based and region-based image segmentation using a GT method. The two segmentation models were the participants in a two-person nonzero-sum non-cooperative game model that they employed. Better image segmentation was produced as a result than either technique could have achieved on its own. The authors provided examples of image segmentation in a variety of image formats, including nuclear medicine single photon emission computed tomography (SPECT) images of the kidney and bladder, a Magnetic Resonance Image (MRI) image of the brain, and an MRI of a canine heart. The original image was shown in each instance, along with a comparison between the segmentations created by each segmentation method alone and those created by using game-theoretic integration. The integrated technique was "more resilient to both noise and bad initialization" (Chakraborty and Duncan 1999) and produced superior results in every case.

7.1.6 Special Considerations in Modern Day Decision-Making

In this section, we describe some important considerations that the modern-day decision-maker needs to understand and correctly assess. The prevalence of complex systems means that emergent behavior will become more likely requiring the necessity and ability to deal with unexpected events and situations. We will examine some of these considerations.

Collective behavior: Collective Behavior is the actions and behaviors of large groups of people that are loosely coupled together. An anthill or a beehive is the most traditional illustration of this. Even though the insects are all individuals, they cooperate because they have some basic drives such as the need for food to thrive (Peppler et al. 2013).

"Collective behavior provides a framework for understanding how the actions and properties of groups emerge from the way individuals generate and share information" (Bak-Coleman et al. 2021). When it comes to connecting collective behavior with complex systems, the relationship is readily apparent. Collective Behavior considers the function, kind, and classification of the person. It also considers how the person interacts with data, information, and objects. However, when people work together as a group, certain patterns become observable that are indicative of emergence, which is mentioned as a distinguishing feature of complex systems.

7.1.6.1 Collective Intelligence

This phrase is again typically thought of in the context of large group dynamics. The internet has made the term "Global Brain" incredibly relevant in today's environment. "A form of universal, distributed intelligence, which arises from the collaboration and competition of many individuals" is how collective intelligence is defined (Salminen 2012).

More people than ever before can be connected thanks to the internet these days. Everyone has a unique history, set of interests, and contributions to society. Within moments, people with similar interests from all over the world may come together to exchange and improve knowledge on a wide range of topics, and so give intelligence its "collective" quality (Yuan et al. 2007). Some truly astonishing and revolutionary ideas will emerge because of this consideration.

Pattern formation: Understanding patterns, where they come from, and how they form is an essential requirement for making good decisions. A variety of methods, mathematics, and models can be used to gain this understanding, and some common approaches are presented below.

Partial differential equations: Any equation with derivatives is a differential equation. The highest order derivative in an equation determines the order of the equation. For example: An automobile can travel a set distance, say 10 miles. From this we get a change of distance (the difference between the starting point of the car and the end destination – exactly 10 miles). If the car accomplishes this in 1 hour, we can determine the first derivative of its position (let x represent its position from its starting point) with respect to time, t. The first derivative of x with respect to time is dx/dt which is the cars change in position (i.e. its distance traveled, 10 miles) divided by the change in time it took to move that distance (1 hour). This rate provides the vehicle's speed (10 miles/1 hour = 10 mph). And what a slow rate it is! This derivative has order 1. If we now take the derivative of the speed with respect to time, or $d/dt(dx/dt) \Rightarrow d^2x/dt^2$, we will take the change in speed over the change in time to provide the acceleration. If the speed is constant, the change in speed is zero and there is no acceleration, and the vehicle moves at constant speed. It is important to distinguish between partial differential equations and ordinary differential equations. Ordinary differential equations lack partial derivatives, whereas partial differential equations have them. What then is meant by a partial derivative?

"When you have function that depends upon several variables, you can differentiate with respect to either variable while holding the other variable constant. This spawns the idea of partial derivatives" (Ivrii 2021). Therefore, to analyze a multivariate system, the system can be divided into several equations, each of which takes the partial derivative with respect to one variable and holds the others constant.

Nonlinear dynamics: In a nutshell, functions and equations that when plotted result in a straight line are linear functions/equations. The simplest example of this is a line with a slope of +5 and a y-intercept of 10, such as $f(x) = y = 5x + 10$. Since most equations do not graph as a straight line, they are classified as nonlinear equations (Clark and Luis 2020). Some examples of nonlinear equations are x^7, $\tanh(3x)$, and $\cos(x)$. Accordingly, nonlinear equations characterize most systems in both science and nature. Finding the best approximation in the form of a linear equation is a best practice when possible because nonlinear equations are renowned for being challenging and difficult to work with. The goal then is to linearize the nonlinear equations where practical. According to Troger and Steindl, "The treatment of nonlinear problems has been either avoided completely or is confined to special courses where a great number of different ad-hoc methods are presented. The wide-spread belief [sic] that no straightforward solution procedures for nonlinear problems are available prevails even today in engineering circles. Though in some courses it is indicated that in principle nonlinear problems are solvable by numerical methods, the treatment of nonlinear problems, more or less, is considered to be an art or an intellectual game" (Troger and Steindl 1991).

Time series analysis: Analyzing time series means precisely what it says. Time moves forward at a continuous pace. A time series is produced if interesting data points are observed and collected at specific intervals. In the case of nonlinear systems, this means collecting as many data points as possible that have an impact on the output over time. "The purpose of time series analysis is generally twofold: to understand or model the stochastic mechanism that gives rise to an observed series and to predict or forecast the future values of a series based on the history of that series and, possibly, other related series or factors" (Cryer and Chan 2008). After the data or time series is captured, we want to understand the data, build models, and provide forecasts and predictions based on all the available data. A simple analytical example is the goodness of fit test, or R^2-value which indicates how well a set of data fits a given model (variations of points around a straight line for a linear model for instance).

Phase space: "More specifically, a phase space is a space (a set with an imposed structure) the elements (phase points) of which (conventionally) represent the states of the system (for example, a phase plane)" (Anosov 2014). Various attributes are input into the system, and one or more outputs are produced. Oftentimes, a 3D graph in a regular "space" is provided with several inputs and outputs. With that in mind, a phase space is a three-dimensional space where the points or planes reflect a certain phase or condition of the system at certain conditions. Figure 7.16 below shows an example of a plot of total energy on the x-axis, density on the y-axis, and gas energy on the z-axis. Each point could represent a certain phase state, and the totality of points in the figure represents the phase space. The same type of plot as in Figure 7.16 could be used as a visual to represent the phase space results in other studies, such as in the case for a mortality study on flour beetles. As an example, a type of flour beetle called Tribolium has a birth and mortality rate that was examined by Cushing et al. (1998). The conditions of the nonlinear Tribolium populations with different mortality rates were captured and displayed with plots like Figure 7.16. Consequently, with x number of pupae, y number of adults, and z number of larvae, a snapshot of the system's state can be provided.

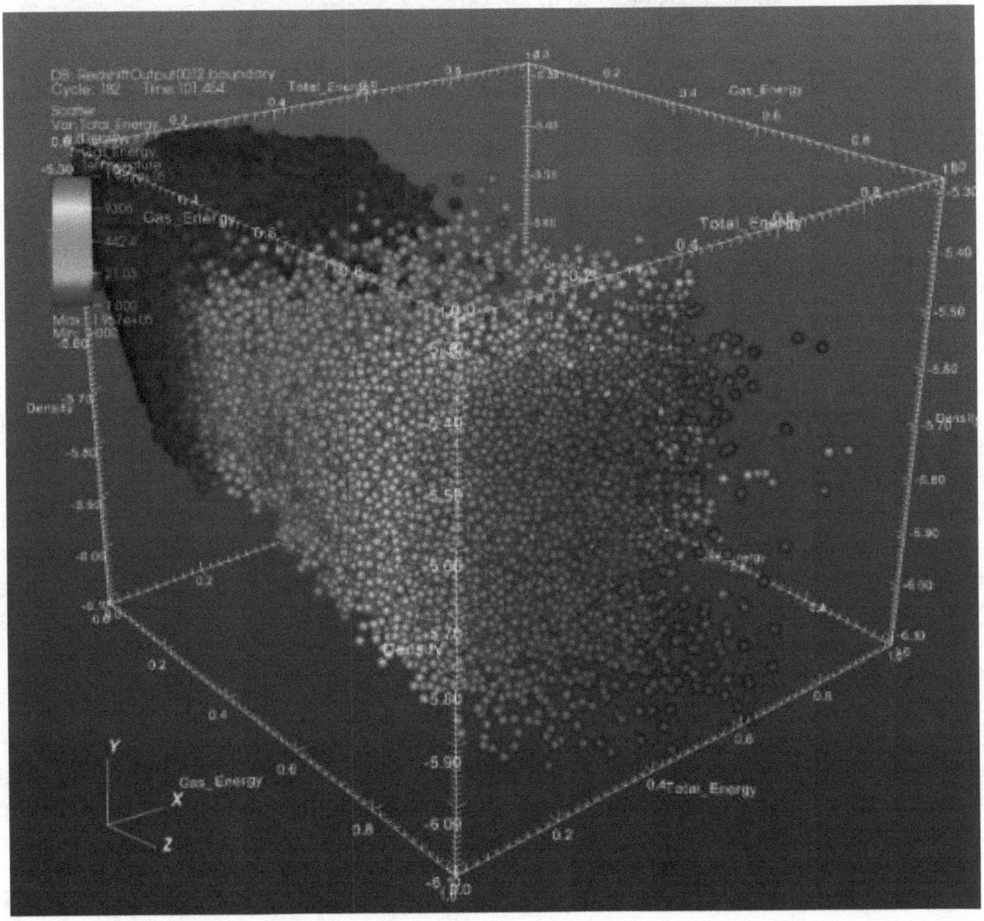

FIGURE 7.16 Representative phase space image. Each point (little sphere) is a phase state, and the entirety of the 3D-plot is the phase space. (By UCRL - Visualizations that have been created with VisIt. at wci.llnl.gov, Public Domain, https://commons.wikimedia.org/w/index.php?curid=4358059).

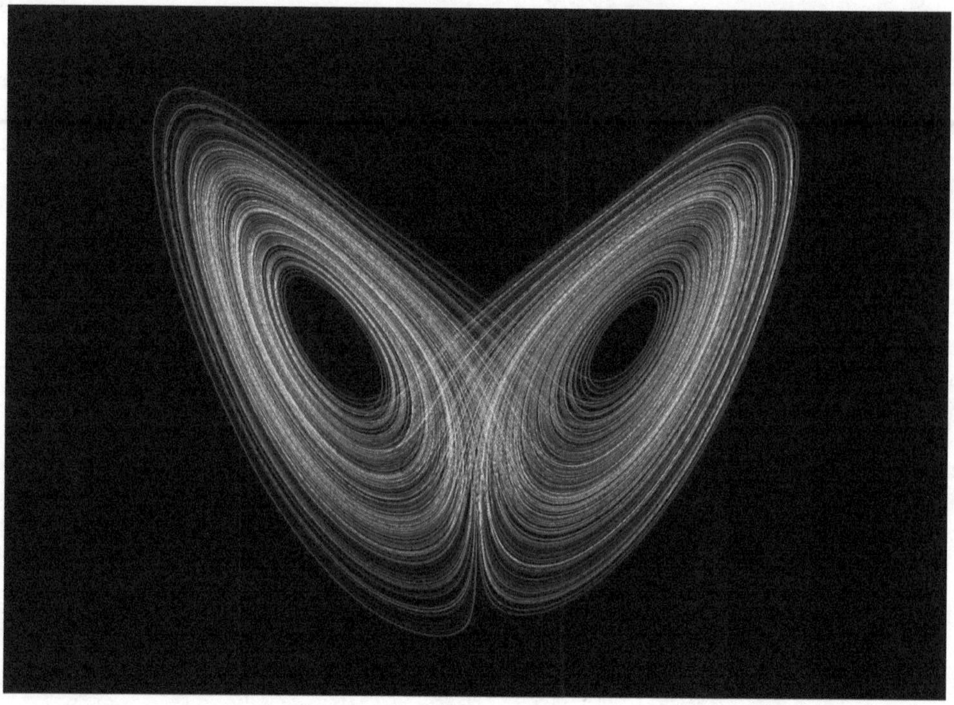

FIGURE 7.17 Computed in Fractint by Wikimol, Public Domain, https://commons.wikimedia.org/w/index.php?curid=165697.

Attractors: Edward Lorenz is likely the first person that springs to mind when one discusses nonlinear attractors. The discussions on nonlinear dynamics can be traced back to his March 1963 publication (Lorenz 1963). Some of the types of attractors that occur in nonlinear systems are, fixed-point attractors, limit-cycle attractors (these are also known as cyclic, periodic, or oscillating attractors), toroidal attractors (a.k.a. quasiperiodic attractors), and chaotic, or strange attractors. For example, the output of a nonlinear system or equation is "pulled" or "attracted" by a fixed-point attractor as input values or parameters of the equation rise or fall. This is best demonstrated by the fluid dynamics "Lorenz system," which is depicted in Figure 7.17.

Doesn't this look fantastic? In the Lorenz system, there are two attractors: one in the forming spiral (right) and the other in the left spiral as it is depicted. This attraction can be thought of as a basin into which a ball is rolling and trending toward the bottom-most point. The Lorenz system's three-dimensional nature is also noteworthy because, although the crossover locations are being observed from a flat, two-dimensional perspective, they are not truly 2D crossovers. The system is seen rotating around the two attractors in a more three-dimensional view in this interesting 3D representation of the same system (Figure 7.18).

Limit cycles/Periodic orbits: The motion that a particle undergoes in limit cycles remains either exactly on fixed orbits or quite near to them. An example is a swinging pendulum in the absence of friction, gravity, or resistive forces. If plotting the motion over time and space, a continuous perpetual loop should be observable.

Quasiperiodic orbits: A quasiperiodic orbit is created when the same system is subjected to two or more periodic orbits with different frequencies. Looking at the standard image of the Lorenz system, we can see this effect since it shows the well-known "Figure 8" loop (seen in some car races) with the crossing in the middle. However, the Lorenz system is also subjected to another orbit on

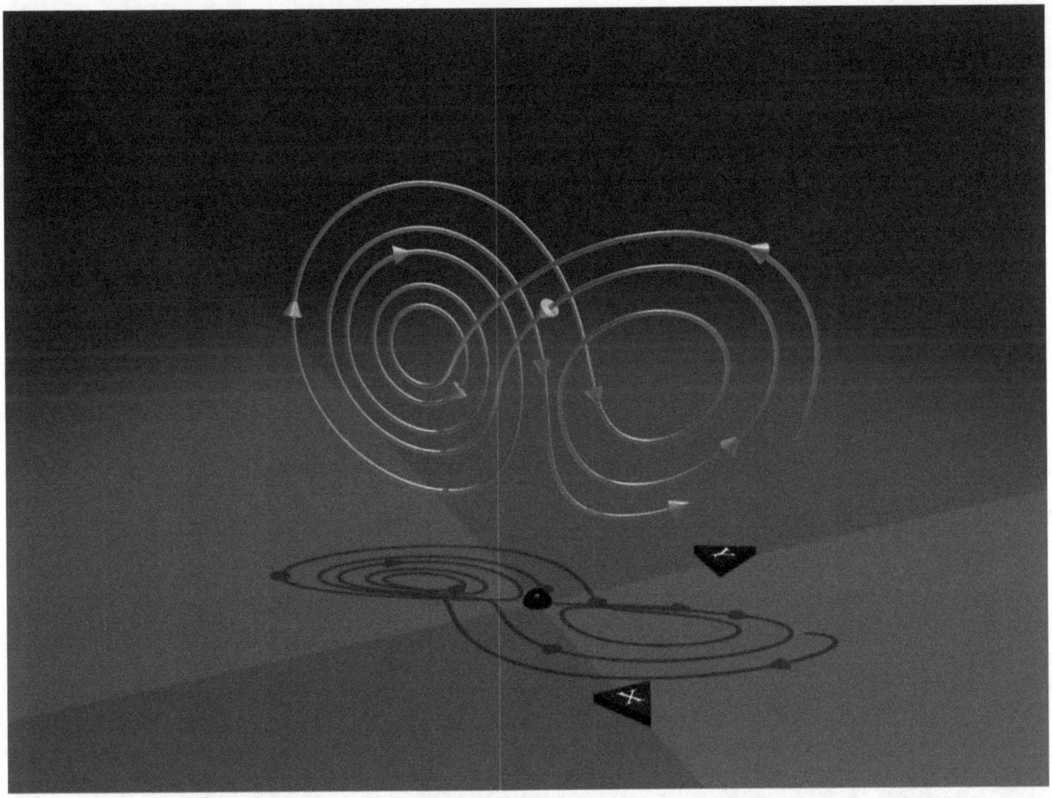

FIGURE 7.18 By original uploader was Mrubel. (At en.wikipedia - Transferred from en.wikipedia, CC BY 2.5, https://commons.wikimedia.org/w/index.php?curid=4050688).

the orthogonal axis, where it is similarly drawn into what is essentially another "Figure 8" pattern. This translates to two distinct "orbits" on perpendicular axes with varying frequencies (Vasilevich and Alexandrovich 2006).

Chaotic attractors: The *R*-values of the Lorenz models displayed above are all higher than a particular threshold. A bifurcation occurs at or above a specific *R*-value. For chaotic Lorenz systems, the *R*-value is the ratio of the Rayleigh number to its critical value. In effect, the bifurcation produces a second attractor in the direction of the values' desired trend. What occurs when the *R*-value (used as an input) is below, but approaching, the bifurcation point? The result is shown in Figure 7.19 which depicts the same Lorenz model from before in Figure 7.17. When contrasting these two images, we see that the former image depicts "chaos," with two distinct fixed points and chaotic attractors. Figure 7.21 illustrates the system's sensitivity to its initial inputs, in this case the *R*-value input. Nonlinear chaotic systems are known for this sensitivity.

Bifurcation: Bifurcation is a property of some nonlinear systems. Within a particular input set, a system may exhibit a distinct order as well as a separate output or group of outputs. Bifurcation occurs when the output values abruptly become "either this or that," with no values in between, because of the input set's values exceeding a certain threshold. More splits occur the further the threshold is exceeded. Troger refers to this phenomenon as "loss of stability," or "post bifurcation" (Troger and Steindl 1991). Figure 7.20 shows a simple illustration. The system exhibits stability for most of the evaluated input values. The branches are the perfect example of bifurcation and are easily visible.

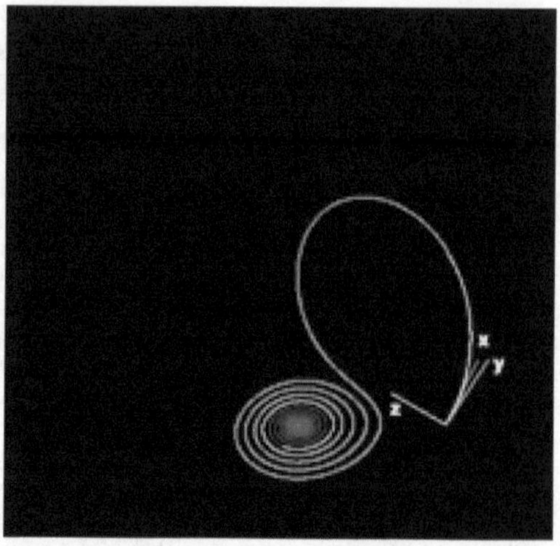

FIGURE 7.19 António Miguel de Campos. (Rescaled by Lulu of the Lotus-Eaters - [1], Public Domain, https://commons.wikimedia.org/w/index.php?curid=2074487).

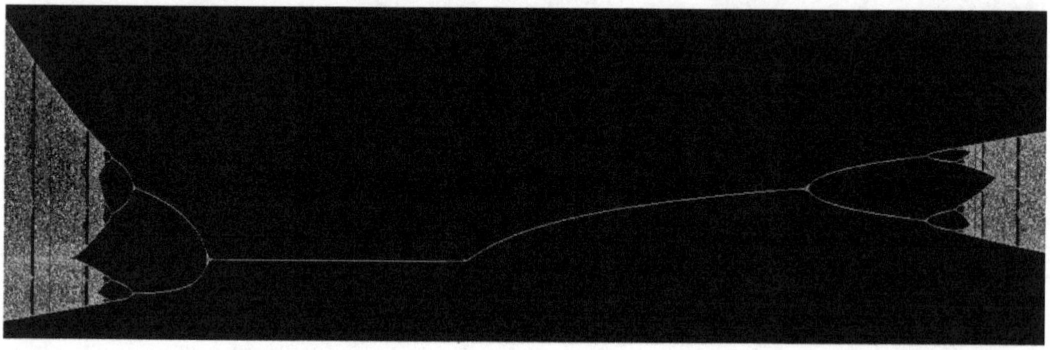

FIGURE 7.20 No machine-readable author provided. (Deeptrivia assumed (based on copyright claims). - No machine-readable source provided. Own work assumed (based on copyright claims), CC BY-SA 3.0, https://commons.wikimedia.org/w/index.php?curid=665066).

Chaos: According to the Encyclopedia Britannica, chaos theory is "the study of apparently random or unpredictable behavior in systems governed by deterministic laws" (Britannica 2024). Chaotic processes can manifest in a variety of common occurrences, such as in observing fluid dynamics, population cycles, and even the weather. Consequently, models based on chaos theory are useful in understanding these types of systems. The combination of "unpredictability or randomness" and "failing to account for all the factors," as in some level of abstraction, makes the study of chaos a fascinating undertaking. When a system behaves in an apparently random manner, it may be responding to some overly sensitive input in a complex way. One fundamental property of chaotic dynamical systems is their profound sensitivity to initial conditions. In essence, if a process is initiated with seemingly identical, but slightly different inputs, the end results could be completely different. An interesting way to think about chaos theory in general is to consider it as describing deterministic randomness. This concept can be drawn from the Encyclopedia Britannica's article on Chaos Theory, wherein the authors state, "In recent decades, however, a diversity of systems have been studied that behave unpredictably despite their seeming simplicity and the fact that the forces involved are governed by well-understood physical laws. The common element in these systems is a very high degree of sensitivity to initial conditions and to the way in which they are set in motion" (Britannica 2024). Notable points in this quote are that the system is extremely sensitive to (1) initial conditions, and also, (2) HOW the system is launched into motion. Figure 7.21 illustrates this concept by plotting the Lorenz system that we discussed previously as a function of a set of parameters

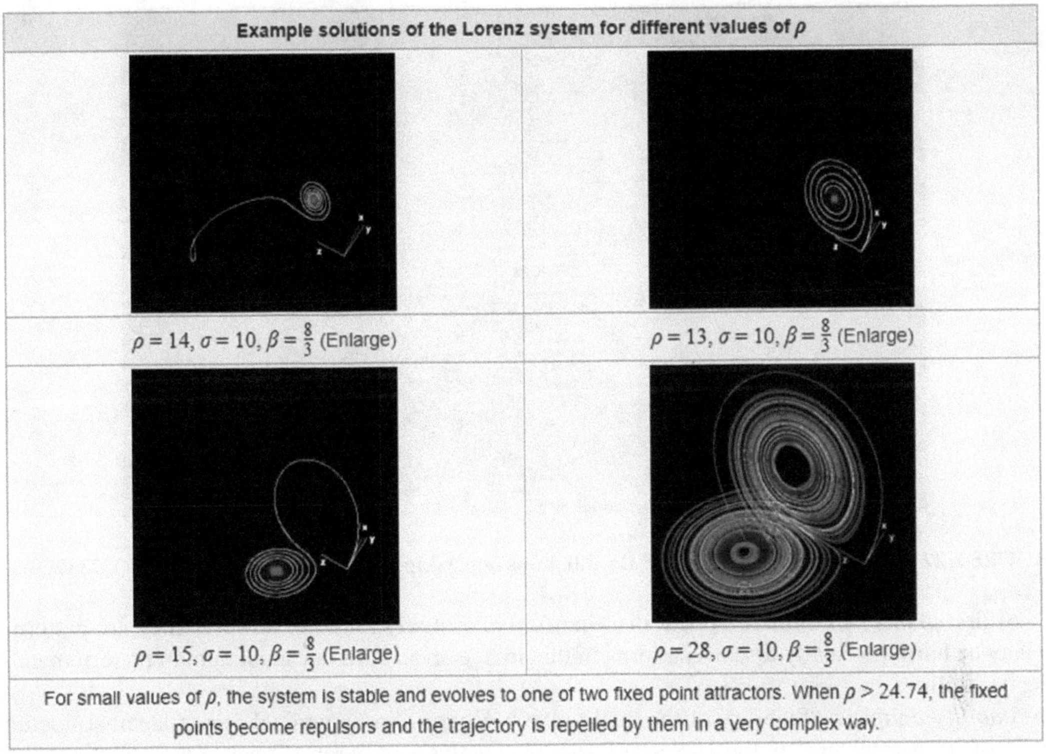

FIGURE 7.21 António Miguel de Campos. (Rescaled by Lulu of the Lotus-Eaters - [1], Public Domain, https://commons.wikimedia.org/w/index.php?curid=2074497).

ρ, σ, and β. To demonstrate the sensitivity of the Lorenz system to input conditions, the input parameters were held constant except for one parameter (ρ) was allowed to vary slightly.

These plots show how different the solutions to the Lorenz system are with only slight changes made to the inputs. The combination of extreme input sensitivity with completely different outcomes are telltale signs that you may be dealing with a chaotic system. The goal of chaos theory is to bring perceived randomness into order. An interesting complementary concept is that of multi-stability wherein multiple stable patterns can exist (Liz and Ruiz-Herrera 2013).

Iterative maps: A function (map) that is applied to an item recursively is called an iterative map (Weisstein 2022; Mikhailov and Loskutov 1996). This involves applying the function/map to a starting parameter (value), then applying the same function to the result, and so on, indefinitely. The way this simulates emergence is what makes this process especially interesting in the context of complex systems. When a system is first launched, it may not have an obvious attractor for the output value; but, after some time, perhaps short or not, an attractor that accurately reflects the system may appear (Cueli and Bordons 2008).

Coupled map lattices: Every element in a coupled map lattice is on a specific dimension. Discrete time (map), discrete space (lattice), and continuous state, characterize the resultant dynamical system (Kaneko and Yanagita 2014). Simplified a little, Coupled Map Lattices are three-dimensional spaces of inputs and outputs with time as a fourth dimension. One of the main applications of Coupled Map Lattices is the analysis and visualization of relationships in partial differential equations. In this sense, Coupled Map Lattices are frequently used to illustrate nonlinear dynamics. Frozen Random Patterns, Traveling Waves, Pattern Selection with Suppression of Chaos, Spatiotemporal Intermittency, and Quasi-Stationary Super-Transients are a few examples of interesting emergent behaviors. Figure 7.22 shows an image of Coupled Map Lattices from Kaneko and Yanagita (2014).

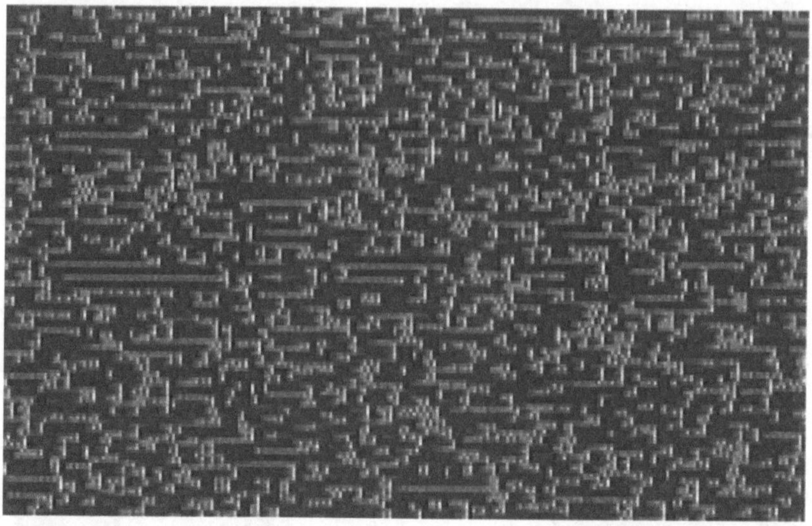

FIGURE 7.22 Coupled map lattices. (CC BY 3.0, https://en.wikipedia.org/w/index.php?curid=22238999).

At first glance, all of these appear disorganized and utterly random, however they are patterns. It may be hard to visually discern patterns in this image, even with our great human pattern matching capabilities.

Stability analysis: "Stability of the qualitative behavior of a dynamical system against fluctuations of its parameters" is the subject of stability analysis research (Cenci and Saavedra 2018). A system, particularly one that is complex, can require many input parameters to provide an output. According to the preceding definition, a stable system should therefore generate an acceptable output (one that meets requirements), despite fluctuations in the input parameters. A representative illustration of this would be a factory that creates engineered wood products such as fence posts. The factory should be able to conduct a stability analysis to demonstrate that the product will meet stakeholder requirements, even with some variation in the wood used as input to the production process. Why? Because wood naturally varies, the factory should expect some variations in its input materials. These variations need to be studied in the stability analysis to ensure the factory can operate successfully over the expected variation of its input parameters.

Population dynamics: The study of population growth, decline, and evolution is known as population dynamics. Most members of these populations are active. They could be bacteria, plants, animals, or insects. For example, in the previously mentioned flour beetle mortality study by Cushing et al., Cushing looked at the flour beetle populations' maturation and mortality rates to create a model to explain these population dynamics over time. The beetle cultures were kept in a well-regulated setting (Cushing et al. 1998). According to Clark and Luis, "As an example, animal population dynamics are suspected to be nonlinear because of the complexity and interrelatedness of variables that influence populations and the regularity in which some animal populations fluctuate" (Clark and Luis 2020). Consequently, great care needs to be taken in these types of studies when generalizing results to make real-world predictions. The Complex, Complicated, and Adaptive Nature of Systems will provide a whole set of wicked problems that the system engineering community needs to successfully engage.

The use of artificial intelligence (AI) in engineering and decision-making: The question of using artificial intelligence in the workplace, especially in engineering circles, has seen great attention recently resulting in spirited and heated discussions and debate, with a wide range of opinions on the matter of the technical, practical, legal, ethical, and even safety aspects and nature of using this revolutionary and disruptive technology. This topic is something that systems developers need to carefully consider and address in developing contemporary, complex systems with emerging behaviors. The potential for AI to help in a systems development effort are boundless but its use and implementation must be carefully evolved. This need for care stems from at least two major considerations: (1) is the AI providing correct advice, and (2) could there be emergent behaviors and conditions that lead to future wicked problems and the possibility of safety and recovery concerns? An example might be that an AI's "advice" or its embedded functionality is determined to be correct over a long period of time, and so confidence is gained in the AI's decision-making capability. After confidence is established in the AI's capabilities, a dramatically incorrect response is given by the AI that puts health and safety at risk for people affected by the system. If the AI is then intrinsically embedded in the system such that it becomes a "wicked problem" to correct and/or remove it, we then have a major problem. If the system with this AI affects society, then we have a major societal problem. Even if there is no manifested technical or safety problem, some questions are already arising. Is it cheating to use AI to write a document, or even use it in helping write a document? What about using AI to generate software code, write requirements, or conduct an analysis? These are all timely questions and are central in current discussions about the use of AI in the workplace.

We looked at the first of the major considerations above and wanted to gain some insights as to how well available AIs can provide the "correct" answer. This investigation was more of a qualitative and subjective effort than a rigorous quantitative study but nevertheless provided some interesting insights. It was also narrowly focused, in that we looked at how well AI can be used to write engineering-oriented, working software code. Although not regarded as an "AI" tool application, we also qualitatively evaluated how well a prevalent paraphrasing tool was able to catch the nuances in engineering writing, such as in writing requirements and nuanced technical content. The idea was just to get a feel for how well the tool was able to accomplish this task. We wanted to start with some initial technical writing that had essentially correct technical information, but not necessarily the best grammar, or organization of the content, and see whether or not the paraphraser could improve the writing and preserve the technical nuances. We used a well-known respected, profession grade writing aid that had a paraphrasing capability and applied the paraphraser to several dozen technically correct but poor to moderately well-written engineering papers and reviewed the results. Our anecdotal findings were that for almost all cases where there was information that required precision of meaning and/or the preservation of nuance, the paraphrased material was inaccurate and required massive amounts of re-writing. We tried several different "styles", and we did see a little improvement on a case-by-case basis, but not enough overall improvement. In this case, it was more work to correct the "help" we received from the paraphraser tool than if we had simply corrected the initial writing ourselves, without the use of the paraphrasing tool. In contrast, for written work that did not require technical precision or the preservation of nuance, the tool did fine (apart from the choice of some quirky words here and there). With regards to capturing and preserving nuance, in rigorous, and precise technical writing, the paraphrasing tool fell short. This may change over time as AI-based tools enter the fray of highly precise and nuanced technical writing and become refined. I fully expect this to be the case in the future, however, for the case of simple (non-AI-based) paraphrasing tools, they aren't there yet.

The second investigation we made was to use Chat-GPT to write software code. We had several teams (four) that used Chat-GPT to write software for simulating computer network architecture problems. Some of the participants had very little software writing experience while others had more experience. A few participants were capable programmers and so we had a broad range of software coding experience in the participants of this AI-based coding experiment. In this case, we had some pleasantly surprising results. Every participant on all teams was able to use Chat-GPT to write a working simulation that achieved their objectives. All participants had to use Chat-GPT to implement a simulation of a computer network layer protocol that simulated data traffic over the network layer including error handling. They had to learn how to repetitively give Chat-GPT instructions for what they wanted it to write, but in the end, the AI provided a working simulation, that compiled and ran, and gave them technically correct, and expected results. Everyone in all groups thought that this was very useful and Chat-GPT immediately provided them with a starting point that they could iterate until they had something they wanted. AI-based technology therefore can be an effective tool in the modern engineering development environment.

In this author's opinion, if AI-based tools are used in this fashion (i.e., providing a base, or starting point for solving problems), and the engineer, author, or DM must check and adapt the AI-provided "advice," then this is not an ethical violation, or cheating. It is just a matter of using available tools to become more efficient in doing standard engineering tasks. The review, revision, and final assessment of the AI-assisted product, whether it is a document, software, or other engineering artifact, remains in the hands of the engineer, author, or DM, and they are responsible for the final content, accuracy, and effectiveness of the AI-assisted work product. However, once AIs become more sophisticated and better at writing technical content with precision, accuracy, and preserving nuance, or are capable of rigorously performing engineering tasks, then these lines become more blurred, and waters of ethics, fairness, intellectual property, and legality become more difficult to navigate. For example, going back to our earlier illustration of an embedded AI that through an emergent property generates a societal wicked problem. Although we are not practically there yet, what if this AI *was* (or *becomes*) a *strong* AI such that it is capable of replicating (or exceeding) human functions like reasoning, planning, problem solving, and establishes human-like behaviors, then ethical and legal questions begin to manifest that need to be addressed (e.g., does human-like artificial intelligence have rights for instance, who is responsible for errors and damage, and what can, should, or must be done if/when things go wrong). Interesting questions all, and questions that must be addressed to practically and successfully integrate this disruptive technology into modern systems development efforts.

7.1.7 Transition to Optical Systems Building Blocks

Throughout this section, we provided an introduction and historical background of risk and decision-making concepts and methods used throughout the systems engineering discipline, including GT. In the next section, we present the optical systems building blocks for understanding how the turbulent atmosphere impacts the imaging system. We will develop methods for including atmospheric turbulence effects in the OTF and show how the OTF can be used to mathematically model, simulate, and ultimately predict the spatial resolution performance of the imaging system. Several different OTFs are mathematically described that are useful in modeling atmospheric turbulence effects, such as the short exposure OTF, the long exposure OTF, the average OTF, and the OTF representation commonly used in ATC. Parameters such as the atmospheric coherence length r_o, and the isoplanatic angle θ_0, are presented as useful metrics in deciding whether atmospheric turbulence compensation will be effective or not. For example, if the diameter of the optical system is larger than r_o, then ATC methods become worthy of consideration since there is a potential benefit in removing the atmospheric turbulence effects from the optical system. If the optical system's entrance pupil plane diameter is equal to (or smaller) than r_o, then ATC methods won't provide an increase in the optical system's spatial resolution and so are not beneficial. Correspondingly, the isoplanatic angle provides a metric for deciding the maximum angular subtense for the attempted atmospheric correction. Expressions for the image spectrum SNR are also presented to help decide if there is sufficient signal present for good detection or not.

7.2 OPTICAL SYSTEMS BUILDING BLOCKS: MODELING ATMOSPHERIC EFFECTS – THE OTF AND SNR

In this section, we explore the effects that atmospheric turbulence has on both coherent and incoherent optical radiation. We present key assumptions and develop the mathematical framework used to understand how the atmosphere degrades the performance of an optical system in terms of its potential spatial resolution, and reduction in its detector signal-to-noise ratio. We introduce the useful OTF which carries the effects of the atmosphere and illustrate several important limiting forms of the OTF. For example, the long exposure and short exposure OTFs encapsulate the atmospheric effects for imaging systems that have long, and short integration times compared to the time scale in which the atmospheric imaging conditions change states (on the order of several milliseconds). Signal-to-noise expressions are presented that allow modeling the expected signal strength at the detector to determine if there is sufficient signal on the detector to capture the image. The SNR expressions are complimentary to the spatial resolution considerations, and both must be satisfied for the imaging system to function as required. Often, the medium through which reflected light waves propagate from an object to an imaging system is affected by atmospheric turbulence resulting in degraded imaging system performance. This causes the received image to have far less spatial resolution than what is possible from a diffraction-limited imaging system. These atmospheric turbulence effects result from the nonuniform heating of the Earth's surface by the Sun and the effects of wind and result in inhomogeneities of the refractive index of the air. These inhomogeneities are ever present and can have devastating on the overall performance of an imaging system that is operating within such an environment. One way to understand and help mitigate these effects is by treating the optical atmospheric aberrations as random processes and determining statistical measures of performance and then using adaptive optics (AO) or atmospheric turbulence compensation methods which will be discussed later on in this book.

Images with resolution much lower than the theoretical diffraction limit are often generated by an imaging system that has illuminating light waves propagating through the atmosphere and imperfect materials. Temperature-induced variations in the atmosphere's refractive index resulting from the Sun's uneven heating of the Earth's surface, is the biggest culprit along with wind, as previously stated. The atmospheric turbulence effects can reduce the spatial resolution performance of a large entrance pupil plane imaging system, like a 1 m diameter telescopic imaging system (or larger), to no better than what can be achieved with an 8 cm store bought telescope. To predict the average performance of such large optical imaging systems, it is necessary to treat the optical atmospheric aberrations as random processes and to determine certain statistical metrics as described below. First, though we need to make some initial assumptions,

1. The source/object radiates incoherently,
2. The scale sizes for observation are much larger than the wavelength of the propagating radiation that is being detected/imaged.

With these assumptions, we develop the incoherent imaging model for determining atmospheric turbulence effects on optical imaging systems. Notice that this also applies for coherently illuminated objects (such as from a laser) that have surface roughness on the scale of the wavelength (common to many person-made, non-mirror-like surfaces).

7.2.1 Incoherent Imaging Model

For incoherent imaging systems, the object is assumed to radiate in a spatially incoherent manner and to be describable by a radiant emittance in the object plane $I_o(\xi,\eta)$ where ξ and η are orthogonal components in the object plane. Lenses L_1 and L_2 have focal lengths f. A blurred and distorted image of the object appears in the (u,v) plane and is described by the irradiance distribution $I_i(u,v)$ The random screen can be represented mathematically by a multiplicative amplitude transmittance $t_s(x,y)$ as seen in Figure 7.23.

7.2.2 Optical Transfer Function

One effective way to describe the spatial frequency response of an incoherent imaging system is in terms of its OTF given by (Goodman 1985; Roggemann and Welsh 1996),

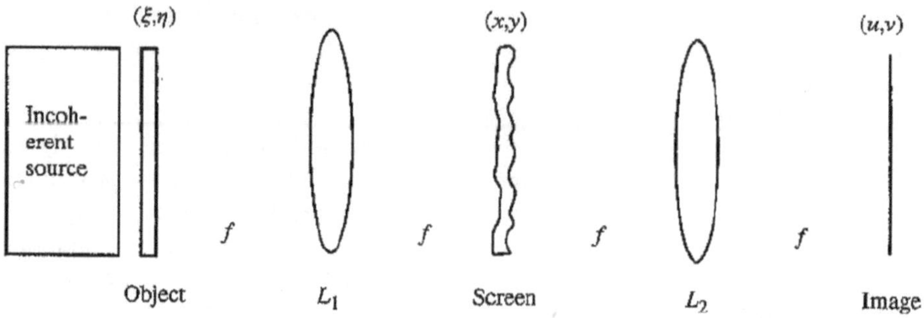

FIGURE 7.23 Single screen model for an incoherent imaging system.

$$\mathcal{H}(\nu_U,\nu_V)=\frac{\iint_{-\infty}^{\infty}\boldsymbol{P}(x,y)\boldsymbol{P}^*\left(x-\bar{\lambda}f\nu_U,y-\bar{\lambda}f\nu_V\right)dxdy}{\iint_{-\infty}^{\infty}\left|\boldsymbol{P}(x,y)\right|^2dxdy}, \tag{7.7}$$

and ν_U and ν_V are spatial frequency variables and $\boldsymbol{P}$ is the generalized pupil function (GPF). The parameter f is the distance from the lens to the object. By defining a screen that has an amplitude transmittance $\boldsymbol{t}_s(x,y)$, and placing this screen in the entrance pupil plane of the imaging system, the generalized pupil function is consequently modified, and a new generalized pupil function $\boldsymbol{P}'(x,y)$ is obtained,

$$\boldsymbol{P}'(x,y)=\boldsymbol{P}(x,y)\boldsymbol{t}_s(x,y). \tag{7.8}$$

Substituting Equation 7.8 into Equation 7.7, the OTF becomes:

$$\mathcal{H}(\nu_U,\nu_V)=\frac{\iint_{-\infty}^{\infty}\boldsymbol{P}(x,y)\boldsymbol{P}^*\left(x-\bar{\lambda}f\nu_U,y-\bar{\lambda}f\nu_V\right)\boldsymbol{t}_s(x,y)\boldsymbol{t}_s^*\left(x-\bar{\lambda}f\nu_U,y-\bar{\lambda}f\nu_V\right)dxdy}{\iint_{-\infty}^{\infty}\left|\boldsymbol{P}(x,y)\right|^2\left|\boldsymbol{t}_s(x,y)\right|^2dxdy}, \tag{7.9}$$

Since, in general, we do not have a priori knowledge of the specific values of $\boldsymbol{t}_s(x,y)$ at each screen coordinate (x,y), it is impractical to proceed with a purely deterministic analysis. Therefore, we will use knowledge of the statistics of $\boldsymbol{t}_s$ to determine some measure of average frequency response of the imaging system. We define the average OTF, $\bar{\mathcal{H}}(\nu_U,\nu_V)$, to be the expected value of the numerator of the right-side expression of Equation 7.9, divided by the expected value of its denominator,

$$\bar{\mathcal{H}}(\nu_U,\nu_V)\triangleq\frac{E[\text{Numerator of the OTF}]}{E[\text{Denominator of the OTF}]}. \tag{7.9}$$

Since the function $\boldsymbol{P}(x,y)$ here is a deterministic aperture function that is defined as one inside the clear aperture of the entrance pupil, and zero outside the clear aperture of the entrance pupil of the imaging system, the statistical expectation is on the amplitude transmittance function as shown in Equation 7.11,

$$\bar{\mathcal{H}}(\nu_U,\nu_V)=\frac{\iint_{-\infty}^{\infty}\boldsymbol{P}(x,y)\boldsymbol{P}^*\left(x-\bar{\lambda}f\nu_U,y-\bar{\lambda}f\nu_V\right)\overline{\boldsymbol{t}_s(x,y)\boldsymbol{t}_s^*\left(x-\bar{\lambda}f\nu_U,y-\bar{\lambda}f\nu_V\right)}dxdy}{\iint_{-\infty}^{\infty}\left|\boldsymbol{P}(x,y)\right|^2\overline{\left|\boldsymbol{t}_s(x,y)\right|^2}dxdy}. \tag{7.11}$$

Making the common assumption that the spatial statistics are wide-sense stationary, the expected value shown in Equation 7.11 is independent of x and y and can be moved outside of the integrals. This results in an average OTF given by

$$\bar{\mathcal{H}}(\nu_U, \nu_V) = \mathcal{H}_o(\nu_U, \nu_V)\bar{\mathcal{H}}_s(\nu_U, \nu_V), \tag{7.12}$$

where $\mathcal{H}_o$ is the OTF of the imaging system when the screen is not present, and $\bar{\mathcal{H}}_s$ represents the average OTF of the screen, given by

$$\bar{\mathcal{H}}_s(\nu_U, \nu_V) \triangleq \frac{\Gamma_t(\bar{\lambda} f\nu_U, \bar{\lambda} f\nu_V)}{\Gamma_t(0,0)}, \tag{7.13}$$

and where Γ_t is the spatial autocorrelation of the transmissive screen,

$$\Gamma_t(\Delta x, \Delta y) \triangleq \overline{\boldsymbol{t}_s(x,y)\boldsymbol{t}_s^*(x-\Delta x, y-\Delta y)}. \tag{7.14}$$

We can see from Equation 7.14 that this is a spatial autocorrelation function that is a function of only spatial differences Δx and Δy.

7.2.3 The Average Point Spread Function

Another interesting and useful imaging system metric is the "average point-spread function (PSF)" of a system. The PSF can be defined as (Goodman 1985),

$$\bar{s}(u,v) \triangleq \mathcal{F}^{-1}\{\bar{\mathcal{H}}(\nu_U, \nu_V)\}, \tag{7.15}$$

where the function $\mathcal{F}^{-1}\{*\}$ is the 2D inverse Fourier Transform of the argument in the brackets.

The average PSF has the property that it is always nonnegative and real valued. Also, an interesting property of $\bar{\mathcal{H}}$ is that, because it has been normalized at its zero frequency components, the related PSF $\bar{s}(u,v)$ will always have unit volume. Because we note from Equation 7.12 that the average OTF is a product of the system's OTF without the screen, and another OTF that is a function of the screen alone, according to Fourier Transform properties, the PSF must be able to be written as a convolution of the PSFs resulting from the OTFs of the system and screen respectively,

$$\bar{s}(u,v) = s_o(u,v) \otimes \bar{s}_s(u,v), \tag{7.16}$$

and the expression for the diffraction-limited PSF is given by

$$s_o(u,v) = \mathcal{F}^{-1}\{\mathcal{H}_o(\nu_U, \nu_V)\}, \tag{7.17}$$

with the expression for the average point spread function (PSF) for the screen given by

$$\bar{s}_s(u,v) = \mathcal{F}^{-1}\{\bar{\mathcal{H}}_s(\nu_U, \nu_V)\}. \tag{7.18}$$

The symbol $\otimes$ is used for 2D convolution. Equation 7.16 stems from the fact that the inverse Fourier Transform of the product of two functions in the spatial frequency domain is the convolution of the inverse Fourier Transform of each of the functions in the spatial domain.

7.2.4 Random Phase Screens

In this section, we look at a very common occurrence of modeling atmospheric aberrations as phase-only aberrations. To do this, we write the screen as an exponential function of only the phase. The phase itself is a function of x and y position coordinates in the clear aperture of the screen and can be modeled as

$$\boldsymbol{t}_s(x,y) = \exp[j\phi(x,y)]. \tag{7.19}$$

Here $\phi(x,y)$ represents the phase due to the atmospheric turbulence and is assigned to the screen at location (x,y). Commonly, a thin screen model is assumed that allows the phase shift $\phi(x,y)$ to be written as

$$\phi(x,y)=\frac{2\pi}{\lambda}\left[L(x,y)-L_0\right]. \tag{7.20}$$

The quantity $L(x,y)$ represents the optical path length at (x,y), L_0 is the mean optical path length pathlength for the screen, and λ is the general wavelength. In the following analysis, the wavelength λ is replaced by the mean wavelength $\bar{\lambda}$. In doing so, we can ignore the wavelength dependence on the phase. This is acceptable if we can assume that narrow-band assumptions apply when using a narrow-band filter at the detector. If the optical path length differences that are introduced by the screen are less than the spatial coherence length of the propagating light, then this narrow-band assumption is valid (Goodman 1985). If we want to determine how a random phase screen affects the performance of an incoherent imaging system, we first need to determine the spatial autocorrelation expression of the amplitude transmittance $\boldsymbol{t}_s$ shown in Equation 7.5:

$$\Gamma_t(x_1,y_1;x_2,y_2)=\overline{\boldsymbol{t}_s(x_1,y_1)\boldsymbol{t}_s^*(x_2,y_2)}. \tag{7.21}$$

If we then substitute the definition for $\boldsymbol{t}_s$ shown in Equation 7.19 into Equation 7.21, we get

$$\Gamma_t(x_1,y_1;x_2,y_2)=\overline{\exp\left[j\phi(x_1,y_1)-j\phi(x_2,y_2)\right]}. \tag{7.22}$$

In Equation 7.22, we can observe that this resembles the second-order characteristic function for the joint random variables $\phi_1=\phi(x_1,y_1)$ and $\phi_2=\phi(x_2,y_2)$ as given by

$$\Gamma_t(x_1,y_1;x_2,y_2)=\boldsymbol{M}_\phi(1,-1). \tag{7.23}$$

In the characteristic function, the radial frequencies ω_1 and ω_2 are assigned (1, –1) respectively, in the general second-order characteristic function expression:

$$\boldsymbol{M}_\phi(\omega_1,\omega_2)=\overline{\exp(j\omega_1\phi_1+j\omega_2\phi_2)}. \tag{7.24}$$

We can also develop an alternate expression using the first-order characteristic function by defining a phase difference as $\Delta\phi=\phi_1-\phi_2$ and assigning the first-order radial frequency as one,

$$\Gamma_t(x_1,y_1;x_2,y_2)=\boldsymbol{M}_{\Delta\phi}(1). \tag{7.25}$$

7.2.5 Gaussian Random Phase Screen

Another interesting result is if we use Gaussian statistics to represent the random phase screen (RPS). If this time we define the random-phase $\phi(x,y)$ as a zero-mean Gaussian random process, we can take advantage of Gaussian statistics to generate some useful theoretical and mathematical models. Recall that Gaussian statistics are applicable in the high photon region where there are lots of independent photons that are being detected. Since in this case, ϕ_1 and ϕ_2 obey Gaussian statistics, then the difference in these random phases $\Delta\phi$ must also obey Gaussian statistics. The mean and variance of the difference of the phases is given by

$$\overline{\Delta\phi}=0, \tag{7.26}$$

and,

$$\sigma_{\Delta\phi}^2 = \overline{(\phi_1 - \phi_2)^2} = D_\phi(x_1, y_1; x_2, y_2), \tag{7.27}$$

where the quantity D_ϕ is commonly known as the structure function for the random process $\phi(x, y)$. We can write the first-order characteristic function for this phase difference as follows:

$$M_{\Delta\phi}(\omega) = \exp\left(-\frac{1}{2}\sigma_{\Delta\theta}^2\omega^2\right). \tag{7.28}$$

The thin screen spatial autocorrelation function upon substituting Equations 7.27 and 7.28 into 7.25 becomes

$$\Gamma_t(x_1, y_1; x_2, y_2) = \exp\left[-\frac{1}{2}D_\phi(x_1, y_1; x_2, y_2)\right]. \tag{7.29}$$

Given that the atmosphere is locally inhomogeneous, we can assume that statistically, the random process is stationary in first increments, and the structure function in Equation 7.29 only depends on the coordinate differences $\Delta x = x_1 - x_2$ and $\Delta y = y_1 - y_2$. This stationary in first increments assumption leads to a very useful expression for the thin phase screen spatial autocorrelation:

$$\Gamma_t(\Delta x, \Delta y) = \exp\left[-\frac{1}{2}D_\phi(\Delta x, \Delta y)\right]. \tag{7.30}$$

Note that this is now solely dependent on the phase structure function. From Equation 7.13, the average OTF of the screen becomes

$$\bar{\mathcal{H}}_s(\nu_U, \nu_V) = \exp\left[-\frac{1}{2}D_\phi\left(\bar{\lambda} f \nu_U, \bar{\lambda} f \nu_V\right)\right]. \tag{7.31}$$

For a wide sense stationary random process assumption, we can express the structure function as a normalized version of the phase autocorrelation function, $\gamma_\phi(\Delta x, \Delta y)$,

$$D_\phi(\Delta x, \Delta y) = 2\sigma_\phi^2\left[1 - \gamma_\phi(\Delta x, \Delta y)\right]. \tag{7.32}$$

By inserting Equation 7.32 into Equation 7.31 and substituting variables, we can get an expression of the average OTF of the screen in terms of the normalized phase autocorrelation,

$$\bar{\mathcal{H}}_s(\nu_U, \nu_V) = \exp\left[-\sigma_\phi^2\left(1 - \gamma_\phi\left(\bar{\lambda} f \nu_U, \bar{\lambda} f \nu_V\right)\right)\right]. \tag{7.33}$$

If we want to understand the behavior of the OTF of an imaging system that has a random thin phase screen, the behavior of the structure function D_ϕ must first be understood. If the phase due to the atmospheric aberrations is wide-sense stationary, then the structure function has the following properties:

$$D_\phi(0, 0) = 0, \tag{7.34}$$

and

$$D_\phi(\infty, \Delta y) = 2\sigma_\phi^2, \quad D_\phi(\Delta x, \infty) = 2\sigma_\phi^2, \tag{7.35}$$

where σ_ϕ^2 is a constant variance, and the term on the right sides of the equals signs in the expressions in Equation 7.35 is reached if either Δx or Δy (or both) grow large enough.

7.2.6 Limiting Forms for the Average OTF and the Average PSF for Large Phase Variance

Given a large phase variance, the average PSF and average OTF can be obtained from the slopes of the random phase function. We can define the partial derivatives of the phase as

$$\phi_X(x,y) = \frac{\partial}{\partial x}\phi(x,y) \quad \phi_Y(x,y) = \frac{\partial}{\partial y}\phi(x,y). \tag{7.36}$$

By assuming that the partial derivatives of the screen phases are jointly stationary random processes, we can get an expression for the spatial autocorrelation function for a thin phase screen of the form

$$\Gamma_t(x_1,y_1;x_2,y_2) = \overline{\exp\left[j\left(\phi(x_1,y_1) - \phi(x_1-\Delta x, y_1-\Delta y)\right)\right]}. \tag{7.37}$$

For a large phase variance, we can expect the spatial autocorrelation of the thin screen Γ_t to drop toward zero when the intervals Δx or Δy become small as compared to the spatial correlation width of $\phi(x,y)$. In this case, we can approximate the spatial phase difference as

$$\phi(x_1,y_1) - \phi(x_1-\Delta x, y_1-\Delta y) \approx \Delta x\frac{\partial}{\partial x_1}\phi(x_1,y_1) + \Delta y\frac{\partial}{\partial y_1}\phi(x_1,y_1). \tag{7.38}$$

Substituting Equation 7.38 into Equation 7.37, we obtain the following expression for the thin screen autocorrelation function

$$\Gamma_t(x_1,y_1;x_1-\Delta x,y_1-\Delta y) \approx \overline{\exp\left[j\Delta x\frac{\partial}{\partial x_1}\phi(x_1,y_1) + j\Delta y\frac{\partial}{\partial y_1}\phi(x_1,y_1)\right]} = \boldsymbol{M}_{\phi_X,\phi_Y}(\Delta x,\Delta y). \tag{7.39}$$

In Equation 7.39, the term $\mathbf{M}_{\phi_X,\phi_Y}$ is the joint characteristic function, this time of the X and Y partial derivatives of the thin screen phase. We can now express the average OTF, with appropriate coordinate scaling, in terms of the spatial, thin screen, autocorrelation function $\Gamma_t(\Delta x,\Delta y)$, or if you prefer the joint characteristic function,

$$\bar{\mathfrak{H}}_s(v_U,v_V) = \mathbf{M}_{\phi_X,\phi_Y}\left(\bar{\lambda}fv_U,\bar{\lambda}fv_V\right). \tag{7.40}$$

By taking the inverse Fourier Transform of this expression, the average PSF of the thin phase screen is given by Goodman (1985)

$$\bar{s}_s(u,v) = \frac{1}{\left(\bar{\lambda}f\right)^2}p_{\phi_X,\phi_Y}\left(\frac{u}{\bar{\lambda}f},\frac{v}{\bar{\lambda}f}\right), \tag{7.41}$$

where p_{ϕ_X,ϕ_Y} is the joint probability density function (PDF) of the phase partial derivatives, ϕ_X and ϕ_Y. Consequently, the energy distribution of the average PSF is determined by the slopes of the random thin screen phases. If we consider the special case of random phase screens that obey Gaussian statistics and have zero mean, the partial derivatives of the phase function are also Gaussian distributed. The joint characteristic function can then be expressed as

$$\boldsymbol{M}_{\phi_X,\phi_Y}(\omega_X,\omega_Y) = \exp\left[-\frac{\sigma_0^2}{2}\left(\omega_X^2+\omega_Y^2+2\sigma_X\sigma_Y\rho\omega_X,\omega_Y\right)\right], \tag{7.42}$$

where σ_X and σ_Y are the standard deviations of ϕ_X and ϕ_Y and ρ is the correlation coefficient. From Equations 7.42 and 7.40, the average OTF for the thin phase screen becomes

$$\bar{\mathcal{H}}_s(\nu_U,\nu_V)=\exp\left[-\frac{\left(\bar{\lambda}f\sigma_0\right)^2}{2}\left(\nu_U^2+\nu_V^2+2\sigma_X\sigma_Y\rho\nu_U\nu_V\right)\right]. \tag{7.43}$$

The average PSF for the thin phase screen is obtained by the inverse Fourier Transform of Equation 7.43

$$\bar{s}_s(u,v)=\frac{1}{2\pi\left(\bar{\lambda}f\right)^2\sigma_X\sigma_Y\sqrt{1-\rho^2}}\exp\left[-\frac{\left(\frac{u}{\sigma_X}\right)^2+\left(\frac{v}{\sigma_X}\right)^2-2\rho\left(\frac{u}{\sigma_X}\right)\left(\frac{v}{\sigma_Y}\right)}{2\left(\bar{\lambda}f\right)^2\left(1-\rho^2\right)}\right]. \tag{7.44}$$

Is also a real-valued function. If we now add the restriction that the partial derivatives of the phase have identical variances σ_0^2, and are not correlated with each other, then we can write the covariance matrix $\underline{C}$ as follows:

$$\underline{C}=\begin{bmatrix}\sigma_0^2 & 0\\ 0 & \sigma_0^2\end{bmatrix}. \tag{7.45}$$

The joint characteristic function now takes on the simple form of

$$M_{\phi_X,\phi_Y}(\omega_X,\omega_Y)=\exp\left[-\frac{\sigma_0^2}{2}\left(\omega_X^2+\omega_Y^2\right)\right], \tag{7.46}$$

and the average OTF for the thin phase screen becomes

$$\bar{\mathcal{H}}_s(\nu_U,\nu_V)=\exp\left[-\frac{\left(\bar{\lambda}f\sigma_0\right)^2}{2}\left(\nu_U^2+\nu_V^2\right)\right]. \tag{7.47}$$

Upon inspection, Equation 7.47 is a circularly symmetric Gaussian function in the (ν_U,ν_V) plane. The average PSF for the thin phase screen is obtained by the inverse Fourier Transform of Equation 7.47

$$\bar{s}_s(u,v)=\frac{1}{2\pi\left(\bar{\lambda}f\sigma_0^2\right)^2}\exp\left[-\frac{u^2+v^2}{2\left(\bar{\lambda}f\sigma_0\right)^2}\right]. \tag{7.48}$$

Note that this is a circularly symmetric function of the radius squared ($r^2=u^2+v^2$).

7.2.7 The Earth's Atmosphere as a Thick Phase Screen

In this section, we relax the thin phase screen assumption and focus on an inhomogeneous atmosphere that we will represent using a thick phase screen. We still focus on an incoherent imaging system where the object is assumed to transmit (or reflect) incoherent light such as is often encountered with natural illumination sources. Recall that laser sources reflecting from natural sources that are rough on the scale of the illuminating wavelength are also well-modeled as incoherent scattering objects. The light propagating from the source to the detector encounters a randomly

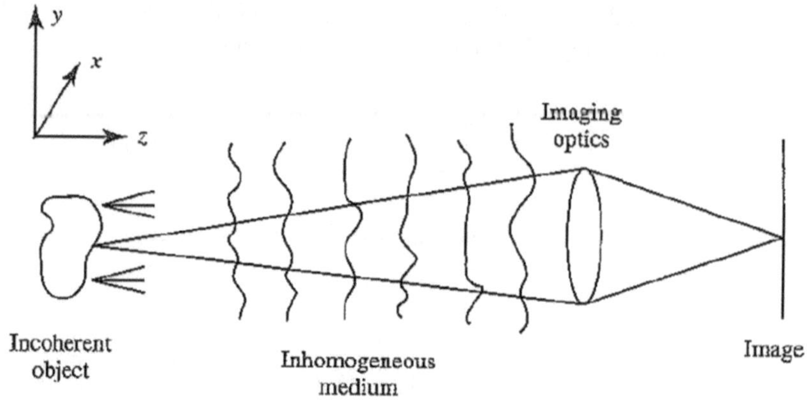

FIGURE 7.24 Conceptual model of a thick phase screen.

inhomogeneous Earth's atmosphere over extended distances. We will develop the mathematics to statistically describe these effects so that we can understand and model the impacts on the incoherent imaging system (Figure 7.24).

A general model for the refractive index of the Earth's atmosphere is expressible in terms of position, time, and wavelength, and can readily be separated into a deterministic term and a random term as follows:

$$n(\vec{r},t,\lambda) = n_0(\vec{r},t,\lambda) + n_1(\vec{r},t,\lambda). \tag{7.49}$$

Here, n_0 represents the deterministic part of the index of refraction and n_1 is random. The n_1 term fluctuates about a mean value of one such that $\bar{n} = n_0 \approx 1$. The deterministic term n_0 has long time scales and so the time parameter t can be suppressed. The cause of the fluctuations of n_1 are the turbulence effects in the Earth's atmosphere. Often, the wavelength dependence can be ignored in the random term n_1 and so Equation 7.49 can be rewritten as

$$n(\vec{r},t,\lambda) = n_0(\vec{r},\lambda) + n_1(\vec{r},t). \tag{7.50}$$

Additionally, because of the speed of light, it typically only takes a fraction of the time of the fluctuations to propagate through the atmosphere. Consequently, the time element of the fluctuating term is often ignored.

7.2.8 Correlation Function and Power Spectral Density of the Refractive Index

In this section, we present some important statistical quantities using the fluctuating part of the Earth's refractive index. We will start with the spatial autocorrelation function and develop a very useful analytical metric, the Power Spectral Density (PSD). We define the spatial autocorrelation function of the refractive index $n_1(\vec{r})$ in the Earth's atmosphere as

$$\Gamma_n(\vec{r}_1,\vec{r}_2) = \overline{n_1(\vec{r}_1)n_1(\vec{r}_2)}. \tag{7.51}$$

Given the assumption that $n_1(\vec{r}_1)$ is locally stationary in all three dimensions, we can say we have a locally homogeneous atmosphere. With this assumption, we get some simplifications of Equation 7.51 allowing us to right the autocorrelation function in a homogeneous medium as

$$\Gamma_n(\vec{r}) = \overline{n_1(\vec{r}_1)n_1(\vec{r}_1-\vec{r})}, \tag{7.52}$$

where $\vec{r} = \vec{r}_1 - \vec{r}_2 = (\Delta x, \Delta y, \Delta z)$. According to the Wiener-Khinchin theorem, we can relate the power spectral density to the 3D spatial Fourier Transform of the spatial autocorrelation of the refractive index as in $\Gamma_n(\vec{r})$ (Roggemann and Welsh 1996)

$$\Phi_n(\vec{\kappa}) = \frac{1}{(2\pi)^3} \iiint_{-\infty}^{\infty} \Gamma_n(\vec{r}) e^{j\vec{\kappa}\cdot\vec{r}} d^3\vec{r}, \tag{7.53}$$

where $\vec{\kappa} = (\kappa_X, \kappa_Y, \kappa_Z)$ is called the wavenumber vector, and the sign on the forward direction of the Fourier Transform kernel in Equation 7.53 is in this case arbitrary but has been chosen to be positive for consistency with Tatarskii's work (1967). This relationship can be reversed and the autocorrelation function of the index of refraction can be obtained from the power spectrum of the index of refraction

$$\Gamma_n(\vec{r}) = \iiint_{-\infty}^{\infty} \Phi_n(\vec{\kappa}) e^{-j\vec{\kappa}\cdot\vec{r}} d^3\vec{\kappa}. \tag{7.54}$$

Notice that Equations 7.53 and 7.54 are a Fourier Transform pair. Given that the spatial autocorrelation function of the refractive index is spherically symmetric, then the fluctuations of the random part of the refractive index $n_1(\vec{r}_1)$ are said to be isotropic. For an isotropic index of refraction fluctuations, the integrals in Equations 7.53 and 7.54 can be reduced to single integrals respectively, given by

$$\Phi_n(\kappa) = \frac{1}{2\pi^2\kappa} \int_0^{\infty} \Gamma_n(r) r \sin(\kappa r) dr, \tag{7.55}$$

and,

$$\Gamma_n(r) = \frac{4\pi}{r} \int_0^{\infty} \Phi_n(\kappa) \kappa \sin(\kappa r) d\kappa, \tag{7.56}$$

where the parameter κ is the magnitude of the wavenumber vector, and r is the magnitude of $\vec{r}$.

7.2.9 Integral Form of the Structure Function

We now want to focus on some details of the structure function for the refractive index and so start with noting that this structure function $D_n(\vec{r}_1, \vec{r}_2)$ is the expected value of the difference of the fluctuating index of refraction at two different spatial coordinates

$$D_n(\vec{r}_1, \vec{r}_2) = \overline{\left[n_1(\vec{r}_1) - n_1(\vec{r}_2)\right]^2}. \tag{7.57}$$

If the atmosphere is homogenous, then the structure function $D_n(\vec{r}_1, \vec{r}_2)$ can be expressed as

$$D_n(\vec{r}) = 2\left[\Gamma_n(0) - \Gamma_n(\vec{r})\right], \tag{7.58}$$

as long as $\Gamma_n(0)$ exists. Equation 7.58 is a function of the autocorrelation of the index of refraction evaluated at zero and at $\vec{r}$. If the definition of the autocorrelation function for the index of refraction in Equation 7.54 is substituted into Equation 7.58, and symmetry properties are applied, the structure function of the index of refraction can be written as

$$D_n(\vec{r}) = 2\iint\int_{-\infty}^{\infty}\left[1-\cos(\vec{\kappa}\cdot\vec{r})\right]\Phi_n(\vec{\kappa})d^3\vec{\kappa}. \tag{7.59}$$

This expression can be simplified to

$$D_n(r) = 8\pi\int_0^{\infty}\Phi_n(\kappa)\kappa^2\left[1-\frac{\sin\kappa r}{\kappa r}\right]d\kappa, \tag{7.60}$$

when the atmospheric statistics are isotropic.

7.2.10 Autocorrelation Function and Power Spectrum of the Refractive Index

Not all instances of the spatial autocorrelation function require the 3D integral form as shown. In some cases, it is necessary to determine the two-dimensional spatial correlation function of the refractive index or the two-dimensional power spectral density. For example, it might become necessary to determine these two-dimensional quantities in a plane that is perpendicular to the direction of propagation of an electromagnetic wave. If we choose the direction of propagation as the z direction, then the plane of interest normal to this direction is the x–y plane. We assume locally homogeneous atmospheric turbulence. For the 2D power spectral density, we use the notation $F_n(\vec{q};z)$ where the parameter $\vec{q} = (\kappa_X, \kappa_Y)$ where κ_X and κ_Y are the magnitude of the wavenumber in x and y directions respectively. The 2D spatial autocorrelation function of the refractive index is represented by $B_n(\vec{p};z)$ where the parameter $\vec{p} = (\Delta x, \Delta y)$ and so is also a function of the x–y plane at position z. The relationship between the two- and three-dimensional power spectral densities is given by

$$F_n(\kappa_X,\kappa_Y;z) = \int_{-\infty}^{\infty}\Phi_n(\kappa_X,\kappa_Y,\kappa_Z;z)d\kappa_Z. \tag{7.61}$$

Like the three-dimensional case, the spatial autocorrelation of the refractive index expression $B_n(\vec{p};z)$ can be related to the expression for the power spectral density $F_n(\vec{q};z)$ by

$$F_n(\vec{q};z) = \frac{1}{(2\pi)^2}\int\int_{-\infty}^{\infty}B_n(\vec{p};z)e^{j\vec{q}\cdot\vec{p}}d^2\vec{p}, \tag{7.62}$$

and,

$$B_n(\vec{p};z) = \int\int_{-\infty}^{\infty}F_n(\vec{q};z)e^{j\vec{q}\cdot\vec{p}}d^2\vec{q}. \tag{7.63}$$

If we now bring in the assumption of locally statistically homogeneous atmospheric fluctuations, we can write the 2D spatial autocorrelation function of the index of refraction as

$$B_n(\vec{p};z) = \overline{n_1(\vec{p}_1;z)n_1(\vec{p}_1-\vec{p};z)}. \tag{7.64}$$

Further if the atmospheric fluctuations are statistically isotropic in the x–y plane, then we can get further simplifications as before. Here, for a constant z both $B_n(\vec{p};z)$ and $F_n(\vec{q};z)$ become circularly symmetric and can be written as

$$F_n(q;z) = \frac{1}{2\pi}\int_0^{\infty}B_n(p;z)J_0(qp)\,pdp, \tag{7.65}$$

and,

$$B_n(p;z) = 2\pi \int_0^\infty F_n(q;z) J_0(qp) q dq, \tag{7.66}$$

where $p = |\vec{p}| = \left[\Delta x^2 + \Delta y^2\right]^{\frac{1}{2}}$, $q = |\vec{q}| = \left[\kappa_X^2 + \kappa_Y^2\right]^{\frac{1}{2}}$, and J_0 is the mutual intensity.

7.2.11 Temperature Inhomogeneities Lead to Refractive Index Inhomogeneities

In this section, we look at the connection between temperature inhomogeneities and refractive index inhomogeneities. An interesting expression for the index of refraction of air at optical frequencies, and as a function of wavelength, pressure and temperature is given by Goodman and others as (1985),

$$n = 1 + 77.7\left(1 + 7.52 \times 10^{-3} \lambda^{-2}\right) \frac{P}{T_k} \times 10^{-6}. \tag{7.67}$$

In Equation 7.67, the units for the wavelength λ is in microns, the pressure P is in millibars, and the temperature T_k is given in degrees Kelvin. Variations in the index of refraction from the pressure are small and can often be ignored. It is the random spatial distribution of temperature that causes the larger variations in the fluctuating part of the index of refraction.

7.2.12 Power Spectral Density of the Refractive Index Fluctuations

Now we want to get some expressions for the power spectral density in terms of the refractive index fluctuations. We first start with some observations about three important regions of the power spectral density $\Phi_n(\kappa)$.

- The first region corresponds to very small wavenumbers κ and so deals with very large-scale turbulence. On this scale, the atmosphere is likely neither isotropic or homogenous.
- The second region is for larger κ with values that are larger than some critical wavenumber value κ_0. For this condition, Kolmogorov developed an expression for the power spectral density given by

$$\Phi_n(\kappa) = 0.033 C_n^2 \kappa^{-\frac{11}{3}}, \tag{7.68}$$

  where the quantity C_n^2 is a very important parameter known as the structure constant.
- The third region of importance for the power spectral density is for κ larger than some other wavenumber κ_m. In this case, Tatarski used the following model to describe the power spectral density (Tatarski 1967),

$$\Phi_n(\kappa) = 0.033 C_n^2 \kappa^{-\frac{11}{3}} \exp\left(-\frac{\kappa^2}{\kappa_m^2}\right). \tag{7.69}$$

For all these cases, there is a non-integrable pole when the wavenumber approaches zero. Also, because the Earth's atmosphere is not infinitely large, the power spectrum cannot get infinitely large when the wavenumber goes to zero. Von Karman developed an alternate expression for the power spectral density that eliminates these concerns given by

$$\Phi_n(\kappa) = \frac{0.033 C_n^2}{\left(\kappa^2 + \kappa_0^2\right)^{\frac{11}{6}}} \exp\left(-\frac{\kappa^2}{\kappa_m^2}\right). \tag{7.70}$$

Since wavenumbers are not all that intuitive for some since they are proportional to inverse distance, a distance metric was developed to help in understanding the bounds where the above power spectral density expressions apply. If the wavenumber κ is very high, then the associated distance scale is very small since,

$$L = \frac{2\pi}{\kappa}, \tag{7.71}$$

represents the scale of the atmospheric turbulence in meters. The atmosphere has a finite length so the wavenumber cannot increase unbounded. Consequently, there is a cutoff wavenumber κ_0 that corresponds to the largest wavenumber that can be contained within the Earth's atmosphere. This large value of κ_0 provides the cutoff boundary for the smallest scale size of atmospheric turbulence, the so-called *outer scale* of atmospheric turbulence, which is given by

$$L_0 = \frac{2\pi}{\kappa_0}. \tag{7.72}$$

In Equation 7.72, L_0 represents the scale size of atmospheric turbulence for which larger values of L (i.e., smaller sizes of the wavenumber κ) cannot be contained by the atmosphere, and so act as an outer bound for the scale size of the atmospheric turbulence. The scale size of the atmospheric turbulence is a function of height, and near the ground, the outer scale l_0 ranges somewhere between 1 and 100 m. On the other end of the spectrum, for large values of κ, the value of L becomes very small until another limit is reached. The quantity κ_m represents the wavenumber upper limit resulting in the smallest scale size known as the inner scale l_m for which smaller atmospheric turbulence cell sizes dissipate away due to viscous forces. This smaller scale size is given by

$$l_m = \frac{2\pi}{\kappa_m}. \tag{7.73}$$

Putting these facts together, the equations for the power spectral density are valid in the wavenumber regions $\kappa_0 \le \kappa \le \kappa_m$, and correspondingly in between the inner and outer scale sizes, $l_m \le L \le L_o$. The inner scale of atmospheric turbulence near ground level has values close to a few millimeters. The range of the wavenumber between its upper and lower bounds leads to multiple different scale sizes for the atmospheric turbulence ranging from millimeters to 100 m or so. For each atmospheric turbulence cell, the index of refraction within that cell is assumed constant for the corresponding scale size. For example, in the upper atmosphere, where the cell sizes are large, there can be large regions of constant index of refraction. Near the ground, where the cell sizes can be on the order of millimeters, there is a lot more variation in the index of refraction. Further, the wind moves these cells around and so the collection of these turbulent cells is known as "turbulent eddies" (a good name for a rock band if I ever heard one)!

7.2.13 The Refractive Index Structure Function

We want to point out another useful simplification when it comes to the structure function $D_n(\vec{r}_1, \vec{r}_2)$. The structure function is a very important and useful atmospheric turbulence metric that is central to many theoretical and mathematical underpinnings and constructs in modeling the physics of atmospheric propagation for optical waves. One advantage in using the structure function is that it has low sensitivity to nonhomogeneous and nonisotropic atmospheric conditions in regions where the wavenumber is low (like approaching the outer scale of the atmospheric turbulence). Said another way, even if the atmosphere has conditions that are inhomogeneous and non-isotropic, the structure function can still be homogeneous and isotropic, a quite interesting and useful fact. If we then use isotropic and homogeneous assumptions and the Kolmogorov spectrum in Equation 7.68

and substitute this into Equation 7.60, we get a radial expression of the structure function that is valid over the range, $\kappa_0 \le \kappa \le \kappa_m$,

$$D_n(r) = C_n^2 r^{\frac{2}{3}}. \tag{7.74}$$

7.2.14 Propagation of the Electromagnetic Wave through the Inhomogeneous Atmosphere

At this point, we want to quickly go over the propagation of the electromagnetic wave, but this time through an inhomogeneous atmosphere. We go back to the basic Maxwell's equations to start. Also, we bring in our earlier model of the index of refraction that we presented which is a function of spatial position and not temperature

$$n(\vec{r}) = n_0 + n_1(\vec{r}). \tag{7.75}$$

Recall that the time dependence of the propagating electromagnetic wave is given by $\exp(-j\omega t)$, where ω is the radial frequency in radians. As usual, the magnetic permeability is given by $\mu$and is assumed constant in the atmosphere. On the other hand, the dielectric constant ϵ is considered space-variant in the atmosphere. With these assumptions, we write Maxwell's equations as

$$\begin{aligned} \nabla \cdot \vec{\mathbf{H}} = 0 \qquad & \nabla \times \vec{\mathbf{H}} = -j\omega\epsilon\vec{\mathbf{E}} \\ \nabla \cdot \left(\epsilon\vec{\mathbf{E}}\right) = 0 \qquad & \nabla \times \vec{\mathbf{E}} = j\omega\mu\vec{\mathbf{H}} \end{aligned}, \tag{7.76}$$

where as usual $\vec{\mathbf{E}}$ represents the electric field, $\vec{\mathbf{H}}$ represents the magnetic field, and the term ∇ has partial derivatives consisting of $\left(\frac{\partial}{\partial x}, \frac{\partial}{\partial y}, \frac{\partial}{\partial z}\right)$. If we apply vector operations like the gradient, curl, and divergence to the individual expression in Equation 7.76, upon simplification we arrive at the wave equation for a general field U (either the electric field or magnetic field) for a source-free region

$$\nabla^2 \mathbf{U} + \frac{\omega^2 n^2}{c^2}\mathbf{U} = 0. \tag{7.77}$$

In Equation 7.77, we note that all components of the field obey the same wave equation so we can drop the vector notation and allow the general field **U** to represent any of the scalar components, E_X, E_Y, or E_Z. The value c is the speed of light in free space.

7.2.15 The Rytov Approximation

Since we know that the fluctuating part of the index of refraction is much smaller than the deterministic part of the index of refraction (i.e., $n_1 \ll n_0$), we can make some interesting observations by expanding the general field about a point $\vec{r}$ such that

$$\mathbf{U}(\vec{r}) = \mathbf{U}_0(\vec{r}) + \mathbf{U}_1(\vec{r}) + \mathbf{U}_2(\vec{r}) + \cdots, \tag{7.78}$$

And that each successive term is much smaller in magnitude than the preceding term (i.e., $|\mathbf{U}_0| \gg |\mathbf{U}_1| \gg |\mathbf{U}_2|$) such as in the case of weak scattering. With the Born approximation, we drop the terms after $\mathbf{U}_1$, effectively eliminating multiple scattering effects, and only including one scattering element of the field, $\mathbf{U}_1$. The term $\mathbf{U}_0$ corresponds to the unscattered part of the field. With these assumptions, we can represent the Rytov approximation as

$$\mathbf{U}(\vec{r}) = \exp\left[\Psi(\vec{r})\right], \tag{7.79}$$

where Ψ is given by $\Psi = \Psi_0 + \Psi_1$ and is known as the complex log amplitude of the wave. We can separate out amplitude and phase portions of the general field by letting the amplitude for the general field **U** be given by A, and the phase of **U** given by S. Similarly, the free-space corresponding quantities for the free-space field $\mathbf{U_0}$ are given by A_0 and S_0, so that the general field **U** and free-space field $\mathbf{U_0}$ are given by

$$\mathbf{U} = A\exp(jS) \qquad \mathbf{U}_0 = A_0\exp(jS_0). \tag{7.80}$$

Consequently, the variations of the complex log amplitude of the wave with respect to the free space wave are given by

$$\Psi_1 = \Psi - \Psi_0 = \ln\frac{A}{A_0} + j(S - S_0). \tag{7.81}$$

If we define χ as the log-amplitude fluctuation

$$\chi \equiv \ln\frac{A}{A_0}, \tag{7.82}$$

and the phase fluctuations as

$$S_\delta \equiv S - S_0, \tag{7.83}$$

We get the expression for the wave fluctuations Ψ_1 in terms of the log amplitude fluctuations and phase fluctuations to be

$$\Psi_1 = \chi + jS_\delta. \tag{7.84}$$

This expression provides a description of the argument of the fluctuating general field shown and can be used with Equation 7.79 to provide both amplitude and phase effects associated with the fluctuating field. Notice that the amplitude portion χ has a natural logarithm term as seen in Equation 7.82 and is known as the log-amplitude.

7.2.16 The Long Exposure OTF

If we need to remove atmospheric turbulence effects from imagery, it becomes necessary to "freeze" the atmospheric turbulence and mitigate time averaging effects. The time scale that the atmosphere changes depends on the velocity of the wind and typically falls somewhere in the range of 1–10 ms. Approximate general exposure ranges for short and long exposure imaging scenarios are,

$$\text{Long Exposure} \quad T \gg 10\ \text{ms},$$

$$\text{Short Exposure} \quad T \ll 10\ \text{ms}.$$

When it comes to long exposure imagery, this approach is used when the object that is being imaged is very dim such as in astronomical objects. These types of objects may require from seconds to hours of integration time on the detector to get sufficient signal to become observable. Using the log-normal and phase fluctuation representation of the optical wave, we can write the instantaneous field distribution incident on the left side of the lens in Figure 7.26 as

$$\mathrm{U}(x,y)=\sqrt{I_0}\exp\left[\chi(x,y)+jS(x,y)\right]. \tag{7.85}$$

If we assume ergodicity, we note that the statistical ensemble average of the OTF is the same as the time averaged, long-exposure OTF. The average OTF can be expressed as the point-wise product of the diffraction-limited OTF and the average long exposure OTF as in,

$$\bar{\mathcal{H}}(\nu_U,\nu_V)=\mathcal{H}_0(\nu_U,\nu_V)\bar{\mathcal{H}}_{LE}(\nu_U,\nu_V). \tag{7.86}$$

The diffraction limited OTF, $\mathcal{H}_0$, is the OTF of the optical system in the absence of turbulence. The long-exposure OTF, $\bar{\mathcal{H}}_{LE}$, can be expressed in terms of the wave fluctuations as

$$\bar{\mathcal{H}}_{LE}(\nu_U,\nu_V)=\frac{\Gamma\left(\bar{\lambda}f\nu_U,\bar{\lambda}f\nu_V\right)}{\Gamma(0,0)}, \tag{7.87}$$

where

$$\Gamma\left(\bar{\lambda}f\nu_U,\bar{\lambda}f\nu_V\right)=\overline{\exp\left[(\chi_1+\chi_2)+j(S_1-S_2)\right]}. \tag{7.88}$$

A simple and useful expression for the average OTF can be obtained by carrying out the expectation at the right of Equation 7.88 under the assumption of a homogenous and isotropic atmosphere

$$\bar{\mathcal{H}}(\nu)=\mathcal{H}_0(\nu)\exp\left[-\frac{1}{2}D\left(\bar{\lambda}f\nu\right)\right], \tag{7.89}$$

where the term $D\left(\bar{\lambda}f\nu\right)$ is the wave structure function given by

$$D\left(\bar{\lambda}f\nu\right)\triangleq D_\chi\left(\bar{\lambda}f\nu\right)+D_S\left(\bar{\lambda}f\nu\right). \tag{7.90}$$

7.2.17 Near-Field Calculation of the Wave Structure Function

At this point, we want to mention that there are times that we want to observe an object that is in the near field of the optical imaging system, and therefore want to determine the impact on the wave structure function for this case. We first make some severe simplifying assumptions that are only sometimes met in practice, to get things started. These simplifying assumptions are as follows:

1. We assume an incoherent radiating object (this one is often true).
2. The distant object that is being imaged is very far away from the entrance pupil of the imaging system, so that the atmosphere affects all parts on the object equally. For this case, the atmosphere is assumed isoplanatic.
3. The atmospheric turbulence is near the imaging aperture over a finite distance z and is also homogeneous and isotropic. Also, the structure constant C_n^2 is assumed constant over the atmospheric turbulence region.
4. There is no significant bending of rays that hit the homogeneous atmospheric turbulence medium because the imaging system is located deep inside the near field of the turbulent eddies that have the greatest effect on the imaging system. The rays of light in the medium just experience a delay.

Although these assumptions, may initially seem overly restrictive, the results found using these assumptions are found to be more generally applicable. Under the assumptions above, the long exposure OTF is found to be,

$$\bar{\mathcal{H}}_{LE}(v) = \exp\left[-57.4C_n^2 \frac{zf^{\frac{5}{3}}}{\bar{\lambda}^{\frac{1}{3}}} v^{\frac{5}{3}}\right], \tag{7.91}$$

where C_n^2 is the structure constant, z is the separation distance of the object of interest and the imaging system, $\bar{\lambda}$ is the mean wavelength, f is the effective focal length of the imaging system, and v is the spatial frequency of interest. $\bar{\mathcal{H}}_{LE}(v)$ is a radially symmetric function. Often, it can be more convenient to write the OTF as a function of frequency $\Omega = fv$ which has dimensions of cycles per radian of arc in the sky instead of cycles per meter,

$$\bar{\mathcal{H}}_{LE}(\Omega) = \exp\left[-57.4C_n^2 \frac{z}{\bar{\lambda}^{\frac{1}{3}}} \Omega^{\frac{5}{3}}\right]. \tag{7.92}$$

The diffraction limited OTF for a circular entrance pupil can similarly be written as a function of Ω, and is given by

$$\bar{\mathbb{H}}_0(\Omega) = \begin{cases} \frac{2}{\pi}\left[\cos^{-1}\left(\frac{\Omega}{\Omega_0}\right) - \frac{\Omega}{\Omega_0}\sqrt{1-\left(\frac{\Omega}{\Omega_0}\right)^2}\right] & \Omega \le \Omega_0 \\ 0 & \text{otherwise} \end{cases}. \tag{7.93}$$

7.2.18 The Atmospheric Coherence Length

The atmospheric coherence length (a.k.a. the Fried parameter: pronounced "Free'd parameter) is a very useful metric in characterizing the effects of the turbulent atmosphere. This metric represents the scale of the atmosphere for which the atmosphere can be considered constant (frozen). As such this metric provides an excellent way of determining if there is any benefit to the imaging system in removing the atmospheric turbulence effects from the imagery. If the diameter of the entrance pupil of the imaging system is larger than the Fried parameter r_0, then the imaging system benefits from removing the atmospheric aberrations. If the diameter of the imaging systems' entrance pupil is equal to or less than r_0, then there is no benefit provided from removing the atmospheric turbulence. The benefit to the imaging system manifests in terms of increased spatial resolution. Another way to think of the atmospheric coherence length is as the effective diameter of a large imaging system that has been degraded by atmospheric turbulence. For example, a large 5 m-class optical telescope will have no better spatial resolution than a 4-in. telescope bought at Walmart if the effects of atmospheric turbulence are not removed. The larger telescope of course will be able to collect signals from dimmer objects. The Fried parameter is given by (Roggemann and Welsh 1996),

$$r_0 \triangleq 0.186\left[\frac{\bar{\lambda}^2}{\int_0^z C_n^2(\xi)d\xi}\right]. \tag{7.94}$$

The Fried parameter is a function of the structure constant that we mentioned before as is the isoplanatic angle that we will discuss in the next section. Another interesting, related metric is the ratio of the entrance pupil plane diameter to the Fried parameter $\frac{D}{r_0}$ which provides a measure of the expected improvement in expected spatial resolution increase by using adaptive optics or atmospheric turbulence compensation methods. For example, for an entrance pupil plane diameter of

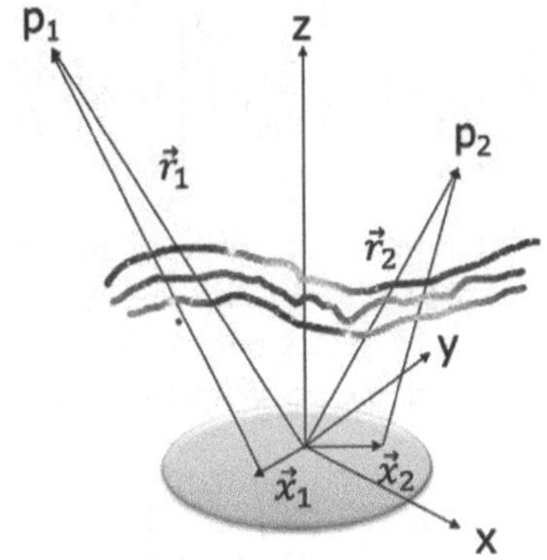

FIGURE 7.25 Geometry for determining the isoplanatic angle.

20 cm and an r_o value of 2 cm, by removing the effects of atmospheric turbulence, we would expect a ten-fold increase in the imaging system's spatial resolution. This marks the atmospheric coherence length as a mighty useful metric indeed! The Fried parameter typically varies from 2 cm under relatively poor imaging conditions such as in ground-to-ground imaging scenarios, and 20 cm for good imaging conditions such as on a mountain-top observatory.

7.2.19 Isoplanatic Angle

The isoplanatic angle represents the maximum angular separation between two points so that the turbulence between the two points is similar. An example would be a bright star beacon and an object being imaged by an adaptive optical system. The isoplanatic angle is the maximum angular separation between the object and the beacon so that the atmosphere sensed by the AO system from the beacon can be used for the object (Figure 7.25).

Fried gives the isoplanatic angle as

$$\theta_0 = 58.1 \times 10^{-3} \lambda^{6/5} \left[\int_0^L dz C_n^2(z) z^{5/3} \right]^{-3/5}. \tag{7.95}$$

7.2.20 The Short Exposure OTF

We now make some observations concerning the differences between the long exposure and short exposure OTFs.

Long Exposure

- PSF is smooth and broad;
- OTF is narrow and smooth.

Short Exposure

- PSF is jagged and narrower;
- OTF has significant fluctuations of both magnitude and phase as a function of frequency;
- Short-exposure image quality is not affected by the tilt component of the Wavefront deformations.

The average short exposure OTF can be written in terms of the atmospheric coherence length r_o and the diffraction-limited structure function as (Goodman 1985),

$$\bar{\mathcal{H}}_{SE}(\nu) = \exp\left[-3.44\left(\frac{\bar{\lambda} f\nu}{r_0}\right)^{\frac{5}{3}}\left[1-\alpha\left(\frac{\bar{\lambda} f\nu}{D_0}\right)^{\frac{1}{3}}\right]\right]. \tag{7.96}$$

However, this expression can be written in a more convenient form by writing the argument in the form of an angular measurement where $\Omega = f\nu$. In this case, the OTF is a function of Ω which represents an angular frequency with units of cycles per radian of arc in the sky instead of the usual spatial frequency with units of cycles per meter. With this argument, the average short exposure OTF is represented as

$$\bar{\mathcal{H}}_{SE}(\Omega) = \exp\left[-3.44\left(\frac{D_0}{r_0}\frac{\Omega}{\Omega_0}\right)^{\frac{5}{3}}\left[1-\alpha\left(\frac{\Omega}{\Omega_0}\right)^{\frac{1}{3}}\right]\right]. \tag{7.97}$$

We now provide an example using the short exposure OTF with the atmospheric turbulence compensation technique known as Stellar Speckle Interferometry. An interesting feature of taking a short exposure image of a point source is that there is a lot of high-frequency structure which is known as "speckle." By averaging many of these short-exposure images together after they have been properly centered, the high-frequency content becomes suppressed. The centering process is used to make sure that the images align with each other. The resulting averaged image agrees with the predictions of the average short exposure image produced by using Equation 7.96 or Equation 7.97. To recover the high-frequency information, the speckle interferometry technique developed by Labeyrie uses the ensemble average of the squared modulus of the OTF, which interestingly retains high-frequency information, whereas the ensemble average of the short-exposure OTF rapidly loses the higher-frequency information (Labeyrie 1970). Consequently, we want to develop an expression for the average squared magnitude of the OTF.

In speckle interferometry, we can use an astronomical telescope to capture many short-exposure images of an object of interest. Say we collect K images in this fashion. If k represents the index of the kth short-exposure image, the Fourier Transform of the kth image is given by

$$\mathfrak{I}_i^{(k)}(\nu_U,\nu_V) = \iint_{-\infty}^{\infty} I_i^{(k)}(u,v)\, e^{j2\pi(u\nu_U+v\nu_V)}\,du\,dv, \tag{7.98}$$

where the subscript i stands for image, and $I_i^{(k)}(u,v)$ is the kth short exposure image irradiance at image coordinates u and v. The quantity $\mathfrak{I}_i^{(k)}(\nu_U,\nu_V)$ is the corresponding irradiance spectrum which can also be written in terms of the kth OTF and the unchanging object spectrum as

$$\mathfrak{I}_i^{(k)}(\nu_U,\nu_V) = \mathcal{H}^{(k)}(\nu_U,\nu_V)\,\mathfrak{I}_o(\nu_U,\nu_V). \tag{7.99}$$

In Equation 7.99, $\mathcal{H}^{(k)}(\nu_U,\nu_V)$ represents the OTF corresponding to the kth image, and $\mathfrak{I}_o(\nu_U,\nu_V)$ is the object spectrum, both at spatial frequency coordinates ν_U and ν_V. If we now take the magnitude squared of Equation 7.99, we get a collection of "energy spectra" given by

$$\mathcal{E}_I^{(k)}(\nu_U,\nu_V) = \left|\mathfrak{I}_i^{(k)}(\nu_U,\nu_V)\right|^2, \tag{7.100}$$

where

$$\overline{\left|\mathfrak{I}_i^{(k)}(\nu_U,\nu_V)\right|^2} = \overline{\left|\mathcal{H}^{(k)}(\nu_U,\nu_V)\right|^2}\left|\mathfrak{I}_o(\nu_U,\nu_V)\right|^2. \tag{7.101}$$

The term $\overline{\left|\mathcal{H}^{(k)}(\nu_U,\nu_V)\right|^2}$ represents the quantity of interest, namely the average of the squared magnitude of the short-exposure OTF corresponding to the kth image. Incidentally, since the magnitude of the OTF is the modulation transfer function (MTF), this can also be interpreted as the average of the squared MTF. As stated before, this term is noteworthy since it retains high-frequency information.

7.2.21 Signal-to-Noise (SNR) Considerations

In this section, we present how to model the effects of the photons detected on the imaging system through the SNR at the detector. Signal-to-noise calculations are a critical complement to the spatial and temporal characteristics of propagation, detection, and characterization of an optical wave. In addition to understanding spatial performance parameters like the spatial resolution of the imaging system, it is necessary to know if sufficient signal is present, compared to the noise effects present in the imaging system, to properly form the image. The model we use is a semi-classical model for the detection of photo-electric light and including the effects of noise. Noise results from several sources such as the random locations where the photons strike the detector and the random arrival rate of the photons themselves. Consequently, we use a doubly stochastic random process to describe the detection statistics. Two of the critical assumptions we make are as follows:

- We have photon noise (shot noise) that stems from two sources of randomness: the arrival rate of the photons on the detector surface and the random location of the photons on the detector surface (two sources of randomness, therefore a doubly stochastic random process);
- We use a semi-classical imaging model where classical propagation methods are used for the electromagnetic field up to the detector. However, at the detector's surface, the interactions of light and the detector use statistical methods.

For low light conditions, which is where SNR effects are most critical, we further use a Poisson probability density function, whereas, for high light levels, we use a Gaussian probability density function due to there being lots of independent photons arriving, and the Central Limit Theorem ensuring that the resulting distribution is Gaussian. We start with the case of low light level detection. For low light levels, the random number of photo-events that are governed by the Poisson probability density function are given by K, and we define, the mean number of photo-events, as

$$\overline{K} = \int_A d\vec{x}\lambda(\vec{x}), \tag{7.102}$$

where the integration is over the arrival area of the photons, and $\lambda(\vec{x})$ is a rate function for the arrival of photons. The units of the rate function are the same as for the irradiance in an imaging system. The second moment of the photo-events is given by

$$\overline{K^2} = \overline{K} + \left(\overline{K}\right)^2, \tag{7.103}$$

and the variance of the photo events is given by

$$\sigma_K^2 = \overline{K^2} - \left(\overline{K}\right)^2 = \overline{K}. \tag{7.104}$$

Note that for the case of low-light illumination, the probability density function obeys Poisson statistics. The model for the arrival locations of the signal and noise photons uses the Kronecker delta function

$$d(\vec{x}) = \sum_{n=1}^{K} \delta(\vec{x} - \vec{x}_n) + \sum_{p=1}^{P} n_p \delta(\vec{x} - \vec{x}_p), \tag{7.105}$$

where $\vec{x}_n$ is the random arrival location of the signal photon, P is the total number of pixels in the detector, $\vec{x}_p$ is the location of the pth pixel, and n_p is the random noise variable in units of photo events that represents the amount of noise present at pixel p. For modeling the noise-free system, the term n_p is set to zero. The random variable n_p is assumed to be (1) zero mean, (2) spatially uncorrelated, (3) one variance can describe the noise variation at each pixel, and (4) is independent from the arrival location random variable $\vec{x}_n$ and the random number of photo events K. If we take the Fourier Transform of Equation 7.105, we get a frequency-based representation of the signal model:

$$D(\vec{f}) = \sum_{n=1}^{K} \exp(-2\pi\vec{f} \cdot \vec{x}_n) + \sum_{p=1}^{P} n_p \exp(-2\pi\vec{f} \cdot \vec{x}_p), \tag{7.106}$$

where we have used the definition of the Fourier Transform between the spatial domain $\vec{x}$ and the spatial frequency domain $\vec{f}$ to be

$$D(\vec{f}) = \int d\vec{x} d(\vec{x}) \exp(-2\pi\vec{f} \cdot \vec{x}). \tag{7.107}$$

In Equation 7.107, the integration is over all space. Often, the frequency-based analytical formalism is the one that is most mathematically tractable and so we define the SNR in frequency space to be

$$\text{SNR}_D(\vec{f}) = \frac{\left| E\{D(\vec{f})\} \right|}{\left[\text{var}\{D(\vec{f})\} \right]^{\frac{1}{2}}}, \tag{7.108}$$

where $E\{*\}$ and $\text{var}\{*\}$ are the *expectation* and *variance* operators on the terms * inside the brackets (in this case $D(\vec{f})$). This signal-to-noise ratio is called the image spectrum signal-to-noise ratio. Carrying out the statistical calculations in Equation 7.108 requires the use of conditional probabilities, and these have been worked out for the case of doubly stochastic random processes by Roggemann and Welsh (1996) as well as others to be

$$\text{SNR}_D(\vec{f}) = \frac{\bar{K} \left| \bar{\mathcal{H}}(\vec{f}) O_n(\vec{f}) \right|}{\sqrt{\bar{K} + (\bar{K})^2 \left| O_n(\vec{f}) \right|^2 \text{var}\{\bar{\mathcal{H}}(\vec{f})\} + P\sigma_n^2}}, \tag{7.109}$$

where σ_n^2 is the noise variance due to the additive noise process and has units of photoelectrons per pixel on a per read-out basis. From before, $\bar{K}$ is the mean number of photo-events given by Equation 7.111, P is the total number of detector pixels, $\bar{\mathcal{H}}(\vec{f})$ is the average OTF at spatial frequency $\vec{f}$, $\mathcal{H}(\vec{f})$ is the random OTF at spatial frequency $\vec{f}$, and $O_n(\vec{f})$ is the normalized object spectrum defined by

$$O_n(\vec{f}) = \frac{O(\vec{f})}{O(\vec{f} = 0)}. \tag{7.110}$$

In Equation 7.110, the numerator is the Fourier Transform of the object radiant emittance distribution in units of Watts/area, and the denominator is the total object radiant emittance. Expression

7.109 captures the random nature of the OTF including the random arrival locations of the incoming photo-events and the random arrival rate of the photons themselves, as well as the additive Gaussian noise effects. Notice also that Equation 7.109 captures details about the object in the normalized object spectrum term $O_n(\vec{f})$ and the atmospheric effects and deterministic optical system entrance pupil plane aperture effects are captured in the random OTF. There are some interesting limiting forms of Equation 7.109 such as in the noise free case where the noise variance σ_n^2 is set to zero

$$\mathrm{SNR}_{D,nf}(\vec{f}) = \frac{\bar{K}\left|\bar{\mathcal{H}}(\vec{f})O_n(\vec{f})\right|}{\sqrt{\bar{K} + (\bar{K})^2\left|O_n(\vec{f})\right|^2 \mathrm{var}\left\{\mathcal{H}(\vec{f})\right\}}}. \tag{7.111}$$

In regions where the second term in the denominator approaches zero, but $\bar{K}$ is still significant (such as in very low regions of the normalized object spectrum or regions where the variance of the OTF approaches zero), the first term in the denominator dominates and the signal to noise ratio in Equation 7.111 scales as the square root of the average number of photo-events. Another interesting case is the high light limit case where the average number of photons approaches infinity (or a very high number compared to the additive optical system noise effects). In this case, we take the limit of the SNR in Equation 7.109 and get

$$\mathrm{SNR}_{\mathcal{H}}(\vec{f}) = \lim_{\bar{K}\to\infty}\left\{\mathrm{SNR}_D(\vec{f})\right\} = \frac{\left|\bar{\mathcal{H}}(\vec{f})\right|}{\sqrt{\mathrm{var}\left\{\mathcal{H}(\vec{f})\right\}}}. \tag{7.112}$$

Note that the square of the mean photons term dominates the expression in Equation 7.109 and $\bar{K}$ cancels in the numerator and denominator. Notice also that this is not a function of object properties and is only a function of the mean OTF and its variance. This SNR is known as the SNR of the random OTF. These signal-to-noise expressions are a useful complement to image formation methods in the sense that it is necessary to know that there is sufficient signal (with respect to the noise) to trigger a successful detection event. A typical rule of thumb is to use a per pixel signal to noise value of 10 or higher to indicate the presence of sufficient signal for successful detection. In practice, this number is conservative, and this author has seen successful detections at lower per pixel detector signal to noise ratios (such as 5 or 6). Nevertheless, the per pixel SNR threshold of 10 or greater is a good initial goal.

In the next section, we apply the concepts learned in Sections 7.1 and 7.2 to the development of an optical system for Homeland Security, border patrol meant to be placed on border patrol owned Predator class UAVs. In this section, we see Futuristic Imaging Technologies engaging in decision and risk analysis activities wherein Game Theory methods are used to develop optimal search patterns at the U.S.-Mexican border. Some sensor analysis activities are also accomplished to determine the effectiveness of FIT's optical sensors.

7.3 INTEGRATED CASE STUDY: DECISION AND RISK ANALYSIS, GT, AND OPTICAL ANALYSIS METHODS IN ASSESSING IMPLEMENTATION ISSUES AND OPTICAL SYSTEM PERFORMANCE

Previously, we covered Decision and Risk Analysis (D&RA) methods such as GT and showed how they can be applied to optical systems as part of the systems engineering development process. In this section, we will see how D&RA methods such as GT can be used in optical development programs/projects to tackle some implementation issues. Additionally, we will illustrate optical system performance estimation using the OTF and signal-to-noise estimation (SNR) methods. As part of our running case study with our fictitious company, Fantastic Imaging Systems (FIT), and their

Department of Homeland Security, Customs and Border Protection (DHS CBP) program, a central focus recently has been on an important upcoming stakeholder TIM, where the stakeholders want to understand the expected performance of the optical system in its operational environment. FIT is now facing some mission and operational planning considerations in how to best deploy the sensor suite that they are developing. Dr. Bill Smith, FIT"s CEO called a meeting with the FIT technical staff to meet with two stakeholder representatives to discuss the issue. At the meeting are Dr. Tom Phelps CTO, Mr. Karl Ben, Senior Systems Engineer, Jennifer O, Systems Engineer, Ron S., new systems engineer, Christian R., Analyst, Tristan R., MBSE Modeler, Aria R., DHS Mission Planner, and Adalind R., DHS CBP UAV Operator.

Bill: "Hello everybody and welcome! I would like to thank our two visitors from the Department of Homeland Security, and also their Customs and Border Protection office, Aria R., and Adalind R. Aria is a mission planner at DHS, and Adalind is providing some operational experience in flying UAVs as an operator at DHS CBP. Both have joined us to give us some operational insights as to how our sensor package will be flown in preparation for the big TIM we will have all our stakeholders later this month. Aria and Adalind, thanks for joining us. Before I turn this over to Karl Ben, our senior systems engineer, is there anything that you would like to see from us?

Aria: "Thanks Bill, and hello everyone. We appreciate that you reached out to us and we hope we can provide the operational details you need. We want to give you a bit of the operational picture for how the missions will typically be flown so that you understand the operational environment in which the sensor package will find itself, and what it needs to accomplish. Our objectives here are that we were hoping to get a bit of an advanced view on how you evaluate the performance of your imaging system. Ultimately, we want to make sure that when we have the TIM with all the stakeholders, we can satisfactorily address all their concerns."

Bill: "Thanks Aria, we will certainly cover our performance assessments as part of this meeting. I think the mission planning insights and operational details of the host platform will be invaluable to our development effort. Would you like to say a few words to kick things off?"

Aria: "Sure, Bill. Thanks! Well, first thing to realize is that we need to think of the operation from the point of view of an adversary who is trying to beat our system. The operational mission most often will not be a cooperative engagement. The folks or items that we need to find, most likely don't want to be found, or seen. So, we may not be able to guarantee optimal detection conditions. The sensing system will need to perform over a range of environmental conditions including day, night, rain, fog, clouds, or even snow. The sensor suite needs to have solutions for all these situations."

Tristan: "We have the latest ConOps and OpsCon documents and graphics in our MBSE system. We can use these to develop Use Cases for each of the relevant missions and then develop scenarios for the different environmental conditions that we expect. We will develop *sunny-day*, *rainy-day*, and *nominal* for each environmental condition based on whether things go right, things go wrong, or what we think will be a typical scenario. Afterwards, we can develop steps for each of the scenarios, and use these to generate requirements for the system. All of this will be linked together in our MBSE environment and used and linked to our modeling effort."

Aria: Okay! Can you show these during our upcoming stakeholder TIM?"

Tristan: "Of course. We can show any of the MBSE products or run our models in our conference center for you all. We can do more than that too. We will be able to link the engineering parameters and operational details from the Use Cases directly to our models and will be able to show results for different parameter variations like height and speed of the platform, platform jitter, look angle, bore-sight jitter, ground sample distance, and a whole host of conditions. We can even have standard parameter sets that can tweak our models for different scenarios of interest."

Aria: "Wow, that's amazing!"

Karl: "That's right Aria, our digital twin of the sensor suite on the host platform is quite sophisticated. We can even change atmospheric conditions to directly simulate the effect on the different sensor system that are in our sensor suite. Amanda, our optical engineer, can give you a better idea of some of the important aspects of our modeling effort when it comes to characterizing our sensor suite performance, but first let's talk a little bit about the general nature of the flight profile during the missions. Aria, you mentioned getting into the head of our adversary. Well, as part of our Optical Systems Artificial Intelligence package, we are incorporating real-time decision and risk methods to be used for mission planning purposes, and for real-time operational decision-making. Our optical sensor suite has its own AI that can make decisions about selecting objects of interest (OoI) and establishing their relative priorities, performing adaptive and dynamic sensor configurations based on real-time environmental conditions and mission profiles and objectives, and can also accomplish real-time data analysis. It also works with the AI on the host platform and provides the sensor inputs needed by the main UAV AI system. So, for example, the optical AI can use information such as the actual instantaneous four vector (x, y, z, for position, and time t) of the UAV platform, its speed, heading, and orientation in real-time, choose the sensor type based on the mission (like the hyperspectral sensor for a concealed target detection mission), and automatically configure the sensor for optimal detection (set the focal length, sensor gain parameters, filter selections, and scan pattern for instance). We did have a question on what types of patterns that the UAV flies when trying to find moving targets at day or night using our high-resolution EO/IR sensors."

Adalind: "I think I might be able to shed some light on that. The type of patterns we fly depends strongly on the mission, the Target-of-Interest (TOI) itself, and the number of other assets we have in the area. If we have multiple A/AI-UAVs, we designate one as the lead and this leader has supervisory responsibility for setting the initial mission parameters, flight patterns, target sets, and safety zones for all the other A/AI-UAVs in the swarm. We call the lead A/AI-UAV the *mother ship* and it chooses and initiates the search patterns of the host platform based on the local terrain, the number of A/AI-UAVs in the swarm and the mission. The mother ship is incredible. It has all sorts of redundant systems and can even function autonomously. The cool thing is that if something happens to the mother ship, another member in the swarm can pick up the responsibilities and becomes the new mother ship. Once the initial swarm flight pattern is assigned to the subordinate drone, it searches the space given by the flight pattern."

Dr. Tom: "It seems that the worst case would be if there was no swarm, and the only available asset was a single A/AI-UAV, right?"

Adalind: "That is correct. The more A/AI-UAVs we have, the more we can divide up the search space and the faster we can cover the necessary terrain. The worst case would be to be limited to only one A/AI-UAV since it must cover all the terrain by itself."

Dr. Tom: "Then we need to plan for the worst case and show that the mission will be successful using a single FIT sensor suite on a single A/AI-UAV host platform. Adalind, can you go over the different types of search patterns that you typically employ?"

Adalind: "Of course! We have a variety of different search patterns when we have any type of a mission that involves search (Schult and Kurucar 2023). Let me bring up my presentation on the overhead. As we said earlier, the type of mission and local terrain are naturally fundamental considerations in setting up the search pattern. For instance, is the mission a cooperative one where the subject of the search has interest in being found? Is the subject neutral such as in an animal, or heard of animals? Or does the subject not want

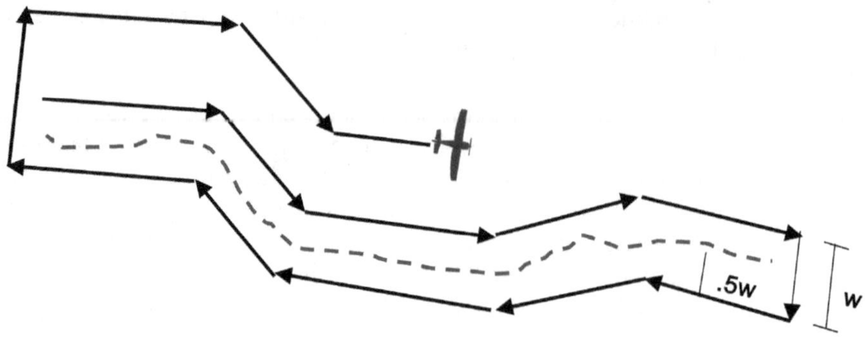

FIGURE 7.26 Track crawl search pattern.

to be seen, such as if we are looking for a baddie. In the latter case, there might even be some anticipated hostility such as gunfire, missile fire, laser blinding techniques, or electromagnetic attack of our assets, in addition to the expected evasive tactics. So, the A/AI-UAV is capable of different types of search patterns, can alter its fight profile such as in increasing altitude, can vary its speed and direction, and engage in evasive maneuvers if necessary. Let me start off with a common search pattern that we call the *track crawl.* This is shown on the overhead (Figure 7.26) and works when we know where our search subject is going. A good example would be a vehicle going down a road or a well-traveled mountain pathway. In the figure, *W* represents the width of the instantaneous field of view (IFOV) of the primary sensor projected along the track. We call this the track width. The A/AI-UAV starts on the left side of the figure at the tail of the middle, right-facing arrow. The IFOV of the A/AI-UAV sensor is centered on the track so that the projected IFOV straddles the track and covers each side of the track equally. The A/AI-UAV flies at half the track width along the track shown by the black arrows. This way, if the subject of the search is on the track (shown by the dashed lines in the center of the figure), they will be found during the first pass along the designated search area along the track. At the end of the track (shown at the right side of the figure) the A/AI-UAV moves one track width to the other side of the track (shown by the downwards arrow at the right side of the figure) and searches the same area on the return path. If the subject of the search is not found at the end of the return path search, the A/AI-UAV moves over by two track widths and covers the adjacent IFOV as shown between the two parallel directional arrows at the top of the figure. Typically, with this pattern, the return cross over is slightly smaller than two widths of the IFOV because it is generally preferable to have a slight overlap between the initial IFOV in the first, round trip, and the IFOV on the subsequent round-trip path. Note that with this pattern there is a decision point that occurs at the end of the prescribed track (shown on the left side of Figure 7.26). Should the subsequent round trip track crawl occur to the top, or bottom of the original track? Local terrain can sometimes be used to help make this decision. Should subsequent track crawls continue in one direction, or bounce back and forth from the top of the original track crawl to the bottom of the original track crawl? The latter case takes more transition time when changing from one track crawl to another. Consequently, this type of pattern is used if one direction can be excluded such as if an impassible ridgeline blocks off the area to the south of the initial track. Of course, the subsequent track crawls are only required if the subject(s) have the capability to go off track which often is not the case for conventional vehicles without off-road capabilities. This track can also be modified to a single pass track that follows a road system, and the A/AI-UAV can be positioned either to the left, right, or directly above the track line shown in the figure. The next type of search pattern is the parallel sweep. You can see this pattern

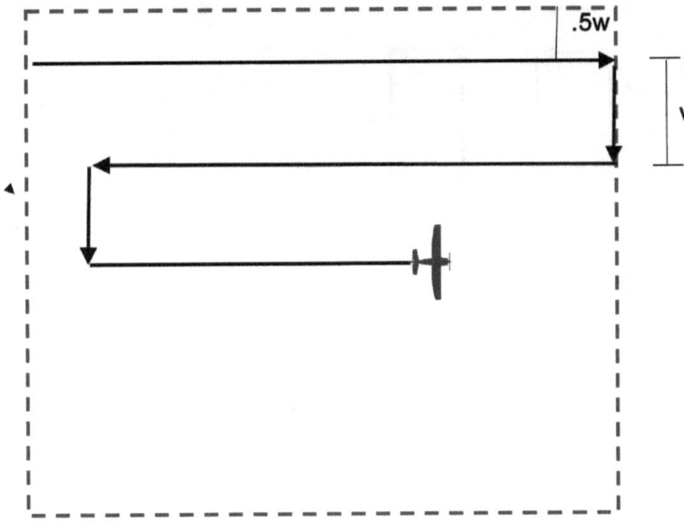

FIGURE 7.27 Parallel sweep search pattern.

in this next slide (Figure 7.27). Notice that this pattern is like the track crawl method we just went over but doesn't have a definitive track that it follows, and this search pattern doesn't repeat the same track on its return path. This type of search pattern is very good for covering large amounts of flat terrain wherein the subject of the search can be anywhere within the rectangular search area and isn't restricted to follow a given track. Examples here would be open plains, wild lands, desert, or tundra. Of course, there may still be terrain limitations like steep ridges, mountains, cliffs, ravines, or large bodies of water that restrict access but generally the target of interest can be anywhere within the search region. Normally, a grid search space is used. In the figure, you can see that the A/AI-UAV starts at the top-left and follows the black arrow pattern as shown. Notice that the track is flown so that the center line of the A/AI-UAV flight path is such that the edge of the IFOV of the sensors on the A/AI-UAV define the dashed region shown in the figure. As before, the A/AI-UAV starts on the top-left of the figure and then doubles back at the end of the track, approximately one, track-width from the original track. At the end of each track, when the A/AI-UAV reverses its position, the IFOV maps out new terrain at a width W in which the A/AI-UAV flies down its center. This pattern looks a bit like a reverse "S" pattern. Any questions or comments yet?"

Karl: "This is quite interesting. It looks like decision trees could be set up to determine which search patterns to use and what the optical system parameters should be. I think we could also use Game Theory to capture the adversarial component that Aria was talking about earlier."

Dr. Bill: "Yes, we can go over our ideas on this shortly but let's hear from Adalind on all the different types of search patterns first, so we have a complete picture. Go ahead Adalind."

Adalind: "Thanks! We also have a variant of the parallel sweep called the *creeping line* search pattern. This one is very similar to the parallel sweep pattern but works well with rectangular search regions wherein one side of the search space is shorter than the other (like in a rectangular search space). We can see on this next slide what this looks like (Figure 7.28). This pattern starts along the shorter edge of the search pattern shown at the top-left of the figure. The pattern proceeds in the reverse-*S* direction from the top-left downwards and then returns one track-width away to start another reverse-*S* pattern. This pattern is useful if the search area is narrower in one dimension than the other and if the TOI is suspected to be in the region closer to where the search starts (the left-half of the

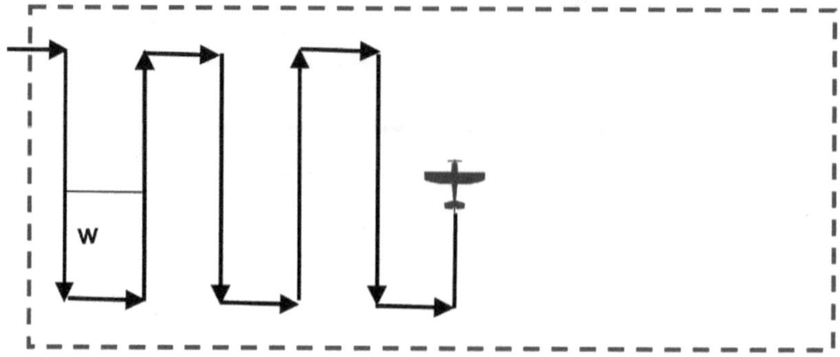

FIGURE 7.28 Creeping line search.

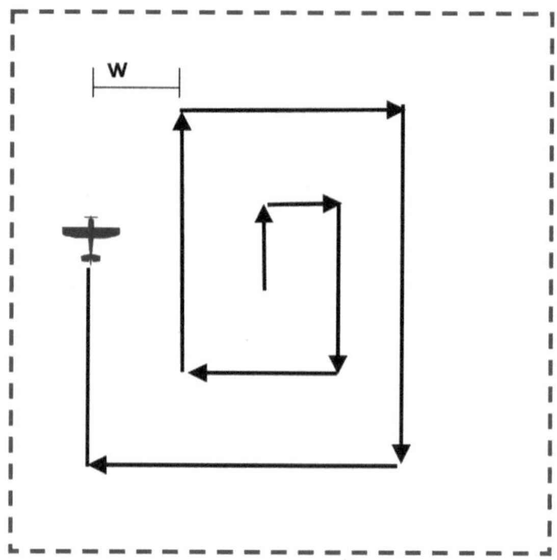

FIGURE 7.29 Expanding square search.

rectangle). If the target is expected to be in the right-half of the rectangle in Figure 7.28, then the search scan should start on the right side of the figure and meander leftward, if possible. Of course, if the TOI happens to be in the opposite half of the rectangle than what is being scanned, it will take longer to find the TOI and depending on the relative speed of the TOI with respect to the speed of the search scan through the area, the TOI may cross the region undetected."

Tristan: "We can use parametric models in the MBSE environment to check the relative speeds of the TOI to the A/AI-UAV regional scan time to see how likely that event is to occur. We may also be able to use this to adjust the flight parameters such as the speed of the A/AI-UAV and the optical system scan pattern and sensor slew rate to prevent this from happening."

Karl: "Let's run some simulations before our meeting with the stakeholders so that we can address this point. Please continue Adalind."

Adalind: "That sounds great. Next, we have a search pattern that is a little more complicated. This search pattern is the *expanding square* and is shown on the next slide (Figure 7.29). This type of search pattern is very good if you suspect where your target is, such as if you have an initial set of coordinates, from their last known, or anticipated location.

This type of pattern is used if you have a small search area such as less than 20 miles (32.2 km) or so. The search pattern starts where the initial TOI is expected to be and then expands in a clockwise direction with an observational track width of *W*. Sometimes, the search space can be further constrained by imposing maximum travel limits based on the maximum speed of the TOI and the terrain conditions. For instance, a person on foot in slightly uneven terrain may only be able to cover 3 miles (4.83 km) in an hour. If the time from the initial location fix has been 6 hours, then the person is likely within 18 miles (28.97 km) of the original fix location and so a radius of 18 miles from the fix location may be established as the primary search space. If we were to search 10 km by 10 km for a search and rescue mission where we start at the last known location of the victim, this type of a search pattern might be a prime candidate. The next search pattern we have is the *sector search*. This type of pattern is shown here [Adalind advances the slide deck to Figure 7.30]. In this type of a search pattern, the A/AI-UAV performs a search pattern much like tracks that follow the pattern of a pizza being sliced. Here, the center part of the search pattern is repetitively crossed over by the searching A/AI-UAV but different sectors away from the center are mapped out. This pattern is useful if there is a high priority target that needs to be monitored (at the center of the search pattern) and avenues towards this asset monitored. As the figure indicates, the A/AI-UAV comes in from the horizontal arrow at the left of the figure and flies over the center until it gets to the edge of the search space. The A/AI-UAV then moves one track width counter-clockwise and follows a new linear track that intersects the center of a circular search area shown as the circular short-dashed line segments. The A/AI-UAV turns again in the same counterclockwise direction by one track-width and continues this pattern. This pattern is like the expanding square method in that there is a reference to a known, or fixed position, in this case located at the center of the circular search region. We might

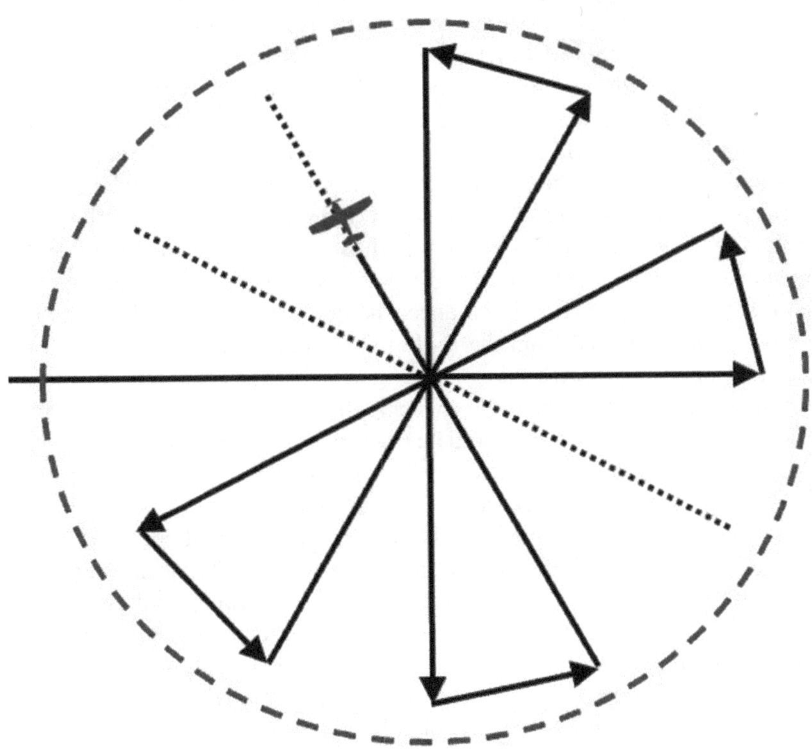

FIGURE 7.30 Sector search pattern.

FIGURE 7.31 Contour search pattern.

use this type of a pattern if there is a high value TOI at fixed coordinates that needs to be monitored and is not likely to move. Another search pattern is the *contour search*. This type of pattern works well were there are obstructions in the flight space such as mountainous terrain and if there are no known tracks to follow. The search pattern starts at an altitude near and above the highest peak and flies concentric ellipses around the ridge. At the completion of an ellipse, the A/AI-UAV drops 500 ft and repeats the pattern as can be seen on the next slide (Figure 7.31). As the A/AI-UAV descends, the ellipses become large and cover more territory within. In this way, the A/AI-UAV follows the local contour lines, search the local terrain along its flight path. This search pattern does not assume a preferred track and so the TOI can be anywhere within search area. There are variants that can be employed with this type of search path such as, when a new searchable region appears, such as in a new peak for an isolated mountain, one complete pass could be made around the peak before continuing with the original ellipsoidal trajectory. The location of the new peak could be updated in the flight path so that subsequent returns to this region would require another revolution about the new peak at the lower height. Once a mountain is completely scanned, it is treated as having the local scan completed, and the completely scanned region is removed from the flight path of the search pattern. In essence, the completely scanned mountain is treated as a no-fly zone, removed from the flight path, and a new optimized flight plan is generated that optimally routes the remaining search space. The optimization is chosen to minimize the time of flight while ensuring the entire remaining search space is covered. This type of a search pattern could be used in search and rescue missions, or for finding TOIs in a mountainous region.

With this type of a search pattern, there are also safety concerns for the A/AI-UAV since it will be flying around obstacles and terrain that could be used to conceal threats, such as those with hostile intent towards the A/AI-UAV. These are some common search patterns that we use. Out of all of these, the parallel sweep or creeping line search patterns are likely the ones that will be most often used along the border regions, when there aren't mountains to contend with. The next slide shows these search patterns with the US – Mexican border as a reference. If it is highly probable that the TOI is close to one side (say the left side), then the creeping line sweep pattern shown on the left side of the slide (Figure 7.32) is likely to be effective. The downwards dash-dot lines are representative paths of border crossing TOIs. Notice that for the creeping line pattern shown on

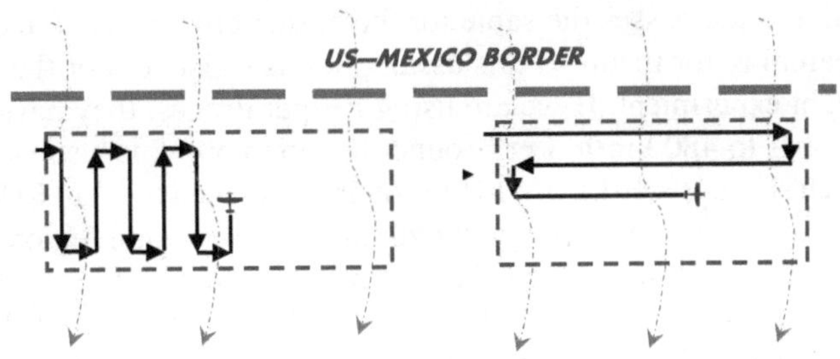

FIGURE 7.32 Creeping line (left) and parallel sweep (right) border crossing scenarios.

the left, the border crosser on the far right (third downward dash-dot line) has the best chance of making it across. If there is no indication of a preferred direction, then the parallel sweep method shown on the right covers the long side of the rectangle most quickly. These search methods can be changed dynamically or mixed and matched if appropriate. For instance, say that initially there is no advanced indication where a border incursion attempt is to occur. Consequently, an initial parallel sweep search pattern may be employed. Say that supporting intel then provides a potentially high probability location for the TOI. The search pattern could be altered to the expanding square or sector search that is centered on the intel provided coordinates. Does this help?"

Tom: "Yes, Adalind, this is great! We can use these search patterns when we are generating the various scenarios for our Use Cases. We can also develop some decision trees to help our AI in selecting the optimal sensor parameters based on the intended track."

Karl: "Yes, and Tristan's MBSE modeling comments are going to be beneficial too. I was just thinking about a comment that Aria made earlier. She said that we had to think like our adversary in some of these situations. Well, this seems like a great application for Game Theory. What do you think Chris?"

Chris: "Oh, definitely! Game theory is all about making the best decision under competitive and potentially adversarial conditions. The decision-making here is perfect for that. In fact, this situation can be modeled as a zero-sum game. We initially want to start off with a basic, simple, scenario and we can then build on it. We can use game theory to set up a zero-sum game where the DHS CBP is one contestant that seeks to detect the border crossers to win. The border crossers represent the opposing player and want to get across the border without being detected. For them, the goal is not to be detected at all. In this kind of an initial, basic game, we need to make some assumptions before setting up the game such as

1. Both players have the same amount of time to complete their play.
2. The time it takes to cross the search area by the border crossers on the paths has to be faster than the entire search space can be searched by the A/AI-UAV in some circumstances (if the A/AI-UAV guesses wrong starts on the wrong side for instance). Otherwise, the border crossers will always be caught, and there isn't much of a game (or a game with the same winner all the time). If one side always does better than the other side, that is a dominant solution for the winning side. If we evolve into a 100% catch scenario based on what we learn from the GT approach, that of course is good for DHS CBP, but we first want to start with a basic game, add to it so it is more realistic, and then optimize our models.
3. The search space is confined to a physical area of finite dimensions.
4. If there are more than one border crossers, they spread out along the track to minimize the liklihood of an all or nothing capture (at least initially).

For the first assumption, time passes the same for the border crossers as it does for the DHS CBP A/AI-UAV that is searching for them, so this assumption is satisfied. For the second assumption, let's do a quick thought experiment. If we are using Reaper drones, they have a speed of roughly 300 mph, which converts to 480 km/h. Let's round this up to 500 km/h so we can do the math in our heads. If the search space is a 10 km by 10 km square area, then the A/AI-UAV would take only 0.02 hours to make a single pass through one length of the observation region, at top speed. If the projected IFOV distance of the sensors on the ground has a width W of 100 m, it will take 100 passes to map out the entire search space area and so it will take the A/AI-UAV 2 hours to map the entire grid (0.02 hours/pass × 100 passes = 2 hours). For a person walking at an average rate of 3 mph through offroad terrain this equates to 4.8 km/h. Again, rounding up to 5 km/h for easy head math, the person, or group of people also takes 2 hours to cross the terrain (10 km travel distance divided by average walking speed of 5 km/h leads to 2 hours of crossing time). Now, the speed used for the Reaper A/AI-UAV is its top speed. It likely, can't image all that well flying as fast as it can, never mind the crazy operational tempo that would set! So, the Reaper might have to fly at a slower speed to get good images and so won't be able to cover the entire grid in those 2 hours. That means there is a "chance" that the terrain could be crossed without being detected. There is also a chance that the A/AI-UAV can catch one or more of the border crossers if the correct scan pattern, and correct entry point into the search grid is chosen (left or right side, and top or bottom of the search grid in Figure 7.32). Consequently, the second assumption seems reasonable as well. The third assumption is satisfied by the 10 km × 10 km search grid, and the fourth assumption certainly is feasible and may be assumed. The game might later be modified to let the crossers choose whether they want to group (all or nothing) or spread out (accept partial losses). We can add some details later if we don't overcomplicate the game. Every new decision adds a new dimension in the game, so three decisions would require a three-dimensional matrix to map out the decision space. If we have too many decisions, the math becomes too hard, and it takes too long to get a solution. If the decision space is large, agent-based and other simulation methods are more practical. We are going to keep things simple to start, and only allow one decision each for the border crossers, and DHS CBP. Since our initial assumptions are reasonable, we can set up our game as follows. For the border crossers, let's use the three optional tracks shown in Figure 7.32 and assume that the search grid is a 10 km × 10 km square grid instead of the rectangle shown in the figure. The game can be set up so that the border crossers can choose to cross on the left, middle, or right side of the grid, and the A/AI-UAV operator, or artificial intelligence circuitry chooses to start at the left, or right side of the grid, but not in the middle. If the border crosser gets across the grid without being detected, they are free and clear (win for border crossers and loss for DHS CBP). If the A/AI-UAV projected IFOV falls on the border crosser, they are detected and assumed caught. To give the border crossers a fair chance, we can add in a safe zone for them so that if they are within a certain distance from the border, they are safe and will escape. This could represent the situation that if the border crossers are less than a fixed distance away from their destination edge, the border patrol responders can't get to the detection location fast enough to apprehend the border crossers. The payoff (gain) for the DHS CBP is the number of border crossers detected/caught whereas this would be a loss for the border crossers. Let's also say that it takes slightly over an hour for a border patrol response team to get to the vicinity of any of the detection spots once alerted by the A/AI-UAV. This means, that if a border crosser gets halfway across before being detected, they are safe and will make it through the search space before DHS CBP can get their ground agents to the intercept point at the end of the path (bottom edge of Figure 7.32). We can now analyze the different patterns that Adalind just presented. For our initial game, we can start with the creeping line pattern. We know straight off that the best chance the border crossers have in making it across is to start at the far end of the sweep (shown as the third path from the top-left example in Figure 7.32). Let's say we have eight border crossers. Say that the farthest path is directly on the rightmost wall of the search grid. If the border crossers space themselves equally by 200 m between themselves along the track while crossing, the A/AI-UAV will find them all in the last 72 seconds of its mission (the last line of its search pattern) using the creeping line pattern (flying at top speed). The first person will be at the departing edge of the search area and the furthest one will be 1.4 km back along the trail. It will take the furthest border crosser back along the trail 16 minutes and 48 seconds to cross the remaining 1.4 km to the edge of the search space, far less time than it takes the border agents to reach them (1 hour plus from detection). Consequently, all the border crossers make it safely across. If instead the border crossers started near the middle path (second downward path from the left in Figure 7.32), then the A/AI-UAV would detect them halfway through the search pattern. The first person would be halfway across and have 1 hour left to go the remaining way. Since the response team takes longer than an hour, the first person makes it across and escapes. How many of the others escape depends on how much longer than an hour the response team takes to get there. If the border crossers attempt to cross on the same side as the starting position of the A/AI-UAV, they will be spotted before they reach the halfway point, and all border crossers will be caught. The GT payoff table for this zero-sum game can be represented as shown in Table 7.12 where the DHS CBP is the column player with only one choice with two options (which side to start, left or right), and the border crossers are the row players with one choice and three options (which crossing point to use, left, middle, or right). Since it is a zero-sum game, there is only one outcome. The payoff can be the number of border crossers that are caught. This payoff is a good thing for the DHS CBP player and a bad thing for the border crossers. Negative numbers are shown for caught border crossers because this is a negative payoff for the border crossers. For each caught border crosser, that would be a corresponding positive payoff for DHS CBP, hence a zero-sum game. If both the DHS CBP's A/AI-UAV and border crossers decide to start and cross on the same side of the search grid, then all eight border crossers are caught. This is shown as the −8 entries in the top-left and bottom-right cells of the payoff matrix shown in Table 7.12. Again, the −8 payoff would be a +8 payoff for DHS CBP because this is a zero-sum game. If the middle path is chosen by the border crossers, the number of border crossers caught depends on the arrival speed of the border agents. If they come very near the hour (say an hour and 10 seconds) then only the first person makes it out and all the others are apprehended. The longer it takes for law enforcement to arrive, the more border crossers successfully make the crossing. Say that law enforcement arrival timing is such that three border crossers are caught. Then the middle crossing entry for both rows will be −3 since the arrival time of law enforcement will be the same regardless of which side the A/AI-UAV starts its search pattern. If the border crossing takes place on the opposite side of the search grid, then all the border crossers make it across and so 0 is shown in the cell entries for the number of apprehended border crossers. There is no dominant strategy in this game for either side. Once we have this table, we can define probabilities of choosing the East, Middle, or Western path for the border crossers, and the left or right-side entry points for the A/AI-UAV by DHS CBP and solve for the strategies to use for each side (e.g., which path to select and at what percentage, versus which side to enter the search space and at what percentage). We can use software to solve this such as Gambit. In fact, let's plug in these numbers and see what we get for this simple example!

[Chris clicks some buttons and enters the matrix into the Gambit GT software.]

TABLE 7.12
Payoff Matrix for Creeping Line Game for Border Crossing Scenario

	East	West
East	−8	0
Middle	−3	−3
West	0	−8

Interesting! Gambit is suggesting a pure strategy for the border crossers to always choose the middle path. Gambit came up with two possible solutions indicating that there are two Nash equilibria for this payoff matrix. Which one ends up being the overall best solution requires a more detailed analysis but let's go with the first row solution from Gambit for now. For DHS CBP, Gambit came up with a mixed strategy wherein the DHS CBP strategy is to have the A/AI-UAV enter the search grid from the left side five-eighth of the time and enter from the right side three-eighth of the time. This will provide the best overall strategy for DHS CBP. If the border crossers had access to Gambit to see what their strategy should be, we would see something interesting. Gambit says that a pure strategy is the best approach for the border crossers, with the border crossers choosing the middle path 100% of the time. Interesting! Doing anything else would end up as a less optimal strategy for the border crossers. If the payoff matrix changes in anyway, such as in the response time changing for the border patrol getting to the interception point, then Gambit would have to be rerun to determine the new optimal strategies.

If we now swap the search mode of the A/AI-UAV to the parallel sweep pattern instead, we can start a new game for the parallel sweep mode and generate a payoff matrix for that situation. Assume the border crossers do not know the entry point of the A/AI-UAV as it enters the search space. They are then just as likely to choose any of the starting points to cross (near, middle, or far crossing locations relative to the A/AI-UAV entry point into the search grid). If the first person starts on the same side of the search grid as the A/AI-UAV (say left-most or West side), the person will be immediately detected if the A/AI-UAV starts its search at the closest line to the border. The person can walk at an average speed of 1.389 m/s and the A/AI-UAV takes 72 seconds to cross the 10 km search grid at top speed. At the initial point, if the search starts on the same side as the border crossers, and the A/AI-UAV starts at the location near the border line, the first person is detected attempting to cross. In the 72 seconds it takes the A/AI-UAV to make an entire pass along the 10 km stretch inside our border, the detection width is 100 m and the person traveling at average speed cannot cover a greater distance than this. In fact, the maximum distance that the person can cover, at an average rate of 5 km/h is exactly 100 m (5,000 m/h × 1 hour/3,600 seconds × 72 seconds = 100 m) before the A/AI-UAV searches an entire single pass through the search grid. If the A/AI-UAV search pattern starts at the top-left of the search area associated with the parallel sweep search pattern in Figure 7.34, then at best, the first person at the furthest path away from the A/AI-UAV will be near the first 100 m detection width edge and will be either detected then, or in the subsequent return sweep of the A/AI-UAV shortly thereafter. We can assume that if you are on the detection edge of the sweep pattern, that this counts as a detection. In this case, none of the border crossers would make it across regardless of the starting location. From a game perspective, this is not really an interesting game since the DHS CBP wins all the time, regardless of the border crosser's choice. This is a dominant position for DHS CBP. Recall, that assumption 4 states all border crossers are spread out on the *same* track. Under this assumption, once the track is identified by the A/AI-UAV in either the first or second pass, DHS CBP just needs to send the border agents to the end of the trail and wait for them to arrive (no challenge, no interesting game). If we modify assumption 4 so that each border crosser chooses which trail to use, then the situation is changed, and parking the border agents at one location doesn't guarantee the capture of all border crossers. Only the first border crosser that was detected by the A/AI-UAV was apprehended. If we further assume that the border agent doesn't wait around once the border crosser is apprehended, and returns the prisoner to the home base, the DHS CBP border agent response time resets for the next encounter. By making these changes, there still is just one decision that the border crossers make, which path to choose. This decision now applies to each border crosser instead of to all of them at once. The single decision that DHS CBP has is where to start the A/AI-UAV search (near the border at the top-left of the parallel sweep example in Figure 7.32), or at the bottom-left location at the end of the search space. Note that the left or right entry points to the search grid don't really matter with the parallel sweep search pattern, combined with the assumption that border crossers are unaware of (decisions are unaffected by) the entry point of the A/AI-UAV. Analyzing this situation to determine the payoff matrix, goes as follows. If the DHS CBP chooses to start the sweep of the A/AI-UAV at the top-left location near the border, the first border crosser will be caught, regardless of which path is chosen, just as was illustrated before. However, the rest of the border crossers all make it since they basically follow the A/AI-UAV through the search grid and are detected by the A/AI-UAV on its return trip too close to the border for the border agents to get there in time. So, one border crosser is caught, and the others all make it. If, on the other hand the initial parallel sweep pattern started at the bottom-left of the search grid shown in Figure 7.32, and moved upwards towards the border, then the first person to attempt crossing the border would be detected approximately at the midpoint (exactly at the midpoint if the middle path was chosen). Since the time for law enforcement to reach the border incursion point is over an hour from the initial detection, and the first border crosser only has 1 hour remaining to reach the safe zone on the other side, the first person will make it. The second and subsequent border crossers will either make it or not based on the relative speed of the response team getting to the intercept point (1 hour plus) versus the time remaining for each individual border crosser (1 hour$+n\times 144$ seconds) to get to the border. Here n is the relative position back from the first border crosser. The 144 seconds is the time it takes a border crosser to move 200 m. For instance, to be consistent with the previous example, if only three border crossers are captured (five make it through), with the first crosser using the middle path, then $n=4$ (four vertical positions back, equals 200 m$\times$4$=$800 m back on an arbitrary path relative to the first border crosser, which after an hour will be safely on the departing edge of the search space). This leads to a crossing time for the first five border crossers of 1 hour+576 seconds. This means the border agents don't get to the intercept locations until slightly over 1 hour, 9 minutes, and 36 seconds, yet before the next border crosser arrives in 144 seconds later. The time to get to the border is longer for each of the remaining border crossers so they all will be caught (three of them to match our first example). Whether or not the border crosser is on the path near to furthest away from the A/AI-UAV as it sweeps the middle row of the search space, makes a slight difference, but not a noticeable one. Say that the A/AI-UAV enters the scan of the center line of the search grid from the left (West) and is flying to the right (East). If the first border crosser entered on the top-left path (say this path was pushed up to the left wall of the search grid), then the A/AI-UAV would detect the border crosser in the first cell of the center line scan during the first second of its 72 second scan. This would be an early detection for the A/AI-UAV since the border crosser is only on the upper (Northmost) detection window edge. The border crosser still has half a cell width to go (50 m) before arriving at the midpoint of the search space and so has 1 hour+36 seconds of time required before reaching the border. If the response time of the border agents is under 1 hour and 36 seconds (recall the assumption that their response time is 1 hour +) then the first border crosser is caught, along with all the rest. If instead, the first border crosser uses the rightmost path (East), then A/AI-UAV detects the border crosser 36 seconds later than the middle crossing point, which means that in this situation the border crosser has 1 hour minus 36 seconds to get to the safe zone, and consequently makes it. This early or late detection ripples through all the other border crossers depending on whether they are near to, or far away from the A/AI-UAV as it enters the line scan path where they get detected. Instead of using East, Middle, and West for the entry path positions, we will use Near, Middle, and Far as the designators for the relative nearness of the border crossers to the A/AI-UAV when the A/AI-UAV enters the scan path where they will be detected. The payoff table for the parallel sweep game is shown in Table 7.13.

TABLE 7.13
Parallel Sweep Game

	Top	Bottom
Near	−1	−3
Middle	−1	−3
Far	−1	−3

This is an interesting result in several ways. Note that with the current scenario and assumptions, there is no way to catch all the border crosses unless the response time is increased to under an hour. Note also that the right column is strongly a function of the response time. The left column shows that nearly all the border crossers get through no matter where they start (how near they are to the starting position of the A/AI-UAV). For the parallel sweep strategy, someone is always caught as opposed to the creeping line strategy where there is the possibility for all to escape. In this case, we see that DHS CBP has a dominant solution wherein they should always choose to enter from the bottom, versus close to the border. This of course makes it very uninteresting and scarry for the border crossers! If DHS CBP can increase their response time, they will catch more and more of the border crossers, all the time.

Once the basic game is set up correctly, we can then add new features and parameters to the game. For example, things get interesting if you allow the speed of the A/AI-UAV and the border crossers to change such as if the A/AI-UAV is conserving fuel (or slowing down to get better imagery), or if the border crossers are allowed to run for short periods or get on vehicles like ATVs or motorbikes. Also, if you relax the border incursion point and allow more crossing points than the three paths we used as examples here, wherein different paths may have impacts on crossing speed (easier or harder to cross) and associated probability of detection. For instance, a rougher path might severely slow down the border crossers but may also provide better cover and so affect the A/AI-UAV sensors by reducing their probability of detecting the crossers. Further, the different Illumination and Sensing System sensors could have different probabilities of detection for different imaging conditions. For instance, if the border crossers are trying to conceal themselves (say with a tarp) while they happen to be in the field of view of the A/AI-UAV IaSS sensors, then if the visible EO sensor was being used, it would have a low probability of detecting the border crossers. On the other hand, if the hyperspectral imaging sensor was employed, it would easily pick up the tarp and so indicate the likelihood that there are border crossers under the tarp. It also takes a certain amount of time for the A/AI-UAV to turn itself around for a return pass. Another interesting feature is to account for the different detection widths W based on the scanning pattern of the optics and the nominal height of the A/AI-UAV above local ground (i.e., allowing for a changing Instantaneous Field-of-View (IFOV) of the sensors on the A/AI-UAV). Finally, adding probabilities of detection depending on how fast the A/AI-UAV is going (i.e., flying faster, there is a higher chance of not "seeing" the border crossers) and let the A/AI-UAV set its search speed, would add another interesting twist to the game. Remember though that for each new decision that is added to the Game, this adds a new dimension in the payoff matrix. For two decisions (with k by l options) as in the current example, a k by l matrix is sufficient for determining the payoffs (2 by 3 matrix in our example above). However, by adding a third decision, this requires a set of matrices. It isn't really feasible to add lots of decisions to the game because this becomes computationally burdensome. If there are a lot of decisions, simulation methods work well such as agent-based modeling methods. All these things could be readily modeled using discrete-event simulation and Monte-Carlo methods allowing for statistical parameter variations to see the overall effect on the number of detections. Then, conduct subsequent optimizations to capture the most border crossers.

One other thing to mention, if we slow down the air speed of the A/AI-UAV to the point where only one-third of the search space can be monitored, and the decision from DHS CBP is which one-third of the search space is to be monitored, then if we also remove the safe zone/agent response time parameter from consideration, we basically end up with the traditional "cat and mouse" game. The border crossers decide which of the three paths to cross, and the DHS CBP decides which of the three paths to scan with the A/AI-UAV. If the border crossers are in the same zone as is being scanned (left, middle, or right sector of the search area, where each sector is one-third of the area of the total search space), then they are caught. If not, they get through. Each border crosser is evaluated independently (i.e., chooses which crossing to take). DHS CBP chooses which sector the A/AI-UAV scans. Long-term results give a 33% chance of getting caught in this well-known game.

To summarize, the idea is to start with a simple GT model, and then adapt it to something more interesting until the math gets too hard (or things take too long to compute) using the GT approach. Then switch to modeling and simulation using agent-based modeling methods, or discrete-event and Monte-Carlo simulation methods and optimize the results.

Aria: "Wow, that is cool. Have you done these kinds of simulations before? I am sure the stakeholders would love to see these results."

Jennifer: "Oh, sure. Building this model is something that we have experience with. We have done this before on other projects and even have some base models to work with. Amanda and Christina are amazing with the optics. We should be able to have some performance predictions when we get to our meeting and some simulations to show. We will need some reality checks on the operational response times, detection parameters, and corresponding expected variances from DHS CBP. Can you get those for us Aria."

Aria: "Sure can! I am looking forward to seeing these results. I will get you some mean response time data along with standard deviations for different hot spots that we are interested in. As you know, the 10km by 10km search grid is a good start. We might want to vary this too if necessary."

Jennifer: "No problem, Aria."

Aria: "This also brings to mind another question; what sensor metrics are you using to give you confidence that your sensors are performing as expected?"

Tom: "Aria, that is an excellent question. Let's take a 10-minute break and we will bring in some members of our optics and sensors team in case you have any direct questions."

[Ten minutes later after Amanda R., Optical Lead, and Christina R., Optical Technician arrive, the meeting proceeds.]

Tom: "Thanks Amanda and Christina for coming. We may have some questions regarding the sensor metrics that we are using to ensure that they are working as expected. Can you tell us a little about that, please?"

Amanda: "No problem! The first thing to note is that we typically solve optics and imaging problems in three stages. First, we determine the imaging properties considering only the spatial aspects of the optical problem without any scaling. In many cases we can use linear systems theory to model and understand the optical system given that some geometrical conditions hold, namely the paraxial imaging conditions. In many cases, the imaging problem can be cast into a linear, shift-invariant system form and we use frequency-based methods to capture the spatial effects, often determining the OTF to capture aperture effects, system effects, and atmospheric turbulence effects. Second, we scale these results using the entrance pupil plane scaling relationship

$$\vec{x} = \lambda f_{\text{eff}} \vec{f}, \tag{7.113}$$

where $\vec{x}$ is a two-dimensional position vector in the entrance pupil of the imaging system, λ is the mean wavelength, f_{eff} is the effective focal length of the imaging system, and $\vec{f}$ is a two-dimensional spatial frequency variable in the entrance pupil of the imaging system. Magnification is then used to map effects into the image plane as required. Third, a SNR analysis is accomplished to make sure there is enough signal for a reliable detection. We can use the OTF to capture the important atmospheric statistics and simulate the spatial characteristics of the imaging system for various atmospheric scenarios (e.g., long exposure imaging, short exposure imaging, or as is common for atmospheric turbulence compensation, a frozen atmospheric state that corresponds to roughly 2 ms of sensor integration time). Recall that the relationship between the object spectrum $O(\vec{f})$ and the image spectrum $I(\vec{f})$ is given by

$$I(\vec{f}) = O(\vec{f})H(\vec{f}), \tag{7.114}$$

where $H(\vec{f})$ is the OTF. For appropriate short-exposure imaging conditions, the OTF can be replaced with the average short exposure OTF

$$\boldsymbol{H}(\vec{f}) = \bar{\boldsymbol{H}}_{os}(\vec{f}) = \boldsymbol{H}_o(\vec{f})\bar{\boldsymbol{H}}_s(\vec{f}), \tag{7.115}$$

where $\boldsymbol{H}_o(\vec{f})$ is the diffraction limited OTF, and $\bar{\boldsymbol{H}}_s(\vec{f})$ is the short exposure OTF at spatial frequency coordinate $\vec{f}$ in the entrance pupil plane of the imaging system. This OTF describes the atmosphere where image snapshots with less than 10 ms of integration time are averaged together after being spatially aligned with each other. A similar expression can be used for long exposure imaging (longer than 10 ms sensor integration time). The expression for the transfer function in Equation 7.115 becomes in this case

$$\boldsymbol{H}(\vec{f}) = \bar{\boldsymbol{H}}_{oL}(\vec{f}) = \boldsymbol{H}_o(\vec{f})\bar{\boldsymbol{H}}_L(\vec{f}), \tag{7.116}$$

where $\boldsymbol{H}_o(\vec{f})$, as before, is the diffraction limited OTF, and $\bar{\boldsymbol{H}}_L(\vec{f})$ is the long exposure OTF at spatial frequency coordinate $\vec{f}$ in the entrance pupil plane of the imaging system. For very short integration times (≤ 2 ms), we *freeze* the atmosphere in accordance with Taylor's frozen flow hypothesis and can use ATC to remove the atmospheric turbulence from the imagery by estimating the OTF, inverting it at every spatial frequency, multiplying Equation 7.105 by the inverse of the estimated OTF, and recovering the object spectrum $O(\vec{f})$. Once the object spectrum has been estimated in this fashion, the 2D Inverse Fourier Transform of $O(\vec{f})$ can be determined to get the object radiant emittance, $o(\vec{x})$, where $\vec{x}$ is a 2D spatial coordinate in the object plane. In this manner, we can simulate the short exposure atmospheric effects, long exposure atmospheric effects, and frozen atmosphere effects and accomplish ATC for the images that meet the Taylor's frozen flow assumptions (i.e., those images with integration times on the order of a few milliseconds). Afterwards, we adjust the scaling and system magnification as mentioned to get actual coordinate distances in the object, entrance pupil, or image plane. Once this is done, we can use the detector signal-to-noise ratio at the detector output to make sure we have enough signal to form a proper image. For example, the image SNR is given by

$$\mathrm{SNR}_D(\vec{f}) = \frac{\bar{K}\left|\bar{\mathcal{H}}(\vec{f})\right|\left|O_n(\vec{f})\right|}{\sqrt{\bar{K} + (\bar{K})^2\left|O_n(\vec{f})\right|^2 \mathrm{var}\left\{\mathcal{H}(\vec{f})\right\} + P\sigma_n^2}}, \tag{7.117}$$

where σ_n^2 is the noise variance due to the additive noise processes and has units of photoelectrons per pixel, on a per read-out basis, and $\bar{K}$ is the mean number of photo-events. Notice that in the high light limit with very low additive noise, the $(\bar{K})^2$ term in the denominator dominates, and Expression 7.117 becomes

$$\mathrm{SNR}_{\mathcal{H}}(\vec{f}) = \lim_{\bar{K}\to\infty}\left\{\mathrm{SNR}_D(\vec{f})\right\} = \frac{\left|\bar{\mathcal{H}}(\vec{f})\right|}{\sqrt{\mathrm{var}\left\{\mathcal{H}(\vec{f})\right\}}}, \tag{7.118}$$

which is only a function of the OTF. The term $|\bar{\mathcal{H}}(\vec{f})|$ also is only a function of the OTF and has the special name, Modulation Transfer Function. This is an important general imaging metric that also directly relates to the high light-level SNR of the image spectrum as seen in Equation 7.118. Other metrics that help us understand the performance of the imaging system are the isoplanatic angle, which determines the angular subtense of a patch of atmosphere that can be considered part of the same frozen atmospheric state. So, when ATC is applied to a region of atmosphere that lies within the isoplanatic angle, it can be corrected. A counter example is that if the region that is to be corrected lies outside of the isoplanatic angle, then there is more than one atmospheric state that is being corrected which generally is not practical. Related to the isoplanatic angle is the Fried parameter, another important imaging performance metric which is the atmospheric coherence length. Like the isoplanatic angle, the Fried parameter, r_o is the length of a wavefront that may be considered coherent (i.e., has less than 1 rad difference in phase across this length). It has some extremely practical applications such as illustrating that entrance pupil plane diameters that are larger than r_o benefit from adaptive optics, or atmospheric turbulence compensation methods, whereas those that are smaller than r_o do not provide any higher spatial resolution by removing atmospheric turbulence effects."

Aria: "So you are saying that you simulate various different atmospheric conditions and apply them to an…um…object spectrum is it…and that allows you to model all different kinds of atmospheric states and detection scenarios?"

Amanda: "Yes, that is the case. We can also remove the turbulence with different types of ATC algorithms or even some AO and hybrid systems. Christina, our optics tech, has done an excellent job in setting these up. Christina, did you want to add anything?"

Christina: "You covered it well, Amanda. The only other thing is that we inject the atmospheric effects in the entrance pupil plane using the generalized pupil function (GPF). We typically model the atmospheric aberrations as phase effects in the entrance pupil and then calculate the OTF using Fourier optics methods on the normalized autocorrelation of the generalized pupil function, which produces the OTF. We use basis function expansions in some approaches to model the phase effects. We can then generate simulations of atmospheric effects on our sensors and apply the AO methods and/or the ATC algorithms to correct the abberated images. We can also model amplitude effects associated with the GPF but there are many instances where the phase effects dominate and so amplitude effects can be neglected in those circumstances. We check our results with experimental data both in the labs and the field so we have high confidence that the sensors will perform as designed."

Aria: "That is great! What about on the hardware side? I know you all are the experts on building the sensors, but what about things like the signal processing elements? Did you all look at that?"

Karl: "Excellent point, Aria. There are critical aspects to consider for both hardware and software. We recently did a trade study evaluating three high-performance digital signal processing chips, and used the Analytical Hierarchy Method (AHP) as our multi-criteria decision analysis approach to select the best chip. Jennifer, do you want to bring up the trade study on the overhead and give Aria a run down?"

Jennifer: "Sure, Karl. The engineering team developed a set of evaluation parameters that we could use to evaluate the different chips against each other. The engineering factors that are important in this evaluation were determined to be: (1) the billions of floating point operations per second (GFLOPS) which is a computational speed metric, (2) the response time of the finite impulse response filter (in nanoseconds, per tap) which relates to the digital filters' latency, (3) the Giga Multiply Accumulates per second (GMAC) metric, which is also a useful computational speed metric, (4) the DSP memory (in Mb), and

(5) failures in time (FIT) metric which is relevant and relatable to reliability considerations. This metric indicates the number of failures in 1 billion hours of operations. Our engineering team then identified three candidate DSP chips that are currently commercially available, satisfied the general processing and packaging requirements, and had a suitably robust supply chain. Since the name of the chip manufacturer is considered sensitive information at this stage, we use generic names for the different chips such as DSP1, DSP2, and DSP3. Table 7.14 show the three candidate chips, along with the evaluation criteria that we use in the AHP method to decide which chip is best for our application. We have three candidate chips that serve as our alternatives, and five criteria that are used in the evaluation process. When using the AHP method, we need to determine the relative importance of the evaluation criteria to each other. This is done by generating a pairwise comparison of the evaluation criteria using an established ranking methodology. A common methodology is to use a five-point preference scale where the number represents how much the row entry is preferred over the column entry.

The convention is to map the criterion from the row entry to the criterion of the column entry. With this mapping strategy in mind, a table can be generated that compares the criteria to each other on a pairwise basis. There are multiple ways to obtain the entry values for this table. For instance, we sat down with the engineering team to determine the ranking based on the importance of the specific criterion on the overall expected impact it has on system performance. For example, the differences in computational speed will likely be more important than the differences in failures in time, based on the extremely low failures over time shown in Table 7.14. We used a Delphi method to systematically capture the relative weights of these preferences. Table 7.15 shows the criteria versus criteria preference ranking using a 1–3 preference rating.

In Table 7.15, the diagonal elements of the table are all ones since they illustrate the preference of the row element with itself. Looking at the top row (GFLOPS), the 3 in the intersecting cell of the first row and second column shows that the GFLOPS criterion is preferred 3 times over the FIR criterion. This is because the engineering team collectively

TABLE 7.14
DSP Evaluation Matrix

	DSP1	DSP2	DSP3
GFLOPS	2.8	2.4	3.6
FIR (ns)	0.97	1.11	0.83
GMAC	2	1.9	4
Memory (Mb)	8	8	12
FIT	3.7	5.9	6.3

TABLE 7.15
DSP Criterion versus Criterion Table

	GFLOPS	FIR	GMAC	Memory	FIT
GFLOPS	1	3	3	1	2
FIR	1/3	1	1	1/3	2/3
GMAC	1/3	1	1	1/3	2/3
Memory	1	3	3	1	2
FIT	1/2	3/2	3/2	1/2	1

determined through the Delphi process that the nanosecond response time of the filter made a smaller impact on the estimated digital signal processing performance than the computational speed of the DSP. Notice that if according to Table 7.15, the GFLOPS criterion is preferred 3 times as much as the GMAC criterion, then the comparison of the GMAC criterion to the GFLOPS criterion must be preferred 1/3 times (inverse relationship). This can be seen from the third row and first column intersecting cell. A similar process can be implemented to generate the set of matrices for the alternatives that are required in the AHP process. Here all three chips are compared to each other from the perspective of each of the criteria. For example, for the first criterion of GFLOPS, a matrix of the three alternative DSP chips is generated comparing the performance of GFLOPS to each other in tabular form. This can be readily and accurately done from the DSP parameter values in Table 7.14 and multiply this by a ratio such that the row element equals the column element. In Table 7.16, we see the set of matrices generated in comparing the DSPs to each other from the perspective of each of the technical evaluation criteria. Once again, the diagonal elements will be one because the comparisons are made between the row element and itself. To demonstrate how these tables were generated by our engineering team, we can use the top-left table as an example. The first-row entry (DSP1) is preferred 7/6 times to DSP2 as seen by the 7/6 in the intersecting cell of the first row and second column. Where did the 7/6 come from? In this case, we have the actual GFLOP numbers for each of the DSPs in Table 7.14. If the GFLOPS of DSP2 is multiplied by 7/6, the GFLOPS for DSP1 is obtained. Note that the GFLOP entry for DSP1 is higher than that for DSP2 and so DSP1 is preferred over DSP2 (higher is better with regards to computational speed). This multiplication factor directly weights the relative preference between DSP1 and DSP2. The remaining entries are generated in a similar fashion. Notice that in comparing DSP2 to DSP1 for GFLOPS, just as in the case for the criteria versus criteria matrix from before, this can be automatically determined from the inverse of the DSP1 versus DSP2 result (6/7 in this case). Therefore, only one side of the diagonal matrix needs to be determined and the other side can be directly computed by inversion. This process is repeated for each of the other criterion and the results are

TABLE 7.16
Alternative versus Alternative Matrices for Each Criterion

	DSP1	DSP2	DSP3
DSP1	1	7/6	7/9
DSP2	6/7	1	2/3
DSP3	9/7	3/2	1

GFLOPS

	DSP1	DSP2	DSP3
DSP1	1	20/19	1/2
DSP2	19/20	1	19/40
DSP3	2	40/19	1

GMAC

	DSP1	DSP2	DSP3
DSP1	1	11/10	4/5
DSP2	10/11	1	8/11
DSP3	5/4	11/8	1

FIR

	DSP1	DSP2	DSP3
DSP1	1	1	3/2
DSP2	1	1	3/2
DSP3	2/3	2/3	1

Memory

	DSP1	DSP2	DSP3
DSP1	1	3/2	7/4
DSP2	2/3	1	7/6
DSP3	4/7	6/7	1

FIT

shown in Table 7.16 above. The next step in the AHP is to normalize these results so that the scale of each column is the same. To do so, we determine the column means for all the columns in all our matrices and divide each column entry by the column sum. After this, the row means are determined for all our matrices. It is interesting to note that the AHP methodology is cast in the form of a linear algebra problem that can be solved through Eigen analysis. The Eigen vector $\vec{X}_e$ is determined by the row means of the criterion versus criterion matrix (factor versus factor matrix). The preference matrix $\boldsymbol{A}$ is determined by sequentially arranging the row means determined from the above process for each of the alternative versus alternative matrices shown in Table 7.16. This arrangement can be seen in the left-most 3×5 matrix in Table 7.17 that has the alternatives (DSP1, DSP2, and DSP3) in the row entry positions, and the evaluation criteria (GFLOPS, FIR, GMAC, Memory, and FIT) in the column entry positions. The column cell entries are the row means determined in the fashion above from Table 7.16 for the alternative versus alternative matrices. The Eigenvector $\vec{X}_e$ is the normalized row means from the criteria versus criteria Table above (Table 7.15), and is shown as the five-entry column vector to the immediate right of the multiplication sign. The result of the vector multiplication is the three-element vector shown to the right of the equal sign. The highest number provides the best choice according to the weights we set in the pair-wise preference process. In this case, DSP1 is our best choice. Since DSP1 and DSP3 are so close, the engineering team took a close look at the choice of weights that were chosen and accomplished a sensitivity analysis to see if the results would change based on uncertainties in the weights. In the end, the sensitivity analysis supported the conclusion that DSP1 was the best choice. Since our preference weights were based on a pair-wise comparison and therefore could have some inconsistencies between the weights, we then wanted to make sure that our chosen weights in the pairwise comparisons were consistent. We did this by calculating the consistency ratio as follows. First, we multiplied the unnormalized criteria versus criteria matrix shown in Table 7.15 by the Eigenvector $\vec{X}_e$ and obtained the product of the Eigenvalue with the Eigenvector shown at the right side of Figure 7.33. If we now divide the vector on the right side of the equal sign pointwise by the corresponding Eigenvector

TABLE 7.17
AHP Formulation for Selecting Best Digital Signal Processor

	GFLOPS	FIR	GMAC	Memory	FIT					
DSP1	0.955	0.950	0.759	1.125	1.340		1.579		5.241	DSP1
DSP2	0.818	0.863	0.722	1.125	0.894	×	0.526	=	4.608	DSP2
DSP3	1.227	1.187	1.519	0.750	0.766		0.526		5.151	DSP3
							1.579			
							0.789			

	GFLOPS	FIR	GMAC	Memory	FIT		$\vec{X}_e$		$\lambda\vec{X}_e$
GFLOPS	1	3	3	1	2		1.579		7.892
FIR	1/3	1	1	1/3	2/3		0.526		2.6307
GMAC	1/3	1	1	1/3	2/3	X	0.526	=	2.6307
Memory	1	3	3	1	2		1.579		7.892
FIT	1/2	3/2	3/2	1/2	1		0.789		3.946

FIGURE 7.33 Matrix calculations in determining the consistency ratio in the AHP.

$\vec{X}_e$ elements, then sum the result of each cell, Dr. Saaty showed that dividing this quantity by the number of criteria (five in our case) gives a measure of the inconsistency of the relative weights assigned in the preference rankings. This number is the consistency ratio and will be greater than or equal to the number of criteria that we selected depending on how consistent our weights were chosen. In this fashion, the consistency ratio metric is an indicator for how consistent we were in our decision-making (assigning the relative preference rating weights). For perfect consistency in assigning the relative preference weights, the consistency ratio is exactly equal to the number of criteria we have chosen. If we were inconsistent anywhere in the assigning of the weights, the consistency ratio would be greater than the number of criteria and thereby indicate the presence of an inconsistency in our decision process in assigning the relative weights in our pairwise comparisons. If inconsistencies are present, then a further step is taken to transform the consistency ratio to the consistency index by normalizing the consistency ratio to account for statistical variations introduced by the pairwise comparison process. In our case, however, we calculated the consistency ratio to have a value of 5.000 which is exactly equal to the number of criteria we have. This indicates that we were perfectly consistent in our decision-making process of assigning our preference weights. With a little thought, this is not unexpected since our choice of preference weights was determined by the exact relationship between the criteria metrics themselves. Consequently, no subjective inconsistency could creep into our pair-wise relative preference rankings, and we ended up having a perfect consistency ratio. So, summing things up, the AHP methodology applied to the selection of the digital signal processing chip ended up with our selection of DSP1, with a confirmation that our decision-making process in choosing the weights used in our evaluation was perfectly consistent."

Aria: "Impressive! Not only did you go through a rigorous process in selecting the chip, but you also have documentation on how you arrived at your decision. Additionally, you have a metric that tells you how consistent you were in making your decision. This means you could revisit the decision by evaluating the weights that you assigned. However, no one can argue with the consistency of the application of the weights because they were fully consistent."

Jennifer: "Exactly! That is one of the big benefits of AHP."

Karl: "I am glad you feel that way Aria. Since you asked about the hardware, I want to share with you how we solved a security issue related to our data using decision analysis methods. I think Ron did this, didn't he, Jen?"

Jennifer: "Yes, Ron did a great job with that. Ron, can you fill us in please?"

Ron: "Sure can and thank you. Well, the issue came up with our requirement to be able to wipe the onboard data if something went terribly wrong with the A/AI-UAV such as if it was compromised somehow and we lost positive control of the onboard data. Normally, you can wipe the data using physical degaussing methods but that isn't practical in this case. So, the engineering team went to work on software data wiping methods, and we came up with three potential solutions. It turned out that none of the software methods were 100% effective, but the three potential solutions were the best of the available options. We needed a decision analysis method that would help us decide which option to choose. We could use AHP as before, but we also had some additional probabilistic information that led us in a different direction. With this probabilistic information, the engineering team decided a decision tree would provide the best result. For the first potential solution (Solution 1) the probability of having no remaining sensitive data was 85%. For the remaining 15%, there were 14 kB of sensitive data that could remain and potentially be retrieved. For Solution 2, there was a 98% chance of destroying everything except 7 kB of data. There was also a 2% chance of leaving only 1 kB of sensitive data. Solution 3 had a 96% chance of leaving 3 kB of data, a 3% chance of leaving 2 kB of sensitive data, and

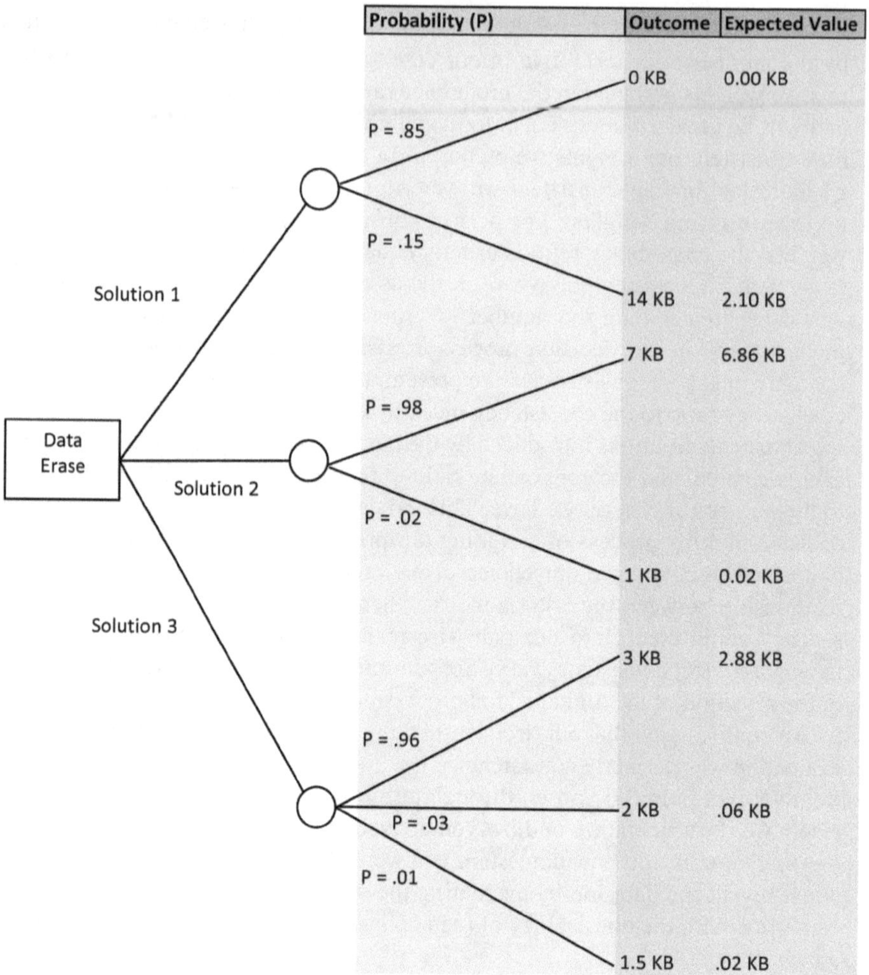

FIGURE 7.34 Data wipe decision tree.

TABLE 7.18
Software-Based Data Wiping Evaluation Results

Alternative	Residual Sensitive Data (kB)
Solution 1	2.10
Solution 2	6.88
Solution 3	2.96

1% chance of leaving 1.5 kB of sensitive data. Figure 7.34 shows the decision tree used to select the best solution. In this decision tree, multiplying the individual branch elements on the decision tree provides the Expected Value shown at the right of the figure. The Expected Value entries are summed for each option and then compared to each other. The sum indicates the total potential data loss for the respective solution so the lowest number wins. Table 7.18 shows the results.

In this case, Solution 1 provides the lowest Expected Value for data loss and so this solution was chosen by the engineering team to implement the software-based data wiping functionality."

Aria: "Fantastic! I think you all have a lot of interesting results to present in our upcoming TIM. I think you all are coming along great. I am sure the TIM will be a huge success."

Bill: "We are looking forward to it! Tom and Karl, can you please get with me early next week with the preliminary slide deck for the TIM? Thanks everyone and keep up the great work!"

Tom: "Sure, Bill. We should have those ready by next Wednesday. We will also want to integrate some of the search pattern information into our models."

Aria: "Awesome!"

7.4 SUMMARY

In this chapter, we presented background information on Risk and Decision Analysis methods that are commonly used in systems engineering. We provided key events regarding the history of decision analysis and provided some historical background on GT. We then showed how risk relates to the fields of program management and systems engineering. We provided a definition of risk and presented ways to assess, manage, and mitigate risk. We also presented concepts in feasibility analysis such as the failure mode effects and criticality analysis, and showed how statistical methods such as the Weibull analysis can be used to assess risk. We also presented fault tree analysis and illustrated some of the software tools that can facilitate risk assessments. Next, we presented some national and international risk related case studies. As part of the international case study, we introduced a probabilistic risk assessment example from NASA to illustrate how the risk assessment framework works at NASA. Afterward, we focused on important aspects of decision-making. We presented different useful decision criteria such as the Maximax criterion, Maximin criterion, and the Expected Value criterion. We provided a short discussion on different types of biases such as the anchoring bias, confirmation bias, focusing illusion, framing bias, overconfidence bias, and the planning fallacy. We then made the connection of decision analysis to GT and presented single criterion, multi-criteria, multiple attribute criteria, and multiple objective criteria decision-making methods (SCDM, MCDM, MADM, and MODM). Ranking methods such as dominance, conjunctive, disjunctive, lexicographic, elimination by aspects, linear assignment, additive weighting, the technique for order preference by similarity to ideal solution (TOPSIS) in addition to maximax, and maximin where introduced. We presented the Analytical Hierarchy Method as a premier method for evaluating alternatives and as a variant of linear assignment. We also presented decision trees to capture probabilistic aspects of decision-making and presented some applications. For GT, we provided insights to understand GT methods. We showed that GT consists of games of chance, games of strategy, and games of skill. We presented concepts such as cooperative games, non-cooperative games, zero-sum games, non-zero-sum games, mixed motive games, games with complete information, or incomplete information. We discussed cooperation versus competition and illustrated games such as the prisoner's dilemma and rational decision-making. We also provided some GT applications. Next, we presented some special considerations in modern decision-making such as collective behavior, collective intelligence, and understanding pattern formation. Insights into patterns can be gained from differential equations, non-linear dynamics, time-series analysis, phase-space analysis, attractors, bifurcation, chaos theory, multi-stability, iterative maps, coupled map lattices, stability analysis, population dynamics, and provided some comments on the role of artificial intelligence in modern engineering practices.

In the second part of the chapter, we presented optical methods for modeling atmospheric turbulence effects such as the OTF and SNR effects at the detector output. We presented the incoherent imaging model and showed the mathematical expression for the OTF. We introduced the concept of the generalized pupil function and related this mathematically to the OTF. We then used statistical aspects of the atmosphere to develop mathematical expressions for the short exposure OTF and the long exposure OTF. The PSF was shown to be the inverse Fourier transform of the OTF. Random phase screens were then presented to capture the statistical aspects of the atmosphere. Gaussian random phase screens were then used to generate the average OTF of the phase screen in terms of the structure function. Then, the limiting forms of the OTF were developed for the case of large phase variance. The Earth's atmosphere was modeled as a thick phase screen and the correlation function and power spectral density of the refractive index were presented. Subsequently, the structure function, autocorrelation function, and power spectrum of the refractive index were illustrated. We then looked at electromagnetic wave propagation through the inhomogeneous atmosphere and used the Rytov approximation to develop the expression for wave fluctuations in terms of the log-amplitude fluctuations and the phase fluctuation. The average long exposure OTF was then written in terms of the diffraction-limited OTF and the wave structure function for an isotropic and homogenous medium. Next, the near-field wave structure function was determined and used to mathematically describe the average long exposure OTF in terms of the structure constant, and two very important metrics known as the atmospheric coherence length, r_o, and the isoplanatic angel, θ_o, were presented. Expressions for the short exposure OTF in terms of the atmospheric coherence length were provided. Various image spectrum SNR expressions that incorporate photon statistics, the modulation transfer function (MTF), and additive noise processes finish this section.

In the last section, we found our fictitious imaging company, Fantastic Imaging Technologies, meeting with some stakeholder representatives to plan for an upcoming TIM. Here we combined and applied the material in the previous sections to show how the principles and methods in those sections fit together in a program development environment. FIT first received a briefing from DHS Customs and Border Protection (DHS CBP) on different search patterns that their A/AI-UAV host platform typically flies for different missions. FIT then showed how GT could be applied to choose which search pattern and sensor combination to use and select where and when to deploy their troops. FIT then demonstrated how the short exposure OTF, the long exposure OTF, and the OTF with an integration time less than or equal to 2 ms can be modeled and simulated to capture spatial effects of an imaging scenario and ultimately remove atmospheric turbulence effects from the imagery. They also showed that the image spectrum SNR complements the spatial resolution information obtained from the OTF and provides information on whether there is sufficient signal to detect the image and apply the ATC methodology. FIT then showed how the Analytical Hierarchy Process (AHP) was used to select the best digital signal processing chip and determined the consistency ratio to prove that they had meaningful results. FIT also used a decision table to choose which type of disk-wiping software to use if the data onboard the A/AI-UAV needs to be wiped.

7.A APPENDIX: ACRONYMS

A/AI-UAV	Autonomous/Artificial Intelligent-Unmanned Aerial Vehicle
AHP	Analytic hierarchy process
ATC	Atmospheric turbulence compensation
CBP	Customs and Border Protection
CI	Consistency index
CONOPS	Concept of operations
CR	Consistency ratio
CTO	Chief technology officer
DCMA	Defense Contract Management Agency
DHS	Department of Homeland Security
EU	Expected Utility
FIT	Fantastic Imaging Technologies
GAO	Government Accountability Office
GPF	Generalized Pupil Function

GT	Game Theory
INCOSE	International Council on Systems Engineering
MADM	Multiple Attribute Decision Making
MBSE	Model-Based Systems Engineering
MODM	Multiple Objective Decision Making
MRL	Manufacturing Readiness Levels
MTBF	Mean Time between Failure
MCDM	Multiple Criteria Decision Making
NIST	National Institute of Standards and Technology
NASA	National Aeronautics and Space Administration
OoI	Object of Interest
OpsCon	Operational Concept
OTF	Optical Transfer Function
PDF	Probability Density Function
PRA	Probabilistic Risk Assessment
PSF	Point Spread Function
RPN	Risk Priority Number
SCDM	Site-Centered Decision Making
SDLC	System Development Lifecycle
SE	Systems engineering
SNR	Signal-to-Noise-Ratio
TIM	Technical Interchange Meeting
TOI	Target of Interest
TOPSIS	Technique for Order Preference by Similarity to Ideal Solution
TRL	Technology Readiness Levels
UAV	Unmanned aerial vehicle
U.S.	United States

REFERENCES

ACQNotes. 2021. Technology readiness level. https://acqnotes.com/acqnote/tasks/technology-readiness-level (accessed March 7, 2022).

Agosto, H.D. 2016. Introduction to Manufacturing Readiness Levels (MRLs), Rev#13. 2016: Defense Contract Management Agency.

Air Force Institute of Technology. 2010. *Global Hawk Systems Engineering Case Study.* Wright Patterson AFB: Air Force Institute of Technology. https://apps.dtic.mil/sti/pdfs/ADA538761.pdf (accessed March 8, 2022).

Anosov, D.V. 2014. Phase space. *Encyclopedia of Mathematics.* April 30. https://encyclopediaofmath.org/wiki/Phase_space (accessed June 4, 2022).

Aumann, R.J. and M. Maschler. 1985. Game theoretic analysis of a bankruptcy problem from the Talmud. *Journal of Economic Theory,* Vol. 36, pp. 195–213.

Bak-Coleman, J.B., M. Alfano, W. Barfuss, C.T. Bergstrom, M.A. Centeno, I.D. Cousin, J.F. Donges, et al. 2021. Stewardship of global collective behavior. *Proceedings of the National Academy of Sciences*, Vol. 118, p. e2025764118. https://doi.org/10.1073/pnas.2025764118

Baron, J. 2006. A decision analysis of consent. *American Journal of Bioethics*, Vol. 6, No. 3, p. 46.

Bearden, D., C. W. Freaner, R. E. Bitten, D. J. Emmons, and T. Coonce,. 2008. Perspectives on NASA Mission Cost and Schedule Performance Trends. (June 3, 1–3 pm). Presentation at the Goddard Space Flight Center, 8800 Greenbelt Road, Glenn Dale, MD USA.

Bertels, C. 2006. Crew maintenance lessons learned from ISS and considerations for future manned missions. *SpaceOps 2006 Conference,* Wessling, Germany: AIAA. 8.

Bierman, H.L., L.E. Fouraker, and R.K. Jaedicke. 1961. *Quantitative Analysis for Business Decisions.* Homewood, IL: Richard D. Irwin, Inc.

Blanchard, B.S. and W.J. Fabrycky. 2010. *Systems Engineering and Analysis,* 5th edn. London, UK: Pearson.

Boeing. 2022. Boeing KC-135 Stratotanker. February 23. https://en.wikipedia.org/w/index.php?title=Boeing_KC-135_Stratotanker&oldid=1073626418 (accessed March 7, 2022).

Brams, S.J. 1980. *Biblical Games: A Strategic Analysis of Stories in the Old Testament*. Cambridge, MA: MIT Press.

Britannica, T. Editors of Encyclopaedia. 2024. Chaos Theory. *Encyclopedia Britannica*. https://www.britannica.com/science/chaos-theory (accessed April 21, 2024).

Buede, D.M. and W.D. Miller. 2016. Decision analysis for design trades. In: Buede, D.M. (eds), *The Engineering Design of Systems: Models and Methods*. Hoboken, NJ: John Wiley & Sons, Inc., pp. 398–444.

Buehler, R., D. Griffin, and M. Ross. 1994. Exploring the "planning fallacy": why people underestimate their task completion times. *Journal of Personality and Social Psychology*, Vol. 67, pp. 366–381.

Cenci, S. and S. Saavedra. 2018. Structural stability of nonlinear population dynamics. *Physical Review*, Vol. 97, p. 012401. American Physical Society (APS). https://doi.org/10.1103/PHYSREVE.97.012401

Chakraborty, A. and J.S. Duncan. 1999. Game-theoretic integration for image segmentation. *IEEE Transactions on Pattern Analysis and Machine Intelligence*, Vol. 21, No. 1, pp. 12–30.

Chalkiadakis, G. 2012. Cooperative game theory. *IEEE Intelligent Systems,* Vol. 27, No. 3, p. 86.

Clark, T.J. and A.D. Luis. 2020. Nonlinear population dynamics are ubiquitous in animals. *Nature Ecology & Evolution*, Vol. 4, pp. 75–81. https://doi.org/10.1038/s41559-019-1052-6

Clemen, R.T. and T. Reilly. 2014. *Making Hard Decisions with DecisionTools*. Mason, IA: South-Western, Cengage Learning.

Cueli, J.R. and C. Bordons. 2008. Iterative nonlinear model predictive control. *Control Engineering Practice,* Vol. 16, pp. 1023–1034.

Cushing, J.M., R.F. Costantino, B. Dennis, R.A. Desharnais, and S.M. Henson. 1998. Nonlinear population dynamics: models, experiments and data. *Journal of Theoretical Biology,* Vol. 194, pp. 1–9. https://www.math.arizona.edu/~cushing/JTB1998.pdf.

Defense Acquisition University. 2003. *Risk Management Guide for DoD Acquisition*. Fort Belvoir, VA: Department of Defense, Defense Acquisition University.

Defense Acquisition University Press. 2001. *Systems Engineering Fundamentals*. Fort Belvoir, VA: Defense Acquisition University Press.

Denœux, T. 2019. Decision-making with belief functions: a review. *International Journal of Approximate Reasoning*, Vol. 109, pp. 87–110.

Dhillon, B.S. (ed). 1999. Maintainability tools. In: *Engineering Maintainability*. Houston, TX: Gulf Professional Publishing, pp. 50–81.

Dionne, G. 2013. Risk management: history, definition, and critique. *Risk Management and Insurance Review*, Vol. 16, pp. 147–166.

Drechsler, M. and M.A. Burgman. 2004. Combining population viability analysis with decision analysis. *Biodiversity and Conservation*, Vol. 13, pp. 115–139.

Drezner, J.A., G. Sommer, and R.S. Leonard. 2007. *Innovative Management in the DARPA High Altitude Endurance Unmanned Aerial Vehicle Program*. Santa Monica, CA: RAND Corporation, pp. 48–49.

Dunbar, B. 2019. *6.4 Technical Risk Management*. Shea, G. (ed). National Aeronautics and Space Administration. December 12. https://www.nasa.gov/seh/6-4-technical-risk-management (accessed March 8, 2022).

Dutta, P.K. 1999. *Strategies and Games: Theory and Practice*. Cambridge, MA: MIT Press.

Ehrlinger, J., W.O. Readinger, and B. Kim. 2016. Decision-making and cognitive biases. In: Friedman, H.S. (ed), *Encyclopedia of Mental Health*. Academic Press. https://www.researchgate.net/publication/301662722_Decision-Making_and_Cognitive_Biases

FAA. 2014. FAA systems engineering manual. https://segoldmine.ppi-int.com/node/72313 (accessed April 6, 2024).

Gates, R. 2007. Memorandum for secretaries of the military departments. May 30.

Goodman, J.W. 1985. *Statistical Optics*. New York: Wiley-Interscience.

Griffin, J.M. 2005. *C-5A Galaxy Systems Engineering Case Study*. Wright Patterson AFB: Air Force Institute of Technology. https://apps.dtic.mil/sti/pdfs/ADA572211.pdf (accessed March 8, 2022).

Hallinan, Jr., A.J. 1993. A review of the Weibull distribution. *Journal of Quality Technology,* Vol. 25, No. 2, pp. 85–93.

Haselton, M.G., D. Nettle, and D.R. Murray. 2015. The evolution of cognitive bias. In: Buss, D.M. (eds), *Handbook of Evolutionary Psychology*. Hoboken, NJ: John Wiley & Sons, Inc., p. 968. https://doi.org/10.1002/9781119125563.evpsych241

Hauert, C. and G. Szabo. 2005. Game theory and physics. *American Journal of Physics*, Vol. 73, No. 5, pp. 405–414.

Heathcote, P.M. 2005. *'A' Level Computing*. Payne-Gallway Publishers. Ipswich, IP1 3DB, UK.

Hofstadter, D.R. 1979. *Gödel, Escher, Bach: An Eternal Golden Braid*. New York: Basic Books.

Howard, R.A. 1966. *Decision Analysis: Applied Decision Theory*. Stanford, CA: Stanford Research Institute.

Howard, R.A. 1980. An assessment of decision analysis. *Operations Research*, Vol. 28, No. 2, pp. 4–27.

Howard, R.A. 2007. The foundations of decision analysis revisited. In: Miles Jr., R. and D. Von Winterfeldt (eds), *Advances in Decision Analysis: From Foundations to Applications*. Cambridge, MA: Cambridge University Press, pp. 32–56.

HQ AFSEC . 2025. Aviation Class A Mishap Summary by Fiscal Year. https://www.safety.af.mil/Portals/71/documents/Aviation/Mishap%20Summaries/USAF_Aviation_Class_A_Summary-Copy.pdf (accessed March 8, 2025).

International Organization for Standardization. 2018. *Risk Management - Guideliness (ISO 31000:2018)*. Geneva: ISO.

Irwin, T. 2009. Risk quantification of systems engineering documents improves probability of DoD project success. https://apps.dtic.mil/sti/pdfs/ADA510447.pdf (Accessed March 8, 2025).

ISO/IEC/IEEE. 2011. Systems and software engineering - Architecture Description ISO/IEC/IEEE 42010.

ISO/IEC/IEEE. 2015. ISO/IEC/IEEE International Standard - Systems and software engineering -- System lifecycle processes. *ISO/IEC/IEEE 15288 First edition 2015-05-15*. May 15. https://doi.org/10.1109/IEEESTD.2015.7106435

ISS. 2022. International space station. March 5. https://en.wikipedia.org/w/index.php?title=International_Space_Station&oldid=1075419968 (accessed March 7, 2022).

Ivrii, V. 2021. *Partial Differential Equations*. Toronto: University of Toronto. https://www.math.toronto.edu/ivrii/PDE-textbook/PDE-textbook.pdf. (Accessed March 8, 2025).

Iyer, R. and M. Mazzara. 2009. *Systems Engineering for Military Ground Vehicle Systems*. Warren: US Army RDECOM-TARDEC.

Kahraman, C. 2008. *Fuzzy Multi-Criteria Decision Making Theory and Applications with Recent Developments*. New York: Springer Science+Business Media, LLC.

Kandiller, L. 2007. *Principles of Mathematics in Operations Research*. New York: Springer Science+Business Media. LLC.

Kaneko, K. and T. Yanagita. 2014. Coupled maps. *Scholarpedia*. http://www.scholarpedia.org/article/Coupled_maps (accessed June 4, 2022).

Kelly, A. 2003. *Decision Making Using Game Theory: An Introduction for Managers*. Cambridge, UK: Cambridge University Press.

Kishk, M. and C. Ukaga. 2008. The impact of effective risk management on project success. https://www.researchgate.net/publication/266065458_The_impact_of_effective_risk_management_on_project_success (accessed March 2022).

Kockler, F., T. Withers, J. Poodiack, and M. Gierman. 1990. *Systems Engineering Management Guide*. Fort Belvoir, VA: Defense Systems Management College.

Koduru, S.D., M.A. Nessim, and P. Gardoni. 2017. *Risk and Reliability Analysis: Theory and Applications: In Honor of Prof. Armen Der Kiureghian*. Cham: Springer International Publishing.

Kossiakoff, A., W.N. Sweet, S.J. Seymour, and S.M. Biemer. 2011. *Systems Engineering Principles and Practice*, 2nd edn. Hoboken, NJ: John Wiley & Sons, Inc.

Labeyrie, A. 1970. Attainment of diffraction limited resolution in large telescopes by Fourier analysing speckle patterns in star images. *Astronomy and Astrophysics*, Vol. 6, p. 85.

Ladkin, P. and K. Loer. 1998. Analysing aviation accidents using WB analysis - an application of multimodal reasoning. *Spring Symposium*. AAAI. Madison, Wisconsin. https://www.aaai.org/Papers/Symposia/Spring/1998/SS-98-04/SS98-04-031.pdf.

Li, S. 2003. The role of expected value illustrated in decision-making under risk: single-play vs multiple play. *Journal of Risk Research*, Vol. 6, p. 113.

Life Data Analysis. 2022. Life Data Analysis. https://www.weibull.com/basics/lifedata.htm (accessed April 27, 2024).

Liz, E. and A. Ruiz-Herrera. 2013. Attractivity, multistability, and bifurcation in delayed Hopfield's model with non-monotonic feedback. *Journal of Differential Equations*, Vol. 255, No. 11, pp. 4244–4266. https://doi.org/10.1016/j.jde.2013.08.007

Lockheed. 2022. Lockheed C-5 Galaxy. February 28. https://en.wikipedia.org/w/index.php?title=Lockheed_C-5_Galaxy&oldid=1074412441 (accessed March 8, 2022).

Long, A.R. 2010. Beauty and the beast - use and abuse of the fault tree as a tool. March 29. https://www.fault-tree.net/papers/long-beauty-and-beast.pdf (accessed March 2022).

Lorenz, E.N. 1963. Deterministic nonperiodic flow. *Journal of the Atmospheric Sciences*, Vol. 20, pp. 130–141. https://math.bu.edu/people/mabeck/Fall14/Lorenz63.pdf.

Macaulay-Brown, Inc., Chrislaghi, D., R. Dyer, and F. Jay. 2009. *KC-135 SIMULATOR*. Case Study, Wright-Patterson AFB: Air Force Center for Systems Engineering. https://scholar.afit.edu/docs/37 (accessed March 7, 2022).

Maschler, M., E. Solan, and S. Zamir. 2013. *Game Theory.* Cambridge, MA: Cambridge University Press.

Mentari, E. and D. Aron. 2006. Decision analysis in endocrinology. In: Montori, V.M. (ed), *Evidence Base Endocrinology*. Totowa, NJ: Humana Press. Mayo Foundation for Educational Research.

Merriam-Webster. 2022. *Risk.* https://www.merriam-webster.com/dictionary/risk (accessed 2022).

Mikhailov, A.S. and A.Y. Loskutov. 1996. *Foundations of Synergetics II.* Berlin, Heidelberg: Springer Verlag.

Modarres, M., M. Kaminskiy, and V. Krivstov. 1999. *Reliability Engineering and Risk Analysis:A Practical Guide*. New York, NY: Marcel Dekker, Inc., pp. 198.

Montibeller, G. and D. von Winterfeldt. 2015. Cognitive and motivational biases in decision and risk analysis. *Risk Analysis*, Vol. 35, pp. 1230–1251.

Moore, D.A. and P.J. Healy. 2008. The trouble with overconfidence. *Psychological Review*, Vol. 115, pp. 502–517.

Mudholkar, G.S. and G.D. Kollia. 1994. Generalized Weibull Family: a structural analysis. *Communication in Statistics - Theory and Methods,* Vol. 23, No. 4, pp. 1149–1171.

Mudholkar, G.S., D.K. Srivastava, and G.D. Kollia. 1996. A generalization of the Weibull distribution with application to the analysis of survival data. *Journal of the American Statistical Association,* Vol. 91, No. 436, pp. 1575–1583.

Murthy, D., N. Prabhakar, M. Bulmer, and J.A. Eccleston. 2004. Weibull model selection for reliability modelling. *Reliability Engineering & System Safety*, Vol. 86, pp. 257–267.

Myerson, R.B. 1997. *Game Theory: Analysis of Conflict.* Cambridge, MA: First Harvard University Press.

NASA. 2016. NASA systems engineering handbook. https://www.nasa.gov/wp-content/uploads/2018/09/nasa_systems_engineering_handbook_0.pdf (accessed April 6, 2024).

NASA. 2021. A guide for aircraft certification by analysis. https://ntrs.nasa.gov/api/citations/20210015404/downloads/NASA-CR-20210015404%20updated.pdf (accessed March 8, 2025).

National Research Council (U.S.). 1982. Committee on Risk and Decision Making. *Risk and Decision Making: Perspectives and Research.* Committee Report, Washington (DC): National Academies Press (US). https://www.ncbi.nlm.nih.gov/books/NBK218911/ (accessed 2022).

National Research Council (U.S.). 2003. *One Step at a Time the Staged Development of Geologic Repositories for High-Level Radioactive Waste.* Washington, DC: National Academies Press. . Committee on Principles and Operational Strategies for Staged Repository Systems Staff Corporate Author.

Nickerson, R.S. 1998. Confirmation bias: a ubiquitous phenomenon in many guises. *Review of General Psychology,* Vol. 2, pp. 175–220.

Nilchiani, R.R. and A. Pugliese. 2016. A complex systems perspective of risk mitigation and modeling in development and acquisition programs. *Proceedings of the Thirteenth Annual Acquisition Research Symposium*, Monterey, CA. p. 62. https://apps.dtic.mil/sti/pdfs/AD1016800.pdf (accessed 2022).

National Institute of Standards and Technology. 2018. NIST Special Publication 800-37 Rev 2: Risk Management Framework for Information Systems and Organizations: A System Life Cycle Approach for Security and Privacy. https://csrc.nist.gov/pubs/sp/800/37/r2/final (accessed March 7, 2025).

Northrop Grumman RQ-4 Global Hawk. 2024. Northrop Grumman RQ-4 Global Hawk. https://en.wikipedia.org/wiki/Northrop_Grumman_RQ-4_Global_Hawk (accessed April 27, 2024).

Olechowski, A., S.D. Eppinger, and N. Joglekar. 2015. Technology readiness levels at 40: a study of state-of-the-art use, challenges, and opportunities. *PICMET 2015: Management of the Technology Age*, Portland, OR, pp. 11.

Ortiz, A. 1992. *System Engineering Concept Demonstration, Process Model.* Long Beach, CA: McDonnel Douglas Corporation.

OSD Manufacturing Technology Program. 2020. Manufacturing readiness level (MRL) deskbook. https://www.dodmrl.com/MRL%20Deskbook%20V2020.pdf (accessed March 8, 2025).

Peppler, K., J.A. Danish, and D. Phelps. 2013. Collaborative gaming. *Simulation & Gaming*, Vol. 44, No. 5. https://doi.org/10.1177/1046878113501462

Pernin, C.G., E. Axelband, J.A. Drezner, B.B. Dille, J. Gordon IV, B. Held, K. Scott McMahon, et al. 2012. *Lessons from the Army's Future Combat Systems Program.* Santa Monica, CA: RAND Corporation.

Perry, W.J. 1995. *DoD News Briefing.* Washington, DC: Office of the Assistance Secretary of Defense (Public Affairs). https://nuke.fas.org/control/ctr/news/95-02-06.html (accessed April 6, 2024).

Peters, H. 2008. *Game Theory a Multilevel Approach.* Berlin, Heidelberg: Springer-Verlag.

Pham, A. and C.-D. Lai. 2007. On recent generalizations of the Weibull distribution. *IEEE Transactions on Reliability,* Vol. 56, pp. 454–458.

Philips, D.T., A. Ravindran, and J.J. Solberg. 1976. *Operations Research Principles and Practice*. New York: John Wiley & Sons, Inc.

QRA Corp. 2025. Automating the INCOSE guide for writing requirements. *QRA Corp.* https://qracorp.com/incose-requirements-guide/ (accessed March 8, 2025).

Rasmusen, E. 1989. *Games and Information: An Introduction to Game Theory*, 2nd edn. Malden, MA: Blackwell Publishers.

Roggemann, M.C. and B. Welsh. 1996. *Imaging through Turbulence*. Boca Raton, FL: CRC Press.

Roy, K., P. Bhattacharya, and C. Suen. 2011. Iris recognition using shape-guided approach and game theory. *Pattern Analysis and Applications*, Vol. 14, No. 4, pp. 329–348.

Roy, K., P. Bhattacharya, and C. Suen. 2012. Iris segmentation using game theory. *Signal, Image and Video Processing*, Vol. 6, No. 2, pp. 301–315.

Roy, K., P. Bhattacharya, C. Suen, and J. You. 2010. Recognition of unideal iris images using region-based active contour model and game theory. *2010 IEEE International Conference on Image Processing*, Hong Kong, China. ISSN 1522–4880, 2010, ISBN 1424479924, pp. 1705–1708.

Saaty, R.W. 1987. The analytic hierarchy process-what it is and how it is used. *Mathematical Modelling,* Vol. 9, pp. 161–176.

Sage, A.P. and W.B. Rouse. 2014. *Handbook of Systems Engineering and Management.* Hoboken, NJ: John Wiley & Sons.

Salminen, J. 2012. Collective intelligence in humans: a literature review. *Proceedings, CI2012*. Lahti, Finland. https://arxiv.org/ftp/arxiv/papers/1204/1204.3401.pdf. (accessed March 8, 2025).

Savage, L.J. 1951. The theory of statistical decision. *Journal of the American Statistical Association*, Vol. 46, pp. 55–67.

Schkade, D.A. and D. Kahneman. 1998. Does living in California make people happy? A focusing illusion in judgments of life satisfaction. *Psychological Science*, Vol. 9, pp. 340–346.

Schuldt, D. and J. Kurucar. 2018. Efficient partitioning of space for multiple UAVs in an unobstructed environment. International Journal of Intelligent Robotics and Applications 2(3), March 2018, DOI: 10.1007/s41315-018-0045-y, pp. 98-109.

Smith, J.E. and W. Detlof. 2004. Decision analysis in management science. *Management Science*, Vol. 50, No. 5, pp. 561–574.

Society for Risk Analysis. 2018. Risk analysis glossary. *Society for Risk Analysis.* https://www.sra.org/wp-content/uploads/2020/04/SRA-Glossary-FINAL.pdf (accessed August 2022).

Stanley, G. 2010. *Manufacturing Readiness Levels Overviews.* Air Force Research Laboratory, AFRL/RXMT,Manufacturing Technology Division,Wright Patterson AFB,OH. Presented at Defense Industrial Base Seminar and Workshops, 15-16 June 2010, Fort McNair, Washington, DC, p. 14.

Stockman, B., J. Boyle, and J. Bacon. 2010. *International Space Station Systems Engineering Case Study.* Wright-Patterson AFB: Air Force Center for Systems Engineering, p. 124. https://www.nasa.gov/wp-content/uploads/2015/05/design_iss_systems_engineering_case_study.pdf (accessed March 7, 2022).

Swedberg, R. 2001. Sociology and game theory: contemporary and historical perspectives. *Theory and Society*, Vol. 30, No. 3, pp. 301–335.

Tatarskii, V.I. 1967. *Wave Propagation in a Turbulent Medium.* New York: Dover Publications.

Taylor, D.E. 2019. *Estimating and Depicting Risk during Large Scale Combat Operations.* Fort Leavenworth: U.S. Army Command and General Staff College.

Troger, H. and A. Steindl. 1991. *Nonlinear Stability and Bifurcation Theory: An Introduction for Engineers and Applied Scientists.* Vienna: Springer Wien, Proquest Ebook Central. https://ebookcentral.proquest.com/lib/fit/detail.action?docID=3099285 (accessed March 8, 2025).

Tversky, A. and D. Kahneman. 1974. Judgment under uncertainty: heuristics and biases. *Science*, Vol. 185, pp. 1124–1131.

Ullman, D.G. and B.P. Spiegel. 2006. Trade studies with uncertain information. *Sixteenth Annual International Symposium of the INCOSE*, Orlando, FL.

Under Secretary of Defense for Acquisition, Technology, and Logistics. 1985. DoD Directive 4245.7-M. *Transition from Development to Production.* September.

U.S. Department of Defense. 1998. MIL-HDBK 338B: *Electronic Reliability Design Handbook.* Reliability modeling and prediction. Pp. 6-19. https://quicksearch.dla.mil/Transient/FE4C04C90023474E87271E10B4A33887.pdf (accessed March 8, 2025).

U.S. Government Accountability Office. 2013. *DOD Financial Management: Ineffective Risk Management Could Impair Progress toward Audit-Ready Financial Statements.* Washington, DC: U.S. Government Accountability Office. https://apps.dtic.mil/sti/pdfs/ADA585260.pdf (accessed March 8, 2022).

U.S. Government Accountability Office. 2020. *Technology Readiness Assessment Guide (GAO-20–48G).* Washington, DC: U.S. Government Printing Office.

Valerdi, R. and R.J. Kohl. 2004. An approach to technology risk management. *Engineering Systems Division Symposium*, Los Angeles.

Vasilevich, S.S. and M.N. Alexandrovich. 2006. *New Methods for Chaotic Dynamics.* Hackensack, NJ: World Scientific. https://doi.org/10.1142/6117

Vasquez, M.E.B. 2015. Analysis of the Effectiveness of F-15E Risk Management during Peacetime. Master's Thesis, Wright Patterson AFB: Air Force Institute of Technology.

Weisstein, E.W. 2022. Iterated map. *Wolfram Mathworld.* June 2. https://mathworld.wolfram.com/IteratedMap.html (accessed June 4, 2022).

West, P.D. and T.E. Trainor. 2006. *Risk-Based Decision Support of Water Resource Management Alternatives.* West Point, NY: The Operations Research Center of Excellence, United States Military Academy.

Whitney, S. 2010. Game theory. *National Jeweler*, Vol. 104, No. 5, p. 4.

Wilson, P.F., L.D. Dell, and G.F. Anderson. 1993. *Root Cause Analaysis: A Tool for Total Quality Management.* Milwaukee, WI: ASQ Quality Press.

Winston, W.L. 1994. Decision making under uncertainty. In: Winston, W.L. (ed), *Operations Research: Applications and Algorithms*. Belmont, CA: Wadsworth Publishing Company, pp. 727–770.

Yentis, S.M. 2006. Decision analysis in anaesthesia: a tool for developing and analysing clinical management plans. *Anaesthesia*, Vol. 61, pp. 651–658.

Yuan, W., Y. Chen, R. Wang, and Z. Du. 2007. Collective Intelligence in Knowledge Management. In: Xu, L.D., Tjoa, A.M., Chaudhry, S.S. (eds), *Research and Practical Issues of Enterprise Information Systems II. IFIP — The International Federation for Information Processing*, vol 254. Springer, Boston, MA, pp. 651–655. https://doi.org/10.1007/978-0-387-75902-9_72.

8 Agile Systems Engineering

Intelligence is the ability to adapt to change.

—**Stephen Hawking**

8.1 AGILE METHODS IN SYSTEMS ENGINEERING

8.1.1 Overview of Agile Methods

Traditional systems engineering development efforts have been document-centric and require huge planning efforts. Because the traditional development methods are requirements-driven, it can become difficult to adapt to change, especially after requirements are baselined. Additionally, to enact change, expensive review and update processes are in place that are often manual and prone to error. Also, the traditional development effort is mostly sequential and with typically long integration, verification, and validation phases, and it is hard to maintain over long periods. Consequently, these established methodologies and associated practices have major constraints throughout the program lifecycle, including problems in making changes or revisions to requirements, delays in reaching program milestones, and difficulties in monitoring progress. In response to these constraints, Agile methods have arisen as an often viable and modern alternative to old practices, marking a major shift in the program management and system development environment. This chapter begins an investigation of the Agile methodology, beginning with an examination of its origins and historical background. Following that, it deconstructs the underlying concept and foundational assumptions that underpin Agile methods. This investigation includes various tools and approaches that help with the adoption of Agile practices in different program circumstances. The final half of this chapter provides an assessment of the usage of Agile methodologies in the domain of systems engineering, backed up by relevant case studies. These case studies highlight the importance and revolutionary potential of Agile methods in this industry. This chapter also emphasizes how Agile methods are more than just a development alternative for improving efficiency, adaptability, and overall program performance in the systems engineering profession. Traditional development approaches have long played an important role in designing and manufacturing systems. However, these traditional methodologies have significant limitations throughout the program lifecycle, including difficulty in amending or revising requirements, delays in meeting deadlines, and difficulties in progress tracking. In many situations, Agile approaches have arisen as a viable alternative to these old practices in response to these limitations.

8.1.2 Introduction

The pace of technology and innovation is steadily increasing, with advancements in artificial intelligence, computers, software, and electronics leading to complex and sophisticated capabilities never thought possible. With these rapid technology evolutions (and sometimes technology revolutions), there is a strong demand for engineering to expand and adapt, embracing modern development practices, and effectively dealing with complexity and change across the systems development lifecycle (SDLC).

To start, what exactly is agile development? According to the IEEE article, "Analysis of the Criteria for Comparing Systems Engineering and Agile Methods," by Humper et al., "In general, agile means *nimble* or *flexible*. This is related to the ability to react swiftly in business" (Humpert et al. 2022). The same source goes on to describe Agile methodologies, which can be considered

DOI: 10.1201/9781003624097-10

an evolution beyond this fundamental statement. Agile methods are systematic approaches that are marked by short feedback loops, adaptability to changing circumstances, a strong focus on client needs, and stringent quality standards. Notably, these techniques emphasize the critical importance of human interaction in the process (Humpert et al. 2022).

Agile methods evolved because of two fundamental issues in the software development arena (Douglass 2015). Initially, there was a clear trend toward implementing stringent restrictions and tight procedures in software development, which frequently included responsibilities unrelated to the actual software creation process. As a result, the software development process became over-organized and extremely bureaucratic, with too many rules, regulations, and procedures that did not always contribute to the efficient development of software. These inefficient methods frequently included activities that had little to do with the essential tasks of coding and building software products. Simultaneously, despite the enforcement of these time-consuming processes, software quality had steadily deteriorated, frequently failing to satisfy the expectations and requirements of its intended users. It was believed that this recurring problem resulted from an overemphasis on processes at the expense of providing software that effectively responded to user demands and satisfied their expectations. In response to these obstacles, Agile methods place a premium on taking a user-centric approach. Their primary goal is to provide software that closely matches the preferences and requirements of its target customers. Agile methods achieve this goal by emphasizing collaboration, adaptability, and iterative development, thereby correcting the perceived flaws in the rigid and relatively ineffective practices that preceded their introduction (Douglass 2015).

Essentially, the Agile method can be defined as a program management and software development methodology that emphasizes flexibility, collaboration, and incremental progress. Agile methods are popular because they are fast paced, focus on team collaboration, and produce the final product faster due to parallel development practices and their continuous feedback loop.

When considering the Agile method, what often comes to mind is its association with software-only development efforts. However, the landscape has changed dramatically; in that, Agile methods are no longer limited to just software development programs, or the software engineering discipline, but have grown to include hardware and system-level development efforts and consequently have been embraced by the general engineering community, including systems engineering.

8.1.3 Historical Background

If we want to understand the Agile methodology and how it relates to system engineering, we must first review a little of its history. From 1970 to 2000, the Waterfall methodology was popular in the software, systems engineering, and military communities due to the discipline it brought to software/systems developments, and the clear path it gave for each stage of the overall systems development lifecycle. *Waterfall* is a technique in which teams define the system needs and requirements, then design a system based on the system's specifications, develop and test the system, and finally deliver the system to its intended users. The system is then supported and eventually retired. This approach lies at the heart of the traditional systems development process. However, a problem with this approach is that the driving requirements can change over the system development lifecycle. Especially now, when technology is moving so quickly that stakeholders have difficulty in articulating their wants and needs considering the potential capabilities that these rapidly changing technologies offer. In turn, developers find it difficult to stick with a given technical approach when new advancements in technology allow for new and better capabilities. During the 1970s, various methods evolved in response to the increased need for flexibility in system development efforts. Agile methods, as defined by Marvin Zelkowitz in his book "Advances in Computers: Advances in Software Engineering," are a heterogeneous group of methodologies and activities that share common values and fundamental ideas. Many of these methodologies, for example, are inspired by iterative improvements, a concept first proposed in 1975 (Zelkowitz 2004, p. 3). Consequently, Agile is a collection of methods and practices that has some of its essential roots stemming from

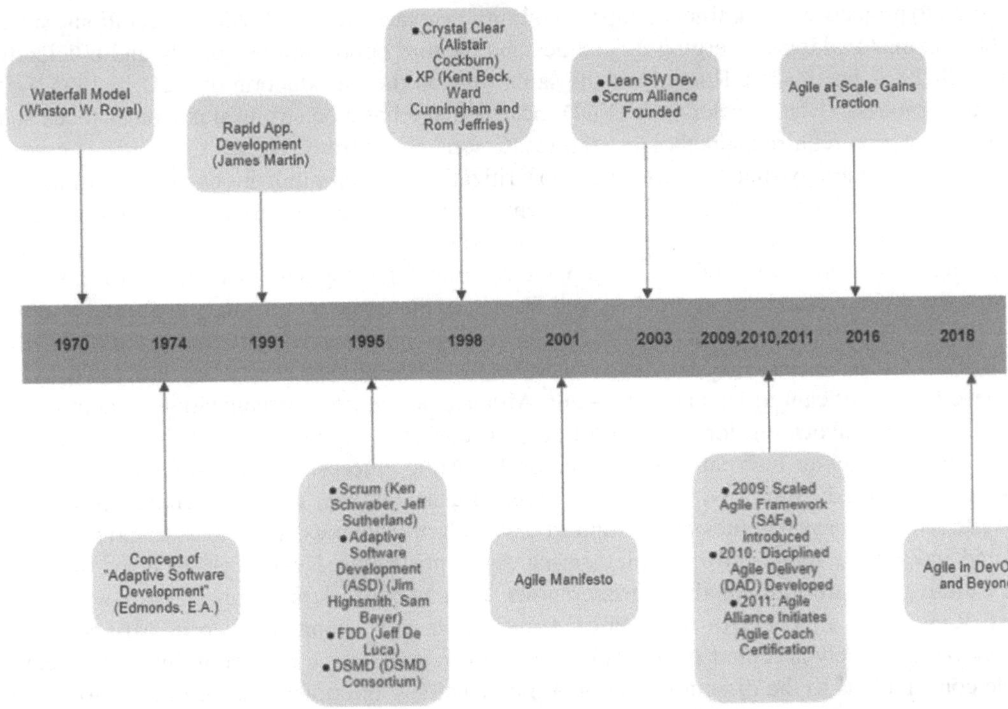

FIGURE 8.1 Key contributions to Agile methods and practices.

lean software and lean manufacturing initiatives to eliminate waste, mistake proofing (poka-yoke), Six Sigma, and Kanban. It is important to note that Agile methods did not inherently introduce revolutionary concepts; rather, they aimed to diverge from the traditional Waterfall methodology. Agile methods are differentiated from conventional development methods by virtue of their fundamental emphasis and core ideals (Zelkowitz 2004, p. 3). Figure 8.1 provides a visual snapshot of the timeline of key contribution that led to our current understanding of Agile.

In 1994, the Dynamic System Development Method (DSDM) developed by the DSDM consortium was introduced to aid the Rapid Application Development (RAD) framework for rapid software prototyping. RAD has four stages that include (1) requirements assessment, (2) prototyping, (3) testing and gathering feedback, and (4) presenting the product. DSDM provided more governance and discipline to the traditionally freeform methods in the RAD framework. It also introduced a prioritization scheme based on MoSCoW (Must have, Should have, Could have, and Would like). The Scrum concept was introduced by Ken Schwaber and Jeff Sutherland, which looked at short (approximately 2 week) bursts of focused developmental activity called sprints with a planning phase at the beginning and a self-evaluation at completion. A backlog tracks required fixes and changes that are worked off in the next sprint. Adaptive Software Development (ASD) is a software development approach developed by James A. Highsmith in 1995 that is frequently associated with the name "Adaptive" but has also been attributed to Jim Edmonds, who invented the notion in 1974. According to James Highsmith's book "Adaptive Software Development: A Collaborative Approach to Managing Complex Systems," adaptive software development is a pliable and iterative technique for software development that prioritizes the ability to adapt to changing [system] dynamics and needs. It emphasizes the significance of encouraging cooperation among team members and stakeholders, supporting constant feedback systems, and maintaining the ability to realign the project's trajectory as new insights become available. Adaptive development approaches, such as Adaptive Software Development (ASD), are focused on incremental value delivery, allowing for frequent assessments and adaption throughout the system's lifecycle. This strategy is particularly well-suited to projects with significant unpredictability and variable stakeholder expectations, since it enables teams to address emerging difficulties and capture opportunities quickly and efficiently (Highsmith 1999, pp. 14–15). Roughly at the same time as the introduction of ASD in 1995, Jeff De Luca's Feature Driven Development (FDD) provided emphasis on strong stakeholder focus and design emphasis on required stakeholder features. Crystal Clear from Alistair Cockburn is part of the Crystal Methodology that focuses on the prioritization of communications and team interactions and team attributes such as skills and talents over processes and tools. It introduces team self-organization to determine useful processes and tools.

Although there were several efforts to supplant traditional development methods, it was not until 2001 that the Agile methodology, which is well framed in the Agile manifesto, was formally introduced. The Agile Alliance, which consisted of 17 technologists and included notable names like Kent Beck, Martin Fowler, Ron Jeffries, and Robert Martin, among others, worked collectively to achieve this significant milestone. The Agile Alliance, a non-profit organization, is committed to advancing and advocating for Agile practices and concepts beyond the software development industry. It serves as an international association for Agile enterprises and professionals, all of which are committed to improving software development, delivery, and maintenance processes. The Agile Alliance provides substantial support, a wealth of resources, and events to aid organizations and individuals interested in implementing Agile principles. Kent Beck, Ron Jeffries, and Ward Cunningham are individuals of great renown within the Agile software development community. Kent Beck, a pioneer in the field of Agile software development, is primarily noted for his work on the widely accepted Agile framework known as Extreme Programming (XP). Beck's notable contributions to the dissemination of Agile principles and practices, include his works on "Test-Driven Development: By Example" and "Extreme Programming Explained," which have solidified his position as a pivotal influencer within the Agile movement. Another significant member of the software development community is Ward Cunningham. One of his most renowned contributions was the development of the first wiki, an online collaborative environment where users can jointly produce and modify web pages. Although he did not initially sign the Agile Manifesto, his contributions to wiki development and participation in early Agile dialogues have established him as a reputable person within the Agile movement. Ron Jeffries, an original signatory of the Agile Manifesto, software developer, consultant, and advocate for Agile development practices, is widely acknowledged for his contributions to Extreme Programming (XP) and his unwavering support for Agile development methodologies. Jeffries has made substantial contributions to the Agile community through his coaching and consulting with firms aiming to implement Agile principles, in addition to his extensive output of books and presentations. In 2009, the Scaled Agile Framework (SAFe) was introduced to scale Agile methods to the enterprise level. SAFe was developed for large enterprises (more than 1,000 individuals) and was developed by Scaled Agile from Boulder Colorado. The Disciplined Agile Delivery (DAD) is an Agile framework focused on IT delivery solutions that puts people first and promotes learning. DAD was developed in 2010 and addresses all facets of the delivery lifecycle and different ways of working. In 2016, Agile at Scale began seeing some recognition. Agile at scale means applying Agile methods at whatever practical scales are required. For instance, Agile can be commonly applied at the team level as part of multiple parallel teams. For Agile at Scale, Agile principles and practices are applied, and results experienced, at various levels of the organization up to and including the enterprise level.

The individuals mentioned above significantly contributed to the development and dissemination of Agile methods, thereby drastically altering the strategy and implementation of software development in enterprises across the globe. One of the signatories, Kent Beck, stated, "The Agile movement does not deny the notion of methodology; on the contrary, a significant number of us strive to reestablish its credibility. We aim to achieve a harmonious equilibrium by endorsing modeling without relegating diagrams to dusty corporate archives, embracing documentation without generating bulky volumes that are seldom utilized, and recognizing the value of planning while also acknowledging its limitations in an ever-evolving environment (Beck 2001)." Considerable advancements in the incorporation of Agile principles into the field of systems engineering were realized in the years 2015 and 2016. The International Council on Systems Engineering (INCOSE) exhibited proactive leadership during this period with the formation of an "Agile Working Group." The principal aim of this collective was to enhance the systems engineering techniques outlined in ISO/IEC/IEEE 15288-2015 through the integration of Agile methods. The objective of this strategic effort was to enhance the synergy between systems engineering methods and Agile principles, acknowledging the criticality of adaptation and flexibility in the ever-changing domains of technology and systems/program management.

8.1.4 Concepts and Principles

The purpose of the Agile methodology, according to Kent Beck and other Agile developers at the agilemanifesto.org, is not antimethodology in the classic sense, but rather to "return credibility to the word methodology." To clarify, Paul Kirvan and Mary Pratt stated that "the [Agile] developers strove to create a balance between the existing modes of development and the new possibilities." They acknowledged to accepting modeling and documentation, but only where it was clearly useful. The Agile developers also stated that, while planning is vital, it is also necessary to realize that plans change, and that flexibility is required for any changes. Overall, the manifesto emphasizes the importance of people and connections over processes and instruments (Kirvan and Pratt 2023).

Accordingly, the Agile Manifesto emphasizes the following four key priorities (Agile Manifesto 2001):

- Individual and interactions over processes and tools;
- Working software over comprehensive documentation;
- Customer collaboration over contract negotiation;
- Responding to change over following a plan.

At first glance, these four priorities appear to contradict each other and to contradict standard software/system engineering development methodologies. Ann Garrison Darrin and William S. Devereux proposed in their article, "The Agile Manifesto, Design Thinking, and Systems Engineering," that the four concepts of the Agile Manifesto appear to "encourage an undisciplined approach." However, rather than being undisciplined, "agile methods have a rigid methodology since they require both team consensus and a high level of discipline to follow and execute the agreed-upon rules and methods" (Darrin and Devereux 2017). Douglass did a comparison of traditional and Agile methodologies in Agile Systems Engineering to have a better grasp of the concepts of the Agile Manifesto in comparison to traditional methods. Traditional methods involve developing work products across multiple program silos at the same time, using a 'breadth-first' strategy. Verification of system/program/project requirements, to ensure that those requirements are met, is frequently postponed until later in the systems development lifecycle, or sometimes even when the system is completed or even later. Agile methods, on the other hand, take an incremental approach, checking work items as they are developed. This continuous verification provides immediate input on quality and conformity with engineering goals, promoting adaptability and efficiency. Agile stresses dynamic, iterative verification, which allows for faster issue detection and resolution, resulting in increased engineering effectiveness (Douglass 2015, p. 42). We can see that the Agile Manifesto's aim is to establish flexibility between engineering and stakeholders and their requirements. The conventional development methodology differs from the Agile method; in that, the Waterfall method is more focused on delivering the system on time, and it is harder to return to prior processes. While using the Agile method, engineers have 'enough room' to go back and create the product, adapting the specs to fit the new stakeholder requirements.

The Agile Manifesto also emphasizes the aspects of the development effort that genuinely matter. The Agile Manifesto represents a significant shift in how we think about creating software or systems.

The most important things are making stakeholders happy and getting the software right. This is more important than strictly adhering to program plans and completing excessive amounts of documentation. The manifesto recognizes that making stakeholders happy and being able to "adjust things as we go" are essential to program success. However, that doesn't mean we should abandon planning and documentation entirely. Instead, there needs to be a balance in which planning, and documentation are important, but not as important as making stakeholders happy and getting the software correct. This new style of thinking has significantly influenced how software, and systems, are created. It makes the systems development team more adaptable, better at cooperating, and better at listening to clients, all of which leads to better and more stakeholder-focused software and systems development processes (Douglas 2015, p. 43). The use of Agile methods emphasizes the importance of the four values stated in the Agile Manifesto. These ideals are intended to improve the software/system development process, with a primary focus on providing a high-quality product that seamlessly corresponds with the demands and expectations of the stakeholders.

The Agile Alliance articulated a comprehensive set of 12 principles that match with the Agile Manifesto's concepts and values. These principles are specifically developed to foster and cultivate a work environment that prioritizes addressing the needs of stakeholders and assuring their satisfaction. Furthermore, they are oriented toward connecting program goals with broader business objectives and developing a responsive and adaptable environment to changing user needs and market dynamics (Kirvan and Pratt 2023). These principles are easily accessible on the Agile Manifesto's official website, agilemanifesto.org and are directly repeated here for convenience (Agile Manifesto 2001):

- "Our highest priority is to satisfy the customer through early and continuous delivery of valuable software."
- "Welcome changing requirements, even late in development. Agile processes harness change for the customer's competitive advantage."
- "Deliver working software frequently, from a couple of weeks to a couple of months, with a preference to the shorter timescale."
- "Businesspeople and developers must work together daily throughout the project."
- "Build projects around motivated individuals. Give them the environment and support they need and trust them to get the job done."
- "The most efficient and effective method of conveying information to and within a development team is face-to-face conversations."
- "Working software is the primary measure of progress."
- "Agile processes promote sustainable development. The sponsors, developers, and users should be able to maintain a constant pace indefinitely."
- "Continuous attention to technical excellence and good design enhances agility."
- "Simplicity – the art of maximizing the amount of work not done – is essential."
- "The best architectures, requirements, and designs emerge from self-organizing teams."
- "At regular intervals, the team reflects on how to become more effective, then tunes and adjusts its behavior accordingly."

The Agile Manifesto has played an important part in the advancement of the Agile methodology over time. Zelkowitz demonstrates how the Manifesto reflects the key values of Agile, distinguishing it from conventional developmental methods, in his book, Advances in Computers: Advances in Software Engineering (Zelkowitz 2004, p. 8).

8.1.5 Benefits of Agile

Agile methods provide significant benefits for project execution. According to Douglass' book, the major purpose of Agile systems engineering is "to establish specifications that may drive downstream engineering to create a system that meets the needs of the customer. It is not to create documentation. It is not to go to meetings. It is not to make schedules. It is not to develop productivity indicators (Douglass 2015, p. 46)." All the tasks mentioned by the author of Agile Systems Engineering can be smoothly integrated into the system's core aim, and their importance is revealed only when the system aligns with client needs. The second critical purpose of any Agile development is to build flexibility, allowing the system development to evolve. In his book, Agile Systems Engineering, Douglass emphasized that certain critical aspects must be in place to ensure the successful evolution and adaptation of a previous system instantiation. To begin, the system's architecture should be developed in such a way that additional features or extensions can be easily added. Second, extensive documentation is required to enable future teams to successfully understand and alter the system. Furthermore, the availability of assistance is critical for recognizing and reducing the risks connected with future development. Finally, a well-established infrastructure for change and configuration management should be in place to allow for the easy integration of system updates and alterations. These parts are synergistic in making the system ready for growth, improvement, and long-term development (Douglas 2015, p. 46).

According to Douglass, the following are the primary benefits of agile methods for system engineering:

- *Improved Engineering Data and Quality:* Douglass claims, "[In] system engineering, this is the most obvious and important benefit of applying Agile methods" (Douglass 2015, p. 46). The reason is, "Agile methods employ practices for early and continuing verification of engineering data and specifications, they will contain fewer and less significant defects" (Douglass 2015, p. 46). Essentially, using Agile methods enhances the overall quality of the developed system and provides for increased accuracy in ways that guarantee the prompt and ongoing validation of technical specifications and engineering data. Consequently, fewer faults with less consequences can rightfully be expected.
- *Better engineering efficiency:* Since there is continuous verification and improvement when employing Agile methods, more problems are found at earlier times, meaning that there is less "rework" required, resulting in less waste. This results in a more efficient engineering operation and saves time and money.
- *Return on investment (ROI) seen earlier:* From Douglass' perspective, "Agile methods provide earlier ROI than traditional approaches because demonstrably correct engineering data and specifications are actionable, whereas data of unknown quality are not" (Douglass 2015, p. 47). Also, when considering the traditional development methods, the waterfall model, the formal testing process starts after production. If flaws or design deviations are discovered at this late stage, a large cost and schedule impact to the program can result.
- *Stakeholder satisfaction:* When all is said and done, the system must satisfactorily address the stakeholders needs and approved wants. There can be a variety of important stakeholders that need to be satisfied, such as funding agencies, the user community, regulation authorities, collaborators, and more. To be successful, the needs and approved wants of the important stakeholders must be addressed. Failing in this capacity can threaten the current and future programs, stain the developer's reputation, and may even have financial and legal consequences.
- *Better project controls:* Dr. Douglass pointed out that various systems projects can find themselves in a state of disarray because of incorrect assumptions and uncertainty in the accuracy of data. When implementing project tracking and control features, in many traditional developmental approaches, the emphasis is placed on sticking with a schedule versus the achievement of project goals. In these traditional developments, the adherence to the plan is paramount. In contrast, Agile methods focus more on objectives in evaluating project progress with continuous assessments being made along the way. In this manner, the developer can get early insights and signals if something is amiss, or if deviations from the intended path are beginning to occur, or even if there is a flaw within the plan itself.

Also, the prevalence of metrics, such as the defect density, failure rate, instantaneous reliability, velocity and more allow for better insight into the project, leading to better project control.

- *Improved response to change:* On an Agile development effort, the continuous assessment and improvement translate into an increased awareness on the current state of the project. Incoming changes can quickly be assessed an integrated into the solution path.
- *Risk reduction benefits:* One of the benefits of an incremental development, especially with Agile methods, is that functionality and capability are incrementally provided. This means that the stakeholder has access to something usable typically much earlier than in a traditional development environment. Having useful desired functionality earlier in the development cycle directly reduces risk; in that, evolutions of capability and features can be tailored and change directions along the development path as opposed to waiting until the system is completed to make any changes. The early and continuous involvement of the stakeholders can mitigate differences in expectations and reduce the risk of delivering an incomplete, or incorrect system.

8.1.6 Agile Methods, Tools, and Techniques

So far, Agile methods have been presented in general terms. But what do we specifically mean when we refer to Agile methods? What Agile tools and techniques are available? How can we use Agile in systems engineering to optimize our chances for success in developing sophisticated contemporary systems? According to Humpert, Mundt, Bretz, and Anacker in their paper, Analysis of the Criteria for Comparing Systems Engineering and Agile Methodologies, many approaches to Agile exist for systems engineering and system/program management. Seven Agile approaches that are all widely used were investigated. These approaches consist of SCRUM, LeSS, SAFe, DevOps, OKR, Kanban, and Agile Product Development (PD) (Humpert, Mundt, Bretz, and Anacker). Agile methods are a set of ideas and practices that guide software/system development and program management, with an emphasis on adaptability, teamwork, and stakeholder satisfaction. Several tools and software solutions are available to support and streamline the implementation of Agile methods across systems development lifecycle. These tools and software help Agile teams strategize, monitor progress, oversee tasks, and help ensure seamless communications. Figure 8.2 shows the popularity of various Agile methods. As the figure shows, there are various Agile methods that are available in an Agile development in practice.

At the bottom of Figure 8.2, we see the interesting statistics in the relative usage of these methods expressed as a percentage. We see that Scrum and Kanban are at the lead in popularity with 89% and 43% usage, respectively. Note that these percentages are not measured against each other but rather reflects their usage in overall Agile programs (i.e., programs can use more than one method). We will not go into detail about each method but instead will use the article Analysis of the Criteria for Comparing Systems Engineering and Agile Methods as a starting point to evaluate each method's shared qualities and distinct attributes. Even though there are a variety of Agile methods as shown in Figure 8.2, each with its own distinct elements, these all adhere to the Agile Manifesto and share the same core principles and assumptions. The basic process common to all these Agile methods follows a cyclical pattern that includes iteration planning, execution, demonstration, review, and retrospection. A primary artifact, sometimes referred to as a backlog, houses a full list of pending tasks and is at the center of these Agile methods. While each Agile method may use different languages and display different degrees of the backlog, the overall purpose and structure of the backlog artifact remains essentially constant. Furthermore, three main roles are present in these Agile methods: the product owner, who controls project priorities; the coach, who is responsible for improving the agile development process; and the non-hierarchically structured developers, who are involved in the actual development work. It is worth noting that under some Agile methods, analogous roles (excluding developers) execute similar but higher-level functions.

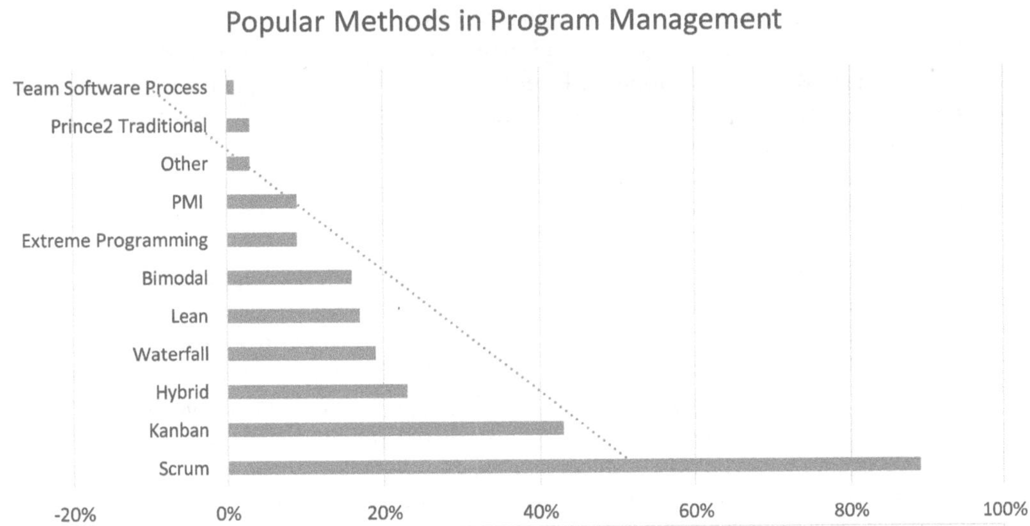

	Scrum	Kanban	Hybrid	Waterfall	Lean	Bimodal	Extreme Programming	PMI	Other	Prince2 Traditional	Team Software Process
Series1	89%	43%	23%	19%	17%	16%	9%	9%	3%	3%	1%

FIGURE 8.2 Popular Agile methods in program management.

8.1.6.1 Agile Tools and Techniques

According to VersionOne's 2014 research, there was a significant growth in the use of Agile methods throughout 2013, with an increase of up to 52%. Concurrently, globally dispersed software projects have shown constant annual growth rates ranging from 10% to 20%. The number of support tools has grown in tandem with the growing number of Agile methods and concepts. Özkan and Mishra, the authors of "Agile Project Management Tools: A Brief Comparative View," claim that software tools facilitate and improve communications in projects (Özkan and Mishra 2019). For example, the usage of groupware systems is said to improve both project communications and end results. We will go over four common Agile software tools.

- Atlassian's JIRA stands out as a widely used project management and problem tracking system. Its versatility has led to its adoption across a wide range of industries and teams, including software development, project management, and business operations. According to Özkan and Mishra, "JIRA is a popular project management tool for software professionals, institutional designers, and users of shared environments." JIRA's customized Scrum boards and flexible Kanban boards are essential features. JIRA is fully compatible with popular environments such as Eclipse. Furthermore, custom filters can be applied using JIRA Query Language (JQL). Other significant JIRA capabilities include customized developer tool connectors, cumulative flow diagrams, reporting, problem, and bug tracking. This technology has an "active user community because it is a mature and proven tool preferred by large organizations worldwide" (Özkan and Mishra 2019). JIRA stands out as a robust and easily customizable tool for project management, issue tracking, and Agile development methodology implementation. Its extensive use and the presence of an active user community make it a popular choice for teams looking to improve collaboration, efficiency, and project monitoring.
- *Agilo for scrum:* Agilo for Scrum is a Scrum and Agile project management solution that assists software development teams in applying Scrum practices and methodologies.

It has several features and capabilities that are well suited for Scrum procedures, making it easier for teams to plan, execute, and manage their projects. According to the article Agile Project Management Tools: A Brief Comparative View, "Agile for Scrum is mostly ideal for distant teams when extensive communication is necessary." This tool provides a real-time workflow, which includes a real-time scrum board, high-level themes, customization, and email integration. Furthermore, it is able to gain an overview of the product's ongoing process, which facilitates sprint planning and sprint estimation (Özkan and Mishra 2019).

- *Microsoft visual studio team services (VSTS):* Microsoft Visual Studio Team Services, formerly known as Azure DevOps Services, is a comprehensive set of cloud-based collaboration capabilities and features aimed at enhancing the entire software development lifecycle. This platform provides as a consolidated base for operations, tools for software development, project management, and DevOps deployment. "One of the powerful agile project management tools, VSTS has almost all the functionalities that are required to support scrum teams: flexible Kanban boards, exceptional traceability through inventory, customizable dashboards, and inherent scrum boards, extended integration, and Git client support. This application automates the handling of code versions, test documentation, and work items such as scenarios, test tasks, problems, and risk. Backlog management, capacity control, and an easy-to-use and appealing interface are also essential elements" (Özkan and Mishra 2019).
- Microsoft Visual Studio Team Services has a high degree of flexibility and adaptability, allowing it to meet the needs of development teams ranging from small startups to large corporations. Its comprehensive set of functions naturally integrates with modern software development processes and DevOps philosophies, facilitating collaborative efforts and accelerating software deployment.
- *VersionOne:* VersionOne, which has subsequently merged and is now known as CollabNet VersionOne, is a comprehensive software platform precisely designed to support Agile and DevOps approaches in corporate settings. This adaptable platform includes a variety of tools and features designed to help with many aspects of software development, project management, and delivery. "VersionOne is valuable for Agile software development experts, especially those working in distributed situations, assisting agile framework, Kanban, and Hybrid approach. The following are the most significant aspects of this tool: end-to-end visibility, team integrity, customizations, user story management, sprint release and planning, storyboard, task board, test board, and test tracking (Özkan and Mishra 2019)." Overall, VersionOne is a versatile platform that supports Agile and DevOps transformations, stimulates collaboration, and provides organizations with the tools they need to efficiently create high-quality software. It is appropriate for a wide range of firms, from small teams to major corporations, that want to incorporate Agile and DevOps principles into their software development processes.

Some other useful resources include *Pivotal Tracker*, which is beneficial for mobile projects. Task management, communication, and collaboration are all made as simple as possible by this cloud-based software. However, its primary purpose is to assist in keeping track of and monitoring tasks on a minute-by-minute basis. Prefix, from Stackify, is a free tool that helps find and fix bugs before they are deployed. Retrace, by Netreo, provides monitoring, error correction, and logging capabilities. Retrace is a user-friendly Software as a Service (SaaS) solution for application monitoring that integrates the following essential components: Application Performance Monitoring (APM) provides application performance visibility at the code level. Tracing at the code level, transaction tracing determines what your code is doing, monitors server applications, and provides custom metrics. There are also tutorials, webinars, and useful publications produced by the International

Council on Systems Engineering (INCOSE). According to Moumita Deb Choudhury (2022), there are a wide variety of certifications related to Agile methods available. Sixteen of these notable certifications are as follows:

- Agile Programme Management (AgilePgM);
- Agile Project Management (AgilePM);
- Professional Scrum Master (PSM);
- Professional Scrum Product Owner (PSPO);
- Certified Agile Project Manager;
- SAFe Agilist;
- SAFe Product Owner/Product Manager;
- SAFe Scrum Master;
- Certified ScrumMaster (CSM);
- Certified Scrum Product Owner (CSPO);
- ICAgile Agile Fundamentals (ICP);
- ICAgile Agile Product Ownership (ICP-APO);
- ICAgile Agile Team Facilitation (ICP-ATF);
- ICAgile Business Agility Foundations (ICP-BAF);
- ICAgile Agile Project and Delivery Management (ICP-APM);
- PMI Agile Certified Practitioner (PMI-ACP).

and a summary of these certifications as well as their costs (as of October 4, 2022) are contained in the referenced article. There is also training and certification available from the Scrum Alliance regarding the SCRUM agile framework and provides coaching, certifications, resources, and global and regional events.

There are many methods that can be used within the Agile framework, such as Flow Analysis. This method, originated in the realm of lean product development, provides useful data for tracking and analyzing performance. According to the authors of the article, Agile Metrics: Agile Progress Monitoring, "the technique focuses on work items that go through a set of state transitions and allows the user to count items in each state and evaluate the time items spend in each state, thereby identifying possible bottlenecks and risky logjams in the process (Hayes et al. 2014)." To understand how this works, we will need to go through four states:

1. *Ready:* Here, a review process is undertaken wherein the story is evaluated by a team that assesses the user story's scope and determines if any changes, details, or clarifications are required. Simultaneously, a review involving the customer stakeholders occurs who may have either written the user story or evaluate it. At the completion of a successful review, the user story is then eligible to be included in an upcoming sprint.
2. *Coding:* The user story is first recognized as one of the most valuable items in the backlog. It then becomes included in a sprint, and the development team is in process to the user stories implementation.
3. *Testing:* The coding for the user story is complete and ready for execution. The user story is currently undergoing testing.
4. *Done:* Everything is complete. The user story has been coded and successfully tested. No further coding or testing is required.

Figure 8.3 depicts a basic Agile lifecycle. There are 30 story points in total. The graph indicates the number of tasks allocated to each timeline's four states (indicated by the different colors in the figure). According to the figure, there are 19 story points that are "Ready" and "Complete," and 16 that are in the process of coding, testing, and review currently.

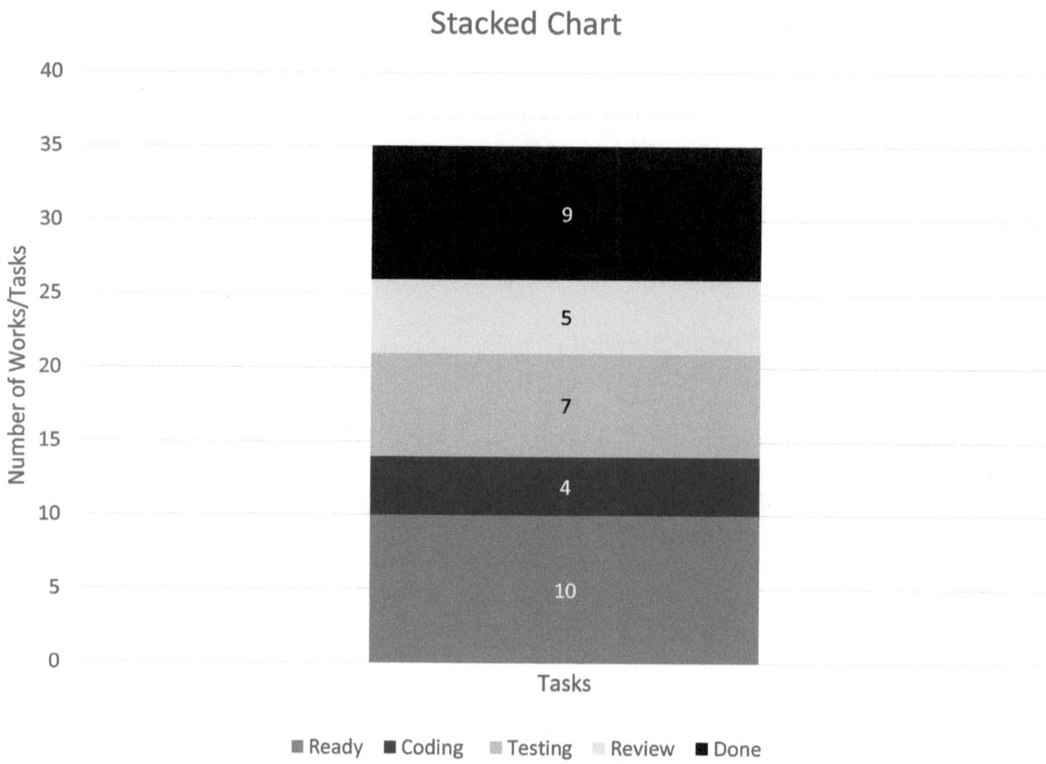

FIGURE 8.3 Project stacked column chart.

8.1.7 Implementation Methodology

Agile, like any other methodology, is not appropriate for every program, and thorough research is always necessary to choose the best approach for each individual development case. When a program's requirements are not entirely visible from the start, or are expected to evolve during the program development, an Agile program development method may be appropriate. An Agile style of development may also be a suitable fit if regular feedback is required or when close alignment of deliverables with stakeholder needs is required. Agile program management is also a good option when speed to market is critical, or when dealing with complicated, or complex programs because Agile's cross-functional, iterative approach allows teams to respond to changes quickly while resolving conflicts as they arise.

On the other hand, Agile methods may not work as intended if the stakeholder is unclear about program goals, the project manager or team is inexperienced, or they do not work well under pressure. Agile promotes developers, program teams, and stakeholder goals throughout the development process, but not necessarily the end user's experience. Agile may not always be easily absorbed within larger, more traditional business structures that can have considerable amounts of rigidity and inflexibility within processes, regulations, or teams, due to Agile's less formal and more flexible processes. An Agile methodology may also encounter issues when used with stakeholders who have similarly rigid processes or operating methods, or when used on programs that are subject to strict regulation and documentation requirements, those with fixed contracts and scope, or environments where stakeholder feedback is unavailable, or feedback participation is likely to encounter resistance.

According to Phyllis Marbach's in "Agile Systems Engineering," the Agile Systems Engineering process executed in each stage frequently consist of the following five steps (SEBOK 2024):

- "Define the highest priority and/or highest risk item to work first, keeping design options open, until the last responsible moment. This produces a list of work items with the highest work item on top. This prioritized list of work items is called a program backlog."
- "During development iterations engineers analyze the requirement, design the solutions to meet those requirements, develop their products, perform tests of that product, and demonstrate that product. For the systems engineering products these records are best retained in tools such as a requirements management tool, a model-based system engineering tool, and a configuration-controlled repository, to name a few. The development iteration is a short period of time, usually from two weeks to four weeks in duration."
- "For large products in development where multiple teams integrate their work items together to show a demonstratable product, several iterations may be needed to get to that point. This multiple iteration period is often referred to as an increment and frequently takes about three months."
- "Prior to starting an increment, all teams working to produce demonstrable products, should meet to plan their work, identify dependencies between the teams and establish commitments to meet the plan. This planning is called different things depending on the framework used by the program. Some call it the big room planning, some call it the Program Increment Planning."
- "At the end of the set of iterations the demonstrable product may be "released" to the stakeholders. Then all members of the program meet to plan the next increment of work."

These five important stages can be categorized as follows:

Forming and prioritizing the backlog: This stage begins with the highest priority task (or the task with the highest risk) being identified for processing while at the same time keeping design options open as long as possible. This results in a carefully crafted task list that commonly called the project *backlog.*

Development iterations: In this stage, the development engineers go through a thorough and complete set of activities, that include fundamental development activities such as analyzing the requirements, designing the solution, creating a product, rigorously testing the product, and presenting their work. With regard to the connection to systems engineering, some useful tools that aid in this stage are requirements management tools, configuration management tools, digital engineering tools, and model-based systems engineering tools and their associated methods. Typically, the duration of these development iterations is kept short and range from 2 to 4 weeks.

Incremental progress: There are some situations in which the development effort is quite substantial and involves the integration of effort from multiple teams to get to a useful deliverable. In this case, there may be several iterations required before reaching a useful result. When several basic development iterations are needed to produce a useful feature or capability, this is called an "increment." The typical duration of an increment is about 3 months.

Planning for increments: Prior to starting an increment, the extended collaborating teams on the increment get together to plan the work, look for and address interdependencies between the teams, and obtain commitments to the plan. The planned work must result in a demonstrable product at the end of the increment. This planning process can go with several different names, such as "big room planning," or "Program Increment Planning."

Release and future planning: Once a set of iterations are completed and the development team has demonstrable product, this can be released to the stakeholders. The stakeholders than can provide feedback to the development team. The development team then begins to strategize and plan for the next work increment.

This description summarizes the essential steps of the Agile Systems Engineering (ASE) process, highlighting the ASE methodology's iterative and collaborative nature. Throughout the systems development lifecycle, the ASE methodology emphasizes the significance of good planning, ongoing communication, and adaptation.

Kanban is one of the Agile method implementations. Kanban is a scheduling technique that has its origination in manufacturing but has now spread to other sectors such as software development, systems and program engineering. Its primary goal is to increase team or organizational productivity by reducing downtime and optimizing workflow processes. According to the authors of the article, Overview on Kanban Methodology and its Implementation, by Rajat Wakode, Laukik Raut, and Pravin Talmale, Kanban "has recently been applied in a number of manufacturing systems, assembly systems, and supply chain systems in a highly effective way" (Wakode et al. 2015). It is a work scheduling technique that increases team productivity by eliminating idle time. Idle time can exist inside any process, workflow, or procedure and is usually traceable back to opportunities within the process. Wakode, Raut, and Talmale also state, "The Kanban system focuses on reducing waste in all aspects, including overproduction, superfluous motion, faults, overprocessing, and waiting." This quotation highlights Kanban's ability in detecting and reducing various types of waste (or any activity or resource that does not directly contribute to generating value to the stakeholder) within a process. Kanban's good practices include:

- Work visualization;
- Provide limits for work in progress;
- Manage workflow;
- Ensure well-documented process policies;
- Provide feedback loops;
- Collaborative Improvement.

Dr. Ian Mitchell's Kanban board is shown in Figure 8.4. This Kanban Board contains multiple tasks. A Kanban system has several fundamental components, such as sections, task cards, Work in Progress (WIP) limits, swim lanes, visual indicators, established protocols, and a Cumulative Flow Diagram (CFD). Sections, such as "To Do" and "Completed," define the various stages of workflow. Task cards indicate work tasks, displaying details such as descriptions and deadlines, and they change horizontally across sections to represent progress. To avoid overburdening, WIP limitations limit the number of tasks that can be assigned to each segment. Swim lanes help to organize duties

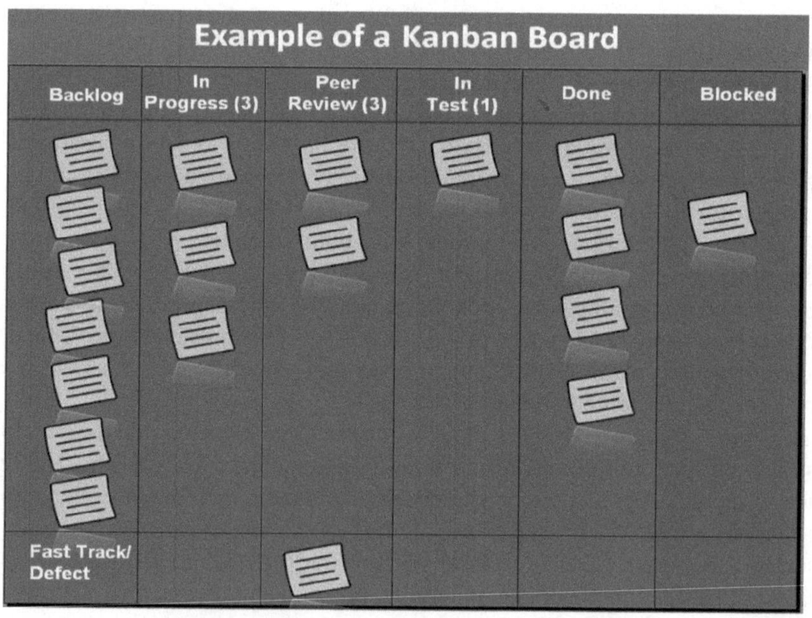

FIGURE 8.4 Kanban Board Template from Dr Ian Mitchell.

by categorizing them. Color-coding and symbols, for example, indicate additional work specifics. Established procedures and guidelines provide the rules of the workflow, promoting consistency. CFDs improve the board by graphically tracking job progression over time, allowing for trend analysis and process improvement.

Dr. Richard Turner and his colleagues at the Stevens Institute of Technology, the Naval Postgraduate School, and the University of Southern California were able to integrate systems engineering and software engineering by developing a scheduling system based on Kanban. "A general Kanban-based Scheduling System (KSS) was defined and coupled with a service-oriented approach for integrating multiple related projects with a resource pool of systems engineers" (Turner et al. 2012). This approach was modeled using several modeling tools to evaluate its expected benefits.

8.1.8 Agile Assessment Metrics

Every development effort requires a set of metrics that provide key insights on how well the effort is progressing, measures important program attributes (e.g., cost, schedule, and defects) and provides assessments on productivity throughout the system's lifecycle phases. These metrics aid in tracking and controlling the quality of a system, as well as assessing team performance. Program completion is not the only goal. As systems engineers, our goal is to provide high-quality systems on schedule and within budget. Accordingly, there are key performance indicators (KPIs) that map to a program's goals and include the success criteria for the developmental effort. For an Agile development method, some useful KPIs are (Radigan 2024) as follows:

- *Sprint burndown:* Dan Radigan, in his article, *Five Agile KPI Metrics you won't hate*, states, "Scrum teams organize development into time boxed sprints. At the outset of the sprint, the team forecasts how much work they can complete during a sprint. A sprint burndown report then tracks the completion of work throughout the sprint. The x-axis represents time, and the y-axis refers to the amount of work left to complete, measured in either story points or hours. The goal is to have all the forecasted work completed by the end of the sprint (Radigan 2024)." Development tasks are arranged using the Scrum method into distinct, time-boxed units called sprints. The development team estimates the amount of work they can finish in the allotted time at the beginning of each sprint. Teams use a sprint burndown report, which shows time on the x-axis and remaining work, measured in story points or hours, on the y-axis, to monitor their progress. Ensuring that all the planned tasks are completed before the end of the sprint is the main goal.

  Below we see a Burndown Chart that visually shows the progression of tasks over time. Created in MS Excel, the Burndown Chart indicates the completion rate of the work indicated by the reduction of the remaining tasks that need to be accomplished. The vertical axis shows the number of remaining tasks, whereas the horizontal axis shows the time within a sprint. As the chart shows, on the first day, there are 30 tasks that need to be completed. As time in the sprint progresses, these tasks are completed, and the remaining tasks are reduced by the completed amount. The process of completing these tasks is known as "burn down." The dashed line shows the linear fit to the reduction in tasks over time. Gridlines are used to make it easier to read and interpret the chart. Trend analysis is used on these charts to determine how well the development team compares to the planned burndown rate. If there are significant deviations from the planned progress, this can be easily seen from the chart. Note that the development team can also predict its completion date for the sprint by projecting the work rate (assuming it doesn't deviate too much daily). Note that in this Burndown Chart, the team started off slow and then sped up toward the middle of the sprint. Toward the end of the sprint, the team slowed down again (Figure 8.5).
- *Velocity:* In the article *Agile Metrics: Progress Monitoring of Agile Contractors* by Will Hayes, Suzanne Miller, Mary Ann Lapham, Eileen Wrubel, and Timothy Chick, Velocity

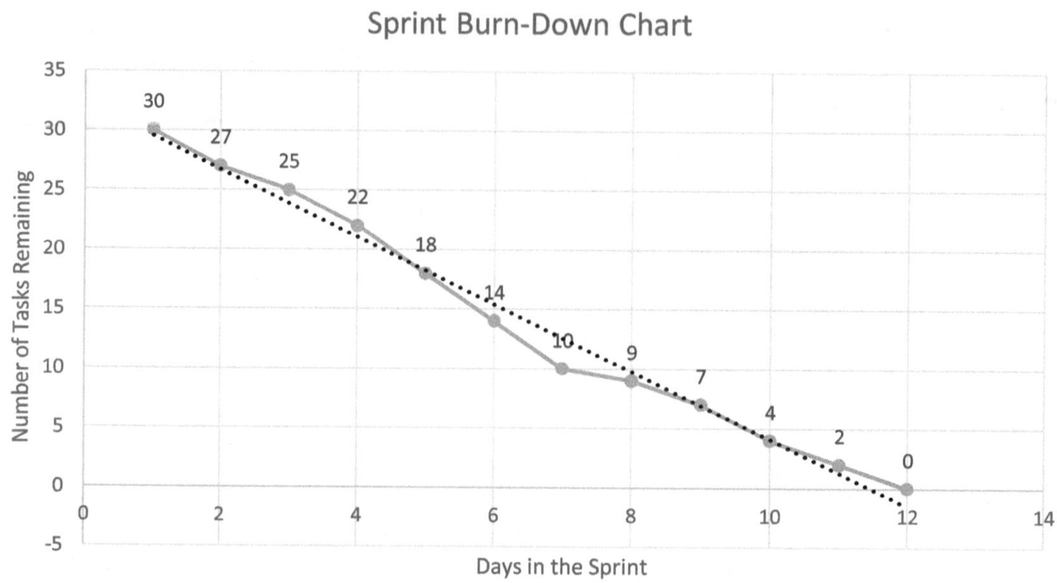

FIGURE 8.5 Sprint burn-down chart.

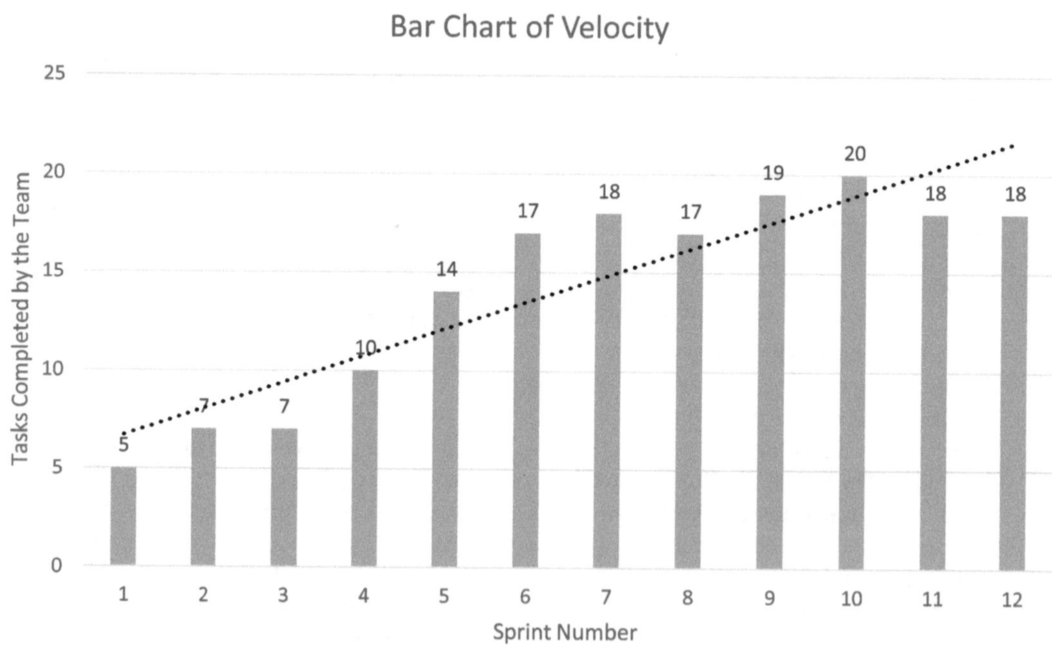

FIGURE 8.6 Bar chart of velocity.

is explained as "the volume of work accomplished in a specified period, by a given team. Typically, this is measured as story points accomplished per sprint" (Hayes et al. 2014). The velocity metric relates to how fast a team can burn down a backlog. This metric is useful in helping a project manager keep track of the project backlog and establish project timelines. This metric needs to be monitored to track how the velocity evolves over time. Figure 8.6 shows a bar chart that provides a representative development teams velocity over several sprints. The development team improved over the course of the sprints, starting with only five tasks completed in the first sprint, and increasing to the neighborhood of 18 tasks for later sprints. Notice that the improvements were gradual. Various reasons could account for this task completion pattern, such as adjusting to being on a newly formed team, with new processes and procedures, or learning to navigate the waters of self-directed work. Getting used to working with user stories and sticking to established completion criteria can also take some getting used to. Velocity is a lagging indicator which means that it is based on work that was done in the past. Predictions of upcoming spikes or dips in activity due to change in scope are not provided by the velocity metric. However, predictions based on past data, with repetitive deterministic and quantifiable data are possible.

- *Release burn-up chart*: "A complementary graphical technique for the sprint burn-down, the release burn-up chart is also commonly used. Many cling to the notion that sprints burn down, and releases burn up — though there is no mathematical principle that governs this choice. With each completed sprint, the delivered functionality grows, and the release burn-up chart depicts this progress in an intuitively logical fashion." (Hayes et al. 2014).
- This view suggests that the release burn-up chart (a useful tool in Agile project management) demonstrates progress as additional features and capabilities are supplied after the completion of period of time, or number of sprints. This provides for a monotonically increasing function that reflects the completion of tasks at the end of each time period or completed features/capabilities at the completion of each sprint. When using the number of sprints as the horizontal axis on a release burn-up chart, the chart provides a clear and logical way to visually represent the growth of implemented features over time. Its purpose is to allow teams and stakeholders to assess the progress of the project toward its release goals by visually representing the gradual growth of features, capabilities, or user stories as sprints are completed. This chart is very beneficial for tracking progress over the whole period of the project and ensuring that the team stays on course in achieving the desired results for the release.

If the burn-up chart uses completed tasks (instead of completed features or capabilities) on the vertical axis to illustrate the cumulative value delivered, then the x-axis is plotted in units of time, such as in days to show tasks completed over days. As the line progresses upward diagonally from the lower left to the higher right, it signifies the progress achieved. The graph includes a line at the highest point, which represents the cumulative story points planned for the period of time. This line helps to illustrate how the project is in alignment with its scope and any development changes made along the way. The release burn-up chart provides clear insight how the project is advancing, captures the impact of scope changes, and helps project managers make data driven decisions.

The release burn-up chart is a dynamic visual representation that shows the progress of a project toward completion. The blue line in Figure 8.7 shows a representative total number of tasks that a team aims to complete. This includes all the tasks within the project scope, which is 200 tasks or 100% of the whole 2-week development effort. Figure 8.7 also shows the estimated trajectory as the linear dashed line (yellow) that is obtained from the initial point of no tasks completed to all 200 tasks completed. The orange line shows how many tasks are completed at the end of each day, and the grey line shows the total number of completed tasks. The team's main goal is to complete all these tasks within a specified 2-week period. During the early phases, progress seems to be sluggish, with only a small number of tasks counted as finished. However, with time, we can see a consistent increase in the number of completed tasks. As the development team approaches the completion of the 200 tasks, it is expected that the rate at which they finish tasks would slow down, indicating that they are getting closer to their goal. The key advantage of this graphical representation is its ability to offer a clear and distinctive portrayal of the project's progress throughout its developmental evolution. It provides project management with a necessary tool to create accurate estimates about the team's capacity to meet deadlines. Project managers can assess the need for adjusting or increasing resources to maintain momentum in achieving the project's stated goals. This chart is a great tool for managing and coordinating project timelines and reaching targeted project goals.

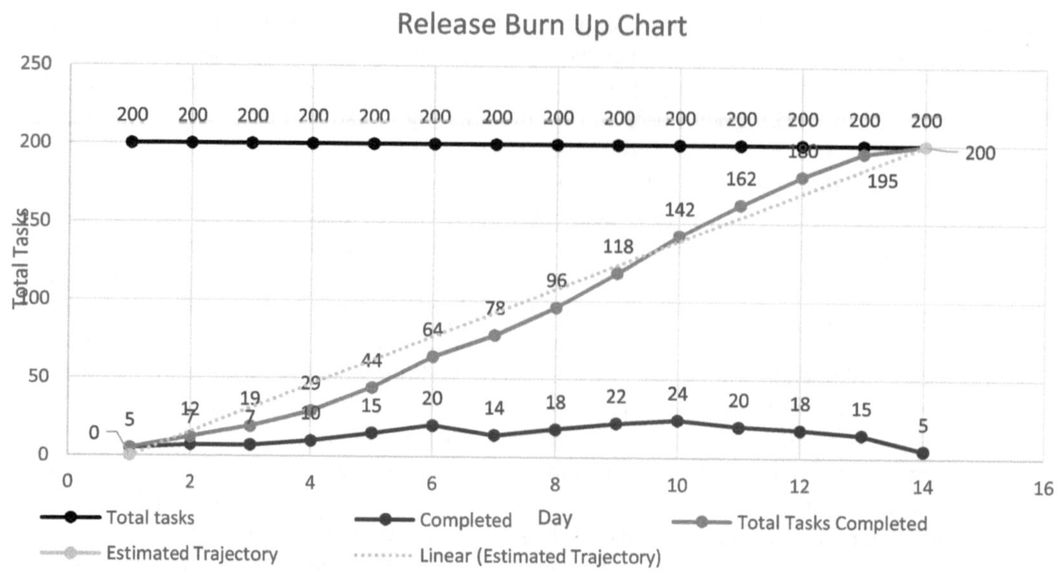

FIGURE 8.7 Release burn-up chart in MS Excel.

Finally, the release burn-up chart is a useful tool in Agile project management since it provides a visual representation of progress toward a project's release targets. It assists teams and stakeholders in tracking scope changes, evaluating progress, and making educated decisions to ensure the project's successful completion. By combining data visualization and historical analysis to troubleshoot problems and design effective solutions, agile metrics provide several benefits, including:

- Determining velocity trends;
- Managing unexpected tasks;
- Effectively handling defects;
- Resolving lags in velocity, and
- Strategically addressing activity spikes.

8.1.9 Agile Scrum Artifacts

Chandler Harris noted in "Agile Scrum Artifacts" that, "Agile scrum artifacts are information that a scrum team and stakeholders use to detail the product being built, actions to produce it, and actions executed during the project. These artifacts include metadata points that provide insight into the performance of a sprint (Harris 2024)."

Agile Scrum artifacts serve as information repositories on which the Scrum team and stakeholders rely to elaborate on the program's development, define the procedures needed to build it, and document program activities. These artifacts, which serve as metadata references, lead to a better understanding of sprint performance. The primary Agile artifacts are the product backlog, sprint backlog, and increments. The artifacts are created during the key activities of a scrum sprint which consist of (Harris 2024):

- Plan work and future goals;
- Create tasks to achieve these goals;
- Organize tasks into sprints based on dependency and priority;"
- Execute the tasks;
- Review and analyze results in order to compare to the goals;
- Repeat these steps.

8.1.10 Applicability of Agile Methods

Agile methods can make a powerful contribution to the success of applicable development efforts. But what do we mean by applicable development efforts? Not all development efforts benefit from adopting an Agile development approach. In fact, under certain conditions, Agile methods can interfere with and hinder the overall success of a systems development effort. So then, what conditions do we need to have in place so that an Agile approach can be expected to provide an overall beneficial result in the systems development effort?

Incremental systems development: The systems development effort must be able to be broken into parallel development efforts, wherein largely independent teams can work in parallel to provide functionality that is integrated at the end of a sprint to provide progressive, incremental capability/functionality. Not all projects/systems development efforts allow for this kind of division of work. The idea is somewhat akin to artistic painting, where initially broad outlines give way to ever increasing detail as the painter proceeds throughout the work. The systems development effort must be able to provide useful results to the stakeholders early on and be structured in such a way so that subsequent sprints add value, functionality, and/or capability through a parallel development process.

Short incremental development (sprint) timelines: The incremental developments need to come in short, rapid, fixed sprints that provide useful functionality/capability. The focus of the sprints is not documentation, but producing a useful function or capability that has demonstrable utility. Functions/capabilities are improved in subsequent short sprints until the final satisfactory result is achieved. Consequently, the function/capability must be able to be evolved. Systems that cannot be incrementally developed in this fashion are not a good fit for the Agile methodology.

Continuous involvement of stakeholders: Since there are useful deliverables at the end of each sprint, the stakeholders have opportunity to provide feedback, guidance, and have input on the evolution of the design. The emphasis is on sprints with short intervals with "something useful to deliver" at the end of the sprint. The stakeholders can evaluate the deliverable and provide useful feedback to the design team and influence the evolution of the design itself. In this fashion, the stakeholders become continuously involved and can provide heading checks, operational advice, and continual guidance throughout the design and development process. This condition becomes even more important and downright necessary if there is volatility in the requirements or if the requirements initially are incomplete, ill-defined, or perhaps even unknown.

Time boxing of the effort: Sprints themselves are given a short, set amount of time in which to perform the work. The idea is to take smaller, incremental steps to reduce risk. It is sometimes easier to add smaller incremental capabilities that progressively lead to what the stakeholders require, than to go for the entire capability all at once. The systems development effort needs to be capable of being split into a series of smaller efforts that each fit into a fixed *time box* and produce a useful output that can be built upon to eventually realize the full required capability.

Self-organizing teams: Teams work on parallel efforts that come together to produce functionality/capability. The teams are allowed to self-organize to take advantage of team strengths and optimally develop the work products of the sprint, whether that is a new function/capability, or burn down of tasks from previous sprints. Teams must be capable of working in self-organizing structure and independently from the other teams.

Excellent internal and external communications: Since there is a large degree of independent work, teams and team members must have excellent communications both inside the team among the individual team members, and between teams. If this condition is not satisfied, such as in highly siloed industries or organizational structures, then great harm can result, and the Agile approach may fail altogether. The consequence could be a design decision made in one team that is not understood and correctly incorporated by another team resulting in incompatible work products and broken functionality or loss of capability.

Rapid change response: The Agile environment is dynamic, and the amount of feedback from stakeholders is increased, and changes to the design are to be expected. These changes must be rapidly addressed and integrated into the design and development effort. Sometimes, the development schedule itself is a primary driver such as short time to market projects like competitive, high-technology projects. A good example is the latest cell phone technology that has time to market pressures and has a fixed viable market lifespan (until the next cell phone technology is introduced by a competitor). Organizations must be capable of having processes in place to facilitate rapid change for Agile methods to be effective.

These are the essential conditions that must be in place to benefit from an Agile approach. The further the departure from these conditions, the higher the risk of failure. It should also be understood that just because Agile methods work for one organization, which does not mean other organizations will have similar results. Each organization has different characteristics and makes unique decisions, and so results can be different. However, the conditions above represent fundamental considerations when evaluating whether the adoption of Agile methods are likely to provide beneficial results. Note also that the adoption of Agile methods does not mitigate issues such as staffing problems and manufacturing delays. Neither does it aid in establishing scheduling priorities. Since Agile prioritizes individuals and interactions over processes and tools, process errors such as missed inputs, checks, approvals, and imprecision can manifest, especially over longer sprints. The priority is also placed on working software/systems over comprehensive documentation which increases program risk in solving downstream errors, or in effectively handling personnel turnover, where this documentation is generally required. The Agile approach favors stakeholder collaboration over what is negotiated in the contract. This is fine if and while there is a friendly and cooperative relationship between the stakeholders and the systems development team. However, if disagreements occur, it is the contract that settles the differences. It is risky to accept unfavorable contract language or relying on in situ contract negotiations for the sake of favoring stakeholder collaboration. Finally, having the ability to rapidly adapt to change favored over following a plan may in some instances introduce risk. This Agile concept can introduce technical, cost, and schedule risk if the introduced changes collectively cause deviations from the expected schedule, costs, and/or long-term technical performance of the system. Without the plan, these deviations may be difficult to spot and correct in a timely and effective manner. Note that these risks increase if the conditions for successfully employing Agile methods are not met. For instance, if there are bad communications between the Agile teams, and there is no emphasized standard plan, then changes made within an Agile team run the risk of not being compatible with changes made in other Agile teams, resulting in incompatibilities and errors during sprint integration efforts. Without an emphasized plan to boot, sorting out what needs to be changed and what direction the collective effort should take can be difficult.

Extension to mixed hardware and software systems and lessons learned: As mentioned earlier, Agile methods came from the software development world for software-based, or software-dominant systems. So, what about the applicability of Agile methods for mixed hardware and software systems, or hardware-dominant systems? Do Agile methods still apply in these cases? The answer depends in part on the developmental organizations' ability to adapt to change, its understanding and proficiency in implementing Agile methods, and the proper conditions just mentioned being in place for the development effort. Both good and bad examples can be found in the literature in applying Agile methods to other than software only development efforts. A good litmus test is to look at the application of Agile methods to Department of Defense programs and projects since these are general systems development oriented and not just software-centric systems. We will look at a representative DoD example that used Agile methods, present some lessons learned, and then make some general observations in connecting Agile methods to large, complex development efforts like in the DoD.

Agile methods and systems engineering: Although, at first blush it doesn't appear so, yet there is compatibility between Agile methods and systems engineering (Allen 2016). One of the key aspects of Agile is iteration. Iteration can already be supported in Systems Engineering with spiral development processes. Larger, linear efforts, with deliberate intent, can be broken into smaller spiral evolutions, each with clear exit criteria. If planning and assessment stages are added to the spiral development, then this looks very much like a sprint in the Agile methodology. Parallelization is already inherent in the general SE process. For instance, different subsystems are largely, independently developed in parallel as part of the normal systems engineering development process. If subsystems can be organized with regard to providing functionality, and requirements generation organized around the provided functionality, then the requirements generation process itself could operate very much like in an Agile development. If the requirements themselves are expected to be volatile, then the Agile methodology offers advantages over the traditional development methodology. According to the *INCOSE Systems Engineering Handbook* (4th Edition),

> The value proposition of an agile SE process is risk management, appropriate when development speed and customer satisfaction are likely to be affected by requirements understandings that evolve during system development.

The relevance and importance of Agile in Systems Engineering has been recognized and addressed in adopted standards. For instance, Section 9.9 of the current INCOSE Systems Engineering Handbook (4th Edition) is entitled, "Agile Systems Engineering" and provides relevant information on Agile in systems engineering including a discussion of the connection of agile to architectural frameworks and metrics. The emphasis on incrementally developing requirements provides for the capability to adapt to changes in mission, scope, or technology, while the corresponding incremental design provides focus on evolving interfaces and introduces some adaptability and resilience into the design process. Some of the Agile metrics such as the burndown rate, and the backlog itself can be useful in assessing program risk. One area that requires some reconciliation is integrating the Agile methodology with the Earned Value Management (EVM) system. The EVM system is used in the DoD to measure and track cost, schedule, and technical performance. This assessment system requires good cost estimation methods and consequently will be addressed in more detail in the lifecycle cost estimation chapter later where it is more directly applicable. With regard to Agile, there is compatibility between the EVM system and Agile methods, but some adaptations are required. According to John S. McGregor, the Deputy Director, for Earned Value Management Performance Assessments and Root Cause Analyses at the Office of the Under Secretary of Defense for Acquisition, Technology, and Logistics (McGregor 2016),

> Agile and EVM are complementary when properly implemented together and help enable a robust overall management process. In order to be effective, Agile must be evaluated for its applicability on a program-specific basis and tailored to best align with programmatic and contractual requirements.

A formal EVM system is required for DOD acquisitions systems exceeding \$20M and require formal validation in accordance with the EVM standard ANSI/EIA-748 for DOD acquisitions greater than \$50M. According to John Allen, senior systems engineer with Zenetex, the EVM system, "tends to fail if costs cannot be predicted with high confidence" (Allen 2016). This can result in multiple adjustments to the acquisition budget (re-baselining), which is highly undesirable. Agile methods can complement the EVM system required by the DOD; however the statements of work for the development effort need to be feature-oriented and capabilities-driven. In Agile, system decomposition is accomplished by decomposing strategic initiatives (themes), which are the largest unit of work in Agile, into Features, then decomposing the Features into Epics, and finally decomposing the Epics into Stories. Both EVM and Agile metrics are useful and complementary and must be incorporated in the Integrated Master Schedule (IMS) for the systems development effort. Spiral, sprint-based planning windows should be accomplished with corresponding frequent updating and baselining of the system along with necessary systems documentation. Allen has the following interesting recommendations with regard to measuring performance in Agile Systems Engineering development efforts:

1. Progress should be tied to scope completion;
2. When reporting performance status, "done" is defined as all or nothing;
3. Dollar amounts should be attached to work/metrics;
4. Agile methods are used to provide more accurate Estimate to Complete (ETC) as work progresses.

EVM system compliance needs to be maintained and so standard terminology needs to be used. Agile methods and metrics can be used to derive and support EVM deliverables and should be set to the same granularity as the reporting level of the EVM. Also, EVM metrics should trace back to Agile metrics.

Agile and testing in systems engineering: Agile methods are also relatable to testing in systems engineering development efforts. A notable Agile testing concept is that of Test-Driven Development (TDD). In TDD, the idea is to avoid defects as opposed to defect identification. The idea behind TDD is to develop, and apply, test cases as the system is developed with corresponding demonstrations that the system evolution is correct. This approach is applicable to complex system development efforts. Small changes are introduced to a developing Use Case during the nanocycle (smallest timescale of evolution; roughly 20–60 minutes). Desired behaviors of the system are identified for the small incremental changes and test cases are generated and executed to demonstrate the correctness of the incremental change. The process is continued until all requirements for the Use Case, and all scenarios have been satisfied. Other similar testing concepts are Acceptance Test Driven Development (ATDD) wherein the TDD unit-level testing concepts are extended to the acceptance testing level. Behavior Driven Development (BDD) is an Agile method where the design, development, and documentation of a system are based on the behavior of the stakeholder. *Exploratory testing*, wherein testers are allowed to freewheel and follow their testing inclinations, is also an effective Agile testing method. Like exploratory testing, *structured testing* has a structure or framework that guides the testing effort. *Session-based* testing is a structured testing and time-based testing method that provides more structure to exploratory testing methods. Agile testing methods follow the best practices from Agile development methods such as continuous testing and supports DevOps. To develop an Agile test strategy, often the test plan is intentionally left incomplete and only partially documented for the sake of adaptability and flexibility. The Agile Test Strategy (ATS) should include a *purpose* which often come in the form of a user story. The ATS requires *objectives* which often take the form of the test cases themselves. There must also be a well-defined *scope* in the ATS which describes what is to be tested. The *methods* that are used in the testing process that describe how the tests are conducted are also part of the ATS. There is a high degree of communication between Agile testers and system developers to make sure that the system development effort has proper requirements verification and proper testing. Benefits of Agile testing include better management of scope, shorter testing cycles and faster, more regular release dates, and improved quality. This testing methodology is also capable of finding and fixing errors faster than traditional testing methods due to the shorter development times of the sprints and the emphasis on continuous testing.

Best practices in Agile systems engineering (ASE): Given the right conditions, we have indicated that the use of Agile methods in systems engineering can be quite beneficial and applicable to software-dominant, hardware-dominant, or mixed systems development efforts. Some of the *best practices* of ASE include the ability to produce *high-fidelity models* in support of model-based engineering (MBE) efforts. Having Use Case-driven *incremental functional analysis* is a best practice wherein functionality is analyzed incrementally. This approach results in a more manageable and understandable decomposition which facilitates more accurate and efficient modeling. Also, this fits hand-in-glove with the best practice of using a *test-driven development approach* in producing the necessary documentation/models such as the system specification. Requirements generation, functional analysis, and incremental, continuous testing effort are accomplished over shorter development time scales. *Incremental traceability* is a best practice that stems directly from the previous incremental best practices. As requirements, functions/features, and/or test elements are incrementally developed, they can also be incrementally traced to each other (e.g., requirements to test, test to functions, requirements to functions, and so on). It is a best practice to use the Agile metrics, and insights gained from the Agile method in the *risk management* program for the systems development effort. Another best practice is to conduct *integrated safety and reliability analysis.* These can be done in conjunction with the incremental functional analysis activity. As functional block diagrams are incrementally generated, safety and reliability analysis can be accomplished in parallel. Two other best practices are to accomplish a *model-based handoff* to downstream engineering efforts and *use model artifacts to automatically generate documents.* Some effort is involved in setting up these capabilities, but the results are well worth it.

8.1.10.1 Agile Example with Lessons Learned: Implementing Agile on Defense Systems and Hardware

This case study illustrates General Dynamics' proactive approach to implementing Agile methods in the development of mixed systems, encompassing both software and hardware components (Smith 2019). Traditionally, only software systems were considered as applicable candidates for Agile methods. However, with the approval of the Department of Defense (DoD) for a project with low risk, General Dynamics was authorized to proceed with their Agile approach, aiming to replace contractual milestone reviews with a more flexible approach.

One of the many challenges that General Dynamics encountered was the assumption that Agile methods "assume a low cost of change." This was a natural assumption since Agile methods stemmed from the software community where changes, compared to hardware changes, were much cheaper. To overcome this difficulty, GD implemented Model-Based Engineering (MBE) tools and methodologies, considering the models as an initial version of the system. This had a number of down-stream advantages. The hardware and software engineers focused on enhancing the models instead of using them as requirements inputs for their set of processes. Additionally, during the Preliminary Design Review (PDR), they presented their models instead of generating PowerPoint slides. The study asserts that this led to a significant increase in stakeholder feedback, which served as an indicator of a superior quality review.

This project used a traditional waterfall methodology. The customer acquisition process used a systems engineering (waterfall) methodology, which GD had experience in handling. The challenge involved achieving standard traditional systems engineering milestones such as SRR, PDR, and CDR, while at the same time using an Agile methodology. The SRR used an Agile methodology for the system engineering efforts, dividing tasks into ½–3-day intervals and conducting scrums. This was successful, but there was a discrepancy between the Earned Value (EV) measures and the Agile Scrum estimates to completion. The EV measurements indicated a delayed finish, while the Agile Scrum projections showed an on-time finish. The Agile estimations proved to be accurate. The team ascribes this to the greater overall productivity of Agile Processing and claimed that the EV (waterfall) and Agile estimation methods are incompatible. Ultimately, this difference resulted in the realization that the Agile stories were centered around specific tasks rather than being focused on values. As a result, the teams had to perform a "backlog rework" in which they realigned the stories to focus on high-value items for large projects. As a result, the duration of the sprints was extended, but there was a notable decrease in mission creep. The addition of extra teams to the project resulted in dependencies becoming an issue. This was resolved with the implementation of a quarterly meeting, during which the teams were able to discuss and address potential dependencies and associated risks.

The most challenging aspect was the integration of the hardware team into the process. They faced obstacles, such as resistance from the team, lead time issues, and the lack of control over service companies like drafting. To address these issues, GD incorporated these uncertainties into their planning process and implemented lengthier sprints lasting 3 weeks. Ultimately, it was discovered that this procedure significantly enhanced the hardware team's ability to detect Long Lead items and improved coordination with the software teams. The team was taken aback by the level of collaboration required with the software teams.

Although commendable, the attempt to integrate Agile methods into a project involving many disciplines and mixed systems (hardware and software), certain aspects that were purportedly addressed by the Agile method should have already been included in their initial program planning efforts. In systems development efforts, it is standard practice to incorporate the identification of Long Lead material/items and find ways to start software initiatives early. Further, several items that were attributed as benefits from the Agile approach were a result of optimizing GD processes. For example, the need for using systems engineering to generate an architecture diagram, which would then be further improved by the next discipline. This was achieved by ensuring that all disciplines used a common tool instead of having each discipline choose a tool of their own. This could have been recognized as a process improvement activity irrespective of using the Agile methodlogy. However, the implementation of Agile methods was the catalyst that brought about the need for this change.

The Agile approach to the systems and hardware design phases was highly effective and well-received by the Department of Defense (DoD). Consequently, General Dynamics is undertaking a complete transformation to this Agile methodology for their Defense Systems. GD recorded 12 lessons learned during the implementation of this Case Study, and they are provided here in their entirety:

1. "*Lesson one*: Agile Systems Engineering provides more accurate predictions for near-term work."
2. "*Lesson two*: Complex backlog functional decomposition and system design may require systems engineers."
3. "*Lesson three*: Teams developing either software or hardware can be functionally cohesive."
4. "*Lesson four*: Agile requires a balance when defining EV packages."
5. "*Lesson five*: This customer appreciates reviewing live work products."
6. "*Lesson six*: Some systems work products require longer sprints (3–4 week) to produce (without splash)."
7. "*Lesson seven*: Stories in the systems domain often represent deliveries to the next development stage, rarely directly to the users themselves."
8. "*Lesson eight*: Including systems engineers on cross-functional software teams can support functional decomposition and verification for that team."
9. "*Lesson nine*: Shared services do not work for agile teams. Invest the time to get specific people assigned to your agile team, even if the person can only work part-time for your agile team."
10. "*Lesson ten*: Working together, systems and hardware engineering can learn to define system design increments (or "slices") faster, reducing long lead times."
11. "*Lesson eleven*: In addition to analyzing system behavior for implementation by software teams, Systems efforts should include "system design analysis" to ensure hardware is in place to support software deployment."
12. "*Lesson twelve*: Our need for external review has not changed. By using actual work products, we still accommodate external review, while spending less time on decks and more time on product development."

We now present some additional case studies that illustrate some of the concepts, principles, and methods we presented above. We have both a national and international case study as examples.

8.1.11 Case Studies

There are a number of useful case studies demonstrating how the Agile method is used in systems/programs. We will give two case studies and their lessons learned for actual Agile system engineering development efforts below. We use as reference the publication, "Review of Agile Case Studies for Applicability to Aircraft Systems Integration," by Robert Carlson and Dr. Richard Turner (2013).

8.1.11.1 Johns Hopkins Multi-Mission Bus Demonstrator (MMBD) CubeSat

Description: The case study of the Johns Hopkins Multi-Mission Bus Demonstrator (MMBD) CubeSat investigates the use of Agile methods in the creation and operation of CubeSat missions. CubeSats are small, low-cost satellites that are used for scientific and technological demonstrations. This case study looks at how Agile principles such as iterative development, stakeholder collaboration, and adaptability may be used to improve the success of CubeSat missions. The MMBD project aims to improve the flexibility, reactivity, and efficiency of satellite development and mission execution through the use of Agile methods, thereby enhancing the capabilities and success rates of CubeSat missions in the fast-expanding area of space exploration.

Background: The case study of the Johns Hopkins Multi-Mission Bus Demonstrator (MMBD) CubeSat provides insight into a key CubeSat project done by the Johns Hopkins University Applied Physics Laboratory (APL). This project aimed to demonstrate the adaptability and versatility of CubeSat platforms for a variety of space missions. CubeSats are small, standardized spacecraft noted for their low cost and flexibility for a wide range of applications. The MMBD project, which began with the goal of demonstrating advanced technologies and capabilities in space science and exploration, centered on the development of a flexible and robust CubeSat bus that could be customized to support scientific, communication, or technology demonstration missions.

The use of Agile methods in the development of the MMBD CubeSat was an essential feature of this case study. The project was deemed technologically advanced because it included new, unproven concepts and system components such as the bus structure and solar array deployment mechanism, all the while meeting NASA's stringent quality assurance and dependability standards.

The case study highlights an exceptional achievement: Johns Hopkins University APL estimated that they were able to create their CubeSat for around 3% of the cost (equal to $10 million) and within 25% of the average schedule (14 months) for a standard satellite project. This was a substantial difference from the traditional satellite development processes, establishing the MMBD CubeSat project as a forerunner in space technology efficiency and creativity (Huang et al. 2012).

Connection to systems engineering (systems test and evaluation): The case study of the Johns Hopkins Multi-Mission Bus Demonstrator (MMBD) CubeSat emphasizes the significance of System Test and Evaluation in complex projects. The MMBD effort placed a heavy emphasis on rigorous evaluation of individual components prior to their integration by using Agile methods and step-by-step testing. This method aided in the early detection and correction of flaws, fitting well with the fundamental concepts of System Test and Evaluation, which emphasize the full evaluation of a system's operational capabilities and overall performance. The MMBD CubeSat project's success in demonstrating cutting-edge space technologies can be attributed in part to its systematic testing processes, emphasizing the critical role of System Test and Evaluation in ensuring the dependability and operational effectiveness of modern systems.

Lesson learned: The Johns Hopkins Multi-Mission Bus Demonstrator (MMBD) CubeSat case study provided important insights into the Agile development of contemporary systems similar to their small, CubeSat space systems development. The Case Study placed an emphasis on the effectiveness of an organizational structure that had been flattened so that functional leaders report to a single empowered program manager. By co-locating teams, the number of necessary meetings was reduced while also fostering effective collaboration and adaptable processes. Early problem resolution and methodical progress were facilitated by incremental testing, which involves evaluating individual components before integrating them. The MMBD CubeSat project is a great example of how Agile methods may improve the development of systems by placing an emphasis on adaptability, efficiency, and the early discovery of problems (Carlson and Turner 2013).

8.1.11.2 Nokia Agile Requirements Development

Description: The Agile Requirements Development case study at Nokia investigates the application of Agile methods in the domain of software and requirements development. This study examines the transition of Nokia from traditional and rigid requirement processes to Agile methods, with an emphasis on collaborative efforts, flexibility, and promptness. The implementation of Agile methods by Nokia was driven by the company's desire to enhance the efficiency of communication, align product development with stakeholder requirements more precisely, and bolster the organization's ability to respond to changing market demands. This case study highlights the benefits and challenges that are linked to the integration of Agile requirements development in a sizable corporate organization such as Nokia. It emphasizes the importance of customer engagement, iterative approaches, and cross-functional collaboration in achieving positive outcomes in industrial settings with swiftly changing technologies.

Background: Nokia intended to shorten their development cycle and increase their agility in responding to changing client demands. They implemented the Scrum methodology as the development approach for addressing stakeholder needs across two divisions of over 500 phone software developers each. Scrum, the most popular Agile method, employs sprints, which are time-boxed development iterations that typically span 30 days. The team maintains a backlog that prioritizes current requirements, and daily meetings are held to guarantee progress reporting and work allocation among team members. The obstacles and issues that Nokia faced, had scalability as being a major concern. They were confronted with the impracticality of managing a staff of 500 people. They established a master backlog file and dispersed it across other team-specific files as a solution. To manage this process, a higher-level scrum team was formed to handle the master backlog file and requirement allocation. Furthermore, Nokia struggled with adequate documentation and communication of choices to other design teams, which caused delays in disseminating architectural changes to affected teams. For example, an increase in battery life required a decrease in power use, which took some time to spread throughout the development team.

Another source of worry was Nokia's use of "frameworks" to generalize the behavior of its product lines. These frameworks did not directly handle user requirements, but instead interacted with programs that met those criteria, which Nokia referred to as derived requirements. The difficulty resulted from differences in the functional scope of these frameworks as well as their proximity to the user experience. Changes to a framework were a complex effort that required a detailed consideration of potential consequences on all other product line members because each framework supported distinct product lines. Due to the complexities of framework implementation, the task was delegated to experienced and senior-level professionals capable of performing in-depth examinations of abstracted needs and evaluating the effects of changing the framework (Savolainen et al. 2010).

Nokia launched several initiatives to improve their Agile framework, with the goal of improving development efficiency and quickly adjusting to changing stakeholder expectations. These changes were made to address typical difficulties connected with Agile methods. Here is an overview of Nokia's activities and their expected outcomes:

1. *Improved scalability*: Nokia had to deal with the difficulty of aligning the activities of two divisions of 500 software developers each. This was dealt with by developing a scrum team that would be in control of a master backlog repository and would allocate tasks to smaller teams. The goal was to improve coordination and provide for effective management of the large workforce.

Result: These organizational changes resulted in better synchronization between the teams and more efficient allocation of tasks across the developing units. This was expected to produce higher productivity, and faster development cycles.

2. *Better communication and documentation*: The importance of proper documentation and effective communications between the design teams was not lost on Nokia. Nokia developed procedures to improve their documentation ensuring that changes to their architecture and decisions were well-documented and well-communicated to the development team members.

   Result: Because of better communications and documentation, the development team experienced improved decision-making with less misunderstandings. Additionally, architectural changes were assimilated faster throughout the entire development team
3. *Optimized architectural changes*: Delays in implementing architectural changes were addressed by Nokia by developing streamlined execution processes and effective communications.

   Result: Architectural changes were implemented faster and the stakeholder response time to evolving stakeholder requirements was improved with a more agile and responsive workflow.
4. *Responsiveness and adaptability were elevated*: The process of Scrum was adopted and used as the primary development method for establishing stakeholder requirements. Additionally, daily meetings were established for better communications and progress updates and to allocate tasks. The intent of Nokia was to become more responsive and agile in addressing stakeholder needs.

   Result: Nokia became more adaptable and responsive to stakeholder needs. This allowed for the provided features to the stakeholder being better aligned with the stakeholder's expectations and quickly changing market dynamics.

Connection to systems engineering: Systems engineering, and the Nokia Agile Needs Development case study are related because they both emphasize iterative and incremental development, stakeholder participation, frequent testing, flexibility in the face of shifting requirements, and a dedication to continuous improvement. Companies like Nokia can better adapt their ever-changing requirements of their software development projects by incorporating Agile methods into their requirements generation processes.

Lesson learned: The major takeaway from Nokia's Agile Requirements Development experience is that while Agile methods can scale effectively, they require a higher-level scrum or coordination mechanism to manage job distribution efficiently. The use of a master backlog file in this context might occasionally result in bottlenecks, necessitating careful schedule integration and management. Furthermore, architectural decisions must be prioritized and documented to ensure they are transmitted down the development chain. This strategy helps to avoid investing resources in areas that may change in the future. Furthermore, Nokia's experience highlights the value of having more seasoned and highly skilled personnel on the development team. This is because Agile processes entail breaking down requirements into physical and logical technical specifications, which requires technical expertise and a thorough understanding of the whole system architecture to ensure successful development and integration (Carlson and Turner 2013).

8.1.12 Summary

Agile project management is a philosophy and development methodology that was initially founded in the 1970s for the development of software systems. Agile developments effectively increase stakeholder satisfaction through faster release, and incremental capability releases. Agile methods have seen widespread adoption throughout various industries. For modern Agile development efforts, the 12 prominent Agile principles that were articulated in the Agile Manifesto need to be at the forefront of developmental considerations. When it comes to satisfaction, stakeholders have the highest priority. Stakeholder satisfaction is achieved by early, rapid, and continuous set of evolving, useful deliverables. "Change" is embraced whether due to evolving requirements, changes in the operational environment, or lessons learned from iterations on the design. This provides stakeholders with a flexible, incremental, and adaptive development environment that leads to a competitive edge. Useful functionality is provided early in the development effort, with early stakeholder involvement benefiting the design evolution. Updates to functionality occur more frequently than in traditional development efforts providing for opportunities to influence functionality and provide useful guidance. There exists a necessary high degree of required collaboration between stakeholders and the development team and with the development team itself. Sometimes daily meetings are required, and face-to-face meetings are often deemed the most efficient and effective method for program success. However, with the effectiveness of modern-day communications, and tools such as Microsoft Teams, Zoom, and the like, effectively involving stakeholders, and/or dynamic remote teammates, or collaborators in essential and frequent online meetings can be highly efficient and productive in the right circumstances. Teammates must be highly motivated and enabled. By enabled, teammates need to have made available necessary tools, organizational support, and trust to act and accomplish their required tasks. The work effort is organized into parallel "sprints" that are designed to be repetitive with regular intervals. The goal for each sprint is to have a useful functional product that evolves with each subsequent sprint with additional useful functionality. At the end of a sprint, fine-tuning behaviors are accomplished to improve efficiency. The final sprint must result in a system, service, product, and/or process that successfully meets the stakeholder's expectations/requirements and is the ultimate measure of success. Ensuring Agile processes provide a constant and continual pace makes for a sustainable development effort. Focusing on technical excellence and simplicity provides agility throughout the design process. Allowing the individual teams in the Agile development to self-organize will optimize the chances in meeting (or exceeding) established stakeholder requirements and producing the best architectures and designs.

In the future, Agile methods are expected to play an important role in both software and hardware dominant developed environments and become an integral part of the systems engineering discipline. In software development, well-known Agile frameworks such as Scrum and Kanban have become established conventions, allowing teams to provide incremental improvements and respond quickly to stakeholder feedback. Similarly, in the realm of systems engineering, the use of Agile principles and methods is on the rise to deal with the complexities inherent in creating large, integrated systems. Numerous industries, ranging from information technology to automotive and aerospace, are increasingly recognizing the benefits of Agile methods. These methods can shorten development times, improve product quality, and increase stakeholder satisfaction. This is especially important in areas where technical improvements are rapid, and there is a constant state of innovation.

Nonetheless, implementing Agile methods comes with its own set of challenges. Agile involves a culture shift within companies, requiring stakeholder participation and a commitment to cross-functional collaboration. Furthermore, integrating Agile methods into traditional systems engineering processes requires a careful balance to guarantee that strict standards and compliance requirements are not jeopardized. To summarize, the prospects for Agile methods in software, systems engineering, and a variety of industries are extremely favorable. As technology continues to advance at a rapid pace, Agile methods are set to play a larger role in enabling enterprises to efficiently create and deliver innovative goods and services and effectively develop complex systems of the future.

8.1.13 Transition to Optical Systems Building Blocks

In this section, we have provided an overview of Agile methods and shown how they relate and contribute to modern-day systems engineering development practices. Agile methods had their roots in the world of software development although their methods, tools, and techniques also apply to the development of complex, contemporary systems that may be software-based, hardware-based, or mixed. In the next section, we present the optical analog to the hardware-based, software-based, and mixed systems (both hardware and software) for high spatial resolution optical systems. Hardware-dominant optical systems that provide high spatial resolution by removing the effects of atmospheric turbulence are called adaptive optics (AO) systems. Optical systems that remove atmospheric turbulence from imaging systems that are software-dominant are called Atmospheric Turbulence Compensation (ATC) systems, and mixed hardware and software systems that provide high spatial resolution are called hybrid systems. We present an overview of these different methods, along with theoretical and mathematical descriptions where necessary, relevant, and practical. As an analog to the Agile metrics presented above, we also present useful statistical metrics, where necessary, to understand the utility of the different high spatial-resolution imaging methods and approaches.

8.2 OPTICAL SYSTEMS BUILDING BLOCKS: ADAPTIVE OPTICS AND ATMOSPHERIC TURBULENCE COMPENSATION METHODS, WITH STATISTICAL METRICS

8.2.1 INTRODUCTION

In this section, we investigate methods to remove atmospheric turbulence from optical imagery using adaptive optics and ATC methods. We also present some deterministic and statistical methods that are useful in analyzing the spatial and temporal characteristics of an optical wave and that also provide useful metrics in assessing the performance of an imaging system. We discuss both passive and active imaging methods but focus predominantly on incoherent imaging methods here and save coherent imagery for the next chapter.

In the field of optical imaging, if a well-designed (low noise) imaging system has an aperture that is larger than the atmospheric coherence length, r_0, the greatest source of aberrations and degradation of spatial resolution performance is the turbulence in the atmosphere. Other causes of loss in spatial resolution include the choice of operating wavelength (the larger the wavelength(s) the lower the attainable spatial resolution), the relative size of the imaging system's entrance pupil diameter, the atmospheric coherence length, r_0, (Fried parameter), the angular instantaneous field of view (IFOV), the isoplanatic angle, θ_i, spatial sampling effects, signal-to-noise issues, and system effects such as imperfect optics, imaging system noise, and platform jitter. The following development stems from an analysis of scalar diffraction theory using the semi-classical model and applies to any system that (1) has an object maximum extent that is small compared to the separation distance of the object from the imaging system, (2) has the property that the dimensions of the diffracting aperture are large compared to the size of the illuminating wavelength.

For an entrance pupil plane aperture, the best possible achievable spatial resolution (without super-resolution methods) is the spatial resolution associated with the classical diffraction limit. For a circular entrance pupil plane aperture, the classical diffraction-limited spatial resolution is given by Equation 8.1 and, for a square entrance pupil plane aperture, given by Equation 8.2:

$$\Delta x_{\text{diff, circle}} = 1.22\frac{\bar{\lambda}}{D}z, \tag{8.1}$$

and

$$\Delta x_{\text{diff,square}} = \frac{\bar{\lambda}}{D}z, \tag{8.2}$$

where $\bar{\lambda}$ is the center wavelength of the illuminating light, D is the diameter of the entrance pupil of the imaging system, and z is the distance between the imaging system's exit pupil and the object to be imaged. In the above equations, it is assumed that the circular aperture is contained within the square aperture (i.e., the square aperture includes the points in the corners that the circular aperture doesn't have, hence the better spatial resolution for the square aperture). For example, a 5-in. (12.7 cm) circular camera lens looking at an object 3 km away at a center wavelength of 500 nm would have a diffraction-limited resolution of 1.44 cm. Essentially, if diffraction-limited imaging conditions are attained, any object features smaller than 1.44 cm will not be resolvable/detectable in the image.

In a situation where the Earth's atmosphere is present in the optical path between the object and the imaging system, the turbulent atmosphere severely degrades the ability to achieve the diffraction-limited spatial resolution indicated in Equations 8.1 and 8.2. Instead, these spatial resolution equations are modified as follows:

$$\Delta x_{\text{atm,circle}} = 1.22\frac{\bar{\lambda}}{r_0}z, \tag{8.3}$$

and

$$\Delta x_{\text{atm,square}} = \frac{\bar{\lambda}}{r_0}z, \tag{8.4}$$

where r_0 is the atmospheric coherence length (Fried Parameter). Given experimental values of 1 and 4 cm for r_0, we would attain corresponding spatial resolutions of 18.3 cm ($r_0 = 1$ cm) and 4.57 cm ($r_0 = 4$ cm) for a circular entrance pupil plane aperture using the example values listed. This represents a "loss" in spatial resolution equal to a factor of 12.71 ($r_0 = 1$ cm) and 3.17 ($r_0 = 4$ cm), respectively. This can be obtained by dividing the circular or square resolutions given by Equation 8.3 or 8.4 by their corresponding Equation 8.1 or 8.2. This results in an improvement metric of D/r_o that represents the attainable increase in spatial resolution, given that the atmospheric turbulence effects are completely removed from the image. Note also that as the size of the entrance pupil plane aperture grows, so too does the potential spatial resolution performance increase by removing the effects of atmospheric turbulence from the imagery. For example, if we use a telescope that has a primary mirror (circular entrance pupil plane aperture) with a diameter of 10 in. (25.4 cm) with a camera attached to view the same object discussed above in the example for Equation 8.1, the diffraction-limited attainable spatial resolution would be 0.9 cm without atmospheric turbulence. When considering the effects of atmospheric turbulence on this 10-in. optical system, substituting our values for r_0 of 1 cm and 4 cm into Equation 8.3 the potential spatial resolution increase becomes 25.4 ($r_0 = 1$ cm) and 4.4 ($r_0 = 4$ cm), respectively. It is evident that as the size of the entrance pupil plane aperture of the imaging system increases, the expected improvement in the optical imaging system's spatial resolution also increases. There are three general methods to estimate and remove the effects of atmospheric turbulence from optical imagery. These methods include designing and implementing (1) adaptive optics (AO) systems, (2) post-processing systems (a.k.a. ATC systems, and (3) hybrid imaging systems. AO systems are largely hardware-dominant; in that, they compensate for atmospheric turbulence using hardware components such as wavefront sensors (WFSs), deformable mirrors (DMs), and high-speed computers to provide real-time correction of atmospheric turbulence capable of providing classical diffraction-limited spatial resolution in the detected imagery. AO imaging systems tend to be bulky, complex, cumbersome, and finicky and are relatively costly and typically relegated to fixed sites. These AO systems are only capable of real-time correction of images taken at the specific site with their hardware. Often, specialized expertise is required to align, calibrate, repair, and maintain the AO system. AO sites require attention placed on vibration and shock isolation and environmental conditioning (such as light/dust shields, detector cooling, and temperature/humidity control). ATC methods, on the other hand, tend to be *software-dominant* in that most of their functionality can be implemented in software, digitally, on a computer. Many of the ATC applications have been applied to incoherent passive imaging systems (systems that detect incoherent light such as objects illuminated by sunlight, moonlight, incandescent lights, thermal radiation, and/or non-coherent sources). Traditional post-processing ATC methods (e.g., Phase Diversity) have been shown to successfully compensate for atmospheric turbulence effects in optical imagery. These methods have often used an iterative 2-D Fourier transform-based process that attempts to minimize an error metric to estimate the aberrations present in an image. Traditionally, these methods have been mostly employed in a post-processing mode where the object reconstruction from the atmospheric turbulence degraded image could take hours, and even days to correct with iterative software methods. The benefit of these post-processing ATC solutions is that they tend to cost less, have more flexibility, and are much easier to upgrade and fine-tune. As will be seen below, there are some exceptions where high-speed ATC methods with and without parallel processing have been developed for real-time applications based on traditional and contemporary ATC methods. These ATC methods can be used to provide real-time, software-dominant, imaging systems that are portable and can be implemented in rugged environments. One specific ATC method, the Well Optimized Linear Finder (WOLF), is a software-dominant ATC method that, unlike many other ATC methods, can be implemented in real time. Hybrid systems are systems that use a combination of adaptive optics and post-processing methods to achieve ATC. This usually involves a less stringent adaptive optics configuration that also uses post-processing methods to further improve the spatial resolution of the imaging system. Consequently, the performance of the hybrid methods lies somewhere between pure AO and pure ATC methods in terms of system complexity, mobility/portability, and speed.

In this chapter, we provide an overview on various AO, ATC, and hybrid imaging methods. We start with a general history of adaptive optics and ATC. We then present AO and ATC systems from a basic component and function viewpoint and provide an overview of how AO/ATC system work. Then, we provide a theoretical and mathematical framework that required to understand how these systems operate and perform, and under what conditions and constraints these systems produce reasonable results. We follow this with a discussion of some of the predominant traditional and contemporary methods and approaches and also provide a detailed look at a promising ATC method along with some simulated results. We begin with a brief history of adaptive optics and ATC.

8.2.2 History of Adaptive Optics and ATC

For ages, it was known that stars twinkle but planets don't. That is until you look at the stars with a large aperture, like a telescope, and notice that in so doing, they cease to twinkle. The reason was known in Isaac Newton's time as the larger aperture combining fluctuating light rays, averaging out their spatial variations due to the atmosphere, so that the eye can't perceive them. It was noticed by Isaac Newton that telescopes in the visible spectrum that are larger than roughly 20 cm cannot achieve diffraction limited performance without some sort of correction (Roggeman and Welsh 1996). This applied even to imaging systems that were well designed. In the 1940s, Kolmogorov made a connection between energy transport from large-scale turbulent motion to small-scale motions and developed a model for the spatial power spectrum of the index of refraction variations (Kolmogorov 1961). It wasn't until the 1950s that photography progressed to the point where short exposure images could be captured. This permitted the filming of speckle patterns in the imagery which were high contrast light and dark regions about the size of the diffraction limit of the optical system. Babcock realized in 1953 that atmospheric turbulence could be compensated using mechanical principles and devices (Babcock 1953). Fried used Zernike polynomials to develop the generally useful atmospheric coherence length parameter r_o that can relate atmospheric turbulence effects to the achievable spatial resolution of an imaging system (Fried 1966). Tatarski took the ideas from Kolmogorov and solved the wave equation for propagation through weak random turbulence (Tatarski 1967). From the 1970s through the 1990s, the seminal results from Kolmogorov, Tatarski, and Fried led to large number of fundamental scientific contributions in the areas of adaptive optics, ATC, and hybrid imaging methods. The Knox-Thompson method was fundamental in phase retrieval methods (Knox and Thompson 1974). Noll used Zernike polynomials to describe atmospheric aberrations (Noll 1976). ATC methods like phase diversity gained ground as post processing methods, but these methods were computationally slow as compared to adaptive optics methods (Gonsalves and Chidlaw 1979). Other ATC methods evolved from the phase diversity idea such as the joint estimation of both object and aberrations (Paxman et al. 1992) and multi-frame blind deconvolution (Schulz 1993). In the early 2000s, some researchers looked at other diversity methods such as wavelength diversity, but early models assumed the same object spectrum at different wavelengths (false assumption). Ingleby and McGaughey developed an error metric for wavelength diversity, but differences in object spectra at different wavelengths and processing restrictions made this approach difficult to implement (Ingleby and McGaughey 2002). In contemporary times, after the 2000s, advances in high-speed parallel processing methods led to real-time capable ATC methods. Boeing implemented the phase diversity ATC method on a Field Programmable Gate Array (FPGA) and demonstrated real-time ATC. Arrasmith developed the Well Optimized Linear Finder (WOLF) correlation-based ATC method in 2013 (Arrasmith 2013) and presented refinements and improvements through 2023 (Arrasmith and He 2023). The WOLF methodology has the advantage of directly estimating the atmospheric phase aberrations and not requiring an approximation through truncated basis function expansions as in many traditional ATC methods. The WOLF method also is several orders of magnitude faster than traditional ATC methods with good scalability for larger images.

8.2.3 Basic Overview of AO and ATC Systems

Optical systems that are embedded in the Earth's atmosphere suffer performance degradations if the entrance pupil plane aperture of the imaging system is larger than the atmospheric coherence length (Fried parameter) r_o. Adaptive optics and ATC methods are ways to estimate and remove atmospheric turbulence effects from images. Optical waves that travel through space outside of the Earth's atmosphere do not suffer these effects which manifest as loss in spatial resolution of the imaging system. The source of the atmospheric aberrations results from the non-uniform heating of the Earth's surface. Different materials have different absorption, reflection, and transmittance properties that result in temperatures of the material. For example, a metal roof on a hot day has a different temperature to a patch of grass in the near vicinity of the roof. The non-uniform heating of the Earth's surface manifests in differences in the index of refraction, which affects the "bending" of the light rays. The index of refraction is a function of temperature and pressure and so affects the propagating optical wave. If a distant object like a star is observed with the naked eye through the Earth's atmosphere, it appears to twinkle. If an extended object is viewed through a telescope, it appears blurry and fuzzy. Whether a point source object like a star, or an extended image, adaptive optics and ATC systems can be used to mitigate the effects of the atmosphere and restore the high spatial-resolution imaging potential of a well-designed optical imaging system. In Figure 8.8, we can see the effect of the atmosphere on the left image that shows a stellar scene with a variety of stars. The image on the right shows the same stellar scene with the Adaptive Optics system turned on. Looking closely at the right image, some stars that previously were not visible become visible (see the bottom right corner of the right image for instance) and the visible objects become more defined and intense. Adaptive Optics imaging systems can provide near diffraction-limited spatial resolution, which is the best possible classical spatial resolution that an ideal optical system can achieve. Unfortunately, system noise and statistical noise inherent in the propagating optical radiation can affect the imaging system and reduce its spatial resolution performance, but near diffraction-limited resolution is achievable and is demonstrable in modern AO imaging systems. So, what makes up an adaptive optics imaging system? Figure 8.9 shows the fundamental components that are required for AO imaging systems. Figure 8.9 shows optical radiation that has propagated through some path to reach the entrance pupil plane aperture of the imaging system (shown as the system pupil at the top right of Figure 8.9).

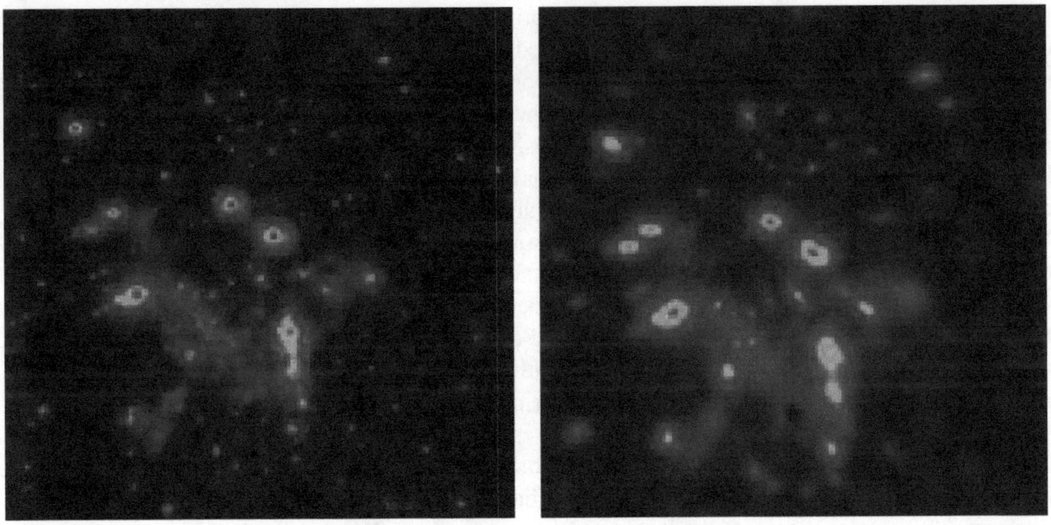

FIGURE 8.8 View of a star system (a) without, and (b) with adaptive optics.

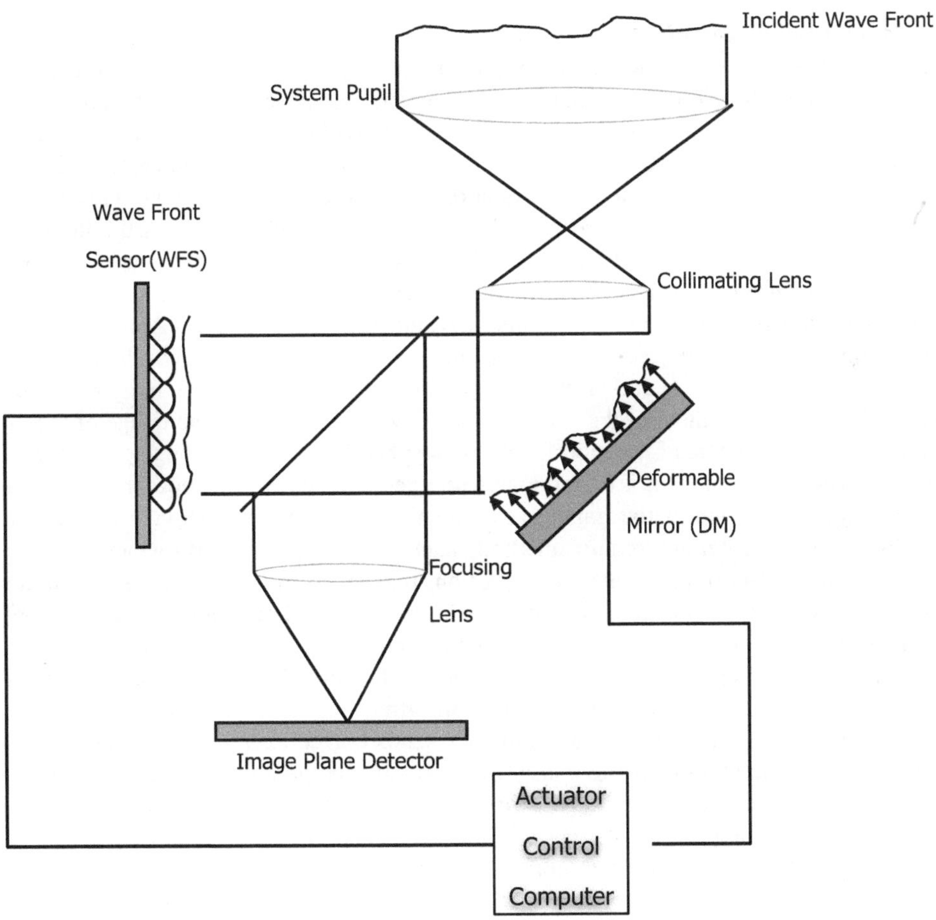

FIGURE 8.9 Basic physical layout of an adaptive optics imaging system.

The light is focused and collected by a collimating lens that projects a column of incident light onto a DM. The DM can change its shape to modify the phase information of the incident optical radiation in two dimensions. The DM is adding or subtracting phase to the incident optical field at various points across the entrance pupil plane aperture so that the atmospheric aberrations, which manifest as localized entrance pupil plane phases, are removed by the DM in real time. How? The light reflected off the surface of the DM is partially sent to a WFS which senses the slope (phase differences of the phase due to the atmospheric aberrations) and sends these slopes to a computer. Part of the light is sent to a detector to form the image, which is still degrade from the phase aberrations of the atmosphere (initially). The wavefront slope information is sent to a high-speed computer which rapidly estimates the atmospheric phase aberrations and sends, in real-time commands to the DM to put the opposite phase onto the incoming optical wave at the sampled entrance pupil plane points. This operation has to occur at every spatially sampled point in the entrance pupil at the DM. Further, this operation has to occur on the order of milliseconds since the atmosphere changes its state on that order. The DM applies the opposite phase locally in the entrance pupil, which cancels the atmospheric aberrations at every point, resulting in a diffraction-limited image detected at the Image Plane Detector. The AO system typically has an outer loop (not shown) that removes tip and tilt from the image and centers the image on the image plane detector. The phase corrections then remove the higher order phase aberrations from the imagery. We will describe some the essential components of the AO system in further detail.

8.2.3.1 Deformable Mirror

DMs are the active devices in the optical system that introduces phase changes on the incoming optical wave causing the minimization of the phase errors that have been introduced by the atmosphere. If an optical wave propagates through free space without any atmospheric effects, if the original source is far enough away so that it is readily modeled by a point source, then the expected phase front in the orthogonal plane of propagation is expected to be constant. This fact is quite useful; in that, if a far distant object in space for instance is viewed from an earthbound telescope (a start for instance), then the optical field outside of the Earth's atmosphere is well-approximated by a plane wave. Consequently, any phase aberration sensed by the AO system is solely because of the Earth's atmosphere. Some AO systems take advantage of this fact by looking at a distant star moments before imaging an object of interest such as a man-made satellite near the Earth. The star can be treated as a point source and provides the point spread function of the imaging system. The trick is to take the measurement of the start within a few milliseconds of the image of the target of interest so that the point spread function for the atmosphere matches the point spread function for the viewed target of interest. This is easier done with fixed objects, such as those in geosynchronous orbit, and using the rotation of the Earth to find a suitable nearby star. Moving targets like Low Earth Orbiting (LEO) satellites require methods such as laser guide starts which produce point source objects in the atmosphere in arbitrary locations that act as artificial stars (bright point-source objects) that provide the direct measurement of the point spread function at spatial coordinates close to an arbitrary target of interest. Once the point spread function is determined, the 2D Fourier Transform produces the optical transfer function for the state of the atmosphere at the time of interest and computer methods can be used to estimate the atmospheric phase aberrations each required sampled entrance pupil plane point. The computer sends corrections to the actuators of the DM which adjusts itself to put an opposite phase (one that cancels the atmospheric phase aberrations) onto the incoming optical wavefront in real-time, thereby mitigating the effects of the atmosphere. There are two different types of DMs: (1) continuous membrane and (2) segmented actuators. An example of continuous membrane actuators is shown in Figure 8.10. The continuous membrane mirrors are made of a very thin sheet of reflective material that can be deformed.

Underneath the thin reflective materials are actuators that are spread out in two dimensions in the entrance pupil plane behind the DM. The spacing of the actuators must conform to spatial sampling requirements of the imaging system as discussed in Chapter 2. When an electrical signal (current or voltage) is applied to the actuators, they change the surface of the DM by contacting or raising the

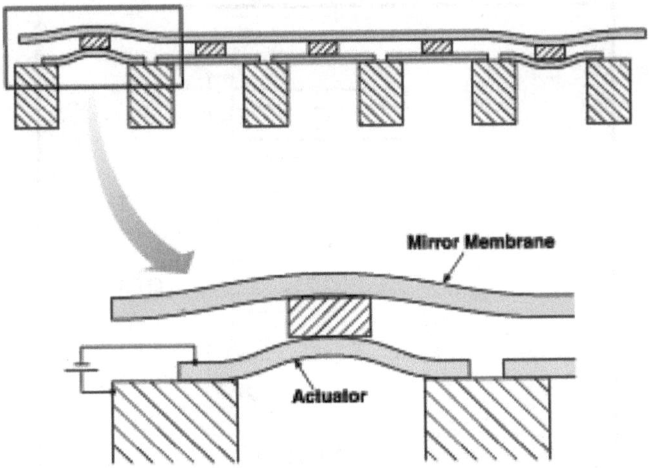

FIGURE 8.10 Schematic of continuous membrane deformable mirror.

FIGURE 8.11 Continuous membrane deformable mirror with 336 permanent magnetic actuators.

surface of the DM to introduce the correct (opposite) phase at that location in the entrance pupil. The intent is to cancel the phase effects introduced by the atmosphere at locations where the actuators touch the surface of the DM. Figure 8.11 gives a nice view of the back side of a DM with all the actuators. This DM is 0.64 m across and has 336 actuating elements to locally change its shape. The surface of the DM is 2 mm thick. Advantage of the continuous membrane mirrors is that there is full spatial coverage of the entrance pupil plane field surface but because of mechanical limitations, the amount of correction that can be locally applied is limited. For situations where these mechanical limitations of the magnitude of the mirror deflection becomes an issue, there are segmented mirrors that can be effective. Segmented mirrors, as the name suggests, are DMs that are made up of different mirror segments that can be independently controlled. Each individual segment may separately provide local tip, tilt, and piston control and, in general, can provide correction over a larger range

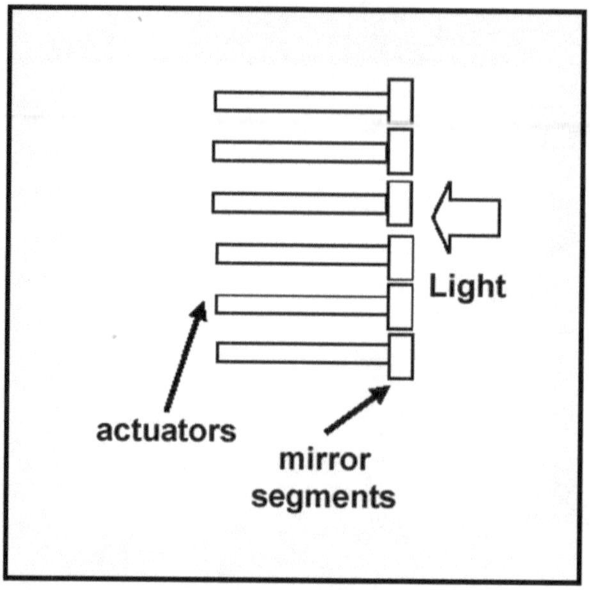

FIGURE 8.12 Schematic of a segmented deformable mirror.

of deflections than a continuous membrane mirror. Figure 8.12 provides a conceptual view of a segmented mirror. Each of the actuators is capable of raising or lowering (piston control) the optical element at the right end of the figure. Many segmented mirrors can also remove the x-tilt and y-tilt to better fit the local wavefront slope as well. The light shown at the right of the figure strikes the segmented DM surface and is reflected back. As shown in Figure 8.12, there is no change to the wavefront in this case due to the change in actuator positions, since all actuators have the same piston, x-tilt, and y-tilt. Figure 8.13 shows several different views of segmented mirror surfaces. One disadvantage of the segmented DM can be seen by the space between actuators where there is no sensing material. Careful alignment between actuators is required to minimize piston errors between actuators. Several of the images show a popular hexagonal structure to the segmented surface elements of the DM. Whether continuous or segmented, the amazing thing about the DM is that it must change its shape completely on the order of a few milliseconds since the atmosphere changes during this time. The closed-loop nature of the AO system is such that the reception, detection, phase estimation, and correction processes must happen continuously on this time scale. To impart the phase corrections on the incoming optical wave, the phase aberrations must first be sensed by the AO system. The phase sensing function is accomplished by the WFS.

8.2.3.2 Wavefront Sensor

We have already alluded to the fact that the expected wavefront for an optical field is expected to have constant phase if the object is located far enough away from the optical system to be in the far-field of the imaging system. The far-field imaging criteria were presented in Chapter 2. The basic way the WFS works is to look at a far-field object that is a point source with respect to the sampling capabilities of the imaging system. If this is the case, the expected field at the entrance pupil plane of the imaging system is well-modeled by a plane wave in the transverse direction of propagation. This means that the phase of the optical field in the transverse plane (x–y plane) to its direction of propagation (z-axis) is constant. Since for single aperture imaging systems, the relative phase and not the absolute phase is important, this constant can be set to zero without any loss of generality. The WFS then can measure the change in position of the phase relative to a constant reference and

FIGURE 8.13 Various views of segmented deformable mirror.

the computer can "stitch" together the phase relative phase estimate in the entrance pupil plane of the imaging associated with the atmospheric turbulence. One way to do this is by sensing the phase difference (or slope) of the incident entrance pupil plane phase locally. Some ways of doing this include interferometers, Shack Hartmann WFS, curvature sensors, and pyramid sensors. Interferometers were already presented in Chapter 2, so we now focus on the more common Shack Hartmann and Curvature WFS and discuss the less common Pyramid WFS approach.

Shack Hartmann wavefront sensor: One of the most common WFS types is the Shack Hartmann WFS, which dates back to the early 1970s. A screen with tiny holes was illuminated by an incoming wavefront, and a detector was used to measure the resulting pattern. Later, lenslets (a series of tiny lenses) were used in place of the holes to collect more light. The detector was placed in the focal plane of the lenslet array. Figure 8.14 shows the fundamental principles of the Shack Hartmann WFS. At the top of Figure 8.14, an incoming wave with no aberrations is shown arriving in the entrance pupil of the imaging system. Since there are no aberrations present, the far-field point source object produces a wavefront in the entrance pupil plane of the imaging system, which we set to zero for convenience, since we are modeling a single aperture imaging system. The image at the top-right shows a series of spots that are centered on the optical axis of each of the lenslets. This corresponds to a zero-phase slope across each lenslet as shown dimensionally by the line at the top left of Figure 8.14. If the wavefront experiences phase aberrations such as a plane wave propagating through the Earth's atmosphere, then Shack Hartmann WFS would see phase changes to the relative phase wherein each lenslet would provide the phase difference (i.e., slope) of the phase. The bottom-left wavefront shows the optical phase front in one direction that is at the entrance pupil of the imaging system wherein the lenslet array is placed. Each lenslet would sense the localized slope of the phase and the detector shown at the bottom right of Figure 8.14 shows the movement of the spots corresponding to each lenslet in the transverse direction of the propagating optical field. The way to measure the slope of the entrance pupil plane phase due to the atmospheric aberrations is to use a quadrature detector in the focus of each of the lenslets that can determine the centroid of the spot. The quadrature detector breaks the spot image formed by each lenslet into four quadrants. If there is no slope in the vertical direction

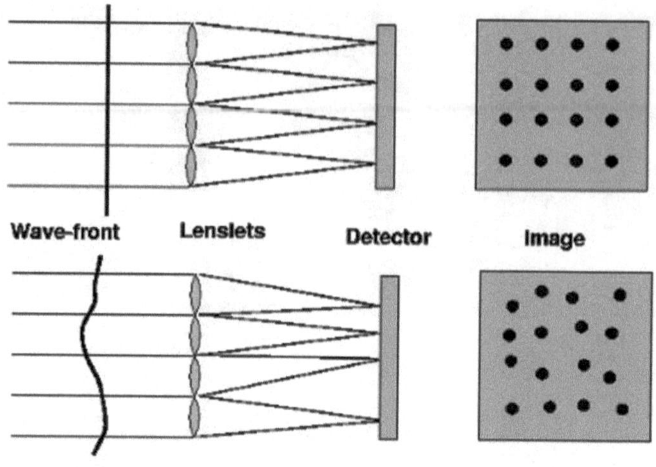

FIGURE 8.14 Depiction of Shack Hartmann wavefront sensor.

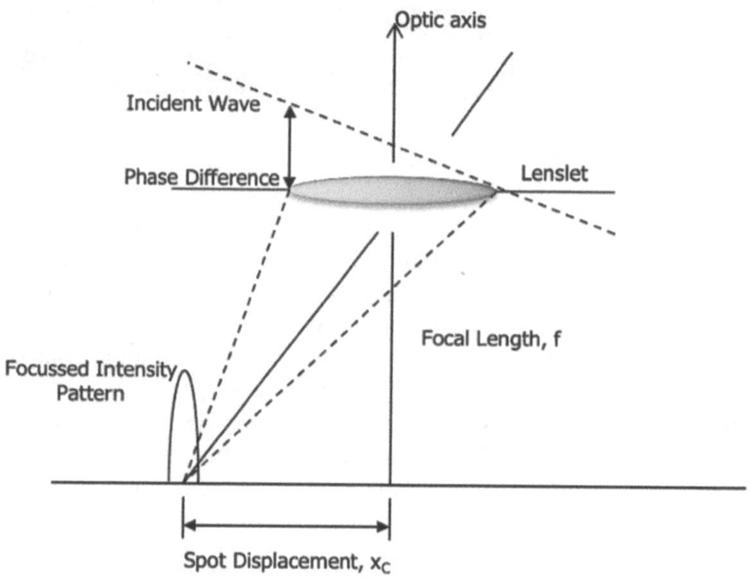

FIGURE 8.15 Spot displacement induced by phase difference across a Shack Harmann WFS lenslet.

of the quadrature sensor, then the difference between the upper half and lower half of the quadrature sensor is zero since the spot is evenly distributed with half its irradiance in the upper quadrant and half the irradiance in the lower quadrant. Figure 8.15 shows this idea. If the phase difference across the Shack Hartmann lenslet is non-zero, this corresponds to a phase tilt across the lenslet as shown. The focused spot on the detector would move in accordance with the amount of tilt in the incoming optical wave. Notice that if there were no tilt present, the focused intensity pattern would straddle the optical axis with half its energy on the left side of the optical axis and an identical amount of optical energy on the right side of the optical axis. The difference between these intensities would be zero indicating a zero phase difference across the lenslet. If the slope is as shown in Figure 8.15, there could be a correspondence between the spot displacement x_c and the slope of the phase. Notice also that the direction of the displacement also gives the direction of the tilt. Figure 8.15 only shows one representative direction, but the shift in the spot could also be measured in the other direction of the detector (into and out of the page in this case). We would need two additional quadrants in and out of the page to measure the slope in that direction (say y_c). In this way, a quadrant detector could provide both the displacements x_c and y_c, and we can infer the x-slope and y-slope from these results. These types of sensors need to be calibrated so that zero centroid measures a zero tilt. The focal length needs to be chosen to provide a reasonable size spot with enough room for displacement to measure a reasonable range of slopes. The actual phase values can be estimated by "guessing" the first phase value and using the slopes to determine the relative phase. This is perfectly legitimate for single aperture imaging systems since they are insensitive to the actual value of phase (piston) and only respond to relative phase. Besides long-range imaging applications, the Shack Hartmann WFS is useful in the medical profession such as measuring the aberrations in the eye. It should be mentioned that this is a simpler application, though no less important, since the aberrations in the eye don't change on the order of milliseconds but remain relatively fixed over the span of the measurement. For medical applications, an optical beam is sent into the eye from a laser or super luminescent diode (SLD) as shown in Figure 8.16. The input beam is projected onto the retina through the lens of the eye and is reflected through the lens. The aberrations in the lens are picked up by the returning optical wave as shown by the squiggly line. The return optical beam is projected through the beam splitter and is collimated as shown by the two lenses. The Shack Hartmann WFS is used to measure the aberrations of the eye with the charge-coupled detector (CCD). The Shack Hartmann WFS is operated the same as was presented above, but a set of orthonormal basis functions are expanded to describe the decomposition of the aberrations of the eye in the so familiar terms we hear at the optometrist, defocus, astigmatism, coma, spherical aberrations and more. These terms come from the Zernike basis functions, which are useful in fully describing not only aberration effects in the eye but also variable aberration effects in the atmosphere. We provide some mathematical insight into Zernike basis functions in the mathematical section below along with a discussion of other useful basis function expansions used to describe phase aberrations (e.g., Seidel basis functions and Karhunen-Loeve expansions). Figure 8.17 shows a comparison of the Shack Hartmann WFS output for a healthy eye (top right) and an aberrated eye (bottom right). As can be seen from the bottom image by inspection, the spots for each lenslet are not spaced uniformly indicating the presence of aberrations. The type and amount of aberrations can be determined by using orthonormal basis function expansions.

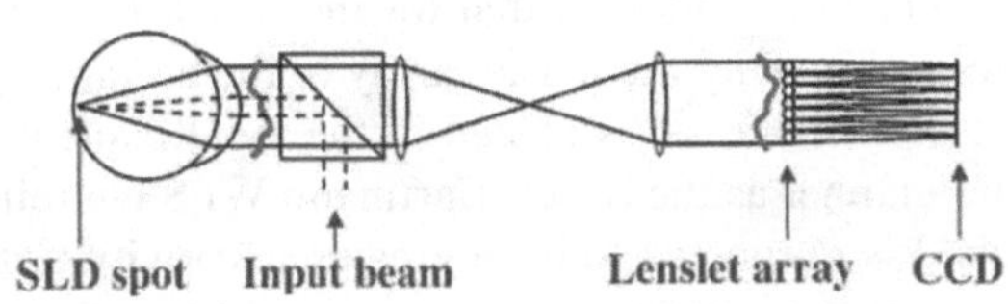

FIGURE 8.16 Shack Hartmann wavefront sensor used for corneal imaging.

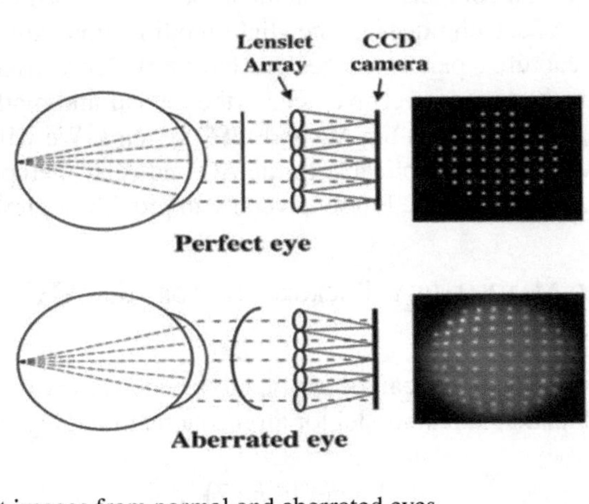

FIGURE 8.17 Wavefront images from normal and aberrated eyes.

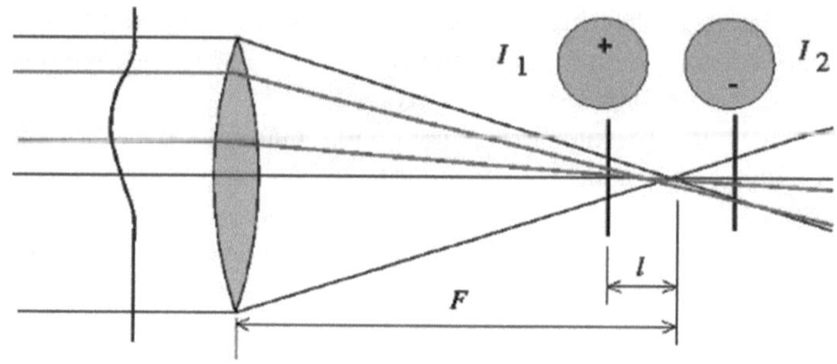

FIGURE 8.18 Depiction of a curvature wavefront sensor.

Curvature wavefront sensor: The idea of a curvature WFS was presented by Francois Roddier in 1988. An interesting property of the curvature sensor is that it senses the derivative of the slope (second derivative of the phase) and not the slope of the phase itself like Shack Hartmann sensors. The principle is to take the difference between two images. Both images are out of focus but are on the opposite sides of the focal plane. Figure 8.18 shows the concept. An electromagnetic field with phase aberrations is shown at the left of Figure 8.18 Image I_1 (shown with a +) is defocused to the left of the focal plane by a distance l, and image I_2 (shown by –) is defocused an equal amount but to the right of the focal plane. The difference of the images is proportional to the derivative of the slope (the Laplacian of the phase of the wavefront). This method has good resolution in the lateral direction but loses information about the phase difference (first derivative). The signal processing of curvature sensors is also more complicated than that of tilt sensors like the Shack Hartmann sensor. These types of sensors are located on some sophisticated sensing platforms like the 8.3 m Japanese telescope (Subaru) that is next to the Keck Observatory in Mauna Kea, Hawaii. The European Southern Observatory also has the Multi Application Curvature Adaptive Optics project which is implementing curvature sensors on a variety of optical instruments.

Pyramid wavefront sensor: Another interesting method for sensing phase aberrations is the pyramid WFS. This method is not as common as that for the Shack Hartmann WFS and Curvature WFS but it is similar. Instead of placing the lenslet array in the entrance pupil of the imaging system, for a pyramid WFS, the lenslet array is placed in the focal plane of the imaging system. As such, it operates in a similar manner as the Shack Hartmann WFS but reimages the aperture of the imaging system. This method is purported to have greater sensitivity to slope measurements than the Shack Hartmann WFS and slope errors which make up 87% of the aberrations introduced by the atmosphere into an imaging system (Chew et al. 2006). For ATC systems, there is very little hardware required apart from a computer to implement the ATC algorithms. For some techniques like the "diversity" based ATC techniques, some slight modifications are required to the imaging system to simultaneously capture a pair of images that have turbulence effects related to each other. Before discussing ATC methods, however, we need a theoretical and mathematical framework to understand the image formation process for AO and ATC systems. We will present the theoretical and mathematical framework that is applicable to the AO and ATC methods and focus on a traditional, well-documented ATC technique, Phase Diversity to provide context.

8.2.4 Theoretical and Mathematical Background for AO, ATC, and Hybrid Methods

A linear, shift invariant imaging model can be applied to a wide variety of incoherent imaging systems. The theoretical and mathematical model for an incoherent imaging system is given by

$$i(\vec{x}) = o(\vec{x}) * |h_i(\vec{x})|^2, \tag{8.5}$$

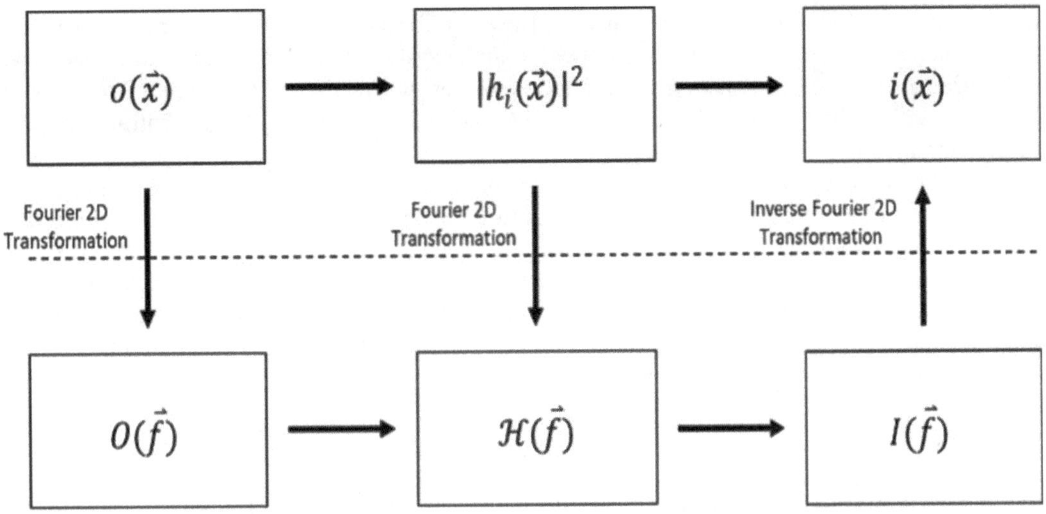

FIGURE 8.19 Linear model for general purpose incoherent imaging system.

where $o(\vec{x})$ is the un-aberrated radiant emittance (object brightness function), $h_i(\vec{x})$ is the impulse response, $|h_i(\vec{x})|^2$ is the imaging system's point spread function (PSF) and $i(\vec{x})$ is the aberrated image due to atmospheric and optical imaging system effect, $(\vec{x})$ is a 2D position vector and the asterisk represents 2D spatial convolution.

Taking the 2D Fourier transform of Equation 8.5 leads to Equation 8.6, which is the frequency space equivalent of Equation 8.5,

$$I(\vec{f}) = O(\vec{f})\mathcal{H}(\vec{f}). \tag{8.6}$$

Here $I(\vec{f})$ is the image spectrum, $O(\vec{f})$ is the object spectrum and $\mathcal{H}(\vec{f})$ represents the optical transfer function (OTF). In Equation 8.6, $(\vec{f})$ is a 2D spatial frequency variable. Equations 8.5 and 8.6 apply at each spatial coordinate $\vec{x}$ and spatial frequency $\vec{f}$. The general relationship between Equations 8.5 and 8.6 is shown in Figure 8.19.

The PSF is related to the OTF by,

$$\mathcal{H}(\vec{f}) = \frac{\mathcal{F}\left[|h_i(\vec{x})|^2\right]}{\mathcal{F}\left[|h_i(\vec{x})|^2\right]_{\vec{f}=0}}, \tag{8.7}$$

where $\mathcal{F}[\cdot]$ represent a 2D Fourier transform of the expression in the brackets. Therefore, the OTF is the 2D Fourier transform of the PSF that is normalized so that the maximum value of the OTF (at its center) is exactly one.

The relationship between the system's impulse response, $h_i(\vec{x})$, and the generalized pupil function (GPF) is given by

$$h_i(\vec{x}) = \mathcal{F}^{-1}\left[\text{GPF}(\vec{x})\right], \tag{8.8}$$

where $\mathcal{F}^{-1}[\cdot]$ represent an inverse 2D Fourier transform of the expression in the brackets, and the GPF is given by

$$\text{GPF}(\vec{x}) = A(\vec{x})e^{j\theta(\vec{x})}. \tag{8.9}$$

In Equation 8.9, $A(\vec{x})$ is an amplitude function. However, for the case of near-field atmospheric turbulence such as in a ground-to-air/space optical imaging system, the amplitude becomes one inside of the clear aperture of the imaging system and zero everywhere else. In Equation 8.9, $\theta(\vec{x})$ is the atmospherically induced phase aberration realized at spatial coordinate $(\vec{x})$ in the entrance pupil of the imaging system.

In traditional post-processing ATC methods, a common method is to insert some sort of known diversity into the system and simultaneously capture the original image and the diversity image. The OTFs of the aberrated and diversity images are then mathematically related to each other based on known physical relationships, and then an error metric is used to select the OTF that produces the lowest local and global error. As an example of a known physical relationship, the well-understood phase-diversity method uses an additive phase term that can be obtained ahead of time in the expression of the phase diversity generalized pupil function

$$\mathrm{GPF}_{\mathrm{pd}}(\vec{x}) = A(\vec{x})e^{j(\theta(\vec{x})+\theta_{\mathrm{pd}}(\vec{x}))}. \tag{8.10}$$

Here $\theta_{pd}(\vec{x})$ represents the introduced phase diversity at spatial position $(\vec{x})$. Note also that the OTF can be calculated directly from the GPF from the following equation

$$\mathcal{H}(\vec{f}) = \frac{\mathrm{GPF}(\vec{x}) \otimes \mathrm{GPF}(\vec{x})}{\mathrm{GPF}(0) \otimes \mathrm{GPF}(\vec{0})}, \tag{8.11}$$

where ⊗, represents the autocorrelation process. The entrance pupil plane spatial position variable is related to the spatial frequency by

$$\vec{x} = \lambda d_i \vec{f}. \tag{8.12}$$

In Equation 8.12, d_i is the distance from the imaging system's exit pupil plane to the focal plane and λ is the wavelength of the illuminating light. Figure 8.20 shows the relationships between the GPF, the impulse response, the PSF, and the OTF.

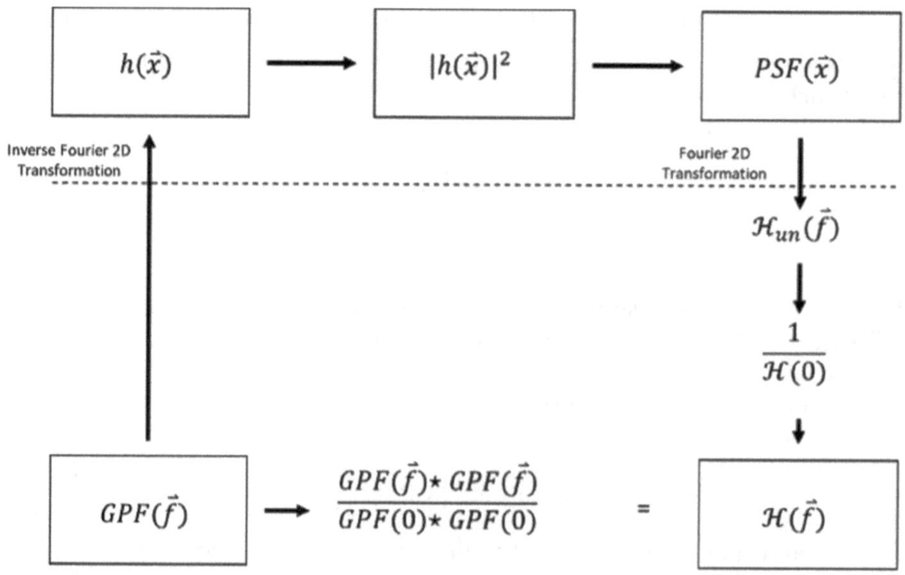

FIGURE 8.20 GPF, impulse response, point spread function, and optical transfer function relationships.

In the above diagram, the important result is the Optical Transfer Function $\mathcal{H}(\vec{f})$. It can be seen from Equation 8.6 that the image is a point-wise multiplication between the un-aberrated object spectrum and the Optical Transfer Function. When **modeling** the effects of atmospheric turbulence, the job is to model the appropriate OTF, pointwise multiply this model with the spectrum of the object, and inverse 2D Fourier Transform the result to obtain the aberrated (blurry) image. To **remove the atmospheric effects** from an image, the reverse process applies. In this case, the task is to estimate the OTF and remove these effects by applying an appropriate inverse filter that inverts the OTF and applies the inverted Optical Transfer Function, $\mathcal{H}^{-1}(\vec{f})$, to every image spectrum point.

To model the object spectrum, OTF, and image spectrum, we use the traditional phasor formalism

$$O(\vec{f}) = \left|O^{\lambda_1}(\vec{f})\right| e^{j\phi_O^{\lambda_1}(\vec{f})}, \tag{8.13}$$

$$\mathcal{H}(\vec{f}) = \left|\mathcal{H}^{\lambda_1}(\vec{f})\right| e^{j\theta_{\mathcal{H}}^{\lambda_1}(\vec{f})}, \tag{8.14}$$

and

$$I(\vec{f}) = \left|O^{\lambda_1}(\vec{f})\right|\left|\mathcal{H}^{\lambda_1}(\vec{f})\right| e^{j\phi_O^{\lambda_1}(\vec{f})} e^{j\theta_{\mathcal{H}}^{\lambda_1}(\vec{f})}, \tag{8.15}$$

where the wavelength superscript λ_1 indicates at which wavelength the operations take place. This vector notation is used throughout the mathematical development below. As can be seen from these three equations, the relationships identified by Equation 8.6 can be mathematically modeled by phasor terms that include a magnitude term and a phase term and that apply at every spatial frequency $\vec{f}$ in the entrance pupil plane of the imaging system. As is conventional, the measurable component of the mathematical expressions throughout our development can be obtained by taking the real part of the mathematical expression whenever desired. Note also that Equation 8.6 is a function of position only and not of time. This interpretation is consistent with the assumption that we are capturing a single image pair (image and diversity image) within the timeframe that the atmosphere remains relatively constant (the so-called Taylor's frozen flow approximation). This means that the integration time of the optical system is no more than a few milliseconds. Since single aperture imaging systems are dependent on relative phase and not absolute phase, the temporal effects can be completely ignored. Hence, we have omitted the temporal effects from our equations and left the mathematical development as strictly functions of position. Also note that for removing atmospheric turbulence in video cameras or a series of camera snapshots, the integration time of each image frame is assumed to be less than 2 ms. This "frozen" state of the atmosphere is what is corrected in our real-time AO, ATC, or hybrid methodology as further discussed below. At this point, it can be seen from Equations 8.6 and 8.13–8.15 that the heart of the atmospheric turbulence detection, estimation, and correction methodology is to determine the OTF magnitude and phase at each spatial frequency coordinate in the entrance pupil plane. Once this is accomplished, the OTF can be inverted and the un-aberrated object spectrum and ultimately the un-aberrated object brightness function can be determined.

8.2.5 Adaptive Optics and Active Imaging Methods

Adaptive optics systems can also be used with high-powered lasers where the lasers generate a point source radiating element in the upper layers of the atmosphere, like the sodium layer. This technique is referred to as a *guide star* technique since the radiating element caused by the laser acts like a far distant point source providing a constant phase wavefront that can be used by the AO system to determine the atmospheric phase aberrations. In this mode of operation, the AO system behaves in

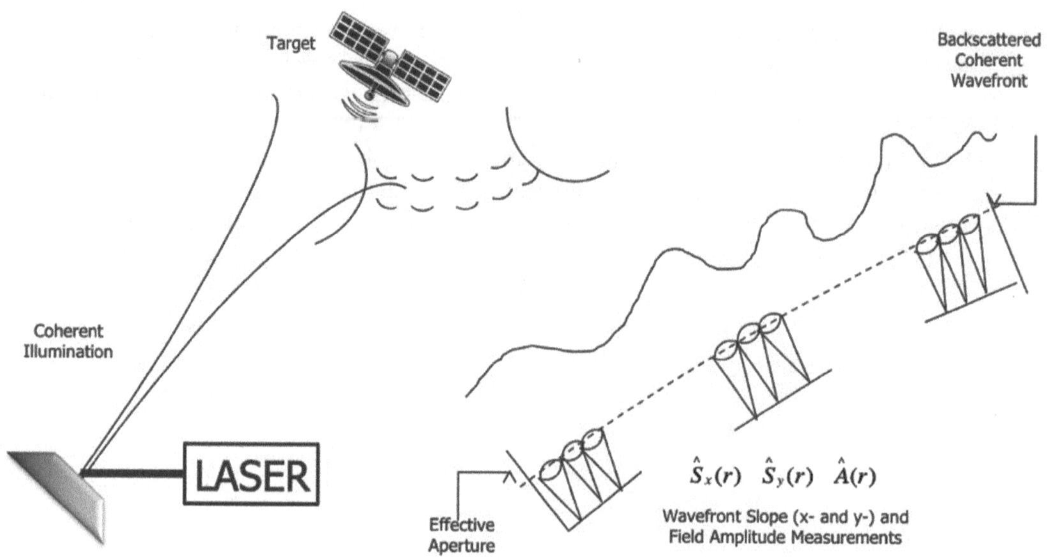

FIGURE 8.21 Sheared coherent interferometric photography (SCIP) method using active illumination.

a similar fashion as previously discussed with the difference being the artificial "star" generated by the laser to provide a reference phase for the propagating optical wave. The advantage of this approach is that the guide star can be generated near a target of interest whereas natural guide star methods require the object to be near an existing, acceptable reference star which often is not the case. Laser properties and physical characteristics are discussed in the next chapter.

Another interesting laser method is the sheared coherent interferometric photography (SCIP) method which conceptually can generate lensless imaging with an arbitrarily large aperture for high spatial resolution (diffraction-limited) imagery (Hutchin 1993). Figure 8.21 illustrates the concept. A distant object shown at the top left of Figure 8.21 is coherently illuminated with an active source such as a high power, short pulse laser. The projected beams are shifted with respect to each other in the vertical y and horizontal x direction and superimposed on the target of interest. The returned light is collected by lenslet array in the entrance pupil plane of the imaging system and the wavefront slope information is used in a digital computer to reconstruct the object radiant emittance. What is measured in this method is amplitude information from the difference of two shifted speckle patterns which preserve the phase. Exponential reconstruction methods developed by David Ehn from ITEK and the Knox Thompson method are used to retrieve phase information. Table 8.1 shows a summary of some entrance pupil plane techniques, along with the type of the first column of Table 8.1 shows the type of measurement that is being made (i.e., coherent, quasi-coherent, or incoherent). The second column shows the specifics field quantity that is measure such as the amplitude of the field (AMPL) and phase (instant phase, differential phase). The third column provides detector options such as Heterodyne detectors, shearing interferometers, Shack-Hartmann sensors, or direct, square law detectors. Some references to these methods are provided in the last column. In the next section, we present hybrid techniques that blend aspects of AO and post-processing methods. The performance and advantages of these methods fall somewhere in between pure AO systems and ATC methods.

8.2.6 Hybrid Techniques

Hybrid imaging systems attempt to relax some of the stringent adaptive optics requirements with regard to hardware requirements and compensate for this with post-processing methods. These methods are not as bulky, expensive, or fast as pure AO systems but are still more hardware-intensive,

TABLE 8.1
Comparison of Entrance Pupil Techniques, Field Quantities Measured, and Detector Types

Type of Measurement	Field Quantity Measured	Detector Possibilities	Image Type
Coherent phase sensitive	AMPL+instant phase (Doppler)	Heterodyne	Coherent
"QUASI"-coherent complex-envelope sensitive (Doppler insensitive)	AMPL+differential phase	Shearing Interf ($\Delta\phi$) Hartmann	Coherent
Incoherent	Amplitude	Direct, square-law det.	Coherent Incoherent (via correlations)

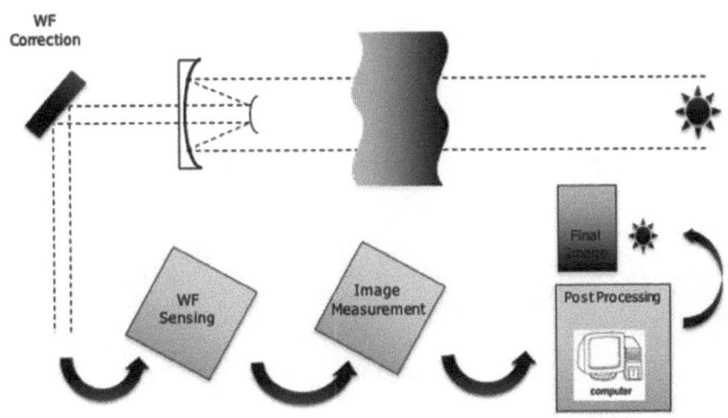

FIGURE 8.22 Components of a hybrid imaging system.

and complex than ATC methods discussed later. Figure 8.22 shows an example of a hybrid imaging method. The object of interest is shown at the top right of the figure. Light propagates through the atmosphere (represented by the gray wavy column at the top center) and is captured by a telescope which focuses and collimates the light onto a wavefront correction device for some initial wavefront correction. The partially corrected light is sent on to a WFS to determine information about the wavefront such as the detected phase differences. The WFS information along with the detected image is passed on to a post-processing station which applies ATC methods to fully remove the atmospheric turbulence from the imagery. This hybrid process is not done in real time. In Figure 8.23, we show a specific hybrid technique called Deconvolution from Wavefront Sensing (DWFS). In this method, both AO and ATC methods are employed. On the AO side, a distant image is observed through the Earth's atmosphere by a telescope shown as the cone on the left side of the figure. A DM is placed in the entrance pupil plane of the imaging system and the light is reflected to a beam splitter shown at the right of the DM. The beam splitter sends part of the light to and imaging sensor which initially captures the aberrated image of the distant object. The beam splitter sends the other part of the optical wave to a WFS which senses the phase properties such as the phase differences of the optical wave. The phase differences are sent to the actuator command computer at the bottom left side of Figure 8.23, which accomplishes partial ATC in real time through the DM. The partially corrected optical wave is sent through the beam splitter to continually update and partially correct the image on the imaging sensor. The WFS information is simultaneously sent to the wavefront computer and the partially corrected image is fully corrected in post-processing using the deconvolution processor. Figure 8.24 shows the modeling process used in generating simulated DWFS imagery. At the top left of the model, system parameters that can be pre-calculated and/or normalized are determined and red into the host computer. A random phase screen is generated

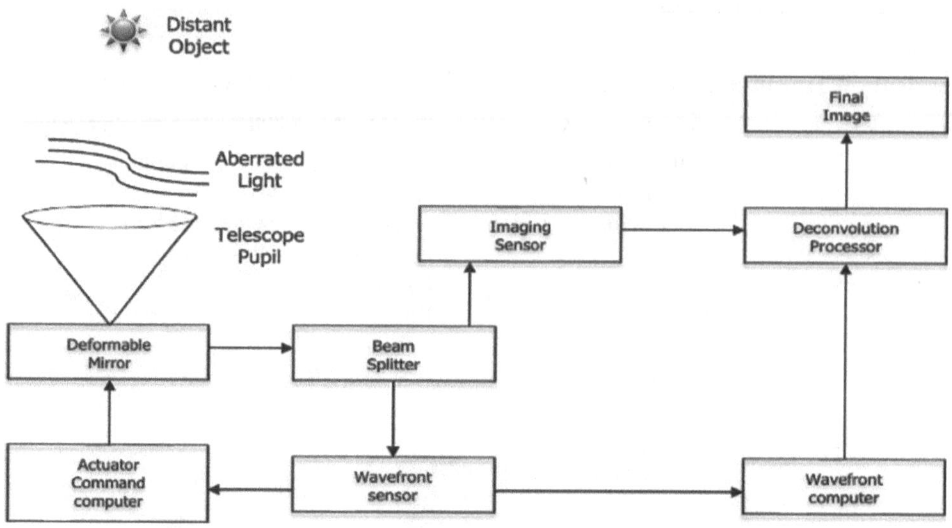

FIGURE 8.23 Deconvolution from wavefront sensing hybrid imaging method.

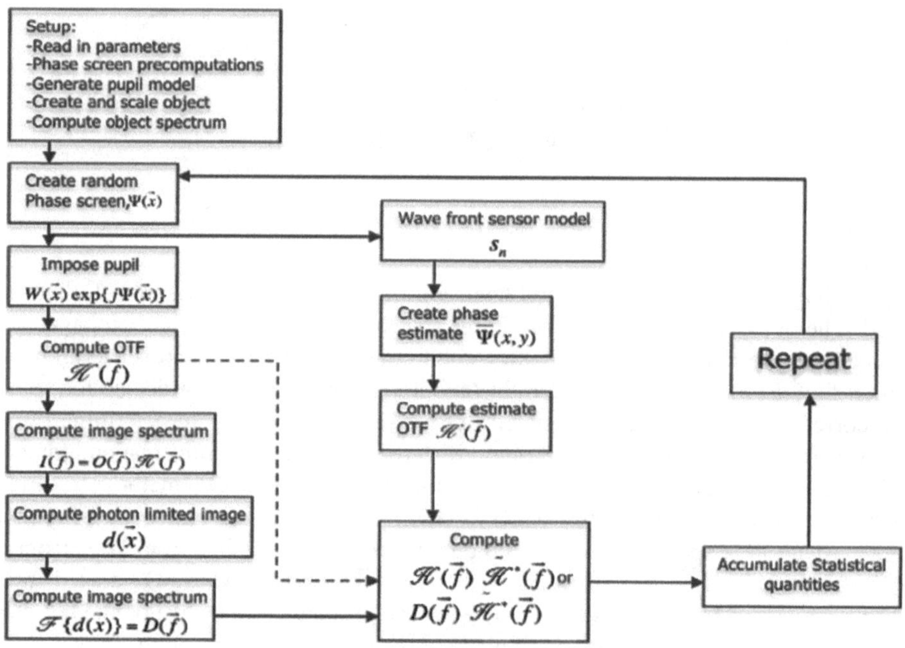

FIGURE 8.24 Process for modeling deconvolution from wavefront sensing.

that models atmospheric phase aberrations with the correct atmospheric statistics and a generalized pupil function is generated with the random phase screen. The phase screen is also sent to the WFS model in the center column of Figure 8.24 shown as S_n. The WFS model then produces the expected phase differences which then are used to generate an estimated phase $\bar{\psi}(x, y)$ where the coordinates are x and y positions in the entrance pupil plane of the imaging system. The phase estimates are used to generate the estimate of the OTF, $\bar{\mathcal{H}}(\vec{f})$, in accordance with Equations 8.6–8.9. In the left column of Figure 8.24, the reference phase (truth phase model) is given by $\psi(x, y)$ and is used to generate the reference generalized pupil function. This reference generalized pupil function then

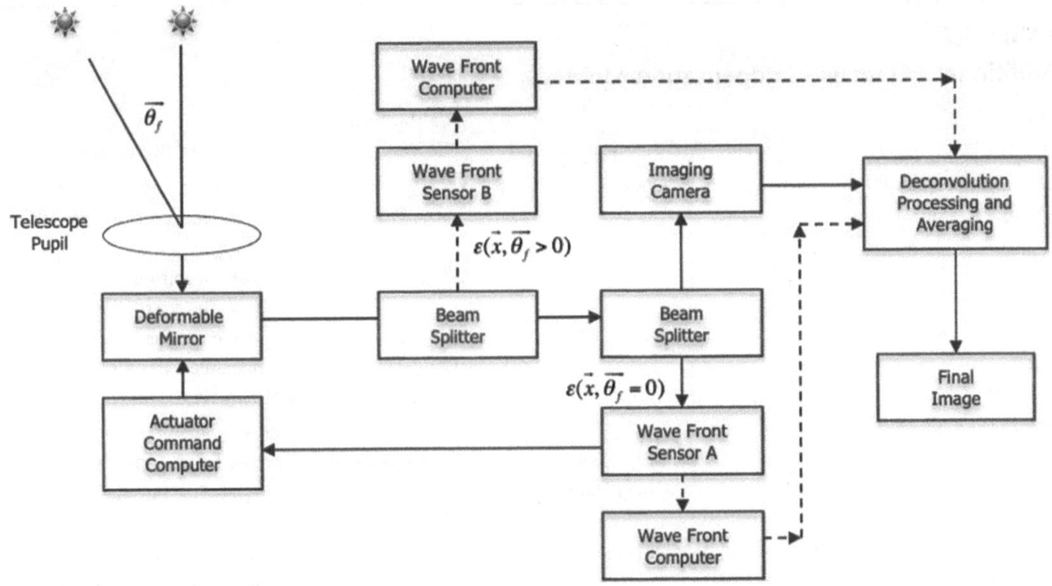

FIGURE 8.25 Compensated deconvolution from wavefront sensing.

generates the reference OTF, $\mathcal{H}(\vec{f})$, and then, based on the chosen object model, is used to provide the reference irradiance spectrum at spatial frequency $\vec{f}$. System noise effects are added and capture in the noisy object spectrum estimate $D(\vec{f})$. Necessary items for the object estimation process are gathered at the bottom center of Figure 8.24 (such as the reference OTF, estimated OTF, conjugate of the OTFs, and noisy irradiance spectrum). Important stored for the current phase screen and the process is repeated for other phase screens as necessary. An issue with the DWF method is that if the viewed object has a different angle of observation than the guide star, then the atmosphere may change the further the observation point moves from the guide star point. Figure 8.25 shows the angular separation between the guide star reference (vertical point) and the observation point at $\vec{\theta}_f$. If the angle of separation is large enough, then the object of interest is viewed through different atmosphere than that which is corrected by the guide star observation. The compensated deconvolution from wavefront sensing method attempts to correct this by performing a second correction for the slant path viewed at the angle $\vec{\theta}_f$ (Roggemann and Welsh 1996). In this method, two WFS estimates are made (one at WFS A and one at WFS B). As shown in Figure 8.25, initially WFS A estimates the wavefront made by the vertical guide star with $\vec{\theta}_f = 0$. The correction commands for the phase adjustments based on the zero-angle guide star are sent to the actuator control computer which then changes the shape of the DM to correct the zero-angle wavefront. If the object is in the vicinity of the second guide star point at angle $\vec{\theta}_f$, then the AO correction applied from the first guide star isn't quite correct and only a partial correction of the atmosphere is obtained. However, since the WFS loop is much faster than the closed loop bandwidth of the adaptive optics system (by an order of 10), the AO system has the capability of sensing the phase much faster than the AO system can correct it with the DM. Consequently, WFS B can be used to *sense* the wavefront phase due to the atmosphere along the slant path $\vec{\theta}_f$ as shown by the estimate $\epsilon(\vec{x}, \vec{\theta}_f > 0)$ where this estimate is everywhere in entrance pupil plane coordinate, $\vec{x}$, at viewing angles $\vec{\theta}_f > 0$. These improved WFS estimates are sent to the WFS computer that uses post-processing deconvolution methods to further compensate and improve the DWS measurement. Table 8.2 shows some categorization of some useful imaging methods. The first column provides the type of illumination source that is used

TABLE 8.2
Additional AO Image Compensation Methods

Illum Source	Measurement Domain	Detection Method	Calibration (Correction)	Name (References)
Solar (and/ or laser)	Focal	Direct	None	Classical imaging (Galileo)
			Non-R/T Star	Astronomical speckle imaging (Labeyrie)
			Reference	Cross/Bi-spectrum (Knox-Thompson, Weight)
			R/T DM	Pre-det. compensated (Hardy)
			R/T+Star	Compensated speckle
			Reference	Direct deconvolution
			NLO crystal	One-way imaging (Yariv)
			WFS calibration	Speckle holography (Gonglewski)
	Pupil	Direct Het Dyne	Phase-closure	Long-baseline interf. (Michelson)
			N/A	IR Het' Dyne interf. (Townes)
Laser	Pupil (or pupil-time)	Direct	None	Self-ref holography
				(Imaging) correlography
			Various	Pulsed laser photography
				Hartmann
	(Pupil)-time	Heterodyne	Various	Coherent pupil-plane
			N/A	NB coherent Lidar
	(Pupil)-time-bandwidth	Heterodyne	N/A	WB coherent-ISAR

(solar and/or laser). The second column shows the measurement domain where the WFS is placed. It also shows whether a series of time measurements and/or bandwidth information is required. The third column indicates whether direct detection methods such as square law detectors are used in the method or whether mixed frequency methods such as in heterodyne detection are required. The fourth column shows calibration methods needed by the various methods (R/T represents Real Time, and NLO represents Non-Linear Optics). The last column provides the name of the AO method along with some references. Note, for further insights into some of these techniques, please see (Idell and Roggemann n.d.). In the next section, we present ATC methods.

8.2.7 ATC Methods

We now provide an overview of some of the more notable ATC imaging methods, both traditional and contemporary. We distinguish ATC methods from AO methods in that ATC methods tend to be largely implementable in software (software dominant) whereas AO methods require hardware such as WFSs and DMs to operate (hardware dominant). Pure AO systems are real-time systems in that the hardware corrects the incoming wavefront in real time. Traditional ATC methods have been mostly relegated to post-processing applications with a few exceptions. With the emergence of high-speed processing environments like field programmable gate arrays (FPGAs), hardware-based neural networks, and Graphical Processing Units (GPUs), some of the traditionally slow post-processing ATC methods are being considered for possible real-time implementation (Boeing 2006) and (Arrasmith 2013). We now provide an overview of some of the more notable ATC imaging methods, both traditional and contemporary. We distinguish ATC methods from AO methods in that ATC methods tend to be largely implementable in software (software dominant) whereas AO methods require hardware such as WFSs and DMs to operate (hardware dominant). Pure AO systems are real-time systems in that the hardware corrects the incoming wavefront in real time. Traditional

ATC methods have been mostly relegated to post-processing applications with a few exceptions. With the emergence of high-speed processing environments like field programmable gate arrays (FPGAs), hardware-based neural networks, and Graphical Processing Units (GPUs), some of the traditionally slow post-processing ATC methods are being considered for possible real-time implementation (Boeing 2006) and (Arrasmith 2013). We start with looking at one of the fundamental traditional ATC methods, Speckle Imaging.

Speckle imaging: Speckle imaging is a traditional ATC method that uses post-processing methods to remove atmospheric turbulence from collected imagery. It requires the capture of a very large number of short exposure images and nearby reference star images with intensive post-processing to produce the ATC image. Figure 8.26 shows the speckle imaging process.

A short exposure image and reference star image are simultaneously collected. The irradiance spectrum is obtained by taking the Fourier Transform of the image. The power spectrum $|I(u,v)|^2$ is determined and the bias is subtracted. To estimate the object's radiant emittance spectrum the magnitude of the object and object phase are required. For the phase estimation process, either the cross spectrum (Knox and Thompson 1974) or the bispectrum (Lohmann et al. 1983) of the image contain extractable phase information and are used in practice to obtain the phase information necessary for the reconstruction of the object radiant emittance. In speckle imaging, the Labeyrie method is used to estimate the object spectrum magnitude by averaging the power spectrum (Labeyrie 1970). The object spectrum phase and magnitude estimates are sent to a reconstruction filter to provide the final corrected image. We now present the Phase Diversity method, which is a well-developed ATC method that requires only a pair of simultaneously captured images to remove atmospheric turbulence.

Phase diversity: The phase diversity imaging method developed by Gonsalvez has been used extensively in removing atmospheric turbulence from imagery (Gonsalvez 1982). This method is software dominant in that it only needs hardware to simultaneously capture an in-focus and out-of-focus image. This traditional ATC method has been implemented on a field programmable gate array to provide real-time ATC (Boeing 2006). Two-simultaneous short exposure images that look through the same atmospheric turbulence are captured, but one of the images has a known change added with respect to the original atmospheric phase aberrations. For example, a slight defocus of one of the images introduces a known quadratic phase change in the entrance pupil plane of the imaging system. The image with the a priori known phase change is called the *diversity image* because of the added phase difference relative to the original image. By knowing these phase differences everywhere in the entrance pupil plane of the imaging system, mathematical expressions

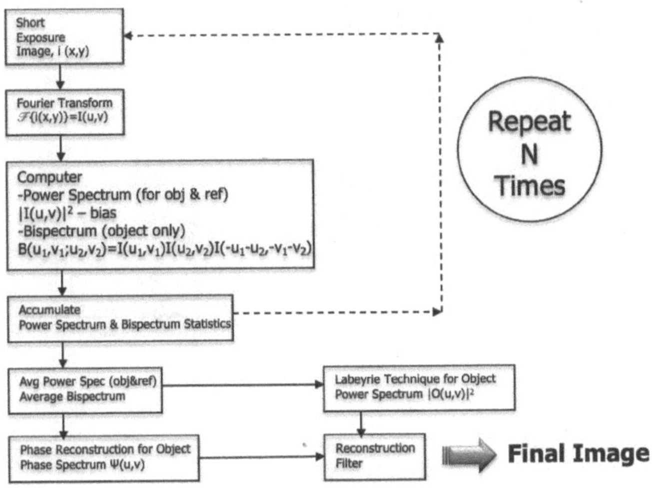

FIGURE 8.26 Process flow for speckle imaging.

for the OTF and diversity OTF can be generated corresponding to the image and diversity image respectively. An error metric developed by Gonsalvez is minimized to choose the correct entrance pupil plane phase corresponding to the atmospheric aberrations permitting the OTF to be estimated. The OTF can be inverted and used to produce the unaberrated object spectrum. Some assumptions and conditions for this method are:

The observation angle is less than the isoplanatic angle θ_a,

- The diameter of the entrance pupil plane must be greater than the atmospheric coherence length r_o,
- The original method is computer-intensive and not real-time.
- Phase aberrations need to dominate (no amplitude effects)

The phase diversity idea goes back to the 1950s when Born and Wolf discussed the general nature around 1951. The expression Phase Diversity was coined by Gonsalvez in 1976. Many researchers contributed to the body of knowledge of phase diversity since then and the method became quite prominent over the years. Some of the notable researchers include Paxman, Fienup, and Miller. Miller applied neural networks to the phase diversity method in 1991 and 1992. There was even an entire conference dedicated to the phase diversity topic in 1992, the Sunspot Conference. The phase diversity approach is shown in Figure 8.27, this method does not require a guide star and works on extended objects. An aberrated image of an extended object is captured by the imaging system and split into two parts. One of the image pair images is captured in the image plane as shown by the bottom In-focus Image in Figure 8.27. The second diversity image has a slightly different optical path length, an so introduces a known defocus in the entrance pupil relative to the original phase aberrations of the In-focus image. A computer implements the phase diversity algorithms, and the corrected image is obtained. The phase diversity algorithms are based on the linear shift-invariant systems methods previously shown in Equations 8.5 and 8.6 which show the image irradiance and irradiance spectrum respectively for an incoherent image. In the Phase Diversity ATC method, the image and image spectrum for the diversity image that has the a priori known phase offsets are simultaneously captured. The pair of simultaneously captured image and irradiance and diversity image irradiance are given by

$$i(\vec{x}) = o(\vec{x}) * |h_i(\vec{x})|^2, \tag{8.16}$$

and

$$i_d(\vec{x}) = o(\vec{x}) * |h_{i;d}(\vec{x})|^2, \tag{8.17}$$

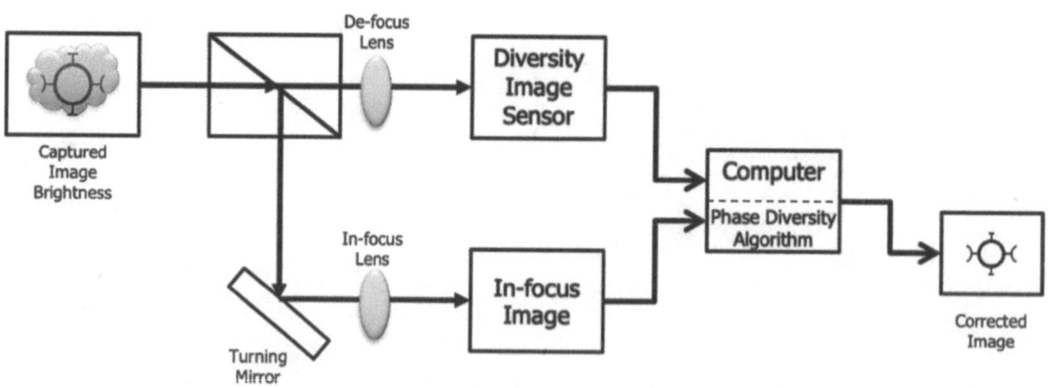

FIGURE 8.27 Phase diversity imaging approach.

where the subscript d refers to the diversity image $i_d(\vec{x})$ and diversity impulse response $h_{i;d}(\vec{x})$ respectively, and $\vec{x}$ is a 2D position vector in the image plane. The point spread functions for the image and diversity image are given by the respective magnitude squared of the impulse responses as follows:

$$s(\vec{x}) = |h_i(\vec{x})|^2, \tag{8.18}$$

and

$$s_d(\vec{x}) = |h_{i;d}(\vec{x})|^2. \tag{8.19}$$

The corresponding spatial frequency relationships for the image spectrum

$$I(\vec{f}) = O(\vec{f})\mathcal{H}(\vec{f}), \tag{8.20}$$

and the diversity image spectrum,

$$I_d(\vec{f}) = O(\vec{f})\mathcal{H}_d(\vec{f}), \tag{8.21}$$

are obtained from the 2D Fourier Transform of the image irradiance and the diversity image irradiance. One common way to proceed is to use the following traditional phase diversity method to accomplish the ATC. In the traditional Phase Diversity-based ATC method, a diversity OTF is generated by using Equations 8.10 and substituting this into Equation 8.11 or, more often, into Equation 8.8 and thereafter substituting the result into Equation 8.7. A common error metric such as the well-known Gonsalvez error metric

$$E(\vec{f}) = \frac{\left|I(\vec{f})\hat{H}_d(\vec{f}) - I_d(\vec{f})\hat{H}(\vec{f})\right|^2}{\left|\hat{H}(\vec{f})\right|^2 + \left|\hat{H}_d(\vec{f})\right|^2}. \tag{8.22}$$

can then be applied at every point in the image spectrum to determine when the OTF estimate is accurate enough. Note that the carat symbol ^ above the OTF and diversity OTF indicate that these quantities are estimated.

Wavelength diversity: Another method for post-processing ATC is wavelength diversity. This method is also intended to be performed within an isoplanatic angle but instead of introducing a known defocus element, it introduces a known diversity in the wavelength of the image spectrum which when combined with the original image spectrum can be used to estimate the un-aberrated object spectrum. However, just like phase diversity, traditional methods of wavelength diversity are also computer-intensive. Figures 8.28 and 8.29 show general models for how wavelength diversity is generally performed. Just like with phase diversity, you start with the measurement of a radiant emittance. From there, the light is sent through a beam splitter where each beam is sent through two different narrow band filters to isolate the target wavelengths. From there, the light is collected on two separate detectors who feed their data into either a computer or field programmable gate array (FPGA) that is running some kind of wavelength diversity algorithm. At the end of this, a corrected image is returned. Figure 8.29 generally acts as a simplification of Figure 8.28, but also shows that if three wavelengths are isolated and measured, say red, green, and blue, you can return a full color image.

Another version of the wavelength diversity model using a sensor capable of simultaneously detecting multiple wavelengths at each pixel is shown in Figure 8.29. With three wavelengths, color imagery can be quickly produced.

Some of the assumptions and attributes for this method include that the observations must be smaller than the isoplanatic angle, the entrance pupil of the imaging system needs to be larger than

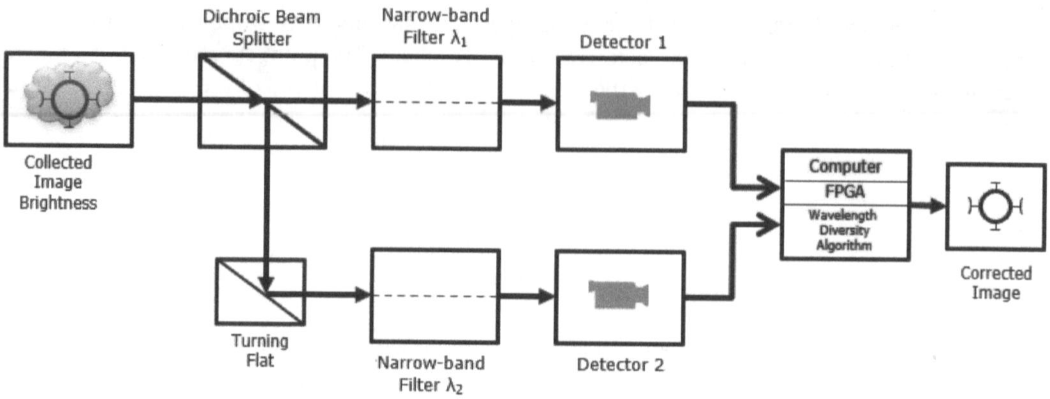

FIGURE 8.28 Wavelength diversity model.

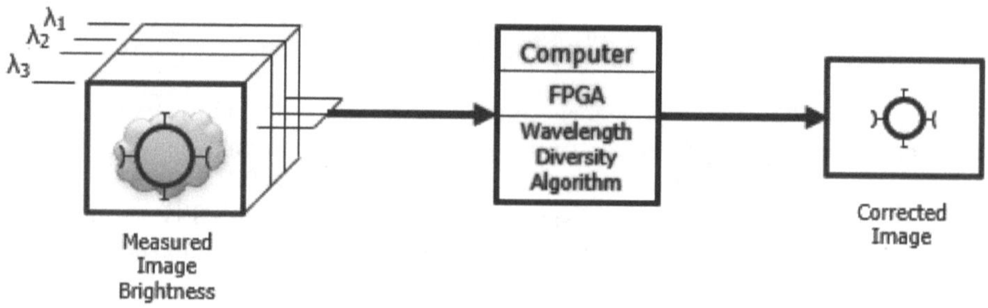

FIGURE 8.29 Alternate wavelength diversity model.

the Fried parameter, a linear (better yet, a linear shift-invariant) systems theory is applicable, and it requires the simultaneous capture of images at a minimum of two different wavelength.

The mathematical fundamentals for this process start with the same fundamentals as phase diversity in terms of the image brightness and the image spectrum (Equations 8.16 – 8.21). However, it should be noted that the object spectrum will be different based on which wavelength is being analyzed, and because of this the Gonsalvez error metric cannot be used.

To get around this, Ingleby and McGaughey developed their own error metric that considers wavelength dependence in object spectra (Ingleby and McGaughey 2002). In this error metric, m represents the index of the wavelength while n represents the index of the image frames. One pitfall of this error metric however is that it is possible to get this error metric to go to zero for any estimated value of the optical transfer function.

$$E_m = \sum_{f}\left[\sum_{n=1}^{N}|I_{mn}|^2 - \frac{\left|\sum_{n=1}^{N}\hat{\mathcal{H}}_{mn}^{*}I_{mn}\right|^2}{\sum_{n=1}^{N}\left|\hat{\mathcal{H}}_{mn}\right|^2}\right]. \tag{8.23}$$

However, a different error metric was developed by Greywolf Technical Services that uses red, green, and blue wavelengths and proprietary constraints (Greywolf Technical Services 2013). This is coupled with a high-speed parallel processing device to recover unaberrated object brightness information.

8.2.7.1 Phase Estimation for ATC Methods

To estimate the entrance pupil plane phase due to the atmospheric aberrations in the generalized pupil functions in Equations 8.9 and 8.10, an orthonormal basis function expansion is used to estimate the phase everywhere in the entrance pupil of the imaging system. A common and practical set of orthonormal basis functions are the Zernike polynomials which are used to model fixed and random aberrations in an imaging system's circular aperture (typically entrance pupil plane). Zernike polynomials are orthonormal on a circle of unit radius. Figure 8.30 is useful in relating the x and y coordinates to the radial components. The x and y coordinates of a vector $\vec{x} = (x, y)$, are related to the

$$\theta = a\tan\left(\frac{y}{x}\right), \quad y = r\sin\theta \quad x = r\cos\theta \quad r = \sqrt{x^2 + y^2},$$

radial and angular coordinates (r,θ) by the four equations shown in Figure 8.30. We can normalize the radial component so that the radius falls somewhere on a circle with radius r to the point of interest within the circle,

$$\rho = \frac{r}{R_{\max}}, \tag{8.24}$$

and $R_{\max}$ is the maximum radial component of the circle shown at the boundary of the circle in Figure 8.30. We can expand the entrance pupil phase in terms of a set of orthonormal basis functions as,

$$\varphi(R\rho,\theta) = \sum_{i=1}^{N} a_i Z_i(\rho,\theta), \tag{8.25}$$

where a_i is the weight on the ith Zernike component at radius ρ and angle θ in the entrance pupil plane of the imaging system. As shown, N Zernike basis functions are used. The Zernike polynomials are given for m not equal to zero ("i" is a single index by Noll that sums over the n and m Zernike indices) by,

$$Z_{i=\text{even}}(\rho,\theta) = \sqrt{n+1}R_n^m(\rho)\cos(m\theta), \tag{8.26}$$

and

$$Z_{i=\text{odd}}(\rho,\theta) = \sqrt{n+1}R_n^m(\rho)\sin(m\theta). \tag{8.27}$$

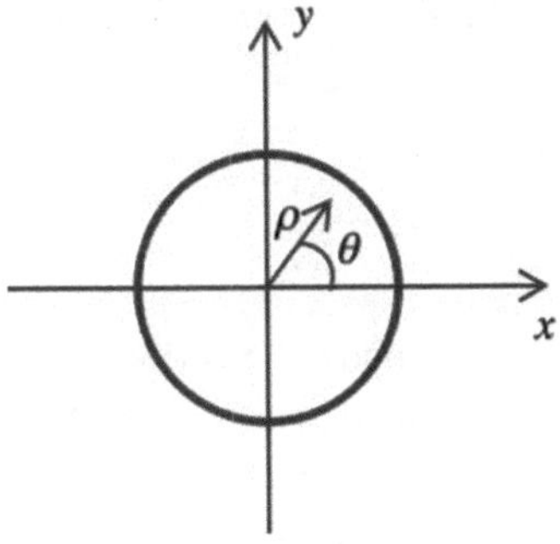

FIGURE 8.30 Radial versus rectilinear coordinates.

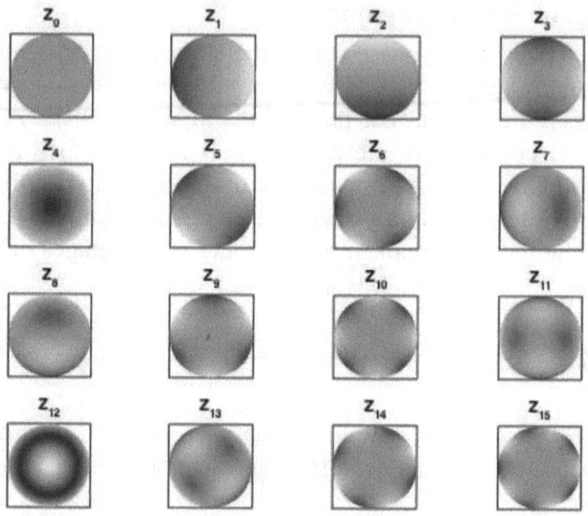

FIGURE 8.31 First 16 Zernike polynomials using the Noll ordering method (Noll 1976).

For $m=0$, the Zernike polynomials are given by,

$$Z_i(\rho,\theta) = R_n^m(\rho) \quad m = 0, \tag{8.28}$$

where

$$R_n^m(\rho) = \sum_{s=0}^{(n-m)/2} \frac{(-1)^s (n-s)!}{s![(n+m)/2 - s]![((n-m)/2 - s]!} \rho^{n-2s}. \tag{8.29}$$

Notice that, as normalized, the Zernike polynomials are defined everywhere on the unit circle and only require the weighted sum of Zernike polynomials in accordance with Equation 8.25 to provide the phase everywhere within the normalized entrance pupil aperture of the imaging system. Figure 8.31 shows the first 16 Zernike polynomials including some of the familiar terms we are used to hearing at the optometrist's office (namely, astigmatism, coma, defocus, and spherical aberration for example). Notice that the fifth Zernike polynomial identified by the Noll ordering as Z_4 is the defocus term which is introduced in the entrance pupil by slightly changing the optical path length between the in-focus image and the defocused image. The amount of defocus that is generated in the entrance pupil (at the edge/maximum as shown) can be related to the optical path difference between the in-focus and image and the slightly out-of-focus diversity image by,

$$\Delta\theta\left(\vec{f}\right) = \frac{2\pi}{\lambda} \cdot \underbrace{A_d}_{\substack{\text{Maximun}\\ \text{Defocus}\\ \text{Waves}}} = \frac{2\pi}{\lambda} \cdot \underbrace{\frac{n}{2}}_{\substack{\text{Index of}\\ \text{Refraction}}} \cdot \Delta z \cdot \frac{1}{4F/\#^2} = \frac{\pi}{2} \frac{\Delta Z}{f_{\text{eff}}^2} \left|\vec{f}\right|^2 \quad \text{for} \quad n = 1, f/\# = \frac{f_{\text{eff}}}{D_p}, \tag{8.30}$$

where A_d is the area of the detector, n is the index of refraction of the atmospheric turbulence, Δz is the offset of the diversity image has from the image plane, $F/\#$ is the f-number of the optical system, f_{eff} is the effective focal length of the imaging system, D_p is the diameter of the entrance pupil plane aperture and λ is the mean wavelength of the optical system. The quantity $\Delta\theta\left(\vec{f}\right)$ is the phase offset produced at spatial frequency $\vec{f}$ in the entrance pupil plane by the difference Δz in the optical path between the image and diversity image. The weights on the Zernike polynomials can be determined by,

$$a_i = \int_0^\infty d\vec{\rho} W(\rho)\psi(R\rho,\theta)Z_i(\rho,\theta), \tag{8.31}$$

where $W(\rho)$ is an aperture function that is one inside the clear aperture (entrance pupil) of the imaging system, and is otherwise zero. The quantity $\psi(R\rho,\theta)$ in Equation 8.31 is the same term as the left side of Equation 8.25. In modeling the atmospheric phase aberrations, the correlation properties of the weights are of paramount importance. For an excellent treatment of the correlation properties of these weights please see the work of Roggemann and Welsh (1996). Other basis function expansions can also be used such as Seidel basis functions and Karhunen-Loeve expansions are also possible.

In traditional diversity-based ATC methods, the process for estimating the OTF (and the diversity OTF by analogy) is to,

1. Use a suitable basis set like the Zernike polynomials to initially generate an entrance pupil plane phase estimate. This is done by just initially guessing the phase - say all phase values are zero.
2. Form the generalized pupil function with this entrance pupil phase "guess" using Equation 8.9. Note: use Equation 8.10 for estimating the diversity OTF.
3. Zero-pack the GPF for sampling reasons in preparation of generating an OTF estimate.
4. Form the impulse response in accordance with Equation 8.6.
5. Determine the PSF estimate by pointwise taking the magnitude squared of the result of step 4 above.
6. Use Equation 8.7 to form the OTF estimate.
7. After forming both the OTF and diversity OTF estimate, apply an error metric such as the Gonsalves error metric given in Equation 8.22 and possibly some constraints to determine the instantaneous error at each spatial location in the OTF.
8. Sum the errors to determine the total summed error due to the initial entrance pupil plane phase estimation error.
9. Change the weights on the Zernike polynomials in a methodical manner to come up with a new entrance pupil plane phase estimate.
10. Repeat steps 2 through 8 to generate a new error estimate.
11. Compare the new error estimate to the old estimate and keep the phase estimates associated with the lowest error.
12. Continue to execute steps 10 and 11 until the error is minimized and the best OTF estimate is obtained.

Once the error minimized OTF estimate is obtained, a Wiener filter (or other digital image processing inverse filter) can be generated that removes the effect of atmospheric turbulence from the collected imagery. Notice that the OTF itself is generated by phase estimates that are due to (1) atmospheric turbulence effects, and (2) aperture effects (e.g. diffraction effects). If the effects of the atmospheric turbulence are mitigated by filtering them out using the Wiener filter (or other digital filter), then the only remaining effect is due to diffraction and so the diffraction-limited result is obtained. To attempt to remove the effects of the aperture, super-resolution methods need to be employed.

After the error minimized OTF estimate is determined, the Wiener filter is given by,

$$H^{-1}(\vec{f}) = \frac{H^*(\vec{f})}{\left(\left|H(\vec{f})\right|^2 + \alpha\right)}, \tag{8.32}$$

where the asterisk on the right side of Equation 8.32 represents the complex conjugation operation. Care must be taken for the case where the denominator of Equation 8.32 approaches zero. A parameter α, that is based on system noise, is sometimes included in the denominator to prevent Equation 8.32 from blowing up as $H(\vec{f})$ approaches zero. As can be seen from Equation 8.6, multiplying the image spectrum by the Wiener filter leads to an un-aberrated object spectrum,

$$O(\vec{f}) = I(\vec{f})H^{-1}(\vec{f}), \tag{8.33}$$

and the atmospheric turbulence free object brightness estimate is simply obtained by taking the 2-D inverse Fourier transform of Equation 8.33. Unfortunately, the steps listed above, while effective, can be very slow and may take hours to days to produce an accurate OTF for images with larger amounts of entrance pupil sampling (say 127 by 127 samples). In the patent, *High-Speed Diversity-Based Imaging Method for Parallel Atmospheric Turbulence Compensation*, Arrasmith introduced a new solution that allowed for parallelization, a specialized phase "stitching" method, a new less computationally intensive error metric, resulting in more accurate, less complex, simplified, and direct diversity-based general transfer function estimation methodology (Arrasmith 2013). This methodology is applicable to any diversity-based transfer function estimation problem across the electromagnetic and acoustic spectrum that can be modeled as a linear system. Non-diversity-based transfer function estimation may also possible but with the added requirement of entrance pupil plane masks (Arrasmith 2023).

8.2.7.2 ATC Using the WOLF Methodology

In this section, we provide some details on the Well-Optimized Linear Finder (WOLF) methodology that can be used for removing atmospheric turbulence from passive imagery (monochromatic, multi-color, image frames, and/or video) in real time, where we consider real-time as 30 Hz. The WOLF ATC method first estimates the entrance pupil plane phases caused by atmospheric turbulence and then calculates the OTF from the entrance pupil plane phase estimates. The OTF is determined at every spatial frequency within its passband and non-zero points of the OTF are inverted using Equation 8.32. The inverted OTF is then used in Equation 8.33 to estimate the unaberrated object spectrum. The final step to determine the atmospheric turbulence compensated image is to take the collective 2D spatial inverse Fourier Transform of the object spectrum $O(\vec{f})$ at every spatial frequency within the OTF's passband. The result, as will be seen below, is a near diffraction-limited *reconstructed image* that mitigates the effects of the atmospheric turbulence and provides a crisp image with high spatial resolution. Before we show a representative example of the WOLF methodology in action, we must first come to understand how the methodology works. In order to do that, we must develop some notation and illustrate some of the fundamental theoretical and mathematical concepts.

We start off with a short discussion of notation to distinguish the important indices and grids necessary to illustrate the critical relationships between the fundamental physical parameters needed to remove atmospheric turbulence from imaging systems. We assume a square (or rectangular) aperture because this type of aperture can circumscribe any other unique specialized aperture, and the methods used by the WOLF algorithm for the square/rectangular aperture work also for any other specialized aperture that can be contained within the bounds of the square or rectangle. This does not mean that the methods presented in this section are the most efficient implementation for any specialized aperture, just that the methods presented below work for the general case. The approach for a specialized aperture is to embed the specialized aperture in a square, or rectangular array of zeros and then apply the methods presented in this section at every relevant point of the square/rectangular array. Note that many computational speed optimizations can be made by tailoring the process illustrated below for the square/rectangular array to the specialized aperture

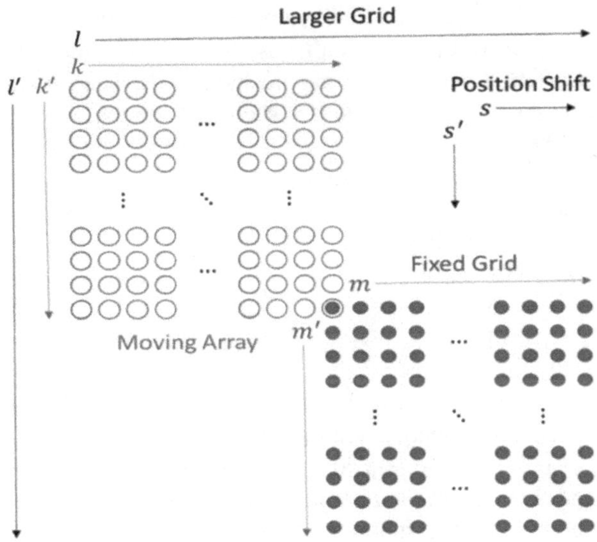

FIGURE 8.32 Index structure for the WOLF ATC methodology.

itself, by employing optimization methods, and by employing parallel processing techniques along with General Purpose Parallel Processing (GPPP) hardware technology. We coin the term applying a "stitching method" to the entrance pupil plane phases that best matches the aperture shape under consideration.

For a general square or rectangular 2D array, that is fundamental to many general imaging applications, we use Figure 8.32 to illustrate the indexing method at the core of the WOLF methodology. It is useful to have a mental model for what these different grids and indices represent. For the larger grid that is represented by the l' and l indices in Figure 8.32, this refers to items such as the object radiant emittance, object spectrum, impulse response of the imaging system, point spread function, OTF, image spectrum, image point, or camera pixels that are indexed by the larger grid coordinates (l', l). Which one of these imaging systems concepts is being addressed is made clear in the context of the discussion. These all share the same coordinate system indicated by the larger grid with indices (l', l). For the smaller grids that are represented by the indices (m', m) and (k', k), these each represent the grid structure for the entrance pupil plane phases wherein the indices (m', m) represent a fixed sampled entrance pupil plane square aperture (shown by the blue array), and the (k', k) indices represent a moving (shifted) entrance pupil plane array (shown as green open circles) with the same phases as in the fixed array. With regards to the indices themselves, we attempt to be consistent with the notation used in Matlab™ in that the top-leftmost array point has the (1,1) coordinate, and the convention is that the horizontal reference position (x) runs from left to right, whereas the vertical reference position (y) runs from top to bottom. Consequently, a matrix coordinate of (1,3) would correspond to the top row, third coordinate to the right (row = 1, column = 3) and, conceptually, this corresponds to a (y, x) ordering scheme. We now present some notational conventions relative to the grid structure shown in Figure 8.32. Boldened and italicized notation refers to the grid under consideration in its totality (includes all elements of the grid). For instance, the larger $\boldsymbol{l}$-grid consists of coordinate elements at specific (but arbitrary) location (l', l) where the italicized l' and l can take on the paired spatial position $\vec{x} = (y, x)$, or spatial frequency $\vec{f} = (f_y, f_x)$ coordinates of **any** element in the larger $\boldsymbol{l}$-grid array. Consequently, the $\boldsymbol{l}$-grid array contains the set of all possible (l', l) index values for the entire grid. Whether-or-not the grid coordinates refer to the spatial position or spatial frequency grid location at the index locations will depend on the context of the problem at hand. If we wish to identify a specific grid-point, we will use subscripts to identify the coordinate of interest.

For instance, on the **l**-grid, if we are interested in a point that is in the second row from the top and fourth column to the right, we would refer to this as (l_2', l_4). These indices without the subscripts refer to a general point on the **l**-grid at arbitrary position (l', l). The maximum number of points in the **l**-grid are given by italicized capital letters such as L' and L, where the prime (or absence of a prime) determines whether the maximum refers to the row or column maximum number of points. In a similar fashion, the **k**-grid refers to the moving entrance pupil plane phase array shown in Figure 8.32 as the green (open) circles. As seen in Figure 8.32 the **k**-grid represents the shifted version of the entrance pupil plane phases relative to the **m**-grid which represents the fixed entrance pupil plane phase array given by the blue (solid circles). The general **k**-grid coordinates are given by (k', k) and specific moving array coordinates are given by (k_d', k_e), where d and e are independent integers in the range $\{1{:}K', 1{:}K\}$, and K' and K are the maximum number of points in the **k**-grids' vertical and horizontal directions, respectively. The fixed grid shown as the solid (blue) circles in Figure 8.32 is referred to as the **m**-grid and has general coordinates (m', m) and specific coordinates (m_d', m_e), where d and e are independent integers in the range $\{1{:}M', 1{:}M\}$, and M' and M are the maximum number of points in the **m**-grids' respective vertical and horizontal directions. The parameters capital S' and S represent the number of overlapping points between the moving array and fixed array in the vertical and horizontal directions respectively. Notice that in Figure 8.32, we have what is called a *one-point overlap* situation where $(S', S)=(1, 1)$, resulting in the moving grid (green grid) having only one point that overlaps with the fixed grid (blue grid). We can now use this index convention to understand some of the fundamental mathematical and theoretical concepts of the WOLF ATC methodology.

We first show the general set of equations that are used by the WOLF method and that are applicable at all 2D points of the larger **l**-grid (i.e., the grid referencing the OTF and other imaging system parameters of the same dimensions such as the object spectrum and image spectrum). Let's assume we want to generate the OTF on the right side of Equation 8.32 to generate its inverse. The WOLF equations presented below have the property that the results of the one-point overlap scenario, as indexed by the grid structures shown in Figure 8.34, are used in all the other overlap scenarios required to generate the WOLF-based OTF estimate. Similarly, the two-point overlap scenario results, obtained by solving the set of equations below with $(S', S)=(1,2)$, or $(S', S)=(2, 1)$ can be used as known constant terms used in determining all the remaining required OTF terms. Once solved, these terms just need to be combined differently in a special way (as detailed by our constant expression) below. This combining can be done by a simple look-up process once the constants have been determined, instead of solving for them again. This attribute of the WOLF method dramatically reduced the required number of computations as compared to traditional ATC methods. Another key feature of the WOLF method is that for any OTF point with an overlap of three or more points (i.e., where $S'+S >= 3$), only a three-point overlap scenario is required, where one of the three points is a determinable constant from previous overlap results. Also, the first and last terms in the overlap scenario contain only the results from the one-point overlap scenario, and two-unknown results. This holds true for all arbitrary number of overlapping points where the condition $S'+S\geq 3$ is met. These properties will be directly evident from the mathematical expressions below. There are also relationships between the fundamental mathematical and theoretical expressions presented below. In effect, the two-point overlap expressions can be obtained from the three-point (or more) overlap expressions by setting the complex constant that is determined by the three-point (or more) overlap process to zero. Also, the one-point overlap expressions can be converted to a two-point equivalent overlap expression and solved by using a virtual point method (Arrasmith 2024).

We now want to introduce the key equations used in the WOLF method. We first want to introduce the key equations needed by the WOLF methodology using the indexing scheme shown in Figure 8.32. The 2D general expression for the irradiance spectrum at spatial frequencies (f_l', f_l) can be represented by the formula

$$I_{s',s}^{\lambda_1}(f_{l'},f_l)=\left|O^{\lambda_1}(f_{l'},f_l)\right|e^{j\phi_0^{\lambda_1}(f_{l'},f_l)}\begin{cases}\displaystyle\sum_{m'=1}^{s'}\sum_{m=1}^{s}\left[\frac{e^{j\left[\theta_{(m',m)}^{\lambda_1}-\theta_{(M'-s'+m',\ M-s+m)}^{\lambda_1}\right]}}{MM'}\right], & \forall\,(l',l)\le(M',M),\\ \displaystyle\sum_{m'=1}^{s'}\sum_{m=1}^{s}\left[\frac{e^{j\left[\theta_{(m',M-s+m)}^{\lambda_1}-\theta_{(M'-s'+m',\ M-s+m)}^{\lambda_1}\right]}}{MM'}\right], & \forall\,(M',M)<(l',l)\le(L',L)\end{cases},\tag{8.34}$$

where the expressions to the right of the single parenthesis in Equation 8.34 depend on the relative position of the center of the moving array to the center of the fixed array. On the right side of Equation 8.34, $\phi_0^{\lambda_1}(f_{l'},f_l)$ represents the object phase at wavelength λ_1, and at vertical and horizontal spatial frequencies at larger grid coordinate (l', l), $O^{\lambda_1}(f_{l'},f_l)$ is the object spectrum at the same wavelength and spatial frequency components, and $\theta_{(a',b)}^{\lambda_1}$ is the entrance pupil plane phase at the fixed grid location a' and b. The $\forall$ symbol means *for all*, and M' is the maximum number of entrance pupil plane sampled grid points in the vertical direction, and M is the corresponding maximum number of entrance pupil plane sampled points in the horizontal direction of the fixed entrance pupil plane grid. On the left side of Equation 8.34, $I_{s',s}^{\lambda_1}(f_{l'},f_l)$ is the irradiance spectrum detected by the imaging system at spatial frequencies associated with the larger grid coordinate (l', l), for fixed and moving entrance pupil plane array overlaps of S' points in the vertical direction and S points in the horizontal direction, and wavelength λ_1. The ***l***-grid coordinates (l', l) are assigned to the center of the moving array shown by the green open circles in Figure 8.34. The vertical direction is governed by the l' index and runs from top to bottom, whereas the horizontal direction is governed by the index l and runs from left to right, starting with the first non-zero overlapping point between the moving and fixed arrays as shown in the figure. The top expression in the single parenthesis of Equation 8.34 applies to the moving grid's horizontal center position (given by l in Equation 8.34) being located to the left, or equal with the horizontal center position of the fixed grid. The bottom expression in the single parenthesis of Equation 8.34 is used when the moving grid's horizontal center l lies to the right of the horizontal center of the fixed grid. Note also that the OTF can be directly determined by inspecting the right side of Equation 8.34 where the sums and the terms in the brackets provide the OTF. In this way, the OTF is seen to be a normalized sum of complex exponential phase differences. It should be mentioned that because of the conjugate symmetry of the OTF, the maximum values of S' does not need to exceed the maximum number of entrance pupil plane points M′ in the vertical direction. Further, the entrance pupil plane phase differences are fixed in relation to each other once the number of elements and the entrance pupil plane geometry have been determined. Equation 8.34 is true for an arbitrary number of overlapping points subject to the maximum point restriction in the vertical direction. This restriction does not pose any limitation since the conjugate symmetry inherent in the OTF determines half the points that are needed to fully remove the atmospheric turbulence from the imagery. This conjugate symmetry property also reduces the number of required points that need to be calculated to fully determine the OTF.

Alternatively, we can take advantage of parallel processing technology to get a factor of two improvement in the number of required entrance pupil plane phase points that need to be determined. Instead of calculating both terms to the right of the single parenthesis in Equation 8.34, we can instead calculate half the points in accordance with Equation 8.34 and using only the top expression to the right of the single parenthesis. The other half of the required points needed to fully determine the OTF is obtained by reversing the horizontal spatial frequency coordinate in Equation 8.46 as follows:

$$I_{s',s}^{\lambda_1}\left(f_{l'},f_{L+1-m}\right)=\left|O^{\lambda_1}\left(f_{l'},f_{L+1-m}\right)\right|e^{j\phi_0^{\lambda_1}\left(f_{l'},f_{L+1-m}\right)}\left[\mathcal{H}_{s',s}^{\lambda_1}\left(f_{l'},f_{L+1-m}\right)\right],\quad M<l\le L, \tag{8.35}$$

where

$$\mathcal{H}_{s',s}^{\lambda_1}\left(f_{l'},f_{L+1-m}\right)=\sum_{m'=1}^{s'}\sum_{m=1}^{s}\left[\frac{e^{j\left[\theta_{(m',M-s+m)}^{\lambda_1}-\theta_{(M'-s'+m',\,m)}^{\lambda_1}\right]}}{MM'}\right]. \tag{8.36}$$

Equation 8.35, with the form of the OTF as shown in Equation 8.36, is effectively identical to Equation 8.34 except that conceptually, the first point in the overlap starts at the top-right side of the fixed array and moves leftwards (instead of the top left of the fixed array and moving toward the right). Consequently, the same WOLF algorithms can be applied to Equations 8.35 and 8.36 with just the horizontal spatial frequency components of the array reversed along with the horizontal components of the entrance pupil plane phases. This also indicates the first opportunity for parallelization in the WOLF methodology since Equations 8.34 and 8.35 can be calculated at the same time. Since Equations 8.34 and 8.35 are essentially equivalent as far as implementation is concerned, we focus our attention on the top part of Equation 8.34, realizing that either the bottom expression after the single parenthesis in Equation 8.34, or Equations 8.35 and 8.36 must be evaluated either sequentially, or in parallel, to fully determine the OTF.

We notice that for any number of overlapping points in the conventional left-to-right moving array shown in Figure 8.34, the first point (top-left point of the fixed array) and last point (bottom-right point of the moving array) are always associated with the two unknown phases that need to be determined. We label these unknown entrance pupil plane phases as $\theta_r^{\lambda_1}$ and $\theta_s^{\lambda_1}$. It is useful and notable that the first point on the fixed array (top-left in this case) is symbolically given as $\theta_i^{\lambda_1}$, and the last point (bottom-right) on the moving array is symbolically given as $\theta_j^{\lambda_1}$, we can substitute these symbolic points into the top expression of Equation 8.34, expand the sums so that the one-point and two-point overlap expressions are specifically shown, to get the general expression

$$I_{s',s}^{\lambda_1}\left(f_{l'},f_l\right)=\left|O^{\lambda_1}\left(f_{l'},f_l\right)\right|e^{j\phi_0^{\lambda_1}\left(f_{l'},f_l\right)}\begin{cases}\dfrac{\left(e^{j\Delta\theta_{i,j}^{\lambda_1}}\right)}{MM'}, & \text{if } m \text{ and } m'=1\\[2ex] \dfrac{\left(e^{j\Delta\theta_{i,s}^{\lambda_1}}+e^{j\Delta\theta_{r,j}^{\lambda_1}}\right)}{MM'}, & \text{if } m\neq m';\forall\ m \text{ and } m'\le 2.\\[2ex] \displaystyle\sum_{m'=1}^{s'}\sum_{m=1}^{s}\left[\frac{e^{j\left[\theta_{(m',m)}^{\lambda_1}-\theta_{(M'-s'+m',\,M-s+m)}^{\lambda_1}\right]}}{MM'}\right], & \forall\ (m+m')>3\end{cases} \tag{8.37}$$

As indicated in Equation 8.34, Equation 8.37 is still restricted to the range $(l',l)\le(M',M)$. In Equation 8.37, the entrance pupil plane phase difference $\Delta\theta_{i,j}^{\lambda_1}=\theta_{(m'=1,m=1)}^{\lambda_1}-\theta_{(k'=M',k=M)}^{\lambda_1}$, where $\theta_i^{\lambda_1}=\theta_{(m'=1,m=1)}^{\lambda_1}$ is always the first point on the fixed grid (blue, solid), and $\theta_j^{\lambda_1}=\theta_{(k'=M',k=M)}^{\lambda_1}$ is the entrance pupil plane phase on the last point of the moving array (green, open circle). It can be seen that the (k', k) indices of the moving array can be related to the (m',m) fixed array by ($k'=M'-s'+m'$, $k=M-s+m$). Apart from the one-point overlap case, given by the top expression of Equation 8.37, the parameters $\theta_r^{\lambda_1}$ and $\theta_s^{\lambda_1}$ are always the unknown entrance pupil plane phases that need to be determined by the WOLF method. Interestingly, $\theta_r^{\lambda_1}$ and $\theta_s^{\lambda_1}$ are always associated with the first and last points of the respective fixed and moving arrays, as are $\theta_i^{\lambda_1}$ and $\theta_j^{\lambda_1}$. The array counting convention that we adopt is that the first point of an array under consideration is at the top-left of the array. The second point is immediately to the right of the first point in the first row. The counting convention proceeds to the right of the second point until the last point in the first row is reached. The next point is the first point of the second row of the array. This counting convention is identical to how we read a page (from left-to-right, and top-to bottom). As an example, for a 7×7 square array of entrance pupil plane phases, the phase at the first point is given by $\theta_{i=1}^{\lambda_1} = \theta_{(m'=1,m=1)}^{\lambda_1}$ and the last point, $\theta_{j=49}^{\lambda_1} = \theta_{(k'=M'=7,k=M=7)}^{\lambda_1}$. For any OTF point, other than the one-point overlap case (which does not have $\theta_r^{\lambda_1}$ and $\theta_s^{\lambda_1}$ as unknown phases), the unknown phase $\theta_s^{\lambda_1}$ is always associated with $\theta_i^{\lambda_1}$, and $\theta_r^{\lambda_1}$ is always associated with $\theta_j^{\lambda_1}$. Also, interestingly, any remaining complex exponential terms in the sums at the bottom of Equation 8.37 are complex constants that are completely determinable from previous overlap results (i.e., the complex constant in the three-point overlap result can be determined from the results of the one-point and two-point overlap results of the WOLF methodology). Keeping this in mind, we can write Equation 8.37 in a slightly more compact and general notation as,

$$I_{s',s}^{\lambda_1}(f_{l'},f_l) = \left|O^{\lambda_1}(f_{l'},f_l)\right| e^{j\phi_0^{\lambda_1}(f_{l'},f_l)} \begin{cases} \dfrac{\left(e^{j\Delta\theta_{i,j}^{\lambda_1}}\right)}{MM'}, & \text{if } m \text{ and } m' = 1 \\[2ex] \dfrac{\left(e^{j\Delta\theta_{i,s}^{\lambda_1}} + e^{j\Delta\theta_{r,j}^{\lambda_1}}\right)}{MM'}, & \text{if } m \neq m';\ \forall\ m \text{ and } m' \leq 2, \\[2ex] \dfrac{\left(e^{j\Delta\theta_{i,s}^{\lambda_1}} + C_{s',s}^{\lambda_1}(m',m) + e^{j\Delta\theta_{r,j}^{\lambda_1}}\right)}{MM'}, & \forall\ (m+m') > 3 \end{cases} \tag{8.38}$$

where

$$C_{s',s}^{\lambda_1}(m',m) = \sum_{m'=1}^{s'} \sum_{m=1}^{s} \left[\frac{e^{j\left[\theta_{(m',m)}^{\lambda_1} - \theta_{(M'-s'+m',\ M-s+m)}^{\lambda_1}\right]}}{MM'}\right], \quad \forall\ (m+m') > 3. \tag{8.39}$$

Again, the range of Equations 8.38 and 8.39 apply for $(l',l) \leq (M',M)$ and the alternate expressions in Equations 8.34 or 8.35 need to be used to generate the values of the OTF for the top-right quadrant of the OTF (and by conjugate symmetry, the bottom-left quadrant). The top expression in Equation 8.38 corresponds to the one-point overlap case for the 2D imaging scenario. The expression to the right of the one-sided parenthesis is the OTF, and the expression to the left of the one-sided parenthesis before the equals sign is the object spectrum. We want to write Equation 8.38 in terms of magnitude and phase and in terms of the sum and differences of the entrance pupil plane phases. We start with the top expression in Equation 8.38 which represents the one-point overlap case as is shown in Figure 8.34. Taking the magnitude of Equation 8.34 with the top term on the right side of the one-sided parenthesis, we get,

$$\left|I_{1;s',s}^{\lambda_1}(f_{l'},f_l)\right| = \frac{\left|O^{\lambda_1}(f_{l'},f_l)\right|}{N_{ep}}, \quad (l',l) \leq (M',M), \tag{8.40}$$

where we substituted the total number of entrance pupil plane samples N_{ep} for the equivalent product of M' with M. Since N_{ep} is known in advance, and $\left|I_{1;s,',s}^{\lambda_1}(f_{l'},f_l)\right|$ is obtainable from the 2D Fourier

Transform of the collected imagery at the spatial frequency location corresponding to the one-point overlap on the l-grid, it is interesting to note that the magnitude of the object spectrum can be directly determined for the one-point overlap case from the measured irradiance spectrum at spatial frequency location $(f_{l'}, f_l)$. We can also obtain an expression for the image spectrum phase from Equation 8.38 for the one-point overlap case as follows:

$$\measuredangle\left(I^{\lambda_1}_{1;s',s}(f_{l'},f_l)\right)=\phi_0^{\lambda_1}(f_{l'},f_l)+\Delta\theta^{\lambda_1}_{i,j}, \quad (l',l)\le(M',M). \tag{8.41}$$

Equation 8.41 already has the OTF phase component written as an entrance pupil plane phase difference so no trigonometric conversions to the entrance pupil plane phase sum and difference formulation are necessary for the one-point overlap case. The two-point overlap general expression for the irradiance spectrum from Equation 8.38 is:

$$I^{\lambda_1}_{2;s',s}(f_{l'},f_l)=\left|O^{\lambda_1}(f_{l'},f_l)\right|e^{j\phi_0^{\lambda_1}(f_{l'},f_l)}\frac{\left(e^{j\Delta\theta^{\lambda_1}_{i,\delta}}+e^{j\Delta\theta^{\lambda_1}_{r,j}}\right)}{N_{ep}}, \quad (s',s)\le(2,2), \quad s'\neq s, \tag{8.42}$$

where we again have substituted N_{ep} for M multiplied by M' and the valid two-point overlap ranges for (l',l) are (1,2) and (2,1). We can convert Equation 8.42 to express the OTF as an amplitude term and a phase term by using trigonometric identities and simplifying to obtain

$$I^{\lambda_1}_{2;s',s}(f_{l'},f_l)=\left|O^{\lambda_1}(f_{l'},f_l)\right|e^{j\phi_0^{\lambda_1}(f_{l'},f_l)}\mathcal{H}^{\lambda_1}_{2;s',s}, \quad (s',s)\le(2,2), \quad s'\neq s, \tag{8.43}$$

where

$$\mathcal{H}^{\lambda_1}_{2;s',s}=\frac{\left[2\cos\left[\frac{1}{2}\left(\Delta\theta^{\lambda_1}_{i,s}-\Delta\theta^{\lambda_1}_{r,j}\right)\right]\right]}{N_{ep}}e^{j\left(\frac{1}{2}\left(\Delta\theta^{\lambda_1}_{i,s}+\Delta\theta^{\lambda_1}_{r,j}\right)\right)}. \tag{8.44}$$

By taking the magnitude and angle of Equation 8.43, we can get a magnitude and phase representation for the two-point overlap irradiance spectrum at spatial frequency $\vec{f}_{(l',l)}=(f_{l'},f_l)$ as follows,

$$\left|I^{\lambda_1}_{2;s',s}(f_{l'},f_l)\right|=\left|O^{\lambda_1}(f_{l'},f_l)\right|\left|\mathcal{H}^{\lambda_1}_{2;s',s}\right|, \quad (s',s)\le(2,2), \quad s'\neq s, \tag{8.45}$$

and the spectral irradiance phase component spatial frequency $\vec{f}_{(l',l)}=(f_{l'},f_l)$ is given by,

$$\measuredangle\left(I^{\lambda_1}_{2;s',s}(f_{l'},f_l)\right)=\phi_0^{\lambda_1}(f_{l'},f_l)+\frac{1}{2}\left(\Delta\theta^{\lambda_1}_{i,s}+\Delta\theta^{\lambda_1}_{r,j}\right). \tag{8.46}$$

We can also express Equations 8.44 and 8.46 in terms of the sum and differences of the unknown phases $\theta_r^{\lambda_1}$ and $\theta_s^{\lambda_1}$. By doing so, this will allow for an optimization of the WOLF algorithm that can improve markedly its speed (by several orders of magnitude in some cases). Using the notation, $\Sigma\theta^{\lambda_n}_{\alpha,\beta}=\theta^{\lambda_n}_{\alpha}+\theta^{\lambda_n}_{\beta}$, and $\Delta\theta^{\lambda_n}_{\alpha,\beta}=\theta^{\lambda_n}_{\alpha}-\theta^{\lambda_n}_{\beta}$ where α and β are variables, and λ_n is the nth narrow-band, mean wavelength, and for the sum and difference of $\theta_i^{\lambda_1}$, $\theta_j^{\lambda_1}$, $\theta_r^{\lambda_1}$ and $\theta_s^{\lambda_1}$, we can rewrite Equations 8.44 and 8.46 as having an OTF of,

$$\mathcal{H}^{\lambda_1}_{2;s',s}=\frac{\left[2\cos\left[\frac{1}{2}\left(\Sigma\theta^{\lambda_1}_{i,j}-\Sigma\theta^{\lambda_1}_{r,s}\right)\right]\right]}{N_{ep}}e^{j\left(\frac{1}{2}\left(\Delta\theta^{\lambda_1}_{i,s}+\Delta\theta^{\lambda_1}_{r,j}\right)\right)}, \tag{8.47}$$

and the irradiance spectrum phase component is given by

$$\measuredangle\left(I^{\lambda_1}_{2;s',s}\left(f_{l'},f_l\right)\right)=\phi^{\lambda_1}_0\left(f_{l'},f_l\right)+\frac{1}{2}\left(\Delta\theta^{\lambda_1}_{i,j}+\Delta\theta^{\lambda_1}_{r,s}\right). \tag{8.48}$$

Equations 8.42 through 8.48 provide the theoretical, conceptual, and mathematical structure needed by the WOLF methodology to estimate the transfer function for the two-dimensional two-point overlap case. We now look at the general expressions for the three-point overlap or more cases to complete the required theoretical and mathematical concepts. As before, we start with the appropriate expression from Equation 8.38 that describes the three-point (or more) overlap case

$$I^{\lambda_1}_{3+;s',s}\left(f_{l'},f_l\right)=\left|O^{\lambda_1}\left(f_{l'},f_l\right)\right|e^{j\phi^{\lambda_1}_0\left(f_{l'},f_l\right)}\frac{\left(e^{j\Delta\theta^{\lambda_1}_{i,s}}+C^{\lambda_1}_{s',s}\left(m',m\right)+e^{j\Delta\theta^{\lambda_1}_{r,j}}\right)}{MM'},\quad\forall\left(m+m'\right)>3, \tag{8.49}$$

where the constant $C^{\lambda_1}_{s',s}\left(m',m\right)$ is given by

$$C^{\lambda_1}_{s',s}\left(m',m\right)=\sum_{m'=1}^{s'}\sum_{m=1}^{s}\left[\frac{e^{j\left[\theta^{\lambda_1}_{(m',m)}-\theta^{\lambda_1}_{(M'-s'+m',\,M-s+m)}\right]}}{MM'}\right],\quad\forall\left(m+m'\right)>3. \tag{8.50}$$

We see from Equation 8.49 that the OTF can be determined by inspection to be the collection of terms that are not related to the object spectrum terms. The OTF is seen to be

$$\mathcal{H}^{\lambda_1}_{3+;s',s}\left(f_{l'},f_l\right)=\frac{\left(e^{j\Delta\theta^{\lambda_1}_{i,s}}+C^{\lambda_1}_{s',s}\left(m',m\right)+e^{j\Delta\theta^{\lambda_1}_{r,j}}\right)}{MM'},\quad\forall\left(m+m'\right)>3, \tag{8.51}$$

which is identical to the previous two-point overlap OTF except for the additive complex constant shown in Equation 8.51. Said another way; by setting the complex constant in Equation 8.51 to zero, we can use this result to also cover the two-point overlap results discussed above. The complex constant term, $C^{\lambda_1}_{s',s}\left(m',m\right)$ in Equation 8.51 is determinable from entrance pupil plane phase results from previous overlap results. For instance, the three-point overlap complex constant can be obtained from the results of the two-point overlap results, and the complex constant for the four-point overlap case can be obtained from the two-point and three-point results and so forth. By combining the two exponential terms inside the square brackets in Equation 8.51 and applying trigonometric identities, we get a simplified expression of the OTF that has similarities with the two-point overlap OTF

$$\mathcal{H}^{\lambda_1}_{3+;s',s}\left(f_{l'},f_l\right)=\frac{\left(2\cos\left[\frac{1}{2}\left(\Delta\theta^{\lambda_1}_{i,s}-\Delta\theta^{\lambda_1}_{r,j}\right)\right]e^{j\left(\frac{1}{2}\left(\Delta\theta^{\lambda_1}_{i,s}+\Delta\theta^{\lambda_1}_{r,j}\right)\right)}+C^{\lambda_1}_{s',s}\left(m',m\right)\right)}{N_{\text{ep}}},\quad\forall\left(m+m'\right)>3. \tag{8.52}$$

Note that Equation 8.52 is entirely a function of entrance pupil plane parameters $\Delta\theta^{\lambda_1}_{i,s}$, $\Delta\theta^{\lambda_1}_{r,j}$, N_{ep}, and $C^{\lambda_1}_{s',s}\left(m',m\right)$ and that the numerical value for the reduced expression on the right side of the equal sign in Equation 8.52 is assigned to the spatial frequency $\left(f_{l'},f_l\right)$ of the OTF. Also, notice, that the expression for the OTF is identical to the OTF for the two-point overlap case, apart from the extra complex constant term that is added in Equation 8.52. To further simplify Equation 8.52 to get the OTF in terms of an expression of magnitude and phase, we need to separate the complex constant and the complex exponential into their respective real and imaginary parts and combine the results. The complex constant in Equation 8.52 written in terms of its real and imaginary part is

$$C_{s',s}^{\lambda_1}(m',m) = C_{s',s}^{\lambda_1;r}(m',m) + jC_{s',s}^{\lambda_1;i}(m',m), \quad \forall\,(m+m') > 3, \tag{8.53}$$

where $C_{s',s}^{\lambda_1;r}(m',m)$ is the real part of the complex constant $C_{s',s}^{\lambda_1}(m',m)$, and $C_{s',s}^{\lambda_1;i}(m',m)$ is the imaginary part of $C_{s',s}^{\lambda_1}(m',m)$. Equations 8.52 and 8.53 apply for any number of overlapping points larger than or equal to three. For instance, for afour-point overlap problem, the expression for the OTF is given by Equation 8.52 with an updated complex constant $C_{s',s}^{\lambda_1}(m',m)$ determined by Equation 8.53 for any given four-point overlap scenario (e.g., $(S', S)=(1, 4), (2, 2), (4, 1)$). It is remarkable that Equations 8.49 through 8.53 apply to all possible overlap scenarios greater than or equal to three, and up to the upper left quadrant of the OTF (and lower right quadrant through conjugate symmetry). Just as was shown previously, the top right quadrant of the OTF (and bottom-left quadrant by conjugate symmetry) can be determined by either the bottom expression of Equation 8.34, or by having the horizontal spatial frequency and entrance pupil plane horizontal phase components reversed like in Equations 8.35 through 8.36. It is also remarkable that the OTF itself is only a function of two known and two unknown entrance pupil plane phases, the total number of entrance pupil plane points N_{ep}, and a determinable complex constant $C_{s',s}^{\lambda_1}(m',m)$. We can combine the real and imaginary parts of Equation 8.52, with Equation 8.53 substituted into Equation 8.52 for $C_{s',s}^{\lambda_1}(m',m)$, to get the OTF magnitude and angle expressions for the three-point (or more) overlap scenario. After some trigonometric substitutions and gathering of terms, for the OTF magnitude we get

$$\left|\mathcal{H}_{3+;s',s}^{\lambda_1}(f_{l'},f_l)\right| = \frac{\left[\begin{array}{l}\left(2\cos\left(\frac{1}{2}\left(\Delta\theta_{i,s}^{\lambda_1}-\Delta\theta_{r,j}^{\lambda_1}\right)\right)\cos\left(\frac{1}{2}\left(\Delta\theta_{i,s}^{\lambda_1}+\Delta\theta_{r,j}^{\lambda_1}\right)\right)+C_{s',s}^{\lambda_1;r}(m',m)\right)^2\\ +\left(2\cos\left(\frac{1}{2}\left(\Delta\theta_{i,s}^{\lambda_1}-\Delta\theta_{r,j}^{\lambda_1}\right)\right)\sin\left(\frac{1}{2}\left(\Delta\theta_{i,s}^{\lambda_1}+\Delta\theta_{r,j}^{\lambda_1}\right)\right)+C_{s',s}^{\lambda_1;i}(m',m)\right)^2\end{array}\right]^{\frac{1}{2}}}{N_{ep}}, \tag{8.54}$$

and for the corresponding OTF phase term at spatial frequency $(f_{l'},f_l)$,

$$\theta_{3+;s',s}^{\lambda_1}(f_{l'},f_l) = \measuredangle\left(\mathcal{H}_{3+;s',s}^{\lambda_1}(f_{l'},f_l)\right) = \operatorname{atan2}\left(\frac{\left(2\cos\left(\frac{1}{2}\left(\Delta\theta_{i,s}^{\lambda_1}-\Delta\theta_{r,j}^{\lambda_1}\right)\right)\sin\left(\frac{1}{2}\left(\Delta\theta_{i,s}^{\lambda_1}+\Delta\theta_{r,j}^{\lambda_1}\right)\right)+C_{s',s}^{\lambda_1;i}(m',m)\right)}{\left(2\cos\left(\frac{1}{2}\left(\Delta\theta_{i,s}^{\lambda_1}-\Delta\theta_{r,j}^{\lambda_1}\right)\right)\cos\left(\frac{1}{2}\left(\Delta\theta_{i,s}^{\lambda_1}+\Delta\theta_{r,j}^{\lambda_1}\right)\right)+C_{s',s}^{\lambda_1;r}(m',m)\right)}\right), \tag{8.55}$$

where the symbol $\measuredangle(*)$ represents taking the angle of the argument inside the parentheses (the asterisk which in this case is $\mathcal{H}_{3+;s',s}^{\lambda_1}(f_{l'},f_l)$), and atan 2(*) is the four-quadrant arctangent of the argument inside the parentheses. The parameters $C_{s',s}^{\lambda_1;r}(m',m)$ and $C_{s',s}^{\lambda_1;i}(m',m)$ are given by Equations 8.50 and 8.53. The sum and difference formulations for Equations 8.54 and 8.55 are obtained by re-arranging the phases inside the trigonometric functions in accordance with the phase sum and difference definitions we presented earlier, namely $\Sigma\theta_{\alpha,\beta}^{\lambda_n} = \theta_\alpha^{\lambda_n} + \theta_\beta^{\lambda_n}$, and $\Delta\theta_{\alpha,\beta}^{\lambda_n} = \theta_\alpha^{\lambda_n} - \theta_\beta^{\lambda_n}$. Note that the complex constant term is already in the form of a sum of known phase differences and so it does not need to be converted into the sum and difference formalism. After re-organizing the phase terms, we get the following expressions for the magnitude of the OTF and the OTF phase

$$\left|\mathcal{H}^{\lambda_1}_{3+;s',s}(f_{l'},f_l)\right| = \frac{\left[\begin{array}{l}\left(2\cos\left(\frac{1}{2}\left(\Sigma\theta^{\lambda_1}_{i,j}-\Sigma\theta^{\lambda_1}_{r,s}\right)\right)\cos\left(\frac{1}{2}\left(\Delta\theta^{\lambda_1}_{i,j}+\Delta\theta^{\lambda_1}_{r,s}\right)\right)+C^{\lambda_1;r}_{s',s}(m',m)\right)^2 \\ +\left(2\cos\left(\frac{1}{2}\left(\Sigma\theta^{\lambda_1}_{i,j}-\Sigma\theta^{\lambda_1}_{r,s}\right)\right)\sin\left(\frac{1}{2}\left(\Delta\theta^{\lambda_1}_{i,j}+\Delta\theta^{\lambda_1}_{r,s}\right)\right)+C^{\lambda_1;i}_{s',s}(m',m)\right)^2\end{array}\right]^{\frac{1}{2}}}{N_{\mathrm{ep}}}, \tag{8.56}$$

and

$$\theta^{\lambda_1}_{3+;s',s}(f_{l'},f_l) = \measuredangle\left(\mathcal{H}^{\lambda_1}_{3+;s',s}(f_{l'},f_l)\right) =$$

$$\operatorname{atan2}\left(\frac{\left(2\cos\left(\frac{1}{2}\left(\Sigma\theta^{\lambda_1}_{i,j}-\Sigma\theta^{\lambda_1}_{r,s}\right)\right)\sin\left(\frac{1}{2}\left(\Delta\theta^{\lambda_1}_{i,j}+\Delta\theta^{\lambda_1}_{r,s}\right)\right)+C^{\lambda_1;i}_{s',s}(m',m)\right)}{\left(2\cos\left(\frac{1}{2}\left(\Sigma\theta^{\lambda_1}_{i,j}-\Sigma\theta^{\lambda_1}_{r,s}\right)\right)\cos\left(\frac{1}{2}\left(\Delta\theta^{\lambda_1}_{i,j}+\Delta\theta^{\lambda_1}_{r,s}\right)\right)+C^{\lambda_1;r}_{s',s}(m',m)\right)}\right). \tag{8.57}$$

Notice that by substituting our results into Equation 8.49, we get the general expression for the irradiance spectrum applicable to the three-point (or more) overlap scenarios

$$I^{\lambda_1}_{3+;s',s}(f_{l'},f_l) = \left|O^{\lambda_1}(\bar{f}_l)\right|\left|\mathcal{H}^{\lambda_1}_{3+;s',s}(f_{l'},f_l)\right| e^{j\phi^{\lambda_1}_0(\bar{f}_l)} e^{j\theta^{\lambda_1}_{3+;s',s}(f_{l'},f_l)}. \tag{8.58}$$

By looking at the individual components of Equation 8.49, we see that the general three-point (and more) expressions also apply to the two-point overlap case. If the constant term $C^{\lambda_1}_{s',s}(m',m)$, both real and imaginary parts, are set to zero in all the expressions above, then these equations reduce to the two-point overlap case. Further, since the solution for the one-point overlap problem involves converting the one-point problem into a two-point problem by using a virtual point method with known entrance pupil plane phase terms, the general expressions for the three-point (or more) overlap conditions apply to all points in the OTF (Arrasmith 2024). This means that the WOLF methodology can take the same approach to solve any of the overlap cases. It also means that, regardless of the number of point overlaps that are being solved, the entrance pupil plane phase estimation problem can be reduced to no more than two unknowns ($\theta^{\lambda_1}_r$ and $\theta^{\lambda_1}_s$) and a determinable constant $C^{\lambda_1}_{s',s}(m',m)$. With Equations 8.49 through 8.58 and the aid of the WOLF methodology, we can solve for the OTF at any required point in the OTF to recover the un-aberrated object spectrum, and subsequently the un-aberrated object radiant emittance. The WOLF methodology uses a suitable error metric like the Gonsalvez error metric and minimizes this error metric to determine the required entrance pupil plane phases caused by atmospheric turbulence. The determined entrance pupil plane phases are then used to determine the OTF and its inverse. The inverse OTF, at all required spatial frequencies, is then used to determine the atmospheric turbulence compensated object spectrum in accordance with Equation 8.33. The atmospheric turbulence compensated image is then obtained by 2D inverse Fourier Transforming the atmospheric turbulence compensated object spectrum. In the next section, we provide an example for how the WOLF methodology removes atmospheric turbulence from imagery.

8.2.7.3 WOLF ATC Simulated Results

Below we show an example of the WOLF ATC method by removing simulated turbulence from an image. The process is to simulate atmospheric turbulence and apply this turbulence to a pristine reference image to get an atmospheric turbulence degraded image. The WOLF method can then be used to remove the atmospheric turbulence using only a diversity image pair as discussed above.

FIGURE 8.33 WOLF algorithm results – truth image (un-aberrated, no noise).

Variable amounts and types of turbulence can be applied using Zernike polynomials as the basis functions in describing the atmospheric turbulence. The correct atmospheric turbulence statistics can also be applied so that the simulated turbulence that is applied to the reference image (pristine image) correctly captures the statistical nature of the atmospheric aberrations. In this example, a camera-to-object distance of 1,500 m and circular entrance pupil with dimensions of 127 by 127 sample points are used in the simulation. From this a 256 by 256 corrected image is recovered that indicates the removal of atmospheric turbulence using the WOLF method. For this example, we start with Figure 8.33, a reference image (or Truth image) of the well-known Lenna image. In this example, the Lenna picture serves as our object radiant emittance and consequently represents our atmospheric turbulence-free reference image.

In applying simulated atmospheric turbulence to the image, we used a Von Karman atmospheric turbulence model to incorporate the correct atmospheric statistics. These statistics are represented in the weights of the Zernike basis function expansions and were implemented using the methods of Roggemann and Welsh (1996). The Von Karman phase model generated entrance pupil plane phases, then properly reflected the statistical nature of the atmosphere. Gaussian noise can optionally be added to the phase to represent imaging system noise effects. These entrance pupil plane phases were then used to generate a Generalized Pupil Function that captures both atmospheric turbulence effects and entrance pupil plane aperture effects. The Generalized Pupil Function was zero-packed and then inverse Fourier Transformed to get the imaging systems impulse response. The square of the magnitude of the impulse response was used to produce the point spread function of the imaging system. The point spread function was then 2D Fourier transformed and normalized so that the maximum value of the result was equal to one, resulting in the OTF of the imaging

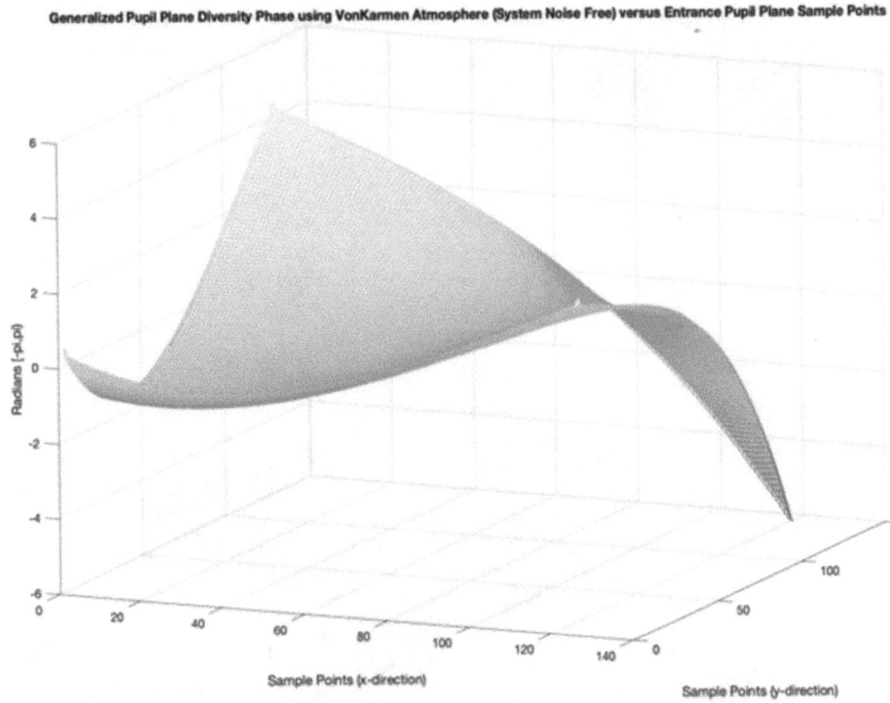

FIGURE 8.34 Simulated Von Karmen atmosphere (no noise).

system. The OTF was then point-wise multiplied with the 2D Fourier Transform of the reference image of Lenna above, and the result was then 2D inverse Fourier Transformed to get the atmospheric turbulence degraded image. Figure 8.34 shows the entrance pupil noise-free diversity phase generated for the Von Karman simulated atmospheric turbulence realization. This entrance pupil plane phase reflects the effects of the atmospheric turbulence as seen in the entrance pupil of the imaging system. This phase underwent the process described above to generate the Optical Transfer Function that reflects the instantiation of the entrance pupil plane phase shown in Figure 8.34. The Optical Transfer Function is a complex function (has real and imaginary parts) and so the Modulation Transfer Function (MTF) is often plotted to illustrate the frequency-dependent attenuation effects of the OTF. The MTF is the magnitude of the OTF and can be seen in Figure 8.35.

Notice how the MTF is sharply attenuated the further away from the center of the MTF. If there were no atmospheric turbulence effects present in the image, the MTF would just be the autocorrelation of a circular aperture with the value of one within the clear aperture of the entrance pupil of the imaging system (circular aperture in this case). In Figure 8.36, we add Gaussian noise to the entrance pupil plane phase shown in Figure 8.34. Here we are showing the entrance pupil plane phase and not the diversity phase as before.

These phases are identical except for a small quadratic phase term that is added to the general entrance pupil plane phase to produce the diversity phase in the imaging system's entrance pupil plane. Gaussian noise is also added to the general entrance pupil plane phase.

We show the MTF of the noisy entrance pupil plane phase in Figure 8.37 and can see the additional attenuating effects of the system noise. For the center point of the MTF, the atmospheric turbulence effects exactly cancel each other out, and so the central MTF value will be one. Slightly off center, this is not the case, hence the evident dramatic off center-axis attenuation. Figure 8.38 shows the results of the simulated atmospheric turbulence applied to our reference model of Lenna. You can see that the original image is now blurry and fuzzy, and there is loss of spatial resolution as compared to the original pristine image of Lenna.

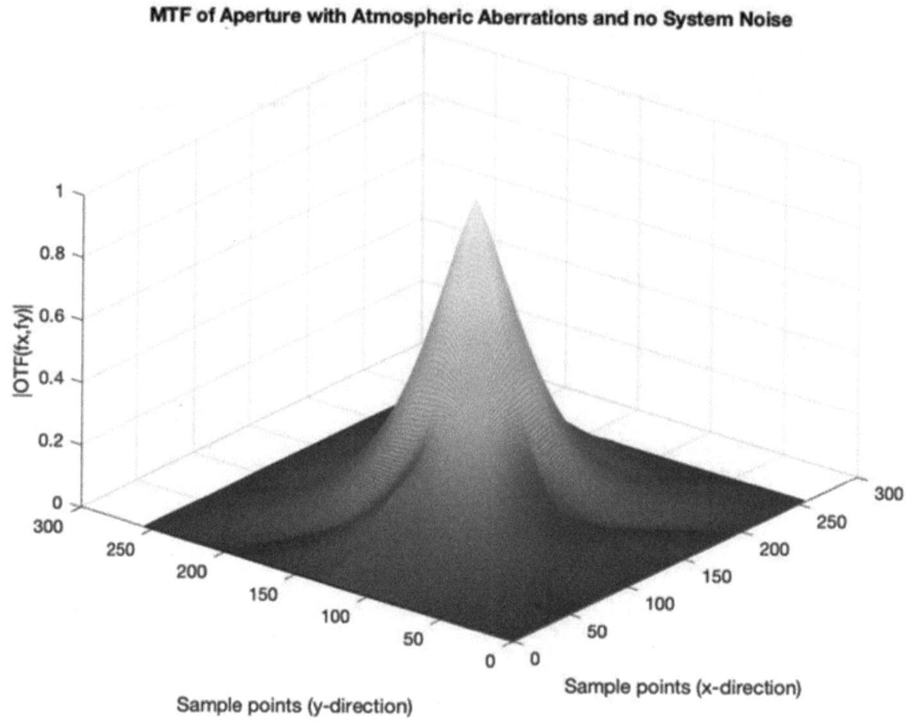

FIGURE 8.35 Simulated MTF (no noise).

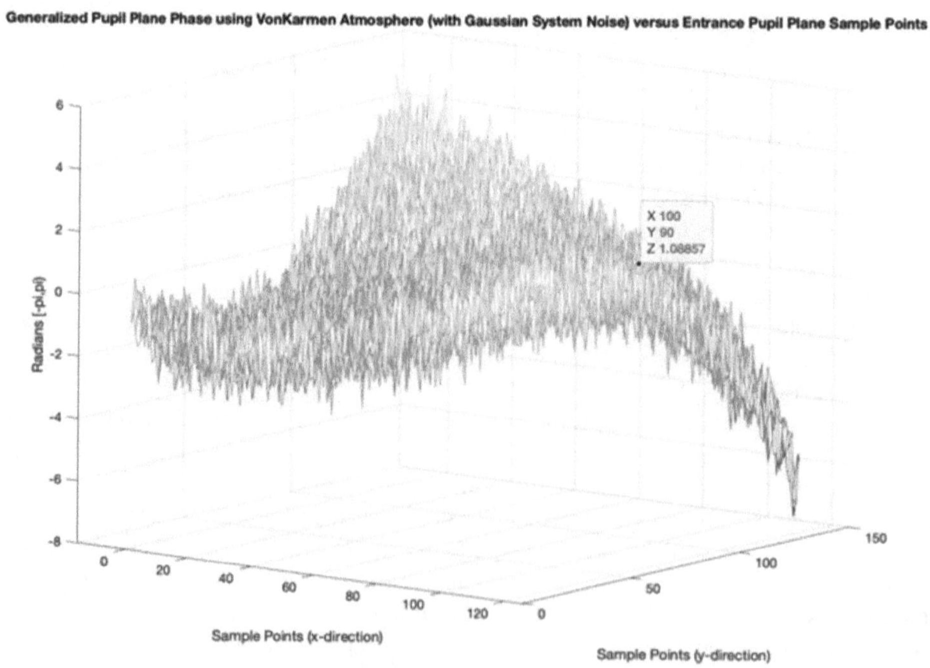

FIGURE 8.36 Simulated Von Karmen atmosphere (with Gaussian noise).

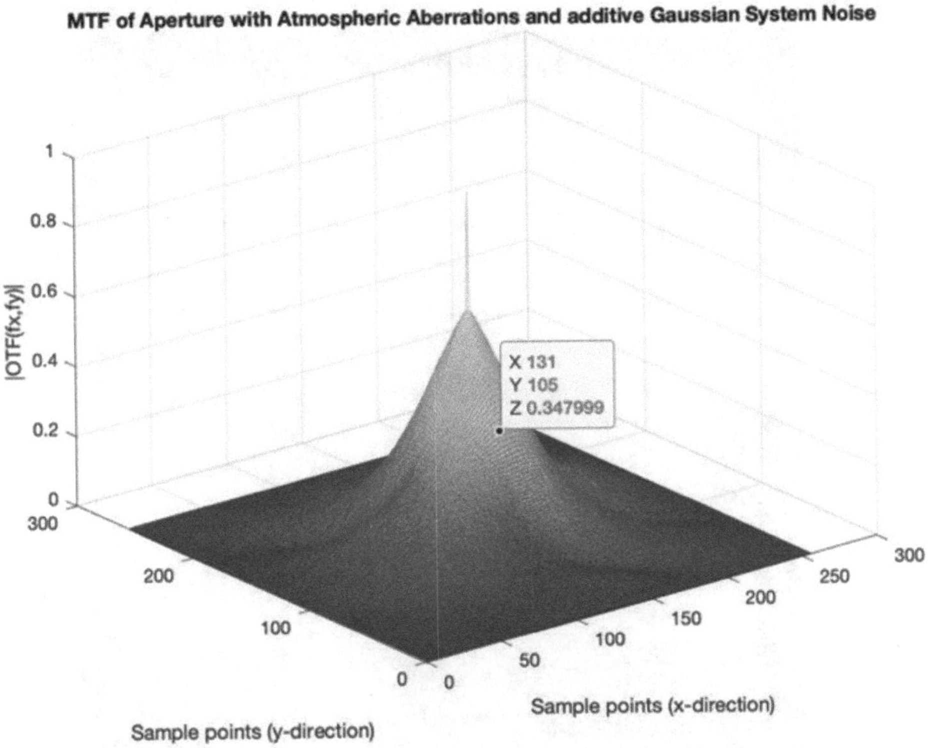

FIGURE 8.37 Simulated MTF (with Gaussian noise).

FIGURE 8.38 WOLF algorithm results – atmospheric turbulence degraded and noisy image of Lenna.

FIGURE 8.39 WOLF algorithm results – reconstructed image.

The WOLF method can now be used to remove the effects of atmospheric turbulence from the image. Figure 8.38 is what an imaging system would produce under similar imaging conditions as those that were simulated. The WOLF method requires only a captured image such as shown in Figure 8.38, along with a simultaneously captured diversity image (that is generated as previously discussed), and nothing more. Figure 8.39 shows the corrected image. Notice that the high spatial resolution is restored with the fine detail that is present on the reference object (Figure 8.35) becoming observable again. The atmospheric turbulence effects have been removed and the original pristine radiant emittance has been restored. To achieve these results, the algorithm took about 8 second while in MATLAB, on a single processor and it is expected that with parallel processing and optimization methods, real-time processing at video rates is possible. Currently, the WOLF algorithm requires the simultaneous capture of the image and its corresponding diversity image but is also able to work for any blind deconvolution or transfer function estimation problem that is linear.

8.2.7.4 Summary of the WOLF Method

The Well-Optimized Linear Finder (WOLF) methodology is a high-speed transfer function estimation method that is generally effective in a wide variety of applications. The current application example is geared toward ATC; however, the WOLF methodology can be used in other circumstances, applications, and CONOPS. The fundamental characteristics of the WOLF method are as follows:

1. Applies to linear systems (especially good for linear, shift-invariant systems),
2. Applies to coherent (e.g., laser-based) and incoherent (passive) imaging systems,
3. ATC capability currently directly implementable on diversity-based imaging systems,

4. Software-dominant method (requires little additional hardware),
5. Real-time capable (equal to or faster than 30 Hz) with the use of general-purpose parallel processor (GPPP) methods,
6. Possibly real-time capable without parallel processing with special optimizations,
7. Applicable to moving targets,
8. Complexity analysis and simulations show baseline WOLF methodology is faster than traditional ATC methods,
9. Works for many kinds of CONOPS,
10. Extendable to spectral imaging systems (e.g., hyperspectral), polarimetric, colored imagery, and video,
11. Extendable to synthetic aperture imaging methods and stereoscopic imaging systems,
12. Requires capability of short integration time images (<2 ms) to freeze the atmosphere,
13. Works with entrance pupil plane diameters (i.e., primary mirror diameters) that exceed the atmospheric coherence length (Fried parameter, r_o),
14. Resilient to additive noise.

When applying the WOLF method, two potential downsides come to mind. First, when using an error metric like the Gonsalvez error metric in Equation 8.22 within the WOLF method, the correct phase difference estimate that truly minimizes the error metric (i.e., finds the global minimum) must be very precise. This results from the error metric plot, as a function of entrance pupil plane phase differences, having a very narrow, but deep well. Potential solutions to this issue could be adaptive sampling methods and/or the use of simultaneous error metrics that have minimums at the same global minimum, but not at the same local minimums for non-solutions. Different error metrics may also be useful. Another issue is that the initial entrance pupil plane phase difference estimate occurs in a region at the edge of the OTF where signal levels are weak. Multiple grid sampling strategies and/or averaging methods may be helpful for this issue. Also, the WOLF method is a point estimator and is resilient to additive noise.

In the next section, we apply the concepts in the first two parts of this chapter to an integrated case study that has our fictitious imaging company, Fantastic Imaging Technologies (FIT) developing a high-end illumination and sensing system for a public customer (and a hidden one too). The program that FIT is undertaking has both systems engineering elements presented in Section 8.1, and atmospheric turbulence compensating sensing system elements as discussed in Section 8.2. These two elements are brought together in this fictitious program development example to show the interplay between the systems engineering driving mechanisms (e.g., the generated requirements and systems architecture), and the evolving, driven elements of the design.

8.3 INTEGRATED CASE STUDY: SOME NEW DIRECTIONS AT FIT!

In this section, we find Fantastic Imaging Technologies in the middle of requirements changes due to stakeholder directed changes in mission parameters. Worldwide tensions, the potential for escalation of hostilities in several international hotspots, and recent technology breakthroughs have driven required adaptations to the existing optical systems development effort FIT has with DHS, U.S. Customs and Border Protection. The CEO of FIT, Bill Smith has called a general staff meeting to bring the FIT team up to speed.

Bill: "Hello everybody. Thanks for coming. I have some news from Wilford Erasmus regarding our SkyNet project for DHS CBP. It turns out that our stakeholders want us to make some changes based on pressures they are receiving from the hill."

Tom (sitting up): "Changes? What changes?"

Bill: "Well, it turns out that Wilford and our special customer are getting pressure to do something more to address the threats arising from recent emerging, worldwide hostilities.

Apparently, our target set has expanded beyond the bi-pedal human traffic to a more inclusive threat target set to include ground and water-based vehicles such as cars, trucks, scooters, motorcycles, boats, All Terrain Vehicles, and even drones."

Tom: "That is quite a change!"

Bill: "And that's not all. The mission has been expanded beyond the search, detection, identification, and tracking mission we had before. They now want us to add capabilities for concealed target detection, low-light target detection, and material characterization with an adaptive, self-learning capability."

Garry: "You are talking about machine learning and artificial intelligence (AI)!"

Bill: "Exactly! They have pre-planned product improvement plans to integrate an AI package into their fleet of UAVs that gives them the ability for autonomous operations and decision making made by the AI. Our optical package needs to support this AI capability. No, that isn't quite right. It needs to be an extension of this AI package. In fact, our sensors need to be able to adapt and learn from the scene and become more efficient in turning data into useful information."

Garry: "Learn and adapt? Like how?"

Bill: "Well, our sensor package needs to provide things requested information from the main UAV AI like position data, time stamps, feature vectors of items of interest, confirmation data, potential targets and associated detection probabilities, material estimates, and things like that. The feature set and interface requirements are still evolving. One thing is sure though, we will need our own sensor AI that is able to handle the UAV AI's data requirements. We will need to add machine learning features and likely several other features. We will need to figure out the impact of these changes on our current baseline."

Karl: "Good thing we have a solid set of Model-Based Systems Engineering (MBSE) artifacts. Also, it helps that we are organized in Agile teams. Do we have any more details about the specifics of these changes?"

Bill: "Karl, why don't you get ahold of your technical contacts in Wilford's group and see what they have. I know they were working on a revised OpsCon so you might start with that."

Arlene: "Bill, these changes are way out of scope from the original contract. How are they going to handle this? Will this be a new contract?"

Bill: "Hah Arlene! I knew I could count on you to pick that up! Yes, they are fully aware that these changes require additional resources, and they are prepared to make them available. We will need to do a full-blown integrated lifecycle cost and schedule estimate (ILCaSE) for this effort. Good news is they said they are okay with modifying our existing contractual effort and adding in these new features/capabilities. We can start with the original ILCaSE and adapt that with these new expectations. Can you lead this effort?"

Arlene: "Sure Bill, happy to do that."

Bill: "Great! Thanks Arlene. As usual, put together your all-star proposal team and work and see Tom and Karl for help with any of the technical details. Let's get this rolling right away."

Arlene: "Consider it done! Marie, can you give me access to the official configuration controlled ILCaSE for the SkyNet program and I will go ahead and assemble the team."

Marie: "Sure, no problem, Arlene."

Bill: "Wonderful! Tom and Karl let's get a technical assessment for these changes and get together in a week. Let me know if you need more time."

Tom: "I can support that if Karl can."

Karl: "That works for me too. I will let you know if I hear anything from my DHS contacts that changes things."

Bill: "Okay then, we will meet on this topic one week from now."

One week later ….

Bill: "Welcome back ladies and gentlemen. Have we made any progress in getting a handle on understanding the technical impacts of the new features and capabilities that need to be incorporated for the SkyNet program?"

Tom: "We sure have. We took a holistic look at the new features and capabilities that are required for the new detection system and have inputs from the software, hardware, optics, production, and support teams, and the news is looking pretty good. Karl, do you want to take it from here?"

Karl: "Glad to! As Tom was saying, after we left the meeting last week, we went back to our Agile teams to get their take on the requirements changes and what was needed for the new features/capabilities. We also formed an Integrated Product Team to bring in production, test, support, and quality. Here's the good news, we have a solid technical base to work from, and can adapt our design and development strategy to accommodate the new features and capabilities. We are already doing an Agile development so the new features and desired capabilities can be planned into our existing workflow."

Bill: "Even the hardware changes?"

Karl: "Yes, even the hardware changes. I will let George talk to the specifics in a moment but even though the Agile development methodology came from the software world, it works just fine on hardware-dominant, or mixed systems development projects that satisfy the right requirements for an Agile development. The idea that Agile only works for software projects is a myth and has been proven false. In fact, as you will see, the type of features and new capabilities that our stakeholder is asking for are right in line with our Agile development strategy, and we are organized well to deal with these changes. Agile is a great strategy to use if you have evolving, volatile, or uncertain requirements, and/or if you have rapid delivery expectations. For this effort, the stakeholder's expectation in feature/requirements evolution, and the schedule pressure from the Hill makes our Agile methodology a good choice."

Bill: "But wait, aren't there some conditions that need to be in effect to make sure that Agile is a good fit?"

Karl: "Yes that is true. As you know, we already addressed these conditions for the current effort, and if you recall, they were a key component in ultimately selecting an Agile development methodology. Since the current effort is already organized for an Agile development, we just need to make sure that the requested new features and capabilities also fit that development model. We gathered the team and looked at the expected features and capabilities under the lens of these conditions and found that an Agile development methodology fits very well with these new features and capabilities too."

Bill: "That is good news. Can you go over the conditions again that convinced you that the Agile methodology was right?"

Karl: "Sure Bill. There are seven conditions that need to be satisfied before considering Agile as a beneficial development methodology. These are,

1. Being able to set up an incremental systems development,
2. Have short development cycles (sprints),
3. Ensure you have continuous stakeholder involvement,
4. Be able to time box the effort,
5. Set up self-organizing teams,
6. Have *excellent internal and external communications,*
7. Be able to have a *rapid change response.*

With respect to the new expected features and capabilities from the stakeholder, we expect all these conditions to be satisfied and in place. Why? There are several reasons. First, the biggest part of this effort is adapting our ATC capability to the new sensing modalities. This is largely a software effort, and Agile came from the software environment as you know. Second, the requirements are still evolving like you said in the meeting last week. Features, capabilities, and requirements may not settle down immediately, so we need a development strategy that accommodates the potential volatility in requirements. We also need to have continuous involvement with the stakeholder as these features and capabilities, and associated requirements evolve. Third, there is likely going to be schedule pressure for these new features and capabilities. Our stakeholders are already feeling pressure from the Hill, and I expect this pressure to be extended to us in the form of a tight schedule. Fourth, we also need to provide functionality rapidly and so need to have an incremental development strategy with short development cycles and tangible results at the end of each cycle. To ensure short development cycles, we need to time box our effort like we have for the current work. The new features and capabilities can be decomposed from *themes*, to *features*, to *epics*, and then to *stories* and we can use parallel sprints with our existing self-organizing teams to simultaneously address the needed functionality. Since we are in continuous involvement with the stakeholder and we are providing needed capabilities in short, time-boxed efforts, we need excellent communications within and without the development teams. We already have this. Collectively, there is good indication that the Agile methodology will be beneficial to this new effort."

Bill: "Nice summary and I agree, our current Agile approach is well suited for this new work if the work itself lends itself to an incremental development. How do we know this is the case?"

Karl: "Good question Bill! I will let the team leads address this. Since the biggest part of the expected work is software related, I will let Phil address this question for the software effort."

Phil: "Thanks Karl. The Integrated Product Team (IPT) we put together to study the technical impact for the proposed changes came up with the AI software add-on, the machine learning, the modifications to the ATC, and the advanced signal processing software being the biggest portion of the new effort. Yes, there likely will be sensor work that needs to be done like adding the new all-weather imaging capability, and we also may need to resurrect the laser illuminator idea we had before, for the low-light capability, but we think that is not where the bulk of the development effort is going to be. We must do a trade study to evaluate the best approach to provide the material characterization, low-light, and expanded target set detection, identification, and tracking capabilities. We have a pretty good idea how to do this, but we still need to do the study. Regardless, of the eventual sensor types and hardware that is selected, George says the hardware development is relatively straight forward. The technology edge will be in the software side in the form of the ATC, machine learning, and AI, in addition to our current software efforts. The good news is that much of this functionality can be parallelized and so different Agile teams can work on it. Also, we can introduce early functionality and then expand on that so Agile is perfect for this development. For instance, the optics folks say the ATC capability can first be provided for single wavelength, for a nominal EO/IR camera. We can look at a functional/feature progression starting with a single wavelength ATC correction on a small image segment like in a particular region of the image in non-real time. The next feature could be to expand the corrected size in non-real time. Next, provide the feature for a small ATC image at a single wavelength in real time (higher than 30 Hz). Then, do a larger image segment at one wavelength at 30 Hz. Afterwards, provide color images at 30 Hz, and keep expanding from there. We can also use as features the expanded functionality of applying the ATC capability to the other sensors that will perform the material characterization capability. Expanding the target set that the ATC is capable of correcting could each be added features, functions, and capabilities. There are useful deliverables at the end of each of these efforts and so an incremental development for the ATC is certainly possible."

Bill: "Well that certainly is good news. Let's get a revised CONOPS and start the trade studies for how to best provide the low-light, material characterization and all-weather imaging capabilities. Let's also run our models on the new target sets and see what develops there. Let's also start thinking about how to organize and decompose these efforts into Themes, Features, Epics, and Stories and begin planning how to provide parallel sprints to incrementally provide these features and capabilities. I know you need the results of the trade studies to determine what sensors and hardware is going to be developed, but please work on that. Let's also develop the plan for reconciling and integrating this new work with our existing tasks so that we can update our Integrated Master Schedule (IMS) when the work is approved and funded. Arlene, please keep me updated on the progress of the development of the Integrated Lifecycle Cost and Schedule Estimate and let me know if you need anything."

Arlene: "Sure thing, Bill."

Bill: "Tom, please stay on top of the technical analyses and schedule a meeting when all this is accomplished."

Tom: "Certainly, my pleasure."

Bill: "Thanks everyone, great job!"

A few days later in a planning session of the IPT, Tom calls the meeting to order.

Tom: "Hi everyone! Let's see where we are at with getting a grip on these new expected capabilities. Let's start with the big picture. Karl, did you close the loop with our special customer and DHS to make sure we are accounting for everything in the CONOPS? We need that to make sure we have the correct conceptual big picture from an operational point of view."

Karl: "We got a response over the secure fax just this morning and the team is already on it. We are comparing the updated CONOPS graphic that is part of the CONOPS document to our current OV-1. We are looking at the changes, assessing the impacts, and prepare for integrating the changes into our modeling framework once we receive the authorization to proceed."

Tom: "Is this it then, or can we expect other changes?"

Karl: "Yes, I talked to both Wilford and Rebecca, and they are both good with this current CONOPS. They also got inputs from their user community, and they think this is it. Hah, at least for now! Let me bring the CONOPS graphic on screen as a point of reference (Figure 8.40)."

Karl (picking up the laser pointer): "As you all can see, there are a lot of similarities from the previous CONOPS graphic that we have been working with. However, there are some significant differences. Notice the expanded target set that includes animals, boats, vehicles, man-made structures (points to the corresponding icons), and even aircraft. After talking with Wilford and Rebecca, we found that by aircraft, they mean low flying aircraft, helicopters, and drones. For vehicles, anything goes from mopeds, bicycles, motorcycles, cars, to large trucks. Notice also that they have the concealed target detection requirement indicated by the concealed person icon, and the low-light detection requirement, indicated here (points to icon with person next to the dark tree). And here is something we didn't expect, an anti-glare capability is needed. Additionally, they want all-weather, day-night capabilities and be operable in all types of terrain (points to terrain icons at the bottom right). It also looks like there will be atmospheric turbulence to deal with for all these capabilities. The artificial intelligence enhanced autonomous unmanned aerial vehicles (AI-AUAV) will be able to fly solo missions or cooperate in a swarm with other AI-AUAVs. They will look like search and rescue drones (pointing to the symbols in the

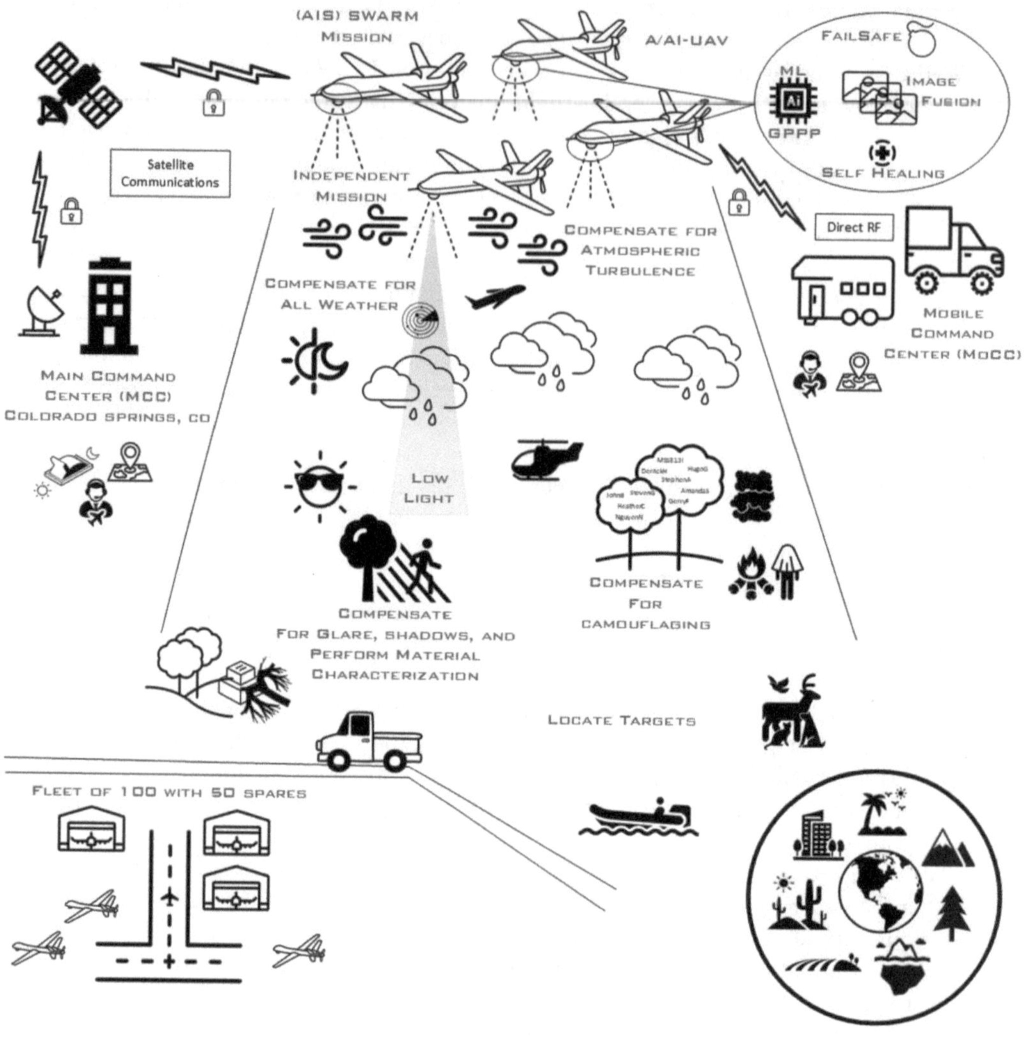

FIGURE 8.40 Concept of operations for SkyNet program.

top-right bubble), have tamper proofing, remote control override function, target detection, tracking, and identification, snapshots and video, and AI functions. We asked about the AI functions and Wilford and Rebecca stated that this included machine learning features, data mining, natural language processing, multi-sensor adaptive feature vector extraction and pattern recognition, and interfacing with the larger, main AI on the AI-AUAV host platforms. Apart from the image and data characterization and analysis functions, the optical AI needs to provide the sensor data and supporting meta data to the main AI-AUAV in actionable form and be able to respond to directions and commands from the main AI on the UAV. Secure comms are used over a satellite link from the Mission Command Center in Colorado, or from a Mobile Command Center that can establish a local command link. According to this CONOPS, 150 of these A/AI-UAVs will be generated."

Amanda: "This is fantastic. There are a lot of similarities with the current project. However, optics can't give you an all-weather capability. You will get degraded performance with rain, and the imaging sensors can't look through clouds in the visible part of the electromagnetic spectrum."

Tom: "Good point Amanda. I think our stakeholders know that a one-sensor solution is out of the question. They will have to have radar and other ground, air, and space assets and sensors like acoustic, or infrasound to get a solid, robust all-weather capability. We will have to confirm with them that they have a solution, or if they want us to provide a non-optical imaging solution. We will check into it."

Amanda: "Ah, good idea! That makes sense. Thanks Tom."

Karl: "Amanda, can you please give Tom the rundown on the analysis your optics group did, based on this new CONOPS?"

Amanda: "Sure Karl. Nice job summarizing the capabilities and features by the way. Based on some previous trade studies we did for a similar airborne platform; we believe that there are some good candidates to address the requested optical features and capabilities. We had a previous environmental characterization project to characterize ground contaminants in farmland and riverways from the air. We used a predator class UAV with a hyperspectral sensor to do the material characterization. We flew the sensor from the payload of the UAV. The hyperspectral sensor could also be used to discriminate man-made objects from natural ones, such as in the stakeholders concealed object detection requirement. By the way, the UAVs that our stakeholders have are generic UAVs. The payload and onboard processing make them autonomous or, with this and other contractor upgrades, capable of hosting an artificial intelligence. The name should be A/AI-UAV, not AI-AUAV since their base UAVs aren't autonomous."

Tom: "Perceptive Amanda. I will get that back to our stakeholders."

Amanda: "It's not that important, but it does get me thinking. For the low-light capability, an eye-safe laser illuminator would fit the bill. We have some experience in using those too on a previous project. For the anti-glare requirement, Polarization sensing is good for that. It can also do some scene characterization and pull things from the background. So, we are looking at a multi-senor platform consisting of an EO imaging sensor that can do grayscale, and color images and video in the visible part of the electromagnetic spectrum, and an IR sensor that can take thermal images and video for nighttime. A hyperspectral imaging system for the concealed target/material characterization requirement, and a polarization imaging sensor for anti-glare. All these sensors can benefit from our ATC methodology and so we should centralize the ATC functionality."

George: "Even the hyperspectral sensor?"

Amanda: "Yes, George. Even the hyperspectral sensor. The hyperspectral sensor should be a hyperspectral imaging sensor (HSI) system that is capable of producing 2D-spatial images at each of its wavelengths. The wavelengths will span the visible and go into the near infrared so atmospheric turbulence will be an issue in the achievable spatial resolution of the hyperspectral images, especially in the visible regime. Also, the atmosphere smears your signal across your focal plane and so the atmosphere reduces the potential signal available at each pixel. This also impacts the spectral analysis since the signals across all the wavelengths are correspondingly spread across adjacent pixels. At a pixel, the signal decreases because of the atmospheric turbulence, yet the pixel noise due to the system remains the same. Consequently, the Signal-to-Noise ratio at each pixel is markedly less than it could be if it had ATC."

Karl: "So, what I am hearing is that the ATC capability is even more important with these new features and capabilities that our stakeholder wants. It is sounding like the bulk of the work will be in the ATC, AI, ATR, and advanced signal processing areas."

George: "Karl, that is sort of true. However, we still have a significant hardware effort in designing the laser illuminator and integrating that with the EO, IR, HSI and polarimetric imaging sensors. The good news is that we have done each of these things before and we have robust processes in place for developing the sensors, optical and computational hardware. We also have reusable or modifiable MBSE and other modeling artifacts, and still have some Hardware-in-the-Loop (HiL) and Software-in-the-Loop (SiL) test structures available that we can capitalize on. In essence, this is a mixed systems development effort, but the bulk of the design innovation and corresponding risk will be on the software side."

Karl: "Phil, do you concur?"

Phil: "Karl, yes, I do. Jennifer, Amanda, George, Kari, and I had a separate meeting yesterday and we laid things out on the whiteboard. The software is going to be the biggest effort. We must consider and provide the autonomous/artificial intelligence capability, centralize the ATC capability so that all the different sensor modalities can access it, and provide the highest spatial resolution, real-time imagery, video, and best possible corrected spectral data as needed. We must implement automatic target recognition, integrated digital filtering, advanced signal processing, machine learning, and data-mining technology and make that available to all the sensors too. Because of the AI requirement, the whole thing needs to be adaptive, evolutionary, independent, and self-correcting. As Steven Hawkings said, *Intelligence is the ability to adapt to change* so this will be a significant part of the overall effort. Also, the individual sensor configurations (Instantaneous Field of View, Transverse and Lateral Magnification, filter settings, timing changes, mask settings, sensor gain, effective focal length,....), Mission Planning Software for Sensors, Sensor Digital Twin, Sensor Post Processing Software, and data preparation for the ATC/Other Signal Processing activities are the purview of the individual sensor elements (i.e., hyperspectral sensor, polarimetric sensor, and EO (visible/daytime) sensor, and IR (thermal/nighttime) sensor) and must also be implemented."

Tom: "That is quite the effort. It sounds like the AI, ATC, ATR, and machine learning, and data mining are the long poles in the tent."

Phil: "Well, the machine learning is considered an application of AI, and the data mining is connected to AI in the sense it is a requirement for AI, so there is a natural bundling of effort there. The ATR can have the bulk of the processing done in the Mission Operations Center and, if desired, the Mobile Command Center so we just need to make sure the data is captured, processed, and disseminated in a format compatible with the ATR algorithms, and that the data gets to the ATR algorithms fast enough. We will be able to upload some high-priority ATR target sets for onboard, immediate processing but the main ATR functionality can be implemented remotely for lower priority targets, or target sets where time is not of the essence. I would say, apart from the main AI function, which we are not developing, the ATC extension to the hyperspectral imaging system would require the biggest effort from us. Yes, we must integrate with the main AI on the UAV and provide and get information from it, but we have the relative luxury of focusing only on AI features and capabilities that are relevant to the optical sensing features and capabilities."

Tom: "Thanks Phil. That gives me a good idea on where we stand. Another consideration is that we might be able to get the laser illuminator, and maybe even the "all weather imager" as Government Furnished Equipment (GFE). These are relatively common and that might help keep our costs down. I know they won't be as integrated into our sensing systems as they would be if we designed them ourselves, but I think that for the initial capability, we may be able to get by with Commercial Off-The-Shelf (COTS), Militarized Off-The-Shelf (MOTS), or Ruggedized Off-The-Shelf (ROTS). What do you think, Karl?"

Karl: "That is a great idea, Tom. We will start looking around and see what is available."

Tom: "Thank you, Karl." Let's start to decompose the planned efforts into manageable units so our Agile teams can incrementally develop these functions, features, and capabilities. We want to break the effort into Themes, Features, Epics, and Stories. Themes are the strategic initiatives for the effort and has the largest block and scale of work that needs to be accomplished. At the strategic level, the new CONOPS and supporting information is a good starting point for us. The Theme for our current Sky Net program is human target detection, identification, and tracking in support of DHS, US Customs and Border Protection (and our special customer's) missions. We are now considering expansions to the existing Theme in not only technical areas, but also in missions, capabilities, features, and even target sets. From some of the high-level, strategic artifacts, like the CONOPS, the Features are readily apparent, as articulated in the required capabilities. For instance, the low-light detection capability, concealed target detection capability, material characterization capability, expanded target detection capability, target characterization, identification and/or tracking capabilities, anti-glare capability, day night capability, 24/7 mission capability, all-terrain capability, are essential elements of the Theme, and all have associated Features that need to be provided by our illumination and sensing system on the host UAV platforms. These Features need to be broken down into large consistent, and coherent chunks of work called Epics, and these Epics then need to be further broken down into User Stories. The User Stories are organized to achieve the Epics, and the Epics organized to achieve the Features, and the Features provide the capabilities needed by the Theme(s). For example, for a Use Case associated with the CONOPS that is describing the capability of human target identification in an open terrain environment during daytime, a possible technical response for providing this Feature could be to use a diversity-based ATC imaging system approach with an EO sensor operating in the visible wavelength regime. The general 24/7 ATC imaging effort for human target detection in an open terrain environment during the day is a good example of a Feature that needs to be provided. Notice, that there would be many features that make up the target detection capability itself including other target sets, other terrain, weather, and temporal conditions. There likely would be other Use Cases related to the CONOPS that would provide the operational and conceptual context for the Users and shed light on the required features. Features are decomposed into Epics, and Epics can be broken down further into a series of User Stories, that have a common goal and can span multiple Sprints. There are roughly 5–15 User Stories in a typical 2-week Sprint. Story Points are blocks of effort or tasks of around 2–4 hours each. For example, if our team can accomplish and maintain three Story Points a day, that would amount to 42 Story Points accomplished in a 2-week Sprint. To get an idea about the scale of work associated with User Stories, the User Stories themselves are articulated from the User's perspective in a short sentence, such as, *As an account owner, I have to be able to set my user name and password.* User Stories are at the lower levels of effort, with higher levels of detail, and are written from the Users perspective. According to Ron Jeffries who co-invented the User Story practice, User Stories should follow the three C's: Card, Conversation, and Confirmation. For the three C's, the *Card* represents writing a simple sentence that reminds everyone what the User Story is about. For *Conversation*, the User Story should be written to promote continuous conversations about the effort permitting refinement of the requirements. The *Confirmation* aspect of the User Story involves ensuring that acceptance criteria are baked into the User Story. For the ATC example above, an Epic could be implementing ATC on a small image segment, for one wavelength, in non-real time. This is a complex, consistent, and coherent large effort that will span multiple 2-week sprints. For our ATC example, User Stories could have cards that read like this,

As an image analyst, I need to get the image and diversity images in digital form.

As an image analyst, I need to have my images in a central digital processing location.

As an image analyst, I must be able to take the 2D Fast Fourier Transform (2D-FFT) of the image and diversity image.

As an image analyst, I need to be able to select a small region from the image spectrum and the diversity image spectrum.

As the image analyst, I need to calculate the One-Point overlap parameters in accordance with the Well Optimized Linear Finder (WOLF) ATC methodology.

Of course, there are many more of these types of statements that are needed to complete the stated Epic of a diversity-based ATC correction of a small image segment, for a single wavelength, in non-real time. We can look at the progression of Epics for the ATC imaging of a human target in an open environment during daytime Feature, by starting with a single wavelength ATC correction on a small image segment like in a particular region of the image in non-real time (our first example Epic above). The next Epic could be to expand the ATC corrected size of the image in non-real time. Next, provide an Epic for a small ATC image at a single wavelength in real time (higher than 30 Hz). Then, an Epic for a larger image segment at one wavelength at 30 Hz or better. Afterward, an Epic for providing ATC color images at 30 Hz, and keep expanding from there until the Feature has been developed. Notice that there are *exit criteria* (Confirmation) for the User Stories and the Epics, and that there is useful functionality provided in the form of deliverables at the User Story level, Epic level, and Feature level.

So then, let's finish the trade studies for the impacts of the new expected capabilities and generate a notional OpsCon graphic and supporting document with the results."

Jennifer: "OpsCon? Isn't that the same thing as a CONOPS?"

Karl: "Thanks Jen, there is a lot of confusion between those two words. The CONOPS is usually developed by the Stakeholders and community to show how the system fits in the larger environment and relates to other systems from an operational perspective. The OpsCon is often developed by the development team, from the perspective of the users, and has the system under development as its focus. Notice that the CONOPS graphic presented the required capabilities themselves and not the developers' solutions for implementing those capabilities. After the trade studies are complete, we can have an OpsCon graphic that shows the system under development that implements the required capabilities. For instance, for the low-light capability, we may have laser system that will illuminate the target of interest and provide that capability. The OpsCon will show the laser system, and not just list the required capability itself. We can then generate Use Cases based on the OpsCon that employs the various sensors that are being developed for different operational uses."

Tom: "That's correct. We can then use the CONOPS, OpsCon and their associated Use Cases, Scenarios, and Scenario Steps for a variety of things such as fleshing out the system architecture, generating requirements, developing system modeling artifacts, and refining the decomposition of work into Themes>Features>Epics>User Stories for our Agile teams."

Jennifer: "I see!"

Tom: "Good! I think we know our next steps. Next time, let's look at a draft version of the OpsCon graphic and start writing the supporting document. Karl, can you please organize this and put the next meeting on the schedule?"

Karl: "Happy to, Tom!"

Tom: "All right, thanks everybody!"

[After the OpsCon was generated, a Technical Interchange Meeting of the IPT was called. Tom starts the meeting ....]

Tom: "Welcome back everyone! Get comfortable and grab some coffee and munchies. We have a lot to go over. I will kick-things off with a quick overview of the sensor Trade Study results and relate our sensor suite selection to the demanded capabilities from our Stakeholders. I will then go over an OpsCon graphic that we generated so that we can effectively communicate with our Stakeholders and provide them some perspective on the operational aspects of the sensor suite from the development side. In a previous IPT TIM, we identified that a significant part of the Sky Net program effort will lie in developing and implementing the ATC capability for all the different sensor modalities, so we ran some simulations that summarize. From the simulations, we found that the ATC HSI system presented the largest risk, and so we will then go over those findings and discuss some strategies for risk mitigation. Finally, I want to go over any updates to the workflow and get your thoughts to see if we missed anything. If we are good to go with everything, then we can tell Bill that we have finished the technical assessment. Any questions or comments before we get going?"

Arlene: "Just a reminder. After this meeting, I will need to get with the technical leads for their inputs for the Integrated Lifecycle Cost and Schedule Estimate (ILCaSE) to integrate this new work into our existing effort. We've started to put the cost models together but need some technical parameter estimates and bounds for our parametric models. Once this technical impact assessment is completed, we will need some help on the cost estimation side."

Tom: "No problem, Arlene. Bill has made this top priority, and you will get whatever help you need. In fact, after we are done here, let's meet with the team leads and get things rolling."

Arlene: "Very much appreciated!"

Tom: "Okay, the good news from the sensor Trade Studies is that there weren't any real surprises with regard to our sensor solution for the new required stakeholder capabilities. For the low-light capability, we will add a laser illuminator package. The main application would be for an illuminator for monochromatic, ATC imagery, but we will also look at laser tunability for color images and video, and possibly aiding the spectral sensors. We will keep the EO/IR ATC image and video approach for the current effort, add a hyperspectral imaging capability (HSI) for the material characterization and concealed target detection capability, and use a polarimetric imaging system for anti-glare and additional scene polarization characterization and imaging. Our WOLF ATC methodology will scale to all these sensor types but as you will see, the HSI system generates lots of data and so we will need to carefully manage expectations when it comes to how many wavelengths of data, and the size of the ATC corrected image segment for real-time processing. The good news is that our current approach for correcting single wavelength image data for removing atmospheric turbulence effects scales directly to hyperspectral and polarimetric imaging modes. Karl, can you bring up the OpsCon graphic and go over it please for reference?"

Karl: "Sure! Here is the OpsCon showing our technical approach (Figure 8.41)."

[Picks up the laser pointer.]

Karl: "You can all see the similarity between this OpsCon graphic and the CONOPS graphic we had before. Notice that instead of showing the capabilities that need to be addressed by the system in its environment, we are showing the system under development from the Stakeholders perspective. Our systems engineering group and technical teams developed this OpsCon to show the operational concepts of the optical system under development in its environment. For example, all the imaging and sensing modalities – EO/IR, low-light imaging, polarimetric imaging for anti-glare, scene characterization, and some concealed target detection (shown by the trees over the man-made structure), and the HSI system for material characterization and concealed target detection are directly shown. The need for ATR for target classification and identification, AI with machine learning to adaptively identify useful patterns in the data, the General-Purpose Parallel Processing element for high speed signal/data/information processing, image/data/

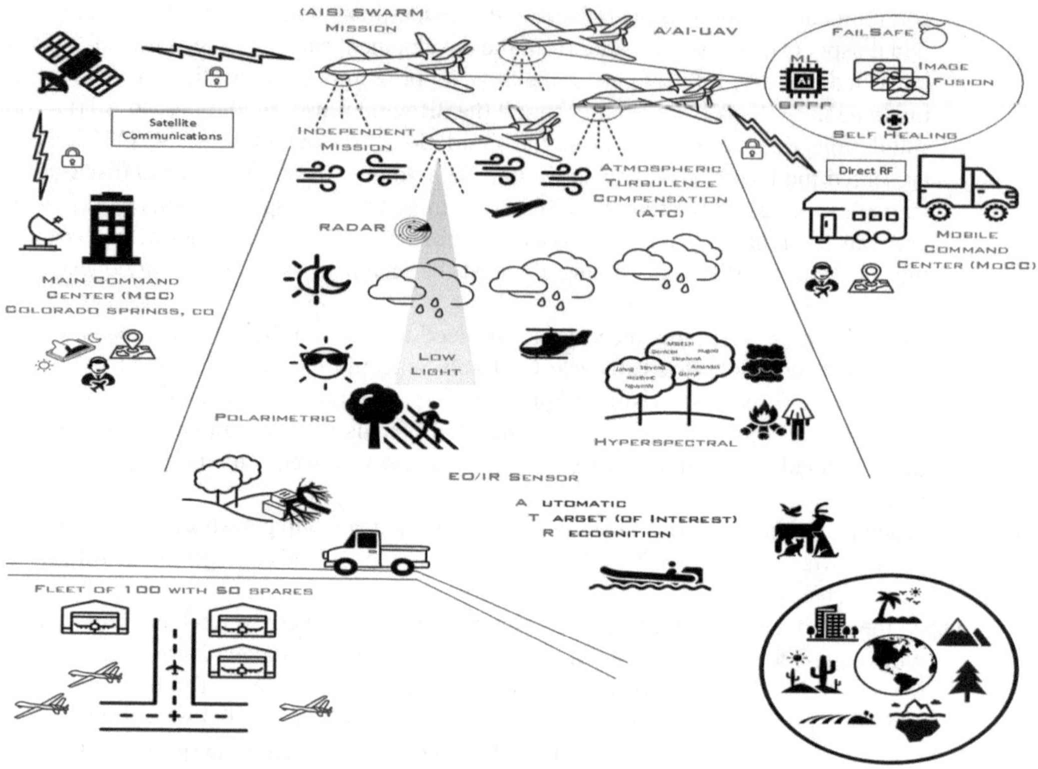

FIGURE 8.41 SkyNet program OpsCon illustrating selected sensor suite.

information fusion, self-healing of the optical system in case of damage, sensor fail safe methods, and the ATC function is shown in the OpsCon. This OpsCon graphic shows how the developed sensor technology fits into the big operational picture to satisfy the capabilities illustrated in the CONOPS graphic."

Tom: "Okay, this is all sounding good. Let's look at what tasks are needed to integrate these new capabilities in our development effort. Karl, do you have an overview?"

Karl: "Sure, Tom. We generated a table with the main tasks that need to be accomplished. Let me bring it up. We broke the required Features into Epics and Tasks. I will show you representative examples of the first set of Epics and Tasks for some of the Features related to the Hyperspectral Imaging approach to the Concealed Target and Material Characterization capabilities. We aligned these Tasks with our Agile team structure so that each Task is approximately equal to about two Sprint of about 2 weeks. Notice that the left column in the table shows significant larger efforts with a common purpose (Epics), and the right column of Table 8.3 shows the associated Tasks with roughly a 4-week granularity."

This table only shows representative starting Epics and associated Tasks supporting a single feature of the concealed target and material identification and characterization capabilities, and that uses the HSI methodology as a solution approach. The first representative Epic above initiates some AI and advanced signal processing methods to potentially expand the number of applicable sensors for which the WOLF ATC approach works. The second Epic focuses on the extension of the WOLF ATC to HSI systems. The third and fourth Epics focus on continuous risk reduction, determine/adapt the collection requirements, and specify/adapt the Key Performance Parameters (KPPs). There are many more

TABLE 8.3
Representative Decomposition of HSI Related Effort into Epics and Supporting Stories

Epics	Stories
1. Investigate the utility of machine learning and/or statistical optics methods to enable the application of the WOLF ATC methodology to current raw data sets within the United States Department of Homeland Security, Customs and Border Protection problem space (US DHS C&BP)	*Story 1:* Generate theoretical and statistical models to represent the statistical atmospheric fluctuations of the argument of the cosine term shown in Equation 8.3 *Story 2:* Conduct Monte Carlo simulations with these atmospheric phase fluctuations to determine the feasibility of estimating the object radiant emittance spectrum magnitude *Story 3:* Integrate Machine Learning and/or Statistical Optics methods in the WOLF methodology to independently estimate the magnitude of the object radiant emittance spectrum *Story 4:* Demonstrate results on simulated imagery
2. Develop the theoretical and simulated extension of the WOLF methodology to hyperspectral imaging systems	*Story 1:* Generate a 3D hyperspectral data model that simulates the collected spatial and spectral components (i.e., data cube) for a representative, generic hyperspectral push-broom imaging system for an arbitrary 256-pixel (spatial) × 256-pixel (wavelength) image segment data set *Story 2:* Generalize the data cube size to an arbitrary N pixel (spatial) × L pixel (wavelength) × M pixel (spatial) data cube *Story 3:* Adapt WOLF methodology to work with the generalized 3D data cube *Story 4:* Generate 3D per pixel signal-to-noise model for the WOLF ATC methodology *Story 5:* Simulate 3D Wolf ATC using Von Karman atmospheric phase aberration realizations and assess performance
3. Identify, assess, and mitigate where possible technology and implementation risks	*Story 1*: Conduct trade study for best parallel processing environment (platform(s), software, and hardware) *Story 2:* Determine best way to protect/secure technology and data
4. Determine data and collection requirements, and Key Performance Parameters (KPPs)	*Story 1:* Determine data and collection requirements for applying the WOLF ATC methodology to currently collected data from existing imaging/video/hyperspectral imaging systems in the US DHS C&BP problems space *Story 2:* Determine data and collection requirements for applying the WOLF ATC methodology to future collected data from existing imaging/video/hyperspectral imaging systems in the US DHS C&BP problem space *Story 3:* Determine data and collection requirements for applying the WOLF ATC methodology to future developed imaging/video/hyperspectral imaging systems *Story 4:* Identify and document additional and/or revised KPPs as appropriate

Features, Epics, and Tasks that require specification (both software and hardware-related) beyond the limited set shown above. In addition to these representative Epics and Tasks, there is an associated schedule with Exit Criteria established as follows,

The First column shows the Epics and General cross-cutting activities needed. The second column shows which organization has lead for the Epic and also which Agile Team or Administrative Unit in parentheses that has responsibility for completing the effort. The third column shows the week of completion and the duration of the Epic in parentheses. The fourth column shows the Exit Criteria for knowing when the Epic is complete (Table 8.4)."

Tom: "That is excellent! Karl, do you have the effort breakdown tables for all the Features, Epics, Stories and Tasks that are needed for the existing work and the effort associated with these new capabilities?"

TABLE 8.4
The Following Schedule Is Associated with the Representative Epics and Stories

Epics and General Activities	Lead (Agile Team)	Completion (Duration)	Exit Criteria
WOLF Parameter Estimation using Machine Learning and/or Statistical Optics Methods	Software (SW-01)	Week 16 (16 Weeks)	Successfully estimated WOLF parameter (from simulation)
WOLF 3D Adaptation and Extension to Hyperspectral Imaging Scenarios	Software (SW-02)	Week 20 (20 weeks)	Successful application of WOLF methodology on 3D hyperspectral data cube (from push-broom simulation)
Risk Identification, Assessment, and Mitigation	IPT (team leads)	Week 36 (36 weeks)	Identified risks have viable documented mitigation strategy
Determine Data and Collection Requirements, and KPPs	Systems (SYS–01)	Week 36 (16 weeks)	Documented data and collection requirements and KPPs
Epic I Documentation, Deliverables, Epic II Planning, and Backlog Update	Systems (SYS-02)	Week 38 (2 weeks)	Epic I cycle complete

Karl: "Tom, we have many of them but are still fleshing some out. We have enough detailed to have a good handle on the work that is required. Bottom line is that we understand what needs to be done and have organized the work so that we can use our Agile structure to accomplish it. I think we understand the technical implications for the new efforts required by our stakeholders and should generate the Integrated Lifecycle Cost and Schedule Estimate (ILCaSE) and submit it to our stakeholders for funding, and once approved, implement our plan."

Tom: "Okay, does anyone else have anything to add?"

Kari: "One thing! Remember that we have a Test-Driven Development (TDD) environment so before the code is written, we need to first develop the test cases for it."

Karl: "Of course, Kari. We have folks from test on the IPT and on the Agile teams and they have been very insistent and helpful on this. Thanks for bringing it up!"

Tom: "Any other comments? No? Okay, I will recommend to Bill that we proceed to focus on the ILCaSE that is currently underway. Thanks everyone! Great job!"

8.A APPENDIX: ACRONYMS

AO	Adaptive Optics
ASD	Adaptive Software Development
ASE	Agile Systems Engineering
ATC	Atmospheric turbulence compensation
ATR	Automatic Target Recognition
BDD	Behavior Driven Development
CBP	Customs and Border Protection
CONOPS	Concept of operations
CTO	Chief technology officer
DAD	Disciplined Agile Delivery
DHS	Department of Homeland Security
DM	Deformable Mirror

DSDM	Dynamic System Development Method
DWFS	Deconvolution from Wavefront Sensing
EVM	Earned Value Management
FDD	Feature Driven Development
FFT	Fast Fourier Transform
FIT	Fantastic Imaging Technologies
FPGA	Field Programmable Gate Array
GEO	Geosynchronous Orbit
GPF	Generalized Pupil Function
GPPP	General Purpose Parallel Processor
HSI	Hyperspectral Imaging
IFOV	Instantaneous Field of View
IMS	Integrated Master Schedule
INCOSE	International Council on Systems Engineering
IPT	Integrated product team
KPI	Key Performance Indicator
KPP	Key Performance Parameter
LEO	Low-Earth Orbiting
MTF	Modulation Transfer Function
NASA	National Aeronautics and Space Administration
NLO	Non-Linear Optics
OpsCon	Operational Concept
OTF	Optical Transfer Function
PSF	Point Spread Function
RAD	Rapid Application Development
SCIP	Sheared Coherent Interferometric Photography
SDLC	Systems Development Lifecycle
SE	Systems engineering
SLD	Super Luminescent Diode
SSPP	System, Service, Product, and/or Process
TDD	Test Driven Development
UAV	Unmanned aerial vehicle
U.S.	United States
WFS	Wavefront Sensor
XP	Extreme Programming

REFERENCES

Agile Manifesto. 2001. Website: agilemanifesto.org (accessed August 19, 2024).

Allen, J. 2016. Using agile methods in a DoD environment: a systems engineering perspective. *INCOSE Orlando Chapter Meeting* (16 June 2016). Orlando, FL. https://www.incose.org/communities/chapters/americas-sector/orlando/events-detail/2016/06/16/default-calendar/orlando-chapter-meeting---june (Accessed March 8, 2025).

Arrasmith, W.W. 2013. High-speed diversity-based imaging method for parallel atmospheric turbulence compensation. US Patent US8,447,129B2, May 21.

Arrasmith, W.W. 2024. Wolf Methodology. Technical Report, v01 (25JUL2024).

Arrasmith, W., and E. He. 2023. Generalizing the well-optimized linear finder (WOLF) atmospheric turbulence compensation (ATC) methodology to non-diversity based imaging systems using entrance pupil plane aperture masks. *Universal Journal of Lasers, Optics, Photonics, and Sensors*, Volume 1, No 2, July 2023. https://www.etextonline.org/articlepdfs/Well-Optimized-Linear-Finder-(WOLF)-Atmospheric-Turbulence-Compensation-(ATC)-computational-speed-improvement-through-the-adoption-of-parallel-pre-calculated-constant-complex-exponential-phase-difference-chains.pdf (Accessed March 8, 2025).

Babcock, H.W. 1953. The possibility of compensating astronomical seeing. *Publications of the Astronomical Society of the Pacific*, Vol. 65, pp. 229–236.

Beck, K., M. Beedle, A. Van Bennekum, A. Cockburn, W. Cunningham, M. Fowler, and D. Thomas. 2001. Manifesto for Agile software development. https://agilemanifesto.org/. (Accessed March 8, 2025).

Boeing. 2006. Closed-loop wavefront sensor using field programmable gate array. US Patent 20060256321A1. S20.

Carlson, R.B. and R. Turner. 2013. Review of Agile case studies for applicability to aircraft systems integration. *Procedia Computer Science,* Vol. 16, pp. 469–474. https://doi.org/10.1016/j.procs.2013.01.049

Chew, T.Y., R.M. Clare, and R.G. Lane. 2006. A comparison of the Shack-Hartmann and pyramid wavefront sensors. *Optics Letters*, Vol. 268, No. 2, pp. 189–195.

Choudhury, M.D. 2022. Online CIO article: top 16 Agile certifications available today. https://www.cio.com/article/230223/7-agile-certifications-to-take-your-career-to-the-next-level.html (accessed November 6, 2023).

Darrin, M.A.G. and W.S. Devereux. 2017. The Agile manifesto, design thinking and systems engineering. *2017 Annual IEEE International Systems Conference (SysCon)*, Montreal, QC. https://doi.org/10.1109/syscon.2017.7934765

Douglass, B.P. 2015. *Agile Systems Engineering*. Waltham, MA: Morgan Kaufmann.

Fried, D.L. 1966. Optical resolution through a randomly inhomogeneous medium for very long and very short exposures. *Journal of the Optical Society of America*, Vol. 56, pp. 1372–1379.

Gonsalvez, R.A. 1982. Phase retrieval and diversity in adaptive optics. *Journal of Optical Engineering*, Vol. 21, p. 5.

Gonsalves, R.A. and R. Chidlaw. 1979. Wavefront sensing by phase retrieval. *Proceedings on Applications of Digital Image Processing III, SPIE*, Vol. 207, pp. 32–39.

Greywolf Technical Services. 2013. Methods and systems for modified wavelength diversity image compensation. U.S. Patent: 20130222615 (August 29, 2013). https://patents.justia.com/inventor/greywolf-technical-services-inc (accessed March 8, 2025).

Harris, C. n.d. Learn about agile scrum artifacts. Atlassian. https://www.atlassian.com/agile/scrum/artifacts#:~:text=The%20main%20agile%20scrum%20artifacts,the%20development%20of%20a%20product (accessed May 27, 20224).

Hayes, W., S. Miller, M.A. Lapham, E. Wrubel, and T.A. Chick. 2014. Agile metrics: progress monitoring of Agile contractors. https://www.researchgate.net/publication/266049888_Agile_Metrics_Progress_Monitoring_of_Agile_Contractors.

Highsmith, J. 1999. *Adaptive Software Development: A Collaborative Approach to Managing Complex Systems*. New York: Dorset House Publishing Co., Inc.

Huang, P.M., M.A.G. Darrin, and A.A. Knuth. 2012. Agile hardware and software system engineering for innovation. *IEEE Aerospace Conference*, Big Sky, MT, 1 March 2012. https://doi.org/10.1109/aero.2012.6187425

Humpert, L., E. Mundt, L. Bretz, and H. Anacker. 2022. Analysis of the criteria for comparing systems engineering and Agile methods. *2022 IEEE International Symposium on Systems Engineering (ISSE)*, Vienna. https://doi.org/10.1109/isse54508.2022.10005556

Hutchin, R. 1993. Sheared coherent interferometric photography: a technique for lensless imaging. *Proceedings of SPIE – The International Society for Optical Engineering*, Vol 2029, pp. 161–168. Digital Image Recovery and Synthesis II, 161.

Idell, P.S. and M.C. Roggemann. n.d. *Unconventional Imaging*. Optical Society of America Short Course.

Ingleby, H.R. and D.R. McGaughey. 2002. Parallel multi-frame blind deconvolution using wavelength diversity. *Proceedings of Image Reconstruction from Incomplete Data III, SPIE*, Vol. 5562, pp. 58–64.

Kolmogorov, A.N. 1961. The local structure of turbulence in incompressible viscous fluid for very large Reynold's numbers. In: Friedlander, S.K. and L. Topper (eds), *Turbulence, Classic Papers on Statistical Theory.* New York: Wiley-Interscience, pp. 150–155.

Kirvan, P. and M.K. Pratt. 2023. Agile manifesto. Tech Target/*CIO*. https://www.techtarget.com/searchcio/definition/Agile-Manifesto#:~:text=4%20values%20of%20Agile,Customer%20collaboration%20over%20contract%20negotiation (accessed March 8, 2025).

Knox, K.T. and B.J. Thompson. 1974. Recovery of images from atmospherically degraded short exposure images. *Astrophysics Journal,* Vol. 195, pp. L45–L48.

Labeyrie, A. 1970. Attainment of diffraction limited resolution in large telescopes by Fourier analyzing speckle patterns in star images. *Astronomical Astrophysics*, Vol. 6, p. 85.

Lohmann, A.W., G. Weigelt, and B. Wirnitzer. 1983. Speckle masking in astronomy: triple correlation theory and applications. *Applied Optics*, Vol. 22, pp. 4028–4037.

McGregor, J.S. 2016. *Agile and Earned Value Management: A Program Manager's Desk Guide.* Office of the Under Secretary of Defense for Acquisitions, Technology, and Logistics (3 March 2016).

Noll, R.J. 1976. Zernike polynomials and atmospheric turbulence. *Journal of the Optical Society of America*, Vol. 66, pp. 207–211.

Özkan, D. and A. Mishra. 2019. Agile project management tools: a brief comparative view. *Cybernetics and Information Technologies*, Vol. 19, No. 4, pp. 17–25. https://doi.org/10.2478/cait-2019-0033

Paxman, R.G., T.J. Schulz, and J.R. Fienup. 1992. Joint estimation of object and aberrations by using phase diversity. *Journal of the Optical Society of America*, Vol. 9, pp. 1072–1085.

Radigan, D. 2024. Five Agile metrics you won't hate. Atlassian. https://www.atlassian.com/agile/project-management/metrics (accessed May 27, 2024).

Roggemann, M.C. and B. Welsh. 1996. *Imaging through Turbulence*. New York: CRC Press.

Savolainen, J., J. Kuusela, and A. Vilavaara. 2010. Transition to Agile development - rediscovery of important requirements engineering practices. *Proceedings of the 2010 18th IEEE International Requirements Engineering Conference*, Sydney, NSW, September 1, 2010. https://doi.org/10.1109/re.2010.41

Schulz, T. 1993. Multi-frame blind deconvolution of astronomical images. *Journal of the Optical Society of America*, Vol. 10, No. 5, pp. 1064–1073.

SEBOK. 2024. Agile systems engineering, frameworks: Agile systems engineering process steps. https://sebokwiki.org/wiki/Agile_Systems_Engineering (accessed May 26, 2024).

Smith, A. 2019. Beyond software: when Agile meets defense systems and hardware. Agile Alliance. https://www.agilealliance.org/resources/experience-reports/beyond-software-when-agile-meets-defense-systems-and-hardware/ (accessed November 9, 2023).

Tatarski, V.I. 1967. *Wave Propagation in a Turbulent Medium*. New York: Dover Publications.

Turner, R., R. Madachy, D. Ingold, and J.A. Lane. 2012. Modeling kanban processes in systems engineering. *International Conference on Software and System Process*, Zurich, Switzerland, June 2, 2012, pp. 23–27. https://apps.dtic.mil/sti/tr/pdf/ADA602568.pdf

Wakode, R.B., L.P. Raut, and P. Talmale. 2015. Overview on kanban methodology and its implementation. *International Journal for Scientific Research and Development,* Vol. 3, No. 2, pp. 2518–2521.

Zelkowitz, M. 2004. *Advances in Computers: Advances in Software Engineering*. Elsevier, Amsterdam, Netherlands.

Appendix

Mathematical Formulas

A.1 TRIGONOMETRIC IDENTITIES

Pythagorean formulas

$$\sin^2 x + \cos^2 x = 1$$

$$1 + \tan^2 x = \sec^2 x$$

$$1 + \cot^2 x = \operatorname{cosec}^2 x.$$

Reciprocal formulas

$$\sin x = \frac{1}{\operatorname{cosec} x} \quad \cos x = \frac{1}{\sec x} \quad \tan x = \frac{1}{\cot x}$$

$$\operatorname{cosec} x = \frac{1}{\sin x} \quad \sec x = \frac{1}{\cos x} \quad \cot x = \frac{1}{\tan x}.$$

Product formulas

$$\sin x = \tan x \cos x \cos x = \cot x \sin x \tan x = \sin x \sec x$$

$$\cot x = \cos x \operatorname{cosec} x \sec x = \csc x \tan x \operatorname{cosec} x = \sec x \cot x,$$

$$\sin\alpha \sin\beta = \frac{1}{2}\cos(\alpha - \beta) - \frac{1}{2}\cos(\alpha + \beta)$$

$$\cos\alpha \cos\beta = \frac{1}{2}\cos(\alpha - \beta) + \frac{1}{2}\cos(\alpha + \beta)$$

$$\sin\alpha \cos\beta = \frac{1}{2}\sin(\alpha + \beta) + \frac{1}{2}\sin(\alpha - \beta)$$

$$\cos\alpha \sin\beta = \frac{1}{2}\sin(\alpha + \beta) - \frac{1}{2}\sin(\alpha - \beta).$$

Angle-sum and angle-difference formulas

$$\sin(\alpha + \beta) = \sin\alpha \cos\beta + \cos\alpha \sin\beta$$

$$\sin(\alpha - \beta) = \sin\alpha \cos\beta - \cos\alpha \sin\beta$$

$$\cos(\alpha + \beta) = \cos\alpha \cos\beta - \sin\alpha \sin\beta$$

$$\cos(\alpha - \beta) = \cos\alpha \cos\beta + \sin\alpha \sin\beta,$$

$$\tan(\alpha+\beta)=\frac{\tan\alpha+\tan\beta}{1-\tan\alpha\,\tan\beta},$$

$$\tan(\alpha-\beta)=\frac{\tan\alpha-\tan\beta}{1+\tan\alpha\tan\beta},$$

$$\sin(\alpha+\beta)\sin(\alpha-\beta)=\sin^2\alpha-\sin^2\beta=\cos^2\beta-\cos^2$$

$$\cos(\alpha+\beta)\cos(\alpha-\beta)=\cos^2\alpha-\sin^2\beta=\cos^2\beta-\cos^2\alpha.$$

Double-angle formulas

$$\sin 2\alpha=2\sin\alpha\cos\alpha=\frac{2\tan\alpha}{1+\tan^2\alpha},$$

$$\cos 2\alpha=\cos^2\alpha-\sin^2\alpha=2\cos^2\alpha-1=1-2\sin^2\alpha=\frac{1-\tan^2\alpha}{1+\tan^2\alpha}$$

$$\tan 2\alpha=\frac{2\tan\alpha}{1-\tan^2\alpha}$$

$$\cot 2\alpha=\frac{\cot^2\alpha-1}{2\cot\alpha}.$$

Power formulas

$$\sin^2\alpha=\frac{1}{2}(1-\cos 2\alpha)\quad \sin^3\alpha=\frac{1}{4}(3\sin\alpha-\sin 3\alpha)\quad \sin^4\alpha=\frac{1}{8}(3-4\cos 2\alpha+\cos 4\alpha)$$

$$\cos^2\alpha=\frac{1}{2}(1+\cos 2\alpha)\quad \cos^3\alpha=\frac{1}{4}(3\cos\alpha+\cos 3\alpha)\quad \cos^4\alpha=\frac{1}{8}(3+4\cos 2\alpha+\cos 4\alpha)$$

$$\tan^2\alpha=\frac{1-\cos 2\alpha}{1+\cos 2\alpha}\quad \cot^2\alpha=\frac{1+\cos 2\alpha}{1-\cos 2\alpha}.$$

Half-angle formulas

$$\sin\frac{\alpha}{2}=\pm\sqrt{\frac{1-\cos\alpha}{2}}\quad \cos\frac{\alpha}{2}=\pm\sqrt{\frac{1+\cos\alpha}{2}}$$

$$\tan\frac{\alpha}{2}=\pm\sqrt{\frac{1-\cos\alpha}{1+\cos\alpha}}=\frac{1-\cos\alpha}{\sin\alpha}=\frac{\sin\alpha}{1+\cos\alpha}$$

$$\cot\frac{\alpha}{2}=\pm\sqrt{\frac{1+\cos\alpha}{1-\cos\alpha}}=\frac{1+\cos\alpha}{\sin\alpha}=\frac{\sin\alpha}{1-\cos\alpha}.$$

Euler's formulas

$$e^{\pm i\alpha}=\cos\alpha\pm i\sin\alpha\quad i=\sqrt{-1}$$

$$\sin\alpha=\frac{e^{i\alpha}-e^{-i\alpha}}{2i}\quad \cos\alpha=\frac{e^{i\alpha}+e^{-i\alpha}}{2}$$

$$\tan\alpha=-i\left(\frac{e^{i\alpha}-e^{-i\alpha}}{e^{i\alpha}+e^{-i\alpha}}\right)=-i\left(\frac{e^{i2\alpha}-1}{e^{i2\alpha}+1}\right).$$

Function-sum and function-difference formulas

$$\sin\alpha + \sin\beta = 2\sin\frac{1}{2}(\alpha+\beta)\cos\frac{1}{2}(\alpha-\beta)$$

$$\sin\alpha - \sin\beta = 2\cos\frac{1}{2}(\alpha+\beta)\sin(\alpha-\beta)$$

$$\cos\alpha + \cos\beta = 2\cos\frac{1}{2}(\alpha+\beta)\cos\frac{1}{2}(\alpha-\beta)$$

$$\cos\alpha - \cos\beta = -2\sin\frac{1}{2}(\alpha+\beta)\sin\frac{1}{2}(\alpha-\beta)$$

$$\tan\alpha + \tan\beta = \frac{\sin(\alpha+\beta)}{\cos\alpha\cos\beta}$$

$$\tan\alpha - \tan\beta = \frac{\sin(\alpha-\beta)}{\cos\alpha\cos\beta}.$$

A.2 POLAR COORDINATES

Fix an origin O and an initial ray from O. Then each point P can be located by assigning to it a polar coordinate pair (r, θ), where r is the directed distance from O to P and θ is the angle between the initial ray and OP

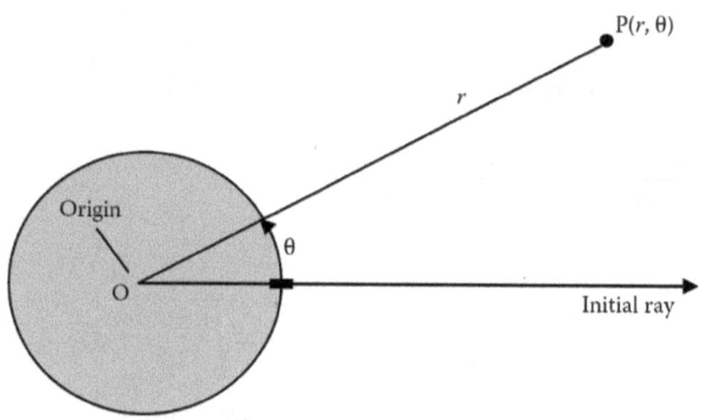

The origin in polar coordinates corresponds to the origin in Cartesian coordinates. We can imagine an x, y plane overlaying the above, with coinciding origins. To change from Cartesian coordinates to polar coordinates, we observe the following identities:

$$x = r\cos\theta,$$

$$y = r\sin\theta,$$

$$\frac{x}{y} = \tan\theta,$$

$$x^2 + y^2 = r^2.$$

A.3 COMPLEX NUMBERS

Since there is no real number that satisfies the equation $x^2+1=0$, we develop a set of complex numbers of the form $a+bi$, where a and b are real, and $i=\sqrt{-1}$. *We* call i the *imaginary unit,* and it has the property that $i^2=-1$. If $z=a+bi$, we say that a is the *real part* of z and b is called *the imaginary part* of z. We denote them Re$\{z\}$ and Im$\{z\}$, respectively.

Two complex numbers $a+bi$ and $c+di$ are *equal* if and only if $a=c$ and $b=d$. We consider the real numbers as a subset of the complex numbers where all $b=0$.

The *complex conjugate,* or merely *conjugate,* of a complex number $a+bi$ is $a-bi$. The conjugate of the complex number z is usually denoted $\bar{z}$.

Fundamental operations with complex numbers are

1. *Addition*: $(a+bi)+(c+di)=a+bi+c+di=(a+c)+(b+d)i$
2. *Subtraction*: $(a+bi)-(c+di)=a+bi-c-di=(a-c)+(b-d)i$
3. *Multiplication*: $(a+bi)(c+di)=ac+adi+cbi+bdi^2=(ac-bd)+(ad+bc)i$
4. *Division*:

$$\frac{a+bi}{c+di}=\left(\frac{a+bi}{c+di}\right)\left(\frac{c-di}{c-di}\right)=\frac{ac-adi+bci-bdi^2}{c^2-d^2i^2}$$

$$=\frac{ac+bd+(bc-ad)i}{c^2+d^2}=\frac{ac+bd}{c^2+d^2}+\frac{bc-ad}{c^2+d^2}i.$$

A.3.1 Absolute Value of Complex Numbers

The absolute value of a complex number $a+bi$ is defined as $|a+bi|=\sqrt{a^2+b^2}$.

If $z_1, z_2, z_3, \ldots, z_n$ are complex numbers, the following properties hold:

1. $|z_1z_2|=|z_1||z_2|$ and, in fact, $|z_1z_2\ldots z_n|=|z_1||z_2|\ldots|z_n|$
2. $\left|\frac{z_1}{z_2}\right|=\frac{|z_1|}{|z_2|}$ if $z_2\neq 0$
3. $|z_1+z_2|\leq|z_1|+|z_2|$ and $|z_1+z_2+\cdots+z_n|\leq|z_1|+|z_2|+\cdots+|z_n|$
4. $|z_1+z_2|\geq|z_1|-|z_2|$ and $|z_1-z_2|\geq|z_1|-|z_2|$

A.3.2 Ordered Pair Form of Complex Numbers

We can define a complex number as an ordered pair (a, b). This allows us to represent the complex numbers graphically in the way we graph real ordered pairs on a Cartesian coordinate plane, where the x-axis becomes the *real* axis and the y-axis becomes the *imaginary* axis. We plot our ordered pairs on the (r, i) axes.

We define equality, sum, and product as follows:

$$(a,b)=(c,d) \text{ if and only if } a=c \text{ and } b=d,$$
$$(a,b)+(c,d)=(a+c,b+d),$$
$$(a,b)(c,d)=(ac-bd,ad+bd),$$
$$m(a,b)=(ma,mb).$$

This is merely a different notation than the previous, and all properties of the complex numbers hold. Further, the complex numbers constitute a field.

A.3.3 Polar Form of Complex Numbers

If P is a point in the complex plane with coordinates (x, y) or $x+iy$, then

$$x = r\cos\theta, \quad y = r\sin\theta,$$

where $r = \sqrt{x^2+y^2} = |x+iy|$ is called the amplitude, modulus, or absolute value of $z=x+iy$ and θ, called the argument of $z=x+iy$, is the angle that OP makes with the positive real axis. It follows that

$$z = x+iy = r(\cos\theta + i\sin\theta).$$

A.3.3.1 De Moivre's Theorem

If $z_1=x_1+iy_1=r_1(\cos\theta_1+i\sin\theta_1)$ and if $z_2=x_2+iy_2=r_2(\cos\theta_2+i\sin\theta_2)$, then

$$z_1z_2 = r_1r_2\left(\cos(\theta_1+\theta_2)+i\left(\sin(\theta_1+\theta_2)\right)\right),$$

$$\frac{z_1}{z_2} = \frac{r_1}{r_2}\left(\cos(\theta_1-\theta_2)+i\sin(\theta_1-\theta_2)\right).$$

We can generalize this to many complex values. Thus, if $z_1=z_2=\cdots=z_n=z$, de Moivre's theorem states that $z^n=(r(\cos\theta+i\sin\theta))^n=r^n(\cos n\theta+i\sin n\theta)$.

A.3.4 Dot and Cross Products on Complex Numbers

Let $z_1=x_1+iy_1$ and $z_2=x_2+iy_2$ be two complex numbers and $\bar{z}$ represents the complex conjugate of z. We define the *dot product* (also called the *scalar product*) of z_1 and z_2 to be

$$z_1 \circ z_2 = |z_1||z_2|\cos\theta = x_1x_2 + y_1y_2 = \mathrm{Re}\{\bar{z}_1z_2\} = \frac{1}{2}\{\bar{z}_1z_2 + z_1\bar{z}_2\},$$

where θ is the angle between z_1 and z_2, $0<\theta<\pi$.

The *cross product* of z_1 and z_2 is defined to be

$$z_1 \times z_2 = |z_1||z_2|\sin\theta = x_1y_2 - y_1x_2 = \mathrm{Im}\{\bar{z}_1z_2\} = \frac{1}{2i}\{\bar{z}_1z_2 - z_1\bar{z}_2\}.$$

Putting these together, we find that $\bar{z}_1z_2 = (z_1\circ z_2)+i(z_1\times z_2) = |z_1||z_2|e^{i\theta}$.

We also have the following:

1. A necessary and sufficient condition that z_1 and z_2 be *perpendicular* is that $z_1 \circ z_2 = 0$.
2. A necessary and sufficient condition that z_1 and z_2 be *parallel* is that $z_1\times z_2=0$.
3. The magnitude of the projection of z_1 on z_2 is $| z_1 \circ z_2 |/| z_2 |$.
4. The area of a parallelogram having size z_1 and z_2 is $|z_1\times z_2|$.

A.4 CHAIN RULE

If $q(x)=u(x)v(x)$, then

$$\frac{d}{dx}q(x) = \frac{d}{dx}(uv) = u\frac{dv}{dx} + v\frac{du}{dx}.$$

If $q(x) = \dfrac{u(x)}{v(x)}$ $v(x)\neq 0$, then

$$\frac{d}{dx}q(x) = \frac{d}{dx}\left(\frac{u}{v}\right) = \frac{v\left(\frac{du}{dx}\right) - u\left(\frac{dv}{dx}\right)}{v^2}.$$

If $(x) = \frac{c}{u(x)}$, $c \in \Re$, then

$$\frac{d}{dx}q(x) = \frac{d}{dx}\left(\frac{c}{u}\right) = c\frac{d}{dx}\left(\frac{1}{u}\right) = -\frac{c}{u^2}\frac{d}{dx}(u).$$

If $q(x) = u^n(x)$, then

$$\frac{d}{dx}q(x) = \frac{d}{dx}\left(u^n\right) = nu^{n-1}\frac{d}{dx}(u).$$

Let $y = f(u)$ and $u = g(x)$. Then we can write y as a function of a function, $y = f(g(x))$. Now, if y is a differentiable function of u, and if u is a differentiable function of x, then $y = f(g(x))$ is a differentiable function of x, and

$$\frac{dy}{dx} = \frac{dy}{du}\frac{du}{dx}.$$

If $z = f(x, y)$ is a continuous function of the variables x and y with continuous partial derivatives $\partial z/\partial x$ and $\partial z/\partial y$, and if x and y are differentiable functions of the variable t, $x = g(t)$ and $y = h(t)$, then z is also a function of the variable t, and dz/dt (called the *total derivative of z with respect to t*) is given by

$$\frac{dz}{dt} = \frac{\partial z}{\partial x}\frac{dx}{dt} + \frac{\partial z}{\partial y}\frac{dy}{dt}.$$

If $z = f(x, y)$ is a continuous function of variables x and y with continuous partial derivatives $\partial z/\partial x$ and $\partial z/\partial y$, and if x and y are continuous functions $x = g(r, s)$ and $y = h(r, s)$ of the independent variables r and s, then z is a function of r and s with

$$\frac{\partial z}{\partial r} = \frac{\partial z}{\partial x}\frac{\partial x}{\partial r} + \frac{\partial z}{\partial y}\frac{\partial y}{\partial r} \quad \frac{\partial z}{\partial s} = \frac{\partial z}{\partial x}\frac{\partial x}{\partial s} + \frac{\partial z}{\partial y}\frac{\partial y}{\partial s}.$$

Suppose that $f(x) = g(x)h(x)$ where both g and h are continuously differentiable scalar functions of the vector $x \in \Re^{n\times1}\left(x = \left[x_1, x_2, \ldots, x_n\right]^T\right)$. Then

$$\nabla_x f(x) = \nabla_x g(x)h(x) + \nabla_x h(x)g(x),$$

where $\nabla_x f(x) = \frac{\partial f(x)}{\partial x} = \left[\frac{\partial f}{\partial x_1}, \frac{\partial f}{\partial x_2}, \ldots, \frac{\partial f}{\partial x_n}\right]^T$ gradient of f with respect to x.

A.5 ALGEBRA

A.5.1 Vectors and Vector Spaces

A vector (linear) space over a field $\mathcal{F}$ is designated as $(\mathcal{L}, \mathcal{F})$ and contains the set of elements $\mathcal{L}$ called vectors (of arbitrary length); a field $\mathcal{F}$; and two operations, which are *scalar multiplication* and *vector addition.*

Two often-used vector spaces are $(\Re,\Re)$ and (C,C). While $(C,\Re)$ is a vector space, $(\Re,C)$ is not because scalar multiplication will not, in general, yield vectors whose elements are real numbers.

A.5.2 Matrices and Matrix Operations

A *matrix* is a rectangular array of numbers. The numbers in the array are called the *entries* in the matrix. The rows of entries in a matrix can be viewed as vectors, as can the columns of the matrix:

$$A = \begin{bmatrix} a_{11} & a_{12} & a_{13} & a_{14} \\ a_{21} & a_{22} & a_{23} & a_{24} \\ a_{31} & a_{32} & a_{33} & a_{34} \end{bmatrix}.$$

The subscripts of a indicate the *row* and *column* location of the entry. We can consider matrix A as a set of row vectors:

$$b_1 = (a_{11}, a_{12}, a_{13}, a_{14}),$$
$$b_2 = (a_{21}, a_{22}, a_{23}, a_{24}),$$
$$b_3 = (a_{31}, a_{32}, a_{33}, a_{34}),$$

or as a set of column vectors:

$$c_1 = \begin{vmatrix} a_{11} \\ a_{21} \\ a_{31} \end{vmatrix}, c_2 = \begin{vmatrix} a_{12} \\ a_{22} \\ a_{32} \end{vmatrix}, c_3 = \begin{vmatrix} a_{13} \\ a_{23} \\ a_{33} \end{vmatrix}, c_4 = \begin{vmatrix} a_{14} \\ a_{24} \\ a_{34} \end{vmatrix}.$$

We define the size of a matrix by the number of rows and columns it contains: matrix A is a 3×4 matrix. We denote the size of a matrix using superscripts: $M^{m,n}$ indicates a matrix M with m rows and n columns. We may say a matrix is an element of the real field $\Re^{m,n}$ or of the complex field $C^{m,n}$.

A matrix with n rows and n columns is called a *square* matrix, and the entries $a_{11}, a_{22}, a_{33}, \ldots, a_{nn}$ are called the *diagonal* of the matrix.

The *sum* of two matrices, $A+B$, is defined *only* if the matrices are the same size and is then obtained by adding together the corresponding entries of the two matrices.

The *product* of a *scalar* c with a matrix A—denoted cA—is the matrix obtained by multiplying each entry of A by c.

The *product* of matrix A with matrix B requires that if A is $m\times r$, then B must be $r\times n$. The entry in row i and column j of matrix AB is calculated as follows:

a. Single out row i of matrix A and column j of matrix B.
b. Multiply corresponding entries by each other.
c. Add all products together. This becomes the ijth entry of AB.

A *null matrix* is one in which all entries are 0s.

A matrix $M^{n,n}$ in which the diagonal entries are all 1s and all other entries are 0s is the *identity matrix,* denoted I_n.

The *transpose* of a matrix M is denoted M^T. To transpose a matrix, the rows and columns of the matrix are interchanged. That is,

$$A^T = \left[a_{ij}\right]_{n\times m}^T = \left[a_{ji}\right]_{m\times n} \in \Re^{m\times n}.$$

The following are properties of the transpose:

1. $(A^T)^T = A$
2. Given $A \in \Re^{n\times m}$, $B \in \Re^{m\times p}$, $(AB)^T = B^T A^T \Re^{p\times n}$
3. $(A+B)^T = A^T + B^T$

The *inverse* of a matrix M is denoted M^{-1}. If a matrix $M \in \Re^{n\times n}$ is *invertible,* it is also called *nonsingular,* and there is a matrix M^{-1} such that $MM^{-1} = M^{-1}M = I_n$.

Let $A \in \Re^{n\times n}$, $B \in \Re^{n\times n}$, $C \in \Re^{m\times m}$, $u \in \Re^{n\times 1}$, $v \in \Re^{n\times 1}$ Then

1. $(A^{-1})I = A^{-1}$
2. $(AB)^{-1} = B^{-1}A^{-1}$
3. $(A^T)^{-1} = (A^{-1})^T = A^{-T}$

A.6 EIGENVALUE PROBLEMS

Let A be a square matrix of order n. A number λ is said to be an eigenvalue of A if there exists a nonzero solution vector Λ of the linear system:

$$A\Lambda = \lambda\Lambda.$$

The solution vector Λ is said to be an eigenvector corresponding to the eigenvalue λ. This is typically called the *standard eigenvalue problem.*

If A is real symmetric, the eigenvalue problem has the following desirable features:

a. The eigenvalues λ are all real.
b. The eigenvectors can be chosen to be mutually orthogonal—that is,
   or, equivalently,

$$Az_i = \lambda_i z_i \quad \text{for} \quad i = 1,2,\ldots,n,$$

where
Λ is a real diagonal matrix whose diagonal elements λ_i· are the eigenvalues
Z is a real orthogonal matrix whose columns z_i· are the eigenvectors

This implies that

$$AZ = Z\Lambda,$$

Further, we can write $A = Z\Lambda Z^T$.

This is known as the eigendecomposition or spectral factorization of A.

If A is real nonsymmetric, it may have complex eigenvalues, occurring as complex conjugate pairs. If x is an eigenvector corresponding to a complex eigenvalue λ, then the complex conjugate vector $\bar{x}$ is the eigenvector corresponding to the complex conjugate eigenvalue $\bar{\lambda}$.

A.7 CONVOLUTION

If functions f and g are piecewise continuous on $[0, \infty)$, then the convolution of f and g, denoted $f * g$, is given by the integral

$$f * g = \int_0^t f(\tau)g(t-\tau)d\tau.$$

Convolution is commutative. That is, $f * g = g * f$.

A.7.1 Convolution Theorem

Let $f(t)$ and $g(t)$ be piecewise continuous on $[0, \infty)$ and of exponential order. Then

$$\mathfrak{F}\{f * g)\} = \Im\{f(t)\}\mathfrak{F}\{g(t)\},$$

where $\Im$ denotes the Fourier transform described in the following texts. That is, convolution in one domain (i.e., time) equals point-wise multiplication in the other domain (i.e., frequency).

A.7.2 Fourier Series

The Fourier series of a function f defined on the interval $(-p, p)$ is given by

$$f(x) = \frac{a_0}{2} + \sum_{n=1}^{\infty}\left(a_n \cos\frac{n\pi}{p}x + b_n \sin\frac{n\pi}{p}\right),$$

where

$$a_0 = \frac{1}{p}\int_{-p}^{p} f(x)dx,$$

$$a_n = \frac{1}{p}\int_{-p}^{p} f(x)\cos\frac{n\pi}{p}xdx,$$

$$b_n = \frac{1}{p}\int_{-p}^{p} f(x)\sin\frac{n\pi}{p}xdx.$$

A.7.3 Fourier Transforms

Fourier transforms occur in pairs. That is, if

$$F(\alpha) = \int_{a}^{b} F(x)K(\alpha,x)dx,$$

is an integral that transforms $f(x)$ into $F(\alpha)$, then the function f is recovered by means of another integral transformation

$$f(x) = \int_{c}^{d} F(\alpha)H(\alpha,x)d\alpha,$$

called the *inversion integral* or *inverse transform*. The functions K and H are called the *kernels* of the transforms.

Some common Fourier transforms are

1. Fourier transform: $\Im\{f(x)\} = \int_{-\infty}^{\infty} f(x)e^{i\alpha x}dx = F(\alpha)$

   Inverse Fourier transform: $\Im^{-1}\{F(\alpha)\} = \frac{1}{2\pi}\int_{-\infty}^{\infty} F(\alpha)e^{-i\alpha x}d\alpha = f(x)$

2. Fourier sine transform: $\mathfrak{I}_s\{f(x)\} = \int_0^\infty f(x)\sin\alpha x dx = F(\alpha)$

Inverse Fourier sine transformation: $\mathfrak{I}_s^{-1}\{F(\alpha)\} = \frac{2}{\pi}\int_0^\infty F(\alpha)\sin ax d\alpha = f(x)$

3. Fourier cosine transform: $\mathfrak{I}_c\{f(x)\} = \int_0^\infty f(x)\cos\alpha x dx = F(\alpha)$

Inverse Fourier cosine transform: $\mathfrak{I}_c^{-1}\{F(\alpha)\} = \frac{2}{\pi}\int_0^\infty F(\alpha)\cos\alpha x d\alpha = f(x).$

We see that the Fourier transform is a generalization of Fourier series and uses complex numbers. Fourier transforms are particularly useful in analyzing the frequency content of continuous signals, in quantifying optical information, and in studying electrical and magnetic fields and their properties. The Fourier transform takes a signal in the time domain with real variables and transforms it into a function of angular frequency with complex variables. See the following figure:

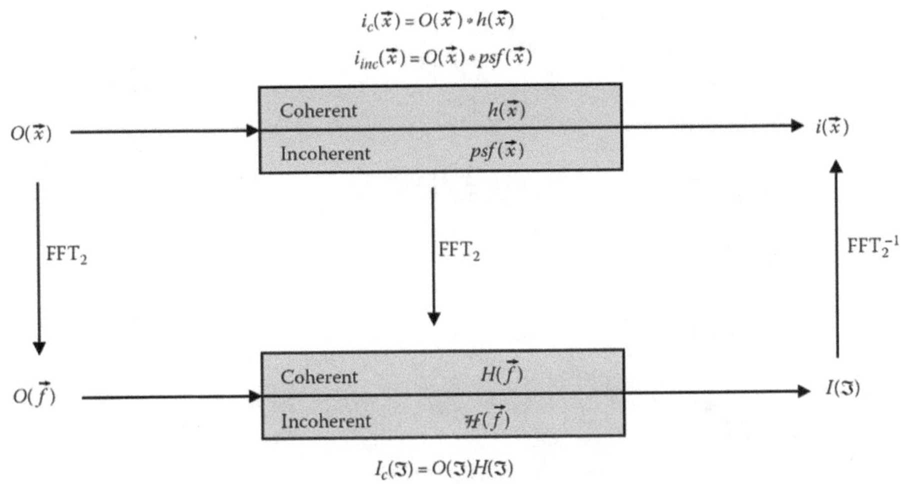

Some common Fourier transforms and properties include the following:

Operation	f(t)	F(ω)
Addition	$f_1(t)+f_2(t)$	$F_1(\omega)+F_2(\omega)$
Scalar multiplication	$kf(t)$	$kF(\omega)$
Complex conjugation	$f^*(t)$	$F^*(-i\omega)$
Time reversal	$f(-t)$	$F(-i\omega)$
Scaling (a is real)	$f(at)$	$\frac{1}{\lvert a\rvert}F\left(\frac{\omega}{a}\right)$
Time shift	$f(t-t_0)$	$F(\omega)e^{-i\omega t_0}$
Frequency shift (ω_0 is real)	$f(t)e^{i\omega_0 t}$	$F(\omega-\omega_0)$
Time convolution	$f_1(t) * f_2(t)$	$F_1(\omega)F_2(\omega)$
Frequency convolution	$f_1(t)f_2(t)$	$\frac{1}{2\pi}F_1(\omega)*F_2(\omega)$

(*Continued*)

(Continued)		
Time differentiation	$\frac{d^n f}{dt^n}$	$(i\omega)^n F(\omega)$
Time integration	$\int_{-\infty}^{t} f(x)dx$	$\frac{F(\omega)}{i\omega} + \pi F(0)\partial(\omega)$

f(t)	F(ω)
$u(t)$	$\pi\partial(\omega) + \frac{1}{i\omega}$
$\mathrm{Sgn}(t)$	$\frac{2}{i\omega}$
$\cos(\omega_0 t)u(t)$	$\frac{\pi}{2}[\partial(\omega-\omega_0)+\partial(\omega+\omega_0)] + \frac{i\omega}{\omega_0^2-\omega^2}$
$\sin(\omega_0 t)u(t)$	$\frac{\pi}{2i}[\partial(\omega-\omega_0)-\partial(\omega+\omega_0)] + \frac{\omega_0}{\omega_0^2-\omega^2}$
$e^{-at}\sin(\omega_0 t)u(t)$	$\frac{\omega_0}{(a+i\omega)^2+\omega_0^2},\quad a>0$
$e^{-at}\cos(\omega_0 t)u(t)$	$\frac{a+i\omega}{(a+i\omega)^2+\omega_0^2},\quad a>0$
$\mathrm{Rect}\left(\frac{t}{\tau}\right)$	$\tau\,\mathrm{sinc}\left(\frac{\omega\tau}{2}\right)$
$\frac{W}{\pi}\mathrm{sinc}(Wt)$	$\mathrm{rect}\left(\frac{\omega}{2W}\right)$
$\Delta\left(\frac{t}{\tau}\right)$	$\frac{\tau}{2}\mathrm{sinc}^2\left(\frac{\omega\tau}{4}\right)$
$\frac{W}{2\pi}\mathrm{sinc}^2\left(\frac{Wt}{2}\right)$	$\Delta\left(\frac{\omega}{2W}\right)$
$\sum_{n=-\infty}^{\infty}\partial(t-nT)$	$\omega_0\sum_{n=-\infty}^{\infty}\partial(\omega-n\omega_0),\quad \omega_0=\frac{2\pi}{T}$
$e^{-t^2/2\sigma^2}$	$\sigma\sqrt{2\pi}e^{-\sigma^2\omega^2/2}$

A.8 LINEAR SYSTEMS THEORY

A physical system that changes over time and whose mathematical model is a linear differential equation is said to be a *linear system.* A linear system has the property that the *superposition principle* holds—that is, the response to a superposition of inputs is a superposition of outputs. A linear system may also be *shift invariant*—that is, a phase shift in input leads to an identical phase shift in output.

For a linear differential equation

$$a_n(t)y^n + a_{n-1}(t)y^{n-1} + \cdots + a_1(t)y' + a_0(t)y = g(t),$$

the values of the variables $y(t), y'(t), \ldots, y^{(n-1)}(t)$ at a specific time t_0 describe the *state* of the system. The function g is called the *input function,* the *forcing function,* or the *excitation function.* A solution $y(t)$ is called the *output* or the *response* of the system.

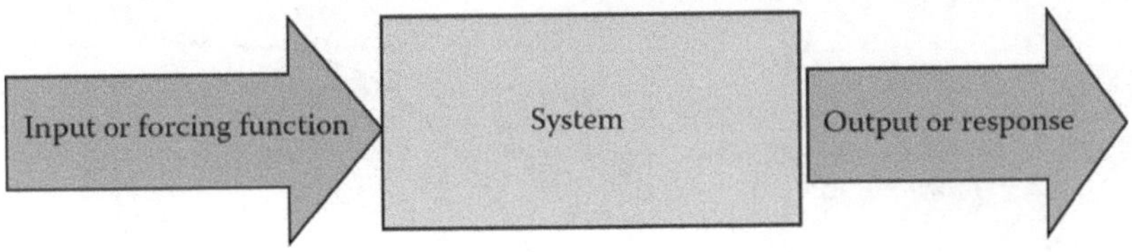

A.9 SUPERPOSITION PRINCIPLE

Simply stated, the superposition principle says that

a. The sum of two or more solutions of a homogeneous linear differential equation is also a solution
b. A constant multiple of a solution of a homogeneous linear differential equation is also a solution

A.10 PROBABILITY AND STATISTICS

A.10.1 Probability Axioms

1. Any probability $P(A)$ obeys $P(A) \geq 0$
2. If S is an event that is certain to occur, then $P(S) = 1$
3. If A_1 and A_2 are events that are mutually exclusive, the probability of the event A_1 or A_2 satisfies $P(A_1 \text{ or } A_2) = P(A_1) + P(A_2)$.

Mutually exclusive: Two events where the occurrence of one guarantees that the other does not occur.

Relative frequency: An experiment is repeated N times and a specific event A is observed n times, the relative frequency of the event A is defined as n/N.

A.10.2 Random Variable Types

Discrete: A random variable where the possible experimental outcomes consist of a discrete set of numbers.

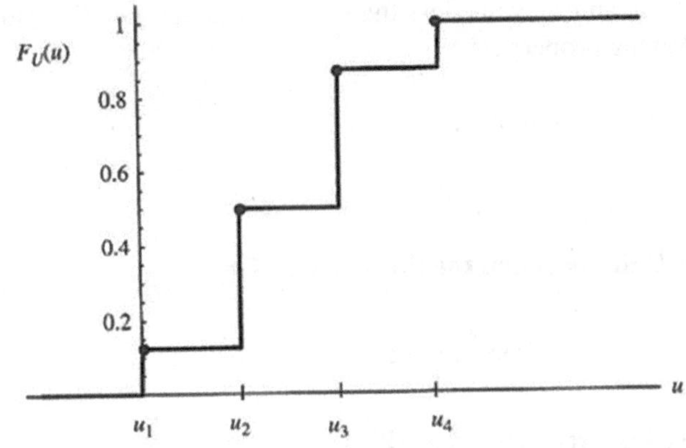

Continuous: A random variable where the possible experimental results can lie anywhere on a continuum of values.

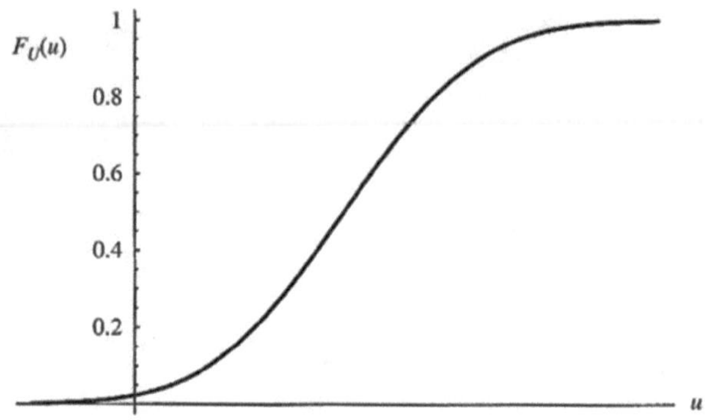

Mixed: A random variable with a combination of discrete and continuous outcomes.

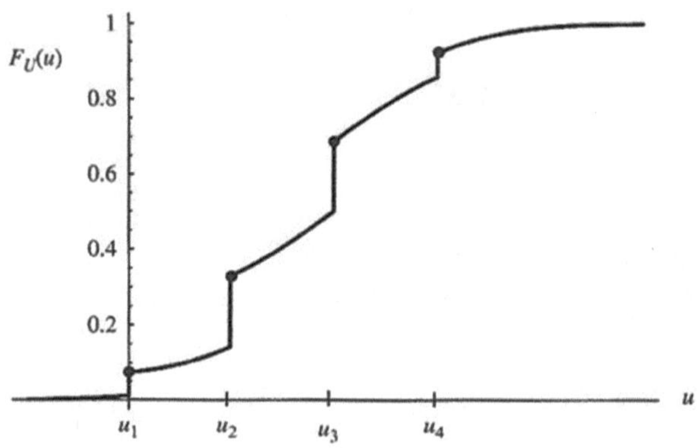

A.10.3 Cumulative Distribution Function (CDF)

Also known as the Cumulative Density Function (CDF). In all cases, it is convenient to describe the random variable U by a probability distribution function $F_U(u)$

$$F_U(u) = \text{Prob}\{U \le u\}.$$

The probability that U assumes a value less than or equal to the specific value u

$F_U(u)$ has the following properties:

1. $F_U(u)$ is non-decreasing to the right
2. $F_U(-\infty) = 0$
3. $F_U(\infty) = 1$

The probability that U lies between the limits $a < U \le b$

$$\text{Prob}\{a < U \le b\} = F_U(b) - F_U(a).$$

A.10.4 Probability Density Function (PDF)

The probability density function $p_U(u)$ is defined as the derivative of the probability distribution function.

$$p_U(u) \triangleq \frac{d}{du} F_U(u).$$

Discrete

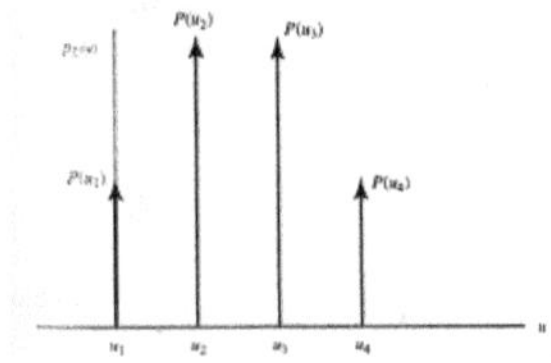

Continuous

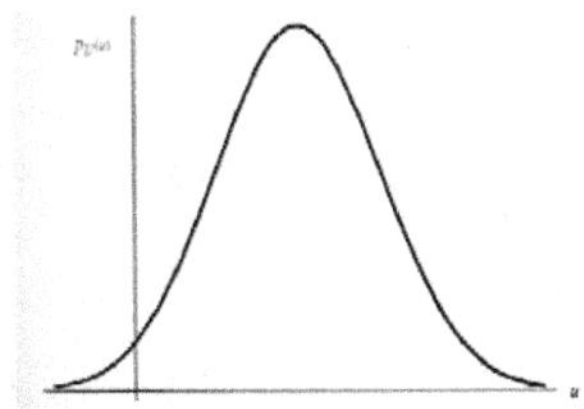

Mixed

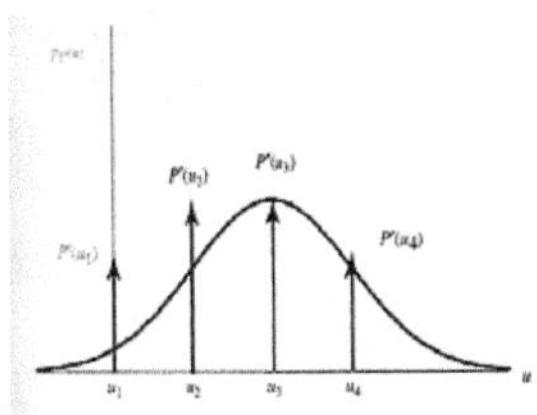

Discrete random variable PDF: For a discrete random variable U, $F_U(u)$ is discontinuous, thus $p_U(u)$ does not exist. By introducing a Dirac delta function δ, we can include this case,

$$\delta(u - u_k) = 0 \quad \text{for} \quad u \neq u_k,$$

$$\int_{-\infty}^{\infty} g_U \delta(u - u_k) du = g(u_{\bar{k}}).$$

Discrete PDF becomes:

$$p_U(u) = \sum_{k=1}^{\infty} P(u_k)\delta(u - u_k),$$

$g(u_{\bar{k}})$ **is the limit of** $g(u)$ **as** u **approaches** u_k **from the left.**

Poisson probability density function is given by:

$$p_U(u) = \sum_{k=0}^{\infty} \frac{k^{-k}}{k!} e^{-\bar{k}} \delta(u - k),$$

$\bar{k}$ represents the mean of the random variable U.

A.10.5 Continuous Random Variables

For a continuous random variable U,

$$p_U(u) = \lim_{\Delta u \to 0} \frac{F_U(u) - F_U(u - \Delta u)}{\Delta u}.$$

If Δu is sufficiently small,

$$p_U(u)\Delta u \approx F_U(u) - F_U(u - \Delta u) = \text{Prob}\{u - \Delta u < U \le u\},$$

implies that $p_U(u)$ has the following properties:

1. $p_U(u) \ge 0$
2. $\int_{-\infty}^{\infty} p_U(u)\,du = 1.$

The probability that U assumes a value between the limits a and b can be expressed in terms of the PDF

$$\text{Prob}\{a < U \le b\} = \int_a^b p_U(u)\,du.$$

Normal Density:

$$p_U(u) = \frac{1}{\sqrt{2\pi}\sigma} \exp\left[-\frac{(u - \bar{u})^2}{2\sigma^2}\right],$$

$\bar{u}$ is the mean of the random variable U
σ^2 is the variance of the random variable U.

A.10.6 Joint Random Variables

Given two random experiments with

- Sets of possible outcome events $\{A\}$ and $\{B\}$
- Set of possible joint outcomes $\{A \times B\}$
- Relative Frequency $\frac{n}{N}$
  - *n = number of times event A occurs jointly with event B*
  - *N = number of joint experimental trials*
- Assign numerical value $u(A)$ to each outcome A
- Assign numerical value $v(B)$ to each outcome B
- Joint random variable UV is the collection of all possible joint numbers (u, v), together with associated measures of their probabilities.

Joint probability distribution function (CDF),

$$F_{UV}(u, v) = \text{Prob}\{U \le u \text{ and } V \le v\}.$$

Joint probability density function (PDF),

$$p_{UV}(u,v) \triangleq \frac{\partial^2}{\partial u \partial v} F_{UV}(u,v).$$

Marginal probability: If the joint probabilities of all specific events A and B are known, the marginal probability of A or B occurring regardless of relation to the other event

$$P(A) = \sum_{\text{all } B} P(A,B),$$

$$P(B) = \sum_{\text{all } A} P(A,B).$$

Marginal PDF,

$$p_U(u) \triangleq \int_{-\infty}^{\infty} p_{UV}(u,v)dv,$$

$$p_V(v) \triangleq \int_{-\infty}^{\infty} p_{UV}(u,v)du.$$

Conditional probability: The probability of observing the event B in one experiment, given the event A has already been observed in another experiment is known as the conditional probability of B given A and is written as $p(B|A)$

$$\text{Relative frequency} = \frac{n}{N} = \frac{n}{m}\frac{m}{N},$$

- n=number of times joint event (A,B) occurs in N trials
- m=number of times A occurs regardless of B
- N=number of joint experimental trials
- $\frac{m}{N}$=marginal frequency of A
- $\frac{n}{m}$=conditional relative frequency of B, given A occurred

Bayes rule: The previous probabilities must satisfy:

$$P(A,B) = P(A)P(B \mid A).$$

Or,

$$P(B|A) = \frac{P(A,B)}{P(A)}.$$

Similarly,

$$P(A|B) = \frac{P(A,B)}{P(B)}.$$

Taken together,

$$P(B/A) = \frac{P(A/B)P(B)}{P(A)}.$$

Conditional PDF: The conditional PDF of U and V is defined by,

$$p_{V|U}(v/u) = \frac{p_{UV}(u,v)}{p_U(u)},$$

and,

$$p_{U|V}(u|v) = \frac{p_{UV}(u,v)}{p_V(v)}.$$

Statistical Independence: Two random variables U and V are called statistically independent if knowledge about the value assumed by one does not influence the probabilities associated with possible outcomes of the second.

For statistically independent random variables U and V, we have

$$p_{V|U}(v|u) = p_V(v).$$

This implies,

$$p_{UV}(u,v) = p_U(u)\,p_{V|U}(v|u) = p_U(u)\,p_V(v).$$

The joint PDF of two independent random variables factors into the product of their marginal density functions.

A.10.7 Statistical Average, Moments, and Characteristic Functions

Statistical Average: Given a function $g(u)$, where for every value of u, $g(\cdot)$ assigns a new real number $g(u)$ then the *Statistical Average* is the mean value (expected value) of $g(u)$ such that,

$$\overline{g} = E[g(u)] \triangleq \int_{-\infty}^{\infty} g(u)\,p_U(u)\,du.$$

For discrete random variables,

$$\overline{g} = \sum_k P(u_k)\,g(u_k).$$

Moments of a Random Variable U,

Given a function $g(u)$ of a random variable U such that,

$$g(u) = u^n,$$

The first moment: Mean Value, Expected Value, Average Value is given by,

$$\overline{u} = \int_{-\infty}^{\infty} u\,p_U(u)\,du.$$

The second moment: Mean-Square Value, is given by,

$$\overline{u^2} = \int_{-\infty}^{\infty} u^2\,p_U(u)\,du.$$

The nth moment of the random variable U is given by,

$$\overline{u^n} = \int_{-\infty}^{\infty} u^n p_U(u)\,du.$$

Central moments are fluctuations of a random variable about its mean, $\bar{u}$ such that,

$$g(u) = (u - \bar{u})^n.$$

The second central moment: Variance, is given by,

$$\sigma^2 = \int_{-\infty}^{\infty} (u - \bar{u})^2 p_U(u)\,du,$$

where σ is the standard deviation of the random variable U.

A.10.7.1 Joint Moments of Random Variables

Let U and V be random variables jointly distributed with PDF $p_{UV}(u, v)$

Joint moments are defined by:

$$\overline{u^n v^m} \triangleq \iint_{-\infty}^{\infty} u^n v^m p_{UV}(u, v)\,du\,dv.$$

Correlation

$$\Gamma_{UV} = \overline{uv} \triangleq \iint_{-\infty}^{\infty} uv p_{UV}(u, v)\,du\,dv.$$

Covariance

$$C_{UV} = \overline{(u - \bar{u})(v - \bar{v})}.$$

Correlation coefficient

$$\rho = \frac{C_{UV}}{\sigma_U \sigma_V}.$$

ρ is the direct measure of the similarity of the fluctuations of U and V

$$0 \le |\rho| \le 1.$$

Perfect Correlation $\rho = 1$

Fluctuations are essentially identical

Anticorrelation $\rho = -1$

Fluctuations are essentially identical but in an opposite sense

Uncorrelated $\rho = 0$

A.10.7.2 Characteristic Functions

The characteristic function of a random variable U is defined as the expected value of $\exp(j\omega u)$ i.e. $e^{j\omega u}$

$$M_U(\omega) = E\left[\exp(j\omega u)\right] \triangleq \int_{-\infty}^{\infty} \exp(j\omega u)\, p_U(u)\, du.$$

Thus the characteristic function is the Fourier Transform of the PDF of U. If this integral exists, the relationship is invertible.

$$p_U(u) = \frac{1}{2\pi} \int_{-\infty}^{\infty} M_U(\omega) \exp(-j\omega u)\, d\omega.$$

A.10.7.3 Negative Random Variables (–*U*)

The characteristic function of $-U$ is the complex conjugate of the characteristic function of U

$$\int_{-\infty}^{\infty} \exp(j\omega u)\, p_U(-u)\, du = \int_{-\infty}^{\infty} \exp(-j\omega u)\, p_U(u)\, du = M_U^*(\omega).$$

Complex conjugate is two complex numbers having their real parts identical and their imaginary parts of equal magnitude but opposite sign.

A.10.7.4 Common Characteristic Functions

Gaussian variable characteristic function

$$M_U(\omega) = \exp\left(-\frac{\sigma^2\omega^2}{2}\right) \exp(j\omega \bar{u}).$$

Poisson variable characteristic function

$$M_U(\omega) = \sum_{k=0}^{\infty} \frac{k^{-k}}{k!} e^{-\bar{k}} \delta(u-k) \exp(j\omega k) = \exp\left\{\bar{k}\left(e^{j\omega} - 1\right)\right\}.$$

A.10.7.5 Joint Characteristic Functions

Joint characteristic function for random variables U and V

$$M_{UV}(\omega_U, \omega_V) = \iint_{-\infty}^{\infty} \exp\left[j(\omega_U u + \omega_V v)\right] p_{UV}(u, v)\, du dv.$$

Joint PDF can be recovered from the joint characteristic function by a two-dimensional Fourier inversion.

Joint moments of U and V are expressible by:

$$\overline{u^n v^m} = \frac{1}{j^{n+m}} \frac{\partial^{n+m}}{\partial \omega_U^n\, \partial \omega_V^m} M_{UV}(\omega_U, \omega_V)\,|_{\omega_U = \omega_V = 0},$$

provided that $\left|\overline{u^n v^m}\right| < \infty$.

A.10.7.6 *K*th-Order Joint Characteristic Functions

Joint characteristic function of random variables $U_1, U_2, \ldots, U_k$

$$M_{\underline{U}}^{(k)}(\omega_1, \omega_2, \ldots, \omega_k) \triangleq E\left\{\exp\left[j(\omega_1 u_1 + \omega_2 u_2 + \cdots + \omega_k u_k)\right]\right\}.$$

Simplified in matrix notation

$$M_U(\underline{\omega}) \triangleq E\left\{\exp\left[j\underline{\omega}^t\,\underline{u}\right]\right\},$$

$$\underline{\omega} = \begin{bmatrix} \omega_1 \\ \omega_2 \\ \vdots \\ \omega_k \end{bmatrix} \quad \underline{u} = \begin{bmatrix} u_1 \\ u_2 \\ \vdots \\ u_k \end{bmatrix},$$

*k*th-order PDF $p_U(\underline{u})$ can be obtained by a *k*th-order Fourier inversion of $M_U(\underline{\omega})$.

Moment-generating function $G_U(\zeta)$ for random variable U is defined similarly to the characteristic function of U, but by means of a two-sided Laplace transform, rather than a Fourier transform,

$$G_U(\zeta) = E\left[e^{\zeta u}\right] = \int_{-\infty}^{\infty} e^{\zeta u} p_U(u)\,du.$$

A.10.8 Sums of Real Random Variables

For two random variables U and V that have joint PDF $p_{UV}(u,v)$, the probability density function of the sum of the random variables,

$$Z = U + V,$$

is given by

$$p_Z(z) = \int_{-\infty}^{\infty} p_{UV}(z - v, v)\,dv.$$

If U and V are independent random variables

$$p_{UV}(z - v, v) = p_U(z - v)\,p_V(v),$$

which yields

$$p_Z(z) = \int_{-\infty}^{\infty} p_U(z - v)\,p_V(v)\,dv.$$

This integral is a convolution of the individual PDFs of U and V,

$$p_Z = p_U \otimes p_V.$$

If $Z = U - V$,

$$p_Z(z) = \int_{-\infty}^{\infty} p_U(z + v)\,p_V(v)\,dv.$$

This integral is an anticorrelation,

$$p_Z = p_U \star p_V.$$

The characteristic function can also be used to derive $p_Z(z)$

$$M_Z(\omega) = \overline{\exp(j\omega z)} = \overline{\exp[j\omega(u+v)]}.$$

If U and V are independent random variables

$$M_Z(\omega) = \overline{\exp(j\omega u)\exp(j\omega v)} = M_U(\omega)M_V(\omega).$$

To find $p_Z(z)$ we must inverse Fourier transform $M_Z(\omega)$.

The inverse Fourier transform of the product of two functions is equal to the convolution of the individual inverse Fourier transforms.

A.10.8.1 The Central Limit Theorem

Let $U_1, U_2, \ldots, U_n$ be independent random variables with arbitrary PDFs with means $\overline{u_1}, \overline{u_2}, \ldots, \overline{u_n}$ and variances $\sigma_1^2, \sigma_2^2, \ldots, \sigma_n^2$.

Z is defined as:

$$Z = \frac{1}{\sqrt{n}} \sum_{i=1}^{n} \frac{U_i - \overline{u_i}}{\sigma_i}.$$

For every n, Z has zero mean and unit standard deviation.

As n approaches infinity, $p_Z(z)$ approaches a Gaussian density.

$$\lim_{n\to\infty} p_Z(z) = \frac{1}{\sqrt{2\pi}} e^{-z^2/2}.$$

The following conditions must be satisfied for all i

$$\sigma_i^2 > p > 0,$$

$$\overline{|u_i - \overline{u_i}|^3} < q.$$

A.10.9 Gaussian Random Variables

A random variable U is called Gaussian (or normal) if its characteristic function is of the form:

$$M_U(\omega) = \exp\left[j\omega\overline{u} - \frac{\sigma^2\omega^2}{2}\right].$$

The nth central moment is found to be

$$\overline{(u-\overline{u})^n} = \begin{cases} 1\cdot 3\cdot 5\cdot \ldots \cdot (n-1)\sigma^n & n \text{ even} \\ 0 & n \text{ odd} \end{cases}.$$

Fourier Inversion of $M_U(\omega)$

$$p_U(u) = \frac{1}{\sqrt{2\pi}\sigma} \exp\left[-\frac{(u-\overline{u})^2}{2\sigma^2}\right].$$

A.10.9.1 Joint Gaussian Random Variables

Joint characteristic function

$$M_U(\underline{\omega}) = \exp\left(j\underline{\bar{u}}^t\,\underline{\omega} - \frac{1}{2}\underline{\omega}^t\underline{C}\underline{\omega}\right),$$

$$\underline{\omega} = \begin{bmatrix} \omega_1 \\ \omega_2 \\ \vdots \\ \omega_n \end{bmatrix} \quad \underline{\bar{u}} = \begin{bmatrix} \bar{u}_1 \\ \bar{u}_2 \\ \vdots \\ \bar{u}_n \end{bmatrix},$$

C is an $n \times n$ covariance matrix with element σ_{ik}^2 in the ith row and kth column defined

$$\sigma_{ik}^2 = E\left[\left(u_i - \bar{u}_i\right)\left(u_k - \bar{u}_k\right)\right].$$

nth-order PDF

$$p_U(u) = \frac{1}{(2\pi)^{n/2}\left|\underline{C}\right|^{1/2}}\exp\left[-\frac{1}{2}\left(\underline{u} - \underline{\bar{u}}\right)^t \underline{C}^{-1}\left(\underline{u} - \underline{\bar{u}}\right)\right].$$

Two jointly distributed Gaussian random variables U and V with zero mean and common variance

$$p_{UV}(u,v) = \frac{1}{2\pi\sigma^2\sqrt{1-\rho^2}}\exp\left[-\frac{u^2 + v^2 - 2\rho uv}{2\left(1-\rho^2\right)\sigma^2}\right],$$

ρ is the normalized correlation coefficient of u and v

$$\rho \triangleq \frac{\overline{uv}}{\sigma^2}.$$

Two uncorrelated jointly Gaussian random variables are also statistically independent.

$$p_{UV}(u,v) = \frac{1}{2\pi\sigma^2}\exp\left(-\frac{u^2 + v^2}{2\sigma^2}\right),$$

$$p_{UV}(u,v) = \frac{1}{\sqrt{2\pi}\sigma}\exp\left[-\frac{u^2}{2\sigma^2}\right]\cdot\frac{1}{\sqrt{2\pi}\sigma}\exp\left[-\frac{v^2}{2\sigma^2}\right],$$

$$p_{UV}(u,v) = p_U(u)\,p_V(v).$$

The sum of two statistically independent jointly Gaussian random variables $Z = U + V$ is itself Gaussian

$$M_U(\omega) = \exp\left[j\omega\bar{u} - \frac{\omega^2\sigma_U^2}{2}\right] \quad M_V(\omega) = \exp\left[j\omega\bar{v} - \frac{\omega^2\sigma_V^2}{2}\right],$$

$$M_Z(\omega) = M_U(\omega)M_V(\omega),$$

$$M_Z(\omega) = \exp\left[j\omega(\bar{u} + \bar{v}) - \frac{\omega^2}{2}\left(\sigma_U^2 + \sigma_V^2\right)\right],$$

Z is a Gaussian random variable with mean $\overline{u}+\overline{v}$ and variance $\sigma_U^2+\sigma_V^2$.

The sum of two dependent (correlated) joint Gaussian Random variables U and V is itself Gaussian

$$\rho=0 \quad \overline{u}=\overline{v}=0 \quad \sigma_U^2=\sigma_V^2=0,$$

$$p_{UV}(u,v)=\frac{1}{2\pi\sigma^2\sqrt{1-\rho^2}}\exp\left[-\frac{u^2+v^2-2\rho uv}{2(1-\rho^2)\sigma^2}\right],$$

$$p_Z(z)=\int_{-\infty}^{\infty}p_{UV}(z-v,v)dv,$$

$$p_Z(z)=\int_{-\infty}^{\infty}\frac{1}{2\pi\sigma^2\sqrt{1-\rho^2}}\exp\left[-\frac{(z-v)^2+v^2-2\rho(z-v)v}{2(1-\rho^2)\sigma^2}\right]dv.$$

Complete the square in the exponent of the integral

$$p_Z(z)=\frac{\exp\left[-\frac{z^2}{4(1+\rho)\sigma^2}\right]}{2\pi\sigma^2\sqrt{1-\rho^2}}\int_{-\infty}^{\infty}\exp\left[-\frac{(v-z/2)^2}{(1-\rho)\sigma^2}\right]dv.$$

Perform the integral

$$p_Z(z)=\frac{1}{\sqrt{4\pi(1+\rho)\sigma^2}}\exp\left[-\frac{z^2}{4(1-\rho)\sigma^2}\right].$$

Z is Gaussian with zero mean and variance

$$\sigma_Z^2=2(1+\rho)\sigma^2,$$

when $\rho\to 0, \sigma_Z^2\to 2\sigma^2$,

when $\rho\to 1, \sigma_Z^2\to 4\sigma^2$,

when $\rho\to -1, \sigma_Z^2\to 0$.

This will also show that any linear combination of jointly Gaussian random variables, dependent or independent, is a Gaussian random variable.

A.10.9.2 Joint Moments of Order Higher than Two

Jointly Gaussian random variables $U_1, U_2, \ldots, U_n$ joint moments of order higher than two can always be expressed in terms of the first- and second-order moments.

A moment in the form $u_1^p u_2^q \ldots u_n^k$ can be obtained by partial differentiation of the n-dimensional characteristic function.

$$\overline{u_1^p u_2^q \ldots u_n^k}=\frac{1}{j^{p+q+\cdots+k}}\frac{\partial^{p+q+\cdots+k}}{\partial\omega_1^p\,\partial\omega_2^q\ldots\partial\omega_n^k}M_U(\underline{\omega})\,|_{\underline{\omega}=\underline{0}}.$$

Since the only parameters that appear in the n-dimensional Gaussian characteristic function are means, variances, and covariances, the $(p+q+\cdots+k)$ th-order moment must be expressible in terms of the first- and second-order moments.

A.10.9.3 Moment Theorem for Gaussian Random Variables

By differentiating the characteristic function for an appropriate number of times, it is possible to prove the following basic properties of zero-mean Gaussian random variables.

$$\overline{u_1u_2\ldots u_{2k+1}}=0,$$

$$\overline{u_1u_2\ldots u_k}=\sum_P\left(\overline{u_ju_m}\;\overline{u_lu_p}\ldots\overline{u_qu_s}\right)\quad j\neq m\quad l\neq p\quad q\neq s,$$

where Σ_P is the summation over all possible distinct grouping of the $2k$ variables in pairs.

There are $(2k)!/(2^k k!)$ such distinct groupings

For $k=2$

$$\overline{u_1u_2u_3u_4}=\overline{u_1u_2}\;\overline{u_3u_4}+\overline{u_1u_3}\;\overline{u_2u_4}+\overline{u_1u_4}\;\overline{u_2u_3}.$$

A.10.10 Complex-Valued Random Variables

Given that $\{A\}$ is a set of events of associated probabilities $P(A)$. A complex number $u(A)$ is assigned to each event. Together with the associated probability measures, we have a complex-valued random variable U.

The real and imaginary parts are described as joint statistical properties.

$$U=R+jI\quad u=r+ji,$$

$$F_U(u)\triangleq F_{RI}(r,i)\triangleq \text{Prob}\{R\leq r\text{ and }I\leq i\},$$

$$p_U(u)\triangleq p_{RI}(r,i)=\frac{\partial^2}{\partial r\,\partial i}F_{RI}(r,i),$$

$$M_U(\omega^r,\omega^i)\triangleq E\left[\exp\left[j(\omega^r r+\omega^i i)\right]\right].$$

A.10.10.1 Joint Complex Random Variables

Random variables $U_1,U_2,\ldots,U_n$ take of values $u_1=r_1+ji_1,u_2=r_2+ji_2,\ldots$

$$F_U(\underline{u})\triangleq \text{Prob}\{R_1\leq r_1,R_2\leq r_2,\ldots,R_n\leq r_n,I_1\leq i,I_2\leq i_2,\ldots,I_n\leq i_n\}.$$

Corresponding to $F_U(\underline{u})$ is a joint PDF of the $2n$ variables $\{r_1,r_2,\ldots,r_n,i_1,i_2,\ldots,i_n\}$

$$p_U(\underline{u})\triangleq\frac{\partial^2}{\partial r_1\ldots\partial r_n\,\partial i_1\ldots\partial i_n}F_U(\underline{u}).$$

Finally, the characteristic function

$$M_U(\underline{\omega})\triangleq E\left[\exp(j\underline{\omega}^t\underline{u})\right],$$

$$\underline{u}=\begin{bmatrix}r_1\\ \vdots\\ r_n\\ i_1\\ \vdots\\ i_n\end{bmatrix}\quad \underline{\omega}=\begin{bmatrix}\omega_1^r\\ \vdots\\ \omega_n^r\\ \omega_1^i\\ \vdots\\ \omega_n^i\end{bmatrix}.$$

Random variables $U_1, U_2, \ldots, U_n$ are said to be jointly Gaussian if their characteristic function is of the form.

$$M_U(\underline{\omega}) = \exp\left(j\underline{\bar{u}}^t\underline{\omega} - \frac{1}{2}\underline{\omega}^t\underline{C}\underline{\omega} \right).$$

$\underline{C}$ is a $2n \times 2n$ covariance matrix

$$\underline{C} = \overline{(\underline{u} - \underline{\bar{u}})(\underline{u} - \underline{\bar{u}})^t}.$$

$2n$-dimensional inverse Fourier transform yields

$$p_U(\underline{u}) = \frac{1}{(2\pi)^n |\underline{C}|^{1/2}} \exp\left[-\frac{1}{2}(\underline{u} - \underline{\bar{u}})^t \underline{C}^{-1} (\underline{u} - \underline{\bar{u}}) \right].$$

Let $\underline{r}$ and $\underline{i}$ be n-element column matrices of the real and imaginary parts of n complex variables. $U_k, (k = 1, 2, \ldots, n)$

$$\underline{r} \triangleq \begin{bmatrix} r_1 \\ r_2 \\ \vdots \\ r_n \end{bmatrix} \quad \underline{i} \triangleq \begin{bmatrix} i_1 \\ i_2 \\ \vdots \\ i_n \end{bmatrix}.$$

Covariance matrix is defined by

$$\underline{C}^{(rr)} \triangleq \overline{(\underline{r} - \underline{\bar{r}})(\underline{r} - \underline{\bar{r}})^t} \quad \underline{C}^{(ii)} \triangleq \overline{(\underline{i} - \underline{\bar{i}})(\underline{i} - \underline{\bar{i}})^t},$$

$$\underline{C}^{(ri)} \triangleq \overline{(\underline{r} - \underline{\bar{r}})(\underline{i} - \underline{\bar{i}})^t} \quad \underline{C}^{(ir)} \triangleq \overline{(\underline{i} - \underline{\bar{i}})(\underline{r} - \underline{\bar{r}})^t},$$

$$\underline{C} = \begin{bmatrix} \underline{C}^{(rr)} & \underline{C}^{(ii)} \\ \underline{C}^{(ri)} & \underline{C}^{(ir)} \end{bmatrix}.$$

A.10.10.2 Jointly Circular Random Variables

The complex variables $U_k, (k = 1, 2, \ldots, n)$ are jointly circular Gaussian variables if the following hold:

$$\underline{\bar{r}} = \begin{bmatrix} 0 \\ 0 \\ \vdots \\ 0 \end{bmatrix} \quad \underline{\bar{i}} = \begin{bmatrix} 0 \\ 0 \\ \vdots \\ 0 \end{bmatrix},$$

$$\underline{C}^{(rr)} = \underline{C}^{(ii)} \quad \underline{C}^{(ri)} = -\underline{C}^{(ir)}.$$

A.10.10.3 Single Circular Complex Random Variable

U is a single circular complex random variable

$$\underline{\bar{r}} = \bar{r} \quad \underline{\bar{i}} = \bar{i},$$

$$\underline{C}^{(rr)} = \sigma_r^2 \quad \underline{C}^{(ii)} = \sigma_i^2,$$

$$\underline{C}^{(ir)} = \sigma_r \sigma_i \rho \quad \underline{C}^{(ri)} = \sigma_r \sigma_i \rho.$$

Imposing the circularity conditions

$$\overline{r} = \overline{i} = 0 \quad \sigma_r^2 = \sigma_i^2 = \sigma^2 \quad \rho = 0,$$

$$\underline{C} = \begin{bmatrix} \sigma^2 & 0 \\ 0 & \sigma^2 \end{bmatrix}.$$

PDF of U becomes

$$p_U(u) = \frac{1}{2\pi\sigma^2} \exp\left(-\frac{r^2 + i^2}{2\sigma^2}\right).$$

A.10.10.4 Complex Gaussian Moment Theorem

Let the complex-values random variables $U_1, U_2, \ldots, U_k, U_{k+1}, U_{k+2}, \ldots, U_r$ have zero means and obey jointly Gaussian statistics

$$\overline{\overset{*}{u_1} \ldots \overset{*}{u_1} u_{k+1} \ldots u_r} = 0 \qquad r \neq k$$

$$\overline{\overset{*}{u_1} \ldots \overset{*}{u_1} u_{k+1} \ldots u_r} = \sum_{\pi} \overline{\overset{*}{u_1} u_p} \, \overline{\overset{*}{u_2} u_q} \ldots \overline{\overset{*}{u_k} u_t} \qquad r = k\text{'}$$

Σ_π denotes a summation over the $k!$ possible permutations $(p, q, \ldots, t)$ of $(k+1, k+2, \ldots, 2k)$.

If $k = 2$

$$\overline{\overset{*}{u_1} \overset{*}{u_2} u_3 u_4} = \overline{\overset{*}{u_1} u_3} \; \overline{\overset{*}{u_2} u_4} + \overline{\overset{*}{u_1} u_4} \; \overline{\overset{*}{u_2} u_3}.$$

See Page 4 of book for higher values of k.

A.10.11 Random Phasor Sums

A phasor is a complex-valued contribution to a complex-valued random variable that represents the amplitude and phase of a monochromatic or nearly monochromatic wave disturbance.

A complex addition of many small independent phasors occurs when we calculate the total complex amplitude of the wave that arises as the result of scattering by a collection of small, independent scatterers. Sums of complex random variables are referred to as random phasor sums.

Consider a sum of a very large number N of complex phasors, the kth phasor having random amplitude $a_k / \sqrt{N}$ and random phase ϕ_k. The resultant phasor, a, with length a and phase θ is defined by

$$a = ae^{j\theta} = \frac{1}{\sqrt{N}} \sum_{k=1}^{N} a_k e^{j\phi_k},$$

1. The amplitude $a_k / \sqrt{N}$ and phase ϕ_k of the kth elementary phasor are statistically independent of each other and of the amplitudes and phases of all other elementary phasors

2. The lengths a_k are identically distrusted for all k, with mean $\overline{a}$ and second moment $\overline{a^2}$
3. The phases ϕ_k are all uniformly distributed on $(-\pi, \pi)$.

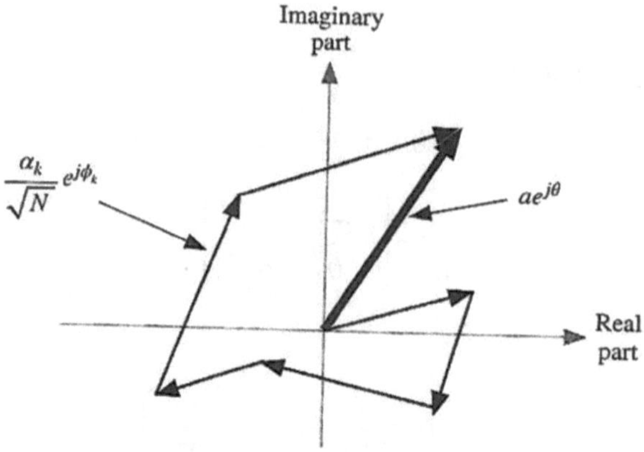

The real and imaginary parts r and i of the resultant phasor are defined as

$$r \triangleq \mathrm{Re}\left\{ae^{j\theta}\right\} = \frac{1}{\sqrt{N}} \sum_{k=1}^{N} a_k \cos\phi_k,$$

$$i \triangleq \mathrm{Im}\left\{ae^{j\theta}\right\} = \frac{1}{\sqrt{N}} \sum_{k=1}^{N} a_k \sin\phi_k.$$

By virtue of the central limit theorem both r and i are asymptotically Gaussian random variables as $N \to \infty$

To determine the joint PDF of r and i, we must first calculate $\overline{r}, \overline{i}, \sigma_r^2, \sigma_i^2$ and their correlation coefficient ρ.

Mean Values of r and i.

$$\overline{r} = \frac{1}{\sqrt{N}} \sum_{k=1}^{N} \overline{a_k \cos\phi_k} = \frac{1}{\sqrt{N}} \sum_{k=1}^{N} \overline{a_k} \, \overline{\cos\phi_k} = \sqrt{N}\,\overline{a}\,\overline{\cos\phi},$$

$$\overline{i} = \frac{1}{\sqrt{N}} \sum_{k=1}^{N} \overline{a_k \sin\phi_k} = \frac{1}{\sqrt{N}} \sum_{k=1}^{N} \overline{a_k} \, \overline{\sin\phi_k} = \sqrt{N}\,\overline{a}\,\overline{\sin\phi}.$$

Since ϕ is uniformly distributed on $(-\pi, \pi)$, then $\overline{\cos\phi} = \overline{\sin\phi} = 0$ and $\overline{r} = \overline{i} = 0$.

Variance of r and i.

To find σ_r^2 and σ_i^2, we will evaluate the second moments $\overline{r^2}$ and $\overline{i^2}$ since the means are zero.

Using the independence of amplitudes and phases

$$\overline{r^2} = \frac{1}{N} \sum_{k=1}^{N} \sum_{n=1}^{N} \overline{a_k a_n \cos\phi_k \cos\phi_n},$$

$$\overline{i^2} = \frac{1}{N} \sum_{k=1}^{N} \sum_{n=1}^{N} \overline{a_k a_n \sin\phi_k \sin\phi_n},$$

$$\overline{\cos\phi_k \cos\phi_n} = \overline{\sin\phi_k \sin\phi_n} = \begin{cases} 0 & k \neq n \\ 1/2 & k = n \end{cases}.$$

Due to the uniform distribution of the phases

$$\overline{r^2} = \overline{i^2} = \frac{\overline{a^2}}{2} \triangleq \sigma^2.$$

Correlation between r and i,

$$\overline{ri} = \frac{1}{N}\sum_{k=1}^{N}\sum_{n=1}^{N} \overline{a_k a_n \cos\phi_k \sin\phi_n}.$$

Since $\cos\phi\sin\phi = 1/2\sin 2\phi$**, we have**

$$\overline{\cos\phi_k \sin\phi_n} = \begin{cases} \overline{\cos\phi_k}\ \overline{\sin\phi_n} = 0 & k \neq n \\ \frac{1}{2}\overline{\sin 2\phi_k} & k = n \end{cases}.$$

Thus, the real and imaginary parts of the resultant phasor are uncorrelated.

Zero mean, equal variance, and lack of correlation are true for any N, finite or infinite.

Joint PDF or r and i,

$$p_{RI}(r,i) = \frac{1}{2\pi\sigma^2}\exp\left(-\frac{r^2+i^2}{2\sigma^2}\right),$$

where

$$\sigma^2 = \frac{\overline{a^2}}{2}.$$

A.10.11.1 Statistics of Length and Phase

In many applications, it is desired to know the statistics of the length a and the phase θ of the random phasor sum.

$$a = \sqrt{r^2+i^2} \qquad \theta = \tan^{-1}\frac{i}{r}.$$

Changing from rectangular to polar coordinates is a one-to-one mapping, thus we can use the methods described in Section 2.5 to find the joint PDF of a and θ

$$r = a\cos\theta \qquad i = a\sin\theta,$$

$$|J| = \left\|\begin{matrix} \partial r/\partial a & \partial r/\partial\theta \\ \partial i/\partial a & \partial i/\partial\theta \end{matrix}\right\| = \left\|\begin{matrix} \cos\theta & -a\sin\theta \\ \sin\theta & \cos\theta \end{matrix}\right\| = a.$$

Thus

$$p_{A\Theta}(a,\theta) = p_{RI}(r = a\cos\theta, i = a\sin\theta)\cdot a,$$

$$p_{A\Theta}(a,\theta) = \begin{cases} \frac{a}{2\pi\sigma^2}\exp\left(-\frac{a^2}{2\sigma^2}\right) & -\pi < \theta \le \pi \text{ and } a > 0 \\ 0 & \text{otherwise} \end{cases}.$$

A.10.11.2 Rayleigh Density Function

The marginal PDF of the length and phase can now be found.

$$p_A(a) = \int_{-\pi}^{\pi} p_{A\Theta}(a,\theta)d\theta = \begin{cases} \dfrac{a}{\sigma^2}\exp\left(-\dfrac{a^2}{2\sigma^2}\right) & a > 0 \\ 0 & \text{otherwise} \end{cases},$$

where

$$\bar{a} = \sqrt{\frac{\pi}{2}}\sigma \quad \sigma_a^2 = \left(2 - \frac{\pi}{2}\right)\sigma^2,$$

while the qth moment is

$$\overline{a^q} = \int_0^{\infty} a^q p_A(a)da = 2^{\frac{q}{2}}\sigma^q\Gamma\left(1 + \frac{q}{2}\right),$$

$\Gamma(\cdot)$ is a gamma function.

PDF of θ,

$$p_\Theta(\theta) = \begin{cases} \dfrac{1}{2\pi}\displaystyle\int_0^{\infty} \frac{a}{\sigma^2}\exp\left(-\frac{a^2}{2\sigma^2}\right)da & -\pi < \theta \le \pi \\ 0 & \text{otherwise} \end{cases}.$$

The integral is precisely the integral of a Rayleigh density function and, hence, must be unity. We conclude that the phase θ of the resultant is uniformly distributed an $(-\pi, \pi)$

$$p_\Theta(\theta) = \begin{cases} \dfrac{1}{2\pi} & -\pi < \theta \le \pi \\ 0 & \text{otherwise} \end{cases}.$$

Note: Since $p_{A\Theta}(a,\theta)$ is equal to the simple product of $p_A(a)$ and $p_\Theta(\theta)$, it follows that A and Θ are independent random variables.

A.10.11.3 Constant Phasor Plus a Random Phasor Sum

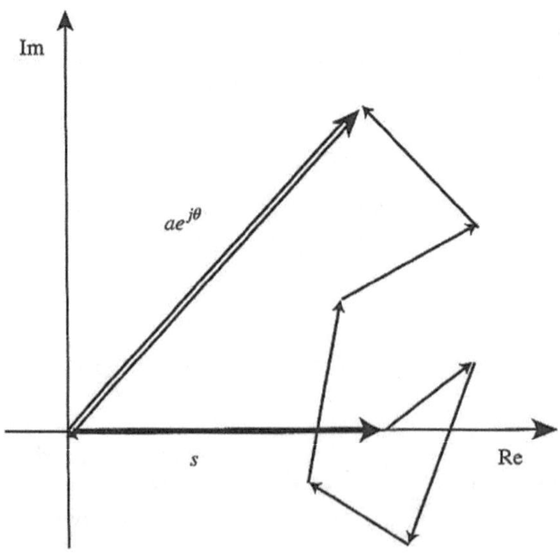

A known phasor is entirely real and positive with length s

$$r = s + \frac{1}{\sqrt{N}} \sum_{k=1}^{N} a_k \cos\phi_k,$$

$$i = \frac{1}{\sqrt{N}} \sum_{k=1}^{N} a_k \sin\phi_k.$$

As $N \to \infty$ the joint statistics of R and I remain Gaussian, but with a modified mean

$$p_{RI}(r,i) = \frac{1}{2\pi\sigma^2} \exp\left(-\frac{(r-s)^2 + i^2}{2\sigma^2}\right),$$

$$p_{A\Theta}(a,\theta) = \begin{cases} \dfrac{a}{2\pi\sigma^2} \exp\left[-\dfrac{(a\cos\theta - 2)^2 + (a\sin\theta)^2}{2\sigma^2}\right] & -\pi < \theta \le \pi \text{ and } a > 0 \\ 0 & \text{otherwise} \end{cases},$$

$$p_A(a) = \int_{-\infty}^{\infty} p_{A\Theta}(a,\theta)\,d\theta = \frac{a}{2\pi\sigma^2} \exp\left(-\frac{a^2 + s^2}{2\sigma^2}\right) \int_{-\pi}^{\pi} \exp\left(\frac{as}{\sigma^2}\cos\theta\right) d\theta,$$

$$p_A(a) = \begin{cases} \dfrac{a}{\sigma^2} \exp\left(-\dfrac{a^2 + s^2}{2\sigma^2}\right) I_0\left(\dfrac{as}{\sigma^2}\right) & a > 0 \\ 0 & \text{otherwise} \end{cases},$$

$I_0(\cdot)$ is modified Bessel function of the first kind, zero order.

This result is known as a Rician density function.

The qth moment of a Rician variate can be shown to be given by

$$\overline{a^q} = \left(2\sigma^2\right)^{\frac{q}{2}} \exp\left(-\frac{s^2}{2\sigma^2}\right) \Gamma\left(1 + \frac{q}{2}\right) {}_1F_1\left(1 + \frac{q}{2}, 1, \frac{s^2}{2\sigma^2}\right),$$

$$= \left(2\sigma^2\right)^{\frac{q}{2}} \Gamma\left(1 + \frac{q}{2}\right) L_{q/2}\left(-\frac{s^2}{2\sigma^2}\right).$$

where $\Gamma(\cdot)$ is a gamma function, ${}_1F_1(\cdot)$ is a confluent hypergeometric function and $L_n(x)$ is a Laguerre polynomial of order n

$$\bar{a} = \frac{1}{2}\sqrt{\frac{\pi}{2\sigma^2}} e^{-s^2/4\sigma^2} \left[\left(s^2 + 2\sigma^2\right) I_0\left(\frac{s^2}{4\sigma^2}\right) + s^2 I_1\left(\frac{s^2}{4\sigma^2}\right)\right],$$

$$\overline{a^2} = s^2 + 2\sigma^2,$$

$I_0(\cdot)$ and $I_1(\cdot)$ are modified Bessel functions of the first kind, zero and first orders.

$$p_\Theta(\theta) = \int_0^{\infty} p_{A\Theta}(a,\theta)\,da,$$

$$p_{\Theta}(\theta)=\frac{\exp\left(-\frac{s^2}{2\sigma^2}\right)}{2\pi}+\sqrt{\frac{1}{2\pi}}\exp-\left(\frac{s^2}{2\sigma^2}\sin^2\theta\right)\frac{1+\operatorname{erf}\left(\frac{s\cos\theta}{\sqrt{2}\sigma}\right)}{2}\cos\theta,$$

$\operatorname{erf}(\cdot)$ is a standard error function

$$\operatorname{erf}(z)=\frac{2}{\sqrt{\pi}}\int_0^z e^{-t^2}dt.$$

A.10.11.4 Strong Constant Phasor Plus a Weak Random Phasor Sum

If the constant phasor is much stronger than the random phasor sum, we can greatly simplify the equations.

For $s \gg \sigma$

$$p_A(a)=\frac{1}{\sqrt{2\pi}}\exp\left[-\frac{(a-s)^2}{2\sigma^2}\right],$$

$$\theta \approx \tan\theta \approx \frac{i}{s},$$

$$p_{\Theta}(\theta)\approx s\ p_I(i=s\theta),$$

Or

$$p_{\Theta}(\theta)\approx\frac{k}{\sqrt{2\pi}}\exp\left[-\frac{k^2\theta^2}{2}\right].$$

A.10.12 Poisson Random Variables

A discrete random variable K is said to be a Poisson variate if the probability of observing integer outcome k is given by

$$p_K(k)=\frac{a^k}{k!}\exp(-a),$$

$$\overline{k}=\sum_{k=0}^{\infty}k\frac{a^k}{k!}\exp(-a)=a\sum_{k=1}^{\infty}\frac{a^{k-1}}{(k-1)!}\exp(-a)=a,$$

$$\overline{k^2}=\sum_{k=0}^{\infty}k^2\frac{a^k}{k!}\exp(-a)=a\sum_{k=1}^{\infty}k\frac{a^{k-1}}{(k-1)!}\exp(-a)=a+a^2.$$

Characteristic function of Poisson Random Variable

$$M_K(j\omega)=\overline{e^{j\omega k}}=\sum_{k=0}^{\infty}\frac{\left(e^{j\omega}\right)^k \overline{k}^{-k}}{k!}e^{-\overline{k}}=\exp\left[\overline{k}\left(e^{j\omega}-1\right)\right].$$

Let K_1 and K_2 be statistically independent Poisson random variables having different means and $K=K_1+K_2$

$$M_K(j\omega)=M_{K_1}(j\omega)M_{K_2}(j\omega)=\exp\left[\left(\overline{k_1}+\overline{k_2}\right)\left(e^{j\omega}-1\right)\right].$$

A.11 RANDOM PROCESSES

A random process is a generalization of the concept of random variables in which the basic unpredictable or random events are functions (Usually of time and space) rather than numbers. $\{A\}$ is a set of possible events and associated probability measures. To define a random variable, a real-valued number $u(A)$ is assigned to each elementary event A. To define a random process, we assign a real-valued function $u(A:t)$, with independent variable t, to each elementary event A. The set of possible sample functions $u(A:t)$, together with their associated probability measures, constitute a random process. The dependence of the random process on $\{A\}$ is not usually noted in the notation

$u(t)$ – A random sample function

$U(t)$ **– Random process i.e. all possible** $u(t)$

Example: Coin Toss – Two Tosses

The elementary events in {A} are A_1 = HH, A_2 = HT, A_3 = TH, and A_4 = TT

Assign a sample function to each event

$$u(A_1:t)=\exp(t) \quad u(A_2:t)=\exp(2t)$$
$$u(A_3:t)=\exp(3t) \quad u(A_4:t)=\exp(4t)$$

For the sample above, each sample function has a probability of ¼. If multiple events generate the same sample function, all possible ways of generating the sample function must be discovered. This becomes difficult and time-consuming and is quite often not possible.

A.11.1 First Order PDF

The parameter t is viewed as fixed. $p_U(u:t)$ is a first-order PDF is specified for the random variable $U(t)$. **Can be used to calculate** $\overline{u}, \overline{u^2}$**, and other moments of** U **for any value of** t

A.11.2 Second Order PDF

The second-order PDF of U **utilizes two parameters** t_1 **and** t_2

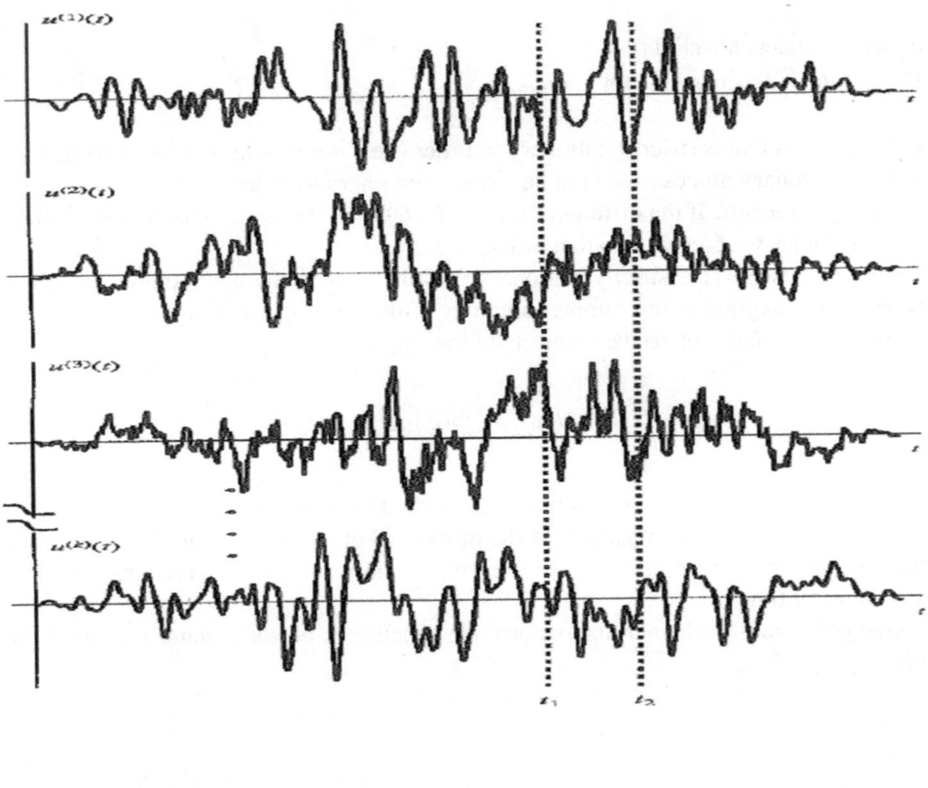

Joint PDF of $U(t_1)$ and $U(t_2)$

$$p_U(u_1,u_2:t_1,t_2).$$

From this description, we can calculate joint moments

$$\overline{u_1u_2} = \iint_{-\infty}^{\infty} u_1u_2p_U(u_1,u_2:t_1,t_2)du_1\,du_2.$$

To completely describe the random process $U(t)$, we specify the kth-order PDF as

$$p_U(u_1,u_2,\ldots,u_k:t_1,t_2,\ldots,t_k).$$

A random process is a mathematical model that is useful to us only before the exact sample function $u(t)$ is determined by measurement.

A.11.3 Stationarity and Ergodicity

Strictly stationary random processes: A random process is called strictly stationary to order k if the kth-order PDF is independent of the choice of time origin

For all T

$$p_U(u_1,u_2,\ldots,u_k:t_1,t_2,\ldots,t_k)= p_U(u_1,u_2,\ldots,u_k:t_1-T,t_2-T,\ldots,t_k-T).$$

For process stationary to order $k\geq 1$, the first-order PDF must be independent of time and hence $p_U(u)$.

For process stationary to order $k\geq 2$, the second-order PDF depends only on the time difference $\tau = t_1-t_2$ and is written as $p_U(u_1,u_2:\tau)$.

A wide-sense stationary process – A process is called wide-sense stationary if the following two conditions are met:

1. $E[u(t)]$ is independent of t
2. $E[u(t_1)u(t_2)]$ depends only on $\tau = t_1-t_2$

Note: Every process that is strictly stationary to order $k\geq 2$ is also wide-sense stationary; however, a wide-sense stationary process need not be strictly stationary to order $k=2$.

Stationary increments: If the difference $U(t_2)-U(t_1)$ is strictly stationary to some order, $U(t)$ is said to have stationary increments to that order.

If random process $\Phi(t)$ is strictly stationary to order k, then the new random process (below) constructed from integrals of the sample functions of $\Phi(t)$ is nonstationary to all orders but does have stationary increments up to the order of stationarity of $\Phi(t)$

$$U(t)=U(t_0)-\int_{t_0}^{t}\Phi(\xi)d\xi \quad (t>t_0).$$

Ergodic random processes: Most restrictive class of random processes

Attention is focused on a comparison of the properties of a single sample function as it evolves along the time axis with the properties of the entire ensemble of sample functions observed at one or more instants of time.

The issue of interest if whether any one sample function is on some sense typical of the entire ensemble.

A sample process is said to be ergodic if every sample function takes on values along the time axes with the same joint relative frequencies observed across the ensemble at any instant or collection of instants.

While a process must be strictly stationary to be ergodic, not all strictly stationary processes are necessarily ergodic (See Page 63 in book for example).

If a random process is ergodic, any average calculated along a sample function must be equal to the same average calculated across the ensemble.

If $g(u)$ is the quantity to be averages, the time average

$$g = \lim_{T\to\infty}\frac{1}{T}\int_{-T/2}^{T/2} g\left[u(t)\right]dt,$$

where $\langle\cdot\rangle$ indicates an infinite time average.

Must be equal to the ensemble average.

$$\bar{g} = \int_{-\infty}^{\infty} g(u)\,p_U(u)\,du.$$

For an ergodic random process, time and ensemble averages are interchangeable.

To establish ergodicity, it is necessary to consider the entire ensemble of sample functions. The ensemble can be said to be ergodic, provided the following:

- The ensemble is strictly stationary
- The ensemble contains no statistically different, strictly stationary sub-ensembles that occur with probability other than 0 or 1

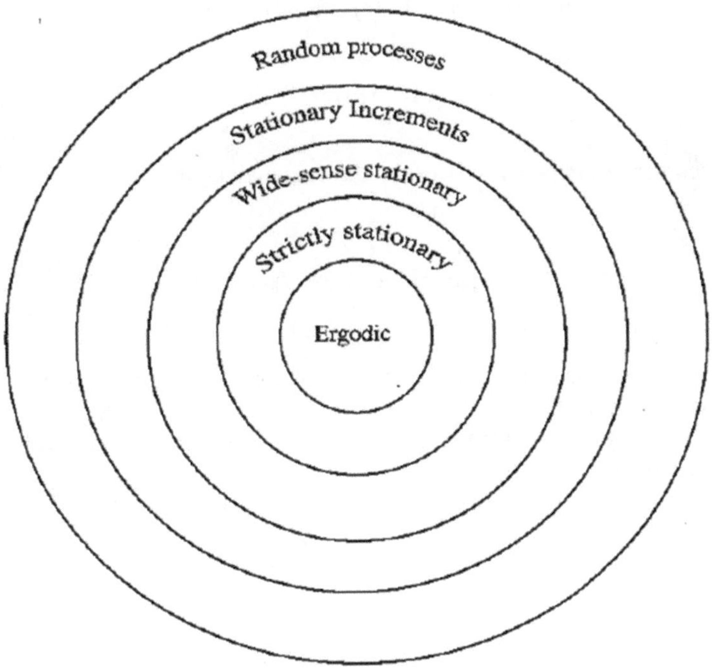

In certain situations, a weaker definition of ergodicity may be used

1. The process may be ergodic only with respect to certain averages, rather than with respect to the full PDF. For example if $x(t)x(t+\tau)=x(t)x(t+\tau)$, the process is ergodic with respect to the autocorrelation function. This type of ergodicity does not require strict stationarity.
2. If the process extends over the infinite time axis, then a change in statistics over a finite interval of time will not change the relative frequencies in time of the values of infinite time averages. Ergodicity can be defined such that intervals of vanishing significance $T \to \infty$ are ignored. Under such a definition, the process need not be strictly stationary to be ergodic.

A.11.4 Spectral Analysis of Random Processes

Spectral density: Let $u(t)$ be a known function of time. If $u(t)$ has the property that

$$\int_{-\infty}^{\infty} |u(t)|\, dt < \infty.$$

We say that $u(t)$ is Fourier transformable. It is important in practice to be able to specify the distribution of energy over frequency. This description is called the energy spectral density of $u(t)$.

On the other hand, it may be that $u(t)$ does not satisfy the first equation, but does satisfy

$$\lim_{T\to\infty} \frac{1}{T} \int_{-T/2}^{T/2} u^2(t)\, dt < \infty.$$

In this case, we say that $u(t)$ has finite average power. It is important in practice to be able to specify the average power over frequency. This description is called the power spectral density of $u(t)$.

Spectral densities of a known function: If $u(t)$ is Fourier-transformable, then

$$\mathcal{U}(\nu) = \int_{-\infty}^{\infty} u(t) e^{j2\pi\nu t}\, dt,$$

always exists. According to Parseval's theorem, the area under $|\mathcal{U}|^2$ is equal to the total energy contained in $u(t)$

$$\int_{-\infty}^{\infty} u^2(t)\, dt = \int_{-\infty}^{\infty} |\mathcal{U}(\nu)|^2\, d\nu.$$

Thus the quantity $\mathcal{E}(\nu) = |\mathcal{U}(\nu)|^2$ has the dimensions of energy per unit frequency. This is referred to as the energy spectral density of $u(t)$

If $u(t)$ is *not* Fourier-transformable but has finite average power, then the integral $\mathcal{U}(\nu)$ does not exist. However, the truncated function

$$u_T(t) = \begin{cases} u(t) & -\dfrac{T}{2} \le t \le \dfrac{T}{2} \\ 0 & \text{otherwise} \end{cases}.$$

Does have a transform, denoted by $\mathcal{U}_T(\nu)$.

$|\mathcal{U}(\nu)|^2$ represents the distribution of energy over frequency for the truncated waveform $u_T(t)$, thus the normalized energy spectrum

$$\mathcal{G}_T(\nu) = \frac{\left|\mathcal{U}_T(\nu)\right|^2}{T}.$$

has the dimensions of power per unit frequency, and we are logically led to define the power spectral density of $u(t)$ by

$$\mathcal{G}_U(\nu) \triangleq \lim_{T\to\infty} \frac{\left|\mathcal{U}_T(\nu)\right|^2}{T}.$$

Spectral densities of a random process: Since we wish to find a spectral distribution that characterizes the entire random process, it is logical to define such quantities in terms of averages over the entire random process ensemble. Accordingly, we define the energy and power spectral densities of a random process by:

$$\mathcal{E}_U(\nu) = E\left[\left|\mathcal{U}(\nu)\right|^2\right],$$

$$\mathcal{G}_U(\nu) = \lim_{T\to\infty} \frac{E\left[\left|\mathcal{U}_T(\nu)\right|^2\right]}{T}.$$

Spectral densities – basic properties: Several basic properties of spectral densities follow directly from the above equations

1. $\mathcal{E}_U(\nu) \ge 0, \mathcal{G}_U(\nu) \ge 0$; Energy and power spectral densities are nonnegative and real valued
2. $\mathcal{E}_U(-\nu) = \mathcal{E}_U(\nu), \mathcal{G}_U(-\nu) = \mathcal{G}_U(\nu)$; energy and power spectral densities are even functions of ν
3. $\int_{-\infty}^{\infty} \mathcal{E}_U(\nu)d\nu = \int_{-\infty}^{\infty} \overline{u^2(t)}dt$

$$\int_{-\infty}^{\infty} \mathcal{G}_U(\nu)d\nu = \begin{cases} \overline{u^2} & \text{if } U(t) \text{ is stationary} \\ \left\langle \overline{u^2(t)} \right\rangle & \text{if } U(t) \text{ is nonstationary} \end{cases}.$$

Spectral Densities for Linearly Filtered Random Processes

Let random process $V(t)$ consist of sample functions that results from passing all sample functions of $U(t)$ through a known linear filter.

If the sample functions of $U(t)$ are Fourier transformable, then the linear filter is time invariant, in which case a single output function $v(t)$ is related to the corresponding input sample function $u(t)$ by a convolution equation.

$$v(t) = \int_{-\infty}^{\infty} h(t-\xi)u(\xi)d\xi,$$

where $h(t)$ represents that known response of the filter at time t to a unit impulse applied at the input at time $t = 0$.

$h(t)$ is called the impulse response.

In the frequency domain, this relationship becomes a simple multiplicative one.

$$\mathcal{V}(v) = \mathcal{H}(v)\mathcal{U}(v),$$

$\mathcal{V}(v)$ is the Fourier transformation of $v(t)$

$\mathcal{U}(v)$ is the Fourier transformation of $u(t)$

$\mathcal{H}(v)$ is the Fourier transformation of $h(t)$ and is known as the transfer function of the linear, time-invariant filter

$$\mathcal{E}_V(v) = E\left[\left|\mathcal{H}(v)\mathcal{U}(v)\right|^2\right] = \left|\mathcal{H}(v)\right|^2 \mathcal{E}_U(v).$$

If the sample functions of $U(t)$ are *not* Fourier transformable, then we must utilize truncated waveforms

$$v_T(t) \approx \int_{-\infty}^{\infty} h(t-\xi)u_T(\xi)d\xi.$$

As T grows, this approximation becomes arbitrarily good.

In the frequency domain, this relationship becomes a simple multiplicative one.

$$\mathcal{V}_T(v) = \mathcal{H}(v)\mathcal{U}_T(v),$$

$$\mathcal{G}_V(v) = \lim_{T\to\infty} \frac{E\left[\left|\mathcal{V}_T(v)\right|^2\right]}{T} = \lim_{T\to\infty} \frac{E\left[\left|\mathcal{H}(v)\mathcal{U}_T(v)\right|^2\right]}{T} = \left|\mathcal{H}(v)\right|^2 \lim_{T\to\infty} \frac{E\left[\left|\mathcal{U}_T(v)\right|^2\right]}{T},$$

$$\mathcal{G}_V(v) = \left|\mathcal{H}(v)\right|^2 \mathcal{G}_U(v).$$

A.11.5 Autocorrelation Functions and the Wiener-Khinchin Theorem

Time autocorrelation function: Given a single real-valued known time function $u(t)$, which may be one sample function of a random process, the time autocorrelation function of $u(t)$ is defined by

$$\tilde{\Gamma}(\tau) \triangleq \langle u(t+\tau)u(t)\rangle = \lim_{T\to\infty} \frac{1}{T} \int_{-T/2}^{T/2} u(t+\tau)u(t)dt,$$

where $\langle\cdot\rangle$ indicates an infinite time average and the ~ denotes that this quantity is defined by a time average rather than a statistical average. This measures the structural similarity of $u(t)$ and $u(t+\tau)$.

Statistical autocorrelation function: The statistical autocorrelation function is defined by

$$\Gamma_U(t_2,t_1) \triangleq \overline{u(t_2)u(t_1)} = \iint_{-\infty}^{\infty} u_2 u_1 p_U(u_1,u_2 : t_2,t_1)du_1 du_2.$$

Thus measures the statistical similarity of $u(t_1)$ and $u(t_2)$.

For an ergodic process $\widetilde{\Gamma_U}(\tau) = \Gamma_U(\tau)$.

The following properties are also important for random processes that are at least wide-sense stationary

1. $\Gamma_U(0) = \overline{u^2}$
2. $\Gamma_U(-\tau) = \Gamma_U(\tau)$
3. $\left|\Gamma_U(\tau)\right| \le \Gamma_U(0)$

Wiener-Khinchin theorem: The autocorrelation function and the power spectral density form a Fourier transformation pair

$$\mathcal{G}_U(\nu) = \int_{-\infty}^{\infty} \Gamma_U(\tau) e^{j2\pi\nu\tau} d\tau,$$

$$\Gamma_U(\tau) = \int_{-\infty}^{\infty} \mathcal{G}_U(\nu) e^{-j2\pi\nu\tau} d\nu.$$

Autocovariance function: The autocovariance function is defined as

$$C_U(t_2,t_1) \triangleq \overline{\left[u(t_2) - \overline{u(t_2)}\right]\left[u(t_1) - \overline{u(t_1)}\right]}.$$

Thus

$$C_U(t_2,t_1) = \Gamma_U(t_2,t_1) - \bar{u}(t_2)\bar{u}(t_1).$$

Structure function: The structure function $D_U(t_2,t_1)$ is defined by

$$D_U(t_2,t_1) \triangleq \overline{\left[u(t_2) - u(t_1)\right]^2}.$$

Thus

$$D_U(t_2,t_1) = \overline{u^2(t_2)} + \overline{u^2(t_1)} - 2\Gamma_U(t_2,t_1).$$

The structure function has the advantage that it depends only on the delay $\tau = t_2 - t_1$.

Cross-correlation function: A natural generalization of the concept of an autocorrelation function is the cross-correlation function of two random processes $U(t)$ and $V(t)$, defined by

Ensemble Average

$$\Gamma_{UV}(t_2,t_1) \triangleq E\left[u(t_2)v(t_1)\right].$$

Time Average

$$\tilde{\Gamma}_{UV}(\tau) \triangleq \langle u(t+\tau)u(t)\rangle.$$

$U(t)$ and $V(t)$ are said to be joint wide-sense stationary if

$$\Gamma_{UV}(t_2,t_1) = \Gamma_{UV}(\tau),$$

and have the following properties

1. $\Gamma_{UV}(0) = \overline{uv}$
2. $\Gamma_{UV}(-\tau) = \Gamma_{VU}(\tau)$
3. $|\Gamma_{UV}(\tau)| \le \left[\Gamma_U(0)\Gamma_V(0)\right]^2$

Cross-spectral density functions: Given,

$$\mathcal{G}_{UV}(\nu) \triangleq \lim_{T\to\infty} \frac{E\left[\mathcal{U}_T(\nu)\mathcal{V}_T^*(\nu)\right]}{T} \qquad \mathcal{G}_{VU}(\nu) \triangleq \lim_{T\to\infty} \frac{E\left[\mathcal{U}_T(\nu)\mathcal{V}_T^*(\nu)\right]}{T}.$$

A normalized, complex-valued correlation coefficient that describes the statistical similarity between U and V at each frequency ν is defined as

$$\rho(\nu) = \frac{\mathcal{G}_{UV}(\nu)}{\left[\mathcal{G}_U(\nu)\mathcal{G}_V(\nu)\right]^{1/2}}.$$

The cross-spectral density has the following basic properties for any real-valued random processes $U(t)$ and $V(t)$

1. $\mathcal{G}_{VU}(\nu) = \mathcal{G}_{UV}^*(\nu)$
2. $\mathcal{G}_{UV}(-\nu) = \mathcal{G}_{UV}^*(\nu)$
3. $\left|\mathcal{G}_{UV}(\nu)\right| \le \sqrt{\mathcal{G}_U(\nu)\mathcal{G}_V(\nu)}$.

The Wiener-Khinchin Theorem also applies to jointly wide-sense stationary random processes

$$\mathcal{G}_{UV}(\nu) = \int_{-\infty}^{\infty} \Gamma_{UV}(\tau) e^{j2\pi\nu\tau} d\tau,$$

$$\Gamma_{UV}(\tau) = \int_{-\infty}^{\infty} \mathcal{G}_{UV}(\nu) e^{-j2\pi\nu\tau} d\nu.$$

When passing through linear, time-invariant filters $\mathcal{H}_n(\nu)$, $U(t)$ becomes $W(t)$ and $V(t)$ becomes $Z(t)$

$$\mathcal{G}_{WZ}(\nu) = \mathcal{H}_1(\nu)\mathcal{H}_2^*(\nu)\mathcal{G}_{UV}(\nu).$$

Sums of sample functions

$$Z(t) = U(t) + V(t).$$

Then

$$\mathcal{G}_Z(\nu) = \mathcal{G}_U(\nu) + \mathcal{G}_V(\nu) + \mathcal{G}_{UV}(\nu) + \mathcal{G}_{VU}(\nu),$$
$$\Gamma_Z(\tau) = \Gamma_U(\tau) + \Gamma_V(\tau) + \Gamma_{UV}(\tau) + \Gamma_{VU}(\tau).$$

A.11.6 Gaussian Random Processes

A random process $U(t)$ is said to be a Gaussian random process if the random variables $U(t_1), U(t_2), \ldots, U(t_n)$ are jointly Gaussian for all finite sets of time instants. For n time instants $t_1, t_2, \ldots, t_n$, the joint PDF of these variables is

$$p_U(\underline{u}) = \frac{1}{(2\pi)^{n/2}\left|\underline{C}\right|^{1/2}} \exp\left[-\frac{1}{2}(\underline{u} - \underline{\bar{u}})^t \underline{C}^{-1} (\underline{u} - \underline{\bar{u}})\right],$$

where

$$\underline{u} = \begin{bmatrix} u(t_1) \\ u(t_2) \\ \vdots \\ u(t_n) \end{bmatrix} \quad \underline{\bar{u}} = \begin{bmatrix} \bar{u}(t_1) \\ \bar{u}(t_2) \\ \vdots \\ \bar{u}(t_n) \end{bmatrix},$$

and $\underline{C}$ is a covariance matrix with element in the ith row and jth column defined by

$$\sigma_{ij}^2 = E\left[\left(u(t_i)-\bar{u}(t_i)\right)\left(u(t_j)-\bar{u}(t_j)\right)\right].$$

The corresponding joint characteristic function is defined as

$$M_U(\underline{\omega}) = \exp\left[j\underline{\omega}'\underline{\bar{u}} - \frac{1}{2}\underline{\omega}'\underline{C}\underline{\omega}\right],$$

where

$$\underline{\omega} = \begin{bmatrix} \omega_1 \\ \omega_2 \\ \vdots \\ \omega_n \end{bmatrix}.$$

A linearly filtered Gaussian random process is also a Gaussian random process.

A.11.7 Poisson Impulse Processes

Consider a random process $U(t)$ with sample function $u(t)$ that consists of a multitude of Dirac delta functions. This random process will be called a Poisson impulse process, or simply a Poisson process, if the following two conditions are satisfied:

1. The probability $P(K;t_1,t_2)$ that K impulses fall with the time interval $(t_1 < t \le t_2)$ is given by:

$$P(K;t_1,t_2) = \frac{\left(\int_{t_1}^{t_2} \lambda(t)\,dt\right)^K}{K!} \exp\left(-\int_{t_1}^{t_2} \lambda(t)\,dt\right),$$

   where $\lambda(t) > 0$ is called the rate function of the process
2. The numbers of impulses falling in any two non-overlapping time intervals are statically independent.

A.11.7.1 Moments

For a given $\lambda(t)$, the mean and second moment of the number of impulses in the time interval $(t_1 < t \le t_2)$ are given by:

$$\bar{K} = \int_{t_1}^{t_2} \lambda(t)\,dt,$$

$$\overline{K^2} = \bar{K} + \left(\bar{K}\right)^2.$$

For higher moments

$$\overline{K(K-1)\ldots(K-q+1)} = \left(\bar{K}\right)^q.$$

A.11.7.2 Energy Spectral Density, Deterministic Rate Function

Let $\lambda(t)$ be a deterministic Fourier-transformable function. A sample function of the corresponding Poisson impulse process and its Fourier transform can be expressed as

$$u(t) = \sum_{k=1}^{K} \sigma(t - t_k) \quad \mathrm{U}(\nu) = \sum_{k=1}^{K} \exp(j2\pi\nu t_k).$$

The energy spectral density for the random process $U(t)$ is the ensemble average of the energy spectral densities of its sample functions.

$$\mathcal{E}_U(\nu) = E\left[\left|\mathcal{U}(\nu)\right|^2\right].$$

Following the steps listed on pages 86 and 87 of the text, we find that

$$\mathcal{E}_U(\nu) = \bar{K} + \left(\bar{K}\right)^2 \widehat{\mathcal{E}_\lambda}(\nu),$$

where $\widehat{\mathcal{E}_\lambda}(\nu) = \dfrac{\mathcal{E}_\lambda(\nu)}{\mathcal{E}_\lambda(0)}$.

Thus the energy spectral density of the Poisson impulse process consists of a constant $\bar{K}$ plus $\left(\bar{K}\right)^2$ times the normalized energy spectrum of the rate function.

A.11.7.3 Power Spectral Density, Deterministic Rate Function

When the known rate function is not Fourier transformable, we once again truncate the random process $U(t)$ and rate function $\lambda(t)$. We find that:

$$\mathcal{G}_U(\nu) = \langle\lambda\rangle + \mathcal{G}_\lambda(\nu),$$

where $\langle\lambda\rangle$ is the time average of $\lambda(t)$

$$\langle\lambda\rangle = \lim_{T\to\infty} \frac{1}{T} \int_{-T/2}^{T/2} \lambda(t)\,dt.$$

A.11.7.4 Doubly Stochastic Poisson Processes

Suppose that $\lambda(t)$ is a simple sample function of a wide-sense stationary random process $\Lambda(t)$. By allowing the T to become arbitrarily large, we obtain

$$\mathcal{E}_U(\nu) = \bar{K} + \mathcal{E}_\Lambda(\nu),$$

$$\mathcal{G}_U(\nu) = \bar{\lambda} + \mathcal{G}_\Lambda(\nu),$$

where $\bar{\lambda} \triangleq \langle E[\lambda(t)]\rangle$.

A.11.7.5 Spectral Densities of Linearly Filtered Poisson Impulse Processes

Consider the case of a linearly filtered Poisson impulse process and the energy r power spectral density of such processes. Assuming that the process consists of pulses of identical shape and area, we find:

$$\mathcal{E}_U(\nu) = \bar{K}|\mathcal{H}(\nu)|^2 + |\mathcal{H}(\nu)|^2 \mathcal{E}_\Lambda(\nu),$$

$$\mathcal{G}_U(\nu) = \bar{\lambda}|\mathcal{H}(\nu)|^2 + |\mathcal{H}(\nu)|^2 \mathcal{G}_\Lambda(\nu),$$

where $\mathcal{H}(\nu)$ is the transfer function of the filter

If pulses of $U(t)$ have random shape and/or area, a modification is necessary.

$$\mathcal{E}_U(\nu) = \bar{K}\overline{|\mathcal{H}(\nu)|^2} + \left[\overline{\mathcal{H}(\nu)}\right]^2 \mathcal{E}_\Lambda(\nu),$$

$$\mathcal{G}_U(\nu) = \bar{\lambda}\overline{|\mathcal{H}(\nu)|^2} + \left[\overline{\mathcal{H}(\nu)}\right]^2 \mathcal{G}_\Lambda(\nu).$$

A.11.8 Random Processes Derived from Analytic Signals

It is common practice in physics and engineering to represent a real-valued signal by a related complex-valued signal. The complex representation is chosen so that its real part is the original real-valued signal; thus, provided that only linear operations are performed on the complex signal, the response to the real-value signal can be found by taking the real part of the complex response.

The reason for this can be traced to the eigenfunctions of linear, time invariant systems, which are complex exponentials of the form $\exp(-j2\pi\nu t)$. Thus, if we represent the linear, time-invariant system by an operator $\mathcal{L}\{\}$, we can show that

$$\mathcal{L}\{\exp(-j2\pi\nu t)\} = \mathcal{H}(\nu)\exp(-j2\pi\nu t).$$

A.11.8.1 Representation of a Monochromatic Signal by a Complex Signal

Consider a monochromatic (single-frequency) real-valued signal $u^{(r)}(t)$ described by

$$u^{(r)}(t) = A\cos(2\pi\nu_0 t - \phi),$$

A – Amplitude,
ν_0 – Frequency,
ϕ – Phase.

The complex representation of this signal is:

$$u(t) = A\exp[-j(2\pi\nu_0 t - \phi)].$$

Related to this complex representation is the phasor amplitude of $u(t)$ defined by the complex constant

$$A \triangleq A\exp(+j\phi),$$

Which represents the amplitude and phase of the monochromatic signal.

The complex representation of a real-valued cosinusoidal function is obtained by doubling the strength of its positive frequency component and removing its negative-frequency component (p. 94).

A.11.8.2 Representation of a Non-Monochromatic Signal by a Complex Signal

Consider a nonmonochromatic real-valued signal $u^{(r)}(t)$ with Fourier Transform $\mathcal{U}(\nu)$. We can represent the complex signal $u(t)$ with the same method as the monochromatic case, doubling the positive frequency components and removing the negative frequency components. Thus:

$$u(t) = 2\int_0^{\infty} \mathcal{U}(\nu)e^{-j2\pi\nu t}d\nu,$$

$u(t)$ is called the analytic signal representation of $u^{(r)}(t)$.

There are three important properties of analytic signals

1. $u^{(r)}(t) = \text{Re}\{u(t)\}$
2. $u^{(i)}(t) = \text{Im}\{u(t)\}$
3. $\mathcal{F}\{u^{(i)}(t)\} = -j\ \text{sgn}\nu\, \mathcal{F}\{u^{(r)}(t)\} = -j\ \text{sgn}\nu\, \mathcal{U}(\nu)$

With the last property, the imaginary part of the analytic signal can be obtained by passing the real part through a linear, time-invariant filter with transfer function

$$\mathcal{H}(\nu) = -j\ \text{sgn}\ \nu.$$

A.11.8.3 Complex Envelopes or Time-Varying Phasors

If the real signal $u^{(r)}(t)$ is a sample function of random process $U(t)$, then $u(t)$ is a sample function of complex valued $U(t)$. If $U(t)$ has zero mean then:

$$\begin{array}{ll} \mathbf{G}_U^{(i,i)}(\nu) = \mathbf{G}_U^{(r,r)}(\nu) & \mathbf{G}_U^{(i,r)}(\nu) = -\mathbf{G}_U^{(r,i)}(\nu) \\ \Gamma_U^{(i,i)}(\tau) = \Gamma_U^{(r,r)}(\tau) & \Gamma_U^{(i,r)}(\tau) = -\Gamma_U^{(r,i)}(\tau) \end{array}.$$

When the real and imaginary parts of $U(t)$ are at least wide-sense stationary, $\Gamma_U(\tau)$ has the following properties

1. $\Gamma_U(0) = \overline{\left[u^{(r)}(t)\right]^2} + \overline{\left[u^{(i)}(t)\right]^2}$
2. $\Gamma_U(-\tau) = \Gamma_U^*(\tau)$
3. $|\Gamma_U(\tau)| \le |\Gamma_U(0)|$.

For joint wide-sense stationary random processes $U(t)$ and $V(t)$

$$\Gamma_{UV}(\tau) = 2\Gamma_{UV}^{(r,r)}(\tau) + j2\Gamma_U^{(i,r)}(\tau).$$

A.11.9 Circularly Complex Gaussian Random Processes

$U(t)$ is called complex Gaussian if the real and imaginary parts are joint Gaussian processes. If the real parts obey Gaussian statistics, so do the imaginary parts. A zero mean complex Gaussian can be shown to have circular Gaussian statistics at any t.

When calculating the fourth order of such an equation of the form $\overline{u^*(t_1)u^*(t_2)u(t_3)u(t_4)}$ of a complex Gaussian, there are methods that can be applied if the following conditions are met:

1. $\overline{u^{(r)}(t_m)} = \overline{u^{(i)}(t_m)} = 0$
2. $\overline{u^{(r)}(t_m)u^{(r)}(t_n)} = \overline{u^{(i)}(t_m)u^{(i)}(t_n)}$
3. $\overline{u^{(r)}(t_m)u^{(i)}(t_n)} = -\overline{u^{(r)}(t_n)u^{(i)}(t_m)}$

$$E\left[u^*(t_1)u^*(t_2)u(t_3)u(t_4)\right] = \Gamma_U(t_3,t_1)\Gamma_U(t_4,t_2) + \Gamma_U(t_3,t_2)\Gamma_U(t_4,t_1).$$

A.11.10 Karhunen-Loève Expansion

Sometimes it is useful to expand $u(t)$ in $U(t)$ as a set of functions that are orthonormal on the interval $(-T/2, T/2)$. Given the orthonormal complex-valued functions $\{\phi_1(t), \phi_2(t), \ldots, \phi_n(t)\}$:

$$u(t) = \sum_{n=0}^{\infty} b_n \phi_n(t), \quad |t| \le \frac{T}{2},$$

where

$$\int_{-T/2}^{T/2} \phi_m(t) \phi_n^*(t)\, dt = \begin{cases} 1 & n = m \\ 0 & n \ne m \end{cases} \qquad b_n = \int_{-T/2}^{T/2} u(t) \phi_n^*(t)\, dt \quad n = 0, 1, 2, \ldots.$$

REFERENCES

Anton, H. 1991. *Elementary Linear Algebra,* 6th edn. Hoboken, NJ: John Wiley & Sons, Inc. https://www.amazon.com/Elementary-Linear-Algebra-Howard-Anton/dp/0471509000

Simmons, G.F. and S.G. Krantz. 2007. *Differential Equations: Theory, Technique, and Practice.* New York: McGraw Hill.

Thomas, G.B. and R.L. Finney. 1990. *Calculus and Analytic Geometry,* 7th edn. Reading, MA: Addison Wesley.

Zill, D.G. and M.R. Cullen. 1986. *Differential Equations with Boundary-Value Problems,* 3rd edn. Boston, MA: PWS-KENT Publishing Co.

Index

FSC
www.fsc.org
MIX
Papier | Fördert gute Waldnutzung
FSC® C083411

Kolibri 360 GmbH
Ferdinand-Jühlke-Straße 7
99095 Erfurt, Deutschland
produktsicherheit@kolibri360.de